Stanford's Geographical Estab.ᵗ London.

Biogeography

THIRD EDITION

Biogeography

THIRD EDITION

Mark V. Lomolino
SUNY College of Environmental Science and Forestry

Brett R. Riddle
University of Nevada, Las Vegas

James H. Brown
University of New Mexico

Sinauer Associates, Inc. • Publishers
Sunderland, Massachusetts

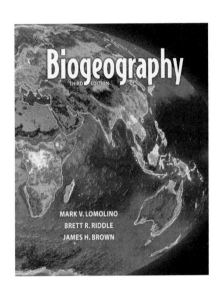

The Cover

Average annual net primary productivity of the Earth during 2002. This map displays carbon metabolism—the rate at which plants absorbed carbon out of the atmosphere to produce organic matter during photosynthesis. Courtesy of NASA's Earth Observatory.

The Endpapers

(Front) Wallace's (1876) scheme of biogeographic regions, which attempts to divide the landmasses into classes reflecting affinities and differences among terrestrial biotas. Numbers identify subregions. (Back) Map of the world illustrating many of the ocean basins, island chains, mountain ranges and other topographic features discussed in this text.

Biogeography, Third Edition

Copyright © 2006 by Sinauer Associates, Inc. All rights reserved.

This book may not be reproduced in whole or in part without permission from the publisher.

For information, address:

Sinauer Associates, Inc., 23 Plumtree Road, Sunderland, MA 01375 U.S.A.

Fax: 413-549-1118

E-mail: publish@sinauer.com

Internet: www.sinauer.com

Library of Congress Cataloging-in-Publication Data

Lomolino, Mark V., 1953-
 Biogeography/Mark V. Lomolino, Brett R. Riddle, James H. Brown.—
3rd ed.
     p. cm.
 Rev. ed. of: Biogeography/James H. Brown, Mark V. Lomolino. 2nd ed.
c1998. Includes bibliographical references (p.   ) and index.
 ISBN-10: 0-87893-062-0
 ISBN-13: 978-0-87893-062-3
 1. Biogeography. I. Riddle, Brett R. II. Brown, James H., 1942 Sept.
25- III. Brown, James H. 1942 Sept. 25- Biogeography. IV. Title.

QH84.B76 2005
578'.09—dc22

                              2005014443

Printed in U.S.A.
5 4 3 2

To our families for their love and patience,
to our colleagues for their insights and inspiration, and
to our students for challenging our ideas and redefining the frontiers of science.

Brief Contents

Contents

UNIT 3

Earth History and Fundamental Biogeographic Processes 139

UNIT 5
Ecological Biogeography 467

UNIT 6
Conservation Biogeography and New Frontiers 641

Preface

Simply defined, biogeography is the geography of life. Biogeographers study an impressive and sometimes overwhelming diversity of patterns in spatial variation of Earth's life forms, from molecular variation among populations of the same species, to geographic variation in productivity and diversity of ecosystems, and distinctiveness of entire biotas among the continents and ocean basins. Today, the field is enjoying a modern renaissance, although its roots extend back to the earliest efforts of humans to understand the geography of nature.

Indeed, biogeography must be one of the most ancient human endeavors. A knowledge of spatial variation in numbers and types of organisms across areas inhabited by our aboriginal ancestors was vital to their survival as they searched for particular kinds of animals and plants to use for food and shelter, and, more generally, as they adapted to a heterogeneous but geographically predictable environment. Explorers and naturalists of the seventeenth and eighteenth centuries greatly expanded our understanding of the geographic variation of the natural world, as their interests and experiences expanded from local and regional scales to a global one. They discovered that organisms and biological communities varied in a highly regular fashion along geographic gradients of latitude, elevation, isolation, and habitat area, and they learned that different regions of the globe—even those with similar climates and environments—were inhabited by different assemblages of species and higher taxa. These very general biogeographic patterns challenged the prevailing views which held that the Earth's diversity was relatively limited and that its continents, oceans, climates, and species were immutable—having changed little since the origin of the planet. Thus, the fascinating geographic signatures of nature would ultimately lead to paradigmatic shifts in our understanding of the origins, spread, diversification, and extinction of life. From these scientific revolutions would emerge entire new disciplines focusing on the dynamics of the Earth (geology and, in particular, plate tectonics theory), interactions among organisms and their environment (ecology and environmental biology), and heritable changes in the characteristics of populations and species (evolutionary biology).

Ironically, despite its long and distinguished history and its central role in development of other disciplines, biogeography was not widely recognized as a major discipline in its own right until the latter decades of the twentieth century. The very reasons responsible for its great insights and promise—its holistic and integrative nature—also posed great challenges. Few scientists could master the many fields relevant to the diversity of patterns and processes biogeographers study. Indeed, up until the 1980s, the most comprehensive works in biogeography were those written many decades earlier by the "fathers" of the field, including those by Alfred Russel Wallace, Charles Darwin, and Joseph Dalton Hooker. Few universities offered a course in biogeography and many of us, including two authors of this textbook, never had the opportunity to take a course in the subject.

In order to address the need for a modern, comprehensive textbook in *Biogeography*, James H. Brown and Arthur Gibson published the First Edition of Biogeography in 1983. For many of us, Brown and Gibson's text was our first comprehensive introduction to the field, and it fundamentally transformed our careers. Soon other authors followed suit, new biogeography journals were launched, more and more universities and colleges began offering courses in the subject, and a growing number of scientists began referring to themselves as "biogeographers."

Recognition of the heuristic interest and applied relevance of the field of biogeography continued to grow at an accelerating rate. By the 1990s it was clear that this textbook required a complete revision. The Second Edition of *Biogeography* emulated the original text in its balanced coverage of the entire breadth of the discipline, its integration of ecological and evolutionary approaches, and its emphasis on general concepts richly illustrated with empirical examples drawn from a wide variety of organisms, ecosystems and geographic regions. We were pleased to see that, in addition to continuing as a popular textbook, the Second Edition became a general reference and frequently cited source in the primary literature.

Although 15 years passed between the publication of the First and Second Edition of *Biogeography*, after only five years it was clear to us that another revision was required. With the addition of Brett Riddle to our team, we again set out to provide a comprehensive and integrative account of the entire field, restructuring the text and updating it with over 1000 new publications from the most exciting frontiers of biogeography. We hope that the Third Edition of *Biogeography* will contribute to this modern scientific renaissance, and that it will soon be necessary once again to update and revise our text with emerging advances in our understanding of the geography of life. That will be a sign that the field is still healthy, relevant, and advancing.

In addition to those who contributed to the previous editions, many additional people—far too many to recognize individually—have helped to revise, update, and improve this book. We are especially indebted to Douglas Kelt for his careful reading and insightful comments on the chapters in this book, and to other colleagues including Rob Channell, Paul Giller, Larry Heaney, David Perault, Christopher Scotese, and Robert J. Whittaker for their comments on topics discussed in particular sections of the book. Many readers of the previous editions provided suggestions and corrections that we have incorporated here, and we thank them for their assistance. We also thank Andy Sinauer, Kathaleen Emerson, Christopher Small, Janice Holabird, David McIntyre, Suzanne Lain, and the other editors and staff at Sinauer Associates for their unfailing professionalism and Herculean efforts to produce a high-quality book on a tight schedule. Finally, we are indebted to all biogeographers and other scientists, from the earliest workers to our contemporaries, for their contributions to the discipline. It is their individual, collaborative, and cumulative contributions that make biogeography so exciting to teach and study.

UNIT ONE

Introduction to the Discipline

CHAPTER *1*

The Science of Biogeography

*L*IFE VARIES FROM PLACE TO PLACE in a highly non-random and predictable manner. This seemingly simplistic observation is none-the-less one of the most fundamental and most important patterns in nature. Even the earliest human societies were aware that as they expanded their search area they would encounter a greater number and greater diversity of plants and animals. Those societies living in mountainous regions could see that vegetation changed in an orderly manner as they moved from the lowlands to the summit. The earliest fishing and seafaring societies learned that their catch varied from place to place and from the shallows to deeper waters of the ocean. These same societies (or their descendants) eventually learned that the numbers and diversity of terrestrial organisms increased as their journeys took them from the tiniest to the largest islands, and from environments of temperate regions to those of the tropics.

Knowledge of geographic variation of life on this planet was thus unavoidable and likely essential to survival of these ancient societies. Thousands of years later, scientists would rediscover these patterns and add many additional insights into the geography of nature during the Age of European Exploration. In order to explain such patterns, these early biogeographers would eventually realize that the Earth, its climate, and its species were dynamic over both space and time. Soon, their explanations for the development and distribution of life would include references to past environments, extinct life forms, and competition and predation among species. Thus the geography of nature became the foundation to entirely new fields of science including geology, meteorology, paleontology, evolution, and ecology.

The early ecologist and evolutionary biologist Ernst Haeckel (1876) once wrote that "the actual value and invincible strength of the Theory of Descent ... [is] ... that it explains *all* biological phenomena, that it makes *all* botanical and zoological series of phenomena intelligible in their relations to one another."Later, Theodosius Dobzansky put it much more succinctly—"Nothing in biology makes sense except in the light of evolution." We do not take issue with either of these visionaries but offer an assertion that is just as bold and important; indeed, it is the fundamental theme of this book and modern biogeography in general. Few patterns in ecology, evolution, conservation biology—and for that matter, most studies of biological diversity—make sense unless viewed in an explicit geographic context.

We know that living things are incredibly diverse. There are probably somewhere between 5 million and 50 million kinds of animals, plants, and microbes living on Earth today. Of these, fewer than 2 million have been formally recognized as species and described in the scientific literature. The remainder are represented by specimens awaiting description in museums, or by individuals awaiting discovery in nature. Additional untold millions, probably billions, of species that lived at some time in the past are now extinct; only a small fraction of them have been preserved as fossils.

Nearly everywhere on Earth from the frozen wastelands of Antarctica to the warm, humid rainforests of the tropics; from the cold, dark abyssal depths of the oceans to the near-boiling waters of hot springs—even in rocks several kilometers beneath the Earth's surface—at least some kinds of organisms can be found. But no single species is able to live in all of these places. In fact, most species are restricted to a small geographic area and a narrow range of environmental conditions. The spatial patterns of global biodiversity are a consequence of the ways in which the limited geographic ranges of the millions of species overlap and replace each other over the Earth's vast surface.

What Is Biogeography?

Definition

Biogeography is the science that attempts to document and understand spatial patterns of biological diversity. Traditionally, it has been defined as the study of distributions of organisms, both past and present. Modern biogeography, however, now includes studies of all patterns of geographic variation in nature—from genes to entire communities and ecosystems—elements of biological diversity that vary across geographic gradients including those of area, isolation, latitude, depth, and elevation.

As with any science, biogeography can be characterized by the kinds of questions its practitioners ask. Some of the questions posed by biogeographers include the following:

1. Why is a species or higher taxonomic group (genus, family, order, and so on) confined to its present range?
2. What enables a species to live where it does, and what prevents it from colonizing other areas?
3. What role does geographic variation in climate, topography, and interactions with other organisms play in limiting the distribution of a species?
4. How do different kinds of organisms replace each other as we go up a mountain or move from a rocky shore to a sandy beach nearby?
5. How does a species come to be confined to its present range?
6. What are a species' closest relatives, and where can they be found? Where did its ancestors live?
7. How have historical events—such as continental drift, Pleistocene glaciation, and recent climatic change—shaped a species' distribution?

8. Why are animals and plants of large, isolated regions—such as Australia, New Caledonia, and Madagascar—so distinctive?

9. Why are some groups of closely related species confined to the same region, while others are found on opposite sides of the world?

10. Why are there so many more species in the tropics than at temperate or arctic latitudes?

11. How are isolated oceanic islands colonized, and why are there nearly always fewer species on islands than in the same kinds of habitats on continents?

The list of possible questions is nearly endless, but in essence we are asking: *How does biological diversity vary over the surface of the Earth?* This is the fundamental question of biogeography. It has always intrigued scientists and laypersons who were curious about nature. Only within the last few decades, however, have scientists begun to call themselves biogeographers and to focus their research primarily on the distributions of living things. Not surprisingly, biogeographers have not yet answered all the questions listed above. They have, however, learned a great deal about where different kinds of organisms are found and why they occur where they do. Much of this progress has been made in just the last few decades, stimulated in large part by exciting new developments in the related fields of ecology, genetics, systematics, paleontology, and geology, as well as by technological developments.

Biogeography is a broad field. To be a complete biogeographer, one must acquire and synthesize a tremendous amount of information. But not all aspects of the discipline are equally interesting to everyone, including biogeographers. Given different biases in their training, their biogeography courses and writings tend to be uneven in coverage. A common specialization is taxonomic—for example, phytogeographers study plants and zoogeographers study animals, and within these categories one finds specialists in groups at all taxonomic levels. Although viruses and bacteria play crucial roles in ecological communities and in human welfare, microbial biogeography is poorly known and rarely discussed. Some biogeographers specialize in **historical biogeography** and attempt to reconstruct the origin, dispersal, and extinction of taxa and biotas. This approach contrasts with **ecological biogeography**, which attempts to account for present distributions and geographic variation in diversity in terms of interactions between organisms and their physical and biotic environments. Paleoecology bridges the gap between these two fields by investigating the relationships between organisms and past environments and using data on both the biotic composition of communities (abundance, distribution, and diversity of species) and abiotic conditions (climate, soils, water quality, etc.) derived from fossilized remains preserved in ancient sediments, ice cores, tree rings, and other places. Different biogeographers have emphasized different methods for understanding distributions: some approaches are primarily descriptive, designed to document the ranges of particular living or extinct organisms, whereas others are mainly conceptual, devoted to building and testing theoretical models to account for distribution patterns. All of these approaches to the subject are valid and valuable, and discounting or overemphasizing any division or specialization is counterproductive and unnecessary. Whereas no researcher or student can become an expert in all areas of biogeography, exposure to a broad spectrum of organisms, methods, and concepts leads to a deeper understanding of the science itself. As we hope to show, the various subdisciplines contribute to, and complement each other, unifying the science.

Relationships to Other Sciences

Biogeography is a synthetic discipline, relying heavily on theory and data from ecology, population biology, systematics, evolutionary biology, and the Earth sciences. Consequently, we do not want to draw sharp lines between biogeography and its related subjects, as some authors have attempted to do. For example, various authors have recommended that paleontology (the study of fossils and extinct organisms) and ecology be divorced from biogeography; this would make biogeography largely a descriptive, mapmaking endeavor. It would deprive biogeography of its central role as a synthetic discipline that not only has its own theoretical and empirical approaches, but also readily incorporates conceptual and factual advances from many other sciences.

Biogeography is a branch of biology and, not surprisingly, a good knowledge of biology is an important starting point. This is why our treatment devotes considerable space to reviewing and developing the ecological and evolutionary concepts that are used throughout the book (Units 2 and 3). In addition, one must be acquainted with the major groups of plants and animals and know something about their physiology, anatomy, development, and evolutionary history. These topics are not the subjects of separate chapters but are integrated throughout the text usually by the device of using different kinds of organisms with distinctive biological characteristics to illustrate biogeographic patterns, processes, and concepts. For example, it is only by knowing several critical features of the biology and evolutionary history of polar bears that we can understand why they are limited to the northern hemisphere, whereas penguins are limited to the southern hemisphere; or why amphibians and freshwater fishes have only rarely crossed even modest stretches of ocean to colonize islands, whereas birds and bats have done so much more frequently.

Naturally, it is important to know some geography and geology. The locations of continents, mountain ranges, deserts, lakes, major islands and archipelagoes (groups of islands), and seas, during the past as well as the present, are indispensable information for biogeographers, as are past and present climatic regimes, ocean currents, and tides. To remind the reader of the major geographic features of the Earth, we have illustrated many of them on the colored maps in the back endpaper of this book. In addition, Unit 2 provides an overview of patterns of geographic variation in environmental conditions—what we call the geographic template (Chapter 3), and basic patterns of distributions of species and entire communities across the globe (Chapters 4 and 5). Unit 3 describes the fundamental processes influencing geographic variation of species and communities, including immigration (Chapter 6), speciation and extinction (Chapter 7), geological processes and dynamics of the continents and ocean basins (Chapter 8), and the nature and biogeographic responses to climate change (Chapter 9). Unit 4 focuses on the evolutionary and geographic history of biotas, including the geography of diversification (Chapter 10), the history of lineages (Chapter 11), and how lineages and biotas have coevolved with changes in the Earth and its geographic template (Chapter 12). Unit 5 explores recent and modern patterns of the geography of nature, including patterns in species diversity (Chapter 13) and assembly and evolution of isolated biotas (Chapter 14). The last chapter in this section (15) explores similar patterns, focusing on continental and oceanic biotas, and introducing and exploring three related subdisciplines—ecogeography, areography, and macroecology. The final section of this book, Unit 6, provides an overview of the status of biological diversity and how the now impressive body of knowledge and tools of biogeographers can be applied to advance our understanding of, and abilities to conserve, the geography of nature.

Philosophy and Basic Principles

Most people have a vague and perhaps misleading impression of what science is, how scientists work, and how major scientific advances come about. Scientists try to understand the natural world by explaining its enormous diversity and complexity in terms of general patterns and basic laws. Philosophers and historians of science, viewing its progress with 20/20 hindsight, often suggest that it is possible to give a recipe for the most effective way to conduct an investigation. Unfortunately, as most practicing scientists know, scientific inquiry is much more like working on a puzzle or being lost in the woods than like baking cookies or following a road map. There are numerous mistakes and frustrations. Luck, timing, and trial and error play crucial roles in even the most important scientific advances. Some important discoveries, such as Alfred Wegener's evidence for continental drift, are long ignored or even totally rejected by other scientists. While the progress of science owes much to such admirable human traits as intelligence, creativity, perseverance, and precision, it is also retarded by equally human but less admirable characteristics such as prejudice, jealousy, short-sightedness, and stupidity. This is not meant to imply that science has not made great advances in our understanding of the natural world. It has. But like most human activities, the progress of science has followed a much more complex and circuitous path than is usually portrayed in textbooks. As we shall see, beginning in the next chapter, biogeography is no exception.

In essence, scientists proceed by investigating the relationships between pattern and process. Pattern can be defined as nonrandom, repetitive organization. The occurrence of pattern in the natural world implies causation by some general process or processes. Science usually advances by the discovery of patterns, then by the development of mechanistic explanations for them, and finally by rigorous testing of those theories until the ones that are necessary and sufficient to account for the patterns become widely accepted.

Traditional treatments of the philosophy of science usually devote considerable space to distinguishing between inductive reasoning—reasoning from specific observations to general principles—and deductive reasoning—reasoning from general constructs to specific cases. Several influential modern philosophers, especially Popper (1968), have strongly advocated so-called hypothetico-deductive reasoning. Any good scientific theory contains logical assumptions and consequences, and if any of these can be proven wrong, then the theory itself must be flawed. The hypothetico-deductive method provides a powerful means of testing a theory by setting up alternative, falsifiable hypotheses. First, an author puts forth a new, tentative idea stated in clear, simple language that can be tested and potentially falsified by means of experiments or observations. After the statement has withstood the severest empirical tests, it can be considered to be supported or corroborated, but by hypothetico-deductive logic a theory can never be proven true, only falsified (see also Kuhn 1970; Meadows 2001; Carpenter 2002; Graham et al. 2002).

New general theories or paradigms are the ultimate source of most major scientific advances, and most of them have been arrived at by inductive methods. The theory of evolution by natural selection, the double helical structure of DNA, and the equilibrium theory of insular biogeography all were derived largely by assembling factual data, recognizing a pattern, and then proposing a general mechanistic explanation. Although it might be safe to say that theories usually arise by inductive methods and are tested by deductive ones, often the actual conduct of scientific research is much more complex. Empirical observations and conceptual models are played back and forth against each other, theories are devised and modified, and understanding of the natural world is acquired slowly and irregularly. This is particularly true of biogeography, in which new empirical observations and theoretical paradigms have frequently overturned existing dogma. For example, although Alfred Wegener's ideas of continental drift were rejected for many decades, by the 1970s the data and theory of tectonic plate movements became overwhelmingly convincing. Biogeographers had to revise historical explanations for distributions that had been based on a static distribution of continents and oceans.

Unlike much of contemporary biology, biogeography usually is not an experimental science. Questions about molecules, cells, and individual organisms typically are most precisely and conveniently answered by artificially manipulating the system. In such experiments, the investigator searches for patterns or tests specific hypotheses by changing the state of the system and comparing the behavior of the altered system with that of an un-manipulated control. It is impractical and often impossible to use these techniques to address many of the important questions in biogeography and the related disciplines of ecology and evolutionary biology. Historical evolution and historical biogeography are, as their names imply, history—they produce no exact predictions about the future.

A few biogeographers have used experimental techniques to manipulate small systems such as tiny islands, sometimes with spectacular success (e.g., Simberloff and Wilson 1969). However, most important questions involve such large spatial and temporal scales that experimentation would be impractical or unethical. This methodological constraint does not diminish the rigor and value of biogeography, but it does pose major challenges. Other sciences, such as astronomy and geology, face the same problems. As Robert MacArthur once observed, Copernicus, Galileo, Kepler, and Newton never moved a planet, but that did not prevent them from advancing our understanding of the motion of celestial bodies (see Brown 1999). Wallace, Darwin, and Hooker used the patterns of animal and plant distributions that they observed in their world travels to develop important new ideas about evolution and biogeography. Islands have had a great influence on these and numerous subsequent biogeographers, ecologists, and evolutionists because they represent natural experiments—replicated natural systems in which many factors are held relatively constant while others vary from island to island. Despite the difficulty of performing artificial experiments, it is possible to develop and rigorously evaluate biogeographic theories by the same logical procedures used by other scientists: search for patterns, formulate theories, and then test assumptions and predictions independently with new observations.

In dealing with historical aspects of their science, most biogeographers make one critical assumption that is virtually impossible to test: they accept the principle of **uniformitarianism** or **actualism**. Uniformitarianism is the assumption that the basic physical and biological processes now operating on the Earth have remained unchanged throughout time because they are manifestations of universal scientific laws. The principle of uniformitarianism is usually attributed to the British geologists Hutton (1795) and Lyell (1830), who realized that the Earth was much older than had been previously supposed and that its surface was constantly being modified by the same geological processes. In this same spirit, one of Darwin's great insights was the recognition that changes in domesticated plants and animals over historical time through selective breeding resulted from the same process as natural changes in organisms over evolutionary time.

As noted by Simpson (1970a), acceptance of uniformitarianism has never been universal in part because some authors have attached additional meanings to the concept. Some have assumed that the term implies that the average intensities of processes have remained approximately constant over time and that both geological and biological changes are always gradual. Neither of these amendments is acceptable. We know that some historical events have been of great magnitude but so infrequent that they have never been observed in recorded human history. For example, contrary to science fiction movies, no humans were living 65 million years ago when dinosaurs roamed the Earth. Consequently, no one observed the collision of the Earth with an asteroid that presumably eliminated the dinosaurs along with roughly half of all other organisms in existence (see Chapter 8). Yet there is abundant evidence, preserved in ancient rocks, for this event. Scientists expect that the intensity of forces will vary from time to time and from place to place; only the nature of the laws that control these processes is timeless and constant.

To avoid these unfortunate connotations associated with uniformitarianism, we will follow Simpson in adopting the term *actualism*, which is conceptually similar to methodological uniformitarianism (Gould 1965). Historical biogeographers in particular use this principle to account for present and past distributions, assuming that the processes of speciation, dispersal, and extinction operated in the past by the same mechanisms that they do today. This premise has become an accepted tool for interpreting the past and predicting the future. The most serious problem in using this principle is discovering which timeless processes have operated to produce different patterns.

The Modern Science

Doing Contemporary Biogeography

Biogeography differs from most of the biological disciplines, and from many other sciences, in several important respects. We have mentioned one of them above: biogeography is, for the most part, a comparative observational science rather than an experimental one because it usually deals with scales of space and time at which experimental manipulation is impossible. Thus most of the inferences about biogeographic processes must come from the study of patterns—from comparisons of the geographic ranges, genetics, and other characteristics of different kinds of organisms or the same kinds of organisms living in different regions, and from observations of differences in the diversity of species, and the composition of communities over the Earth's surface. But while it is difficult for biogeographers to manipulate these systems in controlled experiments, it is possible for them to study the effects of natural and human-caused perturbations (i.e., to learn from both natural and anthropogenic experiments). Thus, for example, we have learned a great deal about the colonization of oceanic islands by monitoring the buildup of biotas after volcanic eruptions, as on Krakatau in the East Indies and Surtsey in the North Atlantic. There is also much to be learned from the changes being caused by modern humans who have altered habitats, connected previously isolated areas and fragmented previously continuous ones, introduced exotic species, and caused the extinctions of native ones.

Another way that biogeography differs from most other sciences is that it is usually dependent on data collected by many individuals working over large areas for long periods of time. Much of biogeography involves studying the geographic ranges of species or the species composition of regional biotas. Doing either of these accurately usually requires reliance on the previous fieldwork of many persons who have collected and identified specimens, deposited them in museum collections, and published the information in the scientific literature. Although being a biogeographer carries the sometimes deserved connotation of adventurous collecting expeditions to exotic places, most practicing biogeographers obtain more of their data from museums and libraries than from their own fieldwork. In this respect biogeography is unlike many other disciplines in which individual scientists or groups of collaborating investigators collect most or all of their primary data themselves.

Finally, as mentioned above, biogeography is typically a synthetic science. Biogeographers work at the interfaces of several traditional disciplines: ecology, systematics, evolutionary biology, geography, paleontology, and the "Earth sciences" of geology, climatology, limnology, and oceanography. Much of their best work comes from putting together data or theories from two or more of these disciplines. This might seem to require such exceptionally broad training as to be intimidating to the beginner. However, while some breadth of knowledge is desirable, the interdisciplinary nature of biogeography also means that individuals with diverse backgrounds and interests can make important contributions—sometimes on their own but more often through collaboration with specialists from other areas. In fact, biogeography is accessible to almost anyone with curiosity and motivation. Large research grants, modern laboratory facilities, and even sophisticated statistical techniques—while often desirable—are not required to do state-of-the-art biogeographic research. All that is required is a good idea and access to a library, museum collection, or other source of data on distributions of organisms. This means that even beginning students can do original research. The authors of this book usually require that each student in their biogeography courses do an original research paper. Typically this involves coming up with an interesting question, finding a source of appropriate data in the literature or a museum, and doing the analyses to try to answer the question. Some of the work discussed in this book and published in distinguished journals began as student research projects in our courses.

Current Status

Biogeography has a long and distinguished history. Many of the greatest scientists of their eras were biogeographers including the great evolutionary biologists of the nineteenth century—Charles Darwin, Alfred Russel Wallace, and Joseph Dalton Hooker, as well as some of the most eminent scientists of the twentieth century—George Gaylord Simpson, Ernest Mayr, Robert MacArthur, and Edward O. Wilson. While these giants did not usually refer to themselves as "biogeographers," their comparative studies of distributions and diversity provided much of the raw material for their theoretical and empirical contributions, which have laid the foundations for much of contemporary evolutionary biology and ecology. In the next chapter, we place the contributions of these and other influential scientists in the context of the historical development of biogeography, evolution, and ecology—disciplines that saw their greatest development during the past three centuries.

It is only recently, however—within the last few decades—that biogeography has emerged as a widely respected science with its own identity. Although there were a few earlier influential works, especially in plant geography (e.g., Wulff 1943; Cain 1944), it was really not until the 1950s and 1960s that major books and papers with the word "geography" in their titles began to appear regularly (e.g., Hesse et al. 1951; Croizat 1952, 1958; Ekman 1953; Simpson 1956, 1965; Dansereau 1957; Darlington 1957, 1965; MacArthur and Wilson 1963, 1967; Nelson 1969; Neil 1969; Udvardy 1969). It was about a decade later, in 1973, that the first issue of the *Journal of Biogeography*, the first scientific periodical devoted exclusively to the discipline, was published. Since then, however, the journal has grown rapidly in size, circulation, and reputation. So great was its success that in 1991 its editors created a second journal, *Global Ecology and Biogeography*, which continued to grow in readership and is now recognized as a leading journal for both biogeography and macroecology. These journals have been joined by others focusing on biogeography including *Diversity and Distributions* and *Ecography*, and many other journals regularly publish biogeography papers. One quantitative measure of the coalescing interest in the field is provided by a computerized search of *BIOSIS* for occurrences of the word "biogeography..." in titles, abstracts, or keywords at 10-year intervals—33 in 1975, 358 in 1985, and 1238 in 1995. Classes and textbooks in biogeography were virtually nonexistent before the 1980s but are now part of the curriculum at most major universities and many smaller institutions.

The emergence of biogeography as a vigorous modern science can be attributed to several coincidental and interacting developments. One is the transformation of the discipline—and of its image in the eyes of other scientists—from a largely descriptive science closely allied to traditional taxonomy to a conceptually oriented discipline concerned with building and testing biogeographic theory. The introduction of new and mathematical theory into ecology, evolution, and systematics has spilled over to trigger major conceptual progress in both ecological and historical biogeography (see Chapters 2, 8, 9, 10, and 11). Contemporary advances in the Earth sciences, especially the theory of plate tectonics (Chapter 8) and a wealth of data from the fossil record, have also been influential.

A second stimulus to the progress of modern biogeography has been the development and application of new technology. Computers have allowed the compilation and manipulation of enormous quantities of data on truly geographic scales: distributional records and other information on organisms, as well as data on climate, landforms, geology, soils, limnological and oceanographic conditions, vegetation, land uses, and other human activities. Advances in computer hardware and software have made possible the development and application of powerful new techniques including simulation modeling, geographic information systems (GIS), and a whole battery of statistical methods such as multivariate analyses and geostatistics. Satellites, submersible vessels, and ground-based data collection systems—usually automated and computerized—have provided a wealth of new information on the environment of the Earth. Other technologies, including radioisotope, stable isotope, and molecular biological techniques have permitted increasingly accurate reconstructions of the history of both the Earth itself and its organisms. Biogeography, being one of the most synthetic sciences, has readily adopted new technologies from many other disciplines and has made rapid use of the new kinds of information that they can provide.

A final stimulus to the emergence of a vigorous modern science of biogeography has come from the need to understand and manage the impact of human beings on the Earth. A major wave of environmental concern and activism developed in the late 1960s and early 1970s. Initially, much of the focus was on local responses to fairly small-scale problems: seeking alternatives to persistent pesticides; alleviating local air, water, and soil pollution; setting aside reserves of relatively unspoiled habitat; and saving local populations of endangered species. By the 1980s, however, it was becoming increasingly apparent that modern humans have been transforming the Earth on regional and global, as well as local, scales. We now know that every place on Earth—from the polar ice caps to the abyssal ocean depths to the upper atmosphere—has been substantially changed by the activities of our ever-growing population and our increasingly industrial and technological society. These changes include increases in atmospheric carbon dioxide and other greenhouse gasses, depletion of stratospheric ozone, extinctions of species and losses of other components of biological diversity, and conversion of natural landscapes and ecosystems to agricultural fields and human habitations. With the realization of the global scale of human influences has come a need for biogeographic expertise to measure and predict the effects of these changes on organisms and ecosystems and to find ways to manage, mitigate, or minimize their most damaging effects (Unit 6). One fortunate result has been increasing respect and funding for biogeographic research, both basic and applied.

In less than two decades, biogeography has gone from being a rather esoteric and unappreciated science, whose practitioners usually called themselves ecologists, systematists, or paleontologists rather than biogeographers, to a vigorous and respected science whose practitioners are using the latest conceptual advances and technological tools to say important things about the present and future state of the Earth and its living inhabitants.

CHAPTER *2*

The History of Biogeography

BIOGEOGRAPHY HAS HAD A LONG HISTORY, one that is inextricably woven into the development of evolutionary biology and ecology. The problems of distribution and variation on geographic scales were matters of primary interest to evolutionary biologists, including the distinguished "fathers" of the field such as Linnaeus, Darwin, and Wallace. The field of ecology, a relatively young offspring of this lineage, grew out of attempts to explain biogeographic patterns in terms of the influence of environmental conditions and interactions among species. But the origin of these three fields—biogeography, evolutionary biology, and ecology—is ancient, dating back well before the Darwinian revolution. Indeed, Aristotle was contemplating many of the questions we ponder today:

> But if rivers come into being and perish and if the same parts of the Earth are not always moist, the sea also must necessarily change correspondingly. And if in places the sea recedes while in others it encroaches, then evidently the same parts of the Earth as a whole are not always sea, nor always mainland, but in process of time all change. (Meteorologica, ca. 355 B.C.)

Aristotle offered this prophetic view of a dynamic Earth in order to explain variation in the natural world over space and time. He was one of the earliest of a long and prestigious line of scientists to ask the same questions: Where did life come from, and how did it diversify and spread across the globe?

Despite Aristotle's impressive insights, answering these questions would require a much more thorough understanding of the geographic and biological character of the Earth. Thus the development of biogeography, evolutionary biology, and ecology was tied to the Age of European Exploration. As we shall see below, the early European explorers and naturalists did far more than just label and catalogue their specimens. They immediately, perhaps irresistibly, took to the task of comparing biotas across regions and developing explanations for the similarities and differences they observed. The comparative

TABLE 2.1 *Persistent themes in biogeography*

1. Classifying geographic regions based on their biotas.
2. Reconstructing the historical development of lineages and biotas, including their origin, spread, and diversification.
3. Explaining the differences in numbers as well as types of species among geographic areas, and along geographic gradients including those of area, isolation, latitude, elevation and depth.
4. Explaining geographic variation in the characteristics of individuals and populations of closely related species, including trends in morphology, behavior, and demography.

method served these early naturalists well, and by the eighteenth century the study of biogeography began to crystallize around fundamental patterns of distribution and geographic variation. Here, we trace the development of biogeography from the Age of Exploration to its current status as a mature and respected science.

Many, if not all, of the themes central to modern biogeography (Table 2.1) have their origins in the pre-Darwinian period. This is not to say that biogeography has not advanced tremendously in the past few decades—only that modern biogeographers owe a great debt to those before them who shared the same fascination with, and asked the same types of questions about, the geography of nature. As Newton once replied when asked how it was that he had developed such unparalleled insights in physics, "If it appears that I have been able to see more than those who came before me, perhaps it is because I stood on the shoulders of giants." Like physics, biogeography has a history of "giants," visionary scientists each building on the collective knowledge of those who came before them.

The Age of Exploration

It is hard for us to appreciate that roughly 250 years ago, biologists had described and classified only 1 percent of all the plant and animal species we know today. Biogeography was essentially founded and rapidly accelerated by world exploration and the accompanying discovery of new kinds of organisms. Biologists and naturalists of the eighteenth century were largely driven by a calling to serve God. The prevailing belief was that the mysteries of creation would be revealed as these naturalists/explorers developed more complete catalogues of the diversity of life. Up until the mid-eighteenth century, the prevailing world view was one of stasis—the Earth, its climate, and its species were immutable. However, as the early biogeographers (then called naturalists or simply geologists) returned with their burgeoning wealth of specimens and accounts, two things became clear. First, biologists needed to develop a standardized and systematic scheme to classify the rapidly growing wealth of specimens. Second, it was becoming increasingly obvious that there were too many species to have been accommodated by the biblical Noah's Ark. It was just as difficult for these early biologists to explain how animals and plants—now isolated and perfectly adapted to dramatically different climates and environments—could have coexisted at the landing site of the Ark before they spread to populate all regions of the globe.

One of the most ambitious and visionary of these eighteenth-century biologists was Carolus Linnaeus (1707–1778). He believed that God spoke most clearly to man through the natural world, and he felt that it was his task to methodically describe and catalogue the collections of this divine museum. Toward that end, Linnaeus developed a scheme to classify all life: the system of binomial nomenclature that we continue to use today. Linnaeus also set his energies to the task of explaining the origin and spread of life. Like his contemporaries, he believed that the Earth and its species were immutable. He realized that the challenge was not just in explaining the number of species, but in explaining patterns of diversity and distribution as well. The rapidly growing list of species included organisms adapted to environments ranging from the moist tropics to deserts, forests, and tundra. Given that species were immutable, how could they have spread from a single site (Paradise and then Noah's landing place) to become perfectly adapted to such diverse environments? Linnaeus's solution, although perhaps naive in retrospect, was logically sound. He hypothesized that all life forms had originally been placed along the slopes of a "Pardisical Mountain" located near the equator where each species was perfectly adapted to the environmental conditions and interacting species of that particular "station" (or habitat) along the mountain slope. According to Linnaeus (1781), because slopes of tropical mountains include a full complement of stations—from tropical forests to alpine tundra—all of nature's species could be accommodated on this one tropical mountain. Linnaeus believed that the Paradisical Mountain was actually an island—the only land that rose above sea level of this primordial period (roughly 6000 years ago). Following these origins, rocks and various depositional materials were spewed up onto the land by the great forces of the sea. The land, and ultimately the nascent continents, expanded as the once pan-global sea contracted and then deepened. And as the terrestrial realm expanded, its species—still immutable and restricted to the same stations—dispersed to colonize newly emergent regions—some on the backs or in the fur, feathers, and bellies of others, and many by virtue of a myriad of dispersal mechanisms richly detailed in the volumes of Linnaeus's treatises. His explanation for the survival and redistribution of these same immutable species following the biblical deluge was analogous to that of his theory of creation. Linnaeus hypothesized that Noah's landing occurred along the slopes of one of the world's tallest mountains—possibly Mount Ararat, which is located near the border of Turkey and Armenia (Figure 2.1). Once the flood receded, these species migrated down from the mountain and spread, to eventually colonize and inhabit their respective environments in different regions of the globe.

Georges-Louis Leclerc, Comte de Buffon (1707–1788) was a contemporary of Linnaeus, but Buffon's studies of living and fossil mammals led him to a very different view of the origin and spread of life. Buffon (1761) noted two problems with Linnaeus's explanation. First, he observed that different portions of the globe—even those portions with the same climatic and environmental conditions—were often inhabited by distinct kinds of plants and animals. The tropics, in particular, contained a great diversity of unusual organisms. Second, Buffon reasoned that Linnaeus's view of the spread of life required that species migrate across inhospitable habitats following the Flood. Species adapted to montane forests, for example, would have had to migrate across deserts before they could colonize deciduous and coniferous forests to the north. If species were immutable, and therefore incapable of adapting to new environments, then their spread would have been blocked by these environmental barriers. Buffon, therefore, hypothesized that life originated not on a Paradisical Moun-

FIGURE 2.1 Two early hypotheses proposed to account for the diversity and distributions of terrestrial organisms. Linnaeus hypothesized that terrestrial plants and animals survived the biblical Flood along the slopes of a mountain such as Mount Ararat, near the present-day border of Turkey and Armenia, and then spread to suitable environments from that point (thin arrows). Buffon, on the other hand, hypothesized that species originated in a region in Northern Europe (circled area) and then spread into the New World and southward, changing ("improving" or "degenerating") as they colonized climatically and ecologically diverse landmasses in both the New and Old Worlds (thick arrows).

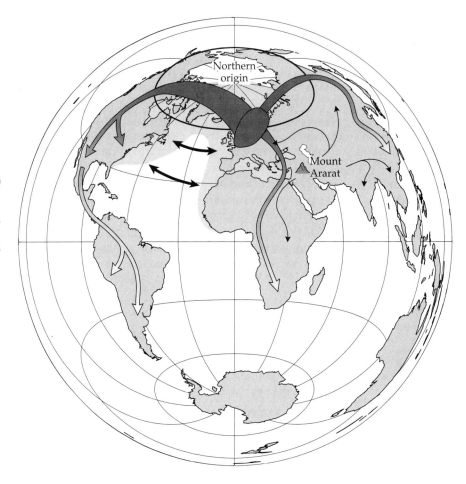

tain in the tropics, but in a region of northwestern Europe during an earlier period when climatic conditions were more equable (see Figure 2.1). He speculated that, when climates cooled, life forms then migrated into the New World (and ultimately south) to colonize the Southern Hemisphere as well. During this migration, the populations of the New and Old Worlds were separated and became increasingly modified until tropical biotas of the New and Old Worlds shared few, if any, forms. According to Buffon,

> Man is totally a production of heaven; But the animals, in many respects, are creatures of the Earth only. Those of one continent are not found in another; or, if there are a few exceptions, the animals are so changed that they are hardly to be recognized.

While Buffon's explanation may now seem fanciful, it provided two key elements of what would become central parts of modern biogeographic theory. Not only was the Earth (both land and sea) and its climate dynamic, but so were its species; they evolved, or in his terms—"improved" or "degenerated"—as they became further isolated in different regions and environments. The results included survival of the improved forms at the expense of those that either could not change, or degenerated (and ultimately died out), a process that Alfred Wallace and Charles Darwin would later call "natural selection." Buffon's view of the process of evolution—degeneration, in particular—were admittedly extreme and betrayed an obvious distaste for the New World and its foundling nations.

In this New World, … there is some combination of elements and other physical causes, something that opposes amplification of Nature. There are obstacles to development, and perhaps to the formation of large germs … [persisting forms that] shrink under the niggardly sky and an unprolific land, thinly peopled with wandering savages …"

One very positive outcome of Buffon's otherwise objectionable view of the New World is that it did spur Thomas Jefferson, then President of the United States, to commission one of the most ambitious biogeographic surveys ever conducted—Lewis and Clark's surveys across the central and northwestern portions of the U. S.

To his credit, Buffon's contributions strongly influenced the development of science during the eighteenth century, not the least of which was his observation that environmentally similar but isolated regions have distinct assemblages of mammals and birds. This became the first principle of biogeography, known as **Buffon's law.**

From 1750 to the early 1800s, many of Buffon's colleagues continued to explore the diversity and geography of nature and to write catalogues and general syntheses of their work. One of the most prominent naturalists/collectors of this period was Sir Joseph Banks, who, during a three-year voyage around the world with Captain James Cook on the *H.M.S. Endeavor* (1768–1771), collected some 3600 plant specimens including over 1000 species not known to science (i.e., in addition to the 6000 species described by Linnaeus in his *Species Plantarum*). The efforts of Banks and many other naturalists/explorers resulted in two important developments. First, they affirmed and generalized Buffon's law. Second, they developed a much more thorough understanding of, and appreciation for, the complexity of the natural world. Banks and his colleagues discovered some interesting exceptions to Buffon's law—specifically, cosmopolitan species. Moreover, they noted other biogeographic patterns, which in their own right would become major themes as biogeography developed.

In the latter part of the eighteenth century, Johann Reinhold Forster (1729–1798) made many important contributions to phytogeography and to biogeography, in general. In his account of his circumnavigation of the globe with Captain Cook (published in 1778), Forster presented one of the first systematic world views of biotic regions, where each is defined by its distinct plant assemblages. He found that Buffon's law applied to plants as well as to mammals and birds and to all regions of the world—not just the tropics. Forster also described the relationship between regional floras and environmental conditions and how animal associations changed with those of plants. He provided some important early insights into what was to become island biogeography and species diversity theory. He noted that insular (or island) communities had fewer plant species than those on the mainland, and that the number of species on islands increased with available resources (island area and variety of habitats). Forster also noted the tendency for plant diversity to decrease from the equator to the poles, a pattern that he attributed to latitudinal trends in surface heat on the Earth.

In 1792, Karl Ludwig Willdenow (1765–1812), another German botanist, wrote a major synthesis of plant geography. He not only described the floristic provinces of Europe, but offered a novel explanation for their origin. Rather than one site of creation (or survival during the biblical deluge), Willdenow suggested that there were many sites of origination—mountains that in ancient times were separated by global seas. Each of these mountain refuges was inhabited by a distinct assemblage of locally created plants. As the Flood receded, these plants spread downward to form the floristic regions of the world.

Ironically, it is not Willdenow but one of his students—Alexander von Humboldt (1769–1859)—who is generally viewed as the father of phytogeography. Humboldt was able to advance the insights of his mentor and add many of his own from his experiences in the New World tropics. He was a great naturalist and was keenly aware that fundamental laws of nature could be discovered through the study of distributions. After studying the works of Latreille on arthropods, and Cuvier on reptiles, Humboldt further generalized Buffon's law to include plants as well as most terrestrial animals. In an essay published in 1805, Humboldt noted that the floristic zonation that Forster described along latitudinal gradients could also be observed at a more local scale along elevational gradients. He studied in great detail climatic associations of plant communities, and he invented the isobar and isotherm (lines joining points of equal atmospheric pressure and temperature) to help map these associations. After conducting many floristic surveys in the Andes—including those along the slopes of 18,000-foot Mount Chimborazo—Humboldt concluded that even within regions, plants were distributed in elevational zones, or **floristic belts,** ranging from equatorial tropical equivalents at low elevations to boreal and arctic equivalents at the summits.

Thus Forster, Willdenow, and Humboldt all observed that plant assemblages were strongly associated with local climate. One of Humboldt's friends and colleagues, Swiss botanist Augustin P. de Candolle (1778–1841), added another fundamentally important insight: not only are organisms influenced by light, heat, and water but they compete for these resources as well. Candolle emphasized the distinction between biotic provinces or regions (which he termed "habitations") and local habitats (or "stations"—a term used much earlier by Linnaeus). Later, he returned to Forster's observations on insular floras adding that, while species number is most strongly influenced by island area, other factors—such as island age, volcanism, climate, and isolation—also influence floristic diversity. Thus, some of the key elements of ecological biogeography and modern ecology were established by the early 1800s. Moreover, in an essay published in 1820 Candolle appears to be the first to write about competition and the struggle for existence, a theme that would prove central to the development of evolutionary and ecological theory.

Biogeography of the Nineteenth Century

By the early 1800s, the first three themes of biogeography were well established. Biogeographers (then called "naturalists" or simply "geologists") were studying the distinctness of regional biotas, their origin and spread, and the factors responsible for differences in the numbers and kinds of species among local and regional biotas. Buffon's observations on mammals from the New and Old World tropics had been generalized to a law applying to most forms of life. But with this increased appreciation of the complexity and geographic variation of nature, the challenges for biogeography were becoming even greater. Scientists had made great progress in describing fundamental biogeographic patterns, including many that we continue to study today, but explanations for those patterns were wanting. How was it, for example, that isolated regions with nearly identical climates shared so few species? Causal explanations would have to include factors accounting for the differences as well as the similarities among isolated regions. Explanations for Buffon's law would also have to account for the exceptions to it.

During the next century, the legacy of Buffon's work would be evidenced by a long and continuing succession of explanations for his law, which would lead from a view of a static Earth populated by immutable, cosmopolitan species to one in which the Earth, its climate, and its species were dynamic. This view would prove essential for classifying biogeographic regions based on their biota (see Table 2.1, Theme 1) and for reconstructing the origin, spread, and diversification of life (see Table 2.1, Theme 2). Other biogeographers of the nineteenth century would focus their energies on fundamental patterns of species richness (see Table 2.1, Theme 3). It was already well established that in nearly all taxa the number of local species (at Candolle's "stations") tended to increase with area and to decrease with distance from the equator or up a mountain. But why?

To summarize, we see that nearly two centuries ago biogeographers had described many of the patterns we study today, explored the generality of those patterns, and developed causal explanations. For the most part, however, they fell short on this last challenge. Their understanding of the mutability of the Earth and its species still lagged far behind their rapidly increasing knowledge of the Earth's geological structure, climatic patterns, and biological diversity. For the field to come of age and to develop more rigorous, testable explanations for its fundamental patterns, it would have to await three important advances of the nineteenth century:

1. A better estimate of the age of the Earth (many early biogeographers were working with an estimate of only a few thousand years)

2. A better understanding of the dynamic nature of the continents and oceans (i.e., continental drift and plate tectonics)

3. A better understanding of the mechanisms involved in the spread and diversification of species—specifically dispersal, vicariance, extinction, and evolution

To make these advances, biogeography had to draw on new discoveries in geology and paleontology. During the nineteenth century, Adolphe Brongniart (1801–1876) and Charles Lyell (1797–1875; Figure 2.2), regarded as the fathers of paleobotany and geology, respectively, concluded that the Earth's climate was highly mutable. Both men used the fossil record to infer conditions of past climates (Ospovat 1977). They found that many life forms adapted to tropical climates had once flourished in the now temperate regions of northern Europe. Lyell also documented that sea levels had changed and that the Earth's surface had been transformed by the lifting up and eroding down of mountains. This, he argued, was the only way to account for the existence of marine fossils on mountain summits. Lyell also provided incontrovertible evidence for the process of extinction. Many fossil forms, once dominant and presumably perfectly adapted to existing climatic conditions, had perished and left no further trace in the fossil record. The causal agent, again, was inferred to be climatic variation and associated changes in sea level. Lyell, however, held firm to the belief that, although extinctions occurred, species were not mutable—at least not to the point that new species arose from existing ones. He also believed that, despite episodes of extinction, the diversity of the Earth on a grand scale remained relatively constant. He resolved this apparent contradiction by suggesting that each episode of extinction was followed by an episode of creation, which established a new set of species perfectly adapted to the altered climatic conditions. Not only were there many sites of creation, but there were many periods of creation as well!

Lyell saw in the layers of rock and their fossils undeniable evidence that the Earth's surface and its biota were dynamic. Moreover, he argued that these great changes resulted from physical processes such as mountain building and erosion—processes that had operated continually throughout the history of

(A)

(B)

FIGURE 2.2 (A) Charles Lyell and (B) Adolphe Brongniart, often regarded as the "fathers of geology and paleobotany," respectively. Lyell strongly influenced the development of biogeography in the nineteenth century largely through his *Principles of Geology*, first published in 1830. (A courtesy of King's College, London; B courtesy of the Council of the Linnaean Society of London.)

the Earth. This theory of **uniformitarianism** replaced earlier catastrophic explanations for the changes in the Earth, its landforms, and its inhabitants. It permitted new thinking about the dynamics of living systems because, after all, their physical and biotic parts are historically inseparable. Moreover, Lyell and other geologists such as James Hutton (1726–1797) realized that, given the gradual nature of these geological processes, the Earth must be much older than just a few thousand years. Only with an ancient Earth could they account for the formation and erosion of mountains, the submergence of ancient landmasses, and the migration or replacement of entire biotas that were documented in the fossil record.

As we shall see, the acceptance of uniformitarianism and the antiquity of the Earth was essential to the theories of both Darwin and Wallace, that organic diversity results from the gradual yet persistent effects of natural selection operating over thousands of generations. Ironically, for most of his long and prestigious career Lyell rejected the idea that species, like the geological features he studied, were the results or "creations" of physical forces that acted throughout time. Despite his insistent denial of what we now regard as the incontrovertible fact that species arise from other species, Lyell played a key role in the development of biogeography largely through his treatise entitled *Principles of Geology*. In this massive work he discussed at great length not only the geological dynamics and antiquity of the Earth but also the geography of terrestrial plants and animals and the distributions of marine algae.

Four British Scientists

Perhaps most prominent among nineteenth-century naturalists were four British scientists: Charles Darwin, Joseph Dalton Hooker, Philip Lutley Sclater, and Alfred Russel Wallace (Figure 2.3). Although many other naturalists produced important works during this period, these four British scientists are responsible for major advances in both biogeography and evolutionary biol-

(A)

(B)

(C)

(D)

FIGURE 2.3 Four British scientists who, in the mid-nineteenth century, revolutionized our understanding of the history of the Earth and the distributions of its organisms: (A) Charles Darwin, (B) Joseph Dalton Hooker, (C) Philip Lutley Sclater, and (D) Alfred Russel Wallace. (A from Darwin 1859; B from Turill 1953; C from Goode 1896; D from McKinney 1972.)

ogy. They all studied the works of Linnaeus, Buffon, Forster, Candolle, and Lyell. They shared similar experiences as naturalists and explorers—traveling to distant archipelagoes, high mountains, and tropical and temperate regions of the New and Old Worlds. They also shared a common goal: to account for the diversity of life including the origin, spread, and diversification of biotas. Therefore, in retrospect, it is not surprising that they featured so prominently in the development of biogeography and evolutionary biology or that they developed great mutual respect and lasting friendships. The correspondence between these four scientists is in many cases as informative and insightful as their formal papers. In their letters to one another they reveal their shared preoccupation with what we now call biodiversity and their conviction that the key to understanding the natural world was to study patterns of distribution. In a letter to Hooker in 1845 Darwin referred to the study of geographic distribution as "that grand subject, that almost keystone of the laws of creation."

With a copy of the first volume of Lyell's *Principles of Geology* in hand, young Charles Darwin set sail in 1831 on a five-year surveying voyage aboard *H.M.S. Beagle* on which he served as a naturalist and gentleman companion for its captain, Robert Fitzroy (Figure 2.4). His travels would take him around the world where he would visit islands of the Atlantic, Pacific, and Indian

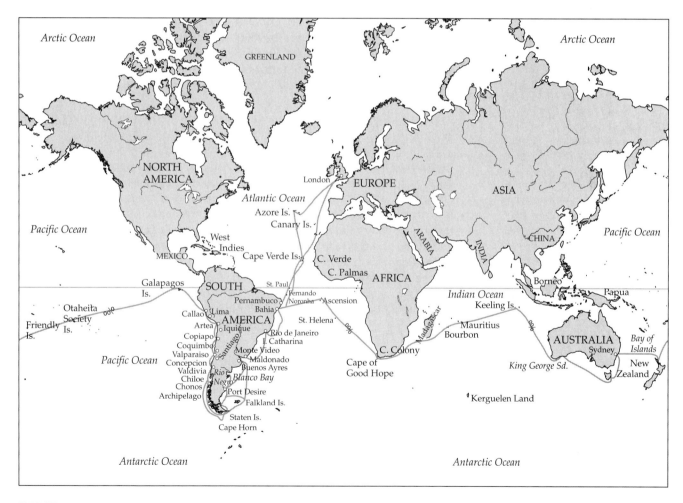

FIGURE 2.4 The route of the *H.M.S. Beagle* (1831–1836). Darwin's voyage on the *H.M.S. Beagle* was instrumental in the development of his theory of natural selection and the origin of species.

Oceans and explore the tropical, pampas, and Andes regions of South America. Darwin studied geology, native plants and animals, indigenous peoples, and domesticated animals in an attempt to understand the order of life. From his diary and collections of specimens, he later published a fascinating account (1839) of his adventures and observations during the voyage of the *H.M.S. Beagle*. Darwin was intrigued and perplexed by the patterns he observed: the fossils of extinct mammals in Argentina, the presence of seashells at high elevations in the Andes, and the occurrence of unique forms of life on islands. The patterns of variability in the Galápagos Archipelago, where different forms of tortoises and finches inhabited different islands, suggested to him the idea that geographic isolation facilitates inherited changes within and between populations. On his return to England, Darwin developed his theory of evolution, invoking natural selection as the primary mechanism by which new forms of life arose and are still arising today. This theory ranks as one of the most important scientific advances of all time and is woven into all aspects of biogeography.

The writing and eventual publication of Darwin's theory of evolution by natural selection is itself an interesting story that has been the subject of much review. Darwin drafted a manuscript on the subject in 1845, but withheld the idea from print for 15 years while he continued to amass evidence to support his theory. He was finally forced to publish when he learned that another brilliant scientist, Alfred Russel Wallace, had written a manuscript that expounded the identical theory—developed independently—but based on similar observations of the natural world. A paper by Darwin and one by Wallace were read together before the Linnaean Society of London in 1858; the following year, Darwin's great book, *The Origin of Species,* was published and became an immediate best seller.

It is difficult to overemphasize Darwin's contribution to the field of biogeography. He, along with Wallace, provided the basis for understanding changes in the adaptations and distributions of organisms across time as well as space. He proposed that the diversification and adaptation of biotas resulted from natural selection, while the spread and eventual isolation and disjunction of biotas resulted from long-distance dispersal. Darwin argued this latter point perhaps more passionately and more convincingly than anyone in the history of biogeography. His arguments were drawn not only from inferences based on the distributions of isolated biotas but also on ingenious "experiments" on dispersal and colonization. Through these studies Darwin was able to show that seemingly unlikely events, such as dispersal of seeds embedded in mud clinging to the feet of birds, were the most likely means by which land plants had colonized oceanic islands.

Darwin's arguments on dispersal threatened to overturn the long-held static view of biogeography. In the mid-nineteenth century this older view was championed by Louis Agassiz (1807–1873), a Swiss-born paleontologist and systematist who trained most of North America's leading zoologists and geologists (Dexter 1978). In two papers—one published in 1848 and the other in 1855—Agassiz argued that not only were species immutable and static, but so were their distributions, with each remaining at or near its site of creation. However, as a result of Darwin's arguments, which were later bolstered by those of Asa Gray and Alfred Wallace, the static view was abandoned by most biogeographers of the nineteenth century (Fichman 1977).

But the battles to be waged by the dispersalist camp had just begun. They were soon challenged by much more formidable adversaries—the "extensionists," whose ranks were no less prestigious than those of the dispersalists and included such respected scientists as Charles Lyell, Edward Forbes, and

Joseph Hooker. Both camps agreed that distributions were dynamic in time and space. The extensionists, however, argued that long-distance dispersal across great and persistent barriers was too unlikely to explain distributional dynamics and related phenomena such as cosmopolitan species and disjunct distributions. Rather, they proposed that species had spread across great, but now submerged, landbridges and ancient continents (Figure 2.5). Despite Darwin's great respect for Lyell, nothing vexed him more than those extensionists who created landbridges "as easy as a cook does pancakes." In a letter to Lyell

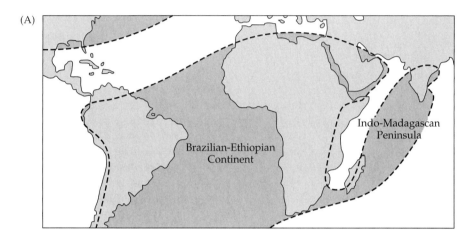

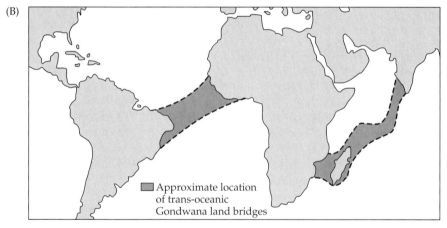

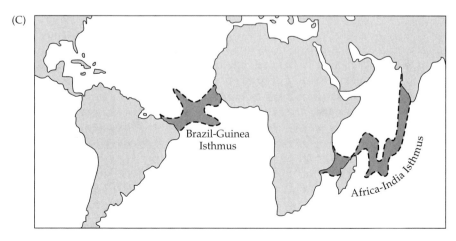

FIGURE 2.5 Hypothetical landbridges proposed by extensionists during the late nineteenth and early twentieth centuries to account for major disjunctions in the distributions of terrestrial organisms. (After Hallam 1967.)

in 1856, Darwin complained of "the geological strides, which many of your disciples are taking. … If you do not stop this, if there be a lower region of punishment of geologists, I believe, my great master, you will go there."

Despite Darwin's persistent and passionate arguments, the extensionist camp would remain a viable and influential force throughout the latter decades of the nineteenth century. One of its greatest proponents, Joseph Dalton Hooker (see Figure 2.3B), was also a good friend and admirer of both Darwin and Wallace. He was a remarkably ambitious plant collector (Turrill 1953). At the age of 22 he became the assistant surgeon and botanist on an expedition to the Southern Hemisphere led by Sir James Clark Ross. During this expedition Hooker studied the floras of many southern landmasses including Australia, Tasmania, Tierra del Fuego, and many archipelagoes in temperate and subantarctic waters. Hooker also traveled to Africa, Syria, India, and North America where he studied the flora of the Rocky Mountain region with Asa Gray. In addition to his own collections, Hooker studied those of other botanists, especially those from the Galápagos Archipelago and Arctic and Antarctic regions. These collections became the raw material for his analyses of the affinities and probable origins of the flora of each of these regions, group by group.

Upon his return to England in 1843, Hooker formed a friendship with Charles Darwin whom Hooker had admired from his account of the voyage of the *H.M.S. Beagle*. Within a year of their meeting, Hooker had read Darwin's manuscript on the theory of natural selection—the only person to see the manuscript before Asa Gray in 1857. Hooker shared with Darwin his ideas on the geographic distribution of plants and was one of the few to encourage Darwin to work on, and later publish, *The Origin of Species*. In Darwin's original introduction to the book, Hooker was the only person singled out for acknowledgment. Hooker also influenced Wallace, who dedicated *Island Life* (1880) to him.

While admitting that long-distance dispersal across open oceans might account for the occurrence on remote islands of plants with easily dispersed seeds and fruits, Hooker contended that, for the most part, long-distance dispersal was an insufficient explanation for the distributions of species. He argued that the biogeographic patterns and peculiarities of the southern floras were not consistent with Darwin's dispersalist hypothesis. Rather, Hooker (1867) believed that the floristic evidence supported "the hypothesis that of all being members of a once more continuous extensive flora, … that once spread over a larger and more continuous tract of land, … which has been broken up by geological and climatic causes."

Hooker was correct about the affinities of Southern Hemisphere plants, although he was mistaken in his explanation. We now know that the ancestors of these plants occurred on a giant southern continent, Gondwanaland, which broke up and whose fragments began to drift apart about 180 million years ago (see Chapter 8). But the nineteenth-century geologists and biogeographers—from Lyell and Hooker to Darwin and Wallace—believed that the relative sizes and positions of the continents and oceans had changed little over geological time. So Hooker hypothesized the emergence and submergence of ancient and undiscovered continents and landbridges to account for the disjunct distributions of closely related plants. Geological evidence for these ancient landbridges never materialized. By the end of the nineteenth century most geologists and biogeographers (including Lyell) had abandoned the extensionist doctrine.

However, we must not discount Hooker's contributions. He is still regarded as the founder of causal historical biogeography. He developed and applied many of the principles of what we now call "vicariance biogeography." In addition to emphasizing the importance of a dynamic Earth and changes in climate, Hooker stressed the importance of studying insular biotas to gain insights into biogeographic processes. He confirmed the earlier observations of Forster and others that insular floras tend to be more depauperate than those on the mainland, and he noted that as island isolation increases, the number of plant species decreases, while the distinctness of the flora increases. He also observed that the most diverse floras tend to be those in lands with the greatest diversity of temperature, light, and other environmental conditions. Hooker drew an analogy between the floras of oceanic islands and those of high mountains, suggesting that both are influenced by the same processes. As we shall see, these observations and principles would be echoed by biogeographers of the twentieth century.

During most of its early history, biogeography was dominated by the contributions of botanists—from Linnaeus and Candolle to Forster and Hooker. It is no great mystery why zoogeography lagged behind phytogeography. There are many more animal species than plant species, and most animals are relatively small and difficult to collect and identify. Over 70 percent of all known species are animals, and most of these are insects. The search for general zoogeographic patterns had to await a better understanding of animal diversity and distributions. By the mid-nineteenth century, however, zoologists had made great strides toward these goals. In his important work on zoology published in 1866, Ernst Haeckel not only introduced the concept of ecology but called for the recognition of biogeography (which he termed "chorology") as a new discipline, one that must incorporate the theory of natural selection.

In addition to Darwin and Hooker, two other British zoologists made major contributions to biogeography: Philip Lutley Sclater and Alfred Russel Wallace. Sclater (see Figure 2.3C), a good friend of Darwin's, was an eminent ornithologist who first described 1067 species, 135 genera, and 2 families of birds. His ambitious career included service as chief executive officer of the Zoological Society of London for over 30 years; he was also a life member of the Royal Geographic Society of London, as well as the first editor of the ornithological journal *Ibis*. Of the hundreds of articles he wrote, only a handful focused on zoogeographic patterns, but they had a major impact on zoogeography and on biogeography, in general.

In 1857 Sclater read a paper before the Linnaean Society of London entitled "On the general distribution of the members of the class Aves." He proposed a scheme that divided the Earth into biogeographic regions that would reflect "the most natural primary ontological divisions of the Earth's surface." As Sclater acknowledged, many others had developed biogeographic schemes, but these earlier systems had given too little regard to the flora and fauna and too much to arbitrary boundaries such as latitude and longitude. Sclater wished to do much more than just categorize each landmass with a list of species. He wanted to develop a system that would reflect the origination and development of distinctive biotas. He assumed that there were many areas of creation and that most species must have been created in the geographic regions where they now occurred. Therefore, the ontological divisions of the globe could be identified by analyzing the similarity and dissimilarity of biotas. Sclater based his scheme on the group he knew best—birds. He acknowledged, however, that given their impressive dispersal abilities birds might not be the best group for such an exercise. Accordingly, he limited his work to passerines, which he believed had higher site fidelity than other birds. The system of biogeographic regions Sclater developed is illustrated in Figure 2.6. While coarse, it formed the basis for the system of six biogeographic regions we continue to use today.

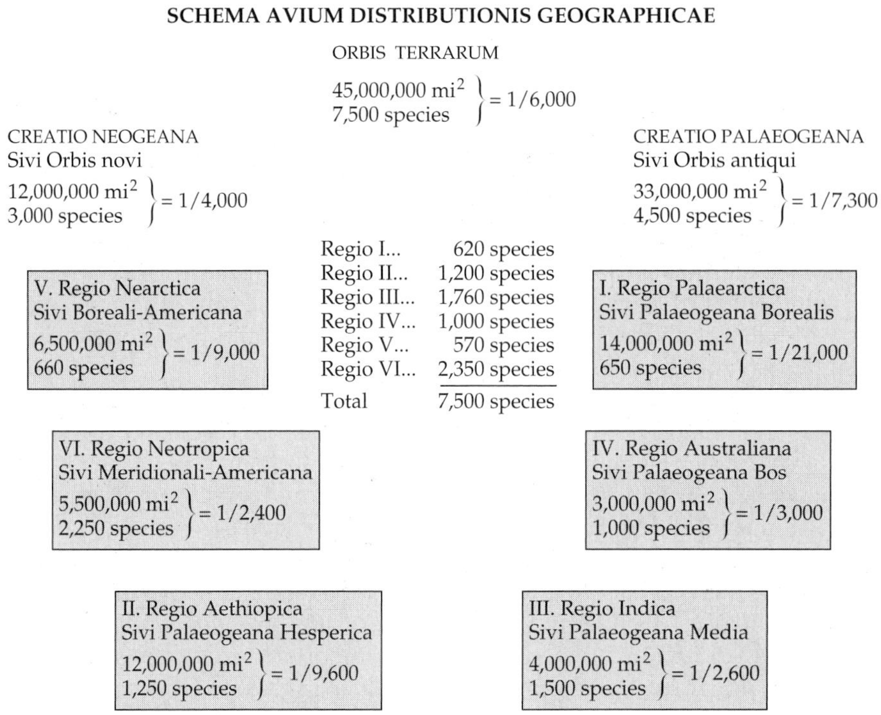

FIGURE 2.6 Sclater's (1858) scheme of terrestrial biogeographic regions based on the distributions of passerine birds.

SCHEMA AVIUM DISTRIBUTIONIS GEOGRAPHICAE

ORBIS TERRARUM

$$\left. \begin{array}{l} 45,000,000 \text{ mi}^2 \\ 7,500 \text{ species} \end{array} \right\} = 1/6,000$$

CREATIO NEOGEANA
Sivi Orbis novi

$$\left. \begin{array}{l} 12,000,000 \text{ mi}^2 \\ 3,000 \text{ species} \end{array} \right\} = 1/4,000$$

CREATIO PALAEOGEANA
Sivi Orbis antiqui

$$\left. \begin{array}{l} 33,000,000 \text{ mi}^2 \\ 4,500 \text{ species} \end{array} \right\} = 1/7,300$$

Regio I...	620 species
Regio II...	1,200 species
Regio III...	1,760 species
Regio IV...	1,000 species
Regio V...	570 species
Regio VI...	2,350 species
Total	7,500 species

V. Regio Nearctica
Sivi Boreali-Americana

$$\left. \begin{array}{l} 6,500,000 \text{ mi}^2 \\ 660 \text{ species} \end{array} \right\} = 1/9,000$$

I. Regio Palaearctica
Sivi Palaeogeana Borealis

$$\left. \begin{array}{l} 14,000,000 \text{ mi}^2 \\ 650 \text{ species} \end{array} \right\} = 1/21,000$$

VI. Regio Neotropica
Sivi Meridionali-Americana

$$\left. \begin{array}{l} 5,500,000 \text{ mi}^2 \\ 2,250 \text{ species} \end{array} \right\} = 1/2,400$$

IV. Regio Australiana
Sivi Palaeogeana Bos

$$\left. \begin{array}{l} 3,000,000 \text{ mi}^2 \\ 1,000 \text{ species} \end{array} \right\} = 1/3,000$$

II. Regio Aethiopica
Sivi Palaeogeana Hesperica

$$\left. \begin{array}{l} 12,000,000 \text{ mi}^2 \\ 1,250 \text{ species} \end{array} \right\} = 1/9,600$$

III. Regio Indica
Sivi Palaeogeana Media

$$\left. \begin{array}{l} 4,000,000 \text{ mi}^2 \\ 1,500 \text{ species} \end{array} \right\} = 1/2,600$$

It is hard to imagine how Sclater, and especially Darwin, could be overshadowed by any other zoogeographer, but they were indeed. While Sclater and Darwin spent much of their energies on other interests, biogeography was Alfred Russel Wallace's life work (see Figure 2.3D). Wallace, who is considered the father of zoogeography, developed many of the basic concepts and tenets of the field, combining the insights of others with his own experiences and his theory of evolution through natural selection. While Darwin and Sclater wrote only a few papers or chapters on biogeography, Wallace amassed his ideas and accounts in three seminal books: *The Malay Archipelago* (1869—dedicated to Darwin), *The Geographical Distribution of Animals* (1876), and *Island Life* (1880—dedicated to Hooker). We have summarized Wallace's contributions to the field in Box 2.1. Many of the concepts enunciated by Wallace were actually introduced by earlier scientists, but Wallace restated, documented, and interpreted them in an evolutionary and biogeographic context. As you read this book, you may wish to refer periodically to this box and note how many of Wallace's ideas are still being investigated by contemporary biogeographers (see also endpapers).

Wallace collected innumerable specimens in the East Indies and made many natural history observations on the biota of that region. He was the first person to analyze faunal regions based on the distributions of multiple groups of terrestrial animals. Wallace's analysis supported—but greatly expanded upon—Sclater's 1858 scheme. His system was based not just on birds, but on vertebrates in general, including nonflying mammals, which because of their limited dispersal abilities should more precisely reflect the natural divisions of the Earth. Wallace developed a detailed and very precise map of the Earth's biogeographic regions. His map includes sharp divisions between regions as well as subregions, along with bathymetric divisions reflecting the isolation of different archipelagoes. A distinctive original contribution was his observation of a sharp faunal gap between the islands of Bali and Lombok in the East Indies, where many species of Southeast Asia reach their distributional limit and are

BOX 2.1 *Biogeographic principles advocated by Alfred Russel Wallace*

■■■ These conclusions are summarized from Wallace's writings and have been verified many times by researchers in the twentieth century.

1. Distance by itself does not determine the degree of biogeographic affinity between two regions; widely separated areas may share many similar taxa at the generic or familial level whereas those very close may show marked differences—even anomalous patterns.

2. Climate has a strong effect on the taxonomic similarity between two regions, but the relationship is not always linear.

3. Prerequisites for determining biogeographic patterns are detailed knowledge of all distributions of organisms throughout the world, a true and natural classification of organisms, acceptance of the theory of evolution, detailed knowledge of extinct forms, and sufficient knowledge of the ocean floor and stratigraphy to reconstruct past geological connections between landmasses.

4. The fossil record is positive evidence for past migrations of organisms.

5. The present biota of an area is strongly influenced by the last series of geological and climatic events; paleoclimatic studies are very important for analyzing extant distribution patterns.

6. Competition, predation, and other biotic factors play determining roles in the distribution, dispersal, and extinction of animals and plants.

7. Discontinuous ranges may come about through extinction in intermediate areas or through the patchiness of habitats.

8. Speciation may occur through geographic isolation of populations that subsequently become adapted to local climate and habitat.

9. Disjunctions of genera show greater antiquity than those of single species, and so forth for higher taxonomic categories.

10. Long-distance dispersal is not only possible, but is also the probable means of colonization of distant islands across ocean barriers; some taxa have a greater capacity to cross such barriers than others.

11. The distributions of organisms not adapted for long-distance dispersal are good evidence of past land connections.

12. In the absence of predation and competition, organisms on isolated landmasses may survive and diversify.

13. When two large landmasses are reunited after a long period of separation, extinctions may occur because many organisms will encounter new competitors.

14. The processes acting today may not be at the same intensity as in the past.

15. The islands of the world can be classified into three major biogeographic categories: continental islands recently set off from the mainland, continental islands that were separated from the mainland in relatively ancient times, and distant oceanic islands of volcanic and coralline origin. The biotas of each island type are intimately related to the island's origin.

16. Studies of island biotas are important because the relationships among distribution, speciation, and adaptation are easier to see and comprehend on islands.

17. To analyze the biota of any particular region one must determine the distributions of its organisms beyond that region as well as the distributions of their closest relatives. ■■■

---

replaced by forms from Australasia (Wallace 1860). This break has been called Wallace's line (see endpapers; Mayr 1944a; Carlquist 1965).

Other Contributions of the Nineteenth Century

Other scientists of the nineteenth century were also looking for and interpreting important patterns in distributional data. Rather than considering only names and numbers of species, some of these pioneering biogeographers began to analyze geographic variation in the characteristics of individuals and populations (see Table 2.1, Theme 4). Chief among these early contributions were the generalized morphogeographic rules of C. Bergmann (1847), J. A. Allen (1878), and D. S. Jordan (Chapter 15). **Bergmann's rule** states that, in endothermic (warm blooded) vertebrates, races from cooler climates tend to have larger body sizes and hence to have smaller surface-to-volume ratios than races of the same species living in warmer climates. The original explanation for this pattern was that a lower surface area per volume ratio helps to conserve body heat in cold environments and, conversely, small size and a relatively large surface area facilitate the dissipation of heat in hot regions. Along this same line of reasoning **Allen's rule** states that, among endothermic species, limbs and other extremities are shorter and more compact in individuals living in colder climates: birds and mammals of polar regions tend to be stout with short limbs. A similar phenomenon was reported by D. S. Jordan in 1891 for ectothermic (cold blooded) teleost fishes inhabiting marine environments. According to **Jordan's law of vertebrae**, as one moves farther from the equator the vertebrae of teleost fishes become smaller and more numerous.

Despite the allure of such simple patterns, the generality and causality of these "rules" have been questioned by many biogeographers. Notwithstanding these problems, these early morpho- or ecogeographic rules were pioneering efforts in the study of geographic variation and adaptation. They stimulated the development of the field of physiological ecology and led to important observations on allometry—that is, how traits scale with body size.

In addition to studying geographic variation in the traits of individuals, biogeographers noted that the demographic characteristics of populations also varied across regions. In 1859 Darwin observed that within most genera, species "which range widely over the world are the most diffused in their own country, and are the most numerous in individuals"—in other words, wide-ranging species tend to occur at relatively high densities. This pattern, while not a major emphasis of early biogeography, was rediscovered and documented for a variety of taxa during the twentieth century. The study of how population-level parameters vary along geographic dimensions is now a central focus of an emerging discipline in biogeography termed **macroecology** (see Chapter 15).

Some early evolutionary "rules" were described by paleontologists who were searching for patterns in the history of life and trying to interpret the fossil record. The theory of **orthogenesis** is an example. This theory states that the evolution of a group continues in only one direction and that this orientation is an intrinsic property of the organism and is not controlled by natural selection. Of course, this theory of evolutionary inertia was used to oppose Darwin's theory of natural selection as the agent of evolutionary change. While few, if any, modern evolutionary biologists believe in orthogenetic trends, evolution is nonetheless conservative and is constrained by the phylogenetic history and preexisting characteristics of lineages (see Chapter 7).

A special type of orthogenesis was **Cope's rule**, which states that the evolution of a group shows a trend toward increased body size. Although there are many exceptions to this rule, it does seem that certain advantages of large size have resulted in repeated increases in size in many animal lineages. Large body size, however, seems to make species susceptible to extinction, so that large forms (such as the dinosaurs and many now extinct groups of giant birds and mammals) die out and are replaced by large representatives of new groups, which in turn evolve to a larger size (see Stanley 1973).

Simultaneously with these advances in zoogeography, phytogeographers continued to make important contributions to the development of biogeography. In 1860 E. W. Hilgard demonstrated that climatic factors and plants are directly responsible for converting parent rock into different kinds of soils varying in pH, mineral composition, texture, and so forth. (These interrelationships are discussed in more detail in Chapters 3 and 4.) Shortly afterward, a Russian scientist named V. V. Dokuchaev recognized that each soil has a characteristic structure. These two contributions led scientists to understand that the soils of a region are governed in large part by climatic patterns, which influence the breakdown of parent materials, the growth of plants, the decomposition of organic materials, and ultimately the kinds of plants and even animals that occur there.

In the late 1800s plant researchers in Europe began to develop novel classifications in which plant taxa were grouped according to their external architectural designs or their tolerance of abiotic stresses such as shortages of water or excesses of salts. By the early twentieth century, contributions by the Danish workers O. Drude (1887) and E. Warming (1898) quickly led to the now famous classification of life forms by C. Raunkiaer (1934), who defined major types of plants based on the positions of their perennating tissues. An ecological rather than a taxonomic approach was also adopted by the great German phytogeographer, A. F. W. Schimper (1898, 1903), who summarized in elaborate detail the forms and habits of plants from around the world. These works later engendered two vital areas of biogeography and ecology: plant physiological ecology, which seeks to understand how various species are adapted to the habitats in which they are found, and phytosociology, a subdiscipline of plant community ecology that attempts to explain why certain combinations of plant species, but not others, co-occur in a given habitat (see Good 1974).

Early botanists such as Candolle and Humboldt were well aware that different types of vegetation occurred at different elevations, but a zoologist, C. Hart Merriam (1894), provided one of the most valuable insights into these broad patterns. Based on his extensive field studies in southwestern North America, Merriam confirmed that elevational changes in vegetation type and plant species composition are generally equivalent to the latitudinal vegetational changes found as one moves toward the poles (Figure 2.7). He called

(A)

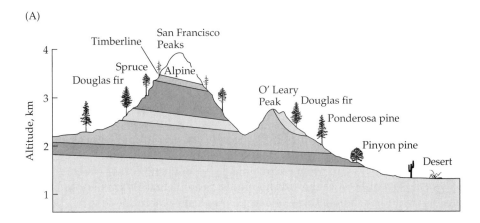

(B)

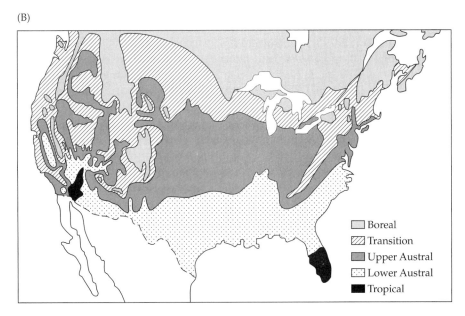

FIGURE 2.7 C. H. Merriam's "life zones," which were based on the relationship between climate and vegetation. (A) Elevational distribution of life zones on the San Francisco Peaks of Arizona as viewed from the southeast. (B) Latitudinal distribution of life zones in North America. (A after Bailey 1996; B after Merriam 1890, 1894.)

these belts of similar vegetation "life zones." Although Merriam was not successful in generalizing his concept of life zones to animals and to other regions of North America, he correctly concluded that elevational zonation of vegetation, like latitudinal zonation, is a response of species and communities to environmental gradients of temperature and rainfall.

Nineteenth century biogeographers continued to develop a much more comprehensive understanding of the dynamics of Earth's climates and how this in turn influenced the dynamics of its biotas. Jean Louis Rodolphe Agassiz studied under Georges Cuvier before emigrating to America. Although, as we noted above, Agassiz refused to accept Darwin's theories on the mutability of species, he developed some of the most seminal insights on the dynamics of Earth's climates and its biotas. While visiting the Alps on a return trip to his homeland in 1836, Agassiz studied the dynamics and ecological impacts of alpine glaciers. From these and other observations on distributions of fossils and extant species across the Northern Hemisphere, Agassiz (1840) developed the first comprehensive theory of the "Ice Age" and its influences on Holarctic biogeography. For these contributions, Agassiz is often recognized as the "father of glaciology," and his theory indeed proved fundamental to understanding biogeographic dynamics of plants and animals throughout the Pleistocene (see Chapter 9).

While Agassiz, Cuvier, and other paleontologists developed a more comprehensive understanding of the geography of nature (past and present) by deciphering the fossil record, other nineteenth century naturalists/explorers began to turn their attention to a new frontier: the oceans and their biotas. As mentioned earlier, in his *Principles of Geology* Charles Lyell (1830) discussed the distributions of marine algae. Edward Forbes (1815–1854) produced the first comprehensive work on marine biogeography (1856), in which he divided the marine world into nine horizontal (latitudinal) regions of similar fauna ("homozoic belts"), which he then subdivided into five zones of depth. Echoing the observations of Linnaeus and Forster on elevational and latitudinal variation of terrestrial communities, Forbes (1844) noted that "parallels in depth are equivalent to parallels in latitude." Similarly, Forbes observed that whereas distributions of terrestrial organisms are primarily influenced by climate, soil, and elevation, those of marine organisms are influenced by climate, water chemistry, and depth.

The American geologist, James Dwight Dana, expanded on some of Forbes's observations while also providing some fundamental insights on volcanism, mountain building, the origins of the continents and the seas, and distributions of corals and crustaceans. Dana showed that distributions of marine organisms could be explained and mapped using "isocrymes" (i.e., lines of equal minimum sea surface temperature). Near the end of the nineteenth century John Murray (1895), G. Pruvot (1896), and Arnold Ortmann (1896) published important general works on marine geography. In 1897 Philip Sclater (primarily known for his ornithological studies) published a paper on the distributions of marine mammals. Much like his earlier scheme for terrestrial zoogeographic regions (see Figure 2.6), Sclater divided the marine realm into six regions based on the distributions of geographically localized genera of marine mammals (Figure 2.8; Sclater 1897).

Despite these early contributions, just as zoogeography lagged behind phytogeography, marine biogeography would not come of age until more explorers focused on this final frontier. The present-day system of marine biogeographic "regions" was not generally accepted until the publication of John Briggs's work in the 1970s, over a century after the terrestrial system of biogeographic regions was established by Sclater and Wallace. While there remained

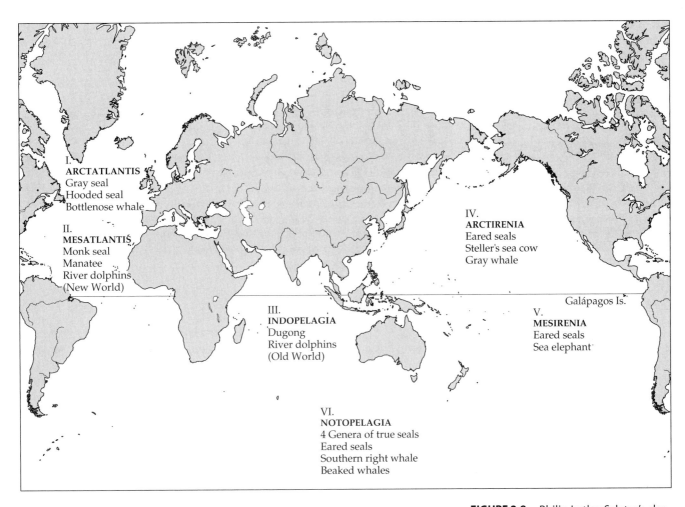

I.
ARCTATLANTIS
Gray seal
Hooded seal
Bottlenose whale

II.
MESATLANTIS
Monk seal
Manatee
River dolphins
(New World)

III.
INDOPELAGIA
Dugong
River dolphins
(Old World)

IV.
ARCTIRENIA
Eared seals
Steller's sea cow
Gray whale

V.
MESIRENIA
Eared seals
Sea elephant

Galápagos Is.

VI.
NOTOPELAGIA
4 Genera of true seals
Eared seals
Southern right whale
Beaked whales

FIGURE 2.8   Philip Lutley Sclater's classification of biogeographic regions in the marine realm. Like his scheme for terrestrial regions, this system included six regions and was based on the distributions of geographically localized genera, in this case marine mammals.

much to learn about the marine frontier, the insights garnered from explorations of the oceans in the twentieth century would not only expand the study of biogeographic patterns to marine life, but would challenge and eventually overturn the long-held view on the permanence of the oceans and continents.

The First Half of the Twentieth Century

From 1900 through the early 1960s, several major trends in research had extraordinary effects on biogeography. Paleontology in particular deserves credit for providing new and fascinating descriptions of faunal changes on each continent. Numerous paleontologists—especially C. Ameghino, W. D. Matthew, G. G. Simpson, E. H. Colbert, A. S. Romer, E. C. Olson, and B. Kurtén—described the origin, dispersal, radiation, and decline of land vertebrates. They showed that new groups increase in number of species, radiate to fill new ecological roles, expand their geographic ranges, and become dominant over and contribute to the extinction of older forms. Thus our present-day continental faunas have extremely long and complex histories that can be understood only by elucidating the **phylogenies** of the groups involved and the history of their movements between landmasses.

As mentioned earlier, explanations for how terrestrial organisms could have spread from one landmass to another were legion. Investigators proposed an incredible number of short-lived landbridges (or island archipelagoes), now vanished; former continents, now sunken; or once-joined continents, now drifted apart. During this period, tempers often flared as investigators debated their explanations for how groups arrived in such places as Australia, the Galápagos, and the Hawaiian Islands. Many such explanations are now rejected or considered unlikely, but these efforts infused phylogeny, paleoclimatology, and geology into biogeographic syntheses.

A common question arising out of such inquiries concerned "centers of origin." Where were the cradles of formation and diversification of various groups or biotas? Biogeographers of the early part of the twentieth century returned to the same challenges tackled by their predecessors, but with much expanded evidence and unparalleled intensity. A common goal of systematic monographs (treatments of the phylogenetic affinities and taxonomic classification of a group) was to propose a probable place of origin for the group and to describe its spread. The fossil record was spotty, so such reconstructions were often made from characteristics that included the geographic distributions of contemporary organisms, using arguments that now seem circular. Some authors were bolder and more dogmatic than others regarding the assumptions that could be used to deduce centers of origin from such limited data (see, for example, Matthew 1915; Willis 1922).

Early in the twentieth century, researchers began to investigate patterns of variation within single species. Following the lead of Bergmann and Allen on ecogeographic patterns, Joseph Grinnell, Lee R. Dice, and B. Rensch demonstrated close relationships between geographic and ecological properties of the environment and patterns of morphological variation within and among species. Subsequently, physiological and genetic variations were related to distributions in nature through the pioneering studies of Theodosius Dobzhansky on fruit flies (*Drosophila*) and J. Clausen, D. Keck, and W. Hiesey on plants. By the early 1940s, evolutionary biologists were building on Darwin's synthesis to investigate patterns of geographic variation and to infer the mechanisms responsible for the origin of new species. A long list of scientists contributed to our understanding of the modes of speciation. Ernst Mayr made major contributions in the fields of systematics, evolution, and historical biogeography. Out of this work arose one unifying theme: the biological species concept, which states that a species is definable as a group of populations that is reproductively isolated from all other such groups (Mayr 1942, 1963). Moreover, Mayr's studies of patterns in the geographic distributions of species and of the underlying evolutionary mechanism now known as allopatric speciation enabled an important new synthesis in evolutionary biology and biogeography (Mayr 1944a,b, 1965a,b, 1969, 1974).

By the middle of the twentieth century, many authors had produced new and more general syntheses of biogeographic patterns for various taxa. These included studies focusing on vertebrates by Phillip J. Darlington (1957) and George Gaylord Simpson (1968); marine zoogeography by Sven Ekman (1953) and J. W. Hedgpeth (1957); vascular plants by Evgenii V. Wulff (1943), Stanley A. Cain (1944), and R. Good (1947); and island biogeography by Sherwin Carlquist (1965). Moreover, an impressive body of literature began to accumulate on ecological biogeography, which was summarized for animals by Hesse, Allee, and Schmidt (1937, 1951), and Niethammer (1958); and for plants by Dansereau (1957). These works—and others too numerous to list here—established biogeography as a respected science that could provide insights for other fields including evolutionary biology, ecology, and conservation biology. In a similar manner biogeographers would continue to draw upon advances in these complementary fields as well as those in geology. This led to a revitalization that would continue throughout the remainder of the twentieth century (see Watts 1979).

Biogeography since the 1950s

Four major developments have revitalized biogeography in the last 40 years: the acceptance of plate tectonics, the development of new phylogenetic methods, new ways of conducting research in ecological biogeography, and investigations of the mechanisms that limit distributions.

Until the 1960s, most biogeographers considered the Earth's crust to be fixed and without lateral movement. A theory of plate tectonics and continental drift was first introduced by Antonio Snider-Pelligrini in 1858. His radical and largely unsubstantiated theory was readily dismissed by his contemporaries. It would take another 60 years before the theory resurfaced with the arguments of a German meteorologist, Alfred Lothar Wegener, and an American geologist, F. B. Taylor (Taylor 1910; Wegener 1912, 1915, 1966). Wegener's theory of continental drift, published between 1912 and 1956, was based on extensive geological and some biological evidence for great movements of the continents. But again, the theory was harshly criticized and rejected by most biogeographers, including distinguished leaders such as Simpson and Darlington. Darlington, for example, argued that it was much easier to move animals than it was to move entire continents.

The theory of continental drift was accepted only in the late 1960s when geological evidence for the process became irrefutable. Once accepted, however, this theory revolutionized historical biogeography and required scientists to rethink their explanations for many distributional patterns. Hooker had not been far off the mark—changes in the relative sizes and positions of landmasses and oceans had indeed resulted in important movements of biotas. The connections he proposed, however, rather than resulting from the emergence of mysterious continents and landbridges, were caused by great lateral movements of existing continents. Whatever one's interpretation of a distribution, the explanation eventually had to be consistent with this geological history of the Earth's surface. We trace the history and modern development of plate tectonics in greater detail in Chapter 8.

Since the 1960s, biologists also have made tremendous strides toward achieving phylogenetic classifications that trace the history and relationships of taxa, thus vastly improving our understanding of how biotas are and have been related. Guidelines for reconstructing phylogenies were already available for traditional systematics (Simpson 1961) and for phylogenetic systematics (or cladistics) (Hennig 1950), but the issues regarding the history of lineages were crystallized in 1966 when the work of Willi Hennig was published in English. Phylogenetic research was transformed from a discipline that compared anatomical and other similarities among taxa, into one in which the historical diversification of a lineage is reconstructed and the evolutionary relationships among species are quantified.

Early in the historical development of the field, biogeographers attempted to use information on geographic distributions of related species to reconstruct the histories of continents and other landmasses. In the mid-1800s, Asa Gray pioneered research on plant *disjunctions*—cases in which two closely related species are widely separated in space. Since then, biogeographers have been fascinated by disjunctions because they can reveal past land or water connections or long-distance dispersal between two regions. Interest in disjunctions as a means of evaluating past land connections has been renewed in the last 30 years, in part through the writings of L. Croizat (1958, 1960, 1964). With the availability of new phylogenetic approaches, the study of disjunct species—now called "*vicariants*"—has taken a central position, particularly in historical research. Because of this renewed interest in vicariants, some of the older phylogenetic and biogeographic reconstructions are being tested and sometimes greatly revised (see Chapters 10 and 11).

Up until the 1960s, the emphasis in biogeography had been primarily evolutionary and historical—emphasizing the phylogenies of groups and their means of dispersing into and surviving in different regions and habitats. By the late 1950s, G. E. Hutchinson began to focus attention on questions about the processes that determine the diversity of life and the number of species that coexist in local areas or habitats. Ecologists began to emphasize the importance of competition, predation, and mutualism in influencing the distributions of species and their coexistence as ecological communities. Of all the work in ecological biogeography, perhaps the most influential was Robert H. MacArthur and Edward O. Wilson's attempt to develop a radically new theory to account for the distributions of species on islands (1963, 1967; see Chapter 13). While Agassiz's static view of distributions had long been abandoned by most biogeographers, early biogeographers still assumed that species distributions within archipelagoes changed only very slowly on an evolutionary time scale. MacArthur and Wilson's equilibrium theory of island biogeography challenged this view and eventually became the new paradigm of the field. Their work changed the direction of ecological biogeography by focusing attention on a new set of questions. They asked abstract questions about patterns of distribution and species diversity, and they suggested that these patterns reflected the operation of two fundamental and opposing processes—immigration and extinction (see Chapters 13 and 14).

Questions about species diversity and coexistence, in addition to dominating the fields of ecology and theoretical biogeography, have spawned other areas of inquiry such as the extent to which the dispersal, establishment, and radiation of a lineage are *stochastic* (i.e., random) or *deterministic* (i.e., predictable if the underlying mechanisms are understood; Raup et al. 1973; Simberloff 1974b; Stanley 1979; Eldredge and Cracraft 1980). Moreover, these abstract questions have stimulated experimental testing of biogeographic concepts (e.g., Simberloff and Wilson 1969) as well as new mathematical ways of quantifying and analyzing observations (e.g., Pielou 1977a, 1979; Upton and Fingleton 1990; Cressie 1991; Manly 1991; Maurer 1994; Gotelli and Graves 1996).

Although some ecologists have emphasized the roles of interspecific interactions in influencing the distribution of species and communities (MacArthur and Connell 1966; MacArthur 1972; Whittaker 1975), others have emphasized the importance of the abiotic environment in limiting the distributions of individuals and populations and, in turn, determining the diversity of species in different regions. Important advances in techniques and instrumentation since the mid-1960s have permitted a flurry of research in this area, which has resulted in a tremendous amount of information that must be selectively integrated with biogeography. Advances in computer technology and related techniques including satellite imagery, geographic information systems (GIS), and spatial statistics (geostatistics) have enabled quantum leaps in our ability to explore and analyze biogeographic patterns from local to global scales.

In summary, many great scientists have contributed to the development of biogeography. All of them, from Linnaeus, Buffon, and Candolle to Simpson, Croizat, MacArthur, and Wilson, have shared a common goal—understanding the origin, spread, and diversification of biotas. In the mid-1700s Buffon gave biogeography its first principle, while others tested its generality and broad-

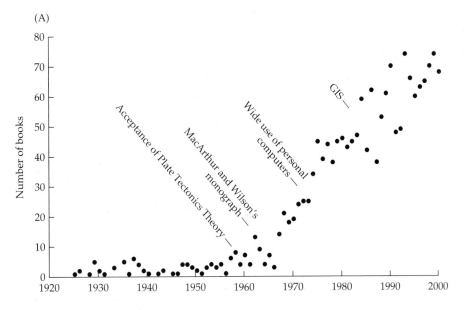

(A)

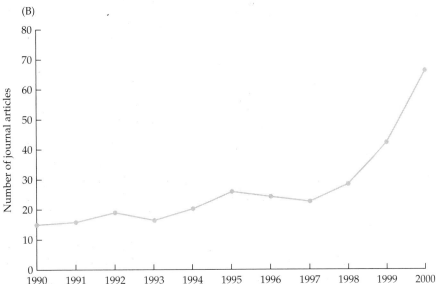

(B)

FIGURE 2.9 The maturity and vitality of biogeography is evidenced by the increasing number of publications in the field, especially over the last three decades, as tallied using OCLC–First Search (Online Computer Library Center, Inc.). (A) Number of books and monographs on "biogeography" found in the WorldCat database. (B) Number of articles on "biogeography" found in the Article 1st database.

ened the field to explore a diversity of other patterns and their causal forces. As the field progressed, theories of natural selection, continental drift, and immigration/extinction dynamics provided new mechanisms to explain long-standing patterns, while new comparative and experimental methods, phylogenetic analyses, and advances in computer science and statistics provided invaluable tools. By the 1960s, biogeography had finally come of age as a rigorous and respected science. It has continued to build on those gains and flourish in subsequent decades. Since 1900, the number of publications on biogeography has increased exponentially, with most of the increase taking place in the past three decades (Figure 2.9). There are at least four biogeography journals and, most recently, a new society—the International Biogeography Society (www.biogeography.org), whose mission is to foster the advancement of all studies of the geography of nature. In addition, the society has contributed to the development of two new books describing both the historical foundations and most recent advances in the field: *Foundations of Biogeography* and *Frontiers of Biogeography*.

The developmental history of biogeography, like that of the natural world itself, is a story of increasing diversification often accompanied by isolation and specialization of its descendant disciplines. The science of Linnaeus, Buffon, Darwin, and Wallace was a holistic, integrative one that included all phenomena that may have influenced the distribution of life. Scientists of the twentieth century became increasingly more specialized, focusing on particular disciplines and subdisciplines including taxonomy, evolutionary biology, and ecology, often restricting themselves to particular taxa and time periods. Fortunately, the frontiers of modern biogeography is now advancing by a reticulation of those descendant and divergent disciplines to develop a much more integrative and, we think, more insightful view of the geography of nature.

Given the long list of biogeography's conceptual achievements—in themselves the seeds of whole disciplines—one can easily comprehend how it has become impossible for one person to understand and follow completely all aspects of the field. Students of biogeography can either be frustrated by their inability to comprehend all the subtleties of this awesome body of knowledge, or they can be challenged and encouraged by the prospect of using biogeography as a focal point to synthesize many separate disciplines and to acquire a unique perspective on the development, diversity, and distribution of life on Earth.

UNIT TWO

The Environmental Setting and Basic Biogeographic Patterns

CHAPTER 3

Physical Setting
The Geographic Template

ORGANISMS CAN BE FOUND almost everywhere on Earth: from the cold, rocky peaks of high mountains to the hot, windswept sand dunes of lowland deserts; from the dark, near-freezing depths of the ocean floor to the steaming waters of hot springs. Some organisms even live around hydrothermal vents in the deep ocean, where temperatures exceed 100° C (but the water does not boil because of the extreme pressure). Yet no single kind of organism lives in all of these places. Each species has a restricted geographic range in which it encounters a limited range of environmental conditions. Polar bears and caribou are confined to the Arctic, whereas palms and corals are rare outside the tropics. There are a few species, such as *Homo sapiens* and the peregrine falcon, that we call cosmopolitan because they are distributed over all continents and over a wide range of latitudes, elevations, climates, and habitats. These species, however, are not only exceptional, but they are also much more limited in distribution than they appear at first glance. Humans and peregrines, for example, are absent from the three-fourths of the Earth that is covered with water; indeed, they are little more than rare visitors to large expanses of the terrestrial realm with extremely harsh environments.

The Geographic Template

As we point out in later chapters, we may need to invoke unique historical events or ecological interactions with other organisms to account for the limited geographic ranges of some species, but the most obvious patterns in the distributions of organisms occur in response to variations in the physical environment. In terrestrial habitats these patterns are largely determined by climate (primarily temperature and precipitation) and soil type. The distributions of aquatic organisms are limited largely by water temperature, salinity, light, and pressure.

As Edward Forbes and other early biogeographers observed (see Chapter 2), climate, soil type, water chemistry, and a long list of other environmental conditions vary in a highly non-random manner across geographic gradients of latitude, elevation, depth, and proximity to major landforms such as coastlines and mountain ranges. In addition, regardless of whether we consider the aquatic or terrestrial realms, the environmental variables tend to exhibit strong spatial autocorrelation, or what is sometimes referred to as **distance-decay.** Simply put, similarity of environmental conditions between sites decreases as we compare more distant sites.

Taken together, these non-random patterns of spatial variation in environmental conditions constitute a multifactor **geographic template**, which forms the foundations of all biogeographic patterns. That is, most biogeographic patterns ultimately derive from this very regular, spatial variation in environmental conditions. For example, as we move from low to high latitudes, from the equator to the poles, or from the ocean through estuaries and then upstream, environmental temperatures tend to cool—either directly or indirectly influencing biotic communities along each of these geographic and environmental gradients (see Chapter 15). Diversity, species composition, and vital processes (e.g., productivity and decomposition) of biotic communities along these gradients change in a highly predictable manner, with similarity among communities being much higher for those located in close proximity along the geographic template.

The conceptual model illustrated in Figure 3.1 may prove useful in understanding a great diversity of biogeographic patterns. Again, the fundamental layer or foundation for all patterns of variation across space is the geographic template. Organisms can then respond to this highly non-random spatial variation through adaptations (both behavioral and physiological), dispersal, or, over many generations, populations may respond through evolutionary changes or suffer local extinction. All of these responses, taken together, ultimately determine the geographic distributions and patterns of variation among populations, species, and communities across geographic gradients. To more fully account for the diversity of patterns biogeographers study, this conceptual model must include at least two additional layers of complexity—both of which may be viewed as feedback or interactions among system components. Not only are species distributions influenced by environmental condi-

FIGURE 3.1 All biogeographic patterns are ultimately influenced by the geographic template. Distributions and other patterns of geographic variation of life forms and communities result from (1) the highly non-random patterns of spatial variation in environmental characteristics across the Earth; (2) the responses of biotas (including adaptation, dispersal, evolution, or extinction) to this variation; (3) interactions among organisms (e.g., competitive exclusion and mutualistic interactions); (4) impacts of particular species—"ecosystem engineers"—such as beavers, prairie dogs, and humans on the geographic template; and (5) the temporal dynamics of the template (including the so-called TECO events—plate Tectonics, Eustatic changes in sea level, Climate change, and Orogeny [mountain building]).

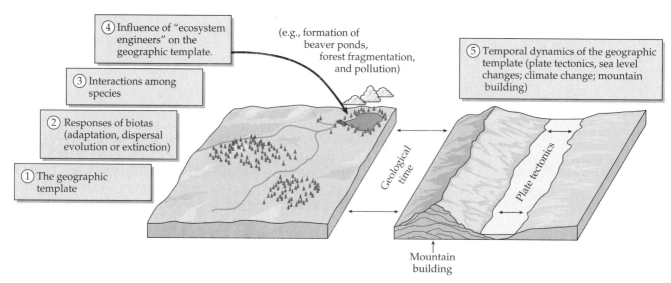

tions, but those distributions and related patterns are also influenced by inter-actions among the species themselves (e.g., mutualism, parasitism, and competitive exclusion—see Chapter 4). Furthermore, some species—often referred to as ecosystem engineers (e.g., beavers, prairie dogs, and humans)—can modify the geographic template itself. Even before our own species rose to become the world's dominant **ecosystem engineer,** microscopic and otherwise "primitive" life forms fundamentally altered Earth's atmosphere thereby increasing its oxygen content and its abilities to store heat, modifying its climate, and eventually molding and recasting the geographic template across both the terrestrial and aquatic realms. We still need to introduce one final but none-the-less fascinating layer of complexity to this conceptual model. The geographic template itself is dynamic—not just across space but across time as well. Over the 3.5 billion-year history of life on Earth, and indeed even over much shorter timescales, climates have swung dramatically from glacial to interglacial episodes of the Pleistocene (including most of the past 2 million years), and from the much earlier periods of the so-called snowball Earth to the global sauna of the mid-Eocene (roughly 50 million years ago; see Chapter 9). Just as fundamental and sometimes driving these major climatic shifts, Earth's crust itself is highly dynamic, emerging from the mantle below to drift, split, and collide in a kaleidoscopic jigsaw puzzle known as plate tectonics and continental drift (see Chapter 8).

All of these dynamics in space and time are themselves driven by two great engines, which are powered by two different sources of energy. The energy stored in the Earth's core at the time the solar system was formed is also supplemented by compressional heating due to gravity. A portion of this energy is gradually but continuously being dissipated through the Earth's mantle and crust and ultimately out into space. This transfer of heat energy moves and shapes the Earth's crust, shifting the positions of the crustal plates containing the continents, thrusting up mountains, and causing earthquakes and volcanic eruptions.

The other great engine is driven by the energy of the sun. Radiant energy emitted by the sun strikes the Earth's surface, where it is absorbed and converted into heat, warming the surface of the land and water and the atmosphere above them. The resulting differences in the temperature and density of air and water cause them to move over the Earth's surface, both horizontally and vertically, creating the Earth's major wind patterns and ocean currents. The heating of surface water also causes evaporation, and the resulting water vapor is carried by the air and re-deposited as rain or snow. These processes, which are responsible for the Earth's climate and for many physical characteristics of its oceans and fresh waters, are the subject of the present chapter.

Climate

Solar Energy and Temperature Regimes

SOLAR RADIATION AND LATITUDE. Sunlight sustains life on Earth. Solar energy not only warms the Earth's surface and makes it habitable, but it also is captured by green plants and converted into chemical forms of energy that power the growth, maintenance, and reproduction of most living things.

According to the principles of thermodynamics, heat is transferred from objects of higher temperature to those of lower temperature by one of three mechanisms: (1) conduction, a direct molecular transfer (especially through solid matter); (2) convection, the mass movement of liquid or gaseous matter; or (3) radiation, the passage of waves through space or matter. Heat flows as

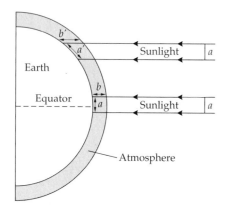

FIGURE 3.2 Average input of solar radiation to the Earth's surface as a function of latitude. Heating is most intense when the sun is directly overhead, so that incoming solar radiation strikes perpendicular to the Earth's surface. The higher latitudes are cooler than the tropics because the same quantity of solar radiation is dispersed over a greater surface area (*a'* as opposed to *a*) and passes through a thicker layer of filtering atmosphere (*b'* as opposed to *b*).

radiant energy from the hot sun across the intervening space to the cooler Earth. When incoming solar radiation strikes matter such as water or soil some of it is absorbed, and the matter is heated. Some solar radiation is initially absorbed by the air, particularly if it contains suspended particles of water or dust (e.g., clouds), but most passes through the sparse matter of the atmosphere and is absorbed by the denser matter of the Earth's surface. This surface is not heated uniformly. Soil, rocks, and plants absorb much of the radiation and may be heated intensely. Water also absorbs much solar radiation, but its heating effect is not confined to as narrow a surface layer as on land. Although air is heated to some extent by absorption of incoming solar radiation, most of the heating of air occurs at the Earth's surface where it is warmed by direct contact with warm land and water, by latent heat released by the condensation of water, and by long-wave infrared radiation emitted from the surfaces of warm objects such as leaves and bare soil.

The angle of incoming radiant energy relative to the Earth's surface affects the quantity of heat absorbed. The most intense heating occurs when the surface is perpendicular to incident solar radiation, for two reasons: (1) the greatest quantity of energy is delivered to the smallest surface area; and (2) a minimal amount of radiation is absorbed or reflected back into space during passage through the atmosphere because the distance it travels through air is minimized (Figure 3.2). This differential heating of surfaces at different angles to the sun explains why it is usually hotter at midday than at dawn or dusk, why average temperatures in the tropics are higher than at the poles, and why south-facing hillsides are warmer than north-facing ones in the Northern Hemisphere (and the reverse in the Southern Hemisphere).

Because the Earth is tilted 23.5° from vertical on its axis with respect to the sun, solar radiation falls perpendicularly on different parts of the Earth during an annual cycle. This differential heating produces the seasons. The seasons are also characterized by different lengths of day and night. Only at the equator are there exactly 12 hours of daylight and darkness every 24 hours throughout the year (Figure 3.3). At the spring and fall equinoxes (March 21 and September 22, respectively) the sun's rays fall perpendicularly on the equator, equatorial latitudes are heated most intensely, and every place on Earth experiences the same day length. At the summer solstice (June 22), sunlight falls directly on the Tropic of Cancer (23.5° N latitude) and the Northern Hemisphere is heated most intensely, experiencing longer days than nights and enjoying summer, while the Southern Hemisphere has winter. At the winter solstice (December 22), the sun shines directly on the Tropic of Capricorn (23.5° S latitude) and the Southern Hemisphere enjoys its summer while the Northern Hemisphere has winter, cold temperatures, and long nights. The seasonality of climate increases with increasing latitude. Therefore, the Tropics of Cancer and Capricorn mark the northern- and southern-most latitudes, respectively, that receive direct sunlight one day each year (i.e., on the solstices). At the Arctic and Antarctic Circles (66.5° latitude), there is one day each year of continuous daylight when the sun never sets and one day of continuous darkness—each at a solstice. Although every location on the Earth theoretically experiences the same amount of daylight and darkness over an annual cycle, the sun is never directly overhead at high latitudes; however, considerable solar radiation is absorbed during the long summer days. Temperatures in excess of 30° C are commonly recorded in Alaska. The warmest days typically are in July (after the summer solstice) because of the time lag required to heat up the Earth's surface.

THE COOLING EFFECT OF ELEVATION. The processes just described account for seasonal and latitudinal variation in temperature, but it remains to be explained why it gets colder as we ascend to higher altitudes. The fact that Mount Kilimanjaro, (nearly on the equator in tropical East Africa) is capped with permanent ice and snow seems to be in conflict with our intuitive expectation. Mountain peaks are nearer the sun, so why are they cooler than nearby lowlands? The answer lies in the thermal properties of air. As a climber moves up a mountain, the length (and the pressure) of the column of air that lies above the climber decreases. Thus, the density and pressure of air decrease with increasing elevation. When air is blown across the Earth's surface and forced upward over mountains, it expands in response to the reduced pressure. Expanding gases undergo what is called **adiabatic cooling**, a process where they lose heat energy as their molecules move farther apart (temperature essentially is a measure of activity and the frequency of collisions of molecules). The same process occurs in a refrigerator as Freon gas expands after leaving the compressor. The rate of adiabatic cooling of dry air is about 10° C per km elevation, as long as no condensation of water vapor and cloud formation occurs.

Higher elevations are also colder because the less dense air allows a higher rate of heat loss by radiation back through the atmosphere. Water vapor and carbon dioxide in the atmosphere retard such radiant heat exchange and produce the so-called **greenhouse effect**. These gases act like the glass in a greenhouse: they allow the short wavelengths of incoming solar radiation to pass through, but trap the longer wavelengths emitted by surfaces that have been warmed by the sun. The resulting warming effect is pronounced in moist lowland areas, where water vapor in the air retards cooling at night. In contrast, mountains and deserts typically experience extreme daily temperature fluctu-

FIGURE 3.3 Seasonal variation in day length with latitude is due to the inclination of the Earth on its axis. At the equinoxes, the sun is directly overhead at the equator and all parts of the Earth experience 12 hours of light and 12 hours of darkness each day. At the summer solstice in the Northern Hemisphere, however, the 23.5° angle of inclination causes the sun to be directly over the Tropic of Cancer, while the Arctic Circle and areas farther north experience 24 hours of continuous daylight; at the same time all regions in the Southern Hemisphere experience less than 12 hours of daylight per day, and the sun never rises south of the Antarctic Circle.

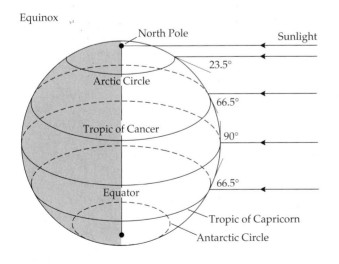

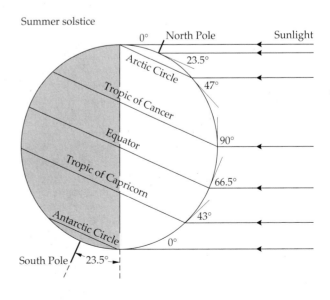

ations, because there is little water vapor in the air to prevent heat loss by radiation to the cold night sky.

Winds and Rainfall

WIND PATTERNS. Differential heating of the Earth's surface also causes the winds that circulate heat and moisture. As we have already seen, the most intense heating is at the equator, especially during the equinoxes when the sun is directly overhead. As this tropical air is heated, it expands, becomes less dense than the surrounding air, and rises. This rising air produces an area of reduced atmospheric pressure over the equator. Denser air from north and south of the equator flows into the area of reduced pressure, resulting in surface winds that blow toward the equator (Figure 3.4). Meanwhile, the rising equatorial air cools adiabatically, becomes denser, is pushed away from the equator by newly warmed rising air, and eventually descends again at about 30° N and S latitude (the *Horse Latitudes*). This vertical circulation of the atmosphere results in three convective cells (Hadley, Ferrel and Polar) in each hemisphere, with warm air ascending at the equator and at about 60° N and S latitude, and cool air descending at about 30° N and S and at the poles. These circulating air masses produce surface winds that typically blow toward the equator between 0° and 30° and toward the poles between 30° and 60°. In the upper atmosphere between the convective cells are the jet streams—high-speed winds blowing approximately parallel to the equator.

The surface winds do not blow exactly in a north-south direction; instead, they appear to be deflected toward the east or west by the **Coriolis effect**. Although the Coriolis effect is often called the *Coriolis force*, it is not a force but a straightforward consequence of the law of conservation of angular momentum. Every point on the Earth's surface makes one revolution every 24 hours. Because the circumference of the Earth is about 40,000 km, a point at the equator moves from west to east at a rate of about 1700 km h^{-1}. But the parallel

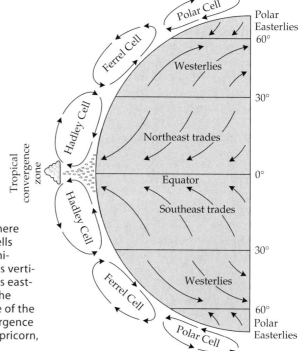

FIGURE 3.4 Relationship between vertical circulation of the atmosphere and wind patterns on the Earth's surface. There are three convective cells (Hadley, Ferrel and Polar) of ascending and descending air in each hemisphere. As the winds move across the Earth's surface in response to this vertical circulation, they are deflected by the Coriolis effect, which produces easterly trade winds in the tropics, and westerlies at temperate latitudes. The latitudinal locations of these cells shift with the seasons as the latitude of the most direct sunlight and most intense heating (i.e., the Tropical Convergence Zone) shifts between 23.5° N and 23.5° S (the Tropics of Cancer and Capricorn, respectively).

lines of latitude become increasingly shorter as we move from the equator to the poles. Therefore, points north or south of the equator travel a shorter distance each 24-hour rotation of the Earth; that is, they move at a slower rate than points closer to the equator. Consider what happens at the equator if you shoot a rocket straight upward. Where does it come down? Right where it was launched; the rocket travels not only up and down but also eastward at a rate of 1700 km h^{-1}, the same rate as the Earth moving beneath it. Now suppose the rocket is propelled northward away from the equator. It continues to travel eastward at 1700 km h^{-1}, but the Earth underneath it moves ever more slowly as the rocket travels farther north and consequently its path appears to be deflected toward the right. The Coriolis effect describes this tendency of moving objects to veer to the right in the Northern Hemisphere and to the left in the Southern Hemisphere. The winds approaching the equator from the horse latitudes appear to be deflected to the west and are therefore called northeast or southeast **trade winds.** (Winds are described based on the direction of their *sources.*) Winds blowing toward the poles between about 30° and 60° N and S latitude are called **westerlies** and are deflected to the east (see Figure 3.4). These winds naturally were very important to commerce in the days of sailing ships when both the westerlies and the trade winds (or *trades*) got their names. Ships coming to the New World from Europe traveled south to the Canary Islands and Azores in tropical latitudes to intercept the trades before heading westward, but they returned to Europe at higher latitudes with the westerlies behind them.

The surface winds, influenced by the Coriolis effect, initiate the major ocean currents. The trade winds push surface water westward at the equator, whereas the westerlies produce eastward-moving currents at higher latitudes. Responding to the Coriolis effect, these water masses are deflected toward the east or west, and the net result is that the ocean currents move in great circular gyres—clockwise in the Northern Hemisphere and counterclockwise in the Southern Hemisphere (Figure 3.5). Warm currents flow from the tropics along eastern continental margins; as these water masses reach high latitudes, they are cooled, producing cold currents down the western margins.

PRECIPITATION PATTERNS. By superimposing these patterns of temperature, winds, and ocean currents we can begin to understand the global distribution of rainfall. We will also need some additional background in physics. As air warms, it can absorb increasing amounts of water vapor evaporated from the land and water. As it cools, it eventually reaches the **dew point,** at which it is saturated with water vapor. Further cooling then results in condensation and the formation of clouds. When the particles of water or ice in clouds become too heavy to remain airborne, rain or snow falls. In the tropics, the cooling of ascending warm air laden with water vapor produces heavy rainfall at low and middle elevations, where rain forests and cloud forests occur. Rainy seasons in the tropics tend to fall when the sun is directly overhead and the most intense heating occurs. The tropical grasslands of Kenya and Tanzania in East Africa, which lie virtually on the equator but at higher elevations than rain forests, experience two rainy seasons each year corresponding approximately to the equinoxes (when the **Tropical Convergence Zone** shifts overhead; see Figure 3.4) and two dry seasons, which correspond to the solstices. In contrast, the area around the Tropic of Cancer in central Mexico has only one principal rainy season—in the summer. Most tropical regions have at least one dry season.

At the Horse Latitudes, where cool air descends from the upper atmosphere, two belts of relatively dry climate encircle the globe. Descending air

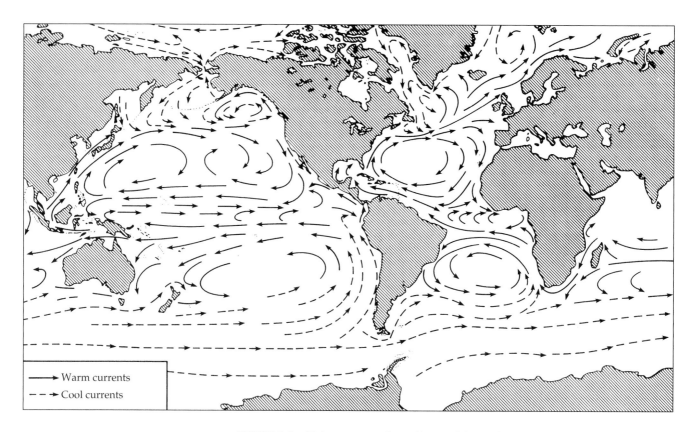

FIGURE 3.5   Main patterns of circulation of the surface currents of the oceans. In each ocean, water moves in great circular gyres, which move clockwise in the Northern Hemisphere and counterclockwise in the Southern Hemisphere. These patterns result in warm currents along the eastern coasts of continents and cold currents along the western coasts. Note the Pacific equatorial countercurrent: the small current along the equator that flows from west to east opposite to the gyres, and which strengthens in some years to cause the El Niño phenomenon.

warms and can therefore absorb more moisture, drying the land. In these belts lie most of the Earth's great deserts (including the Mojave, Sonoran, and Chihuahuan in southwestern North America; the Sahara in North Africa; and the Arid Zone in central Australia), and adjacent to these deserts are regions of semiarid climates and grassy or shrubby vegetation. Within these belts, the seasonality of climate is very marked on the western sides of continents, which experience **Mediterranean climates.** Parts of coastal California, Chile, the Mediterranean region in Europe, southwestern Australia, and southernmost Africa have dry, usually hot summers and mild, rainy winters (see Figure 1.2). In winter, when the land tends to be cooler than the ocean water, the westerly winds bring ashore moisture-laden air, condensation occurs, and fog and rain result. In summer, when the land is warmer than the ocean, the westerlies blowing inland from the cold offshore currents are warmed, able to hold more water vapor, and result in dry conditions on land. The effects of cold currents are even more pronounced in localized regions of western South America and southwestern Africa, where they contribute to the formation of coastal deserts that are the driest areas on Earth (Amiran and Wilson 1973).

Several of the deserts between 30° and 40° N and S latitude are located not only on the western sides of continents, but also on the eastern sides of major

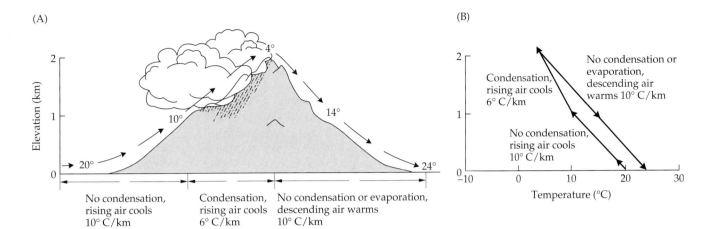

FIGURE 3.6 Factors causing rain shadow deserts. (A) Air blowing over a mountain cools as it rises, water vapor condenses, and the air loses much of its moisture as rain on the windward side, so that the leeward side experiences warm, dry winds. (B) The rate of change in air temperature with elevation is greater for drier air, resulting in warmer, drier conditions on the leeward side than at the same elevation on the windward side. (After Flohn 1969.)

mountain ranges. As westerly winds blow over the mountains, they are cooled until eventually the dew point is reached and clouds begin to form. Condensation releases heat—the latent heat of evaporation—so that wet air cools adiabatically at a slower rate than dry air: 6° C per km of elevation, as opposed to 10° C per km for dry air. As the air continues to rise and cool, most of its moisture falls as precipitation on the western side of the mountain range. When the air passes over the crest and begins to descend, the remaining clouds quickly evaporate, and the dry air warms at the higher rate. This **rain shadow** effect causes the warm, dry climates found on the leeward sides of temperate mountains (Figure 3.6). Thus, for example, the Sierra Nevada in California has lush, wet forests of giant sequoias and other conifers on its western slopes, but arid woodlands of piñons and junipers on its eastern (leeward) slopes; a bit farther east with an elevation below sea level lies Death Valley—the driest place on the North American continent. Similarly, the Monte Desert of South America is in the rain shadow on the eastern side of the Andes in Argentina.

These global patterns of temperature and precipitation frequently are summarized in climatic maps like that in Figure 3.7. Such maps are useful, but can be misleading because they fail to show the small-scale patterns of spatial and temporal variation that influence the abundance and distribution of organisms.

SMALL-SCALE SPATIAL AND TEMPORAL VARIATION. The same processes that we have just described on a global scale can also produce great climatic variation on a local scale. The effect of mountains is particularly great, as we can illustrate with several examples. From Tucson, Arizona it is only 25 km by a paved road to the top of Mount Lemmon (2800 m elevation) in the Santa Catalina Mountains. But the climate and the plants at the summit are far more similar to those in northern California and Oregon—1500 km to the north—than to those in the desert just below (Table 3.1). Similarly, the spruce-fir forests on the summit of the Great Smokey Mountains in Tennessee are more similar to the

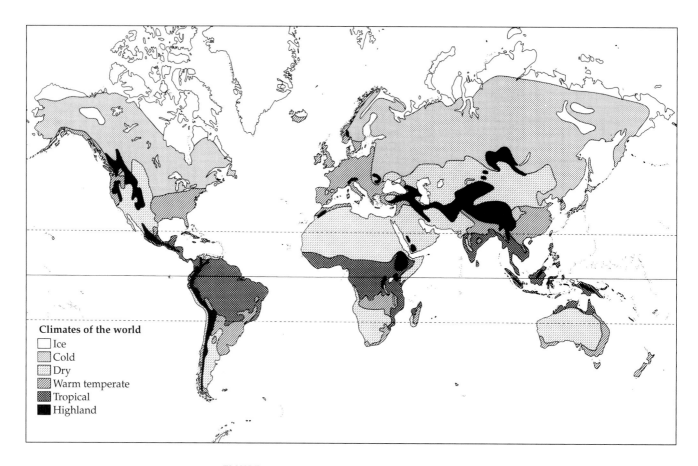

FIGURE 3.7 Major climatic regions of the world. Note that these regions occur in distinct patterns with respect to latitude and the positions of continents, oceans, and mountain ranges. (After Strahler 1973.)

TABLE 3.1 *The influence of elevation on climate*

Site	Elevation (m)	Temperature (°C)				Mean annual precipitation (cm)
		Mean January	Mean July	Lowest	Highest	
Tuscon, Arizona	745	10.8	30.7	−9.4	46.1	27.3
Mt. Lemmon, Arizona	2791	2.3	17.8	−21.7	32.8	70.0
Salem, Oregon	60	3.2	19.2	−24.4	40.0	104.3

Source: Data from U.S. Weather Bureau.

Note: Two of the sites are near one another in Arizona; the third site is in Oregon. Note that the climate of the high-elevation site in Arizona, Mt. Lemmon, is much more similar to that of Salem, Oregon, 1700 km to the north, than to that of Tucson, only 25 km away but 2000 m lower in elevation.

boreal forests of northern Canada than to the deciduous forests in the valleys below. Puerto Rico, which lies in the Caribbean Sea at 18° N latitude, is about 150 km long and 50 km wide, and has a central mountainous backbone rising to about 1000 m. The lowlands on the northern and eastern sides are lush and tropical, but much more rain falls at higher elevations on the northeastern slopes, and this is where the best-developed rain forests are found. So much moisture is lost as the northeast trade winds traverse the mountains that the southwestern corner of Puerto Rico is extremely dry; the cacti and shrubby vegetation that occur there remind a visitor of the deserts and tropical thorn forests of western Mexico (Figure 3.8). Even more dramatic are the combined effects of the cold Humboldt Current and the rain shadow cast by the westward flowing southeast trades over subtropical regions of the Andes in Peru and Chile. Here, up to 10 m of precipitation per year drenches the tropical rain forests on the eastern slope, while there may be several years in succession with no rain at all in the Atacama Desert on the western slope. Note that because they are located in regions where the prevailing winds come from different directions, the Atacama Desert (10°–15° S, with westward flowing southeast trades) and the Monte Desert (around 30° S, with eastward flowing westerlies) are located on opposite sides of the Andes.

There are also year-to-year and longer-term temporal variations in climate. The entire global system of moving air masses, ocean currents, and patterns of precipitation fluctuates on a five- to seven-year cycle. The fluctuations appear to be initiated by events in the vast tropical Pacific Ocean (although similar events occur in the tropical Atlantic). This pattern is called the **El Niño Southern Oscillation**, or **ENSO** for short. We are uncertain about its initial cause—perhaps a variation in the output of solar radiation or intrinsic fluctuations in the atmosphere-ocean system. But whatever the ultimate cause, the pattern of tropical ocean circulation changes. While the primary ocean currents are the great hemispheric gyres mentioned above, close inspection of Figure 3.5 will

(A)

(B)

FIGURE 3.8 Comparison of vegetation on opposite sides of the central mountain range on the tropical island of Puerto Rico. (A) On the northeastern side, which receives the moisture-laden trade winds, lush rain forests occur. (B) In marked contrast, the southwestern side lies in a rain shadow, has a hot and dry climate, and has cacti and other plants typical of desert regions. (A courtesy of E. Orians; B courtesy of A. Kodric-Brown.)

show a small current running west to east right along the equator. It is called the **equatorial countercurrent** because it runs in the opposite direction to the gyres. It is usually small, as the figure suggests, but in some years it becomes much stronger and pushes warm water away from the equator up the coasts of North and South America. As the westerly winds pass over this warm water they pick up moisture and carry it onto the adjacent continents causing heavy precipitation in the winter when the land is colder than the offshore waters. This phenomenon is called **El Niño** (literally, "little boy" in Spanish), because the resulting rains tend to fall around Christmas when Hispanic cultures celebrate the birth of the Christ child. El Niño years are the only times that it rains in the extremely arid coastal deserts of South America. The seemingly lifeless Atacama Desert bursts into bloom as plants that have survived as seeds (dormant in the soil) germinate, grow, and reproduce. In contrast, ENSO events also are characterized by reduced coastal upwelling that leads to dramatic reductions in nutrients, impacting the entire food chain of these regions. Seabirds and other marine organisms along the Pacific coast and in the Galápagos Islands suffer wholesale reproductive failure and mortality due to the unusual rain and reduced upwelling.

Other kinds of temporal variation can also have important biogeographic consequences. For example, a hurricane may pass over a Caribbean island only once in a century on average, yet these rare, unpredictable storms wreak incredible devastation. Hurricanes probably are one of the primary causes of disturbance on Caribbean islands. Such large, infrequent storms may increase or decrease biodiversity; while they can inundate tiny islets, causing extinction of some terrestrial animals and plants, they also clear space in forests and coral reefs, which facilitates the continued existence of competitively inferior species (Spiller et al. 1998). The more general lesson for biogeographers cannot be overstated: seemingly unpredictable and extremely rare events such as hurricanes, cataclysmic volcanic eruptions, long-distance chance dispersal to isolated oceanic islands, or collision with a wayward asteroid can fundamentally transform the development and distributions of life on Earth.

Soils

Primary Succession

Except for the polar ice caps and the perpetually frozen peaks of the tallest mountains, almost all terrestrial environments on Earth can and do support life. Areas of bare rock and other sterile substrates created by volcanic eruptions or other geological events are gradually transformed into habitats capable of supporting living ecological communities by a process called **primary succession**. This process involves the formation of soil, the development of vegetation, and the assembly of a complement of microbial, plant, and animal species.

We cannot understand the distribution of soils without a knowledge of the role of climate and organisms in successional processes. The type of vegetation covering a region depends primarily on three ingredients: climate, type of soil, and history of disturbance. For example, three distinct vegetation types (temperate deciduous forest, pine barrens, and salt marsh) occur in northern New Jersey in close proximity to one another but on different soil types (Forman 1979). Moreover, if a mature stand of deciduous forest is destroyed, such as at the hands of humans or by natural fire, it is not reestablished immediately. Instead, certain plant species colonize the area and are in turn replaced by later colonists beginning with weedy pioneer species and continuing until the mature or **climax** vegetation is reestablished. This process is called **secondary succession**. Throughout this process both the microclimate and the soil of the site also change, becoming more favorable for some species and less favorable for others.

Soil formation is both a chemical and a biological process resulting from weathering of rock and the accumulation of organic material from dead and decaying organisms. The process by which new soil is formed from mineral substrates is usually long and complicated. Physical processes such as freezing and thawing, and water and wind erosion, break down the parent rock material. Organisms also play key roles: lichens hasten the weathering of rock; decaying corpses of plants, animals, and microbes add organic material; the activities of roots and microbes alter the chemical composition of the soil; and burrowing animals mix and aerate it. Totally organic soils (or **histosols**) such as peat, form in certain unusual environments.

The rate of soil formation varies widely depending largely on the nature of the parent material and the climatic setting. The formation of shallow soils may take thousands of years in Arctic and desert regions, where temperature and moisture regimes are extreme (e.g., McAuliffe 1994). For example, soils only a few centimeters deep cover much of eastern Canada where the retreat of the last Pleistocene ice sheets left bare rock only about 10,000 years ago. In other cases, especially when soils are formed from sand, lava, or **alluvial** materials in regions with warm, moist climates, primary succession can be amazingly rapid. In 1883, the small tropical island of Krakatau in Indonesia experienced an explosive volcanic eruption that exterminated all living things and left only sterile volcanic rock and ash. Organisms rapidly recolonized Krakatau from the large neighboring islands of Java and Sumatra, and by 1934—only 50 years after the eruption—35 cm of soil had been formed and a lush tropical rain forest containing almost 300 plant species was rapidly developing (van Leeuwen 1936; Thornton 1996; Whittaker 1998).

Formation of Major Soil Types

Anything we write about soils must be a gross oversimplification because both the classification and the distributions of soils are very complex, even controversial. Visit the vast flat plains of the United States or the Ukraine and you will find just one or a few soil types distributed as far as the eye can see, but in other geographic regions—especially mountainous areas—soil and geological maps are mosaics that look like complicated abstract paintings. Great Britain has a series of unusual organic soil types formed in cold, wet environments as well as soils overlaid onto a complex geological foundation and greatly modified by centuries of human activities (Figure 3.9).

We can begin to appreciate the diversity and distribution of soils by studying the four major processes that produce the primary (or *zonal*) soil types. These so-called **pedogenic regimes** are those that typically occur in habitats characterized by temperate deciduous and coniferous forests (**podzolization**), tropical forests (**laterization**), arid grasslands and shrublands (**calcification**), and waterlogged tundra (**gleization**).

Podzolization occurs at temperate and subarctic latitudes and at high elevations where temperatures are cool and precipitation is abundant. In such climates plant growth may be substantial, but the low temperatures inhibit microbial activity so organic matter, called **humus**, accumulates. As the humus decays, organic acids are released and carried downward (**leached**) through the soil profile by percolating water. The hydrogen ions of these acids tend to

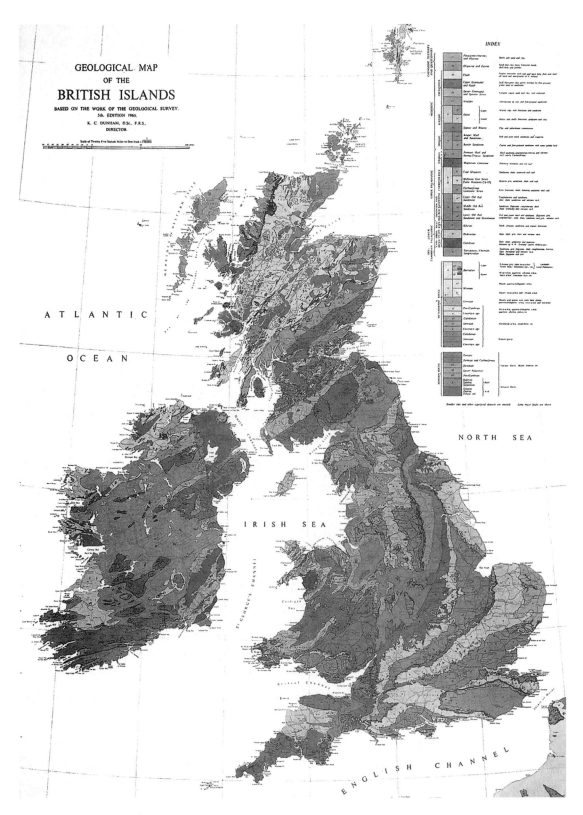

FIGURE 3.9 William Smith's classic *Geological Map of the British Islands* is sometimes referred to as "the map that changed the world" (see Winchester 2001). This was the first comprehensive geologic map of a region, one that described the successive layers and development of rock, fossils, and soils and, thus, revealed the temporal development of Earth's geographic template.

replace cations that are important for plant growth, such as calcium, potassium, magnesium, and sodium, which are removed by leaching from the soil (Figure 3.10A). This process leaves behind a silica-rich upper soil containing oxidized iron and aluminum compounds, but few cations. Coniferous forests, which thrive in such acidic conditions, are a characteristic vegetation type on podzolic soils.

In the humid tropics, which experience high temperatures and heavy rainfall, microbes and other organisms rapidly break down dead organic material, so little humus can accumulate. In the absence of organic acids, oxides of iron and aluminum precipitate to form red clay or a bricklike layer (laterite). The heavy rainfall causes silica and many cations such as potassium, sodium, and calcium to be leached out of the soil (Figure 3.10B), leaving behind a firm and porous soil with very low fertility. In some areas, if the tropical forest cover is removed, the organic material and its bound nutrients are easily lost and the intense equatorial sun bakes the exposed lateritic soils hard, retarding secondary succession and making the area unsuitable for agriculture.

Calcareous soils typically occur in arid and semiarid environments, particularly in regions where thick layers of calcium carbonate were deposited beneath ancient shallow tropical seas. Where rainfall is relatively low, so that evaporation and transpiration exceed precipitation, cations are generally not leached out. Instead they are carried downward through the soil profile to the depth of greatest water penetration where they precipitate and form a layer rich in calcium carbonate (Figure 3.10C). In desert soils, the scanty rainfall penetrates only a short distance below the surface, where it leaves behind a rocklike layer of calcium carbonate called **caliche**. In regions where precipitation is higher, water and roots penetrate deeper into the soil profile, leading to the formation of deep, fertile soils rich in organic material and essential nutrients such as potassium, nitrogen, and calcium. Such soils are typical of tallgrass and shortgrass prairie habitats, although little of the former remains because these soils are so highly prized for agriculture.

In cold and wet polar regions, gleization is the typical process of soil formation. At the permanently wet (or frozen) surface, where the low temperatures and waterlogged conditions prevent decomposition, acidic organic matter builds up, sometimes forming a layer of peat that can be several meters thick (Figure 3.10D). Below this organic upper layer an inorganic layer of grayish clay, containing iron in a partially reduced form, typically accumulates. While few nutrients are lost through leaching, the highly acidic conditions cause nutrients to be bound up in chemical compounds that cannot be used by

FIGURE 3.10   Schematic representations of the four major pedogenic regimes showing the resulting soil profiles: (A) podzolization, (B) laterization, (C) calcification, and (D) gleization. (After Strahler 1975.)

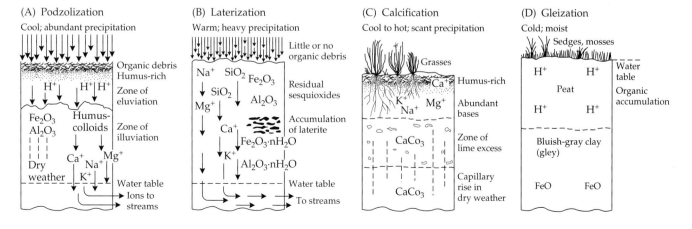

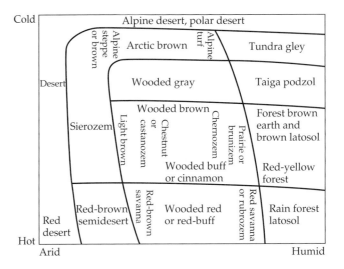

FIGURE 3.11   Schematic diagram depicting the relationships between major soil types and climate, showing that different combinations of temperature and precipitation cause the formation of distinctive soil types. (After Whitaker 1975.)

plants. Thus gley soils typically support a sparse vegetation of acid-tolerant species.

The above descriptions represent four idealized cases of soil formation processes. Given the complex variation in parent material and climate over the Earth's surface, pedogenic regimes vary in complex but predictable ways. The processes of soil formation and soil types described above occur where the chemical composition of the parent material is typical of the common rock types: sandstone, shale, granite, gneiss, and slate. The soils that are derived from these "typical" rocks are called **zonal soils.** A simplified summary of the relationship between climate and zonal soil type is given in Figure 3.11. The global distribution of zonal soil types (Figure 3.12) can also be compared with the global climate map (see Figure 3.7) to demonstrate the close relationship between soils and climate.

Unusual Soil Types Requiring Special Adaptations

In addition to such zonal soils, there are unusual soil types derived from parent material of unusual chemical composition. Certain rock types such as gypsum, serpentine, and limestone contain unusually high amounts of some compounds and little of others. Serpentine, for example, is particularly deficient in calcium, and gypsum contains an excess of sulfate. Few plant species can tolerate such azonal soils, and the low-diversity plant communities that do grow on such soils have special physiological adaptations for dealing with their unusual chemical composition.

One example of a soil type that requires special adaptations by plants is **halomorphic soil,** which contains very high concentrations of sodium, chlorides, and sulfates. Halomorphic soil typically occurs near the ocean in estuaries and salt marshes, and in arid inland basins where shallow water accumulates and evaporates, leaving behind high concentrations of salts. A small number of specialized **halophytic** (salt-loving) plant species grow in such areas. They include a variety of taxonomic and functional groups, each of which has special adaptations for dealing with the problem of maintaining osmotic and ionic balance in these environments. Some species of mangroves

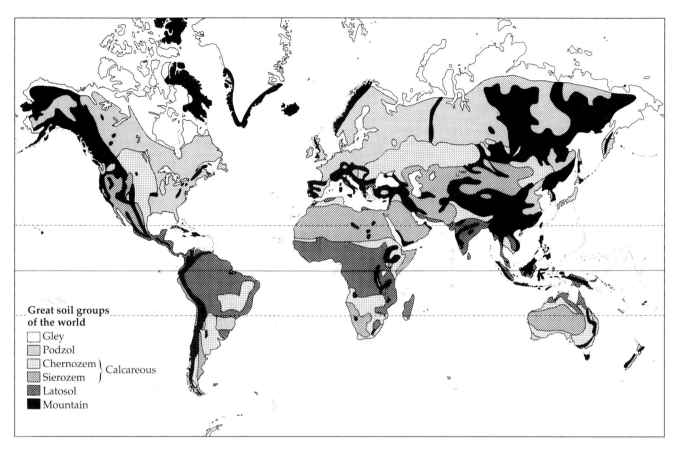

FIGURE 3.12 World distribution of major soil types. Note the close correlation of these soil types with the climatic zones shown in Figure 3.7, which reflects the influence of temperature and precipitation on soil formation.

and grasses excrete salts from specialized cells in their leaves, whereas pickleweeds and ice plants store salts in special cells in their succulent leaves.

As mentioned above, highly acidic soil conditions cause essential nutrients—especially nitrogen and phosphorus—to be bound in compounds that plants cannot use. Pitcher plants, sundews, Venus's flytraps, and other insectivorous plants can grow in highly acidic soils or other environments where nutrients are severely limited. These plants obtain their nitrogen and phosphorus by capturing living insects, digesting them, and assimilating the nutrients. A less spectacular adaptation to acidic and other nutrient-poor soils is evergreen vegetation (Beadle 1966). Because nutrients are lost when leaves are dropped, and because more minerals must then be taken up by the roots to produce new leaves, plants can use limited nutrients more efficiently by retaining their leaves for longer periods. In mesic temperate climates, where the predominant vegetation is usually deciduous forest, it is common to find evergreens growing on acidic and nutrient-poor soils. Examples are the pine barrens of the eastern United States and the eucalyptus forests of Australia (Daubenmire 1978; Beadle 1981).

In addition to their chemical composition, the physical structure of soils can influence the distribution of plant species and the nature of vegetation. In arid regions, for example, the size and porosity of soil particles affect the availability of the limited moisture to plants by affecting the runoff, infiltration, penetration, and binding of water. Thus, even within a small region of uniform climate, differences in soil texture can cause large differences in vegetation. A striking example is provided by the **bajadas** (or alluvial fans) of desert regions (Figure 3.13). These interesting geological formations are made up of sedi-

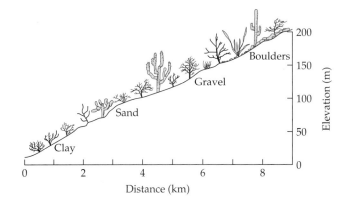

FIGURE 3.13 Schematic representation of the local elevational distribution of soil particle size and vegetation on a desert bajada on the Sonoran coast of the Gulf of California (Sea of Cortez). At the upper end of the alluvial fan, where large boulders have been deposited, the vegetation is dominated by cacti and other succulents that can take up water rapidly before it percolates below the root zone. At the lower end, where water infiltration is poor and the existing water is tightly bound by fine clay particles, the vegetation consists of sparse, shallowly rooted shrubs. The greatest water availability, productivity, and species diversity occur at intermediate elevations, where the soils are sandy, infiltration is high, and water is not tightly bound by soil particles.

ments carried out of mountains by infrequent but heavy flooding of the canyons. As the floodwater gradually loses energy it deposits sediments in a gradient—dropping large, heavy rocks at the mouth of the canyon and small sand- and clay-sized particles at the bottom of the fan. The resulting bajada shows a corresponding gradient in water availability and vegetation (Bowers and Lowe 1986). Cacti predominate on the coarse, rocky, well-drained soils high on the bajada, where water is available only for short periods during and after rains. These succulents can take up water rapidly through their extensive shallow roots and store it in their expandable tissues. Shrubs and grasses are much more common farther down the bajada, where their roots can extract the water held on and among the smaller soil particles.

A somewhat similar situation occurs along the coast of the Gulf of Mexico in the southeastern United States. The uplands have coarse, sandy, well-drained soils that support a drought-tolerant coniferous woodland/savanna vegetation. In contrast, the lowlands have accumulated fine, water-retaining soils, and this—as well as their proximity to the water table—allows them to support a much more mesic vegetation. Thus a person interested in the factors influencing plant distributions and community composition at this local to regional scale must pay particular attention to how subtle characteristics of soil structure affect the runoff, infiltration, and retention of rainwater.

Although we have concentrated here on the relationship between soils and vegetation, soils also affect the distributions of animals, both indirectly—by controlling which plant species are present, and directly—through the effects of the chemical and physical environment on their life cycles. Many kinds of mammals, reptiles, and invertebrates are restricted to particular types of soils that meet their specialized requirements for burrowing and locomotion. For example, in North American deserts, lizards of the genus *Uma*, the kangaroo rat *Dipodomys deserti*, and the kangaroo mouse *Microdipodops pallidus* are all restricted to dunes and similar patches of deep, sandy soil. Another set of species, including chuckwallas (*Sauromalus obesus*), collared lizards (*Crotophytus collaris*), and rock pocket mice (*Chaetodipus intermedius*) show just the opposite habitat requirement, that of being restricted to rocky hillsides and boulder fields.

Aquatic Environments

As anyone who has ever tried to keep tropical fish knows, warm and relatively stable temperatures are essential for their survival and reproduction. Salinity, light, inorganic nutrients, pH, and pressure also play key roles in the distributions of aquatic organisms. Like terrestrial climates, the physical characteristics of water often exhibit predictable patterns along geographic gradients, which can be understood with a basic background in physics.

Stratification

THERMAL STRATIFICATION.   When solar radiation strikes water, some is reflected but most penetrates the surface and is ultimately absorbed. Although water may appear transparent it is much denser than air and its absorption of radiation is rapid. Even in exceptionally clear water, 99% of the incident solar radiation is absorbed in the upper 50 to 100 m, and this absorption occurs even more rapidly if many organisms or colloidal substances are suspended in the water column. Longer wavelengths of light are absorbed first; the shorter wavelengths—which have more energy—penetrate farther, giving the depths their characteristic blue color.

This rapid absorption of sunlight by water has two important consequences. First, it means that photosynthesis can occur only in surface waters where the light intensity is sufficiently high (the **photic zone**). Virtually all of the primary production that supports the rich life of oceans and lakes comes from plants living in the upper 10 to 30 m of water. Along shores and in very shallow bodies of water, some species such as kelp are rooted in the substrate. These plants may attain considerable size and structural complexity, and may support diverse communities of organisms. In the open waters that cover much of the globe, however, the primary producers are tiny, often unicellular algae (called **phytoplankton**), which are suspended in the water column. **Zooplankton**, tiny crustaceans and other invertebrates that feed on phytoplankton, migrate vertically on a daily cycle: up into the surface waters at night to feed and down into the dark, deeper waters during the day to escape predatory fish that rely on light to detect prey.

Second, the rapid absorption of sunlight by water means that only surface water is heated. Any heat that reaches deeper water must be transferred by convection or by currents. Consequently, deep waters are characteristically cold, even in the tropics. The density of pure water is greatest at 4° C and declines as its temperature rises above or falls below this point. This unusual property of water is significant for the survival of many temperate and polar organisms because it means that ice floats. Ice provides an insulating layer on the surface that prevents many bodies of water from freezing solid. The presence of salts in water lowers its freezing point, and some organisms are therefore able to exist in unfrozen water below 0° C (de Vries 1971).

A more general consequence of the relationship between water density and temperature is that water tends to acquire stable thermal stratification. When solar radiation heats the water surface above 4° C, the warm surface water becomes lighter than the cool deeper water, and so tends to remain on the surface where it may be heated further and become even less dense. In tropical areas and in temperate climates during the summer, the surfaces of oceans and lakes are usually covered by a thin layer of warm water. Unless these bodies of water are shallow, the deep water below this layer is much colder (some-

FIGURE 3.14 Vertical temperature profiles of Lake Mendota, Wisconsin, at different dates from summer through winter showing the loss of thermal stratification as the lake cools. In July the thermocline is pronounced and shallow; in September it is less pronounced and deeper; and by November it has disappeared, allowing surface and deep waters to mix during the fall overturn. Because the density of water increases as it cools down to 4° C, but then decreases as water molecules begin to form ice crystals, thermal stratification is typically reversed in winter (February). (After Birge and Juday 1911.)

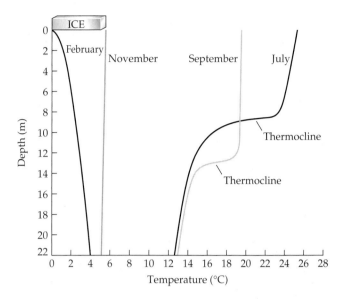

times near 4° C). The change in temperature between the surface layer and deeper water is called a **thermocline** (Figure 3.14). Mixing of the surface water by wave action determines the depth of the thermocline and maintains relatively constant temperatures in the water above it. In small temperate ponds and lakes that do not experience high winds and heavy waves, the thermocline is often so abrupt and shallow that swimmers can feel it by letting their feet dangle a short distance. In large lakes and oceans, where there is more mixing of surface waters, the thermocline is usually deeper and less abrupt.

Tropical lakes and oceans show pronounced permanent stratification of their physical properties, with warm, well-oxygenated, and lighted surface water giving way to frigid, nearly anaerobic, and dark (aphotic) deep water. Oxygen cannot be replenished at great depths where there are no photosynthetic organisms to produce it, and the stable thermal stratification prevents mixing and reoxygenation by surface water. Only a relatively few, but fascinating menagerie of organisms can exist in these extreme conditions. The feces and dead bodies of organisms living in the surface waters sink to the depths taking their mineral nutrients with them. The lack of vertical circulation thus limits the supply of nutrients to the phytoplankton in the photic zone. Consequently, deep tropical lakes are often relatively unproductive and depend on continued input from streams for the nutrients required to support life.

OVERTURN IN TEMPERATE LAKES. The situation is somewhat different in temperate and polar waters. Deep lakes, in particular, undergo dramatic seasonal changes: they develop warm surface temperatures and a pronounced thermocline in summer, but freeze over in winter. Twice each year, in spring and fall, the entire water column attains equal temperature and equal density, the temperature/density stratification is eliminated, and moderate winds may then generate waves that mix deep and shallow water, producing what is called overturn (see Figure 3.14). This semiannual mixing carries oxygen downward and returns inorganic nutrients to the surface. Phosphorus and other mineral nutrients may be depleted during the summer, when warm temperatures allow algae to grow and reproduce at high rates; overturn replenishes these nutrients by stimulating the growth of phytoplankton. Temperate lakes, such as the Great Lakes of North America are often quite productive and support abundant plant and animal life, including valuable commercial fisheries.

However, abnormally high nutrient inputs—often due to runoff from agricultural fields and discharges of inadequately treated sewage—can cause excessive production, rapid algal growth, depletion of oxygen, fish kills, and other environmental problems.

Oceanic Circulation

The vertical and horizontal circulation of oceans is more complicated than that of lakes, in part because oceans are so vast, extending through many climatic zones, and in part because salinity affects the density of water. Salts are dissolved solids carried into the oceans by streams and concentrated by evaporation over millions of years. The presence of salts in water increases its density, causing swimmers to experience greater buoyancy in the ocean than in fresh water. Varying salinity and density have important effects on ocean circulation. Rivers and precipitation continually supply fresh water to the surface of the ocean, and this lighter water tends to remain at the surface. If you have ever flown over the mouth of a large, muddy river such as the Mississippi, Thames, or Nile, you may have noticed that its water remains relatively intact, flowing over the denser ocean water for many kilometers out to sea. In polar regions, the input of fresh water to the ocean from rivers and precipitation generally exceeds losses from evaporation, but the reverse is true in the tropics. This pattern creates a somewhat confusing situation because warm tropical surface water tends to become concentrated by evaporation and to increase in density, counteracting to some extent stratification owing to temperature. Conversely, cold polar water—which would be expected to show little stratification—may become somewhat stabilized as low-density fresh water accumulates on the surface.

Vertical circulation occurs in oceans, but the rates of water movement are so slow that a water mass may take hundreds or even thousands of years to travel from the surface to the bottom and back again. Areas of descending water tend to occur at the convergence of warm and cold currents in polar regions, where the colder, denser water sinks under the warmer, lighter water. Areas of rising water, called **upwelling**, are found where ocean currents pass along the steep margins of continents. This happens, for example, along the western coast of North and South America, where there is little continental shelf and the land drops sharply offshore. As the Pacific gyres sweep toward the equator along these shores, the Coriolis effect and, in tropical latitudes, the easterly trade winds, tend to deflect the surface water offshore, and water wells up from the depths to replace it. Because upwelling, like overturn in lakes, returns nutrients to the surface, productivity tends to be high in areas of upwelling (see global productivity map on the cover of this text). Probably the greatest commercial fishery in the world is located in the zone of upwelling off the coasts of Chile and Peru, making episodic ENSO events discussed earlier in this chapter particularly devastating to local economies.

Surface currents, such as the great hemispheric gyres (see Figure 3.5), are relatively shallow and rapidly moving so that they tend to form discrete water masses, each of which has a characteristic salinity and temperature profile distinct from those of neighboring water masses. Some organisms with limited capacity for locomotion may drift in currents for long distances without leaving a single uniform water mass. Organisms that can move actively to overcome the currents must also be able to tolerate the contrasting physical environments in different water masses.

Although oceanographers have recognized the existence of distinct water masses within the oceans for many years, modern technology has revealed the extent of spatial heterogeneity in shallow ocean waters. For example, investigators from the Woods Hole Oceanographic Institution have studied the physical environment and the biota of Gulf Stream rings (Wiebe 1976, 1982; Lai and Richardson 1977; Katsman et al., 2003). These rings are small masses of cold or warm water that have broken away from the southern or northern edges of the Gulf Stream to drift through water of contrasting temperature in the North Atlantic. They can be readily seen on infrared satellite images that show sea surface temperatures (Figure 3.15). These rings not only have physical environments that are strikingly different from their surroundings, but they also contain a unique biota that can persist in these special conditions far from its normal distribution in the Gulf Stream. The possible roles of these floating warm- or cold-water eddies in trans-Atlantic dispersal—both now and in the past—are intriguing subjects for future biogeography research.

Pressure and Salinity

Pressure and salinity vary greatly among aquatic habitats. These variations have major effects on the distributions of organisms because special physiological adaptations are necessary to tolerate the extremes. As every scuba diver knows, water pressure increases rapidly with depth. It becomes a major problem for organisms in the ocean, where the deepest areas are up to 6 kilometers below the surface. Pressure increases at a rate of about one atmosphere (about 1.5 mega-Pascals) for every 10 m of depth. In the abyssal depths, pressures are more than 200 times greater than at the surface. Organisms adapted to living in surface waters cannot withstand the pressures of the deep sea, and vice versa.

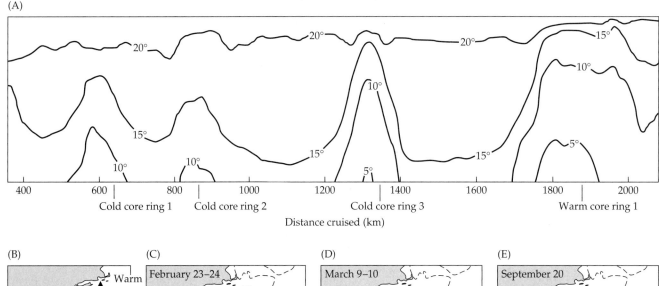

(A)

Cold core ring 1   Cold core ring 2   Cold core ring 3   Warm core ring 1

Distance cruised (km)

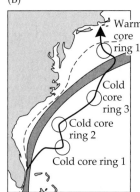

(B) Warm core ring 1 / Cold core ring 3 / Cold core ring 2 / Cold core ring 1

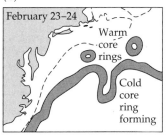

(C) February 23–24   Warm core rings   Cold core ring forming

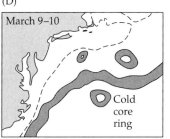

(D) March 9–10   Cold core ring

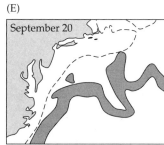

(E) September 20

FIGURE 3.15 Small-scale spatial and temporal heterogeneity of surface waters in the North Atlantic Ocean is caused by small water masses called "rings" that split off from the Gulf Stream (dark shaded line). (A) Temperature/depth profile recorded by an oceanographic vessel that traveled through several rings, as indicated by the line on map (B). (C–E) Changes in water surface temperatures as mapped by infrared satellite imagery showing the formation, movement, and disappearance of rings. (After Wiebe 1982.)

Variation in salinity is relatively discontinuous. The vast majority of the Earth's water is in the oceans and is therefore highly saline (greater than 34 parts per thousand). In contrast, freshwater lakes, marshes, and rivers, which account for less than 1% of the Earth's waters, contain very few dissolved salts. Habitats of intermediate or fluctuating salinity, such as salt marshes and estuaries, constitute only a tiny fraction of the Earth's aquatic habitats. Consequently, most aquatic organisms are physiologically adapted and geographically restricted either to fresh water, where the physiological problem is obtaining sufficient salts to maintain osmotic balance, or to salt water, where the problem can be eliminating excess salt. Only a few widely tolerant (**euryhaline**) organisms have the special physiological mechanisms required to survive in the widely fluctuating salinities of estuaries and salt marshes.

Tides and the Intertidal Zone

We can learn a great deal about the factors determining the distributions of organisms by studying environmental gradients: both gradual changes such as variation in light and pressure with depth in lakes and oceans, and rapid changes such as the variation in temperature in the cooling outflow of a hot spring. One of the steepest, best-studied, and most interesting environmental gradients occurs where the ocean meets the land. Along the shore is a narrow region that is alternately covered and uncovered by seawater. It is called the **intertidal zone** because it experiences a regular pattern of inundation and exposure caused by tides.

Sir Isaac Newton explained how the gravitational influences of the moon and sun interact to cause the global fluctuations in sea level that we call tides. The entire story is complicated but the main pattern and its mechanism are simple. The tides are flows of surface waters. They occur in response to a net tidal force, which reflects a balance between the centrifugal force of the Earth and moon revolving around their common center of mass, and the gravitational forces of the moon and sun (Figure 3.16). Because the gravitational force exerted by one object on another is equal to its mass divided by the square of the distance between those objects, the smaller but nearer moon has a greater effect than the sun.

Most shores typically experience both a daily and a monthly tidal cycle: there are two high and two low tides every 24 hours, and there are two periods of extreme tides each month, corresponding to the new and full moons (Figure 3.17). During these periods, the moon and sun are in the same plane as the Earth, and their gravitational effects are additive, causing high-amplitude or **spring tides,** with the highs occurring at dawn and dusk and the lows near noon and midnight. During the quarter moons, the sun and moon are at right angles to each other (from the perspective of Earth), and their gravitational effects tend to cancel each other, resulting in low-amplitude or **neap tides**.

(A)

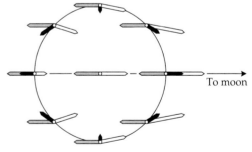

(B)

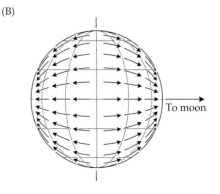

FIGURE 3.16 (A) Schematic representation of how the centrifugal force of the spinning earth and the gravitational force of the moon cause the tides. On the side of the Earth closest to the moon, the gravitational force (white) is stronger than the centrifugal force (gray), and the net tidal force (black) tends to draw surface water toward the moon. On the side opposite the moon, the gravitational force is weaker than the centrifugal force, and the net tidal force tends to draw water away from the moon. In between these extremes, the gravitational and centrifugal forces are balanced, and there is essentially no net tidal force. (B) The movement of surface waters in response to these tidal forces.

FIGURE 3.17   A tide calendar for the northern Gulf of California (Sea of Cortez) showing the typical pattern of tides due to the gravitational influences of the moon and sun. Note that there are two high and two low tides each day. There are also two periods of low-amplitude (neap) and high-amplitude (spring) tides each month; the latter correspond to the times of the new and full moons when the gravitational forces of moon and sun are aligned. (Courtesy of D. A. Thomson.)

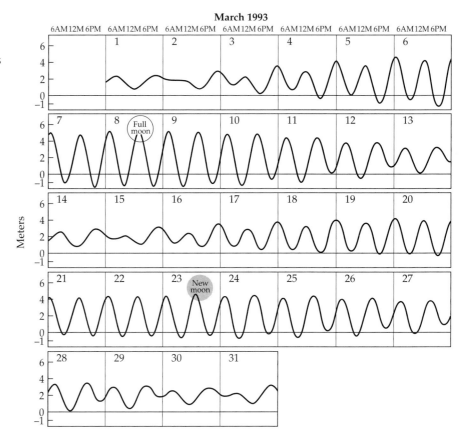

A distinct community of plant and animal species lives in the intertidal zone. Nearly all aspects of the lives of these organisms are dictated by the cyclical pattern of inundation by seawater at high tide and exposure to desiccating conditions at low tide. Most species are confined to a very narrow zone of tidal exposure, so that their distributions form thin bands running horizontally along the shore. As we shall see in Chapter 4, the narrow ranges of species in the intertidal zone (typically only a few centimeters or meters) and the ease with which critical environmental conditions can be manipulated experimentally, have produced a wealth of information about the factors limiting the distributions and regulating the diversity of species.

Microenvironments

Small-Scale Environmental Variation

It would be misleading to end here without a word of qualification. In this chapter we have been concerned primarily with the geographic template and global patterns of variation in abiotic environments that influence the distributions of organisms. Upon closer inspection, however, these patterns tell us surprisingly little about the actual conditions experienced by an organism living in a particular region. This point was best stated by the observer who noted that the climatic data recorded by the U. S. National Weather Service measures accurately only the climate experienced by the spiders living in the shelters that house the recording instruments!

Microenvironmental variation is less of a problem in aquatic habitats than on land because the physical properties of water tend to prevent the occurrence of abrupt small-scale changes. But even there, changes can be very abrupt and conditions can go from favorable to intolerable in distances of just a few centimeters or meters. Examples include the rapid changes in temperature at thermoclines and around hydrothermal vents (see Chapter 8 and 16), and in salinity in estuaries where rivers enter the ocean.

In terrestrial habitats, the climates of small places—called **microclimates**—may bear little relationship to large-scale climatic patterns. On one hand, two organisms living only a few centimeters apart may live in radically different physical environments: humid or arid, hot or cold, windy or protected. On the other hand, by selecting appropriate microenvironments, individuals can be distributed over a wide range of latitudes and elevations and still experience virtually identical physical conditions (Figure 3.18). Examples of both situations abound. Lizards are conspicuous elements of most desert faunas because they are active during the day and are able to tolerate the hot, dry conditions. The same deserts, however, may also be inhabited by frogs and toads, which spend most of their lives buried in the cool, relatively moist soil, emerge to feed only on rainy or humid nights, and possess adaptations for breeding in ephemeral ponds that form after occasional heavy rains. Perhaps the best examples of organisms that live in similar physical environments over a wide geographic range are internal parasites and microbial symbionts of birds and mammals. The same species may occur in tropical rain forests and arctic tundras, but still live in virtually identical, homoeostatically regulated environments within the bodies of their hosts.

The most distinctive microenvironments are typically small and widely dispersed sites. The capacity of organisms to exploit specific microenvironments depends largely on their mobility or vagility (regardless of whether they are actively or passively transported), body size, special physiological properties, and behavioral selectivity. We can readily imagine how mobile animals can seek out and settle in a particular habitat, but we should keep in mind that plants also may have adaptations that result in effective microhabitat selection. For example, many species have seeds that are attractive to certain kinds of animals, which disperse them to favorable microsites. Many seeds also require specific cues for germination that indicate the presence of favorable environmental conditions.

Colonizing Suitable Microenvironments

In order to live in isolated localities and microclimates, organisms must be able to get to them. Many plants, invertebrate animals, and microbes accomplish this during special dispersal stages of their life cycles, being carried long distances as seeds, eggs, or spores that can tolerate extreme environments while in transit. Often, however, their arrival at a suitable microsite is largely a matter of chance. In contrast, many animals are able to use their sophisticated sensory and locomotor systems to seek out particular isolated microenvironments. To demonstrate both passive and active dispersal it is only necessary to create a small artificial pond and observe the rapidity with which it is colonized both by zooplankton (copepods and other small crustaceans) that disperse passively as resistant eggs, and by large insects (diving beetles and dragonflies) that fly long distances to actively seek out suitable sites for colonization.

Some distinctive microenvironments are so isolated that the specialized organisms that inhabit them cannot disperse among them. How, then, were they originally colonized? Some organisms colonized them in the past when bridges of suitable habitat existed between them, or the intervening areas

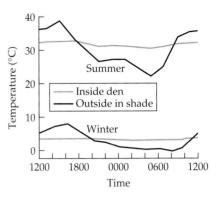

FIGURE 3.18 Temperatures inside and outside the den of a bushy-tailed woodrat (*Neotoma cinerea*) and a deep crack between large boulders in the high desert of southeastern Utah during midsummer and midwinter. Because the den (where the animal spends most of its time) experiences much less variation than the macroclimate outside, it affords vital protection from stressfully high and low temperatures in summer and winter, respectively. (After Brown 1968.)

were at least not so extensive and inhospitable. Examples include the fishes of isolated lakes, which require freshwater connections in order to disperse. Still other microenvironments are so inaccessible that their biotas include many unique species that have evolved in situ, diverging from ancestral forms that occurred in neighboring habitats. Examples include many of the highly differentiated, blind, unpigmented cave animals that have evolved in each cave system from surface-living ancestors, and the so-called "extremophiles"—microbial organisms that are adapted to the extremely hot (~80°C) and acidic (pH~3) waters of the world's isolated, geothermal springs (e.g., those of Kamchatka Peninsula, Russia; Yellowstone and Lassen Volcanic National Parks, North America; and Western Iceland; see Whitaker et al. 2003).

The fact that many organisms are found only in particular microenvironments has important consequences for our understanding of geographic patterns. On one hand, it means that some species may have much broader geographic ranges than we would have predicted from a cursory comparison of their physical tolerances and climatic patterns. On the other hand, careful studies have shown that the local distributions of many organisms are highly patchy, because within their geographic ranges they are confined to microsites that provide very specific environmental conditions. In the next chapter, we consider in more detail how different kinds of environmental conditions and ecological interactions limit the local distributions and geographic ranges of individual species.

CHAPTER 4

Distributions of Species
Ecological Foundations

THE PROPOSITION THAT EACH SPECIES HAS A unique geographic range is central to all of biogeography. Biogeographers study many phenomena—locations of occurrence of individual organisms, shifts in the local or regional distribution of a population, present and past distributions of higher taxa or clades (lineages of species descended from a common ancestor; see Chapter 11), clines of variation in morphology, physiology, behavior, and diversity across geographic gradients—but the ecological processes and historical events that have shaped the ranges of species are directly relevant to nearly all of them.

Before exploring the factors and processes influencing this fundamental unit of biogeography—the geographic range—we first discuss an important, albeit often overlooked methodological challenge: representing distributions of species across Earth's curved surface on the simplified, two-dimensional surfaces of maps. We then proceed in a hierarchical fashion, first considering the various factors influencing the distributions of individuals, then discussing the additional factors and processes influencing the ranges of populations and entire species. While it seems natural to consider these phenomena within the realm of ecological biogeography, it is important to remember that geographic ranges have been influenced by responses of individuals, populations, and species to environmental conditions throughout their evolutionary history.

Methodological Issues

Projections and Geographic Coordinate Systems

Before we can describe and analyze distributional patterns and search for explanations for those patterns, it seems appropriate that we review the basis for visualizing these patterns (i.e., maps, which are simplifications and, quite often, distortions of the true geographic template). The distortion arises from two challenges: (1) describing and

FIGURE 4.1 Although map projections are created using various mathematical formulae to transform spatial data from a curved three-dimensional surface to a flat two-dimensional map, the basic approach is not difficult to understand. If the features of interest (e.g., continental outlines or geographic ranges) are drawn onto a transparent model of the globe, those features can be transformed ("projected") by a light source that passes through the globe and then onto a piece of paper located either inside or outside the globe. By folding the paper (e.g., to form a plane, cylinder, or cone) and by moving it to different positions around the globe, projections can be adjusted to better preserve one or more of the key geographic features (i.e., shape, area, distance, or direction).

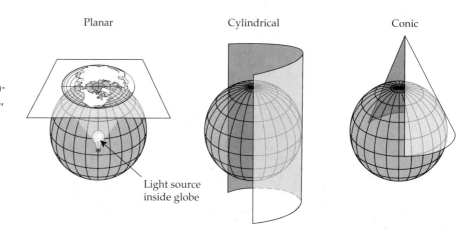

Planar   Cylindrical   Conic

Light source inside globe

somehow transforming three-dimensional data that describe patterns across the curved surface of the Earth onto a two-dimensional plane of a map, and (2) locating particular points or areas of interest on that plane. Fortunately, cartographers have been tackling these challenges for centuries, providing us with a wide array of **projections** and **geographic coordinate systems** to address each of these challenges, respectively. As Figure 4.1 illustrates, developing a map projection is relatively simple in theory, but in practice it turns out to be impossible to accurately represent a three-dimensional entity on a two-dimensional surface without distortion. For relatively small areas, across which the curvature of the Earth is negligible, map distortions are so minor that they can be ignored for most applications. At larger scales, however, map projections can cause substantial distortions of shape, area, distance, and direction. Fortunately, cartographers have developed a diversity of projections with known properties such that we can choose the projection most appropriate to our question (i.e., conformal, equal-area, equidistant and true directional—preserving the shape, area, distances between points and directions, respectively).

Once we have decided on an appropriate projection, we still cannot locate particular points or areas of interest without a geographic coordinate system. The two most commonly used types of coordinate systems include latitude and longitude, and the Universal Transverse Mercator (UTM) system. Locations in the UTM system are expressed as number of meters east of 1 of 60 arbitrary reference meridians ("false eastings") and, in the Northern Hemisphere, number of meters north of the equator. In the Southern Hemisphere, "eastings" are expressed in the same manner, while "northings" are expressed as number of meters north of an imaginary "false origin" located 10,000 meters south of the equator. With the addition of elevation, each of these systems can better approximate true complexities of the Earth's surface.

Mapping and Measuring the Range

How do we define and measure a geographic range? At first glance, this seems straightforward. Field guides and more technical systematic publications on regional floras or faunas are filled with **range maps.** These maps are seemingly easy for researchers to prepare, and equally easy for other scientists to use as sources of data for biogeographic studies. Before we start using range maps to illustrate biogeographic patterns and processes, however, we should critically consider just what they tell us.

There are three basic kinds of range maps: outline, dot, and contour. **Outline maps** usually depict a range as an irregular area—often shaded or col-

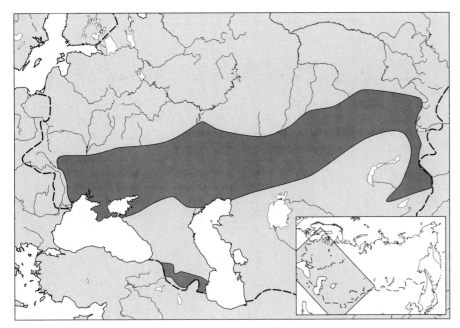

FIGURE 4.2 An example of an outline map of the geographic range of a species—in this case the endangered sooty orange-tip butterfly (*Zegris eupheme*) which occurs in southwestern Asia. The outer boundary has been drawn by hand to include localities where the species is known to occur. (After Borodin et al. 1984.)

ored—within a hand-drawn boundary (Figure 4.2). A boundary line presumably defines the limits of the known distribution of a species, but its accuracy can vary widely depending on how well the distribution is actually known and how precisely the author has incorporated this information into the map. Often the author will use his or her knowledge of the organism to make educated guesses about the probable distributional limits when adequate data are not available.

Dot maps plot points on a map where a species has been recorded (Figure 4.3). Dot maps are often prepared as part of a taxonomic study of a species, and the dots show localities where verified museum specimens have been collected. Such maps convey both more and less information than outline maps. On one hand, they accurately depict known records of a species' distribution. On the other hand, locations of specimens or other records such as sightings of bird species can represent only an infinitesimal fraction of the actual places where individuals of most species live at present or occurred in the past. While a small minority of species with tiny ranges are known to be restricted to just one or a small number of highly localized sites, documented records of occurrence of most species represent only a small sample of their actual distribution. So a disadvantage of dot maps is that they do not extrapolate beyond the relatively few sampled locations to make inferences about the potential distribution of a species. Sometimes, however, the author draws a free-form line around peripheral location records, creating a combination dot and outline map (e.g., Figure 4.4).

Recently, investigators have obtained sufficient information to produce **contour maps** on biotic variables such as abundance within the geographic ranges of some species (Figure 4.5). Because they use contour lines or other graphical techniques to indicate variation in density, these maps convey much more information than either outline or dot maps. Contour maps should be interpreted with caution, however, because information on abundance usually is available for only a limited number of fairly widely separated localities. Typically, computer programs and a statistical technique called **kriging** are used to interpolate between data points and generate a three-dimensional

FIGURE 4.3   An example of a dot map of the geographic range of a species—in this case the Sonoran Desert canyon ragweed (*Ambrosia ambrosioides*). Each circle represents a locality where someone has documented the presence of the species by collecting a voucher specimen and depositing it in an herbarium. Each cross represents an additional record based on a sighting and identification of the plant in the field. (After Turner et al. 1995.)

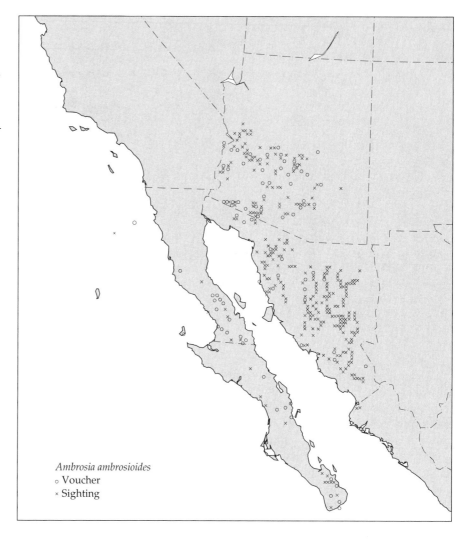

Ambrosia ambrosioides
∘ Voucher
× Sighting

FIGURE 4.4   An example of a combination dot and outline map of the geographic range of a species—in this case the endangered southern festoon butterfly (*Zerynthia polyxena*), which is restricted to a small area north of the Black Sea in southern Eurasia. Each dot represents a locality where the species has been recorded. A line has been drawn by hand to include the outermost dots, thereby enclosing the known geographic range. (After Borodin et al. 1984.)

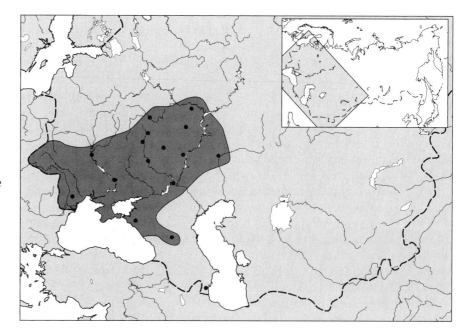

(A)

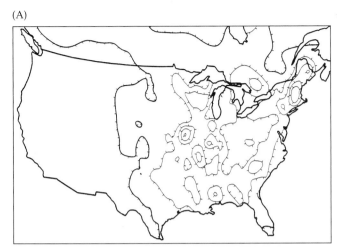

(B)

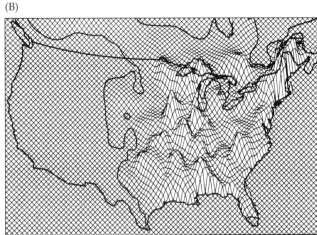

FIGURE 4.5 An example of a contour map of the geographic range of a species—in this case the winter range of the blue jay (*Cyanocitta cristata*), showing geographic variation in abundance. (A) Each contour line (or isocline) indicates a 20th-percentile class of relative abundance. (B) A three-dimensional landscape depicting relative abundance. Data on abundance come from North America Christmas Bird Counts. Raw data from these census counts (number of birds seen per hour per field party) were entered into a computer program that averaged and smoothed them to estimate abundance between actual census localities in order to draw the maps. (From Root 1988a.)

landscape depicting variation in abundance within a range. So let the user beware: even contour maps are highly oversimplified depictions of the complex and dynamic structure of ranges.

Nevertheless, despite the limitations of all three kinds of range maps, they are invaluable summaries of biogeographic information. Although their precision may be limited, they usually provide a reasonably accurate and unbiased, large-scale picture of the geographic range of a species. They can be compiled, georeferenced and displayed (using software such as geographic information systems [GIS]), and then analyzed quantitatively. It is relatively straightforward, for example, to measure such variables as area, northernmost and southernmost latitude, and easternmost and westernmost longitude of ranges from maps and to use such data in comparative biogeographic studies. You will see many such applications in this book.

The Distribution of Individuals

Even the best map can convey only a highly simplified and abstract picture of the geographic distribution of a species. Real units of distribution are the locations of all the individuals of a species. A map depicting these locations would be impossible to prepare for most kinds of organisms, but we can get some idea of what one would look like from aerial photographs in which we can identify individuals of certain conspicuous species (Figure 4.6). Rapoport (1983) prepared a map of the distribution of a distinct, easily recognizable palm tree (*Copernicia alba*) from aerial photographs along a transect through part of its range in Argentina (Figure 4.7). As we can see from both the sample aerial photograph and Rapoport's data, distribution is complex, with individual plants occurring in clumps separated by gaps. As the edge of the range is approached, individuals tend to be more sparsely distributed, clumps smaller, and gaps between them larger.

The clumpy-gappy distribution of individuals across a landscape means that any kind of map of a species range is not only an abstraction, it is a scale-dependent abstraction. Imagine that we superimposed a grid on an aerial photograph such as Figure 4.6. Whether an individual will be found in an individual grid cell will depend on the size of that cell: the larger the cell, the higher the probability of its containing at least one individual. This exercise reveals that the edge of a range is also a scale-dependent abstraction. Al-

FIGURE 4.6 An aerial photograph near the edge of the local distribution of the juniper tree (*Juniperous osteosperma*) in eastern Nevada. Individual trees, which are recognizable as dark spots, generally decrease in both size and abundance as elevation decreases from left to right. Note three things: (1) the overall complexity of the pattern of abundance and the difficulty of defining a precise range boundary; (2) the relatively uniform distribution of plants along an alluvial outwash plain at the top of the photograph; and (3) the patchy distribution of plants on southeast-facing slopes of small hills toward the bottom of the photograph. (Photograph courtesy of U.S. Forest Service.)

FIGURE 4.7 Abundance of the palm tree (*Copernicia alba*) along a 3 km wide transect from west to east through the edge of its geographic range. Data were taken from aerial photographs on which individual palms were easily recognized by their distinctive shapes. Note that near the edge of the range, abundance tends to be low and the distribution of trees tends to be patchy (as indicated by values of zero abundance.) (After Rapoport 1983.)

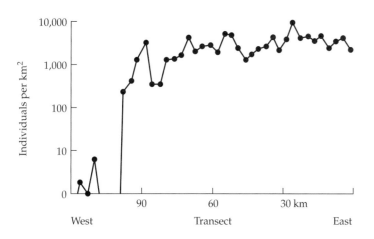

though in Figure 4.6 there are no individuals in the lower right corner, the exact definition of the distributional boundary will depend on the scale at which we connect the locations of individuals to draw an edge. Furthermore, in addition to relatively obvious range boundaries, there are "holes" within the range where no individuals occur. Therefore, Rapoport (1983) has likened the range to a slice of Swiss cheese. But even this is an oversimplification, because the sizes and locations of the areas where no individuals are considered to occur will also depend on the spatial scale of analysis. Perhaps the best representation of the effect of spatial scale on the perceived distribution of a species is still Erickson's (1945) classic distribution maps of the shrub *Clematis fremontii* (Figure 4.8).

Even an aerial photograph—thought to be a faithful depiction of the distribution of a species—fails to capture another critical feature of the geographic range because it represents only a snapshot in time. The distribution of any species is dynamic, and any accurate depiction of its range should in theory be constantly updated to reflect the changes that occur as individuals are born, move, and die, and as populations colonize new areas and go locally extinct in parts of their former range. Andrewartha and Birch (1957), for example, documented large shifts in the geographic ranges of several species of insects in Australia. They showed diagrammatically how the apparent range boundary varies as local populations episodically go extinct and then are recolonized from other areas (Figure 4.9). Despite the seemingly static nature of most published range maps, such expansions and contractions are always occurring in response to both natural environmental variation and human activities (see below and Chapters 16 and 17).

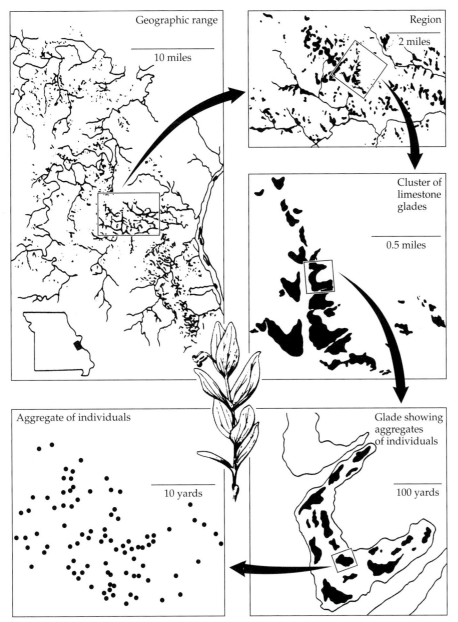

FIGURE 4.8 Erickson's classic depiction of the distribution of Fremont's leather flower (*Clematis fremontii*) within the state of Missouri in the central United States, on different spatial scales. The largest scale shows the geographic range based on known collecting localities. Successively smaller scales show distributions of populations. The smallest scale shows dispersion of individual plants within a single local population. Note that at all scales the distribution is patchy and that areas where plants are found are separated by uninhabited areas. (After Erickson 1945.)

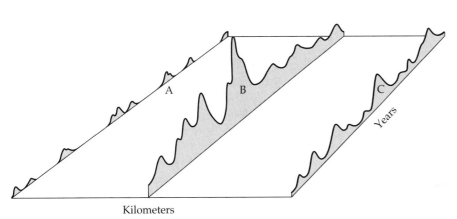

FIGURE 4.9 A schematic diagram showing how abundance and distribution of a hypothetical organism might vary in time and space. Shown are fluctuations in abundance over many years at three different localities (A–C) separated by distances of several kilometers. Note that all three populations fluctuate over time. At locality A, which is presumably at the margin of the local or geographic range of the species, only a few individuals are intermittently present, indicating repeated episodes of local extinction and recolonization. (After Andrewartha and Birch 1954.)

The Distribution of Populations

Population Growth and Demography

The size of a range, location of its boundaries, and shifting patterns of abundance within those boundaries reflect the influence of environmental conditions on the survival, reproduction, and dispersal of individuals, and the dynamics of populations. In 1798, in his *Essay on the Principle of Population*, Thomas Malthus showed that all kinds of organisms have the inherent potential to increase their numbers exponentially. A population increases when the combined rates of birth and immigration exceed the combined rates of death and emigration. We can express this concept mathematically as

$$r = b + i - d - e \quad (4.1)$$

where r is the per capita rate of population growth (if r is positive, the population increases; if r is negative, it decreases); b and d are per capita birth and death rates, respectively; and i and e are the respective per capita rates of immigration from and emigration to other populations. Given unlimited resources and favorable environmental conditions, a population will grow continuously at its maximum possible r. It will increase its numbers as described by the equation

$$dN/dt = rN \quad (4.2)$$

where dN/dt is the rate of change in numbers of individuals, N, with respect to time, t; and r is the population growth rate, as above. We call this **exponential growth,** and we can describe its rate in terms of the time interval required for a population to double its numbers. If it kept growing exponentially, any species would eventually cover the Earth with its own kind. The time required would depend on r: bacteria and houseflies would require only a few years; whereas slowly reproducing trees and elephants, with their lower values of r, would take a few thousand years. The global human population has been growing at a nearly exponential rate for the last several thousand years (Figure 4.10). Malthus

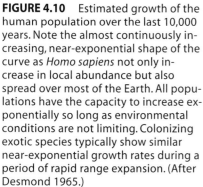

FIGURE 4.10 Estimated growth of the human population over the last 10,000 years. Note the almost continuously increasing, near-exponential shape of the curve as *Homo sapiens* not only increase in local abundance but also spread over most of the Earth. All populations have the capacity to increase exponentially so long as environmental conditions are not limiting. Colonizing exotic species typically show similar near-exponential growth rates during a period of rapid range expansion. (After Desmond 1965.)

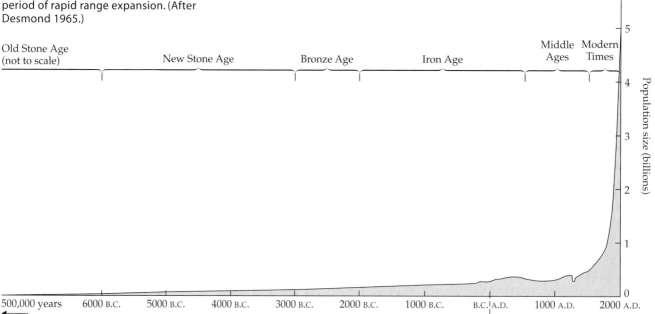

recognized, however, that because resources ultimately limit growth, and because many environments are unsuitable, no organisms actually continue to increase indefinitely at such exponential rates.

Hutchinson's Multidimensional Niche Concept

In 1957 Evelyn Hutchinson developed the concept of the multidimensional **ecological niche** in order to conceptualize how environmental conditions limit abundance and distribution. Hutchinson's view of the niche was a modification of the earlier niche concepts of Grinnell (1917) and Elton (1927; see also James et al. 1984; Schoener 1988). Hutchinson realized that over a period of time and over its geographic distribution, every species is limited by a number of environmental factors. He conceptualized a species' environment as a multidimensional space or "hypervolume" in which different axes or dimensions represent different environmental variables. The niche of a species represents the combinations of these variables that allow individuals to survive and reproduce, and populations to maintain their numbers.

This concept sounds intimidating and it is indeed a bit abstract, but the basic idea is very simple (Figure 4.11). Imagine the effects of just two environmental variables—say, temperature and salinity—on some aquatic species. If all other conditions are favorable, individual performance and population growth will be limited by the joint effects of these two variables. We can plot out the range of conditions under which population growth will be negative, zero, or positive; the space inside the zero growth contour, or isocline, represents the niche space. In reality, it is almost certain that our aquatic organism would also be limited by other variables such as dissolved oxygen concentration, a competing species, presence of a predator, or all three. Each of these variables would represent another dimension of the niche, causing the niche space to be a multidimensional volume, which is hard to visualize or draw but not too hard to imagine. It is easy to see that the niche of every species is unique because each species differs at least slightly—sometimes greatly—from all others in the combinations of environmental conditions required for the survival and reproduction of its individuals and the growth of its populations.

The Geographic Range as a Reflection of the Niche

The geographic range of a species can be viewed as a spatial reflection of its niche, along with characteristics of the geographic template and the species that influence colonization potential. If species were unlimited by geographic barriers and colonization abilities, they might be expected to occur wherever environmental conditions (including interactions with other species) are suitable. The boundaries of the range, and the pattern of abundance within these boundaries, constantly shift as local populations grow, decline, colonize, and go extinct in response to changing environmental conditions. In addition, just as interspecific interactions can limit a species to its **realized niche**, which is only a subset of its entire **fundamental niche**, over broader spatial and temporal scales these interactions along with limited time and opportunities for dispersal can restrict a species' **realized geographic range** to a small subset of its **fundamental geographic range**.

One of the earliest studies of the ecological niche of a species is still one of the most complete. Joseph Connell (1961) studied the environmental factors that limit the range of a barnacle (*Chthamalus stellatus*) on the rocky coast of the Isle of Cumbrae in Scotland. As mentioned in Chapter 3, the intertidal zone is that narrow strip of coastline that is alternately inundated by seawater and then exposed between tides. Species are typically restricted to a narrow range of exposures within the intertidal zone. Connell used some elegantly simple

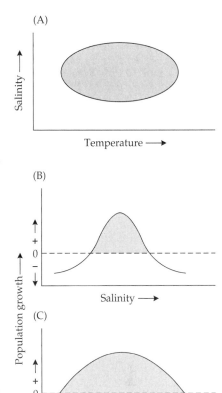

FIGURE 4.11 (A) A diagram representing two dimensions—temperature and salinity—of the niche of a hypothetical aquatic species. The shaded area represents the combinations of these two variables—a broad range of temperatures but a narrow range of salinities—under which individuals can survive and reproduce or a population can increase. In (B) and (C), population growth rate is plotted as a function of each variable, with the horizontal dashed line showing the zero value used to plot the elliptical niche space in (A).

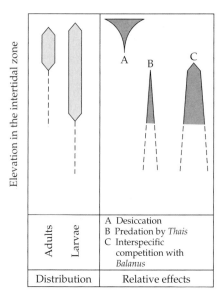

Elevation in the intertidal zone

Adults	Larvae	A Desiccation B Predation by *Thais* C Interspecific competition with *Balanus*
Distribution		Relative effects

FIGURE 4.12 Diagrammatic representation of the effects of interspecific competition and other factors on the distribution of the barnacle (*Chthamalus stellatus*) in the intertidal zone on rocky shores in Scotland. Diagrams on the left show species distribution; the width of each bar indicates population density at that elevation. Larvae settle over a wide range, but many die before reaching maturity, leaving adults confined to a much narrower zone. Diagrams on the right indicate the effects of three mortality-causing factors: desiccation between tides, A, which sets the upper limit of distribution; predation by the snail (*Thais*), B; and competition from the barnacle (*Balanus balanoides*), C, which together with A and B set the lower limit. The width of each bar shows the strength of each effect at that elevation. (After Connell 1961.)

field experiments to characterize important variables of the niche of *C. stellatus*, and to explain its distribution in the uppermost portion of the intertidal zone (Figure 4.12). He showed that the upper edge of the species' range is set by the ability of the barnacles to tolerate the physiological stress of desiccation while exposed during low tides. The lower edge of the species' range is set by interactions with other intertidal organisms, primarily by competition with another barnacle species (*Balanus balanoides*) and secondarily via predation by a snail (*Thais lapillus*). The power of the experimental method is illustrated by the effect of removing *B. balanoides* from small patches of shore—on those plots where its competitor was absent, *C. stellatus* extended its range lower into the intertidal zone.

Connell's study pioneered the use of field experiments in ecology. It is also a classic for demonstrating how three niche variables—exposure to desiccation, competition with another barnacle species, and predation by a snail—can largely explain the limited distribution of *C. stellatus* on the rocky shores of the Isle of Cumbrae. Note, however, that these are almost certainly not the only niche variables affecting the distribution of this barnacle: for example, some other factor(s) presumably account for the absence of *C. stellatus* from the sandy and muddy substrates that occur only a short distance from Connell's study site.

While the multidimensional environmental niche provides a conceptual framework for understanding how environmental limiting factors influence both geographic ranges and local population densities of a species, niche variables alone are inadequate to account for all patterns of distribution and abundance. Three complications will be mentioned briefly here, and then considered further in later chapters. First, it is too simplistic to assume that environmental conditions are equally favorable for a species at all localities where it occurs. Some localities may be so favorable that birth rates exceed death rates; these localities can serve as **"source habitats,"** producing surplus individuals that migrate out to other areas (Figure 4.13; Pulliam 1988). Some other localities may be so unfavorable that death rates exceed birth rates, but they may still be inhabited if they act as **"sink habitats"** and receive a sufficient supply of immigrants to maintain a local population (refer to Equation 4.1 and note the terms *i* and *e*, representing the contributions of immigration

Habitat 1 (Source population)

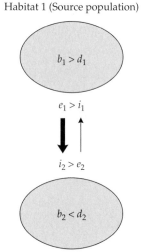

$b_1 > d_1$

$e_1 > i_1$

$i_2 > e_2$

$b_2 < d_2$

Habitat 2 (Sink population)

FIGURE 4.13 Diagrammatic representation of source and sink habitats showing the relative magnitude of the four processes that determine growth and persistence of their populations. In the source population (Habitat 1), the birth rate (b_1) is greater than the death rate (d_1), but the population does not increase. Instead, the "excess" individuals disperse, resulting in a higher rate of emigration (e_1) than immigration (i_1). The opposite situation results in the sink population (Habitat 2), which is able to persist despite having a lower birth (b_2) than death rate (d_2) because the rate of immigration (i_2) is sufficiently higher than the rate of emigration (e_2).

and emigration to the population growth rate). One might expect that some of the areas near the border of a species' range would be sink habitats and that their environmental conditions would be so marginal that they would not be able to sustain populations in the absence of immigration. The sea rocket (*Cakile edentula*), a broadly distributed annual plant that lives on coastal sand dunes, provides an example of this pattern. The small proportion of individuals growing in exposed seaward sites have the highest growth rates and produce the majority of seeds, but most plants occur in unfavorable inland sites, where storms have carried large numbers of dispersing fruits (Keddy 1982).

Second, just as there may be sites where environmental conditions are unfavorable but are nonetheless inhabited, there may also be favorable localities that are uninhabited. Some ecologists would say that there is an unfilled niche for the species in such places, but most prefer to define the niche as a characteristic of organisms (i.e., of species) rather than of places. As mentioned in Chapter 3, a species is likely to be absent from many places where it could live. Often this is because such favorable sites are isolated from inhabited areas by some combination of distance and intervening areas with unfavorable environmental conditions, so that individuals have not been able to disperse to these locations. This situation is common, as demonstrated by the large number of exotic species that have been successful in becoming established in new habitats when humans have transported them over intervening barriers (see Chapters 16 and 17). Biogeographers often invoke "history" to account for situations in which species are absent from apparently suitable areas. Indeed, to understand why a species occurs where it does and not elsewhere, it is necessary to understand both the history of apparently favorable places and the barriers between them, and the history of the species itself. We will return to consider these problems in Chapters 6, 10, and 11.

Finally, some places may be inhabited only intermittently. Local populations increase and decrease—sometimes to local extinction—as environmental conditions fluctuate or as stochastic events affect their growth. If the habitat is patchy, a species may consist of a series of isolated populations separated by uninhabited areas. When a species population is subdivided in this way it is said to be a **metapopulation,** comprised of multiple **subpopulations**. In such cases, some of the subpopulations are likely to go extinct intermittently, especially if they are sink populations. Even source populations in favorable patches of habitat, however, may go extinct by chance. New subpopulations are also likely to be founded by immigration to uninhabited patches including those vacated by previous local extinction events (e.g., Gilpin and Hanski 1991). We will return to consider the implications of such metapopulation and source/sink dynamics in Chapters 6, 13, and 16. For the moment it is sufficient to stress that they are especially likely to occur on the periphery of a species' range and to contribute to dynamic shifts in range boundaries.

The Relationship between Distribution and Abundance

The contour maps discussed earlier (see Figure 4.5) imply that there is considerable variation in abundance within a species' range. In fact, most published contour maps underestimate the magnitude of this variation because of the statistical and graphical methods used to interpolate between data points and construct the maps. The real spatial abundance patterns of nearly all species are extremely heterogeneous—what statisticians would call **clumped** or **aggregated** (Brown et al. 1995). That is, compared with a random distribution of individuals across a landscape, some places have many more individuals and others have many fewer or none at all.

(A) Red-eyed vireo

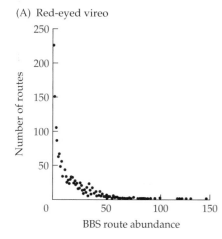

(B) Carolina wren

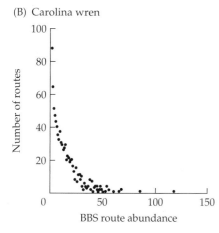

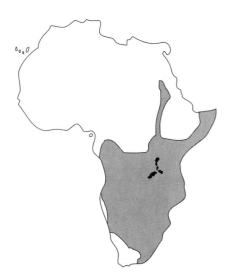

FIGURE 4.14   Variation in the abundance of two common songbird species, (A) the red-eyed vireo (*Vireo olivaceus*), and (B) the Carolina wren (*Thryothurus ludovicianus*), among hundreds of census routes distributed throughout their respective geographic ranges in eastern North America. Numbers of birds counted on each standardized census route of the North American Breeding Bird Survey are plotted in rank order, so that routes with one bird are on the left and routes with the maximum number of birds recorded on any route, are on the right. Note that for both species, fewer than 5 birds were recorded on the vast majority of census routes, but more than 100 birds were counted on at least six routes. This highly clumped pattern of abundance is characteristic of most birds as well as many other organisms. (After Brown et al. 1995.)

Although the complications of source/sink dynamics and history discussed above are relevant here, most of this spatial variation in abundance presumably reflects the extent to which the local environment meets the niche requirements of a species. Each species tends to be most abundant where all niche parameters are in the favorable range, and to be rare or absent where one or more environmental factors are strongly limiting (Brown 1984; Hengeveld 1990; Lawton et al. 1994; Brown et al. 1995).

Common species are typically several orders of magnitude more abundant at some sites than at others (Figure 4.14). Thus, for example, only a single red-eyed vireo (*Vireo olivaceus*) was recorded on more than 200 of the standardized census routes of the North American Breeding Bird Survey, but more than 100 individuals were counted on 6 routes. One consequence of this highly clumped distribution pattern is that the majority of individuals of a species actually occur in a very small proportion of its geographic range. For the majority of common songbirds in eastern North America, more than half of the individuals occurred at fewer than 20 percent of the sites within their geographic ranges. Of course, rare species may be uncommon throughout their ranges (Rabinowitz et al. 1986; Gaston 1994), but they too typically have patchy distributions and are absent from many, presumably unfavorable, localities.

There is also variation in abundance and distribution over time, and most of it presumably reflects temporal variation in niche parameters. The fluctuations of Australian insect populations in response to climatic variation documented by Andrewartha and Birch (1954) are excellent examples (see Figure 4.9). Migratory locusts of the Old World provide additional, perhaps even more dramatic, examples (Waloff 1966; Albrecht 1967; White 1976). Source populations of these grasshoppers persist in limited regions called outbreak sites, where conditions are suitable for their continued survival and reproduction. During periods when weather and food supplies are particularly favorable, these populations increase fantastically, change their morphology and behavior, aggregate into huge swarms, and migrate outward from the outbreak sites to forage over an enormous area. Such plagues of both the African migratory locust (*Locusta migratoria*) and the red locust (*Nomadacris septemfasciata*) have occurred two or three times in the last century, sweeping over most of southern Africa, an area consisting of millions of square kilometers—more than 1000 times the size of the outbreak area from which the locusts originated (Figure 4.15).

FIGURE 4.15   The temporally shifting range of the red locust (*Nomadacris septemfasciata*) in Africa. The small black areas are source habitats at the core of the range. These outbreak areas are the only places known to sustain permanent populations. Enormous population increases and geographic expansions begin in these areas, and in favorable years the locusts can spread into sink habitats throughout the invasion area (gray area), which includes about half of the continent. (After Albrecht 1967.)

Snowy owl

White-winged crossbill

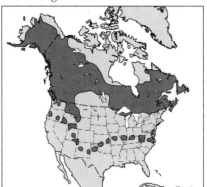

Common redpoll

FIGURE 4.16 Winter ranges of three bird species: snowy owl (*Nyctea scandiaca*), white-winged crossbill (*Loxia leucoptera*), and common redpoll (*Carduelis flammea*), that normally winter at high latitudes (dark gray area), but in years of food shortage disperse far to the south (to the dotted line), greatly expanding their ranges.

Similar fluctuations occur in insects inhabiting other regions. In the tundra and taiga (coniferous forest; see Chapter 5) of northern North America, Europe, and Asia, several species of voles and lemmings (mouse-like rodents, subfamily Arvicolinae, formerly Microtinae) fluctuate by several orders of magnitude in abundance over a 3- to 4-year period (see Finerty 1980; Lidicker 1988; Stenseth and Ims 1993). Unlike locusts, however, these rodents have very limited dispersal abilities, so their geographic ranges do not change very much during these cyclical fluctuations; however, their local patterns of habitat use may shift considerably. Some northern birds, such as snowy owls that feed on these rodents, and crossbills that feed on conifer seeds, also show wide fluctuations in abundance. Associated with population dynamics of these birds are large shifts in their winter ranges—sometimes hundreds of kilometers to the south—in years of low food availability (Figure 4.16; Bock and Lepthian 1976).

Superimposed on all of this spatial and temporal variation in abundance are certain spatial patterns. We will illustrate and comment briefly on three of them (Figure 4.17). First, like many other variables, abundance is autocorrelated in space—which is a technical way of saying that abundances tend to be more similar at sites that are closer together. This is just what we would expect if abundance reflects the suitability of the local environment and if niche variables also exhibit spatial autocorrelation across the geographic template. Second, abundance often varies systematically over the geographic range, from relatively high numbers near the center, to zero at the boundaries. In particular, localities near the edges of the range tend to have consistently low populations, whereas sites near the center of the range can have a wide range of abundances, from very low to very high. This pattern is consistent with the idea that boundaries of ranges occur where one or more niche factors become unfavorable. Third, an exception to the previous pattern is that abundance may be high near an edge of a range that is set by a coastline. Again, this is what would be expected if a range boundary is determined by just one niche variable that abruptly becomes unfavorable.

Range Boundaries

So far we have talked about niche dimensions and limiting factors only in abstract terms. What are these environmental variables that cause the boundaries of a species' range by limiting that species' abundance and distribution? Our previous discussion of the multidimensional environmental niche, spatial and temporal variation in abundance, and Connell's study of barnacle distributions in the intertidal zone would suggest that locations of range bound-

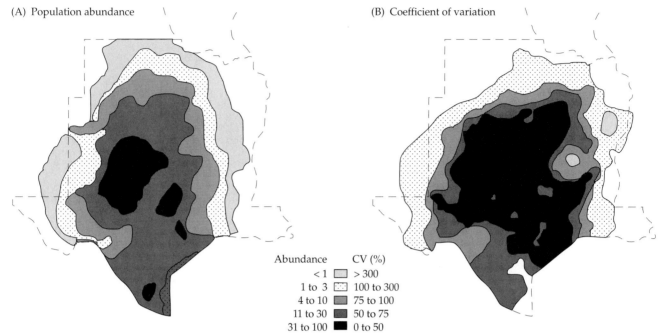

(A) Population abundance

(B) Coefficient of variation

Abundance	CV (%)
< 1	> 300
1 to 3	100 to 300
4 to 10	75 to 100
11 to 30	50 to 75
31 to 100	0 to 50

FIGURE 4.17 Abundance (A) and population variability (B) for the scissor-tailed flycatcher (*Muscivora forficata*) in its breeding range of Texas and Oklahoma. These maps illustrate three general tendencies: (1) autocorrelation, such that nearby areas tend to have similar abundances; (2) maximal abundances in one or more areas, usually located near the center of the range; and (3) generally low abundances and relatively high variability at the edge of the range, except where the boundary is set by a coastline. Abundance = average count in Breeding Bird Surveys; CV = coefficient of variation of abundance over a 25-year period (compare with Figures 4.5 and 4.7). (From D. Certain, unpublished report.)

aries are set by multiple environmental factors, and that these can include both physical conditions (**abiotic factors**) and effects of other organisms (**biotic factors**).

Unfortunately, there have been few studies as comprehensive as Connell's. Most investigators have studied the effects of just one or a few factors in a small area near one edge of a geographic range. This is understandable because geographic ranges of most species are large, and the scientific expertise of most ecologists is limited (i.e., physiological ecologists study the effects of different kinds of abiotic stresses, and community ecologists study the different kinds of interactions among species). Nevertheless, we are left with a literature that provides many examples of single limiting factors, but little overview and synthesis (but see Gaston 1994; Brown et al. 1996). So first let's examine some of these examples, and then try to see whether they reveal any general patterns and processes.

Physical Limiting Factors

Many widespread species appear to be limited in at least part of their geographic ranges by physical factors such as temperature regime, water availability, and soil and water chemistry. For example, many Northern Hemisphere plants and animals become increasingly restricted to low elevations and south-facing exposures as they approach the northern limits of their ranges, suggesting that their distributions are determined by ambient temperature. An excellent example is afforded by the giant saguaro cactus (*Carnegiea gigantea*) as it approaches the northern edge of its range in the Sonoran Desert of southern Arizona (see Figure 4.18 and the more detailed discussion below). Such correlations provide only circumstantial evidence, however, and do not necessarily indicate direct causal relationships. The species in question might, for example, be limited not by its inability to tolerate low temperatures but by competition from other species that are superior competitors in cold climates.

It might seem an easy matter to investigate the distributional limits of a particular species in order to identify the limiting factor and discover its mechanism of action on the organism. In 1840 Justus von Liebig suggested that biological processes are limited by that single factor that is in shortest supply relative to demand, or for which the organism has the least tolerance. At one time ecologists accepted this idea so completely that they called it **Liebig's law of the minimum** and tried to identify *the* single factor that limited the growth of each population. Now, however, as implied by Hutchinson's multidimensional niche concept and demonstrated by numerous empirical studies, we realize that Liebig's concept is too simplistic. Abundance and distribution are influenced by multiple and interacting environmental variables.

For example, many temperate and arctic birds and mammals appear to be limited by their inability to tolerate cold temperature regimes in the winter. Often this is not because they simply cannot survive at such low temperatures (Dawson and Carey 1976; Root 1988b,c), but rather because these temperature regimes increase the energy requirement for thermoregulation beyond the food supply available in the environment. In this case, food supply and temperature interact to limit distributions; increasing the food supply enables populations to inhabit colder climates. A number of bird species including the cardinal, tufted titmouse, and mockingbird, have expanded their ranges far northward in eastern North America (e.g., Boyd and Nunneley 1964). These birds are year-round residents, and a new reliable winter food supply provided by backyard bird feeders has almost certainly contributed to their expansion into colder environments.

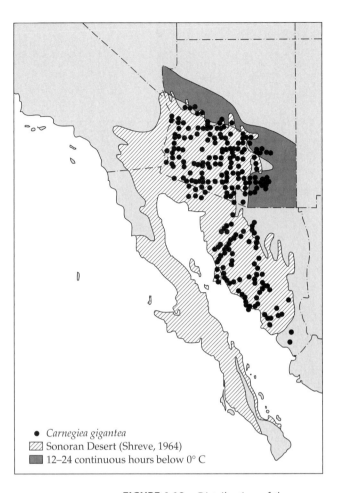

● *Carnegiea gigantea*
▨ Sonoran Desert (Shreve, 1964)
▮ 12–24 continuous hours below 0° C

FIGURE 4.18 Distribution of the saguaro cactus (*Carnegiea gigantea*) in relation to winter temperature regime. This cactus, like many other Sonoran Desert plants, is intolerant of prolonged freezing. Note the close correspondence between the northern limit of the saguaro, the northern boundary of the Sonoran Desert, and the region where temperatures remain below 0° C for more than 12 hours. (Data from Hastings and Turner 1965; Hastings et al. 1972).

Another problem in determining the causes of distributional limits is the difficulty of identifying the mechanisms by which environmental factors affect the growth of populations. Cold, for example, is not a single variable, and different aspects of low temperature regimes limit different populations in different ways. Adults of some plant species may be killed by critically low short-term temperatures such as those experienced on a single, exceptionally cold winter night. Other species may be more susceptible to damage from prolonged freezing. Still other species may be limited by cold climates—not because they cannot withstand low winter temperatures, but because the summer growing season is too short to allow for sufficient growth and reproduction.

As mentioned above, one of the best-documented examples of cold temperatures limiting the upper elevational and latitudinal distribution of a species is provided by the saguaro cactus (Figure 4.18). This giant multi-armed columnar cactus, which may reach 15 m in height and grow to be 200 years of age, is a conspicuous part of the landscape in much of the Sonoran Desert of southern Arizona and northwestern Mexico. Although it lives where winter nighttime frosts are not infrequent, the saguaro is extremely sensitive to temperatures below –7° C and to prolonged freezing. Individual cacti are often killed by frost damage to their tissues, especially by destruction of the growing shoot tips. Young saguaros are more susceptible to frost damage than are adults, but seedlings typically become established under the canopy of small desert trees, which provide young cacti with a protective microclimate for the first few decades of their lives (Nobel 1980b). These "nurse trees" shield young saguaros from the cold night sky and prevent their freezing in much the same way that frost damage to tomato plants can be prevented by covering them at night with paper or plastic—the loss of heat by infrared radiation to the sky is retarded by these nurse trees. Before they reach reproductive age, saguaros grow above their nurse trees, often killing the trees in the process; but by then they are large enough not to be affected by overnight frosts. Nobel (1978, 1980a) has studied the thermal relations of stems and shoot apices using computer simulations and direct field measurements. Results show that the large stem diameter of the saguaro enables it to maintain higher minimal temperatures of its apical buds, and thus it has a more northern distribution than related species of columnar cacti.

Steenburgh and Lowe (1976, 1977) studied populations of saguaros at Saguaro National Monument outside Tucson, Arizona, near the northeastern and upper elevational limit of the species range. Extensive mortality of both young and adult plants occurred as a result of exceptionally low temperatures in January of 1937, 1962, 1971, and 1978. The 1971 freeze killed about 10 percent of individual cacti and severely injured an additional 30 percent; many of the injured cacti died during the next few years as a result of microbial infections that started at the site of the frost damage (Figure 4.19). These observations of episodic winter kill, together with a close correspondence between the northern and eastern boundaries of the species range and areas that experience below-freezing temperatures for more than 12 hours at a time (see Figure 4.18), suggest that low temperatures directly limit the distribution of saguaros in this region.

Distributions of many plant species appear to be limited by low temperatures interacting with other environmental conditions such as water availability and soil chemistry. Hocker (1956) studied the distribution of loblolly pine (*Pinus taeda*) in the southeastern United States and concluded that the northern and western edges of the pine's range were set by low temperatures in concert with low soil moisture. He suggested that this resulted from the inability of the pine's roots to take up sufficient water to replace quantities lost by

FIGURE 4.19 Matched photographs of a stand of saguaro cacti near Redington, Arizona, near the upper elevational and northern edge of the species' range. (A) In 1961. (B) In 1966, showing the loss of one large individual (center foreground) and scars (white patches near tips of arms) on several other cacti as a result of severe frost in 1962. (C) In 1979, showing much additional mortality due to severe frosts in 1971 and 1978; several of the individual cacti still standing are dead or dying. (A and B courtesy of J. R. Hastings; C courtesy of R. M. Turner.)

(A)

(B)

(C)

FIGURE 4.20 Relationship of timberline to elevation and latitude in three different major mountain chains in North America. In general, timberline increases in elevation with decreasing latitude reflecting the influence of increasing temperature. Note, however, that the relationship is different in each mountain chain due to other factors such as length of the summer growing season. Note also that along the primary cordillera (chain extending from the northern Rocky Mountains to Panama) there is essentially no change in the elevation of timberline within the tropics and subtropics between about 35° N latitude and the equator. (After Daubenmire 1978.)

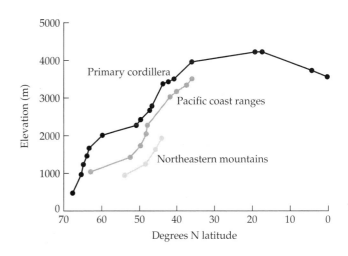

evaporation when environmental temperatures were low. Shreve (1922) noted that upper elevational limits of many desert plant species appear to be determined largely by temperature because extensive mortality occurs in populations at the highest elevations during exceptionally cold winters. However, many of these same species such as the ocotillo (*Fouquieria splendens*) occur at much higher elevations and tolerate substantially colder winter temperatures on limestone soils than on granitic and other zonal soil types (Shreve 1922; Whittaker and Niering 1968).

Across the world's mountainous regions, many investigators have attempted to determine the cause of **timberline**, the upper elevational limit of trees on mountains. The large-scale geographic position of timberline seems to be related to the mean or maximum temperature during warm months of the growing season, which vary with latitude and elevation (Figure 4.20). Timberline is influenced locally by other factors such as wind, snow depth, and energy balance (Daubenmire 1978; Stevens and Fox 1991). Trees at timberline of windswept alpine regions are sometimes called krummholz (crooked timber), referring to their dwarfed nature, contorted shape and lack of branches on their windward side. At timberline, established trees often live for a long time, but their growth is very slow, and successful reproduction and seedling establishment are rare. Bristlecone pines (*Pinus longaeva*), which often grow just below timberline on arid mountains in the southwestern United States, are the oldest known living things, some individuals being more than 3000 years old. Bristlecone pines grow slowly and produce exceptionally hard, dense wood that is highly resistant to decay (Figure 4.21). The annual growth rings of living and dead bristlecones provide a valuable record of the climatic history of this region because the width of each ring is proportional to the suitability of the growing season in the year it was laid down (Fritts 1976). In some places dead bristlecones form a "fossil timberline" above the timberline of living trees. Apparently this area was once favorable for tree growth, but climatic changes during the last few thousand years killed the trees and caused a contraction in their elevational range (La Marche 1973, 1978).

Unlike plants and sessile animals (such as the barnacles discussed earlier), most animals can move to seek out favorable microenvironments and thus avoid the most stressful abiotic conditions. Nevertheless, even highly mobile animals such as fishes can be limited directly by physical factors such as envi-

FIGURE 4.21 A bristlecone pine (*Pinus longaeva*) growing near timberline in the White Mountains of California. As is typical of individuals at the upper elevational edge of the range, this one has much dead wood and a highly contorted growth form. (Courtesy of A. Kodric-Brown.)

ronmental temperatures. Pupfishes of the genus *Cyprinodon* are extremely eurythermal and euryhaline; that is, they can tolerate a wide range of temperatures and salinities (Brown 1971c; Brown and Feldmeth 1971; Soltz and Naiman 1978; Varela-Romero et al. 1998). Species of this genus occur in rigorous physical environments including shallow streams and marshes in deserts, small pools in tidal flats, estuaries, and mangrove swamps, where temperature and salinity may fluctuate widely. Some populations even inhabit hot springs, although they generally cannot tolerate temperatures in excess of about 43° C (which is still amazingly high for a fish). Some pupfishes occur in the cooler outlets of hot springs whose temperatures exceed lethal limits. The local distribution of one such population of *C. nevadensis* near Death Valley, California, is limited directly by temperature (Brown 1971c). Fish occur in all waters cooler than about 42° C, including small pools only a few centimeters away from much hotter water (Figure 4.22). Occasionally individuals stray or are frightened into lethally hot water and die instantly—a clear demonstration of the thermal niche axis! Less frequently, rapidly changing weather conditions cause lethally hot water to flow far downstream from the source, trapping thousands of fish in pools where they are killed when the temperatures rise to over 43° C.

As in many organisms, adult pupfishes are tolerant of a wider range of physical conditions than are pupfishes in early developmental stages. Eggs of *C. nevadensis* develop normally only in water between 20° and 36° C, although adults of this species can withstand temperatures between 0° and 42° C (Shrode 1975). Eggs are also less tolerant than adults of extreme salinity. Con-

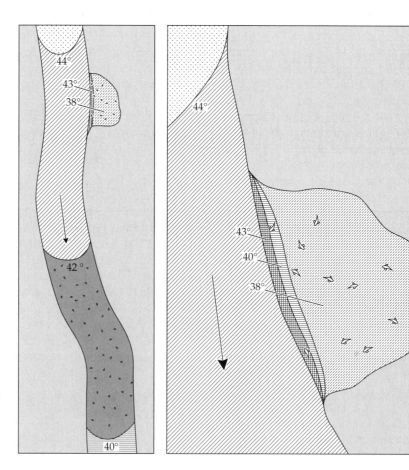

FIGURE 4.22 Temperature limits the local distribution of a desert pupfish (*Cyprinodon nevadensis*) in the outflow of a hot spring near Death Valley, California. The fast-flowing main channel is above the lethal temperature of 43° C; fish are trapped, but are able to survive in the cooler side pool (enlarged at right). (After Brown 1971c.)

sequently, adult pupfishes can be found in sink habitats where reproduction is impossible, so long as they can immigrate from nearby source microhabitats that are suitable for egg development. On the other hand, species of *Cyprinodon* are conspicuously absent from some cold springs and other habitats where the adults can grow and survive, but where there are no microclimates suitable for earlier stages.

In addition to temperature, other physical and chemical factors, such as moisture, light, oxygen, pH, salinity, and soil and water elements, limit the distributions of animals—either singly or in interaction with one another. A simple example of such an interaction is the effect of temperature and oxygen concentration on many fishes and aquatic invertebrates. As water increases in temperature, its capacity to hold oxygen and other dissolved gases decreases. The resulting combination of high temperature and low oxygen concentration is very stressful to many fishes and aquatic invertebrates because high temperatures cause elevated metabolic rates and increased demand for oxygen.

A few additional examples will suffice to illustrate distributional boundaries that can be attributed, at least in part, to such factors as moisture, salinity, and soil chemistry. Many terrestrial plants are limited by low soil moisture at the drier edges of their ranges, just as they are by low temperatures at the colder margins. In nearly all vascular plants, photosynthetic rates decline as soil moisture decreases. Plants can compensate for decreased water uptake through the roots by closing their stomates and reducing transpiration from the leaves, but rates of photosynthesis are reduced concomitantly.

Plants have diverse anatomical, physiological, and phenological adaptations that enable them to grow in a wide range of temperature, moisture, and light regimes (Bazzaz 1996). For example, species that can grow in full sunlight on dry soils (**xerophytes**) show many specialized mechanisms for keeping their stomates open despite low levels of water in their leaves. In contrast, species from wetter and more shaded environments (**mesophytes**) typically close their stomates when subjected to drought and temperature stress; without evaporative cooling, their leaves suffer high—often fatal—heat loads. On the other hand, xerophytes typically have relatively low rates of photosynthesis when abundant water is available, and they are intolerant of shade. A consequence of this trade-off is that mesophytes are physiologically incapable of growing on dry soils, whereas xerophytes can grow where there is little moisture, but are competitively excluded from wetter soils by shading from mesophytes that have higher growth rates (see Odening et al. 1974) (Figure 4.23) . These physiological findings provide a mechanistic basis for the conclusion of Shreve (1922) and other early plant ecologists that in gradients of increasing aridity, the limits of plant distributions are determined largely by an inability to tolerate

FIGURE 4.23 Trade-offs between growth rate and drought tolerance in two species of desert shrubs: creosote bush (*Larrea tridentate*), which grows in some of the driest North American deserts; and desert willow (*Chilopsis linearis*), which has an overlapping geographic range, but is more mesophytic, occurring in microhabitats along watercourses where the soil is permanently moist. Note that under relatively high drought stress (light gray region) creosote bush has the higher net photosynthetic rate and is able to grow faster, shade, and competitively exclude desert willow. (After Odening et al. 1974.)

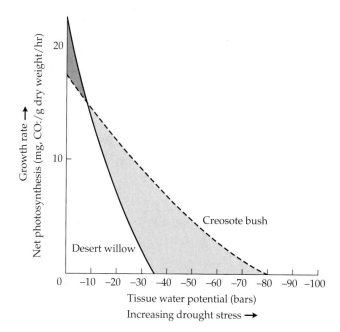

low soil moisture. Widespread diebacks in drought years are commonly observed in local populations at the margins of a species range (e.g., Sinclair 1964; Westing 1966). Similar kinds of trade-offs between photosynthetic rate and ability to tolerate low nutrient levels, high salinity, extreme pH, or high concentrations of toxic minerals, probably account, at least in part, for the failure of many otherwise widespread plant species to occur locally on soils with these characteristics, while species with special adaptations to these soil types are often restricted to them (Whittaker 1975).

These kinds of physical and chemical factors also limit animal distributions. Because of their osmoregulatory physiology, the vast majority of freshwater fishes and invertebrates are intolerant of salinities even approaching the concentration of seawater, whereas marine species cannot survive in fresh water. One consequence is that salt marshes and estuaries are inhabited by neither freshwater nor marine species, but rather by specialized euryhaline species that can tolerate great—often daily—fluctuations in salinity caused by tides, floods, and storms.

Only a few kinds of specialized organisms occur in those lakes and springs that are even saltier than seawater. Great Salt Lake in Utah, for example, has a salinity approximately seven times that of seawater. It contains no fish and only two macroscopic invertebrates: the pelagic brine shrimp (*Artemia salina*), and the benthic brine fly (*Ephydra cinerea*). Several fish species and numerous invertebrates inhabit the freshwater streams and marshes that empty into the lake. These animals have abundant opportunities to extend their ranges and colonize the lake, but they are prevented from doing so by their inability to tolerate its high salinity. This is dramatically illustrated by the widespread mortality that occurs when occasional windstorms push salt water from the lake into the adjacent marshes.

Disturbance

Another class of factors that influences both local and geographic distributions of many organisms includes fires, hurricanes, volcanic eruptions, and other agents of sudden widespread disturbance and destruction. These natural disasters are capable of completely destroying habitats and their inhabitants. In regions where such disturbances occur with some frequency, however, they are a natural part of the environment (Figure 4.24). Many species are dependent on these periodic disturbances for their continued existence, and there is a regular pattern of colonization and replacement of species following a disturbance, which ecologists call secondary succession. For example, jack pines (*Pinus banksiana*) in eastern North America and lodgepole pines (*P. contorta*) in the Rocky Mountains have closed cones that require the heat of periodic forest fires to release their seeds. Somewhat similar cyclical successional processes occur in the chaparral shrublands of coastal California and the coniferous forests of the Rocky Mountain region, which are swept by periodic fires; on the forested islands and offshore coral reefs of the Caribbean region, which are occasionally but inevitably decimated by hurricanes; and in intertidal habitats throughout the world, which are subjected to heavy wave action and sand scouring during major storms.

On the other hand, periodic natural disturbances may be sufficiently severe and frequent to prevent the expansion of some species into areas where they could otherwise survive. In most grasslands, lightning-caused fires are a natural part of the environment. At forest/grassland boundaries, where soil moisture is high enough for woody vegetation to become established, frequent fires may prevent trees and shrubs from extending their ranges (see Figure 4.24). Artificial fire suppression within the last 200 years has contributed to the

(A)

(B)

(C)

FIGURE 4.24 Effect of a natural lightning-caused wildfire on savanna vegetation near Elgin, Arizona. (A) A few days after that fire in June 1975. (B) Several months later, September 1975. (C) Eleven years later, February 1986. Note that while the live oak trees lost most of their leaves as a result of the fire, only a few, such as the one in the center foreground, were killed or suffered major, lasting damage. Fires kill invading shrubs and some trees, maintaining this open, grassy habitat. (Photographs courtesy of R. M. Turner.)

expansion of forest and shrubland at the expense of prairie and other grassland habitats along the eastern margin of the Great Plains of central North America (e.g., Beilmann and Brenner 1951; Hartnett et al. 1996). A similar phenomenon occurs along the drier margins of the Great Plains in southwestern North America, where fire influences the boundary between arid grassland and desert shrubland. Fire suppression, along with livestock grazing, has played a

FIGURE 4.25 Contraction of grassland habitat in southern Texas between (A) 1860 and (B) 1960. These changes, reconstructed from historical records, are due largely to fire suppression, livestock grazing, and invasion of woody vegetation, especially mesquite (*Prosopis*). (After Johnston 1963.)

(A)

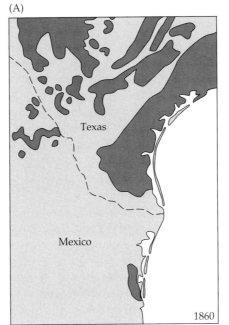

1860

(B)

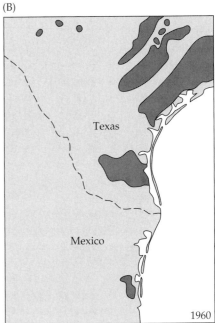

1960

major role in **desertification** and degradation of grasslands in the southwestern United States, northern Mexico, and other arid regions throughout the world (Figure 4.25; Johnston 1963; Bahre 1995). In many of these places where natural fires once burned unchecked over enormous areas, prescribed burns must now be used as a management tool to preserve grassland and prevent the invasion of woody vegetation and exotic species. The tiny reserves of tallgrass prairie in the north central United States, for example, are too small to experience a natural frequency of lightning-caused fires. Now, prescribed burns are essential to prevent local extinction of native prairie plant species.

At smaller scales, other causes of disturbance—especially biological ones—can be equally important. Typically, such small-scale disturbance has the effect of removing dominant plants or sessile animals, and allowing fast-growing but competitively inferior species to colonize. This creation and filling of gaps results in a patchwork of microsuccessional stages that, in total, supports many species. In many tropical and temperate forests, gaps in the canopy caused by the falling of single trees, or even large limbs, create a sunny microclimate on the forest floor that is essential for the existence of certain understory species and for the establishment of seedlings of some canopy trees (e.g., tulip tree (*Liriodendron tulipifera*): Picket and White 1985). On rocky intertidal shores, such as those of northwestern Washington, the predatory starfish (*Pisaster ocraceous*) removes the competitively dominant mussel (*Mytilus californianus*) creating gaps in the otherwise continuous mussel beds that can then be colonized by rapidly growing algae and invertebrates (Paine 1966). On some exposed shores, logs banging against the mussel beds can have a similar effect (Dayton 1971). In North American tallgrass and shortgrass prairies, soil disturbance caused by the activities of mammals may be as important as fire for the maintenance of a diverse grassland community. Wallows of bison and mounds of pocket gophers all provide patches of bare, sunny soil that are essential for the establishment and persistence of certain plant species (e.g., Platt 1975; Reichman and Smith 1985; Inouye 1987). Similarly, grazing by prairie dogs (genus *Cynomys*) maintains a dynamic mosaic of shortgrass prairies across broad expanses of the Great Plains Region, providing habitat for hundreds of species of plants and terrestrial vertebrates, while the extensive burrow system of the prairie dogs creates microhabitats and refugia for many other animals and plants (Miller et al. 1990, 1994; Stapp 1998; Lomolino and Smith 2003, 2004; Smith and Lomolino 2004).

Interactions with Other Organisms

In many cases, geographic distributions are not limited directly by physical factors. Botanical gardens and zoos provide perhaps the most dramatic evidence that individuals can survive, grow, and even reproduce under a much wider range of physical conditions than they encounter anywhere in their natural geographic range. The fact that many plants can thrive in botanical gardens, suburban landscapes, or agricultural fields—but only if they are protected from competing plants, animal herbivores, and microbial pathogens—demonstrates the importance of interspecific interactions in limiting distributions.

There are three major classes of interspecific interactions: competition, predation, and mutualism. All of these can influence the dynamics of populations and limit the geographic ranges of species.

COMPETITION. **Competition** is a mutually detrimental interaction between individuals. Organisms that share requirements for the same essential resources necessarily compete with each other and suffer reduced growth, survival, and reproduction if the resources are in sufficiently short supply. Plants may compete for light, water, nutrients, pollinators, or physical space, whereas animals most frequently compete for food, but also for shelter, nesting sites, mates, or for living space that contains these resources. These interactions may be purely **exploitative** so that individuals use up resources and make them unavailable to others. Alternatively, competition may involve some form of **interference** in which aggressive dominance or active inhibition is used to deny other individuals access to resources. For example, some plants such as the black walnut (*Juglans nigra*), and some sessile marine animals such as bryozoans and corals, use a form of chemical warfare called **allelopathy** to defend space from competitors (see Jaksic and Fuentes 1980).

There is much circumstantial evidence that competition limits geographic ranges. There are many examples of ecologically similar, closely related species that occupy adjacent but nonoverlapping geographic ranges. Five species of large kangaroo rats (*Dipodomys*) are found in desert and arid grassland habitats in the southwestern United States and northern Mexico, but their geographic ranges do not overlap (Figure 4.26). Two species, *D. ingens* and *D. elator*, have isolated (or disjunct) ranges, but *D. spectabilis* shares an extensive border with *D. deserti* in the west and with *D. nelsoni* in the south.

These three species, however, segregate their realized niches at a more local scale: *D. deserti* occurs only in sandy habitats, which lack *D. spectablis*; *D. nelsoni* indhaibts the core of the Chihuahuan desert, while *D. spectablis* is limited to arid grasslands of this region. Although such cases strongly suggest that competition limits distributions by preventing coexistence, they are subject to alternative explanations, and often there is no direct evidence of competitive interactions occurring on the boundaries.

Better evidence for the limiting effects of competition comes from "natural experiments" in which one species, simply by chance, is absent from regions that are apparently suitable for it. If a second species has expanded its range to include habitat types that are normally occupied by the first species, this implies that in other areas where the first species is present, the second will be limited by competition. Forests and shrublands of the western United States are inhabited by more than 20 species of chipmunks of the genus *Neotamias*

FIGURE 4.26   Non-overlapping geographic ranges of five species of large kangaroo rats (*Dipodomys*) in southwestern North America. These rodents are similar in their ecology, and the fact that their ranges frequently come into contact—but rarely overlap—suggests that they limit one another's distributions through competition. (After Bowers and Brown 1982.)

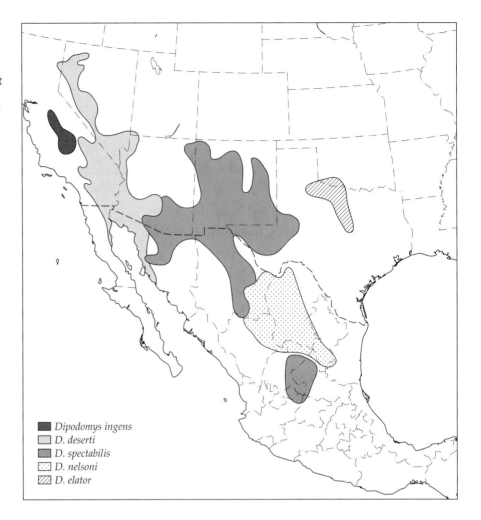

Dipodomys ingens
D. deserti
D. spectabilis
D. nelsoni
D. elator

(formerly *Eutamias*). In mountains of the Southwest, two or three species typically occur in woodlands and forests, but they are segregated by habitat and elevation. *Neotamias dorsalis* inhabits open, xeric woodlands at lower elevations, and is replaced in the denser coniferous forests at higher elevations by a species of the *N. quadrivittatus* group. A third species, *N. minimus*, is sometimes present in spruce and fir forests and above-timberline habitats on the highest peaks. There are at least 24 isolated desert mountain ranges where appropriate habitats seem to be present, but one of these species is absent, apparently because it either never colonized or became extinct sometime in the past. In every case, regardless of which species is absent, the remaining species has expanded its range to include all forested habitats from the edge of the desert to the timberline (Patterson 1980, 1981; Figure 4.27). Such examples of niche and range expansion are particularly convincing evidence of competition when, as in the present case, similar distributional shifts have occurred independently in several different places.

The mechanisms of competitive interaction among these chipmunk species have been investigated in some detail. Brown (1971a) placed feeding stations and observed behavioral interactions in the narrow zone where the ranges of *N. dorsalis* and *N. umbrinus* come into contact. He concluded that *N. dorsalis*, the more aggressive and terrestrial species, was able to exclude *N. umbrinus* from open woodlands by aggressively defending patchy food resources on the

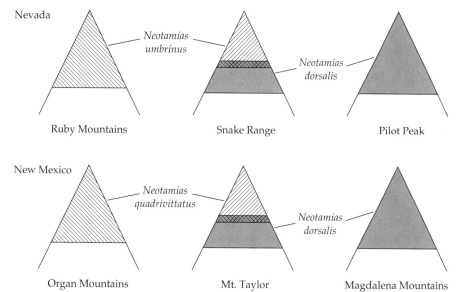

FIGURE 4.27 Diagrammatic representation of elevational distributions of chipmunks (*Neotamias*) on mountain ranges in the southwestern United States. On most ranges two species are present and their ranges overlap only slightly. On some ranges, however, only a single species occurs, in which case its range has expanded to include nearly all habitats and elevations normally occupied by both species. Such natural experiments are good evidence of competitive exclusion, which in this case has been confirmed by field studies.

ground, where chipmunks have to do most of their traveling. However, in denser forests, where food is harder to defend because it is more abundant and chipmunks can travel through the trees, *N. umbrinus* wins out because *N. dorsalis* wastes excessive time and energy on fruitless chases. Chappell (1978) studied a more complex situation where the ranges of several species come into contact in the Sierra Nevada of California. He also found that these mutually exclusive distributions could be attributed to a combination of differences in thermal tolerance and body size among species, with larger species able to coopt the more optimal, forest and woodland habitats.

Such "natural experiments" implicating competition in limiting geographic ranges are by no means confined to small mammals. Diamond (1975) gives several examples of niche and range expansion by birds in the absence of competing species on isolated mountain ranges within New Guinea and on nearby islands. Schoener (1970) and Roughgarden (1996; see also Roughgarden et al. 1983; Pacala and Roughgarden 1985) describe several cases among *Anolis* lizards on islands of the West Indies. Perhaps the best example in plants is the segregation of two species of cattails along a gradient of increasing water depth in marshes in central North America. In an elegant series of reciprocal transplant experiments, Grace and Wetzel (1981) showed that *Typha latifolia* was little affected by its congener, *T. angustifolia*. In the absence of its competitor, however, *T. angustifolia* could grow over the entire gradient.

Competition need not involve similar species to limit distributions. In fact, the kinds of pairwise interactions between closely related, ecologically similar species described above probably account for only a small part of the competition experienced by most species. For example, distantly related plants of different growth forms often experience intense, asymmetrical competition (Johansson and Keddy 1991). We described above how stressful physical conditions and fire limit the distributions of many plant species and vegetation types. The other side of the story is that the plants that are the least tolerant of abiotic stresses (drought, fire, and flooding) are usually superior competitors (Keddy and MacLelan 1990). For example, while some combination of fire, grazing, and drought may prevent trees and shrubs from invading grasslands, these woody plants are superior competitors where these conditions are not too severe. They not only can prevent establishment by grassland species, but during the last two centuries, as fire frequency has been reduced and grazing regimes altered, these woody plants have aggressively invaded grasslands and replaced herbaceous species (Steinauer and Collins 1996) (see Figure 4.25). Such asymmetrical niche relationships, in which one species is limited by competition while the other is restricted by its inability to withstand disturbance or physical stress, appear to be extremely common among many kinds of organisms—recall Connell's study of barnacles.

PREDATION.   **Predation** can be defined as any interaction between two species in which one benefits and the other suffers. According to this definition, relationships between herbivores and their food plants (see Harper 1969), parasites and their hosts, and Batesian mimics and their models, would also be classified as predation. Predator/prey interactions can limit the distribution of either participant because, on one hand, predators may depend on particular prey for food or other benefits necessary to support their own populations, whereas on the other hand predators may limit prey populations by killing or damaging individuals.

When predators are highly specific, it is obvious that their distributions must depend in part on the availability of appropriate prey. It is hardly surprising that the geographic ranges of many specific parasites and herbivores correspond almost precisely with those of their animal or plant hosts. Thus, the distribution of the checkerspot butterfly (*Euphydryas editha*), in coastal California is limited to the immediate vicinity of patches of serpentine soil to which its host plant (*Plantago hookeriana*), is restricted (Ehrlich 1961, 1965). Of course, even highly specific predators may range less widely than their hosts because in some areas their populations are limited by factors other than the availability of suitable prey.

It is much more difficult to document cases in which the distributions of prey populations are limited by their predators. Some of the best examples are artificial in that they involve introductions by humans of predators into regions where they did not originally occur. In some cases these introductions were deliberate attempts to control pest species. Two conspicuously successful examples of such biological control involve drastic reductions in plant populations by introduced herbivores. The prickly pear cactus (*Opuntia stricta*) was introduced into Australia in the mid-1800s to serve as an ornamental garden plant. By the early 1900s the cactus had escaped from cultivation and had become a serious pest on grazing lands. In 1926 Australian scientists introduced a moth (*Cactoblastis cactorum*) whose larva is a specific feeder on *Opuntia* in its native Argentina. By 1940 *O. stricta* had been effectively checked as a pest species in eastern Australia, although small patches of cacti and local populations of the moth remain (Dodd 1959). Unfortunately, many other species of exotic plants are invading Australian habitats, and the search for suitable agents of biological control is ongoing.

Similarly, Klamath weed (*Hypericum perforatum*) was introduced into northwestern North America from Eurasia in about 1900 and became an agricultural pest, but was subsequently controlled by the introduction of a specific leaf-eating beetle (*Chrysolina quadrigemina*) from its native habitat. In the southern part of its range, Klamath weed persists only in small populations along roadsides and in shady areas, but beetle populations do not do well in colder climates, and the weed is more widely distributed in British Columbia (Huffaker and Kennett 1959; Harris et al. 1969). In the cases of both prickly pear cactus and Klamath weed, specific herbivores have drastically reduced plant populations, resulting in very limited local distributions, but they have not greatly altered the distributions of the weeds on a larger geographic scale.

Perhaps the best examples of complete elimination of prey populations from parts of their geographic ranges by voracious but nonspecific predators are provided by artificial introductions of large predatory fishes into certain freshwater habitats. Many of the small native fishes of the southwestern United States have suffered great reductions in their geographic ranges and complete extinction of local populations as a result of the introduction of large predatory game fishes (especially largemouth black bass [*Micropterus salmoides*]) into their habitats (Miller 1961a; see also Braba et al. 1996; Whittier and Kincaid 1999; White and Harvey 2001; Winfield and Hollingworth 2001). Native fishes are not adapted to large generalist predators because they have evolved in isolated lakes, streams, and springs for thousands of years. Zaret and Paine (1973) documented a similar example: the extinction of at least seven species of native fishes in Lake Gatun, Panama, following the introduction of the predatory fish *Cichla ocellaris*. Many of the approximately 300 endemic species of cichlid fishes have declined following the introduction of the Nile perch (*Lates niloticus*) into enormous Lake Victoria in central Africa. The lake trout (*Salvelinus namaycush*) is widely distributed over northern North America, but it is not adapted to the presence of lampreys of the genera *Petromyzon* and *Entosphenus*, which are voracious predators. Niagara Falls formerly prevented *Petromyzon* from entering the upper Great Lakes, which supported large populations of lake trout. Construction of the Welland Canal, however, enabled the lamprey to colonize these lakes. The result has been a precipitous decline of lake trout in the upper Great Lakes despite a major effort to control the lamprey and save this valuable commercial fishery.

In the previous examples, the effects of predators on prey populations are particularly clear because we have been able to observe responses to artificial introductions of predators. Without this historical perspective, however, we would be hard put to infer the extent to which prey are limited by their predators. Today, most remaining patches of *Opuntia* in Australia are not infested with *Cactoblastis*. Similarly, it is difficult to observe black bass preying on pupfishes of the southwestern United States because the native fishes have already been extirpated from most waters where bass are present. It is likely that many prey populations are limited at least in part by their predators, but it is difficult to obtain convincing evidence without performing manipulative experiments.

While the effects of predators have been studied for decades, the effects of parasites and disease-causing microbes received much less attention from ecologists and biogeographers until quite recently. It is apparent, however, that they can also limit both local abundances and geographic distributions. There are an increasing number of examples in which domesticated plants and animals have been able to expand their ranges into previously uninhabitable areas following elimination or control of their parasites or pathogens. Likewise, there are examples of range contraction following the human-aided invasion of new parasites or diseases. One example of the latter is the influence of avian malaria on native Hawaiian birds. Ever since the first humans arrived on the islands, this spectacular endemic avifauna has been suffering extinctions, range contractions, and population reductions (see Chapters 16 and 17). An initial wave of extinctions, presumably due primarily to hunting and habitat destruction, followed the arrival of the Polynesians about 1000 years ago. More extinctions followed European settlement, which began less than 200 years ago. Europeans not only caused additional habitat destruction and brought in exotic avian competitors and mammalian predators, but also initiated biological warfare. Along with the exotic birds they brought in as domesticated fowl, game birds, and pets—and sometimes deliberately released—came avian malaria (*Plasmodium*). This parasite has become pandemic at lower elevations, where its persistence and transmission is favored by warm temperatures, availability of mosquito carriers, and relatively resistant exotic avian hosts (e.g., game birds and domesticated fowl). While it is difficult to tease out the multiple interacting influences of all the factors affecting native bird species, avian malaria has clearly played a major role in the extirpation of most endemic Hawaiian birds from the lower elevations, even in areas where relatively undisturbed habitat remains (Werner 1968; van Riper et al. 1986).

MUTUALISM. The third class of interspecific interactions is **mutualism**, in which each species benefits the other. Examples of mutualistic associations are provided by plants and their animal pollinators and seed dispersers, corals and the photosynthetic zooxanthellae (algae) that live in their tissues, ants and aphids, and cleaner fishes and their hosts. Compared with competition and predation, mutualism has been little studied by ecologists and biogeographers (Boucher et al. 1984; Boucher 1985; Fleming and Estrada 1993). Much remains to be learned about the effects of these mutually beneficial associations on the abundance and distribution of participating populations.

When mutualistic relationships are obligate for at least one—and especially for both—of the partners, then the interaction must have a major influence on distributions. Some plant populations are dependent on the services of specific pollinators for sexual reproduction. For example, red clover (*Trifolium pratense*) did poorly after being introduced into New Zealand until its pollinator, the bumblebee (*Bombus spp.),* was also introduced (Cumber 1953; Free 1970). Janzen (1966) studied the association between ants of the genus *Pseudomyrmex* and trees and shrubs of the genus *Acacia* in the New World tropics. Although many species of *Acacia* have no ants associated with them, and some species of *Pseudomyrmex* may not be dependent on acacias, the relationship is apparently obligate for numerous species. The trees provide the ants with enlarged thorns in which to build their nests and with specialized foods rich in sugars, oils, and proteins. In return, the ants attack herbivorous insects and vertebrates that attempt to feed on the trees, and clear away surrounding vegetation, reducing competition from other plants. Such coevolved specializations apparently have made these two mutualists so dependent on each other that they have virtually identical ranges.

These examples of close ecological and biogeographic associations may be exceptional, however. Often, relationships among mutualists, at least at the species level, are not so obligate, and the influence of this interaction on geographic distributions is not so apparent. Usually at least one of the partners does not absolutely require the mutualistic service of the other, or several different species can supply the service. Thus, most plant/pollinator and plant/seed disperser relationships are not obligately species-specific—at least outside the tropics (Waser et al. 1996).

Figure 4.28 shows the relationship between geographic ranges in North America of eight hummingbird species and some plant species with red tubular flowers that they feed upon and pollinate. There is no correspondence between the ranges of particular pairs of hummingbird and plant species, but the distributions are such that, no matter where they occur, the hummingbirds have flowers to feed upon and the plants have hummingbirds to pollinate them (Kodric-Brown and Brown 1979). A similar relationship is seen in the ranges of clownfishes (*Amphiprion spp.)* and the sea anemones with which

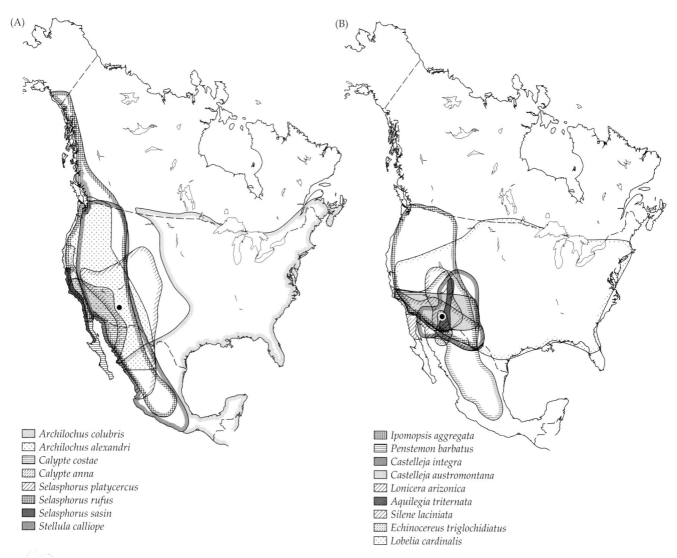

(A)

Archilochus colubris
Archilochus alexandri
Calypte costae
Calypte anna
Selasphorus platycercus
Selasphorus rufus
Selasphorus sasin
Stellula calliope

(B)

Ipomopsis aggregata
Penstemon barbatus
Castelleja integra
Castelleja austromontana
Lonicera arizonica
Aquilegia triternata
Silene laciniata
Echinocereus triglochidiatus
Lobelia cardinalis

FIGURE 4.28 Distributions of temperate North American hummingbirds and some of the plants they pollinate while foraging for nectar. (A) Breeding ranges of the eight species of hummingbirds that occur substantially north of the United States/Mexican border. (B) Geographic ranges of the nine species of red tubular flowers commonly visited by hummingbirds at just one site in Arizona (circle). It would be too confusing to plot ranges of the approximately 130 species of flowers used by these hummingbirds throughout their breeding ranges. Note that despite their close mutualistic relationships, there is little relationship between geographic ranges of the specific plants and their pollinators. (From Kodric-Brown and Brown 1979.)

they are associated (Figure 4.29). While all clownfishes appear to require the protection of anemones, and some kinds of anemones may be unable to exist without clownfishes, the particular species are much more interchangeable.

MULTIPLE INTERACTIONS. In addition to those cases in which it is possible to isolate the limiting effect of one species on the distribution of another, there are undoubtedly many situations in which ranges are structured by more diffuse biotic interactions. Such limits may be the result of different, interacting effects of several species (see earlier discussion of competition between woody and herbaceous plants). MacArthur (1972) noted that the southern limits of the ranges of many North American bird species apparently could be attributed neither to climate (because climate becomes more equable at lower latitudes), nor to habitat (although it might be important for some species), nor to any particular species of competitor or predator. He suggested that for many of these species, the limiting factor must be **diffuse competition** from an increasing number of tropical species (see also Diamond 1975). He noted, for example, that of 202 land bird species that breed in Texas, only 29 also breed in Panama, but that Panama has a total of 564 breeding land bird species. One of the few species that breeds in both the United States and Panama is the yellow warbler (*Dendroica petechia*), a small, insectivorous foliage gleaner. In the

FIGURE 4.29 Distribution of the Indo-Pacific clownfish (*Amphiprion clarkii*) and two sea anemones (*Heteractis malu* and *Macrodactyla doreensis*) that serve as its hosts. *Amphiprion clarkii* is the only fish that is mutualistic with *H. malu,* but it uses other anemones, including *M. doreensis.* As with the hummingbirds and flowers shown in Figure 4.28, even relatively closely coevolved mutualists may be able to switch partners, and therefore do not necessarily have coincident ranges. (Unpublished data courtesy of D. Dunn.)

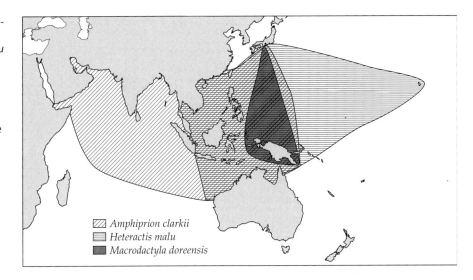

United States the yellow warbler is abundant in a wide variety of shrubby and forested habitats, but in Panama it is restricted to mangrove swamps and small offshore islands (Figure 4.30). Because the forests of Panama contain many species of highly specialized foliage-gleaning birds, whereas mangroves and islands have few such species, MacArthur attributed the distribution of the yellow warbler to diffuse competition—the combined negative effects of competition with many other bird species.

In such isolated cases, however, there is little hard evidence that competition alone is responsible for observed distributional patterns. Other factors such as predators, parasites, diseases, and mutualists, could—and very likely do—play major roles. Recall the example of avian malaria limiting the lower elevational distributions of several native bird species in the Hawaiian Islands. In the absence of historical and epidemiological evidence for the role of this parasite, it would be easy to mistakenly attribute restricted distributions of native species solely to competition with introduced birds. Bertness (1989, 1991, 1993) suggests that in abiotically stressful environments such as salt marshes and intertidal zones, mutualistic interactions may have at least as much influence on the abundance and distribution of species as competition and predation.

Careful studies of highly coevolved species at different trophic levels such as parasites and their hosts or plants and their pollinators, reveal many instances in which such pairs of species have, as we might expect, virtually identical ranges. Most of the species in a community, however, do not interact strongly with each other. The number of possible pairwise interactions between species rapidly increases with the number of species (S) as $(S^2 - S)/2$. If a community contained only 50 species, each species could interact with each of the 49 others, resulting in a total of 1225 possible direct pairwise interactions. Clearly each species cannot be finely adapted to all of the other species with which it coexists. On the other hand, an organism may be involved with just a few other species in strong competitive, predator/prey, or mutualistic interactions that have important influences on its abundance and distribution. In some cases, a single **keystone species** may strongly influence community structure through its direct as well as indirect effects (Paine 1974; Terborgh 1986; Cox et al. 1994; Menge et al. 1994; Paine 1995; Power et al. 1996; Gill 2000).

Interactions among three or more species also may be quite common, and may sometimes be as influential as two-way interactions. For example, many species may share a common limiting resource. Even where competition is

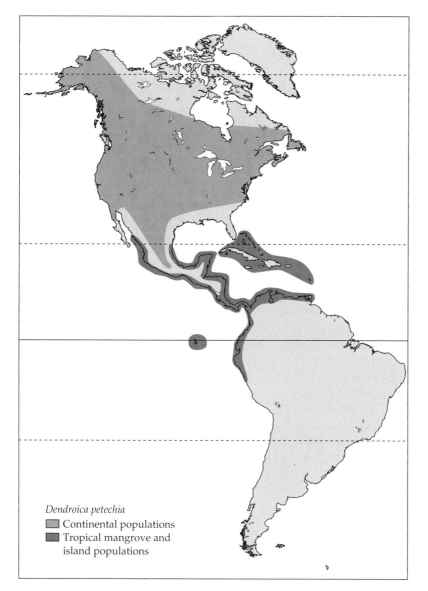

FIGURE 4.30 Geographic range of the yellow warbler (*Dendroica petechia*). This passerine bird is widely distributed over temperate North America, but in the tropics it is restricted to coastal mangrove swamps and islands. MacArthur (1972) suggested that diffuse competition from the many species of small insectivorous birds in most habitats on the mainland of tropical America accounts for the limited distribution of yellow warbler in the tropics.

Dendroica petechia
  Continental populations
  Tropical mangrove and
  island populations

intense, predators can promote the coexistence of otherwise incompatible competitors by preying most heavily on the most abundant prey species (a phenomenon termed **predator mediated coexistence;** Holling 1965; Murdoch and Owen 1975; Roughgarden and Feldman 1975; Caswell 1978; Hus 1981; Holt and Tilman 1994). Changes in the population densities of such predators and other keystone species may have pervasive effects that cascade through the community to affect species across many trophic levels (Carpenter et al. 1985, 1987; Schmitt 1987; Carpenter 1988; Vanni and Findlay 1990; Strauss 1991; Wootton 1992; Holt 1994; Jones and Lawton 1994; Menge et al. 1994; Leibold 1996). In response to the extirpation of sea otters from some regions of their historical range, populations of sea urchins (invertebrate herbivores) increased to the point at which they overgrazed aquatic plants such as kelp, reducing them to a fraction of their original biomass (see Estes and Duggins 1995). On land, the extirpation of wolves and other large predators in Eurasia and North America has also resulted in the ecological release of their herbivorous prey. Populations of deer (*Odocoileus spp.*) and other browsing mammals have increased to the extent that they have become pests. As well as entering suburbs and cities, these animals have denuded the understory of temperate forests, removing important cover and nesting materials, and decreasing the breeding success of native songbirds.

While there is every reason to suspect that these kinds of multispecies interactions also influence distributions and diversity of species on geographic scales, rigorous documentation is usually lacking. Increased emphasis on larger-scale phenomena and nonexperimental methods should lead to a better understanding of the biogeographic consequences of these complex interactions.

Synthesis

Having presented numerous examples of the kinds of abiotic and biotic factors that limit the geographic ranges of species, it is appropriate for us to draw some general conclusions. First, however, a note of caution is in order: Despite all of the examples given above, it may not be productive to search for single limiting factors and simple explanations for the geographic distributions of most species. Not only may a single species be limited by different factors in different parts of its range, but even in one local area, several factors may interact in complex ways to prevent expansion of populations.

Given this complexity, however, there is a hint of one pervasive pattern, which has been alluded to above: Many species appear to be limited on one range margin by abiotic stress and on the other by biotic interactions. This pattern is particularly apparent in studies of distributions along environmental gradients, as of sessile organisms in the intertidal zone and plants along terrestrial gradients of temperature and moisture. An interesting feature of many such examples is that on the margin limited by physical stress there are relatively few other species, whereas on the margin limited by biotic interactions the diversity of other organisms is much higher. This pattern suggests an interesting relationship between species diversity and geographic range limitation: Where conditions are physically stressful for one species, they are likely to be unfavorable for other organisms as well, and vice versa at the other end of the gradient where abiotic conditions are more favorable. This is not surprising. What is more interesting is the suggestion that biotic interactions tend to become limiting when physical conditions are less severe, and that the higher the diversity of other species in the environment, the more likely that some of these will be sufficiently powerful enemies—whether competitors, predators, parasites, or diseases—to limit the range by excluding the species from areas where it otherwise could occur.

In 1853, after developing a map that separated the surface waters of the world's oceans into zones based on isocrymes (lines of mean minimum temperature), James Dana (1853) made the prescient observation that "the cause which limits the distribution of species northward or southward from the equator is the cold of winter rather then the heat of summer or even the mean temperature of the year." Later, Theodosius Dobzhansky (1950) and Robert MacArthur (1972) offered a similar hypothesis for the terrestrial realm: Biotic interactions are more likely to limit abundances and distributions in the tropics, while abiotic stresses are more likely to be limiting at higher latitudes. We will consider the implications of these relationships in Chapter 15 when we discuss the correlates and causes of geographic gradients of diversity across oceanic and continental regions.

In the next chapter we scale up the ecological hierarchy to describe patterns of geographic variation of biotic communities and ecosystems.

The Geography of Communities

N O LIVING THING IS SO INDEPENDENT that its abundance and distribution are unaffected by other species. In Chapter 3, we observed that not only must species respond to environmental variation across the geographic template, but many also modify it and, in turn, significantly influence distributions of other species. The physical and chemical composition of the Earth's surface has been drastically altered by the activities of organisms, which, among other things, have created the oxygen in the atmosphere and contributed to the development of soil. Organisms vary greatly, however, in the extent to which they are dependent on other organisms. Some autotrophic organisms, such as certain kinds of algae and lichens, not only make their own food from sunlight, carbon dioxide, water, and minerals, but also inhabit extremely rigorous physical environments, such as hot springs and bare rocks, where they may encounter and interact directly with few, if any, other species. Heterotrophic organisms, which include all animals, fungi, and some non-photosynthetic vascular plants, cannot make their own food and are dependent on autotrophs for usable energy. Even many photosynthetic plants are directly dependent on specific kinds of organisms, such as nitrogen-fixing bacteria and mycorrhizal fungi, (which make available essential mineral nutrients), and insects and vertebrates, (which pollinate flowers and disperse fruits and seeds).

Historical and Biogeographic Perspectives

Species occur together in complex associations called ecological communities. In the early twentieth century, as the new science of ecology was becoming firmly established, some of its foremost practitioners debated the nature of relationships among species that determine the organization of communities. F. E. Clements (1916) suggested that a community could be regarded as a type of superorganism with its own life and structure as well as its own spatial and temporal limits. According to this view, individual organisms and species could be analogized to the cells and tissues of an organism, and the process of secondary succession could be likened to the growth and development of an individual. In contrast to Clements's concept of the community as a discrete and highly integrated unit, H. L. Gleason (1917, 1926) viewed a community as merely the coexistence of relatively independent individuals and species in the same place at the same time. Focusing primarily on plants, Gleason pointed out that the occurrence of species in an area depends primarily on their individual capacities to immigrate to, and to grow in, the local environment.

As is often the case with such debates, both sides made some important points. Communities do have certain properties, (analogous to those of individuals), that can be measured and studied. These properties include photosynthesis (total **primary productivity**) and metabolism (community **respiration**), as well as more complex processes associated with the transfer and use of energy and nutrients and with changes in species composition and habitat during succession (e.g., Odum 1969, 1971). Some species are interdependent because they require others as mutualists, prey, or hosts (see Chapter 4). On the other hand, plant ecologists such as J. T. Curtis (1959) and R. H. Whittaker (1967, 1975; Whittaker and Niering 1965) amassed much data showing that the abundances and distributions of most vascular plant species vary in time and space as if they were independent of those of other members of their communities (McIntosh 1967; Palmer et al. 1997; Davis et al. 1999; Menge and Lubchenco 2001; Parker 2002).

A question that is of particular interest to biogeographers is: To what extent are species distributed together as interdependent communities as opposed to being distributed essentially independently of one another? If communities are integrated sets of species that are adapted to one another and tolerant of similar physical environments, then we might expect communities to be distributed as discrete units.

Even the casual observer will notice that certain kinds of plants tend to occur together in particular climates to create distinctive vegetation types. Ecologists and biogeographers refer to these as **life zones** and **biomes**—or, in more recent years, **ecoregions**—and recognize that specific kinds of animals and microorganisms are associated with these vegetation formations. For example, a broad band of coniferous forest, sometimes referred to facetiously as the "spruce-moose biome," extends around the world at high latitudes in the Northern Hemisphere. Not only similar vegetation, but also many of the same species and genera of plants, animals, and microbes, are distributed over the Old World from Scandinavia to Siberia and across the New World from Alaska to Nova Scotia and Newfoundland. These organisms generally are adapted to living with one another in similar physical environments. Their ranges extend southward together where mountain ranges provide appropriate habitats at high elevations. In North America these coniferous spruce-fir forests extend southward along the Appalachian Mountains in the east and along the Cascades, Sierra Nevada, and Rocky Mountains in the west. Several kinds of organisms are restricted to these coniferous forest habitats and have similar geographic distributions (Figure 5.1). We should be cautious, however, not to infer from such patterns that communities represent discrete, highly integrated ecological and biogeographic units, because there are alternative explanations, including historical ones (see Chapters 10, 11, and 12).

Communities and Ecosystems

Definitions

Communities and ecosystems—the highest levels of biotic and ecological organization—are rather arbitrarily defined. As stated above, a community

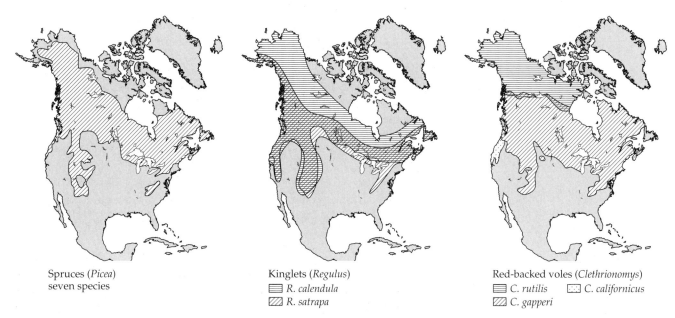

Spruces (*Picea*)
seven species

Kinglets (*Regulus*)
▤ *R. calendula*
▨ *R. satrapa*

Red-backed voles (*Clethrionomys*)
▤ *C. rutilis*   ⬚ *C. californicus*
▨ *C. gapperi*

FIGURE 5.1   Three groups of organisms inhabiting coniferous forests exhibit similar geographic ranges in North America. Seven species of trees (not shown individually), two species of birds, and three species of mammals are typical inhabitants of the coniferous forests that spread across the northern part of the continent and extend southward at high elevations in the mountains.

consists of those species that live together in the same place. A member species can be defined either taxonomically or on the basis of more functional ecological criteria, such as life form or diet. Ecologists study two fundamental properties of communities: community structure (which refers to static properties that include diversity, composition, and biomass of species) and community function (which includes all of the dynamic properties and activities that affect energy flow and nutrient cycling such as photosynthesis, interspecific interactions, and decomposition). The place where member species occur can be designated either by arbitrary boundaries or on the basis of natural topographic features. Thus we can speak of the fish community in some ecologist's 2-hectare study area on a coral reef on the north shore of Jamaica; of the grazing community (which would include representatives of several invertebrate phyla as well as some fish species) on that entire reef (which might extend uninterrupted over many square kilometers); or of the entire community of all coral reef-inhabiting organisms in the Caribbean Sea.

An **ecosystem** includes not only all the species inhabiting a place, but also all of the features of that place's physical environment. Because ecosystem ecologists are interested primarily in the exchange of energy, gases, water, and minerals among the biotic and abiotic components of a particular ecosystem, they often try to study naturally confined areas where input and output of energy and materials are restricted and hence easier to control or monitor (e.g., Odum 1957; Teal 1957, 1962; Hutchinson 1957, 1967, 1975, 1993). Small, relatively self-contained ecosystems are often called **microcosms** because they represent miniature systems in which most of the ecological processes characteristic of larger ecosystems operate on a reduced scale. A sealed terrarium is a good example of an artificial microcosm, and a small pond is an example of a natural one.

Few ecosystems are as isolated and independent as they might appear. The largest and only completely independent ecosystem is the **biosphere**, which encompasses the entire Earth. The interdependence of all ecosystems is vividly demonstrated by the widespread impact of human activities. Acid rain falls on virgin forests and is carried into pristine lakes far from the source of pollution. Deforestation—especially the cutting of tropical forests—and the burning of fossil fuels are changing the composition of the atmosphere, and perhaps altering the climate, throughout the world.

Many ecologists are actively investigating the structure and function of communities and ecosystems. As is usually the case in a rapidly developing field, many important questions remain unanswered, and some tentative conclusions are highly controversial. Because many of these unresolved issues are relevant to biogeography, this situation can be both challenging and frustrating for those trying to account for plant and animal distributions. On one hand, biogeographers have an opportunity to use their own methods and data to make important advances in community ecology. Researchers with a broad biogeographic perspective are in a position to make unique contributions to the unraveling of interacting ecological processes that influence the numbers and kinds of species that live together in different parts of the Earth. On the other hand, until these ecological processes are better understood, biogeographers will be unable to integrate them with the effects of historical events in order to provide a solid conceptual basis for interpreting and predicting distribution patterns.

Community Organization: Energetic Considerations

In Chapter 4 we emphasized that each species—the fundamental unit of an ecosystem—has a unique ecological niche that reflects the biotic and abiotic environmental conditions necessary for its survival. How are these individual niches organized to produce complex associations of many species? This is the ultimate question of community ecology, and one for which we can advance only tentative answers. Two characteristics of species that strongly influence their effects on community organization are their body mass and trophic status (i.e., whether they are primary producers, herbivores, carnivores, or decomposers.) In fact, if you sought to take a single measurement that would provide the most information on the biology of an organism, you would be best advised to measure its body mass. The larger an organism, the more energy it requires for maintenance, growth, and reproduction. The rate of energy uptake and expenditure of animals at rest [or the **basal metabolic rate** (m)] varies with body mass (M) according to the relationship $m = cM^{0.75}$. Although the value of the constant (c) varies somewhat among taxonomic groups, the exponent (0.75) is very general (Hemmingsen 1960; Peters 1983; Calder 1984; Figure 5.2). The ecological consequences of this simple formula are profound. Because the exponent is positive, total energy requirements increase with body size. It takes about 14,000 times more energy to maintain a 5000 kg elephant than a 15 g mouse. The same applies to all organisms: a mature redwood tree uses much more energy than a strawberry plant. On the other hand, because the exponent is less than 1, the metabolic rate *per unit of mass* of a small organism is greater than that of a larger one. It requires about 25 times more energy to maintain a gram of mouse than a gram of elephant. Apparently for this reason, almost all rate processes (e.g., cardiac rate and respiration frequency in vertebrates) are accelerated in small organisms, which are more active and have higher reproductive rates and shorter life spans than do large organisms.

Although metabolic rate and energy requirements increase as less than linear functions of body mass, storage capacities (e.g., energy stored as fat, water volume, mineral stores in bone tissue) increase in direct proportion to mass. Therefore, all else being equal, larger organisms have greater capacities to withstand prolonged stresses such as starvation, dehydration, and subfreezing temperatures. Body size also has important ecological consequences because it influences the scale at which organisms use the environment. Because small organisms require fewer resources per individual than large ones, they can use smaller areas, be more specialized, and still maintain population densities

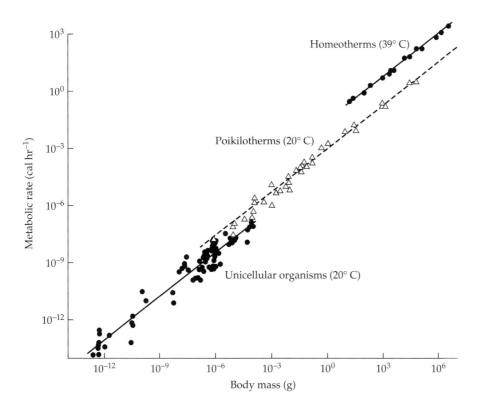

FIGURE 5.2 Relationship between metabolic rate (m) and body mass (M) for a wide variety of organisms, ranging from unicellular forms to poikilothermic ("cold-blooded") animals to homeothermic birds and mammals. Note that the axes are on a logarithmic scale, so that the relationship is described by a power function of the form $m = cM^{0.75}$, where the constant (c) varies slightly among the three different groups, but the exponent or slope (0.75) is remarkably constant. (After Hemmingsen 1960.)

high enough to avoid extinction. Small organisms are better able than large ones to respond to what Hutchinson (1959) termed the mosaic nature of the environment—the spatial heterogeneity or patchiness of the environment that is most pronounced on a small scale. Consequently, small organisms have been able to divide the environment more finely, so that any geographic area contains a greater number of small-bodied species than large ones (Van Valen 1973a; May 1978; Brown and Maurer 1986; Figure 5.3). Consider the tremendous diversity of insects as compared with terrestrial vertebrates (about 750,000 versus 24,000 known species, respectively). A biogeographic correlate of this pattern is that, at least among animals, large organisms are constrained to having broad geographic ranges. Reasons for this should be obvious: because large individuals require more space (McNab 1963; Schoener 1968b; Kelt and Van Vuren 1999, 2001), carrying capacity of the environment for these animals is low, and only large areas can support sufficient numbers of individuals to maintain the species over evolutionary time. In contrast, small areas can support large populations of small organisms, which consequently have a low probability of extinction. Small species may or may not have very restricted ranges (Figure 5.4; see also Chapter 15).

The **trophic status** of organisms (or how they acquire energy) also influences the role they play in community structure. Outside of a handful of fascinating exceptions (e.g., chemosynthetic communities of geothermal vents on the ocean floor), all of the energy used by living things ultimately comes from

FIGURE 5.3 Frequency distribution of body size among species for several different kinds of organisms: (A) terrestrial mammals of the world; (B) land birds of the world; (C) beetles of Great Britain; and (D) all terrestrial animals of the world (approximately). Note that the axes are on a logarithmic scale, so that small species are much more numerous than large ones. This pattern is very general, and accounts for the obvious fact that insects are much more numerous (in numbers of species as well as individuals) than vertebrates in terrestrial environments. (After May 1978.)

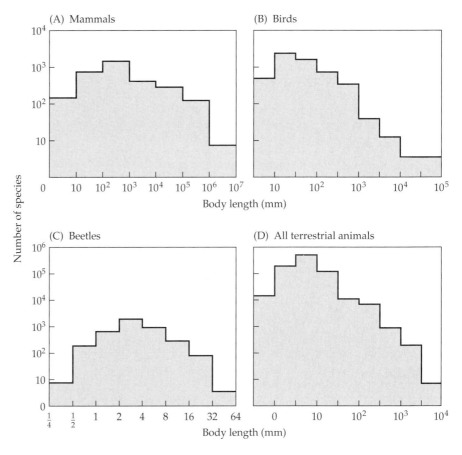

FIGURE 5.4 Relationship between area of geographic range and body mass among North American terrestrial mammals. Note that although there is much variation, the areas of the smallest ranges increase with increasing body mass and are smaller for herbivores (A) than for carnivores (B). Open circles indicate species that actually have larger ranges than indicated because they range into Central America (only the North American area of the distribution has been measured). (After Brown 1981.)

the sun. Autotrophic green plants use solar radiation, carbon dioxide, water, and minerals to synthesize organic compounds, and produce oxygen as a by-product. These organic compounds are not only used by plants for making structures and fueling basic metabolism, but are also the sole source of energy for heterotrophic organisms in the community. Solar energy, trapped in organic molecules, is transferred from one species to another and gradually used up as herbivores eat plants, carnivores eat herbivores, and so on. These heterotrophic species oxidize organic compounds to obtain usable energy, and in the process consume oxygen and release carbon dioxide.

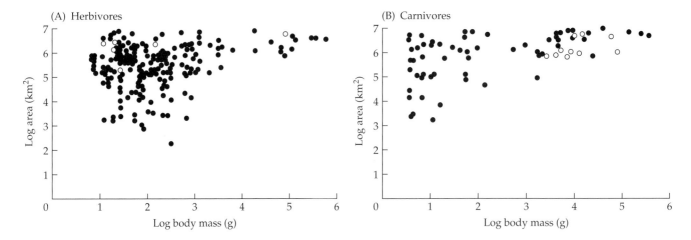

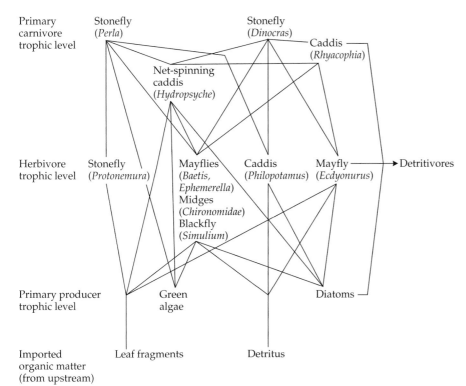

Primary
carnivore
trophic level

Herbivore
trophic level

Primary producer
trophic level

Imported
organic matter
(from upstream)

FIGURE 5.5 Part of the food web for an aquatic community inhabiting a small stream in Wales. The diagram shows interconnections of food chains to form food webs. There are three trophic levels, but some organisms, such as *Hydropsyche*, which feeds on both plant and animal material, occupy intermediate positions. The many individual species of plants, green algae, and diatoms that make up the primary producer trophic level are not shown individually. (After Jones 1949.)

Unidirectional paths of energy that flow between species and through communities are termed **food chains**. Different links in a food chain are called **trophic levels** (Figure 5.5). The first level contains green plants, or **primary producers**; the second, **herbivores**, or **primary consumers**; the third, **carnivores**, or secondary consumers; and so on. At the ends of food chains are decomposers, or **detritivores**, consisting mostly of bacteria and fungi, which break down remaining organic matter and release inorganic minerals into the soil or water, where they can be recycled and again taken up by plants.

Although each tiny packet of energy follows a linear path along a food chain, actual trophic relationships among species are complex. Because most consumers feed on several species and are in turn consumed by several kinds of predators and decomposers, the aggregate paths of energy that flow through a community form interconnected branching patterns called food webs (see Figure 5.5; see also Schoener 1989; Winemiller 1990; Pimm 1991; Morin and Lawler 1995; Winemiller et al. 2001; Dunne et al. 2002). As energy is distributed through a community in this fashion it obeys the laws of thermodynamics, which have important implications for community organization. Simply stated, the **first law of thermodynamics** says that energy is neither created nor destroyed, but can be converted from one form to another. During photosynthesis, for example, solar energy is converted into high-energy bonds in the organic molecules of plants, which in turn can be consumed by animals and transformed into ATP to support muscular contractions. The **second law of thermodynamics** states that as energy is converted into different forms, its capacity to perform useful work diminishes, and disorder (entropy) of the system increases. Organisms use energy stored in organic molecules to perform the work of moving, growing, and reproducing. As organic compounds are oxidized to provide energy for these activities, much of the energy is dissipated as heat. In fact, most organisms are very inefficient. Most animals are able to incorporate only 0.1% to 10% of the energy

FIGURE 5.6 Ecological pyramids of energy flow, biomass, and number of individuals for several different communities. Note that three of the pyramids (B–D) are regular, with each successive trophic level reduced compared with the one below, but two of the pyramids (A, E) are not. The pyramid of individuals for the temperate forest in Great Britain (A) and the pyramid of biomass for the English Channel (E) have fewer individuals and a lower biomass of primary producers than herbivores. P, primary producers; H, herbivores; C_1, primary carnivores; C_2, secondary carnivores. (After Odum 1971.)

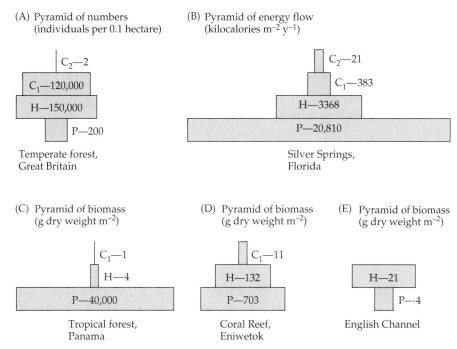

(A) Pyramid of numbers (individuals per 0.1 hectare)

C_2—2
C_1—120,000
H—150,000
P—200

Temperate forest, Great Britain

(B) Pyramid of energy flow (kilocalories $m^{-2} y^{-1}$)

C_2—21
C_1—383
H—3368
P—20,810

Silver Springs, Florida

(C) Pyramid of biomass (g dry weight m^{-2})

C_1—1
H—4
P—40,000

Tropical forest, Panama

(D) Pyramid of biomass (g dry weight m^{-2})

C_1—11
H—132
P—703

Coral Reef, Eniwetok

(E) Pyramid of biomass (g dry weight m^{-2})

H—21
P—4

English Channel

they ingest into the organic molecules of their own bodies; the remaining 90% to over 99% is "lost" as heat. [Note, however, that in endotherms (birds and mammals) and an intriguing selection of other animals, heat lost from cellular metabolism plays a critical role in maintaining relatively warm body temperatures.]

One ecological consequence of these thermodynamic properties of organisms is that substantially smaller quantities of energy are available to successively higher trophic levels. This can be shown diagrammatically by portraying the community as an ecological pyramid (Figure 5.6). In any community, green plants have the highest rates of energy uptake, and they usually comprise the largest number of individuals, the greatest number of species, and the greatest **biomass** (the total quantity of living organic material). Successively higher trophic levels tend to have less than 10% of the rate of energy uptake of the level below them, and usually contain a proportionately lower biomass as well as fewer individuals and species (Lindeman 1942).

Availability of energy must decrease substantially with each successive trophic level in accordance with the laws of thermodynamics, but there are exceptions to the other patterns. Distribution of biomass, individuals, and species among trophic levels depends on how energy is acquired and used by species in the community. In some communities, such as those inhabiting caves (Poulson and White 1969) and the dark depths of ocean trenches, there are no autotrophic organisms, and all energy is imported in organic form, usually as dead material called **detritus**. Deciduous forests of eastern North America and western Europe may have inverted pyramids of abundance and species richness—that is, fewer individuals and species of primary producers than of primary consumers. Here, most of the energy used by plants is monopolized by a relatively small number of large trees, whereas energy used by herbivores is allocated among many individuals and species of small insects (Elton 1966; Varley 1970) (see Figure 5.6A). Inverted pyramids of biomass also are possible and, in fact, quite common in some aquatic ecosystems (see Figure 5.6E). In the open ocean, photosynthetic rates of phytoplankton may be so high that these tiny producers can support a biomass of consumers far exceeding their own. This is possible because the consumers depend on rates of energy transfer (i.e., molecules metabolized per unit of time) and not simply biomass per se. Phytoplankton populations are so productive that they can replace themselves many times per day and thus support high rates of herbivory by zooplankton. However, consistent with the first law of thermodynamics, pyramids of energy flow can never be inverted.

Because the **carrying capacity** (measured in units of usable energy) of any area is lower for successively higher trophic levels, organisms occupying these levels exhibit predictable characteristics that affect their ecological roles and geographic distributions. Not only are there fewer species of carnivores than of herbivores and plants, but the carnivores also tend to be larger and more generalized than herbivores. Carnivores usually have to be large enough to overpower their prey. They also tend to feed on several prey species, and to have relatively broad habitat requirements and wide geographic distributions (see Figure 5.4; Rosenzweig 1966; Van Valen 1973a; Wilson 1975; King and Moors 1979; Zaret 1980; Brown 1981; Gittleman 1985). The cougar or mountain lion (*Puma concolor*), for example, is one of the top carnivores in the New World. It weighs 50 to 100 kg and takes a variety of prey, mostly mammals ranging in size from rabbits to elk. The cougar also has the widest geographic distribution of any American mammal. Its range extends from Alaska to the southern tip of South America and, before European settlement, from the Atlantic to the Pacific coasts on both continents. The cougar inhabits tropical rain forests, coniferous and deciduous hardwood forests, shrublands, and deserts. Other top carnivores, such as the wolf (*Canis lupus*) and jaguar (*Panthera onca*), also have broad geographic and habitat distributions, whereas herbivores of comparable size tend to have more restricted ranges.

Parasites are an exception to these patterns, but they still illustrate the fundamental importance of energetic relationships in community organization. Parasites are usually much smaller and more highly specialized than their hosts. Consequently, individual parasites need not consume large numbers of host individuals to meet their energy requirements, and parasites can be more numerous than their hosts. One or more parasites usually inhabit a host for an extended period, often for the entire life span of either parasite or host. Parasites are often highly specialized to infect only one or a few, often taxonomically related, host species. Although these adaptations have allowed parasites to maintain geographic ranges and numbers of individuals roughly comparable to those of their hosts in the trophic level below them, many species have had to solve the problem of finding and infecting sparsely distributed hosts. The elaborate, highly specialized life cycles of many parasites appear to be largely adaptations for locating and infecting appropriate hosts. Price (1980), Hawkins (1994), and Bush and his colleagues (Bush et al. 2001) provide interesting accounts of some special features of parasite ecology, biogeography, and evolution.

These kinds of energetic considerations lead to the prediction that capacities of habitats or geographic areas that support many individuals and diverse species of organisms should ultimately depend on their total productivity. Everything else being equal, the higher the fixation rate of sunlight into organic material, the more usable energy should be available to be subdivided among individuals, species, and trophic levels. As we shall see, productivity varies greatly among different habitats, depending on such factors as climate, soil type, water availability, and the influence of human activity (Rosenzweig

1968; Jordan 1971; Lieth 1973; Whittaker and Likens 1973). In general, the predicted relationship between productivity and diversity is observed (see Chapter 15). Widespread, highly productive habitats such as tropical rain forests and coral reefs are renowned for their great diversity of specialized species. In contrast, small, isolated areas, such as small islands, and widespread, unproductive habitats, such as boreal forests and tundra, contain fewer and, for the most part, more generalized species. We will examine such patterns of species diversity in more detail in Chapters 13, 14, and 15.

Distribution of Communities in Space and Time

Spatial Patterns

Because species often are adapted to the same physical conditions as other, ecologically similar species, we might expect biotic communities to be distributed as discrete units, with rapid turnover of species as we move along an environmental gradient from one community to the next. On the other hand, because of competition, we might expect ecologically similar species to adapt to different niches, and thus exhibit exclusive, or non-overlapping distributions.

It is obvious that abrupt changes in the environment, such as those found at the shore of a lake or the edge of a forest, will usually be accompanied by rapid transitions between two different associations of species. If the environmental discontinuity is rapid and severe, most of the species living on each side will be limited almost simultaneously when they encounter inhospitable conditions at the border. If the two habitat types are not too dissimilar, however, the edge (or **ecotone**) may actually contain more species than either pure habitat type (Holland et al. 1991). Species from either side of the boundary may be able to mix and occur together in the narrow area where the two environments meet. If the transition zone is fairly productive and not too narrow, it may even support its own assemblage of organisms uniquely adapted to live there. This is especially likely if the boundary is not simply an ecotone, with conditions intermediate between those on either side, but an environment in which conditions are unique or fluctuate back and forth between those found on either side. The most dramatic example of such a community is provided by the intertidal zone. Special adaptations are required to withstand periodic inundation in seawater followed by exposure to a desiccating terrestrial environment. Along a narrow strip of shore live entire communities of such plants and animals, which show their own small-scale patterns of association and segregation within the vertical gradient of tidal exposure (e.g., Connell 1961, 1975; Menge and Sutherland 1976; Lubchenco and Menge 1978; Lubchenco 1980; Souza et al. 1981; Peterson 1991; Menge et al. 1994).

How are species distributed along environmental gradients, such as the relatively abrupt gradient in exposure within the intertidal zone or the more gradual gradient in climatic conditions, caused by latitudinal or elevational changes? Some of the most thorough studies of such patterns called **coenoclines** have been those of Robert H. Whittaker and his colleagues on the distribution of tree species on mountainsides in the United States. Their data can be used to evaluate five alternative hypotheses (see Whittaker 1975; Figure 5.7):

1. Groups of species exhibit similar ranges along the gradient and are distributed as discrete communities ("superorganisms" in the Clementsian sense) with sharp boundaries between them. This pattern could be caused by competitive exclusion between dominant species and by other species evolving to coexist with the dominants and with one another (see Figure 5.7A).

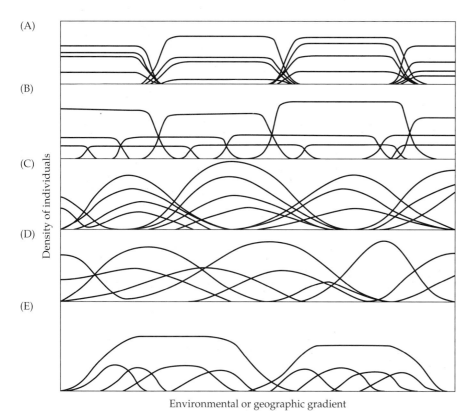

(A)

(B)

(C)

Density of individuals

(D)

(E)

Environmental or geographic gradient

FIGURE 5.7 Five hypothetical coenoclines—patterns of distributions of species along an environmental or geographic gradient. (A) Species are distributed as discrete communities that replace each other abruptly. (B) Species are not segregated into communities, but some sets replace each other abruptly. (C) Species are distributed as discrete communities, which gradually replace each other. (D) Species behave as if they are independent of each other, neither associating in discrete communities nor replacing each other abruptly at transition zones, or "ecotones." (E) Most species are nested within the ranges of a few dominant species, but otherwise occur independently of each other. (After R. H. Whittaker 1975.)

2. Individual species abruptly exclude one another along sharp boundaries, but most species are not closely associated with others to form discrete communities (see Figure 5.7B).

3. Much as in hypothesis 1, species form discrete communities, but replacement of communities along the gradient is gradual. This could happen if groups of species evolved to coexist with one another, but competitive exclusion did not cause rapid replacement of species along the gradient (see Figure 5.7C).

4. Individual species gradually appear and disappear, seemingly independent of the presence or absence of other species. Species neither competitively exclude each other nor associate to form discrete communities, and species replacement along the gradient is random (see Figure 5.7D).

5. Ranges of most species are nested within the ranges of a few dominant species that are over-dispersed (non-overlapping) along the gradient. Thus, species distributions may appear highly ordered at geographic scales (with little overlap among assemblages), but random at local scales (see Figure 5.7E).

Whittaker and his associates (e.g., Whittaker 1956, 1960; Whittaker and Niering 1965) sampled several sites at varying elevations on mountains. Sample sites were chosen so as to keep soil type, slope, and exposure as constant as possible, but to allow natural variation in temperature and rainfall. The results of two such studies (shown in Figure 5.8) clearly support the hypothesis that species are distributed as if they were independent of one another, showing no evidence of either abrupt replacements that might be attributed to competitive exclusion or association of species to form discrete communities.

FIGURE 5.8 Coenoclines illustrating the distributions of tree species along two moisture gradients: (A) in the Siskiyou Mountains, Oregon, 760 to 1070 m elevation; and (B) in the Santa Catalina Mountains, Arizona, 1830 to 2140 m elevation. Note that the species replace one another gradually and seemingly independently of one another. The species have narrower elevational ranges along the steeper moisture gradient in Arizona. (After Whittaker 1967.)

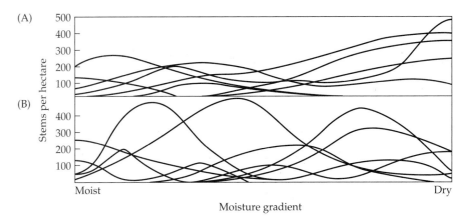

It should be pointed out, however, that Whittaker's methods of collecting and analyzing data may have contributed to his obtaining the patterns shown in Figure 5.8. By censusing relatively large plots and then averaging the results across several mountain slopes, Whittaker would have tended to miss abrupt replacements of species owing to competitive exclusion and mediated by the distribution of microsites in a spatially heterogeneous environment (see Davis et al. 1999; Parker 2002). Yeaton (e.g., Yeaton et al. 1980; Yeaton 1981) has made careful analyses of the elevational distributions of pines (*Pinus*) in western North America. He found many instances of abrupt replacements by species of similar growth form, apparently as a result of interspecific competition (Figure 5.9; see also Shipley and Keddy 1987). For example, gray pine (*P. sabiniana*)

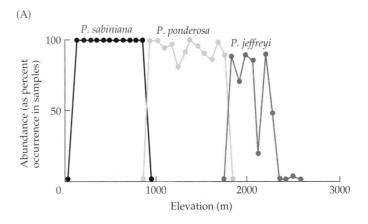

FIGURE 5.9 Elevational distributions of pine (*Pinus* spp.) trees on western slopes of the Sierra Nevada in California, including (A) species of three-needled pines, and (B) species of five-needled pines. The species with the same number of needles are morphologically and ecologically similar, and they overlap little in elevation on sites with similar slope, exposure, and soil type. Note also that distributions of these pines (three- and five-needled species, combined) suggests some degree of community organization, with *P. ponderosa* and *P. lambertiana* occupying similar elevations, as do *P. jeffreyi* and *P. monticola*, while *P. sabiniana* and *P. albicaulis* exhibit exclusive distributions (at the lowest, and highest elevations, respectively). (After Yeaton 1981.)

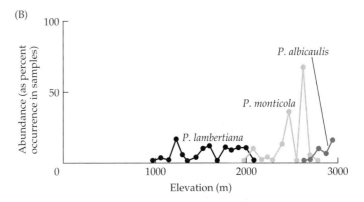

reaches its upper elevational limit and is abruptly replaced by ponderosa pine (*P. ponderosa*) at approximately 840 m on southeastern-facing slopes of the Sierra Nevada near Yosemite National Park. Gray pine is at its greatest density and growth rate in mesic environments at higher elevations just before it is supplanted by ponderosa pine, suggesting that it is competitive exclusion, not inability to tolerate the physical environment, that determines the upper limit of its range. Thus, although Whittaker's results accurately reflect the typical distribution of many plant species along environmental gradients, abrupt replacements by competing species can undoubtedly occur in many cases in which ecologically similar or closely related species come into contact (e.g., chipmunks along elevational gradients; see Figure 4.27).

Temporal Patterns

ECOLOGICAL SUCCESSION.   Individual species and entire ecological communities replace each other over time as well as in space. These changes are of great interest both to ecologists, who study relatively short time scales, and to paleontologists, who study very long time scales. Many ecologists, especially those studying sessile organisms, have devoted much effort to the study of succession (Cowles 1899, 1901; Cooper 1913; Clements 1916; Beckwith 1954; Monk 1968; Odum 1969; Drury and Nisbet 1973; Horn 1974, 1975, 1976, 1981; Connell and Slayter 1977; Usher 1979; McIntosh 1981; Miles 1987; Facelli and Pickett 1990; Farrell 1991; McCook 1994). Succession, as we saw in Chapter 3, is progressive change in community structure and function over ecological time. Succession is a normal process in any ecosystem in which disturbances repeatedly eliminate entire communities from patches of local habitat. If all soil and organic material is removed—as by a glacier or a volcanic eruption—the process is termed primary succession. On the other hand, if a disturbance such as a fire or storm removes most of the living organisms but leaves the soil intact, the process is termed secondary succession.

Despite decades of research, succession theory remains a hotly debated topic among ecologists (McIntosh 1999; Walker and Moral 2003). Much of the controversy echoes the Clementsian-Gleasonian debate over community organization; indeed, much of Clements's work was devoted to the study of succession. Here, the question is whether communities replace each other as interdependent units or as collections of independent species over *temporal* gradients (rather than spatial gradients, as in Figure 5.7). Clements's view, which dominated successional theory for most of the twentieth century, held that succession was deterministic, predictable, and convergent. Pioneering species colonize a site and then modify it to the point that another, better adapted, set of species can invade, outcompete, and replace the first. This process is then repeated through a progressive sequence of communities, or **sere**, until a relatively stable "climax" community is established. According to the Clementsian school of thought, successional seres are characterized by an increase in biomass, complexity, and stability (Odum 1969). The climax community and its particular sere are largely determined by climate and soil conditions, and therefore are predictable and characteristic of a given region (or biome).

In contrast to this orderly, deterministic view of succession, the Gleasonian camp observed an impressive diversity of successional seres and alternative climaxes, even for a given region. These ecologists argued that succession may not be the orderly process envisioned by Clements. Rather than being driven largely by species interactions and autogenic modification of local conditions, succession may simply reflect the idiosyncratic capacities of independent species to disperse, establish themselves, and survive at a local site with a particular combination of environmental conditions.

This debate, while nearly a century old, is not likely to be resolved soon. Until recently, succession was largely regarded as a botanical phenomenon. Whether deterministic or stochastic, it was viewed as a sequence of progressive changes in plant communities. Plant ecologists acknowledged that animals exhibited similar successional dynamics, but felt that they did so only because of their dependence on plants. This view, of course, is problematic. Grazers, granivores, pollinators, and decomposers—whether animals, bacteria, fungi, or protists—all influence plant community structure, and animal communities also exhibit progressive changes in species composition during primary and secondary succession (e.g., see McCook and Chapman 1997). Indeed, the search for the driving force, or forces, of ecological succession may require ecologists to look outside their taxonomic biases and study communities defined by functional—rather than taxonomic—criteria.

As with all debates between two such extreme points of view, it is likely that the answer to this one lies somewhere in between. Both the purely deterministic Clementsian camp and the stochastic Gleasonian camp agree that within a region of similar soil and climate, succession is a directional process that tends to result in communities characterized by a predictable set of dominant species.

PALEOECOLOGICAL PERSPECTIVES. Paleoecologists and historical biogeographers also study temporal changes in communities and biotas, albeit over larger geographic areas and longer time periods. Regional biotas are strongly influenced by the major shifts in climate and soil conditions that have occurred throughout history. Paleobotanists have often used analyses of fossil pollen to document the reestablishment of eastern deciduous forest communities in response to climatic and geological changes that followed the retreat of the last glaciers in North America (Figure 5.10; see Davis 1969, 1976; see also Bernabo and Webb 1977; Grayson 1993; Peng et al. 1995). Julio Betancourt and his colleagues (e.g., see Betancourt et al. 1990; Betancourt 2004) analyzed pack rat middens (plant material collected by these rodents and preserved in caves and in rock crevices) to reconstruct the dynamics of vegetation in arid regions of western North America over the past 40,000 years. Such studies have revealed that the retreat of glaciers and associated climatic changes resulted in major shifts in terrestrial vegetation far removed from the ice sheets. Communities, however, did not behave as perfectly interdependent units. Apparently, rates of invasion depended in large part on mechanisms of seed dispersal and other aspects of the life history that differed markedly among species. While some groups of species shifted in concert, others became associated with different communities, while still others were unable to adapt and went extinct (see Jackson 2004, and Chapters 9 through 12).

Paleoecologists such as those cited above study the dynamics of biotas over the span of millennia. Other paleobiologists focus on major upheavals that have occurred on an evolutionary, or geological, time scale (i.e., on the order of millions of years). Taxonomic specialization of most paleontologists and the fragmentary nature of the fossil record pose formidable challenges for those attempting to reconstruct changes in communities that have occurred over geological time (Gray et al. 1981; Behrensmeyer et al. 1992). Much attention has been focused on **mass extinctions**—episodes of relatively abrupt (on the span of a few million years) replacement of virtually entire biotas. Although the exact causes of these catastrophic changes are still hotly debated, many, if not all, must have been triggered by drastic environmental perturbations. In many ways these biotic upheavals are comparable to the succession that occurs after ecological disturbances, or to abrupt spatial replacement of com-

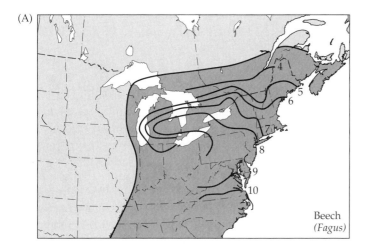

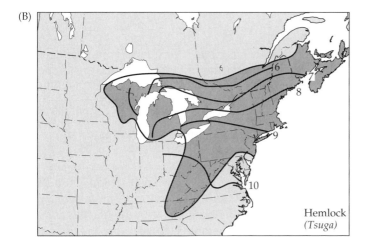

FIGURE 5.10 Reconstruction from fossil pollen records of the recolonization of North America by two tree species, (A) beech (*Fagus*) and (B) hemlock (*Tsuga*), after the last Pleistocene glaciation. Numbered lines indicate the fronts of each species' range at 1000-year intervals B.P., showing the progressive northern migration of each species. Note that the migration of these two tree species was quite different, although the northern borders of their present ranges (darker areas) are virtually identical. (After Bernabo and Webb 1977.)

munities across major environmental discontinuities. These catastrophic events were spectacular, but they were the exceptions—these short pulses of extinction and evolutionary change were interspersed with extremely long spans of more gradual changes in regional biotas.

Throughout geological time, many species have colonized, speciated, and become extinct relatively independently of one another in a pattern much like that of the spatial replacement of many contemporary species observed along gradual physical gradients. As evidence of this pattern we can point to the existence in modern communities of recently evolved species alongside "living fossils" (e.g., the ginkgo, horseshoe crab, and coelacanth), forms that have survived virtually unchanged for hundreds of millions of years. However, we must caution against sweeping generalizations. While much of the fossil record suggests independent shifts of species, other evidence may reflect interdependence among species. Mutualism, widespread today (Janzen 1985), must also have been common in the distant past. We know from research on contemporary communities that the loss of a keystone species, or the invasion and establishment of an exotic species, can cause major changes and even wholesale reorganization of communities. It is therefore likely that the evolution, range shifts, and extinctions of such keystone species contributed to the dynamics and upheavals of regional biotas in the past (e.g., see Petuch 1995). The fossil record, riddled as it is with gaps and mysteries, also may hold many insights into the forces structuring ecological communities, both past and present.

Before delving any further into the past, we now present an overview of contemporary terrestrial biomes and aquatic communities.

Terrestrial Biomes

The fact that communities do not represent perfectly discrete associations of species in either time or space obviously complicates any attempt to classify the communities of species that occur in particular environments or geographic regions. Where physical and geographic changes are abrupt, it is relatively easy to recognize distinct community types, but where environmental variation is gradual, we are faced with the problem of dividing the essentially continuous variation in species composition, life form, and other community traits into a discrete number of arbitrary categories. The human mind seems to require, or at least to depend heavily upon, such categories (e.g., the stages of mitosis, the geological periods, the biogeographic regions) even when we are well aware that they represent artificial divisions of continuous processes or variables.

In the last few decades, mathematicians and biologists have developed sophisticated multivariate statistical techniques for quantifying the degree of similarity (or difference) between two samples based on a large number of variables. Many plant and animal ecologists have applied these methods to ecological and biogeographic data (see Pielou 1975, 1979; Omi et al. 1979; Robinove 1979; Rowe 1980; Smith 1983; McCoy et al. 1986; Birks 1987; Cornelius and Reynolds 1991; Bailey 1996). If comparable measurements are available for a large number of communities, these techniques can be used to group them into hierarchical clusters reflecting their similarities in species composition, life form, or other attributes of interest. As we shall see in Chapters 11 and 12, these statistical methods are extremely useful for detecting quantitative patterns of floral and faunal resemblance that may suggest the influence of historical geological events or contemporary ecological conditions.

Beginning with the pioneering classifications of Schouw (1823) and the early phytogeographers, continuing with Merriam's (1894) life zones (see Figure 2.7) and Herbertson's (1905) natural regions, and extending to the contemporary concept of biomes, ecologists and biogeographers have almost without exception classified terrestrial communities on the basis of the structure (or physiognomy) of the vegetation (Figure 5.11). Implicit in all of these classifications is the recognition that life forms of individual plants, and the resultant three-dimensional architecture of plant communities, reflect the predominant influence of climate and soil on the kinds of plants that occur in a region. Some authors, such as Holdridge (1947) and Dansereau (1957), have attempted to depict these relationships more quantitatively, showing fairly tight relationships between ranges of climatic variables (such as temperature and precipitation) and specific vegetation types (Figure 5.12) (Whitaker 1975; see also Leith 1956). Similar climatic regimes do tend to support structurally and functionally similar vegetation in disjunct areas throughout the world. Often these similarities result from convergence—that is, unrelated plant species in geographically isolated regions having evolved similar forms and similar ecological roles under the influence of similar selective pressures (see the discussion on convergence of geographically isolated communities in Chapter 10).

There are almost as many different classifications of vegetation types as there are textbooks in ecology and phytogeography. In general, most biogeographers recognize six major forms of terrestrial vegetation:

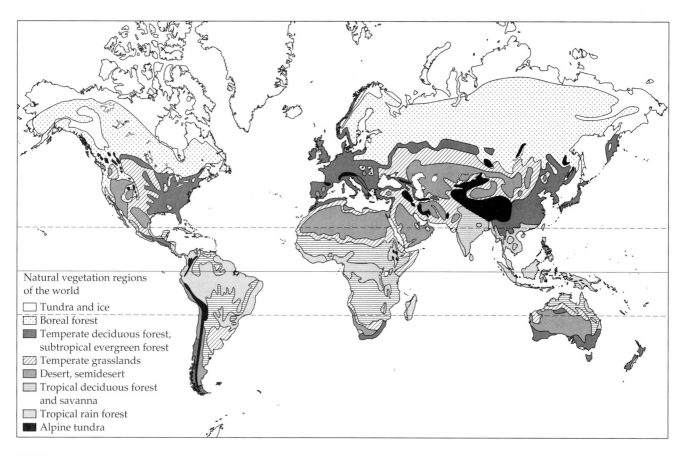

FIGURE 5.11 World distribution of major terrestrial biomes. Note that the locations of these vegetation types correspond closely to the distribution of climatic regimes and soil types (see Figures 3.6 and 3.11). In some cases, several different vegetation types (e.g., tropical deciduous forest and savanna) have been grouped together so that the general zonal pattern of biomes can be observed.

1. **Forest:** a tree-dominated assemblage with a fairly continuous canopy
2. **Woodland:** a tree-dominated assemblage in which individuals are widely spaced, often with grassy areas or low undergrowth between them
3. **Shrubland:** a fairly continuous layer of shrubs, up to several meters high
4. **Grassland:** an assemblage in which grasses and forbs predominate
5. **Scrub:** a mostly shrubby assemblage in which individuals are discrete or widely spaced
6. **Desert:** an assemblage with very sparse plant cover in which most of the ground is bare

We recognize 12 common terrestrial biomes, whose geographic distributions are mapped in Figure 5.11. Note that the occurrence of these biomes corresponds approximately to the distribution of climatic zones (see Figure 3.5) and soil types (see Figure 3.12, Table 5.1). These latitudinal and elevational patterns reflect the fact that vegetation is highly dependent not only on local climate and underlying soil, but also on the influence of regional climate and topography on soil formation (see Chapter 3). We now consider characteristics of the principal biomes, and then conclude this chapter with a global compar-

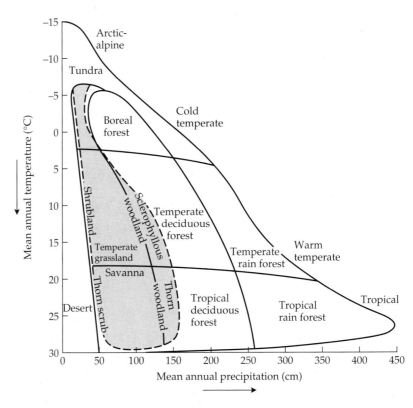

FIGURE 5.12   A climograph, which is a simple diagram quantifying some aspects of the relationships between climate and vegetation types. (After Whittaker 1975.)

TABLE 5.1 *Relationship between climatic zone, soil type, and vegetation communities*

Zonal soil type[a]	Zonal vegetation
Latisols (Oxisols)	Evergreen tropical rain forest (selva)
Latisols (Oxisols)	Tropical deciduous forest or savanna
Chestnut, brown soils, and sierozem (Mollisols, aridisols)	Shortgrass
Desert (Aridisols)	Shrubs or sparse grasses
Mediterranean brown earth	Sclerophyllous woodlands
Red and yellow podzolic (Ultisols)	Coniferous and mixed coniferous-deciduous forest
Brown forest and gray-brown podzolic (Alfisols)	Coniferous forest
Gray-brown podzolic (Alfisols)	Deciduous and mixed coniferous-deciduous forest
Podzolic (Spodosols and associated histosols)	Boreal forest
Tundra humus soils with solifluction (Entisols, inceptisols, and associated histosols)	Tundra (treeless)

Source: Bailey (1996).

[a]Names in parentheses are soil taxonomy orders (USDA Soil Conservation Service 1975).

ison of their salient features and a brief overview of recent advances in ecosystem geography.

Tropical Rain Forest

Tropical rain forests (Figure 5.13) are the richest and most productive of the Earth's terrestrial biomes, covering just 6% of the Earth's surface, but harboring about 50% of its species. Hundreds of tree species may occur in just a few hectares of tropical rain forest, and here one also finds the world's highest diversity of arboreal insects and other invertebrates. The diversity of terrestrial and flying vertebrates is no less impressive. This enormous diversity of species sets the stage for an incredible complexity of biotic interactions.

Tropical rain forests are found at low elevations along tropical latitudes (chiefly 10° N to 10° S) where rainfall is abundant (over 180 cm annually; Figure 5.14). Although most tropical rain forests receive some precipitation throughout the year, rainfall tends to be seasonal. Temperatures are nearly uniform year round (typically more than 18° C), and vary less seasonally than diurnally. The dominant plants are large evergreen trees that form a closed canopy at 30 to 50 m. The architecture of the trees is often convergent, featuring buttressed bases and smooth, straight trunks, but the height and shape of the crowns can be highly variable. The evergreen leaves also tend to be convergent in form, robust and broad with smooth edges. There may be several levels of trees below the uppermost canopy, and palms and other distinctive plants typically occur in the understory. Growing in the upper layers of trees are numerous **lianas** (woody vines) and **epiphytes** (orchids, ferns, and in the New World—bromeliads), and the leaves of these trees may be covered with

FIGURE 5.13 Tropical rain forest, Equador. (Courtesy of A. Sinauer.)

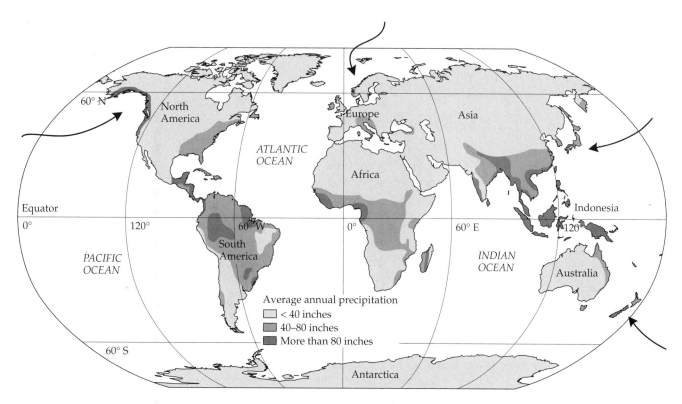

FIGURE 5.14 Global distribution of tropical and temperate rain forests. Temperate rain forests (indicated by arrows) are limited to regions along the west coasts of continents and large islands at the mid-latitudes, whereas tropical rain forests are much more broadly distributed between 10° N and 10° S latitudes.

FIGURE 5.15 Tropical deciduous forest in the dry season when most plants are leafless. Santa Rosa National Park, Costa Rica. (Courtesy of E. Orians.)

epiphylls (thin layers of mosses, lichens, and algae). Very little light penetrates the dense, multilayered canopy to reach the forest floor, which is usually surprisingly open and devoid of vegetation. Annual plants are conspicuously absent.

Lowland tropical rain forests are the most diverse and productive of the major terrestrial biomes. As a result of high temperatures and high humidity, decomposition of dead organic matter occurs so rapidly that little litter accumulates on the forest floor or in the soils. Many trees have adapted to this environment by developing extensive horizontal root mats—mostly within the upper 20 cm of soil—to capture nutrients released when detritus decomposes. Mycorrhizal fungi, which facilitate the uptake of nutrients, are also closely associated with these root mats. These adaptations, combined with the rapid leaching rates characteristic of tropical soils, are responsible for the paradox of tropical rain forests: one of the world's most productive systems grows on some of the world's poorest soils (Richards 1996).

Tropical Deciduous Forest

Tropical deciduous forests (Figure 5.15) usually occur in hot lowlands outside the equatorial zone (between 10° and 30° latitude) where rainfall is more seasonal, and the dry season more pronounced and more extensive, than in regions of tropical rain forest. Compared with a tropical rain forest, the canopy of a tropical deciduous forest is lower and more open, and because more light reaches the ground, there is often more understory vegetation. To conserve water, many of the trees and understory plants shed their leaves during the long dry season, although much flowering and fruit maturation may occur at this time.

The dominant vegetation is often called **rain-green forest** because forest trees leaf out during the first heavy rains following the dry season. The most luxuriant form is the **monsoon forest**, the layperson's "jungle." Monsoon forest, especially well developed in southern Asia, has many large leaves and dense undergrowth rich in bamboos. These areas are frequently drenched with torrential rainstorms, and some are among the rainiest, but most seasonal, habitats in the world.

Thorn Woodland

Tropical and subtropical thorn woodlands (Figure 5.16) are low arborescent vegetation types that grow in hot semiarid lowlands. Dominant plants are small spiny or thorny shrubs and trees. Members of the genus *Acacia* and other legumes (Fabaceae) are common in these biomes on all continents. Succulents, such as cacti (Cactaceae) in the New World and convergent forms of the genus *Euphorbia* (Euphorbiaceae) in Africa, are often abundant. Most plants lose their leaves during the prolonged dry season, but during the rainy season trees leaf out and a dense herbaceous understory develops.

Thorn woodlands are often found on drier sites adjacent to tropical deciduous forests. As the climate becomes even drier along a gradient, thorn woodlands give way to **thorn scrub**. A minimum of 30 cm of annual rainfall usually is necessary to establish a thorn scrub, and there is usually a 6-month dry season with virtually no rainfall.

Tropical Savanna

Tropical savannas (Figure 5.17) are biomes dominated by a nearly continuous layer of xerophytic perennial grasses and sedges scattered amongst fire-resistant trees or shrubs. Savannas usually occur at low to intermediate elevations along intertropical latitudes (primarily between 25° N and 25° S). Savannas

FIGURE 5.16   Tropical woodland near Lake Chircro, Zimbabwe. (Courtesy of E. Orians.)

are characterized by marked seasonality of precipitation, with one or two rainy seasons followed by intense droughts. These weather patterns are largely driven by seasonal shifts in the intertropical convergence zone, the zone of most intense solar radiation and convergence of trade winds from the north and south (see Figure 3.4). As the sun shifts between tropical latitudes, the intertropical convergence zone shifts as well, passing twice over the equator, but just once over higher latitudes of the tropics. As a result, equatorial

FIGURE 5.17   Tropical Savanna, Champagne Ridge, Buffalo Springs National Reserve, Kenya. (Courtesy of E. Orians.)

savannas experience two rainy seasons, while those near the limits of the tropics experience a single, longer rainy season. Thus, annual rainfall varies from 30 to 160 cm. Most savannas, however, are strongly influenced by three common factors: seasonally intense precipitation, fire during the dry season, and migratory or seasonal grazing. These dynamic forces combine to make the savanna one of the most spatially and temporally heterogeneous biomes on Earth.

The most extensive savannas are found in intertropical Africa, where they support the most abundant and diverse community of large grazing mammals in the world, as well as a variety of large carnivorous mammals. While easily overlooked, the diversity of smaller herbivores and carnivores living on savannas is no less impressive—typically many times that of their larger counterparts. Savannas have also played an important role in the development of human civilizations. Native peoples inhabiting the savannas of Africa and other continents followed the natural migrations of ungulates, which they depended on for food and clothing. Eventually these early "pastoralists" domesticated some species, such as cattle, horses, donkeys, and camels, and moved their herds along with the natural grazers to exploit seasonal shifts in the productivity of the savanna's native communities.

Desert

Hot deserts and semideserts (Figure 5.18) occur around the world at low to intermediate elevations, especially within the belts of dry climate from 30° to

FIGURE 5.18   Desert vegetation near Puertecitos in northern Baja California, Mexico. Note that the vegetation, which consists almost entirely of small shrubs, is extremely sparse. A few small trees of palo verde (*Cercidium microphyllum*) grow in the sandy bottom of the dry watercourse in the center, where the plants in general are larger and denser than on the surrounding rocky hillsides. (Photograph by J. R. Hastings.)

40° N and S (the horse latitudes) between the humid tropics and the mesic temperate biomes. Rainfall is not only scanty (often less than 25 cm per year) and seasonal but also highly unpredictable. The evaporative potential of hot desert climates is so strong that most plants have acquired special adaptations that enable them to take up, store, and prevent the loss of water. The key feature common to deserts is that the amount of rainfall is far less than the evaporative potential besides being highly unpredictable within, as well as among, years. Even regions with well over 30 cm of average annual precipitation may be dominated by desert vegetation under these circumstances. Some extremely arid regions may not experience any rainfall for several years in a row, and therefore have little or no perennial vegetation. In less arid regions, the dominant vegetation consists of widely scattered low shrubs, sometimes interspersed with succulents (cacti, euphorbs, yuccas, and agaves). Where shrubs predominate, the vegetation is usually called desert scrub.

Desert plants possess a variety of adaptations to withstand periods of drought and capitalize on the short, unpredictable, but sometimes heavy, rains. Following these brief periods of rains, ephemeral forbs and grasses may grow rapidly to carpet the normally bare ground. Many plants, especially succulents, are able to swell and store water, and often possess extensive shallow root systems, which act like inverted umbrellas to capture rainfall before it penetrates the soil. Animals also exhibit a variety of adaptations to arid environments, including an ability to derive all the water they need from seeds and other foods, or to remain seasonally or diurnally inactive until environmental conditions become more equable. Some large desert mammals are able to store substantial amounts of water in their tissues, while birds and bats can avoid the most stressful seasons by practicing migration.

Despite their hardy appearance, desert ecosystems are quite fragile. They show very poor resilience and, once disturbed, may take centuries to recover. On the other hand, overgrazing by livestock and diversion of water for agriculture and other uses have converted otherwise more mesic systems into deserts, albeit atypical ones.

Sclerophyllous Woodland

Sclerophyllous woodlands and chaparral (Figure 5.19) occur in mild temperate climates with moderate winter precipitation but long, usually hot, dry summers. Sclerophyllous woodlands may also occur in regions with moderate precipitation, but whose sandy soils have little water-holding capacity. This biome includes a broad variety of xeric woodlands ranging from piñon-juniper woodlands and pine barrens to sandhill pine woodlands, sandpine scrub, and pine flatwoods. The dominant plants have sclerophyllous (hard, tough, evergreen) leaves. Sclerophyllous woodlands can be tall, open forests that receive over 100 cm of annual rainfall, like the eucalypt woodlands of southwestern Australia, or shorter woodlands that experience less rainfall, like the oak and conifer woodlands of western North America (with evergreen *Quercus, Juniperus,* and *Pinus* species).

Those areas that receive less than 60 cm of rainfall per year tend to have low, shrubby vegetation. Sclerophyllous scrublands—called **chaparral, matorral, maquis, fynbos,** or **macchia**—are characteristic of Mediterranean climates (see Chapter 3). Much of their land surface is covered with a dense and almost impenetrable mass of evergreen vegetation only a few meters high. Fires frequently sweep through these habitats, burning off the aboveground biomass and apparently playing a major role in preventing the establishment of trees. Shrubs then resprout from their root crowns to reestablish the vegetation.

FIGURE 5.19 Sclerophyllous woodland. (A) Chaparral, Mendocino County, California. (B) Fynbos, Cape of Good Hope, South Africa. (Courtesy of E. Orians.)

(A)

(B)

Subtropical Evergreen Forest

Subtropical evergreen forests (Figure 5.20), some of which have been called oak-laurel forests or montane forests, are common in subtropical mountains at intermediate elevations. These broad-leaved forests cover extensive areas of China and Japan, disjunct areas in the Southern Hemisphere, and much of the southeastern United States. These areas may receive as much as 150 cm of annual rainfall evenly distributed throughout the year, but subtropical evergreen forests cannot occur where the mean annual temperature is much below 13° C or where severe frosts occur (Wolfe 1979).

Most of the dominant species consist of dicotyledons with broad, sclerophyllous evergreen leaves, such as laurels (Lauraceae), oaks (*Quercus*, Fagaceae), and magnolias (Magnoliaceae) in the Northern Hemisphere and southern beeches (*Nothofagus*) in the Southern Hemisphere. The canopy is not usually well stratified, and understory plants, especially mosses, may be exceedingly common. A number of temperate broad-leaved deciduous trees occur in these forests, and as the climate becomes colder, broad-leaved evergreens are gradually replaced by deciduous trees or conifers.

Temperate Deciduous Forest

Temperate deciduous forests (Figure 5.21) grow throughout temperate latitudes almost wherever there is sufficient water to support the growth of large trees. They are also called summer-green deciduous forests because they have a definite annual rhythm—the trees are dormant and leafless in the cold and snowy winter and leaf out in the spring. Temperate broad-leaved deciduous forests are extremely variable in their structure and composition across eastern North America, western Europe, and parts of eastern Asia. In otherwise arid southwestern North America, similar vegetation occurs along permanent watercourses (riparian deciduous woodland). The height and density of the canopy and the importance and composition of the understory vary greatly depending on local climate, soil type, and the frequency of fires. The diversity and coverage of understory plants can be quite high, especially during spring before the trees leaf out. As a result of their extensive accumulation of organic matter and the high capacity of their soils to hold water, temperate deciduous rain forests are much less prone to fire than many other biomes.

In many parts of the Northern Hemisphere, temperate deciduous forests are located next to other arborescent communities, especially temperate evergreen forests, sclerophyllous woodlands, and coniferous forests. Consequently, many phytogeographers recognize a long list of hybrid associations between these communities, such as mixed evergreen-deciduous forest. The trees of temperate forest climax communities grow slowly, and most forests have been significantly affected by logging over the last few centuries.

FIGURE 5.20 Subtropical evergreen forest, Queensland, Australia. (Courtesy of E. Orians.)

FIGURE 5.21 Temperate deciduous forest, near Moscow, Russia. (Courtesy of E. Orians.)

Temperate Rain Forest

Temperate rain forest (Figure 5.22) is an uncommon but interesting biome found along the western coasts of continents where precipitation exceeds 150 cm per year and falls during at least 10 months out of each year (see Figure 5.14). Cool temperatures predominate year-round, but these regions are always above freezing, and they experience much fog and high humidity, which permits the growth of large evergreen trees. A moderately dry season during the summer inhibits the growth and dominance of deciduous trees. Cool temperatures account for the absence of any true tropical plants, such as palms, and the relatively low number of tree species. Temperate rain forests do not have many kinds of lianas, but the epiphyte diversity is high, consisting of mosses, lichens, epiphyllous fungi, and some ferns. These cool and moist forests are largely uninfluenced by fire. Therefore, while growth rates are relatively slow, temperate rain forests are renowned for possessing some of the world's oldest and largest trees. Their canopies tend to be closed, with many dead standing trees, or "snags," while the humid understory is often covered with lush mats of mosses, ferns, and lichens. Decomposition rates tend to be slow, again due to the relatively low temperatures. Combined with slow growth rates of the dominant tree species, this means that forest development requires many centuries of relatively stable climatic conditions.

The best examples of a temperate rain forest include those of New Zealand, Chile, and the Pacific Northwest region of North America. Along these coastal regions, maritime airmasses flow inland and then ascend the coastal mountain ranges, releasing 100 to 300 cm of water in the form of rain and fog. Although highly varied among regions, these forests are dominated by immense, but slow-growing trees including *Agathis*, *Eucalyptus*, *Nothofagus*, and *Podocarpus* in the Southern Hemisphere and *Picea* (spruce), *Abies* and *Pseudotsuga* (firs) in the Northern Hemisphere. The understories of these forests are typically covered with lush and diverse mats of mosses, epiphytes, fungi, and ferns.

FIGURE 5.22 Temperate rain forest of the Olympic Peninsula, Washington State. (A) Riparian corridor of Olympic National Forest. (B) The world's largest sitka spruce tree, which has a circumference of 17.6 m (58 ft) and is approximately 1000 years old. (Photographs by Mark V. Lomolino.)

(A)

(B)

Temperate Grassland

Temperate grasslands (Figure 5.23) are situated both geographically and climatically between deserts and temperate forests. While broadly distributed between 30° and 60° latitude, temperate grasslands are most extensive in the interior plains of the Northern Hemisphere. The climate of these regions is markedly seasonal, with substantial annual variation in both temperature and rainfall. The vegetation is confined to a single stratum, which is dominated by grasses, sedges, and other herbaceous plants. Vegetation height tends to vary directly with precipitation, both factors decreasing from tall grasslands, or prairie (veldt of South Africa, puszta of Hungary, tallgrass prairie of North America, or pampas of Argentina and Uruguay), to shortgrass plains, or steppes, in colder latitudes and desert grasslands adjacent to warm arid regions. Even in relatively moist tallgrass prairies, drought, fire, and heavy grazing pressures combine to prevent the establishment of woody plants while favoring the dominance of herbaceous plants.

The dominant grasses are perennials with **basal meristems** (growth tissue located in the soil), which make them tolerant to defoliation; indeed, in these species, vegetative growth is stimulated by fire and grazing. Although typically dominated by just a few grass species, temperate grasslands actually harbor a surprising diversity of both plants and animals. Grasses, which may account for over 90% of the biomass, typically constitute less than 25% of the plant species in grassland ecosystems. Despite the variable and sometimes luxuriant layer of vegetation above the surface, most grassland biomass lies belowground in the extensive root systems of perennial plants. The ratio of belowground to aboveground biomass varies with annual precipitation, ranging from less than 2:1 for arid grasslands to 13:1 for tallgrass prairies (Wiegert and Owen 1971). Accordingly, grassland soils tend to have high accumulations of organic material, which in turn supports a rich diversity of soil invertebrates and microbial decomposers. Many vertebrate species, especially some rodents, are completely **fossorial** (burrow-dwelling), while others are cursorial grazers, consuming as much as two-thirds of the aboveground production.

Because of their deep, fertile soils, many temperate grasslands have been converted to agricultural uses. As a result of cultivation or desertification, natural grasslands, which once covered approximately 40% of the Earth's surface, have now been reduced to about half of their presettlement range.

FIGURE 5.23   Temperate grassland, the Tallgrass Prairie Preserve in Northern Oklahoma. (Courtesy of Bruce Hoagland.)

FIGURE 5.24 Boreal coniferous forest, near Eagle Trail, Alaska. (Courtesy of E. Orians.)

Boreal Forest

Boreal forests (also called taiga or "swamp forest") occur in a broad band across northern North America, Europe, and Asia in regions with cold climate and adequate moisture (Figure 5.24). At high elevations, this biome also extends well southward into temperate latitudes. For example, boreal forests extend down the cordilleras of western North America all the way to southern Mexico; indeed, much of highland Mexico is covered by boreal forest.

Boreal forests, although often thick, are typically dominated by just a few species of coniferous trees, such as spruce (*Picea*), fir (*Abies*), and larch (*Larix*). Because of cool temperatures and waterlogged soils, decomposition rates are relatively slow, resulting in an accumulation of peat and humic acids, which render many soil nutrients unavailable for plant growth. Acidic soils combine with relatively cool temperatures to limit diversity and productivity of the few tree species able to survive these stressful conditions. Often the canopy is not dense, and a well-developed understory of acid-tolerant shrubs, mosses, and lichens may be present in the most mesic sites.

Tundra

Tundra (Figure 5.25) is a treeless biome found between the boreal forest and the polar ice cap and at high elevations on tall mountains (alpine tundra). Even more than the boreal forest, tundra is characterized by stressful environmental conditions. Temperatures remain below freezing for at least 7 months of the year, precipitation is often less than in many hot deserts, and tundra soils tend to be even more nutrient-limited than those of the boreal forest. Soils are also saturated with water because of slow evaporation rates and the presence of permafrost (a frozen, impermeable layer of soil that lies at a depth of a meter or less in the summer). Consequently, primary productivity, biomass, and diversity of the tundra are lower than that of almost all other terrestrial biomes.

FIGURE 5.25 Tundra, Denali National Park, Alaska. (Courtesy of E. Orians.)

Arctic, antarctic, and alpine tundra are all covered with a single dense stratum of vegetation, usually only a few centimeters or decimeters in height. The dominant plants tend to be dwarf perennial shrubs, sedges, grasses, mosses, and lichens. Despite generally low productivity during the rest of the year, tundra plants exhibit bursts of productivity during the short growing season. The lush vegetation is then heavily grazed by migratory or nomadic ungulates, including caribou (*Rangifer tarandus*), muskoxen (*Ovibos moschatus*), and Dall sheep (*Ovis dalli*). Other important herbivores of the tundra include geese (*Branta* spp.), ptarmigan (*Lagopus* spp.), and small mammals (voles and lemmings), whose populations fluctuate dramatically in a complex interaction with the plant community.

Found above the timberline on mountaintops in the equatorial zone is **tropical alpine scrubland**, with vegetation taller than that of the arctic tundra. The dominant plants are tussock grasses and bizarre, erect rosette perennials with thick stems. These vegetation types are found at elevations above 3300 m in the Andes (**páramo**) in South America, on the upper slopes of the highest mountains in East Africa, and on mountaintops in New Guinea.

Just like its hot desert counterpart, the tundra is a fragile system. Oil exploration and other human activities are major threats to this delicate environment. Once these activities disturb the permafrost, natural communities may take many decades to recover.

Aquatic Communities

Marine and freshwater ecologists and biogeographers do not classify aquatic communities into categories analogous to those used for terrestrial biomes. For one thing, the relatively simple arrangement of sessile plants growing on a land surface is not comparable to the three-dimensional diversity of the water column. Terrestrial habitats are essentially two-dimensional, in the sense that organisms do not remain permanently suspended above the soil surface. To the extent that a third dimension is present, it is formed by the vertical growth of sessile plants and by arboreal and flying animals. The three-dimensional organization of aquatic communities is very different. On one hand, a well-developed structure of attached, vertically growing organisms is absent from most aquatic habitats, although there are obvious exceptions, such as kelp forests, coral reefs, and the submersed vegetation of lakeshores. On the other hand, many aquatic organisms spend much or all of their lives suspended in the third dimension, either drifting passively or swimming actively in the water column.

The physical factors that vary in time and space to affect the abundance and distribution of aquatic organisms are also quite different from those that determine both the terrestrial climate and the organization of terrestrial communities. Because of the high specific heat of water, temperature varies less on a daily, seasonal, and latitudinal basis in aquatic environments than in terrestrial ones. On the other hand, variations in pressure, salinity, and light are important in aquatic systems. Tidal cycles, which fluctuate bimonthly with the phases of the moon, are more important to many marine shore communities than are daily or seasonal cycles (see Figure 3.16).

Oceanographers, limnologists, and aquatic ecologists have developed classification systems for marine and freshwater communities. Like the division of terrestrial communities into biomes, these systems use arbitrary groupings that break up a continuous spectrum of biological associations into a number of convenient categories. Salinity, depth, water movement, and nature of the substrate are physical characteristics that most influence the abundance and distribution of aquatic organisms and are most often used in classifying aquatic communities.

The first major division of aquatic systems is into marine and freshwater communities. On biological as well as geographic grounds, the Earth's bodies of water can be divided into the oceans, which form a huge interconnected water mass covering over 70% of the Earth's surface, and the comparatively tiny, highly fragmented lakes, ponds, rivers, and streams, which together cover only a small fraction of the remaining surface of the Earth. The oceans and these various bodies of fresh water differ greatly in salinity. Salt concentration of the oceans varies slightly around 35 parts per thousand, whereas even the hardest fresh waters have salinities of less than 0.5 parts per thousand. As stressed in Chapter 4, this difference in salinity can have dramatic effects on species distributions. Only a tiny fraction of aquatic organisms can live in both salt and fresh water, so salinity effectively divides aquatic communities into two non-overlapping groups. Because of this strong dichotomy, marine and freshwater ecosystems have largely been studied independently by different groups of ecologists (**oceanographers** and **limnologists**, respectively), who have developed different classifications of communities. These classification schemes are best considered separately.

Marine Communities

Compared with terrestrial and freshwater environments, the ocean is immense and essentially continuous, but far from homogeneous in terms of its environments and biotic communities. It also comprises the lion's share of the world's ecosystems, with marine waters dominating the Earth's surface and providing a three-dimensional habitat across its entire 1.37 billion cubic kilometer volume (Duxbury 2000).

Organisms live everywhere in the ocean, but the abundances and kinds of life vary greatly depending on the local physical environment. Perhaps the most important features are light, temperature, pressure, and substrate. The current system of biogeographic regions of the marine realm was not generally accepted until the early 1970s (Figure 5.26A), roughly a century after Sclater (1858) and Wallace (1876) proposed the system of terrestrial biogeographic regions that we continue to use today. This system is primarily based on water temperature; therefore, biogeographic regions of the marine realm encompass broad latitudinal zones and tend to be elliptical due to circular oceanic currents (compare Figures 5.26A and 5.26B).

Within each of these regions, the ocean can be divided into two vertical zones—the **photic zone** and the **aphotic zone**—based on penetration of sunlight (Figure 5.27; see Chapter 3). Because sunlight is gradually absorbed by water with increasing depth, the boundary between these zones is somewhat arbitrary, but is usually set where light penetration is reduced to between 1% and 10% of incident sunlight. The depth of the photic zone increases from coastal waters, where light rarely penetrates more than 30 m because of organisms and inanimate particles suspended in the water column, to the open ocean, where it may extend to a depth of 100 m or more.

The significance of this zonation by light is, of course, that photosynthesis can occur only in the photic zone. Essentially all of the organic energy that

FIGURE 5.26   Biogeographic and climatic regions of the world's oceans. (A) Biogeographic regions: 1, arctic; 2, subarctic; 3, northern temperate; 4, northern subtropical; 5, tropical; 6, southern subtropical ; 7, southern temperate; 8, subantarctic; and 9, antarctic. (B) Climatic regions based on mean monthly water temperatures: A, arctic; NB, northern boreal; SB, southern boreal; T, tropical waters; E, equatorial region; NN, northern notal; SN, southern notal; and ANT, antarctic. (After Rass 1986.)   ▶

(A)

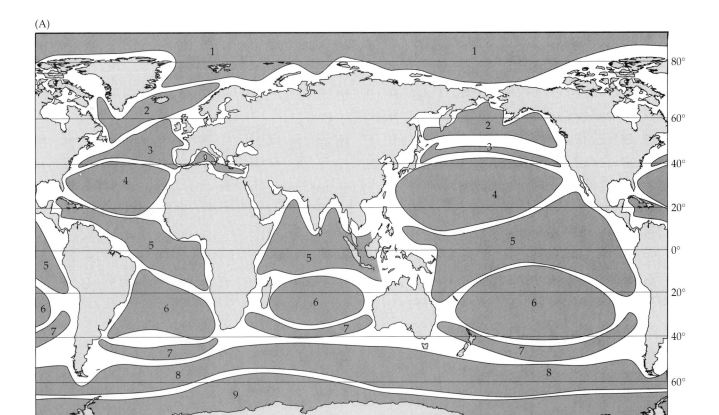

(B)

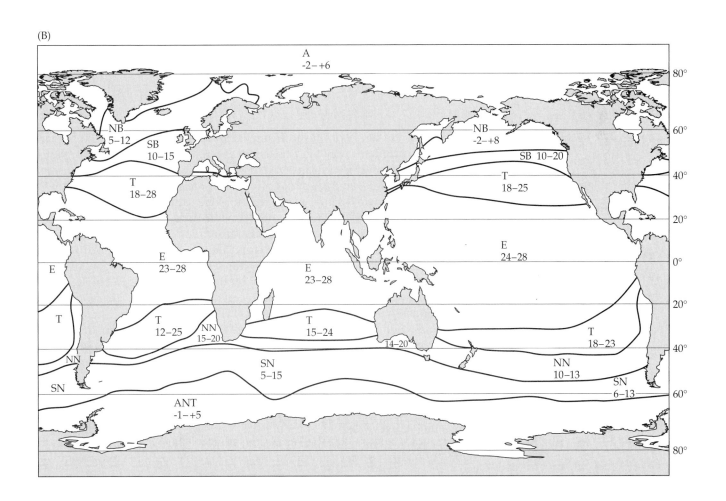

FIGURE 5.27 Division of marine communities into major zones. Note that these zones are based primarily on water depth, light levels, and relationships between organisms and substrates.

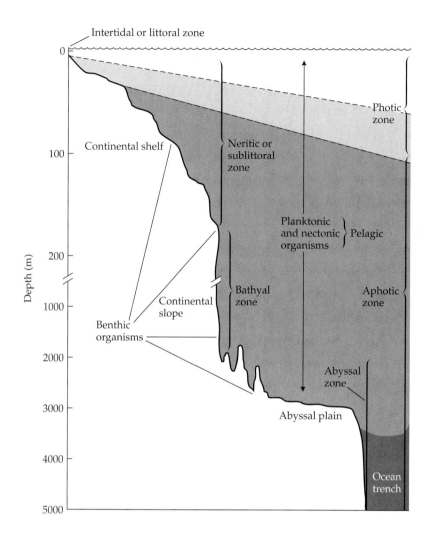

sustains marine life is produced in this shallow surface layer of the ocean. Most organisms in the aphotic zone obtain their energy by consuming organic material that is produced in the photic zone and reaches deep water in the forms of feces and dead bodies that sink to the bottom. In the late 1970s, however, oceanographers discovered entire flourishing communities of organisms, including unique kinds of worms, mussels, and crabs, that do not depend on material from the photic zone. These communities live in highly localized areas such as those on the otherwise barren slopes of the Galápagos rift (eastern tropical Pacific) where submarine hot springs emit hydrogen sulfide. Chemosynthetic bacteria obtain energy by oxidizing hydrogen sulfide, and these unusual autotrophs serve as the base of the food chain for these communities (Jannasch and Wirsen 1980; Karl et al. 1980).

Marine communities are also classified into another set of zones on the basis of **bathymetry** (i.e., depth and configuration of the ocean bottom) (see Figure 5.27). The shallowest zone is the intertidal or **littoral zone**, which occurs on the shore where sea meets land. Although it is inhabited almost exclusively by marine organisms, the intertidal zone is actually an ecotone between land and ocean. Beyond the intertidal zone is the **neritic** or **sublittoral zone**, which encompasses waters of a few meters to about 200 m deep that cover the continental shelves. At the edges of the continental plates is a region of highly varied relief, called the **bathyal zone**, with the marine equivalent of mountainsides and canyons, in which the waters rapidly drop away to the great ocean depths. The **abyssal zone**, which constitutes most of the ocean, covers extensive areas in which the water ranges in depth from 2000 to more than 6000 m. Deep ocean waters provide some of the most constant physical environments; they are continually dark, cold ($4°$ C), subject to enormous pressures, and virtually unchanging in chemical composition.

Organisms that inhabit the oceans are often classified as either **benthic** or **pelagic**, depending on whether they are closely associated with the substrate or distributed higher in the water column (see Figure 5.27). Benthic communities vary greatly in composition depending on the nature of the substrate. On hard substrates, attached benthic organisms often form a three-dimensional structure that varies in complexity from low crusts and turfs of algae and sessile invertebrates to tall "forests" of kelp and coral. On soft sandy or muddy substrates, there is often a comparable three-dimensional complexity, but it is formed by burrowing invertebrates that live beneath the surface. Pelagic (open water) organisms are usually divided into two groups, **plankton** and **nekton**. The former consists of primarily microscopic organisms that float in the water column. The plankton typically includes many simple plants (**phytoplankton**) such as diatoms, and tiny animals, or small crustaceans and the larvae of many invertebrates and fishes (**zooplankton**). Nekton is comprised of actively swimming animals, including fishes, whales, and some large invertebrates, which usually occupy higher trophic levels than planktonic organisms.

Freshwater Communities

Freshwater communities are widely distributed as small, isolated lakes, ponds, and marshes, sometimes connected by long, branching streams and rivers. These environments are usually divided into two categories: lotic or running-water habitats, such as springs, streams, and rivers; and lentic or standing-water habitats, such as lakes and ponds. Lotic habitats are often divided into rapids (or "riffles") and pools. In the former, water velocity is sufficient to keep the water well oxygenated and the substrate clear of silt. Stream rapids are usually inhabited by organisms that live on the surface of the rocky substrate or swim strongly in the current. Pools are characterized by deep, slowly moving water and silty, often poorly oxygenated, bottoms. Swimming animals are common in stream pools, and many benthic species burrow into the substrate. Although some organic material in streams is manufactured in place by benthic plants or phytoplankton, most of it is washed in from the surrounding watershed.

Lentic habitats are often divided into zones reminiscent of those of the oceans, although somewhat different terms and meanings are applied (Figure 5.28). The littoral zone consists of shallow waters where light penetrates to the bottom and rooted aquatic vegetation may be present. Offshore waters are divided into a surface **limnetic zone**, where light penetrates sufficiently for photosynthesis to occur, and a deep **profundal zone** (beyond the depth of effective light penetration). Lakes can be highly productive, supporting extensive food webs based on photosynthesis by attached vegetation in the littoral zone and phytoplankton in the limnetic zone. Productivity is limited largely by the availability of inorganic nutrients, such as phosphorus, which wash in from surrounding watersheds and, in temperate and subarctic lakes, are seasonally replenished from organic material on the lake bottom when thermal stratification disappears and the waters overturn (see Chapter 3). Temperate lakes are often classified as either **eutrophic** or **oligotrophic**. Eutrophic lakes are shallow and highly productive because light penetrates almost to the bot-

FIGURE 5.28 Division of freshwater lentic habitats into major zones. This classification scheme is similar to that used for marine environments because it is based on variation in light penetration, water depth, and relationships between organisms and substrates; nevertheless, somewhat different terms are used.

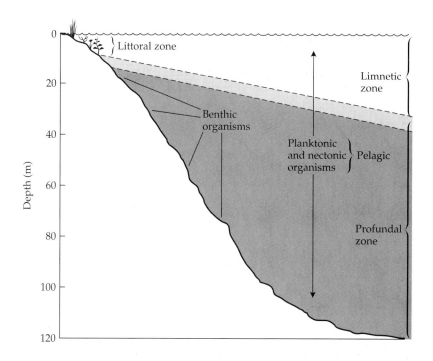

tom and vertical circulation of the water column occurs each spring and fall, returning limiting nutrients to the surface and oxygen to the depths. Oligotrophic lakes are characterized by low nutrient input and tend to be deeper than eutrophic lakes. In fact, many oligotrophic lakes are so deep that little or no vertical circulation occurs. As a result, productivity is relatively low, despite the high water clarity of many oligotrophic lakes.

The preceding classification scheme omits a number of freshwater communities. **Swamps** (**marls**) and **marshes** (**moors**) are two very common types of wetlands that tend to develop on mineral soils and are distinguished primarily by whether their dominant vegetation is woody or herbaceous, respectively. Some other freshwater communities are much rarer, but interesting because they represent atypical physical environments that pose special problems for the few organisms that are able to live there. Examples of such harsh environments are hypersaline lakes (undrained lakes, such as Great Salt Lake in North America, the Dead Sea in the Middle East, and the Aral Sea in Central Asia, all of which are much more saline than seawater), caves (which admit no light and contain communities supported entirely by imported organic matter), and hot springs (among the most physically rigorous of all environments).

Another important group of communities occurs in those areas where fresh water, the sea, and the land meet. The **estuaries**, **salt marshes**, and (in tropical regions) **mangrove swamps** that occur at these sites are highly productive ecosystems. They usually contain only a few species that can tolerate the physical rigors of successive exposure to fresh water, seawater, and the terrestrial climate. On the other hand, the circulation provided by tides and the input of nutrients from rivers—and ultimately, land—often permits these habitats to support high biomasses and densities of individuals.

A Global Comparison of Biomes and Communities

In concluding this discussion of biomes, we must sound a note of caution. The biomes described above, and their distributions as shown on the map in Figure 5.11, indicate only the general kind of climax vegetation that we would expect to find in a region, based primarily on its climate. However, someone visiting many of the areas on the map might have difficulty finding good stands of the vegetation typical of these biomes, and might find some unexpected vegetation types. In some cases, this might be the result of secondary succession occurring in response to natural disturbances. More often, it is caused by human destruction of the original vegetation and modification of the landscape. For example, tallgrass prairie—the most productive temperate grassland—once covered much of Illinois and Iowa and stretched from Saskatchewan to Texas. Now most of this area has been converted to agriculture, and only a few stands of native prairie—totaling less than 5% of its original area—remain, mostly in a small number of preserves (see Steinauer and Collins 1996). Perhaps the most significant change now occurring is the rapid destruction of virgin lowland tropical rain forests, whose boundaries are ever-shrinking, giving way to successional communities of reduced biotic diversity and diminished economic value.

In some places, local variations in topography or soil type cause types of vegetation to be found in regions where one would not predict their presence based on the general map. For example, throughout temperate grasslands—sclerophyllous scrublands, tropical thorn scrub, and deserts—there are galleries of riparian forest vegetation along permanent streams. Such diversity of vegetation types contributes greatly to the overall biotic richness of a region, because distributions of many other plants and numerous animal species are strongly influenced by the dominant vegetation. Bird species distributions, and the diversity of bird communities, for example, are highly dependent on vegetation structure (MacArthur and MacArthur 1961), and the riparian deciduous woodlands in the deserts of the southwestern United States support exceptionally high bird species diversity (Carothers and Johnson 1974).

Given the great diversity of world biomes and communities, it may be instructive to conclude this overview with a global comparison of their salient features (Table 5.2). **Net primary productivity** (NPP) is a measure of the rate at which solar energy is converted to plant tissue, typically expressed as mass produced per unit of surface area (e.g., $g\ m^{-2}\ yr^{-1}$). NPP is one of the most fundamental and important measures of community function, as it represents the energy available to maintain biomass and diversity of almost all forms of life. As we have seen, biomes and communities vary markedly in temperature, precipitation, nutrient availability, and many other factors that influence primary productivity. Consequently, the world's biomes vary markedly in NPP, biomass, and diversity. Tropical rain forests are renowned for their high productivity, and indeed, they are clearly the most productive of the terrestrial biomes. However, as we see in Figure 5.29A, some aquatic systems rival tropical forests in NPP. In aquatic systems, productivity tends to be highest in shallow-water environments, where high photosynthetic rates are favored by relatively high levels of sunlight and nutrients.

These trends in NPP among biomes are paralleled by trends in biomass (Figure 5.29B). Again, the most productive biomes and communities tend to support the highest density of living tissue. Aquatic communities, however, exhibit consistently lower biomass than their terrestrial counterparts. Phytoplankton, the smallest plants, account for approximately 90% of NPP in aquatic communities. Unlike terrestrial macrophytes, more than a third of whose biomass is photosynthetically inactive tissue, phytoplankton are extremely efficient. Under optimal conditions, solar energy is rapidly converted into phytoplankton tissue and in turn made available to aquatic consumers and decomposers. As mentioned earlier, this high turnover rate, which supports extensive food webs, explains why pyramids of biomass are inverted for some lake and ocean ecosystems (see Figure 5.6E).

FIGURE 5.29 Comparisons of (A) net primary productivity, and (B) biomass among terrestrial and aquatic communities (see Table 5.2). Despite their relatively low biomass, aquatic communities often rival tropical rain forests in productivity on a per area basis.

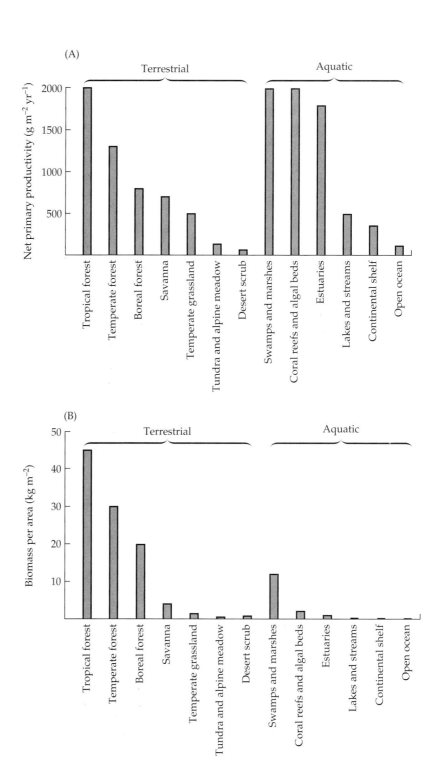

We caution that the above estimates of productivity and biomass are averages over space and time. At finer scales, each of these biomes and communities is remarkably heterogeneous. Each is composed of a collection of successional and disturbance stages, all influenced to varying degrees by seasonal changes. Temperate grasslands, meadows, and deserts, for example, can exhibit impressive bursts of productivity during the growing season.

We should also bear in mind that up to now we have been comparing biomes and communities on a productivity per area basis. Let us now study

TABLE 5.2 *Net primary production and biomass of major kinds of biomes and marine communities*

Biome/Community	Area (10^6 km²)	Net primary production per unit area (g m⁻² yr⁻¹)	Total net primary production (10⁹ MT yr⁻¹)	Mean biomass per unit area (kg m⁻²)
Terrestrial and Freshwater				
Tropical rain forest	17.0	2000.0	34.00	44.00
Tropical deciduous forest	7.5	1500.0	11.30	36.00
Temperate rain forest	5.0	1300.0	6.40	36.00
Temperate deciduous forest	7.0	1200.0	8.40	30.00
Boreal forest	12.0	800.0	9.50	20.00
Savanna	15.0	700.0	10.40	4.00
Cultivated land	14.0	644.0	9.10	1.10
Woodland and shrubland	8.0	600.0	4.90	6.80
Temperate grassland	9.0	500.0	4.40	1.60
Tundra and alpine meadow	8.0	144.0	1.10	0.67
Desert scrub	18.0	71.0	1.30	0.67
Rock, ice, and sand	24.0	3.3	0.09	0.02
Swamp and marsh	2.0	2500.0	4.90	15.00
Lake and stream	2.5	500.0	1.30	0.02
TOTAL TERRESTRIAL AND FRESHWATER	149.0	720.0	107.09	12.30
Marine				
Coral reefs and Algal beds	0.6	2000.0	1.10	2.00
Estuaries	1.4	1800.0	2.40	1.00
Upwelling zones	0.4	500.0	0.22	0.02
Continental shelf	26.6	360.0	9.60	0.01
Open ocean	332.0	127.0	42.00	0.003
TOTAL MARINE	361.0	153.0	55.32	0.01
WORLD TOTAL	510.0	320.0	162.41	3.62

Source: After Whittaker and Likens (1973).

these systems from a different perspective—a global view. If an astronaut were able to view the Earth through a biologically sensitive lens—one that could distinguish productivity levels—it would look something like the map on this text's cover. Both terrestrial and aquatic ecosystems exhibit pronounced latitudinal effects, but other patterns differ markedly between them. On land, NPP tends to be strongly correlated with precipitation and temperature. Thus, except for arid regions along the horse latitudes, terrestrial productivity tends to be highest in tropical and subtropical regions and to decrease as we move toward the poles. In the oceans, however, phytoplankton productivity tends to be most strongly limited by the availability of dissolved nutrients, especially phosphorus and nitrogen. Consequently, hotspots of marine productivity coincide with areas of nutrient input from the discharge of large rivers along continental shelves or upwellings from the nutrient-rich depths of the ocean. As we can see from the map of global productivity, ocean upwelling tends to occur along the west coasts of continents and at higher southern latitudes.

Reviewing Figure 5.11 makes it quite clear that world distribution of biomes and communities is far from uniform. In fact, the Earth's surface is dom-

FIGURE 5.30 Comparisons among total surface area (A) and total NPP (B) of the world's biomes. Tropical rain forests are so efficient (on a per area basis) at transforming solar energy into plant tissue that, even though they cover just 4% of the Earth's surface, that they account for roughly one-fourth of the Earth's total primary production. On the other hand, even though open ocean communities are relatively inefficient at fixing solar energy, because they cover nearly three-fourths of the Earth's surface, they rival tropical rain forests in total primary production.

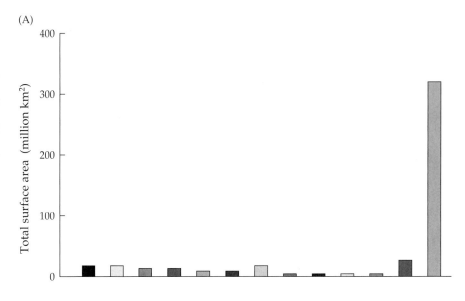

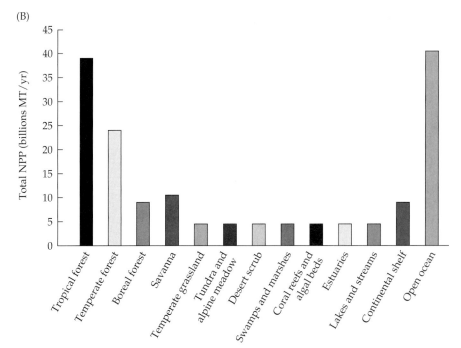

inated by open ocean, one of the least productive ecosystems per unit of surface area (see Figure 5.29A). Open oceans, however, cover nearly three-fourths of the Earth's surface (Figure 5.30A). Consequently, when all open oceans are totaled, we find that they account for over one-fourth of the Earth's primary production—a proportion equal to or perhaps slightly exceeding that of all tropical forests (Figure 5.30B). On the other hand, some of the most productive communities on a per area basis, such as coral reefs and estuaries, contribute only a minor fraction to Earth's total primary production.

Ecosystem Geography

Primary production, predation, and the great diversity of other processes involving energy flow and nutrient cycling among species, and between species and the abiotic components of their environment are, of course, not just community properties, but properties of entire ecosystems. **Ecosystem geography** is the branch of science that focuses on how these processes have influenced the distributions of ecosystems. Given the open nature of ecosystems and the need to study them at a broad range of spatial and temporal scales, ecosystem geographers have developed a hierarchical scheme of ecological regions, or **ecoregions.** This hierarchical, or nested nature, and the focus on processes at a telescoping range of scales from local sites, to landscape mosaics, and to regional and continental (or ocean basin) scales, are what distinguishes the ecoregion approach from traditional schemes based on biomes (Figure 5.31). In addition, the ecoregion scheme is truly global, including a nested scheme of ecological regions for both terrestrial and aquatic realms (Figure 5.32). Nested subdividisions of ecoregions (ecoregions forming

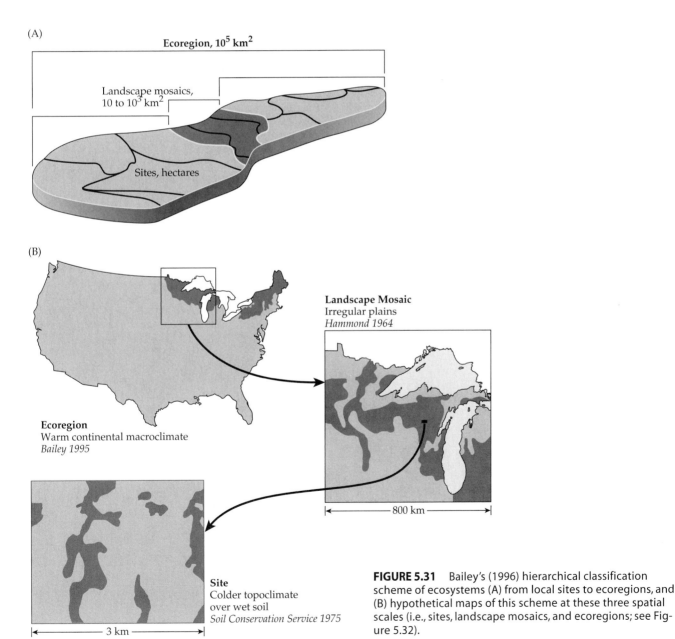

FIGURE 5.31 Bailey's (1996) hierarchical classification scheme of ecosystems (A) from local sites to ecoregions, and (B) hypothetical maps of this scheme at these three spatial scales (i.e., sites, landscape mosaics, and ecoregions; see Figure 5.32).

(A)

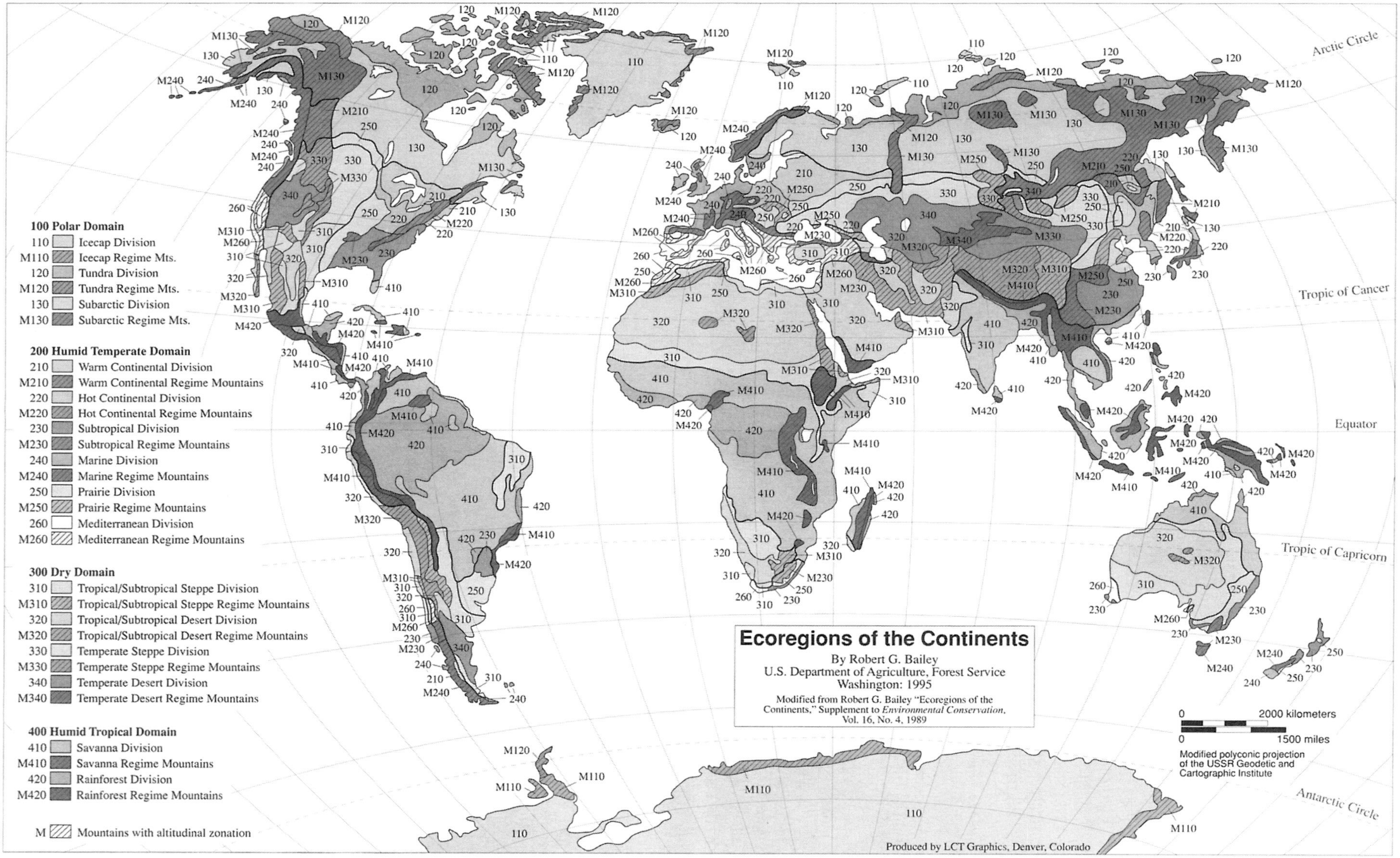
Ecoregions of the Continents
By Robert G. Bailey
U.S. Department of Agriculture, Forest Service
Washington: 1995
Modified from Robert G. Bailey "Ecoregions of the Continents," Supplement to Environmental Conservation, Vol. 16, No. 4, 1989
100 Polar Domain
110 Icecap Division
M110 Icecap Regime Mts.
120 Tundra Division
M120 Tundra Regime Mts.
130 Subarctic Division
M130 Subarctic Regime Mts.
200 Humid Temperate Domain
210 Warm Continental Division
M210 Warm Continental Regime Mountains
220 Hot Continental Division
M220 Hot Continental Regime Mountains
230 Subtropical Division
M230 Subtropical Regime Mountains
240 Marine Division
M240 Marine Regime Mountains
250 Prairie Division
M250 Prairie Regime Mountains
260 Mediterranean Division
M260 Mediterranean Regime Mountains
300 Dry Domain
310 Tropical/Subtropical Steppe Division
M310 Tropical/Subtropical Steppe Regime Mountains
320 Tropical/Subtropical Desert Division
M320 Tropical/Subtropical Desert Regime Mountains
330 Temperate Steppe Division
M330 Temperate Steppe Regime Mountains
340 Temperate Desert Division
M340 Temperate Desert Regime Mountains
400 Humid Tropical Domain
410 Savanna Division
M410 Savanna Regime Mountains
420 Rainforest Division
M420 Rainforest Regime Mountains
M Mountains with altitudinal zonation
Arctic Circle
Tropic of Cancer
Equator
Tropic of Capricorn
Antarctic Circle
0 2000 kilometers
0 1500 miles
Modified polyconic projection of the USSR Geodetic and Cartographic Institute
Produced by LCT Graphics, Denver, Colorado

FIGURE 5.32 Ecoregions of the terrestrial (A) and aquatic (B) realms (From Bailey 1996).

(B)

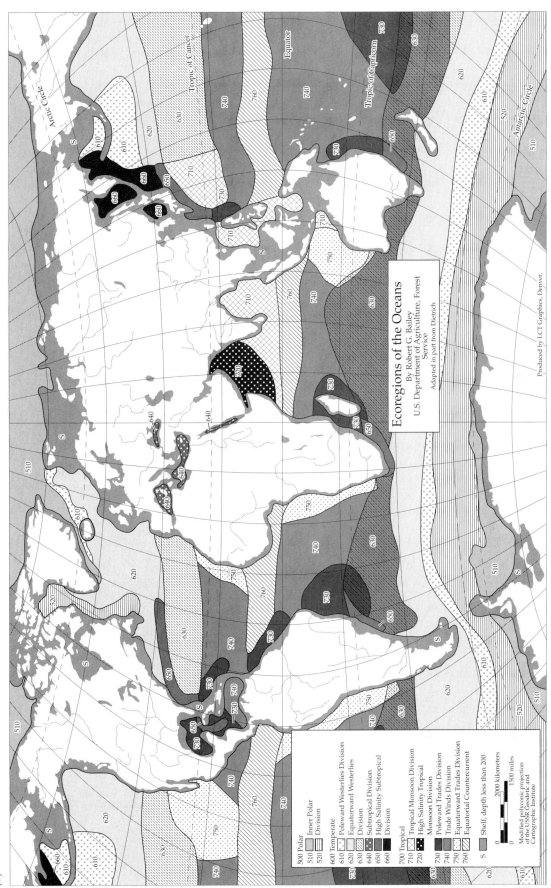

Ecoregions of the Oceans
By Robert G. Bailey
U.S. Department of Agriculture, Forest
Service
Adapted in part from Dietrich

Produced by LCT Graphics, Denver,

500 Polar
510   Inner Polar
        Division
520   Division

600 Temperate
610   Poleward Westerlies Division
620   Equatorward Westerlies
        Division
630   Subtropical Division
640   High Salinity Subtropical
650   Division
660

700 Tropical
710   Tropical Monsoon Division
720   High Salinity Tropical
        Monsoon Division
730   Poleward Trades Division
740   Trade Winds Division
750   Equatorward Trades Division
760   Equatorial Countercurrent

S   Shelf, depth less than 200

0   2000 kilometers
0   1500 miles

Modified polyconic projection
of the USSR Geodetic and
Cartographic Institute

provinces, provinces within divisions, divisions within domains) each reflect finer scale differences in climates, which in turn fundamentally influence development of soils and structure and function of ecological communities. In the aquatic realm, knowledge of water currents (surface and vertical currents, which influence water temperature, water chemistry, and salinity in the aquatic realm), are also used to classify and delineate ecoregions.

Ecoregions, however, are just that—ecological divisions of the Earth based on responses of organisms to climatic conditions. Therefore, unlike the system of biogeographic regions first developed by Sclater and Wallace, ecoregions were not developed to reflect the evolutionary history of regions and their biotas. Nonetheless, an increasing number of biogeographers, ecologists, and conservation biologists are coming to rely on the ecoregion approach and its hierarchical maps of the Earth's ecosystems (see Figures 5.31 and 5.32) as valuable tools for understanding and conserving the geography of nature. There is little doubt that the accuracy, resolution, and utility of these maps—and of ecosystem geography, in general—will continue to increase with future advances in satellite imagery, remote sensing, and other means of interpreting spatial variation in climates, oceanic currents, and biotic communities.

Finally, our current global view of nature, while generally accurate and informative, is really just a snapshot in time. The distribution of biomes and the productivity profile of the Earth have changed dramatically throughout geological time. Since the breakup of Pangaea approximately 180 million years ago, continental drift has caused major changes in global temperatures, precipitation patterns, prevailing winds, and ocean currents. Positions of the continents with respect to latitude and solar radiation have changed substantially as they have drifted, sometimes from one hemisphere to the other (see Chapter 8). These environmental changes have resulted in major shifts and transformations of communities and ecosystems, including extinction of once dominant ancient biomes and communities, and the creation of novel ones (i.e., those without any analogs in the paleoecological record; Behrensmeyer et al. 1992; Erwin 1993; Betancourt 2004; Jackson 2004). Even as recently as 18,000 years ago—only an instant ago on the geological time scale—most of the northern landmasses were covered by glaciers, often a kilometer or more thick (see Chapter 9). Warm deserts and some other biomes, once relatively rare and limited to lower latitudes and lower elevations, have since expanded at the expense of others. As we shall see in subsequent chapters, these historical changes often leave a lasting imprint on biogeographic patterns (see Chapters 10, 11, and 12).

We can learn much from this history lesson and, just as important, apply it to the future. Even within ecological time—decades to hundreds of years—human activities have caused significant changes in global climate and, in turn, fundamentally altered the distribution of biomes. Given recent advances in our ability to monitor and model these changes, we can apply some of these lessons from the past to development of a prospective view of tomorrow's biogeography. This is the focus of the final chapters.

UNIT THREE

Earth History and Fundamental Biogeographic Processes

CHAPTER 6

Dispersal and Immigration

Nature has been defined as a principle of motion and change, and it is the subject of our inquiry. We must therefore see that we understand the meaning of motion for if it were unknown, the meaning of nature too would be unknown. (Aristotle, Physics, III, 1)

THERE ARE THREE FUNDAMENTAL PROCESSES in biogeography: evolution, extinction, and dispersal. These are the means by which biotas respond to spatial and temporal dynamics of the geographic template. Thus, all of the biogeographic patterns that we study derive from the effects of these processes. While Aristotle may have been unaware of the importance, or even existence, of evolution and extinction, his writings clearly speak to the role of movements.

Throughout the development of biogeography and its maturation as a respected discipline, the relative importance of movement, or **dispersal** (inclusive of **immigration**), has been a subject of great debate. As described in Chapter 2, Charles Darwin was one of the most passionate and persuasive champions of the importance of dispersal. He and other dispersalists—including Alfred Russel Wallace and Asa Gray—argued that species disjunctions can best be explained by long-distance dispersal across existing barriers. Their antagonists in this debate were the extensionists, including Charles Lyell, Edward Forbes, and Joseph Hooker, who argued that disjunctions were the result of movements along ancient corridors that have subsequently disappeared. The point of contention was not whether dispersal occurred, but whether range expansion and disjunction required the crossing of extensive barriers.

Evidence for the extensionists' transoceanic landbridges never surfaced, but a new mechanism of dispersal emerged in the early twentieth century: continental drift. No longer was it necessary to propose such rare and unlikely events as long-distance dispersal to account for disjunctions. Species could simply ride on the continents as they split and dispersed across the surface of the Earth. This splitting of once continuous populations would serve as a vicariant event and promote evolutionary divergence of the now isolated populations. The dispersalist/extensionist debate had thus been replaced by one that was no less heated and contentious: a debate between dispersalists and vicariance biogeographers (see Chapters 10, 11, and 12). The earlier debate was transformed into one that focused on the relative importance of long-distance dispersal versus plate tectonics, orogeny, and other vicariant events in shaping biogeographic patterns; in other words, did dispersal occur before or after barriers formed?

Finally, some biogeographers have questioned whether dispersal—over any time frame—has had any lasting influence on biogeographic patterns, either past or present. This view, in perhaps its most extreme form, is summarized by **Bejerinck's Law**: "Everything is everywhere, but the environment selects" (Sauer 1988, but see also Whitaker et al. 2003). Proponents of this theory suggest that, given enough time, dispersal is inevitable for many, if not most, species. Therefore, regardless of their differences in dispersal ability, all species exhibit geographically limited ranges simply because they differ in their abilities to adapt to different environments. This view may apply to some life forms (especially microbes, which Bejerinck studied), but it seems to fly in the face of some very general biogeographic patterns including **Buffon's Law.** Bejerinck's view may have been derived from a misinterpretation of G. G. Simpson's suggestion that even exceedingly unlikely events become more likely if enough trials are performed. Increased likelihood, however, does not mean eventual certainty of colonization by all species to all points across the globe. Of course, the other extreme view—that dispersal is the only force influencing isolated communities—is just as problematic (Box 6.1).

BOX 6.1 *Dispersal versus vicariance: No contest by C. A. Stace*

▇▇▇ It is true that few new ideas or approaches are not controversial, and that controversy is a healthy and possibly essential part of scientific progress. Nevertheless, it is most unfortunate and not in the least beneficial that the advocacy of cladistic methods in taxonomy and biogeography has been so bitterly and vindictively pursued and opposed, and that criticism on both sides has so often been aimed as much at individuals as at ideas. As usual when such emotions surface, polarization of views and extremist stances emerge. It is a fair bet in such cases that neither stance is fully sustainable, and I believe that 'cladistic biogeography' provides an excellent example of such an outcome.

At the center of the argument is whether a disjunct distribution of a taxon was caused by a *dispersal* event (migration by the taxon across a barrier from A to B) or by a *vicariance* event (erection of a barrier between A and B, both of which were already occupied by the taxon)—in other words, whether the occupation of both areas followed or preceded erection of the barrier. The most obvious and strongest

spatial barriers are mountain ranges and oceans, the formation of which are today reasonably well understood and can usually be at least approximately dated. The idea of continental drift—first forwarded in 1912—was not generally accepted until the 1960s, and so it was not until after that date that vicariance was widely championed as the major causative element in disjunction, despite the publication of Croizat's views in the 1950s (Nelson and Platnick 1981).

The most widely used argument against dispersal solutions to distribution patterns 'is that they lack a testable definition of relationship', and so 'they are irrefutable' (Humphries and Parenti 1986). The corollary, of course, is that they are also unprovable. Both lines of reasoning are true, and it is indeed a severe handicap to any scientific hypothesis that it is not capable of objective testing—preferably by using it to predict outcomes whose reality can be observed. It would be preferable to adapt the hypothesis in some way so as to make it testable, or it might be better to abandon it altogether. *But it still could be the*

case that it is the correct hypothesis. Lack of means of proving that an idea is true is not an indication that it is false, nor even a reason to believe that it might be false. It is, however, a reason to consider alternative ideas that might be more testable [italics added].

There can surely be no doubt that long-range dispersal has played—and still plays—a large part in determining distribution patterns. This is nowhere better demonstrated than in the colonization of new oceanic volcanic islands that *de novo* were barren but that have become vegetated. However, there are many other dispersal events that explain distribution patterns far better than do vicariance events. The colonization of northern Europe after glaciation has been much debated, especially that into Britain and Ireland. Indeed, some of the geological facts as well as the biological events are still disputed (Devoy 1985). Rose (1972) demonstrated that most species arrived in Britain before the Channel formed c. 7500 years B.P., probably along the Somme Valley to the Sussex/Dorset axis. The rare continental species of Kent are mainly light-seeded Orchidaceae

This debate over the biogeographic relevance of dispersal and immigration is, of course, an important one. However, as with most debates, the truth lies somewhere between the extremes. The relative importance of dispersal, extinction, and evolution probably varies from one biotic group and region to another. As Simpson (1980) put it, "A reasonable biogeographer is neither a vicarist nor a dispersalist, but an eclecticist." Many of today's historical biogeographers acknowledge that biotas may have reticulate phylogenies—that is, biotas may have repeatedly fragmented and merged as dispersal barriers appeared and disappeared through time. Relatively new advances, such as dispersal-vicariance analysis (DIVA) can be used to evaluate the relative importance of vicariance and dispersal in shaping distribution patterns (e.g., see Sanmartin and Ronquist 2002; Voelker 2002; Sanmartin 2003; see also Chapters 10, 11, and 12). Again, all biogeographic patterns result from the combined effects of these responses to environmental variation across the geographic template. Before we can assess the relative importance of these processes, however, it is essential that we more fully explore the nature of dispersal.

What Is Dispersal?

All organisms have some capacity to move from their birthplaces to new sites. The movement of offspring away from their parents is a normal part of the life

---

BOX 6.1 (*continued*)

and Orobanchaceae that are probably more recent arrivals across what is now the narrowest disjunction. In other words, most British/French disjunctions are explicable by a vicariance event, but some may be accounted for by a dispersal event.

The Irish/North American disjunction has been no less debated. The small group of species that are considered amphi-Atlantic natives (e.g. *Eriocaulon aquaticum*) can be equally argued as being the result of either dispersal or of vicariance events. Other species are more doubtfully native, but some are certainly aliens (e.g. *Juncus tenuis*, probably *Hypericum canadense*) whose arrival in western Ireland has been by an as yet unknown means. Thanks to the increasingly large numbers of amateur bird-watchers on the look-out for rare species, we now know that many birds arrive in the British Isles direct from America each year. Lest anyone doubt the likelihood of dispersal across the Atlantic, let them consider the case of *Juncus planifolius*, quite recently found in natural habitats in western Ireland with its nearest native stations in Chile. Whatever the numbers of amphi-Atlantic species explicable

by vicariance events, some are *only* explicable by dispersal events.

As a third example, the tropical family Combretaceae (c. 100 species in America, c. 250 species in Africa) contains only three amphi-Atlantic species, all of which are coastal (two of them mangroves). Similarly the single species of the family common to Africa and Asia is also a mangrove. The most likely explanation is that these maritime species have been able to occupy both sides of the Atlantic due to their dispersal across it—not that these three alone of hundreds of species have remained indistinguishable for 180 million years.

I think the evidence is overwhelming that both dispersal and vicariance have played major roles in producing disjunct distribution. To deny either is futile. In each case the evidence needs to be weighed by whatever means are available, and it is to be expected that the relative importance of the two types of events will vary from situation to situation, for example from the Hawaiian Islands to Australia/South Africa/South America. Nor is it certain that the two processes are always so different. Geographic disjunctions mostly do not

arise overnight, but gradually, and there is no reason to believe that plants did not migrate to fresh areas at about the same time that these areas were becoming isolated. Post-glacial recolonization might well come into this category. This phenomenon—the wholesale migration of complete floras and faunas—also demonstrates how preposterous is the notion held by some cladistic biogeographers that if many different groups of organisms show the same patterns of disjunction, then dispersal events are not likely to be the cause.

We have a very long way to go before we can be confident of our understanding of disjunct distributions and the interpretation they provide for biogeographic phenomena. When we do get close to the truth I feel absolutely certain that the explanations will be as diverse as the phenomena, and that it will be quite impossible to consider either dispersal events or vicariance events insignificant. ▮▮▮

Source: Stace, C. A. (1989). *Journal of Biogeography* 16: 200–201. Reprinted with permission of Blackwell Science, Ltd.

cycle of virtually all plants and animals. Often dispersal is confined to a particular stage of a plant or animal's life history. Higher plants and some aquatic animals are sessile as adults, but in their earlier developmental stages they are capable of traveling long distances from their natal sites. Mobile animals can shift their locations at any time during their lives, but many settle in one place and confine their activities to a limited home range for long periods of time.

Dispersal should not be confused with **dispersion**, an ecological term referring to the spatial distribution of individual organisms within a local population.

Dispersal as an Ecological Process

Plants and animals have evolved an incredibly diverse array of dispersal mechanisms (see Gadow 1913; Udvardy 1969; Pijl 1972; Carlquist 1974, 1981; Den Boer 1977; Krebs 1978; Sauer 1988; Stenseth and Lidicker 1992; Thornton 1996; Desbruyeres et al. 2000; Bilton et al. 2001; Bullock and Kenward 2001; Clobert 2001; Figuerola and Green 2002; Jones 2003). Yet in all cases dispersal is basically an ecological process that is an adaptive part of the life history of every species. Natural selection favors individuals that move a modest distance from their natal site. A new location is always likely to be more favorable than an individual's exact birthplace, in part because intraspecific competition between parent and offspring and among siblings is reduced, and in part because the environment—and hence the quality—of the natal site is always changing. On the other hand, most environmental variables exhibit spatial autocorrelation. Thus, as distance from the natal site increases, habitats become more dissimilar and, as a result, would-be colonists are less likely to be well adapted to their new habitats.

Dispersal as a Historical Biogeographic Event

The role of dispersal in biogeography is very different from its role as a demographic phenomenon. Although dispersal occurs continually in all species, most of it does not result in any significant change in their geographic distributions. As pointed out in Chapter 4, the geographic ranges of most species are limited by environmental factors and remain relatively constant over ecological time. Biogeographers are concerned primarily with the following exceptions: those rare instances in which species shift their ranges by moving over long distances. This occurs so infrequently that we seldom see it happening, and even less often are we able to study it. Usually biogeographers must look at dispersal as a historical process, and must infer the nature and timing of past long-distance movements from indirect evidence, such as distributions of living and fossil forms. Making such inferences is a monumental task. The distribution of every taxon reflects a history of local origin, dispersal, and local extinction extending back to the very origin of life. Patterns of endemism, provincialism, and disjunction of geographic ranges (see Chapter 10) indicate that the dispersal of some groups has been so limited that their histories are indeed reflected in the distributions of their living and fossil representatives. However, reconstructing dispersal history usually requires that we deal with a process operating so infrequently that its effects may often appear to be highly random and idiosyncratic.

The problem of dealing with rare but important events is not unique to biogeography. In many ways the role of successful dispersal in biogeography is analogous to that of beneficial mutations in evolution. Beneficial mutations provide the raw material for evolutionary change, but essentially they occur randomly and so infrequently that they are difficult to observe and study (see Elena et al. 1996). Every inherited feature of an organism has a history comprising a series of unique genetic changes. Trying to reconstruct the history of the mutations that resulted, for example, in a reptilian scale evolving into a feather is conceptually similar in many ways to attempting to reconstruct the biogeographic events that led to the present disjunct distribution of large predaceous birds (e.g., ospreys, hawks, kestrels, and owls) in Africa, Australasia, New Zealand, and South America.

We draw this analogy to make two points. First, long-distance dispersal events may be infrequent and somewhat stochastic, but that does not mean that they are unimportant. On the contrary, these movements are among the most important of the events that have shaped present distributions. Second, we cannot afford to ignore the role of dispersal in biogeography just because it is difficult to study. One of the great challenges in understanding the geography and evolution of life is to develop ways of evaluating the influence of rare but important events such as beneficial mutations, asteroid impacts, and long-distance dispersal to archipelagoes as remote as the Galápagos.

Dispersal and Range Expansion

In order to expand its range, a species must be able to (1) travel to a new area; (2) withstand potentially unfavorable conditions during its passage; and (3) establish a viable population upon its arrival. Biogeographers often distinguish among three kinds of dispersal events that can have this result, primarily by the relative rates of movement and range expansion involved (e.g., Platnick and Nelson 1978; Pielou 1979). These three mechanisms of range expansion are called jump dispersal, diffusion, and secular migration (Pielou 1979).

Jump Dispersal

There is abundant evidence that many species have undergone long-distance dispersal, also known as **jump dispersal**. Anyone who has built a small pond in the backyard cannot help but be impressed by the rate at which populations of aquatic insects, snails, other invertebrates, vascular plants, and algae become established there. The same process of colonization occurs on a much larger geographic scale. When a volcanic eruption blasted the Indonesian island of Krakatau in 1883, it covered both what was left of Krakatau and the neighboring island of Sertung ("Verlaten") with a deep blanket of ash, obliterating all life. Biological surveys, primarily of birds and plants, documented the rapidity with which new populations of organisms became established on the island (Figure 6.1; van Leeuwen 1936; Dammermann 1948; Rawlinson et al. 1992; Bush and Whittaker 1993; Whittaker and Jones 1994; Thornton 1996; Whittaker et al. 1997, 2000; see also Brown et al. 1919). By 1933—only 50 years after the eruption—Krakatau was once again covered with a dense tropical rain forest; indeed, 271 plant species and 31 kinds of birds, as well as numerous invertebrates, were recorded on the island. Where did these organisms come from, and how did they get there? In the case of Krakatau, the answer to the first question is relatively clear: they dispersed across the water gap from the large islands of Java and Sumatra, which lie 40 and 80 km, respectively, from Krakatau (see discussion of the colonization of Krakatau in Chapter 13).

The case of Krakatau is unusual only in that the creation of completely virgin habitat was so dramatic that the sources of colonizing organisms could be easily identified and their immigration carefully documented. The same processes have occurred over much longer distances and time periods for other archipelagos. The Galápagos and Hawaiian archipelagoes lie far out in the Pacific Ocean—800 km west of Ecuador and 4000 km west of Mexico, respectively. These oceanic islands are actually the tops of volcanoes that arose

FIGURE 6.1 Rapid recolonization of the Krakatau Islands by (A) plants of the three Krakatau Islands of Panjang, Rakata, and Sertung ("Verlaten"); and (B) birds of Rakata, after all life on the island was destroyed by a volcanic eruption in 1883. Several biological surveys recorded the arrival of colonizing species, which probably traveled across at least 40 km of ocean. (After MacArthur and Wilson 1967; see also Whittaker 1998.)

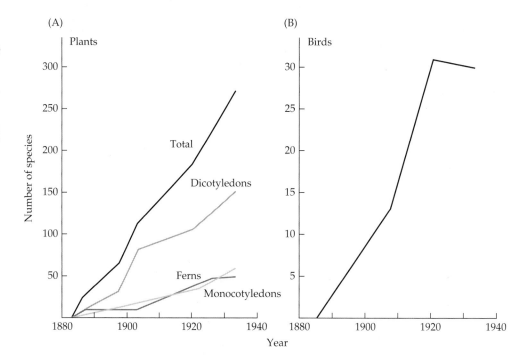

from the ocean floor. Although they have never been much nearer to continents than they are today, the Galápagos have acquired diverse biotas from **propagules** that dispersed across the ocean.

Some authors continue to downplay the importance of long-distance dispersal in favor of alternative explanations (e.g., the ancient landbridges of the extensionists). Nevertheless, there is undeniable evidence that many groups of organisms reached distant islands by traveling through the air or floating on the sea (Ridley 1930; Vagvolygi 1975; Baur and Bengtsson 1987; Enckell et al. 1987; J. M. B. Smith et al. 1990; J. M. B. Smith 1994). We also have evidence—both from recent occurrences and from historical patterns—that other species have traveled equally long distances over continental areas (McAtee 1947; Cruden 1966). Moreover, long-distance dispersal may have a strong selective component. Those species that have special adaptations for long-distance travel, or that send out more than one immigrant to found a population, have been especially successful at colonizing islands (see Chapters 13 and 14). Both the Galápagos and the Hawaiian Islands have land snails and bats as well as numerous species of trees, insects, and birds. In addition, giant tortoises and native rats inhabit the Galápagos. Noteworthy in their absence from these islands are **nonvolant** (non-flying) mammals, amphibians, freshwater fishes, and other forms poorly adapted for dispersal across open oceans.

Long-distance dispersal offers at least three important consequences for biogeographers. First, it can be used judiciously to explain the wide and often discontinuous distributions of many taxa of animals, plants, and microbes. Second, it accounts in part for both the similarities and the differences among biotas inhabiting similar environments in different geographic areas. As we have seen, the ability to disperse successfully over distance and over habitat barriers varies in a predictable manner among different kinds of organisms (e.g., bats and birds vs. amphibians and nonvolant mammals). However, because chance plays such an important role in the successful dispersal and establishment of colonists, there is a certain degree of taxonomic randomness (or **stochasticity**) in the composition of biotas. Finally, it emphasizes the importance of the many changes that have occurred as expanding human civilizations have aided the long-distance transport of species to the most remote points of the globe. We shall return to the biogeographic and ecological effects of anthropogenic dispersal in Chapters 16 and 17.

Diffusion

In comparison to jump dispersal, **diffusion** is a much slower form of range expansion that involves not just individuals, but populations. Whereas jump dispersal can be accomplished by just one or a few individuals within a short period of their life span, diffusion typically is accomplished over generations by individuals gradually spreading out from the margins of a species' range. However, these two mechanisms of range expansion are closely related, as diffusion often follows the jump dispersal of a species into a distant—but uncolonized—region of hospitable habitat.

An excellent example of both kinds of range expansion is provided by the cattle egret, *Bubulcus ibis* (Crosby 1972). This small heron originally was native to Africa, where it still inhabits tropical and subtropical grasslands in association with large herbivorous mammals, foraging for insects and other small animals that are flushed out by grazing herbivores. In the late 1800s cattle egrets colonized eastern South America, having dispersed under their own power across the South Atlantic Ocean (Figure 6.2). During succeeding decades the immigrants and their descendants thrived and spread throughout much of the New World, finding abundant food and habitat as a result of the

FIGURE 6.2 Colonization of the New World by the cattle egret, (*Bubulcus ibis*). This heron crossed the South Atlantic from Africa under its own power, becoming established in northeastern South America by the late 1800s. From there it dispersed rapidly, and it is now one of the most widespread and abundant herons in the New World. (After Smith 1974.)

FIGURE 6.3 Range expansions of selected species of animals: (A) European starling (*Sturnus vulgaris*) in North America; (B) house sparrow (*Passer domesticus*) in North America (note that the current populations derive from at least three introductions, including those of 1852 in New York, 1871 in California, and 1873 in Utah; (C) American muskrat (*Ondatra zibethica*) in Europe since its 1905 introduction near Prague; (D) nine-banded armadillo (*Dasypus novemcinctus*) in the southern United States; (E) European rabbit (*Oryctalagus cuniculus*) in Australia; and (F) red fox (*Vulpes vulpes*) in Australia. (A compiled from various sources; B after Lowther and Cink 1992; C after Van den Bosch et al. 1992; D after Taulman and Robbins 1996; E after Stoddart and Parer 1988; and F after Dickman 1996.)

clearing of tropical forests, primarily for grazing livestock. The cattle egret has now expanded its breeding range northward to the southern United States and has colonized all of the major Caribbean Islands.

Other well-known examples of diffusion involving bird species include the range expansions of starlings (*Sturnus vulgaris*) and house sparrows (*Passer domesticus*) after their intentional introductions into North America (Figure 6.3A–B). The American muskrat (*Ondatra zibethica*, see Figure 6.3C) rapidly expanded its range after introduction into Europe, and opossums (*Didelphis virginiana*) and armadillos (*Dasypus novemcinctus*, see Figure 6.3D)—both natives of South America—continue to expand their ranges northward through North America. Similarly, the European rabbit (*Oryctalagus cuniculus*;

(A)

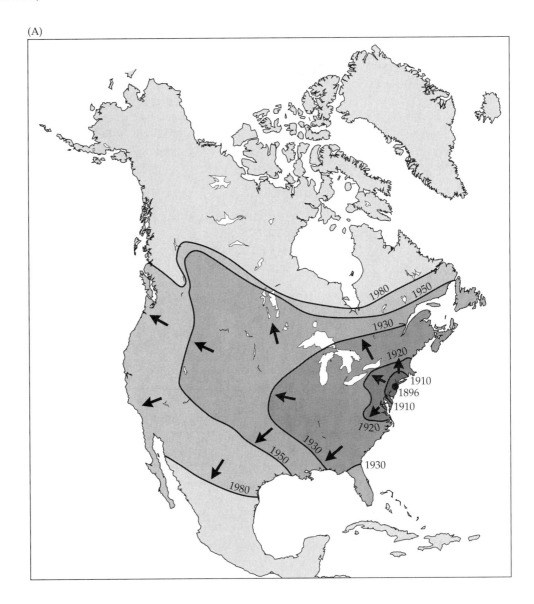

(B)

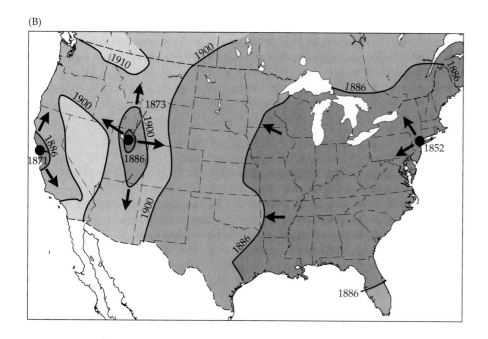

(C)

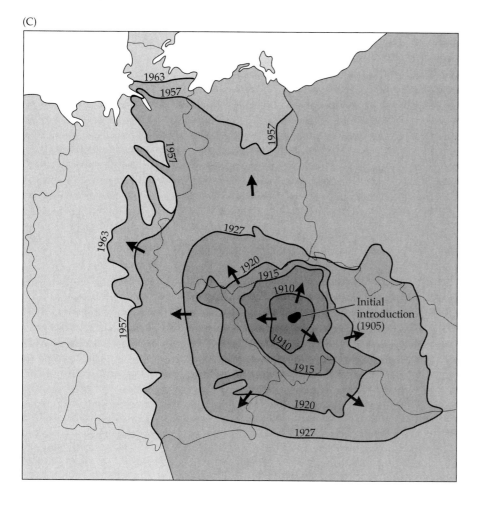

(D)

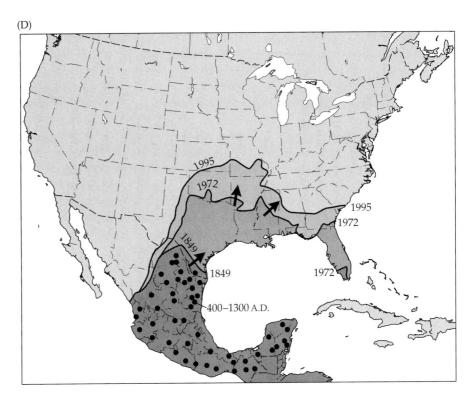

(E)

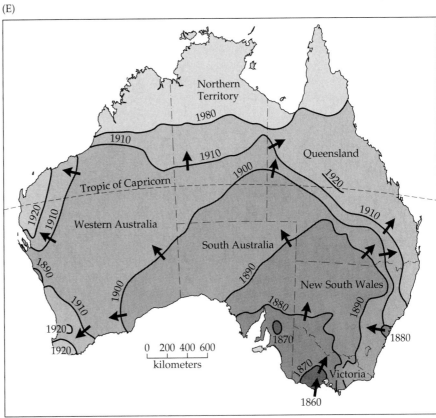

see Figure 6.3E) and red fox (*Vulpes vulpes*; see Figure 6.3F), both purposely introduced into Australia in the late 1800s, over the next century subsequently expanded their exotic ranges to occupy most of the island continent. Cases of

(F)

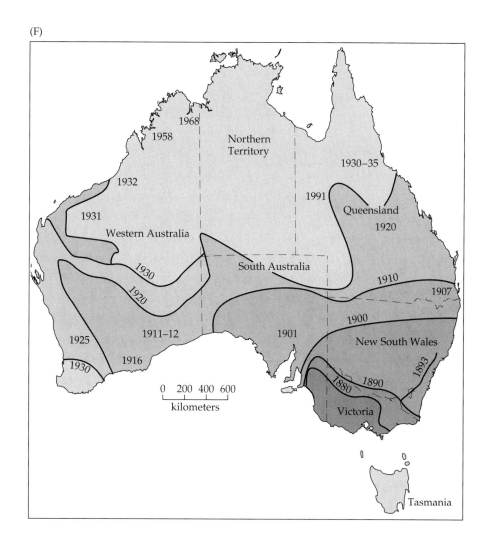

diffusion of invertebrates are perhaps all too common, but among them, the most notorious and problematic is the spread of Africanized "killer" bees (*Apis mellifera*) northward following their introduction into Brazil, and of fire ants (*Solenopsis invicta*), which are continuing their relatively rapid expansion into the southern portion of the United States. In aquatic ecosystems around the globe, many "weeds," such as water hyacinths (*Eichhornia crassipes*), water fern (*Salvinia molesta*), and Hydrilla (*Hydrilla verticilatta;*), as well as many animal pests, including zebra mussels (*Dreissena polymorpha*) have exhibited rapid range expansion following their accidental introduction into lakes and rivers outside of their original range (Figure 6.4).

Although diffusion is an inherently complex process, it typically proceeds in three stages. Initially, invasion and range expansion may be very slow, and they may require repeated dispersal events and adaptations to the characteristics of the ecosystems being invaded. Once an invasive species becomes established, however, its geographic range often expands at an exponential rate and in a roughly symmetrical fashion (Van den Bosch et al. 1992). Eventually, range expansion slows when the species encounters physical, climatic, or ecological barriers, which also tend to distort the shape of the species' geographic range. At this stage, geographic ranges may remain relatively stable unless environmental conditions become substantially altered or the species somehow crosses the barrier and reinitiates the invasion sequence.

FIGURE 6.4 Range expansion in selected plants. Maps show the spread of (A) "Fertile Crescent" crops across western Eurasia; (B) oaks (*Quercus* spp.) in Great Britain (numbers indicate years B.P.); (C) elm (*Ulmus* spp.) in Great Britain (numbers indicate years B.P.); (D) purple loosestrife (*Lythrum salicaria*) in North America. (A after Diamond 1997; B and C after Birks 1989; and D after Thompson et al. 1987.)

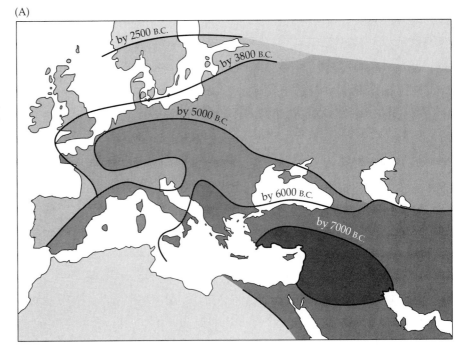

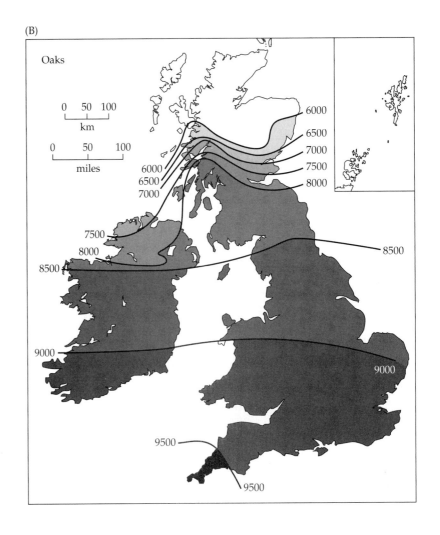

(C)

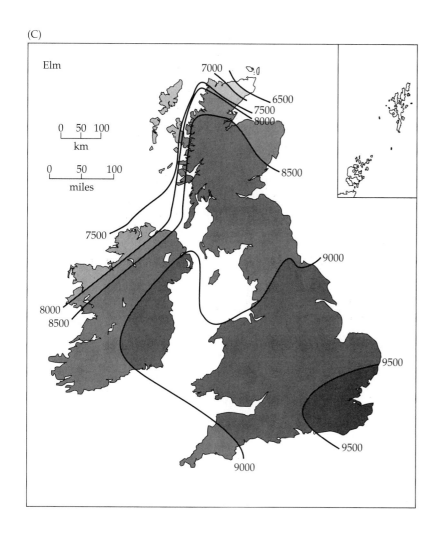

(D)

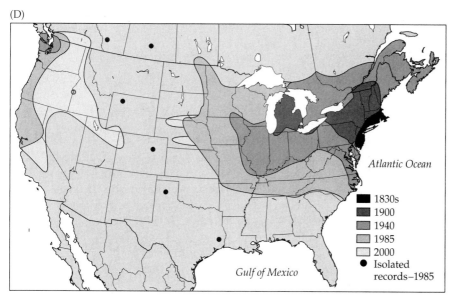

Secular Migration

In contrast to these relatively rapid forms of range expansion, **secular migration** occurs so slowly—on the order of hundreds of generations—that species have ample opportunity to evolve en route. Although Herbert Louis Mason appears to have coined the term in 1954, the concept of secular migration is an old one, dating back to the early development of biogeography (Pielou 1979). During the eighteenth century, Buffon hypothesized that most life forms originated in the northern regions of the Old World, but then spread to the New World, apparently at a time when the northern continents were much less isolated (see Chapter 2). From there, they migrated southward through the New World and, once isolated, evolved ("degenerated" in Buffon's terms) and became modified such that the biotas of the New and Old World tropics share very few forms (i.e., Buffon's Law).

While Buffon's explanation for the spread and diversification of terrestrial life forms (including mammoths and humans!) now seems fanciful, the fossil record provides convincing evidence of evolutionary divergence during range expansion. For example, some now extinct forms of camels spread southward through North America and eventually across the newly formed Central American landbridge during the Pliocene epoch. Extant products of this secular migration include the guanaco (*Lama guanicoe*) and vicuña (*Vicugna vicugna*) of South America. Similarly, camels and horses spread from the Nearctic region into the Old World. In both cases, the descendants (including camels, zebras, Przwalski's horse, and Asian wild ass) persisted although their ancestors were extirpated from their North American homeland.

In Chapter 4 we described several cases in which a species had gradually expanded its geographic range by colonizing new regions at the boundaries. Most such cases that have been documented in written records involve shifts in response to habitat modification by humans, possibly accompanied by adaptations of the species to the new niches. Presumably the same processes—habitat change and adaptation to new conditions—were responsible for historical expansions that cannot be attributed to human influence, such as the invasion and subsequent adaptive radiations of South America by many groups of North American mammals during the Pliocene (see discussion of the Great American Interchange in Chapter 10).

Mechanisms of Movement

Active Dispersal

Organisms can disperse either actively (moving under their own power) or passively (being carried by a physical agent, such as wind or water, or by other organisms). The terms **vagility** and **pagility** are sometimes used to denote the ability of organisms to disperse by active or passive means, respectively. Only a few animals have the capacity to travel long distances under their own power. Of these, strong fliers, such as many birds, bats, and large insects (e.g., dragonflies, some lepidopterans, beetles, and bugs), have the greatest capability for active, long-distance dispersal. Many of these animals regularly travel hundreds or thousands of kilometers during their seasonal migrations, which are a normal part of their annual life cycles. When stressed or aided by favorable winds, some of these same animals can cover comparable distances during a single flight.

A few examples of normal migratory routes demonstrate the potential of dispersal by flight. The golden plover (*Pluvialis dominica*), a medium-sized shorebird weighing about 150 g, breeds in the Arctic and winters in southern

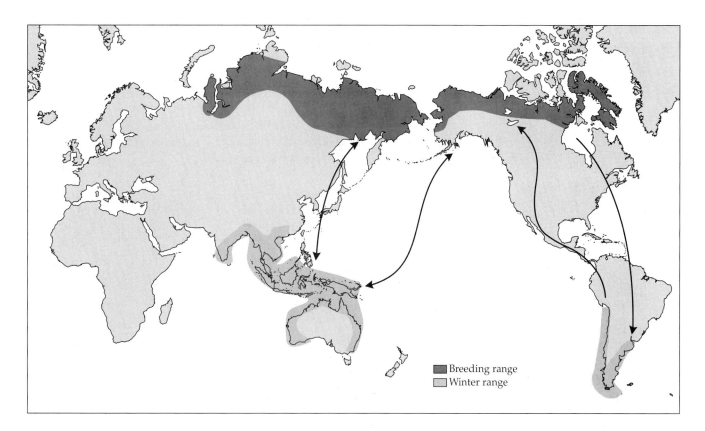

FIGURE 6.5 Migratory routes of the golden plover (*Pluvialis dominica*), a shorebird that breeds in the Arctic and winters in temperate regions of the Southern Hemisphere. Each year these birds fly prodigious distances, often crossing huge expanses of open ocean.

South America, southern Asia, Australia, and the islands of the Pacific. Migrating individuals regularly fly nonstop from Alaska to Hawaii, a distance of 4000 km (Figure 6.5) (Dorst 1962). The ruby-throated hummingbird (*Archilochus colubris*), one of the smallest birds at 3.5 g, regularly commutes twice a year nonstop across the Gulf of Mexico en route between its breeding grounds in the eastern United States and its wintering grounds in southern Mexico (Dorst 1962), a distance of 800 km.

In one of his most insightful yet often overlooked papers, Joseph Grinnell (1922) discussed "accidental" (extralimital) occurrences of birds. In addition to noting that these seemingly rare events are actually commonplace for many species, Grinnell concluded that "the continual wide dissemination of so-called accidentals … provided the mechanism by which each species spreads [i.e., diffusion] or by which it travels from place to place when this is necessitated by shifting barriers [i.e., jump dispersal]." Extralimital sightings are well documented for many taxa. Every year a few individuals of bird species native to Europe are seen in eastern North America, and vice versa, usually following severe North Atlantic storms. Hoary bats (*Lasiurus cinereus*, body mass 10–35 g) migrate from northern North America to winter in the Neotropics (Findley and Jones 1964). A species of this bat that is native to the Hawaiian Islands undoubtedly is derived from migrating individuals that went astray.

Monarch butterflies (*Danaus plexipus*) and some dragonflies (*Anax* spp.) migrate distances comparable to those flown by many songbirds—from southern Canada and the northern United States to the southern United States and central Mexico (Figure 6.6). Individual monarch butterflies may fly as far as 375 km in four days and 4000 km during their lifetimes (Urquhart 1960; Brower 1977; Brower and Malcolm 1991). Occurrences of individuals in Cuba

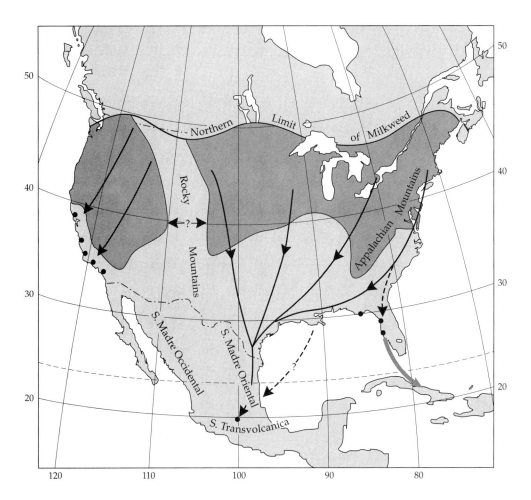

FIGURE 6.6 Fall migration routes of eastern and western populations of the monarch butterfly (*Danaus plexippus*). Most, if not all, overwintering individuals of the eastern populations gather in spectacular winter congregations in montane forests of the Transvolcanic region of Mexico. During the summer, the northern limits of this species are coincident with the northern limits of its host plants, milkweeds (*Asclepias*). (After Brower and Malcolm 1991.)

and other regions of the Caribbean may indeed represent fatal "accidents," but they might also be the equivalent of genetic "hopeful monsters." These misguided few, or their descendants, may someday found a new race or species of monarchs (see gray arrow in Figure 6.6).

In contrast to these volant species, only a few nonvolant animals, such as some of the larger mammals, reptiles, and fishes, are able to disperse substantial distances by walking or swimming. Active dispersal by these means is generally less effective than flight because the animals are forced to swim or walk through unfavorable intervening habitats, whereas flying animals can simply vault barriers. Nevertheless, some large animals—especially aquatic ones such as whales, sharks, predaceous fishes, and sea turtles—have wide, albeit often discontinuous geographic distributions produced in part by active dispersal.

Stories about swimming terrestrial vertebrates have appeared repeatedly in the literature, and as absurd as some reports may seem, we cannot deny that such incidents occasionally occur. One well-documented case involves elephants, which appear to enjoy being in water. However, who would think that an elephant would be found swimming in the ocean, with its trunk serving as a snorkel? Odd as it may seem, such observations are well authenticated. A cow and her calf not only were photographed voluntarily swimming in fairly deep ocean water up to 50 km to an island off the coast of Sri Lanka, but they were also followed and timed, showing that they made a return trip to Sri Lanka (Johnson 1978, 1980, 1981). Johnson and others believe that elephants are able to traverse narrow straits between islands and mainlands in search of new food supplies and, therefore, that their presence is not a reliable indicator of a solid land connection. Other large vertebrates, such as tigers and terrestrial snakes, have been spotted over a kilometer from land, but in many of these cases the animals were probably not there by choice, having been washed out to sea during a torrential storm.

Passive Dispersal

As impressive as active dispersal may seem, the vast majority of organisms disperse largely or solely by passive means. In any plant community, for example, we can easily observe the movement of **diaspores** (seeds, spores, fruits, or other plant propagules) away from the parent plant (Figure 6.7). Wind carries seeds and fruits that have attached wings, hairs, or inflated processes. Birds and mammals consume fleshy and dry diaspores, scattering some of them during feeding and distributing others later with their feces.

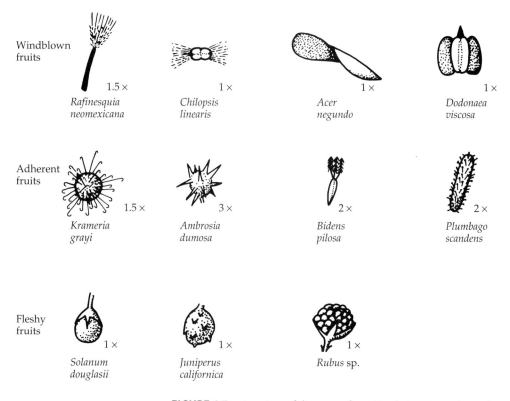

FIGURE 6.7 A variety of diaspores from North American plants designed to enable seeds to disperse from the mother plant. Seeds that are carried by wind often bear hairlike (*Rafinesquia, Chilopsis*) or papery wings (*Acer, Dodonaea*). Seeds that hitchhike on the bodies of animals have barbs (*Krameria, Bidens*), sharp spines (*Ambrosia*), or sticky glands (*Plumbago*). Seeds meant to be eaten and later defecated by birds and mammals are enclosed in fleshy fruits (*Solanum, Juniperus, Rubus*). In an unusual case, seeds of the dwarf mistletoe (*Arceuthobium*) are explosively discharged from a fleshy base.

Seeds, fruits, and spores of some plants become attached to the feathers or fur of animals and ride as hitchhikers until they are dislodged accidentally or by grooming. Some diaspores are explosively released over short distances, whereas others simply fall to the ground at the base of the parent plant. On the ground, ants, rodents, and birds compete for the fallen seeds and may carry them considerable distances—in some cases many meters. Finally, flowing water, tides, and ocean currents may displace diaspores and carry them long distances (Stebbins 1971a; Pijl 1972).

These means of passive dispersal are used not only by plants, but also by invertebrates, fungi, and microbes. Some of them are obviously more effective than others. Of course, the distance passively traveled in a generation by a seed, spore, or other disseminule depends on the velocity, distance, and direction of movement by the dispersal agent. Aerial dispersal by winds—or by birds and bats—is especially successful in moving disseminules over long distances. Animal transporters may be nomadic or migratory, leaving their original habitats and covering great distances, or they may be territorial residents that rarely leave their home ranges. Wind tends to have local effects except in violent storms, during which organisms can be widely disseminated. Jet streams are also alleged to be important dispersal agents, especially for tiny spores (Gressitt 1963; Clagg 1966). Such so-called **aerial plankton** also includes tiny mites, spiders, and insects that do not necessarily have special stages adapted for dispersal so, while their lifeless exoskeletons may drift over great distances, they will not contribute to range expansion.

Even passive mechanisms for dispersal are not necessarily completely haphazard or random. Aerial invertebrates, for example, occupy different strata above the land surface, so it is likely that they differ in their tendency to be wind dispersed (Figure 6.8). In addition—all else being equal—small organisms tend to be more effectively transported by wind than large organisms. Land snail faunas of many Pacific Islands are derived from a disproportionate diversity of *microsnails* (i.e., species so small that they drifted as aerial plankton to colonize even the most isolated islands; see Vagvolygi 1975) But all else is not equal. Structural features that promote long-distance dispersal differ markedly among species and taxonomic groups. For bacteria, yeast, fungi, mosses, and ferns, dispersal is effected by tiny spores that not only are readily transported by wind and water, but also are highly resistant to extreme physical environments. They are light enough to float in water or remain aloft in the air for long periods.

Many invertebrates also have small propagules. Often, fertilized eggs or other life history stages encyst to form thick-walled structures that are metabolically inactive and capable of withstanding long periods of desiccation and wide ranges of temperature. Some protozoans, rotifers, tardigrades, worms, and crustaceans disperse over long distances by such means. For example, the brine shrimp, *Artemia*, lives in highly saline pools and lakes throughout the world. When these bodies of water dry up, the shrimp can survive for months or years as encysted fertilized eggs. When the pools refill, the brine shrimp resume their life cycle. These dry eggs can be purchased at pet stores, and one need only add them to salt water to rear a new generation of *Artemia*. The tiny encysted eggs are also picked up by the wind and dispersed great distances. It is apparently because of this long-distance dispersal ability that *Artemia* is found in isolated localities such as Great Salt Lake in the western United States and the Dead Sea in Israel (see Browne and MacDonald 1982; Stephens and Gardner 1999; Abatzopoulos 2002).

Larger propagules are obviously heavier, and therefore require surface features to keep them aloft in wind currents. We mentioned earlier the adapta-

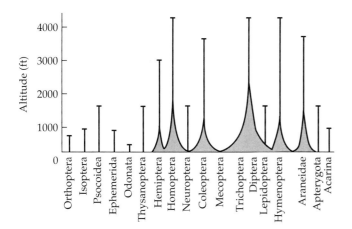

FIGURE 6.8 Aerial distribution of insects, spiders, and mites. Provided they can survive the rigors of high altitudes (especially low oxygen concentrations and temperatures), species found at higher altitudes should have greater dispersal abilities. (After Muller 1974.)

tions of plants for aerial transport (see Figure 6.7); some invertebrates have equally innovative designs, including gossamer parachutes spun by certain spiders (Araneae, Arachnida) and the long dorsal filaments of mealybugs (Hemiptera).

Most marine organisms have free-living juvenile stages that drift near the surface of the water. These tiny planktonic propagules—invertebrate eggs and larvae, fish larvae, algal cells, and the like—move passively with ocean currents. Distances traveled by small versus large forms may differ, but all are controlled ultimately by circulation patterns of the oceans (see Chapter 3). Strong currents may move propagules quickly from one locality to another in a definite direction (see Roughgarden et al. 1987). The wind that drives oceanic circulation also disperses some species, such as the jellyfish, *Vellela* (Cnidaria), which is equipped with "sails" that are oriented to carry individuals in somewhat predictable directions (Figure 6.9).

Although terrestrial vertebrates generally are too large to be carried far by the wind, some can be transported over surprising distances by water, often as passengers on mats of drifting vegetation or other debris, called "rafts." Rafts containing entire trees and carrying a large variety of organisms have been recorded (Ridley 1930; Carlquist 1965). Such large rafts are washed out to sea from large tropical rivers and may not break up for extremely long distances.

Sail

Tentacles

Left

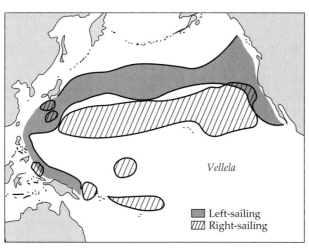

Vellela

■ Left-sailing
▨ Right-sailing

Right

FIGURE 6.9 Like their Cnidarian relative the Portuguese man of war, individuals of the jellyfish *Vellela* possess "sails," or floats, that are oriented so that they tend to drift either to the right or to the left. Each form is restricted to certain regions of the ocean, and it is thought that by sailing in the wind, these animals counteract their tendency to be carried off by ocean currents, thus remaining within a limited area. (After Savilov 1961.)

Rodents—and especially lizards—apparently have managed to colonize many isolated oceanic islands by this means. Although the chance of a particular raft going ashore on a given island is admittedly very small, over sufficiently long periods of evolutionary time there is a high probability of at least a few such events occurring. As we noted earlier, it is through the chance occurrence of such individually unlikely episodes of passive long-distance dispersal that oceanic islands and many other isolated habitats acquire much of their distinctive biotas. A less haphazard means of dispersal by seawater is found among plants—especially those inhabiting beaches and mangrove swamps—some of which have fruits, seeds, or vegetative parts that float in salt water and remain viable even when submerged for long periods of time (Carlquist 1965, 1974; MacArthur 1972; J. M. B. Smith et al. 1990; J. M. B. Smith 1994).

A surprisingly large number of organisms are transported over long distances by other organisms. Many examples of this process, called **phoresy**, are obvious. Parasites, for example, are carried to any new areas colonized by their hosts. As Europeans explored and colonized the Earth, they spread a variety of bacterial and viral diseases. In some isolated areas, the native human populations had never been exposed to these microbial pathogens and had evolved no immunity to them. Consequently, diseases such as smallpox, syphilis, and measles had devastating effects when introduced among American Indians and Pacific Islanders (Marks and Beatty 1976; McNeill 1976).

Some plants have sticky or barbed seeds or fruits that adhere to mobile animals (see Figure 6.7). Aquatic birds may transport small invertebrates in their feathers—or, as Charles Darwin demonstrated, trapped in mud on their feet—between widely separated lakes and ponds (see also Figuerola and Green 2002). Other seeds—usually those found in sweet, fleshy fruits—are adapted to pass through the digestive tracts of animals. Because they are resistant to digestive juices, they are still viable when contained in dropped feces; in fact, the germination of certain seeds is enhanced when they are exposed to animal digestive chemicals. Such internal transport is extremely common and is important not only in the colonization of isolated oceanic islands, but also in dispersal within continental regions, from one habitat to another. Birds are the usual agents, but mammals and reptiles also serve as internal seed dispersers (Janzen and Martin 1982; but see also Howe 1985). Similarly, insects and mycophagus mammals are important agents of dispersal of fungal spores.

A few animals are obligatory hitchhikers. Colwell (1973, 1979) and his associates discovered a fascinating example of this in tiny mites that live in certain flowers (see also Poinar et al. 1990; Brown and Wilson 1992; Zeh et al. 1992; Athias-Binche 1993). These mites do not have a highly resistant stage in their life cycle and seem poorly adapted for long-distance dispersal. However, the mites occur in flowers regularly visited and pollinated by hummingbirds, and they are transported between flowers on the birds' beaks. They crawl into the nasal openings, whose protected microclimate enables them to survive and be carried for long distances. At least two species of these specialized flower mites occur on the Greater and Lesser Antilles. Presumably, they originally colonized these islands from the continental mainland of tropical America, riding hundreds of kilometers on the beaks of dispersing hummingbirds.

The Nature of Barriers

Successful long-distance dispersal usually requires that organisms survive for significant periods of time in environments very different from their usual habitats. These unusual environments constitute physical and biological barriers that successful colonists must cross. The effectiveness of such barriers in preventing dispersal depends not only on the nature of the environment, but also on characteristics of the organisms themselves. Of course, these characteristics vary from one taxonomic group to another, so that particular barriers may not affect all residents of a habitat equally. Thus, barriers are species-specific phenomena. Two examples will illustrate this point.

Most freshwater zooplankton have resistant stages that facilitate long-distance dispersal. Consequently, the same species may be found in widely separated localities wherever environmental conditions are similar, such as in the cold temperate lakes of northern North America, Europe, and Asia. On the other hand, fishes inhabiting these same lakes appear to be unable to disperse across terrestrial and oceanic systems. Consequently, the same fish species inhabit only those lakes that have been connected by fresh water at some time in the past. Many isolated alpine lakes that have never had such connections, support a diverse plankton fauna but have no fish at all (unless they have been recently introduced by humans).

Similarly, many of the isolated mountains of arid regions in the southwestern United States are essentially islands of cool, moist forest in a sea of hot, dry desert. The desert is not a barrier to most birds and bats, which rapidly and repeatedly fly over it to colonize the montane forests. On the other hand, for many small terrestrial mammals, reptiles, and amphibians, which must disperse much more slowly on foot, the desert represents a formidable barrier. It is likely that these animals colonized the most isolated mountains when they were connected by bridges of suitable forest and woodland habitat during the Pleistocene (Figure 6.10; see Findley 1969; J. Brown 1978; Patterson 1980; Lomolino et al. 1989).

Because the effectiveness of a particular kind of barrier depends on both the physical and biotic challenges it poses and the biological characteristics of the organisms attempting to cross it, it is difficult to make sweeping generalizations about the nature of barriers. It is usually true, however, that organisms that inhabit temporary or highly fluctuating environments are much more tolerant of extreme or unusual physical and biotic conditions than are species that are confined to permanent or stable habitats. Plants and animals from fluctuating environments are also more likely to have resistant life history stages, which not only enable them to survive periods of unfavorable conditions, but also can serve as effective propagules. Consequently, the "weedy" species that inhabit temporary or fluctuating environments are likely to be better dispersers and less limited by any kind of barrier than are species from more permanent or constant habitats.

A comparable example is provided by the pupfish, *Cyprinodon variegatus*, an extremely **euryhaline** and **eurythermal** species that inhabits estuaries, tidal flats, and mangrove swamps along the eastern coast of North America from Cape Cod to Yucatán. This little fish has managed to cross hundreds of kilometers of ocean to colonize comparable habitats on many of the Caribbean Islands. In contrast, the strictly freshwater sunfishes (*Lepomis*) and basses (*Micropterus*) that are so abundant and diverse in rivers, lakes, and streams of eastern North America, are not native to the Caribbean. The successful introduction of these species by humans indicates that saltwater barriers, rather than any lack of suitable habitats on the islands, had prevented their colonization.

Daniel Janzen (1967) made a related point in his insightful paper that addressed the restricted distributions of many species in mountainous areas of the tropics. He pointed out that mountain passes of a given elevation are effectively "higher" in the tropics than in temperate zones. Because temperate environments experience seasonal temperature variations, lowland plants and

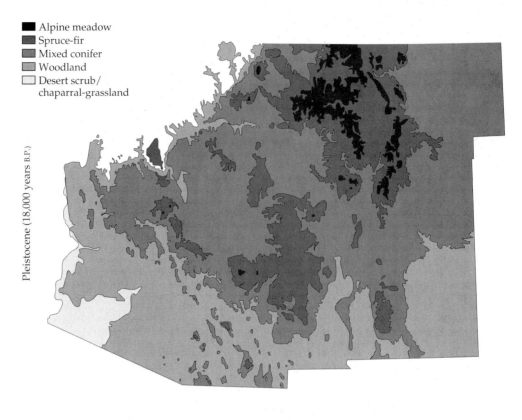

Alpine meadow
Spruce-fir
Mixed conifer
Woodland
Desert scrub/
chaparral-grassland

Pleistocene (18,000 years B.P.)

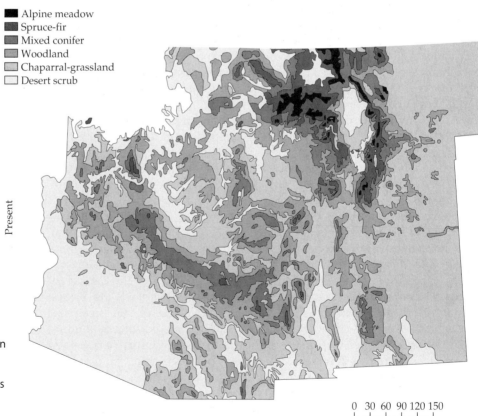

Alpine meadow
Spruce-fir
Mixed conifer
Woodland
Chaparral-grassland
Desert scrub

Present

FIGURE 6.10 Geographic shifts in vegetation zones of the American Southwest since the most recent glacial maximum. (Elevational shifts in these vegetation zones are depicted in Figure 9.22.) (From Lomolino et al. 1986.)

0 30 60 90 120 150
kilometers

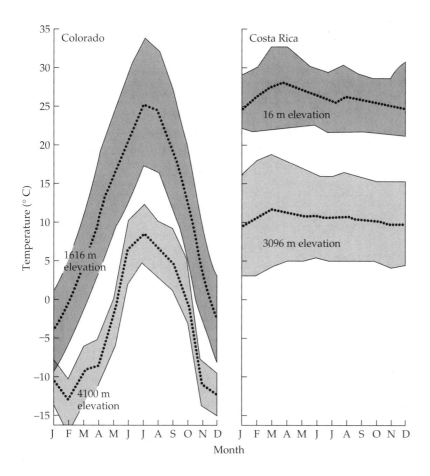

FIGURE 6.11 A given change in elevation tends to be a greater barrier to dispersal in the tropics than at higher latitudes, as shown by these temperature profiles. In tropical regions, sites separated by several thousand meters of elevation usually experience no overlap in temperature, whereas in the temperate zone, winter temperatures at low elevations broadly overlap summer temperatures at much higher sites. (After Janzen 1967.)

animals must be able to tolerate a wide range of environmental conditions, including those they would encounter at higher elevations during the summer (Figure 6.11). In contrast, because of the thermal constancy of the tropics, a tropical lowland species would never experience—and, therefore, would not be adapted to withstand—the temperature regimes it would encounter if it were to travel over a high mountain pass from one lowland area to another.

Physiological Barriers

Probably the most severe barriers are presented by physical environments so far outside the range an organism normally encounters that it cannot survive long enough to disperse across them. The vast majority of aquatic organisms live either in the oceans or in fresh water, and they cannot regulate their water and salt balance sufficiently to survive more than a brief exposure to the other environment. Very few freshwater fishes have successfully colonized across the oceans. The same is true of amphibians, which have permeable skins and are quite intolerant of exposure to salt water. Likewise, many terrestrial plants are unable to withstand prolonged exposure to seawater at any stage of the life cycle, including the seed. Hnatiuk (1979), however, reported that 56 of the 69 terrestrial plant species native to Aldabra Atoll in the western Indian Ocean tolerated total immersion of their seeds in seawater for 8 weeks with no inhibition of germination. This tolerance would not only facilitate the survival of these species on tiny islands subject to wave splash during heavy storms, but would also greatly increase the probability of these species surviving immersion during transport across ocean barriers, thus perhaps explaining how many of them managed to colonize the islands in the first place.

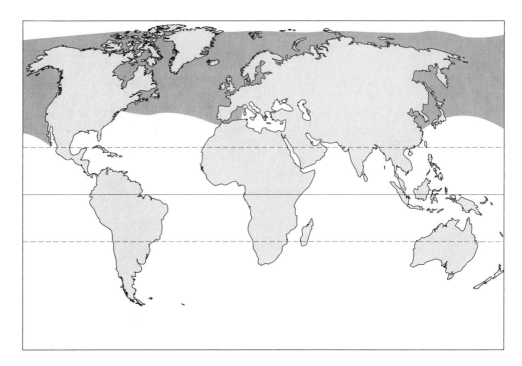

FIGURE 6.12 The avian family Alcidae (auks, puffins, and murres) is restricted to cooler regions of the Northern Hemisphere (shaded area), even though all of these birds are winged and most are strong fliers. The tropics appear to constitute a major barrier to the dispersal of this group, although the presence of distantly related but ecologically convergent seabirds in the Antarctic region may also prevent their colonization of the Southern Hemisphere. (After Shuntov 1974.)

Environmental temperature regimes can also serve as physiological barriers. We have already mentioned the case of mountain passes in the tropics. A similar barrier is created by the tropics themselves, which form a band of high temperatures around the equator, isolating the cooler temperate and arctic areas toward either pole. For many cold-adapted organisms—both terrestrial and aquatic—these warm tropical climates are a major barrier to dispersal, resulting in **amphitropical** (sometimes referred to as "bipolar" or "antitropical") distributions of many species (see Figures 9.25, 10.19, 10.26, and Box 8.2 Figure A). Some groups, such as the Nearctic avian family Alcidae (auks, puffins, guillemots, and murres), are good dispersers within their climatic zone and are broadly distributed throughout one hemisphere (North or South), but have not managed to cross the tropics to colonize the other hemisphere (Figure 6.12). Thus, the two hemispheres are inhabited by different forms with convergently similar morphology and ecology, such as penguins (Spheniscidae) and diving petrels (Pelecanoididae), or skuas (*Catharacta* spp.) and jaegers (*Stercorarius* spp.) of the Southern and Northern Hemispheres, respectively.

The nature of barriers may also vary dramatically with the seasons. For example, in temperate regions of North America, large bodies of water serve as effective barriers to the movement of many terrestrial species, including nonvolant mammals. However, during winter these waters freeze, providing a seasonal avenue for the dispersal of winter-active species to islands or adjacent mainland sites (Figure 6.13) (Jackson 1919; Banfield 1954; Lomolino 1984, 1988, 1989, 1993; Tegelstrom and Hansson 1987).

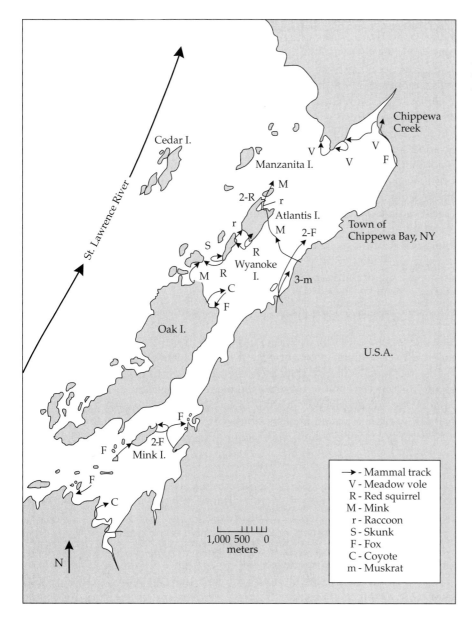

FIGURE 6.13 Movements of terrestrial mammals across the ice of the St. Lawrence River within the vicinity of Chippewa Bay, New York. Arrows indicate the direction of mammal tracks observed within a 24-hour period (March 1, 1979). (After Lomolino 1983, 1988.)

Ecological and Psychological Barriers

Dispersing organisms must be able to survive not only the physiological stresses imposed by the environments they traverse, but also the ecological hazards. Just as predation and competition can limit the local abundances of species, they can also prevent successful dispersal. One would expect biotic interactions to be particularly important components of barriers for animals that neither move very rapidly nor disperse at a resistant stage. Although it is likely that competition and predation limit the dispersal of such organisms, we are not aware of any well-documented examples.

Surprising as it may seem, behavioral or psychological barriers appear to play a major role in preventing the long-distance dispersal of some organisms. Most organisms appear to possess mechanisms of **habitat selection**, which is the ability to recognize and respond appropriately to favorable environments. In some animals these traits are so well developed that they strongly inhibit active dispersal. For example, some forest birds that seem perfectly capable of flying long distances are apparently unwilling to disperse across open habitats, including open waters. Willis (1974) has described species of antbirds (Formicariidae) that have become extinct on Barro Colorado Island and have not recolonized, even though they would have to fly only a few hundred meters across Gatun Lake. MacArthur et al. (1972), Diamond (1975b), and others have documented groups of tropical birds, such as New World cotingas and toucans and Old World barbets and pittas, that are strong fliers, but that are repeatedly absent or poorly represented on oceanic islands, even where good habitat appears to be abundant. Often species restricted to virgin rain forest are particularly sedentary, whereas other, even closely related, species that are characteristic of second growth habitats or edges are good dispersers. Species adapted to successional habitats must continually disperse to new environments as their habitat patches undergo succession and become unsuitable. Such species are therefore much more likely to strike out across barriers and disperse successfully over long distances than are species that are restricted to very stable environments.

Biotic Exchange and Dispersal Routes

As we emphasized in the preceding section, barriers are species-specific phenomena. Even within a group of related species, a barrier to one is not likely to be a barrier to all. Now we shall broaden our perspective to look at entire biotas and the process of biotic exchange between biogeographic regions. Biogeographers often distinguish three kinds of dispersal routes based on their effects on biotic exchange. Listed in order of increasing resistance to biotic exchange, these include corridors, filters, and sweepstakes routes (Simpson 1940).

Corridors

The term **corridor** refers to a dispersal route that permits the movement of many or most taxa from one region to another (Simpson 1936, 1940; see also Udvardy 1969). By definition, a corridor does not selectively discriminate against any form, and must therefore provide an environment similar to that of the two source areas. A corridor therefore allows a taxonomically balanced assemblage of plants and animals to cross from one large source area to another, so that both areas obtain organisms that are representative of the other. Any differences between biotas on either side of the corridor result either from chance, or from ecological and evolutionary responses to different environments on either end of the corridor. In Simpson's example, the Pre-Columbian landscape between Florida and New Mexico served as a corridor for terrestrial vertebrates of this region (Simpson 1940). He attributed differences between biotas to differences between subtropical environments of Florida and those of the arid lands of New Mexico.

As we will discuss in Chapters 8 and 9, the geological record is replete with evidence of great pulses of biotic exchange across ancient marine and terrestrial corridors. The ancient Tethyan Seaway, a circum-equatorial marine system, extended all the way from the Orient and Malaysia to westernmost Europe, separating Africa from Eurasia (see Figure 8.21B). For nearly 500 million years the Tethyan Seaway served as a dispersal highway for marine organisms, including both benthic and pelagic forms (Adams and Ager 1967). Exchange of taxa within the Tethys was not always uniform, and may have been strongly influenced by its predominantly westward flowing currents. This exchange was disrupted about 60 million years ago when Africa became connected with Asia through Arabia, and India drifted northward to join Asia. Now the Mediterranean Sea has a vastly different fauna than the Indian Ocean.

Many ancient, transient connections have served as corridors for exchange among terrestrial biotas. Most recently, a great diversity of terrestrial plants and animals dispersed across the landbridges of Beringia, the Sunda Shelf, and the Tasman and Arafura Basins during the glacial maxima of the Pleistocene (see Figure 9.9). The Bering landbridge between Alaska and Siberia permitted passage in both directions, but it was probably a corridor only during the first half of the Cenozoic, when a mild climate prevailed. Even then, biotic exchange across Beringia tended to be asymmetrical, with more species moving from west to east, (i.e., from the relatively diverse regions of eastern Asia and Siberia into Alaska).

Filters

As its name implies, a **filter** is a dispersal route that is more restrictive than a corridor. It selectively blocks the passage of certain forms while allowing those able to tolerate the conditions of the barrier to migrate freely. As a result, colonists tend to represent a biased subset of their respective species pools. Thus the biotas on the two sides of a filter share many of the same taxonomic or functional groups, but some taxa are conspicuously absent in each.

The Arabian subcontinent is a harsh filter that permits the dispersal of only a limited number of mammals, reptiles, nonpasserine ground birds, invertebrates, and xerophytic plants between northern Africa and Central Asia. Organisms that need abundant water, such as freshwater fishes, most amphibians, and forest dwellers of many taxa, are stopped by the deserts in this region. This has not always been true. As mentioned above, for millions of years Arabia and Asia were widely separated by the Tethyan Seaway—a formidable barrier to terrestrial organisms. When the Arabian Peninsula became emergent and contacted Persia, the region intermittently served as a landbridge for movements of Asian taxa into Africa and vice versa, as in the Upper Miocene, when many African forms appeared in Asia for the first time (Cooke 1972). Thus, today's filter has been both more of a corridor and a more formidable barrier in the past, and it might become either in the future.

Filters may be produced by abiotic or biotic factors. They are generally easy to identify because the number of species in certain taxa decreases with distance from the source area. Filters often form transition zones between two biogeographic regions. Figure 6.14 depicts such a transition zone: a two-way filter formed by the Lesser Sunda Islands between Java and New Guinea. Reptile species of Oriental origin decrease proportionately as one moves eastward,

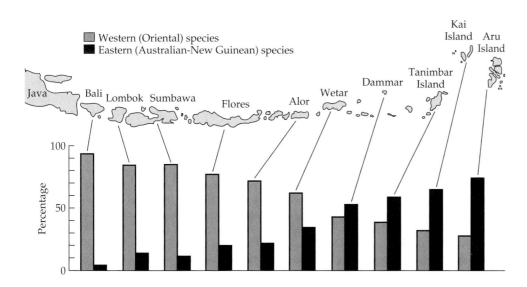

FIGURE 6.14 The Lesser Sunda Islands between Java and New Guinea serve as a two-way filter for the reptilian faunas of southeastern Asia and Australia. Bars quantify the decline in Oriental species and the increase in Australian species going from west to east down the island chain. (After Carlquist 1965.)

(A)

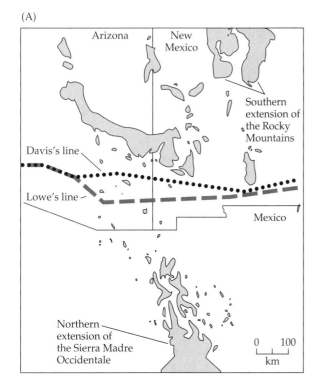

(B)

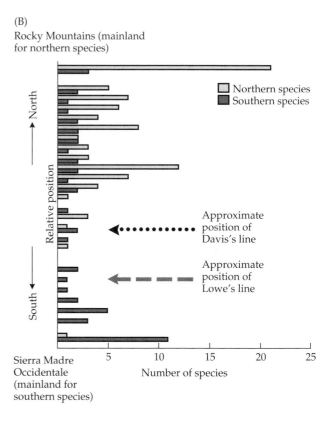

FIGURE 6.15 The desert "sea" of the American Southwest may serve as a transition zone, or two-way filter for forest and woodland species between the Rocky Mountain "mainland" to the north and the Sierra Madre "mainland" to the south. (A) Two biogeographic lines—Lowe's line and Davis's line—have been proposed to mark the divisions between these regional faunas. (B) Bars quantify the decline in species of nonvolant mammals as one moves away from their respective source regions. (A after Davis 1996; B after Lomolino and Davis 1997.)

where Australasian groups become dominant; of course, the same decreasing trend is found for Australasian groups as one moves westward through Wallacea (i.e., the transition zone bisected by Wallace's line).

On the opposite side of the globe, biogeographers have discovered a terrestrial version of Wallacea. For most of the Pleistocene, assemblages of forest-dwelling terrestrial vertebrates in western North America have been isolated in two major "mainland" regions—forests of the Rocky Mountains to the north and the Sierra Madre to the south. However, intervening habitats have allowed some mixing among these regional biotas, especially during glacial maxima, when forests and woodlands expanded at the expense of more xeric habitats (see Figures 6.10 and 9.22). Thus, the xeric landscapes of the American Southwest serve as a two-way filter or transition zone between the two faunal regions (Figure 6.15; see Davis 1996; Lomolino and Davis 1997).

In Chapter 10 we discuss the biotic exchange across another terrestrial filter—the Central American landbridge. This event, termed the Great American Interchange (G. G. Simpson 1969, 1978), represents perhaps an unparalleled case study of the combined effects of dispersal, evolution, and extinction. As many have observed, however, the formation of this landbridge created a barrier to the dispersal of marine organisms. Construction of the Panama Canal reunited the Caribbean Sea and the Pacific Ocean, but biotic exchange between them has been extremely limited. Low-salinity waters flowing from Gatun Lake serve as an effective physiological barrier to most marine animals (Figure 6.16) (Hildebrand 1939; Woodring 1966; Rubinoff and Rubinoff 1977; Abele and Kim 1989).

Sweepstakes Routes

On September 5, 1995, Hurricane Luis ripped through the Lesser Antilles region of the Caribbean. Its torrential rains and winds, which were in excess of

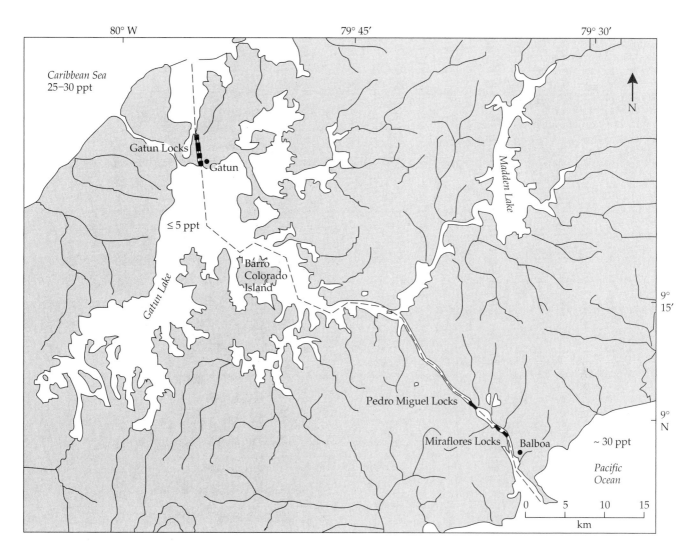

FIGURE 6.16 Map of the Panama Canal and associated waterways, showing salinity levels. Low-salinity waters of Gatun Lake serve as an effective physiological barrier to most marine animals. (After Abele and Kim 1986.)

140 miles per hour, were powerful enough to cause massive landslides and send large tangled mats of vegetation adrift in the sea. Rafting along on one of the largest mats—comprised of hundreds of logs and trees torn from the island of Guadeloupe— were at least 15 green iguanas (*Iguana iguana*). Incredibly, these unwitting and unwilling castaways not only survived Hurricane Luis and Hurricane Marilyn—which followed closely on its heals—but they also clung to and survived on their raft for a month until it washed ashore onto the shores of Anquilla—some 200 miles distant (Figure 6.17; Censky et al. 1998; see also Raxworthy et al. 2002; Rieppel 2002). The saga of these rafting lizards once again speaks to the importance of events which, although seemingly rare and unlikely in ecological time, are highly likely over longer time periods. Again, not everything is going to get everywhere, but things will definitely end up in surprising places and, as a result, such rare events strongly influence the geography of nature.

The distinguished paleontologist, George G. Simpson, coined the term **sweepstakes route** to refer to barriers that must be crossed by such rare, chance interchanges (i.e., by long-distance, jump dispersals). When people enter a sweepstakes, they do so knowing that although many individuals enter the contest, only one or a handful of lucky ones will win prizes. Yet,

FIGURE 6.17 Tracks of Hurricane Luis and Hurricane Marilyn across the Lesser Antilles during September and October of 1995, apparently carried at least 15 green iguanas (*Iguana iguana*) on an immense natural raft of logs and trees that broke off from the island of Guadeloupe during the storms, depositing the unintentional immigrants on the shores of Anguilla, some 200 miles distant (Raxworthy et al. 2002). (After Censky et al. 1998.)

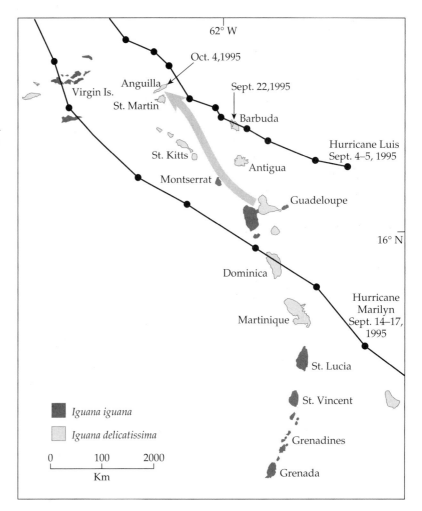

because winners are so few and far between, prizes can be very high indeed. In the natural world, propagules continually disperse from established populations into new areas, but only a small fraction are ever successful in founding a new population. Yet, their arrival in such isolated and ecologically simple environments, if successful, is often followed by adaptive radiations (see Chapter 7).

The seemingly chance colonization of isolated oceanic islands by lizards and any other terrestrial animals is the classic case of interchange across a sweepstakes route. This process is not as random as it may seem, however. Even when organisms cross barriers independently of one another, they may still use the same dispersal routes (e. g., sweepstakes dispersal of green iguanas followed the hurricane tracks through the Lesser Antilles). Insular biotas, for example, usually have taxonomic affinities with the organisms inhabiting the nearest continent or other nearby landmass, which serves as a source of propagules. Similarly, as we discussed earlier, most of the species inhabiting islands of the South Pacific have Indo-Malayan and, to a lesser extent, Australasian affinities, and the number of taxa shared with these landmasses decreases with increasing distance out into the Pacific (see Figures 6.14 and 6.18) (Wilson 1959, 1961; Solem 1981). Mainland ancestors of most species inhabiting these remote islands were probably pre-adapted for long-distance

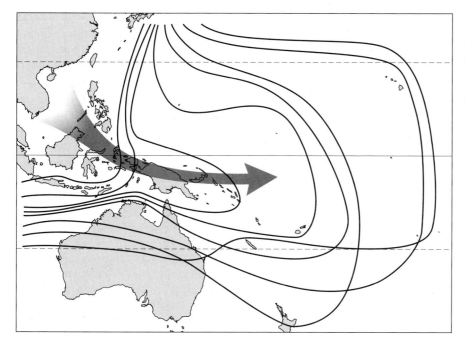

FIGURE 6.18 A classic example of a sweepstakes route. Lines show limits of the distributions of eight different families of land snails in Australasia and the South Pacific. Each of these groups originated in Southeast Asia and then spread southward and eastward to a different extent. (After Solem 1981).

dispersal—so-called waif species. Thus, dispersal across these sweepstakes routes is random only in that it is impossible to predict when it will occur and which of these waif species will actually make it to any given island.

Other Dispersal Routes

When a landmass is shifted from one place to another by seafloor spreading, it carries a biota on board. These species can thus be transferred directly without crossing major barriers. India, which moved from its ancient connection with southern Africa to Asia, has been called a "Noah's Ark" by McKenna (1973) because an assemblage of organisms rafted on it en masse to a new environment. There are other fragments of tectonic plates in southern Europe, western North and South America, and eastern Asia that may have served as arks for their respective biotas. These plate fragments also carried fossil beds associated with their former locations. Movements of these beds of long-extinct fossils, sometimes termed "Viking funeral ships," may complicate biogeographic analyses by mixing the remains of biotas that may never have occurred together in time.

Biologists are now eager to help Earth scientists find evidence of past land connections and physical barriers. There was a time early in the twentieth century, however, when biogeographers could have made substantial contributions toward establishing the occurrence, mechanisms, and timing of continental drift, especially if our understanding of the phylogeny of organisms had been more advanced. But this did not happen, perhaps because most biogeographers were too conservative (see Chapter 8). Now we must design our reconstructions with a realistic picture of the Earth's history in mind. Biogeography still can make some original contributions in these areas, however, especially by identifying barriers other than those created by drifting continents. Biological distributions generally cannot be used as proof of land connections in lieu of evidence from rocks, but their analysis can corroborate or contradict geological data and can provide valuable clues to past events overlooked by geologists. The analysis of dispersal routes is best done separately by geologists and biogeographers, with both keeping watchful eyes out for conflicting

FIGURE 6.19 (A) An idealized lognormal frequency distribution of dispersal potential among individuals or species. (B) Dispersal curves describe the expected decline in immigration rate as a function of increasing distance from a source region. If dispersal is a passive and random process with a constant probability of organisms dropping out with each increment of distance, then the dispersal curve should take the form of a negative exponential. On the other hand, if dispersal is active or if organisms are carried on rafts with normally or lognormally distributed persistence times, then the dispersal curve should take the form of a "normal" function. (C and D) Seed dimorphism in the composite, *Heterotheca latifolia*, affects dispersal curves of two types of fruits: disk achenes, which have an attached parachute-like structure, and ray achenes that lack this dispersal structure, but have thicker fruit walls and can survive in the soil. (A and B after MacArthur and Wilson 1967; C and D courtesy of L. Venable.)

results. Meanwhile, the burden of determining positions of landforms and dates of tectonic changes still falls on geologists, whereas reconstructions of biotas, paleoecology, and paleoclimatology are projects for biogeographers (Chapter 12).

Dispersal Curves within and among Species

For nearly all organisms, immigration rates (i.e., numbers of individuals or species arriving at a site per unit of time) tend to decline with increasing distance from the parent or source. Depending on the mechanism of dispersal, this relationship between immigration rate and distance often takes one of two general forms (Figure 6.19). If dispersal is a purely random process with a constant probability of organisms "dropping out" with each increment of distance, then the dispersal curve should take the form of a negative exponential (see Figure 6.19B). This pattern is often found among wind-borne propagules and other forms that are passively dispersed.

On the other hand, for actively dispersing organisms or those dispersing on logs or rafts with normally distributed persistence times, dispersal abilities may approximate a normal or lognormal distribution. The precise form of this frequency distribution is not critical, only the fact that most individuals (or species) can disperse beyond some minimal distance, and that few can travel

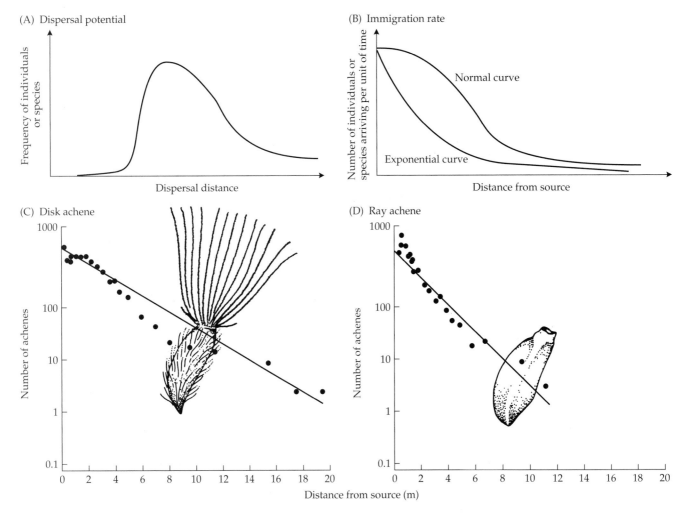

beyond some greater distance. In such cases, immigration curves should approximate what MacArthur and Wilson (1967) termed a "normal" or **Gaussian function** (upper curve in Figure 6.19B). Immigration rates should thus remain relatively high until the distance exceeds the dispersal abilities of the least vagile individuals, decline rapidly until the modal dispersal ability is approached, and then slow to asymptotically approach zero for the most distant sites.

Because dispersal abilities of related species result from a combination of physiological, behavioral, and morphological traits—many of which are distributed as normal or lognormal functions—we expect the same patterns to hold for dispersal curves compiled across species. That is, for a given taxonomic group or pool of species, dispersal abilities should be normally or lognormally distributed, and immigration rates should also decline as a normal or lognormal function of distance.

We will return to the biogeographic relevance of dispersal curves in our chapters on island biogeography. It is important to emphasize here, however, that actual rates of dispersal or arrival at distant sites are influenced by the nature of dispersal barriers as well as by characteristics of the species in question.

Establishing a Colony

If dispersal is to be of biogeographic significance, an organism not only must be able to travel long distances and cross barriers, but it also must be able to establish a viable population upon its arrival at a new site. As Sherwin Carlquist (1965) has stated, "Getting there is half the problem." Obviously, a successful colonist is one that survives and reproduces. Several factors, including interspecific interactions, habitat selection, and reproductive strategies, may play little role in dispersal per se, but may be of great importance in determining the fate of an immigrant once its journey is over.

Habitat Selection

All organisms exhibit some form of habitat selection. Highly mobile organisms actively seek out favorable environments, which they recognize either instinctively or from having learned the characteristics of their place of origin. Many passively dispersed organisms, such as the planktonic larvae of sessile marine invertebrates, will not settle unless they perceive certain sensory cues indicating a substrate that is suitable for establishment. Even the seeds and spores of plants and the cysts of invertebrates have some capacity for habitat selection. Although these resistant structures are passively dispersed, certain specific conditions of temperature, moisture, light, and other factors are usually required to break their dormancy and initiate growth and development. Unless cues indicating favorable conditions are received, the diaspore remains in the resistant stage and is capable of further dispersal.

Stanley Wecker (1963, 1964; see also Harris 1952) performed a classic study of habitat selection in deer mice (*Peromyscus maniculatus*). He worked with two subspecies that inhabit the north central and northeastern regions of the United States: *Peromyscus maniculatus gracilis*, a form primarily restricted to mature mesic forests, and *P. maniculatus bairdi*, which inhabits grasslands. The latter subspecies is of particular interest because it is an animal that originally inhabited the prairies of central North America, but extended its range eastward by colonizing agricultural cropland and old fields as the forests were cleared by Europeans (Hooper 1942; Baker 1968). In a series of experiments in which he reared mice in different environments and then tested them in a

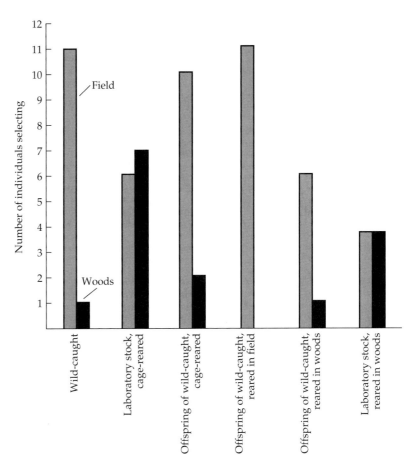

FIGURE 6.20   Habitat selection by the field (grassland) subspecies of the deer mouse, (*Peromyscus maniculatus bairdi*), depends on both inherited ability and early experience. Field-caught individuals and their offspring—especially if they were exposed to field habitat at the time of weaning—spent most of their time in the field end of a large enclosure that covered a field/forest boundary. Many generations in the laboratory or early experience in the woods diminished this preference. (After Wecker 1963.)

large outdoor enclosure that was half old field and half forest, Wecker showed that each subspecies tended to wander and explore until it found its appropriate habitat, whereupon it would establish residence (Figure 6.20). The mice used genetically inherited information, supplemented by early experience in their natal habitat, to select the correct environment. Because forest and field habitats form a temporally fluctuating patchy mosaic throughout the northeastern United States, this precise habitat selection has probably been important to the successful dispersal of deer mice into these successional habitats.

What Constitutes a Propagule?

In addition to finding the right kind of habitat upon arrival, successful colonists must be able to reproduce and establish a viable new population. The size and composition of the dispersing unit are of critical importance in this regard. MacArthur and Wilson (1967) used the term propagule to define the unit necessary to found a new colony.

Colonization is a major problem for the many higher plants and animals that reproduce sexually, because reproduction normally requires the participation of two individuals of different sexes or mating types. Since long-distance dispersal is a rare event, it is relatively unlikely that potential mates will arrive sufficiently close to each other in space and time to find each other and reproduce. In contrast, asexually reproducing organisms are often successful colonists, because a single individual can found an entire population by fission, budding, or some other asexual means (see Browne and MacDonald 1982). Although many microbes and lower plants and animals are capable of sexual reproduction, many of them can also reproduce asexually. Many of the crustaceans that constitute the freshwater zooplankton can go through numerous generations of asexual reproduction after emerging from encysted zygotes (which are produced by a sexual generation that occurs periodically during the life cycle). Many plants also are capable of asexual reproduction by vegetative growth via rhizomes, stolons, root crowns, or bulbs.

Even among the normally obligately sexual higher plants and animals, there are some species with reproductive patterns that make them much better colonists than others. A large number of plant species are both hermaphroditic and self-compatible; that is, the same individual has both male and female flower parts and can produce viable seeds by self-fertilization. This is true of many species, such as tumbleweed (*Salsola iberica*), that we normally think of as weeds. Dandelion (*Taraxacum*), another common weed, is **apomictic:** it can produce viable seeds from unfertilized ovules. In all of these kinds of weeds a single seed can serve as a propagule. Animals also exhibit a marvelous diversity of reproductive strategies. A few species of fish, amphibians, and reptiles are parthenogenetic: the entire population of these species consists solely of females that produce female offspring asexually (White 1973; Cuellar 1977). Some fishes can also change their sex, depending on the sex of other members of their local population.

Although these reproductive patterns do not necessarily mean that the organisms will be good colonists, they appear to facilitate successful colonization in some species. At least three genera of gecko lizards, for example, contain parthenogenetic species that are distributed primarily on oceanic islands (Cuellar 1977). It takes only a single individual, drifting ashore on a floating log or arriving in an islander's canoe, to establish an entire new population. On the other hand, sexuality, if it is associated with aggressive behavior, may confer a competitive advantage to some invading lizard species over asexual natives (Petren et al. 1993; Case et al. 1994).

Some sexual species have reproductive cycles and social behaviors that facilitate simultaneous colonization by several individuals. In animals with internal fertilization, a propagule can consist of a single female with stored sperm or developing embryos. In plants in which several seeds are normally retained and dispersed in a fruit, the fruit can be the unit of dispersal even in self-incompatible, obligately sexual species. Some birds, which have virtually none of the characteristics listed above, are good dispersers and successful colonists not only because they can fly long distances, but also because they normally travel in flocks, which obviously increases their chances of establishment in new locations. Thus we find that fruit pigeons (*Ducula*)—medium-sized, highly social birds that often travel in flocks far out to sea—are widely distributed in the tropical Pacific (Diamond 1975b), and were even more widespread before extinctions caused by aboriginal humans (see Chapters 16 and 17; Steadman 1995).

Survival in a New Habitat

Because long-distance dispersal almost inevitably means colonizing a habitat that is not exactly like the one in which the colonists originated, successful colonists must be able to survive physical stresses and ecological challenges of their new environments. Thus, organisms from highly fluctuating and unpredictable environments probably tend to be good dispersers not only because they can tolerate stressful conditions encountered en route, but also because they are well prepared to meet the unknown physical and biotic challenges they must face after they arrive. Biological hazards posed by competitors and predators (including parasites and diseases) should not be underestimated. Organisms from large, continuous areas with diverse ecological communities, such as productive continental habitats, tend to be relatively successful in establishing populations in small, isolated habitats containing few species, such as oceanic islands. Insular species, however, are rarely successful in invading mainland habitats, presumably because they are not adapted to cope with the variety of threats posed by a diverse array of competitors and predators.

When small numbers of individuals are involved in colonization, as is almost always the case in long-distance dispersal, chance becomes very important in determining the course of events. Colonies may fail, for example, because a freak storm or an unlikely predator kills just one or two individuals. Thus, despite this incredible diversity of pre-adaptations for long-distance dispersal, even if a viable propagule arrives in an apparently suitable environment, successful colonization is not assured: in fact, most colonizations fail. Most human-assisted introductions of exotics also fail to establish breeding populations. Therefore, successful long-distance dispersal typically is a very rare phenomenon.

Yet rare does not mean unimportant or intractable. Continuing advances in fields such as molecular biology, genetics, and the analysis of stable isotopes hold great promise for providing invaluable clues on the nature and importance of these rare events. In the latter case, because the levels and combinations of stable isotopes (e.g., those of carbon [^{13}C], hydrogen [^{2}H], sulfur [^{34}S], and strontium [^{87}Sr]) have characteristic geographical distributions, assays for these isotopes in, for example, the feathers of birds on their winter range can be used to identify the location of their natal range (Hobson 2002, 2005). These and other, perhaps unanticipated advances in our abilities to investigate rare events such as dispersal across sweepstakes routes will provide invaluable insights into the nature of dispersal and the ability of organisms to respond to the temporal and spatial dynamics of the geographic template (see Nathan 2005). Before focusing on the latter subject (i.e., the dynamics of the Earth and its environments, Chapters 8 and 9), we review two other, fundamental biogeographic processes—speciation and extinction—and their influences on the geographic variation of nature.

CHAPTER *7*

Speciation and Extinction

THE THREE FUNDAMENTAL PROCESSES of biogeography are speciation, extinction, and immigration. In this chapter we focus on the two former processes—speciation and extinction—but both of these processes are strongly influenced by immigration, and all three have a strong geographic context. Speciation (evolution) occurs across space as well as across time, and it often requires geographic isolation, while extinction is the final stage of the long-term, large-scale process of geographic range collapse.

Species have histories that are somewhat analogous to the lives of individuals. New species originate through the multiplication of old species (**speciation**); they survive for varying periods of time, during which they may or may not leave descendants; and they die (**extinction**). There are, however, important differences between the histories of individuals and the histories of species. Whereas the births and deaths of most individuals are discrete, easily recognizable events, it is often difficult to determine when a new species has come into existence, and when the last individual of a species has died.

Species are comprised of groups of populations. Even if a local population goes extinct, the species it belongs to does not (unless that population represented the entire species—e.g., the Devil's Hole pupfish; Figure 10.3). As such, a species is considered to have temporal and spatial **cohesion**, which could be maintained genetically in species with different sexes, or perhaps ecologically through adaptation to a specific set of niches. Whatever the mechanisms involved, it is this property of spatial and temporal cohesion that makes a species an evolving entity with both a discrete beginning and end, and thus an important entity in evolutionary biology and biogeography (note for example that the title of Darwin's most influential work began with "*On the Origin of Species…*").

As evolving entities, species have a unique set of properties. In sexual organisms, a species includes groups of individuals held together genetically through the process of reproduction, whereas each "clone" in an asexual species remains genetically distinct from one another

FIGURE 7.1   Contrasting modes of inheritance between (A) sexual and (B) asexual organisms in species level lineages. In sexual organisms, reproduction produces a complex network of relationships among individuals, whereas "clones" in an asexual species must be held together by some process other than reproduction. Circles represent individuals; different colors represent different sexes; and solid lines represent parent-offspring reproductive connections. (After de Queiroz 1998.)

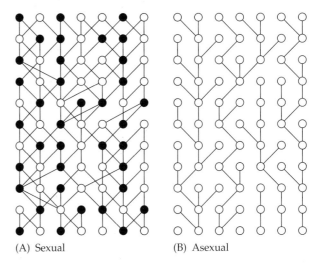

(A) Sexual   (B) Asexual

(Figure 7.1). Individuals within a sexual species are said to have a **tokogenetic** or **reticulate** (netlike or interwoven) set of relationships. On the other hand, because each mating represents a joining of different parental lineages, the process of speciation represents the splitting (**cladogenesis**) of an ancestral species into two or more daughter species—each with its own evolutionary trajectory—the species represents the boundary between tokogenetic evolutionary processes among individuals and **cladogenetic** evolution *among* different species (Figure 7.2). This demarcation is relatively straightforward, it is also possible for different species to exchange genes through **hybridization**. Sometimes hybridization is restricted to an occasional cross-species mating at a narrow hybrid zone, but other times it can produce a hybrid species with its own evolutionary trajectory, ecological niche, and geographic structure. Such **reticulate species** are particularly common in plants (Grant 1981), but also occur in animals (e.g., *Rana esculenta* in Europe; Spolsky and Uzzell 1986).

Like related individuals, related species possess particular combinations of characters that have been inherited from their common ancestors (**homologies**). Even if a character changes along the way (Darwin's "descent with modification"), it is still recognized as being homologous if it was inherited along an **evolutionary lineage** connecting descendant taxa with their common ancestor. Many characters inherited by descendants from ancestors might have arisen through an evolutionary force other than natural selection. However, some characters that make a lineage distinct represent innovative solutions (**adaptations**) to a problem that may have limited survival and reproduction in other lineages. These sorts of characters often are acquired independently in unrelated species that are faced with similar environmental challenges (e.g., the wings of birds, bats, and butterflies that are required for powered flight), and are called **homoplasies**, or **convergent** characters.

Further, although a particular lineage may flourish for a period of time and give rise to a diverse group of species, this success is usually ephemeral on a geological time scale. Eventually, the rate of extinction exceeds the rate of speciation, the diversity of the group decreases, and all or nearly all of its representatives go extinct. While the vast majority of species that were at one time alive went extinct without leaving descendants, it's obvious that some lineages have continued to survive in order to produce more species. The vast variety of living and extinct organisms on Earth have been produced through

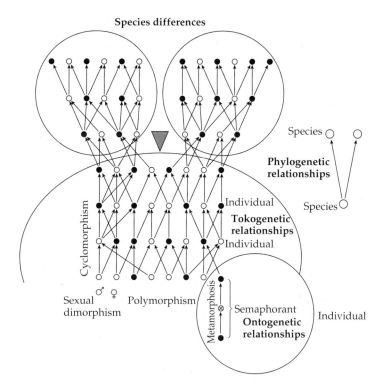

FIGURE 7.2 Hennig's original depiction of the demarcation between *ontogenetic* (within individuals of a sexual species), *tokogenetic* (among individuals within a sexual species), and *phylogenetic* (between different species) relationships. (After Hennig 1966.)

repeated cycles of speciation and extinction, often through rapid bursts of speciation following an abrupt mass extinction, and all set against a background of changes in Earth's geographic template (see Chapters 8 and 9).

Systematics

The science of biology that addresses patterns and processes associated with biological diversity and diversification (**systematics**) consists of two interrelated but distinct subdisciplines, **taxonomy** and **phylogenetics**. Taxonomy is the discipline that assigns names to organisms and classifies biological diversity using Linnaeus's binomials (species and genus) and a hierarchical classification scheme (genera within families, families within orders, etc.). Phylogenetics is the discipline that reconstructs the history of evolutionary relationships between organisms—the study of **cladistics** is a subdiscipline of phylogenetics that studies cladogenesis (i.e, splitting but not reticulation) that results in separate evolutionary lineages. A number of the **species concepts** summarized below rely on explicit hypotheses of phylogenetic relationships to delineate separate species. However, apart from the few definitions needed to proceed through this discussion, we will defer a formal development of phylogenetic theory and practice until Chapter 11.

What Are Species?

If the fundamental unit of biogeography is the geographic range of a species, then what is a species, and what criteria shall we use to recognize different species? Biologists have long wrestled with the problem of how to recognize closely related but distinct species (Mayr 1982), and the debate is not over (e.g., Mayden 1997; de Queiroz 1998; Wheeler and Meier 2000). Recent reviews have analyzed and summarized the large number of currently debated **species concepts** (e.g., Bisby 1995; Mayden 1997; de Queiroz 1998; Avise 2000).

TABLE 7.1 *Twenty-two species concepts and standardized abbreviations*

1. Agamospecies (ASC)	14. Morphological (MSC)
2. Biological (BSC)	15. Non-dimensional (NDSC)
3. Cohesion (CSC)	16. Phenetic (PhSC)
4. Cladistic (CISC)	17. Phylogenetic (PSC)
5. Composite (CpSC)	a. Diagnosable Version (PSC$_1$)
6. Ecological (EcSC)	b. Monophyly Version (PSC$_2$)
7. Evolutionarily Significant Unit (ESU)	c. Diagnosable and Monophyly Version (PSC$_3$)
8. Evolutionary (ESC)	18. Polythetic (PtSC)
9. Genealogical Concordance (GCC)	19. Recognition (RSC)
10. Genetic (GSC)	20. Reproductive Competition (RCC)
11. Genotypic Cluster Definition (GCD)	21. Successional (SSC)
12. Hennigian (HSC)	22. Taxonomic (TSC)
13. Internodal (ISC)	

Source: Mayden 1998.

Each of a large number of modern species concepts (Table 7.1) has its own strengths and weaknesses. Some may be more applicable to a given kind of organism than to others (e.g., sexual vs. asexual; plant vs. animal; fossil vs. extant species), and emphasize a particular aspect of what may turn out to be a more general theoretical framework (Mayden 1997; de Queiroz 1998), and some are considered more difficult to implement in practice than others (Bisby 1995).

Mayden (1997) argued that only the **evolutionary species concept** (ESC) qualifies as a primary definition of species—all others are secondary concepts that serve as operational criteria for delineating species, but are not sufficiently general to serve as a primary concept. Although the ESC was originally developed by Simpson (1961), the most modern definition is "…an entity composed of organisms that maintains its identity from other such entities through time and over space and that has its own independent evolutionary fate and historical tendencies" (Wiley and Mayden 2000, p. 73). Although seemingly counterintuitive, Mayden (1997) also suggested that the ESC is the only concept that is not operational, thereby requiring maintenance of an array of secondary concepts to employ in practice—a few of the more important of these are discussed briefly below.

Under the **morphological species concept** (MSC), "Species are the smallest groups that are consistently and persistently distinct, and distinguishable by ordinary means" (Cronquist 1978, p. 15). This is also known as the **classical species concept** because it has, since the beginning of taxonomy, been the main operational criterion used to delineate different species. The main assumption is that morphological differences between species have an underlying genetic basis that is heritable, and it has been applied to asexual as well as sexual species. This criterion fails when speciation occurs without detectable morphological change (**sibling** or **cryptic species**), or when morphological traits are highly plastic (e.g., subject to rapid change—with or without genetic changes—as ecological conditions shift) within a single species.

The **biological species concept** (BSC) defines a species as "…a group of interbreeding natural populations that is reproductively isolated from other such groups" (Mayr and Ashlock 1991, p. 26). This influential species concept arose from the Modern Synthesis in evolutionary biology that developed in the early twentieth century. Mayr (1963; 1970) considers reproductive isolation as the most fundamental attribute delineating closely related species because it protects a reproductive community of harmonious genotypes that have been molded by evolution to work within a particular ecological context.

As straightforward as it may sound, the criterion of reproductive isolation is difficult to employ. It cannot be applied to fossil organisms. The best one can do is to determine whether the morphological gaps between specimens are as large as—or larger than—those between living species that are reproductively isolated (i.e., reproductive isolation is inferred from an MSC perspective). The BSC is unwieldy for extant groups as well. The reproductive isolation criterion cannot be applied to organisms that are exclusively asexual and do not exchange genetic information with one another (see Figure 7.1). While asexual reproduction is the hallmark of prokaryotes (bacteria, archea), it is also common throughout the fungal, plant, and animal world. Even for sexual species, application of the BSC can be difficult or in some cases impossible to apply. For instance, we know of many examples of **hybridization:** interbreeding that produces viable and fertile offspring, but which occurs between populations that remain distinct genetic and evolutionary units. This is especially problematic in vascular plants, in which hybridization is common between species and genera and has been known to occur even between plants in different families (Levin 1979; Grant 1981; Briggs and Walters 1984; Wyatt 1992). Recently, biologists have argued that hybridization may frequently produce novel evolutionary lineages with equal or higher fitness than the parental lineages (Arnold and Emms 1998), which runs counter to Mayr's original emphasis on the importance of isolating mechanisms for protecting co-adapted gene complexes. Finally, spatially isolated populations cannot interbreed under natural conditions, and so their potential to interbreed must be investigated under artificial laboratory conditions. Only in a few cases have such mating tests actually been conducted, because of the time and expense involved and the fact that many organisms cannot easily be reared in the laboratory. Usually, as in the case of fossil species, it has traditionally been left to a systematist to use a surrogate such as the MSC to make an educated guess about whether individuals would have interbred sufficiently freely to form a single population if the isolated populations had come back into contact under natural conditions. Mayden and Wood (1995) discussed ten criteria under which the BSC fails as a general and operational species concept.

There are several forms of a **phylogenetic species concept** (PSC). Each of these has a common goal of defining species based on evolutionary lineages (clades) derived through the method of phylogenetic systematics (see Chapter 11). The concept of a phylogenetic species was originally proposed in the early 1980s (Cracraft 1983), and as elucidated by McKitrick and Zink (1988) has three components. It must be

1. a **monophyletic** lineage;
2. derived through an evolutionary process of descent from an ancestral lineage;
3. diagnosable through examination of character state transformations.

Those derived characters that uniquely define a monophyletic lineage are called **apomorphic** characters. This definition considers the reproductive isolation criterion that is fundamental to the BSC as irrelevant, even misleading, in the delineation of species (McKitrick and Zink 1988; Mayden and Wood 1995), and is applicable to both sexual and asexual species.

One criticism of the PSC is that it might lead to an inflation of species diversity if species designations are based on as few as a single apomorphic character. Another is that, at the molecular level, phylogenies derived from independent parts of an organism's genome (e.g., mitochondrial vs. nuclear genes) will not always be exactly concordant in their delineation of species—the so-called gene tree versus species tree problem (Nei 1987), outlined in greater detail in Chapter 11. Both of these issues were considered under the **genealogical concordance concept** (GCC), which proposes that "…population subdivisions concordantly identified by multiple independent genetic traits should constitute the population units worthy of recognition as phylogenetic taxa" (Avise and Ball 1990, p. 52). In practice, most advocates of the PSC do recognize the gene tree versus species tree problem, and normally do incorporate information from more than a single genetic trait into decisions about species boundaries.

Perhaps of greatest relevance to asexual organisms is the **cohesion species concept** (CSC), which defines species as "…the most inclusive population of individuals having the potential for phenotypic cohesion through intrinsic cohesion mechanisms [genetic and/or demographic exchangeability]" (Templeton 1989, p. 12). Thus, cohesion within a species can be maintained through the exchange of genetic information in sexual species, but also through the occupation of a particular ecological niche even when gene exchange is not possible as in asexual lineages. Cohan (2001) has argued that different bacterial "ecotypes"—populations of cells that might overlap spatially but differ in their ecological niches—qualify as different species under an ESC definition based on the demographic exchangeability criterion of the CSC.

Brown (1995) suggested that the difficulties of defining species reflect fundamental problems that stem from the way in which organisms vary in their traits. Variation among organisms in most traits is **discontinuous:** there are clumps separated by gaps. In other words, some individuals are very similar to one another but are considerably different from other such clumps of individuals. Figure 7.3 illustrates this pattern using variation in beak size among Darwin's finches that occur on several islands of the Galápagos. This "clumpy-gappy" organization of biological diversity is apparent in nearly all characters of organisms, not only in morphology and genetics, but also in physiology, ecology, and behavior. It reflects something very basic in the genetic and evolutionary processes through which both individuals and populations inherit characteristics from their ancestors. At the individual level, these processes are essentially the same in all organisms, but at the population level, as we saw above (see Figure 7.1), they are fundamentally different in sexual and asexual organisms: the former have two parents and thus reticulate patterns of inheritance, whereas the latter have only one parent and thus only a simple branching pattern of ancestry and descent.

The essence of most species concepts is that they either explicitly or implicitly attempt to identify the points along an evolutionary trajectory at which a particular level of clumpy-gappiness exists because populations have crossed the line between genetically reticulate (tokogenetic) to phylogenetically divergent associations (see Figure 7.2). The concepts differ from one another in the sorts of clumps they seek to recognize—morphological (MSC), genealogical (GCC), ecological (CSC), phylogenetic (PSC), or reproductive communities (BSC). While the same clumps are often recognizable on the basis of distinctive morphological, genetic, physiological, behavioral, and ecological characteristics, sometimes they differ in some of these features, but not others. Nowhere is this discrepancy more evident than at the level of gene trees nested within species trees. For very good theoretical reasons, we should often

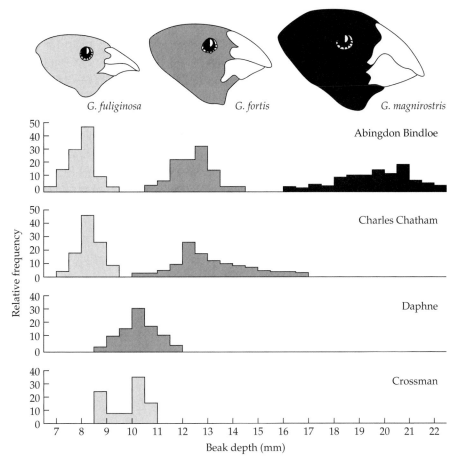

FIGURE 7.3 Distribution of one trait, beak depth, in several populations of Darwin's finches inhabiting different islands of the Galápagos. This discontinuous variation, especially within a single island, illustrates the "clumpy-gappy" distributions of traits that most species concepts try to capture. The pattern of interisland differences also illustrates *character displacement;* that is, the fact that species tend to be more different from one another where they occur together—presumably reflecting adaptive divergence to avoid competition—than where they do not coexist with closely related forms. (After Lack 1947.)

expect that following the cessation of reproductive exchanges between populations, mitochondrial DNA (mtDNA) gene trees in animals will normally cross the tokogenetic-phylogenetic border much faster than will nuclear genes (Moore 1995; see Chapter 11).

The best that can probably be done is to utilize a particular criterion or set of criteria for recognizing the discontinuous organization of biological diversity in a fashion that is practical to use (operational) and consistently associates major clumps with a conceptually sound primary species concept (e.g., the ESC). It is important to recognize that the variety of species recognition criteria (secondary species concepts) have been invented by humans as useful surrogates to simplify and classify biological diversity in a fashion that, hopefully, will increasingly be consistent with an evolutionary lineage-based concept of species. Several investigators have argued that different species concepts are most appropriately applied at different stages in the process of speciation (Harrison 1998; Brooks and McLennan 2002). Nevertheless, the process of evolutionary change includes such a variety of dynamic interactions among populations that the task of species recognition will always include some degree of subjectivity, although some effort has been made to develop empirical methods for delimiting species in nature (Sites and Marshall 2003). However, we can take some comfort from the fact that specialists on most taxonomic groups have traditionally recognized separate species at a scale of gappiness that is similar to those discovered through more recent surveys of molecular-level variation (Avise and Walker 1999).

SUBSPECIES, ECOTYPES, PHYLOGROUPS, AND EVOLUTIONARILY SIGNIFICANT UNITS. Taxonomists, evolutionary biologists, and ecologists recognize that some well-differentiated populations probably do not warrant recognition as distinct species using any species definition. These populations have traditionally been variously called subspecies, geographic races, varieties, or ecotypes. Subspecies is a formal term usually reserved for populations that are morphologically, and presumably genetically, distinct—a genealogical concordance definition introduced by Avise and Ball (1990). Typically, a subspecies is given a trinomial Latin name, such as *Peromyscus maniculatus gracilis*, in which the last part is the subspecies designation. The other terms are applied more loosely and informally. Plant ecologists use the term ecotype to refer to a distinct population that occurs in a particular habitat type (note that Cohan [2001] equated bacterial ecotypes with species). Unlike subspecies and geographic races, which normally have nonoverlapping geographic ranges, two or more ecotypes may occur together in the same local area so long as they are restricted to different habitats. Typically, ecotypes have distinctive morphological and physiological traits, which may reflect genetic adaptations to their unique environments.

More recently, with the advent of molecular approaches to examining biological diversity, populations that are genetically diagnosable have been called **phylogroups** or **evolutionarily significant units** (ESUs). The former has a fairly precise meaning as a reference to distinct, genetically delineated clades (reciprocally monophyletic gene lineages), often defined using mtDNA (Avise 2000); whereas ESU is defined in several different ways—one that is popular is as "historically isolated sets of populations for which a stringent and qualitative criterion is reciprocal monophyly for mitochondrial DNA (mtDNA) combined with significant divergence in frequencies of nuclear alleles" (Moritz et al. 1995, p. 249). We will revisit phylogroups in Chapter 11.

Higher Classifications

Having considered the problem of how species are defined, let's briefly turn to the equally vexing problem of how species are arranged in a taxonomic classification scheme. The great eighteenth-century naturalist, Carolus Linnaeus, developed the hierarchical scheme for classifying organisms that is still used today. In the Linnaean classification scheme, species are grouped into genera, genera into families, families into orders, and so on (Table 7.2). In Linnaeus's time, when species were thought to have been individually created by God, this classification was simply meant to reflect the fact that species differ in the degree to which they resemble one another.

Now that we know that all living things are descended—through a complex historical pattern of branching lineages—from a common ancestor, most taxonomists want their hierarchical classification schemes to reflect this history of ancestor-descendant relationships. The discipline of **phylogenetic systematics** has had considerable success in developing both a rigorous conceptual framework for recognizing evolutionary lineages with a shared history (Hennig 1966), and methods for reconstructing the evolutionary history of lineages (see Chapter 11). These developments have led a large group of phylogenetic systematists to propose a radical new system of taxonomic classification called the **PhyloCode** (http://www.ohiou.edu/phylocode/) that gets rid of the Linnaean hierarchical ranking system and instead ties nomenclature explicitly to the phylogenies generated using modern concepts and methods of phylogenetic reconstruction. However, while exciting, this proposal remains controversial and, at least for now, it is still necessary to describe some of the attributes of the traditional rank-based system.

TABLE 7.2 *The primary units in the Linnaean system of hierarchical taxonomic classification of living things[a]*

Kingdom	Animalia
Phylum	Chordata
Class	Mammalia
Order	Primates
Family	Hominidae
Genus	*Homo*
Species	*sapiens*

---

[a]Illustrated by the classification of the human species, *Homo sapiens*.

Major levels of classification are given in Table 7.2. Additional levels are sometimes recognized by taxonomists working on particular groups—examples include superspecies, subgenera, infraorders, and so on (for a good example, see the classification scheme for mammals by McKenna and Bell [1997]). It is usually easy to figure out from their names where these levels are located in the hierarchy.

The hierarchical classification scheme implies that the taxa grouped together at each level are most closely related to one another. Thus, for example, species within the same genus are more closely related to one another than to species in any other genus, as are genera within the same family, and so on. By *more closely related*, we mean that they more recently shared a common ancestor (i.e., all included taxa comprise a **monophyletic clade**). One thing that the classification scheme does not imply, however, is that all of the taxa at the same level of classification are equally closely related—there is no explicit reference to phylogenetic relationships among clades nested within a particular genus, family, etc. Thus, if three families are placed in an order, and two have split from their common ancestor more recently than the third, this particular hierarchical evolutionary relationship will not be reflected in the classification. In some groups, such as certain vascular plants and vertebrates, species in the same genus may have diverged only within the last few million years, whereas congeners in other groups may have split tens of millions of years ago. In Chapter 11, we will consider in some detail how systematists reconstruct phylogenetic relationships, including estimates of times of divergence, and what these genealogies imply. In Chapter 12, we will examine why these phylogenies are important in biogeography.

While this rank-based system (the Linnaean classification scheme) has worked well for over 250 years, its weaknesses (in addition to those mentioned above) are evident now that we have a rapidly expanding, phylogenetically based framework for understanding the "tree of life" and reconstructing its biogeographic history (Cracraft and Donoghue 2004). First, it does not provide for a clade, once given a name, to retain that name should future analyses indicate that it needs to be repositioned on a phylogenetic tree (in a rank-based system, for example, if a genus-level clade is elevated to a family level rank, then the name of the clade must also be changed). Second, it requires that, should a species be transferred to a different genus, the formal name of the species (which includes the genus name) must also be changed (e.g., the desert pocket mouse, formerly called *Perognathus penicillatus*, is now recognized as *Chaetodipus penicillatus*). While the PhyloCode system seeks to eliminate the conventions of nomenclature that lead to such taxonomic instability, it is quite controversial, and we will not know for another few years if and how rapidly it will gain popularity and be adopted by the systematics community.

Macroevolution

Species delineate two sometimes very different approaches to investigating evolutionary patterns and processes—**macroevolution** above, and **microevolution** below—the species level. Recent studies of changes in organisms observed in the fossil record have given us new insights into the patterns and processes of evolution. Until the 1970s, most biologists assumed that evolution usually proceeds gradually through successive incorporation of relatively small genetic changes as populations respond to environmental change in space and time. However, many eminent evolutionary biologists (e.g., Simpson 1944; Mayr 1963) emphasized that evolutionary rates can vary widely, even within the same evolutionary lineage. Simpson categorized evolutionary lineages into **bradytelic** (slower-evolving than typical), **tachytelic** (faster-evolving than typical), or **horotelic** (typical) rate classes. Mayr and others also suggested that major, rapid changes can occur during the process of speciation.

The prevailing view of evolution was the product of the Modern Synthesis that dominated evolutionary thought for most of the twentieth century. The new synthesis was concerned primarily with how the characteristics of populations change as a result of the genetic mechanisms of natural selection, mutation, genetic drift, and gene flow (i.e., the processes that comprised the arena of microevolution). Microevolutionists have traditionally studied contemporary organisms, either observing experimentally induced changes in laboratory populations of organisms such as microbes and fruit flies, or drawing inferences from comparative studies of populations in the field. A long running debate amongst evolutionary biologists has been whether—and to what degree—these microevolutionary processes can adequately explain a variety of evolutionary changes observed above the species level, including the origins of evolutionary novelties, the differential rates of evolution and success of lineages through time, and the influences of a dynamic Earth history (see Chapters 8 and 9) on patterns of diversification (for an excellent review of emerging syntheses between micro- and macroevolution, see Jablonski 2000).

Evolution in the Fossil Record

In the 1970s, a group of young paleontologists that included Niles Eldredge, the late Stephen J. Gould, and Steven M. Stanley began to emphasize a variety of kinds of evolutionary changes recorded in fossil remains that are not easily explained through simple extrapolation of microevolutionary processes. Because of the long time frame typically involved, these changes tend to be large, and include speciation and extinction events as well as fundamental changes in morphology. The paleontologists thus resurrected the concept of macroevolution that had come to be deemphasized by most of the neontologists of the time. But Eldredge and his colleagues suggested that macroevolution occurs primarily as a result of two processes, which they called punctuated equilibrium and species selection (Eldredge and Gould 1972; Stanley 1979).

The fossil record indicates that the evolution of a lineage usually consists of long periods of virtually no change (**stasis**) interspersed with relatively brief periods of rapid change, which often appear to be associated with speciation events. These bursts of rapid change are often accompanied by morphological innovations, geographic range expansion, and exploitation of new areas of ecological niche space. One of the most profound events in the history of life was the Cambrian Explosion that defined the beginning of the Paleozoic Era, and marked the origination of most major body plans in modern oceans and the extinction of many earlier groups. This pattern of highly variable rates of evolution is referred to as **punctuated equilibrium**. It is supported by data from the fossil record—especially that of marine and freshwater invertebrates—in which new fossil species often appear abruptly and then persist for millions of years with virtually no morphological change. These bursts of rapid change often followed an episode of **mass extinction**, suggesting that the removal of prior occupants from ecological niche space provides a stimulus for the **adaptive radiation** of new forms to fill that space (Raup 1994; Jablonski 2002). Evidence of rapid change associated with speciation events comes, for example, from P. G. Williamson's (1981) work on the molluscan fauna of Lake Turkana Basin in eastern Africa. In a stratigraphic sequence representing several million years of fossil deposition, several lineages of mollusks showed long periods of virtually no change in shell morphology, punctuated by rapid shifts (Figure 7.4). These changes took place within a time interval of less than 50,000 years—so rapidly with respect to the fossil record that often no shells with intermediate characteristics were found. Many of these shifts were accompanied by the splitting of lineages, and they often occurred in conjunction with rapid, climatically mediated changes in lake level. A high frequency of punctuated equilibrium is also detectable in benthic (e.g., the bryozoans *Metrarabdotos* and *Stylopoma*) and planktonic (e.g., the foraminiferan *Globorotalia*) lineages in marine and brackish water environments during the Neogene of the last 25 million years (Jackson and Cheetham 1999).

The other macroevolutionary process evidenced in the fossil record has been called **species selection.** The idea is simple—just as natural selection is a process of evolutionary change caused by differential survival and reproduction of individual organisms with certain heritable characteristics, so species selection is a process of evolutionary change caused by the differential survival and speciation of *species* with particular heritable traits. The demonstration, however, of species selection acting independently of selective processes operating at the level of the individual organism has proven difficult. Even so, macroevolutionists have shown that major episodes of extinction and speciation in the fossil record have been selective with respect to the characteristics of species. For example, the asteroid that struck the Earth at the end of the Cretaceous Period about 65 million years B.P. (resulting in the Chicxulub Crater off the coast of the Yucatan Peninsula) caused wholesale death and destruction, including the extinction of many species and higher taxa. But far from being random, these extinctions were highly selective. Entire lineages of large animals (e.g., dinosaurs, several other groups of giant terrestrial and marine reptiles, and several kinds of large invertebrates, including ammonites) were completely exterminated, while many lineages of small animals (e.g., insects, small mammals, and teleost fishes) and vascular plants survived. The selective nature of this extinction event apparently had a strong geographic bias as well—especially immediately following the impact—it was particularly pronounced to the north owing to the fact that the meteor hit the Earth at a low trajectory, travelling over the Antarctic and the Southern Hemisphere before

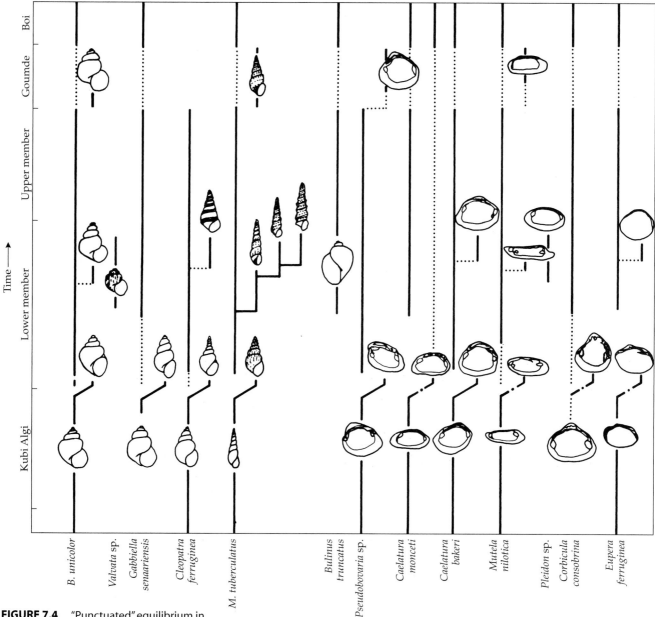

FIGURE 7.4 "Punctuated" equilibrium in the evolution of fossil mollusks in Lake Turkana Basin in eastern Africa. The diagram depicts the reconstructed history of shell morphology in several genera. Dotted lines indicate inferred changes during periods when no fossils have been recovered. Note that most lineages exhibited long periods of virtual stasis followed by rapid, substantial change. The latter, punctuational events often occurred either (1) virtually simultaneously in several different lineages, suggesting major environmental changes, such as shifts in lake level owing to climatic change (confirmed by other evidence); or (2) in association with speciation events, which often left one species virtually unchanged while the other species diverged substantially. (After Williamson 1981.)

impact, thus sending the incendiary heat and a tsunami northward along the epicontinental sea that separated the eastern and western part of North America (Flannery 2001). Similarly, in major episodes of speciation, some species with certain morphological and life history characteristics have produced many more descendant species than have other ancestral species with contrasting characteristics.

Micro- and Macroevolution

When they first aired their ideas in the 1970s, macroevolutionists appeared to be using data from the fossil record to challenge the traditional view of microevolution that had come from the new synthesis. But as is often the case in such apparent controversies, neither side has been proved right or wrong; rather, we have discovered that, for the most part, these apparent antagonists were actually talking about different things. In this case, the primary difference is one of scale. Microevolutionists are concerned with evolutionary changes within populations that occur as a result of the differential births, deaths, and movements of individuals with certain heritable characteristics. Such changes can occur over short periods of time and can cause geographic variation within a species (see examples of industrial melanism, house sparrows, and clines below). Such changes are by no means always slow or uniformly continuous; however, they are rarely of sufficient magnitude to result in the formation of a new species or the extinction of an existing one. Macroevolutionists, on the other hand, are concerned with evolutionary changes in the morphological forms that they can recognize in the fossil record. These changes are often caused by the differential proliferation and extinction of species, so that they substantially alter the diversity of lineages and the composition of biotas. By comparison with the modest scale of microevolutionary change, these changes are often large and, in relation to the eons of evolutionary time over which they occur, they often appear abrupt or "punctuated."

Microevolutionary and macroevolutionary perspectives are complementary. Both are necessary for understanding the influence of geography on evolutionary processes and the influence of evolutionary processes on the distributions of organisms. Several newer subdisciplines of evolutionary biology offer promising approaches for developing new and innovative syntheses across micro- and macroevolutionary arenas (Avise 2000; Goodman and Coughlin 2000; Jablonski 2000). We will illustrate the need for complementary perspectives in the remainder of this chapter, with emphasis on the "macroevolutionary" processes of speciation and extinction. Yet, we will show that a deep understanding of these processes also requires a microevolutionary perspective, focusing on how populations respond to environmental change in both space and time.

Speciation

Although the process of speciation was not widely recognized by that term until the 1930s, a number of researchers prior to the Modern Synthesis were interested in the formation of new species and explored issues that are still being addressed today. They asked whether geographic isolation is required, and they explored the roles of natural selection and genetic drift (Berlocher 1998; Coyne and Orr 2004). Today, speciation is usually seen as a branching process (cladogenesis) in which new kinds of organisms originate from a single ancestral species. However, new species might also arise through hybridization (**reticulate speciation**), and some investigators include the non-branching (anagenetic) process of **phyletic speciation**—the transformation of one ancestral species into a single descendant species—with each major stage in such an evolutionary sequence being called a **chronospecies**. Many examples of phyletic speciation have been cited in the literature, but relatively few have been studied carefully enough that we can be confident that the chronospecies represent an unbroken anagenetic series—and not, in fact, products of cladogenetic events that were followed by rapid extinction of all but one lineage.

The magnitude of this process is staggering, especially when we consider that all species of green land plants that have ever lived ultimately share a common ancestor in a simple green alga that lived over 500 million years ago; that all vertebrates are traceable to some ancient chordate; and that millions of species of insects have evolved in the 400 million years since their ancestor first invaded land. Most of the species in all of these lineages are no longer around—they lived in the past and went extinct. How did this diversity come to be?

Mechanisms of Genetic Differentiation

Regardless of the geographical mode of speciation (reviewed below), the divergence of an ancestral species into two or more daughter species requires genetic changes among populations, ranging from those that are simply used to diagnose different species (e.g., under the PSC) to those that actually cause reproductive isolation between an ancestrally single reproductive community (e.g., under the BSC). Population geneticists have long recognized four primary microevolutionary processes by which fully or partially isolated populations can diverge (or converge)—mutation, genetic drift, natural selection, and gene flow—in genetic composition. Different forms of a gene at the same locus, known as **alleles**, are responsible for differences among individuals in their heritable traits. New alleles arise by mutation, and changes in the frequencies of alleles in populations occur primarily through genetic drift, natural selection, and gene flow.

MUTATION. Changes in DNA are called **mutations** and include four kinds of change: **substitution** of single nucleotides; and **deletion**, **insertion**, or **inversion** of one or more nucleotides. Deletions and insertions are often facilitated by **transposable elements** or **horizontal gene transfer**. Any of these events can cause a change in a protein-coding gene, RNA-coding gene, or gene regulatory region of the DNA. New gene copies in multigene families (e.g., globins) can be generated through **unequal crossover** at **recombination,** and damaged genes can be repaired through **gene conversion.** The importance of mutations in the genetic differentiation of populations is that they can be incorporated into the genetic architecture of a population through genetic drift or natural selection. Gene flow will distribute new mutations among populations, whereas geographic or reproductive isolation will prevent them from happening.

GENETIC DRIFT. **Genetic drift** is a relatively weak force because it involves changes in the genetic constitution of a population caused solely by chance. Given sufficient time—which usually means many generations—the frequencies of alleles in a population tend to change randomly as different individuals happen to survive, mate, and produce offspring. Genetic drift has relatively little effect in large populations, but can have important influences on the evolution of small populations. How small is small is a matter of considerable discussion in evolutionary biology, but we should mention that when population geneticists discuss population size, they are usually more interested in the **effective population size**—the actual number of breeding individuals in a population—which can differ substantially from the **census population size**—the total number of individuals in a population. Factors that can make the effective population much smaller than the census population include social structure, age structure, and genes being transmitted from one generation to the next through a single sex (e.g., female transmission of mtDNA in bisexual organisms). If new species start from small founding populations—as many island forms undoubtedly do—then genetic drift may play an important role in their initial differentiation, as we shall see below.

NATURAL SELECTION. Natural selection, on the other hand, can be a potent force for evolutionary change in both large and small populations. **Natural selection** is the change in a population that occurs because individuals express genetic traits that alter their interactions with their environment so as to enhance their survival and reproduction relative to other individuals in the population. Over many generations, alleles for such adaptive traits tend to increase in frequency at the expense of alleles that confer less fitness. Populations tend to diverge if there is sufficient variation in the environment to select for different characteristics to deal with different environmental conditions. If a new mutation has a strongly positive fitness effect—a **selective sweep**—the rapid fixation of a favorable new mutation, as well as linked alleles that "hitchhike" with it, might trigger a rapid differentiation of populations—such sweeps being of particular importance in the evolution of asexual populations (Cohan 2001).

This fine-tuning of phenotypes to environmental heterogeneity can be readily documented. A classic example involves industrial melanism in moths. Within the past 150 years, moths in industrialized areas around cities have evolved dark color patterns to match their sooty backgrounds and are thus less visible to their avian predators (Kettlewell 1961). Rapid evolution by natural selection has also occurred in house sparrows. Since being introduced to North America from Europe less than 200 years ago, these enormously successful birds have not only spread to colonize most of the continent, but have also evolved distinct geographic races (Johnston and Sealander 1964, 1971; Johnston and Klitz 1977). Other examples include changes in body size in response to climatic shifts (Hadly 1997) and habitat fragmentation (Schmidt and Jensen 2003; see Chapters 14, 16, and 17).

GENE FLOW.   Migration, or **gene flow**, often tends to act counter to genetic drift and natural selection in order to retard genetic divergence. Individuals that migrate to a new area carry their genes with them, and if they subsequently reproduce successfully, their genes are introduced into the local population. Such migration, therefore, tends to have a homogenizing influence, preventing or at least retarding the development of geographically isolated and genetically differentiated populations.

ADAPTATION AND GENE FLOW.   The niche of a species is not constant in either space or time. Since natural selection is a universal process that tends to increase the capacity of individuals to survive and reproduce, we would expect populations to adapt to their local environments. Such adaptation might be evidenced by geographic variation within species in physiology, morphology, or behavior. It might also be reflected in adaptive changes in niche relationships over time. We might expect peripheral populations to be able to adapt to the environmental factors that limit their ranges, increase in density, and colonize adjacent areas (see Baker and Stebbins 1965). Lewontin and Birch (1966) describe an apparent example in the Queensland fruit fly, *Bactrocera* (*Dacus*) *tryoni*. During the last century this species has expanded its range several hundred kilometers to the south along the eastern edge of the continent. These flies are limited by low temperatures, and their expansion has been accompanied by adaptation of the peripheral populations for increasing cold tolerance. But clearly not all species are increasing their ranges in this fashion. A historical perspective suggests that throughout most of the Cenozoic Era, the distributions of some species and higher taxa have indeed increased, but that these expansions have been almost equally matched by contractions in the ranges of other organisms. As a result, there appears to have been little change in global biodiversity during this period.

Why don't the peripheral populations of all species adapt to local conditions, resulting in a continual expansion of the range on all margins? In some cases, the reason is obvious: There are fundamental constraints that cannot be easily overcome by local adaptation. Terrestrial life that forms along coastlines, for example, cannot simply evolve adaptations for aquatic life and invade the marine realm. This kind of adaptation did happen in the past—as terrestrial ancestors of penguins and whales invaded the sea—but it took millions of years and required that these animals give up their capacity for living on land. But such cases hardly explain why species do not expand their ranges along ecological gradients. Why don't local populations adapt to deal with a limiting physical stress or biological enemy? Why don't they expand the species range by evolving to tolerate just a bit more cold, aridity, predation, or competition?

Genetic, ecological, and biogeographic processes seem to interact to limit the capacity of peripheral populations to adapt to local conditions. Exchange of genes among populations through dispersal and gene flow may prevent local populations from acquiring and maintaining the combinations of genes necessary for continual adaptation. Gene flow is caused by the dispersal of individuals or gametes between populations, which bring with them genes for traits that have been selected for different local environments. Sufficiently high gene flow can swamp a local population with genes from outside, effectively working against increasing adaptation to local conditions.

The critical question in any particular case of local adaptation is whether gene flow is high enough to overwhelm natural selection, prevent the continual adaptation of peripheral populations, and thereby preclude expansion into new areas of ever more extreme environments. The answer to this question for many organisms is not clear, and it has been the subject of much debate among population geneticists and evolutionary biologists. Certainly, many species show adaptive genetic changes in response to variation in their environments over their geographic ranges. The deer mouse (*Peromyscus maniculatus*), whose enormous geographic range encompasses most of the North American continent, has been the subject of many genetic and ecological studies. There is a great deal of geographic variation in the color and shape of deer mice, which is reflected by the subdivision of the species into many geographic races or subspecies (Figure 7.5). Classic studies by Lee Dice and his students showed that much of this variation reflects adaptation to local environments. Coat color tends to match the local soil color because of strong selection by owls and other predators against contrastingly colored individuals (Dice and Blossom 1937, 1947). Compared with individuals from desert and grassland habitats, animals from forest populations tend to have longer feet and tails because they use these appendages in climbing (Horner 1954). Populations in regions where contrasting habitat types come into close proximity, however, show lower levels of such morphological and behavioral specialization because of the diluting influence of gene flow (Thompson 1990).

These and similar studies of other organisms support theoretical models that predict that even modest levels of gene flow may be sufficient to preclude local adaptation (Wright 1978). If gene flow could be reduced or eliminated, however, local adaptation could proceed, and further range expansion might be possible. Evidence in support of this conjecture comes from studies of the genetics of Old World burrowing rodents of the genus *Spalax*. Differences in chromosome number among populations of *S. ehrenbergi* appear to have been

FIGURE 7.5 Geographic variation in the deer mouse (*Peromyscus maniculatus*) is in- ▶ dicated by the subdivision of the species into 50 formally recognized subspecies. Each of these geographic races has distinct characteristics—including dorsal coat color, which resembles background color and provides camouflage, and tail and hindfoot length, which are related to climbing ability and habitat structure. (After Hall 1981.)

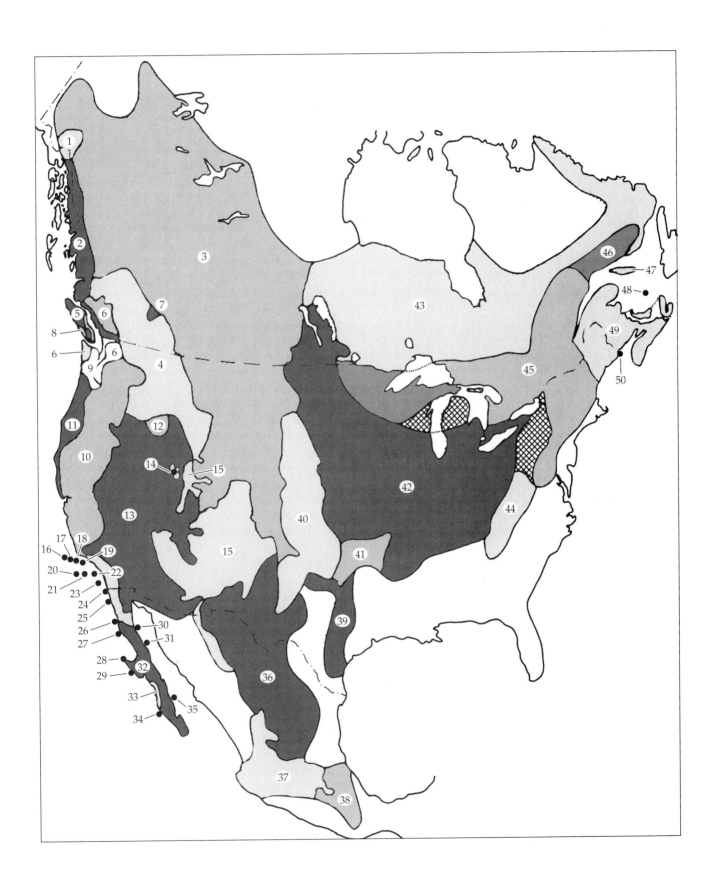

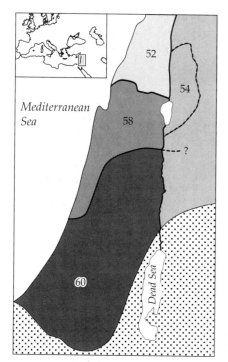

FIGURE 7.6 Distribution of karyotypes of the burrowing rodent, the Palestine mole rat (*Nannospalax ehrenbergi*). Number of chromosomes increases along a gradient of increasing aridity from north to south (and also from west to east) in the Middle East. Note that these "chromosomal races" replace each other with virtually no overlap, suggesting that reduced gene flow between them has facilitated adaptation and expansion into increasingly arid environments. (After Nevo and Bar-El 1976.)

important in facilitating their colonization of the most arid deserts of the Middle East by reducing gene flow from populations adapted to more mesic areas (Wharmann et al. 1969; Nevo and Bar-El 1976; Figure 7.6). Gene flow between populations with different chromosome numbers and configurations is reduced because when individuals with different karyotypes breed, problems with pairing and separation of chromosomes reduce the viability and fertility of their offspring. Patton (e.g., 1969, 1972, 1985) describes other examples of how local and regional ecological conditions, natural selection, and genetic barriers to gene flow interact to influence the distributions of small mammals.

Chromosomal rearrangements also have been implicated in limiting the distributions of plants. Populations at the geographic margins of a species range, as well as those inhabiting extreme soil types or climates, are often characterized by major chromosomal changes, especially polyploidy (Stebbins 1971b). Although these rearrangements of genetic material may themselves confer specific adaptations, their general effect is to reduce the frequency of crossing with individuals from other populations and thereby reduce—or completely block—gene flow, permitting adaptation to the local environment and facilitating colonization of new areas.

Ultimately, of course, the capacity of populations to adapt to new environments and to colonize new habitats is limited. Although some species are more widely distributed and more tolerant of varying conditions than others, there are no superorganisms that occur everywhere. Adaptation inevitably involves trade-offs and compromises. In order to tolerate the physical conditions and deal with the biotic interactions in some environments, a species must sacrifice its ability to do well in other habitats. Using a combination of "common garden" and breeding experiments on yarrow (*Achillea millefolium*), Clausen et al. (1940, 1947, 1948) elegantly demonstrated the interacting roles of genetic isolation, local selection, and trade-offs in determining the degree of adaptive differentiation of local populations and allowing this species to have a wide distribution along an elevational gradient in the Sierra Nevada Mountains of California (Figure 7.7). Gene flow can prevent populations from adapting to different local environments, but when gene flow between populations is interrupted—as it is during the process of speciation—then over evolutionary time, populations can and often do diverge in response to natural selection, adapt to widely different conditions, and expand their ranges to occupy new habitats and geographic areas.

GEOGRAPHIC VARIATION. There is often a geographic component to genetic divergence, because both genetic drift and natural selection are facilitated, and gene flow is retarded, by geographic isolation. Genetic drift can be an important force in small, isolated populations, such as those that inhabit small outlying patches at the periphery of a species range, or those that have recently been founded by long-distance colonization. A population started by a few colonizing individuals usually contains only a small random sample of the alleles present in a much larger ancestral population. Genetic drift that occurs during such a **population bottleneck** has been termed the **founder effect.** Mayr (1942) suggested that this process accounts for the apparently random differences often found among bird populations on different islands (Figure 7.8), each of which was probably derived from a few successful colonists. Arguments that the founder effect can play a major role in speciation by converting this initially random genetic sampling into different species through both genetic drift and natural selection of small colonizing populations (e.g., Carson 1971; Templeton 1980a; Carson and Templeton 1984) have been challenged by theoretical arguments (Barton and Charlesworth 1984).

FIGURE 7.7 Morphological differentiation of the plant, *Achillea millefolium*, along an elevational gradient in the Sierra Nevada Mountains of California. Note the distinctive characteristics of different populations of the plant from different places along the gradient. Since these plants were all grown in the identical environments of a "common garden," we can infer that the differences among them are genetic. Presumably, these differences reflect local adaptation due to a combination of natural selection and reduced gene flow. (After Clausen et al. 1948.)

Whether or not coupled to founder events, geographic separation of populations facilitates genetic differentiation by natural selection. Different environmental regimes tend to select for different traits, and spatially isolated populations are likely to occur in different environments. Under these conditions, natural selection can facilitate a process termed **ecological speciation** (Schluter 1998). The effects of natural selection associated with spatial environmental heterogeneity are exemplified by data on small mammals. Coat color of the

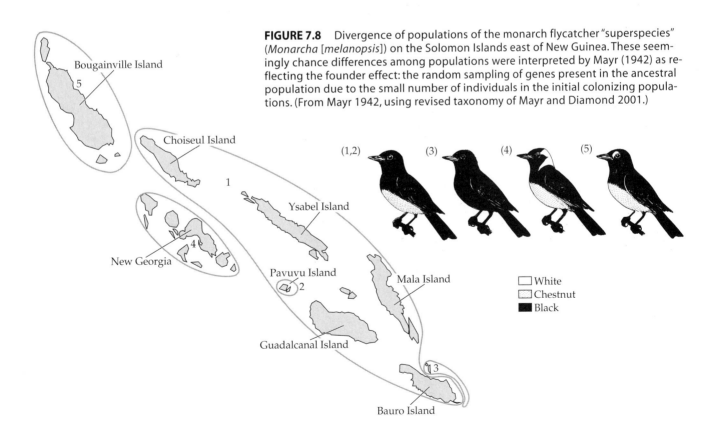

FIGURE 7.8 Divergence of populations of the monarch flycatcher "superspecies" (*Monarcha* [*melanopsis*]) on the Solomon Islands east of New Guinea. These seemingly chance differences among populations were interpreted by Mayr (1942) as reflecting the founder effect: the random sampling of genes present in the ancestral population due to the small number of individuals in the initial colonizing populations. (From Mayr 1942, using revised taxonomy of Mayr and Diamond 2001.)

deer mouse (*Peromyscus maniculatus*) varies greatly over the wide geographic range of the species (see Figure 7.5). The color of the dorsal fur closely matches the color of the soil or other substrate on which the animal is likely to be active, reflecting the consequences of selection by predators (e.g., classic experiments by Dice [1947] showing that owls selectively captured mice that contrasted with their background). More recently, mutations at specific coat color genes have been linked with dramatic coat color differences selected to match substrates in adjacent populations of the rock pocket mouse (*Chaetodipus intermedius*) that inhabit either lava bed or lighter soil substrates (Hoekstra and Nachman 2003; Nachman et al. 2003).

Patterns of geographic variation can take many forms (see Chapter 15). The term **cline** is used to describe a gradual change in one or more features along a single environmental or other geographic gradient. Many birds and mammals exhibit clinal variation in clutch or litter sizes with latitude and elevation (e.g., for *Peromyscus*, see Dunmire 1960; Lord 1960; Smith and McGinnis 1968; Spencer and Steinhoff 1968) (Figure 7.9). Such variation presumably reflects the adaptation of life history traits to environments that differ in temperature, seasonality, productivity, and other factors. Clines can also develop in a **hybrid zone**, which marks a relatively narrow geographic zone where one or more phenotypes or alleles exhibit rapid shifts in frequencies between otherwise isolated populations or species (Butlin 1998).

Allopatric Speciation

Geographic isolation has historically been considered the most frequent, if not necessary, first step in the speciation process (Mayr 1942, 1963), because it is the simplest way to cut off gene flow between populations and ultimately lead to formation of reproductive barriers. Several possible modes of **allopatric** (or **geographic**) **speciation**, along with two alternatives to complete geographic isolation, are described below.

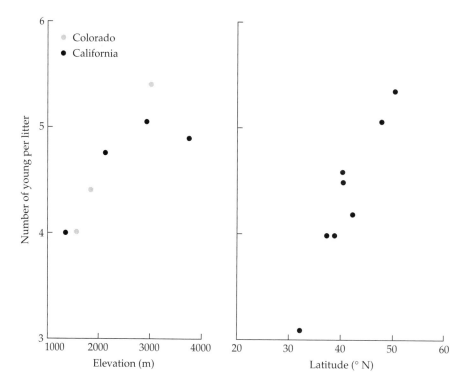

FIGURE 7.9 Latitudinal and elevational variation in average litter size of the deer mouse (*Peromyscus maniculatus*). Note that number of young per litter increases with both latitude and elevation. Presumably, large litters are advantageous in colder climates, perhaps because the shorter growing seasons favor fewer litters per year with more young per litter (see also Chapter 15).

If the environment is heterogeneous, a geographically widespread ancestral population will tend to develop regional genetic differences in response to either natural selection or genetic drift. Because of barriers that limit dispersal, free gene flow from one end of the range to the other rarely occurs. Thus, populations living in environmentally distinct and geographically separated regions tend to become somewhat differentiated from one another, but some gene flow maintains the genetic cohesiveness of the species. Regional races of house sparrows and deer mice, and the clinal variation in litter and clutch sizes mentioned above, are examples of such differentiation.

If, however, the regional populations become sufficiently isolated that the cohesive gene flow between them is cut off or drastically reduced, then they become independent evolutionary units. Without dispersal and gene flow, isolated populations tend to diverge. Divergence proceeds more rapidly when effective population sizes are small, or when substantially different environments subject isolates to different selective pressures. For example, Darwin called attention to the morphological differences among the giant tortoises of the Galápagos, which are obviously descended from a common ancestor but presently occur on different islands. On some of the most arid islands, where treelike cacti are abundant, the endemic tortoises have evolved long necks and forelimbs and distinctively shaped shells that allow them to reach up high to feed on these plants. On wetter islands, where the tortoises feed mostly on lower vegetation, they have more generalized body forms. We can distinguish two general means by which such isolation can occur.

ALLOPATRIC SPECIATION MODE I: VICARIANCE. At one extreme, some environmental change can create a barrier to dispersal somewhere within the range of an ancestral species, isolating previously connected and interbreeding populations. Rising sea levels, for example, can isolate an island on a continental shelf; tectonic events can cause part of a continent to split off and drift away; or, conversely, landmasses can drift together to isolate formerly continuous oceans. Such changes are called **vicariant events**, and they usually isolate relatively large sets of populations (Figure 7.10). We will see later that these sorts of events are important in biogeography because they also tend to isolate many different, co-distributed species at the same time.

EXAMPLE: GONDWANAN VICARIANCE. Biogeographers have long considered the break up of Gondwanaland (see Chapter 8) to have led to speciation caused by vicariance in any ancestral taxon that might have been distributed across two or more of the fragments of the former supercontinent. The fragment that initially included India, Madagascar, and the Seychelles Islands broke away from the other Gondwanan landmasses about 130 million years ago. Subsequently, Madagascar's connection with India-Seychelles was severed about 88 million years ago—the latter colliding with Eurasia about 56 million years ago. Bossuyt and Milinkovitch (2001) presented a molecular phylogeny of

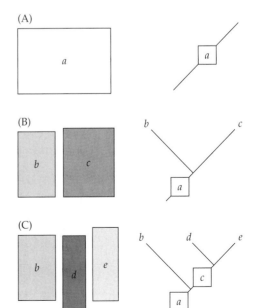

FIGURE 7.10 An illustration of the process of allopatric speciation mode I, and resulting phylogenetic patterns. (A) An ancestral species, *a*, widely distributed throughout a geographical area, is subdivided by the formation of a geographic barrier (vicariance). (B) With gene flow prevented, populations diverge from one another over time and eventually can be recognized as separate species, *b* and *c*. (C) Another geographic barrier subsequently subdivides species *c* and the isolated populations diverge in geographic isolation to form species *d* and *e*. (After Brooks and McLennan 2002.)

FIGURE 7.11 Molecular phylogeny for subfamilies of the anuran family Ranidae. Node with asterisk represents the well-dated geological separation of Madagascar from the India-Seychelles landmass about 88 million years ago. This node also represents the separation of the Mantellinae—endemic to Madagascar—and its sister-clade the Rhacophorinae, which was most likely a member of the ancestral India-Seychelles biota as it drifted northward. The close association between calibrated times of lineage divergence (see discussion of molecular clocks in Chapter 11) and times of separation of land masses, provides an example of vicariance-induced speciation. Diagrams A–D below the phylogeny, and the shaded area of the phylogeny, represent configurations of several Gondwanan landmasses between the time of separation of Madagascar-India-Seychelles about 130 million years ago (A), to the time of India's collision with Eurasia about 56 million years ago (D). (After Bossuyt and Milinkovitch 2001.)

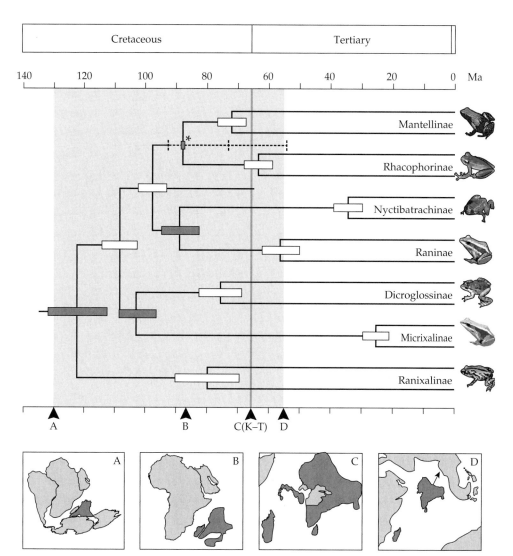

subfamilies of frogs within the family Ranidae (Figure 7.11), with estimated divergence times calibrated from the well-dated separation of Madagascar from India-Seychelles. This study provided a convincing case of at least one instance of vicariance-induced speciation. The subfamily, Mantellinae, is entirely endemic to Madagascar, whereas its sister clade, Rhacophorinae, has its greatest diversity in the Oriental region and most likely was a member of the ancestral India-Seychelles biota prior to the suturing of India and Eurasia early in the Cenozoic.

ALLOPATRIC SPECIATION MODE II: PERIPHERAL ISOLATES, OR FOUNDER EVENTS. At the other extreme, individuals may disperse across an existing barrier to colonize a previously uninhabited region. Such occasional and random events, called **jump dispersal** (see Chapter 6) or **founder events**, are the way in which oceanic islands and many other isolated patches of habitat come to be colonized. Typically involved is a small initial population, sometimes only one or a few individuals (Figure 7.12). Unlike vicariance, these events tend not to happen simultaneously across multiple species. As indicated above, mechanisms and rates of initial genetic divergence may differ depending on the mode of isolation. Genetic drift may play a greater role relative to selection, and divergence may be more rapid, at least initially, in allopatric mode II relative to mode I speciation (Carson 1971; Bush 1975; Templeton 1980a, 1981).

EXAMPLE: ALLOPATRIC SPECIATION IN THE GALÁPAGOS ARCHIPELAGO. Despite uncertainties about the details, there can be little doubt that geographic isolation has been a common mode of speciation in many groups. The finches and giant tortoises that Darwin observed in the Galápagos archipelago provide examples of populations in different stages of the process. As we have seen, there are morphological differences among populations of tortoises on each of the large islands. These populations have diverged from a single ancestor that

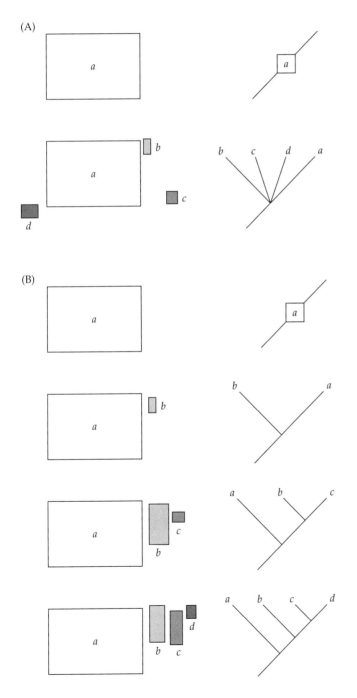

FIGURE 7.12 Illustration of the process of allopatric speciation mode II, and resulting phylogenetic patterns. (A) Random dispersal of members of the geographically widespread species *a* to peripheral areas initiates new, geographically isolated populations. Over time, each new peripherally isolated population will diverge to form new species *b*, *c*, and *d*. (B) Peripheral isolates are again formed by founding events, but now the process of dispersal is temporally and spatially sequential such that the first population isolated to become new species *b* is adjacent to ancestral species *a*; then members of species *b* disperse to the next peripheral area and found the next new species *c*; and finally, dispersing members of species *c* found new species *d*. (After Brooks and McLennan 2002.)

originally colonized the older islands in the archipelago (represented today by Espanola and San Cristóbal), probably from South America about 2 to 3 million years ago (Caccone et al. 2002). Their descendants subsequently dispersed, most likely by drifting along with the prevailing northwesterly ocean currents, to establish populations on the different islands as the younger islands formed (Figure 7.13). Two of the larger islands, Isabela and Santa Cruz, were each colonized more than once at different times and from different source islands, and there appear to be no cases of younger populations on younger islands re-colonizing older islands. Cases in which islands have been colonized more than once appear to have resulted in some islands having more than one named taxon, although taxonomists have not yet determined whether morphological and molecular differences among different populations of Galápagos tortoises justify considering some of them to be distinct species, or whether all are best considered subspecies within a single species (*Geochelone nigra* or *G. elephantopus*).

On the other hand, in Darwin's finches, the final stages of speciation are represented. Here again, all of the species are believed to be derived from a single ancestral population that colonized the archipelago between 2 and 3 million years ago, probably—somewhat surprisingly—from Caribbean islands (Burns et al. 2002). In the case of the finches, however, after diverging in isolation on different islands or in different habitats, some populations have successfully reinvaded already inhabited areas, so that several species (as many as ten on certain islands) now coexist on a single island. These species have not only evolved specific mating behaviors that prevent widespread interspecific hybridization, but have also diverged in morphology and behavior to exploit different ecological niches (Figure 7.14). The process of speciation is an ongoing one, and so there are also documented cases of hybridization between forms that are not yet completely reproductively isolated (Grant and Grant 1998).

CONTACT AND REINFORCEMENT. Once isolates have formed and differentiated, there are several possible outcomes. First, they may come back into contact

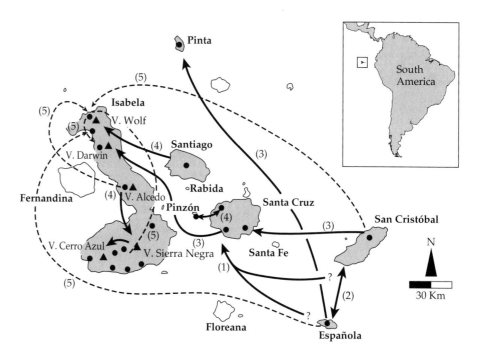

FIGURE 7.13 Proposed phylogeographic history of Galápagos tortoises on the larger islands of the Galápagos archipelago. Mainland progenitors from South America first colonized the older islands of San Cristóbal and Española. Arrows represent likely colonization events, numbers indicate approximate temporal order of colonization, solid arrows represent natural colonization events, and dashed arrows are possible translocations caused by humans. Note that there were at least two independent natural colonizations of the islands of Santa Cruz and Isabela. (After Caccone et al. 2002.)

with one another. This may occur either through the disappearance of a geographic barrier or by dispersal back across it. If the populations reestablish contact, there are three possible outcomes: (1) the two populations may not interbreed because they have developed prezygotic isolating mechanisms, or they may fail to produce fertile offspring if they do interbreed (postzygotic isolating mechanisms), in which case reproductive isolation is complete, speciation has

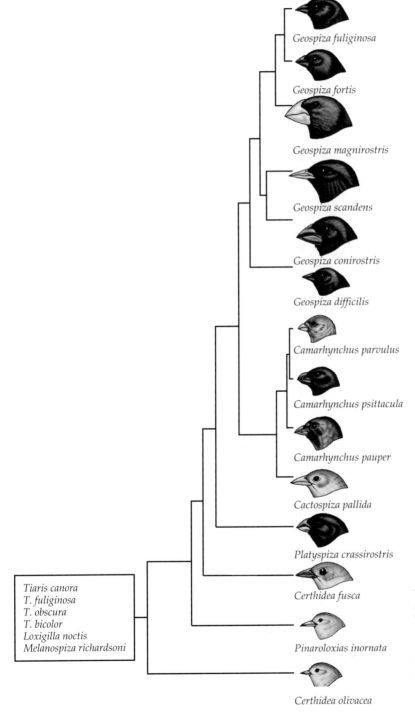

FIGURE 7.14 Adaptive radiation in Galápagos finches, showing the diversity of beak shapes and diets. A single ancestor from within a clade that includes six species in three genera colonized the archipelago. Subsequent allopatric speciation events, due to repeated episodes of colonization and divergence, produced five genera and fourteen species (the mangrove finch, *Cactospiza heliobates,* not shown) as well as the Cocos Island finch, *Pinaroloxias inornata.*

occurred, and they might begin to share the same geographic range but as two separate species; (2) the two populations may interbreed extensively, producing fertile hybrids that then backcross to parental populations to the extent that the populations gradually merge genetically and their differentiation breaks down; or (3) the two populations may not have developed prezygotic isolating mechanisms and do in fact hybridize, but the hybrids may be less fit beyond a narrow hybrid zone than the offspring of within-population matings. In this last case, selection favoring those individuals that choose mates from within their own population may be sufficiently strong to lead to reproductive isolation and completion of the speciation process. This process of selection for prezygotic mechanisms that promotes within-population matings is termed **reinforcement** (Dobzhansky 1937). A final scenario is one in which the isolated populations never do come into contact. In this case, reproductive isolation may take a long time to develop, and, as mentioned above, it may be difficult for a systematist to apply a biological species concept when deciding whether they should be considered different species.

While there is broad agreement that speciation often occurs through geographic isolation, there is much less consensus about the details of the process (see papers in Howard and Berlocher 1998). How much do mechanisms of speciation vary among different kinds of organisms, and even among different speciation events within lineages of closely related organisms? What are the relative frequencies of vicariant events—with isolation between initially large populations less likely to experience rapid genetic change—and founder events—with initial isolates composed of only a few individuals and thus likely to diverge rapidly due to the founder effect and genetic drift? When vicariant events are involved, how often are the isolates small, peripheral populations (**microvicariance**) as opposed to large fragments of a once continuous population? How important is gene flow in maintaining cohesion and preventing differentiation among populations? How important is the genetic inertia of large populations and the influence of similar selective environments in retarding divergence? Conversely, what are the roles of genetic isolation and founder events relative to divergent selective pressures in promoting divergence of isolated populations? To what extent is there active selection for "reinforcement" of prezygotic isolating mechanisms so that individuals actively avoid interbreeding with members of closely related species, and to what extent might we expect these traits to move away from a hybrid zone back into ancestral populations? To what extent does sexual selection for mates with exaggerated traits facilitate the evolution of prezygotic reproductive isolation? There is much disagreement among evolutionary and systematic biologists on the answers to these questions, but progress is being made (see Howard and Berlocher 1998; Coyne and Orr 2004). Rather than searching for one universal process of speciation, we must recognize the enormous variation in the process due to the special characteristics of different kinds of organisms and the different historical and environmental contexts in which speciation occurs.

Sympatric and Parapatric Speciation

Although many evolutionary biologists once maintained that speciation in most organisms has occurred primarily or solely as a result of geographic isolation, most now accept that speciation can—and often does—occur within spatially overlapping populations. This process is called sympatric speciation if the geographic overlap of populations of the ancestral species is extensive (Figure 7.15A), and parapatric speciation if the populations are largely allopatric but overlap in a narrow zone where depressed fitness of inter-pop-

(A)

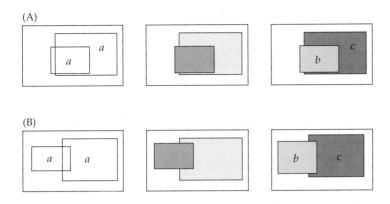

(B)

FIGURE 7.15 Illustration of sympatric and parapatric speciation. In sympatric speciation (A), two populations of ancestral species *a* overlap extensively when population differentiation begins and maintain extensive contact throughout the speciation process until new species (*b* and *c*) are recognized. Alternatively, in parapatric speciation (B), overlap occurs only along a narrow zone of contact between populations of ancestral species *a* and throughout the course of the speciation process. (After Brooks and McLennan 2002.)

ulation matings leads to selection for isolating mechanisms (Figure 7.15B). Two mechanisms of sympatric speciation have been proposed.

DISRUPTIVE SELECTION. If strong selective pressures cause a population to adapt to two or more different environmental regimes or niches, they can progressively pull the population apart and eventually result in speciation. Bush (1975), Price (1980), and others (see papers in Howard and Berlocher 1998) have argued that sympatric speciation by **disruptive selection** may be common in certain groups of phytophagous (herbivorous) insects and animal parasites that are highly specialized for specific host species. Among these organisms, successful colonization of a new kind of host must be a rare event, but when it occurs, the colonists are immediately subjected to selection for the ability to survive and reproduce in a drastically different environment. Usually this selection pressure is intensified by counter-evolution of the host to escape from or reject the parasite. Selection to meet the challenge of a new host could potentially lead to the rapid differentiation of totally sympatric populations. Initially, matings between insect or parasite host-specific races would generate offspring that have reduced fitness in either host niche, which would favor selection for evolution of a **prezygotic isolating mechanism** and completion of the speciation process. A parapatric model of speciation proposed by Endler (1977; see also Slatkin 1973; Rosenzweig 1978) suggested that disruptive selection, acting along an environmental gradient, could gradually sharpen clinal variation until a single ancestral population fragmented into two or more species. Disruptive selection might also underlie sympatric speciation, but in the form of sexual selection rather than natural selection, as has been suggested for the African cichlid radiations described below (Dominey 1984).

CHROMOSOMAL CHANGES. Sympatric or parapatric (also called **stasipatric** by White 1978) speciation can also occur through chromosomal changes. Chance rearrangements of the genetic material of a parent during meiosis, or of an embryo during fertilization or early development, can sometimes change the number of chromosomes or the sequence of genes on chromosomes. Changes in chromosome number are of two kinds: **aneuploidy**, in which a single chromosome breaks or fuses with another to change the total number by plus or minus one; or **polyploidy**, in which an entire additional set of chromosomes is passed on, changing the number by some multiple (e.g., a doubling or tripling). In other cases, the chromosome number remains the same, but some of the genetic material is either rearranged within a chromosome (**inversion**) or transferred to another chromosome (**translocation**).

In diploid organisms, precise pairing during meiosis of the genes and chromosomes inherited from each parent usually is necessary to ensure the transmission of a complete set of genes to each gamete, and hence to produce viable offspring. Consequently, mutant individuals with new chromosomal arrangements often have impaired fertility when they mate with an individual having the original chromosomal arrangement, and may be able to reproduce only by mating with another individual having the new arrangement. For this reason, it is obviously difficult for a population with a new arrangement to become established. However, once established—especially in a small, isolated, inbred population—the new type is genetically isolated from its parental population and can diverge rapidly as a new species. Navarro and Barton (2003) have recently incorporated this ephemeral isolation step into an updated and more realistic "allo-parapatric" model.

Sympatric speciation by way of polyploidy appears to have occurred frequently in some groups of organisms, especially plants (e.g., Stebbins 1971b; de Wet 1979; Lewis 1979; Briggs and Walters 1984). There are many documented ways in which polyploidy has been achieved in plants, but one such process is considered very common (de Wet 1979). Diploid ($2N$) organisms have two sets of chromosomes and produce haploid (N) gametes by meiosis. Occasionally, a female gamete is formed without undergoing meiosis, and remains diploid ($2N$). This unreduced gamete can then fuse with a haploid pollen grain (N) to produce a triploid ($3N$) plant; this will then produce triploid gametes because of complications during meiosis. In the next generation, a triploid female gamete can fuse with a haploid pollen grain to yield a tetraploid ($4N$) zygote, which can survive and produce fertile offspring from diploid ($2N$) gametes, either by self-fertilization or by crossing with other rare tetraploids in the population. The resulting tetraploid population is immediately genetically isolated from the diploid population. Interestingly, tetraploidy has recently been discovered in a mammal—the red viscacha rat, *Tympanoctomys barrerae* (Octodontidae)—in Argentina (Gallardo et al. 1999).

Polyploidy can occur either within a population, called **autopolyploidy,** or as a result of hybridization between different but usually closely related populations or species, called **allopolyploidy.** Allopolyploidy is thought by many researchers to be more common. Because chromosomes from different species may not pair and segregate properly, interspecific hybridization often results in abnormalities in the meiotic process that can facilitate the process described above. Allopolyploids may not only arise more frequently than autopolyploids, but may also be more likely to become established. In addition to possessing a larger genome than either of the parental species, allopolyploids tend to be intermediate in their characteristics, which enables them to be superior competitors in certain habitats.

EXAMPLES: SYMPATRIC SPECIATION IN ISOLATED LAKES. In the decades immediately following the new evolutionary synthesis, there was a tendency to regard allopatric speciation as the general process and sympatric speciation as the exception: something that undoubtedly occurs in some plants and possibly in a few other kinds of organisms, such as phytophagous insects and parasites, but is relatively rare. This assumption was largely due to the enormous influence and powerful arguments of Ernst Mayr (e.g., 1942, 1963), who argued for the near universality of allopatric speciation. Recently, however, the pendulum has begun to swing back as increasing evidence for sympatric speciation has accumulated. For example, as many as 70 to 80% of angiosperm plant species are now thought to be of polyploid origin (Briggs and Walters 1984).

Some of the most convincing cases of sympatric speciation in animals involve the divergence and adaptive radiation of fishes in isolated lakes

(Echelle and Kornfield 1984). There are many examples: cichlids in the Great Lakes of central Africa, and in the giant Cuatro Cienegas spring system in northern Mexico; whitefishes in the Great Lakes of eastern North America; sculpins in Lake Baikal in Siberia; herrings in Scandanavian lakes; pupfishes in Lago Chichancanab in Yucatán; and sticklebacks in lakes in British Columbia. These cases differ in when the isolation occurred and how much differentiation has taken place. At one extreme are the cichlids in Lakes Victoria, Malawi, and Tanganyika of the African Rift Valley. These fishes have been isolated in large lakes for tens of thousands to millions of years, and flocks of hundreds of species have been produced from a few founding lineages of cichlids (e.g., Fryer and Iles 1972; Greenwood 1974, 1984; Kaufman and Ochumba 1993; Kornfield and Smith 2000). The resulting species exhibit enormous variation in morphology, much of it related to specialization for different diets and feeding modes; and in color patterns and behavior, much of it related to courtship and mating displays (Figures 7.16 and 7.17). Recent evi-

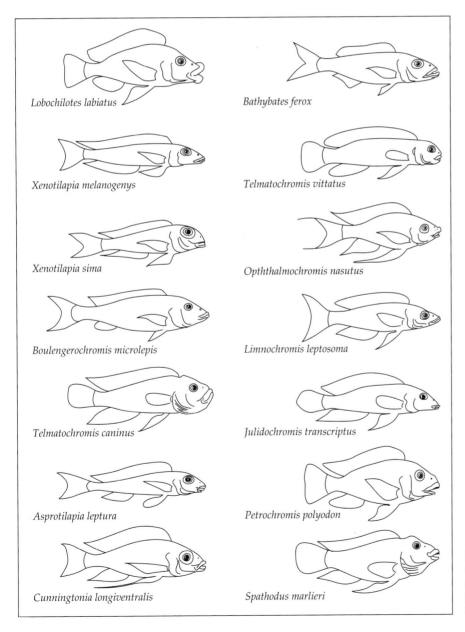

Lobochilotes labiatus

Bathybates ferox

Xenotilapia melanogenys

Telmatochromis vittatus

Xenotilapia sima

Opththalmochromis nasutus

Boulengerochromis microlepis

Limnochromis leptosoma

Telmatochromis caninus

Julidochromis transcriptus

Asprotilapia leptura

Petrochromis polyodon

Cunningtonia longiventralis

Spathodus marlieri

FIGURE 7.16 Examples of the variety of body forms resulting from adaptive radiation of cichlid fishes in Lake Tanganyika in eastern Africa. (After Fryer and Iles 1972.)

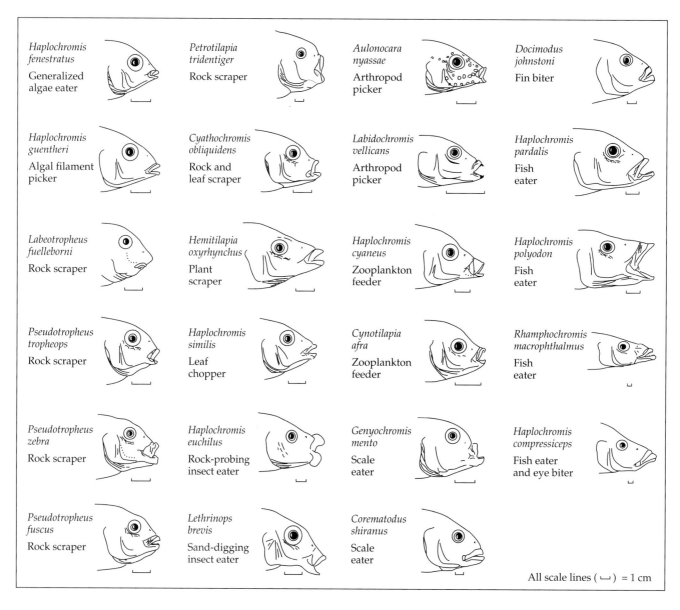

FIGURE 7.17 Examples of the variety of head shapes, mouthparts, and feeding habits resulting from adaptive radiation of cichlid fishes in Lake Malawi in eastern Africa. This amazing variation reflects specialization in diet due to natural selection to reduce competition and exploit ecological opportunities. (After Fryer and Iles 1972.)

dence indicates that shallow Lake Victoria may have been completely dry in the late Pleistocene (until only about 12,000 years ago; Johnson et al. 1996). Even if complete desiccation did not occur, this lake is probably no more than 0.5 to 1 million years old (Martens et al. 1994) and so the fact that it now supports a fauna of more than 300 species of endemic cichlids suggests that under appropriate conditions, speciation and adaptive radiation can be extremely rapid.

At the opposite extreme are the pupfishes of Yucatan's Lago Chichancanab and the sticklebacks of British Columbia, which have been isolated in small lakes for only a few thousand years as a result of changing sea and lake levels since the Pleistocene (see Chapter 9). Lago Chichancanab contains five species of *Cyprinodon*, which have distinctive morphologies related to diet and also exhibit some degree of genetic and behavioral reproductive isolation (Figure 7.18) (Humphries and Miller 1981; Humphries 1984; Strecker et al. 1996). Small lakes in coastal British Columbia typically contain two forms of sticklebacks—

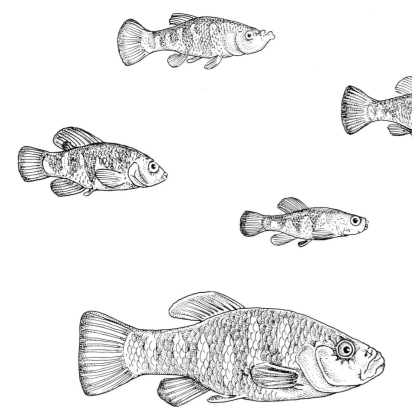

FIGURE 7.18 Morphology of five forms of pupfishes (*Cyprinodon*) in Lago Chichancanab on the Yucatán Peninsula of Mexico. These forms have diverged sympatrically within the last few thousand years since the isolation of the saline lake from the sea. Genetic and behavioral studies suggest that the speciation process is not complete, and that some of these forms still interbreed to some extent. Differences in size and shape reflect dietary differences, and suggest that strong selection for trophic differentiation is driving sympatric speciation. (After U. Strecker 1996.)

benthic and pelagic—which exhibit morphological adaptations for swimming and feeding along the bottom or in open water, respectively (McPhail 1994). While these pupfishes and sticklebacks have diverged much less than the African cichlids, they nevertheless provide an excellent opportunity to study the roles of ecological and evolutionary processes in speciation and adaptation (e.g., Schluter 2000; see discussion of taxon cycles in Chapter 14).

While hypotheses of allopatric speciation, which typically involve changing lake levels forming isolated refugia in the deeper basins, are not entirely excluded as viable hypotheses to explain speciation among the African cichlids (Kornfield and Smith 2000), the simplest explanation is that differentiation has occurred sympatrically within each of the lakes. We do know at this point that each of the larger lakes (Vicotria, Malawi, and Tanganyika) contains its own monophyletic assemblage of cichlids (Meyer et al. 1990; Kocher et al. 1993), which provides support for sympatric speciation hypotheses. While the ultimate mechanism might be disruptive selection to exploit different food resources, a compelling case can be made for speciation being driven by sexual selection, with female choice based on male coloration (Dominey 1984).

These cases suggest that sympatric speciation may be more common than most evolutionary biologists have suspected. They also suggest that either disruptive selection to exploit different ecological niches, or strong sexual selection, may often be sufficiently powerful forces to produce speciation, even in the absence of geographic isolation. Nevertheless, well-documented cases of sympatric speciation are still scarce (Coyne and Orr 2004), and a recent analysis that examined the geography of speciation in a phylogenetic context within several clades of birds, fishes, and insects (Barraclough and Vogler 2000) concluded that the most predominant form of speciation was an allopatric mode.

Diversification

Ecological Differentiation

Once new species have formed, what happens to them? Immediately after a speciation event, the resulting species are often quite similar to each other. Above, we pointed out that they are most likely to diverge if they are subjected to different environments with different selective regimes. Ecological differentiation increases the likelihood that the species will be able to occupy overlapping geographic ranges. Thus, it facilitates the buildup within a region of a biota of closely related, sympatric species. Gause (1934) showed in laboratory experiments with protozoans of the genus *Paramecium* that two species with identical resource requirements could not persist in the same environment: one species was eventually outcompeted and driven to extinction. Ecologists have generalized this phenomenon and termed it the principle of **competitive exclusion** (see Hardin 1960; Miller 1967; Hutchinson 1978).

According to this principle, two or more *resource limited* species cannot coexist in a *stable environment* without segregating their *realized niches*. The important point is not simply that identical species cannot coexist. Rather, the competitive exclusion principle describes how species *can* coexist—they may be limited by something other than resources (e.g., by predators or parasites); they may inhabit unstable environments where floods, fires, or other disturbances limit populations of competitors before they can exclude each other; or they may inhabit different regions of their shared, fundamental geographic range (see Chapter 4).

A biogeographic corollary of the principle of competitive exclusion is that species that are extremely similar in their niches tend to have nonoverlapping geographic distributions, whereas species that coexist in the same area and habitat tend to differ substantially in their resource use. Perhaps the most striking example of this phenomenon is provided by so-called **sibling** or **cryptic species**: species that are genetically distinct, but extremely similar in their morphology and ecology. Examples include species that have recently formed through chromosomal changes as well as many examples that have only recently been detected using modern molecular methods (examples given in Avise 2000)—they may be particularly abundant in the sea (Knowlton 1993). Sibling species often exhibit almost perfectly abutting, but not strongly overlapping (i.e., parapatric), geographic ranges, as demonstrated by numerous examples in both plants and animals (Figure 7.19). Even related species that are no longer extremely similar may competitively exclude each other from local habitats or extensive geographic areas. Examples include the kangaroo rats and chipmunks mentioned in Chapter 4 (see Figures 4.26 and 4.27). Several species of pocket gophers—fossorial rodents of the genera *Thomomys* and *Geomys*—have ranges in western North America that come into contact, but do not overlap. Even though the burrows of parapatric species may be only a few meters apart, the species remain separate (Figure 7.20). There is good evidence that this pattern is maintained by competitive exclusion (e.g., Miller 1964; Vaughan and Hansen 1964; Vaughan 1967).

Conversely, numerous field studies show that when closely related species do coexist in nature, they often differ substantially in their use of limiting resources (e.g., MacArthur 1958, 1972; Cody and Diamond 1975). Often these niche differences are reflected in pronounced morphological, physiological, or behavioral differences. Galápagos finches provide excellent examples of this phenomenon. As the finches have reinvaded inhabited islands after speciating, they have diverged morphologically, behaviorally, and ecologically from other sympatric species. This process, called **character displacement** (Brown

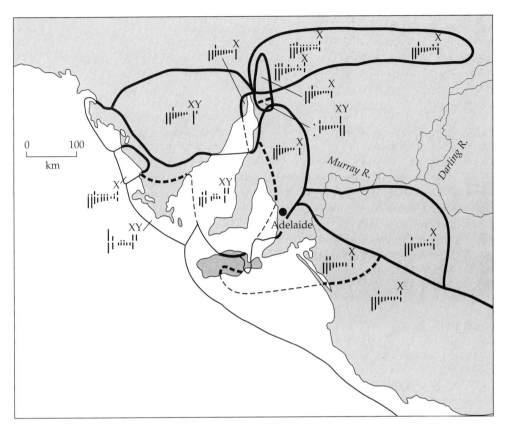

FIGURE 7.19 Ranges of different chromosomal forms of the morabine grasshopper (*Vandiermonetta* spp.) in southern Australia. These grasshoppers are extremely similar in morphology and ecology, but their different chromosomal arrangements effectively prevent hybridization. Their contiguously allopatric, or parapatric (i.e., touching but not overlapping), distributions are typical of those of sibling species formed as a result of chromosomal changes (represented by "X" and "Y" labels and associated bar patterns). (After White 1978.)

and Wilson 1956), has resulted in species being more different where they coexist than where they live allopatrically (see Figure 7.3). In these finches, character displacement is most apparent in the size of the beak, which enables coexisting forms to specialize on different kinds of foods (Lack 1947; Abbott et al. 1977; Schluter and Grant 1984; Schluter et al. 1985; Grant 1986; Grant and

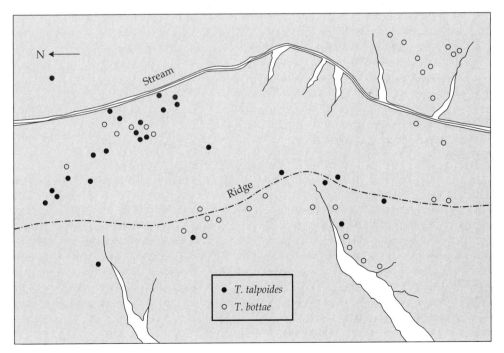

FIGURE 7.20 Small-scale distribution of two species of pocket gophers (*Thomomys talpoides* and *T. bottae*), as indicated by the locations of burrows of individual animals. Map encompasses approximately 1 km of a valley in the Rocky Mountains of Colorado. Note that these morphologically and ecologically similar species come into extremely close proximity, but do not overlap at this scale. (After Vaughan 1967.)

Grant 1989). Some intermediate phenotypes do occur, but strong selection against interspecific hybrids and other deviant phenotypes during periods of food shortage maintains the clumpy-gappy distribution of distinctive sympatric species (see Figure 7.3).

Adaptive Radiation

Adaptive radiation is the diversification of species to fill a wide variety of ecological niches. It occurs when a single ancestral species gives rise, through repeated episodes of speciation, to numerous kinds of descendants that become or remain sympatric. Often, these bursts of speciation are thought to be initiated by the development of a **key innovation** in the ancestral species. These coexisting species tend to diverge in their use of ecological resources in order to reduce interspecific competition. Such character displacement in response to competition obviously cannot occur if closely related species remain divided by physical barriers (i.e., vicariant), but some differentiation will still tend to occur as the allopatric species adapt to different environments.

Today, looking at the variety of living things, we can find numerous examples of successful lineages that have radiated to produce diversity at many levels. We can, for example, consider the adaptive radiation of Hawaiian honeycreepers (Drepanidinae), a group of small perching birds that probably colonized the archipelago within the last few million years (Amadon 1950; Raikow 1976; Tarr and Fleischer 1995). Or, we can examine the major radiation of placental mammals that produced many of the existing mammalian orders (e.g., Lilligraven 1972), which occurred during the Cenozoic era after the mass extinction event that eliminated most of the giant reptiles. In all such cases, the basic ecological opportunities are created either by an adaptive (key) innovation or an environmental change, such as colonization of a new area or the extinction of competing species. These opportunities are exploited as an ancestral form initiates repeated speciation events producing specialists that fill numerous ecological niches (Schluter 2000).

One of the most dramatic examples of adaptive radiation, mentioned above, is provided by the cichlid fishes of the lakes of East Africa, and is described in a fascinating book by Fryer and Iles (1972; see also Greenwood 1974, 1984; Kaufman and Ochumba 1993; Johnson et al. 1996; Kornfield and Smith 2000). These cichlids have diversified morphologically and have specialized behaviorally and ecologically to fill many different niches (see Figures 7.16 and 7.17). Lake Malawi alone contains over 300 species in a group of rockfishes called the "mbuna," and many of these have ecological counterparts in completely separate adaptive radiations in Lake Victoria (the "mbipi") and Lake Tanganyika. There are herbivores and carnivores; species with mouths and teeth adapted for catching tiny zooplankton, crushing snails, and eating other fishes whole; even forms specialized to feed just on the fins, scales, or eyeballs of other fishes.

Islands and archipelagoes provide many other examples of adaptive radiation. Madagascar, with its long history of isolation from Africa and other southern continents (see Chapter 8), has been the site of several spectacular radiations. These include not only the well-publicized lemurs (Primates), but the less well known tenrecs (a morphologically, behaviorally, and ecologically diverse endemic family, Tenrecidae) and vanga shrikes (a similarly diverse group of perching birds belonging to the endemic family, Vangidae). Among the many examples in the Hawaiian Islands (see also Chapter 12; Wagner and Funk 1995) are Hawaiian honeycreepers of the endemic subfamily, Drepanidinae (Pratt 2004); the Hawaiian silversword alliance (Baldwin and Robichaux 1995); and the picture-winged *Drosophila* (Carson and Kaneshiro 1976). In the honeycreepers, there were about 33 species of these birds when Europeans first visited Hawaii, although at least ten have since gone extinct and most of the others are endangered. They are descended from a single common ancestor, now thought to be a cardueline finch that colonized from North America, probably less than 5 million years ago (Tarr and Fleischer 1995). The lineage radiated to produce an amazing variety of species, which differ conspicuously in the sizes and shapes of their beaks (Figure 7.21) and to a lesser extent in color pattern. As in Darwin's finches, ecological differentiation to exploit dif-

FIGURE 7.21 Variety of beak shapes resulting from adaptive radiation of Hawaiian honeycreepers (Drepanidinae). This schematic should be viewed only as a generalized picture of progressive specialization of morphological types, but does not imply a well supported phylogenetic hypothesis of ancestor and descendant relationships among these species. Compare this group with the examples of Galápagos finches (Figure 7.14) and African cichlid fishes (Figures 7.16 and 7.17). (From Primack 1998.)

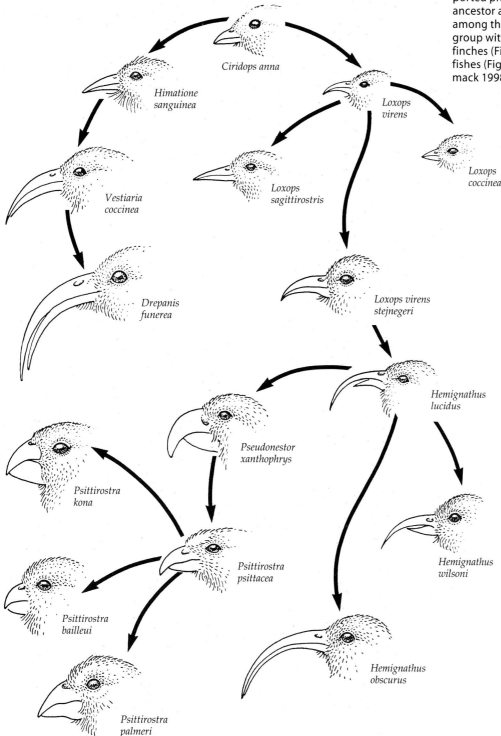

Ciridops anna

Himatione sanguinea

Loxops virens

Loxops sagittirostris

Loxops coccinea

Vestiaria coccinea

Drepanis funerea

Loxops virens stejnegeri

Hemignathus lucidus

Pseudonestor xanthophrys

Psittirostra kona

Hemignathus wilsoni

Psittirostra bailleui

Psittirostra psittacea

Hemignathus obscurus

Psittirostra palmeri

ferent niches—especially food sources—seems to have been the primary process that led to radiation of the honeycreepers (Amadon 1950; Raikow 1976; Taur and Fleischer 1995).

While many examples come from isolated habitats, such as islands and lakes, comparable adaptive radiations have occurred in other groups in oceans and on continents. They have also taken place on many different temporal and spatial scales. An ancient radiation occurred in the marine realm in the cryptically colored marine anglerfishes (Lophiiformes), producing the shallow-water frogfishes (*Antennarius* and *Antennatus*), the open-ocean sargassum fish (*Histrio histrio*), the stingray-like batfishes (Ogcocephalidae), and a variety of deep-water bioluminescent forms (suborder Ceratoidea). In plants, a group that has radiated over the past 20 million years on the North American continent is the phlox family (Polemoniaceae). Within this single family, flower form and color have become amazingly variable as different species have adapted to be pollinated by hawkmoths, bees, butterflies, flies, bats, or hummingbirds, and some have become specialized for self-pollination (Figure 7.22). Australia, the most isolated continent, is the site of many spectacular radiations: marsupial mammals, lizards of the genera *Ctenotus* and *Varanus*, and plants of the genera *Eucalyptus*, *Melaleuca*, and *Acacia*.

Whereas many genera and families exhibit a fascinating variety of ways of life, many others are extremely monotonous, differing mainly in minute structural details. Taxa that have changed only slightly over evolutionary time apparently are good at what they do and have not had the genetic flexibility or ecological opportunity to shift adaptive strategies. The wide variation in rates and degrees of divergence makes it difficult to generalize about the diversification of higher taxa and the influence of geography on adaptive radiation. The many examples of spectacular adaptive radiations on islands (including Madagascar and Australia) seem to show the importance of long

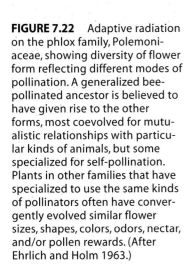

FIGURE 7.22 Adaptive radiation on the phlox family, Polemoniaceae, showing diversity of flower form reflecting different modes of pollination. A generalized bee-pollinated ancestor is believed to have given rise to the other forms, most coevolved for mutualistic relationships with particular kinds of animals, but some specialized for self-pollination. Plants in other families that have specialized to use the same kinds of pollinators often have convergently evolved similar flower sizes, shapes, colors, odors, nectar, and/or pollen rewards. (After Ehrlich and Holm 1963.)

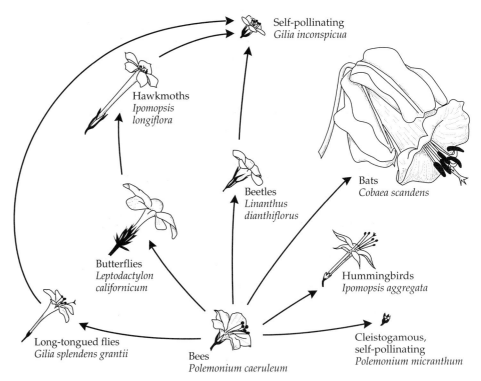

Self-pollinating
Gilia inconspicua

Hawkmoths
Ipomopsis longiflora

Beetles
Linanthus dianthiflorus

Bats
Cobaea scandens

Butterflies
Leptodactylon californicum

Hummingbirds
Ipomopsis aggregata

Long-tongued flies
Gilia splendens grantii

Bees
Polemonium caeruleum

Cleistogamous, self-pollinating
Polemonium micranthum

historical isolation in these cases. Equally impressive radiations in less isolated settings, however, such as those of desert rodents of the family Heteromyidae in southwestern North America; sigmodontine rodents in South America; darters of the fish genus *Etheostoma* in the Mississippi River drainage; lizards in the *Liolaemus darwinii* complex in South America; and the spiny, succulent plants of the family Cactaceae in the arid regions of North and South America, suggest that long episodes of geological and ecological transformation may induce adaptive radiation on less geographically isolated landscapes. Comparative studies that are attempting to assess the influence of phylogenetic relationships and evolutionary constraints (Felsenstein 1985; Harvey and Pagel 1991; Brooks and McLennan 2002) are contributing importantly to our understanding of the enormous differences in rates of speciation and adaptive differentiation among lineages.

Extinction

Ecological Processes

Although all living organisms represent a continuous evolutionary lineage extending billions of years back to the origin of life, the ultimate fate of every species is extinction. This can be appreciated by taking a brief glance at the fossil record. Earth was teeming with life 100 million years ago. Both terrestrial and aquatic habitats were occupied by diverse biotas that formed complex ecological communities. However, the species and genera—and many of the families and orders—that were dominant then have been eliminated or drastically reduced by extinction, and have been replaced by new lineages. On land, dinosaurs and other reptilian groups have been replaced by birds and mammals, while ferns and gymnosperms have been largely supplanted by angiosperms. In the oceans, cephalopod mollusks have been supplanted by teleost fishes, while icthyosaurs and mesosaurs (reptiles) have been replaced by dolphins, whales, seals, and sea lions (mammals). Extinctions have apparently occurred continuously throughout the history of life, although the fossil record also catalogues occasional episodes of widespread disaster when much of the Earth's biota was wiped out, apparently by rapid and drastic environmental change (Raup and Sepkoski 1982). Not surprisingly, extinctions have had a major influence not only on the kinds of organisms in existence at any given time, but also on the geographic distributions of those now extinct forms and the contemporary lineages that are descended from them.

Several authors have likened the evolutionary history of life to a continual race with no winners, only losers—those species that become extinct. This view is probably best expressed in Van Valen's (1973b) **Red Queen hypothesis**, named for the Red Queen in Lewis Carroll's *Through the Looking Glass*, who said "It takes all the running you can do to keep in the same place." The idea is that a species must continually evolve in order to keep pace with an environment that is perpetually changing, not just because abiotic conditions are shifting, but also because all the other species are evolving, altering the availability of resources, and the patterns and processes of biotic interactions. Those species that cannot keep up with the changes become extinct, but others do well temporarily and speciate to produce new forms.

Van Valen points out that the probability of a species becoming extinct appears to be independent of its evolutionary age, but not of its taxonomic and ecological status. Certain taxonomic and ecological groups have consistently higher rates of extinction than others. For example, apparently due to their lower extinction rates, small and herbivorous mammals are found on more and smaller islands than large or carnivorous species (Brown 1971b; Heaney 1986; Lawlor 1986). This appears to be a general pattern (see Van Valen 1973a). Somewhat similarly, differences in diversity and duration in the fossil record among lineages of marine invertebrates are correlated with characteristics of their life history, as we shall see below.

While we speak of extinction as the end of a species, the processes that begin a species on the path to extinction typically occur at the scale of individual populations. Some researchers have developed mathematical models to predict a population's vulnerability to extinction based on its demographic characteristics (MacArthur and Wilson 1967; Richter-Dyn and Goel 1972; Leigh 1981; Gilpin and Hanski 1991). All populations experience fluctuations in size as a result of variations in environmental conditions and the activities of their enemies. When populations become very small, however, purely chance factors, such as random variations in the sex ratio, can also affect their abundance. Mathematical models show that, in general, the smaller a population becomes, the lower its ratio of births to deaths (see Equation 4.1), and the longer it remains at low numbers, the more vulnerable it is to extinction. A population with a low birth rate, especially when coupled with a high death rate, cannot recover rapidly from a temporary reduction in numbers. Long-term overall population size is probably the most important factor. Models suggest that probability of extinction increases nonlinearly as population size decreases, and becomes very high when population size becomes, and remains, very low—say, fewer than 100 individuals (Figure 7.23). These mathematical models are based on the intrinsic demographic characteristics of populations, but it is important to recognize that changes in extrinsic environmen-

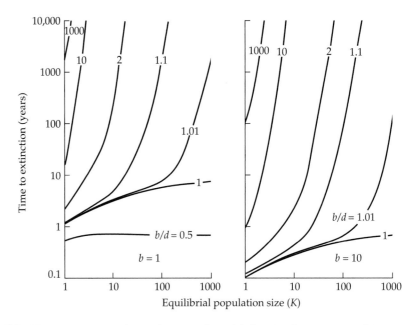

FIGURE 7.23 Output of a mathematical model showing how estimated time to extinction depends on two demographic characteristics of a population: equilibrial population density, or carrying capacity (K), and the ratio of birth rate (b) to death rate (d). The graphs show that the probability of extinction is high (expected time to extinction is low) when populations are small and birth rates are low relative to death rates, but the probability of extinction decreases rapidly as these demographic parameters increase. (After MacArthur and Wilson 1967.)

tal conditions are likely to be the cause of changes in those characteristics that ultimately lead to extinction.

Information on how intrinsic demographic and extrinsic ecological factors interact to cause extinction is difficult to obtain and interpret because extinctions of species, with the exception of those caused by humans, are rarely observed. Researchers have gained valuable insights, however, by studying the effects of these factors on the turnover of small, isolated subpopulations within a metapopulation (Gilpin and Hanski 1991; see also Chapters 4 and 16). One particularly well-documented example is provided by the work of Andrew Smith (1974, 1980) on the American pika (*Ochotona princeps*). The American pika is a small relative of rabbits; it lives in rock slides and boulder fields in the mountainous regions of western North America. Smith carefully monitored pikas that had colonized the rock piles left by a mining operation in the Sierra Nevada Mountains of California. These mine tailings functioned as habitat islands for the pikas in a sea of sagebrush habitat. Smith was able to document extinctions of subpopulations by censusing the rock piles initially in 1972, and then again in 1977. The repeat censuses documented both extinctions of previously existing subpopulations and colonizations of previously uninhabited rock piles. From data on the birth and death rates of pikas and the sizes and degree of isolation of the rock piles, Smith and colleagues developed a stochastic model that quite accurately predicted the frequency of extinction and colonization events as a function of these factors (Moilanen et al. 1998; see also Beever et al. 2003) (Table 7.3). Somewhat similarly, Lomolino (1984, 1993) and Crowell (1986) documented the extinction and immigration of small mammals on islands in the St. Lawrence River, in Lake Huron, and in the Gulf of Maine. Even though these studies are concerned with the turnover of small, isolated subpopulations rather than extinctions of entire species, they illustrate how both intrinsic demographic characteristics and extrinsic environmental conditions influence the extinction process.

Recent Extinctions

Over the last 200 years, humans have caused the extinctions of thousands of species; moreover, we are undoubtedly unaware of many more species of

TABLE 7.3 *Estimated time to extinction for pika subpopulations*

Population size (number of individuals) K	Birth rate (per capita per year) b	Death rate (per capita per year) d	Time to extinction (years) E
1	0.35	0.00	2.9
2	0.35	1.63	6.9
3	0.35	2.44	46.2
4	0.35	2.84	405.1
5	0.35	3.15	3751.5

Source: Data from Smith 1974, 1980.

Note: The populations represented here occupy habitat islands of varying size (see Figure 7.17). Per capita birth and death rates were determined from the age structure of this population. A mathematical model, based on the model of MacArthur and Wilson (1967), predicts that only very small populations ($K < 3$ individuals) should have measurable rates of extinction owing to ordinary random demographic fluctuations. This prediction was confirmed by Smith's subsequent measurement of population turnover on these same islands during a 5-year period.

microbes and small animals and plants that have disappeared. On the other hand, the demise of some larger, more spectacular organisms is well documented. Recent reductions in populations, contractions of geographic ranges, and extinctions of species caused by humans will be considered in more detail in Chapter 16. We will present a few examples here, however, to illustrate some of the processes involved in extinctions, because we have much more information about some of these recent extinctions than about most of those evidenced in the fossil record.

The passenger pigeon (*Ectopistes migratorius*) was incredibly abundant in eastern North America when the first European colonists arrived. Estimates of the total population size are in the millions, perhaps billions. Because the pigeons traveled in dense flocks numbering in the thousands, feeding on beechnuts, acorns, and other abundant seeds and fruits, and nested in huge aggregations, they were very vulnerable to humans, who hunted them for food. In the 1870s, 2000 to 3000 birds were often taken in one net in a single day, and 100 barrels of pigeons per day were shipped to New York City for weeks on end. By 1890, the birds had virtually disappeared. In 1914, the last known passenger pigeon died in the Cincinnati Zoo (Pearson 1936). Similar stories could be told about the demise of the great auk, Carolina parakeet, and Stellar's sea cow. Only a combination of luck and belated conservation action has prevented the whooping crane, trumpeter swan, bison, sea otter, gray whale, northern elephant seal, and black-footed ferret from suffering similar fates.

A different lesson can be drawn from the demise of the American chestnut tree (*Castanea dentata*). Along with beech, maple, oak, and hickory, the chestnut was one of the most abundant trees in the deciduous forests of eastern North America. In 1904, a pathogenic fungus (*Endothia parasitica*) was accidentally introduced, apparently from Asia, where it attacks a related but less susceptible species of chestnut. The disease spread rapidly (Metcalf and Collins 1911), and within 40 years, mature chestnuts were virtually eliminated from their entire range (see map in Figure 17.39). In some areas, scattered small trees that have sprouted from surviving root stock can still be found. Unfortunately, these are usually attacked by the disease and killed before they can reach reproductive size, so it is questionable whether the chestnut can much longer avoid absolute extinction. Similar cases of biological warfare due to introduced pathogens are the near extinction of the American elm due to Dutch elm disease, of native Hawaiian birds due to avian malaria (see Chapter 4), and of native peoples on oceanic islands, and perhaps even on the American continents, due to smallpox, measles, diphtheria, and other diseases brought by European invaders.

One of the best-documented cases of local extinction of multiple species is the loss of bird species from Barro Colorado Island in Panama (see Figure 6.16) Prior to the early 1900s, the island did not exist; it was simply a hill in a tract of tropical lowland forest. During the construction of the Panama Canal, the Chagras River was dammed, and the rising waters of Gatun Lake covered the adjacent lowland areas, creating Barro Colorado, an island of about 16 km^2. Despite its relatively large size, Barro Colorado no longer contains many of its original bird species. Because the Smithsonian Institution has operated a long-term biological research station on the island, its biota is well known, and some of the extinctions are well documented. Using these records, Willis (1974) calculated conservatively that at least 17 species had been lost from a total land bird fauna in excess of 150 species. More recently, by comparing the avifauna of the island with that of a mainland site of similar size and habitat, Karr (1982) concluded that at least 50 species have become extinct on the island since its isolation. Although habitat changes may have contributed to the demise of some forms, most of the missing species were those normally occurring at low densities, and it appears that when the small populations present on the island died out, they were not replaced by colonists from across the water gap.

The above examples illustrate the vulnerability of many species to environmental changes caused by modern humans (see also Chapters 16 and 17). However, humans do not need to have guns, agriculture, or European culture to cause extinctions. Aboriginal humans almost certainly played a major role in many extinctions documented in the fossil record, as we will see in Chapter 9. On the other hand, the capacity of some species to recover from low numbers is amazing. Two species of marine mammals that inhabit the Pacific coast of North America—the northern elephant seal (*Mirouga angustirostris*) and the sea otter (*Enhydra lutris*)—were hunted almost to extinction for their fat and fur, respectively. Tiny populations managed to escape detection, however, and once protected, increased rapidly to produce large, healthy populations. Another Pacific marine mammal, the gray whale (*Eschrichtius robustus*), has been steadily increasing following the protection of its breeding grounds off Baja California and the cessation of hunting (Rice and Wolman 1971), and the gray whale was recently removed from the U.S. Endangered Species List. Several bird species that were hunted to near extinction in the late 1800s for feathers to supply the millinery industry, including trumpeter swans, sandhill cranes, and great and snowy egrets, are again abundant.

Extinctions in the Fossil Record

The fossil record provides abundant evidence of extinctions, but often provides only tantalizing clues to their causes. In many cases, the fossil record is complete enough to show that many diverse species became extinct virtually simultaneously over relatively short periods of geological time. We can infer that such **mass extinctions** were caused by some drastic, widespread environmental change, but the exact nature of the perturbation and its effect on the organisms concerned is often difficult to deduce clearly. Long-standing, vigorous debates about the causes of mass extinctions mark paleontological literature. The six largest mass extinctions recognized in the fossil record have consequences beyond the elimination of large portions of the standing biological diversity. First, they have often been considered important precursors to new cycles of adaptive radiation by creating new opportunities for speciation of surviving lineages to fill the array of unoccupied ecological niches following a mass extinction episode. Second, it often appears that, even if a few species in a group that was extremely successful prior to a mass extinction event happen to survive, they seldom will give rise to another equally successful radiation within that group—a phenomenon called "dead clade walking" by Jablonski (2002).

One of the most recent episodes of mass extinction was the disappearance of the Pleistocene megafauna of North and South America between 15,000 and 8000 years ago. The cause of this dramatic event is still the subject of controversy, as we will see in Chapter 9. Originally, paleontologists favored hypotheses invoking climatic change. The idea that aboriginal humans played a major role, advanced by Paul Martin in the 1960s and known as the "overkill hypothesis," was initially rejected by most paleontologists. Multiple lines of accumulating evidence, however, have shifted the balance, causing many scientists to conclude that humans almost certainly played a significant—and perhaps pivotal—role in the extinctions of megafauna (e.g., Martin and Klein 1984; Owen-Smith 1987, 1989; Barnosky et al. 2004). Examples of additional

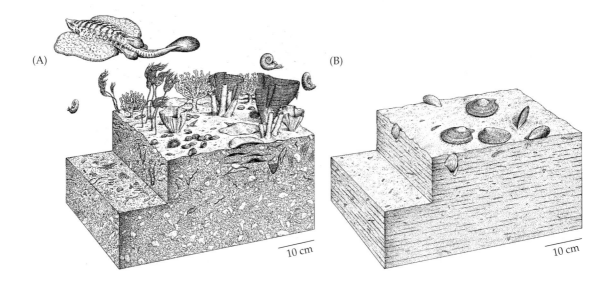

(A)

(B)

10 cm

10 cm

FIGURE 7.24   Reconstructions of the ancient seabed in southern China (A) before, and (B) after the Permo-Triassic mass extinction. Before the crisis, the seabed included more than 100 species of reef-dwellers and burrowing infauna. Species richness was reduced to four or five species following the event. (From Benton and Twitchett 2003; artwork © J. Sibbick.)

extinctions caused or threatened by modern humans will be discussed in detail in Chapter 16.

Causes of other mass extinctions that occurred even longer ago have also been highly controversial. Two of the most dramatic episodes saw the disappearance of dinosaurs and many other groups of terrestrial and marine organisms at the end of the Cretaceous period, about 65 million years ago, and of even more species and higher taxa—particularly marine organisms—at the boundary between the Permian and the Triassic periods, about 250 million years ago (see Raup 1979; Raup and Sepkoski 1982; Jablonski 1991; Benton and Twitchett 2003). Raup (1979) and Jablonski (1991) estimated that 96 percent of all marine species then in existence went extinct at the end of the Permian (Figure 7.24). Causes most often invoked for these and other mass extinctions are large, rapid changes in the climate of the Earth: either global cooling, with resulting continental glaciation and exposure of continental shelves; or global warming, with associated rising sea levels and inundation of continental shelves (e.g., Stanley 1984). Some evidence points to mass volcanic activity with associated global warming as a cause of the Permo-Triassic event (Benton and Twitchett 2003). But other explanations, including continental drift (e.g., Schopf 1974) and reorganization of interaction in ecological communities (Bak 1996), have had their proponents.

While the causes of the Permo-Triassic event remain poorly understood, there is now little doubt about the cause of the Cretaceous-Tertiary event. Its discovery is one of the most exciting detective stories in recent Earth sciences, rivaling the development of plate tectonic theory. It has long been apparent that the end of the Cretaceous was a time of major change. The Cretaceous-Tertiary (K-T) boundary was defined based on dramatic changes in the fossil record. Not only did major groups of previously dominant organisms (such as dinosaurs on land and ammonites in the oceans) go extinct, but there were also major changes in the abundance and species diversity of the lineages that survived (e.g., dramatic increases in birds and mammals on land, and teleost fishes in the oceans; and decreases in marine cephalopod mollusks). Until the early 1980s, these changes were generally thought to have occurred over several million years and to have been caused by climatic change.

In 1980, three scientists at the University of California at Berkeley suggested a radically new hypothesis: an asteroid impact (Alvarez et al. 1980, 1984). Initially, this sounded like an idea from a science fiction movie and was ridiculed by many established Earth scientists. Nevertheless, the group, led by Walter Alvarez, presented intriguing data and a testable hypothesis. The asteroid impact hypothesis was based on the occurrence in rock strata that spanned the K-T boundary of a layer highly enriched in the element iridium (Ir). Iridium is rare in the Earth's crust, but is often present in high concentrations in materials of extraterrestrial origin (meteorites and asteroids). The Alvarez group hypothesized that the iridium-enriched layer was produced by the dust injected into the atmosphere and circulated around the Earth following the impact of a large asteroid. The dust presumably blocked solar radiation and resulted in rapid global cooling, which in turn caused the extinction of dinosaurs and other lineages.

A strength of the Alvarez hypothesis was that it made testable predictions (Alvarez et al. 1984). First, it predicted that when an iridium-enriched layer occurred near the K-T boundary, it would always separate Cretaceous rocks below from Tertiary strata above. This prediction led to efforts to find stratigraphic sequences spanning the K-T boundary, to search for an iridium-enriched stratum, and to date the underlying and overlying rocks. Many independent studies of localities throughout the world confirmed the prediction: when present, the iridium-enriched layer marked the K-T boundary. Second, the hypothesis predicted that the vast majority of extinctions should have occurred right at the K-T boundary. Previous interpretations of fossil remains had suggested that some taxa disappeared millions of years before or after the K-T boundary. More accurate dating of fossil remains, however—especially with reference to the telltale iridium-enriched layer, when present—increasingly supported the hypothesis: extinctions were concentrated during a very brief period. Third, an asteroid impact and resulting global cooling would be expected to lead to differential extinction or survival of different kinds of organisms depending on their life histories and ecologies. And, indeed, the K-T extinctions were highly selective. Those groups with capacities for body temperature regulation, such as endothermic birds and mammals, and those with life history stages that would be resistant to a brief but intense cold stress, such as seed plants, insects, and freshwater invertebrates, suffered fewer extinctions than many other kinds of organisms.

Finally, the hypothesis predicted that the crater formed by the asteroid impact might actually be discovered, and suggested that the thickness of the iridium-enriched stratum should provide clues to its location. Obtaining evidence bearing directly on this prediction initially seemed to be a long shot. No known meteorite craters had the appropriate combination of size and age to be good candidates. Further, since most of the Earth's surface is ocean and most of the seafloor is young, there was a high probability that the impact occurred in the ocean and that the crater had long since been subducted into an ocean trench (see Chapter 8). However, mapping of increasing data on the iridium-enriched stratum showed a clear geographic pattern, with the thickest layers in the Caribbean region. Then came the definitive find: the discovery of a crater—of the right size and age—on the continental shelf off the coast of Yucatán, termed the Chicxulub crater (Swisher et al. 1992).

So we now know that an asteroid did strike Earth about 65 million years ago. It appears to have caused the extinction of dinosaurs, ammonites, and many other organisms, and in so doing, to have opened the way for the adaptive radiation of mammals, teleost fishes, and other lineages to rise to their present dominance. While the major features of this event are now well documented, many pieces of the story are not. How much was the Earth cooled, and how long did the cold period last? Beyond the differences in life history and ecology mentioned above, what determined which species and clades died out and which survived? Clearly, even such an unprecedented and unpredictable event as the impact of an extraterrestrial body did not cause "random" extinctions. Instead, it acted as a severe filter, eliminating many lineages, allowing others to pass through, and causing revolutionary changes in the composition of the Earth's biota that have endured for 65 million years.

Species Selection

Processes of Species Selection

The historical pattern of speciation and extinction has had a major influence on the taxonomic diversity and geographic distribution of living things. Episodic events such as mass extinctions followed by explosive adaptive radiations have been particularly important. The aftermath of the K-T extinction event makes a point that is also illustrated by other parts of the fossil record: lineages of organisms have not been equally successful, and even the most successful lineages have been dominant during different periods of Earth history and have prospered for different lengths of time. Some lineages have radiated rapidly to leave many descendants; others have survived virtually unchanged for millions of years; still others have disappeared quickly, leaving no descendants. Paleontologists and biologists have long recognized that certain lineages appear to possess particular traits that result in high speciation rates or low extinction rates, often leading to episodes of adaptive radiation and relative evolutionary success compared with those groups that die out or barely maintain themselves.

As mentioned above in the discussion on macroevolution, the differential survival and proliferation of species over geological time has come to be termed **species selection** (Stanley 1979), by analogy to the differential survival and reproduction of individuals (*individual selection*), which has traditionally been considered the primary mechanism of evolutionary change by natural selection. Species selection and individual selection should be viewed not as totally different biological processes, but as the consequences of generally similar ecological and genetic processes operating at different levels of biological organization. The same conditions (e.g., rapid environmental changes) that cause large differences in the birth and death rates of individuals of different genotypes, and thus result in rapid evolution by individual selection, are also likely to lead to differential multiplication and survival of species and thus to result in rapid evolution by species selection. The most important requirement for species selection to operate is that species-level traits are heritable. So, for example, if two unrelated marine benthic invertebrate lineages shared a trait such as having pelagic larvae (whereas other lineages within each of their clades did not), and we discovered that those lineages speciated more rapidly than other members of each clade, then we might suspect that selection favoring the trait "pelagic larvae" led to the differential production of species having that trait in two evolutionarily independent lineages.

Examples of Species Selection

To some extent the fate of evolutionary lineages is a matter of chance and opportunity—it depends on a species with particular traits being in a favorable place at an opportune time. As emphasized above, many adaptive radiations begin when a population either colonizes a new area or evolves a key innovation that substantially increases its fitness in a particular environment (Simpson 1952a). In either case, the species is suddenly presented with new ecological opportunities, which stimulate further rapid evolution and speciation. The Cenozoic radiations of mammals illustrate the influence of both of the above factors. During the early Mesozoic, ancestors of modern mammals developed several innovations, including: precise articulation between jaws and crania, and specialized teeth—both traits conferring sophisticated efficiency in processing food; an erect posture increasing efficiency in locomotion; a large brain; and probably endothermic temperature regulation. During the late Mesozoic, ancestors of modern placental mammals acquired several new traits, including: higher metabolic rates; improved temperature regulation; highly developed tactile, auditory, and olfactory systems; and new mechanisms of nourishing their developing young. These traits represented major advances over their reptilian ancestors and over other kinds of primitive mammals, including monotremes and some now extinct groups. Extinction of the dinosaurs and other vertebrate groups at the end of the Cretaceous eliminated their rivals and provided them with new ecological opportunities. Immediately afterward, in the early Cenozoic, a relatively few number of lineages of placental mammals radiated explosively (Figure 7.25). They not only "replaced" the extinct reptilian groups, but also diversified at the expense of

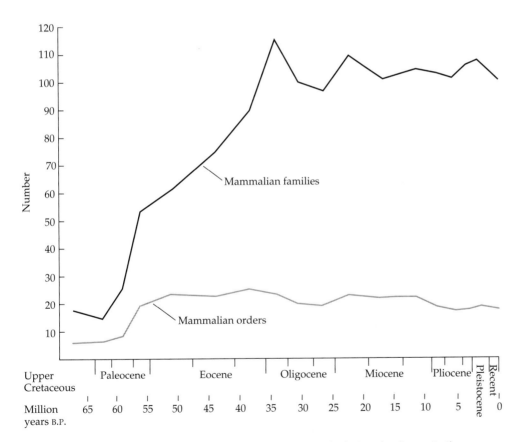

FIGURE 7.25 The "explosive" radiation of placental mammals during the Cenozoic, illustrated by the rapid increase in number of families. This radiation occurred after the K-T mass extinction event as mammals diverged and specialized to take advantage of ecological opportunities presented by the extinction of dinosaurs and other groups of previously dominant reptiles. (After Lillegraven 1972.)

the other mammalian lineages, such as the "primitive" docodonts and the rodentlike multituberculates, that had survived the K-T mass extinction.

Radiation of placental mammals occurred throughout the world except in Australia. On this isolated continent, surrounded by ocean barriers, egg-laying monotremes and pouched marsupials were the norm. In the absence of dominant reptiles, marsupials underwent their own adaptive radiations. South America had a mixture of marsupial and basal placental groups for much of the Cenozoic, but was subsequently colonized by many other groups of placental mammals from North America (and perhaps, Africa), and many of the original South American taxa became extinct (see Chapter 10). Australia has remained isolated right up to the present. Two groups of egg-laying monotremes (duck-billed platypus and echidna) survived, and marsupials underwent an extensive radiation, giving rise to many families and diverse morphological and ecological types that are amazingly convergent with placental forms (but no flying or marine forms, likely due to functional morphological constraints) (see Figure 10.35). These include a sand-burrowing "mole" (Notoryctidae); small to medium-sized insectivores and carnivores—some called marsupial "mice," "cats," and "wolves" (Dasyuridae); numbats (Myrmecobiidae) that converge on anteaters and aardvarks, possums, and gliders; koalas (Phalangeridae); bandicoots (Paramelidae); wombats (Phascolomiidae); and a wide variety of wallabies and kangaroos (Macropodidae). Prior to the Pleistocene and the arrival of humans, dogs (dingoes), and other mammals that the humans imported, only two orders of placental land mammals—rodents and bats—had managed to colonize Australia from Asia, presumably across water barriers. The rodents diversified dramatically, after at least two different colonization events, to produce about 80 species, most belonging to endemic genera. These include water rats (*Hydromys*), desert hopping mice (*Notomys*), and stick nest builders (*Leporillus*), which are ecologically and morphologically similar to North American muskrats (*Ondatra*), kangaroo rats (*Dipodomys*), and packrats (*Neotoma*), respectively (Keast 1972a,b,c).

In the absence of a historical event that eliminated competitors or provided access to a new area, radiations often were slower, and the buildup of diversity took longer. Many such radiations apparently occurred when some evolutionary innovation gave an ancestral species an advantage over coexisting organisms. Despite its apparent superiority, however, it usually took considerable time for the founding lineage to speciate, diversify ecologically, and supplant other groups of organisms that were already present. An example is provided by the neogastropods, a group of specialized, predatory marine snails that originated in the Cretaceous 130 million years ago, which have gradually diversified to become a dominant invertebrate group. Another is provided by angiosperm plants, which also originated in the Cretaceous. They possessed a number of structural and functional innovations in reproductive biology that, among other things, allowed them to evolve mutualistic associations with the animals that pollinate their flowers and disperse their seeds. Despite these advantages, however, it took until the mid-Cenozoic—about 100 million years—for angiosperms to largely supplant the previously dominant ferns and gymnosperms (Figure 7.26; Niklas et al. 1983; Knoll 1986; Taylor 1996).

The importance of evolutionary innovations, or at least particular combinations of adaptive traits, is illustrated by macroevolutionary and biogeographic patterns in mollusks. There is an interesting relationship between speciation rate, geographic distribution, and mode of larval dispersal among these groups (Jackson 1974; Hansen 1980; Jablonski and Lutz 1980; Jablonski 1982; Jablonski 1987). Some groups have planktotrophic larvae—juvenile stages that drift passively in the ocean—whereas others brood their offspring or have

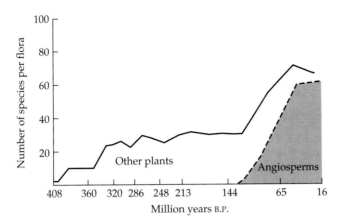

FIGURE 7.26 The relatively gradual radiation of angiosperms as illustrated by changing composition of fossil floras representing past communities. Although innovations in reproductive biology apparently gave angiosperms advantages over the previously dominant gymnosperms, the angiosperms rise to dominance took over 100 million years. Angiosperms have maintained much higher diversity than their gymnosperm progenitors, although some gymnosperms, such as conifers and cycads, still survive today. (After Knoll 1986.)

other specializations that do not involve their young being dispersed in the plankton. Mode of larval dispersal can be determined for fossil as well as for recent forms because an individual's original larval shell is preserved as part of its adult shell, which is readily fossilized. Molluscan species with planktotrophic larvae tend to have larger geographic ranges (Figure 7.27A) and

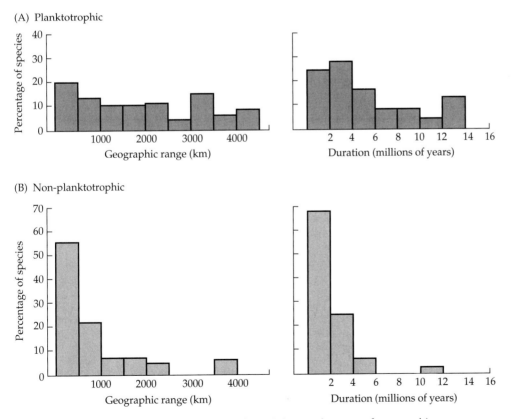

FIGURE 7.27 Relationship between mode of larval dispersal, extent of geographic range, and survival time in the fossil record of late Cretaceous mollusks on the east coast of North America. Note that species whose larvae can disperse long distances in the plankton (A) tended to have larger geographic ranges and lower extinction rates (longer survival times) than those species whose larvae settle close to their parents (B). (After Jablonski 1982.)

lower speciation rates than related species whose young rarely disperse far from their parents. Limited dispersal of offspring presumably reduces gene flow, enables populations to adapt to local conditions, and thus facilitates genetic differentiation and allopatric speciation. On the other hand, because small, specialized populations are also particularly vulnerable to extinction, nonplanktotrophic species tend to persist in the fossil record for shorter periods of time than do those with planktotrophic larvae (see Figure 7.27B).

As implied by this simple example, patterns of species extinction, which depend in part on the characteristics of the organisms themselves, can influence the evolutionary histories of lineages by species selection. Over a period of 600 million years, one group of marine bottom-dwelling invertebrates, the clams (Mollusca, class Bivalvia), has been replacing a phylum, the brachiopods (Brachiopoda; Figure 7.28). Because these two groups have superficially similar morphology, feeding habits, and habitat requirements, it had long been thought that the clams were supplanting the brachiopods by competitive exclusion (Elliot 1951). In fact, species of the two groups may compete significantly where they occur together, but careful examination of the fossil record indicates that the increase of clams and the dramatic decline of brachiopods were also facilitated by mass extinctions that eliminated many more taxa of brachiopods than of clams (Gould and Calloway 1980). Valentine and Jablonski (1982) suggest that brachiopods are differentially susceptible to catastrophic extinction because they do not have planktotrophic larvae, but instead brood their young or produce larvae that spend only a brief time in the plankton. As noted above, marine invertebrates with nonplanktotrophic larvae tend to have restricted geographic distributions and high extinction rates. Brachiopods have largely been eliminated from shallow-water habitats, where they are exposed to environmental fluctuations, but some forms persist in deep waters, where their mode of reproduction is less disadvantageous.

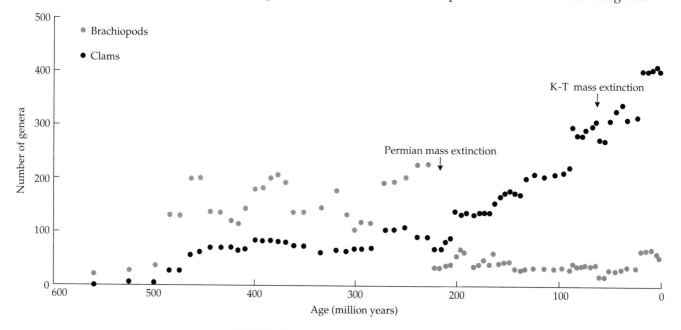

FIGURE 7.28 The "replacement," over a 600-million-year period, of brachiopods (phylum Brachiopoda) by clams (phylum Mollusca, class Bivalvia). This shift in dominance has often been attributed to competition because the two groups are generally similar in morphology and ecology. While competition may have played an important role, so did several mass extinction events: after both the Permian and the K-T mass extinctions, brachiopods declined in diversity, while clams increased. (After Gould and Calloway 1980.)

The fossil record documents the origination and extinction of lineages at all taxonomic levels, from species to phyla. Many of these lineages radiated to produce considerable morphological and ecological diversity, and some spread throughout the world. But success was almost invariably fleeting: most groups declined in diversity and ultimately went extinct. Sometimes their fates were due primarily to chance events, such as random extinctions of small populations or mass extinctions caused by the collision of an asteroid with Earth. Often, however, their biological attributes, shaped by the microevolutionary force of natural selection and the macroevolutionary force of species selection, played key roles. The more we learn about the history of life, the more we appreciate the interplay between the biological and abiotic environmental processes that have shaped the patterns of speciation and extinction. In the next two chapters we describe how this arena for speciation and extinction has been influenced by fundamental changes in Earth's geographic and geologic template.

CHAPTER 8

The Changing Earth

THROUGHOUT THE HISTORY OF LIFE on Earth, countless generations of organisms either adapted to the temporal and spatial dynamics of the geographic template (via evolution or dispersal) or they suffered extinction. Paleontologists refer to the sum total of Earth's dynamic history as TECO events; namely plate Tectonics, Eustasis (global fluctuations in sea levels), Climate change, and Orogeny (mountain building). As shown in this and the next chapter, these processes are interdependent, with drifting plates often causing major changes in global sea level and climates, and creating volcanic islands and mountain chains. The relatively recent acceptance of plate tectonics theory (i.e., that Earth's landmasses and ocean basins are in a constant state of flux) is one of science's most important advances—providing a much more comprehensive understanding of the "history of place"—and knowledge that is essential for understanding the biogeography and evolutionary history of all life forms (see Chapter 2, and Chapters 10 to 12).

The Geological Time Scale

Anyone studying historical biogeography needs to be familiar with the time scale used to date the history of the Earth (Table 8.1). From revelations and painstaking reconstructions of William Smith and his "Geological Map of the British Islands" ("the map that changed the world," see Figure 3.9; see also Winchester 2001), early geologists recognized that each layer in a stratigraphic column contains a unique assemblage of fossils characteristic of a single time span. These assemblages could, therefore, be used to correlate ages of rock strata in one locality with those in distant localities. The most reliable fossils for use in such correlations were wide-ranging species whose lifestyles were largely independent of small-scale patchiness in the environment—especially those that were freely dispersed among marine habitats by currents. These forms are called index or guide fossils. Examples include planktonic, calcareous foraminiferans, and siliceous radiolari-

TABLE 8.1 *The geological time scale (all numbers are in millions of years B.P.)*

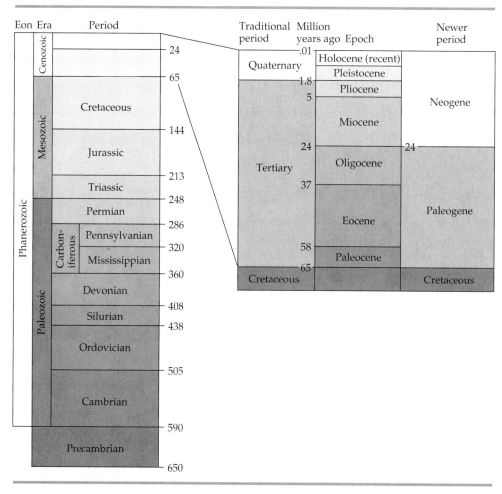

ans (phylum Protozoa) of the Cenozoic era; chitinous colonial graptolite (phylum Hemichordata)—floating animals of the Ordovician and Silurian periods; and swimming ammonoid cephalopods (phylum Mollusca) of the Mesozoic era, whose buoyant calcareous shells probably drifted with currents after their death. On continents, the most widespread fossils are found in coal beds of the Carboniferous period, which are characterized by numerous fossils of distinctive vascular plants.

At first, correlations of fossils around the world gave only relative estimates of the ages of various rocks and fossils; early scientists had no specific knowledge of the actual dates of rocks. Carolus Linnaeus and his eighteenth century colleagues accepted the biblical doctrine that the Earth was just a few thousand years old. One century later, Alfred Russel Wallace (1880) used an estimate of 400 million years for the absolute age of the Earth. In the twentieth century, the discovery of radioactive materials finally led to more exact dating procedures. Radioactive elements are unstable, and they decay forming stable atoms through a series of intermediate unstable products; during this disintegration process, atomic particles are released. The rate of decay can be quantified and expressed as a half-life—the amount of time needed for half of the radioactive material to decay to the stable element. By calculating the ratio of radioactive element to stable end product, one can determine, within limits, the age of a sample. Thus, by using radioactive isotopes of uranium and thorium, whose stable end product is lead, scientists have pushed back the estimate of Earth's age to 4.6 billion years. The oldest known fossils have been dated at 3.5 billion years before the present (B.P.).

The potassium-argon method is another valuable technique for dating Phanerozoic rocks. Radioactive potassium ($^{40}K^{19}$) decays to stable calcium ($^{40}Ca^{20}$) and the inert gas argon ($^{40}Ar^{18}$); half-life of potassium 40 is 1.31 billion years. A major problem with this technique is that argon gas will escape from rock heated above 300° C—the temperature that would be reached during metamorphism—therefore making this method not wholly reliable. Measurement of the decay of rubidium 87 (^{87}Rb) to strontium 86 (^{86}Sr) is used mainly for dating rocks older than 100 million years. For very recent material, radiocarbon dating is extensively used. Carbon 14 decays to carbon 12 at a fairly rapid rate (half-life is 5730 ± 30 years). After 50,000 years, so little radiocarbon remains that detection is very difficult. Other advances, including Accelerator Mass Spectrometry dating (which requires only minute amounts of organic carbon; i.e., < 2 mg), and optically stimulated luminescence (which uses light to stimulate luminescence of material with very low concentrations of uranium, thorium, and potassium) are now often used to provide high quality and high precision estimates of material age within a range of 100 to 200,000 years (see Jackson 2004, and references therein).

Because many trees form an annual ring during each growing season, analysis of tree growth rings in temperate latitudes is a reliable method for dating fossils formed within the last 10,000 years, and the results compare closely with carbon 14 values. In fact, tree rings can be used to calibrate radiocarbon dating. Paleoclimatic reconstructions are also possible because the width of a fossil growth ring is correlated with the length of the growing season and the availability of water (Fritts 1976).

In combination, these methods have been used to estimate the times of major geological and evolutionary events during the past 600 million years, a period known as the Phanerozoic eon (age of obvious, or multicellular, life; see Table 8.1). The geological time scale is hierarchical, with each division among eons, eras, periods, or epochs marking transitions among geological strata and embedded fossil assemblages. Given the great difficulty of dating such ancient events, it should not be surprising that precise dates are not universally accepted. For example, the scale accepted in 1971 divided the Cretaceous period into 12 equal epochs of 6 million years apiece, but radioisotope dating has revealed that some epochs were longer and others quite short (Baldwin et al. 1974). Reexamination of Triassic deposits may eventually lead to a drastic shortening of that period and to increased time spans for the Jurassic and Permian. Accurate dates within the Mesozoic are crucial because the early evolution and dispersal of major lineages of land vertebrates and seed plants, as well as the extinctions of certain marine groups and radiations of others, occurred during that era.

Traditionally, the Cenozoic era was divided into the Tertiary and Quaternary periods. However, the Tertiary covered some 63 million years, while the Quaternary lasted just 1.8 million years. This traditional scheme has now been replaced by a newer one that divides the Cenozoic into two periods of more similar duration: the Paleogene (65 to 24 million years B.P.) and the Neogene (24 to 0.01 million years B.P.; see Table 8.1).

The Theory of Continental Drift

No contribution to biogeography has had more of an impact than the theory of continental drift. This theory developed from a highly speculative idea in the early 1900s to a well-established fact by the 1960s (see Briggs 1987; Strahler 1998; Scotese 2004). Simply defined, the theory of continental drift states that continents and portions of continents have rafted across the surface of the globe on the weak, viscous upper mantle beneath the Earth's crust. Thus the Earth's crust is not composed of fixed ocean basins and continents, as was assumed by an overwhelming consensus of scientists up until the 1950s, but instead the Earth's crust is a changing landscape in which once-distant lands are now in juxtaposition, and others, once attached, are now widely separated.

Evidence in favor of crustal movements is conclusive, and within the last few decades the theory of continental drift has given rise to a respected science. Today's more comprehensive theory—referred to as plate tectonics—explains the origin and destruction of Earth's plates as well as their lateral movement or drift. However, throughout the long history of the field, biogeographers from Linnaeus to Lyell, Darwin, and Wallace, insisted on the fixity of the continents and great oceans. Lyell, Cuvier, and others had long ago established that the land and sea exhibited great vertical fluctuations throughout the geological record. But the notion that great masses of the Earth's crust could drift, collide, and separate like ice on a partially frozen river must have been a tough sell indeed.

Scholars have searched diligently to determine who first proposed the theory of continental drift. In 1596, the Flemish geographer and cartographer Abraham Ortelius (1527–1598) speculated in his Thesaurus Geographicus that the Americas were "torn away from Europe and Africa . . . by earthquakes and floods…The vestiges of the rupture reveal themselves, if someone brings forward a map of the world and considers carefully the coasts of the three [continents]." While some authors have attributed the germ of the idea to Sir Francis Bacon (1620), it appears that he may have never really discussed the subject, beyond the apparent jig-saw-like fit of the continents noted by Ortelius centuries earlier. In 1782, Benjamin Franklin postulated that the Earth's crust was like a shell floating on a fluid interior, and that broken pieces of the shell could move about as the fluid flowed. In the early decades of the nineteenth century, Sir Charles Lyell realized that fossils found in Europe indicated that a tropical climate had once prevailed in that region, and he suggested that the Earth experienced cycles of global climatic change (Lyell 1834). According to Lyell, these changes were triggered by vertical shifts in the Earth's crust, which rose above sea level in one region of the world while sinking in another. Like Franklin, Lyell also postulated that the Earth's crust was comprised of one rigid shell, which at times rotated independently of the fluid mass of molten matter below. When the crust rotated such that most of Earth's landmasses were concentrated near the equator, global warming occurred. During other "great seasons of the Earth," landmasses were concentrated near the poles, resulting in what Lyell called Earth's "great winter," which was accentuated by glacial episodes. Lyell's theory of geoclimatic cycles, albeit novel, had little if any impact on the field perhaps largely because he insisted that, while the Earth's crust shifted vertically and at times rotated as one rigid shell during these great cycles, the continents (plates) changed little in size, shape, or position relative to one another (Figure 8.1). Lyell, one of the most persuasive and revered scientists of his time, never made his theory of Earth's great seasons one of his crusades. As we shall see, his model bore only a trivial resemblance to the modern theory of continental drift and plate tectonics.

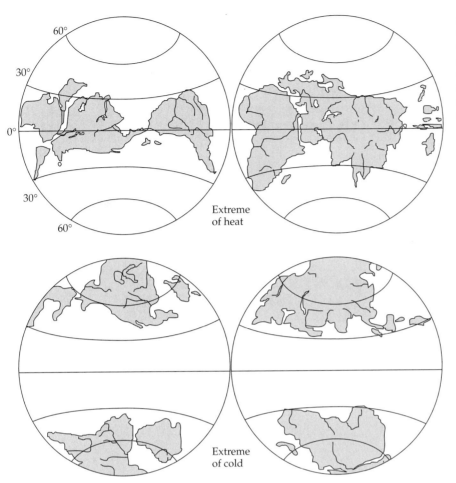

FIGURE 8.1 Lyell's theory of cycles of global climatic change. In the third edition of his *Principles of Geology*, Lyell suggested that although they did not change in shape or position relative to one another, the continents shifted in concert across the globe. This, he argued, could account for the paleontological evidence of major shifts in climates.

On the other hand, one of Lyell's contemporaries, Antonio Snider-Pelligrini (1858), may have been the first to forcefully demonstrate the geometric fit of the coastlines of continents on opposite sides of the Atlantic Ocean and to argue cogently that they once formed a supercontinent that subsequently split apart. Snider-Pelligrini's theory, however, was quite rudimentary and included the suggestion that separation of the continents was somehow caused by Noah's flood.

Throughout the remainder of the nineteenth century and indeed into the early decades of the twentieth century, geologists had little understanding of the past relationships of continents. In 1908, Frank Bursley Taylor, an American geologist and astronomer, presented a detailed model (privately published in 1910) in which the continents were hypothesized to move, distorting crustal materials into mountain ranges and island chains (Figure 8.2). Most of Taylor's research focused on glacial history of the Great Lakes region during the Pleistocene, but he was also strongly interested in the formation of mountain chains, island archipelagoes, and the solar system. Taylor postulated that recent glacial periods were caused by movements and massing of the continents near the poles and, during earlier periods, mountain chains and island arcs formed along the forward margins of moving continents, which also opened ocean basins behind them (Taylor 1928). Although Taylor's ideas were innovative he did not correctly perceive the directions of continental movements and, not surprising given his interests in astronomy, he wrongly attributed these movements to tidal and rotation forces of the Earth, moon, and sun.

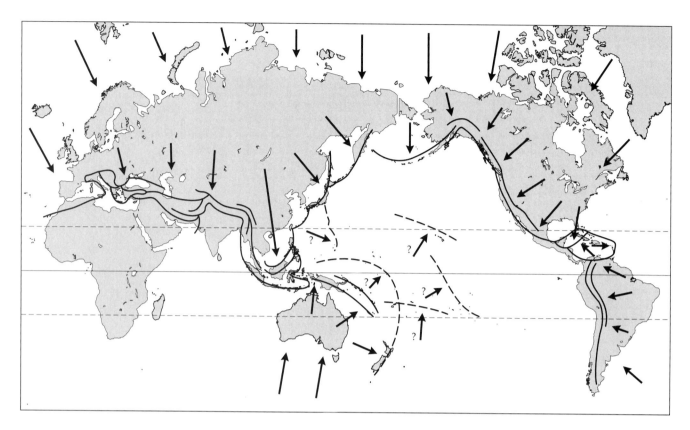

FIGURE 8.2 An early model of crustal movement proposed by Taylor (1910). This scenario suggests a general drift of major landmasses towards the equator.

Later research would indeed confirm that continental movement, crustal upheaval, and the formation of ocean basins are indeed intimately related.

Wegener's Theory

Alfred Lothar Wegener (Figure 8.3), a German meteorologist, conceived and championed the theory of continental drift, which he presented with admittedly scanty evidence at first, but which anticipated much of our current knowledge. Wegener developed his ideas on continental displacement in 1910, independently of Taylor, after visiting tectonically active areas of Greenland as well as observing on a world map the congruence of opposite coastlines across the Atlantic. Unlike Ortelius, Snider-Pelligrini, and others who also commented on the fit of the continents, Wegener dedicated the remainder of his life to advancing what would eventually become the theory of continental drift. In January 1912, Wegener unveiled his working hypothesis, with supporting evidence, in two oral reports, published later that year (1912a,b). These observations were expanded into his classic book, *Die Entstehung der Kontinente und Ozeane* (*The Origin of Continents and Oceans*) (1915), which he continued to expand on and revise over the following two decades.

Wegener's theory on horizontal continental movements not only discussed all of the continents, but also synthesized evidence from many disciplines:

FIGURE 8.3 Alfred Lothar Wegener (1880–1930), who developed the ideas leading to the modern science of plate tectonics and its confirmation of continental drift. (From Schwarzbach 1980; photo courtesy of Deutsches Museum, Munich.)

geology, geophysics, paleoclimatology, paleontology, and biogeography. The strongest attribute of the theory was its integration of these many types of phenomena for the first time. In addition to the geometric fit of the continents, Wegener noted the alignment of mountain belts and rock strata on opposite sides of the Atlantic. As Lyell and others had observed much earlier, coal beds of North America and Europe indicated that both of these landmasses were once situated over the tropical latitudes. Similarly, Wegener observed that glacial deposits, or "**tillites**," in what are now subtropical Africa and South America suggest that they were once displaced poleward. In addition, the tillites of North America seemed continuous with those of Europe. Many anomalous biogeographic patterns, such as the highly disjunct occurrence of extant marsupials in South America and Australia, were easily explained by the theory that these continents were connected during the Permian period.

Like many revolutionary ideas, Wegener's were ignored by most scientists and ridiculed by others. At first, his conclusions were accepted by just a few geologists and biogeographers, most notably those in the Southern Hemisphere. After all, the idea of ancient connections and biotic affinities among the southern biotas was not new to students of the southern biotas. Joseph Dalton Hooker had suggested this in 1853 after reviewing distribution patterns of plants among the southern continents and archipelagoes:

> Enough is here given to show that many of the peculiarities of each of these the three great areas of land in the southern latitudes…is agreeable with the hypothesis of all being members of a once more extensive flora, which has been broken up by geological and climatic causes. (1853)

As we now know, Hooker postulated intermittent connections and dispersal across transoceanic landbridges, but geological evidence for his extensionist theory never materialized. Then again, evidence for Wegener's theory of continental drift, while forthcoming, would require the efforts of generations of scientists. The second edition (1920) of Wegener's treatise received some attention, albeit negative, when it was criticized by several prominent geologists, but wide knowledge of the theory came only after the third edition (1922) was translated into five languages, including English (1924). A fourth edition (1929), the one generally used now, contained much more information, but throughout the various editions, the substance of Wegener's ideas remained the same. The following are some of his pertinent conclusions:

1. Continental rocks, called sial (composed largely of silicon and aluminum), are fundamentally different, less dense, thicker, and less magnetized than those of the ocean floor (basaltic rocks, called sima—consisting primarily of silicon and magnesium). The lighter sialic blocks—the continents—float on a layer of viscous, fluid mantle.

2. Major landmasses of the Earth were once united as a single supercontinent, Pangaea. Pangaea broke into smaller continental plates, which moved apart as they floated on the mantle. Breakup of Pangaea began in the Mesozoic, but North America was still connected with Europe in the north until the late Tertiary or even the Quaternary (Figure 8.4).

3. Breakup of Pangaea began as a rift valley, which gradually widened into an ocean, apparently by adding materials to the continental margins. Mid-oceanic ridges mark where opposite continents were once joined, and ocean trenches formed as the continental blocks moved. The distributions of major earthquake centers and regions of active volcanism and orogeny (mountain building) are related to movements of these blocks.

Upper Carboniferous

Eocene

Lower Quaternary

FIGURE 8.4 Wegener's (1929) model of continental drift. Wegener envisioned the continents—initially united in one giant landmass—to have moved apart during the Mesozoic and early Tertiary. In Wegener's time, the geological epochs and periods were thought to have been more recent than has been indicated by modern dating methods. Nevertheless, comparison with Figure 8.15E and Figure A in Box 8.2 shows that Wegener's view was very similar to current reconstructions of continental movement. (After Wegener 1968.)

4. The continental blocks have essentially retained their initial outlines, except in regions of mountain building, so the manner in which the continents were once joined can be seen by matching up their present margins. When this is done, similarities in the stratigraphy, fossils, and reconstructed paleoclimates of now-distant landmasses demonstrate that those blocks were once united. These patterns are inconsistent with any explanation that assumes fixed positions of continents and ocean basins.

5. Rates of movement for certain landmasses range between 0.3 and 36 meters per year—the fastest being Greenland, which may have separated from Europe only 50,000 to 100,000 years ago.

6. Radioactive heating in the mantle may be a primary cause of block movement, but other forces are probably involved. Whatever the causal processes, they are gradual and not catastrophic.

Early Opposition to Continental Drift

Wegener's ideas were clearly prophetic, and formed the basis of our modern theory of continental drift and the more comprehensive theory—plate tectonics (see Briggs 1987). Yet, despite his cogent and persistent arguments, continental drift was not generally accepted until the early 1960s, some 50 years after he and Taylor first published their ideas. Why was the theory resisted for so long? The history of this debate serves as an important lesson in the nature of scientific revolutions. Strong criticism of Wegener's theory arose as soon as translated volumes made it available to most geologists and biogeographers in the mid-1920s. Some scientists resisted the new idea because it conflicted with their preconceived ideas of fixed continents and a solid Earth, and because it was proposed by a German meteorologist—a man who was not part of the geological establishment of western Europe.

Other scientists opposed Wegener's theory on much more objective and defensible grounds. Although it would eventually prove to be one of science's most important paradigms, Wegener's theory suffered from at least four shortcomings. First, the nature of scientific revolutions is such that new theories that challenge long-held paradigms are resisted until evidence clearly shows them to be more parsimonious. The prevailing attitude in biogeography and paleontology was first expressed by Alexander du Toit: "Geological evidence *almost entirely* must decide the probability of this hypothesis" (1927, p. 118). Wegener's theory included too many assumptions that remained unsubstantiated by available geological evidence. Paleontologists in particular were unconvinced by the biogeographic and fossil evidence marshaled to support the continental drift model. For example, in 1943, G. G. Simpson published an analysis of past and present mammalian distributions to show how these data fit alternative scenarios of past intercontinental connections. After correcting prevalent errors in the literature concerning these distributions, he pointed out that most known patterns of distributions during the Cenozoic Era could be explained without invoking continental drift. It was difficult, using paleontological data, to overturn long-accepted theories based on ancient landbridges and periodic floods. Although Wegener's thesis was plausible, the theory of continental drift would not be accepted until enough unambiguous new evidence was collected to make it the most parsimonious explanation for geological and biogeographic patterns.

A second shortcoming of Wegener's theory is that it contained many factual errors. The theory aroused skeptical interest in many fields, and scientists in each discipline recognized major factual errors in Wegener's presentations.

These inconsistencies had to be resolved. Even du Toit—Wegener's strongest proponent, who published two books (1927, 1937) in favor of the theory—had to concede unquestionable errors by Wegener. For example, Wegener proposed that plates move at an incredibly high rate—as rapidly as 36 meters per year. His theory might have been much more palatable if his estimates were closer to what we now believe to be true—that is, rates of 2 to 12 cm per year. Wegener's error, however, is understandable, as he and his colleagues were working with what they believed to be a much younger Earth than we now assume, and were limited to very crude methods of measuring displacement rates. Third, in addition to correcting these errors, Wegener and his followers would have to gather much more evidence from a variety of disciplines in biogeography and geology (especially marine geology) and to actually test their model.

Finally—and this was perhaps more critical than any of the other shortcomings—Wegener's theory was criticized for lacking a plausible mechanism. How could the plates—rigid, enormous masses of rock—move about, and what force or forces could drive such movements? Wegener's insights in this area, however, are largely overlooked by science historians. In fact, Wegener did indeed discuss three potential driving or "displacing forces" (see Wegener 1929, pp. 167–179). Not surprisingly, given that he was a meteorologist and an astronomer, two of these forces involved celestial phenomena (effects of centrifugal forces on the Earth's surface, and combined effects of gravitational fields of the Earth, moon, and sun). The third force Wegener discussed—convective currents of molten rock beneath the Earth's crust—is indeed the same ultimate force assumed by today's model of plate tectonics. Wegener admitted that his ideas on causal mechanisms were speculative and that "the problem of forces which have produced and are producing continental drift is still in its infancy" (p. 179). Speculation, however, may be a normal and perhaps an essential part of scientific revolutions. As Darwin once wrote in a letter to Wallace (1857), "I am a firm believer that, without speculation, there is no good and original observation."

Soon after completing the fourth and final revision of his book, Wegener set out on an expedition in 1930 and returned to Greenland to document its purportedly rapid movement and, by so doing, to verify his theory of continental drift. In fact, his view of a dynamic Earth with rapidly drifting landmasses was largely stimulated by his earlier expeditions to Greenland in 1906 and 1912. It is one of science's great and tragic ironies that, as Pascual Jordan observed, Wegener "died in a snowstorm on the very Greenland expedition he undertook to verify his theory."

Evidence for Continental Drift

Providing convincing evidence for Wegener's theory was to take nearly five decades of research conducted by many teams of geologists, paleontologists, and biogeographers. From the first time Wegener proposed the supercontinent of Pangaea, his opponents criticized the liberties he took to achieve a "good fit." A good reconstruction was finally achieved when S. W. Carey (1955, 1958b), an Australian geologist, used plasticene shapes of landmasses sliding over a globe. Nonetheless, the fit was not widely accepted until three researchers (Bullard et al. 1965) combined computer mapping techniques and statistical analyses to test continental fits. Their analyses showed that the continents do fit together if one uses submarine contours of the continental shelves to delineate the margins of continental plates (see also Hallam 1970).

Other stratigraphic, paleoclimatic, and paleontological evidence gradually accumulated to support the theory of continental drift. Along with these important discoveries (summarized in Box 8.1), some of the most compelling evidence in favor of the theory of continental drift was provided by marine geologists.

MARINE GEOLOGY. After World War II, a second generation of scientists, who had not been directly involved in the initial debates, made some important discoveries about ocean basins and rock magnetism that encouraged reexamination of the ideas and evidence advanced by Wegener and du Toit (see Briggs 1987). When Wegener proposed his ideas, very little was known about the structure of the ocean floor. On the basis of loose samples obtained by dredging, geologists suspected that the ocean floor was composed of basalt (sima), but no one had actually taken core samples of the deep basins. Sialic continental rocks, however, were well known. Echo soundings from several transoceanic expeditions had portrayed ocean bottoms as smooth structures (abyssal plains) lying 4 to 6 km beneath the ocean surface. A midoceanic ridge was known only in the Atlantic Ocean. Finally, deep cuts in the ocean floor (known as trenches) had been found on the ocean sides of island arcs, and were known to display unusual gravitational properties.

Oceanographic research was just beginning to accelerate before the outbreak of World War II, when charting ocean topography became a practical goal. During the war, Herman H. Hess—a marine geologist who was using an echo sounder as he sailed aboard a U.S. troop transport—discovered some flat-topped submarine volcanoes 3000 to 4000 m high. Peaked submarine volcanoes, called **seamounts**, had been previously identified. The new structures, which Hess later named **guyots** in honor of a Princeton geologist, were thought to be volcanic islands that had formed above the ocean surface, later became truncated by wave action, and finally sunk to 1 or 2 km beneath the waves (Figure 8.5). Guyots are common in the northern and western Pacific Ocean, and as we shall see shortly, they figured prominently in the development of models of continental drift.

Following the war, marine exploration blossomed due to generous funding by Allied navies. Important discoveries were made using new deep-sediment piston corers and explosive charges. From samples obtained using these new techniques, geologists learned that under recent sediments, all ocean floors are

FIGURE 8.5 A highly simplified model of seafloor spreading that depicts how oceanic plates are pushed apart at the spreading center by the upwelling of magma from the mantle, which causes the plates to slide away from midoceanic ridges over the viscous asthenosphere. Magma may also produce volcanic islands near the spreading center, but as a point on the plate is displaced from the ridge, it also descends to 4–6 km below sea level, and the islands become submerged. These submerged volcanic structures (seamounts or guyots) eventually disappear into an oceanic trench where the oceanic plate meets another plate. In the case illustrated, the heavier oceanic plate descends beneath the lighter continental plate, which causes the metamorphosis of the surface material on the oceanic plate into ophiolites and their deposition on the continent, the consumption of volcanic islands, and eventually the remelting of the plate itself. Asterisks indicate epicenters of earthquakes resulting from contact of two plates (the Benioff zone).

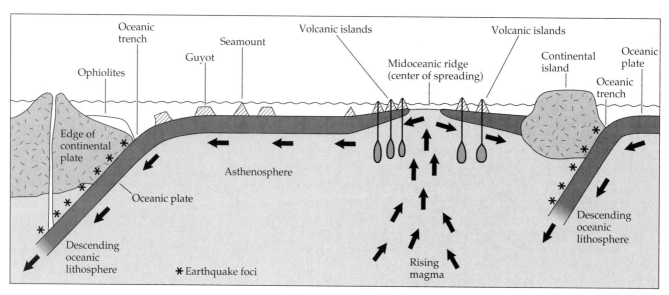

BOX 8.1 *Stratigraphic, paleoclimatic, and paleontological discoveries that contributed to the acceptance of the theory of continental drift*

■■■ By the 1970s, stratigraphic, paleoclimatic, and paleontological data provided strong support for the theory of continental drift. In particular, many pieces of evidence came together to support the matchup of continents—particularly the southern continents—in their former positions as part of Pangaea.

Stratigraphic evidence. Topographic features, including mountains, oceanic ridges, and island chains, along with specific rock strata (e.g., Precambrian shields and flood basalts) and fossil deposits, were found to be aligned along Wegener's hypothesized connections of the now fragmented portions of Gondwanaland (Figure A) (Hurley 1968; Hurley and Rand 1969). In addition, on each of the now isolated southern continents, rocks from the late Paleozoic and early Mesozoic contained the same stratigraphic sequence (see Allard and Hurst 1969): glacial sediments, coal beds, and sand dunes and other desert deposits, all overlain by a layer of volcanic rock.

Paleoclimatic evidence. All of the continents in the Southern Hemisphere have late Paleozoic glacial deposits (tillites) in their southernmost regions. Moreover, as glaciers move, they scour the underlying rocks, leaving deep scratches that mark the direction of their movements. If we plot these glacial lines on a map with the southern continents in their current positions, the patterns appear quite confounding (Figure BI). Not only do the glaciers appear to have been situated in what are now some relatively warm latitudes, but many appear to have risen out of the sea. This perplexing anomaly disappears when the same glacial lines are plotted on a reconstruction of Gondwanaland as it was during the Permian period (Figure BII).

Paleontological evidence. The late Paleozoic glacial deposits of the southern continents are covered with Permian rocks bearing the so-called *Glossopteris* flora (Schopf 1970a,b). These arborescent gymnosperms are presumed to have been adapted to (and therefore indicative of) temperate climates because they had deciduous leaves and conspicuous growth rings in their wood (Schopf 1976). When the occurrence of the *Glossopteris* flora is plotted on a map of Pangaea (Figure B), the points circumscribe a discrete region of

Figure A (I) Distributions of Precambrian shields (hatched areas), illustrating how they match up if the continents are reassembled as they were before the breakup of Pangaea. (After P. M. Hurley and J. R. Rand.) (II) Although substantially isolated today, flood basalts also serve as evidence of the previous connections of the southern continents in the former Gondwanaland. (After Storey 1995.)

(I)

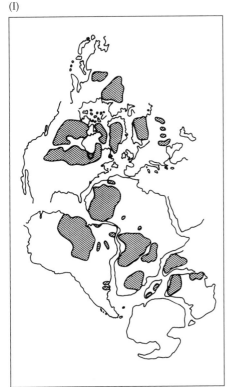

(II)

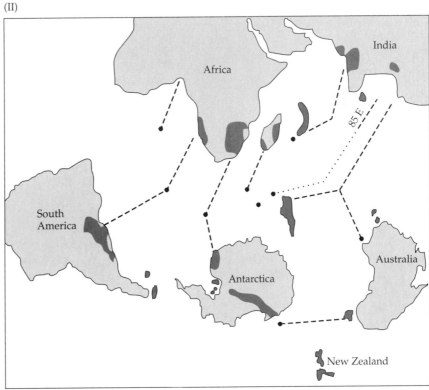

BOX 8.1 *(continued)*

Wegener's Gondwanaland, thought to be correlated with the margins of the glaciers.

In 1969 D. H. Elliot and E. H. Colbert unearthed the first tetrapod fossils found in Antarctica (Elliot et al. 1970) from mudstone and volcanic sandstone of the early Triassic. Additional finds provided convinc-ing evidence that many of these bones belonged *to Lystrosaurus*, a mammal-like reptile also found in rocks of similar age in the Karoo of southern Africa and the Panchet Formation of southern India. The reconstruction of Gondwanaland explained many such biogeographic anomalies in the distributions of extant as well as fossil assemblages. Many vertebrates underwent major radiations during the Permian, when the proximity of the continents allowed their rapid spread across what are now isolated landmasses. Like basaltic rocks, glacial tillites, and fossil assemblages, many of these extant forms exhibit disjunct distributions. The extensive breaks in their current ranges are consistent with the former connections of Pangaea and its subcontinents (Figure C). ▮▮▮

(I)

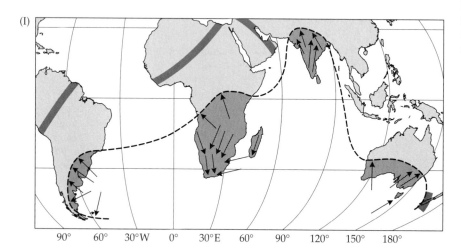

(II)

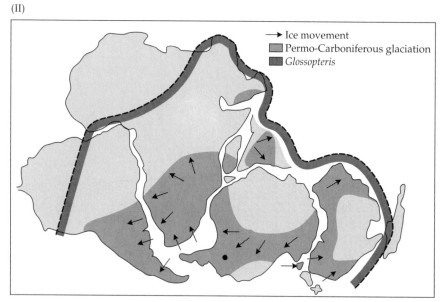

→ Ice movement
▨ Permo-Carboniferous glaciation
▨ *Glossopteris*

Figure B   Two lines of paleontological evidence for continental drift found on the southern continents. Glaciers carved lines in the underlying rock material, marking their location and direction of movement (arrows). The *Glossopteris* flora (or "southern beeches") included several groups of plants that grew along the margins of the glaciers. (I) The origin and directions of glacial movement (shaded area with arrows) and the distributions of *Glossopteris* fossils (darker shading) are difficult to explain based on the current positions of the southern continents because they imply that glaciers moved from oceans onto land. (II) These patterns, however, are consistent with reconstructions of Gondwanaland as it was during the Permian period. (I after Stanley 1987; II after Windley 1977.)

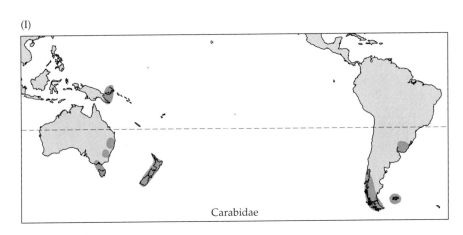

(I)

Carabidae

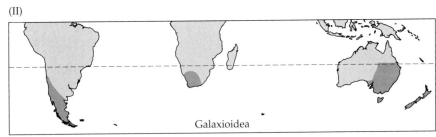

(II)

Galaxioidea

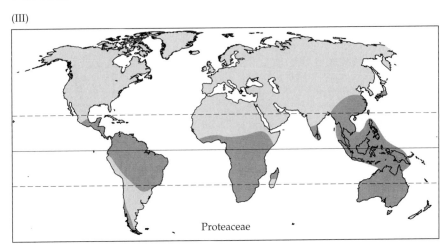

(III)

Proteaceae

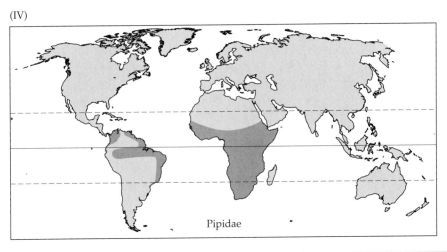

(IV)

Pipidae

Figure C   The disjunct distributions of some living taxa suggest that their ancestral forms radiated across Gondwanaland in the Permian period. (I) Southern temperate beetles of the tribe Migadopini of the family Carabidae. (II) Fishes of the superfamily Galaxioidea. These fishes are restricted to nontropical waters in the Southern Hemisphere. (III) Plants of the family Proteaceae. This group is found on all of the southern continents, but barely reaches the Northern Hemisphere. (IV) Clawed aquatic frogs of the family Pipidae. This family is comprised of two subfamilies, the Pipinae in tropical South America and the Xenopinae in tropical Africa, suggesting a common ancestor that was once distributed in western Gondwanaland. (I after Darlington 1965; II after Berra 1981; III after Johnson and Briggs 1975; IV after Savage 1973.)

composed of basalt, and that this basement is young, at a maximum dating back only to the Jurassic (150 million years B.P.). Thus the oceans are considerably younger than the continents, whose ancient foundations, called **cratons** or **Precambrian shields**, are older than 1 billion years!

By the mid-1950s a team of scientists had recognized that the submarine mountain ranges that bisect the oceans are really segments of a continuous global system 65,000 km long (Figure 8.6). This system is marked by a central rift valley, which is closely associated with a zone of frequent shallow earthquakes. At that time, new instruments measured remarkably high temperatures in these rifts, suggesting that molten mantle material was being released there. As anticipated by Wegener, geologists of the 1950s began to interpret midoceanic ridges as zones where the oceans expand, establishing the concept of **seafloor spreading**.

Located far from the ridges, oceanic **trenches** are so deep that until the 1950s, most knowledge of them was obtained by taking soundings. Oceanic trenches are V-shaped troughs about 10 km deep (see Figure 8.5). Through the use of seismic refraction techniques, marine geologists learned that the Earth's crust is extremely thin in trenches, and that the heat flow beneath the trenches is half that found in the abyssal plain, implying that heat is consumed in trenches. Gravity measurements in the trenches are lower than in any other place on Earth. Geologists therefore postulated that it is here, in the oceanic trenches, where the crust is pulled downward and its material reincorporated into the mantle.

Surrounding the Pacific Ocean is a belt of volcanism and earthquake activity known as the "Ring of Fire." It was proposed that the subduction (down-

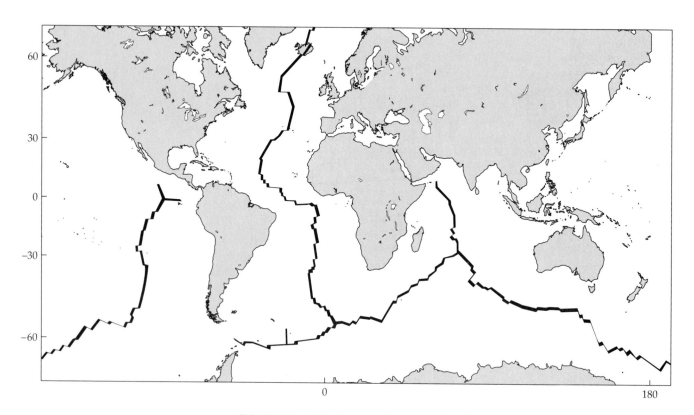

FIGURE 8.6 Global system of midoceanic ridges, which mark regions of seafloor spreading. (After Scotese et al. 1988.)

ward movement) of crustal materials in the trenches was the direct cause of these violent geological events. Hugo Benioff (1954) provided the first convincing evidence for this hypothesis. By plotting positions and depths of earthquake epicenters in the vicinity of the trenches, he demonstrated that the epicenters closest to trenches are shallow and those farther away are progressively deeper. Epicenters are aligned along a zone dipping downward at about 45° behind a trench, indicating that earthquakes are caused as the cold, rigid crustal slab descends into the mantle (see also Calvert et al. 1995). These zones are now termed **Benioff zones** (see asterisks in Figure 8.5).

PALEOMAGNETISM AND THE EMERGENCE OF A MECHANISM. Studies of **paleomagnetism** provided additional evidence for seafloor spreading. Convective flows of molten material from inside the Earth's core and through its mantle generates magnetic fields that permeate the planet (Rai et al. 2002; Dehant et al. 2003). Paleomagnetism refers to the orientation of magnetized crystals at the time of mineral formation (i.e., when molten rock solidifies). Rocks containing iron and titanium oxides become magnetized as they solidify and cool, and this magnetization is reflected in their crystalline structure, which remains "frozen" in the rock, oriented as a fossil compass in the direction and **declination** of the then-prevailing magnetic field. This high-temperature magnetization, referred to as **remnant magnetism**, is very stable unless the rock is reheated to extremely high temperatures (the **Curie point**). Hence, by measuring the direction and declination of remnant magnetism in cooled lavas, it is possible to determine the relationship of any landmass to the magnetic poles at the time the rock was formed; and by using triangulation and computer techniques, it is possible to reconstruct the positions of landmasses relative to the equator and to one another (Figure 8.7).

In the early 1950s the British physicist P. M. S. Blackett invented a new supersensitive magnetometer (magnetic detector) that could be used to determine continental orientation throughout geological history. First, the magnetometer was used to show that the British Isles had rotated 34° clockwise since

FIGURE 8.7 (A) The Earth acts as a great bar magnet. Because its magnetic fields are oriented toward its core (as well as poleward), latitudinal position can be read as declination in a compass needle. The same phenomenon also influences the orientation of crystals during the formation of magnetically active rock, thus recording the latitudinal position of the rock when it was formed. (B) Such paleomagnetic information can be used to reconstruct the positions and movements of the continents, such as the movements of Gondwanaland relative to the South Pole during the Paleozoic era. Gondwanaland (including some present-day equatorial regions) drifted over the South Pole twice—in the late Ordovician and in the late Devonian. (After Stanley 1987.)

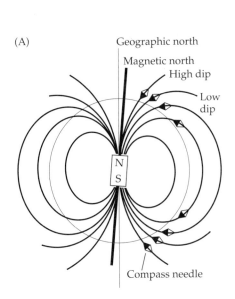

(A)

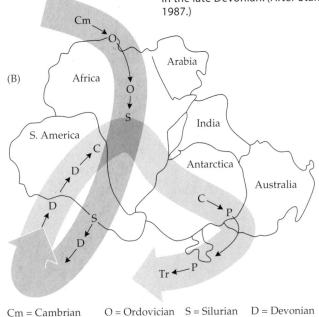

(B)

Cm = Cambrian O = Ordovician S = Silurian D = Devonian
C = Carboniferous P = Permian Tr = Triassic

FIGURE 8.8   Shifts in the orientation and latitudinal positions of Labrador, Africa, and Australia between the Triassic (200 million years B.P., dashed outline) and the present (solid outline) are revealed by paleomagnetism. (After Pielou 1979.)

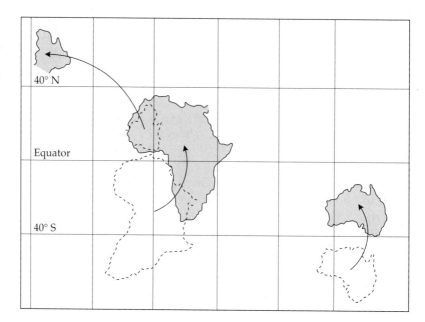

the Triassic (Clegg et al. 1954). A major breakthrough came when other British scientists (Creer et al. 1954, 1957; Runcorn 1956) analyzed geological strata in Europe and North America and provided strong evidence that the two continents had once been joined, but had later drifted apart. Subsequent studies around the world reaffirmed the necessity of continental movement to explain existing paleomagnetic patterns (Irving 1956, 1959; Runcorn 1962; Figure 8.8).

At the beginning of the twentieth century in central France, Bernard Bruhnes (1906) first discovered **magnetic reversals** (anomalies) when he found lavas that were magnetized in a direction opposite that in recently formed ones. Such patterns reflect reversals of the Earth's magnetic field, which occur every 10^4 to 10^6 years and appear to be generated by changes in magma flows through the mantle (Buffett 2000). Since Bruhnes' discovery, many investigators have found additional geological evidence of these reversals. On the ocean floor, alternating patterns of normally and reversely magnetized basalt appear as **magnetic stripes**, which retain their spacing and shapes for long distances (Figure 8.9A; see Cox 1973).

Marine geologists were the first to perceive the significance of magnetic stripes for continental drift theory. Two seminal insights were provided in the early 1960s by Frederick Vine and Drummond Matthews, and by Herman Hess. Vine and Matthews (1963) discovered several important properties of ocean floors:

1. Basaltic rocks at the midoceanic ridges have normal field (present-day) magnetic properties.

2. Widths of alternating magnetic stripes on opposite sides of a ridge are often roughly symmetrical, and the stripes are generally parallel to the long axis of the ridge (Figure 8.9).

3. The banding pattern of any one ocean closely matches that of others, and ocean patterns correspond approximately to reversal timetables from terrestrial lava flows.

Herman Hess proposed the first model of seafloor spreading to account for major tectonic events (see Figure 8.5). He presented his synthesis at Princeton

(A)

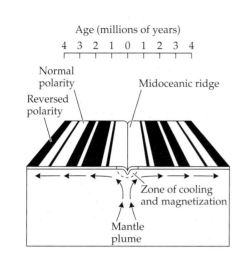

FIGURE 8.9 (A) During seafloor spreading, reversals in the Earth's magnetic field are recorded as the magnetically sensitive, iron-rich crust cools. Differences in the widths of the magnetic stripes reveal differences in the duration of these polarity episodes and in rates of seafloor spreading over time and among regions. (B) Cross-sectional view of the Earth and its major layers. (A after Stanley 1987.)

(B)

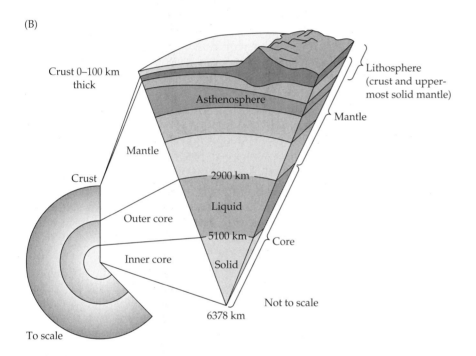

University in 1960, but it was not published for general readership until 1962. By that time, R. S. Dietz (1961) had published similar but less detailed accounts of global continental movements. Hess and Dietz hypothesized that the oceans are formed by addition of material and spreading at the mid-oceanic ridges. Moving away from the ridges, basalt along the ocean floor increases in age and is marked by magnetic stripes, which record the polarity of the prevailing magnetic field. The stripes tend to be highly symmetrical on opposite sides of a ridge. In contrast, differences in stripe widths indicate that the rate of seafloor spreading varies over time, and that it is not uniform across different oceans, or even different parts of the same ocean.

With this model of seafloor spreading, Wegener's theory finally had a plausible mechanism. The initial theory of continental drift was now included in a more general theory, plate tectonics, which included not just lateral movements of the continents and ocean basins, but their origin and destruction as well.

In Great Britain, most geologists who studied global tectonics were converted to the theory of continental drift by 1964, convinced especially by the soundness of Hess' model and the new synthesis. Acceptance in North America, however, lagged behind by several years. Meanwhile, widely circulated articles appearing in *Scientific American*, *Science*, and *Nature* did much to make the entire scientific community aware of the rebirth of Wegenerism. Young scientists became aware of the latest evidence and helped to create a wave of acceptance following the mid-1960s. The theory of plate tectonics is now firmly established as a unifying paradigm for much of geology, paleontology, and biogeography.

The Current Model

The theory of plate tectonics remains an active and exciting area of research (see Briggs 1987; Oreskes 2001; Bercovici 2003; Scotese 2004). Following its general acceptance, geologists, paleontologists, and biogeographers were free to focus their attention on more intriguing questions relating to the temporal and spatial patterns of plate dynamics and potential causal mechanisms.

Lateral movements of the plates result from a complex interaction among the Earth's crust, its underlying mantle, and its core—which is the site of intense heat that drives plate movement (Figure 8.10). This immense heat includes heat stored from when the Earth first formed, plus heat generated by gravitational compression and nuclear reactions (primarily from radioactive potassium, uranium, and thorium) deep in the core. Plates are roughly 100 km thick and are composed of a relatively thin, rigid layer of crust, which adheres to the upper layer of the mantle, together comprising the **lithosphere** (see Figure 8.9B). The mantle also includes a deeper, more fluid layer, the **asthenosphere**, which is composed primarily of molten material. There is an emerging consensus suggesting that plate movements are caused by a combination of forces, including **ridge push**, **mantle drag**, and **slab pull** (Kerr 1995; Deplus 2001; Price 2001; Conrad and Lithgow-Berteiloni 2002; Bercovici 2003). Ultimately, all of these forces are generated by heat and convective forces deep inside the Earth. Ridge push occurs at the midoceanic ridges, where **magma** (molten rock) upwells from the asthenosphere to the surface. It is believed that the parent rock of the mantle is partially melted and the basaltic portion is then brought to the surface. The addition of basaltic magma at the center of the ridge causes the older rocks on either side to spread—to be literally pushed apart. Thus, ridge push is the cause of seafloor spreading in Hess' model (see Figure 8.6).

Mantle upwelling is part of a convective cell system that also includes lateral flow of the mantle beneath the plates and downward flow of cooler rock toward the Earth's core (see Figure 8.10). Lateral flow and friction between the mantle and the overlying plate create a dragging force much like that of a conveyor belt (i.e., mantle drag). However, much, if not most, of the drifting force may be generated at **subduction zones**, where dense oceanic plates are pulled deep into the magma, eventually contributing to the convective gyre of molten rock. As the leading edge of the subducted plate descends, it pulls the rest of the plate laterally toward the subduction zone (see Kerr 1995, and references therein).

The relative importance of these three forces—ridge push, mantle drag, and slab pull—appears to vary markedly across tectonic regions and geological periods (Deplus 2001; Bercovic 2002; Conrad and Lithgow-Berteiloni 2002). Computer modeling conducted by Lithgow-Berteiloni and Richards (1995) suggests that slab pull may at times account for over 90% of net tectonic

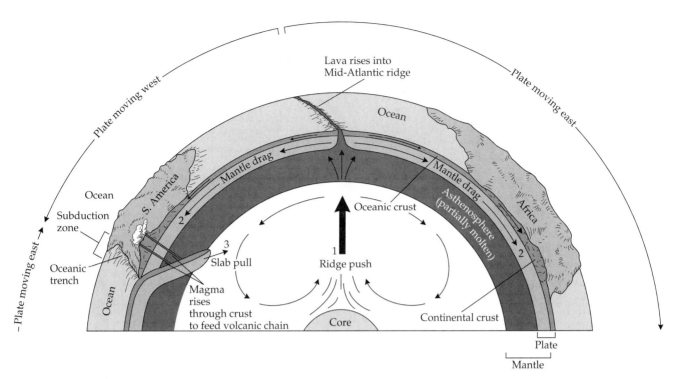

Lava rises into
Mid-Atlantic ridge

Ocean

Plate moving west

Plate moving east

Mantle drag

Mantle drag

Ocean

S. America

Africa

Subduction
zone

2

Asthenosphere
(partially molten)

Plate moving east

Oceanic
trench

3

Slab pull

1

Ridge push

Oceanic crust

2

Ocean

Magma
rises
through crust
to feed volcanic chain

Core

Continental crust

Plate moving east

Plate

Mantle

FIGURE 8.10 The current model of plate tectonics includes the possibility that at least three forces may be responsible for crustal movements: (1) ridge push, or the force generated by molten rock rising from the earth's core through the mantle at the midoceanic ridges; (2) mantle drag, the tendency of the crust to ride the mantle much like boxes on a conveyor belt; and (3) slab pull, the force generated as subducting crust tends to pull trailing crust after it along the surface. (After Stanley 1987.)

forces. On the other hand, slab pull cannot account for movement of those continental plates, such as the South American Plate, that lack subduction zones (actually, the Pacific Plate is being subducted under the western edge of the South American Plate). Obviously, this central issue—the underlying mechanisms of plate tectonics—will remain an active and hotly debated question for some time (Box 8.2, Figure B).

Despite remaining uncertainties about their ultimate causal mechanisms, geologists have developed a sound understanding of plate configurations, plate movements and interactions, and related phenomena, including earthquakes and volcanism. While the number of plates has varied continually throughout geological time, 16 major plates are currently recognized (Figure 8.11). These plates range in size from the Gorda Plate (roughly 750 km^2) to the Pacific Plate (with an estimated area of over 100 million km^2). Rate of lateral movement also varies markedly among the plates, with some, such as western portions of the Pacific Plate, drifting as rapidly as 5 cm per year, while others appear fixed.

The biogeographic relevance of plate tectonics becomes immediately evident when we compare plate configurations with Wallace's map of biogeographic regions (compare Figures 8.11 and the endpapers): biogeographic regions are defined by biotas and portions of the Earth's crust that share both evolutionary and tectonic histories, with representative assemblages on each plate evolving in isolation from those on other plates. In Chapter 12 we explore in depth the relationships between history of the Earth and that of its biotas.

While geological history of the Earth's plates must be profoundly complex, plate boundaries take three basic forms: **spreading zones**, **collision zones**, and **transform zones**. Other tectonic phenomena, including earthquakes, volcanism, and the formation of mountains and island arcs, are closely associated with these interactions among plates. For example, the Himalaya Mountains formed 60 million years ago when the Indian Plate drifted across the equator to collide with the Eurasian Plate (see Briggs 2003). These remain the world's

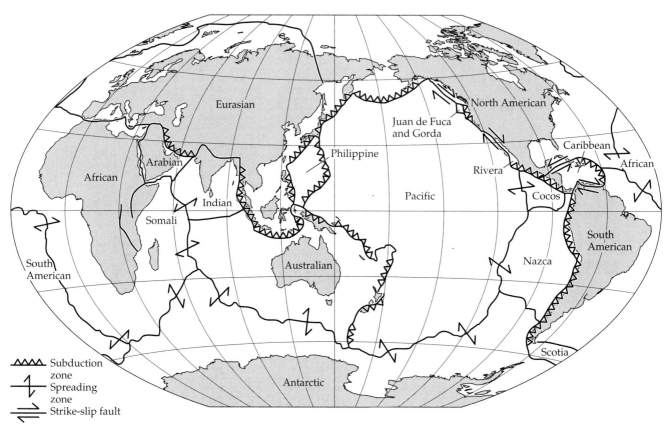

FIGURE 8.11 Earth's major tectonic plates.

tallest mountains. On the other hand, older mountain ranges, such as the Appalachians of eastern North America, which formed over 300 million years ago when North America collided with northwestern Africa, have been reduced by long-term erosion (in this case, such that the tallest mountains in the Appalachians are less than 2000 m; see Scotese 2004).

As noted above, midoceanic ridges mark the sites where two plates are drifting apart. Spreading zones, however, are not confined to oceanic plates. On continents, plates also diverge to form **rift zones**, which in the past created the Red Sea and the great, deep lakes of the Baikal rift zone and the East African Rift Valley. The latter region remains a tectonically active area marked by frequent earthquakes. Rather than continually and gradually sliding away from spreading centers, plates tend to resist moving until tectonic forces finally exceed some threshold, creating powerful bursts of movement, interspersed with relatively long periods of stasis. Spreading zones also are marked by volcanic activity where magma rises to the surface (see Figure 8.5). Along ridge systems, volcanoes may become emergent as oceanic islands. Once formed, such an island is eventually carried away from the ridge and down a slope to the abyssal plain. This movement, along with accumulation of what often amounts to many tons of coral and carbonates, decreases elevation of the island relative to sea level and eventually draws it beneath the surface, making it into a submarine seamount. In classic models, wave action wears an island flat to form a guyot; however, some evidence suggests that flat-topped guyots may actually be formed that way without ever having been emergent (see Figure 8.5). If the seafloor spreading model is correct, the ocean floor and its associated chains of volcanic islands, seamounts, and guyots

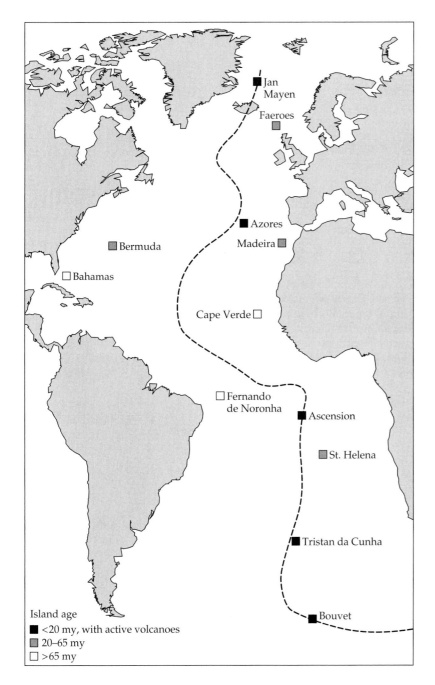

FIGURE 8.12 Because new crust is added along the spreading zones of midoceanic ridges, the age of islands and other crustal features tends to increase with increasing distance from these ridges, as demonstrated by islands at different distances from the mid-Atlantic ridge (dashed line). (After Pielou 1979.)

Island age
■ <20 my, with active volcanoes
▨ 20–65 my
□ >65 my

should be youngest at ridges and oldest—and sink deeper—below sea level as it approaches subduction trenches. Beginning with Tarling (1962) and Wilson (1963b), various researchers have shown that these predictions are correct (Figure 8.12).

Far away from the spreading zone, along a plate's leading margin, it will often collide with another plate. If the two plates are of roughly equal density (i.e., two oceanic or two continental plates), their collision will cause violent uplifting and the formation of mountain ranges along the plate boundary. This is what occurred when the Indian Plate collided with the Eurasian Plate to form the Himalayas. More often, relatively dense oceanic plates collide with and sink beneath lighter continental plates to form a subduction zone and a

BOX 8.2 *Expanding Earth's envelope*

■■▮ The theory of continental drift and plate tectonics has weathered some nine decades of skepticism, criticism, and at times even ridicule, to finally become embraced as a unifying paradigm of geology, paleontology, and biogeography. While most scientists now accept the general tenets of the theory, a few have proposed modifications or alternatives, some of which seem quite radical. Here, we highlight just three of these alternative views.

At least as early as the mid-1970s, geologists and biogeographers hypothesized the existence of an ancient continent, "Pacifica," which lay east of New Zealand and Australia before the Permian (see Nur and Ben-Avraham 1977; Melville 1981; Nelson and Platnick 1981). During the Mesozoic, this continent supposedly fragmented and dispersed, and its remnants became embedded along the growing margins of other continents. The terranes around the current margins of the Pacific Ocean (see Figure 8.14) were thought to be remnants of Pacifica. However, subsequent analyses indicated that these terranes were not of continental origin, but rather accretions of island chains, guyots, and seamounts. Further, reconstructions of seafloor spreading and crustal movements over the past 180 million years indicate that the purported remnants of Pacifica could not have diverged from a single "homeland."

Based primarily on biogeographic evidence, including disjunct distributions of many plant species, Humphries and Parenti (1986) suggested a more substantial modi-

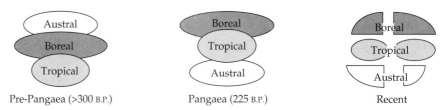

Pre-Pangaea (>300 B.P.)   Pangaea (225 B.P.)   Recent

Figure A A possible explanation for the amphitropical distributions of many plants (see Figure 10.26). This hypothesis suggests a pre-Pangaean supercontinent in which austral and boreal forms were located adjacent to one another. These landmasses may then have separated and drifted toward opposite poles before the formation of Pangaea. Finally, Pangaea split north to south to provide the modern pattern of amphitropical disjunctions. (After Cox 1990.)

fication of the current model of plate tectonics. In their view, many disjunctions, including **amphitropical distributions** (see Figure 10.26), in which species are restricted to boreal and austral regions, are consistent with the existence of a global, pre-Pangaean supercontinent during the early or mid-Paleozoic. The arrangement of landmasses of that supercontinent may have been quite different from that of Pangaea, with today's austral landmasses occupying northern latitudes adjacent to the (then) equatorial, (now) boreal landmasses (see Figure A). This supercontinent then broke up, and its landmasses

migrated to approximate their current relative positions during the late Carboniferous. While this hypothesis would explain many biogeographic anomalies, existing geological and paleomagnetic data are not consistent with the hypothesized movements of this pre-Pangaean landmass (see Cox 1990).

Perhaps the most radical alternative to the current model of plate tectonics is the expanding Earth theory. This theory was first proposed in the 1950s to account for the same features addressed by the Wegenerians: the fit of the continents, their breakup since the Permian, the relative

Figure B Terella model illustrating the expanding earth hypothesis. (I) According to this alternative theory of plate tectonics, for most of earth's 4.5 billion year history it remained relatively small and its surface was covered with a densely packed mass of continental plates (forming Pangaea)—devoid of oceanic plates or a world ocean (Panthalassa). (II) Earth's expansion and the concomitant development and expansion of oceanic basins is hypothesized to occur sometime during the Carboniferous and Permian periods (360–248 million years b.p.), thus initiating the breakup of Pangaea and its subcontinents. (III–V) The hypothesized expansion of the earth continued throughout the late Mesozoic and early Cenozoic Eras, with oceanic expansion occurring at the mid-oceanic ridges (dotted lines in IV and V)—i.e., just as hypothesized for the currently accepted model of plate tectonics.

I	II	III	IV	V
~1.5 billion B.P.	250–360 million B.P.	~1.50 million B.P.	~100 million B.P.	present

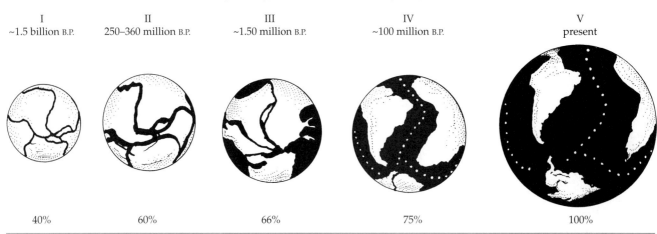

40%   60%   66%   75%   100%

ages of continental and oceanic crust, and the alignment of various geological and paleontological features across the oceans (Egyed 1956, 1957; Carey 1958). Expansionists proposed that the Earth's volume is not constant, but has increased—perhaps even doubled—since the Permian (see also Egyed 1956; Kremp 1992). As a result of this expansion, the Earth's crust—all continental at that time—was ripped apart, and new, oceanic crust filled the expanding gaps. Perhaps the most controversial, yet intriguing, feature of the expansion theory is that it implies that the gravitational "constant" is not—that is, that expansion is a result of weakening of the

Earth's gravitational field. Not surprisingly, these provocative ideas have attracted attention from physicists (see Jordan 1971). Figure B summarizes Kremp's (1992) reconstruction of major geological features associated with Earth's hypothesized expansion since the Permian period (see also Shields 1990; Carey 1996; Maxlow 1996; McCarthy 2003).

Taken together, the above hypotheses lie far outside the mainstream views of plate tectonics and related phenomena. In most cases, they are not supported by the available information, are less parsimonious than accepted paradigm, and are often championed by scientists outside

the primary fields (see Briggs 2004). This may sound familiar: when the theory of continental drift was first proposed by Wegener, a meteorologist, it received much the same notoriety, and many decades passed before what at first seemed a very radical theory was embraced by the scientific establishment.

Thus, while the Earth's plates are unlikely to make great progress in their movements during the next century, views of our ancient Earth and its tectonic machinery may experience great revolutions, perhaps embracing some of what we now view as radical hypotheses, including some yet to be described. ▮▮▮

---

deep oceanic trench (see Figures 8.10 and 8.13). Again, the plates tend to undergo long periods of stasis followed by violent episodes of movement and resulting earthquakes. As an oceanic plate is drawn deep into the molten layer of the mantle, its accumulation of relatively wet sediments is heated, causing mantle plumes to rise to the surface and form a ring of volcanoes far upstream from the superficial layers of the subduction zone. Extensive subduction zones are found along western margins of North and South America. As oceanic plates are subducted over geological time, seamounts and guyots are scraped

FIGURE 8.13 A subduction zone, where a relatively heavy oceanic plate slides beneath a lighter continental plate. Here, seafloor spreading causes the Gorda Plate to be subducted beneath the continental crust of the Pacific Northwest region of North America. This, in turn, results in faults and terranes (accumulations of oceanic material; see Figure 8.14) at the subduction zone and volcanic activity deep beneath the continental plate, forming the Cascade Range.

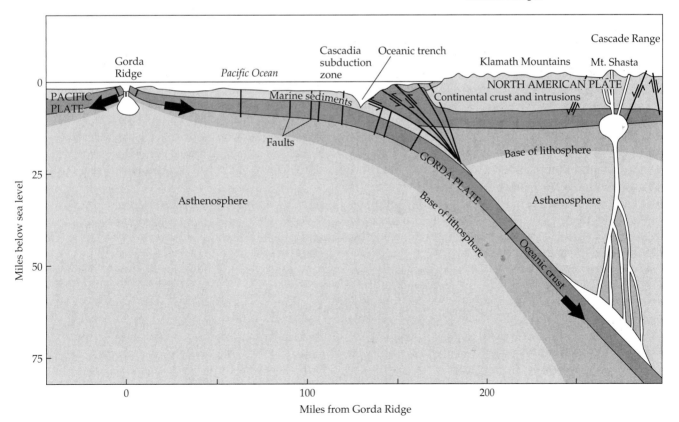

onto continental plates, forming coastal mountain ranges such as the Olympics of Washington State. Thus, subduction zones are marked with parallel bands of earthquakes and volcanism near their active edges, and mountain ranges and accumulations of marine sediments, or **terranes**, which increase in age as we move upstream from the active zone (Figure 8.14; see also Coney et al. 1980; Calvert et al. 1995).

Finally, plates of roughly equal density may slide and grind against each other, without either subducting, to form a **transform zone**. Not surprisingly, transform zones are characterized by high seismic activity as plates grind against each other. In southern California, the North American and Pacific Plates grind past each other along the San Andreas fault. While this may be the most notorious example, transform zones are quite common. Some are closely associated with major plate boundaries (e.g., those along the margins of the Scotia and Antarctic Plates, along the southern tip of South America), while others are found at great distances from plate boundaries (e.g., the Altyn Tagh fault of western China).

Earth's Tectonic History

The tectonic processes described above have recurred throughout Earth's 4.5-billion-year history, and will certainly continue to modify the geographic template as well as subsequent distributions of future biotas (Briggs 1987; Richards 2000; Scotese 2004). It is hard to overemphasize the effects of plate tectonics on biogeographic patterns of virtually all organisms. The origin, spread, and radiation of many taxa took place when the Earth's surface was dramatically different from its current profile. These processes—along with the many extinctions evidenced in the fossil record—were often associated with the collision, separation, or destruction of continents or oceanic basins, which altered opportunities for biotic exchange or affected climate on regional to global scales (Harris 2002).

In the following sections, we summarize the current view of Earth's tectonic history, focusing on the dynamics of its continents and marine basins during the Phanerozoic. As you can imagine, the detective work involved in this 650-million-year reconstruction, while fascinating, is a great challenge. It is without doubt one of the most fundamental and insightful advances in our understanding of the natural world. Over the past four decades, many teams of geologists and biogeographers have taken up this challenge and developed paleogeographic reconstructions of the Earth's dynamic history (Figure 8.15) drawing on four lines of evidence (Scotese 2004; see also Burke et al. 1977; Dewey 1977).

1. Paleomagnetic declinations: as discussed earlier, magnetic declinations can be used to determine latitude of sites where iron- and titanium-bearing rocks cooled and crystallized (see Figure 8.7 A).

2. Symmetrical magnetic stripes ("anomalies") of ocean basins: these magnetic reversals recorded in rocks on either side of a midoceanic ridge (see Figure 8.9A) can be used to "match-up" ancient spreading zones and, with the aid of fossil evidence and radio-isotope techniques, we can virtually "re-zip" the ocean floors back together and back into the mantle, and thus reconstruct the extent, direction, and timing of sea-floor spreading (and drifting of landmasses on opposite sides of that ocean basin).

3. Detailed topographic and bathymetric maps: recent advances in remote sensing, GIS, and spatial analyses have provided geologists and geographers with maps of Earth's land and ocean basins with unprecedented accuracy and precision. Paleogeographic reconstructions are now

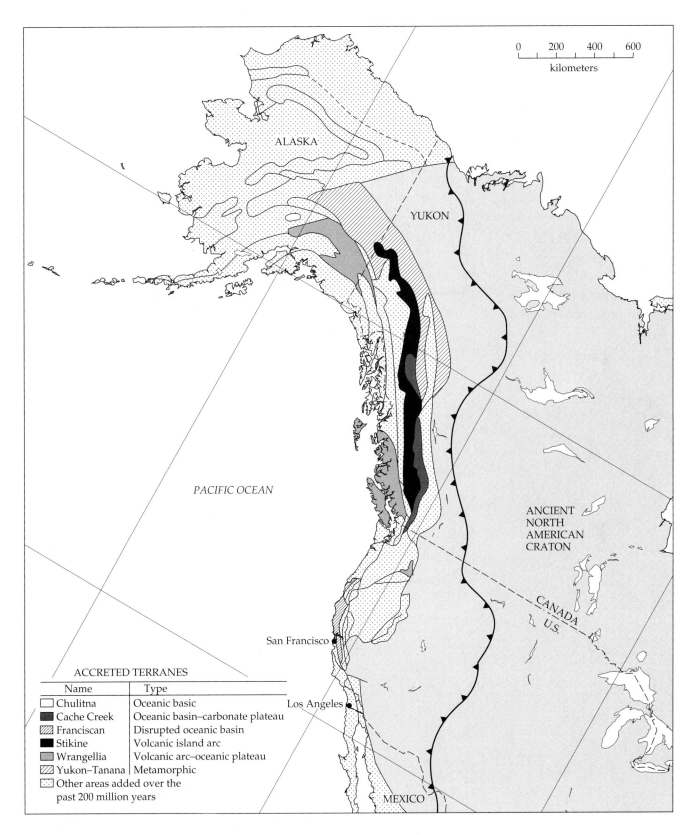

FIGURE 8.14 Terranes, or accumulations of marine sediments, mark the locations of earlier subduction zones along the west coast of North America. Over the past 200 million years, oceanic plates slipped beneath the relatively buoyant continental plates, but islands, seamounts, and other superficial features were scraped off and added to the continental plate. (After Jones et al. 1982.)

(A) Precambrian (660 million years B.P.)

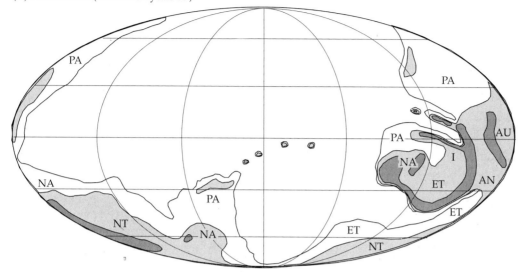

(B) Cambrian (520 million years B.P.)

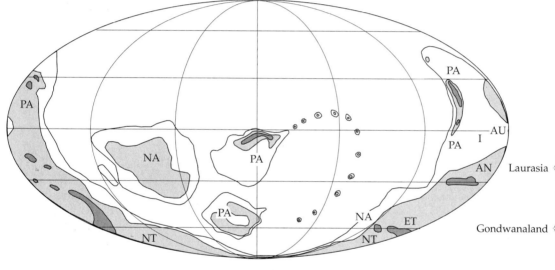

Laurasia $\begin{cases} \text{PA = Palearctic} \\ \text{NA = Nearctic} \end{cases}$

Gondwanaland $\begin{cases} \text{NT = Neotropical} \\ \text{ET = Ethiopian} \\ \text{I = Indian} \\ \text{AN = Antarctic} \\ \text{AU = Australian} \\ \text{M = Madagascar} \end{cases}$

(C) Late Silurian (425 million years B.P.)

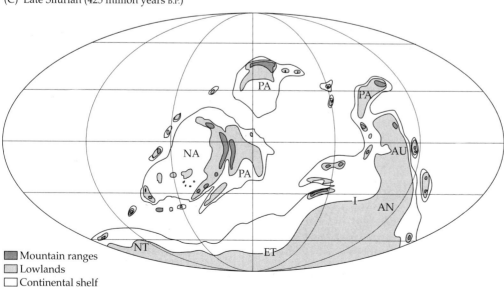

■ Mountain ranges
▨ Lowlands
☐ Continental shelf

FIGURE 8.15 Continental drift and paleotopography of Earth's landmasses from the early Precambrian (650 million years B.P.) to the present and future. [After Chris Scotese of the PALEOMAP project, University of Texas at Arlington; for information on digital images and other PALEOMAP products, contact via internet (www.scotese.com), email (chris@scotese.com), or toll free (888) 288-0160.]

(D) Earliest Triassic (240 million years B.P.)

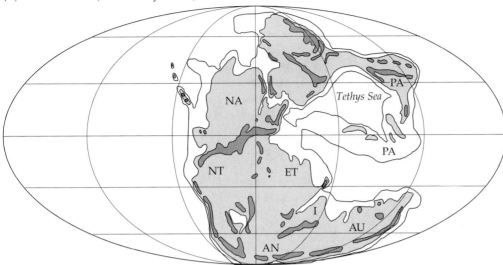

(E) Early/Middle Jurassic (180 million years B.P.)

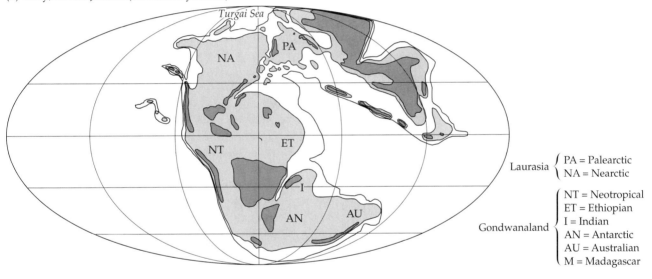

Laurasia $\begin{cases} \text{PA = Palearctic} \\ \text{NA = Nearctic} \end{cases}$

Gondwanaland $\begin{cases} \text{NT = Neotropical} \\ \text{ET = Ethiopian} \\ \text{I = Indian} \\ \text{AN = Antarctic} \\ \text{AU = Australian} \\ \text{M = Madagascar} \end{cases}$

(F) Early Cretaceous (120 million years B.P.)

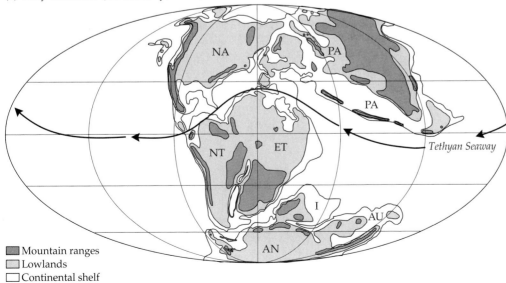

■ Mountain ranges
▨ Lowlands
☐ Continental shelf

(G) Late Cretaceous (80 million years B.P.)

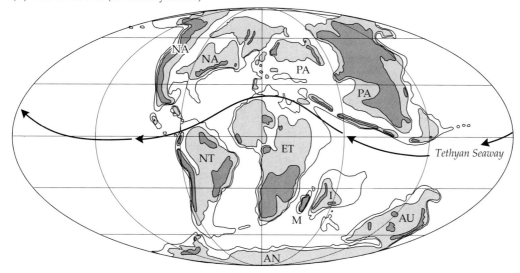

Tethyan Seaway

(H) Middle Paleocene (60 million years B.P.)

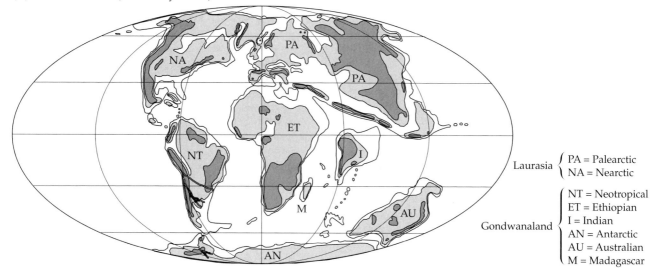

$$\text{Laurasia} \begin{cases} \text{PA = Palearctic} \\ \text{NA = Nearctic} \end{cases}$$

$$\text{Gondwanaland} \begin{cases} \text{NT = Neotropical} \\ \text{ET = Ethiopian} \\ \text{I = Indian} \\ \text{AN = Antarctic} \\ \text{AU = Australian} \\ \text{M = Madagascar} \end{cases}$$

(I) Early Oligocene (30 million years B.P.)

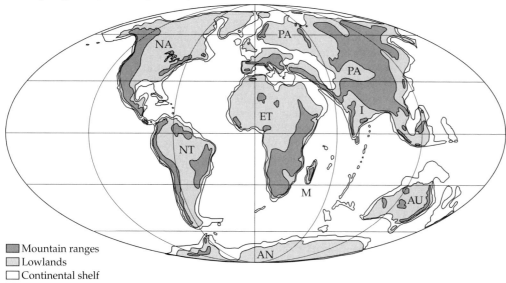

■ Mountain ranges
▨ Lowlands
□ Continental shelf

(J) Late Miocene (10 million years B.P.)

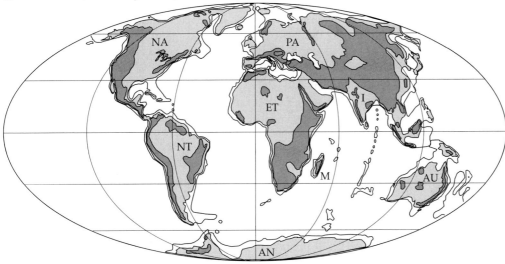

(K) Present day

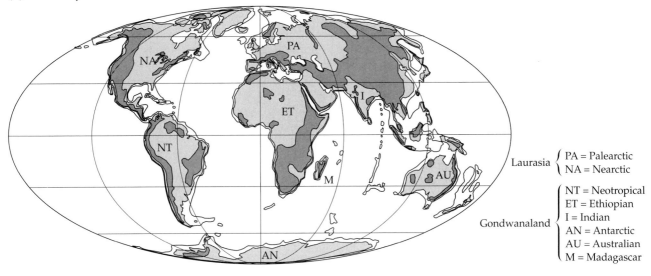

Laurasia $\begin{cases} \text{PA = Palearctic} \\ \text{NA = Nearctic} \end{cases}$

Gondwanaland $\begin{cases} \text{NT = Neotropical} \\ \text{ET = Ethiopian} \\ \text{I = Indian} \\ \text{AN = Antarctic} \\ \text{AU = Australian} \\ \text{M = Madagascar} \end{cases}$

(L) Future (+250 million years)

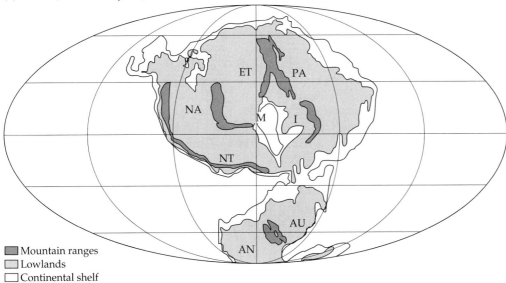

■ Mountain ranges
▨ Lowlands
□ Continental shelf

informed by detailed information on the locations of seamounts and guyots, midoceanic ridges and extinct spreading zones, deep sea trenches, linear fracture zones (indicating directions of drift of previously joined landmasses), mountain chains, linear archipelagoes (e.g., Emperor and Hawaiian Archipelagoes; see Figure 8.20), and tracks of flood basalts (magma from hotspot plumes that burn through the lithosphere and pour onto the continents).

4. Lithologic indicators of climate: as discussed in Chapter 5, Earth's soil and rock formations are strongly influenced by climatic conditions; coals forming where bauxite occurs in wet and warm (i.e., tropical) climates, tillites where cool and wet conditions prevailed (especially along the edges of glaciers), and evaporates and calcretes under warm and dry conditions. Thus, even though continental plates may drift from pole to pole, their rock formations have recorded their previous climates and, indirectly, their general latitudinal positions when they formed. Again, these rocks and associated tectonic events can be aged based on fossils and radiometric techniques.

Tectonic History of the Continents

Unlike the oceanic lithosphere, which is continually, albeit slowly, recycled into great gyres of molten material of the mantle, continental lithosphere is less dense and, therefore, not readily subducted. Thus, whereas the age of ocean basins rarely exceeds 150 million years, continental lithosphere is truly ancient—some as much as 3.8 billion years old (Scotese 2004). As a result, reconstructions of the tectonic history of the continents can be much more extensive than that of ocean basins. In the following sections we will summarize paleogeographic history of continental plates and terranes starting with the formation of the most recent supercontinent—Pangaea, realizing, however, that this is just a small portion of the long and complex tectonic history of these landmasses.

GONDWANALAND, LAURASIA, AND THE FORMATION OF PANGAEA. Early supporters of the theory of continental drift originally deduced that prior to the Mesozoic, all landmasses had been united in a supercontinent—Pangaea, which occupied one-third of the Earth's surface. Now paleogeographers have not only confirmed this, but tell us that other, global-scale supercontinents existed prior to this (see Figure 8.15A), and that Pangaea was a relatively temporary structure that probably existed as a single unit only during late Paleozoic and early Mesozoic times (see Figure 8.15D). Its northern half, **Laurasia**, had a very complex early history that was substantially different from that of the southern half, **Gondwanaland**.

Gondwanaland included the foundations of present-day South America, Africa, Madagascar, Arabia, India, Australia, Tasmania, New Guinea, New Zealand, New Caledonia, and Antarctica. It was, by far, the most ancient of the Pangaean landmasses, forming some 650 million years B.P. during the Precambrian era (see Figure 8.15A). From this time through the middle Ordovician period (about 475 million years B.P.), Gondwanaland remained a relatively continuous supercontinent in the Southern Hemisphere, although it drifted substantially between the equator and the South Pole (see Figure 8.7B–C). If one examines the history of Queensland, Australia—now at 12° S latitude—one discovers that it was located near the North Pole in the Proterozoic (1 billion years B.P.), at the equator in the Silurian, and at 40° S in the Mesozoic (Embleton 1973). The great antiquity of Gondwanaland accounts in part for the similarity of the now-isolated biotas of the Southern Hemisphere, patterns that strongly influenced historical biogeography since the early writings of Joseph Dalton Hooker (see Chapter 2).

In contrast to its ancient southern counterpart, the history of Laurasia is much more recent. Pre-Laurasian landmasses remained isolated until the early Devonian (about 400 million years B.P.), when the precursors of Northern Europe, which once occupied subantarctic latitudes, drifted northward across the equator to collide with the precursors of North America and the northern part of Siberia, forming the "Old Sandstone Continent" (Figure 8.16). The precursors of China and southern Siberia remained isolated in subtropical latitudes of the Northern Hemisphere, far from Gondwanaland, which was then situated near the South Pole. During the Carboniferous period, Gondwanaland began drifting northward, and eventually united with the Old Sandstone Continent about 300 million years B.P. Later, during the early Permian (about 270 million years B.P.), western Asia collided with Europe to form the Ural Mountains and unite the bulk of Pangaea. At this time, eastern Asia (largely derived from accreted remnants of Gondwanaland; Metcalfe 1999) still remained isolated as several large islands, which partially divided the world's oceans and encircled the Tethys Sea along the eastern margin of Pangaea (see Figure 8.15D). By the late Permian (250 million years B.P.), these landmasses consolidated to form Pangaea, a great, continuous landmass stretching from pole to pole.

Although this fact is often overlooked, coincident with the formation of Pangaea, marine basins were united to form one great ocean, **Panthalassa**. Thus, the Permian was a time of great connectivity and exchange among both terrestrial and marine biotas, many of which had evolved in isolation for hundreds of millions of years. On the other hand, while this was a time of interchange among long-isolated biotas, Triassic landscapes (and likely, seascapes

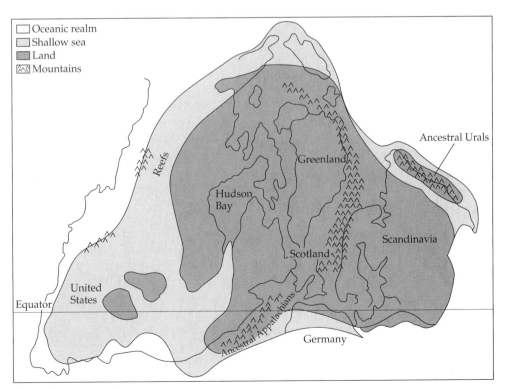

FIGURE 8.16 The landmasses that constituted the Old Sandstone Continent of the late Devonian (365 million years B.P.). Note that reefs covered most of the region that eventually formed North America, and that regions that now lie in temperate zones (such as Germany and the Appalachians) once lay near the equator. After Stanley 1987.)

as well) were not without barriers and ancient, biogeographic divisions. Formidable barriers to biotic interchange during the Triassic included extensive mountain ranges (e.g., the Central Pangaean Mountain Range of the Early Triassic) and the subtropical Pangaean Desert, which at times stretched from central North America to the Amazon Basin and west-central Africa.

THE BREAKUP OF PANGAEA. Great evolutionary radiations of most terrestrial families and many orders, especially among vertebrates, seed plants, and insects, occurred after the breakup of Pangaea and Panthalassa. In fact, the diversification of many taxa may have been a direct outcome of this breakup and the resultant isolation of terrestrial and marine biotas. As continents fragmented and drifted toward different latitudes and climates, reduced gene flow and altered selective pressures allowed rapid speciation and radiation of their rafting biotas. Again, the history of tectonic events provides invaluable insights for interpreting biogeographic patterns, both past and present.

THE BREAKUP OF LAURASIA AND ITS RIFTING FROM GONDWANALAND. The breakup of Pangaea was actually initiated in the early to mid-Jurassic (about 180 million years B.P.) when the Turgai Sea (Figure 8.15E) expanded southward from the Arctic to split Asia from Euramerica (i.e., eastern from western Laurasia). By the late-Jurassic, a shallow "Atlantic Sea" developed and began to separate North America from Europe (see Figure 8.15E–G and Figure 8.17). By this time, active seafloor spreading had begun to open the Atlantic, and the remaining narrow landbridges between Laurasia and Gondwanaland (near Central America and North Africa) were broken, transforming the Tethys Sea into a circum-equatorial seaway. This drifting also marked the origin of the Gulf of Mexico as North and South America separated, while extensive volcanic activity along the eastern margins of the African plate formed the initial basins of the Indian Ocean. Pangaea had fragmented into three great landmasses (Laurasia in the Northern Hemisphere, and Eastern and Western Gondwanaland to the South), and these continued to split into smaller and increasingly more isolated continents as the numerous, incipient ocean basins continued to expand throughout the Jurassic to mid-Cretaceous Periods (roughly 215 to 100 million B.P.).

By the late Cretaceous (75 million years B.P.), a land connection between Laurasian landmasses had been reestablished across the Bering Strait (see Figure 8.15G). Thus, at about the same time that one land connection between eastern North America and western Europe was severed, a new one formed between western North America and Asia, permitting continued exchange of organisms. This connection, Beringia, persisted through most of the Cenozoic, although an uninterrupted landbridge was not continually present (see Chapter 9). Transient connections among the remnants of Laurasia (between North America and Europe) also existed across the Greenland-Scotland Ridge during the early Tertiary (i.e., the Thulian and DeGeer routes). Paleontological evidence indicates that these connections across the North Atlantic were important corridors for the migration of animals and plants (McKenna 1972a,b; Sclater and Tapscott 1979; Barron et al. 1981). Earlier, biotic exchange also occurred across transient land connections that often linked the British Isles with Greenland and Canada during the early Cretaceous (Hallam and Sellwood 1976). As a result of this long history of exchange between the Nearctic and Palearctic regions, they share many species of plants and animals, and have often been coupled as a superregion, the Holarctic.

Other areas of the Nearctic have a much more recent history. As North America drifted away from Europe, its southeastern coast, including land

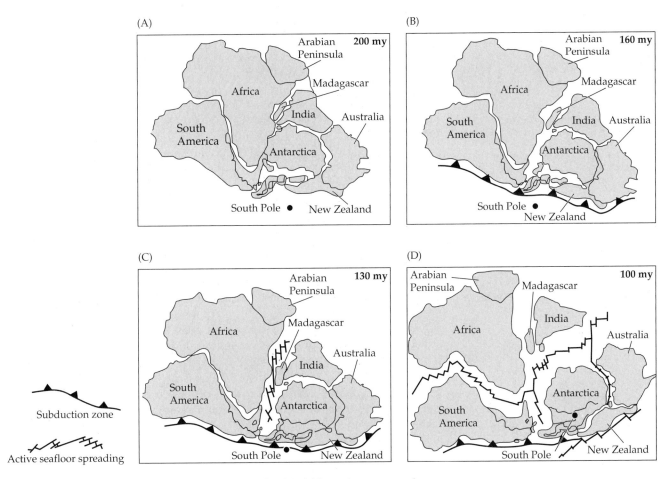

FIGURE 8.17 Stages in the breakup of Gondwanaland. Bold lines show zones of seafloor spreading. (After Storey 1995.)

from New Jersey to Yucatán, Mexico, became submerged to form a great sedimentary basin. The Atlantic Gulf Coast was under water from the late Jurassic until the Eocene, so all of the forms living there today are recent colonists. Similarly, it now appears that the Caribbean Plate formed and then expanded as the North American and South American Plates separated during the late-Jurassic (about 150 million years B.P.; see Figure 8.15E–F). The origins of the Greater and Lesser Antilles continue to be debated, but it appears that they formed de novo as a result of tectonic and volcanic activity, and not as portions of an ancient Central American Archipelago or a portion of the mainlands of North or South America. Therefore, it appears that most of the biota of these islands, including a large number of vertebrates that went extinct during the Holocene, colonized these islands during multiple episodes of over-water immigrations (see Briggs 1995; Meschede and Frisch 1998; Hedges 1996, 2001).

The final punctuation mark to the tectonic development of this region of the New World was the emergence of the Central American Landbridge, which allowed biotic exchange between the long-isolated Nearctic and Neotropical Regions, something the distinguished paleontologist George Gaylord Simpson termed **The Great American Interchange**. The origins, spread, diversification and extinction of species across these two regions and throughout the Caribbean is indeed a complex, but fascinating account of the relationships between the dynamics of the Earth, its climate, oceans, and biotas, and so we further explore the current understanding of these events in Chapter 10.

THE BREAKUP OF GONDWANALAND. Roughly simultaneously with the initial breakup of Laurasia, the landmasses of Gondwanaland began to separate. According to the review by Storey (1995), separation of Gondwanaland proceeded in three stages (see Figure 8.17). First, during the early Jurassic (about 180 million years B.P.), rifting opened the Somali, Mozambique, and Weddell Seas between eastern and western Gondwanaland. This initial rifting was followed by seafloor spreading, which united these relatively narrow seas and split Africa and South America from the remainder of Gondwanaland (Madagascar, India, Australia, Antarctica, and New Zealand) about 160 million years B.P. At this time, Madagascar and India were adjacent to the northeastern margin of Africa, while Australia was located to the south along the eastern margin of Antarctica in a position rotated approximately 90° clockwise from its present position (see Figure 8.17B).

During the second stage of the breakup, Madagascar remained joined to India as this landmass separated from the Antarctica-Australia landmass by a narrow seaway (about 155 million years B.P.; Briggs 2003). Active seafloor spreading had begun by about 130 million years B.P. as India–Madagascar separated from its connection to Antarctica (see Figure 8.17C). Also at about this time, South America and Africa rifted apart, thus forming four Gondwanan landmasses (South America, Africa, Madagascar-India, and Antarctica–Australia–New Zealand). Madagascar then separated from India and the Seychelles landmass approximately 90 million years B.P. This process of global-scale fragmentation of Gondwanaland continued during the late-Creataceous and throughout the Paleocene (100 to 58 million years B.P.), when oceanic basins such as the Mascarene continued to expand, which in this case separated Madagascar and the Seychelles archipelagoes from India (see Figure 8.15F–H). On the other hand, extensive drifting of continents also resulted in collisions and mixing of biotas that had been isolated since the initial rifting of Pangaea. India's migration northward was very rapid indeed, averaging 15 to 20 cm per year, and by about 50 million years B.P., it had collided with Eurasia (Scotese 2004). The period of biotic interchange may, however, have been quite punctuated since the re-connection of these remnants of Pangaea was soon followed by formation of one of the world's most formidable barriers to interchange among terrestrial biotas—the Himalayas—when India collided with the Eurasian plate around 60 million years B.P. (see Figure 8.15H; Butler 1995; Briggs 2003).

Much farther to the south, Australia and New Zealand rifted from Antarctica about 100 million years ago, and extensive seafloor spreading continued to disperse these and other remnants of Gondwanaland during the late-Cretaceous Period. New Zealand retained a connection to Antarctica and what is now the eastern margin of Australia until about 90 million years B.P. New Zealand broke away, first from Australia and then from Antarctica, about 80 million years B.P., drifting farther east of Australia during development of the Tasman Sea. While other remnants of Gondwanaland were drifting toward the lower latitudes, Antarctica continued its slow poleward journey. By the Miocene, (24 million years B.P.) Antarctica was situated over the South Pole, triggering the development of the great polar icecap.

Thus, these reconstructions of the tectonic history of Gondwanaland go a long way toward explaining both the similarities among biotas of now isolated remnants (e.g., those of India, the Seychelles, and Madagascar as well as those of New Zealand and Australia) along with the distinctiveness and endemicity of these biotas.

Throughout the Cretaceous Period (65 to 2 million years B.P.), extensive drifting and collisions of long-isolated continental landmasses continued, frequently resulting in lithospheric uplift and formation of extensive mountain ranges. While new ocean basins expanded, others were subducted and ancient seaways, such as the circum-equatorial Tethys, were transformed into isolated seas. In addition to uplift of the Himalayas (discussed above) collision of drifting landmasses during the mid- and late-Cretaceous formed the Pyrenees Mountains (when landmasses now occupied by France and Spain collided), the Alps (formed by the collision of Italy with France and Switzerland), the Hellenide and Dinaride Mountains (formed by the collision of Greece and Turkey with the Balkan States), and the Zagros Mountains (formed by the collision of Arabia with Iran). Again, each of these collisions of continental plates were likely followed by waves of relatively rapid biotic interchange, which likely decreased as uplifting proceeded to replace an oceanic barrier with a mountainous one.

HISTORY OF CENTRAL AMERICA AND THE ANTILLES.   No matter what brand of historical biogeography you practice, the history of Mesoamerica and the Caribbean region is a complex and intriguing case study. The reasons for this are obvious. First, biogeographers need to know how much opportunity there was for the exchange of organisms between the Nearctic and Neotropical regions following the breakup of Pangaea. Second, we need to know whether the rich biota of the Caribbean islands, especially the Greater Antilles, arrived there exclusively by long-distance dispersal, as suggested by Darlington (1938) and many others, or whether the islands are remnants of an ancient landmass that at one time was continuous with the mainland (i.e., either North or South America).

Following the formation of Pangaea, the precursors of North and South America remained connected until about 160 million years B.P. The ancient connection between Mexico proper and South America was apparently severed by the late Jurassic (approximately 150 million years B.P.). During the early Cretaceous (120 to 140 million years B.P.), a chain of volcanic islands began to emerge along the eastern edge of the Caribbean Plate. By the late Cretaceous, this plate had drifted eastward so that this archipelago, called the Proto-Antilles, was in a position to serve as a stepping-stone route for limited exchange of some terrestrial organisms between the Nearctic and Neotropical regions (Figure 8.18). The Proto-Antilles continued to drift eastward along the leading edge of the Caribbean Plate to eventually form the Greater Antilles, the Caribbean Mountains of Venezuela, and the ABC (Aruba, Bonaire, and Curaçao) islands off the coast of Venezuela. Some small fragments of the Proto-Antilles may also have been displaced to the region of the Lesser Antilles. The core of the Greater Antilles had achieved their present positions by Eocene times (58 million years B.P.), and the Caribbean Plate collided with the Bahamas Platform by 55 million years B.P. (Figure 8.15H; Scotese 2004).

Formation of the current landbridge between North and South America (Nearctic and Palearctic Regions) was an important event for New World biogeography. Waves of migration across this landbridge, called the **Great American Interchange**, profoundly affected the diversity and composition of the Nearctic and Neotropical faunas (see Chapter 10). Like the Proto-Antilles, the Central American landbridge may have first emerged during the late Cretaceous (80 to 65 million years B.P.) as a chain of islands far west of its current position. It then drifted eastward with the boundary between the Cocos and

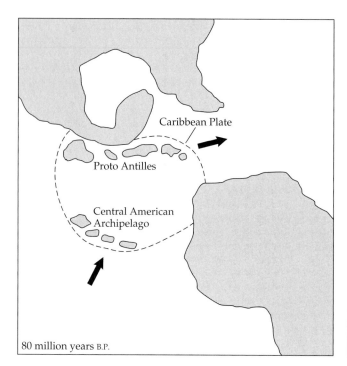
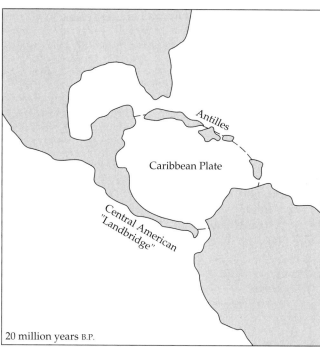

80 million years B.P.

20 million years B.P.

FIGURE 8.18 One possible reconstruction of tectonic events in the Caribbean. Central America first formed as an archipelago in the Pacific Ocean during the early Cretaceous (120 to 140 million years B.P.), then continued to drift eastward along with the Caribbean Plate and the Proto-Antilles. The Central American Archipelago eventually drifted to its position between the two continents by the mid-Miocene (20 million years B.P.), but did not form a complete landbridge until the late Pliocene (around 3.5 million years B.P.). (After Briggs 1994.)

Caribbean Plates (Briggs 1995; see Figure 8.18). At the end of the Cretaceous (65 million years B.P.), the sea level fell substantially to expose a continuous, albeit narrow, landbridge. This scenario remains hypothetical, but it is consistent with available biogeographic evidence, which suggests isolation of marine invertebrates on either side of the Central American rise simultaneous with the Paleocene exchange of some mammals, lizards, frogs, and freshwater fishes between Neotropical and Nearctic regions (see Gayet et al. 1992; Briggs 1994). This hypothesized Cretaceous-Paleocene landbridge, if it ever existed as a continuous strip of land, must have quickly submerged beneath rising waters during the late Paleocene.

The final emergence of the Central American landbridge in the Neogene was produced by convergence of the Cocos, Nazca, and Caribbean Plates (see Figures 10.21 and 10.33). By the late Miocene (10 to 5 million years B.P.), the Central American archipelago provided a stepping-stone route for the dispersal of a variety of terrestrial organisms. Approximately 3.5 million years B.P., the archipelago finally fused to form the current Central American landbridge between the Nearctic and Neotropical regions (see Robinson and Lewis 1971; Coney 1982; Stehli and Webb 1985; Briggs 1994, 1995).

Tectonic Development of Marine Basins and Island Chains

We commented earlier on the unfortunate tendency of biogeographers to ignore tectonics of the marine realm, except as it relates to biogeography and paleontology of terrestrial biotas. Yet the marine realm, in addition to creating major barriers to dispersal for terrestrial organisms and influencing climates on continents and islands, constitutes most of the biosphere. As noted earlier, when the Earth's landmasses were consolidated to form Pangaea, its marine basins also were joined to form one world ocean—Panthalassa. As Pangaean landmasses split apart, seafloor spreading created new marine basins at the expense of Panthalassa, which was fragmented and isolated by the drifting continents. Just as terrestrial climates are strongly influenced by adjacent oceans, changes in the positions of continents changed paleocurrents dramatically, strongly influencing the temperatures and basic chemistry of marine waters. Often throughout the Earth's history, these climatic changes have produced glacial cycles, which lower sea levels during their peak phases and raise sea levels to flood the continents during interglacial periods.

Again, a sound understanding of its unique tectonic history is essential for anyone studying biogeography of the marine realm. While continental plates have drifted, broken apart, and united, their total expanse has changed very little over the past 650 million years. In contrast, the ocean floor is continually created at the midoceanic ridges and consumed at the trenches. Consequently, most of the present ocean floor is less than 50 million years old. In the following sections we summarize some of the major events associated with the formation and disintegration of marine basins and shallow, epeiric seas, and then discuss associated phenomena, including the formation of island chains and the dynamics of marine currents.

EPEIRIC SEAS.   Epeiric seas, also called epicontinental seas, are formed when sea levels rise and oceans flood continental plates. We happen to be living today in one of the driest periods in the history of the Earth. Anyone who has hunted for fossils in the interior of the United States or Eurasia, however, knows that vast seas covered these areas many times, leaving the fossilized remains of marine organisms and thick deposits of limestone, formed from the calcium carbonate of ancient coral reefs.

Epeiric seas usually act as barriers to terrestrial organisms, subdividing a landmass into smaller emergent regions. In the early Cretaceous (about 50 million years B.P.), for example, epeiric seas divided Australia into three subcontinents. Such events had important biogeographic consequences for biotic differentiation and extinction (see Chapters 10, 11, and 12). Some of the more interesting examples since the Paleozoic are marine transgression in North America, which extended from the Gulf of Mexico as far west as Arizona, and the subdivision of mainland Africa (Cooke 1972) (see Figure 8.15G–H). The mid-American seaway dried up by the Paleocene (about 65 million years B.P.), while the Turgai Sea (which isolated Europe from Asia), dried up during the Oligocene (about 30 million years B.P.), again allowing biotic exchange among terrestrial biotas of Europe and Asia. India is one of the few large regions that have had relatively little change in its coastal outline throughout the Phanerozoic.

On a smaller scale, we can find important changes in river systems. In South America, the Amazon had a complex history, changing its drainage several times, from westward in the early Cenozoic to its present eastward drainage into the Atlantic following uplift of the Andes. River systems such as the Mississippi and the St. Lawrence in the United States also have changed course many times. In the western United States, the Great Basin, which includes much of Utah, Nevada, and eastern California, now has no drainage to the ocean—only numerous alkali sinks and dry lake and river beds. However, during wetter periods over the last million years, there were connections to both the Colorado River in the south and the Columbia River (via the Snake River) in the north.

FORMATION OF THE MEDITERRANEAN AND RED SEAS.   After being freed from North America and South America, Africa began to swing counterclockwise toward

Eurasia and it eventually closed the formerly extensive Tethyan Seaway. A bridge was formed between Asia and Africa through Arabia following their collision in the middle Tertiary (35 million years B.P.; Jolivet and Faccenna 2000), which not only created the Zagros Mountains of Iran, but also confined the Mediterranean Sea along its eastern margin. As Africa approached southern Europe, a number of deformations were initiated, including the counterclockwise rotation of Italy, which produced its characteristic diagonal orientation (McElhinny 1973a; see also Hsü 1972; Biju-Duval and Montadert 1977; Bureau de Recherches Géologiques et Minières 1980a,b; Ghebreab 1998). By the early Miocene (24 million years B.P.), the Straits of Gibraltar had closed, causing desiccation and a 6% increase in salinity in the Mediterranean (see Figure 8.15I–J). Seafloor spreading then accelerated during the early Miocene (21 to 25 million years B.P.). The Straits of Gibraltar rifted apart during the late Miocene to open the Mediterranean Sea to the Atlantic. Rifting proceeded more rapidly in the south, resulting in an angular separation of the African and Arabian Plates hinged along their northern juncture. This created a large basin continuous with the Indian Ocean to the south, but isolated from the Mediterranean.

While the Mediterranean formed through the closure of a once expansive, circum-equatorial seaway, the Red Sea formed de novo through updoming and rifting of continental landmasses (Wegener 1929; Coleman 1993), which began approximately 30 million years B.P. (Figure 8.19; Ghebreab 1998). The rift system that created the Red Sea seems continuous with Africa's Great Rift Valley. If current trends continue, the tectonic events creating the Red Sea may be repeated in eastern Africa, with its great lakes eventually connecting to form an extensive inland seaway continuous with the Indian Ocean. The biota of eastern Africa would then drift in isolation, much like that of India during its northward migration from Gondwanaland, providing a fascinating, natural biogeographic experiment for future biogeographers.

DYNAMICS OF THE PACIFIC OCEAN. Early articles on continental drift focused mainly on continental plates, but soon considerable attention was focused on oceanic plates in the Pacific basin, which contain relatively little emergent land. It may surprise you to learn that our largest ocean, the Pacific, has been getting smaller; it was once part of the continuous global ocean, Panthalassa, which covered two-thirds of the globe when Pangaea existed.

HOTSPOTS AND TRIPLE JUNCTIONS. In the basic model of seafloor spreading, volcanic islands are formed either at midoceanic ridges in files perpendicular to the ridge, or behind oceanic trenches as parallel arcs of islands (see Figure 8.6). In the Pacific Ocean, however, there is no single central ridge, and many of the islands are not associated with ridges or trenches. The Hawaiian Islands are a perfect example: this narrow, linear island chain is located far from any ridge or trench. The oldest emergent island is Midway, and the youngest is Hawaii, which remains volcanically active. J. Tuzo Wilson (1963a) proposed a unique mechanism to explain this pattern, which he called a hotspot: a fixed weak spot in the mantle at which magma is released. As the oceanic plate passes over this hotspot, volcanoes are produced at the surface, causing the formation of islands (Figure 8.20). The spot is very narrow, so the islands form in a chain, which is linear as long as the plate moves strictly in one direction. The youngest islands in the chain are emergent, and the progressively older ones gradually become submerged (see also Jarrard and Clague 1977; Wagner and Funk 1995: 14–29).

One nice feature of Wilson's model was that it explained the occurrence of most islands in the eastern Pacific. The Hawaiian Islands—many of which are

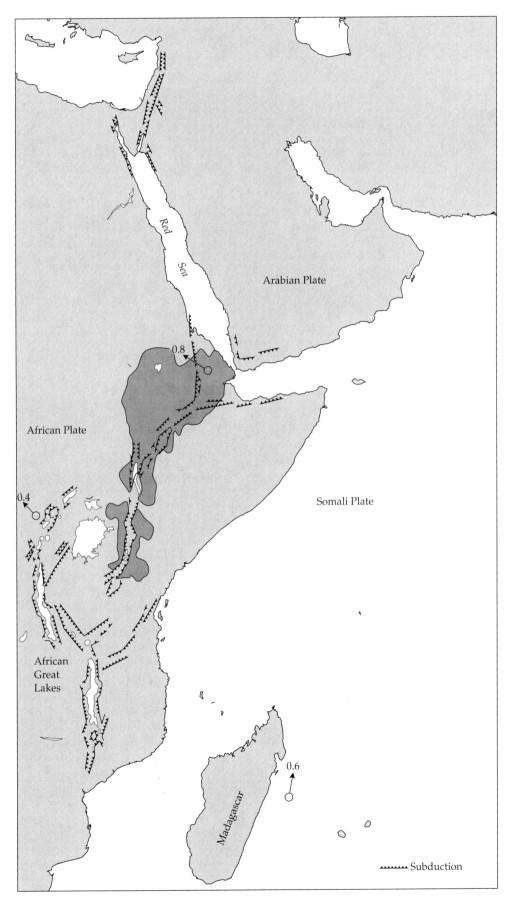

FIGURE 8.19 Tectonic activity in the Great Rift Valley system of eastern Africa. Arrows indicate directions of plate movements; rates of movements are in cm per year. Dark gray areas indicate lava flows. The Rift Valley and its Great Lakes formed as the African Plate separated and rotated away from the Somali Plate. (From Jordan 1971, after Goguel 1952.)

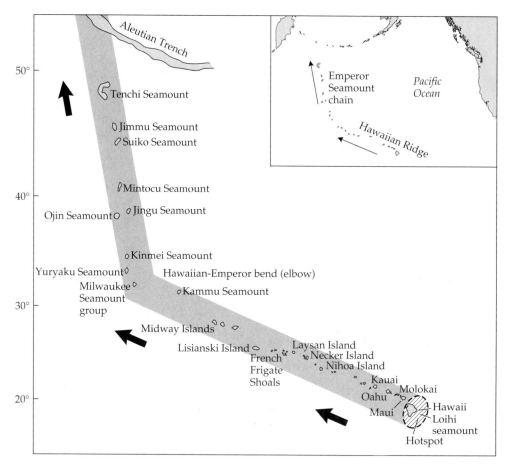

FIGURE 8.20 A hotspot has produced the volcanic Hawaiian Islands and the Emperor Seamount chain in the Central Pacific. The hotspot, presently located near the Hawaiian Deep, is presumably responsible for the intense volcanic activity on the island of Hawaii. Several authors agree that these islands were created as their locations on the Pacific Plate successively drifted over the hotspot. The older islands to the north were formed first; they have disappeared beneath the surface to become seamounts, and the oldest have been subducted into the Aleutian Trench. The sharp bend between the Emperor and Hawaiian chains occurred when the drifting oceanic plate changed direction about 44 million years B.P. (After Jackson et al. 1972.)

still emergent—are really only the most recent part of a long series extending to the Aleutian Trench, including the submerged Emperor seamount chain (Morgan 1972a; Dalrymple et al. 1973). The seamount closest to the trench is naturally the oldest (about 80 million years B.P.). Morgan (1972a,b) noted that three very similar chains of emergent and then submerged volcanoes occur in the Pacific basin: the Tuamoto Archipelago (including Easter Island) and the Line Islands chain; the Austral chain and the Marshall–Gilbert–Turalu island chain; and a chain from the Cobb Seamount opposite Washington through the Gulf of Alaska. Another interesting quality of three of these four chains is that the recent islands are all roughly parallel; then there is a sharp bend (elbow), so that all the older sections are parallel again (see Figure 8.20). This is the pattern you would expect if a rigid plate started moving in a different direction as it passed over a stationary hotspot. The Hawaiian–Emperor elbow is dated at 44 million years B.P. Consequently, we can use the distance and orientation of the chains to reconstruct their movements through geological time.

Volcanic islands may also be formed at a **triple junction**, a place where three plates rest against each other to form a complex of trenches (McKenzie and Morgan 1969; see also Henstock and Levander 2003). Volcanoes are often associated with triple junction. Because each plate has its own rate of subduction into a trench, the triple junction and its associated volcanic activity can shift position through time, forming an island arc, or archipelago. An example of an archipelago formed in this manner is the Galápagos, which originated near the intersection of the Nazca, South America, and Cocos Plates.

INFERRED HISTORY OF THE PACIFIC BASIN. Using the data on hotspots and triple junctions, many researchers have attempted to reconstruct the early history of the Pacific basin. Most such reconstructions recognize six ancient plates. The Phoenix Plate was located in the southern Pacific and disappeared due to the enlargement of the Pacific Plate, and the Kula Plate has disappeared under Alaska and Siberia. A triple junction existed at the intersection of the Kula, Pacific, and Farallon Plates, which migrated to come into contact with the North American Plate; at 150 million years B.P.; this junction was just east of the Hawaiian hotspot. Most of the Farallon Plate has been subducted under western North America, and is responsible for much of the mountain building there, but small pieces of this plate remain in the vicinity of the Pacific Northwest (the Juan de Fuca Plate and its southern extension, the Gorda Plate) and opposite central mainland Mexico (called the Rivera Plate). The Pacific Plate also became fused with a section of mainland Mexico to become Baja California, while its northward movement began the formation of the Gulf of California. This gulf has been spreading over the last 4 million years. The convergence of the Cocos and Caribbean Plates has produced the present mountainous backbone of Central America, and the meeting of the Nazca and South America Plates induced the formation of the Andean Cordillera, mostly since the Oligocene (24 million years B.P.).

PALEOCLIMATES AND PALEOCIRCULATIONS. As discussed in Chapter 3, global circulation patterns of wind and ocean currents are controlled not only by the equatorial to polar gradients of solar radiation and temperature, but also by the relative proportions and distributions of land and water. Given great shifts, splittings, collisions, submergences, and emergences of landmasses that occurred, oceanic currents—and thus terrestrial climates—must have changed dramatically through the Phanerozoic Eon, strongly affecting both marine and terrestrial biotas (Harris 2002; Table 8.2).

During the Permian, surface currents and gyres in the Tethys Sea and throughout Panthalassa were replaced by circulation through a warm, tropical, circum-equatorial Tethyan Seaway, which formed during the late Cretaceous (Figure 8.21A–B). Later, as Australia separated from Antarctica during the late Eocene (about 40 million years B.P.), cold currents became established between those two continents, deflecting the relatively warm South Pacific gyre, which had until then moderated the water temperatures and coastal environments of southeastern Australia and northeastern Antarctica (Figure 8.21C). As it continued to separate from Australia during the Oligocene, Antarctica drifted farther poleward, and cold currents strengthened. Warm circum-equatorial currents of the late Cretaceous were thus interrupted by drifting continental landmasses and replaced by cold circum-Antarctic currents, setting the stage for regional extinctions of thermally intolerant marine invertebrates and for global cycles of glaciation and climatic change.

(A)

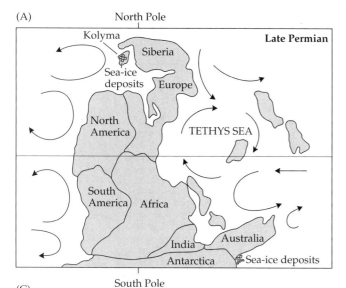

(C)

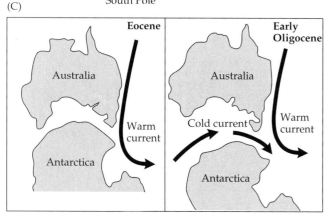

(B)

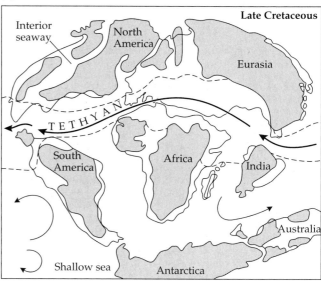

FIGURE 8.21 Continental drift strongly influenced paleocirculations and paleoclimates (see also Table 8.3). (A–B) As a result of the breakup of Pangaea, surface currents throughout Panthalassa (the world ocean) were replaced by flow through the warm, circum-equatorial Tethyan Seaway. (C) In a similar manner, the separation of Australia from Antarctica between the Eocene and early Oligocene epochs established the cool, circum-polar Antarctic seaway. These and many other shifts in paleocirculations dramatically altered the thermal regimes of marine and terrestrial environments, and are often linked to heavy extinctions of stenothermal species. (After Stanley 1987.)

TABLE 8.2 *Key tectonic events of the Cenozoic and their effects on marine waters and paleocurrents*

Event	Effects
Physical Isolation of Antarctica	
Early Eocene (50 million years B.P.)—full deep water separation of South Tasman Rise	Almost complete deep-water isolation of Antarctica. First indications of global cooling at 50–40 million years B.P. and significant 2° C temperature drops in both the late Middle Eocene and Middle-late Eocene boundary. Further isolation of Antarctic marine biota.
Eocene–Oligocene boundary (37 million years B.P.)	Major cooling of both surface and bottom waters by 5° C in 75–100 thousand years B.P. Onset of widespread Antarctic glaciation.
Opening of Drake Passage (~36–23 million years B.P.)	Precise time of opening uncertain, but deep-water connections probably not established until 23–28 million years B.P. Almost complete isolation of Antarctic marine biota.
Middle Miocene (15 million years B.P.)—full establishment of Antarctic Circumpolar Current.	Development of Polar Frontal Zone. Latitudinal temperature gradient similar to that of today.
Closure of Tethyan Seaway in the Middle East	
End of the Cretaceous period (~75–65 million years B.P.)—vast circumpolar--equatorial tropical ocean.	Major westerly flowing equatorial current system. Some faunal differentiation in this vast ocean, but no indications of clear high-diversity foci.
Paleogene (65–23 million years B.P.)—continuity of the tropical Tethyan Ocean.	Largely homogeneous tropical fauna. Major pulse in coral reef development at end of Oligocene (~23 million years B.P.) but even then there were marked similarities between western Tethys (Mediterranean) and Caribbean/Gulf of Mexico.

TABLE 8.2 *(continued)*

Event	Effects
Early Miocene (20 million years B.P.)—closure of Tethyan Seaway by the northward movement of Africa/Arabia landmass.	Westerly flowing tropical current drastically curtailed. Mediterranean Sea excluded from reef belt. Caribbean and eastern Pacific regions become progressively isolated. This marks the beginning of the distinction between Indo-West Pacific (IWP) and Atlantic Caribbean-East Pacific (CEP) foci. Further development through the Neogene (<20 million years B.P.) sees relative impoverishment of ACEP and enrichment of IWP.
Collision of Australia/New Guinea with Southeast Asia	
Beginning of Cenozoic era (~65 million years B.P.)—Australia/New Guinea separated from S E Asia by deep-water gateway (~3000 km wide).	Single tropical (Tethyan) ocean; no differentiation in Indian and Pacific Oceans.
Paleogene (65–23 million years B.P.)—progressive closure of Indo-Pacific gateway; northward subduction of Indian-Australian lithosphere beneath the Sunda-Java-Sulawesi arcs.	
Mid-Oligocene (30 million years B.P.)—gap narrowed substantially but still a clear deep water passage formed by oceanic crust.	
Latest Oligocene (25 million years B.P.)—major changes in plate boundaries when New Guinea passive margin collides with leading edge of eastern Philippines—Halmahera-New Guinea arc system.	
Early Miocene (20 million years B.P.)—continent-arc collision closes deep water passage between Indian and Pacific Oceans.	Major reorganization of tropical current systems; new shallow water habitats appear in Indonesian region.
Early-late Miocene (20–10 million years B.P.)—continued northward movement of Australia/New Guinea; rotation of several plate boundaries and formation of tectonic provinces that are recognizable today.	Widespread growth of coral reefs in IWP region; huge rise in the numbers of reef and reef-associated taxa; many modern genera and species evolve.
Later Neogene (10 million years B.P.–present)—Australia/New Guinea block continues to move north. In Early Pliocene (4 million years B.P.) a critical point is reached where close contact is made with the island of Halmahara.	Warm south Pacific waters deflected eastwards at the Halmahera eddy to form Northern Equatorial Countercurrent. Thus warm waters in Indonesian through-flow are replaced by relatively cold ones from the north Pacific. These changes affect heat balance between east and west Pacific and help promote onset of northern hemisphere glaciation.
Uplift of Central American Isthmus (CAI)	
Mid-Miocene (~15–13 million years B.P.)—sedimentary evidence of earliest phases of shallowing.	Still a deep water connection through CAI.
Latest Miocene—Early Pliocene (6–4 million years B.P.)—continued shallowing to <100m depth.	Major effect on oceanic circulation; Gulf Stream begins to deflect warm, shallow waters northward along eastern US seaboard. Initially leads to some warming in northern mid to high latitudes; however, this water is slightly hypersaline and sinks when it reaches the highest latitudes to form the North Atlantic Deep Water. This then spreads into both the South Atlantic and Pacific Oceans to initiate a major 'conveyor belt' of deep ocean circulation.
Mid-Pliocene (3.6 million years B.P.)—further closure of CAI.	North Atlantic thermohaline circulation system intensifies; Arctic Ocean effectively isolated from warm Atlantic. This leads eventually to the onset of northern hemisphere glaciation at 2.5 million years B.P.
Late Pliocene (~3 million years B.P.)—complete closure of CAI.	Final separation of shallow water Atlantic/Caribbean and eastern Pacific provinces. A further important effect of the closure of CAI is the reverse flow of water through the Bering Strait; this in turn influences Pliocene–Pleistocene patterns of thermohaline circulation in Arctic and North Atlantic Oceans.

Source: After Crame 2004.

Climatic and Biogeographic Consequences of Plate Tectonics

Plate tectonics, perhaps more than any other phenomenon, has had profound effects on the biogeographic patterns of both terrestrial and marine biotas. Over geological time, new plates have arisen and expanded at the expense of more ancient ones, which have been subducted and consumed within the deeper layers of the Earth's mantle. Plates have varied dramatically in shape and size and, during their existence, have split apart, slid against each other, or collided to feed the mantle's convective cycle.

Surface features of the plates have also varied tremendously over time and space. Many plates were often divided by extensive shallow seas, which harbored a great diversity of marine life while isolating terrestrial biotas. Also, as illustrated in the series of maps in Figure 8.15, extensive mountain ranges often served as barriers to lowland species or as corridors for those adapted to montane habitats. Even during the late Permian, when most landmasses were united, extensive mountain ranges along the convergence zone between Gondwanaland and the Old Sandstone Continent of Laurasia may have effectively isolated their respective biotas. As noted earlier, the Himalayas and the Urals also formed as a result of collisions between continental plates, effectively isolating many species on either side of this topographic barrier. Other mountain ranges formed when oceanic plates were subducted beneath continental plates, which caused great, upward buckling of the land surface (as in the Andes) or the welding of seamounts and guyots onto the continental plate (as in the Olympic Mountains). Finally, other mountain ranges, including island chains, formed as a result of volcanic activity associated with subduction zones or as plates drifted over hotspots in the mantle (as in the Hawaiian Island chain; see Figure 8.20).

Biogeographic consequences of these tectonic events are many-fold, but most stem from the effects of plate movements on the area of landmasses and marine basins, on isolation and opportunities for biotic exchange—or more indirectly—on regional and global climates (Chapters 9 and 12). Since the early development of the field, biogeographers have been well aware of the effects of area, isolation, and latitudinal position on diversity. Across a great variety of taxa, time periods, and ecosystems, diversity increases with area, while diversity and similarity among biotas decrease with increasing isolation. As Hallam (1983) has shown, the diversity of ammonites—a once dominant but now extinct group of marine invertebrates—was strongly correlated with the area of epeiric seas (Figure 8.22). Other paleontological work with marine invertebrates shows that when shallow-water marine communities on either side of the North Atlantic drifted apart, their similarity (i.e., the proportion of shared species) decreased (Figure 8.23) (Fallaw 1979).

Precipitation patterns, paleocurrents, wind, and temperature also must have varied tremendously as continents split, converged, or drifted between the poles. On a global scale, the proportion of landmasses at different latitudes has been nearly reversed since the Silurian (425 B.P.), when most landmasses were situated south of the equator (Figure 8.24). When landmasses drifted over the poles, they triggered cycles of glaciation and changes in sea level (Table 8.3). Glacial cycles of the Pleistocene are now familiar to most biologists, but these events were dwarfed by the magnitude of some earlier glaciations and fluctuations in sea level. Similarly, the formation of Pangaea isolated most of its interior land area from the buffering effects of the ocean, increasing temperature fluctuations and aridity and, in turn, causing the drying of shallow seas.

We have tried to summarize the major tectonic, climatic (including currents and sea levels), and biotic events of the Phanerozoic in Tables 8.2 and 8.3.

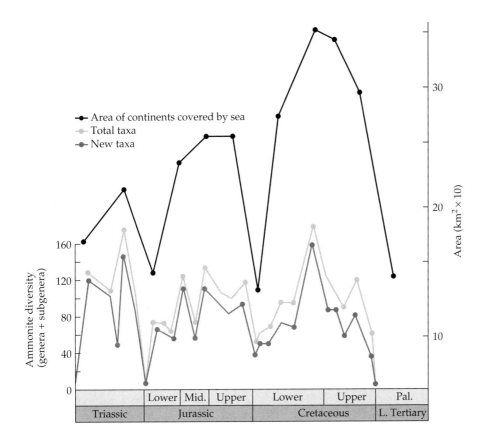

FIGURE 8.22   Many extinction events in the marine realm may be linked to changes in sea level, which dramatically altered the area of the continents covered by shallow epeiric seas—the source for the majority of marine fossils. The diversity of ammonites, once dominant marine invertebrates, appears to be directly correlated with the area of Earth's epeiric seas. (After Hallam 1983.)

Admittedly, this account is far from complete or precise, but it should be instructive. While the mass extinction at the end of the Cretaceous was caused by an asteroid impact, most other extinctions appear to have been associated with tectonic events (see also Ward et al. 2001). The great Permian event (which marked the extinction of over 90% of Earth's species) is associated with the formation of Pangaea, the concurrent drop in sea level, and the drying of shallow seas. The challenge of establishing cause and effect across geological time remains a daunting one indeed, but we should note that Cuvier himself attributed mass extinctions to great changes in relative sea level. We should, however, add a note of caution: Mass extinctions, as spectacular and intriguing as they appear, tend to attract a disproportionate—and perhaps unwarranted—amount of attention. As Stanley (1987) noted, victims of the major mass extinctions include only a small minority of all extinct species:

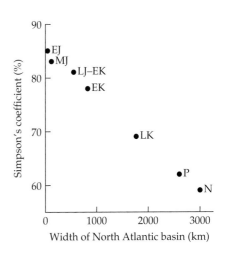

FIGURE 8.23   Seafloor spreading in the North Atlantic caused separation and divergence of shallow-water marine invertebrates during the Mesozoic and Cenozoic. As the North Atlantic widened and the coastal environments on either side became more isolated, the taxonomic similarity of their biotas decreased. Simpson's coefficient of similarity = 100 × (the number of taxa in common/the number in the sample with the smallest number of taxa). EJ, early Jurassic; MJ, mid-Jurassic; LJ–EK, late Jurassic to early Cretaceous; EK, early Cretaceous; LK, late Cretaceous; P, Paleogene; N, Neogene. (After Hallam 1994.)

TABLE 8.3 *Summary of the tectonic, climatic, and biotic events of the Phanerozoic*

MY B.P.	Period	Tectonic events	Climatic/marine events
	Neogene	Central American Archipelago emerges Red Sea forms	Repeated (about 20) glacial events Aridity peak
24	Paleogene	Africa collides with Eurasia	Climate cooling, drop in sea level [Tethys seaway closed, circum-antarctic seaway opens]
65		India collides with Eurasia, Himalayas uplifting Madagascar and Seychelles separate from India Beringia forms	Warming of global climate Tethys circum-equatorial seaway opens
	Cretaceous	Separation of Gondwanaland is complete India and Madagascar drift from Antarctica-Australia Proto-Antilles emerge between Mexico and Colombia	Global transgression of shallow seas Shallow seas recede
144	Jurassic	N. and S. America separate E. and W. Gondwanaland rifting	Global transgression of shallow seas
213	Triassic	Laurasia rifting Formation of Pangaea and Panthalassa	Minor glacial event, sea levels drop, shallow seas dry
248	Permian		Glaciers wane in Gondwanaland
286	Pennsylvanian (Carboniferous)		
	Mississippian (Carboniferous)		
360	Devonian	Old Red Sandstone Continent of Pre-Laurasia forms	Glacial event in Gondwanaland
408	Silurian	Australia near the equator	
438	Ordovician	Most landmasses located in southern or equatorial latitudes	Glacial event in Gondwanaland Glaciation Circum-arctic current established
505	Cambrian		Lowering of sea level
590	Precambrian		
650		Gondwanaland forms	Glacial event

Source: Sea levels after Hallam, 1997.

TABLE 8.3 *(continued)*

Global sea level *(shading marks cooler periods)*	Biotic events
	Extinctions: plankton, marine invertebrates, land mammals Mass biotic exchange between Nearctic and Neotropics *Extinctions:* plankton, marine invertebrates, land mammals *Extinctions:* plankton, marine invertebrates, marine reptiles; dinosaurs Flowering plants become dominant on land *Extinctions:* marine invertebrates *Extinctions:* marine invertebrates, dinosaurs Archaeopteryx *Extinctions:* marine invertebrates ⎤ Earliest birds Proto-Aves Earliest mammals evolve *Extinctions:* sea-floor protozoans, marine invertebrates mammal-like reptiles Mammal-like reptiles evolve Earliest reptiles evolve Vertebrates invade land *Extinctions:* plankton, marine invertebrates, primitive fishes *Extinctions:* marine invertebrates, including reef builders Earliest fishes *Extinctions:* marine invertebrates (trilobites and brachipods) Great diversification of marine life First multicellular organisms

FIGURE 8.24 Continental drift has substantially altered the latitudinal distribution of landmasses and, in turn, has caused dramatic changes in regional and global climates. (After Rosenzweig 1995.)

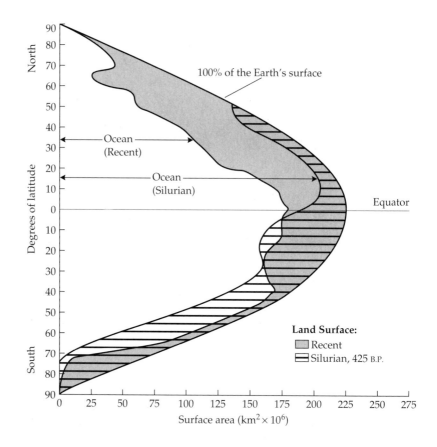

"most of the global turnover of species was very gradual, not catastrophic." On the other hand, tectonic events also set the stage for major radiations of new species, which are evident throughout the fossil record. As noted earlier, many terrestrial vertebrates evolved and spread across the relatively continuous landmasses of Pangaea during the late Permian. However, their major radiations did not take place until Pangaea split, isolating its biotas, which were then free (if they survived) to diversify under different selective pressures and in the absence of gene flow. Few contemporary scientists doubt that tectonic, climatic, sea level, and biogeographic phenomena are causally related (see Fischer 1984).

Our extensive discussions of plate tectonics, ecology, and evolution in the early chapters of this book are recognition of their fundamental importance to biogeography. Again, all of these disciplines are intricately related, and any distinction we attempt to make is surely an arbitrary one. In the next chapter we take a closer look at the relationship between climate and biogeographic events during a relatively thin slice of time—the Pleistocene Epoch (roughly, the past two million years). In Chapters 10, 11, and 12, we return to the theme that is central to both historical biogeography and evolutionary biology—that the Earth and its life forms have evolved in concert (see Lieberman 2000), and that the best approach to understanding the geography of nature—past and present—is to develop a better understanding of the histories of place and of life.

Glaciation and Biogeographic Dynamics of the Pleistocene

HAPTER 8 EXPLORED THE BIOGEOGRAPHIC consequences of events occurring across a broad span of the geological record—the entire 590 million years of the Phanerozoic Eon. As we saw, plate tectonics strongly influenced Earth's climate at all spatial scales, sometimes triggering glacial events even more extensive than those of the recent "ice ages" of the Pleistocene Epoch. For example, glacial events such as those during the Cenozoic Era likely were triggered by a combination of tectonic events that transformed oceanic circulation from predominantly equatorial currents flowing through the Tethyan Seaway to strongly meridional (i.e., north-south) currents following the closure of this circum-equatorial seaway (Chapter 8; see Crame and Rosen 2002; Mitrovica and Vermeersen 2002).

Here we focus on a thin slice of the geologic record—the Pleistocene and Holocene Epochs (i.e., the late Neogene), which include the past 2 million years of Earth's history. While Earth's tectonic plates were relatively stable during this short period, global and regional climates experienced many upheavals, which in turn had profound effects on most biotas. Because these events were so recent, biogeographers and paleoecologists can capitalize on a variety of paleontological methods and data not available to those studying more ancient periods. These methods include analyses of tree rings, pack rat middens, pollen deposits, accumulations of thermally sensitive indicator species in ice cores (e.g., foraminiferans, radiolarians, and coccoliths), lithological evidence (including tillites and evaporites), and the depths of fossilized coral reefs, as well as radiocarbon dating and optically stimulated luminescence. The resulting high-resolution reconstructions of the late Neogene provide us with an intriguing opportunity to explore the relationships between climate and biogeographic patterns of Earth's biotas.

By devoting a full chapter to the Pleistocene Epoch, we recognize that events during that time had profound effects on today's distributions of plants and animals. By making this the second of two chapters on Earth history, however, we are reminding the reader that these more recent changes have not erased the patterns resulting from earlier events. We first review some of the major climatic events of the Pleistocene, then discuss their biogeographic consequences, including range shifts and biotic exchanges, adaptations and evolution, and extinctions. In addition to elucidating the interrelations between historical and ecological biogeography, studies of the climatic and biotic dynamics of the Pleistocene may also provide invaluable insights for predicting biogeographic responses of contemporary biotas to anticipated anthropogenic changes in global climates (see Chapters 16 and 17).

Extent and Causes of Glaciation

As noted in the previous chapter, glacial events were often associated with the positioning of large landmasses over or near the poles (Harris 2002). Thus, as a result of the drifting of continents during the Phanerozoic, the Earth has undergone several periods of extensive glaciation and at times nearly the entire planet may have been covered with ice and snow (i.e., the hypothesized periods of "snowball Earth" [Hyde et al. 2000; Kerr 2000; Lubick 2002]; see Table 8.2). Throughout much of the Mesozoic and early Cenozoic Eras, however, the global climate was relatively warm and equable, with little variation across seasons or latitudes. In contrast, the glacial/interglacial cycles that characterized the late Neogene were dramatic episodes in the biogeographic and evolutionary history of Cenozoic biotas, many of which developed and radiated during an age known as the mid-Eocene "sauna" (Beard 2002; Bowen et al. 2002).

The origins of Pleistocene biogeography may be traced back to Swiss-borne (Jean) Louis Rodolphe Agassiz. After studying comparative anatomy under the renowned paleontologist and systematist Georges Cuvier, Agassiz emigrated to America, but returned to his homeland in 1836 to study glaciers along the slopes of the Alps. Agassiz was struck, not just by the immense mass of the glaciers, but also by their dynamics—sometimes nearly imperceptible, sometimes cataclysmic. From this, he inferred that glaciers are, over long periods of time, highly dynamic—constantly growing at the higher altitudes and sliding down slopes to erode soils and strongly influence vegetation at lower elevations. By combining these and related observations, Agassiz later developed the first comprehensive theory of glaciation, which included the original insight that, like alpine glaciers, those forming Earth's polar ice-caps also are highly dynamic. During an earlier period, which he called the Ice Age, glaciers extended far into subarctic and temperate latitudes of the Northern hemisphere, strongly influencing distributions of plant and animal communities (Agassiz 1840). Generations of paleontologists and climatologists have, of course, confirmed Agassiz's prescient inferences and the fact that a thorough understanding of the climatic dynamics of the Pleistocene is central to understanding the biogeographic dynamics of its biotas.

Although Agassiz hypothesized one "Ice Age," it is now clear that the Earth experienced numerous **glacial-interglacial cycles** during the Pleistocene. These glaciers were incredibly massive sheets of ice. They often were 2 to 3 km thick, and their mass was so great that it deformed the underlying lithosphere by 200 to 300 m. At their maximum extent, these ice sheets covered up to a third of the Earth's land surface. At the height of the most recent glacial period, ice sheets in the Northern Hemisphere extended from the Arctic southward to cover most of North America and Central Asia to approximately 45° N latitude (see Figure 9.1). During these periods of glacial maxima, prevailing winds shifted, sometimes blowing wet oceanic air masses deep into

FIGURE 9.1 Glaciation of the Northern Hemisphere during the most recent glacial maximum (18,000 years B.P.). North-flowing waters of the Gulf Stream tended to warm the North Atlantic, but this same oceanic gyre cooled coastal regions of southern Europe and Africa as it flowed southward. (After Stanley 1989.)

Glacier

Sea ice

Gulf Stream

Tundra

the interior of some continents; therefore, glacial maxima in now arid regions are also referred to as **glacio-pluvial** (ice-rain) periods. In contrast, tropical regions tended to be drier during glacial maxima. Glacial periods alternated with **interglacial** periods, when climate warmed and ice sheets retreated. As the recent paleontological record tells us, these events had profound effects on both terrestrial and marine biotas, including those far removed from the glaciers.

Most authors focus on Pleistocene glaciation episodes in the Northern Hemisphere because over 80% of glacial ice occurred there (see Figures 9.1 and 9.2). In the Southern Hemisphere, where much less land occurs in temperate and subarctic regions, glaciation was mostly confined to high elevations at high southern latitudes, such as the Central Plateau of Tasmania and the New Zealand Alps (see Flint 1971). The Andean Cordillera was glaciated, but the greatest ice coverage was in Chile and Argentina. Mainland Australia was unglaciated except for the Victorian Alps, and Africa lacked glaciation except in the Atlas Mountains of the extreme northwestern corner and in the highest mountains of eastern Africa.

Before turning to the biogeographic effects of these climatic changes, we first focus on their causes. The glacial events of the Pleistocene, unlike those of much earlier periods, were not caused by the drifting of continents. In the past, scientists attributed these events to changes in the output of solar radiation. However, while solar output has varied by as much as 25–30% during Earth's 4.5-billion-year history, it has changed little during the Phanerozoic (the last 590

18,000 years B.P.

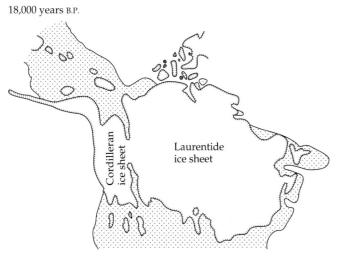

Cordilleran
ice sheet

Laurentide
ice sheet

13,000 years B.P.

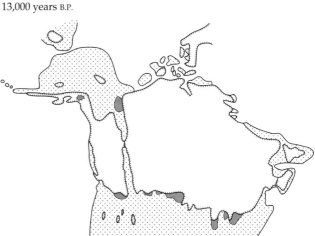

10,000 years B.P.

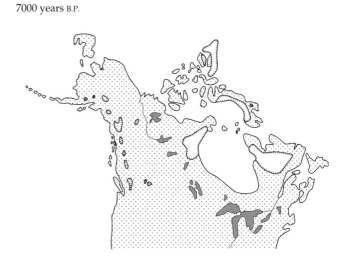

7000 years B.P.

Present

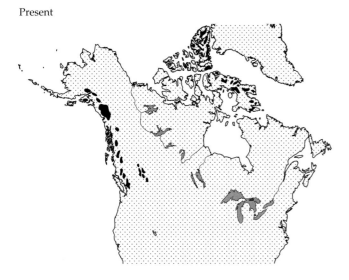

FIGURE 9.2 Sequence of glacial recession in North America since the most recent glacial maximum (18,000 years B.P. to present). White areas with etched borders represent glaciers; stippled areas, unglaciated land surfaces; shaded areas, lakes. Black areas in the present map represent land surfaces currently covered with glaciers. (After Pielou 1991.)

million years: Gates 1993). Instead, the climatic reversals of the Pleistocene were caused by changes in the Earth's orbit, which affected the amount of solar radiation reaching the biosphere (Figure 9.3) (Gates 1993; Muller and MacDonald 1995). Three characteristics of the Earth's orbit around the sun change over time, each with a characteristic periodicity. These changes are called **Milankovitch cycles** after their discoverer, Serbian astrophysicist—Milutin Milankovitch. First, Earth's orbit is not perfectly circular, but varies in ellipticity, or "**eccentricity**," with a period of 100,000 years. Second, the tilt of the Earth on its axis—its **obliquity**—varies from 22.1° to

(A) Eccentricity (ellipticity of orbit)
   Cyclic period = 100,000

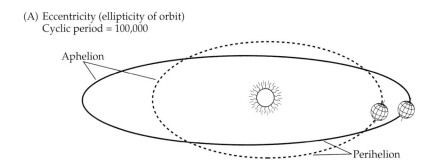

Aphelion

Perihelion

FIGURE 9.3 Milankovitch cycles are periodic changes in the eccentricity, obliquity, and precession of Earth's orbit. Each of these changes influences Earth's interception of solar radiation; therefore, these cycles may have been largely responsible for the glacial cycles of the Pleistocene. (After Gates 1993.)

(B) Obliquity (orbit tilt)
   Cyclic period = 41,000

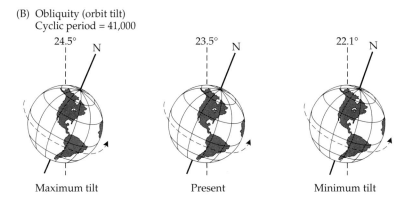

24.5°   N   23.5°   N   22.1°   N

Maximum tilt   Present   Minimum tilt

(C) Precession (pole wandering)
   Cyclic period = 22,000

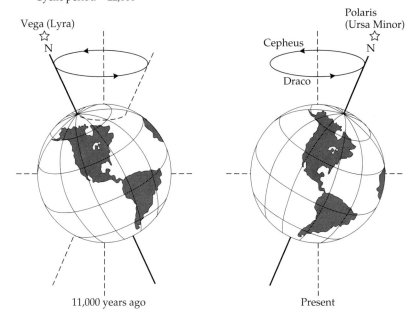

Vega (Lyra)
☆
N

Polaris
(Ursa Minor)
☆
N

Cepheus

Draco

11,000 years ago   Present

24.5° with a 41,000-year period. Finally, Earth's orientation, or **precession**, wanders, with the axis of the North Pole shifting from one "north star" (presently Polaris of Ursa Minor) to another (Vega of Lyra) with a periodicity of approximately 22,000 years. The combined effects of these cyclical changes in the Earth's orbit cause substantial fluctuations in the amount and seasonal variation of solar energy striking the Earth, ultimately combining in a complex fashion to create the glacial-interglacial cycles and climatic reversals of the Pleistocene (Figure 9.4; see Rutherford and D'Hondt 2000).

FIGURE 9.4 Past and predicted changes in Milankovitch cycles. Note that while the periodicities of these orbital phenomena remain relatively constant, their amplitudes vary substantially. (After Gates 1993.)

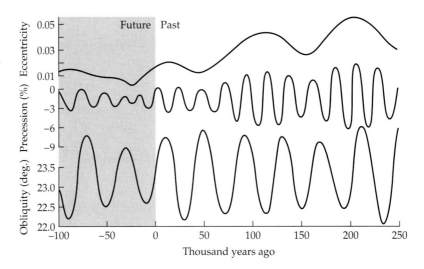

To further complicate this picture, it seems that transitions between glacial and interglacial periods were influenced by some remarkable feedback effects. During the initial stages of glaciation, for example, developing fields of ice and snow increase reflectance (**albedo**) over increasingly larger areas of the Earth. This phenomenon reduced effective solar heating of the planet and further increased cooling rates, causing more rapid glaciation. The reverse process, deglaciation, also seems to be accelerated by feedback and secondary driving forces, including changes in oceanic currents and composition of the atmosphere. It appears that glaciers melted much more quickly than they formed—too quickly to be accounted for solely by the relatively gradual changes in Earth's orbit. Instead, glacial melt seems to have been accelerated by global changes in the concentration of haze particles, water vapor, and especially carbon dioxide and methane. Apparently, global warming accelerated the natural biogenic release of these "greenhouse" gases into the atmosphere. In fact, this natural biogenic warming may have accounted for as much as half of the 4.5° C temperature rise since the last glaciation (Figure 9.5; Gates 1993; Adams et al. 1999).

Effects on Nonglaciated Areas

The steep thermal latitudinal gradient from frozen poles to warm equator that now characterizes our global climate is a relatively recent phenomenon. Throughout most of the Phanerozoic Eon, equatorial climates were tropical, as

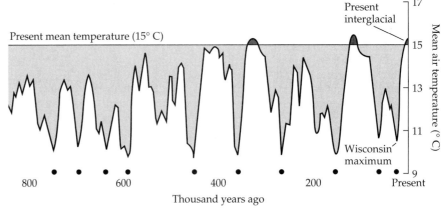

FIGURE 9.5 Estimated average global air temperature over the past 850,000 years, inferred from oxygen isotope measurements from ice cores. Note that the current interglacial represents one of the warmest periods in this record, and that interglacials, in general, tend to be very limited in duration. Circles mark glacial maxima. (After Gates 1993.)

FIGURE 9.6 Records of global temperatures over three temporal scales reveal that climatic change can occur very rapidly. (A) Mean global ocean temperatures over the last 140,000 years show that the last two shifts from glacial maxima to interglacials (4° to 5° C increases in temperature) occurred over a span of just a few thousand years. (B) Air temperatures from Eastern Europe during the past 10,000 years, and (C) mid-latitude air temperatures from the Northern Hemisphere over the past 1000 years show that more recent thermal shifts, while not as dramatic, were also quite rapid, sometimes shifting a degree or two in just a few centuries or even decades. (After Gates 1993.)

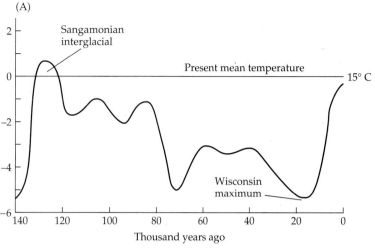

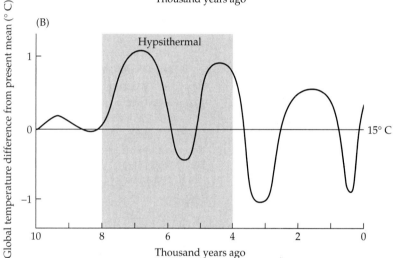

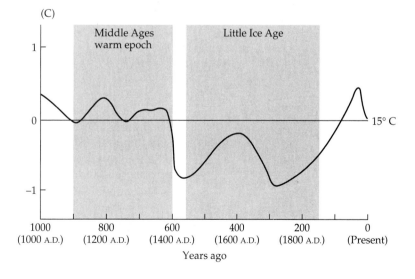

they are today, but latitudinal gradients in temperature were much less pronounced, and terrestrial climates were generally more equable. Beginning in the Miocene, global climates gradually began to cool and become drier. Extension of the Antarctic ice sheet and intensification of oceanic and atmospheric circulation during the mid-Miocene (15 million years B.P.) established a strong latitudinal thermal gradient that intensified during interglacial periods. During glacial maxima of the Pleistocene, average air temperatures were 4° to 8° C cooler than during interglacial periods (Figure 9.6), and cooling appears to have been more pronounced in the lower latitudes.

Due to the high heat capacity of water, global temperature of the oceans varied by just 2° or 3° C between glacial and interglacial periods. On the other hand, atmospheric and oceanic circulation changed dramatically, and climatic zones shifted substantially in latitude and altitude with each climatic flush. In mountainous regions, snow lines shifted as much as 1000 m in elevation between glacial and interglacial periods (Behrensmeyer 1992). Before exploring some of their biogeographic and evolutionary consequences, we summarize some of the major climatic changes associated with glacial-interglacial cycles.

Temperature

Investigators can estimate paleotemperatures by several qualitative methods or, more quantitatively, by determining oxygen isotope ratios in marine fossils or ice cores. Most marine exoskeletons are composed of calcite ($CaCO_3$), made by combining water and carbon dioxide. In water, the two common isotopes of oxygen are ^{16}O and the heavier ^{18}O.

The lighter isotope (^{16}O) evaporates more rapidly, especially during warmer periods. Thus, with some sophisticated calibrations and analysis, the ratio of ^{18}O to ^{16}O locked in the shells of marine organisms or in ice cores can serve as a paleothermometer.

The resultant reconstruction of global temperatures during the past 2 million years reveals that the Earth underwent at least twenty major glacial periods (i.e., periods during which global temperatures were at least 4° C below the current, interglacial temperature). At these times, glaciers in the Arctic expanded into the lowlands down to 40° N latitude, and even farther south along the mountain chains of Eurasia and North America (see Figure 9.1).

As Figure 9.5 indicates, glacial conditions prevailed throughout most of the Pleistocene, with intermittent interglacials accounting for less than 10% of this epoch. A closer look at the thermal record for the Pleistocene and Holocene confirms that climatic change can be remarkably rapid (see Figure 9.6; Jackson 2004). Since the peak of the most recent glaciation—the "Wisconsin" Glaciation in North America or the "Würm" in Europe—global temperatures increased by approximately 4.5° C, with most of that increase occurring within a span of just 3000 years (17,000 to 14,000; see Figure 9.6A). Using higher-resolution reconstructions (see Figure 9.6B), we see that temperatures have varied substantially even during the Holocene. Again, remarkably rapid cooling and warming events ("stadials" and "interstadials," respectively) took place—sometimes occurring over a span of just decades due to positive feedback mechanisms in the global climate system (Adams et al. 1999). The Younger Dryas (12,800 to 11,600) was a period marked by an extremely rapid plunge to cooler temperatures, followed by equally rapid warming to the current interglacial—all of this occurring within just 1200 years (Alley and Clark 1999). Global temperatures from about 8000 to 4000 years B.P.—a period called the hypsithermal—were often 0.5° to 1° C warmer than current temperatures. Following this relatively warm period, global temperatures dropped by 2° C in just 500 years. Finally, global temperatures have varied significantly even over the past millennium (see Figure 9.6C). After the relatively warm Middle Ages (1100 to 1400 A.D.), temperatures cooled rapidly, and much of the world experienced "Little Ice Ages," the most recent occurring between 1650 and 1850 A.D. (Davis 1986; Gates 1993).

Geographic Shifts in Climatic Zones

The glacial-interglacial cycles of the Pleistocene involved shifts not just in temperatures, but in entire climatic regimes. During the most recent glacial period, most nonglaciated regions experienced declines in air temperatures ranging from 4° to 8° C (Figure 9.7). In general, climatic zones shifted toward the equator during glacial periods and poleward during interglacials. The pattern of shifts, however, was complicated by the configurations and positions of land, large bodies of water, and glaciers on atmospheric and oceanic circulation patterns.

Glaciers were so massive that they greatly reduced the flow of cool polar air masses to nonglaciated regions. Even air masses that were not effectively blocked by the glaciers would have undergone substantial adiabatic warming as they descended 2 to 3 km, creating steep thermal gradients at the glacial margins. Thus, despite the generally cooler conditions, glacial winters were less severe, while glacial summers were cooler and less subject to the heat waves that characterize contemporary climates—especially in temperate regions. Recent studies suggest that this pattern of cooler and more equable climates and less pronounced thermal latitudinal gradients during glacial maxima was true of marine environments as well (see D'Hondt and Arthur 1996; Crame and Rosen 2002; Crame 2004).

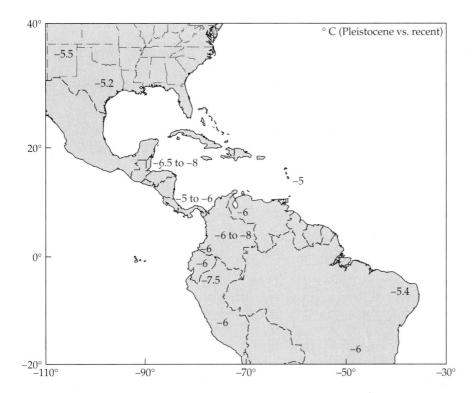

FIGURE 9.7 Glacial cycles of the Pleistocene influenced regional climates far from the edges of the glaciers. Temperatures over much of North and South America, for example, ranged from 4° to 8° C cooler during the Wisconsin. (After Stute et al. 1995.)

An important, but often overlooked, effect of glacial cycles is that as thermal zones shifted, novel combinations of temperature, prevailing winds, ocean currents, and precipitation created climatic and edaphic (soil) zones that lack any contemporary analogues (Jackson 2004). The glaciers caused dramatic changes in prevailing winds and ocean currents, which in turn strongly influenced regional climates. During the Wisconsin, the jet stream of North America was split and diverged around the glacier, and an anticyclonic (clockwise) circulation pattern was established over the Laurentide Ice Sheet (Figure 9.8). As a result, while average global temperature of the oceans cooled by just 2° to 3° C, surface temperatures of waters in the North Atlantic cooled by as much as 10° C (see also Figure 9.12). Along the southwestern edge of the glacier, relatively dry, easterly winds from the interior caused desiccation of lakes in the American Northwest. While cold and dry conditions characterized the glacial climates of other temperate regions, including Europe, other regions were wetter during the glacio-pluvial periods (see Van der Hammen et al. 1971; Wells 1979; Spaulding and Graumlich 1986). Farther south of North America's glaciers, prevailing westerly winds brought saturated oceanic air masses, causing elevated water levels in lakes of the American Southwest (see Figure 9.8A).

Paleoclimatic data also reveal a strong association between glacial events and monsoonal circulation and precipitation. **Monsoons** are driven by differential solar heating of adjacent land and sea during hot summers in tropical and subtropical latitudes. Because of its lower heat capacity, land heats much more rapidly than water. Ascending air masses over land create convective cycles that draw oceanic air masses inland, causing heavy summer precipitation—monsoon rains. Pollen records and global circulation models indicate that the strongest monsoonal events of the Pleistocene coincided with periods of peak summer solar radiation (roughly 126,000, 104,000, 82,000, and 10,000 years B.P.; Gates 1993). Conversely, monsoons were weakest during glacial maxima, often resulting in aridification of otherwise moist tropical regions.

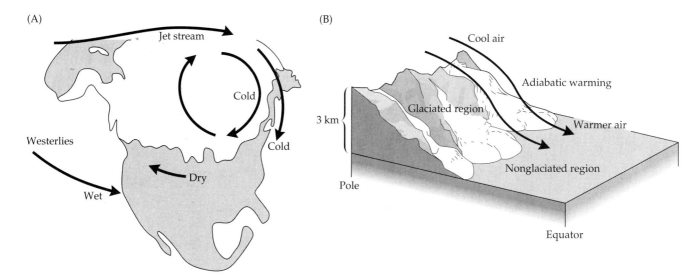

FIGURE 9.8 (A) The Laurentide ice sheet (18,000–9000 years B.P.) caused major changes in prevailing winds, including the jet stream and westerlies over North America, which strongly influenced regional climates. (B) As relatively cool winds descended the 2 to 3 km high faces of the glaciers, they were adiabatically warmed, thus moderating climatic conditions adjacent to the glaciers. (A after Gates 1993.)

In summary, the world we see outside our windows is not typical of the conditions that animals and plants lived and evolved under for the past 2 million years. The current very warm period was perhaps equaled during only 10% of the Pleistocene. While climatic zones tended to shift in latitude and altitude with glacial cycles, the specific nature of these zones varied markedly, often resulting in novel combinations of temperature, precipitation, and atmospheric or oceanic circulation. Not only were climatic zones altered in both character and location, but they frequently were displaced to occupy new soil regimes. Finally, while most climatic zones shifted toward the equator, tropical climates had little opportunity for geographic shifts. The inevitable reduction of what we now view as tropical climatic zones (i.e., the "tropical squeeze") raises an intriguing question: How is it that the biotas adapted to these zones not only survived the cooler climates of glacial cycles, but many maintained or developed a diversity unparalleled by that of any other region on Earth? Later in this chapter and in Chapter 15, we return to this fascinating paradox, along with other biogeographic consequences of glacial events. First, however, we need to focus on one of the most pervasive and most important events associated with glacial cycles: regional and global changes in sea level.

Sea Level Changes in the Pleistocene

Throughout the Pleistocene, sea levels have fluctuated dramatically on both global and regional scales. **Eustatic** changes are global fluctuations in sea level resulting from the freezing or melting of great masses of sea ice, thus decreasing or increasing the global volume of liquid water, respectively. In contrast, **isostatic** changes in sea level occur when portions of the Earth's crust rise or sink into the less buoyant asthenosphere, causing local or relative changes in sea level even when global (eustatic) levels remain unchanged. In some areas of the high northern latitudes, the sheer mass of 2 to 3 km high glaciers caused crustal downwarping by over 300 m. As we shall see, both eustatic and isostatic changes during the Pleistocene strongly influenced the distributions and diversity of biotas.

During the most recent glacial maximum, nearly one-third of all land in the Northern Hemisphere was covered with thick glaciers, and great fields of sea ice occurred in both polar regions (see Figure 9.1). In consequence, a large volume of water was removed from the ocean, probably exceeding the equivalent of 50 million km^3 of ice—more than the present-day ice sheets in Antarctica and Greenland combined. However, the resulting 100 m drop in sea level during the Wisconsin Glaciation was relatively mild in comparison to some earlier glacial maxima, when sea levels dropped by well over 160 m. Similarly, the current interglacial is not the warmest one on record. Sea levels during some of the most intense earlier interglacials may have been as much as 70 m higher than current levels. Thus, the total amplitude of sea level changes during the Pleistocene may have exceeded 230 m. Despite the relative stability of the tectonic plates during this thin slice of the geological record, Earth's biogeographic profile may have been transformed as much as during any period in its history. For example, the lowering of sea level by 100 m during the early Holocene exposed a considerable area of the continental shelf and allowed landbridges to form between biogeographic regions (Figure 9.9). The biogeographic relevance

(A)

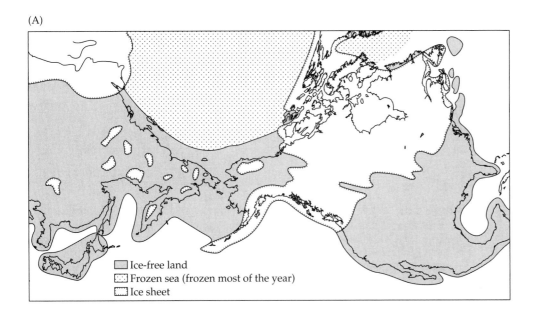

☐ Ice-free land
☷ Frozen sea (frozen most of the year)
☐ Ice sheet

(B)

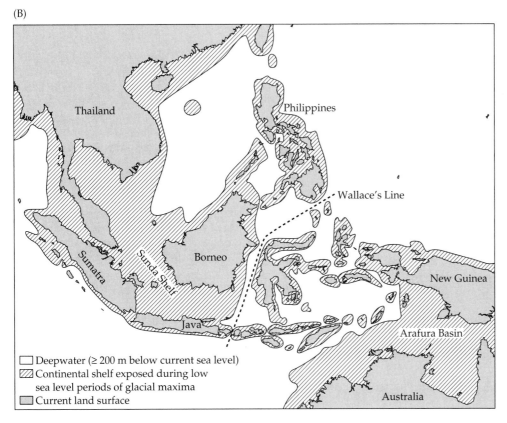

☐ Deepwater (≥ 200 m below current sea level)
▨ Continental shelf exposed during low
   sea level periods of glacial maxima
☐ Current land surface

FIGURE 9.9 Glaciation during the Pleistocene resulted in the lowering of sea levels by 100 m to as much as 160 m below their current levels. As a result, many terrestrial regions and associated biotas now isolated by oceanic barriers were connected during glacial maxima. (A) Beringia connected North America and Asia. (B) Many islands of Indonesia were connected to mainland Asia and Australia, respectively. Wallace's line, marking a division between the biotas of Southeast Asia and Australia, coincides with the division between these glacial landmasses. (A after Pielou 1991; B after Heawey 1991, 2004.)

(A)

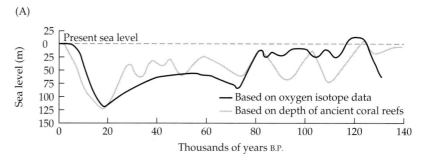

(B)

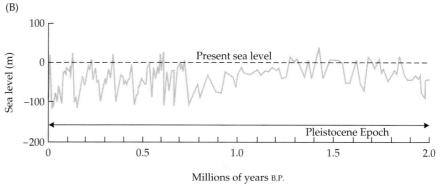

FIGURE 9.10   (A) Changes in global sea level during the past 140,000 years based on oxygen-isotope data (black line) from analysis of benthic foraminifers found in deep sea cores of the Caribbean, and raised coral reefs (gray line) in New Guinea. In addition to these global (or eustatic) changes in sea levels, regional sea levels may vary substantially as Earth's crust rises and sinks in the asthenosphere. Such isostatic fluctuations in sea level can occur even when global levels remain unchanged. (B) Changes in global sea level throughout the Pleistocene (i.e., the past 1.8 million years). This figure illustrates the rapid transitions between full glacial and interglacial periods. (A after Hopkins et al. 1982; B after Dyer 1986.)

of these events is difficult to overstate. Wallace's line, for example, which separates the Oriental and Australian biogeographic regions, coincides with the division between the glacial landbridges of the Sunda Shelf and Australia-New Guinea-Tasmania.

In Figure 9.10 we summarize global, or eustatic, changes in sea level. For many regions, however, isostatic changes have been just as large, with one important difference: isostatic changes tend to lag far behind eustatic changes. It takes many centuries for downwarped crust to rebound after glacial recession. Consequently, as temperatures rise and glacial meltwater accumulates, rising seas often spill over onto the downwarped regions of continents, creating extensive shallow seas. During the early Holocene (about 12,000 to 10,000 years B.P.), for example, the Saint Lawrence River Valley and Great Lakes of North America were inundated with marine waters from the Atlantic (Figure 9.11A). This saltwater corridor provided a dispersal avenue for shoreline flora of the Atlantic coast (e.g., beach plants including *Ammophila breviligulata*, *Cakile edentula*, and *Euphorbia polygonifolia*; coastal bog species such as *Xyris caroliniana*; aquatic macrophytes such as *Utricularia purpurea*: Pielou 1991). Subsequent rebound of Earth's crust by as much as 275 m drained this corridor and transformed it into the freshwater Saint Lawrence River, which drains the Great Lakes. Events such as this may account for the disjunct ranges of many species adapted to coastal environments (Fig 9.11B).

(A)

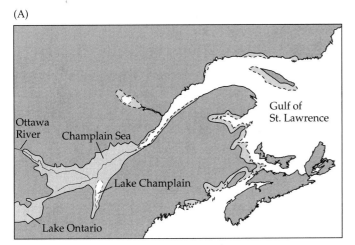

(B)

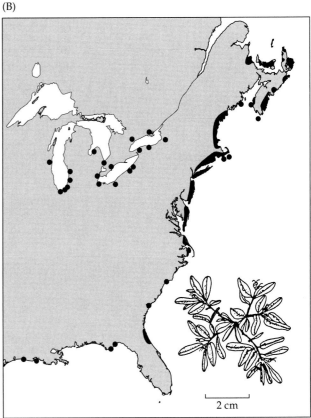

FIGURE 9.11 During the most recent glacial retreat, sea levels rose relatively rapidly, while previously glaciated land surfaces were slow to rebound from the downwarping force of the massive glaciers. As a result, marine waters spilled over onto these low-lying regions of the continents, creating shallow seas and providing saltwater corridors for the migration of plants and animals. (A) About 10,000 years B.P., shallow seas covered the Saint Lawrence River Valley, Lake Champlain, and the Ottawa River, forming the Champlain Sea (dashed line depicts modern coastline; shading represents area covered by shallow seas). (B) The existence of such transient corridors explains the disjunct distributions of many shallow-water, marine, and estuarine species, such as the seaside spurge (*Euphorbia polygonifolia*; distribution indicated by dark areas), along the coastlines of eastern North America and the Great Lakes. (After Pielou 1979.)

In summary, glacial events and associated changes in sea level altered dispersal routes among non-glaciated regions far from the ice sheets and long after the glaciers receded. During glacial maxima, the once continuous ecosystems of northern regions were often isolated across widely scattered pockets of unglaciated land, while other, long-isolated biotas mixed across formerly submerged landbridges. Although these landbridges provided dispersal routes for terrestrial biotas, they created barriers for marine life. Glacial cycles thus created great, alternating waves of biotic exchange and isolation (vicariance) among both terrestrial and marine biotas. Before exploring biogeographic responses to these changes in the geographic template, we re-emphasize that, as Stephen Jackson (2004, page 52) puts it, "… the environment of each glacial period and interglacial period was unique … there is no modal condition for the past century, the past millennium, the Holocene, or the Quaternary."

Biogeographic Responses to Glaciation

The biogeographic dynamics of Pleistocene biotas were triggered by three fundamental changes in their environments:

1. Changes in the location, extent, and configuration of their prime habitats
2. Changes in the nature of climatic and environmental zones
3. Formation and dissolution of dispersal routes

The responses of biotas, long adapted to relatively stable and equable climates, also were of three types:

1. Some species were able to "float" with their optimal habitat as it shifted across latitude or altitude.
2. Other species remained where they were and adapted to altered local environments.
3. Still other species underwent range reduction and eventual extinction.

As we shall see, the environmental stresses of the glacial-interglacial cycles triggered all three of these responses. The specific nature of biogeographic responses varied considerably among species and geographic regions, but we have summarized some of their salient features in Box 9.1. We emphasize that,

---

BOX 9.1   *Biogeographic responses to climatic cycles of the Pleistocene*

1. The gradual period of cooling during the mid-Cenozoic was followed by repeated and dramatic climatic reversals during the Pleistocene.

2. Communities and coevolved assemblages of plants and animals that may have persisted for tens of millions of years during the equable Mesozoic were disrupted, with many species responding independently of one another based on their particular physiological tolerances, life history strategies, and dispersal abilities.

3. Many species were able to track the geographic shifts of their prime climates and habitats, but they typically lagged behind, often by centuries, sometimes by millennia.

4. Vegetation zones tended to shift toward the equator (or lower elevations) during glacial periods and toward the poles (or higher elevations) during interglacials, but the shifts were complicated and strongly influenced by geographic features (e.g., mountains, ocean basins, prevailing winds, and proximity to the ice sheet).

5. In general, open-canopied biomes (tundra, savannas, grasslands, and prairies) expanded during glacial maxima at the expense of closed biomes (i.e., forests). These trends were reversed during periods of global warming, but again, rates of shifts varied substantially among biomes, as did the particular species composition of each biome and community.

6. Despite substantial variation among regions, glacial climates tended to be dry as well as cool. On the other hand, glacial warming resulted in flooding of coastlines, submergence of land-bridges, introgression of marine waters onto land, and formation of extensive shallow seas and great, post-glacial lakes and rivers.

7. On land, climatic zones changed dramatically, not only in location and areal coverage, but also in their characteristic nature (i.e., combinations of temperature, seasonality, precipitation patterns, and soil conditions). As a result, Pleistocene events created novel environments, fostering development of novel communities, while other communities disappeared.

8. Although there was much variation within taxonomic groups, plants tended to shift more slowly than animals. The geographic dynamics of species during the Pleistocene created many isolated populations, in some cases promoting evolutionary divergence and diversification of certain biotas.

9. Many plants and animals that were unable to track their shifting environments were able to remain in situ by adapting to altered conditions.

10. The remaining species, unable to shift or adapt, went extinct. During the initial cycles of climatic reversals, extinctions were much more common among plants than animals. This may have been a consequence of the com-

paratively limited ability of plants to disperse and the decoupling of associations among plants and between plants and animals that served as pollinators, parasites, and herbivores.

11. In contrast, until the most recent glacial cycles, animal extinctions were relatively few, and many groups, especially the large herbivores and carnivores, underwent major radiations.

12. The tables were turned, however, during the more recent glacial cycles, which witnessed waves of extinctions of many animals, especially larger ones, while comparatively few plants suffered extinctions. It appears that the initial climatic reversals may have "weeded out" most of the intolerant plants, leaving behind those more capable of dispersing with, or adapting to, climatic reversals.

13. During the most recent glacial cycle, large mammals may have become too specialized on the now waning glacial habitats (especially steppes and savannas). Alternatively, these "megafaunal" extinctions may have resulted from biotic exchanges associated with glacial events (including invasions by humans during periods of low sea levels), which caused mass anthropogenic extinctions of the naïve, native fauna (see Chapters 16 and 17).

while such a synopsis may be useful, a richer understanding of these truly complex and fascinating events can be developed only by thoroughly exploring individual case studies (see chapters in Unit 5; see also Lieberman 2000).

Biogeographic Responses of Terrestrial Biotas

Many of the trends summarized in Box 9.1 are illustrated in the vegetative dynamics of the most recent deglaciation (see Figures 9.13–9.19). Climatic changes associated with the glacial maximum caused the general expansion of steppes, savannas, and other open-canopied terrestrial ecosystems at the expense of closed ecosystems, especially tropical rain forests. In general, biomes have shifted from 10° to 20° in latitude between glacial and interglacial periods, yet have tended to occupy the same relative positions across latitudes or elevations (i.e., a sequence from tundra and boreal forest to savanna and tropical rain forest). This is because the climatic belts of the Earth (see Figure 3.7), then as now, created zonal patterns of vegetation, although the zones were compressed during glacial episodes. In the marine realm, latitudinal shifts in isotherms and biogeographic patterns have tended to be substantial in the mid-latitudes (35° to 55°), but relatively minor at lower latitudes (Figure 9.12).

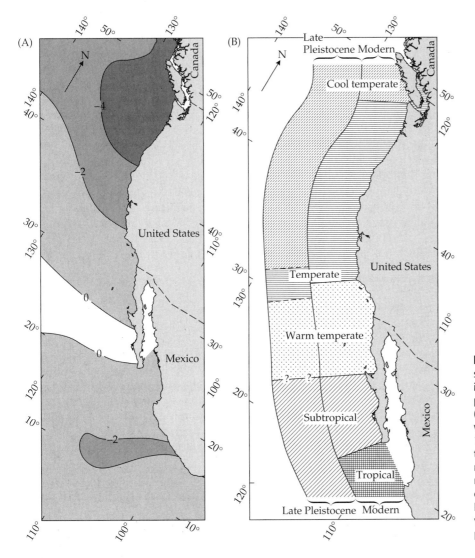

FIGURE 9.12 Shifts in (A) February sea surface temperatures (temperatures are in °C; minus sign indicates cooler temperatures during the Pleistocene), and (B) marine biotic provinces along the west coast of North America between 18,000 years B.P. and the present. Note that sea surface temperatures and nearshore marine provinces have remained relatively stable in the waters off southern California and northern Baja, while those in other regions have varied substantially during the same period. (After Fields et al. 1993.)

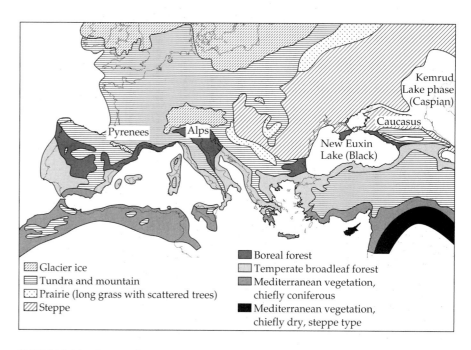

FIGURE 9.13 Zones of vegetation in Europe during the Würm glacial maximum (18,000 years B.P.). Major vegetation types were shifted southward of their present locations by 10° to 20° of latitude, and the precursors of the Black Sea and Caspian Sea were interconnected. Mountain ranges that probably blocked latitudinal shifts of biomes are shown. (After Flint 1971.)

Shifts in climatic zones and biomes, however, are complicated by currents and topographic features, including mountain ranges, large rivers, and other bodies of water. In Europe, the southward shift of some biomes and biotas during the most recent glacial maximum was blocked by the Alps, Pyrenees, and Mediterranean Sea (see Figure 9.13). In contrast, the north-south-running rivers and mountain ranges of North America facilitated extensions of high-latitude biomes deep into subtemperate and subtropical latitudes. During the Wisconsin maximum (about 18,000 years B.P.), boreal forests and tundra penetrated deep into the interior of the continent along both the Mississippi River Valley and the Appalachian Mountains (Figure 9.14). Similar extensions of boreal forests and otherwise high-altitude biomes occurred along the Rocky Mountains of western North America, the Carpathian, Ural and Atlay Mountains of Eurasia, the Great Divide Range of Australia, and the Andes of South America.

In Figure 9.15, pollen profiles of the Andes in Colombia show how each of the vegetation zones has shifted upward since the most recent glacial maximum (see Flenley 1979b). The lower tropical, sub-Andean, and Andean forest types (including the tropical rain forest) have much wider elevational amplitudes now than they did 14,000 years ago. Subpáramo has remained about the same, but **páramo** and superpáramo are elevationally compressed. Because upper vegetation types occur closer to narrow mountain peaks, the total area covered by these biomes has fluctuated greatly, causing extinctions when populations were isolated in restricted montane areas. In these mountainous regions, elevational shifts ranged from 150 to 1500 m between glacial and interglacial periods, and typically were much more rapid than latitudinal shifts.

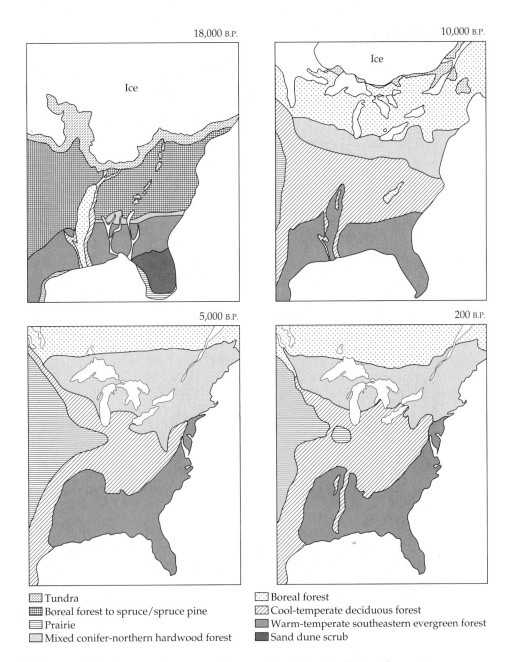

18,000 B.P.

10,000 B.P.

5,000 B.P.

200 B.P.

Tundra
Boreal forest to spruce/spruce pine
Prairie
Mixed conifer-northern hardwood forest

Boreal forest
Cool-temperate deciduous forest
Warm-temperate southeastern evergreen forest
Sand dune scrub

FIGURE 9.14   Shifts in vegetation zones of eastern North America during the most recent deglaciation. Note the great southern expansion of boreal forest and tundra along the Mississippi Valley and Appalachian Mountains during the Wisconsin (18,000 B.P.). (After Gates 1993).

Figure 9.16 tracks the probable upper elevational limit of subtropical forests in three equatorial regions (East Africa, New Guinea, and South America). Each graph suggests several drastic elevational shifts for these forests during the last 34,000 years. The curves are not identical, but in each case the upper elevational limit of tropical forest began to decrease gradually beginning about 29,000 to 27,000 years B.P., but then reversed and increased sharply about 16,000 to 15,000 years B.P. Flenley (1979a, b) also noted that in Africa, South America, Indo-

FIGURE 9.15 Elevational shifts in vegetation zones in the eastern Cordillera of the Andes in Colombia in response to climatic change following the most recent glacial maximum. Note that while all zones tended to shift in concert, the upper zones became narrower as they shifted upward in response to global warming. (After Flenley 1979a.)

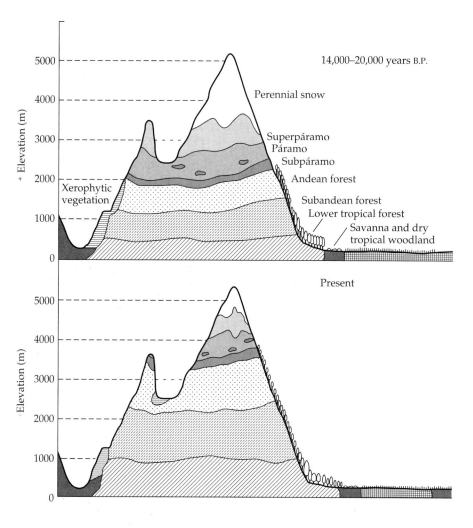

Malaya, and New Guinea, montane vegetational zones were depressed by about the same amounts (1000 to 1500 m) at roughly the same times.

The area just south of New Guinea, Queensland, in northeastern Australia, exhibited an exceptional and rapid change from sclerophyllous woodland to rain forest between the late Pleistocene and 6000 years B.P. (Figure 9.17). The

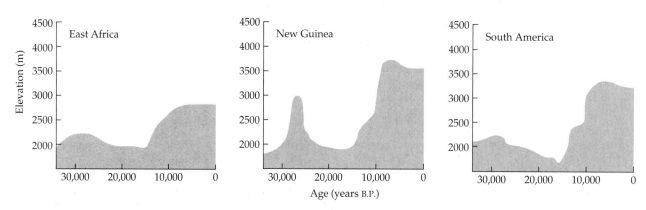

FIGURE 9.16 Shifts in the upper elevational limits of tropical forests in three different regions (South America, East Africa, and New Guinea) during the past 33,000 years. Note that these elevational shifts were roughly synchronous, although the exact upper elevational boundary was somewhat different in each region. (After Flenley 1979a.)

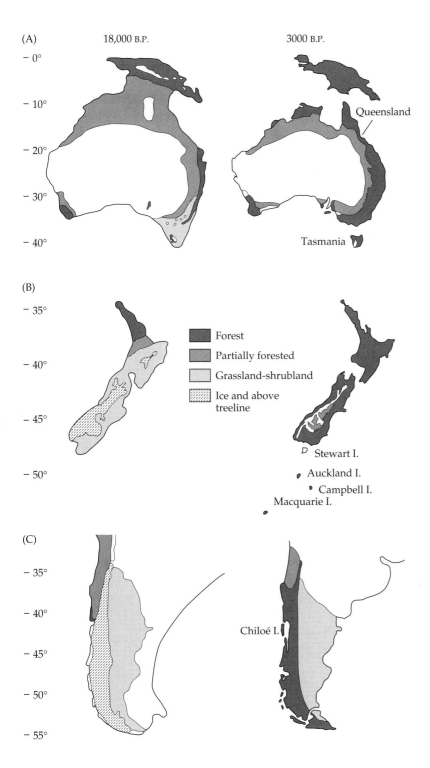

FIGURE 9.17 Comparisons of distributions of vegetation zones during the most recent glacial maximum (18,000 years B.P.) and 3,000 years B.P. (i.e., before significant disturbance by humans) for three regions of the Southern Hemisphere: (A) Australia; (B) New Zealand; and (C) southern South America. (After Markgraf et al. 1995.)

latter point helps to illustrate how difficult it is to extrapolate trends in terrestrial paleoclimatology and to understand vegetational history merely by studying present-day vegetation. Again, the responses to glacial conditions were not uniform across the globe. While African and Amazonian rain forests contracted in response to glacial aridity, those of Sumatra remained intact (see Flenley 1979a; Maloney 1980; see also Figure 9.18).

Geographic shifts in response to climatic changes of the Pleistocene were even more complex for individual species. Rather than responding as discrete units to shifts in climates, biomes and communities often disintegrated, with

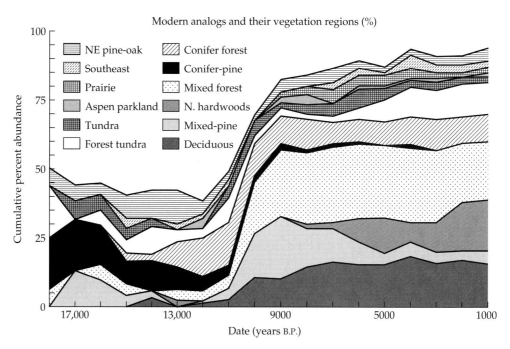

FIGURE 9.18   Variation in the relative abundance of different types of vegetative communities in North America since the most recent glacial maximum (based on samples of fossil pollen). Note that the most rapid changes occurred between 14,000 and 10,000 years B.P. (After Shane and Cushing 1991.)

many, if not most, species responding in an individualistic manner (i.e., consistent with Gleasonian views of community organization; see Chapter 5). Shifts in species ranges were influenced by both extrinsic and intrinsic factors. Extrinsic factors included climate, soils, prevailing winds, ocean currents, and topographic features such as those discussed above. Colonization of deglaciated areas of North America by beech provides an exemplary case study of the influence of these factors. According to Margaret Davis (1986), after beech became established along the southeastern shore of Lake Michigan (about 7000 years B.P.), it took another 1000 to 2000 years to reach the far side of the lake. Birds, including the now extinct passenger pigeon (*Ectopistes migratorius*), may have played an important role in carrying the seeds of beech and many other species across such geographic barriers. The potential for range shifts also may have been strongly influenced by other interspecific interactions, especially competition, predation, and parasitism.

Individualistic responses to climatic change among species derive largely from their considerable differences in ability to respond to these extrinsic factors. Physiological tolerances, behavior, life history strategies, intrinsic rates of increase, and dispersal abilities all vary markedly among species, even those occupying the same community. Pollen analyses and other paleontological records document considerable interspecific variation among tree species in rates of migration (Table 9.1). Average rates of range extension during the Holocene in North American trees varied from 100 to 400 meters per year. As a result, some species now recognized as dominants may be relatively recent arrivals in contemporary communities. The American chestnut (*Castanea dentata*), for example, was a dominant species in oak-chestnut forests of the Appalachian Mountains for over 5000 years, but it was a relatively recent invader of similar forests in Connecticut (arriving about 2000 years B.P.). In west-

(A)

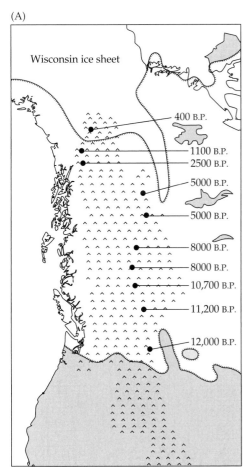

(B)

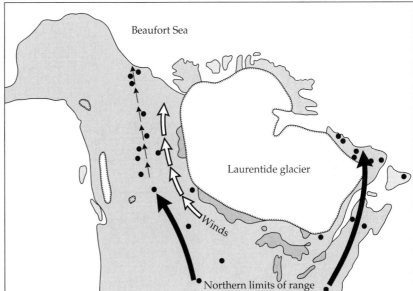

FIGURE 9.19 Rates of migration of trees following glacial recession varied substantially among species, and were strongly influenced by extrinsic factors such as topographic features and prevailing winds. (A) Inland varieties of the lodgepole pine (*Pinus contorta* var. *latifolia*) have expanded their range northward over the past 12,000 years, and may be continuing this range expansion in modern times. Dots indicate northern range boundaries at various times based on pollen samples. (B) Northward range expansion of the white spruce (*Picea glauca*) during retreat of the Laurentide glacier was strongly influenced by prevailing winds (hollow arrows). The species moved rapidly once it reached the western edge of the glacier, where north-flowing winds aided its range expansion. Dots indicate locations of sample points. Long arrows indicate slow migration; short arrows, rapid migration. (After Pielou 1991.)

ern North America, lodgepole pine (*Pinus contorta* var. *latifolia*) has extended its range at about 200 meters per year, reaching southern Alaska just a few centuries ago, and may still be expanding its range northward (Figure 9.19A).

Rates of range expansion can vary considerably within, as well as among, species. At the end of the Wisconsin in North America, for example, white spruce (*Picea glauca*) migrated northward along both the eastern and western margins of the retreating Laurentide glacier. From 14,000 to 7000 years B.P., it migrated along the eastern edge of the ice sheet at about 300 m per year. In contrast, its migration along the western edge of the ice sheet, where north-flowing winds aided its dispersal, was nearly an order of magnitude faster (see Figure 9.19B).

Species varied not only in their rates of range expansion, but in direction as well, with many species shifting as much in longitude as in latitude. For example, the geographic range of the eastern chipmunk (*Tamias striatus*) shifted to the northeast, that of the northern pocket gopher shifted to the west (Figure 9.20), and the range of the papershell pinyon (*Pinus remota*) contracted 300 km to the south (Graham 1986; Graham et al. 1996; Lanner and Van Devender 1998). In a similar manner, many plants and animals inhabiting mountainous regions exhibited individualistic shifts in elevation (e.g., see Jackson and Whitehead 1991; Jackson 2004). The overall result of these individualistic responses was a great reshuffling of communities from one climatic reversal to the next. Again, animals tended to shift much more rapidly than

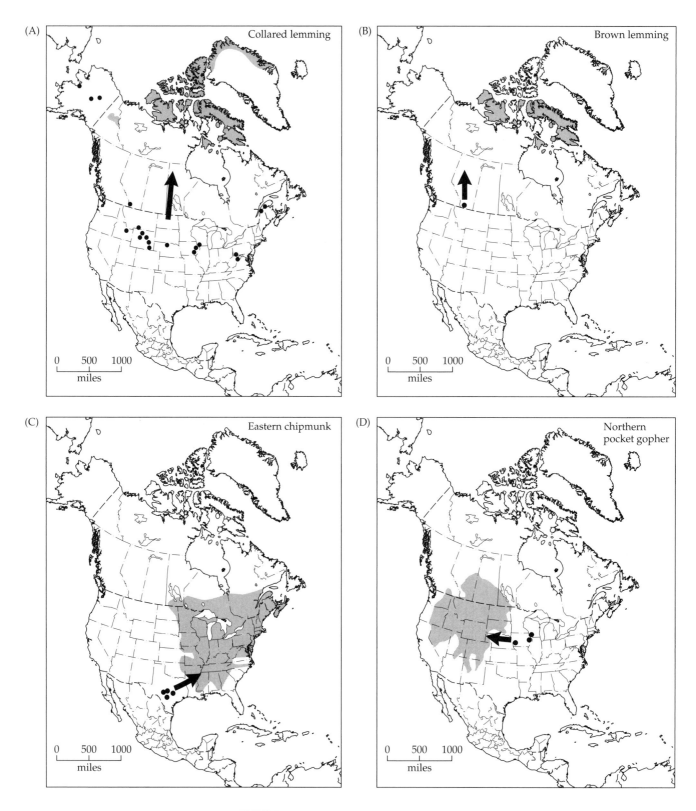

FIGURE 9.20 Geographic range shifts in four species of rodents during the Holocene. Shaded areas represent present range; dots indicate locations of late Pleistocene fossils. Note that these range shifts differed in extent, with collared lemmings (*Dicrostonyx*) (A) shifting farther northward than brown lemmings (*Lemmus*) (B), and in direction, with the above species shifting northward, while eastern chipmunks (*Tamias striatus*) (C) shifted toward the northeast, and northern pocket gophers (*Thomomys talpoides*) (D) shifted toward the west. (After Graham 1986.)

TABLE 9.1 *Average rates of Holocene range extensions of trees (m/yr) following glacial recession*

Species	North America	European mainland	British Isles
Pines	300–400	1500	100–700
Oak	350	150–500	350–500 (50 near northern limit)
Elm	250	500–1000	550 (100 near northern limit)
Beech	200	200–300	100–200
Hazelnut	—	1500	500
Alder	—	500–2000	500–600 (50–150 near northern limit)
Basswood	—	300–500	450–500 (50–100 near northern limit)
Ash	—	200–500	50–200
Spruce, Larch, Balsam Fir, Maples, Hemlock and Hickory	200–250	—	—
Chestnut	100	—	—

Source: North American data after Davis 1981; European mainland and British Isles data after Huntley and Birks 1983.

Note: For most of these species, the rate of range extension slowed as they approached their northern limits.

plants, apparently influenced more strongly by vegetative structure than by abundances of particular plant species. Range shifts of some and perhaps many plant species may have been slowed by relatively limited dispersal abilities of symbiotic fungi—in particular, large-spored, below-ground fruiting mycorrhizal fungi (Wilkinson 1998). Thus, glacial reversals often caused decoupling of animal and plant communities. In the sub-temperate latitudes of the American West (30° to 40° N latitude), for example, glacial and contemporary communities typically share fewer than one-third of their plant species. While biomes shifted in a predictable manner, the species composition of each system varied significantly, often with communities of one glacial stage having no comparable analogue in the next (Figure 9.21).

Not only were species decoupled and reshuffled among communities, but communities and biomes may never have reached equilibrium with their Pleistocene climates (Davis 1986; but see Wright, 1976; Webb 1987). Animals, and especially plants, lagged far behind climatic zones, which shifted at least an order of magnitude more rapidly (Gates 1993). In fact, plant communities of the Holocene often are viewed as ephemeral collections of species that repeatedly disassociated and reassociated with each other as climates changed (Behrensmeyer et al. 1992).

Dynamics of Plant Communities in the Aridlands of North and South America

Many of the biogeographic dynamics discussed above can be illustrated by detailed accounts of vegetative shifts in arid regions of the American Southwest. Because deserts of the Northern Hemisphere lie between about 30° and 40° N latitude (see Chapter 3), they were not covered by ice sheets or mountain glaciers. Yet these and other regions far removed from the glaciers were still strongly influenced by glacial cycles of the Pleistocene. In fact, if it were possi-

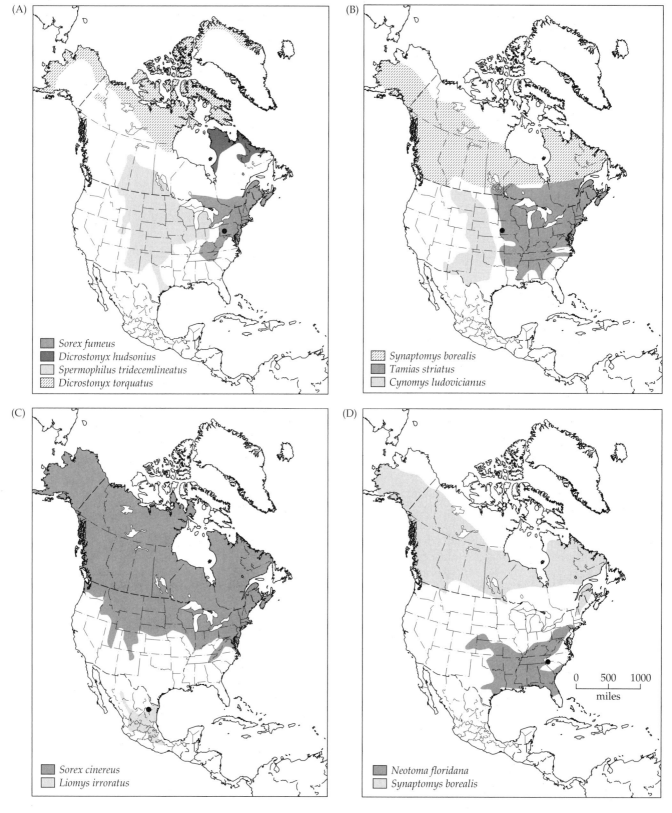

FIGURE 9.21 As a result of the differences in extent and direction of range shifts during the Holocene, species that co-occurred during the most recent glacial maximum often exhibit disjunct ranges today. In each map, black dots indicate coincident occurrences of the named species during the late Pleistocene, while shaded areas indicate their current ranges. (After Graham 1986.)

ble to visit these "desert" regions 12,000 to 18,000 years ago, you would be surprised to find lush, lowland forests in what today are seas of deserts and dry woodlands.

Because deserts are notoriously poor environments for fossilization, it has been difficult to reconstruct the biotic history of deserts. Traditionally, the history of deserts has been inferred mainly from theoretical paleoclimatological models, reinforced wherever possible by fossils, especially pollen records from the sediments of pluvial lakes. Neither provides a very reliable account of the small-scale history of deserts.

In recent decades, a new type of fossil data has been used to reconstruct the vegetational history of one complex region: the desert zones of the semiarid and arid southwestern United States (see Wells and Jorgensen 1964; Betancourt et al. 1990; Grayson 1993; Rhode 2001; Betancourt 2004). Pack rats (*Neotoma*) are abundant rodents in these xeric habitats. They hoard plant materials in large caches, which sometimes are protected in caves or under rock ledges. Remains of these caches become solid structures, called **middens**, and persist for thousands of years if kept dry. They are excellent sources of plant fossils because they provide a relatively complete and often quite continuous sample of the plants growing within roughly 100 m of the rats' den during its occupation. By collecting pack rat middens at different elevations and locations and then dating the materials using radiocarbon methods, researchers can reconstruct the shifts in elevation and composition of vegetation types and deduce past climatic regimes (i.e., those back to 40,000 years B.P.).

Data from pack rat middens of the American Southwest are continuing to be accumulated and analyzed (Van Devender 1977; Van Devender and Spaulding 1979; Wells 1979; Betancourt et al. 1990; Betancourt 2004; see also Fritts 1976). In addition, paleoecologists have recently begun to analyze rodent middens from South America (Latorre et al. 2002), Africa and Australia (Pearson and Betancourt 2002), and will soon explore the midden potential of arid-land regions of central Asia and the Middle East. While midden analyses in these latter areas are still in the preliminary stages, analysis and dating of over 3000 middens from North and South America have provided a rich picture of the dynamics of plant and animal communities over the past 40,000 years (Betancourt 2004).

There have been major changes in the vegetation at all elevations throughout the arid lands of the American Southwest over the last 20,000 years. As explained earlier, climates in this region were substantially cooler and wetter during the most recent (Wisconsin) glacial maximum. During this period, vegetation zones were displaced as much as 500 to 1000 m below their present limits (Figure 9.22), with most of the shift occurring within the last 8000 to 12,000 years. Analyses of middens from the Atacama Desert along the Pacific slope of the Andes (18° to 27° S latitude) indicates similar elevational shifts in plant communities between 14,000 and 10,500 B.P. (Latorre et al. 2002). It appears, however, that vegetative responses to climatic dynamics of the Pleistocene were not universal. Recent studies of rodent middens east of the Andes in Argentina's Monte Desert reveals surprising stability, with little latitudinal displacement of floras during the early Holocene (Betancourt 2004).

Even in those regions with significant shifts of biomes, plant communities did not simply move as entire, integrated entities up and down the mountains. Rather, they changed dramatically in composition. During the most recent glacial maximum (21,000 to 15,000 years B.P.), most plant species in North America's Grand Canyon occurred 600 to 1000 m lower than at the present time. This finding indicates a cooler, wetter climate (Cole 1982). Many of the species that inhabited areas along the rim of the Grand Canyon, however, are no longer found in similar communities in the same region today. In fact, the community

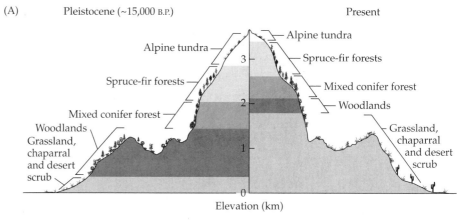

(A) Pleistocene (~15,000 B.P.)   Present

Alpine tundra

Spruce-fir forests

Mixed conifer forest

Woodlands
Grassland,
chaparral
and desert
scrub

Alpine tundra

Spruce-fir forests

Mixed conifer forest

Woodlands

Grassland,
chaparral
and desert
scrub

Elevation (km)

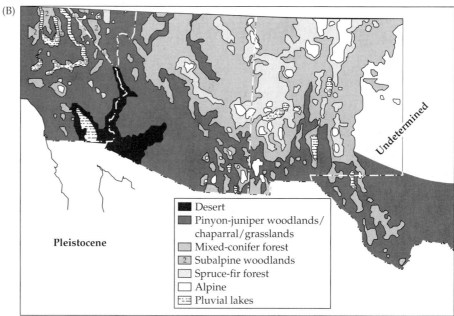

Desert

Pinyon-juniper woodlands/
chaparral/grasslands

Mixed-conifer forest

Subalpine woodlands

Spruce-fir forest

Alpine

Pluvial lakes

Pleistocene

Undetermined

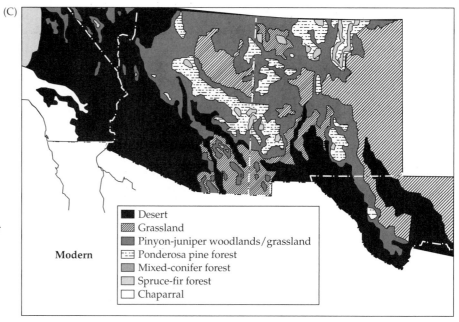

Desert

Grassland

Pinyon-juniper woodlands/grassland

Ponderosa pine forest

Mixed-conifer forest

Spruce-fir forest

Chaparral

Modern

FIGURE 9.22 (A) Elevational shifts in vegetation zones in the mountainous region of the American Southwest near southern Arizona. (B) Vegetation of the southwestern U.S. during last glacial period compared to (C) modern. (A after Lomolino et al. 1989, Merriam 1890; B,C from Betancourt, after Swetnam et al. 1999.)

contained several species that are presently found in northeastern Nevada and northwestern Utah, at least 500 km to the north (Cole 1982).

Similarly, many of the plants now dominant in the coniferous forests of the intermountain West (e.g., ponderosa pine and piñon pine) were rare and restricted in glacial times. On the other hand, species that were much more widespread 10,000 to 30,000 years ago are now narrowly distributed or no longer occur in this region. Results of pollen analyses over the same time periods also reveal individualistic responses to climatic fluxes of the Pleistocene, resulting in what is sometimes referred to as disequilibrium and disruption of vegetative communities that once characterized glacial maxima (Davis 1986, 1994; Jackson et al. 1997; see also Overpeck et al 1992; Williams et al. 2003).

Aquatic Systems: Postglacial and Pluvial Lakes

GLACIAL (CRYOGENIC) LAKES. No other force of lake formation can compare with the glacial activity of the Pleistocene (Hutchinson 1957; Wetzel 1975). As glaciers melted, many shallow-water marine systems were decimated by rising waters. Throughout the world, great volumes of meltwater created rivers and lakes of such magnitude that they dwarf their contemporary analogues. Modern Lake Superior, the world's largest freshwater lake, is less than a fourth the size (surface area) of post-glacial Lake Agassiz (Figure 9.23A), appropriately named after the scientist who developed the first, comprehensive theory of the Ice Ages—(Jean) Louis Rodolphe Agassiz. Lake Agassiz covered approximately 350,000 km^2 of North America during the early Holocene, influencing regional climates and possibly contributing to the very rapid cooling of the Younger Dryas period (12,800 to 11,600 B.P.; see Hostettler et al. 2000). Like Lake Agassiz, most other postglacial lakes developed between 12,000 and 11,000 years B.P. and peaked in extent about 10,000 to 9000 years B.P. While many of these lakes persisted for several millenia, their demise was often catastrophic and spectacular. Even during their northerly retreat, the great mass of glaciers continued to downwarp the lithosphere, causing glacial meltwater to flow toward, and be blocked by, remnants of the retreating glacier (i.e., the great ice dams; see Figure 9.23A, B). As climates continued to warm during the Holocene, ice dams eventually gave way in explosive outbursts. For example, postglacial Lake Missoula, equivalent in size to modern-day Lake Ontario, emptied its 2,000 km^3 volume of water in less than two weeks (McPhail and Lindsey 1986; cited in Pielou 1991). As cataclysmic as this must have been, it appears relatively minor in comparison to the breaching of Lake Agassiz's ice dam, which occurred around 8000 B.P. and poured some 163,000 km^3 of freshwater (about seven times the total volume of the modern Great Lakes) into the Tyrrel Sea (Hudson Bay) and the Atlantic Ocean (see Figure 9.23A; Teller et al. 2002; Leverington and Teller 2002, 2003).

Some of today's lake basins were created when retreating glaciers left great blocks of ice in their wake, which often persisted for many centuries, leaving their imprints as prairie potholes in temperate regions of the Holarctic. The largest blocks of ice created deep and persistent depressions that later filled to form **kettle lakes** (see Figure 9.23C). The legacy of glacial recession can also be seen in the relatively deep, roughly circular lakes called **plunge pools**, which were formed by glacial meltwaters that flowed over the surface of the glacier before plunging off its edge to carve a basin in the Earth some 2 to 3 km below (e.g., Green Lake near Syracuse, New York; see Figure 9.23D). Countless other lakes were formed by the scouring action of glaciers and meltwater, which formed basins, many of them dammed by the deposition of rock debris along glacial moraines (e.g., Lake Mendota and Devil's Lake of Wisconsin, and the Finger Lakes of New York).

FIGURE 9.23 Many of today's lakes were formed by glacial activity. (A) Glacial Lake Agassiz, approximately 9000 B.P., with its outflows (arrows), including its catastrophic release (gray arrows) of some 163,000 km³ of freshwater into the Tyrrel Sea (Hudson Bay) and the Atlantic Ocean once its ice dam burst around 8000 B.P. (B) Shyok Ice Lake along the Western margin of the Himalaya Mountains, an example of an extant, postglacial lake formed when retreating glaciers acted as dams, with glacial meltwater accumulating in valleys carved by previous glacial activity. (C) Formation of kettle lakes: 1, retreating ice sheet leaves stagnant ice blocks on the outwash plain; 2, lakes are formed in the outwash and in the till by melting blocks of ice; 3, a large, partly buried ice block melts, causing irregular sliding of the outwash originally covering the sides of the block. (D) Formation of a glacial plunge pool lake. Glacial meltwater sometimes formed large rivers that flowed over the surface of the retreating glacier, then plunged as much as 3 km down the face of the glacier to carve out a roughly circular lake basin below. (A after Teller et al. 2000; B and C after Hutchinson 1959.)

(A)

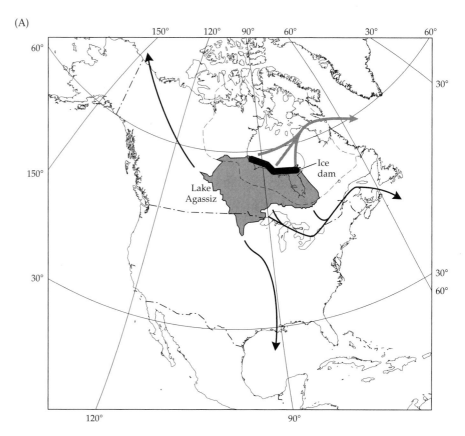

(B)

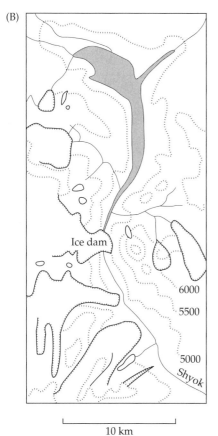

10 km

(C)

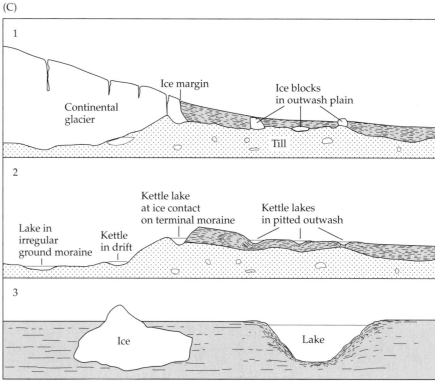

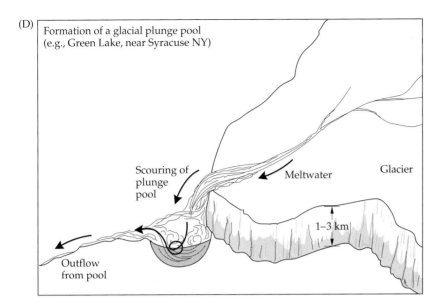

(D) Formation of a glacial plunge pool (e.g., Green Lake, near Syracuse NY)

Scouring of plunge pool

Meltwater

Glacier

1–3 km

Outflow from pool

PLUVIAL LAKES IN ARID REGIONS.   Glacial cycles affected plant and animal communities in regions far withdrawn from glaciers, including those that are now dominated by deserts. During pluvial periods, large freshwater or saline lakes formed in these regions because of a combination of low evaporation rates (primarily due to lower temperatures) and high precipitation rates ("pluvial" means "rainfall"). Few biogeographers appreciate the size and number of pluvial lakes that existed in places that now have desert climates. Aridlands of Nevada and adjacent areas have a **basin-and-range topography**, in which many low, flat areas (**basins**) are interrupted by isolated mountain ranges. During pluvial times, many of these basins filled with water (Figure 9.24). The largest such water body was Lake Bonneville in Utah and parts of Nevada and Idaho. At times it contained fresh water and drained northward into the Snake and Columbia Rivers. In the Middle Wisconsin, this lake was 330 m deep, had an area exceeding 50,000 km^2 (slightly smaller than present-day Lake Michigan), and supported a diverse freshwater community, including cutthroat trout (*Salmo clarki*). The present Great Salt Lake is a small remnant of Lake Bonneville. Nevada, southern Oregon, eastern California, southeastern Arizona, and southwestern New Mexico had numerous large and small pluvial lakes. Yet most of these lakes had evaporated by 10,000 years B.P. One of these lake basins, Death Valley, with the lowest elevation in North America (93 m *below* sea level), now contains perhaps the most extreme desert on the continent.

Pluvial lakes were also present in what are now deserts on other continents. Their remnants—saline lakes and dry lake beds—are abundant in the Atacama Desert of northern Chile, the Monte of Argentina, many areas in interior Australia, the region of the Dead Sea in the Middle East (ancient Lake Lisan), the Kalahari Desert of southern Africa, and many places in arid and semiarid Asia (Flint 1971). The western portion of the Sahara Desert was still relatively mesic until 5000 years ago. Perhaps the most remarkable of these remnants of glacial-pluvial periods is Lake Chad. This lake, now just 16,000 km^2, was once 950 km long and covered over 300,000 km^2 (the present size of the Caspian Sea), including a significant portion of the southern Sahara Desert. Lake Chad remained at this maximum size from 22,000 to 8500 years B.P.

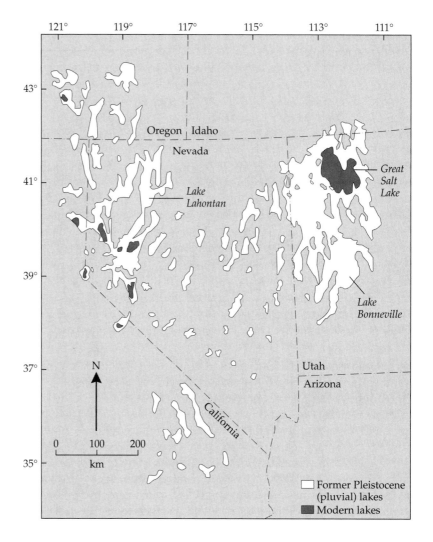

FIGURE 9.24 Distribution of pluvial lakes in western North America during the most recent (Wisconsin) glacial maximum. During glacio-pluvial periods most of the arid region of the continent experienced a wetter and cooler climate, and lakes and marshes filled what are now desert valleys. (After Benson and Thompson 1989; Elias 1997.)

The disappearance of pluvial lakes during the Holocene had several profound biogeographic effects. It caused the wholesale extinction of many plants and animals living in or around these bodies of water. In addition, the dissection of large lakes into smaller, isolated units led either to local extinctions or to the vicariant speciation of surviving forms, as with the pupfishes (*Cyprinodon*) of the southwestern United States (Miller 1961b; Smith 1981).

Biotic Exchange and Glacial Cycles

Figures 9.15 and 9.22 demonstrate how downward shifts of high-altitude vegetation (primarily forests) during glacial maxima could have created avenues of dispersal: such a lowering would allow a plant or animal species, previously isolated in montane forests of individual peaks, to cross ridges and migrate along mountain ranges and mesic lowlands. This mechanism is precisely the one invoked by many authors to account for the intracontinental dispersal of vegetation types and associated animals. When followed by isolation as mesic biomes once again contracted toward higher elevations and higher latitudes, it could result in taxonomic disjunctions (Simpson 1975; Vuilleumier and Simberloff 1980).

Glacial cycles may have had similar effects on the distributions of many marine organisms (Figure 9.25). In the case of cold-water stenothermal species, relatively warm tropical waters serve as an effective physiological barrier to dispersal, limiting their distributions to the middle or high latitudes of the Northern or Southern Hemisphere. During glacial maxima, however, cooling of marine waters could have allowed range expansion into the lower latitudes. Subsequent rewarming of tropical waters during interglacials could have again caused range contraction and possibly bipolar distributions of these species (see Figure 9.25C).

As noted earlier, eustatic and isostatic changes in sea level greatly altered opportunities for biotic exchange for both terrestrial and marine biotas (see Figure 9.9). Greatly reduced sea levels during the Wisconsin created extensive landbridges, such as Beringia (connecting Siberia and North America), the Sunda Shelf (connecting Malaysia and Indonesia), and the Arafura Sea and Bass Straits (connecting New Guinea, Australia, and Tasmania). While these landbridges served as important dispersal corridors for terrestrial organisms, they simultaneously eliminated and fragmented marine biotas. In most cases, however, biotic exchange was asymmetrical, with more species migrating from larger (species-rich) to smaller areas than vice versa (e.g., from Siberia to

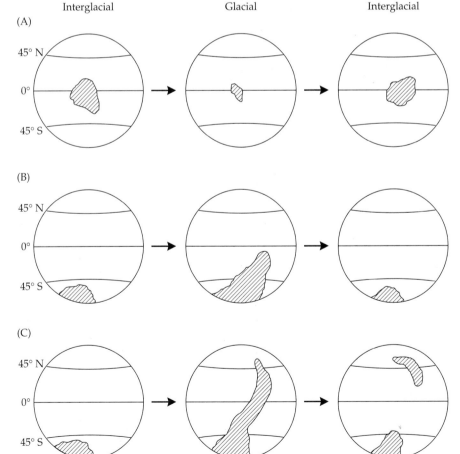

FIGURE 9.25 Potential range expansions and contractions of hypothetical warm stenothermal (A) and cool stenothermal (B and C) marine taxa. During a glacial period, cool stenothermal taxa could penetrate from one hemisphere to the other (middle diagram, C) and could possibly achieve bipolar distributions when waters rewarm and their ranges contract toward the poles. (After Crame 1993.)

Alaska; from Southeast Asia to the "islands" of the Sunda Shelf; from Australia to Tasmania). Biotic exchange of terrestrial organisms across Beringia (see below) contributed significantly to the similarity between Nearctic and Palearctic biotas. Our own species used this glacial landbridge to colonize North America from Siberia.

Similarly, biotic exchange of marine organisms often tended to be asymmetrical, depending on the size and diversity of each species pool and the ocean currents and other factors influencing dispersal (see Briggs 2004; Vermeij 2004). While Beringia served as a dispersal corridor for terrestrial organisms during glacial maxima, the Bering Strait was an important corridor for the dispersal of marine life during interglacial periods. Again, more species migrated from the larger, more species-rich region—that is, from the Pacific basin northward. During the late Cenozoic, 125 species of marine invertebrates invaded the Arctic-Atlantic region from the Pacific, while no more than 16 species colonized in the reverse direction (Durham and MacNeil 1967; see also Vermeij 1991).

In Chapter 6 we discussed the tectonic events that ultimately formed a Central American landbridge between North and South America (about 3.5 million years B.P.). The resultant waves of biotic exchange between Nearctic and Neotropical biotas, referred to as the **Great American Interchange**, were made possible not just by tectonic events, but by eustatic changes and vegetative shifts associated with glacial cycles. During glacial maxima, the lowering of sea levels increased both the area and elevation of the Central American landbridge. Perhaps just as important, the relatively dry conditions that prevailed during glacial maxima caused savannas to expand toward the equator and form a continuous habitat corridor for the migration of many species adapted to these open habitats (savannas and shortgrass prairie; see Webb 1991). We shall return to the profound effects of these waves of biotic interchange in Chapter 10.

Evolutionary Responses and Pleistocene Refugia

As we saw in Chapter 7, evolutionary divergence is intimately related to biogeographic dynamics and the resultant mixing or isolation of gene pools. Given the repeated shifting and fragmentation of species ranges during the Pleistocene, glacial cycles may have had profound evolutionary effects. Here we focus on the biogeographic and evolutionary responses of Pleistocene biotas confined to two types of hypothesized refugia: Neotropical forest refugia and glacial refugia, or nunataks. In Chapter 12, we discuss changes in body size and other evolutionary responses to climatic reversals of the Pleistocene and earlier periods.

NEOTROPICAL PLEISTOCENE REFUGIA. Lowland tropical rain forests (see Chapter 5) are awe-inspiring structures, teeming with an amazing diversity of life. Traditionally such places were regarded by many investigators as stable refuges that had persisted virtually unchanged for many millions of years while climates became harsh and fluctuating elsewhere. However, more recent lines of evidence suggest that these rain forests have experienced marked climatic changes, so that they have varied greatly in distribution within the last 40,000 years, sometimes being much more restricted than at present.

Within the Amazon basin, which contains the most extensive continuous rain forest in the world today, researchers have found centers of high species richness and relatively high endemism for both plants and animals. Haffer (1969, 1974, 1978, 1981), for example, identified six principal areas to which

(A)

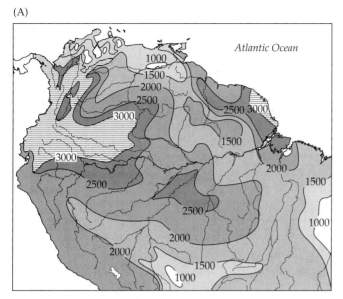

(B)

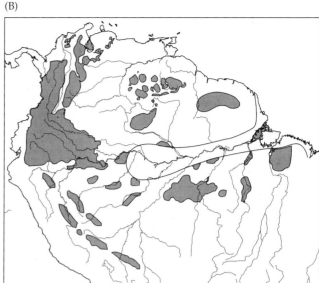

FIGURE 9.26 Haffer's Pleistocene refugium hypothesis was proposed to account for the relatively high species richness and endemicity of Amazonian plants and animals. (A) Contemporary patterns of rainfall were used to identify locations of Pleistocene rain forest refugia (numbers indicate rainfall in mm per annum). (B) Regions receiving more than 2500 mm of rainfall annually were postulated to have served as forested refugia during the Pleistocene. (C) Limited geographic distributions of a number of species groups (such as the toucanets, *Selenidera* spp., in this example) seemed consistent with the postulated distributions of Pleistocene refugia, and with the hypothesis that these refugia served as speciation pumps. (After Terborgh 1992.)

(C)

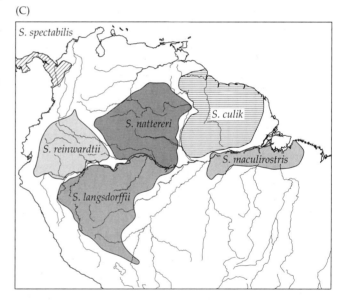

about 150 species of birds are narrowly restricted (Figure 9.26). Frequently these areas are inhabited by several subspecies of the same species, or by several similar species of the same superspecies. This pattern is repeated in families of reptiles, amphibians (Vanzolini and Williams 1970; Dixon 1979; Lynch 1979), woody plants (Vuillemier 1970; Prance 1973, 1978, 1982), and butterflies (Brown, K. S. Jr. 1982), although a greater number of centers can be identified for plants and butterflies.

These insular patterns of species richness and endemism in what had seemed to be homogeneous habitats suggested to Haffer and others that these centers were **refugia**—islands of lowland rain forest that persisted during glacial maxima. Haffer and his colleagues developed a model of **cyclical vicariance**—also known as the **speciation pump model**—to attempt to account for the amazing diversity of tropical rain forests. According to their model, populations of wide-ranging species dependent on tropical forests became isolated as their prime habitat became fragmented during glacial maxima (see Figure 9.26). These relatively dry glacial periods, which lasted up to 100,000 years, were thought to be long enough for significant evolutionary divergence among isolated sister populations in different refugia. As glaciers melted, these isolated rain forests expanded and eventually reconnected to form the continuous forest of Amazonia. Populations from the various refugia likewise expanded their ranges and came into contact with related populations from adjacent refugia. Haffer postulated that their divergence during isolation may have been reinforced by mechanisms promoting reproductive isolation in these zones of secondary contact. However, complete reproductive isolation between many close relatives apparently was not achieved during the most recent glacial cycle. For insects, lizards, and birds, zones of secondary contact are regions of frequent hybridization between closely related forms.

Although Haffer's refugia hypothesis provided an interesting explanation for the high diversity and relatively high endemism of tropical rain forests, over the past two decades or so, many scientists have questioned its validity based on the following objections (see Prance 1982). First, Haffer and his colleagues may have greatly overestimated the reduction of tropical rain forests during glacial maxima. Because cooling was relatively uniform across Amazonia, Neotropical rainforests and their associated populations were not as fragmented and isolated during glacial periods as hypothesized (see Colinvaux 1989; Colinvaux et al. 1996; Colinvaux et al. 2000; Colinvaux and De Oliveira 2002). Second, molecular data and other genetic and phenotypic data indicate that many of the endemic forms are much older than the hypothesized refugia; they appear to have speciated well before the formation of the putative refugia (see Cracraft and Prum 1988; Hackett and Rosenburg 1990; Hackett 1993; Marshall and Lundberg 1996). Third, there appears to be little overlap in centers of endemism among different taxa (see Beven et al. 1984). Finally, as interesting and provocative as it was, Haffer's hypothesis is not the most parsimonious explanation for the high diversity and endemism of today's relatively continuous tropical rain forest. Actually, the high diversity of tropical forest ecosystems may derive, at least in part, from the fact that they are much more spatially heterogeneous than once assumed (see Chapter 15).

Despite these objections, Haffer's hypothesis served an important purpose by stimulating an increased emphasis on understanding the effects of glacial cycles on biotas far removed from the glaciers. Although Haffer's original model now seems tenuous, it appears that diversification of those tropical forest biotas may indeed have developed during periods of climate-driven fragmentation and isolation of these forests (Willis and Whittaker 2000). Rather than forest refugia created by expanding savanna during glacial maxima, however, it appears that fragmentation, isolation, and divergence of populations in mid- to high-elevation forests occurred during interglacial periods. At these times, global sea levels rose by over 100 m, flooding the Amazon Basin and creating a complex and expansive archipelago of tropical forest islands—conditions similar to those of the Philippines, Hawaiian Islands, and other oceanic hotspots of diversity (see Patton 1998; Nores 1999; Gascon et al 2000; Heaney 2004).

NUNATAKS: GLACIAL REFUGIA. **Nunataks**, refugia that persisted within, or adjacent to, ice sheets, are a popular topic in Pleistocene phytogeography. Nunataks could have developed along the continental periphery of a glacier, internally as isolated pockets along a mountain range (e.g., on the tops of the highest peaks), along a coastline where bluffs are too steep for glaciers, or between adjacent ice sheets. Geologists have identified several possible "internal" nunataks. The Laurentide Ice Sheet covered most of northeastern and north central North America (see Figure 9.2). The most famous non-glaciated region within this ice sheet is the so-called **driftless area** in southern Wisconsin and adjacent Illinois and Iowa, an elliptical area that was bypassed by the glacial front. The Cordilleran glacier complex occurred in westernmost Canada, extending from the Canadian Rockies and adjacent southern Alaska (through the Aleutians) to the coast, and including small portions of northern Washington, Idaho, and Montana. A number of narrow nunataks may have occurred between the Laurentide and Cordilleran ice sheets and in intermountain valleys (Figure 9.27).

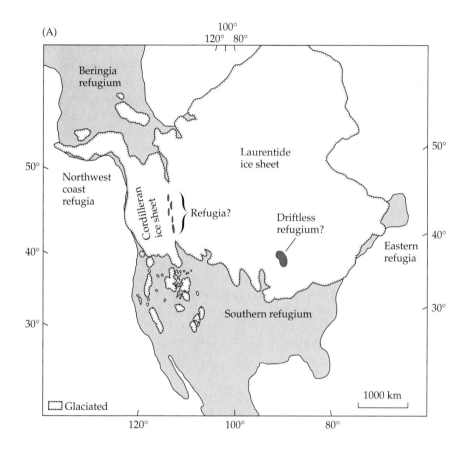

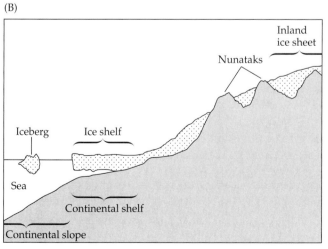

FIGURE 9.27 (A) Even during the Wisconsin glacial maximum, ice-free refugia may have occurred between the Laurentide and Cordilleran ice sheets and in a region called the driftless area. (B, C) Ice-free areas in mountainous regions along the Pacific coast may also have served as refugia—and possibly as migration corridors—for plants and animals, during full glacial conditions. (A after Rogers et al. 1991; B and C after Pielou 1991.)

Numerous phytogeographers have used plant distributions as evidence for other northern nunataks, but the response to these hypothesized refugia, especially from geologists, has not always been favorable. Reminiscent of the hypothesized Neotropical refugia, these nunataks are alleged to have occurred where investigators have found small, isolated pockets of arctic-alpine vegetation that are especially rich in species, including rare and endemic forms. Examples include two alleged nunataks in Norway, which contain some interesting disjunctions. The Lapland rosebay (*Rhododendron lapponicum*), for example, occurs in Greenland and in the Arctic and alpine regions of the New World and Asia, but nowhere else in Europe besides these two areas. Pielou (1979) has discussed the many problems in interpreting these areas as nunataks, the most serious being that it is difficult to demonstrate that they were not areas of recolonization following the retreat of the ice sheets at the end of the Pleistocene.

In contrast to the questionable importance of these relatively small, internal nunataks, those along the periphery of glaciers may well have provided refugia for a great many species. For example, at least three large refugia persisted north of the glaciers during the Wisconsin in North America: the expansive iceless areas of Beringia, coastal regions of the Pacific Northwest, and much of Nova Scotia (see Figure 9.27). Many species that were unable to disperse with their shifting habitats persisted, and in some cases diversified, in these refugia. Thus, many Holarctic biogeographic patterns bear witness to the importance of these glacial refugia.

The thesis that Beringia was a Pleistocene refugium was clearly presented by the phytogeographer E. Hultén (1937) (Figure 9.28). Hultén's work on this subject is classic, and one of the best efforts to document plant distributions on a broad scale. His exact distribution maps, which stimulated an important creative synthesis, should be required study for all biogeographers. Although Hultén's idea was ahead of its time, several lines of evidence helped to con-

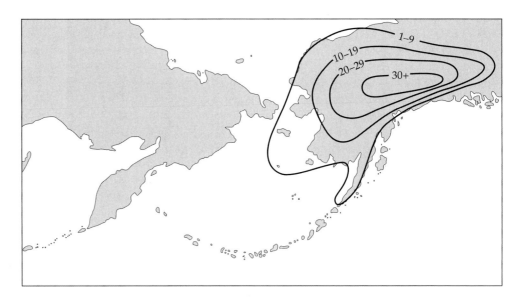

FIGURE 9.28 The high frequency of endemic plant species in central Alaska was used by Hultén (1937) to advance the idea that a large part of Beringia remained unglaciated and served as a refugium for arctic forms during the Pleistocene. Each contour line indicates the number of narrow endemics present in that region.

firm that Beringia was vegetated during the Wisconsin glacial maximum. First, Colinvaux (1981, 1996) and others analyzed fossil pollen records and showed that mesic tundra was widespread in the northern and central portions of the landbridge at the height of the Wisconsin glacial period (see Elias et al. 1996). Second, Hopkins and Smith (1981) presented rather strong evidence for the occurrence of deciduous dicotyledonous trees and larch (*Larix*, Pinaceae) in the Yukon at this same time. Finally, Weber et al. (1981) and others excavated late Pleistocene mammalian fossils (aged from 40,000 years B.P.) in the interior of Alaska, finding a great diversity of large ungulates (woolly mammoth, woolly rhino, barren-ground and forest musk ox, mountain sheep, steppe antelope, reindeer, horse, and camel; see also Hopkins et al. 1982; Guthrie 1990). Paradoxically, these grazing mammals occupied Beringia during a period when its vegetative cover and the productivity of their preferred forage were much lower than in other contemporary communities. The waves of extinctions that followed the Wisconsin glacial maximum, when ungulate forage was increasing, remain an equally controversial and intriguing paradox, and are attributed, perhaps justifiably, to humans, as we shall see below. While it may take many years to solve this mystery, it is clear that the Bering landbridge, at times 1000 km wide, served as an important refuge and dispersal corridor for many Holarctic plants and animals, including humans. Smaller refugia, including those along the coast of the Pacific Northwest (see Figure 9.27), supported fewer species. In general, however, regions formerly within glacial refugia still tend to harbor a higher diversity of animals than those in formerly glaciated regions.

In the absence of gene flow among refugia, populations diverged genetically, often to the specific level in small mammals or to the subspecific level in larger species. Examples of such divergence between populations in Beringia and those in regions south of the glaciers include northern and southern red-backed voles, tundra and arctic shrews, and arctic and Columbian ground squirrels (see Hoffman 1971; Nadler et al. 1973; Nadler et al. 1978). Examples of large mammals include the divergence of moose (*Alces alces gigas*) and Beringian Dall's sheep from their southern counterparts (Peterson 1955; Korobytsina et al. 1974). In contrast, smaller refugia of the Pacific Northwest not only supported fewer species, but their populations underwent less divergence than those inhabiting Beringia. Shrews, deer mice, voles, marmots, chipmunks, brown bears, and ermine have exhibited modest divergence, especially those isolated on islands of the Queen Charlotte Archipelago (see Banfield 1961; Hoffmann et al. 1979; Hoffman 1981; Riddle 1996; Paetkau et al. 1998). Genetic diversity of many European species, including genetically distinct geographic races of mammals, amphibians, grasshoppers and trees, occurred when southward expanding glaciers isolated their ancestral populations in three peninsulas of southern Europe—the Iberian Peninsula (Spain), Italy, and the Balkans (Hewitt 1999; Willis and Whittaker 2000).

Finally, anthropological data also bears strong witness to the legacy of glacial refugia. Biogeographic patterns in the dental traits, genetic markers, blood proteins, and linguistic diversity of native Americans appear to parallel those found in other mammals. Diversity in these traits (e.g., number of languages per unit of area) tends to remain highest where refugia occurred, and higher in larger refugia (i.e., higher south of the ice sheets than in Beringia, and higher in Beringia than in the Northwestern refugia; see Rogers et al. 1991).

In summary, Pleistocene refugia—especially the largest and most isolated ones—provided important opportunities for diversification of regional biotas. Few of their endemic forms, however, spread as rapidly into deglaciated areas as their southern counterparts (i.e., those persisting south of the ice sheets). On the other hand, the high endemicity of former glacial refugia may derive, at least in part, from the inability of many other species to invade these sites.

Glacial Cycles and Extinctions

As noted earlier, most late Neogene plant extinctions occurred during initial glacial cycles—between 5 and 0.7 million years B.P. Those plants that survived these glacial periods did so by virtue of their ability to disperse along with their climatic zones or to adapt to new zones and new environments. Some groups, such as those of the arid land regions of North America, underwent extensive radiations during glacio-pluvial periods. It appears that by the later glacial cycles of the Pleistocene, vegetative communities had come to be dominated by plants that were pre-adapted to climatic changes. Thus, while vegetative communities were reshuffled and shifted substantially in latitude and elevation, relatively few plant extinctions occurred during the most recent glacial cycles.

Mass extinctions of marine invertebrates during the late Neogene also appear to be associated with glacial cycles. As with plants, bivalve extinctions in the Mediterranean and North Seas were heaviest during the earlier glacial cycles, with most occurring in the Pliocene between 3.2 and 3.0 million years B.P.—that is, coincident with the earliest evidence of glaciation in those regions. Additional pulses of extinctions continued through the early Pleistocene, but not into the mid- or late-Pleistocene. It appears that the victims were primarily stenothermal species. By the mid-Pleistocene, these species had been "weeded out" and, as a result, marine biotas became dominated by eurythermal species, which were little affected by fluctuations in water temperatures during subsequent glacial cycles (see Raffi et al. 1985).

The pattern of terrestrial vertebrate extinctions during the Pleistocene is quite different. On display in American natural history museums are skeletons and models of many large mammals that dominated American faunas until as recently as 13,000 years ago, but are now extinct (Figure 9.29). Gone from North America are most of its large herbivores, such as mastodons, mammoths, camels, llamas, horses, tapirs, ground sloths, and cave bears, as well as species of ungulates related to contemporary deer, bison, and pronghorn antelope (Martin and Wright 1967; Kurtén and Anderson 1980). Gone also are many of the large predators that hunted those herbivores, including hyenas, dire wolves and other canids, saber-toothed tigers, and even lions. Large birds, especially raptors and scavengers, were also disproportionately subject to extinctions during this period. These casualties included the teratorns of North America, believed to be the largest flying birds that ever lived (wingspan of *Teratornis incredibilis* was approximately 5 m, as compared with 3 m in today's California and Andean condors; see Figure 9.29E; see Steadman and Martin 1984; Steadman 1987).

So many of these fossils have been unearthed from Pleistocene beds that paleontologists could not help but be impressed by the disappearance of this North American megafauna at the end of the Pleistocene. The record of this mass extinction prompted scientists in the nineteenth and twentieth centuries to search for a specific cause for the wholesale destruction of the North American megafauna. Were the extinctions sudden or gradual? Did other landmasses experience similar waves of extinction, and if so, were they synchronous across regions? Did small animals and plants become extinct at the same time? Were the Pleistocene extinctions caused by climatic and geological changes, or did intense hunting by humans result in the extirpation of these

(A)

(B)

(C)

(D)

(E)

10 feet

17 feet

Condor

Teratorn

FIGURE 9.29 The mass extinction of terrestrial vertebrates that occurred in North America during the late Pleistocene and early Holocene included the loss of a highly disproportionate number of large mammals and birds, often referred to as the Pleistocene megafauna. (A) Mammoths (*Mammuthus* spp.); (B) ground sloths (*Megalonyx* spp.); (C) sabertooth cats (*Smilodon* spp.); (D) giant bison (*Bison latifrons*); and (E) teratorns (*Teratornis* spp.). These groups are either without modern day analogues in North America (mammoths and ground sloths) or they have been replaced by much more diminutive forms (sabertooth cats replaced by mountain lions and smaller felines; giant bison by American bison; teratorns by condors and vultures).

large beasts? These are questions that must be answered if we are to understand faunal change and biogeography in the Pleistocene.

The Overkill Hypothesis

In 1967, Paul S. Martin proposed a controversial, but insightful explanation for the collapse of North America's **megafauna**—the diverse, but now extinct assemblage of large mammals and birds that roamed these lands during much of the Holocene (Martin 1967, 1973, 1984, 1990, and 1995). The prehistoric, or Pleistocene, **overkill hypothesis** states that humans were responsible for the mass extirpation of large herbivorous mammals (over 50 kg), and the carnivores and scavengers dependent upon them, after the Wisconsin glaciers had retreated. This is an old hypothesis, anticipated by both Darwin and Wallace, and it has been applied to explain waves of megafaunal extinctions across other continents as well as oceanic archipelagoes. It is also one that has been clearly presented as a straightforward explanation with potentially falsifiable assumptions and predictions. The Pleistocene overkill model could be tested and possibly falsified by showing that many different types of animals and plants became extinct at the same time, that extinctions were under way before humans arrived, that aggressive human hunters coexisted with large mammals for long periods, that human populations were never at high densities, or that comparable extinctions on other continents did not correspond with an invasion by ecologically significant human societies.

Let us consider this hypothesis in detail for comparison with alternative explanations. The original overkill model suggests that a population of aggressive and skillful human hunters entered North America during the late Wisconsin by crossing Beringia from Asia. Once these hunters colonized the New World, they spread southward and eastward through North America and into South America, killing large animals as they went (Figure 9.30). As human populations expanded and encountered new game, they also adapted and advanced their hunting skills, taking more game and increasing their populations to, and beyond, the carrying capacity of their megafaunal prey. The native animals had never been exposed to such (human) predators, and therefore lacked adequate defensive and avoidance behaviors. The initially abundant food supplies obtained from their hunts permitted human populations to remain high and in constant need of expanding their ranges to exploit new and massive food sources. Behind this trail of carnage, there were no additional waves of megafaunal immigrants from Asia to replace those species that became extinct. Most of the large mammals that somehow managed to survive during this period were those that had spread to the New World from the Old World since the evolution of Pleistocene humans, so they were presumably already adapted to human hunters.

Evidence supporting this scenario is of several types. First, fossil evidence shows that prehistoric humans and large mammals co-occurred (albeit for a limited period) in the Americas, and that these people hunted the now extinct herbivores. Arrow points in carcasses and remains of massive animal kills in box canyons, ravines, and beneath cliffs are clearly demonstrated. Second, late Wisconsin extinctions in North America were nonrandom in that a highly disproportionate number of large mammals became extinct during the period from 12,000 to 10,000 years B.P. (Figure 9.31). Third, as noted above, immigrants from Beringia and Eurasia, including caribou, moose, deer, and Dall's and bighorn sheep, fared much better than native species (Figure 9.32; Kurtén and Anderson 1980). Fourth, extinctions of large mammals appear to have begun in the north and proceeded rapidly and systematically southward

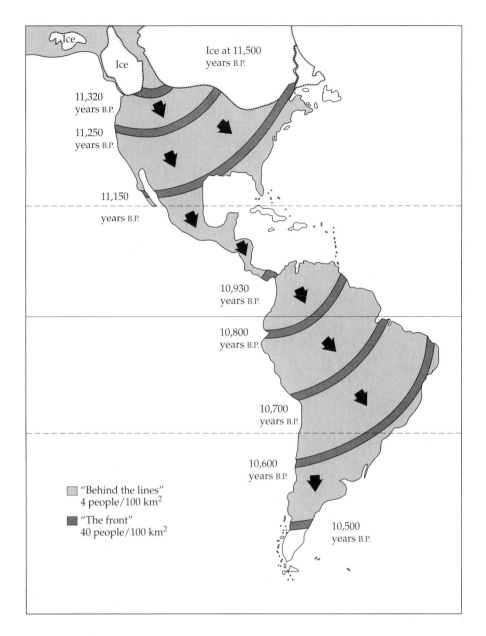

FIGURE 9.30 The temporal sequence of advancing populations of human big-game hunters correlates well with the progressive extinction of large Pleistocene mammal species. According to the Pleistocene overkill hypothesis, sophisticated hunters crossed Beringia and expanded southward, maintaining a relatively dense population by subsisting on large mammals. Human populations may have colonized the Americas well ahead of these dates, but their population densities, technology, and ability to cause significant ecological disturbance were very limited in comparison to the more sophisticated hunting societies that followed. (After Martin 1973.)

(compare to Figure 9.30). Finally, when the dates of the last known occurrences of species are compared with computer simulations of human migration and hunting pressures, the two appear to coincide rather closely (Mosiman and Martin 1975; Alroy 1999; Fiedel 1999; but see also Choquenot and Bowman 1998).

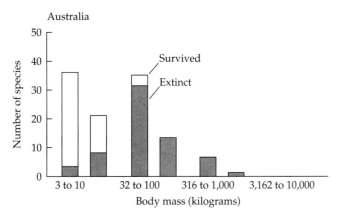

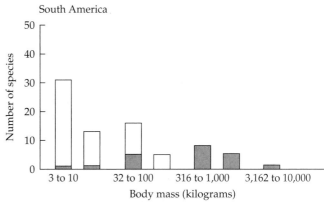

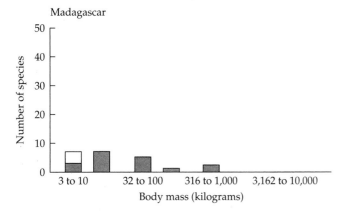

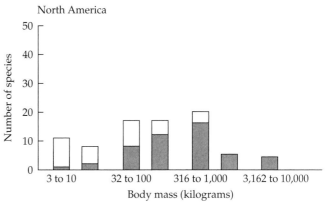

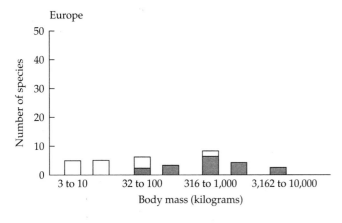

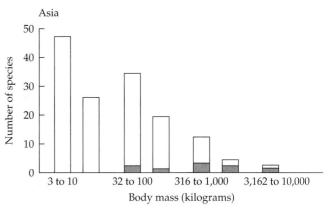

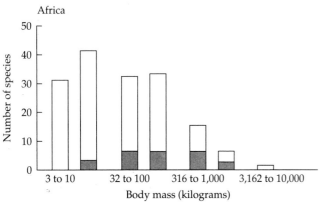

FIGURE 9.31 Extinctions of mammals during the Quaternary Period were highly non-random, varying in timing and magnitude across landmasses but in all cases disproportionately affecting the largest taxa. For example, diversity of medium to large mammals (i.e., those > 5 kg) during the early Quaternary was relatively high, especially for the largest landmasses (compare total height of bars for mammals of Asia, North America, and Africa vs. those for Madagascar and Europe). However, each of these mammalian faunas suffered a disproportionate loss of their largest species during the late Quaternary, albeit at different times on different continents. Grey bars indicate surviving species, and black bars indicate those going extinct during this period; note that mass is graphed along a logarithmic scale. (After Brook and Bowman 2004.)

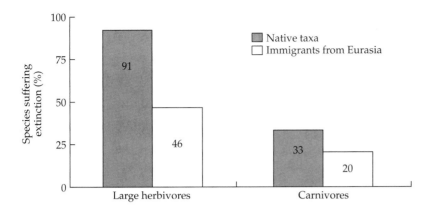

FIGURE 9.32   Extinction rates among native and immigrant large herbivores and carnivores in North America during the late Wisconsin. (After Kurtén and Anderson 1980.)

Alternative Explanations for Pleistocene Extinctions

As with any controversial theory in biogeography, there are several alternative explanations to account for Pleistocene extinctions of mammals. The overkill hypothesis, if correct, paints a rather brutal and disparaging picture of the early human pioneers in North America. Some authors feel that these colonists have been given a "bum rap"—they may have been instrumental in reducing prey population sizes, but extinctions may have already been underway in response to climatic shifts at the end of the last Ice Age. They point to other groups of organisms, such as raptors and large scavenging birds, that experienced high rates of extinction at roughly the same time as large herbivores (Grayson 1977). However, as noted earlier, such extinctions are entirely consistent with the overkill hypothesis, which predicts a loss not just of the megafauna, but of dependent predators and scavengers as well (see Steadman and Martin 1984; Owen-Smith 1988).

One observation that has always been puzzling is that the North American fauna did not disappear until long after the Wisconsin glaciers retreated. Hence the late Pleistocene extinctions cannot be related directly to glaciation, cold climate, or any other catastrophic geological event, such as flooding. Nevertheless, many researchers contend that climatic changes were the direct cause of the extinctions, either through increased aridity (Guilday 1967) or by decreased equability (Axelrod 1967; Slaughter 1967).

An excellent discussion of Pleistocene extinctions by Kurtén and Anderson (1980) explains why many paleontologists preferred a climatic explanation. Pleistocene extinctions of mammals (albeit mostly smaller species) were not restricted to the period of 12,000 to 10,000 years B.P., but were part of a fairly continuous series of episodes during the latest Cenozoic (Table 9.2). The Pliocene Blancan extinction (mainly between 3.3 and 2.4 million years B.P.) resulted in the disappearance from North America of at least 125 mammalian species, of which three-fourths were animals smaller than 1 kg in body mass. During that time, aridity increased, grasslands replaced forests, and many forest dwellers and browsers died out. Following that depletion of this fauna, surviving grazers and rodents underwent evolutionary radiation, and small carnivores also increased in abundance and diversity. In the Irvingtonian extinction of the Pleistocene (1.8 to 0.7 million years B.P.), only 89 taxa disappeared—80% of which were small or medium-sized (< 180 kg). Extinction rates during the Pleistocene were fairly low and constant until the late Wiscon-

TABLE 9.2 *Extinction of North American mammals during the last 4 million years*

Animal size	Blancan (3.5–1.8)	Irvingtonian (1.8–0.7)	Rancholabrean (0.7–0.01)	Extinction total	Surviving species	Percent extinct
Small (1–907 g)	97	55	29	181	166	52%
Medium (908 g–181 kg)	31	25	33	89	50	64%
Large (182–1730 kg)	5	12	35	52	16	76%
Very large (> 1730 kg)	1	2	6	9	1	90%

Source: After Kurtén and Anderson (1980).

Note: Duration of periods is in million years B.P.

sin, when many small, medium, and large animals disappeared. However, as stated earlier, a highly disproportionate number of large mammals became extinct in the late Wisconsin, but not during the 20 or so earlier climatic reversals of the Pleistocene.

The cause and geographic dynamics of the late Pleistocene extinctions remains one of the most important and intriguing mysteries of our field. In a wonderful essay on the nature and causes of historic extinctions, Jared Diamond (1984) called on one of the world's greatest detective minds to help solve the mystery: Sir Arthur Conan Doyle—alias Sherlock Holmes. In "Silver Blaze," Holmes called attention to "the curious incident of the dog in the night-time." When the dim-witted Inspector Gregory observed that "the dog did nothing in the night-time," Holmes pounced on him—"that was the curious incident"—for it indicated that the stables pet was familiar with the "intruder." The decisive clues to the causes of Pleistocene extinctions also may be the "dogs that did nothing in the night-time," namely the species and biotas that survived while others became extinct. Champions of the overkill hypothesis call our attention to four such "curious incidents." First, they ask, why did the North American megafauna diversify when climatic conditions seemed least favorable during the Wisconsin, only to suffer so many extinctions when climates warmed and environmental conditions became more favorable? Second, why did the megafaunal extinctions not occur during an earlier glaciation? Third, why were most groups of small animals spared from this wave of mass extinctions while their larger counterparts were devastated? Finally, why didn't the megafaunal species of Africa suffer extinctions comparable to those among biotas on other large continents (Figure 9.33)?

Some have argued that familiarity is the key to the last question: *Homo sapiens* had such a long history in Africa that, at least until modern times, native African societies have had little impact on the megafauna that coevolved with them. Megafauna likely became ecologically and possibly evolutionarily adapted to the truly "native" populations of humans in their African homeland. Proponents of the overkill hypothesis note that Africa may be the exception that proves the rule—in Holmes' words, 'the dog that did not bark in the night.'

Although megafauna of most other landmasses also suffered mass extinctions during the Pleistocene, they were far from synchronous. A mass extinction of the Australian megafauna for example, occurred roughly 45,000 years B.P. (Figure 9.34), while mass extinctions on Madagascar, New Zealand, The Mascarenes, and the Galápagos Islands have occurred within the last 1000 years (Figure 9.35). These chronologies are difficult to explain according to the climate-based hypothesis, since climatic reversals were synchronous across the globe. The timing of these waves of megafaunal extinctions, however, is coin-

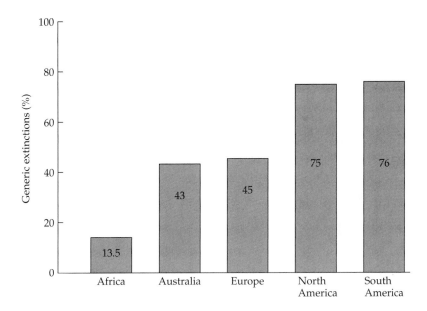

FIGURE 9.33 Extinction rates among mammalian herbivore genera with medium to large body sizes (> 5 kg) on different continents during the late Pleistocene. (After Owen-Smith 1988.)

cident with and at the heels of invasions by ecologically significant humans (Steadman and Martin 1984; Steadman 1989, 1993, 1995; Martin 1990; Flannery, 1994; Holdaway 1996, 1999; Holdaway and Jacomb 2000; Roberts et al. 2001; Worthy and Holdaway 2002; Murray 2003).

In its current version, the "overkill" hypothesis actually encompasses a combination of anthropogenic threats including the use of fire for habitat management (Flannery 1994), competition for seeds and other resources (Neuman 1985), introduction of exotics (Holdaway 1996, 1999), accidental introduction of a "hyperdisease" (MacPhee and Marx 1997), as well as hunting. Its assumptions are relatively simple: the native fauna was naïve to the powers of colonizing humans, and the rate of increase in human populations and their advancements as ecosystem engineers outstripped the reproductive abilities of the native fauna and their ability to adapt to what Darwin termed the "stranger's craft of power." Under these conditions, native faunas should and did collapse, with the most susceptible species being those with relatively high resource requirements and ecological naïveté (in particular, lack of predator avoidance), and low reproductive potential (i.e., the megafauna). Again, given the fact that the waves of megafaunal extinctions were asynchronous, occurring at very different periods and under very different climatic conditions—but following invasions by ecologically significant humans—Paul Martin's controversial hypothesis is attracting support from a growing consensus of scientists (Allroy 1999, 2001; MacPhee 1999; Miller et al. 1999; Dayton 2001; Roberts et al. 2001; Johnson 2002; Kerr 2003; Murray 2003; but see Webb and Barnosky 1989; Choquenot and Bowman 1998; Grayson 2001; Brook and Bowman 2002; Grayson and Meltzer 2002, 2003; Barnosky et al. 2004).

Despite continuing debate on the specific causes of the Pleistocene extinctions, it is clear that elimination of the majority of large mammals, birds, and reptiles has been one of the most important events in the relatively recent history of these terrestrial biotas. Not only did these animals themselves experi-

FIGURE 9.34 Selective extinctions of large, megafaunal mammals in Australia during the Pleistocene and Holocene. Shown are outline drawings of species known to occur in these regions when they were colonized by aboriginal humans. Species suffering extinction during the Pleistocene and early Holocene are shown in black, while those that became extinct or endangered following European colonization are shaded (open outlines indicate extant, non-endangered species). (Information courtesy of T. F. Flannery; art based on original drawings by Tish Ennis, Australian Museum, Sydney.)

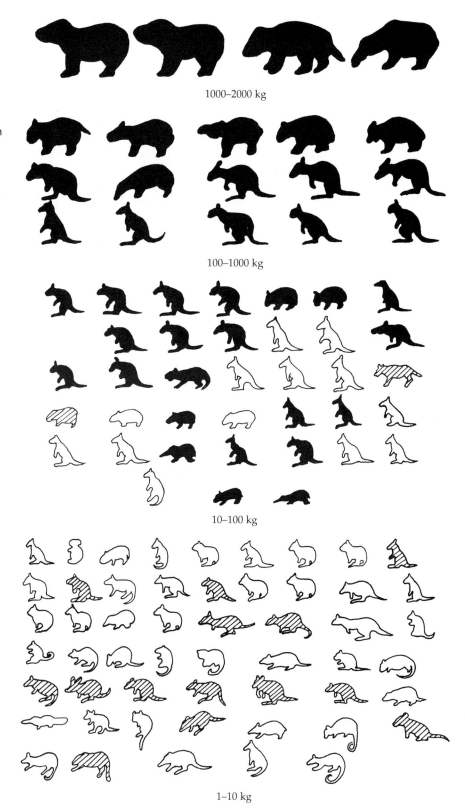

1000–2000 kg

100–1000 kg

10–100 kg

1–10 kg

ence reductions in their populations and geographic ranges ending in their extinction, but their disappearances may also have had important effects on other species. We know from fossil feces (**coprolites**) that the extinct large herbivores consumed great quantities of certain extant plant species. To what

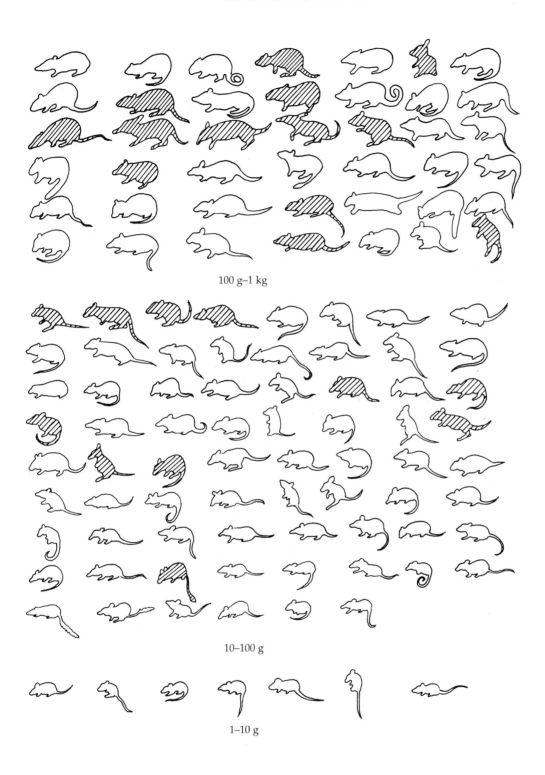

100 g–1 kg

10–100 g

1–10 g

extent did release from such herbivory contribute to the shifts in plant species ranges and the changes in the distribution of vegetation types that are known to have occurred within the last 20,000 years? Herbivores that did survive were faced with fewer potential competitors (paving the way for overgrazing of native plants and for competitive exclusion among native herbivores), and surviving carnivores had to make do with fewer prey. Parasites, mutualists, and scavengers of the extinct species either switched to new associates or became extinct themselves (see Owen-Smith 1988). Accordingly, Steadman

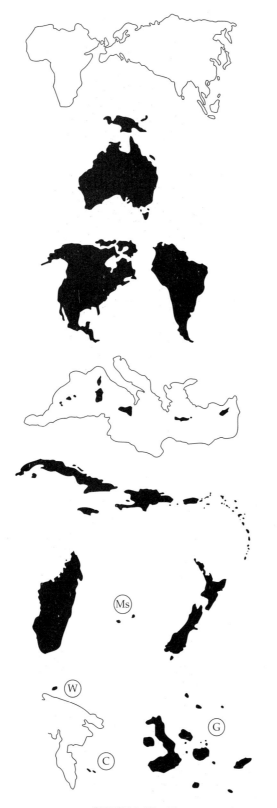

Afroeurasia: No major episodes of extinction during past 100,000 years, although some losses occurred.

Meganesia: Humans arrive around 56, B.P.; mass extinctions recorded between 51,000 and 40,000 B.P.

Americas: Ecologically significant human populations arrive 12,500 years B.P.; major extinction episode terminates circa 10,500 years B.P., few extinctions thereafter.

Mediterranea: Humans arrive 10,000 years B.P.; major extinction episode follows and terminates circa 4,000 years B.P., few extinctions thereafter.

Antillea: Humans arrive 7000 years B.P.; major episode of extinction follows, but extends to circa A.D. 1600 (or later?).

Madagascar: Humans arrive 2000 years B.P.; major episode of extinction follows and terminates circa A.D. 1500, few extinctions thereafter.
Mascarenes (Ms): Humans arrive A.D. 1600; major episode of extinction follows and terminates circa A.D. 1900.
New Zealand: Humans arrive 800–1000 years B.P.; major episode of extinction follows and terminates circa A.D. 1500.

Commander Islands (C): Humans arrive A.D. 1741; Steller's sea cow extinct by A.D. 1768.
Wrangel Island (W): Humans arrive ?; mammoths survive to 4000 years B.P.
Galápagos Islands (G): Humans arrive A.D. 1535; modern-era extinctions only.

FIGURE 9.35 The geography and chronology of Pleistocene extinctions may be correlated with major episodes of human colonization. Extinction episodes during the Pleistocene were relatively minor in regions with a long history of human occupation, but severe and coincident with colonization by ecologically significant human cultures elsewhere (see also Chapter 17). (After MacPhee and Marx 1997; based on drawings by Clare Flemming.)

and Martin (1984) record high extinction rates among carrion-feeding birds at the end of the Pleistocene, including eagles, vultures, teratorns, and condors.

In the following chapters (Unit 4), we dig deeper into the paleontologic record to reconstruct the geologic, climatic, and evolutionary processes that influenced the origins, diversification, and distributions of life on Earth. To paraphrase George Gaylord Simpson, our ability to understand and effectively curtail the ongoing wave of extinctions may well depend on our ability to learn the lessons of these prehistoric extinctions. Finally and just as important, ecologists, conservation biologists, and others studying extant biotas should bear in mind that the many waves of megafaunal extinctions during the Pleistocene may have fundamentally changed the geography of nature (see Chapters 16 and 17). Given the diversity, numbers, and presumed ecological dominance of the thousands of now vanished species, many if not most of the patterns and processes we study are those of highly altered biotas.

UNIT FOUR

Evolutionary History of Lineages and Biotas

CHAPTER *10*

The Geography of Diversification

*I*N UNIT 3, WE LEARNED THAT the evolutionary fates and distributions of species have been dynamic throughout the history of life on Earth. New species have continuously evolved from ancestral species, and occasionally, large numbers of species have been eliminated in an episode of mass extinction followed by the evolution of entirely new sorts of species. Species have often expanded or shifted their distributions following the dispersal of founding populations across a barrier or along with shifting habitats. Some of the most profound changes in distributions have come when entire groups of species (biotas) have crossed from one biogeographic region to another following the erosion of a barrier. Additionally, we learned that the geology of the Earth itself changes continuously through time, and that at various times and particularly over the past several million years, climatic cycles have produced an exceedingly dynamic arena in which, in order to avoid extinction, species and biotas must respond to frequent and unpredictable changes in habitat structure and distribution through dispersal or adaptation. There are thus two kinds of "histories" that have shaped the characteristics of contemporary organisms, which we will call the history of place and the history of lineage (Brown 1995).

The history of place is the history of the Earth itself: the changes in geography, geology, climate, and other environmental characteristics that are extrinsic to the particular organism or group of organisms being studied. We like to think of the history of place as the environmental, geographic template that each organism has experienced during its own unique evolutionary history. Thus, the history of place includes the geological history of the Earth; changes in soils, climate, oceanographic and limnological conditions; and the varying composition of biotic communities. Past environments, by influencing the abundance, distribution, and adaptive evolutionary changes of its ancestors, may have influenced a contemporary species or lineage in many ways. Even the most distantly related kinds of organisms that lived together in the past at a particular place—and hence, experienced some features of a common environment—share some history of place. Sharing a history of place does not depend on sharing ancestors.

The history of lineage is the series of evolutionary changes that have occurred in the intrinsic characteristics of an individual organism, species, or higher taxon. These are changes in heritable characteristics, derived from and constrained by the characteristics of the organisms' ancestors. Only to the extent that organisms share common ancestors do they share a history of lineage. The history of place strongly influences the history of lineage, because characteristics of past environments (e.g., geology, climate, and biotic composition) undoubtedly influenced the survival, distribution, and diversification of all lineages that occurred in those places. But the converse is less true. The history of lineage influences the history of place only to the extent that the activities of particular organisms altered the past environment for themselves and other lineages. While many kinds of organisms substantially modify their environments—the building of reefs and atolls by corals, and the modification of the climate of the Amazon basin by rain forest trees, are just two examples—usually these influences are diffuse and cannot be attributed to just one lineage or taxonomic group.

In this chapter, we describe the fundamental biogeographic patterns that emerge from a history of diversification, dispersal, and extinction of lineages and biotas on a geologically and climatically dynamic Earth. We first consider general patterns of endemism, provincialism, and disjunction, and the abiotic and biotic features that function to maintain geographically separate biotas. We then consider what happens when barriers between previously separate biotas erode, opening corridors for biotic interchange. We finish this chapter with a discussion of several general evolutionary trends that tend to arise when lineages and biotas have long been separated by geographic barriers. In chapter 11, we present an overview of approaches to reconstructing the history of lineages, and then in Chapter 12 we build on that discussion by presenting a number of approaches to reconstructing historical relationships between lineages, biotas, and places.

The Fundamental Geographic Patterns

The most pervasive feature of geographic distributions is the fact that they have limits. No species is completely **cosmopolitan** (organisms that are widely distributed throughout the world), and most species and genera, and even many families and orders, are confined to restricted regions, such as a single continent or ocean—the term **endemic** means occurring in one geographic place and nowhere else. Many distributions are extremely limited. The minute, redfinned blue-eye (*Scaturiginichthys vermeilipinnis*), the sole species in the fish genus *Scaturiginichthys*, lives in a handful of tiny springs in arid western Queensland, Australia (Ivantsoff et al. 1991), which it shares with another equally restricted species, the Edgbaston goby (*Chlamydogobius squamigenus*; Larson 1995). The avian family, Opisthocomidae, contains only a single species, the weird hoatzin (*Opisthocomus hoazin*), a nearly flightless, leaf-eating bird that occurs only in a small region of northern South America. Even some diverse taxa have very restricted distributions. We have already discussed the extreme diversity of cichlid fishes in the Great Lakes of the African Rift Valley (Chapter 7). The avian family, Furnariidae (ovenbirds), with more than 50 genera and 200 species, is nearly confined to South and Central America plus several neighboring islands (one species extends north to northern Mexico). Several large fish families, including catfish eels (Plotosidae), are con-

TABLE 10.1 *Families and subfamilies of angiosperms endemic (or nearly so) to the Neotropical regions*

Aextoxicaceae	Luzuriagoideae of Liliaceae
Agdestioideae of Phytolaccaceae	Malesherbiaceae
Alstroemerioideae of Liliaceae	Marcgraviaceae
Alvaradoroideae of Simaroubaceae	Mayacaceae[a]
Asteranthoideae of Lecythidaceae	Misodendraceae
Bixaceae	Morkilliodeae of Zygophyllaceae
Brunelliaceae	Neotessmannioideae of Tiliaceae
Calyceraceae	Nolanoideae of Solanaceae
Caryocaraceae	Pakaraimoideae of Dipterocarpaceae
Catopherioideae of Lamiaceae	Pellicieroideae of Theaceae
Columellioideae of Saxifragaceae	Peridiscaceae
Cyclanthoideae of Cyclanthaceae	Phytelphantoideae of Arecaceae
Cyrillaceae	Picramnoideae of Simaroubaceae
Dictyolomatoideae of Rutaceae	Plocospermatoideae of Loganiaceae
Duckeodendroideae of Solanaceae	Quillajeoideae of Rosaceae
Eremolepidaceae	Quiinaceae
Goetzeaceae	Rapateaceae[a]
Gomortegaceae	Rhabdodendroideae of Rutaceae
Goupioideae of Celastraceae	Siparunoideae of Monimiaceae
Gyrocarpoideae of Hernandiaceae (except *Gyrocarpus americanus*)	Stegnospermataceae
Halophytaceae	Styloceratoideae of Buxaceae
Henriquezioideae of Rubiaceae	Tetralicoideae of Tiliaceae
Houmiriaceae[a]	Theophrastoideae of Myrsinaceae
Lactoridaceae (Juan Fernández Island)	Thurnioideae of Juncaceae
Lecythidoideae of Lecythidaceae	Tovarioideae of Capparaceae
Ledocarpaceae	Tropaeolaceae
Lennoaceae	Vellozioideae of Velloziaceae
Leonioideae of Violaceae	Vivianiaceae
Lithophytoideae of Verbenaceae	Vochysiaceae[a]
Lophophytoideae of Balanophoraceae	

[a]Taxa with African taxa.

fined to the Indo-Pacific region. In South and Central America, there are over 50 families and subfamilies of flowering plants that occur nowhere else (Table 10.1); many of these are small taxa, but they also include several extremely diverse families. Cacti (Cactaceae) and bromeliads (Bromeliaceae) would be endemic to the New World if one species of each had not recently crossed the ocean by natural means (long-distance dispersal) and become established in Africa.

Endemics are not distributed randomly, but tend to be concentrated in certain regions, a phenomenon that is called **provincialism.** Australia, southern Africa, Madagascar, New Zealand, and New Caledonia, for example, are hot spots of endemism, containing a large percentage of endemic species and numerous endemic higher taxa, whereas other regions, such as Europe, northern North America, and the southern Atlantic Ocean, share much of their biotas with other areas. Often, different groups of plants and animals occur not

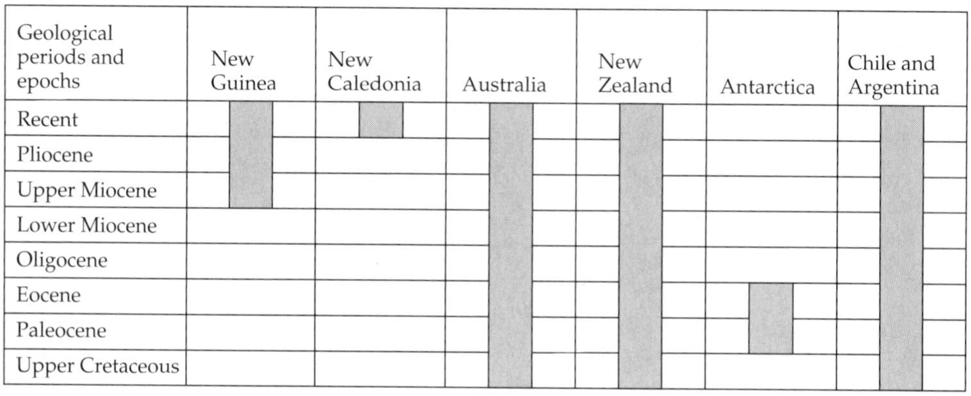

Geological periods and epochs	New Guinea	New Caledonia	Australia	New Zealand	Antarctica	Chile and Argentina
Recent	▓	▓	▓	▓		▓
Pliocene	▓		▓	▓		▓
Upper Miocene	▓		▓	▓		▓
Lower Miocene			▓	▓		▓
Oligocene			▓	▓		▓
Eocene			▓	▓	▓	▓
Paleocene			▓	▓	▓	▓
Upper Cretaceous			▓	▓		▓

FIGURE 10.1 Recent and fossil distribution of the Southern Hemisphere beeches, *Nothofagus* (Fagaceae). This distribution is both relictual and disjunct, with representatives of the genus presently occurring on many of the widely dispersed fragments of the ancient southern continent of Gondwanaland. Also, note that the fossil record of the genus in some of these places extends back to the Mesozoic, and that fossil localities also include Antarctica. As with many other Gondwanan relicts, no representatives of the genus presently occur on Africa, Madagascar, or India. (After Schlinger 1974.)

only in the same ocean or on the same continent or island, but also in the same localities and habitats within those regions (recall the redfinned blue-eye and the Edgbaston goby examples given above). These coincident distributions of endemics often do not coincide precisely to the present boundaries of continents and oceans, and they certainly do not always coincide with obvious characteristics of abiotic and biotic environments. As pointed out in Chapters 4 and 16, the many successful introductions of species by humans demonstrate that species can thrive in regions far from their native habitats. These introductions emphasize the unique influences of historical events in determining where organisms occur today. In many cases, the spread of taxa from regions in which they evolved has been blocked by barriers to dispersal. In other cases, one species of a once widespread group has persisted in a limited area after its representatives in other regions have become extinct.

Disjunctions are those cases in which populations, closely related species, or higher taxa occur in widely separated regions but are absent from intervening areas. Disjunct distributions reflect past events: the disjunct forms dispersed long distances over geographic barriers, were carried to distant sites aboard crustal plates as they drifted apart, or are the surviving remnants of a once widespread taxon. Usually the disjuncts are morphologically similar and inhabit similar environments. The evergreen southern beeches (*Nothofagus*), for example, grow in wet, cool temperate forests in South America, New Zealand, New Caledonia, New Guinea, New Britain, and Australia (Figure 10.1). Disjuncts that are phylogenetically related are not always biologically similar. The flightless ratite birds (superorder Paleognathae) have a disjunct Southern Hemisphere distribution similar to that of the southern beeches and other groups (both living and known only from fossils) that once occurred on Gondwanaland. Most ratites, such as the ostriches of Africa, emus of Australia, and rheas of South America, inhabit open, relatively arid habitats. The cassowaries of New Guinea and northeastern Australia, however, occur in tropical rain forests. The ecologically and morphologically most divergent of the ratites are the strange, "mammal-like" kiwis (family Apterygidae), which live in wet temperate forests of New Zealand, are nocturnal, feed on earthworms, and nest in burrows.

Endemism and Cosmopolitanism

Because the term endemic simply means occurring nowhere else, organisms can be endemic to a geographic location on a variety of spatial scales and at different taxonomic levels. Organisms can be endemic to a location for three different reasons: because they originated in that place and never dispersed, because their entire range has shifted in locality subsequent to origination, or because they now survive in only a small part of their former range.

Taxonomic categories are hierarchical (see Table 7.2), so the distributions of lower taxa within that of a higher taxon are also organized in a hierarchical fashion. An order, for example, contains a nested set of families, genera, and species that represent the historical phylogenetic branching pattern of a single evolutionary lineage. Similarly, the geographic range of an order contains within its boundaries the ranges of all of its families, genera, and species in a cumulative series. For this reason, the lowest taxonomic categories, species and genera, tend to be more narrowly endemic than the higher taxa, such as families and orders, of which they are members. Figure 10.2 provides an example. The rodent family, Heteromyidae, containing the pocket mice, kangaroo rats, and kangaroo mice, is endemic to western North America, Central America, and northernmost South America. Each of the five extant genera of heteromyids have more restricted ranges; the genus *Microdipodops* (kangaroo mice), for example, occurs only in the Great Basin of the western United States. The genus *Dipodomys* (kangaroo rats) has a much wider distribution, and its species vary greatly in their ranges, from *D. ordii*, which occurs in most of the cold desert and arid grassland regions of the western United States, to the endangered *D. ingens*, which is endemic to an area of a few thousand square kilometers in the San Joaquin Valley of California. We note here that non-taxonomic units such as evolutionarily significant units and phylogroups—defined in Chapter 7 and discussed in more detail in Chapter 11—can also be described in the context of their pattern of endemism.

Many species and genera, and even some families and orders, are entirely restricted to tiny islands or equally small patches of habitat. The entire native

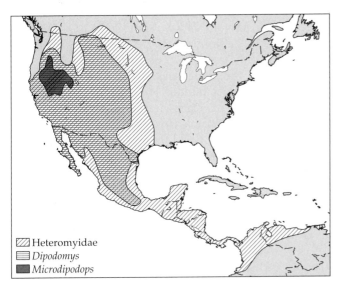

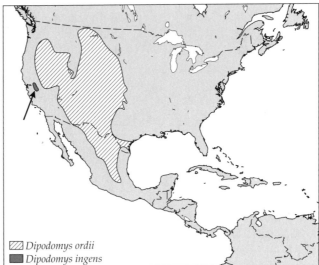

FIGURE 10.2 Hierarchical patterns of endemism in the rodent family Heteromyidae, which includes kangaroo rats and pocket mice. The entire family is endemic to the New World, ranging from southern Canada to northern South America. Distributions of its five genera vary in extent, from the kangaroo rats (*Dipodomys*), which range over most of western North America, to the kangaroo mice (*Microdipodops*), which are endemic to the Great Basin. Similarly, the range of just one species, *D. ordii*, encompasses most of the range of its genus, whereas *D. ingens* is endemic to the San Joaquin Valley of California.

FIGURE 10.3 Native habitat of the Devil's Hole pupfish, (*Cyprinodon diabolis*) near Death Valley in the Mojave Desert, USA. This pool, connected to a large underground water basin and having surface dimensions of about 20 × 3 m, represents the entire native habitat of the species. Photo courtesy of Lawrence Naiman.

population of the Devil's Hole pupfish (*Cyprinodon diabolis*) numbers fewer than 600 individuals, which are confined to a spring pool measuring 20 × 3 m in the Mojave Desert of Nevada, just east of Death Valley (Figure 10.3). Some remarkable plant endemics occur in California, especially on the Channel Islands off the southern coast. The only population of the distinctive shrub, *Munzothamnus* (Asteraceae), lives on the tiny island of San Clemente. The entire known population of a species of mountain mahogany, *Cercocarpus traskae* (Rosaceae), consists of four individuals growing in one small canyon on Santa Catalina Island. Also on Santa Catalina, San Clemente, Santa Rosa, and Santa Cruz islands lives the Catalina ironwood (*Lyonothamnus floribundus asplenifolius*; Rosaceae), an elegant evergreen tree whose fossils are known from the mainland, including some from Death Valley, during the Miocene. Even some flying organisms can have narrow ranges. The todies (Todidae), for example, are a family of tiny birds entirely restricted to a few West Indian islands; and the highly specialized Mexican fishing bat, *Myotis vivesi* (Vespertilionidae), is almost entirely restricted to quiet lagoons on islands and along mainland coasts in the Sea of Cortés of Mexico.

Of course, isolated islands, such as Madagascar and New Zealand, are famous for their endemics. Some of the endemic groups that have radiated on Madagascar are mentioned in Chapter 7. Perhaps the most famous endemic reptile is the tuatara (*Sphenodon punctatus*) of New Zealand, a lizard-like species that is the sole surviving representative of the order Rhynchocephalia, the sister-clade of all other extant lizards and snakes. Widespread on continents in the Mesozoic, the tuatara persists only on a few small islands that have not been reached by introduced rats and cats. Not so lucky was the Stephen Island wren (*Xenicus lyalli*), which occurred only on one small islet in the strait between North and South Islands of New Zealand. The original size of its population is unknown, because the only individuals known to science were "collected" by the lighthouse keeper's cat, which apparently single-handedly exterminated the entire species. Other peculiar New Zealand endemics indicating a long history of isolation of these islands from the rest of the world include the previously mentioned kiwis, the now-extinct moas, and the New Zealand short-tailed bats (Mystacinidae).

In contrast to such narrow endemics are cosmopolitan taxa, organisms that are widely distributed throughout the world. No genus, species, or family is truly cosmopolitan, although our own species, *Homo sapiens*, and certain microorganisms, come close. Some plant, animal, and microbe species are now widely distributed because humans have intentionally or inadvertently introduced them throughout the world. Terrestrial organisms that have achieved nearly worldwide distributions by natural means include the peregrine falcon (*Falco peregrinus*), the diverse plant genus *Senecio* (groundsel), and the bat family Vespertilionidae (Figure 10.4). Although they do not occur on Antarctica and on some remote islands, these and other exceptionally widespread taxa are often said to be cosmopolitan. Numerous minute animals and plants, such as protozoans, algae, and fungi, also have extremely broad ranges because their resistant life stages are dispersed widely by water or wind (see Wilkinson 2001; Finlay 2002).

Even in the sea—where any organism could, theoretically, swim around the world unimpeded by land barriers—there are relatively few species or genera that are actually found in all oceans. The most notable exceptions are certain whales and some invertebrates that have been widely dispersed by the whales or by ships. On the other hand, some kinds of freshwater organisms are surprisingly widely distributed, especially considering that other groups, such as fish and mollusks, contain some of the most narrowly distributed endemics. Most freshwater plant families, and even genera and species, tend to have broad ranges. Some species, including many duckweeds (Lemnaceae), some aquatic ferns (*Azolia, Salvinia, Marsilea*), water milfoil (*Myriophyllum*), and hornwort (*Ceratophyllum*), are nearly cosmopolitan. Similarly, several genera and even species of freshwater zooplankton (e.g., the water flea, *Daphnia*, and other tiny invertebrates, such as rotifers and tardigrades) are distributed

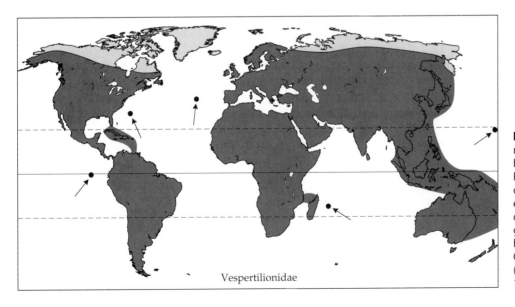

FIGURE 10.4 The nearly cosmopolitan distribution of the bat family, Vespertilionidae. Representatives of this group occur on all of the continents except Antarctica and have colonized isolated archipelagoes (indicated by arrows and black dots) such as Hawaii, the Galápagos, and the Azores. (After Koopman and Jones 1970.)

Vespertilionidae

worldwide, or nearly so. The key to these wide distributions appears to be dispersal (see Chapter 6). Freshwater plants are dispersed largely by wading and swimming birds, which carry the seeds or plantlets from one pool or lake to another. The widely distributed freshwater zooplankters have resistant life history stages, often fertilized eggs that are capable of surviving long periods of desiccation as they are transported by water birds or blown like dust in the wind.

Few species are truly cosmopolitan. However, many families and genera with exceptionally wide distributions contain at least some species that also have very large geographic ranges, indicative of their broad ecological tolerances and their capacities for dispersing long distances with or without human assistance. On the other hand, many high-ranking taxa, such as orders and classes, are essentially cosmopolitan, because the ecological diversity within these groups is broad enough to include forms that can exist in most terrestrial or aquatic habitats, and also because these groups are old enough to either have had historical opportunities to colonize most parts of the world, or to have occupied ancestral landmasses (e.g., Gondwanaland or Laurasia) prior to their fragmentation into the modern land masses.

The Origins of Endemics

The origins of endemics are indicated by a variety of terms. An **autochthonous endemic** is one that differentiated where it is found today. These taxa are particularly important in the analysis of vicariance because in order to infer that a particular geographic feature became a barrier to dispersal within the range of an ancestral species, one must have confidence that current geographic distributions of sister-taxa on either side of that barrier represent their original areas of differentiation (see allopatric speciation mode I and Figure 7.10 in Chapter 7, and a more detailed discussion in Chapter 12). A geographic area that contains two or more non-related autochthonous endemics is formally defined as an **area of endemism**, a concept of central importance to much of modern historical biogeography that will be developed formally in Chapter 12. Alternatively, an **allochthonous endemic** is one that originated in a different location from where it currently survives. Prime examples of allochthonous endemics include a number of relics, organisms such as the tuatara that were once widespread but are now confined to a very small region.

There are two kinds of relics, taxonomic and biogeographic. **Taxonomic relics** are the sole survivors of once diverse taxonomic groups, whereas **biogeographic relics** are the narrowly endemic descendants of once widespread taxa. Often the two categories coincide, which is often true for organisms called **living fossils**. One example, already mentioned, is the tuatara of New Zealand. Another is the monito del monte (*Dromiciops gliroides*), a primitive marsupial that superficially resembles some opossums and is restricted to southern beech forests of Chile and Argentina. Still another is the coelacanth (*Latimeria*), known only from the deep waters of the tropical Indian Ocean. This "primitive" fish is the only living member of the lobe-finned fishes, the crossopterygians, a group that was widely distributed in freshwater habitats as well as in oceans and shallow epicontinental seas in the Paleozoic. An example of a plant relict is the ginkgo (*Ginkgo biloba*, Ginkgoales), a gymnosperm native to a small region in eastern China, the sole survivor of a group of primitive conifers that was quite diverse in the Mesozoic. Now the ginkgo is a widely distributed ornamental tree, valued for its unusual, aesthetically pleasing form; ability to tolerate drought, poor soil, and air pollution; and supposed therapeutic qualities.

The terms **paleoendemic** and **neoendemic** are used to identify old and recently formed endemic species, respectively. One can see immediately that the use of such terms requires judgments, usually subjective ones, about the origins of endemics. In the previous paragraph we mentioned several examples of ancient relicts, all of which could also be considered as paleoendemics.

Also easy to identify are some very recent endemics, those of Quaternary Period (1.8 to 0.1 million years B.P.). As pointed out in Chapter 9, the Pleistocene was a time of great climatic and biogeographic change. Within just the last 10,000 years, many species ranges have shifted dramatically in response to the warming climate and retreating ice sheets. The ranges of many once widespread species, especially those of cool mountain climates, have contracted, so that now only small, isolated populations are found. An example is the bristlecone pine (*Pinus longaeva*), which is now restricted to arid, rocky microenvironments just below timberline on a few isolated mountains in the Great Basin of California and Nevada, but was widely distributed at lower elevations during the most recent glacial period (Figure 10.5, see also Figure 4.21). The saltbush genus *Atriplex* is widely distributed in western North America, but several endemic polyploid forms occur in distinctive habitats surrounding Great Salt Lake in Utah. Since they occur in areas that were covered by the water of

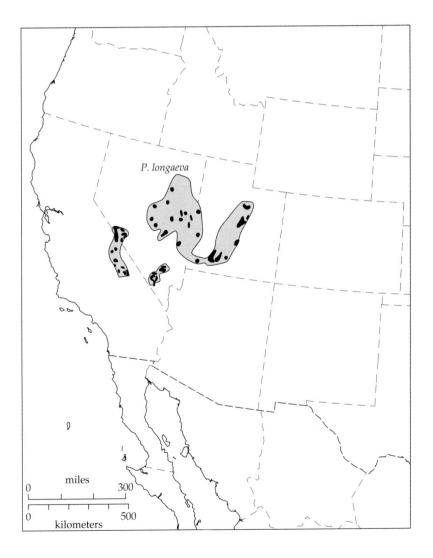

FIGURE 10.5 Current (black) and late Pleistocene (shaded) distributions of the bristlecone pine, *Pinus longaeva*. This species is now restricted to a few high mountain localities in the Great Basin, but only a little more than 10,000 years ago it occurred at lower elevations and was much more widespread. For a photograph of this species, see Figure 4.21. (After Wells 1983.)

Pleistocene Lake Bonneville—of which the Great Salt Lake is a small remnant—they have almost certainly formed within the last few thousand years (Stutz 1978). On the other hand, many desert-dwelling species of plants and animals have expanded their distributions greatly since the beginning of the Holocene—often into valleys that were previously covered with large lakes. At the same time, boreal forest and tundra species have reinvaded large areas of northern North America and Eurasia that were previously glaciated. Species restricted to these newly colonized areas, whether their ranges are large or small, most certainly speciated very recently—including the sticklebacks and lake whitefish in lakes across Canada (Bernatchez and Wilson 1998; Schluter 2000)—and thus are good examples of neoendemics.

As pointed out in Chapter 4, the restriction of a taxon to a particular geographic range is a consequence of both historical events and ecological processes. Ecological explanations must be invoked to explain the present range limits of an endemic. The survival and extinction of its local populations affect its ability to persist in particular localities, and dispersal processes affect its ability to colonize (or re-colonize after local extinction) favorable but isolated localities (see also Chapter 6). Abiotic and biotic limiting factors determine the distribution of habitable places within the geographic range and prevent the species from expanding at the periphery of its range. Together, dispersal ability and limiting factors determine the nature of the barriers that prevent the species from occurring in favorable but distant areas.

One the other hand, historical events must be invoked to explain how the taxon became confined to its present range and to reconstruct the geographic origin, spread, and contraction of the taxon. As one looks for the influence of historical events, such as the formation of barriers by drifting continents, changing sea levels, uplifting mountains, and glaciation (i.e., TECO events), it must also be kept in mind that there were other taxon-specific events occurring at the same time. Further, geological, climatic, and other environmental changes cause the expansion and contraction of the ranges of many different species, allowing new combinations of organisms to come into contact and to limit each other's distributions through biotic interactions. Thus, while investigators may search for satisfying, simplistic explanations for the origins of endemics, what they often find is a complex picture, explained in part by Earth history and in part by past and present ecological processes—sometimes mixed in ways that are hard to disentangle, although our ability to do so is getting better all the time (see Chapter 12).

Provincialism

Terrestrial Regions and Provinces

When the ranges of organisms are examined closely, it can be seen that endemic forms are neither randomly nor uniformly distributed across the Earth, but instead are clumped in particular regions. Three patterns are apparent. First, the most closely related species tend to have overlapping or adjacent ranges within restricted parts of continents or oceans. Second, completely unrelated higher taxa—for example, certain plant and animal orders and classes—often show similar patterns of endemism. Third, a small but significant number of taxa have markedly disjunct ranges, with species living in widely separated areas on different continents or islands. The first two patterns make it possible for biogeographers to identify circumscribed regions of the Earth's surface that share common, taxonomically distinctive biotas. The third pattern encourages us to search for historical explanations for how these organisms came to be distributed among such widely separated regions.

Provincialism was one of the first general features of land plant and animal distributions noted by such famous nineteenth-century biologists as the phytogeographers Schouw (1823), Hooker (1853), and de Candolle (1855); and the zoogeographers Sclater (1858) and Wallace (1876). As soon as biologists traveled among different continents, they were impressed by the differences in their biotas (e.g., "Buffon's law" described in Chapter 2). The limited distributions of distinctive endemic forms suggested a history of local origin and limited dispersal. As discussed in Chapter 2, much biogeographic research has been devoted to identifying centers of origin—areas where one or more groups of organisms originated and began their initial diversification (e.g., Udvardy 1969; Nelson and Platnick 1981; Funk 2004; see also Chapter 12). There have been complementary efforts to find evidence of historical barriers that blocked the exchange of organisms between adjacent regions, and of historical corridors that allowed dispersal between currently isolated regions. The result has been a division of the Earth into a hierarchy of regions reflecting patterns of faunal and floral similarities. In order of decreasing size, the common subdivisions are usually referred to as **realms** or **regions**, **subregions**, **provinces**, and **districts**.

The largest units are the most general. As mentioned in Chapter 2, the division of the Earth into biogeographic regions (see Figure 2.6, and endpapers) recognizes that these areas contain endemic and closely related taxa in many different groups. These large units have often been divided into hierarchically nested subunits on the basis of distinctive endemic biota. A classic example of a biogeographic region, and one that was recognized by Sclater and Wallace, is the Australian Region, whose terrestrial fauna is typified by marsupials (kangaroos, koalas, wombats, possums, etc.), although native placental mammals are also present, and whose flora contains unique eucalypts (Myrtaceae) and proteads (Proteaceae). Within Australia, biogeographers have classically recognized two major subregions (Figure 10.6). One, the Eyrean Subregion, named for the giant Lake Eyre Basin, encompasses the entire central two-thirds of the continent. It is now a vast arid and semiarid area without major mountain ranges or other internal barriers. The other subregion comprises the wetter fringe of Australia, and is further subdivided into three parts. The northern portion, called the Torresian Province, is a warm tropical belt that contains animals and plants with close affinities to those in New Guinea and sometimes also Southeast Asia. Their isolation is relatively recent, often only about 10,000 years old. During glacial episodes of the Pleistocene, when low sea levels created a landbridge across the present Torres Strait, many of these organisms ranged continuously from Australia to New Guinea. The southeastern portion of Australia, including Victoria, Tasmania, and some of the surrounding islands, is called the Bassian Province, and is inhabited by animals and plants adapted to cool, mesic temperature climates, including the southern beech forests, which contain *Nothofagus* and other relicts dating back to the ancient southern continent of Gondwanaland. The other province, called Westralia, includes the southwestern corner of the continent, where great numbers of endemic forms reside. Many of the taxa shared between the Mediterranean climatic regions of Australia and South Africa, such as certain groups in the plant family Proteaceae, occur in Westralia.

The Australian Region is one of the six large units used by Sclater to describe the world distributions of bird families and genera (see Figure 2.6). These were essentially the same six zoogeographic regions later adopted by Wallace in his major treatise on zoogeography (see endpapers). In the North-

☐ Eyrean
☐ Westralia
■ Torresian
■ Bassian

FIGURE 10.6 Biogeographic provinces of Australia. Australia is typically divided into two subregions: the Eyrean Subregion, which includes the entire arid center of the continent; and the strip of moister, mostly forested environments around the northern, eastern, and southwestern periphery. The latter is divided into three provinces, designated the Torresian, the Bassian, and Westralia, respectively. All three provinces are rich in endemics and in groups with representatives on other southern continents. The Bassian province shares southern beeches (*Nothofagus*) with New Zealand, New Caledonia, and South America, while Westralia shares several disjunct plant groups with the Mediterranean climatic region of South Africa.

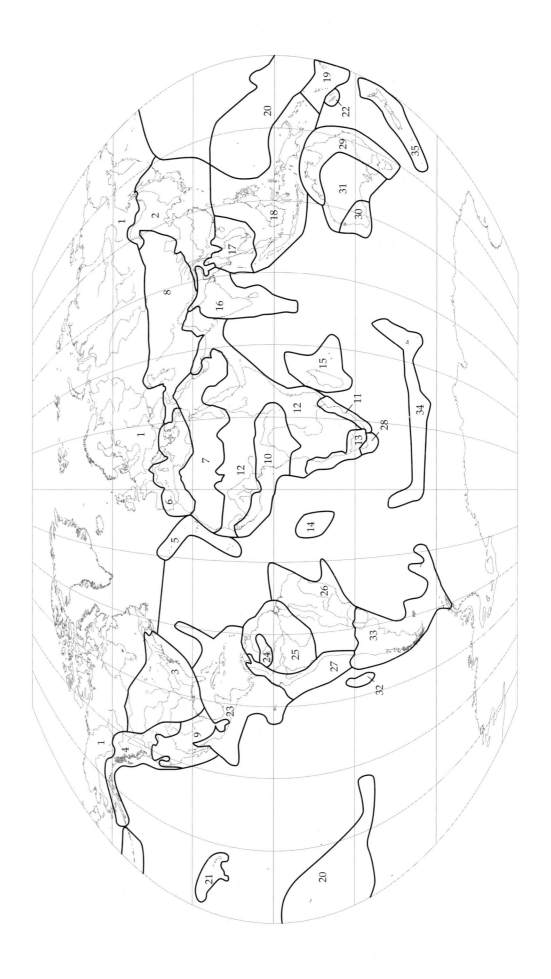

◀ **FIGURE 10.7** Division of the world into biogeographic regions based on the distributions of land plants. While the major regions have been subdivided to a greater extent, comparison with Figure 2.6 and with Wallace's map in the endpapers shows the close correspondence between the divisions used for plants and animals. (After Takhtajan 1986.)

ern Hemisphere, landmasses north of the tropical zone are called the **Holarctic**, which is composed of the **Nearctic** (North America) and **Palearctic** (Eurasia and northernmost Africa) regions. The remaining regions are primarily tropical: the **Neotropical** (Central and South America, and the West Indies), the **Ethiopian** (Africa, south of the Sahara; and Madagascar), and the **Oriental** (Southeast Asia, and the adjacent continental islands). The many oceanic islands positioned between the Sunda and Sahul shelves harbor a large number of endemic species and are often referred to collectively as Wallacea. The remaining faunas of the oceanic islands in the Pacific basin are anomalous in this classification scheme because they either contain a small number of taxa from adjacent continents and have relatively few unique groups, or have their own unique sets of adaptive radiations (e.g., the Philippine and Hawaiian archipelagoes).

The majority of articles and books on the distributions of organisms now use these six regions as fundamental descriptors of animal distributions, and this classification scheme has become one of the primary empirical foundations of biogeography. This is a tribute to the early zoogeographers who created it, who were working with woefully incomplete information on distributions, and who were using classifications that did not accurately reflect phylogenetic relationships. Moreover, these pioneering scientists were working without an understanding of the underlying events in Earth history that have created barriers to dispersal or corridors for biotic movements that explain many of the similarities and differences that exist between the great biogeographic regions.

The Sclater-Wallace classification scheme is basically similar to those used for plants, but there are a number of significant differences (Figure 10.7). Disregarding for the moment the fact that phytogeographers call the provinces of vertebrates "regions," and the districts "provinces," there are other important differences that reflect the distinctive characteristics of plants. Vegetation types are tightly restricted by abiotic factors, such as temperature and rainfall. Consequently, regions of endemism are more clearly defined by climatic and other physical barriers for plants than for animals, which are often able to surmount climatic barriers by dispersal or by physiological and behavioral adaptations to tolerate stressful conditions. Hence, the tip of South Africa, which has a Mediterranean climate, is a very distinctive phytogeographic region. This small area has an exceedingly rich flora, and about 90% of its species are endemic (Dyer 1975; Goldblatt 1978). In contrast, most animal groups of southernmost South Africa are not restricted to that zone and are not spectacularly diverse (Werger and van Bruggen 1978). South America has been divided into numerous provinces based primarily on the distributions of plant taxa. As elsewhere, these provinces are characterized by particular combinations of temperature, precipitation, and soil. However, Cabrera and Willink (1973), who defined the provinces shown in Figure 10.8, pointed out that distinctive animal species also are largely restricted to or exceptionally abundant in those regions. Biogeographers continue to work towards dividing continents into more or less natural biotic provinces based on a common evolution-

FIGURE 10.8 Division of South America into biogeographic provinces based primarily on the distributions of land plants. Note that the configuration of the provinces corresponds closely to geological and climatic features, such as the Andes, showing the influence of climate and soil on the distributions of plant groups and vegetation types. (After Cabrera and Willink 1973.)

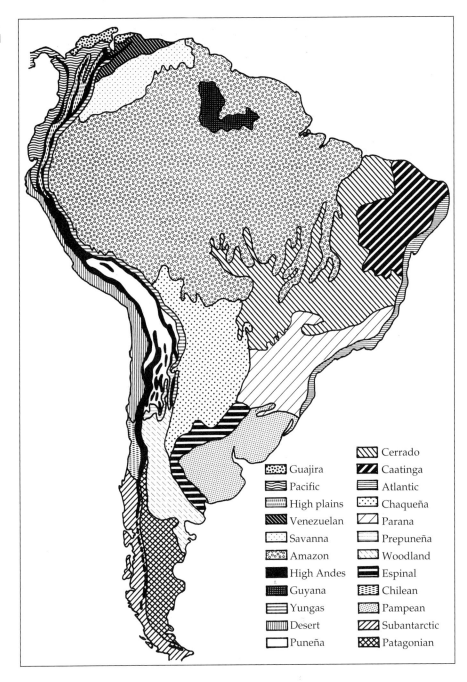

	Guajira		Cerrado
	Pacific		Caatinga
	High plains		Atlantic
	Venezuelan		Chaqueña
	Savanna		Parana
	Amazon		Prepuneña
	High Andes		Woodland
	Guyana		Espinal
	Yungas		Chilean
	Desert		Pampean
	Puneña		Subantarctic
			Patagonian

ary history shared by flora and fauna (e.g., Figure 10.9 shows a recent example for North America).

As implied above, the distinctive floras and faunas that are used to characterize regions and provinces have been shaped by both ecological and historical factors. Continental taxa have been stranded on certain landmasses because their ranges have been limited by both abiotic and biotic factors (see

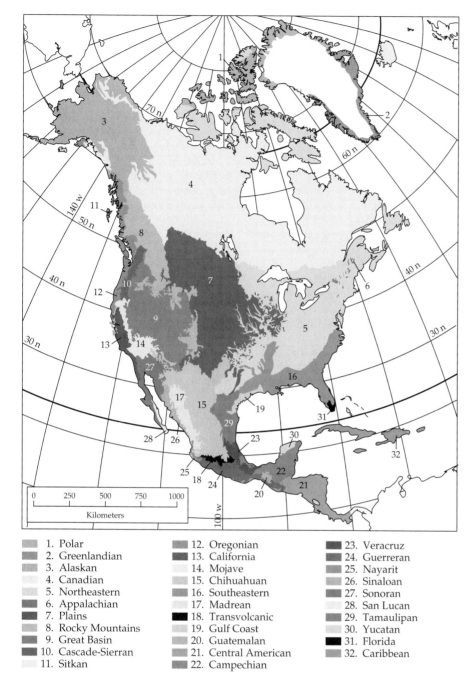

FIGURE 10.9 Biogeographic provinces for North America based on combined distributions of a number of plant and animal taxa. Note that, as for South America (Figure 10.8), the province boundaries largely conform to geological and climatic features. As such, there are more provinces in the western half of North America, with its extreme topographic and climatic complexity, than in the eastern, more homogeneous, half of the continent. Compare with Figure 10.10 and notice that the boundaries of the desert provinces are in close accord, but not identical, with one another. (After D. E. Brown et al. 1998.)

1. Polar
2. Greenlandian
3. Alaskan
4. Canadian
5. Northeastern
6. Appalachian
7. Plains
8. Rocky Mountains
9. Great Basin
10. Cascade-Sierran
11. Sitkan
12. Oregonian
13. California
14. Mojave
15. Chihuahuan
16. Southeastern
17. Madrean
18. Transvolcanic
19. Gulf Coast
20. Guatemalan
21. Central American
22. Campechian
23. Veracruz
24. Guerreran
25. Nayarit
26. Sinaloan
27. Sonoran
28. San Lucan
29. Tamaulipan
30. Yucatan
31. Florida
32. Caribbean

Chapter 4), and because major barriers to dispersal have restricted them to those landmasses for millions of years (see Chapters 8 and 12). To avoid complicating our general discussion here, we shall illustrate patterns of endemism in two groups (land birds and angiosperm plants) in two regions—Australia and South America (Box 10.1). Despite the enormous differences in lifestyle

BOX 10.1 *Endemic birds and plants of South America and Australia*

▌▌▌ South America and Australia share some important features of earth history and biogeography. Both were parts of the giant southern-hemisphere continent of Gondwanaland that drifted apart in the Mesozoic, and both were completely isolated island continents for most of the Tertiary. By the late Tertiary, however, continental drift had carried the northern, tropical parts of both South America and Australia into close proximity to large northern continents that were once part of the giant continent of Laurasia. South America joined with North America at the very narrow Central American Isthmus about 3.5 million years B.P. Australia and New Guinea, while part of the same plate and joined to each other by land connections for most of their history, drifted close to southeastern Asia (especially when the Sunda Shelf was exposed in the Pleistocene), but never established land connections with it.

Both South America and Australia contain many endemic taxa. Here we use patterns of endemism in angiosperm plants and birds to illustrate some features of their biogeography that reflect the long histories of isolation of these continents. Much attention has been given to the few groups, such as southern beeches (*Nothofagus*) and giant flightless birds (ratites, superorder Paleognathae), that are shared between the two continents and are probably of Gondwanan origin. Many of the endemic groups, however, are greatly differentiated, and their nearest presumed relatives do not necessarily occur on other southern continents. At the level of families, approximately 27 plant taxa are endemic, or nearly so, to South and Central America, and more than 18 are endemic, or nearly so, to Australia and New Guinea. The "nearly endemic" taxa have one or a few close relatives that occur in a different biogeographic region, apparently as a result of fairly recent long-distance dispersal: to Africa in the case of the South American Mayacaceae, Rapateaceae, and Vochysiaceae; and to Melanesian or Polynesian islands in the case of the Australian Balonpaceae, Corynocarpaceae, Davidsoniaceae, and Trimeniaceae.

Despite large differences in their capacities for dispersal and other traits, similar patterns of endemism occur in birds. If we consider only landbirds that are considered endemic as breeders, about 20 families containing a total of about 3121 species are endemic, or nearly so, to South and Central America. This compares with approximately 18 families containing more than 1415 species that are endemic breeders, or nearly so, in Australasia. Like plants, while no South American lineages of birds apparently have recently dispersed across the Atlantic to Africa, several groups of Australian birds, including the Psittacidae (lorikeets), Pachycephalinae (whistlers), Meliphagidae (honeyeaters), and Artamidae (wood swallows), have colonized Pacific islands.

Another pattern of endemism seen in both plants and birds is exhibited by a large number of additional taxa that are New World endemics. Many of the families included in the above analysis are not restricted to South America, but have spread northward to colonize at least some of the Central American Isthmus. In addition, many more taxa are most diverse in South America, and almost certainly originated there, but include several species or genera that have spread varying distances into subtropical and even temperate North America. These include the plants (such as those whose ranges are shown in Figure 10.26) that are arid habitat disjuncts, and the main groups of birds (e.g., flycatchers, vireos, wood warblers, and tanagers: see text) that breed in temperate North America but migrate to winter in the tropics of Central and South America. Thus, South American and Australian plants and birds illustrate the point that often (but not always: see text and Chapter 12) very different groups of organisms show strikingly similar patterns of endemism. ▌▌▌

and dispersal ability between birds and plants, it is readily apparent that these historically isolated southern continents are centers of endemism for both groups.

By itself, however, the observation that these regions contain endemic taxa is of limited interest. Even randomly selected areas would be expected to contain endemics, given the fact that most taxa have limited distributions. For example, Beadle (1981) has observed that Australia has about one-fifth as many endemic plant families as South America, but that it is also about one-fifth the size. The recognition of formal biogeographic provinces implies not only that each province contains distinctive endemic taxa, but also that the biota within each province is more homogeneous than between adjacent areas. While this is what would be expected if the provinces have long been isolated by some combination of unique ecological conditions and barriers to dispersal, most biogeographic units have not been quantitatively analyzed to show that they do indeed contain relatively homogeneous assemblages. Much quantitative characterization of patterns of endemism still remains to be done, although several studies demonstrating how this might be accomplished are available (e.g., Flessa et al. 1979; Conran 1995). Another factor that has, until relatively recently, impeded a better quantification of endemism within bio-geographic provinces, is the quality of phylogenetic data used to identify distinct taxa. Consider, for example, passerine birds in the Australian Region. Until recently, many species were considered to be closely related to species elsewhere in the world. Based on a study using DNA-DNA hybridization techniques (Sibley and Ahlquist 1985), we now know that these species instead are distantly related to passerines of the northern continents, and probably represent a number of separate and old lineages that originated on the Australian plate after separation from Gondwanaland (Barker et al. 2002). The striking levels of morphological similarity to non-Australian passerines represent either convergent evolution, or retention of ancestral characteristics in the latter group.

Regardless of difficulties in the quantitative and phylogenetic characterization of biogeographic regions, at least four lines of evidence suggest that many of them do indeed indicate the pervasive influence of geography, geology, and climate on the historical origin, diversification, and spread of lineages—and that the legacy of this history is preserved in the distributions of contemporary taxa.

CONGRUENCE BETWEEN TAXONOMIC AND BIOGEOGRAPHIC HIERARCHIES. A convincing line of evidence is the correspondence between the hierarchy of taxonomic categories and the hierarchy of biogeographic regions, with the clearest patterns occurring at the largest scales. Thus, higher taxa (orders and families) tend to have fairly broad distributions within a continental landmass, presumably reflecting the ancient origin and long confinement of major lineages, while progressively less differentiated forms (genera and species) tend to be confined to small areas nested within those regions. As pointed out earlier, nested patterns are required by the hierarchical classifications of both organisms and biogeographic regions, so this apparent pattern could be an artifact if based merely on taxonomic classifications. Nevertheless, the relationships between the two hierarchies suggest that the progressively finer biogeographic subdivisions reflect a history of less ancient and less severe barriers between more recently differentiated lineages (more on this in Chapter 12).

LINEAGE CONGRUENCE. The second line of evidence that biogeographic subdivisions reflect the long-standing influence of geological events and climatic patterns is that the locations of provinces and their boundaries, determined independently for different groups of organisms, tend to coincide. We have already described, above and in Chapter 2, how the division of the Earth into the six major terrestrial biogeographic regions was originally developed for birds by Sclater, then generalized to terrestrial mammals by Wallace, and finally extended to land plants and, by implication, all terrestrial organisms. The same kind of generality across different taxa often applies to the finer divisions as well. The North America deserts, for example, have long been divided into four provinces—the Great Basin, Mojave, Sonoran, and Chihuahuan (Figure 10.10)—based originally on the pioneering floristic studies of Forest Shreve (1942). This division reflects the distributions of many endemic plants and animals, which are typically restricted to only one or a subset of the four deserts but are fairly wide-ranging within the desert(s) where they occur. Indeed, recent molecular-based studies (e.g., Riddle et al. 2000; Zink et al. 2001; Nason et al. 2002) have revealed even finer subdivisions (e.g., the Baja Peninsula—the southern half, in particular, which harbors a high number of endemic lineages not occurring elsewhere within the Sonoran Desert). These patterns of endemism can in turn be attributed to the unique histories and

FIGURE 10.10 The division of arid North America into four desert provinces: Chihuahuan, Sonoran, Mojave, and Great Basin. The first three are relatively hot, low-elevation deserts, distinguished primarily by an east-west gradient of seasonality of precipitation: rain falls in summer in the Chihuahuan, in both summer and winter in the Sonoran, and in winter in the Mojave. The Great Basin, due to its higher latitude and elevation, is a cold desert, or shrub-steppe. Each province has distinctive endemic plants and animals.

☐ Great Basin
■ Mojave
☐ Sonoran
■ Chihuahuan

environments of the deserts. On one hand, during the glacial periods of the Pleistocene, the climate of southwestern North America was cooler and wetter than at present, and the deserts contracted to lowland basins isolated by barriers of grassland and woodland at higher elevations (Figure 10.11). On the other hand, even during the interglacial periods, when these barriers were removed, the distinctive precipitation and temperature regime of each desert limited distributions and prevented wholesale mixing of the desert biotas. Prior to the Pleistocene, geological events during the Miocene and Pliocene transformed the landscape in southwestern North America into the complex basins and plateaus bracketed by mountain ranges and seas that delineate each of the deserts today. The result is that each desert has distinctive endemic plants and animals. For example, the saguaro, *Carnegia gigantea*, the Sonoran Desert toad (*Bufo alvarius*), the regal horned lizard (*Phrynosoma solare*), Bendire's thrasher (*Toxostoma bendirei*), and Bailey's pocket mouse (*Chaetodipus baileyi*), are but a few of the many plant and animal species that are endemic to the Sonoran Desert of southern Arizona and northwestern Mexico. Such congruence in patterns of endemism across different lineages of organisms suggests that they have responded similarly to geographic variation in important geological, topographic, and climatic features of the Earth (we will return to this topic in Chapter 12).

Freshwater organisms include forms with both some of the most cosmopolitan, and some of the most narrowly endemic, distributions. The former, as mentioned above, tend to have life history stages that can withstand desicca-

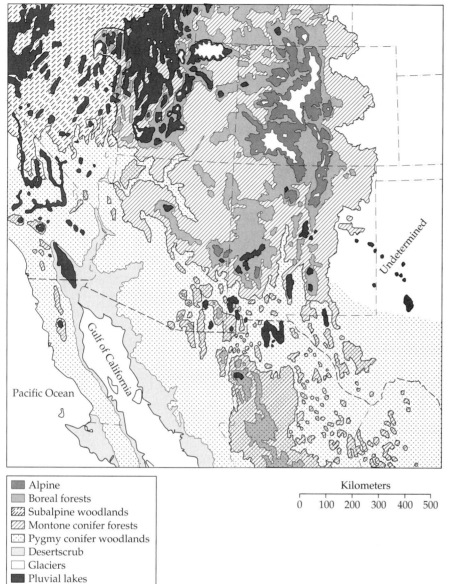

FIGURE 10.11 Reconstruction of late Pleistocene vegetation in southwestern North America. Note, in particular, the restricted distribution of desertscrub vegetation (contrast with distribution of modern deserts shown in Figure 10.10). This reconstruction is based on plant macrofossil records extracted from woodrat (genus *Neotoma*) middens in rocky terrain, and therefore could underestimate distribution of desertscrub habitats outside the home-ranges of the woodrats. (After Betancourt et al. 1990.)

Undetermined

Gulf of California

Pacific Ocean

Alpine
Boreal forests
Subalpine woodlands
Montone conifer forests
Pygmy conifer woodlands
Desertscrub
Glaciers
Pluvial lakes

Kilometers

0   100   200   300   400   500

tion and are readily dispersed. The latter tend to lack such resistant stages and to be equally intolerant of exposure to air and to seawater. As a consequence of their limited dispersal abilities, some freshwater groups, such as fishes and mollusks, show clear patterns of endemism that correspond to the Sclater-Wallace biogeographic scheme. Furthermore, these same taxa typically show pronounced patterns of provincialism and endemism within continents as well. Each major river drainage or lake basin has distinctive, well-differentiated endemic forms (e.g., for fishes, see Figure 10.12; Hocutt and Wiley 1986; Mayden 1992). Thus, each of the world's great river systems, such as the Amazon, Congo, Volga, and Mississippi, has a diverse fish fauna that contains many endemic species and genera. And such endemism is found even at much smaller spatial scales. Each of the Great Lakes of the Rift Valley of Africa, such as Victoria, Tanganyika, and Malawi, has been a major center of endemic speciation (see Chapter 7), and supports unique lineages containing many closely related species of both fish and mollusks. These same groups are spectacularly

FIGURE 10.12 Biogeographic provinces for North American freshwater fishes. Note that most of these provinces correspond to major drainage basins, such as the Great Lakes, and the Mississippi and Colorado Rivers. The two numbers given are the numbers of families and species, respectively, and the values in parentheses are the percentages of the species that are endemic to each province. (After Burr and Mayden 1992.)

diverse not only in the Mississippi River itself, but also in many of its tributaries. Especially in the Ozark and Appalachian highlands, even the smaller drainages contain endemic forms of darters (*Etheostoma*), shiners (*Notropis*), and clams. Giant Lake Baikal in Siberia contains not only endemic fishes and invertebrates, but also the unique Baikal seal (*Phoca sibirica*) that colonized sometime during the Pleistocene when the now landlocked lake drained into the Arctic Ocean. In the Mojave Desert in the vicinity of Death Valley, thermal springs within the drainage of the Pleistocene Amargosa River contain a diverse array of endemic snail lineages (family Hydrobiidae) in the genera *Tryonia* (Hershler et al. 1999) and *Pyrgulopsis* (Liu et al. 2003), and fish in the genus *Cyprinodon* (Echelle and Dowling 1992). Because of the inability of most freshwater fishes and certain invertebrates to disperse across either land or sea, their distributions probably reflect historical events more faithfully than do those of most other organisms (see Berra 2001). These groups offer fruitful systems for research on speciation and diversification, as well as on many aspects of historical biogeography.

BIOGEOGRAPHIC LINES. Another observation that reflects the pervasiveness of provincialism is the rapid turnover of taxa at the boundaries between regions. Traditionally, biogeographers have drawn lines to fairly precisely define the limits of regional biotas. Although these biogeographic lines are usually derived for one taxon at a time, they often coincide with geological or climatic barriers that have prevented the dispersal of many kinds of organisms. A prime example of such a sharp boundary is Wallace's line, drawn between the Indonesian islands of Borneo and Sulawesi (Celebes) on the west, and Bali and Lombok on the east to mark the boundary between the Oriental and Australian Regions. Not all taxa show distributional boundaries corresponding precisely to Wallace's line, and other lines have been described to accommodate them (Figure 10.13). Nevertheless, two things about Wallace's line are noteworthy. First, most of the other lines deviate only slightly from the boundary described by Wallace. Second, Wallace's line corresponds almost exactly with the outer limit of the Sunda Shelf, the part of the continental shelf of Southeast Asia that was intermittently exposed by lowered sea levels during the Pleistocene (Hall 1998). Thus, although Wallace did not realize this, his line corresponds to the deep water marking the limit of historical land connections

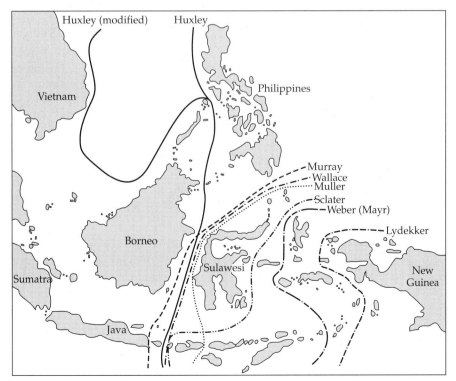

FIGURE 10.13 The biogeographic lines drawn by Wallace (and later biogeographers) to mark the boundary between the Oriental and Australian regions. The locations of the various lines indicate that different taxa have dispersed to different degrees into the islands of the East Indies from their continents of origin. The approximate limits of the continental shelves are given by Lydekker's line in the east and Huxley's modified line in the west, although the Philippines have a complex tectonic history (see Figure 10.22).

among the major East Indian islands, and between them and the Southeast Asian mainland. Lydekker's line (see Figure 10.13), just west of New Guinea, corresponds to the edge of the Sahul Shelf, which was exposed in the Pleistocene as part of the supercontinent that included both Australia and New Guinea. It marks the northwestern limit of the ranges of many species of the Australian Region. In between these two lines are islands surrounded by deep water that were not connected by Pleistocene landbridges to either Southeast Asia or Australia-New Guinea, and which contain an extraordinarily rich mixture of endemic species derived from ancestors from both regions that managed to disperse across the water barriers.

Boundaries between other biogeographic regions are not as distinct as Wallace's line. Such boundaries usually reflect a combination of two interrelated phenomena. First, the historical, geological, and climatic isolation of the regions was either originally not so discrete, or became less discrete following a major geological transformation allowing more dispersal between regions, causing the apparent boundary to be blurred. Second, different groups of organisms responded somewhat differently to the same historical events and climatic patterns, resulting in a lack of consistency across taxa. These two factors have conspired to make it difficult to draw a single line defining the boundary between the Nearctic and Neotropical regions. The efforts undertaken to draw this boundary illustrates the influence of both historical and ecological factors on the phenomenon of provincialism.

The Isthmus of Panama is the location of the historical separation between the two regions. It marks where the last marine barrier, separating the formerly isolated continents of North and South America, closed about 3.5 million years ago. When the Central American landbridge became available, many organisms took advantage of ecological opportunities and dispersed along it, an event called the Great American Interchange (see Biotic Interchange later in this chapter, and Chapter 12). Because the climate and habitats of Central America and southern Mexico are tropical, there was much biotic exchange between this area and tropical regions of northern South America. Thus, even though Central America and southern Mexico have many endemic forms, they are usually recognized as a subregion of the Neotropics. But the problem of where to draw the line remains. Often it is placed at the Isthmus of Tehuantepec, which marks both the approximate northernmost extension of tropical rain forest and a major lowland gap in the mountain cordilleras running the length of western North America. By no means do all Neotropical taxa have their northern limit there, however. Figures 10.14 and 10.15, which plot the northernmost and southernmost limits of the ranges of terrestrial mammals and freshwater fishes, respectively, illustrate the difficulty of defining the line separating the Nearctic and Neotropical regions. Each family has colonized southward or northward from its historical origin in North or South America in a somewhat different pattern, reflecting some combination of its capacity for dispersal and its tolerance for environmental conditions.

CONGRUENCE BETWEEN BIOTIC AND EARTH HISTORY. While the tendency of range boundaries of species and higher taxa to be coincident facilitates the objective recognition of biogeographic provinces, two more interesting questions can be asked: To what extent are the distributions of taxa a reflection of the history of the landmasses or water masses in which these organisms now live? And, to what extent can they be used to reconstruct the history of geographic changes on the Earth's surface? We will consider the relationships between phylogenetic, biogeographic, and Earth history in much more detail in Chapter 12, but finish our overview of terrestrial biogeographic regions in this chapter by pre-

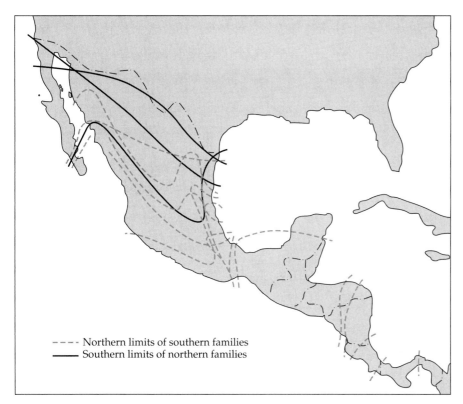

FIGURE 10.14 Northern limits of the ranges of Neotropical mammal families (dashed lines), and southern limits of the ranges of Nearctic families (solid lines). Note that the transition between these historically isolated mammal faunas occurs over a wide area, making it difficult to draw a definitive line separating the two biogeographic regions. Similarly, complex patterns occur in nearly all taxa (e.g., see Figure 10.12 for fishes).

- - - - Northern limits of southern families
———— Southern limits of northern families

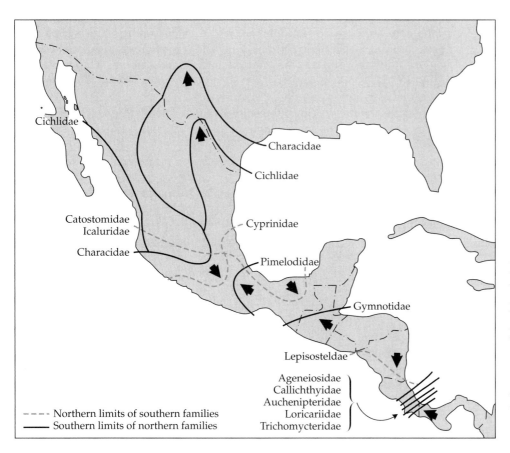

Cichlidae

Characidae

Cichlidae

Catostomidae
Icaluridae

Cyprinidae

Characidae

Pimelodidae

Gymnotidae

Lepisosteldae

Ageneiosidae
Callichthyidae
Auchenipteridae
Loricariidae
Trichomycteridae

- - - - Northern limits of southern families
———— Southern limits of northern families

FIGURE 10.15 Distributional limits of freshwater fish families of South American (dashed lines) and North American (solid lines) origin. Only two species of obligately freshwater fishes of South American origin have reached the United States, and North American forms extend no farther south than Costa Rica. Note that, as in the case of the mammals shown in Figure 10.14, no single line can be drawn to separate unambiguously the Neotropical and Nearctic faunas. (After Miller 1966.)

FIGURE 10.16 Global biogeographic kingdoms and regions based on modern biogeographic analyses. 1–2, Holarctic kingdom (= Laurasia): 1, Nearctic region; 2, Palaearctic region; 3–6, Holotropical kingdom (= eastern Gondwana): 3, Neotropical region; 4, Afrotropical region; 5, Oriental region; 6, Australotropical region. 7–12, Austral kingdom (= western Gondwanaland): 7, Andean region; 8, Cape or Afrotemperate region; 9, Antarctic region; 10, Neoguinean region; 11, Australotemperate region; 12, Neozelandic region. (After Morrone 2002.)

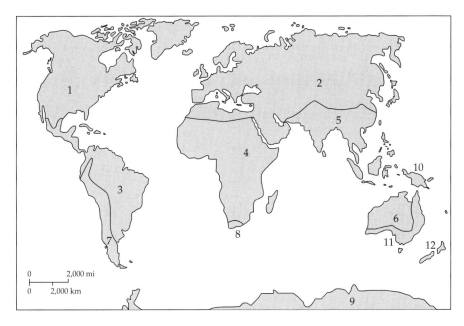

senting a recent update by Morrone (2002) that incorporates information from cladistic phylogenetic analyses and plate tectonics to create a new view of global provinciality (Figure 10.16). While this depiction looks very similar to the original Sclater-Wallace regions, it recognizes that several continents—in particular, South America and Africa—include a mixture of species with very different biogeographic and tectonic histories. For example, the South American biota can be split into a "Laurasic" component with biogeographic connections to the northern landmasses, and a "Gondwanic" component with connections to the southern end of the world. Thus, the original Neotropical Region is split into separate Neotropical and Andean regions, respectively, in Figure 10.16.

That an association should exist between plate tectonics and the development of global patterns of provinciality might be fairly obvious. However, the history of provinciality on a single continent is more difficult to ascertain because the timing and magnitude of geological events are often more difficult to reconstruct and until recently their influence on biotic distributions could only be inferred indirectly from a geographically and taxonomically detailed fossil record (see, for example, Figure 10.17). Much progress has been made in the past two decades, however, by exploiting a different kind of data—the molecular phylogenies of populations within species and across related species—to infer the roles of intracontinental, geological, and climatic histories on the biogeographic and evolutionary histories of extant lineages (see Chapters 11 and 12).

Marine Regions and Provinces

Numerous biogeographers have attempted to define regions and provinces in the oceans. Some of these classifications, however, are really ecological characterizations based on such criteria as water temperature, depth, and substrate. Such classifications often emphasize vertical rather than horizontal divisions of the three-dimensional marine realm—and understandably so, because there is much less difference among the marine biotas in different oceans than among terrestrial biotas on different continents. Not only are the oceans more

(A)

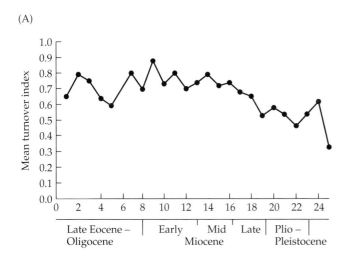

FIGURE 10.17 An example of changes in provinciality through time. Mean turnover index (A) in large mammalian carnivore and in herbivore species between eight sampling regions (B) in western North America from 44 million years ago (beginning of sampling interval 0–1) to the Recent (end of sampling interval 24–25). Changes indicate extent of provinciality, which remains more or less constant from Late-Oligocene through Late-Miocene time, followed by a general trend towards decreasing provinciality (e.g., a greater proportion of species that range across two or more sampling regions). (After Van Valkenburgh and Janis 1998.)

(B)

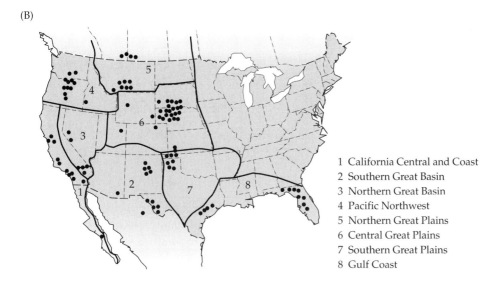

1 California Central and Coast
2 Southern Great Basin
3 Northern Great Basin
4 Pacific Northwest
5 Northern Great Plains
6 Central Great Plains
7 Southern Great Plains
8 Gulf Coast

interconnected than the continents, but many marine forms have tiny planktonic eggs or larvae that are passively and widely dispersed by ocean currents. Most marine plants and animals are cosmopolitan at the familial level, and many even at the generic level. Consequently, it is difficult to detect patterns in the distributions of marine taxa that clearly reflect the histories of the present water masses. This is especially true of open-water, or pelagic, organisms. Benthic plants and animals that live close to, on the surface of, or buried in, marine substrates tend to have more restricted distributions, and consequently, to exhibit more provincialism.

Part of the problem in reconstructing the histories of oceanic biotas is the very long and complicated history of most oceans. By studying fossil marine invertebrates of the Paleozoic and Mesozoic, we can see that during certain periods there was very little provincialism in some, if not the majority, of taxa. For example, in the Silurian, many genera of even benthic forms, such as brachiopods, gastropods (Boucot and Johnson 1973), and graptolites (Berry 1973), had nearly cosmopolitan distributions. On the other hand, the majority of phyla contain distinctive taxa that, at least during some periods, were

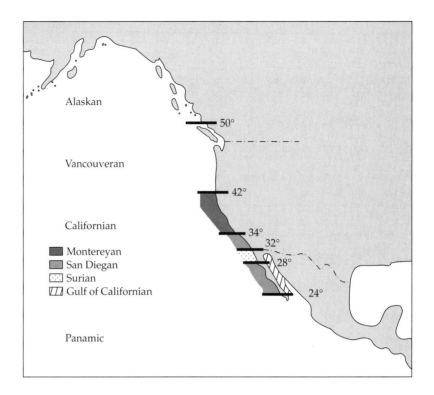

FIGURE 10.18 Map of the Pacific coast of western North America, showing the latitudinal boundaries of the marine provinces that have been designated for shallow-water benthic marine organisms. (After Southern California Coastal Water Research Project 1973.)

restricted in distribution to some early ocean, such as the Tethys (Hallam 1973a; Gray and Boucot 1979). So, marine invertebrates show an accordion effect, with ranges expanding and contracting, to produce alternating stages of provincialism and cosmopolitanism. This fact should cause us to be cautious in searching for legacies of ancient (i.e., pre-Cenozoic) Earth history in the distributions of present-day marine organisms.

Some history is preserved, however, in the distributions of certain taxa, especially those benthic forms with limited dispersal ability. An obvious horizontal pattern is latitudinal variation in species diversity and composition (Chapter 15). Several biogeographers have described marine provinces distributed latitudinally along continental coastlines (Figure 10.18; e.g., Valentine 1966; Pielou 1977b; Horn and Allen 1978) that are separated by zones of rapidly changing species composition. However, these latitudinal patterns appear to be primarily the result of environmental gradients in temperature and correlated abiotic limiting factors, rather than a legacy of unique historical events (see Chapter 7). Thus, the locations of the boundaries between provinces have shifted latitudinally with changing ocean temperatures during the Pleistocene (Valentine 1961, 1989; Valentine and Jablonski 1992), and the taxonomic composition of local assemblages has changed as species have expanded and contracted their ranges (Enquist et al. 1995).

There is, however, marine provincialism that more clearly reflects the influence of tectonic and oceanographic history on the origin and distribution of lineages. Warm tropical oceans have served as significant barriers to dispersal for many cold-adapted temperate and arctic organisms. Thus, the high-latitude seas represent centers of endemism and speciation for groups such as fishes,

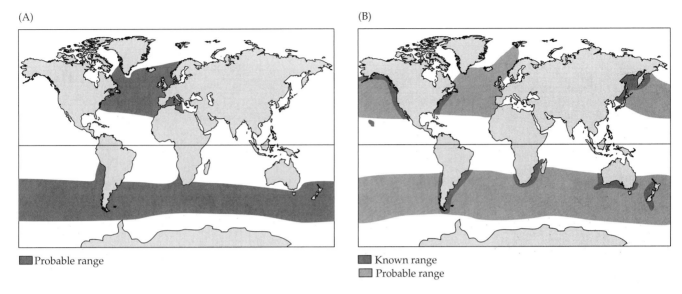

(A)

(B)

Probable range

Known range
Probable range

FIGURE 10.19   Amphitropical distributions of the long-finned pilot whale (*Globi-cephala melas*) (A), and the right whale (*Eubalaena glacialis*) (B). Warm tropical waters are powerful barriers to dispersal, isolating populations of these cetaceans in the Northern and Southern Hemispheres, and apparently preventing the pilot whales from colonizing the North Pacific. (After Martin 1990.)

seabirds (penguins and auks), and marine mammals (whales and seals). Other polar groups do not exhibit such provincialism, apparently because they have frequently dispersed across the equator in deep water, where temperatures are also very cold (an obvious impossibility for air-breathing marine birds and mammals). Some marine organisms exhibit amphitropical disjunctions: ranges that include the cold waters of both the Northern and Southern Hemispheres, but not the warm tropical oceans in between. Figure 10.19 shows two such examples for whales, but similar patterns are also found in fishes and invertebrates. Within the vicinity of Wallacea, several recent molecular studies have shown a surprising historical subdivision between Pacific and Indian Ocean species, including butterfly fish in the genus *Chaetodon* (McMillan and Palumbi 1995) the mantis shrimp, *Haptosquilla pulchella* (Barber et al. 2000), and seahorses, *Hippocampus* (Lourie et al. 2005).

Some shallow-water and coral reef organisms show significant differentiation among different oceans or coastal regions. Coral reef fish faunas are quite variable in composition, and certain areas in the Indo-Pacific and the Caribbean not only are centers of high species diversity, but also contain a number of endemic species and genera (Briggs 1974; Mora et al. 2003; Briggs 2004). There is also significant endemism at the generic level among the corals and mollusks. Fossils indicate that many genera of these invertebrate groups were widely distributed in the Mesozoic, which means that many of today's narrowly restricted forms are relics of once circumtropical forms (Vermeij 1978). Recent information suggests that many shallow-water marine organisms, even some of those with planktonic larval stages, do not disperse as far as once believed (Gaines and Bertness 1992; Palumbi 1997).

On the other hand, we are just beginning to learn about the biology and distribution of deep-water benthic organisms, including those inhabiting hydrothermal vents, cold water seeps, and the vast abyssal plains that constitute the majority of the ocean floor. At least some of these organisms seem to

FIGURE 10.20 Depiction of biogeographic provinces at deep sea hydrothermal vents based on distributions and relationships among specialized vent biotas. (After Van Dover et al. 2002.)

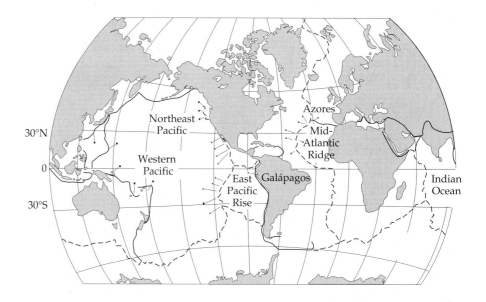

show remarkably little differentiation or endemism, suggesting that they may have enormous powers of dispersal, although the vent and seep faunas appear to exhibit greater levels of differentiation, providing for the delineation of distinct hydrothermal vent provinces (Figure 10.20; see also Figure 17.10). Clearly, the relationship between marine biogeography and historical tectonic events will be a fruitful area for future research as we explore one of the Earth's last frontiers.

Classifying Islands

Since Wallace (1876, 1880), biologists have attempted to classify islands as either continental or oceanic (see Table 10.2). Initially, such classifications were made on the basis of the composition of the insular biotas. As we shall see in Chapter 13, both kinds of islands typically have substantially fewer species than are found in comparable habitats on nearby continents. Continental islands, as their name implies, support plants and animals that are closely related to forms on the nearby mainland; these forms are often so little differentiated as to be considered populations of the same species. Further, continental islands usually have kinds of organisms that are poor over-water dispersers, such as large-seeded plants, amphibians, and nonvolant terrestrial mammals. Thus, they tend to have representative samples of the floras and faunas of nearby continents. These two features suggest recent land connections between continental islands and the mainland.

In contrast, oceanic islands typically have biotas of lower taxonomic richness, not only at the species level but at higher taxonomic levels as well. Often these forms are insular endemics that are well differentiated from their apparent relatives on nearby continents. An oceanic island, or an archipelago of many such islands, typically supports a group of species obviously more closely related to one another than to any continental form. This suggests that not only evolutionary divergence, but also speciation, has occurred within the island or archipelago. Further, the limited taxonomic richness of oceanic islands is strongly biased in favor of the kinds of organisms that are relatively good over-water dispersers, such as flying animals and plants with small wind- or bird-dispersed seeds. Some endemic island forms have become so

---

TABLE 10.2 *Some biogeographically interesting islands classified according to their modes of origin*

FULLY OCEANIC ISLANDS

Totally volcanic islands of fairly recent origin that have emerged from the ocean floor and have never been connected to any continent by a landbridge.

Midoceanic island chains or clusters formed from hot spots (HS) or along fracture zones (FZ) within an oceanic plate

Austral-Cook Island chain (HS or FZ); Carolines (HS); Clipperton Island (FZ); Galápagos Islands (FZ, Carnegie Ridge); Hawaiian Islands (HS); Kodiac Bowie Island chain (HS); Marquesas (HS); Society-Phoenix Island chain (HS, but some contribution by the Tonga Trench)

Island arcs formed in association with trenches

Aleutians (may have been part of the Bering Landbridge in the Cenozoic); Lesser Antilles; Lesser Sunda Islands; Marianas; New Hebrides; Ryukyus (may have been associated with neighboring islands); Solomons; Tonga and Kermadec

Islands formed at presently spreading midoceanic ridges

Ascension Island; Azores (some islands have continental rocks); Faeroes; Gough Island; Tristan da Cunha

CONTINENTAL ISLANDS

Formed as part of a continent and subsequently separated from the landmass. Some of these have added oceanic material since they were formed.

Islands permanently separated from the mainland since the split was initiated (time of final separation in parenthesis)

Greater Antilles (80 million years B.P.); Kerguelen Island (Upper Cretaceous); Madagascar (ca. 100 million years B.P.); New Caledonia (ca. 50 million years B.P.); New Zealand (80–90 million years B.P.); Seychelles (65 million years B.P.); South Georgia (45 million years B.P.)

Island groups with connections of some islands, but not others, to the mainland

Canary Islands

Islands most recently connected to some mainland in the Pleistocene by land or an ice sheet

British Isles; Ceylon; Falklands; Greater Sunda Islands; Japan and Sakhalin; Newfoundland and Greenland; New Guinea (with Australia); Taiwan; Tasmania

---

differentiated that they have lost their structures and capacities for dispersal (e.g., parachutes, hooks, and awns have been lost from seeds; reduction and loss of wings have occurred in insects and birds; see Chapter 14, and Carlquist 1965, 1974). Their phylogenetic affinities, however, suggest that these organisms lost their dispersal abilities secondarily, after their ancestors had colonized the islands. Together, these characteristics of their biotas suggest that oceanic islands have had a long history of isolation, perhaps without any previous land connections to continents.

From Wallace's time until the 1960s, this division between continental and oceanic islands, based on their biotic composition, appeared to work well. It was presumed to reflect geological history in a straightforward way: continental islands, located on the continental shelves, were presumed to have had recent land connections to the mainland during periods of lowered sea levels; whereas oceanic islands, located in deeper waters, were presumed never to have been connected to the mainland, but to have been pushed up from the sea floor, usually as a result of volcanism. As knowledge of plate tectonics and insular geology accumulated, however, this clear dichotomy between continental and oceanic islands was called into question. Some islands, isolated from all continents by oceans hundreds of kilometers wide and hundreds of fathoms deep and long thought to be oceanic, were found to contain continental (sialic or andesitic) rocks. And, as tectonic plate movements were reconstructed, it was realized that these islands had once been part of continental plates. Madagascar, New Zealand, and New Caledonia, for example, were recognized as fragments of the ancient supercontinent of Gondwanaland, and the islands of the Greater Antilles were possibly connected via a continuous land-bridge (along with the Aves Ridge), with northern South American (Figure 10.21; GAARlandia model of Iturralde-Vinent and MacPhee 1999). In the wake of this accumulating geological evidence for ancient continental connections, the biological and biogeographic evidence of isolation was reexamined for these islands (as was the case for continents; see Chapter 8). Many elements of their floras and faunas may indeed be descendants of over-water colonists (e.g., Lack 1976). Often, however, their biotas contain telltale elements, ignored or misinterpreted by earlier biogeographers, that have until quite recently been assumed to more likely be relictual survivors of ancient continental floras and faunas. Examples include the southern beeches (*Nothofagus*) and fresh-water galaxioid fishes of New Zealand and New Caledonia, the elephant birds of Madagascar, and the moas and kiwis of New Zealand, all generally recognized as remnants of the Gondwanan biota that owe their disjunct distributions to vicariance (but see discussion in Chapter 12 of recent findings supporting a significant role as well for dispersal across continents and islands of a former Gondwanaland).

There are also examples of the converse situation for presumably continental islands. On one hand, increasing information on bathymetry has permitted much more accurate estimation of the timing and extent of past land connections. Of particular importance is the depth of the water between island and continent in relation to Pleistocene fluctuations in sea levels. The 100 m depth contour is now known to mark the approximate sea level during the most recent glacial period (see Chapter 9). Islands isolated by shallower depths can usually be assumed to have had their continental land connections severed within the last 10,000 years. This assumption is generally consistent with the biogeographic evidence, such as the occurrence of a diverse and only moderately differentiated terrestrial mammal fauna on the islands of Great Britain and of the Sunda Shelf (e.g., Java and Sumatra), which had extensive land connections with Europe and Southeast Asia, respectively, during the latest Pleistocene.

On the other hand, we cannot simply assume that about 10,000 years ago, all continental islands had their mainland connections severed, and that this stopped all biotic interchange. Just as with oceanic islands, both the geological and biogeographic histories of continental islands can be complex. One example concerns the more isolated islands of the East Indies, especially Borneo and the Philippines (Heaney 1985, 1986, 1990). Both geological and biogeographic evidence suggest that Borneo and a subset of the Philippine islands had past land connections to the islands of the Sunda Shelf and to the mainland of Southeast Asia, but not during the latest Pleistocene (Figure 10.22). Thus, these islands contain much more differentiated endemic land plants, terrestrial mammals, and other organisms, than other islands on the Sunda Shelf. Within the Phillipine archipelago, many current islands were joined during the late Pleistocene into several much larger landmasses (Figure 10.22), whereas a number of smaller islands of volcanic origin have never had such connections.

Also, just because most continental islands have not had land connections with the mainland within the last 10,000 years does not necessarily mean that there has not been more recent biotic interchange. Forms with well-developed capabilities for over-water dispersal, such as bats and some birds, tend to be especially well represented, even on tiny islands, and to exhibit virtually no evidence of evolutionary divergence. For example, the continental island of

Latest Eocene-Early Oligocene (35–32 million years B.P.)

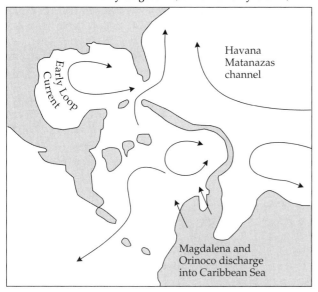

Uplift of Greater Antilles-Aves Ridge (GAARLandia)

Late Oligocene-Middle Miocene (30–14 million years B.P.)

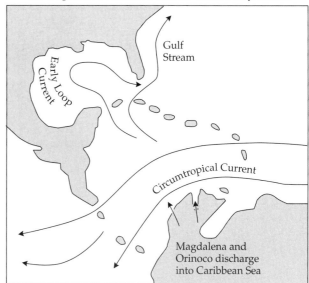

Drowning of Greater Antilles-Aves Ridge

Late Miocene (5–6 million years B.P.)

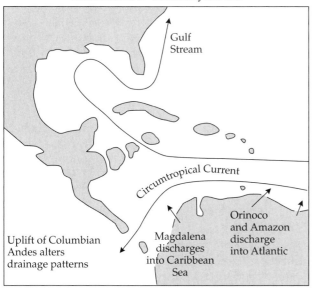

Closure of Havana-Matanzas Channel

Pliocene-Holocene (4–0 million years B.P.)

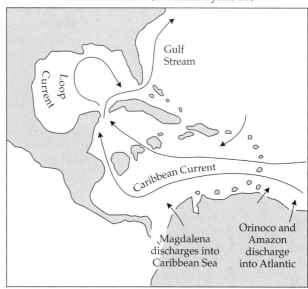

Closure of the isthmus of Panama

FIGURE 10.21 Paleoceanographic reconstructions of latest Eocene to Recent land configurations and connections of the Caribbean region. Note the latest Eocene-Early Oligocene continuous land connection extending from the northern coast of South America to at least a portion of Cuba, and perhaps including a portion of Jamaica (see Figure 10.33), known as the Greater Antilles-Aves Ridge (GAARlandia). The eventual drowning of GAARlandia, leaving a series of disjunct islands, would represent an opportunity for vicariance rather than long-distance dispersal, and make it difficult to classify islands of the Greater Antilles as either oceanic or continental in origin. Notice also the influence of land configurations on surface-current patterns, including the disruption of the Circumtropical Current by the closure of the Isthmus of Panama in the Late Pliocene. (After Iturralde-Vinent and MacPhee 1999.)

FIGURE 10.22 Map of the eastern Sunda Shelf and the Philippines, showing the apparent extent of land above sea level in the late Pleistocene. Note that while many of the currently isolated islands (lighter shading) were connected and the straits between others were much narrower than at present, there were still no land connections between the Philippines and Borneo (to the southwest) or the rest of the exposed Sunda Shelf, and thus to the Southeast Asian mainland. (After Heaney 1986.)

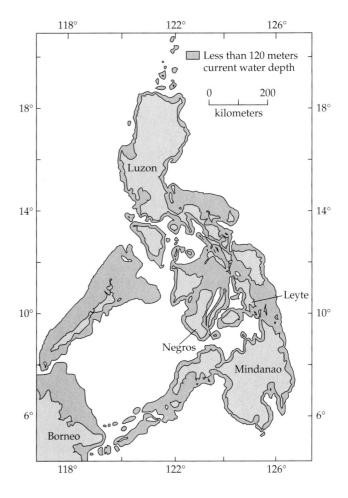

Trinidad, separated from Venezuela by two shallow straits, both less than 25 km wide, is famous in birdwatching circles for its scarlet ibises. These hardly constitute an isolated population, however, because many of the birds commute on a daily basis to feed on the South American mainland. While this is an extreme example, many kinds of organisms have continued to disperse from the mainland after the isolation of continental islands by rising sea levels at the end of the Pleistocene. The rescue effect of such over-water immigration (see Chapter 13) has tended both to recolonize the islands after extinction events, and to prevent extinction and divergence of the insular populations.

Quantifying Similarity among Biotas

Early biogeographers, and many of their successors, defined biogeographic provinces and regions subjectively, based on their intuition about how to interpret the geographic patterns. This does not necessarily mean that their classifications are unreliable; in fact, the human brain has an exceptional capacity for recognizing patterns. The biogeographic regions first defined subjectively by Sclater almost certainly include real patterns that could be defined objectively by modern quantitative analyses. During the last several decades, however, quantitative techniques have been applied increasingly to systematics and biogeography in order to make the process of classifying organisms and biogeographic regions more rigorous, objective, and repeatable.

In principle, quantitative techniques are simple. First, the items to be classified are described using objective criteria. In the case of biogeographic studies,

BOX 10.2 *Simple similarity indexes used by various authors to estimate biotic similarities*

Jaccard $\dfrac{C}{N_1 + N_2 - C}$

Second Kulczynski $\dfrac{C(N_1 + N_2)}{2(N_1 N_2)}$

Braun-Blanquet $\dfrac{C}{N_2}$

Simple matching $\dfrac{C + A}{N_1 + N_2 - C + A}$

Otsuka $\dfrac{C}{\sqrt{N_1 N_2}}$

Fager $\dfrac{C}{\sqrt{N_1 N_2}} - \dfrac{1}{2\sqrt{N_2}}$

Dice $\dfrac{2C}{N_1 + N_2}$

Correlation ratio $\dfrac{C^2}{N_1 N_2}$

First Kulczynski $\dfrac{C}{N_1 + N_2 - 2C}$

Simpson $\dfrac{C}{N_1}$

Note: A, absent in both units compared; C, present in both units; N_1, total present in the first unit; N_2, total present in the second unit (when the first unit contains the fewer taxa). (After Cheatham and Hazel 1969.)

the data usually consist of a complete list of the relevant taxa that occur at a specific site or within a given area. Sometimes other data, such as average abundance or area of geographic range, are available for each taxon for each region. Several mathematical techniques can then be used to quantify the similarity between each pair of biotas. The most commonly used measures are the similarity indexes (Box 10.2) because they all can be computed from simple presence-absence data. These indexes differ primarily in the extent to which they incorporate taxa that are present in both regions, in the range of values they can assume, and how they behave mathematically (i.e., how variations in presence-absence patterns are combined to produce a single number). The Jaccard and Simpson similarity indexes are the two that have probably been most frequently used in biogeographic analyses. Similarity values for each pair of biotas can be conveniently expressed in matrix form (Table 10.3).

TABLE 10.3 *A matrix of similarity coefficients (Simpson index %) between the mammalian faunas of various regions*

	North America	West Indies	South America	Africa	Madagascar	Eurasia	SE Asian Islands	Philippines	New Guinea	Australia
North America		40	55	8	9	19	8	9	6	6
West Indies	67		33	11	9	11	7	7	9	11
South America	81	73		3	7	4	7	6	4	3
Africa	31	27	25		30	25	21	27	17	12
Madagascar	38	27	35	65		32	26	22	22	17
Eurasia	48	27	36	80	69		75	64	25	14
SE Asian islands	37	20	32	82	63	92		73	30	18
Philippines	40	20	32	88	50	96	100		26	18
New Guinea	36	21	36	64	50	64	79	64		46
Australia	22	20	22	67	38	50	61	50	93	

Source: After Flessa et al. 1979. Courtesy of The Geological Society of America.

Note: Values above the diagonal are similarities at the generic level; values below the diagonal are similarities at the familial level.

Once the similarity between each pair of biotas has been computed, a quantitative phenetic clustering method is typically used to divide the biotas into groups that reflect a hierarchy of distinctiveness. There are several such clustering techniques, each of which has unique procedures for making mathematical computations, and each of which can give somewhat different results. Once the biotas have been clustered, all that remains is to decide what levels of similarity to use for designating different biogeographic ranks, such as provinces and subprovinces, so that the results can be plotted with exact boundaries on a map. Given a suitable data set, the similarity indexes can be calculated and the clustering procedure performed rapidly using a computer.

Most applications of quantitative methods to define biogeographic provinces have been limited to a single group of organisms within a limited geographic region. Examples include the division of Australia into ten provinces based on avian distributions (Kikkawa and Pearse 1969); analyses of similarities among mammalian faunas on the different landmasses (Flessa et al. 1979; Flessa 1980, 1981); and the division of the North American deserts into a hierarchy of biogeographic regions based on plant distributions (McLaughlin 1989, 1992). Such studies illustrate the practicalities of applying quantitative methods to biogeographic problems, but are so limited in taxonomic scope that their results may not be general.

Independent analyses of the distributions of several groups in the same region are potentially more informative. Connor and Simberloff (1978) compared the similarities among the various Galápagos Islands for both land birds and land plants, and found both similarities and differences between the two groups that presumably reflect the influences of such factors as the geological histories and ecological environments of the islands and the dispersal capabilities and population biologies of the organisms. An ambitious and early use of quantitative methods was Holloway and Jardine's (1968) analysis of the distributions of butterflies, birds, and bats in the Indo-Australian area (Figure 10.23). They measured similarities among different islands at the species level, and made inferences about past trends of dispersal and speciation. They concluded that all three of these groups of flying organisms dispersed predominantly eastward out of Southeast Asia, but that bats showed a different pattern of faunal differentiation than birds and butterflies. Subsequently, Nelson and Platnick (1981) reanalyzed the original data of Holloway and Jardine, applied the alternative approach of cladistic biogeographic analysis (see Chapter 12), and made somewhat different interpretations.

Interestingly, the use of similarity indexes and clustering methods to define biogeographic provinces has declined since the 1970s. There appear to be two reasons for this. First, as indicated above, at levels finer than the Wallace-Sclater regions, the provinces defined in this way for one group usually do not generalize well to other taxa. This should not be surprising. Different kinds of organisms originated and spread during different historical periods, when the Earth's geography and climate were different, and their patterns of speciation and dispersal have been affected by their different intrinsic characteristics, such as dispersal modes and life histories, and by different environmental factors, such as biogeographic corridors and barriers. Second, efforts to more rigorously characterize and interpret the legacy of biogeographic history have seen methods based on biotic similarity measures largely supplanted by methods based explicitly on the distributions of organisms with respect to their phylogenetic relationships. In doing so, it has become clear that many, if not all, biogeographic provinces are composite or reticulate areas, comprised of a mixture of taxa with evolutionary connections to more than one other province (see South American example above). Indexes provide a single value

(A)

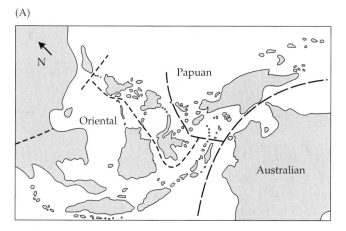

(B)

(C)

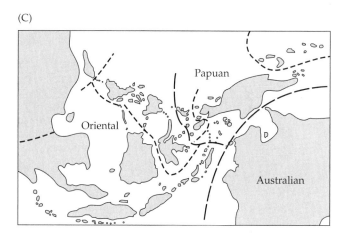

FIGURE 10.23   Biogeographic regions and subregions in the Indo-Australasian area defined for (A) birds, (B) bats, and (C) butterflies, based on the quantitative analyses of Holloway and Jardine (1968). Note that the divisions for birds and butterflies are virtually identical, but those for bats are quite different. Also, compare the lines separating these provinces with those drawn to separate the Oriental and Australian regions in Figure 10.13. (After Holloway and Jardine 1968.)

summarizing overall similarity among areas, and are thus of limited utility in revealing the details of a complex reticulated biogeographic history. We will examine the phylogenetically based approaches in the next two chapters.

Disjunction

Patterns

Disjunctions are those distributions in which closely related organisms live in widely separated areas. There are many classic examples. Many related forms occur on some combination of the Southern Hemisphere landmasses of Australia, Africa, South America, and sometimes New Zealand and Madagascar: these disjuncts include southern beeches (*Nothofagus*), trees and shrubs of the genus *Acacia*, several groups of mayflies, lungfishes, galaxoid fishes, clawed frogs (family Pipidae), ratite birds (emus, cassowaries, ostriches, rheas, tinamous, kiwis, and extinct moas and elephant birds; Figure 10.24), and marsupial (pouched) mammals. Lungless salamanders (family Plethodontidae) are broadly distributed in the wetter parts of the New World; the only representative in the Old World, the genus *Hydromantes*, occurs in southern Europe (France, Italy, and Sardinia), but also occurs in California (Figure 10.25). Many of the common genera and even species of plants in the deserts of southwestern North America, including creosote bush (*Larrea*), mesquite (*Prosopis*), and paloverde (*Cercidium*), also occur in the desert regions of southern South

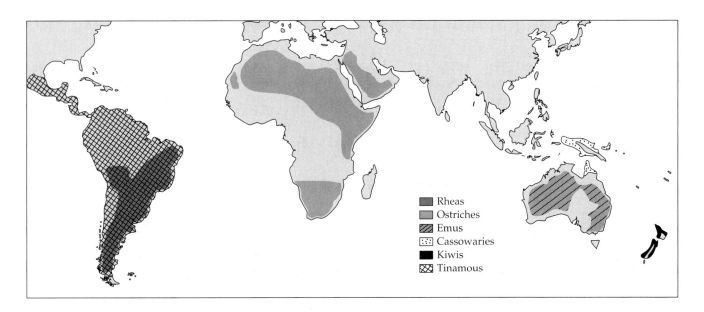

FIGURE 10.24 The disjunct distribution of the surviving members of the bird lineage that includes the tinamous and flightless ratites: ostriches, emus, cassowaries, kiwis, and rheas. The current widely disjunct distribution of this group reflects their original occurrence on Gondwanaland and their subsequent isolation as the ancient southern continent fragmented and drifted apart. Additional evidence of the historical Gondwanan distribution of this group comes from fossils of extinct giant flightless moas on New Zealand, and elephant birds on Madagascar.

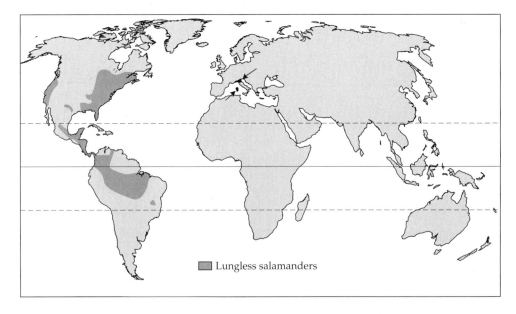

FIGURE 10.25 The disjunct distribution of lungless salamanders (Plethedontidae). Note that this ancient group has many isolated relicts in North America, and even one in the Mediterranean region of France, Italy, and Sardinia. (After Wake 1966.)

America, but not in the extensive tropical and montane areas in between. Other examples of amphitropical distributions are shown in Figure 10.26.

There are disjuncts on smaller scales as well. Plants provide some of the best examples, perhaps because they often have very specific environmental requirements and are able to persist for long periods in small patches as long as conditions are suitable. The sweet gum tree, *Liquidambar*, occurs in the

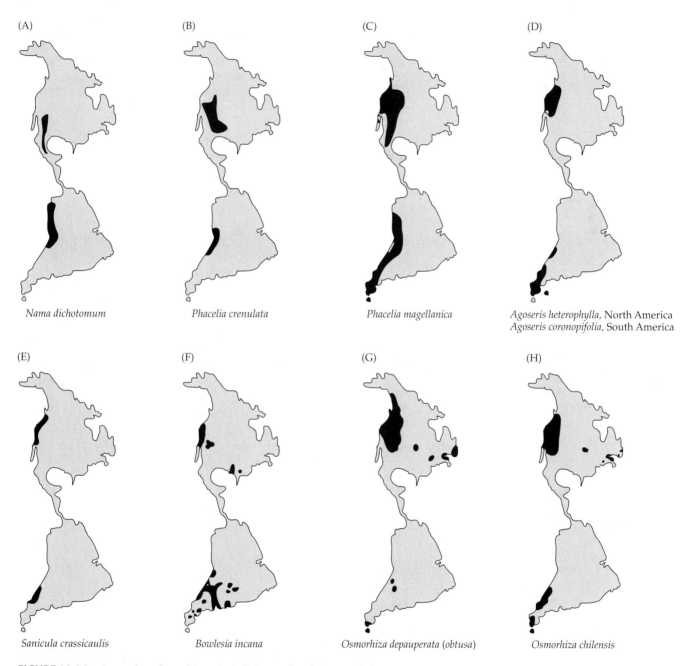

(A) *Nama dichotomum*

(B) *Phacelia crenulata*

(C) *Phacelia magellanica*

(D) *Agoseris heterophylla*, North America
Agoseris coronopifolia, South America

(E) *Sanicula crassicaulis*

(F) *Bowlesia incana*

(G) *Osmorhiza depauperata (obtusa)*

(H) *Osmorhiza chilensis*

FIGURE 10.26 Examples of amphitropical, disjunct distributions of plant species in North and South America. These examples represent only a small fraction of the disjunct distributions of closely related plant species, mostly in arid regions, on the two continents. These disjunct distributions raise interesting questions about the historical, geological, and climatic events that have allowed these plants to disperse across the tropics.

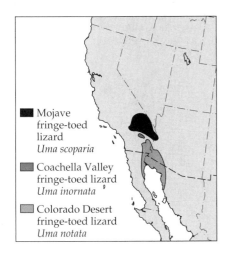

FIGURE 10.27 The disjunct distribution of the lizard genus *Uma*, which is restricted to dunes of wind-blown sand in the desert region of southwestern North America. Many other plants and animals that are restricted to specific kinds of spatially isolated habitats, such as mountaintops, lakes and springs, and patches of unusual soil types, show similar patterns of small-scale disjunct distributions.

■ Mojave
fringe-toed
lizard
Uma scoparia

■ Coachella Valley
fringe-toed lizard
Uma inornata

□ Colorado Desert
fringe-toed lizard
Uma notata

deciduous forests of the eastern United States and also in a series of isolated localities in central and southern Mexico. The Black Belt and Jackson Prairies in the southeastern United States are isolated grassland habitats that occur on unusual soils. They contain a number of disjunct plants whose nearest populations occur far to the west on the Great Plains (Schuster and McDaniel 1973). Some wetland plants found along the Atlantic coastal plain of eastern Canada also occur as isolated populations well inland in central Ontario (Keddy and Wishen 1989).

Animals also exhibit small-scale disjunctions. Like plants, many kinds of insects have been able to disperse to, and persist in, widely separated areas that meet their narrow niche requirements. Vertebrate examples are the fringe-toed lizard genus *Uma* and the Desert kangaroo rat (*Dipodomys deserti*), both of which occur almost exclusively on isolated sand dunes in the southwestern North American deserts and not in the intervening habitats (Figure 10.27). The eastern collared lizard (*Crotaphytus collaris*) is widespread throughout the Chihuahuan and Colorado Plateau deserts, but also has a set of highly disjunct populations in arid, rocky habitat patches in the Missouri Ozarks. The checkerspot butterfly, *Euphydryas editha*, occurs in many isolated populations widely dispersed across western North America (Ehrlich et al. 1975), but many of the isolates near the southern and lower elevational limits of the species' range have gone extinct in the last few decades, possibly due to global warming (Parmesan 1996).

Processes

Since the early nineteenth century, explaining disjunct distributions has been a major goal of biogeography. These distributions raise an obvious challenge: If the organisms are indeed closely related, then how did they get from where their common ancestor occurred, to their widely separated ranges? There can be only three possible answers: (1) disjunction by tectonics—their ancestors occurred on pieces of the Earth's crust that were once united, but have subsequently split and drifted apart; (2) disjunction by intervening extinction—their ancestors were once broadly distributed, but populations in the intervening areas have gone extinct, leaving isolated relics; or (3) disjunction by dispersal—at least one lineage has dispersed a long distance from the area where its ancestor(s) originally occurred.

Examples of all three mechanisms have been found. The Southern Hemisphere disjuncts mentioned above have until recently been considered as mostly vicariant relics of forms that once occurred on Gondwanaland and were isolated when that ancient supercontinent broke up and its fragments drifted apart (but see Sanmartín and Ronquist 2004; de Queiroz 2005). Examples of disjunctions caused by the extinction of intervening populations include, at an intercontinental scale, the camels and tapirs, which now occur only in South America and Eurasia, but are found in North America as fossils, which clearly indicate that they had a much wider and apparently continuous range during the Pleistocene. Many of the organisms that exhibit smaller-scale disjunctions, including the collared lizards mentioned above, are relicts left by the extinction of intervening populations due to changes in climate or other environmental conditions since the Pleistocene. Many fishes found in now isolated desert springs, and mammals restricted to the cool, mesic habitats on isolated mountain ranges, had wider, more continuous ranges in the Pleistocene when their habitats were more continuous (see Chapter 13). Some disjuncts are the result of long-distance dispersal, either by natural means, such as the cattle egrets that a few decades ago crossed the South Atlantic Ocean to colonize South America from Africa; or by human transport, such as the exotic European house sparrows and rabbits that are now established in such outposts as Australia, New Zealand, and southern South America.

There are also examples of forms once thought to be disjuncts that are now known not to be close relatives at all. Thus, "porcupines," rodents characterized by hairs modified to form distinctive defensive quills and by certain features of the skull, were once thought to be disjuncts between the New and Old World. These animals occur in Africa, southern Europe and Asia, and North and South America, but are absent from all of temperate Eurasia. Subsequent systematic research showed, however, that New and Old World porcupines are not closely related; they arose independently from primitive ratlike rodents, retaining unspecialized features of the skull and evolving quills by convergence. Similarly, the anteaters of the tropical New World and the pangolins (or scaly anteaters) of tropical Africa and Asia were once thought to be related disjuncts and were placed in the same order, Edentata. These animals share several characteristics, especially anatomical features of the skull and tongue that are related to their exclusively insectivorous diet, but these have arisen by convergence. These "anteaters" actually are unrelated members of two different orders, Xenarthra (which includes the New World anteaters, armadillos, and sloths) and Pholidota (which includes only the Old World pangolins).

The revelation that some organisms with seemingly disjunct distributions are not in fact close relatives underscores the need for good systematics. Patterns of cosmopolitanism, provincialism, and disjunction raise important questions in historical biogeography. We want to know how the origin, diversification, and spread of a particular kind of organism was related to and influenced by the geographic, geological, climatic, and biotic features of the Earth at the times when those processes were occurring. But unless we have an accurate assessment of the phylogenetic relationships among species and higher taxa, as well as phylogenetic and population genetic relationships among populations within species, we do not even know which biogeographic questions to ask. Those are the subjects of the next chapter. We will finish this chapter with three final questions: What abiotic and biotic attributes serve to keep distinct biotas separated from one another over time? What happens when barriers erode between previously separated biotas? And, what are some of the evolutionary patterns we see following long-term isolation of species and biotas?

Maintenance of Distinct Biotas

Here we consider briefly how distinctness between biotas is maintained. Given the present landbridges connecting Africa and Eurasia, as well as North and South America, and the frequent Pleistocene connections between North America and Eurasia, why hasn't biotic interchange been more complete? What processes are responsible for the preservation of biogeographic provincialism, especially in organisms that are good dispersers?

At one level, the answer is fairly straightforward. Since biotic interchange is due to dispersal and subsequent ecological success, the maintenance of provincialism must be due largely to some combination of continued isolation and resistance to invasion. Both factors are important.

Barriers between Biogeographic Regions

Most biogeographic regions are isolated in the sense that they are separated by ecological barriers to dispersal. For example, inspection of Wallace's map of biogeographic regions (see endpapers) immediately reveals that even where the terrestrial biogeographic realms are connected by landbridges, these are either narrow isthmuses, harsh deserts, or high mountains. Furthermore, the geographic ranges of the majority of species within a region do not extend to the boundary and, therefore, fall short of the landbridge. In order to move between regions, the majority of species, including distinctive endemic forms, would have to disperse long distances through unfavorable habitats. Consequently, the opportunity for interchange between regions is limited to a small fraction of the biota. For example, even though the Bering landbridge was exposed by the lowered sea levels that prevailed during most of the last two million years, it allowed only limited exchange between North America and Eurasia. It did provide a dispersal corridor, but only for those organisms whose ranges extended into steppe, coniferous forest, and tundra habitats at high latitudes. Some species, such as bears, wolves, ermines, caribou, moose, and lemmings among mammals, and ravens, hawks, owls, ptarmigan, siskins, and crossbills among birds, dispersed freely, and now occur on both continents. In contrast, many groups of North American and Eurasian amphibians, reptiles, birds, and mammals that did not inhabit these high-latitude environments did not disperse across the Bering landbridge.

Resistance to Invasion

Although it is tricky to document, it also appears that the biotas of large landmasses are relatively resistant to invasion. When faced with a diverse native biota that has evolved to tolerate an abiotic environment and coevolved to withstand existing biotic interactions, it is difficult for invading species to become established. Evidence in support of this hypothesis comes from the fate of exotic species (for individual case histories, see Chapter 16, and Elton 1958; Udvardy 1969; Drake et al. 1989; Hengeveld 1989). Another factor involved in resistance may be one or more forms of biotic intertia—the retention of native species such as long-lived forest trees, or of soil chemistries produced by allelopathic plants, long after changes of environmental conditions that favor the invader have ocurred (Von Holle et al. 2003).

Numerous Old World plant and animal species have become locally abundant and geographically widespread in North America within the last four centuries. Although the success of these invaders is impressive, they represent only a small fraction of the species that have been intentionally or accidentally introduced. The vast majority of introduced populations have gone extinct, and many others have not spread far beyond the site of their introduction. An example is the European tree sparrow (*Passer montanus*): more than a century after its establishment, it is still confined to a small area near St. Louis, Missouri.

Furthermore, of the hundreds of Eurasian plants established in North America, most can be classified as weeds, species that occur in successional habitats created primarily by human disturbance. Of the many introduced insects, the majority of successful species are crop pests, associated with introduced plant species, or confined largely to disturbed habitats. This pattern holds even for vertebrates. Many Eurasian birds have been introduced into North America, but only a few have become established. The two amazingly successful introduced bird species that have spread to cover most of the North American continent—the house sparrow (*Passer domesticus*), and the starling (*Sturnus vulgaris*; see Figure 6.3A,B)—are largely commensal with humans, using artificial structures for nesting sites, and urban and agricultural habitats for food resources. Mammalian examples include the house mouse (*Mus musculus*), the Norway rat (*Rattus norvegicus*), and the roof rat (*Rattus rattus*). The success of these Eurasian exotics in North America and the much lower success of New World species in the Old World suggests that most of the Eurasian forms are adapted to occupy niches that are dependent on human activity, and exploit similar niches in North America. As evidence of this, we note not only that the exotics have been generally unsuccessful at invading undisturbed native habitats, but also that it is difficult to point with confidence to the extinction of a native continental species owing to replacement in its niche by an introduced competitor.

A quantitative macroecological analysis of resistance to invasion was performed for North American freshwater fishes by Gido and Brown (see Brown 1995). They plotted the number of colonizing exotic species as a function of the number of native species for 135 watersheds in temperate North America. The result is a triangular-shaped distribution (Figure 10.28). Watersheds with few native species show wide variation in the number of exotics present. Some of them have been colonized by few invaders, perhaps because few species have been introduced or because stressful abiotic conditions have prevented their establishment, but other watersheds have been invaded by scores of exotic species. In contrast, watersheds with large numbers of native species have uniformly low numbers of exotics. This pattern is consistent with Elton's

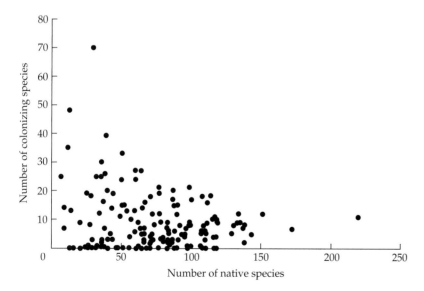

FIGURE 10.28 Number of exotic fish species that have become established as a function of the number of native species for 135 large watersheds in temperate North America. Note that all of the points fall within a roughly triangular space, showing that watersheds with diverse native fish faunas have been relatively resistant to invasion, whereas some of the watersheds with few native species have been colonized by many invading species. (Courtesy of K. Gido.)

(1958) suggestion that diverse biotas that have radiated to fill many niches may indeed be relatively resistant to invasion.

Avian Migration and Provincialism

Birds, with their capacity to disperse by flight over great distances and formidable geographic barriers, would seem to be one of the groups of organisms least likely to exhibit biogeographic provincialism. Indeed, a few bird species, such as the peregrine falcon, osprey, and barn owl, are virtually cosmopolitan, and birds have managed to colonize even the most remote of oceanic islands (see Chapter 14). Remember, however, that Sclater (1858) originally divided the world into biogeographic provinces based on the distributions of passerine birds. This seems counterintuitive.

An examination of avian distributions on a global scale reveals that observed provincialism is due largely to two very different "functional groups" of birds. On one hand, birds with very limited powers of dispersal and specialized adaptations for particular environments often exhibit high degrees of endemism. Such birds include the flightless ratites (ostriches, emus, cassowaries, kiwis, and rheas) of the southern continents; the weakly flying tinamous, hoatzins, and curassows of tropical America; secretary birds and turacos of Africa; megapodes, lyrebirds, bowerbirds, and birds of paradise of Australia; and turkeys of North America. They also include highly specialized birds such as nectar-feeding hummingbirds of the New World, sunbirds of Africa and southern Asia, and honeyeaters of Australasia; as well as the wood creepers, ovenbirds, antbirds, and manakins of South America; the honeyguides, wood hoopoes, and weaver finches of Africa; and the lorikeets, cockatoos, wood swallows, whistlers, and logrunners of Australia. Two additional observations are worth making. First, many of these groups are endemic to one of the southern continents, presumably reflecting their long history of geographic isolation. Second, other groups of birds that might seem to be equally poor fliers or equally specialized, are surprisingly widespread. Examples include the large, rarely flying bustards (Africa, Europe, Asia, and Australia) and the marsh-living, rarely flying rails (cosmopolitan).

The other "functional group" that contributes importantly to biogeographic provincialism consists of small land birds that are long-distance migrants (Bohning-Gaese et al. 1998). These include the hummingbirds, tyrant flycatchers, vireos, wood warblers, tanagers, orioles, blackbirds, and emberizine buntings and sparrows, all families or subfamilies endemic to the New World; and also the true flycatchers and warblers, groups restricted to the Old World. These subfamilies and families of birds include many species that migrate twice each year between widely separated breeding grounds in the subarctic and temperate regions of either North America or Eurasia, and wintering grounds in tropical areas of South and Central America or Africa, southern Asia, and Australia (Figure 10.29). With very few exceptions, representatives of these groups do not occur in both the New and Old Worlds. The exceptions prove the rule: a few species of Old World warblers breed in the Aleutian Islands and even on the Alaskan mainland, but migrate to winter in tropical Asia or Australasia (Figure 10.30).

Also surprising is the fact that similarly small land birds that are either nonmigratory or only short-distance migrants show exactly the opposite pattern: many families and subfamilies contain genera that occur in both North America and Eurasia. These groups include woodpeckers, nuthatches, creepers,

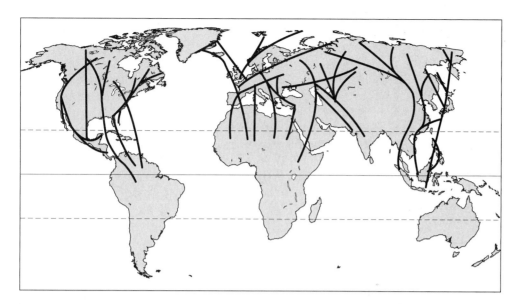

FIGURE 10.29 Major avian migratory flyways. As the plotted routes suggest, in both the New and Old Worlds, many bird species travel hundreds and even thousands of kilometers twice each year, commuting between breeding grounds in Arctic and temperate regions to wintering grounds that are located mostly in the tropics. Very few of these birds, however, cross between the New and Old Worlds. (After McClure 1974, and Baker 1978.)

wrens, crows, tits, finches, thrushes, and waxwings (Figure 10.31). These relatively sedentary avian taxa have distributions very similar to those of land mammals, insects, and both coniferous and angiosperm plants: closely related species occur across Arctic, subarctic, and cool temperate regions of both Eurasia and North America. Apparently such Holarctic distributions reflect a long history of close proximity, including repeated land connections and resulting biotic exchanges, most recently across the Bering landbridge during the Pleistocene (see above, and Chapters 9 and 12).

Why are the long-distance migrants so different? What has prevented interchange between the New and Old Worlds? At least two interrelated factors seem to be involved. First, traits associated with long-distance migration may actually make it difficult for migrants to colonize a new continent. An initially small founding population not only must become established in a suitable breeding area, it must also find a new, distant wintering ground, and a route there and back. The difficulty of developing new migratory routes appears to be so severe as to make exchange between New and Old Worlds highly improbable. Support for this hypothesis comes from the observation that the land bird faunas of isolated islands are constituted primarily of species derived from nonmigrants and short-distance migrants, rather than long-distance migratory ancestors (L. Gonzalez-Guzman, pers. comm.). Long-distance migrants can and do reach distant places, but they are unlikely to stay there or find their way back to breed.

Second, this pattern seems to have a historical component. Lineages that contain most of the long-distance migrants appear to be of tropical origin. Migration seems to be an adaptation that allowed highly mobile birds to rear their young using the seasonal pulse of productivity that is available for only

(A) Arctic Warbler (*Phylloscopus borealis*)

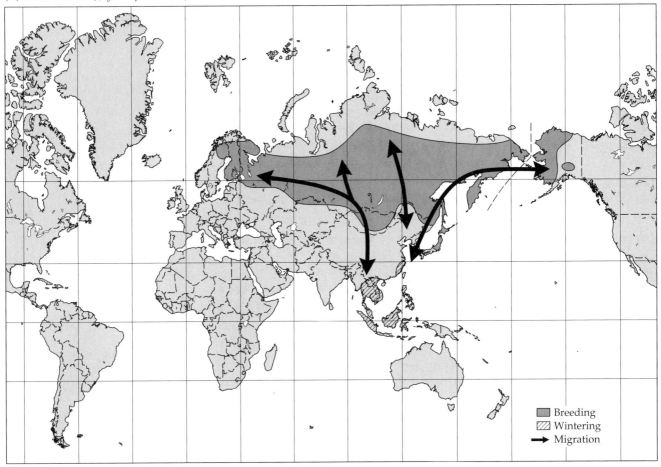

Breeding
Wintering
Migration

FIGURE 10.30 Maps showing the breeding ranges (dark shaded areas), winter ranges (hatched areas), and migratory routes (arrows) of two passerine bird species: (A) the arctic warbler (*Phylloscopus borealis*) which breeds in western Alaska; and (B) the northern wheatear (*Oenanthe oenanthe*) which breeds in Alaska, northern Canada, and Greenland. Both of these species have colonized the New World, but show their Eurasian origins by migrating to winter in the Old World tropics.

a few months at higher latitudes. Thus, most of the long-distance migrants feed their nestlings on insects that are active, abundant, and accessible only during the warm months. This is especially true in the New World, where, as noted above, the families and subfamilies containing most of the long-distance migrants are of South American origin.

The special case of birds illustrates how evolutionary constraints and ecological factors can interact to maintain the historical legacy of biogeographic provincialism. Even though the continents may be connected by landbridges that would seem to provide dispersal routes, and even though organisms may possess traits that would seem to permit long-distance dispersal, the actual interchange and mixing of long-isolated biotas has been relatively limited. The result is that the influence of ancient Earth history, especially of tectonic events, is preserved in the distributions of contemporary forms, even such vagile ones as migratory birds.

(B) Northern Wheatear (*Oenanthe oenanthe*)

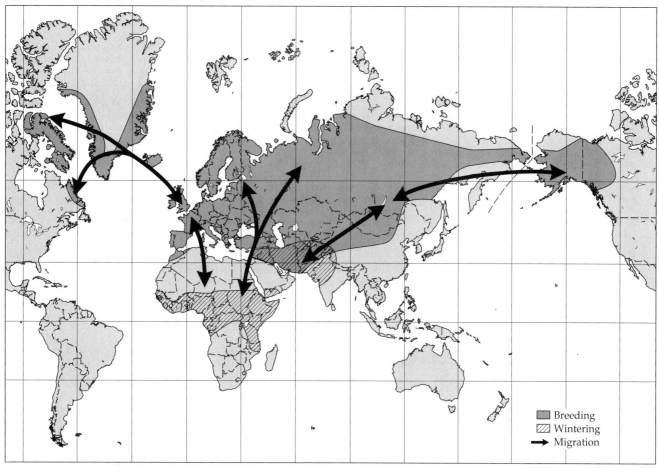

Breeding
Wintering
Migration

Biotic Interchange

The opposite of biotic segregation or biogeographic provincialism is biotic interchange. The fate of artificially introduced species provides some limited insight into the kinds of interactions that can occur when representatives of one formerly isolated biota come into contact with those of another. This has happened naturally in the past, not only when individual species have managed to disperse across barriers, but also when the barriers themselves have been reduced or abolished, bringing two distinct, previously isolated biotas into direct contact. Such contacts have occurred many times as continents have drifted over the Earth and new land and water connections have been formed (Vermeij 1978, 1991; Figure 10.32). Unfortunately, the record of their results is often poor, because it must be pieced together largely from limited fossil evidence. For example, when the Indian plate collided with southern Asia, it presumably brought a distinct Gondwanan biota into contact with a large Eurasian biota. You may recall from our discussion of taxa that were originally distributed on Gondwanaland that little mention was made of relictual Gondwanan populations in India. Was the Indian biota derived from Gondwanaland eliminated by interactions with the more diverse biota of the larger Eurasian landmass? This is a plausible explanation, but unfortunately, the fossil inhabitants of the Indian subcontinent are so poorly known that

(A) Species

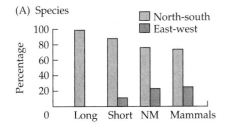

(B) Genera

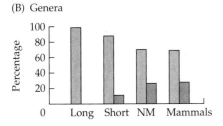

(C) Families

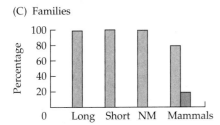

FIGURE 10.31 Comparison of the distributions between hemispheres of long-distance migrant (long), short-distance migrant (short), and nonmigratory (NM) birds and terrestrial nonvolant mammals at three levels of taxonomic classification: (A) species, (B) genera, and (C) families. North-south (light gray bars) indicates that the geographic range of the taxon includes either both North America and South America in the New World, or both Europe and Africa in the Old World; east-west (dark gray bars) indicates that the range of the taxon includes both North America and Eurasia in the Northern Hemisphere. Note that the long-distance migrant birds have exclusively north-south ranges, indicating that they have not colonized between the Old World and the New World; in contrast, resident birds resemble nonvolant mammals in having a substantial number of taxa with east-west distributions, indicating colonization between North America and Eurasia. (After Bohning-Gaese et al. 1998.)

there is little direct evidence to support or refute this hypothesis. In fact, recent molecular phylogenetic evidence, with its increased accuracy in tracing evolutionary histories and, often, good estimates of divergence times, offers the intriguing possibility that some of what we have assumed to be lineages of northern origin instead originated on Gondwanaland and subsequently have diversified after reaching the northern landmasses (see Fig. 7.11). One of the better documented events is the much more recent interchange that took place when North and South America joined, and is the focus of the following discussion.

The Great American Interchange

During the age of mammals, the Cenozoic in geological terms, South America was an island continent. Its land mammals then evolved in almost complete isolation, as in an experiment with a closed population. To make the experiment even more instructive, the isolation was not quite complete, and while it continued two alien groups of mammals were nevertheless introduced, as might be done to study perturbation in a laboratory experiment. Finally, to top the experiment off the isolation was ended, and there was extensive mixture and interaction between what had previously been quite different populations, each with its own ecological variety and balance. (G. G. Simpson 1980.)

The ecological experiment that Simpson was referring to was the exchange of mammals between North and South America following the formation of the Central American landbridge about 3.5 million years ago. This event, known as the Great American Interchange, provides a fascinating case study of the combined effects of dispersal, interspecific interaction, extinction, and evolution on biological diversity. Most of the available information concerns mammals, in part because they left a rich fossil record. Interestingly, it seems that the effects of the interchange were quite different in other terrestrial organisms, such as amphibians, reptiles, birds, and plants.

HISTORICAL BACKGROUND. Mammals evolved from reptilian ancestors during the Mesozoic, beginning about 220 million years ago, when the continents were united in the single great landmass of Pangaea. By the time Pangaea began to break up, mammals had spread over the continents and diversified into the modern lineages (monotremes, marsupials, placentals). Compared with the ruling reptiles, however, they remained a small component of the biota until the mass extinctions at the end of the Cretaceous, about 65 million years B.P. This biotic upheaval, which saw the extinction of many terrestrial and marine lineages (see Chapter 8), including the dinosaurs; it was followed by rapid diversification of birds, mammals, and some other surviving groups.

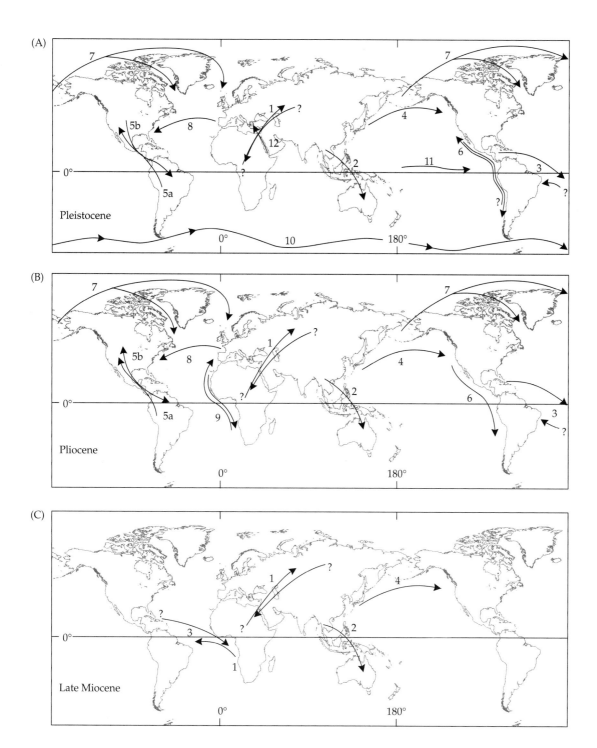

FIGURE 10.32 Major episodes of terrestrial and marine biotic interchange during three time intervals:(A) Pleistocene–Recent, (B) Pliocene, and (C) late Miocene. Note that, in several cases, interchange occurred along similar routes during more than one timeframe. Arrows depict predominant direction of dispersal during interchange, and question marks indicate uncertainty about directionality. 1, Terrestrial, between Africa and Asia; 2, Terrestrial, from southeast Asia to Australia and New Guinea; 3, Marine, across tropical Atlantic; 4, Marine, across North Pacific; 5a, Great American Interchange for lowland rainforest species; 5b, Great American Interchange for savanna and upland species; 6, Marine, transequatorial in east Pacific; 7, Marine, trans-Arctic; 8, Marine, across North Atlantic; 9, Marine, transequatorial eastern Atlantic; 10, Marine, circum-Antarctic; 11, Marine, tropical Pacific; 12, Lessepsian, trans-Suez. (After Vermeij 1991.)

TABLE 10.4 *Summary of tectonic, climatic, and biotic events associated with the Great American Interchange*

220–160 million years B.P.	South America remains connected the rest of Gondwanaland, and to North America. The origin, spread and diversification of mammals and birds across the landmasses.
140–75 million years B.P.	South America becomes isolated. Its biota evolves and diverges on a separate landmass.
140–120 million years B.P.	Proto-Antilles serves as a transient stepping stone, sweepstakes route. Limited biotic exchange among Nearctic and Neotropical regions.
10–5 million years B.P.	Central American Archipelago serves as a transient stepping stone, sweepstakes route. Limited biotic exchange among Nearctic and Neotropical regions.
~3.5 million years B.P.	Emergence of the Central American Landbridge (closure of the Bolivar Trench). Provides a filter dispersal route for terrestrial forms, but a barrier for marine organisms.
2–0 million years B.P.	Lowering of sea-level and extension of savanna and other open-habitat biomes during glacial maxima opens a corridor or filter route for biotic exchange. Subsequent invasions and diversification of invaders, extinctions of invaders and natives.

The most important geological, climatic, and biotic events associated with the history of South American mammals are summarized in Table 10.4 (see also, Stehli and Webb 1985). South America was a part of Gondwanaland until about 160 million years B.P., but then began drifting apart and remained a giant island continent until about 3.5 million years B.P. During this period of what Simpson called "splendid isolation," a distinct endemic land mammal fauna evolved. At least one monotreme (a platypus), several groups of marsupials, and at least one lineage of eutherian mammals speciated and differentiated to produce a morphologically and ecologically diverse fauna that included large carnivores (even a marsupial saber-toothed "tiger"), and giant herbivores. This radiation was in some ways comparable to, but even more diverse than, that which occurred on the other island continents formed by the breakup of Gondwanaland—namely, Australia and Madagascar.

The biogeographic isolation of South America was not complete, however. As Simpson noted in the above quote, it was probably interrupted by one or two transient periods of limited exchange, although presence and configurations of former archipelagoes and landbridges in the Caribbean region become difficult to reconstruct accurately prior to the last several million years (Figure 10.33; Iturralde-Vinent and MacPhee 1999). During these periods, the ancestors of the present South American primates, edentates (armadillos, sloths, and anteaters), caviomorph rodents (porcupines, capybaras, pacas, agoutis, guinea pigs, chinchillas, and other forms), and possibly sigmodontine rodents (New World rats and mice), colonized the continent. In each of these groups, speciation and adaptive radiation subsequently produced multiple families and genera from a few founding lineages.

The isolation of South America ended dramatically about 3.5 million years B.P. when the Central American uplift formed the land connection to North America that still exists today. Since its formation, however, the Central American landbridge has served more as a filter than as a highway (Figure 10.34). David Webb (1991) has suggested that dispersal across the landbridge was strongly influenced by the climatic cycles of the Pleistocene (see Chapter 9). Interchange was greater during glacial periods, when savanna habitats expanded to cover much of Central and northern South America, than during interglacial periods, such as the present, when most of the landbridge was covered with tropical forest.

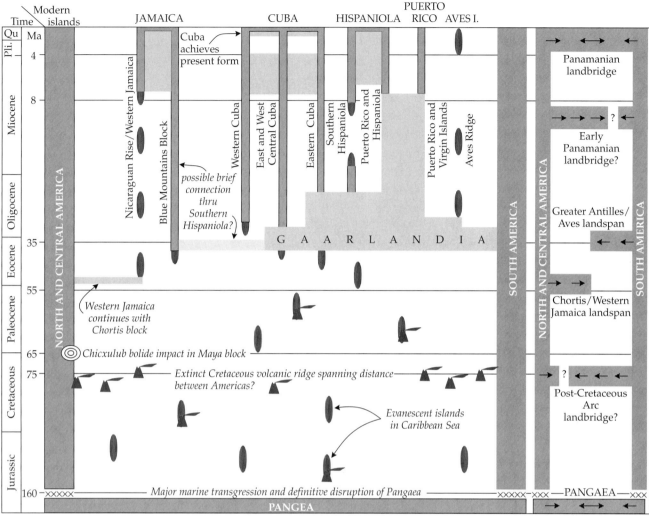

FIGURE 10.33 A paleogeographical scenario for the Caribbean region spanning a middle Mesozoic to late Cenozoic timeframe. Locations and numbers of the "evanescent islands" are intended to be only conjectural, and illustrate the increasing difficulty in reconstructing accurate depictions of paleogeography at older times. Light shading represents times of land connection between two or more terrestrial blocks, including the Greater Antilles-Aves Ridge (GAARlandia; see Figure 10.21) that now form the modern islands of Jamaica, Cuba, Hispaniola, Puerto Rico, and Aves Island. A summary of known (e.g., the Late-Pliocene Panamanian landbridge that resulted in the Great American Interchange) and conjectured (identified with question marks) interconnections across the Caribbean between North and South America are shown on the right. Dark vertical bars that span the entire time frame indicate, from left to right: North and Central America; South America; North and Central America; South America. (After Iturralde-Vinent and MacPhee 1999.)

PATTERNS AND CONSEQUENCES OF MAMMALIAN FAUNAL EXCHANGE. The fossil records of North and South America over the last ten million years reveal the magnitude of the Great American Interchange. The diversity of the North American fauna, at the generic level, remained virtually unchanged despite the invasion of some South American forms such as the extant porcupines, opossums, and armadillos, and the now extinct hippo-sized glyptodonts and bear-sized giant ground sloths. The overall diversity of the South American fauna increased

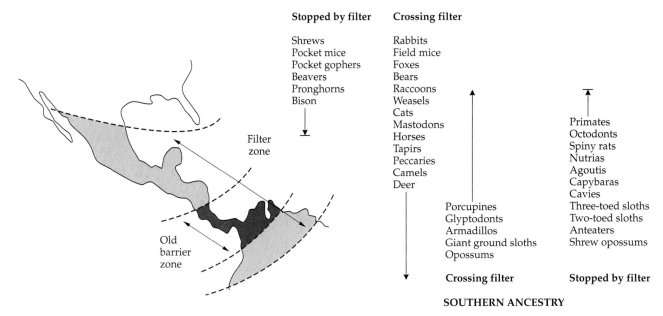

NORTHERN ANCESTRY

Stopped by filter	Crossing filter
Shrews	Rabbits
Pocket mice	Field mice
Pocket gophers	Foxes
Beavers	Bears
Pronghorns	Raccoons
Bison	Weasels
	Cats
	Mastodons
	Horses
	Tapirs
	Peccaries
	Camels
	Deer

Filter zone

Old barrier zone

		Primates
		Octodonts
		Spiny rats
		Nutrias
		Agoutis
		Capybaras
		Cavies
	Porcupines	Three-toed sloths
	Glyptodonts	Two-toed sloths
	Armadillos	Anteaters
	Giant ground sloths	Shrew opossums
	Opossums	

Crossing filter	Stopped by filter

SOUTHERN ANCESTRY

FIGURE 10.34 Map showing the location of the Central American landbridge, with lists of the mammalian families of both North and South American origin that either crossed through the filter of tropical lowland habitats in Central America during the Great American Interchange to colonize temperate regions of the other continent, or were stopped in or near the filter. Note the asymmetry, with more groups of North American origin passing through the filter and more families of South American origin stopped by the filter.

due to the invasion and establishment of lineages from North America. The many North American groups that crossed the Isthmus of Panama and became established in South America included not only the extant shrews, rabbits, squirrels, dogs, bears, raccoons, weasels, cats, deer, peccaries, tapirs, and camels, but also the now extinct mastodons and horses. The magnitude of change was even greater than the final figures for generic or familial diversity suggest, however, because several endemic South American forms went extinct. These included several kinds of large marsupial carnivores and even larger eutherian herbivores.

The imbalance of the interchange can be seen in the fact that about half of contemporary South American species are derived from North American ancestors, whereas only about 10 percent of North American species are of South American origin (see Figure 10.34). The northern forms appear to have had three advantages that contributed to this imbalance:

1. *They were better migrators.* While many northern forms (listed above, and in Figure 10.34) crossed the landbridge and invaded deep into South America, only three species (porcupine, opossum, and armadillo) of ancient South American ancestry managed to colonize and persist to the present in temperate North America. Interestingly, however, several other groups of South American ancestry (primates, sloths, anteaters, and other kinds of marsupials and caviomorph rodents) have colonized northward across the Isthmus of Panama, but extend no farther than the tropical forests of southern Mexico.

2. *They were better survivors and speciators.* Most of the northern forms that invaded South America not only survived (mastodons and horses were exceptions), but also speciated and diversified there, giving rise to such distinctive forms as kinkajous, coatimundis, giant otters, swamp deer, guanacos, and vicuñas. The mouselike sigmodontine rodents, which probably colonized South America by island hopping before the completion of the landbridge, are now the most diverse group of mammals on

the continent, comprising about 61 genera and 299 species (Marshall 1979; Webb and Marshall 1982; Reig 1989; Engel et al. 1998). This diversity is all the more impressive when viewed from two perspectives. First, while the ancestral sigmodontine may have originated in Mexico or Central America, the vast majority of extant diversity in the South American sigmodontines arose within South America following dispersal of one or a handful of ancestral species to that continent. Second, the neotomine-peromyscine rodents of North and Central America, sister group to the South American sigmodontines, represent a highly successful radiation in their own right, yet, with 18 genera and 124 species, are far less diverse than the sigmodontines. Furthermore, over the last 3.5 million years, while lineages of North American ancestry have been diversifying, the ancient South American lineages have been dwindling away because speciation has not kept pace with extinction.

3. *They were better competitors*. In the face of differential colonization, survival, and speciation of North American forms, it is hard to avoid the conclusion that they have been, on average, superior competitors. Clearly, the North American forms not only have increased in generic and species diversity at the expense of their more ancient South American counterparts, but also have radiated to usurp their ecological roles. For example, all of the large carnivores and herbivores, and the vast majority of the mouselike rodents in South America today, are descendants of northern invaders.

Some authors have questioned whether such differential ecological and evolutionary success can be attributed to "competition." Indeed, not all of the interactions are strictly competitive. Predation, parasitism, and disease probably play a significant role. Furthermore, much of the competition that does occur may be diffuse and involve many species rather than simple pairwise interactions. It is also clear that not all mammals of South American ancestry are competitively inferior: witness the success of porcupines, opossums, and armadillos in invading North America. The latter two species have actually been expanding their ranges northward in recent decades, as seen in Figure 6.3D for armadillos.

Elizabeth Vrba (1990) presented a novel alternative to the competition model, which she termed the "habitat-theory." Briefly, her hypothesis combines evidence of long-term climate change (a general cooling trend over the past 2.5 million years) and associated shrinkage of tropical forests in favor of more open and arid savanna habitats extending between northern South America and Central America, with ecological data on the sorts of mammals that crossed the landbridge, to make some predictions for the observed immigration imbalance without invoking competetive displacement. The first prediction is that, because the northern biota contained a greater number of savanna habitat specialists, one would expect a net vector of greater migration success southward based solely on the fact that savanna habitats were more continuously distributed across the landbridge. The second prediction is that the extinction imbalance toward South American natives could be attributed to a higher extinction rate among the closed forest specialists following the climatically induced fragmentation of forest habitats in favor of a greater amount of open savanna habitat in northern South America. The final prediction is that the greater speciation rates of northern invaders into South America was due primarily to their containing a greater proportion of habitat specialists, prone to frequent cycles of habitat vicariance over the past 2.5 million years. Some evidence for this prediction comes from the great diversity of sigmodontine rodent species that appear to have speciated in areas with high topographic diversity, within the higher valleys and plains of the Andes, and the deserts and savannas of the southwestern part of the continent (see Figure 10.8).

Returning to Simpson's analogy, the Great American Interchange constituted a vast natural experiment. By the movement of continents, the mammalian fauna of South America were first allowed to evolve in "splendid isolation," and then brought into contact with a more diverse fauna—one that had evolved on the larger North American landmass and had had frequent interchanges with the fauna of the Old World. The outcome of the experiment is clear. Taken as a whole, the North American mammalian fauna proved superior, and differentially "replaced" much, though by no means all, of the original South American fauna.

INTER-AMERICAN EXCHANGE IN OTHER VERTEBRATES. Information on historical exchanges between North and South America in vertebrate groups other than nonvolant mammals is scanty, because the fossil record of these groups is relatively poor. It is also more difficult to interpret, because some of these groups, such as reptiles—and especially birds and bats—are better over-water colonists than nonvolant mammals. This means that some level of faunal interchange between the two continents probably occurred continuously throughout the Cenozoic, rather than being concentrated in the last 3.5 million years following the completion of the Central American landbridge (Vuilleumier 1984, 1985; Estes and Baez 1985; Vanzolini and Heyer 1985).

Two patterns are noteworthy. First, there is not a clear dichotomy between an ancient South American fauna, which dates back to the isolation of Gondwanaland, and relatively recent invaders. Instead, the South American herpetofauna and avifauna appear to have been assembled more gradually. They are made up of a mixture of ancient Gondwanan forms, lineages that colonized across water during the early and mid-Cenozoic, and forms that crossed the Central American landbridge (or the island chain that preceded it) during the last several million years. Thus, at least for terrestrial reptiles, amphibians, and birds, it is difficult to distinguish easily between South American "natives" and North American "invaders."

Second, the late Tertiary invasion of South America by North American forms, so clearly seen in mammals, simply did not occur in many other groups. The exchange of birds appears to have been much more balanced than that of mammals. While some groups, such as pigeons, owls, woodpeckers, and jays, colonized South America from the north; others, such as hummingbirds, tyrant flycatchers, vireos, wood warblers, blackbirds, orioles, tanagers, and emberizine buntings (grosbeaks and sparrows), moved in the opposite direction. An interesting feature of the North American avifauna is that the Neotropical migrants, which make up the majority of breeding passerines in most temperate habitats, are virtually all of South American ancestry.

The dominance of the North American fauna by lineages of South American origin is even more pronounced in reptiles and amphibians. Thus, South America has no salamanders and perhaps only one species of frog that colonized from North America. In contrast, of the North American frog fauna, only three families (Ascaphidae, Pelobatidae, and Ranidae) are of northern origin, whereas four (Bufonidae, Hylidae, Leptodactylidae, and Microhylidae) are of South American origin. Similarly, in reptiles, the predominant movement has been from South America to North America, although this pattern has sometimes been complicated by secondary centers of speciation and diversification in Central America and Mexico. Thus, for example, the two large families of New World lizards, Iguanidae and Teidae, and most of the snakes, have dispersed northward from South America to North America.

The completion of the Central American landbridge resulted in some limited interchange of freshwater fishes (Miller 1966; Rosen 1975; Bussing 1985; Bermingham and Martin 1998). Again, the predominant direction of dispersal was from south to north. Because of their requirement for freshwater connections in order to disperse, however, the invasion of primary freshwater forms has been limited. Several South American lineages have reached as far north as central Mexico (Miller 1966), whereas no North American group has made it farther south than Costa Rica (Bussing 1985).

To summarize, the Great American Interchange, that wonderful natural experiment so thoroughly documented by Simpson and later researchers, shows a clear pattern: terrestrial mammals of North American ancestry were better dispersers, survivors, and speciators. Their differential success was due to some currently unknown combination of the contingency of habitat distributions across the landbridge and competetive superiority. Over the last 3.5 million years, since the completion of the Central American landbridge, they have invaded South America and largely supplanted the ancient South American groups. Completely different patterns, but usually also with unbalanced exchange, occurred in other vertebrates.

The Lessepsian Exchange: The Suez Canal

Continuing with Simpson's analogy of biotic interchange as a natural experiment, humans have unintentionally performed one such manipulation by bringing into contact two previously long-isolated marine biotas. This experiment began in 1869 with the completion of the Suez Canal between the Mediterranean Sea and the Red Sea. The resulting biotic exchange has been called the Lessepsian migration or exchange, named in honor of the chief engineer of this project, Ferdinand de Lesseps (Por 1971, 1977). Although the hypersaline lakes through which the canal passes constitute an impassable barrier for many marine forms, an increasing number of taxa have been able to disperse between the two seas. Again, the biotic exchange has been very unbalanced. More than 50 species of fish, 20 species of decapod crustaceans, and 40 species of mollusks have colonized the eastern Mediterranean from the Red Sea, but it is hard to document cases of migration in the opposite direction.

Three explanations for this unidirectional dispersal have been proposed, and all three factors probably contribute to successful colonization by Red Sea forms. First, the Gulf of Suez, at the southern end of the canal, is itself more saline than ocean water, so the species that occur there may have been preadapted to cross the barriers formed by the hypersaline lakes. Second, most of the successful migrants inhabit shallow sandy or muddy bottom habitats in the Indian Ocean. Such habitat affinities also appear to be preadaptations, because they facilitate movement through similar habitats in the Suez Canal. Finally, species of the Indo-Pacific biota may be more resistant to predation than their Mediterranean counterparts, competitively superior to them, or both. The Red Sea is an arm of the Indian Ocean, which contains a far more diverse biota than the Mediterranean. Biotic interactions are implicated by observations of declining populations of some endemic Mediterranean species in the eastern part of the sea (e.g., along the coast of Israel) where Red Sea colonists have become well established. (For a much more complete discussion of the history and results of the Lessepsian exchange, see Por 1971, 1975,

1977, 1978; Vermeij 1978, 1991b; Golani 1993; Galil 2000; Gofas and Zenetos 2003; Braidai et al. 2004; Shefer et al. 2004.)

One final point is warranted. The differential extinctions of South American mammals in the face of North American invaders and the unbalanced exchange of marine forms through the Suez Canal, are both consistent with a general pattern noted by many biogeographers. As early as 1915, W. D. Matthew (see also, Willis 1922) noted that organisms from diverse biotas on large landmasses are best able to successfully invade smaller areas and replace the native organisms. Darlington (1957, 1959) reemphasized this point, although he argued that the successful forms usually originated in tropical regions, whereas Matthew had thought they came from temperate climates. This point may be difficult to resolve because the climates of regions that are now temperate or tropical have changed greatly over their geological history (see Chapters 8 and 9). Nevertheless, the success of organisms from large, diverse biotas in colonizing small, isolated regions containing fewer native species seems to be a consistent phenomenon in biogeography. The interactions that have been seen on a continental scale fit the pattern of invasions of the island continent of Australia from larger continents, as well as the natural and human-assisted colonization of many islands by continental forms (see Chapters 13, 14, and 16).

The Divergence and Convergence of Isolated Biotas

We have considered what happens when long-isolated biotas are brought into contact. Now let's examine the opposite situation. Are there detectable general evolutionary trends that tend to arise when lineages and biotas have been long separated by geographic barriers?

Divergence

When the Central American landbridge was formed, it not only permitted the interchange of formerly isolated terrestrial organisms between North and South America, but also created a barrier that completely isolated the formerly continuous tropical Atlantic and Pacific Oceans. This event was as important for marine organisms as it was for the terrestrial forms we discussed above. It is especially important in view of the fragile nature of the isthmus as a barrier.

This fragility became particularly apparent in the 1960s, when the possibility of constructing a sea-level canal across the Isthmus of Panama was being seriously considered. The present Panama Canal, constructed in the early part of the twentieth century, incorporates a large body of fresh water, Gatun Lake, and uses a series of locks to raise and lower ships as they traverse the Isthmus. The fresh water effectively prevents interchange between most elements of the Pacific and Caribbean tropical marine biotas. The construction of a sea-level canal would constitute a biogeographic experiment of gigantic proportions. Its effect on marine biotas would be analogous to the influence of the original establishment of the Isthmus on terrestrial forms, with the exception that the Caribbean and Pacific biotas have been isolated in entirety for only about 3.5 million years, whereas North and South America had been separated for at least 135 million years.

Controversy surrounding the possible ecological effects of an interchange of species as a result of a sea-level canal stimulated much research on the similarities and differences between the Caribbean and eastern tropical Pacific biotas. These studies revealed major differences, especially in species richness, between the marine faunas of the two regions (Briggs 1968, 1974; Rubinoff 1968; Porter 1972, 1974; Vermeij 1978). Most groups are more diverse in the Pacific; examples include most major taxa of mollusks, crabs, and echinoderms. There are exceptions, however. The Caribbean has about 900 species of shallow-water and coral reef fish, compared with only about 650 species in the eastern Pacific. Sea grasses and their specialized animal fauna are abundant, widespread, and diverse in the Caribbean, but virtually absent from the eastern Pacific, where suitable, highly productive shallow-water habitats are not extensive.

In many groups, including fishes, sea urchins, gastropods, crabs, and isopods, there are closely related, so-called geminate sister species on either side of the Isthmus (Jordan 1908; Vermeij 1978). The rates of divergence of some of these forms have been of considerable interest to systematists and evolutionists (e.g., Rubinoff and Rubinoff 1971; Lessios 1998). The time of isolation is often presumed to be the same for all groups, although there might be a broader range of times of isolation than would be expected if all species-pairs were isolated during the final stages of closure of the Isthmus (Lessios 1998). Although there has been some differentiation, most of these species pairs remain similar in morphology and, presumably, in their ecological niches. This suggests that competition between such species pairs might prevent much interchange across a sea-level canal. On the other hand, if some forms were superior competitors and able to invade the other ocean, competitive exclusion might result in the extinction of some sister species without greatly affecting the diversity of the biotas on each side of the Isthmus.

The absence of complementary species in a few exceptional groups has caused more concern. Two Pacific taxa in particular that do not have close relatives or obvious ecological counterparts in the Caribbean are the yellow-bellied sea snake, *Pelamis platurus*, and the crown of thorns starfish, *Acanthaster planci*. The former is highly venomous, whereas the latter feeds voraciously on certain corals and occasionally devastates reefs in its native region (Chester 1969). It is possible that *Acanthaster* might wreak even greater havoc if it were able to colonize the rich Caribbean reefs, where the coral species have had no opportunity to evolve resistance to it (Porter 1972; Vermeij 1978).

Convergence

Most of what we have said thus far would cause one to expect that geographically isolated organisms should diverge. Much of phylogenetic systematics and vicariance biogeography is based on the assumption that genetically and geographically isolated lineages tend to become more different as they evolve independently of each other. This is certainly true of many attributes of organisms, including most of those that are used in classification, and necessarily must be true for characters used in phylogenetic analysis. It need not be true of ecological characteristics, however. If groups are isolated in regions of different area, geology, and climate, these differences in the physical environment will tend to promote ecological divergence. If the physical environments are similar, however, distantly related organisms may independently evolve similar adaptations. This phenomenon is called **convergent evolution**, and it can occur on many levels. Within species, it may be restricted to a few traits, or it can involve essentially the entire organism, resulting in convergence in morphology, physiology, and behavior as unrelated forms specialize for similar niches. Convergence can also occur at the level of entire biotas, resulting in geographically isolated ecological communities with similar structures and functions.

CONVERGENCE AT THE SPECIES LEVEL. It is possible to find examples of geographically isolated, distantly related pairs of species that are spectacularly similar. Some of the best examples occur in plants, especially those living in regions where similarly stressful abiotic environments have selected for similar form and function. In Chapter 3 we discussed the Mediterranean climates that occur at about 30° latitude where cold ocean currents flow down the west coasts of continents, in the Mediterranean region, southwestern Africa, southwestern Australia, Chile, and California. The latter two regions were the sites of extensive comparative ecological studies by the International Biological Program (IBP) in the early 1970s (Mooney 1977). Analyses of the anatomy and physiology of the woody plants revealed many similarities. The dominant plants in both matorral habitat in Chile and chaparral in California are shrubs with small to medium-sized, thick (sclerophylous), evergreen leaves. It is possible to identify many pairs of species that are extremely similar in growth form and leaf morphology (Figure 10.35). In addition, these species usually have similar physiological and life history adaptations to photosynthesize and grow during the cool winter rainy season, to minimize water loss and survive through the hot summer dry season, and to regenerate rapidly—usually from vegetative sprouts from the stumps, but sometimes by seed—after frequent wildfires (Mooney et al. 1977).

Plants from desert regions throughout the world provide equally spectacular examples of convergent form and function. Examples are exhibited in several botanical gardens, including the Arizona Sonora Desert Museum near Tucson, Arizona, and Kew Gardens outside London, England. Different genera of succulent thorny cacti (family Cactaceae) from North and South America are remarkably similar not only to each other, but also to the much more distantly related euphorbias (family Ephorbiaceae) of Africa. Another convergent theme is succulents with whorls of tough, pointed leaves: agaves in North America, terrestrial bromeliads in South America, and aloes in Africa (each representing a different family). And, the similarity does not stop at superficial resemblances among life forms, but extends to details of cellular anatomy, physiology, and biochemistry. For example, these desert succulents all share a special form of photosynthesis, called crassulacean acid metabolism (CAM). This adaptation enables them to conserve water by opening their stomates at night when temperatures and rates of evaporative water loss are low, taking up CO_2, storing carbon in the form of organic acids, and then completing photosynthesis with their stomates closed during the day when the sun shines.

These examples of convergence are convincing. Some geographically isolated plant species in different families that live in similar environments are much more similar to one another than to more closely related species that occur on the same continents, but in different environments. Furthermore, their similarities clearly represent evolutionary adaptations to deal in similar ways with similar kinds of abiotic environmental conditions. It should be emphasized, however, that these convergent plants are not strikingly similar in all of their characteristics. Naturally, they each retain distinctive traits indicating their divergent ancestry. In addition, they often differ considerably in reproductive biology, exhibiting divergent forms and functions of flowers, fruits, and seeds that reflect adaptations for different agents of pollination and seed dispersal.

Examples of convergence among geographically isolated animal species are equally spectacular. As in plants, many come from desert regions. The mammalian order Rodentia has a virtually cosmopolitan distribution, but many rodent families have much more restricted distributions. Each of the biogeo-

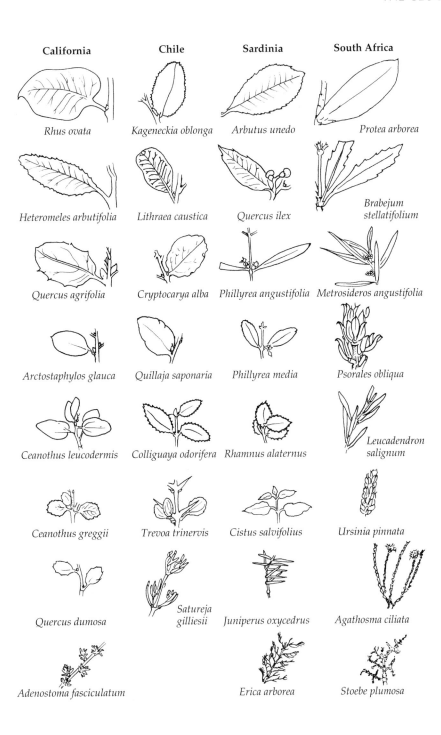

California	Chile	Sardinia	South Africa
Rhus ovata	*Kageneckia oblonga*	*Arbutus unedo*	*Protea arborea*
Heteromeles arbutifolia	*Lithraea caustica*	*Quercus ilex*	*Brabejum stellatifolium*
Quercus agrifolia	*Cryptocarya alba*	*Phillyrea angustifolia*	*Metrosideros angustifolia*
Arctostaphylos glauca	*Quillaja saponaria*	*Phillyrea media*	*Psorales obliqua*
Ceanothus leucodermis	*Colliguaya odorifera*	*Rhamnus alaternus*	*Leucadendron salignum*
Ceanothus greggii	*Trevoa trinervis*	*Cistus salvifolius*	*Ursinia pinnata*
Quercus dumosa	*Satureja gilliesii*	*Juniperus oxycedrus*	*Agathosma ciliata*
Adenostoma fasciculatum		*Erica arborea*	*Stoebe plumosa*

FIGURE 10.35 Leaf morphology of distantly related plant species from evergreen shrub habitats in Mediterranean climates in four widely separated regions: California, Chile, Sardinia (Mediterranean), and South Africa. Presumably, the similarities—not only in the sizes and shapes of the leaves shown here, but also in their physiological characteristics—reflect convergence: the independent evolution of similar traits in response to natural selection for similar adaptations to similar environments. (After Cody and Mooney 1978.)

graphic realms has large areas of desert and semiarid habitat, and each of these has a distinct, highly specialized desert rodent fauna made up of different families that have independently evolved to fill a variety of niches. The most striking case of convergence is perhaps the independent evolution of forms with elongated hind legs; long, tufted tails; and bipedal hopping (saltatorial or ricochetal) locomotion in different families on several continents (Figure 10.36; Mares 1976, 1993a). Some of these rodents also share other adaptations, including light-colored pelages to match their backgrounds; enlarged ear cavities (auditory bullae), perhaps to detect predators; short, long-clawed

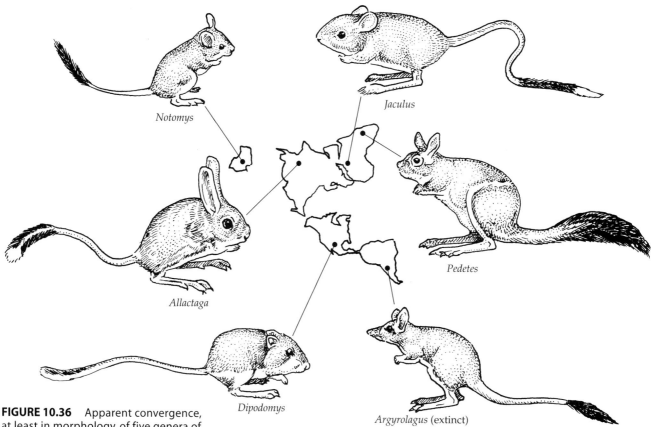

FIGURE 10.36 Apparent convergence, at least in morphology, of five genera of rodents and one extinct marsupial from deserts throughout the world. All of the rodents are derived from unspecialized mouse-like ancestors, and have independently evolved long hind limbs, short forelimbs, long tufted tails, light brown dorsal and white ventral pelage, and bipedal hopping locomotion. Some of them share other morphological, physiological, and behavioral characteristics, but some of them also differ conspicuously in body mass, ear length, diet, and other characteristics. (After Mares 1993a.)

front legs for digging burrows and collecting food; and urine-concentrating kidneys for maintaining water balance on a dry diet. On the other hand, as in the plants, their convergence does not extend to all traits. Thus, the North American kangaroo rats and kangaroo mice are highly specialized seed eaters, whereas the similar-looking hopping mice in Australia and jerboas in North Africa, the Middle East, and central Asia eat some seeds, but also include substantial amounts of other items, such as insects, tubers, and leaves, in their diets. Furthermore, in these different deserts, some of the rodent species that are not very similar in external morphology are much more similar in diet and, as a consequence, in digestive and excretory physiology (Mares 1976; Mares and Lacher 1987; Mares 1993a,b; Kelt et al. 1999).

Similar generalizations could be drawn for other taxa, such as the superficially similar snakes and lizards from arid regions of North America, Africa, Asia, and Australia. The pronghorn or "antelope" of the plains of North America is the sole representative of an endemic family (Antilocapridae) that is convergent in morphology and behavior with the true antelopes (family Bovidae) of the grasslands of Africa. Several species of toucans (family Ramphastidae) of the New World tropics are superficially similar to some species of hornbills (family Bucerotidae) of the Old World. Representatives of both families have striking black and white plumage and very large, light, keel-shaped bills adapted for feeding on fruit. Toucans and hornbills differ conspicuously, however, in their reproductive biology, with only hornbills showing the distinctive behavior of males building mud walls to imprison their mates in a nesting cavity. The European fire salamander (*Salamandra salamandra*) and North American tiger salamander (*Ambystoma tigrinum*) are similar in their large size, robust body shape, poison glands in the skin, and striking yellow on black warning coloration, but they differ in other aspects of their ecology and behavior.

CONVERGENCE OF ENTIRE ASSEMBLAGES?   While such convergence between isolated species and groups may be extremely precise, some investigators have suggested that there can be at least as much convergence at the level of entire biotas. In the literature, there are many illustrations purporting to show pairs of similar species that make up the biotas of isolated regions with similar environments. The mammals of Australia and North America, shown in Figure 10.37, are perhaps the most frequently cited examples, but others are the mammals inhabiting tropical forests of Africa (or Asia) and South America (e.g., Bourliere 1973; Eisenberg 1981), and the plants and birds living in semiarid (including Mediterranean) or desert climates on different continents (e.g., Cody 1968, 1973; Solbrig 1972; Mooney 1977).

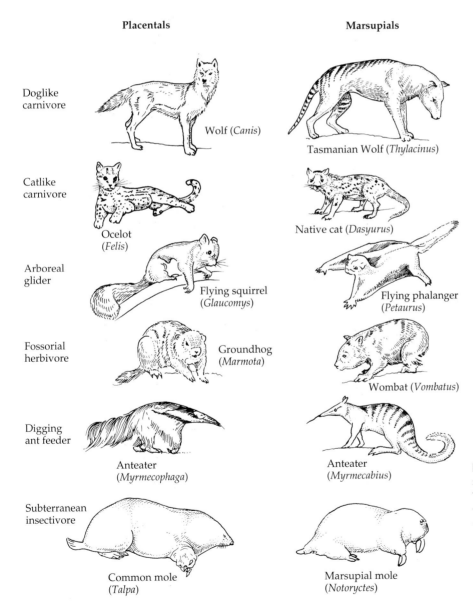

FIGURE 10.37 Drawings of pairs of species of North American and Australian mammals purporting to show convergence. Figures such as this one are somewhat misleading. As described in more detail in the text, the species paired here are often not drawn to the same scale, and some of those that look alike do not have similar ecological niches. (After Begon et al. 1986.)

Such illustrations are not intended to be misleading. Nevertheless, they often exaggerate the degree of overall resemblance between the biotas. For one thing, the pairs of species that are depicted in the drawings may not be as similar in body size, physiology, behavior, and ecology as in general body form. Thus, for example, in Figure 10.37, the marsupial "cat" is smaller, more insectivorous, and less arboreal than the ocelot; the flying phalanger is much larger than, and differs in diet from, the flying squirrel; the wombat is much larger than the groundhog; the numbat or marsupial anteater is much smaller than the giant anteater; and the marsupial "mole" inhabits sandy deserts, unlike any true mole. In addition, such figures rarely show the majority of the species in the two biotas that are not similar. Australia, for example, has no close ecological equivalents of many North American mammals, including mountain lions, bison, weasels, skunks, prairie dogs, and beavers; similarly, North America has no species that are really similar to duck-billed platypuses, spiny anteaters, bandicoots, koalas, and the numerous small and medium-sized wallabies. And, finally, other elements of the North American and Australian biotas exhibit even less convergence than the mammals. North America has no real equivalents of Australia's parrots and emus among birds, or the eucalypts and acacias among plants, and its lizard fauna is much less diverse (Pianka 1986). Australia has no arid-zone plants that are similar to the cacti and agaves of the North American deserts. In fact, after visiting both North America and Australia, many naturalists begin to question the dogma of convergence.

It is not really surprising that the differences in these two biotas far outweigh their similarities. Comparison of the geography, ecology, and geology of the two continents reveals that their environments are in fact very different in several respects. First, as illustrated in Figure 10.38, Australia is much more tropical than North America. The Tropic of Capricorn passes through the very center of Australia just north of Alice Springs, whereas the Tropic of Cancer passes just north of Mazatlan, Mexico. Compared with North America, Australia has extensive older geological formations (Paleozoic and Mesozoic rather than Cenozoic), little recent tectonic activity, more eroded landscapes, and hence much less elevational relief (the highest mountain is only 2229 m compared with 6194 m in North America). As a result of their long exposure to erosion and leaching, Australian soils are extremely low in nutrients. The low availability of essential resources has strongly influenced many aspects of the ecology and evolution of Australian plants, and the plants in turn have affected the animals. Given these major differences in their geological and geographic histories and current environments, it seems naive of biogeographers, evolutionary biologists, and ecologists to expect to find any substantial degree of convergence between the biotas of these two continents. Similar caution should be exercised before uncritically accepting other examples of expected or claimed convergence at the level of whole floras or faunas (e.g., Cody and Mooney 1978; Lomolino 1993a; Kelt et al. 1999).

Overview

So, we end by emphasizing the complexity evident in the diversification and assembly of biotas. On one hand, as noted by such early biogeographers as Buffon, de Candolle, Sclater, and Wallace, and confirmed by many more recent investigators, the biotas of the major landmasses are each made up of distinct kinds of plants and animals. To a large extent these differences reflect geological histories of isolation and the presence of different lineages, each with their unique constraints and potentials to diversify and adapt to abiotic and biotic environments. We are beginning to discover, for example, the surprisingly

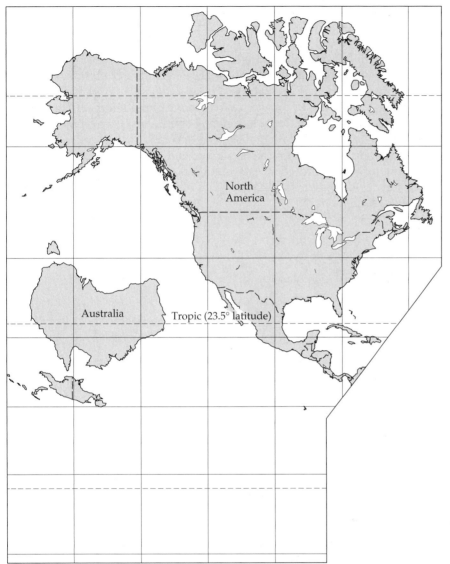

North
America

Australia Tropic (23.5° latitude)

FIGURE 10.38 A map in which the locations of Australia and North America have been juxtaposed while maintaining their relative latitudinal positions. Note that, while Australia is about the same size as the United States, its latitudinal position overlaps more with that of Mexico. The locations of the arid habitats that are often suggested to contain convergent species or ecological communities are, on average, much more tropical in Australia than in North America. (Australia has been inverted in order to maintain its orientation relative to the equator.)

fundamental role of Gondwanaland in the early diversification of many now-common northern hemisphere lineages (e.g., amphibians, Bossuyt and Milinkovitch 2001; modern birds, Cracraft 2001; placental mammals, Eizirik et al. 2001; and grasses, Bremer 2002; see Figure 7.11). The pattern and process of diversification has been additionally influenced by differences, both small and large, in the environmental settings, which have favored the differential diversification of some lineages and the evolution of certain traits.

On the other hand, the biotas of the major continents are also similar in some respects. They share some taxonomic groups—and even some species are cosmopolitan, or nearly so. Some of these shared lineages have persisted from the time before the continents were isolated. Others have spread more recently, either colonizing across long-standing barriers or dispersing at times in the past when bridges of habitat permitted biotic interchange. Similar environments on different continents have facilitated the colonization and persistence of closely related organisms with similar requirements, as well as the convergent evolution of distantly related forms to use similar environments in similar ways. The same processes can be extended, albeit at some notably different spatial and temporal scales, to marine environments (Wainright et al. 2002; Johannesson 2003; Fratini et al. 2005).

The diversity of life on Earth reflects the outcome of opposing forces promoting both convergence and divergence among biotas. Evolutionary conservatism, phylogenetic constraints, gene flow, and similar environments limit the rates and directions of diversification, and thus tend to maintain similarities among biotas. Different environmental conditions, geographic isolation, evolutionary innovations, adaptation, and speciation enhance the rates and directions of divergence and diversification, and thus tend to promote differences among biotas. Nowhere is the complex interplay of these opposing forces more evident than in the diversity of the plants and animals inhabiting biogeographic regions.

Reconstructing the History of Lineages

A T THE BEGINNING OF THE PREVIOUS CHAPTER, we introduced the basic differences between the "history of lineage" and the "history of place." Each of these can, and has often been, investigated without reference to the other, but an important goal of historical biogeography is to discover causal relationships between lineage and place histories. Our ability to reconstruct such an association can only be accurate to the extent that the underlying hypotheses of history are accurate. We have described in detail many of the important events in the history of place in Chapters 8 and 9. We have also introduced the basic concepts of systematics in Chapter 7, but left until now a detailed discussion about how we reconstruct and interpret the histories of lineages.

The importance of phylogenetic systematics to biogeography has grown rapidly in the past several decades, in part because of the increasing availability of high quality phylogenetic trees, in part because of the increasing sophistication of phylogenetically based analyses employed to assess the relative influence of vicariance and dispersal events in biogeographic history, and in part because of an expanded technical capacity to address biogeographic history at relatively recent timeframes (i.e., during the past two million years). Finally, while fossils have always been used in the reconstruction of lineage histories, they assume a renewed importance, particularly in the estimation of times of lineage divergence, when used in combination with molecular approaches to constructing phylogenetic trees. This chapter reviews the basic principles of phylogenetic systematics and phylogeography—the approach in biogeography that studies the geographic distributions of genealogical lineages within species and among closely related species—after a brief review of the development of modern systematics during the last third of the twentieth century. The histories of lineages and places will then be considered together in our formal discussion of modern historical biogeography in Chapter 12.

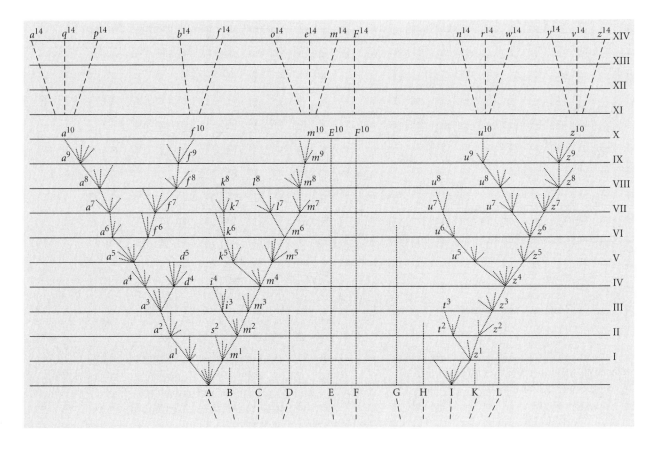

FIGURE 11.1 A hypothetical phylogeny, the only illustration Darwin included in *The Origin of Species*. Darwin used this diagram to illustrate the principal of descent with modification from a common ancestor and his view that a Natural System of classification was one in which "propinquity of descent…is the bond, hidden as it is by various degrees of modification, which is partially revealed to us by our classifications." (After Darwin 1859.)

Classifying Biodiversity and Inferring Evolutionary Relationships

The hierarchical system of biological classification used traditionally to classify biodiversity dates back to the eighteenth-century Swedish naturalist Linneaus, but he believed that each kind of plant and animal had been specially created by God. In the latter half of the nineteenth century, Darwin placed Linneaus' scheme of classification within an evolutionary framework, arguing that species are nested within genera, genera within families, and so forth because of a history of descent with modification within evolutionary lineages—"…the view that an arrangement is only so far natural as it is genealogical" (Darwin 1859). In fact, the only illustration in *The Origin of Species* was a hypothetical phylogeny (Figure 11.1). But while Darwin's observation would form the conceptual foundation for the modern science of systematics, this discipline has wrestled over the past century with basic questions about how to define and diagnose natural groups, and what sorts of characters and procedures can best be used to reconstruct them. Fortunately, much of the heated controversy that infused this discipline throughout the 1960s and into the 1980s has subsided, and tremendous progress continues in both the quality and quantity of phylogenetic hypotheses that are becoming available for use in biogeography. Developments in four principal areas have had a profound impact on our ability to reconstruct the history of lineages: (1) general acceptance and application of the concepts of phylogenetic systematics established by Willi Hennig (1966); (2) utilization of abundant characters within a DNA sequence; (3) continual improvements in our understanding of the tempo and mode of evolution of those characters; and (4) bridging the threshold between the interspecific and intraspecific boundary by using phylogenetic and population genetic approaches to investigate the phylogeography of populations within species.

Systematics

As pointed out in Chapter 2, humans have long recognized that some kinds of organisms are more similar to each other than they are to others, and we have tried to capture these patterns of similarity by naming and classifying living things. Aboriginal peoples have names for the plants and animals in their environments, and these names typically recognize degrees of difference, with classifications incorporating functional information based on the roles each play in the lives of the humans, but nevertheless arriving at a taxonomy surprisingly similar to one developed by trained taxonomists (e.g., Patton et al. 1982). While Darwin provided the conceptual foundation for linking the history of lineages into a nested hierarchy of ancestor-descendant evolutionary relationships, early systematists classified biodiversity for another century without developing a clearly defined methodology for doing so. Beginning in the 1950s, three very different "schools" of classification emerged—**evolutionary systematics**, **numerical phenetics**, and **phylogenetic systematics** or **cladistics** (Mayr 1988). Each school claimed to have developed a robust set of methods for constructing a natural classification, but their underlying conceptual principles differed greatly and resulted in heated controversies and debates throughout the 1960s and into the 1980s. Although the vast majority of systematists now employ some form of phylogenetic systematics, it is worthwhile to review briefly the first two methods, both because each enjoyed great influence in systematics at various times during the 1960s and 1970s, and also because some of the methods that arose originally within a phenetics framework are continued into modern systematics within a phylogenetic framework.

Evolutionary Systematics

Recall from Chapter 7 that modern systematics has two related goals—to classify biological diversity (taxonomy) and to recover patterns of ancestor-descendant relationships (phylogenetics). Prior to the widespread acceptance of the methods and logic formalized by Willi Hennig into cladistics, two founders of the Modern Synthesis—Ernst Mayr (1942) and George Gaylord Simpson (1945, 1961a), among others (Dobzhansky 1951; Rensch 1960)—were leading advocates for the use of two kinds of evolutionary information in the classification of organisms. On one hand, evolutionary classifications were consistent with Hennigian logic in using the best estimates of branching sequences representing the hierarchy of ancestor-descendant relationships. However, they went a step further and sought to include information on degree of divergence between taxa, as well. In the language of phylogenetics, these classifications therefore had a tendency to produce paraphyletic groups—those that contain a common ancestor and some, but not all, of its descendants. Many long-recognized taxa are actually **paraphyletic** and therefore considered artificial rather than natural groups under phylogenetic systematics. Consider, for example, the amniotic vertebrates (reptiles, birds, and mammals). While evolutionary classifications have long placed reptiles, birds, and mammals into separate classes (Reptilia, Aves, Mammalia; Figure 11.2), detailed character analyses—primarily from morphological studies of fossil and extant specimens, but supported by molecular analyses—have generally provided strong support for a basal amniote divergence into two mono-

FIGURE 11.2 An evolutionary classification and reconstructed phylogeny of living vertebrates from the classic work of A. S. Romer. The thickness of the branches indicates the approximate diversity of the groups through time. The degree (rate) of evolutionary divergence from a common ancestor as well as the time of the split are taken into account in grouping organisms into taxa. Note that the cartilaginous fishes (Chondrichthyes), bony fishes (Osteichthyes), amphibians, reptiles, birds (Aves), and mammals are all recognized as equivalent units and given class rank.(After Romer 1966.)

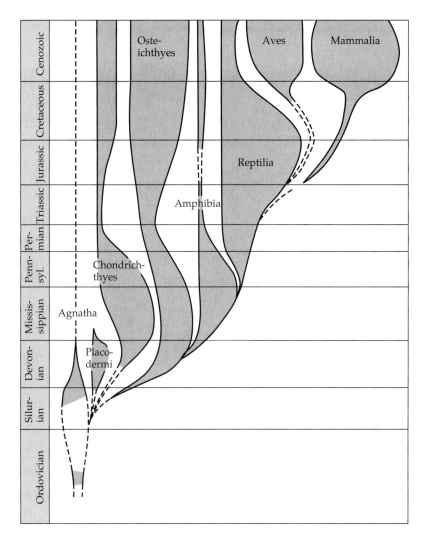

phyletic clades. The first branch is a "sauropsid" clade that includes living reptiles, birds, and the extinct dinosaurs (and nested within this clade, is an "archosaur" clade that contains just the dinosaurs, crocodilians, and birds); the second branch is a "synapsid" clade that includes mammals and the extinct mammal-like reptiles (compare Figure 11.3 with Figure 11.2).

Numerical Phenetics

Another problem with evolutionary classifications was that they tended to be highly subjective. Different systematists, assigning different importance to branch points versus degree of evolutionary divergence and using their own schemes for choosing and weighting characters, produced different classifications for the same groups—and then argued about which one reflected the "true" evolutionary relationships. In response, some biologists questioned the scientific legitimacy and relevance of systematics (Ehrlich 1964), and others attempted to develop a more objective and quantitative process of classification called "numerical phenetics" or "numerical taxonomy" (Sneath and Sokal 1973). As an alternative to attempting to construct explicitly evolutionary classifications, numerical taxonomists advocated purely "phenetic" classifications, which attempted to use *degree of overall similarity* derived through quantitative

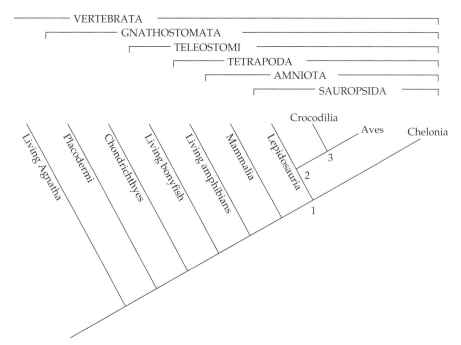

FIGURE 11.3 A cladogram showing nested monophyletic groups within the Vertebrata. Note that reptiles (Lepidosauria, Crocodilia, and Chelonia) would become a paraphyletic group if Aves were removed from the Sauropsida and placed in its own category. Shared-derived (synapomorphic) characters that support the inclusion of Aves in a particular group include: (1) loss of the medial centrale bone of the pes; (2) dorsal and lateral temporal openings and suborbital fenestra; (3) hooked fifth metatarsal and foot directed forward for much of the stride. (Modified from Carroll 1988.)

analyses of phenotypic traits, to cluster taxa into hierarchical units. Although the practitioners of this approach had hoped to be able to arrive at natural classifications through use of many characters, these classifications frequently grouped non-related taxa into the same taxon due to an undue influence of traits that were similar because of convergent or parallel evolution, or because rates of evolution differed among taxa. As with evolutionary systematics, the original form of numerical phenetics has been largely replaced by phylogenetic systematics. However, traces of this school are still found in modern systematic procedures that use quantitative measures of similarity to produce distance matrices that are analyzed within a phylogenetic framework. An important legacy of numerical phenetics survives in disciplines outside of systematics, such as ecology and sociology, that employ a variety of procedures based on ordination and clustering methods to examine attributes and similarities among communities and populations.

Phylogenetic Systematics

While Darwin provided great insight by defining a "natural" classification as one that grouped taxa according to their evolutionary affinities, the systematics community did not have a straightforward procedure for objectively coding this fundamental property into phylogenetic trees until the mid-twentieth century. Willi Hennig, a German entomologist, solved this problem with a landmark work first published in German (1950), but having much greater impact after its translation into English (1966). Hennig's genius was to recognize the logical consequences of a feature of the evolutionary process that had been known for a long time. Organisms are very complex systems; they comprise many parts and processes that must interact in many ways to produce a functional individual that can survive and reproduce. Thus, the kinds of changes that can occur over generations of evolution—and also during the ontogenetic development of individual organisms—are highly constrained. New structures and functions are almost never created de novo. Instead, they are obtained by modifying already existing structures and functions. The history of these changes is recorded in the similarities and differences in the complex characteristics of related organisms—in the extent to which the characteristics of their common ancestors have been modified by subsequent additions, losses, and transformations. Hennig showed how one could deduce a history of ancestral connections between descendant taxa by ordering the transformation of ancestral to derived states of a **character** (any heritable and observable part or attribute of an organism) into a **transformation series**. From this simple insight, Hennig developed an elegantly simple yet logical method for reconstructing the history of a lineage and assessing the nested hierarchy of relationships among its taxa. It is this pattern that is captured within a cladogram, and the construction of cladograms that reflect the true history of cladogenesis is the basic goal of phylogenetic systematics.

ANATOMY OF PHYLOGENETIC SYSTEMATICS. The essence of Hennig's logic is easy to understand. First, a natural taxon is defined as a **monophyletic group** (a clade), which includes all descendant taxa and their common ancestor (Figure 11.4A and B). Any other grouping of taxa comprises an artificial group. For example, as we introduced above, a group that includes an ancestor and some—but not all—descendant taxa, is paraphyletic (Figure 11.4C and D); while a **polyphyletic** group includes taxa that trace back through two or more separate ancestors before reaching a common ancestor (Figure 11.4E and F).

FIGURE 11.4 Cladograms illustrating monophyletic, paraphyletic, and polyphyletic groups. Capital letters represent species and each group of species included within a box represent a taxonomically recognized group. The groups in the top two diagrams are monophyletic because they include an ancestor and all of its descendants. For example, on tree A one group includes Ancestor 2 + species T + R; another group includes Ancestor 3 + species I + L + O + B. Groups in the center diagrams are paraphyletic because each contains an ancestor, but not all of its descendants. For example, on tree C the large group is paraphyletic because it includes the common Ancestor 1 + species R + I + L + O + B, but excludes species T—if the group were revised to include species T, it would become a monophyletic group. Groups in the bottom diagrams are polyphyletic because they need to pass through one or more ancestors before arriving at a common ancestor for the group. For example, on tree E the group includes Ancestor 1 + species R + I, but we need to count back through Ancestor 2 and Ancestor 3 before arriving at the ancestor shared by species R + I. (After Brooks and McLennan 2002.)

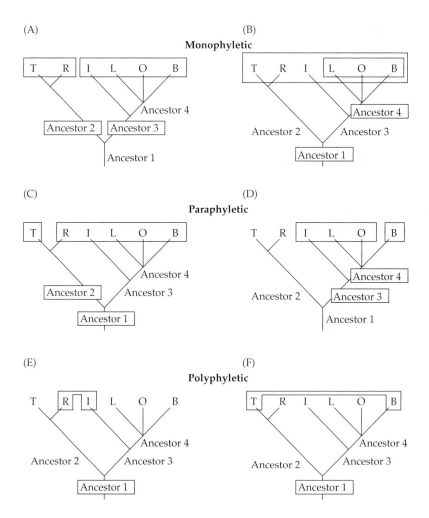

FIGURE 11.5   A simple phylogenetic tree illustrating the concepts of ingroup, outgroups, and the special outgroup that is called the sister group to the ingroup. The ingroup is the focal set of taxa within a monophyletic group, whereas one or more outgroups are typically included in an analysis in order to root the tree and provide an ancestral-descendant orientation to the character transformation series. (After Brooks and McLennan 2002.)

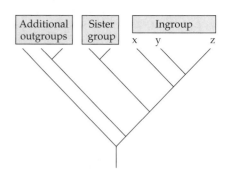

Second, a monophyletic group is supportable through evidence derived in an analysis of the pattern of distribution of different states of a character across taxa. A character that exists in a particular state in two or more taxa is said to be **homologous** if it is shared because of its inheritance from a common ancestor; if it is shared because of convergent or parallel evolution, it is considered a **homoplasy**. Only a homologous character can be used to support a monophyletic group, but not all homologous characters are equally valid for doing so, and in order to understand which are valid, we first need to introduce two more terms. As a character evolves between ancestral and descendant taxa, its form in the ancestor is said to be **plesiomorphic** (primitive), while the transformed state in the descendants is called an **apomorphic** (derived) character. Now, we can define those "special homologies" that can be used to diagnose monophyletic groups—these are called **synapomorphic** (shared-derived) characters, while homologous characters that diagnose a group of taxa more inclusive than the focal monophyletic group are called **symplesiomorphic** (shared-primitive) relative to the focal group.

How do we infer which state of a character is plesiomoporphic, and which is apomorphic (i.e., elucidate the correct order of changes in a transformation series)? The most straightforward approach is to inspect its transformation within a group of taxa considered to comprise a monophyletic group (an **ingroup**) with reference to one or more distantly related **outgroups** (Figure 11.5). The most useful outgroup for this purpose is the next most closely related group to the ingroup, and is called the **sister group**. The logic here, again, is simple—that state of the character that occurs in both the sister group (or another closely related outgroup, if the sister group is unknown) and one or more ingroup taxa is inferred to be the plesiomorphic state. Obviously, this will not always be the correct inference—there will always be a chance that character-states shared by ingroup and outgroup taxa are homoplasies. However, systematists typically include many characters in an analysis, recognize that some of those characters will likely be homoplasies, and employ methods that account for homoplasy while producing the "best" phylogenetic trees possible. In the simplest case, the best tree is the one that requires the least number of character-state changes to group all ingroup taxa into a phylogeny—this is called the most **parsimonious** tree.

Overall, there are only a few basic principles and rules that one needs to know in order to understand the basic logic for building trees under the principles and rules of Hennigian argumentation (Box 11.1). Once the basic terminology, principles, and rules are understood, one can begin to understand the relationship between cladogenesis and character evolution that forms the basis of Hennigian logic (Box 11.2).

METHODS IN PHYLOGENETIC SYSTEMATICS.   While the logic of Hennig's paradigm is quite simple, the current array of competing methods and computer programs available to produce and analyze phylogenetic trees can be intimidating to beginning and experienced systematists alike. A short list of the more popu-

BOX 11.1 *The principles and rules of Hennigian logic*

■■▮ Hennig's Auxiliary Principle: Never presume convergent or parallel evolution; always presume homology in the absence of contrary evidence.

Significance: This principle establishes a testable hypothesis about character evolution. It does not state that convergent evolution, which produces homoplasious characters, is non-existent, or even particularly uncommon. It simply establishes the logical basis of employing putatively homologous characters in phylogenetic analysis.

Relative Apomorphy Rule (outgroup comparison): Homologous characters found within the members of a monophyletic group that are shared with members of the sister group are plesiomorphic, while homologous characters found only in the ingroup are apomorphic.

Significance: This rule establishes the basis, through outgroup comparison (with the sister group being the most impor-tant outgroup), of polarizing the transformation series of character evolution in the ingroup. This rule forms the foundation for stating the following rule.

Grouping Rule: Only synapomorphies (shared special homologies) provide evidence of common ancestry relationships. Symplesiomorphies (shared general homologies), autapomorphies (unique homologies), and homoplasies (convergences and parallelisms) are useless in this procedure.

Significance: A homologous character that is a synapomorphy in reference to a particular clade will become a symplesiomorphy in reference to clades nested within that clade.

Inclusion / Exclusion Rule: The information from two transformation series can be combined into a single hypothesis of relationship if that information allows for the complete inclusion (or the complete exclusion) of groups which were formed by the separate transformation series. Partial overlap of groupings necessarily generates two or more hypotheses of relationship.

Significance: This rule establishes the logic behind being able to say that character-states from one transformation series either are, or are not, completely consistent with one another regarding the groupings of taxa that are supported by synapomorphies. More often than not, phylogenetic analyses will produce more than one tree because conflicts between relationships supported by different characters do not allow them to be combined on a single phylogenetic tree. One must either add more homologous characters to the analysis to further assess degree of support for the alternative hypotheses, or collapse conflicting branches into unresolved polychotomous nodes on a tree that represents a consensus between the conflicting hypotheses.

(Adapted from Brooks and McLennan 2002.)■▮■

lar approaches, each implemented in multiple readily available software packages, includes: distance matrix methods; and the character-based methods consisting of maximum parsimony, maximum likelihood, and Bayesian inference. Distance matrix methods first produce a quantitative estimate of difference between each pair of operational taxonomic units (OTUs), then use one or more clustering methods (e.g., UPGMA, Neighbor-joining) to generate a tree called a phenogram. In contrast, the maximum parsimony method uses discrete character states directly to find the tree, or set of trees, that require the least number of character-state changes to represent relationships among OTUs. Both maximum likelihood and Bayesian methods allow users to choose an explicit, and often rather complex, model of the evolutionary process—the former then searches for the tree that maximizes the probability of observing the data given the tree, whereas the latter searches for the most probable tree given the data under a posterior probabilities framework. These methods are popular in molecular studies, where it is often possible to understand with a fair degree of precision the different rates of evolution in different "partitions" of a genome (see discussion below). A good resource for entry into the available methods of phylogenetic analysis, including a large array of methods for statistical analysis and for comparing trees for degrees of congruence, can be found at www.evolution.genetics.washington.edu/phylip/software.html. Good tutorials on phylogenetic systematics include Wiley et al. (1991), Skelton and Smith (2002), Hall (2004), and the online resource provided by the Museum of Paleontology at the University of California Berkeley at www.ucmp.berkeley.edu/clad/clad4.html.

BOX 11.2 *The basis of Hennig's paradigm: A hypothetical example of cladogenesis and cladogram construction*

■■▮ The following example and diagrams will illustrate Hennig's insight into how the branching pattern of phylogeny leaves its legacy in the derived characteristics of contemporary forms. We begin with a group of four species, numbered 1–4; and six characters, lettered A–F, each of which can exhibit one of two states. The first thing we need to do is to determine which of the character states are ancestral (or plesiomorphic), and which are derived (or apomorphic). To do this, we need to identify an outgroup, a species closely related to species 1–4 but whose lineage diverged at an earlier time (preferably, the sister group to the ingroup). In this case, we choose species 0 to be the outgroup. Each character has two discrete states, A or A', and for simplicity in coding we assign each state that occurs in the outgroup the non-prime value, which, through application of the Relative Apomorphy Rule (Box 11.1) makes those states plesiomorphic and those with prime signs apomorphic (i.e., the transformation series is A → A'). The matrix showing the distribution of character states among the five species is shown below. Note that we will adhere to Hennig's Auxiliary Principle until the Inclusion / Exclusion Rule gives us reason to consider alternative hypotheses.

Now let's look at the first character, for which all ingroup species share the derived state. The tree for the ingroup based on this character is:

Likewise, the tree depicting change in character B is:

Proceeding in a similar way to characters C and D, we get the tree:

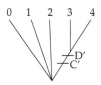

It should be apparent that we can apply the Inclusion / Exclusion Rule at this point to derive the following tree, considering the first four characters all together:

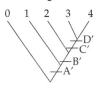

Character state E' for the next character occurs in only one ingroup taxon (Species 2) and is therefore an autapomorphy and therefore uninformative with regard to sister-lineage relationships according to the Grouping Rule.

Finally, when we get to the last character, we find that the derived state specifies the following tree:

According to the Inclusion / Exclusion Rule, this tree is not combinable with the preceding one, and, given the otherwise strong character support for that tree, the simplest way to explain this discrepancy is that F' represents a convergence or homoplasy. Under that hypothesis, we arrive at the final tree using all characters:

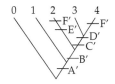

Note that a particular character state can be either a synapomorphy or a symplesiomorphy, depending on the focal group in question. For example, B' is a synapomorphy for the group: (Species 2, Species 3, Species 4), but is a symplesiomorphy for the group: (Species 3, Species 4).

This example is highly simplified. It considers only a few characters, and, although they change from plesiomorphic to apomorphic states, there is only one case of reversal or independent evolution (homoplasy) of character states. ■■▮

Taxon	Character transformation series					
Species 0 (outgroup)	A	B	C	D	E	F
Species 1	A'	B	C	D	E	F
Species 2	A'	B'	C	D	E'	F'
Species 3	A'	B'	C'	D'	E	F
Species 4	A'	B'	C'	D'	E	F'

We turn now to a more detailed discussion of two recent revolutions that have had a profound impact on biogeography—molecular systematics and phylogeography.

Molecular Systematics

In 1953, Watson and Crick elucidated the molecular structure of DNA, the molecule that encodes and transmits genetic information. Systematists soon realized that just as the structural and functional characteristics of organisms could be used to reconstruct phylogenetic relationships, the molecules that code for those characteristics also could be used in phylogenetic inference. The utilization of molecular data, primarily in the form of DNA sequences, within the theoretical framework of molecular evolution, is often treated as a separate subdiscipline in systematics called **molecular phylogenetics** or **molecular systematics**.

EVOLUTION OF METHODS IN MOLECULAR SYSTEMATICS.   Several early approaches to estimating changes in the structure of DNA are indirect. One of these methods estimates evolutionary distances between taxa from *immunological assays* (e.g., Sarich and Wilson 1967). A protein such as albumin is used to assay the strength of antibody-antigen binding with a group of "complement" proteins between a reference and test species in a procedure called micro-complement fixation. The amount of complement "fixed" is related to the number of amino acid replacement substitutions between proteins from different taxa and so can be used as a measure of phylogenetic relatedness. One of the most popular of the early methods that is still used is *protein electrophoresis*, which takes advantage of differences in the mobility of soluble proteins migrating across a gel medium in an electrical gradient to infer amino acid substitutions. Those amino acid changes that cause differences in protein charge lead to changes in the rates of migration of the protein across the gradient and can thus be scored as character-state changes (i.e., those that can be analyzed using either distance or parsimony approaches; e.g., Selander et al. 1971). If a protein is inherited in a Mendelian fashion, this method can also be used to examine attributes of genetic structure within and among populations and species, and test for hybridization between species at contact zones. Some of the more compelling phylogenetic surprises, particularly with regard to relationships across higher taxa (e.g., families or orders of birds), have come from *DNA-DNA hybridization*, which uses the thermal melting profiles of "heteroduplex" double-strand DNA built from DNAs from different taxa that are allowed to form a double-stranded pair as an index of relatedness. The more different the strands are, the lower the temperatures at which bonds are broken between heteroduplex double-stranded DNA (e.g., Sibley and Ahlquist 1990).

Many studies in the 1980s and 1990s used *restriction endonuclease* enzymes produced from recombinant bacterial DNA that "cut" the DNA at a specific (usually four, five, or six base) sequence (e.g., the restriction enzyme *Eco*RI recognizes the sequence 5'-GAATTC-3' and cuts a double-strand DNA everywhere this sequence occurs). If a point mutation occurs at any of the six nucleotide bases in a recognition sequence, the restriction enzyme no longer recognizes it as a site to cut. The DNA fragments are passed through a gel medium in an electrical gradient, and length of fragment determines the rate of migration through the gel. The DNA fragment profiles are then used to infer character-state changes, and, as with the isozymes produced in protein electrophoresis, these profiles can be analyzed with either distance or parsimony approaches in phylogenetic trees and population genetic studies.

The vast majority of biogeographic work using restriction sites was done using the small, circular genome found in animal mitochondria (plant chloroplast DNA was also employed, but not as successfully until direct sequencing methods became easier to use; Avise 2004). This was also the approach that motivated the development of the discipline of phylogeography in the 1980s, as described below. Most recently, an enormous advance in technology in the late 1980s—the polymerase chain reaction (PCR)—motivated a new generation of phylogenetic and population genetic studies. Using PCR, an investigation can begin with very small amounts of DNA, and DNA that has been degraded into small fragments (as happens, for example, in the DNA of museum specimens and fossils), because the method uses a process of cycling through multiple rounds of amplifying the original DNA into many thousands of copies (see Avis 2004). A target region of DNA, and even whole mitochondrial genomes, that have been amplified through PCR can be sequenced directly, for tens or hundreds of individuals, within a few weeks of laboratory work. The vast majority of studies today—whether they examine changes in a DNA sequence directly, or use a newer generation of indirect methods (e.g., with restriction enzymes or denaturing polyacrylamide gels)—begin with PCR amplification of the DNA.

MOLECULAR VS. MORPHOLOGICAL CHARACTERS. From the beginning of the utilization of molecular characters in systematics several decades ago, there have been controversies over the relative merits of molecular versus morphological data in reconstructing evolutionary relationships. Certainly, paleontologists are almost always limited to using morphological characters (Lieberman 2000) although the ability to successfully sequence so-called *ancient DNA* has become much better in recent years (e.g., Cooper et al. 2001; Haddrath and Baker 2001; Kuch et al. 2002), and new fossil finds continue to transform our knowledge of the ages and geography of early diversification in modern groups (Figure 11.6). Nevertheless, as evidenced by the growth in popularity of several journals devoted to publishing molecular-based studies (e.g., *Molecular Phylogenetics and Evolution*; *Molecular Ecology*), and recent papers reviewing the future of historical biogeography arguing for more sophisticated uses of molecular phylogenies (Donoghue and Moore 2003), there is really no question that molecular phylogenetics has assumed an enormous importance in modern biology. The vast number of characters available (even the very small

FIGURE 11.6 Morphological phylogeny placing the 66–68 million-year old Cretaceous fossil bird, *Vegavis iiai*, from western Antarctica, within the waterfowl (Anseriformes); it is most closely related to the living lineage that includes the true ducks (Anatidae). Other living waterfowl lineages represented are the magpie goose (*Anseranas*), two screamers (*Anhimus* and *Chauna*), chicken-like birds (Galliformes), and tinamous (Tinamiformes). Given that at least four nodes leading to modern lineages subtend the node uniting *Vegavis* with Anatidae and *Presbyornis* (another extinct wading bird), at least five divergence events leading to modern bird lineages are inferred here to have occurred prior to the end of the Cretaceous. Note that while previous fossil evidence had supported the idea of an explosive evolution of modern bird lineages following the Cretaceous-Tertiary mass extinction episode (e.g., Feduccia 2003), accumulating molecular evidence, with calibrated molecular clock estimates of divergence times, have consistently indicated a Cretaceous timeframe for the origination of many modern lineages (e.g., Cooper and Penny 1997; van Tuinen and Hedges 2001; Harrison et al. 2004). (After Clarke et al. 2005.)

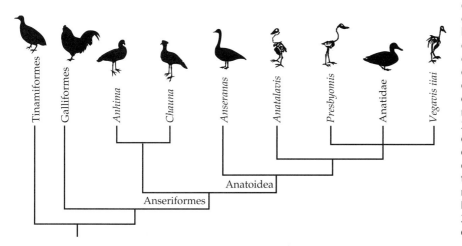

TABLE 11.1 *Representative rates of molecular evolution of an array of different DNA markers[a]*

DNA markers	Organism	Average of divergence (%/Myr)	Reference
Nuclear DNA			
Nonsynonymous sites	Mammals	0.15	Li (1997)
	Drosophila	0.38	Li (1997)
	Plant (monocot)	0.014	Li (1997)
Synonymous (silent) sites	Mammals	0.7	Li (1997)
	Drosophila	3.12	Li (1997)
	Plant (monocot)	0.114	Li (1997)
Intron	Mammals	0.7	Li (1997); Li and Graur (1991)
Chloroplast DNA			
Nonsynonymous sites	Plant (Angiosperm)	0.004–0.01	Li (1997)
Synonymous (silent) sites	Plant (Angiosperm)	0.024–0.116	Li (1997)
Mitochondrial DNA			
Protein-coding region	Mammals	2.0	Brown et al (1979); Pesole et al. (1999)
	Drosophila	2.0	DeSalle et al. (1987)
COI	*Alpheus* (shrimps)	1.4	Knowlton and Weigt (1998)
D-loop	Human	14	Horai et al. (1995)
	Human	17.5	Tamura and Nei (1993)
	Human	23.6	Stoneking et al. (1992)
	Human	260	Howell et al. (1998)
	Human	270	Parsons and Holland (1998)
Nonsynonymous sites	Plant (Angiosperm)	0.004–0.008	Li (1997)
Synonymous (silent) sites	Plant (Angiosperm)	0.01–0.042	Li (1997)

Source: Hewitt 2001.

[a]Note the wide discrepancy of rates, which provides an opportunity to choose a marker that is likely to be most informative within the temporal window of a particular biogeographic study.

animal mitochondrial genome contains over 16,000 nucleotide bases) provides the opportunity to include from hundreds to thousands of characters in a phylogenetic study. Many of those characters are likely to be evolving in a neutral or near-neutral fashion, without the strong natural selection component that could result in convergent evolution. Another attractive feature is that characters can be grouped into different categories or "partitions" that differ in their rates of evolution (see Table 11.1). Thus, slowly evolving characters can be used to reconstruct older events in a phylogeny, whereas faster-evolving characters can be used to address more recent phylogenies and population histories (Figure 11.7). Some of the character partitions could also be evolving in a clock-like fashion, an important attribute that has been exploited to estimate both absolute and relative times of divergence in a phylogeny.

Yet, even given this long list of potential advantages of molecular characters, systematists have not abandoned morphological characters entirely, recognizing that good phylogenetic characters (those with little homoplasy) do not rest entirely within the molecular realm—notice in Figure 11.3 that synapomorphic characters that diagnose major vertebrate clades are morpho-

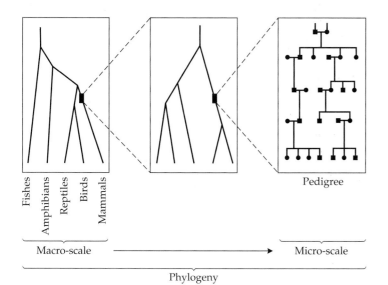

Fishes
Amphibians
Reptiles
Birds
Mammals

Macro-scale

Micro-scale

Pedigree

Phylogeny

FIGURE 11.7 An illustration of the hierarchical nature of phylogeny. Embedded within the phylogeny of the traditional classes of vertebrates are any number of possible phylogenies for groups within each class—shown here is one possible phylogeny across some set of mammalian groups, but we could imagine those groups as anything from taxonomic orders to species and even groups of populations within a single species. Shown here also is the extended genetic pedigree of a familial lineage, completing the continuum of genetic connectivity, from macro- to micro-scale. (After Avise 2004.)

logical. Sometimes systematists employ the **total evidence** approach, incorporating both molecular and morphological characters into a phylogenetic analysis. In other cases, phylogenies are constructed from molecular characters and then morphological characters are "mapped" onto the molecular phylogeny as a means of testing hypotheses of morphological evolution. Either way, there is little question that morphological characters will continue to play an important role in phylogenetic systematics well into the foreseeable future.

MOLECULAR CHARACTERS AND PROPERTIES OF MOLECULAR EVOLUTION. Mechanisms of genetic differentiation were reviewed briefly in Chapter 7. The most important form of mutational change for use in phylogenetic analysis is *nucleotide base substitution* (A = adenine, G = guanine, C = cytosine, T = thymine) at a single position in a DNA sequence, although the *insertion* or *deletion* of one or more bases can also be used as characters. Should a substitution occur in a *protein-coding* sequence, the change in the three-base *codon* that translates DNA information into an amino acid will often not lead to a change in the amino acid (a *synonymous* or *silent* mutation) because of the "redundancy" built within the genetic code, although sometimes it will result in an amino acid change (a *nonsynonymous* or *amino acid replacement* mutation). Within a codon, the rates of DNA substitution differ according to position, with third-base substitutions being generally more frequent than first or second base position substitutions because the genetic code dictates that third base substitutions are less likely to alter the amino acid. Often, because a substitution, insertion, or deletion in a *non-coding* segment of DNA is less likely to lead to a major change in a phenotype (and therefore under less selective constraint), the fixation of mutations in this type of DNA are more frequent than in coding DNA. There is also a bias in substitution rates related to the chemical composition of the bases—substitutions between purines (A–G) and pyrimidines (C–T) are called *transition* substitutions and are generally more frequent than those that substitute a purine with a pyrimidine or vice versa, which are called *transversion* substitutions.

The simplest model of nucleotide base substitution assumes that a mutation at any particular base pair along a DNA sequence is equally as likely as one at any other base pair, which forms the basis for the common maximum parsimony procedure where all characters are given the same "weight" in a phylogenetic analysis. However, variation in rates of evolution across the various partitions in a genome offer the prospect of being able to develop models—ranging from fairly simple to quite complex—of the evolutionary process. One popular approach is to use a maximum likelihood approach to search for the model that best describes the number of partitions that are evolving at different rates (e.g., Posada and Crandall 1998), and then use a maximum likelihood or Bayesian approach to construct phylogenetic trees under the chosen model. Accessible discussions of the various models, their assumptions, advantages, and limitations, are provided in the tutorial by Hall (2004).

The variation in substitution rates across different partitions of a DNA sequence and different genomes results in some distinct advantages to using molecular systematics in biogeography. In particular, it has allowed investigators to target a partition or genome that is most likely to be informative within a given time frame of biogeographic history. Thus, molecular systematic studies can address questions ranging from relationships among taxa originally isolated during the fragmentation of Gondwanaland millions of years ago (e.g., Ortí and Meyer 1997; Eizirik et al. 2001) to the isolation of populations resulting from Late Pleistocene habitat fragmentation only a few thousand years ago (e.g., Knowles 2001).

MOLECULAR CLOCKS AND ESTIMATING TIMES OF DIVERGENCE. The idea that proteins and DNA could be evolving at a sufficiently constant rate to provide a metric for dating branching events in the history of lineages was first proposed by Zuckerkandl and Pauling (1965). The notion that evolutionary events could be dated even when fossils are unavailable captured the imagination of biologists and, in particular, was attractive to those interested in addressing questions about biogeographic histories. Some of the estimates of divergence times were often surprising. For example, Maxson and Roberts (1984) used a molecular clock for albumin evolution to postulate that speciation in several Australian frog lineages occurred through Miocene and Pliocene vicariance (roughly between 12 and 4 million years B.P., respectively), millions of years earlier than required by the then prevailing model of multiple eastern to western episodes of dispersal during Pleistocene glaciations.

Molecular clocks have always been controversial (e.g., Graur and Martin 2004), but they continue to play an important role in modern biogeography for several reasons. First, we have increasingly robust methods for assessing whether rates of evolution vary across lineages. If a fossil record exists for a group of interest, and if those fossils provide good dates for one or more divergence points in a molecular phylogeny, one can calibrate an "absolute rate" of molecular evolution (Figure 11.8). However, even without the possibility of calibration with fossils, the **relative rate test** (Sarich and Wilson 1973) provides a means of testing for rate heterogeneity among lineages in a phylogeny. If a relative rate test indicated a homogeneous rate of evolution in the group, we could employ an estimate of absolute rate, perhaps derived from a related group that did have a good fossil record, to estimate actual time of divergence—with the added assumption, of course, that rates were similar between the focal and related groups. Second, recent studies (reviewed in Arbogast et al. 2002) have shown how important it is to address the often considerable rate variation that occurs among different classes of nucleotide sites in a DNA sequence when estimating rates, and this can be accomplished by employing maximum likelihood methods of choosing a model of molecular evolution that takes into account the appropriate pattern of rate heterogeneity across character partitions. Third, several procedures are now available for

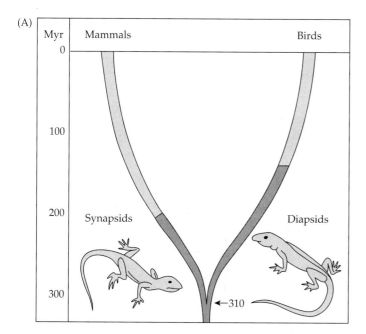

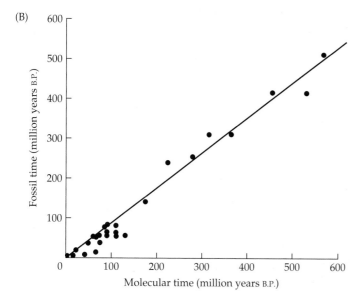

FIGURE 11.8 (A) Calibration of molecular divergence times from the fossil record requires one or more well-dated and diagnostic fossils that can be used to "fix" the time of a cladogenetic divergence event. In this case, synapsids—the stem amniote lineage (darker shading) leading to mammals (lighter shading), and diapsids—the stem amniote lineage (darker shading) leading to birds (lighter shading) have diagnostic cranial morphologies that allow paleontologists to infer a "first appearance" of each in the fossil record at 310 million years ago. (B) This point was then used to estimate rates of molecular divergence across 658 different nuclear genes (details of estimation methods provided in original article), and the composite estimates derived from all genes agreed remarkably well with estimates derived directly from the fossil record. (After Kumar and Hedges 1998.)

estimating absolute rates and times of divergence, even when among-lineage rate heterogeneity does in fact occur in a group (e.g., Sanderson 2002; Figure 11.9). Fourth, often for purposes of testing alternative biogeographic hypotheses, it might not be necessary for estimated rates of divergence to be particularly precise. For example, the corridor known as the Bering Land Bridge opened and closed dispersal routes between the Palearctic and Nearctic as early as 20 million years ago (Chapter 9), and at least two more times before the Pleistocene, and then continued to do so during glacial climatic periods throughout the Pleistocene up until about 10,000 years ago (see Figure 12.15). If one is interested in asking whether Beringian dispersal and subsequent divergence of a group occurred several millions of years ago versus one dating to the Late Pleistocene only a few thousands of years ago, a molecular

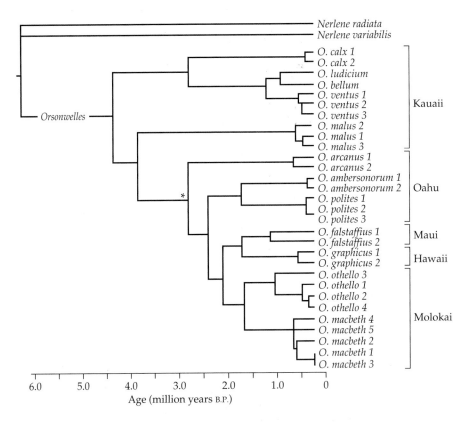

FIGURE 11.9 Phylogeny and estimated divergence times derived from molecular data for 12 of the 13 known species of the endemic Hawaiian linyphiid spider genus *Orsonwelles* (see Figure 12.13 for distributions). Branch lengths differed significantly in the original phylogeny such that a simple molecular clock could not be used to estimate divergence times. Therefore, a non-parametric rate smoothing procedure (Sanderson 1997) was used to estimate divergence times after calibrating the rates of divergence with the divergence event at the node marked by an asterisk, a geologically well-dated age of origination of the Koolau range in Oahu, and the point of separation of species on Kauaii from those on all other islands. (After Hormiga et al. 2003.)

clock would not need to be particularly precise to rule out one or the other of these alternative divergence times.

Phylogeography

The other revolution of the late 1980s took full advantage of the potential of molecular phylogenetics and population genetics, but differed from the traditional realm of phylogenetic systematics in a very important way. While the vast majority of phylogenetically based studies in biogeography had been conducted among taxa at or above the species level, the new discipline of phylogeography was defined explicitly as "a field of study concerned with the principles and processes governing the geographic distributions of genealogical lineages, especially those within and among closely related species" (Avise 2000). In other words, whereas biogeography has typically concentrated on species as the most fundamental units of analysis, and phylogenetic trees were built among species or higher taxa, the newly developing molecular approaches were providing a basis for mapping the spread of lineages (**gene trees**) during their development (Figure 11.10).

In animals, the vast majority of phylogeographic studies have used a genome with some particularly favorable properties for biogeographic questions—the small (about 16,500 base pairs in vertebrates), closed circular genome found in the mitochondrion—the intracellular organelle responsible for cellular respiration. Perhaps the most important property of animal mitochondrial DNA (mtDNA) was the discovery that its rate of evolution is considerably faster than the average rate of sequence evolution of nuclear DNA (Brown et al. 1979; see Table 11.1). Animal mtDNA also has several other important properties that increase its utility for intraspecific studies: it is usually passed clonally, without recombination, from one generation to the next—thus, making it possible to treat the entire molecule—protein coding, tRNA, and rRNA genes, and non-coding sequences—as a single "locus"; it exists as a haploid rather than diploid genome—thus, different genotypes are typically called **haplotypes**; and it is usually transmitted between generations exclusively through a female (matrilineal) pedigree. These properties combine to give mtDNA a much smaller effective population size than nuclear DNA. One consequence of this reduced effective population size is that within a population, as DNA is transmitted from parents to offspring across multiple generations, ancestral mtDNA diversity will be lost more rapidly than nuclear DNA through a process called **lineage sorting**, the stochastic extinction of ancestral haplotype lineages within a population. Thus, following an event that isolates populations previously connected through gene flow (see Figure 11.10), the mtDNA diversity in each now isolated population will **coalesce** to a single mtDNA haplotype much faster than will nuclear DNA. In fact, following the subdivision of an ancestral population into geographically separated populations, these populations are theoretically expected to go from a polyphyletic stage, through a paraphyletic stage, and eventually to a **reciprocally monophyletic** stage with respect to their matriarchical ancestry (compare Figure 11.10 with Figure 11.4, and refer to the definitions of these terms in the glossary). So, with a high rate of mutation rapidly generating new haplotypes, and with a rapid rate of lineage sorting (extinction) of ancestral haplotypes in a population, animal mtDNA will develop a signal of divergence—ultimately reciprocal monophyly—among isolated populations within a much shorter span of time following their initial isolation than is the case for nuclear DNA (Moore 1995). In plants, most phylogeographic studies have used chloroplast (cp) DNA rather than mtDNA, and while the overall rate of evolution is much slower than in animal mtDNA, several gene regions are evolving at a relatively rapid rate; thus, the number of studies demonstrating phylogeographic structure in cpDNA (e.g., Soltis et al. 1997) have increased steadily since the advent of PCR-based sequencing methods.

The Dual Nature of Phylogeography

An interesting feature of phylogeography is that it lies at the intersection between two traditionally separate sets of questions in evolutionary biology. As discussed in Chapter 7, macroevolution is typically concerned with evolutionary patterns at and above the species level, and it is generally approached using the tools of phylogenetics and paleontology. Alternatively, in a microevolution framework, evolutionary questions focus on individuals and populations within a species and are often addressed using the tools of population genetics. Phylogeography incorporates a variety of phylogenetic as well as population genetic approaches and methods to address questions that span the traditional boundaries between the macroevolutionary and microevolutionary arenas (Figure 11.11). Drawing from this variety of methods allows phylogeographers to address a very broad range of possible patterns of phy-

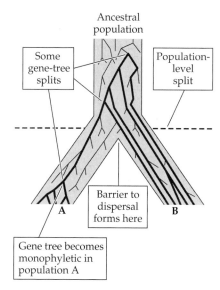

FIGURE 11.10 Gene trees embedded within population trees. Note that the gene tree divergence between populations will not always be concordant with the time of population divergence, and that the rate of ancestral lineage sorting can occur at very different rates in different populations. In this case, the gene tree in population A is monophyletic with respect to population B, but did not become so until well after the time of the population split. Population B is still paraphyletic with respect to population A. At any point in time prior to the coalescence of the gene tree in population 1 to monophyly, populations A and B would have been polyphyletic. (After Avise 2000.)

FIGURE 11.11 Depiction of phylogeography as a bridge between traditionally separate concerns of microevolution—addressed using population genetics methods, and macroevolution—addressed using historical biogeographic approaches. (After Riddle and Hafner 2004.)

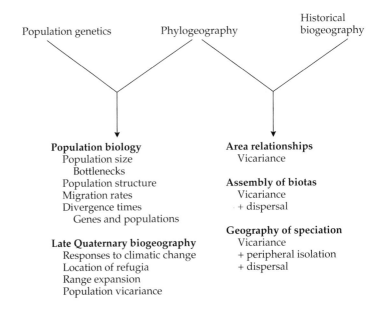

Population genetics   Phylogeography   Historical biogeography

Population biology
Population size
Bottlenecks
Population structure
Migration rates
Divergence times
Genes and populations

Late Quaternary biogeography
Responses to climatic change
Location of refugia
Range expansion
Population vicariance

Area relationships
Vicariance

Assembly of biotas
Vicariance
+ dispersal

Geography of speciation
Vicariance
+ peripheral isolation
+ dispersal

FIGURE 11.12 Illustration of five possible categories of phylogeographic pattern. Ovals and circles encompass mtDNA haplotypes (denoted by letters) or groups of closely related haplotypes, separated from other such groups within a phylogenetic network. The slashes on lines connecting groups of haplotypes into a network each represent a single mutation step, lines without slashes represent a single mutational step between connected groups. Category I occurs when haplotype groups are strongly divergent and geographically separated from one another. Category II demonstrates similar levels of divergence but now without geographic separation, as might occur with dispersal following erosion of a long-term barrier to dispersal between populations. Categories III and IV are similar to I and II, respectively, but occur within a much shallower timeframe and possibly result from life history rather than biogeographic attributes, or in a species with a much slower rate of mtDNA evolution. Category V represents a species with historically continuous gene flow among populations, but with some rare haplotypes that are contained within a single population which might be new mutations that have yet to spread beyond their population of origin. (From Avise 2000.)

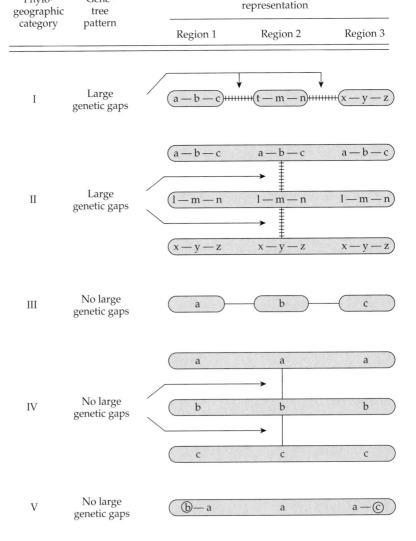

Phylogeographic category	Gene-tree pattern	Geographic representation		
		Region 1	Region 2	Region 3
I	Large genetic gaps			
II	Large genetic gaps			
III	No large genetic gaps			
IV	No large genetic gaps			
V	No large genetic gaps			

logeographic architecture (Figure 11.12), which is expected to arise over a broad range of timeframes and as a consequence of a variety of possible vicariance and dispersal histories.

Phylogeographic patterns exhibiting larger genetic gaps between **phylogroups** (groups of closely related haplotypes separated from other phylogroups by relatively large genetic gaps), such as those representing categories I and II in Figure 11.12, should be amenable to a phylogenetic analysis using methods described previously in this chapter. This is a simple matter of treating each unique haplotype in a dataset as an operational taxonomic unit (OTU) and employing one or more of the many phylogenetic algorithms—either distance matrix or discrete character based—capable of producing a phylogenetic tree of haplotype relationships (Figure 11.13). When variable

(A)

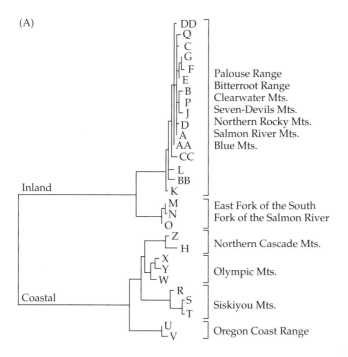

Palouse Range
Bitterroot Range
Clearwater Mts.
Seven-Devils Mts.
Northern Rocky Mts.
Salmon River Mts.
Blue Mts.

East Fork of the South
Fork of the Salmon River

Northern Cascade Mts.

Olympic Mts.

Siskiyou Mts.

Oregon Coast Range

(B)

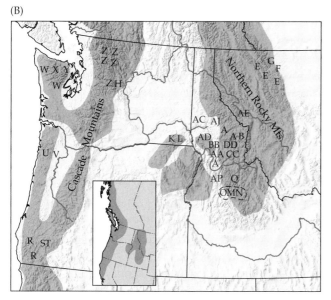

FIGURE 11.13 (A) Mitochondrial DNA phylogenetic tree for mtDNA haplotypes (capital letters) drawn from populations of the Tailed frog, *Ascaphus truei*, from the Pacific Northwest, USA. Haplotypes form two deeply divergent, reciprocally monophyletic and geographically disjunct clades—an Inland group and a Coastal group (B). With exception of the split between haplotype group M + N + O and all other Inland haplotypes, the Coastal clade has generally deeper points of coalescence among subclades, suggesting older splits between populations from different mountain ranges than is shown for the Inland clade. This phylogeny was used to propose that the Inland and Coastal populations represent two separate, morphologically "cryptic" species. (From Nielson et al. 2001.)

FIGURE 11.14 Three unrooted haplo-type networks within and among two species of African (genus *Loxodonta*) and one species of Asian (*Elephas*) elephant. Each network represents a separate gene from the nuclear DNA. The haplotypes from Asian elephants are shown as white circles, those from African savanna elephants are dark gray circles, and those from African forest elephants are light gray circles. Slashes across lines connecting haplotypes each represent a single mutational step (lines without a slash represent one mutational step between haplotypes). Note that haplotypes from each species cluster together in all cases except one (for *PLP*, the two savanna elephant individuals at the arrow carried a haplotype that was otherwise common in the forest species). The pattern of haplotype clustering provides evidence of genetic divergence between the three species. (After Roca et al. 2005.)

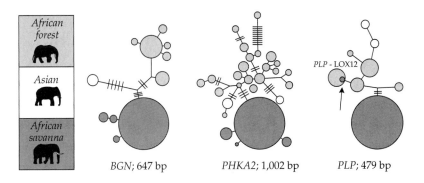

haplotypes are closely related and groups of populations are not clustered within the clearly reciprocally monophyletic phylogroups shown in Figure 11.12, one can still often depict an informative picture of evolutionary history among populations or closely related species (e.g., categories III, IV, V, in Figure 11.12) by using an unrooted phylogenetic network to summarize haplotype relationships (Figure 11.14).

There are, however, several important caveats that arise when using gene trees to infer the evolutionary and biogeographic history of populations or closely related species. First, a gene tree embedded within a population tree can, under a variety of historical demographic scenarios (e.g., number of breeding individuals in a population), be expected to trace to a most recent common ancestral haplotype at a time well before the divergence time for the populations themselves (see Figure 11.10) such that the gene tree might overestimate population divergence time considerably. Second, because of the stochastic nature of the lineage sorting process, and the strong influence of demographic factors on it, the confidence intervals around an estimate of population divergence time from a gene tree can be so large as to be unable to exclude just about any realistic scenario for times of population divergence (Edwards and Beerli 2000). Third, again because of the stochastic nature of the lineage sorting process, a gene tree will not always trace the true historical sequence of population differentiation (Figure 11.15), and an incorrect trace can happen if lineage sorting does not produce reciprocal monophyly in the period of time separating sequential branching events, the likelihood of which increases with decreasing time between branching events.

We will return to a more focused discussion of the applications of phylogeography in reconstructing biogeographic histories in Chapter 12.

The Fossil Record

The task of reconstructing the evolutionary histories of organisms has by long tradition been carried out by two different groups of scientists. Most cladograms are produced by systematists, and many of those are based solely on molecular data and include only contemporary organisms, but with expanding possibilities for incorporating extinct taxa through ancient DNA sequencing. At the same time, paleobiologists have been studying the evolutionary histories of organisms directly from the fossil record, often producing cladograms based solely on fossil evidence. Fossils, the preserved remains of organisms that lived at various times in the past, provide the most direct factual evidence of evolution, but the fossil record has even greater importance for modern systematics and biogeography. We have already seen above that fossils provide an important means of calibrating absolute rates of molecular evo-

FIGURE 11.15 An illustration of a gene tree that differs topologically from the "true" phylogenetic relationships among species. In the top figure, species D and E are allied with C, but the gene tree in D and E is allied with F. This is because population B retains ancestral polymorphism between the time interval t_1 to t_2 and across the speciation event between C and D + E. As this interval becomes shorter (i.e., time between consecutive branching events becomes reduced), the probability of a discordance between species and gene tree topologies increases. The bottom figure shows that once such a discordance arises, it becomes permanently frozen into place, so must always be considered a possible source of error when using gene trees to infer species histories from sampling of extant lineages (G, H, and I). (After Avise 2002.)

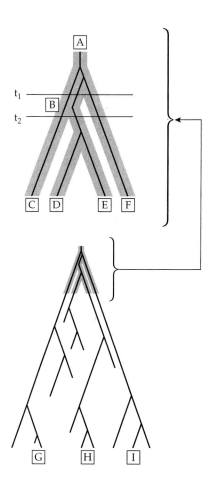

lution so that molecular phylogenies can be used to estimate divergence times. Fossils can also either corroborate or contradict estimates of divergence times based on molecular trees (see Figure 11.6). Long before the advent of molecular systematics, fossils provided a window into evolutionary history that could never come from examination of relationships among extant organisms alone. For example, the history of diversification is accompanied by a history of extinction of lineages (Figure 11.16). Regardless of how accurate a phylogeny based only on extant lineages might be, it cannot address directly the history of lineage extinction in a group without examination of the fossil record. Furthermore, simulation studies have shown that biogeographic infer-

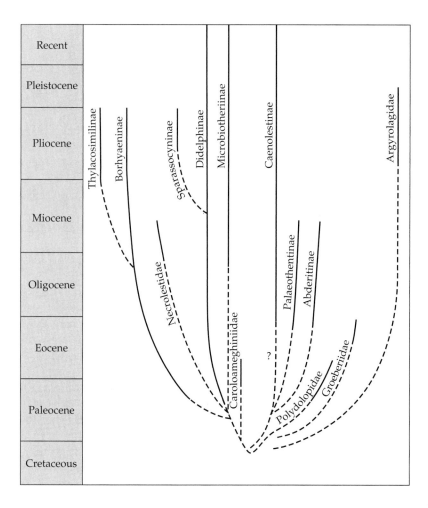

FIGURE 11.16 A phylogenetic hypothesis for the family-level relationships and evolutionary history of South American marsupials. The fossil record shows that this group had a wide radiation on the isolated South American continent during the early Cenozoic, but only three families have survived to the present. The solid lines show the known time spans of the families in the fossil record; the dashed lines show their inferred durations and relationships to other families. This hypothetical reconstruction uses a modification of cladistic methods, based on detailed information on the living forms supplemented by limited data on morphology and dates obtained from the fossils. A classification that employed molecular and strict cladistic techniques would show only the relationships among the three extant families, missing much of the rich history of this groups of South American mammals. (After Patterson and Pascual 1972.)

ences that rely entirely on cladograms produced using only extant taxa can frequently be inaccurate (Lieberman 2002).

Limitations of the Fossil Record

Although it might seem more straightforward to rely on the fossil record than on cladograms constructed only from extant organisms, the fossil record has its own set of practical and logical problems. The known fossil record is incomplete, and its interpretation can be difficult. There are many uncertainties in dating fossils, assigning them to taxonomic groups, understanding how the living organisms were structured and how they functioned, and reconstructing their paleoenvironments.

Only a minute fraction of all the organisms and species that have ever lived have been found as fossils. **Taphonomy**, a subdiscipline of paleontology, is concerned with the processes by which remains of living things become fossilized, and the ways in which these processes can bias the fossil record or cause problems of interpretation (e.g., Behrensmeyer et al. 1992). Animals lacking hard tissues and plants without durable chemicals in their cell walls are poorly represented in the fossil record because they are readily decomposed. Some organisms are known only from their distinctive chemical remains, such as the limestone formed from coral reefs and the coal and oil formed from undecomposed biomass in ancient swamps and shallow seas. Some animals, especially soft-bodied forms, are known only from their tracks or burrows, called **trace fossils**. Of those fossils that are formed, many are destroyed by erosion or tectonic processes, and many others are inaccessible, hidden away in deep strata. The relatively recent fossil record is the most complete (Figure 11.17), one reason why so much more is known about organisms and their environments during the Pleistocene (see Chapter 9) than during earlier periods of Earth history (see Chapter 8).

Even when organisms have been fossilized and those fossils have been discovered, problems of interpretation remain. Most fossils are deposited under water and preserved in sediments. Remains may have been transported by wind or by the currents of streams, rivers, lakes, and oceans, and deposited far from where the organisms originally lived. A striking example is provided by the remains of recently deceased fish that are frequently washed up on the shores of Great Salt Lake. A naive observer might easily conclude that these fish live in the lake. In fact, the reverse is true: the fish die in the lake. These fish are killed when storms push the hypersaline water of the lake into the freshwater marshes and streams where they live, and their bodies float into the lake and are eventually washed ashore.

One consequence of such long-distance transport of remains before fossilization, therefore, is that species that occur together in fossil deposits may not have actually coexisted locally in the same communities (e.g., the "Viking funeral ships" mentioned in Chapter 6). Pleistocene fossils of small vertebrates from river bottom and cave deposits often represent more species than occur together in single habitats today, probably in part because they have been collected from a larger area. It is easy to imagine how a river could collect sediments from a large basin, but it might be thought that a dry cave would contain fossils only of the animals from its immediate vicinity. Many of the small vertebrate fossils preserved in caves, however, were brought in by owls, which then regurgitated the bones of prey captured in their territories, which territories may have been several square kilometers in area and included a variety of habitat types. So, the take-home message seems to be a disappointing one: the fossil record is biased in favor of easily preserved organisms; it is

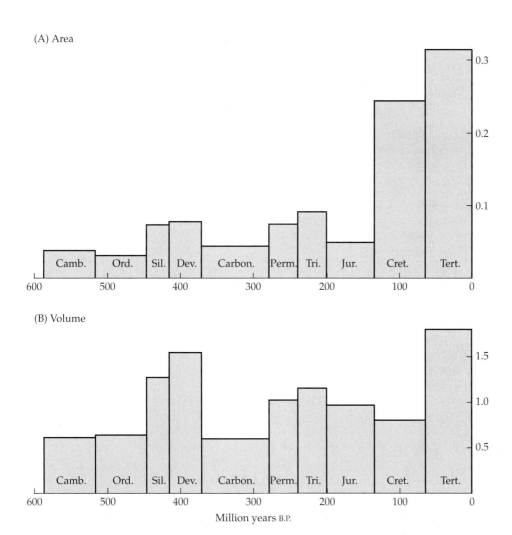

FIGURE 11.17   Estimated area (A) and volume (B) of potentially fossil-bearing sedimentary geological strata of varying ages (here measured as area or volume of rock per year of the past preserved). Note that the area of exposed rocks decreases with age, introducing an important sampling bias into the fossil record. Although some ancient strata are extensive and rich in fossils, in general there is simply less opportunity to find fossils from older geological periods. (After Raup 1976.)

temporally and geographically uneven, still greatly unexplored, and sometimes difficult to interpret.

Biogeographic Implications of Fossils

Despite these limitations, a great deal of information about the history of life on Earth can only or best be obtained from fossils. Only fossils record with certainty the kinds of organisms that occurred at particular times and places in the past. Fossils provide the ages when different taxonomic lineages lived (Table 11.2) and when they occurred in particular areas. Because of the incomplete nature of the record, however, the organisms may have been present for unknown lengths of time before and after their known fossils were preserved. Fossils document some of the enormous diversity of prehistoric life, and some of the important events in the histories of lineages: speciations and radiations, expansions and contractions of ranges, and extinctions. Finally, inference

TABLE 11.2 *Oldest known fossils of selected taxa*

Taxon	Earliest undisputed fossil	Period	Million years B.P.
Vascular land plants	*Cooksonia*	L. Silurian	428
Land animals	*Pneumodesmus*	L. Silurian	428
Insects	*Rhyniognatha*	U. Silurian	412
Lungfish	*Youngolepis* and *Diabolepis*	U. Siluarian	412
Amphibians	*Acanthostega*	U. Devonian	375
Conifers		Mississippian	330
Reptiles	*Hylonomus*	Pennsylvanian	320
Flying insects (and natural flight)		Pennsylvanian	320
Dinosaurs	*Eoraptor*	M. Triassic	228
Turtles	*Proganochelys*	L. Jurassic	210
Mammals	*Morganucodon*	L. Jurassic	200
Birds	*Archaeopteryx*	U. Jurassic	150
Angiosperms (flowering plants)	*Archaefructus*	L. Cretaceous	142
Monotremes (egg-laying mammals)	*Steropodon* and *Kollikodon*	L. Cretaceous	100
Bats	*Icaronycteris*	L. Ecocene	50
Hominids	*Australopithecus*	L. Pliocene	4

Source: Stratigraphic Commission; 2003 Time Chart (www.stratigraphy.org).

about the sorts of environments that fossils could have—and perhaps, more importantly, could not have—occupied can inform paleogeologists about the geographic locations of land masses at the time of fossilization.

FOSSIL HISTORIES OF LINEAGES. Fossils can provide insights into the history of a lineage that cannot be obtained from cladistic reconstructions based solely on recent material. They can supply data on the characteristics and character states of earlier representatives of the lineage, some of which may have been the ancestors of contemporary forms. But even when the fossils are not on the direct branches leading to living representatives, they can show the unique combinations of traits that these organisms possessed. Sometimes the traits present in extinct lineages can be spectacular and unpredicted until a new fossil is discovered. For example, a bird-like dinosaur in the genus *Microraptor*, recently discovered in China, had both forelimb and hindlimb "wings" (Xu et al. 2003). While recent analyses suggest that *Archaeopteryx* may not have been, as was once thought, a direct ancestor of living birds (Feduccia 1996), its exquisitely preserved Jurassic fossils show the special combination of avian and reptilian traits found in a bird that lived 150 million years B.P. One of the most exquisite mammalian fossils is of the earliest known bat, *Icaronycteris index*, which shows unequivocally that the lineage that includes all extant species of bats had already evolved the capacity for powered flight by at least 50 million years B.P. (Figure 11.18). Knowing the characteristics of earlier representatives permits tests of cladistic assumptions about the kinds and directions of change in character states that occurred during the evolution of a lineage. For example, in a recent study it has been possible to include *Icaronyc-*

teris, as well as several other fossil and extant taxa, into a phylogenetic analysis of bat relationships (Simmons and Geisler 1998) that addressed the origins of echolocation and foraging patterns.

In addition, fossils can be dated to provide fairly precise information on when prehistoric representatives of a lineage showing particular degrees of differentiation lived on Earth. As we discussed previously in this chapter, such dating allows cladograms to be calibrated with respect to real time. When fossils are used to calibrate rates of molecular evolution, this calibration is invaluable, because the timing and rates of diversification have been very different among different plant and animal groups, and even among different subclades within the same group. Such information can have important implications for biogeography. For example, in a fossil deposit dated about 15 million years B.P. from the mid-Cenozoic of the Magdalena River Basin in Colombia, South America, Lundberg et al. (1986) found the exquisitely preserved remains of a fish that was identified as a living species, *Colossoma macropomum*, which currently occurs in the Orinoco and Amazon basins. The discovery of fossils of this and other fishes on the western side of the Andes, where they no longer occur, allowed minimum dates to be assigned to events of speciation and extinction as well as to dispersal between now isolated watersheds (see also Lundberg and Chernoff 1992).

PAST DISTRIBUTIONS. Another unique biogeographic contribution of fossils is their ability to document localities of past occurrence. There are countless examples of fossil representatives of lineages being found far outside the current geographic range of the group, and often such occurrences provide key answers to interesting biogeographic puzzles. Of course, it must be borne in mind that when the fossils were deposited, the environment may have been very different than it is today. Some fossils have been transported on drifting crustal plates ("Viking funeral ships") to locations far from where the organisms originally lived. Thus, for example, the discovery in Antarctica of an ever-growing number of fossils, including the plants *Nothofagus* and *Glossopteris* (Schopf 1970a, 1970b, 1976), the reptiles *Lystrosaurus*, *Thrinaxodon*, and *Procolophodon* (Elliot et al. 1970), and the marsupials (Woodburne and Zinmeister 1982; Goin and Carlini 1995; Woodburne and Case 1996), not only provides increasing evidence of the rich biota that continent once shared with other continents of Gondwanaland, but also indicates that Antarctica did not always occur at polar latitudes and was once much warmer than it is today (see Figure 8.7B).

Fossils show that many now narrowly endemic forms are relicts of much more diverse and widely distributed lineages. Examples of such fossils include a member of the avian family Todidae, which is now restricted to the Greater Antilles, but inhabited Wyoming during the Oligocene (~30 million B.P.) (Olson 1976); and a platypus (a representative of the order of egg-laying mammals, Monotremata) whose sole surviving species is now restricted to Australia, although its ancestors once occurred in Argentina during the Paleocene (~60 million years B.P.; Pascual et al. 1992). Cenozoic fossils of tapirs and camels from temperate North America show how intervening populations went extinct to create the present disjunct distributions between South America and Asia (Figure 11.19). The extent of the historic range contractions that can be documented using the fossil record can be impressive. We know that some "living fossils," such as the superficially lizard-like tuatara (*Sphenodon punctatus*), now found only on a few small islands of New Zealand (see Chapter 10), are members of once widespread groups. Another spectacular example is the fish family, Ceratodontidae, now represented by a single species, *Neocer-*

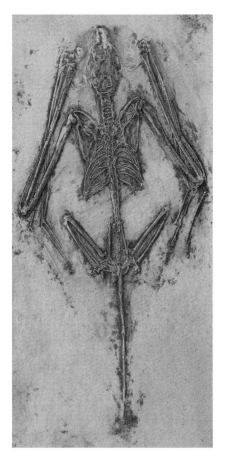

FIGURE 11.18 The earliest known bat fossil, *Icaronycteris index*, from the Green River shale deposits in southwestern Wyoming, USA.

FIGURE 11.19   Fossil localities of camels (family Camelidae) from the Pleistocene in western North America. The records show that these mammals were widely distributed south of the continental ice sheet (shaded area) until they went extinct abruptly—probably due to human influences—less than 20,000 years ago. The extinctions left a disjunct distribution of camelids with guanacos and vicuñas in South America and dromedary and bactrian camels in the Old World. (After Hay 1927.)

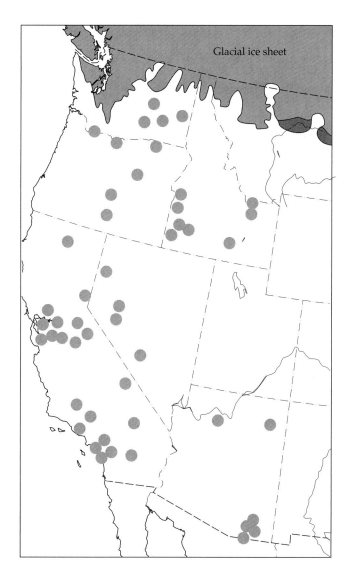

Glacial ice sheet

atodus fosteri, in Queensland, Australia, but known from fossils found on all of the continents except Antarctica (Figure 11.20). Many cladistic or phylogenetic biogeographers are still attempting to reconstruct past geographic distributions of lineages without using fossil evidence (see Chapter 12), but as new fossil finds are made, they promise to provide strong evidence to support or reject these hypotheses (Lieberman 2002; 2003).

PALEOECOLOGY AND PALEOCLIMATOLOGY.   Finally, fossils and the other materials that are preserved with them provide invaluable information on the nature of past environments. Some of this information comes from the physical and chemical composition of fossil-bearing rocks. Their particle size and structure indicates the nature of the substrate and the surrounding geological formations where the remains were preserved. Their chemical composition can indicate whether the environment was marine, freshwater, or terrestrial; anoxic or oxygen-rich; and can often give some idea of the temperature. For example, the fossilized remains of ancient coral reefs form distinctive limestone rocks, and since reef-building corals require sunlight for their symbiotic algae and water temperatures above about 20° C, these strata indicate the past occur-

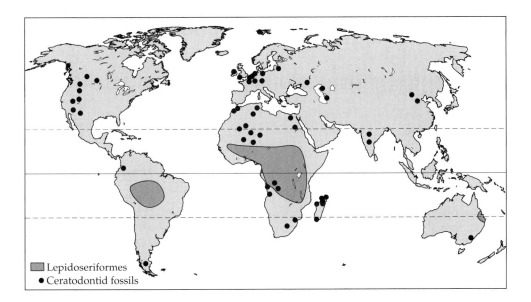

FIGURE 11.20   Information from the fossil record is essential for interpreting biogeographic history. The shaded areas show the current relictual distribution of the lungfishes (Lepidoseriformes) on three southern continents. The circles show fossil localities for one of the three lungfish families, Ceratodontidae. This family contains just one surviving species (*Neoceratodus fosteri*), which is restricted to Queensland, in northeastern Australia, but was distributed nearly worldwide in the Mesozoic. Accurate dating of the fossils is obviously crucial to determining how tectonic events, such as the breakup of Gondwanaland, affected the distribution of the lungfishes. (After Sterba 1966; Keast 1977a.)

rence of shallow tropical and subtropical seas (Figure 11.21). The orientation of crystals due to paleomagnetism indicates the latitude of liquid volcanic rocks when they were cooling (Chapter 8). Mapping of the past positions of the drifting continents and oceans relative to the equator has enabled paleobiologists to determine that the latitudinal gradient of species diversity is an ancient feature of the biogeography of the planet, dating back at least 100 million years (Chapter 15; see also Stehli 1968; Stehli et al. 1969; Stehli and Wells 1971; Crane and Lidgard 1989; Crame 2004).

Remains of organisms preserved together in the same strata can often provide abundant information on the climate, vegetation, and biotic communities when these organisms were alive. Perhaps the fossils most informative about paleoecology are **catastrophic death assemblages**. The lives and environment of the people inhabiting the Roman cities of Pompeii and Herculaneum are frozen in time, exquisitely preserved by the ash spewed out by the eruption of Mount Vesuvius in 79 A.D. Similarly, many soft-bottom marine assemblages were preserved in situ when they were covered by large quantities of sediment. Individuals were trapped in their exact locations, providing an invaluable record of the depth and orientation of burrows, the size and age structure of populations, and the species composition and spatial relations of communities. Extensive, exquisitely preserved fossil deposits in the northeastern corner of Nebraska preserve a catastrophic death assemblage of the terrestrial mammals that inhabited the area during the Miocene, and died over a three- to five-week period about 12 million years ago as a consequence of volcanic ash that drifted eastward from a giant volcano far to the west in the state of Idaho. Along the Florida Gulf Coast, a catastrophic event—possibly a red tide—

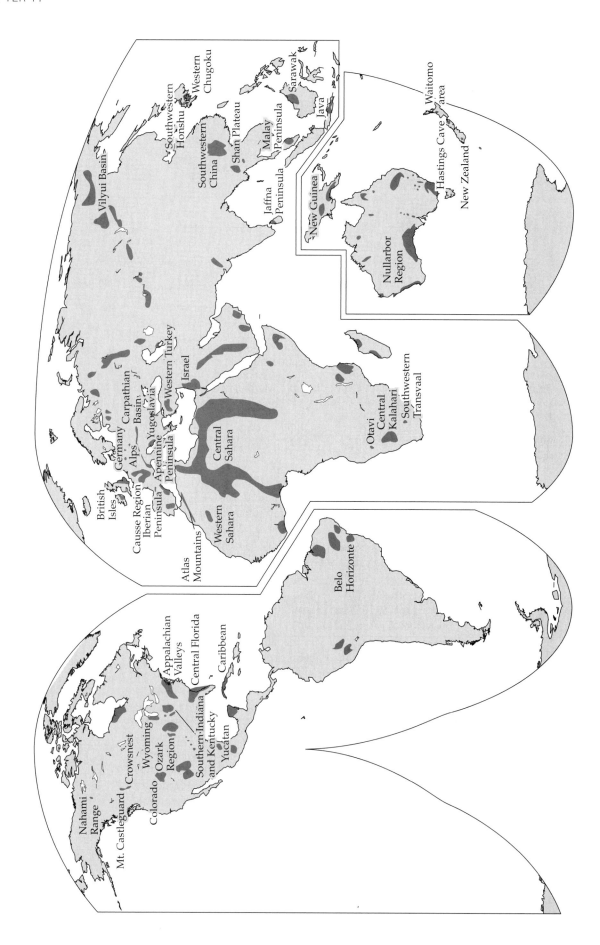

◀ **FIGURE 11.21**   Global distribution of limestone, indicating past location of shallow marine waters and coral reefs. Note that many of these limestone formations occur far from contemporary coastlines, sometimes near the centers of continents. (After Snead 1980.)

killed and deposited over a hundred skeletons of a single species of extinct cormorant during the late Pliocene (Emslie and Morgan 1994).

Pleistocene assemblages provide a wealth of paleoecological information. Not only is there a large quantity of well-preserved fossil material, but the fact that most of the plants and animals were very similar to their living descendants enables us to make strong inferences about their behavior and ecology. Not only do data on oxygen isotope ratios in ancient ice and fossils provide information on global, regional, and local temperatures (see Chapter 9), but the kinds of organisms and their characteristics also allow for reconstructions of past climates. Fossil pollen from lakes and bogs and plant fragments from woodrat middens provide so much information on the occurrences and abundances of still extant plant species that it is possible to reconstruct the structure and composition of Pleistocene plant communities and the climatic regimes in which they lived. Recently, the diets of extinct Shasta ground sloths, *Nothrotheriops shastensis*—large ground-dwelling herbivores that went extinct in North America about 11,000 years ago—have been reconstructed by sequencing ancient plant DNA from fossilized fecal remains (Poinar et al. 1998). From plant and animal fossil deposits, it is also possible to document the shifts in geographic ranges and changes in composition of communities as a consequence of the advances and retreats of glaciers and the associated changes in climate during the Pleistocene. For example, since the maximum of the Wisconsin glacial period about 20,000 years ago, some small mammals in North America have shifted their ranges hundreds of kilometers, and not always consistently in a northward direction, as would be expected if the climate had simply warmed up uniformly. As a consequence, Pleistocene communities contained combinations of species strikingly different from contemporary communities (see Chapter 9). In South America, the post-Pleistocene range shift of a species of leaf-eared mouse, *Phyllotis limatus*, has been inferred by sequencing ancient DNA from hairs left in the midden currently occupied by another species of rodent, demonstrating its former occurrence in a region where it no longer occurs (Figure 11.22).

Fossils can provide convincing evidence of the role of biotic interactions in the past. Many researchers have used shifts in dominance over time between certain functionally similar but distantly related groups of organisms to infer that competition has played a major role in the history of these lineages (e.g., comprehensive dominance shifts in brachiopods and mollusks; see Elliot 1951; but see also Gould and Calloway 1980), gymnosperm and angiosperm plants (Knoll 1986), and multituberculate and rodent mammalian herbivores (Simpson 1953; Lilligraven 1972). Pollen profiles and woodrat middens provide evidence of successional dynamics and reorganization of past plant communities in response to shifts in climate (e.g., Delcourt and Delcourt 1977, 1994, 1996; Davis 1986, 1994; Woods and Davis 1989).

In the marine realm, Vermeij (1974, 1978) has used patterns of co-occurrence of different groups of carnivorous and herbivorous mollusks and evolutionary trends in shell form to document the history of the coevolutionary arms race between predators and prey throughout most of the Phanerozoic. But such coevolution proceeds slowly, at least in mollusks. The drill holes made by

FIGURE 11.22 (A) Geographic distribution, and (B) mtDNA phylogeny for a group of closely related species of leaf-eared mouse, genus *Phyllotis,* that are distributed throughout the Andean region of western South America. The operational taxonomic units (OTUs) labeled "midden consensus" and "midden clone" were derived from rodent feces preserved in an 11,700 year-old rodent midden within the boxed area on the map, located in the Atacama Desert. Note that from the phylogeny, we can infer that (1) *P. limatus* was the species that occupied this region 11,700 years ago (and probably created the midden); and (2), that the current southern limit of its distribution is about 100 km to the north of this location. (After Kuch et al. 2002.)

(A)

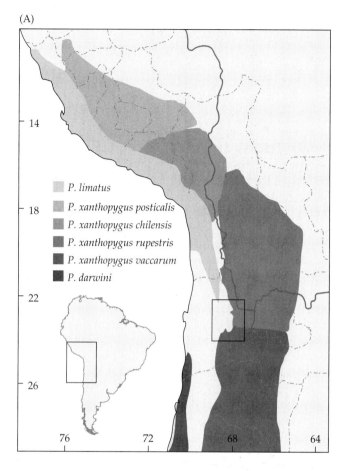

(B)

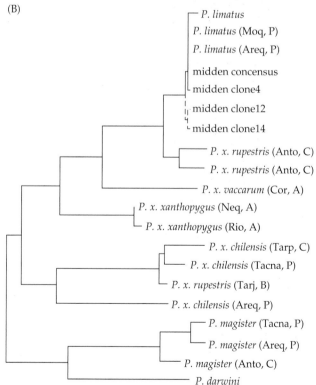

predatory snails in the shells of their prey are clearly visible on well-preserved fossils. The sizes and locations of 100,000-year-old drill marks are identical to those found on contemporary shells, indicating that there has been no detectable change in predatory behavior over that period (Tull and Bohning-Gaese 1993). Fossils are also providing evidence of the role of predation by aboriginal humans in the extinctions of other vertebrates (Barnosky et al. 2004). The correspondence between the time of arrival of humans and whole-sale extinctions of native birds and mammals on islands and continents shows the effect that an invading predator can have on its prey, and demonstrates that humans have been having a major influence on global biodiversity for tens of thousands of years (see Chapter 9).

The Emerging Synthesis

We concluded this chapter in the previous edition of this book by noting that knowledge of the history of lineages was being advanced rapidly by research in several areas. Systematists continue to improve on the methods of phyloge-netic reconstruction and to exploit advances in molecular genetics to recon-struct robust phylogenetic trees, often being able to provide reasonable esti-mates of times of evolutionary divergence for a group. Phylogeography adds the approaches and methods of population genetics to reconstruct the history of gene trees in closely related lineages, and thus opens the door to exploring biogeographic histories within a time frame of the past thousands to several millions of years. Paleobiologists and paleoecologists have been using new finds and better interpretations of fossil remains to document what kinds of organisms lived in particular places at different times in the history of the Earth, along with their ecological associations and the habitats they occupied; and quantitative methods are being developed to assess the incompleteness and accuracy of a fossil record (Kidwell and Holland 2002; Foote 2003). Until quite recently, however, there had been far too little communication between these groups of scientists, and too little synthesis of their findings, even though the need to do so has been recognized for some time (e.g., Grande 1985; Lieberman 2003). In part, this was because the scientists themselves his-torically came from different backgrounds—systematists and phylogeogra-phers from biology, and paleontologists from geology—and because they pre-sented their results in different journals and meetings. But in part, it was because they had often treated each other as rivals and with suspicion, rather than as mutualists who can bring different backgrounds, techniques, and per-spectives to a common endeavor: the formidable task of reconstructing the history of life on Earth. Nowhere has this suspicion been more apparent than when molecular systematists suggest times of divergence that are at odds with the known fossil record.

Fortunately, there are some very positive signs that a mutualism between disciplines is beginning to emerge. First, paleobiogeographers have begun to actively employ modern phylogenetic methods and theory in reconstructions of evolutionary and biogeographic histories (Lieberman 2003). Second, a new bridge between molecular systematics and paleontology is emerging in the form of combining ancient and modern DNA sequences in biogeographic analyses (e.g., Haddrath and Baker 2000; Cooper et al. 2001; Kuch et al. 2002). Third, as discussed previously in this chapter, fossils are increasingly sought after by molecular systematists for calibration of molecular clocks. Fourth, while molecular estimates of divergence times still continue to often be at odds with the known fossil record, many of the more sophisticated estimates appear to be quite consistent with the fossil record (see Figure 11.8). Finally, phylogeography is conceptually positioned to build bridges between modern and paleoecological perspectives on the assembly and disassembly of biotas during late Neogene time frames (Riddle and Hafner 2004). Given the importance of the formidable task of reconstructing the history of life on Earth, it is indeed refreshing to report significant progress in the seven years since the last edition of this book in the direction of more mutualism and less competition, and more synthesis and less parochialism, than in the past.

Reconstructing the History of Biotas

A S WE DISCUSSED IN CHAPTER 2, de Candolle first differentiated the historical from the ecological determinants of geographic distributions, but not until the 1960s were these attributes turned into two generally separate research agendas. As a result, historical and ecological biogeography subsequently developed along very separate routes and concentrated on largely, but not exclusively, separate kinds of questions. For much of the twentieth century, the prevailing paradigm in historical biogeography was one in which certain regions of the Earth were considered to be particularly important centers of origination of new species and, from these centers, new forms would disperse to far corners of the globe (see, for example, Figure 2.1).

Beginning in the latter half of the twentieth century, the center of origin paradigm was largely replaced with one in which historical biogeography became increasingly more narrowly focused on how the distribution and diversification of lineages changed through tectonic events and vicariance. A complementary endeavor was the development of hypotheses about Earth's tectonic history based on the dynamic geography of organisms.

Historical biogeography is again undergoing a rapid transformation with the infusion of new approaches and methods, some of which we reviewed in part in Chapter 11 and others that we present new in this chapter. Most importantly, new data on phylogeny, distribution, and diversity—largely based on DNA sequencing—are appearing at a rapid rate and offering previously unattainable opportunities to analyze biogeographic structure across multiple, co-distributed lineages.

This chapter is divided into four parts. First, we trace the last century of development of concepts and approaches in historical biogeography. Second, we review some of the many currently available approaches and methods in historical biogeography—some that have been around for awhile and some that have emerged more recently. Third, we feature applications of these methods that provide some fascinating insights into the biogeographic histories of terrestrial and marine biotas. Finally, we preview the relatively recent reintegration of divergent lineages and the development of a new synthesis in historical biogeography.

Origins of Modern Historical Biogeography

Historical biogeography serves as an interesting case study in scientific revolutions—in the questions that biogeographers believe can and should be addressed, in the mechanisms underlying changes in biotic distributions and diversification, and in the approaches and methods available (Funk 2004). In the previous chapter, we reviewed recent developments in reconstructing the patterns and timing of diversification in lineages using phylogenetic (and particularly molecular) systematics and paleobiology. These and other recent advances allow us to address questions of historical biogeography at more recent temporal and smaller spatial scales.

We have discussed in several previous chapters the profound influence that advances in plate tectonics theory of the 1950s and 1960s had on biogeographers—no longer was it necessary to construct scenarios of biogeographic history (e.g., ancient landbridges between continents) under an assumption that landmasses had largely remained static over the course of Earth history. If the climate of the Earth and the positions of the continents had remained fixed over time, this would drastically limit the kinds of hypotheses that could be advanced about the causes of endemism, provincialism, and disjunction. For example, there would be only one likely explanation for the disjunct distribution of a terrestrial organism between South America and Australia: that the ancestors originally occurred on one of those continents, and individuals dispersed over long distances across extensive oceanic barriers to colonize the other continent. On a dynamic Earth, however, it was now possible to invoke an alternative hypothesis: that the ancestors originally inhabited Gondwanaland, and vicariance accounted for disjunct distributions and the diversification of distinct biotas on the previously connected but now isolated landmasses. Of course, there are more complicated versions of both hypotheses, perhaps involving past occurrences of the ancestral lineage on the northern continents and extinction of those populations with some combination of over-water dispersal and continental drift being required to explain the current distribution. Using information from phylogenetic reconstruction, plate tectonics, and the fossil record, it should be possible to erect and evaluate an array of alternative hypotheses.

Not all historical biogeographers "think alike" about the questions to address and appropriate methods to employ in a biogeographic study. The late nineteenth and twentieth centuries were times of great change in philosophical and methodological approaches in historical biogeography. Early in the twentieth century, W. D. Matthew developed the *center of origin – dispersal model* from Darwin and Wallace's **evolutionary biogeography** and their original concept of "dispersal over a permanent geography." Throughout the next four decades, this model continued as the ruling paradigm of historical biogeography, largely owing to the many insights of George Gaylord Simpson (1940), Philip Darlington (1957), and their colleagues.

Early Efforts: Determining Centers of Origin and Directions of Dispersal

Since the time of Buffon and de Candolle (see Chapter 2), biogeographers have realized that distributions of organisms have shifted over time. This realization led naturally to efforts to determine the birthplace—or center of origin—of each taxon. Two general questions motivated the search for centers of origin. First, investigators wanted to know whether certain geographic regions served as cradles for the evolution of new kinds of organisms, and of the special features that allowed these lineages to be successful. Second, biogeographers wanted to understand how biotas have been assembled: where taxa started from, what routes they followed to disperse around the world, and what factors produced present patterns of endemism, provincialism, disjunction, and diversity. These are still among the goals of historical biogeography, but fortunately, the methods used to infer the origins and movements of lineages have improved greatly, and they have been enhanced through incorporation of detailed models of Earth history (e.g., plate tectonics) and its influence on dynamic biogeography and diversification of lineages.

Initially, biogeographers tried to develop simple rules for determining the center of origin for a taxon. For example, Adams (1902, 1909) listed ten criteria that could help to identify such centers. Some authors insisted categorically on using a single criterion to the exclusion of others. Matthew (1915), for instance, followed Buffon in believing that centers of mammalian origin were in the Holarctic (Figure 12.1; see also Figure 2.1). Matthew argued that new, successful forms arose in response to the challenges of temperate climates; these forms eventually supplanted their progenitors and other lineages, forcing them into peripheral habitats, then to lower latitudes, and eventually into the Southern Hemisphere. This "boreal superiority" notion was considered by Simpson (1940) as finding some support in the asymmetric effects of the Great American Interchange between northern and southern continents (see Chapter 10).

Matthew (1915) thought that the center of origin is where the most derived forms reside. This is in direct opposition to Hennig's (1996) **progression rule** which holds that an ancestral population remains at or near the point of origin, and progressively more derived forms are found at farther distances away from the center, in a "stepping stone" pattern of sequential dispersal and speciation events. In reality, either explanation might be correct for different groups, because displacement of one form by another depends in part on how

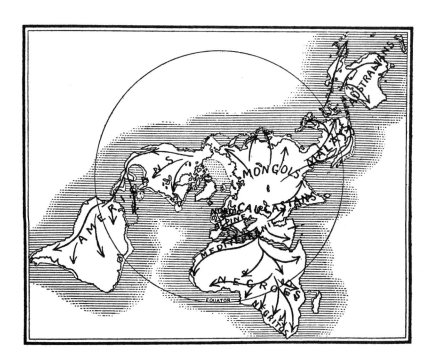

FIGURE 12.1 An example of Matthew's arguments for a center of origin for mammalian speciation in the Holarctic region of the Northern Hemisphere, with subsequent dispersal around the world. This figure shows his scenario for the principal races of humans as they were understood in the early 1900s. In his 1915 book *Climate and Evolution*, Matthew presented a scenario in which consecutive waves of more advanced forms would evolve in the Northern Hemisphere and displace more primitive forms as they dispersed away from the center of origin. (From Matthew 1915.)

they disperse, speciate, and interact with their biotic and abiotic environments. An important difference between the two ideas, however, is that Matthew's was based on a model of evolutionary asymmetry in which environmental challenges at one geographic area drove the improvement of superior traits in new species that then dispersed outward and drove more primitive forms to extinction. In contrast, Hennig's model was based solely on a model of dispersal, isolation, and speciation in peripheral populations, without presumption of evolutionary asymmetry, and thus was testable through phylogenetic analysis.

A thorough evaluation of the challenge of determining centers of origin was made by the phytogeographer Stanley Cain (1944). He listed the thirteen criteria in use at the time (Table 12.1), and argued convincingly that none by itself could reliably identify the center of origin. Some authors had claimed, for example, that the center of origin should be where the greatest number of species in a group reside. This assumption is, of course, invalid if the majority of forms inhabit a region of secondary radiation, as do the heaths of South Africa (*Erica*), with 605 species in the Cape Region (Baker and Oliver 1967), or the hundreds of species of *Drosophila* on the Hawaiian Islands (see Chapter 14). Cain also argued against using the location of a primitive form, or the earliest fossil, as an absolute criterion. Primitive forms often survive in isolated regions located far from their original ranges and containing few competing species (e.g., paleoendemics such as the tuatara in New Zealand; other examples of such relicts are described in Chapter 10).

AN EXAMPLE: SEA SNAKES. To give the flavor of the sort of center of origin scenario that was popular prior to conceptual advances beginning in the 1970s and 1980s, let's examine one case in which a number of Cain's criteria point to

---

TABLE 12.1 *Criteria used and abused for indicating center of origin of a taxon*

1. Location of greatest differentiation of a type (greatest number of species)
2. Location of dominance or greatest abundance of individuals (most successful area)
3. Location of synthetic or closely related forms (primitive and closely related forms)
4. Location of maximum size of individuals
5. Location of greatest productiveness and relative stability (of crops)
6. Continuity and convergence of lines of dispersal (lines of migration that converge on a single point)
7. Location of least dependence on a restricted habitat (generalist)
8. Continuity and directness of individual variation or modifications radiating from the center of origin along highways of dispersal (clines)
9. Direction indicated by geographic affinities (e.g., all Southern Hemisphere)
10. Direction indicated by the anuual migration routes of birds
11. Direction indicated by seasonal appearance (i.e., seasonal preferences are historically conserved)
12. Increase in the number of dominant genes toward the centers of origin
13. Center indicated by the concentricity of progressive equiformal areas (i.e., numerous groups are concentrated in centers, and numbers decrease gradually outward)

Source: After Cain (1994).

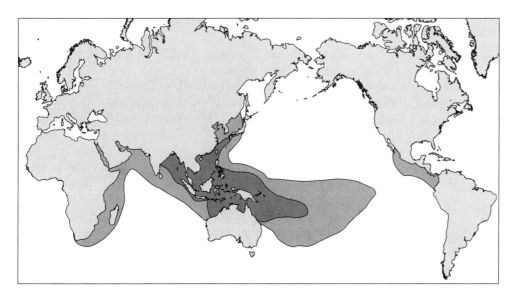

FIGURE 12.2 Map showing the current distribution of sea snakes (family Hydrophidae). All but one species are confined to the densely shaded area. The exceptional, very wide-ranging species, *Pelamis platurus*, occurs in this region, but also in the more lightly shaded areas. The concentration of species and lineages in the Indo-Pacific oceans and the occurrence of the seemingly most closely related outgroup in swamps in Australia led Cogger (1975) and others to hypothesize that sea snakes spread from a center of origin in the Australia-New Guinea region. (After Heatwole 1987.)

a similar conclusion. Consider the true sea snakes (tribe Hydrophiini) of the Indo-Pacific region, a group whose fossil record is unknown. They are venomous marine predators and members of the Family Elapidae, which includes many Australasian snakes as well as cobras, kraits, coral snakes, and mambas. At least 52 species of true sea snakes in 16 genera are known from tropical and subtropical coastal habitats along reefs in the western Pacific and Indian oceans, with related species inhabiting brackish inlets, rivers, and terrestrial habitats in Australia. Several related species have entered and adapted to freshwater habitats in the Philippines and the Solomon Islands (Figure 12.2; Dunson 1975; Heatwole 1987). The species with the widest distribution is the pelagic form *Pelamis platurus* (see Figures 12.2, 12.3), which covers the geographic range of the entire family, and includes the tropical Pacific to the west coast of the Americas, from Mexico to Ecuador. Postulating a center of origin for sea snakes was made easier for earlier investigators because these marine reptiles possess several useful attributes: (1) their inability to tolerate cold water (below 20° C) has confined them to the Indian and Pacific oceans; (2) they appeared to be a young tribe (of Cenozoic age); (3) their closest relatives—the terrestrial members of the subfamily Hydrophiinae—overlap with them in distribution in the Australasian region; and (4) the lines of morphological specialization appear to be relatively unbroken, thus seemingly reducing problems of tracing geographic patterns obscured by major extinctions.

Cogger (1975) concluded that the Hydrophiini originated in the Australia-New Guinea region—the region of highest hydrophiin diversity, with more than 30 species. Most genera of Hydrophiini are classified in the *Hydrophis* group, and most of these taxa are Australian (Cain's first criterion, Table 12.1). Here, too, live what Cogger believed to be their putative elapid ancestors (*Rhinoplocephalus* and *Drepanodontis*—subsequently synonymized with *Hemi-*

(A)

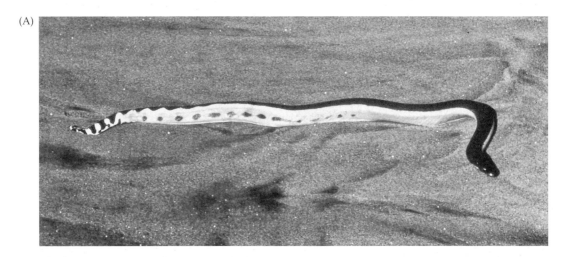

(B)

FIGURE 12.3 Two species of sea snakes. (A) *Pelamis platurus* is pelagic, and is the most widespread species in the group. This individual was washed ashore on a beach in western Mexico during a storm. (B) *Laticauda laticauda* is representative of a distinctive group often classified as a separate tribe or subfamily. This specimen was photographed in shallow water on the island of Fiji. (A courtesy of R. E. Brown; B courtesy of W. A. Dunson.)

apsis), swamp-inhabiting Australian snakes, as well as the most primitive and unspecialized hydrophiins such as *Ephalophis*, *Hydrelaps*, and *Parahydrophis*. These putative basal genera in the hydrophiini radiation share several traits with the elapids and are notably weak swimmers (McDowell 1969, 1972, 1974; Dunson 1975; Cain's third criterion). Genera that range to the northwest of Australia often have at least one representative in the Australian coral reefs as well. Finally, the sea krait genus *Laticauda* (see Figure 12.3), was considered by Cogger (1975) and Voris (1977) to be an early offshoot (perhaps a separate tribe), but still closely related to true sea snakes—as such, it's occurrence around New Guinea provides yet another degree of support for this being the center of origin of true sea snakes under Cain's third criterion.

Scenarios like the above are just that: stories concocted to explain most or all of the known facts about a group that might be relevant to its history. In a sense, such scenarios are hypotheses, because many specific parts of the story can potentially be falsified. For example, all it would take to reject the above scenario would be the discovery of a fossil of appropriate age and with plesiomorphic characters indicating hydrophiin origination along the west coast of Africa. More importantly, this scenario allows us to make a critical observation about the importance of good phylogenetic data in historical biogeography. For example, recent studies of Hydrophiinae evolution (Keogh et al. 1998; Scanlon and Lee 2004) have demonstrated that sea kraits are the sister taxon of the entire clade that includes the very diverse terrestrial radiation of hydrophiinae in Australasia, as well as true sea snakes—the sister group to true sea snakes appears to be a viviparous (live-bearing), terrestrial Australian snake. Second, a recent molecular phylogeny for true sea snakes (S. Keogh, pers. comm.) clarifies relationships among species and genera—the three above-mentioned, morphologically primitive genera are not basal to all remaining sea snakes but rather sister to the most diverse and most recent adaptive radiation, the "Hydrophis" group. Both of these examples demonstrate the value of a robust phylogeny, and specifically demonstrate the weaknesses associated with earlier uses of Cain's third criterion. And so, we conclude that, with no reference to a rigorous phylogenetic analysis or to fossil evidence, and not being linked to any specific events in Earth history, earlier authors had precious little direct evidence to support the above center of origin scenario for sea snakes. This is not to say, however, that as more and better evidence becomes available, all or parts of this scenario will not be supported; for example, Scanlon and Lee (2004) used an updated phylogenetic analysis to conclude that hydrophiin sea snakes had "their origin…almost certainly somewhere along the Australian coast."

CRITICAL ISSUES. Why did we go into so much detail about the center of origin concept? First, inferring centers of origin using some or all of the criteria in Table 12.1 was common practice in historical biogeography up until the 1970s and 1980s—and it is still practiced, at least implicitly. When other evidence is not available, and a group is diverse and widely distributed in one region but has a few disjunct species in another, the former region is often assumed to have been inhabited for a long period and the latter region to have been more recently colonized. For example, several clades of small birds, including hummingbirds (Trochilidae), flycatchers (Tyrannidae), wrens (Troglodytidae), tanagers (Thraupinae), blackbirds (Icterinae), and sparrows (Emberizinae) that occur in temperate North America but are much more diverse in the Neotropics, are often assumed to have originated in South America and to have subsequently colonized North America as the two continents drifted into proximity during the Cenozoic.

Second, scenarios about centers of origin and directions of subsequent dispersal have historically been used to make sweeping generalizations about the innovations that triggered the success of lineages, and about global patterns of diversity. We have mentioned that first Buffon (1761) and then Matthew (1915) argued that new lineages and new innovations arose predominantly as adaptive responses to the climatic rigors of the temperate zones and then spread southward to thrive and diversify in the milder tropics. The exact opposite position was advanced by Darlington (1959b), who argued that new lineages and new innovations were more likely to have originated in the tropics and spread toward the poles. He suggested that the greater diversity of species and higher taxa at low latitudes meant an increased probability that some of these "evolutionary experiments" would be successful (see Chapter 15).

These questions about the origins of lineages and their unique attributes remain of great inherent interest. Biogeographers, systematists, and ecologists still debate whether the tropics are the source or the last refuge of the distinctive combinations of traits that characterize lineages (Gaston and Blackburn 1996; Wiens and Donoghue 2004). A related example is the discussion among marine paleobiologists of whether the rates of origination of new lineages and extinction of existing ones have been higher in the more abiotically stressful onshore environments or in the more benign ones offshore (e.g., Jablonski et al. 1983; Miller 1989; Jablonski and Bottjer 1990, 1991).

Third, we now have much better methods and a greater wealth of data to use in reconstructing biogeographic histories. Currently there are signs that the emerging, more integrative approaches in historical biogeography are replacing the last few decades of discord and controversy (e.g., see Nelson and Rosen 1981; but see also Ebach and Humphries 2002; Van Veller et al. 2003).

From Center of Origin-Dispersal to Vicariance

To appreciate the evolution of the ideas that have been put forward during these contentious times, it is helpful to see what the proponents were reacting against. It is clear that historical biogeography, based largely on studies of single taxa and plagued by flawed concepts of centers of origin and dogmatic ideas about dispersal (see Chapters 2 and 6), was in need of more rigorous approaches.

CROIZAT'S PANBIOGEOGRAPHY. As is often the case with revolutions, the rapid changes in historical biogeography that took place largely in the 1970s were triggered by a few individuals with radical ideas. And like the bizarre evolutionary changes in organisms that have occurred on islands, novel and divergent ideas often seem to arise in isolated circumstances. Thus it was that Leon Croizat, an unorthodox plant biologist who labored for most of his career in relative obscurity in Venezuela, provided the spark that ignited the revolution in historical biogeography. Croizat (1952, 1958, 1960, 1964), like many phytogeographers before him, recognized that the present limited distributions of narrowly endemic, disjunct taxa are the relicts of ancestral taxa—and often of entire floras—that were more broadly distributed in the past. Unlike most previous researchers, who typically studied just one taxon, Croizat compared the distributions of different groups. He realized that very distantly related kinds of organisms often exhibited similar disjunctions, and that these distributional patterns were legacies of historical events that influenced many different lineages in the same way. Amassing data on disjunctions from all over the world, Croizat developed an approach that he termed **panbiogeography**.

Croizat argued that each pattern of multiple disjunctions reflected the fragmentation of a biota that originally inhabited interconnected regions. He plotted the ranges of narrowly endemic species on a map and then drew lines, called **tracks**, connecting the distributions of the most closely related taxa. Croizat then superimposed the maps for multiple taxa and noted where the positions of their tracks coincided. He called the resulting coincident tracks **generalized tracks** (Figure 12.4). He inferred that such tracks indicate historical connections—that they were pathways connecting the isolated fragments of a formerly continuous biota. By plotting all the generalized tracks on a map, he could theoretically recreate the way in which regional biotas had developed in time and over space.

When first published, Croizat's ideas received little favorable response. For one thing, the basic concept of historical subdivision and isolation of formerly widespread biotas was hardly new. For another, Croizat arrogantly dismissed alternative views while zealously promoting his own. In addition, some of his ideas and examples were so extreme that many of his contemporaries did not

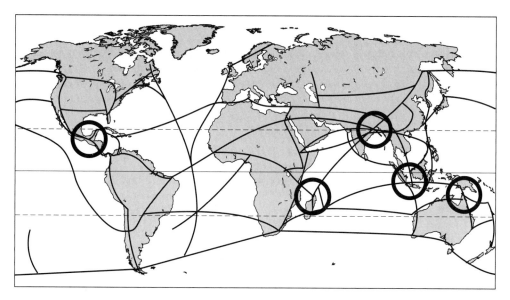

FIGURE 12.4 A composite drawing of the major and moderately important "tracks" drawn by Croizat (1952, 1958, 1960, 1964) to show hypothesized past connections between the distributions of terrestrial organisms that currently occur in widely separated regions. These tracks were derived by plotting the distributions of endemic species within several taxa and then drawing lines to connect regions that share disjunct endemics in several taxonomic groups. Circles identify five major "biogeographic nodes"—intersections between two or more tracks. The tracks radiating out from Madagascar, for example, show that its endemics have close affinities with disjuncts in Africa, India, Australia, and the East Indies. While many of Croizat's tracks correspond to past land connections, others are highly problematic. Note, for example, the lines connecting northeastern North America and northwestern Europe with islands and continents in the South Atlantic Ocean.

take his work seriously. His books contained many technical errors, and some of his purported disjunctions were based on questionable systematics. He categorically rejected dispersal over long distances and across barriers to account for the generalized tracks because he reasoned that "dispersal must be orderly and continuous in time and space" (1958). He argued that long-distance dispersal could not have been a major mechanism of geographic range expansion, even for the colonization of distant oceanic islands. And while continental drift would seemingly have provided a mechanism to explain many of the observed disjunctions, Croizat—at least in his earlier writings—denounced Wegenerism (see Chapter 8). Rather than moving continents and oceans, he kept their locations fixed and invoked ephemeral land and water bridges to explain most of the tracks connecting disjunctions, such as those among the southern continents.

Today, except for a few seemingly zealous disciples of panbiogeography (e.g., Craw 1982, 1988; Heads 1984; Craw et al. 1999), most biogeographers recognize Croizat's approach as being flawed by its unrealistic and idiosyncratic assumptions (Nelson and Platnick 1981; Parenti 1981; Patterson 1981; Humphries and Parenti 1986; Page 1990; Morrone and Crisci 1995) and, most importantly, by the fact that he never used a phylogeny. An influential paper senior-authored by Croizat (Croizat et al. 1974) attempted to conjoin Croizat's track analysis with Hennig's rigorous phylogenetic methods into a new approach that the authors called **"vicariance biogeography"** (see below). Later, Croizat would write a scathing denial of his contribution to this paper, and criticized Nelson at length for blending phylogenetics and track analysis (Croizat

1982). Nevertheless, Croizat made an important contribution to the field. He emphasized the need for formalized methods of inferring past distributions from present ones. His search for distributional patterns that hold across many different kinds of organisms was a major departure from much previous work, which had been content to confine itself to a single taxon and to propose ad hoc events and taxon-specific mechanisms to account for its distribution.

While he went to the extreme of insisting that his methods be applied uniformly across all organisms without regard for their dispersal mechanisms and times of divergence, Croizat was reacting to the lack of rigor in the methods that many of his predecessors and contemporaries had been using to infer centers of origin, mechanisms of dispersal, and directions of range expansion, and thus to develop historical scenarios to explain contemporary distributions. Croizat rightly pointed out that even though these narratives may be correct, many of them are ad hoc, untested constructs of human imagination. To materially advance the science, they needed to be framed and tested as rigorous hypotheses.

BRUNDIN'S PHYLOGENETIC BIOGEOGRAPHY. Croizat criticized evolutionary biogeographers for lacking an explicit method and for the "story telling" flavor of their explanations, but by denying the importance of robust phylogenies in historical biogeographic reconstruction, fell short of developing a general method for biogeography. Instead, it was left to Lars Brundin to introduce the power of Hennig's phylogenetics to a new generation of biogeographers. Long before Brundin, Joseph Dalton Hooker had developed an interest in explaining the disjunct distributions of temperate floras on southern continents, including South America, New Zealand, and Australia-Tasmania (Chapters 2 and 6). Although Hooker (1867) had considered these distributions to represent a separate center of evolution in the Southern Hemisphere, the scenarios of dispersal from northern to southern regions of the world favored by evolutionary biogeographers, including Darwin, Matthew, and Darlington, were still a prevailing theme in historical biogeography into the 1960s.

Brundin clearly was showing his frustration with one of the last of the evolutionary biogeographers, Phillip Darlington, when he accused him of being "… largely unsuccessful because he has misunderstood the method of approach in biogeography. It is plainly meaningless to speculate…without knowledge of the reliability and meaning of the basic data and by perpetual neglect of the principles of phylogenetic reasoning" (Brundin 1966). In his classic 1966 monograph, Brundin argued that a phylogenetic analysis based on Hennigian logic (see Boxes 11.1 and 11.2) was a fundamental component of historical biogeographic analysis. His study involved a detailed phylogenetic analysis of the chironomid midges that occupied high elevation cold water streams throughout the world, including many species in southern temperate forests that occur on many of the landmasses in the Southern Hemisphere.

Brundin produced detailed phylogenetic trees for several tribes of midges, with an example shown in Figure 12.5A. Multiple clades showed a basal split between New Zealand, which was then followed by a South American versus Australia-Tasmania split, leading Brundin to conclude decisively that the southern end of the world was indeed, as predicted by Hooker, an important center of evolution rather than just a receiver of dispersers from the north. Brundin interpreted this history by imagining Antarctica to have played a dual role (Figure 12.5B)—both as an earlier "dispersal" route between New Zealand and South America via west Antarctica, and as a more recent route between South America and Australia-Tasmania via east Antarctica. Note that even though Australia-Tasmania is now geographically much closer to New

(A)

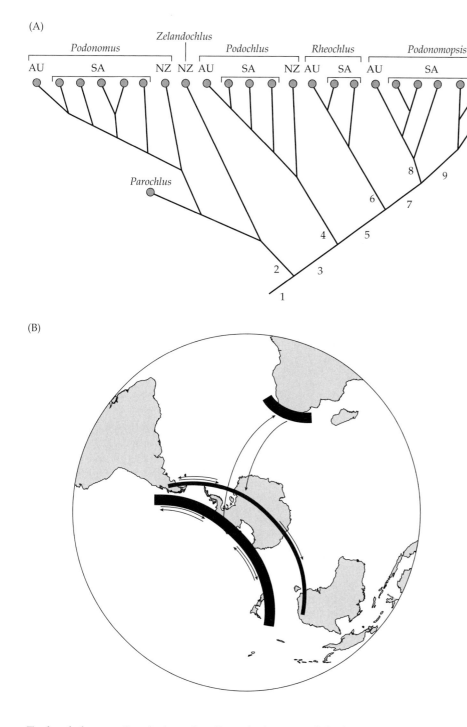

FIGURE 12.5 One of the first uses of Hennig's phylogenetic methods in historical biogeography was in Brundin's (1966) analysis of the chironomid midges of the southern temperate forests. (A) Summary phylogeny of the tribe Podomini, with the species indicated by circles, their grouping into genera indicated above the circles, and their distributions among Southern Hemisphere landmasses indicated as follows: AU = Australia-Tasmania; SA = South America; NZ = New Zealand. Note that there are multiple cases of the species in New Zealand being a sister group to species in South America and Australia-Tasmania; and that the Australia-Tasmania species are always the sister group to an array of South American species (which are always a paraphyletic rather than monophyletic group, if Australian species are not included). (B) Diagram Brundin used to summarize his full set of analyses of several midge tribes. These three landmasses are shown as having two general tracks across Antarctica—a bolder, more dominant track across west Antarctica between New Zealand and South America (with "dispersal" events going in both directions); and a lesser track across east Antarctica between South America and Australia-Tasmania (with dispersal in only one direction into the latter landmass). Phylogenetic trees also supported a track connecting these landmasses with South Africa. (B after Brundin 1966.)

Zealand than to South America, Brundin's powerful phylogenetic methods allowed him to make a bold and robust conclusion that its temperate forest biota had a closer historical affinity with the latter continent. In fact, Brundin was one of the first biogeographers to take full advantage of the plate tectonics revolution in geology of the 1960s (see Chapter 8) to interpret biogeographic histories, thereby initiating a renaissance in the discipline.

NELSON AND PLATNICK'S VICARIANCE BIOGEOGRAPHY. Beginning in the early 1970s, Gareth Nelson and Norman Platnick of the American Museum of Natural History in New York developed an approach that combined elements of both

Croizat's and Hennig's methods, but differed from Brundin, who had used Hennig's Progression Rule to infer historical scenarios for the midges. These biogeographers became ardent advocates of the point of view that dispersal histories—by virtue of being idiosyncratic, lineage-specific events—could not be addressed within a rigorous, hypothetico-deductive framework. Nelson (1974) set the tone for the dominant role of vicariance in biogeography throughout the next several decades: "In summary, I reject as aprioristic all 'clues' or 'rules' used to resolve centers of origin and dispersal without reference to general patterns of vicariance and sympatry … I include as a rejectable apriorism Hennig's 'Progression Rule'…Unencumbered by aprioristic dispersal, historical biogeography is the discovery and interpretation, with reference to causal geographic factors, of the vicariance shown by the monophyletic groups resolved by phylogenetic ('cladistic') systematics." The vicariance method was described in 1978 by Platnick and Nelson (1978), the same year that Donn Rosen published his classic vicariance biogeography study of the poeciliid fish genera *Heterandria* and *Xiphophorus* (Rosen 1978). Other descriptions of the method included those by Nelson and Platnick (1981) and, with some modifications, Wiley (1981, 1988).

The basic reasoning behind vicariance biogeography was simple. Platnick and Nelson formalized the observations made earlier by Brundin that historical explanations for disjunct distributions of related organisms fall into two classes: dispersal hypotheses (which they called **dispersal biogeography**), in which organisms are assumed to have migrated across preexisting barriers, and vicariance hypotheses (vicariance biogeography), in which the formation of new barriers is assumed to have fragmented the ranges of once continuously distributed taxa. Extant distributions are usually inadequate not only to determine whether a given barrier was in existence before or after the migration of a particular taxon to its present disjunct areas, but also to reconstruct the direction of the migration or the sequence of barrier formation. If, however, the organisms inhabit three or more disjunct areas, techniques of cladistic systematics (see Chapter 11) can be used to determine the sequence of branching in the lineage (Figure 12.6).

If we can assume that the speciation events in a lineage were caused by geographic isolation, then the phylogenetic relationships within the lineage also indicate the relative times of initial spatial separation of the now disjunct groups. If we can further assume that these ancient geographic separations have been preserved, and are reflected in the current distributions of the species of the lineage, then the cladistic phylogeny provides not only a phylogenetic hypothesis about the historical ancestor-descendant relationships among the taxa, but also a biogeographic hypothesis about the historical relationships among geographic localities. Both phylogenetic and biogeographic reconstructions are hypotheses about the historical branching of a taxonomic lineage in time and space. The biogeographic reconstruction is called an area cladogram because of its precise logical analogy to a cladistic reconstruction of phylogenetic relationships.

Figure 12.6 shows a hypothetical example of an area cladogram. Three taxa are distributed in three disjunct areas. The arrangement of the diagram indicates the relationships among the taxa as revealed by cladistic analysis: taxa 2 and 3 are more closely related to each other than either is to taxon 1. From this we can infer that taxa 2 and 3 share not only a more recent common ancestor, but also a more recent common geographic range than either taxon does with taxon 1. We can now advance two kinds of biogeographic hypotheses that are consistent with these relationships. A dispersal hypothesis would have the ancestral population inhabiting area A, with a propagule dispersing to area B

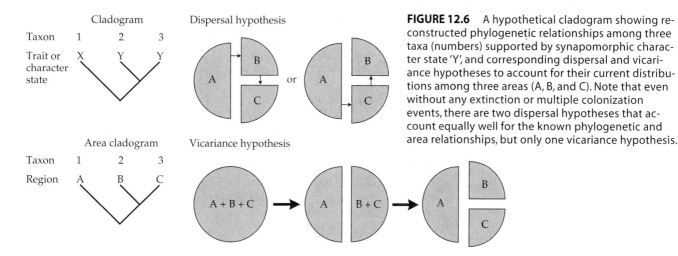

FIGURE 12.6 A hypothetical cladogram showing reconstructed phylogenetic relationships among three taxa (numbers) supported by synapomorphic character state 'Y', and corresponding dispersal and vicariance hypotheses to account for their current distributions among three areas (A, B, and C). Note that even without any extinction or multiple colonization events, there are two dispersal hypotheses that account equally well for the known phylogenetic and area relationships, but only one vicariance hypothesis.

and then, after some additional time, another propagule dispersing to area C. Because taxa 2 and 3 are symmetrically related to each other, the colonization order could also have been A → C → B, but dispersal from either B or C to A is inconsistent with the phylogenetic diagram. A vicariance hypothesis would have the ancestral taxon inhabiting all three areas—A, B, and C—followed by the formation of a barrier to isolate B and C (still interconnected) from A, and finally by the formation of a barrier to isolate B from C.

How would we test these hypotheses? Initially, the American Museum group wanted, like Croizat, to reject out of hand any hypothesis that called for long-distance dispersal. Platnick and Nelson (1978) and others tried to justify this seemingly arbitrary procedure by suggesting that dispersal hypotheses are extremely difficult to falsify, even with information from the fossil record. They argued that an episode of long-distance, or barrier-crossing dispersal is an event of low probability and predictability. Each episode is likely to represent an independent event in which a propagule of one taxon crosses the barrier and colonizes the isolated region; several different kinds of organisms are unlikely to disperse together or simultaneously. Further, if a species could disperse across a barrier once, presumably it could have done so repeatedly, and this would invalidate the assumption that the historical separation created by the original allopatric speciation event has been preserved. It is easy to show that if we allow for the possibility of repeated episodes of colonization and extinction, a wide variety of histories can produce the same area cladogram, and most of them will not preserve the geographic history of past speciation events (Figure 12.7).

The American Museum group argued that a vicariance hypothesis is easier to falsify than a dispersal hypothesis. Provided that barriers—once formed by a vicariance event—cannot be crossed, a vicariance hypothesis makes explicit predictions about the historical connections among areas. In particular, the formation of a new barrier is likely to isolate populations of many different kinds of organisms at approximately the same time. Therefore, a vicariance hypothesis would be supported if multiple taxa exhibit similar patterns of endemism, as Croizat emphasized. A vicariance hypothesis would be falsified if it was incompatible with data on the geological and climatic history of the Earth; with information from the fossil record on past distributions of the lineage(s); or if the phylogenies of two or more co-distributed taxa each generated different area cladograms.

FIGURE 12.7 Hypothesized sequences of colonization and extinction events that are all consistent with a cladogram in which taxa B and C are more closely related to each other than either is to A. The numbers show the sequence of events, arrows indicate colonizations, and crosses indicate extinctions. In the sequence at the lower left, for example, (1) the ancestor colonizes the center, (2) a population disperses to the left, (3) the original population in the center goes extinct, (4) dispersal from the left recolonizes the center, and (5) dispersal from the center colonizes the right. The four sequences shown here, only a subset of the possible sequences that are consistent with the cladogram, demonstrate how repeated colonization and extinction events can erase the evidence of earlier biogeographic history.

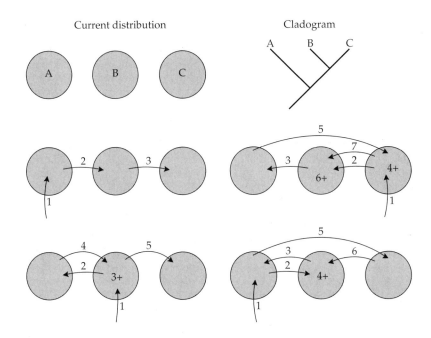
Current distribution   Cladogram

In the case of our hypothetical example in Figure 12.6, the vicariance hypothesis would be falsified if clear geological evidence showed that the barriers between the areas were in existence before the taxa occurred there. The vicariance hypothesis also would be falsified by the occurrence of a fossil of taxon 2 or 3 in area A, or one of 3 in B or of 2 in C, or if a molecular phylogeny indicated that divergence dates between taxa were more recent than the ages of barrier formation. Perhaps the strongest test, however, could be conducted by constructing area cladograms for other extant groups with endemic taxa in areas A, B, and C. If all the lineages were in existence at the same time, one would expect them to have responded similarly to the sequential fragmentation of their ranges. Therefore, they should exhibit similar or congruent area cladograms, reflecting similar responses to the same sequence of barrier-forming geological and/or climatic events (Figure 12.8; Box 12.1).

Of course, there is some probability that area cladograms will be congruent by chance alone. These probabilities can be determined, however, because they depend on the number of areas and taxa being considered. As the number of areas, and in particular, the number of independent taxonomic groups included in the analysis—increases, the probability of observing congruent cladograms simply by chance diminishes rapidly. Calculating these probabilities may not be as simple as it may seem, but the long-proposed possibility of

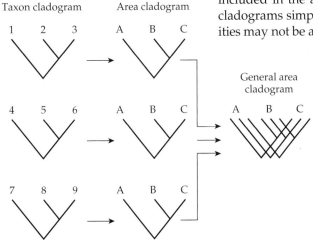
Taxon cladogram   Area cladogram

General area cladogram

FIGURE 12.8 The main steps in several historical biogeography approaches listed in Table 12.2 and discussed in text, including cladistic biogeography, phylogenetic biogeography II, and comparative phylogeography. Here, the general area cladogram reflects a congruent sequence of speciation events among three co-distributed taxa across three areas of endemism, which is more likely to have resulted from a history of vicariance than from a series of unique jump dispersal events in each taxon. Perfect congruence across area cladograms such as that shown here rarely is found in real datasets, even if vicariance did play a role in the biogeographic history of a biota (see Box 10.2). (After Crisci et al. 2003.)

BOX 12.1 *Defining and delineating areas of endemism*

■■I In Chapter 10, we discussed the concept of endemism in detail and its importance in biogeography. In historical biogeography, an **area of endemism** is generally considered as the fundamental unit of analysis in cladistic-based approaches. Several decades of historical biogeography have been developed upon the premise that "the most elementary questions of historical biogeography concern areas of endemism and their relationships" (Nelson and Platnick 1981). Clearly, the importance of understanding relationships among areas, based on the taxa that occupied them, was associated with the idea that vicariance, followed by "allopatric speciation mode I" (see Figure 7.10) would produce congruent relationships across co-distributed taxa (see Figure 12.8).

Delimiting areas of endemism would seem to be an easy thing to do. After all, at the simplest level, they merely represent geographic areas where two or more endemic taxa share overlapping, or congruent, distributions. But we know that it is rare that the distributions of two or more taxa overlap exactly, except in cases where distributional limits are set by very discreet abiotic boundaries (e.g., lakes or islands), and biogeographers still are debating how to define and delineate them. How much or how little overlap in ranges, or sympatry, should we accept in order to delimit an area of endemism? Or should some criterion other than sympatry be applied? Recently proposed definitions emphasize one of three criteria to define an area of endemism: (1) degree of distributional overlap, or sympatry; (2) barriers between separate areas resulting from vicariance; and, (3) as an operational extension of the latter, phylogenetic congruence between co-distributed taxa and their sister-taxa in the area on the other side of the barrier. For example:

Platnick's defintion. "At the minimum, it would seem that an area of endemism can be defined by the congruent distributional limits of two or more species. Obviously 'congruent' in this context does not demand complete agreement on those limits, at all possible scales of mapping, but *relatively extensive sympatry* (italics added) at some scale must surely be the fundamental requirement" (Platnick 1991; see also Morrone and Crisci [1995]; Linder [2001]).

Hausdorf's definition. "Areas of endemism can be defined as areas *delimited by barriers* (italics added), the appearance of which entails the formation of species restricted by these barriers" (Hausdorf 2002).

Harold and Mooi's definition. An area of endemism is "a geographic region comprising the distribution of *two or more taxa that exhibit a phylogenetic and distributional congruence* and having their respective relatives occurring in other such-defined regions" (Harold and Mooi 1994).

So, the definition of an area of endemism can range from requiring extensive sympatry (Platnick's definition) to little or none, with the main criterion being derived from a vicariance model of barrier formation and subsequent speciation (Hausdorf's and Harold and Mooi's definitions). The approaches to delineating areas of endemism are equally diverse, ranging from strongly geopolitical (e.g., "historically persistent Gondwanan land-masses according to paleogeographic reconstructions" [Sanmartin and Ronquist 2004] see figure below), to the quadrat approach of Morrone (1994) using Parsimony Analysis of Endemicity (PAE) to delineate areas of endemism based on the distributions of taxa within a region.

In the Gondwanan example, the approach to delimiting areas would be an example of using Hausdorf's definition, because the appearance of barriers following the fragmentation of Gondwana into separate landmasses is a more important criterion than the "extensive sympatry" of any taxa at smaller scales within each landmass. The areas delimited based on this criterion are:

- Africa, south of the Sahara
- Madagascar, including several Indian Ocean islands
- India, including Nepal, Tibet, and Sri Lanka
- Australia and Tasmania
- New Zealand, including subantarctic islands on the same continental block
- New Caledonia
- New Guinea, including the Solomon and New Hebrides islands
- Southern South America
- Northern South America

These areas of endemism are informative to historical biogeographers because they have arisen from a well-understood sequence of historical fragmentation of Gondwana landmasses (see figure), which forms a basis for addressing the relative importance of vicariance and dispersal in the biogeographic history of these regions.

In many cases, however, the physical discreetness between areas is not so clear-cut, and so other methods need to be employed to delineate areas of endemism. A number of approaches and methods have been proposed (as examples, see Morrone 1994; Linder 2001), including

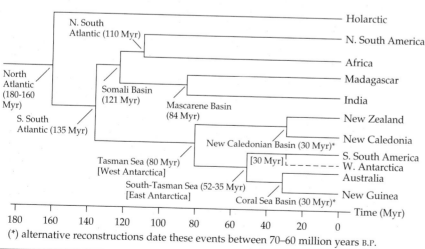

(*) alternative reconstructions date these events between 70–60 million years B.P.

(continued on next page)

BOX 12.1 *Defining and delineating areas of endemism* (continued)

some that employ sophisticated statistical or simulation methods (Hausdorf and Hennig 2003; Mast and Nyffeler 2003; Szumik and Goloboff 2004).

Example: Morrone's (1994) Quadrat-PAE approach to delimiting areas of endemism. This method relies on the use of Parsimony Analysis of Endemicity (PAE; B. Rosen 1988) to delimit areas of endemism. There are five steps in the analysis, explained in greater detail in Morrone (1994) and Crisci et al. (2003):

1. Lay a grid of quadrats across the study area. The actual sizes of the grids are determined by the investigator and depend on the scale of desired resolution:

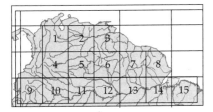

2. Determine the geographic distributions of taxa across the study area. Any number of unrelated taxa, at any taxonomic hierarchy, can be used, constrained by the members of each taxon being a monophyletic group.
3. Construct a data matrix summarizing distributions of taxa × quadrats, assigning a quadrat "1" if a taxon is present in

it, and "0" if a taxon is absent. Construct a hypothetical quadrat with all taxa absent (coded as "0" in all quadrats) as an outgroup:

	Species				
	1	2	3	n
Quadrats 1	1	0	0	-	1
2	1	0	0	-	1
3	1	1	0	-	1
-	-	-	-	-	-
15	0	0	1	-	0

4. Perform a maximum parsimony cladistic analysis on the matrix and construct a tree:

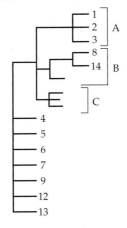

5. Select areas of endemism based on results of the cladistic analysis. In this case, three quadrats (1 + 2 + 3) are sup-

ported by several taxa as a monophyletic clade, and are selected to represent area of endemism 'A'. The same reasoning leads to recognition of areas 'B' and 'C'. This analysis results in the delineation of three areas of endemism.

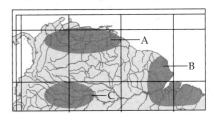

Note that seven of the original quadrats were not grouped into any of the areas, which might happen for several reasons. First, quadrats might not contain any of the species in the data set—these regions might represent current dispersal barriers between areas of endemism. Second, a quadrat might contain only a single endemic species found nowhere else and therefore not share any synapomorphies with other quadrats. Finally, one or more quadrats might have a mixture of some species that otherwise occur in one area of endemism and other species that occur in another area—as might happen with dispersal out of an area following the erosion of a previous dispersal barrier. ▮▮▮

---

testing the degree of congruence of area cladograms in a rigorous statistical framework (Simberloff et al. 1981a,b) has been implemented in several of the more recent methods (e.g., Page 2003).

Beyond Vicariance Biogeography and Simple Vicariance

Fundamental Questions and Issues in Modern Historical Biogeography

The vicariance biogeography advocated by the American Museum group represented an important advance over Croizat's panbiogeography by demanding that historical biogeography be conducted within a rigorous hypothesis testing framework, but it had some of the same—and some other, unique—weaknesses. In its quest for analytical rigor, it created some highly restrictive and simplifying rules or assumptions that reduced its relevance in a world full of complexity:

1. Their approach assumed that long-distance dispersal cannot be incorporated into the cladistic/vicariant framework and is sufficiently infrequent to not overwhelm the signal of vicariance in a general area cladogram. However, we presented many examples of long-distance dispersal in Chapter 6, and will review later in this chapter the emerging evidence for its profound influence throughout the historical assembly of modern biotas.

2. They also assumed that geographic isolation and speciation arising through vicariance is preserved in the spatial configurations of current geographic ranges. But note the cautionary message we must take from the magnitude of relatively recent changes in distributions described in Chapters 9 and 17, and later in this chapter.

3. Finally, they argued that dispersal is nearly always a single-lineage, idiosyncratic process and, therefore, not amenable to an analytical and hypothesis-testing approach. But recent advances in paleobiogeography and phylogeography have shown how the process of **geo-dispersal** can also produce biogeographic congruence, and thus is testable in a cladistic framework (Lieberman 2004).

The vicariance biogeography of the 1970s and early 1980s was renamed **cladistic biogeography** in the mid-1980s and is thoroughly reviewed by Chris Humphries and Lynn Parenti in their 1999 book. With the growth of power in computers in the 1980s, a quantitative implementation of the original method of **component analysis** introduced by Nelson and Platnick (1981) was introduced by Rod Page (1989) in his software, COMPONENT 1.5. Efforts to develop better approaches to resolving the general vicariance structure (see Figure 12.8) shared by two or more co-distributed lineages with widespread taxa, redundant distributions, or missing areas (Box 12.2) continued into the 1990s with Nelson and Ladiges' (1992; 1995) development of a method called **three item statement analysis** (TAS, or TASS), and then one called **paralogy-free subtrees** (Nelson and Ladiges 1996), also implemented using TASS software. These and other methods are addressed thoroughly, with examples, by Crisci et al. (2003).

While much of the original vicariance biogeography has largely been replaced by an array of different approaches and methods (Table 12.2; Crisci et al. 2003), most of them retain three of its key features: (1) its emphasis on rigorous logic and hypothesis testing; (2) its reliance on robust phylogenetic hypotheses; and (3) its use of taxon-area cladograms based on the delineation of areas of endemism (see Box 12.1). Many of these approaches depart from vicariance biogeography largely in their willingness to address a wider range of hypotheses and mechanisms, including long-distance dispersal, geo-dispersal, and pseudo-congruence. Dan Brooks and his colleagues (van Veller and Brooks 2001; van Veller et al. 2003; Brooks 2004) have argued that what traditionally was considered a single research program in cladistic biogeography actually included two different goals. One of these was what we described above as vicariance, or cladistic, biogeography, the goal being to identify the causal association between changes in the configuration and connectivity of areas of the Earth and the emergence of differentiated biotas as a result of that Earth history. In this vein, cladistic biogeographers treat all area relationships on a cladogram that resulted from some process other than vicariance as "noise" that needed to be removed from the analysis (see Box 12.2). They proposed to do so by creating a set of a priori assumptions—called Assumptions 1, 2 (Nelson and Platnick 1981), and 0 (Wiley 1987, 1988; Zandee and Roos 1987)—each based on the processes that were assumed to generate the noise,

BOX 12.2   *Processes that reduce the generality of the general area cladogram*

■■▮ The main goal of cladistic biogeography, and one of the goals of several other approaches in Table 12.2, including phylogenetic biogeography II, is to derive a general area cladogram that summarizes congruent area relationships among a set of co-distributed lineages. In a world in which only vicariance produced disjunct distributions, all speciation resulted from vicariance, and there was no extinction, general area cladograms would be simple to interpret—the area cladograms of all co-distributed taxa would be perfectly congruent, as shown in Figure 12.8. We know, however, that this idealistic world does not actually exist, and instead, species sometimes do not speciate when a vicariant event leads to allopatric speciation in other co-distributed lineages; that organisms do sometimes disperse across barriers and establish populations on the other side, which may then speciate; and that species do go extinct. Adding to this complexity, areas of endemism are often difficult to delineate with precision (see Box 12.1) and could include lineages that have undergone episodes of sympatric or embedded allopatric speciation prior to a vicariant event. Biogeographers recognize three categories of patterns in taxon-area cladograms that reduce their congruence with an underlying general area cladogram.

Suppose the general area cladogram summarized the vicariant history across a set of five areas, A–E, as shown below:

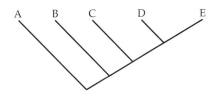

There are three ways in which a taxon cladogram can be less than perfectly congruent with this general area cladogram:

A taxon-area cladogram with a *widespread species* that occurred in both areas A and B, but without a species endemic to either area alone, would look like this:

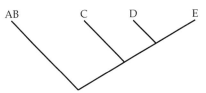

This pattern introduces a source of ambiguity into the interpretation of biogeographic history. One explanation might be that the widespread species did not speciate at the vicariant event that separated area 1 from ancestral area (B + C + D + E). Another is that the widespread species did, in fact, speciate through vicariance but then dispersed to the other area without undergoing yet another round of speciation. Many biogeographers suggest that we cannot know from only this cladogram which of these explanations might be valid, and so we need to make one of three *assumptions*, called Assumptions 0, 1, and 2, that each provide a set of rules about how to remove the ambiguities (see van Veller et al. 1999; Crisci et al. 2003) so that the general pattern of vicariance emerges. Other biogeographers prefer to analyze the widespread lineages using methods that allow them to infer ancestral distributions and dispersal histories in addition to the general vicariance pattern (see discussion in text).

Another source of ambiguity is introduced through the presence of *redundant distributions*, where an area occurs more than once on an area cladogram. The area cladogram below shows two cases of redundant distributions—area B appears once at a more basal—and again, at a more derived—position; and area D is repeated serially. For area B, we might explain the repetition by a dispersal of the taxon that inhabited area E at the time of

vicariance into area B followed by a subsequent speciation event. Alternatively, for area D, we might explain the repetition by invoking a sympatric speciation event in area D prior to vicariance between areas D and E, as shown above in the general area cladogram. Again, biogeographers differ about whether to remove or analyze redundant distributions. Note here that area B has two different relationships with the other areas. For example, with regard to area E, in one case two vicariant events separate areas B and E. In the other case, areas B and E share a sister area relationship. This is an example of a *reticulate history*, where one area has historical connections with more than one other area (also sometimes called *composite areas* by biogeographers).

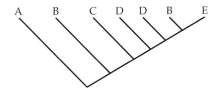

Finally, *missing areas* describe the situation where not all areas on the general area cladogram are occupied by a member of a particular taxon, illustrated in the taxon-area cladogram below where no member of the group occurs in area B.

This source of ambiguity could be explained either as *primitively absent*—the ancestral taxon never got to the area, and so was not a member of the biota in that area at the time of vicariance—or it could be explained as *extinction* sometime subsequent to vicariance. ▮■■

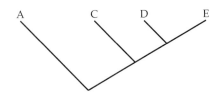

including dispersal, sympatric speciation, and extinction. Implementing one or more of these assumptions would allow investigators to "find" the vicariance structure embedded in a general area cladogram (see Figure 12.8) that also contained non-vicariance relationships. A detailed treatment of the rather complex

TABLE 12.2 *A selection of the more historically important or currently popular approaches and methods in historical biogeography*[a]

Approaches	Goal and selected methods	Original authors and general references
Descriptive biogeography	Comparing species lists	Sclater 1858
		Hooker 1844–60
Evolutionary biogeography	Center of origin-dispersal	Matthew 1915; Cain 1944
Phylogenetic biogeography I	Phylogenetic systematics	Hennig 1966; Brundin 1966
Ancestral areas analysis	Area(s) of origin prior to dispersal	Bremer 1992; 1995
	Weighted ancestral areas analysis	Hausdorf 1998
Panbiogeography	Generalized tracks on a dynamic Earth	Croizat 1958
	Track analysis	Croizat 1958
Cladistic (Vicariance) biogeography	Vicariance on a dynamic Earth	Nelson 1974
	Reduced area cladogram	Rosen 1978
	Component analysis (CA)	Nelson and Platnick 1981; Humphries and Parenti 1999
	Three-area statement (TASS)	Nelson and Ladiges 1992
	Paralogy-free subtrees	Nelson and Ladiges 1996
Phylogenetic biogeography II	Vicariance, dispersal, geography of speciation	Wiley 1980
	Brooks parsimony analysis (BPA)	Wiley 1980
	Primary and secondary BPA	van Veller and Brooks 2000; Brooks et al. 2001
Parsimony analysis of endemicity (PAE)	Natural distribution patterns of taxa	B. Rosen 1988
	Areas of endemism	Craw 1988a; Morrone 1994
Event-based methods	Benefit/cost modeling of events	Ronquist and Nylin 1990
	Dispersal-vicariance analysis (DIVA)	Ronquist 1997
	Parsimony-based tree fitting	Page 1994; Ronquist 2002
Phylogeography	Geography of genealogical lineages	Avise et al. 1987; Avise 2000
	Phylogeny of gene trees	various
	Nested clade analysis	Templeton et al. 1995; Templeton 2004
	Coalescent-based approaches	various; Knowles 2003
	Comparative phylogeography	Zink 1996; Arbogast and Kenagy 2001

Source: After Crisci et al. 2003.

[a] This list differs primarily from Crisci et al. (2003) by distinguishing older and newer uses of "phylogenetic biogeography" as I and II, respectively, and separating the latter from cladistic biogeography.

literature attempting to explain, understand, and implement Assumptions 0, 1, and 2 would take more time than we want to spend on the topic here, but good places to explore this topic include van Veller et al. (1999) and Crisci et al. (2003).

The other goal embedded within cladistic biogeography, according to Brooks et al. (2004), was to try to understand the geography of speciation and historical assembly of biotas. Here, rather than trying to remove the "noise" of taxa incongruent with the vicariance-driven general area cladogram, events including post-speciation dispersal, peripheral isolates speciation (Figure 7.12), sympatric speciation (Figure 7.15), and extinction were considered parts of the whole of biotic histories and, therefore, worth trying to understand if historical biogeographers wanted to contribute significantly to evolutionary biology. These investigators proposed that this latter research program be decoupled from cladistic biogeography and called "*phylogenetic biogeography.*" Notice that this is also the name of Brundin's earlier approach, described above, but the methods being currently developed under the rubric are quite different from those employed by Brundin. Therefore, to reduce confusion, we will call Brundin's approach "**phylogenetic biogeography I**" (see Table 12.2), and that of Brooks et al., "**phylogenetic biogeography II.**"

Three additional examples demonstrate how far modern historical biogeography has drifted away from the strict vicariance biogeography of the American Museum group. First, Bremer's (1992:440) ancestral area analysis is based on assumptions similar to those of dispersal biogeography, "namely that each group originated in a more or less small ancestral area, *the center of origin* [italics added], and that the distribution areas have subsequently expanded by dispersal" (Hausdorf 1998:445). In other words, here is a quantitative, phylogenetically based set of methods that have explicitly resurrected the concept of center of origin-dispersal so loathed by the vicariance biogeographers! Of course, Bremer's method was based on a much more rigorous footing than Matthew, Simpson, and Darlington's approach—including cladistic procedures and delineation of areas of endemism. In a second example of how historical biogeography has diversified in the past decade, one of the so-called *event-based methods* (see Table 12.2) is actually called "**dispersal-vicariance analysis (DIVA),**" which we explain in more detail below. Finally, many of the popular methods in phylogeography are designed to sort allopatric fragmentation from range expansion (i.e., *dispersal*) histories among intraspecific populations (Crandall and Templeton 1996). Clearly, modern historical biogeography has become a much broader endeavor than was advocated by the earlier vicariance biogeographers, although not without a certain level of remaining discord over its appropriate role in evolutionary biology (e.g., Humphries 2000; Ebach and Humphries 2002).

DIFFERENT APPROACHES TO THE SAME QUESTION OR DIFFERENT QUESTIONS? Crisci et al. (2003), in their introduction to historical biogeography, point out that a variety of distinct questions often have unknowingly been embedded within the seemingly single discipline of historical biogeography, leading to confusion in choosing the appropriate methods of analysis. They suggest that approaches and methods have developed along two separate lines. Several are designed to reconstruct the history of lineages across areas at a single lineage at a time, an approach they call **taxon biogeography**. The cladogram of Brundin's midges in Figure 12.5A provides a good example of the early use of this approach. Ancestral areas analysis and event-based methods are often used in this fashion. Many studies using phylogeographic methods are concerned with a single taxon. The ultimate goal, of course, could certainly be to compare separate lineage histories with one another or with a geological area cladogram and assess the level of congruence across them for evidence of vicariance or dispersal histories (Figure 12.9).

Alternatively, an investigator doing **area biogeography** might wish to produce a general area cladogram (see Box 12.2) because it is the historical relationships among the areas and the biotas inhabiting those areas that are of interest to the biogeographers. Here an investigator might use a component analysis to produce a **consensus area cladogram** depicting vicariant history; a **reconciled tree** approach (Page 1994) estimating the number of vicariance, dispersal, and extinction events; or a **Brooks parsimony analysis** (BPA; see below) to address the geography of speciation.

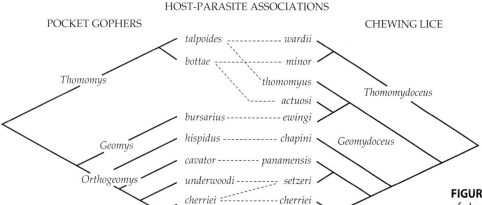

HOST-PARASITE ASSOCIATIONS

POCKET GOPHERS

CHEWING LICE

FIGURE 12.9 Cladograms for a group of chewing lice (right) and their pocket gopher hosts (left) show a substantial degree of congruence. While similarities in the branching patterns suggest that speciation may occur coincidentally in both lineages, differences in the pattern imply that the parasites are also able to colonize distantly related hosts (e.g., such switching seems necessary to account for the occurrence of three closely related louse species on two distantly related gopher genera: *Geomydoceus actuosi* and *G. thomomyus* on *Thomomys*, and *G. ewingi* on *Geomys*). Since the chewing lice are obligate, specific parasites on these rodents, the two cladograms have important historic biogeographic implications: for example, parasite and host must live in the same place for cospeciation or host switching to occur. (After Hafner and Nadler 1991.)

In summary, the questions in historical biogeography are extremely diverse—ranging from the reconstruction of an ancient Gondwanan biota, through colonization and diversification on oceanic island archipelagoes, to other questions that concern biotic responses to the influences of late Pleistocene glaciation. As such, it perhaps is not surprising that the available approaches and methods are equally diverse (see Table 12.2; Crisci et al. 2003). Even so, we probably can find a great deal of common ground in the ultimate goals motivating biogeographers despite their use of alternative—and, we hope, complementary—approaches to similar questions.

HOW COMPLEX, REALLY, ARE BIOTAS? If the relationships among biotas in different areas of the world were mostly the product of either a simple vicariance or predictable dispersal history (e.g., via the progression rule), most of the many available methods would find the same general area cladogram and, given accurate phylogenies, it would not be particularly difficult to reconstruct the history of biotas. Let's look in some detail at an example of a system with a very simple geological history—the Hawaiian Archipelago. Waren Wagner and Vicki Funk (1995) edited an excellent book that brought together much of the phylogenetic and biogeographic research on the biota of the Hawaiian Islands, although a number of more recent studies of this biota are available as well (e.g., see Figure 11.9). Each chapter presented a cladistic and biogeographic analysis of a different group of organisms. In the final chapter, Funk and Wagner provided synthetic analysis of the biogeographic history of over 20 Hawaiian lineages, including such diverse groups as terrestrial invertebrates (insects and spiders), birds (honeycreepers), and flowering plants.

As we saw in Chapter 8, the Hawaiian Archipelago has a dramatically simple geological history resulting from the Pacific Plate drifting over a hot spot now located at the southeastern end of the island chain, currently beneath the island of Hawaii and another volcano (Loihi Seamount) to the southeast, which is growing but still submerged below the ocean (see Figure 8.20). The formation of the islands began 75 to 80 million years ago, but there may have been times since then when there was little or no emergent land. The oldest of the present major islands is Kauai, the northwestern-most island, which was formed about 5.1 million years ago. The ages of the islands decrease down the chain to the southeast: Kauai, Oahu, Molokai, Maui, and Hawaii.

Given this simple geological history, we can develop a relatively straightforward geological area cladogram (Figure 12.10) to use as a hypothesis of

FIGURE 12.10 The area cladogram predicted for the Hawaiian Islands based on a simple progression rule. This is the pattern of phylogenetic relationships that would be expected if a lineage colonized each island in turn as it was formed by the emergence of a volcano from the sea.

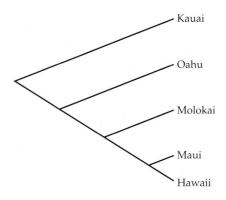

taxon relationships under the following assumptions: that a taxon originally colonized either Kauai, the oldest present island, or an older island to the northwest of Kauai that is now submerged; and that, as newer islands were formed, each was colonized from an ancestral population on the adjacent and older island up the chain. This prediction should look very much like the progression rule we discussed earlier in this chapter (and shown in Figure 7.12B), and is one process that would lead to a high level of congruence across the co-distributed lineages analyzed by Funk and Wagner. Deviations from this predicted pattern could, however, result from back-colonization from younger to already inhabited and older islands, from colonization of non-adjacent islands (which could also be interpreted as extinctions of populations on islands in between), from in situ speciation on a single island (which might occur via sympatric or microallopatric modes), or from more recent colonization of the archipelago itself.

So what do the results of the phylogenetic and biogeographic analyses show? The bottom line is that the simple prediction generated from the geological area cladogram is not supported as a general rule. Instead, the distributions of the lineages studied show a rich variety of relationships with respect to their phylogenetic histories. We can present only some of this variety in a simplified form here (for the full story, read Wagner and Funk 1995 and more recent literature, including an interesting analysis of geological and colonization histories by Price and Clague 2002).

Some clades (e.g., *Drosophila* fruit flies, other invertebrates, and certain plants) do indeed show a more or less clear progression rule, with basal taxa on older islands and progressively more derived forms on younger islands (Figure 12.11A,B; Figure 11.9). But other clades show very different patterns. One variation, found in the closely related plants *Schiedea* and *Alsinidendron*, is a series of subclades, each exhibiting its own progression rule with multiple waves of dispersal from older to younger islands (see Figure 12.11C). A clear exception to any progression rule is found in the plant genus *Tetramolopium* (see Figure 12.11D), in which the ancestral species clearly colonized one of the younger islands (either Hawaii or Maui) and subsequently dispersed to older ones (Oahu and Molokai), but apparently never reached the oldest (Kauai). Several other cladograms show complex patterns that clearly do not support a progression rule (e.g., the honeycreepers in Figure 12.12A, which have sister taxa on islands of contrasting age: Kauai and Hawaii), while still others are not easily resolved, and could be interpreted to suggest two or more very different histories.

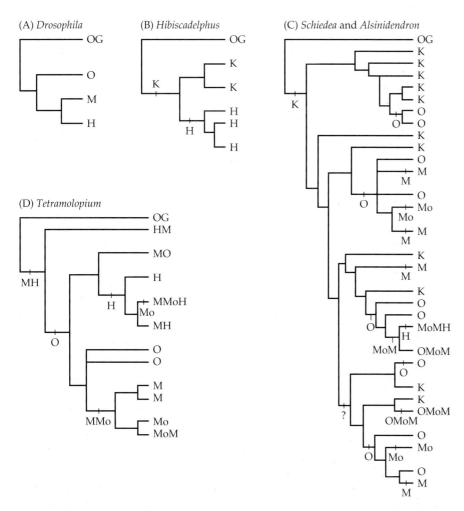

FIGURE 12.11   Area cladograms for four groups of Hawaiian organisms, simplified to include only the outgroup and those taxa found on the five largest islands. OG = outgroup; K= Kauai; O = Oahu; Mo = Molokai; M = Maui; H = Hawaii. Letters on the terminal branches (right) indicate present distributions; letters placed on the tree indicate over-water dispersal to colonize new islands; and multiple letters for the same island without multiple colonization events indicate within-island speciation events. (A) A group of *Drosophila* fruit flies shows a progression rule, with the more derived forms occurring on progressively younger islands. (B) The endemic plant genus *Hibiscadelphus* shows a highly modified progression rule, with the more derived taxa occurring on the youngest island (Hawaii) and the ancestral taxa on the oldest island (Kauai), but with multiple speciation events within these two islands, and no occurrences on the islands of intermediate age. (C) The closely related endemic plant genera *Schiedea* and *Alsinidendron*. This group comprises four subclades, each of which shows a general progression rule. Note, however, that there have also been multiple independent colonizations of the same island (e.g., Oahu, six times) and speciation events within islands (e.g., especially on Kauai and Oahu). (D) The plant genus *Tetramolopium,* which is probably a fairly recent immigrant to the archipelago, shows no evidence of a progression rule. It originally colonized either Maui or Hawaii, and subsequently dispersed to Molokai and Oahu, but apparently never got established on the oldest island, Kauai. (After Funk and Wagner 1995.)

(A) Drepanidinae

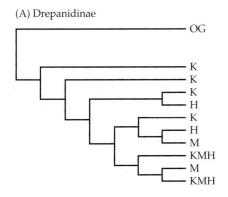

(B) *Prognathogryllus*

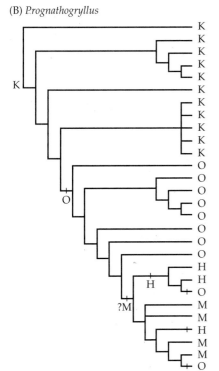

FIGURE 12.12 Area cladograms for two lineages of animal groups endemic to the Hawaiian archipelago and showing contrasting patterns of colonization and speciation. Localities and colonization events are coded as in Figure 12.11. (A) The Hawaiian honeycreepers of the avian subfamily Drepanidinae show many episodes of inter-island colonization followed by speciation in isolation on the different islands. The direction of colonization is not known for the honeycreepers because there have been so many colonization events that the direction of dispersal often cannot be resolved from patterns of phylogenetic relatedness. (B) The cricket genus *Prognathogryllus* shows relatively few interisland colonization events, but each such event has been followed by multiple episodes of within-island speciation. (After Funk and Wagner 1995.)

The cladograms also reveal many different patterns of colonization and speciation. The examples of progression rules mentioned above illustrate cases of dispersal from older to younger islands. But there are also many cases of colonization of older islands from younger ones. One example, also mentioned above, is the genus *Tetramolopium*. Another is the Hawaiian honeycreepers, in which several recently derived taxa occur on the oldest island, Kauai (see Figure 12.12A). Again, it is important to mention that, for several clades, it is difficult to pinpoint the island that was first colonized, and therefore it is equally difficult to determine the direction of subsequent dispersal events that resulted in the colonization of additional islands. Such problems may be due to difficulty in resolving the cladograms, but they may also be due to unresolvable complications in the biogeographic history. For example, branches of lineages that went extinct on islands at different times in the past and did or did not colonize other islands and leave descendants there, can make it difficult to reconstruct the biogeographic history even though the phylogenetic reconstruction may be well resolved and accurate.

With respect to speciation, the cladograms do show examples of allopatric speciation presumably caused by dispersal to—and differentiation, on—different islands. Perhaps the best example is that of the honeycreepers (see Figure 12.12A). In general, the most closely related pairs of species occur on different islands, and often these islands are far from each other (e.g., Kauai and Hawaii, at opposite ends of the archipelago). This pattern of lineage diversification as a result of repeated episodes of colonization and speciation fits well with that seen in other groups of birds in other archipelagoes, such as the Galápagos and East Indies. On the other hand, the predominant pattern, seen in many clades of Hawaiian arthropods and plants, is one of extensive speciation and radiation within islands (see Figure 12.12B). This is very similar to the pattern observed in groups of fishes and mollusks in lakes such as those of Africa's Rift Valley (see Chapter 7). Further, since we know the ages of the Hawaiian Islands, we can estimate the minimum times for various speciation events. Clearly, all of the within-island speciation occurred within the last 5 million years, and some of it probably occurred (e.g., on Hawaii) within the last 500,000 years (Table 12.3).

It is important to note, however, that just because speciation occurred *within* an island and not just among islands, geographic isolation still may have played an important role in the differentiation of the populations. All of the large Hawaiian islands have a great deal of topographic relief and habitat heterogeneity including mountain ranges, large rivers, and other land features that may serve as barriers to dispersal. For organisms that disperse as poorly as some insular plants and invertebrates, this topographic heterogeneity promotes **microallopatric** speciation and rapid divergence and adaptive radiation among populations inhabiting the diversity of environments found on these

TABLE 12.3 *Estimated ages of the most recent common ancestors (MRCA) of some Hawaiian lineages*[a]

Lineage	Type of organism	Number of species	Age (Ma)	Source
Hawaiian fruitflies (Drosophilidiae)	Insect	ca. 1000	26	Russo et al. (1995)
Hawaiian lobelioids (Campanulaceae)	Plant	125	15	Givnish et al. (1996)
Megalagrion damselflies (Coenagrionidae)	Insect	23	9.6	Jordan et al. (2003)
Silversword Alliance (Asteraceae)	Plant	28	5.1	Baldwin & Sanderson (1998)
Laysan duck, *Anas laysanensis* (Anatidae)	Bird	1	< 5	Fleischer & McIntosh (2001)
Hawaiian crows, *Corvus hawaiiensis* + other spp.? (Corvidae)	Bird	1 + ?	< 4.2	Fleischer & McIntosh (2001)
Hawaiian honeycreepers, Drepanidinae (Fringillidae)	Bird	ca. 50	4–5	Fleischer et al. (1998)
Viola spp. (Violaceae)	Plant	6	3.7	Ballard & Sytsma (2000)
Flightless Anseriformes, 'moa-nalos' (Anatidae)	Bird	4	< 3.6	Sorenson et al. (1999)
Hawaiian thrushes, *Myadestes* spp. (Muscicapidae)	Bird	5	< 3.35	Fleischer & McIntosh (2001)
Kokia spp. (Malvaceae)	Plant	4	< 3	Seelanan et al. (1997)
Flightless rails, *Porzana sandwicensis* + other spp.? (Rallidae)	Bird	1 + ?	< 2.95	Fleischer & McIntosh (2001)
Geranium spp. (Geraniaceae)	Plant	6	2	Funk & Wagner (1995)
Hesperomannia spp. (Asteraceae)	Plant	4	1.81–4.91	Kim et al. (1998)
Flightless ibises, *Apteribis* spp. (Plataleidae)	Bird	2	< 1.6	Fleischer & McIntosh (2001)
Hawaiian duck, *Anas wyvilliana* (Anatidae)	Bird	1	< 1.5	Fleischer & McIntosh (2001)
Flightless rails, *Porzana palmeri* + other spp.? (Rallidae)	Bird	1 + ?	< 1.05	Fleischer & McIntosh (2001)
Hawaiian geese, *Branta* spp. (Anatidae)	Bird	3	< 1	Fleischer & McIntosh (2001)
Hawaiian black-necked stilt, *Himantous mexicanus knudsenii* (Recurvirostridae)	Bird	1	< 0.75	Fleischer & McIntosh (2001)
Hawaiian hawk, *Buteo solitarius* (Accipitridae)	Bird	1	< 0.7	Fleischer & McIntosh (2001)
Tetramolopium spp. (Asteraceae)	Plant	11	0.6 – 0.7	Lowrey (1995)
Metrosideros spp. (Myrtaceae)	Plant	5	0.5 – 1.0	Wright et al. (2001)

[a] Refer to Price and Clague 2002 for methods used to calculate divergence times. Note that several lineages are older than the oldest present large island (Kauai, 5.1 million years old), suggesting an initial colonization of an older, now submerged island. The MRCA of other lineages is considerably younger, implying relatively recent colonization events.

large islands (Figure 12.13). Nevertheless, the high frequency of speciation within islands, like that within lakes, raises other important questions about the role of ecological and genetic processes in speciation—and especially about the relative importance of geographic isolation and divergent selection pressures.

THE PERILS OF IGNORING TIME. We have seen in the Hawaiian example that, even in a region with a comparatively simple geological history, phylogenies for co-distributed lineages can generate a diversity of incongruent area cladograms, owing to a complex history of dispersal, vicariance, extinction, and speciation

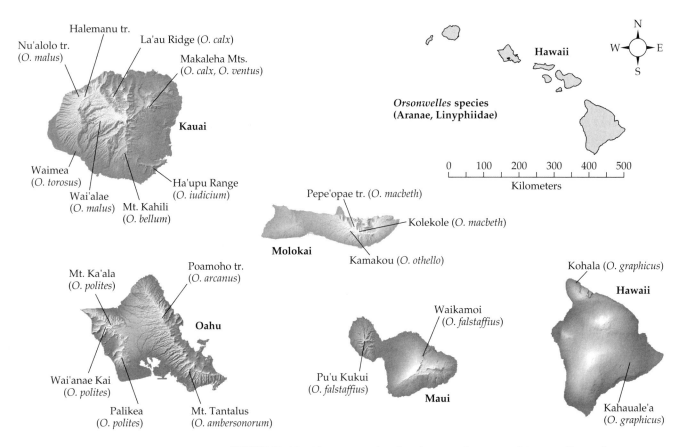

FIGURE 12.13 The geographic distribution of species of *Orsonwelles* spiders across the Hawaiian Islands. Although phylogenetic reconstructions suggest that a number of speciation events occurred within a single island, there is only one case of a completely sympatric distribution (*O. calx* and *O. ventus* in the Makaleha Mountains of Kauai). Even here, the phylogeny is indecisive about whether this is a result of a sympatric mode of speciation (followed by dispersal and divergence between *O. ventus* and the ancestor of *O. bellum* + *O. ludicium*), or of dispersal in the opposite direction resulting in secondary sympatry. Because of the allopatric distributions of all other species on different mountains, a microallopatric speciation mode is generally more likely than sympatric speciation. (After Hormiga et al. 2003.)

within and among areas. Yet, even if we found perfect congruence across a set of area cladograms, leading us to the provisional conclusion that all lineages shared a single history of simple vicariance, we may still have arrived at the wrong conclusion. Comparisons of area cladograms such as those in Figure 12.14 may appear to be relatively straightforward, but only for those clades that diversified at roughly the same times (i.e., the left side of Figure 12.14). Either the divergence events are both topologically and temporally congruent (upper-left box), indicating that the two lineages share a single biogeographic history; or they are geographically incongruent (lower-left box) indicating that they do not share the same history. However, the patterns shown in the boxes in the right of this figure (i.e., those for clades that diversified at different times) are likely to lead us to the wrong conclusions, unless of course our approach explicitly incorporates this asynchrony in clade diversification. Note that the two clades in the upper-right comparison appear to exhibit perfect topological congruence, yet this is an artifact of not incorporating differences in divergence times for these lineages. Without accurate information on tem-

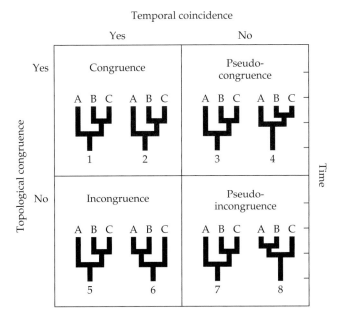

FIGURE 12.14 Four hypothetical sets of area cladograms for two lineages distributed across three areas (A, B, and C) in each comparison. See text for discussion. (After Donoghue and Moore 2003.)

poral coincidence of different clades, such patterns of cryptic biogeographic incongruence (called **pseudo-congruence**; Cunningham and Collins 1994) may generate erroneous conclusions about colonization history and evolution of these lineages (Hunn and Upchurch 2001). On the other hand, as more and more molecular phylogenies incorporate robust estimates of divergence times (see the discussion on molecular clocks in Chapter 11), we may discover that pseudo-congruent patterns are relatively common in parts of the world that have experienced temporally layered cycles of formation and erosion of dispersal barriers—first isolating, then allowing movement of a succession of lineages between the same set of areas (Donoghue and Moore 2003; see Figure 12.15).

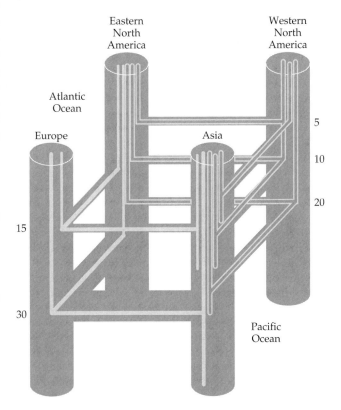

FIGURE 12.15 A depiction of area cladograms that summarizes historical tracks inferred from phylogenetic analyses of seven plant lineages distributed across four Northern Hemisphere areas of endemism. Each track traces one of two postulated intercontinental dispersal routes: either a Beringian route across the Pacific Ocean—with evidence presented here of having dispersal routes forming at three different timeframes (numbers are in millions of years); or a North Atlantic route—with two different timeframes for dispersal shown here. The temporal component of this complex biogeographic history was inferred by estimating divergence times from molecular phylogenies, and demonstrates "pseudo-congruence" embedded within topologically congruent sets of area cladograms. (After Donoghue and Moore 2003.)

RETICULATE AREA RELATIONSHIPS: THE END RESULT OF COMPLEX HISTORIES. As we continue to develop new tools to reconstruct the history of biotas—including molecular-based phylogenies that can provide estimates of the timing of historical vicariance and dispersal events, improved reconstructions of geological histories, and newer and more sophisticated analytical approaches (see examples in next section), we are approaching an empirically broad-based, new synthesis—one that holds that biotic histories are often complex rather than simple. It is becoming increasingly more clear that two or more co-distributed lineages often have sister taxa in an array of different areas (indicating a variety of dispersal as well as vicariant histories in the assemblage of that biota), or even in the same area (suggesting sympatric or embedded vicariant speciation within an area). Thus, even when the phylogenies of two co-distributed lineages suggest a shared history of area relationships (see Figure 12.8), we should use caution in concluding this to be the true history unless we have estimates of the timeframe of isolation and divergence.

Next, we explore in greater detail two distinctly different methods developed to reconstruct complex biogeographic histories—**primary BPA** and **secondary BPA** versus dispersal-vicariance analysis—and then return to where we left off in Chapter 11 in our discussion of phylogeography.

Two Approaches to Unraveling Reticulate Area Relationships

Within the past decade, there have been a number of creative attempts—several mentioned above and listed in Table 12.2—to resolve conceptual and methodological issues that became apparent in response to the strict form of vicariance biogeography advocated by the American Museum group. Our intent is not to provide an exhaustive review of these approaches and methods (for that, see Crisci et al. 2003) but, rather, to compare two innovative but very different approaches that are currently being employed by historical biogeographers. Indeed, even as we write this, new methods continue to appear so frequently that one of those considered in detail here (BPA) is already being replaced by a conceptually related method called **parsimony analysis for comparing trees** (PACT; Wojcicki and Brooks 2004, 2005). Even so, the logic behind the approach is unchanged and can be explored by looking at the BPA implementation.

Brooks parsimony analysis shares a goal with the methods listed in Table 12.2 under "cladistic biogeography": of producing a general area cladogram to describe the vicariant or geo-dispersal "backbone" in the biotic relationships among areas of endemism (see Lieberman 2003, 2004). BPA parts ways with these methods, however, in providing insights on the unique, lineage-specific patterns that arise from processes such as post-vicariance dispersal among areas, peripheral isolates speciation, sympatric speciation, and extinction (van Veller and Brooks 2001; Brooks and McLennan 2002; van Veller et al. 2003; Brooks 2004). It does so by implementing a two-step procedure: primary BPA and secondary BPA (Box 12.3). The end result is a general area cladogram that contains, in addition to the general area cladogram produced through primary BPA analysis, some number of duplicated areas generated through secondary BPA that describe those area relationships that conflicted with the general area cladogram.

In contrast, dispersal-vicariance analysis departs from BPA and the cladistic biogeography methods by not starting with a general area cladogram (Ronquist 1997). Indeed, even in a "worst-case scenario" of reticulated areas with no discernable hierarchical relationships (Figure 12.16), DIVA can be used to estimate episodes of vicariance and dispersal. Although taxon distributions among areas of endemism are still used, as are phylogenies, each taxon-area

BOX 12.3 *Primary and secondary Brooks parsimony analysis*

▨▮▮ Brooks parsimony analysis starts, as with any of the cladistic biogeographic methods, with a phylogeny for each lineage. We do not show the details of the phylogenetic analysis here and assume that it is an accurate depiction of relationships within each lineage. The next step is to summarize the distributions of each taxon within a set of areas of endemism.

Below, we show hypothetical phylogenies for three genera (*A.,* *B., C.*), each containing four species, and their distributions in areas *W, X, Y,* and *Z.* This is a highly simplified example, where each species occupies only one area, all areas are represented only once on each phylogeny, and the topologies of the three taxon-area cladograms are all perfectly congruent. As such, the general area cladogram summarizes all of the information contained in each separate taxon-area cladogram, with no incongruence, an hypothesis of a simple vicariant history. The numbers on each taxon phylogeny represent "characters" coding the phylogenies for the *primary BPA.* Once each phylogeny is coded as such, they are all compiled into a single binary matrix (1 = character present; 0 = character absent; ? = taxon absent from area [does not apply in this example]):

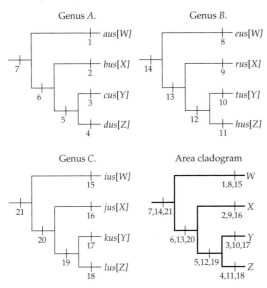

Area	Characters
	2
	1 2 3 4 5 6 7 8 9 . . . 1
H	0 0 0 0 0 0 0 0 0 . . . 0
W	1 0 0 0 0 0 1 1 0 . . . 1
X	0 1 0 0 0 1 1 0 1 . . . 1
Y	0 0 1 0 1 1 1 0 0 . . . 1
Z	0 0 0 1 1 1 1 0 0 . . . 1

Source: After Bouchard and Brooks 2004.

This matrix includes an additional "hypothetical area" *H,* which includes all zeroes, and which is a methodological necessity in order to root the tree with an outgroup for phylogenetic analy-

sis—the next step in the procedure. If all character-state changes for all taxon-area cladograms conform to a single general area cladogram with no incongruence—that is, no *homoplasy* (see Chapter 11)—then, the analysis is completed.

However, this will rarely be the case in real biotas with complex biogeographic histories. *Secondary BPA* was therefore developed as an approach to resolving homoplasy on the general area cladogram produced in the primary BPA. Suppose we add a fourth lineage, genus *D.,* to the biogeographic analysis, and that it matches closely the pattern of area-relationships shown by all of the first three genera, but it is not quite a perfect match. Perhaps *D. nus* crossed a barrier between areas *X* and *W* and speciated (*peripheral isolate speciation,* taxon-area cladogram below):

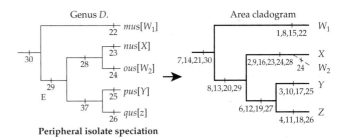

Peripheral isolate speciation

In this case, a new binary matrix could be constructed that treats area W, which now has a component that conflicts with, and will produce homoplasy in, the general area cladogram shown above, as *two separate areas,* W_1 and W_2. This is called the *area duplication convention,* which simply states that areas causing homoplasy on the general area cladogram are to be duplicated until the homoplasy is removed (explained more formally in Brooks and McLennan 2002; Brooks 2004). Parsimony analysis of this new matrix will produce the new general area cladogram shown above on the right. Note that it preserves the information suggesting a "vicariant backbone" to the historical relationships among areas, but now also postulates a subsequent peripheral isolate speciation event between areas *X* and duplicated area *W* (W_2).

As we discussed in the text, several other processes can lead to incongruence between any one of the taxon-area cladograms and the general area cladogram. Below, we show the case of a *sympatric speciation* event in area *X,* resulting in the sister species of *D.* *nus* being *D. ous.* This event does not result in itself in incongruence with the general area cladogram, although note the addition of characters on the branch leading to *X*:

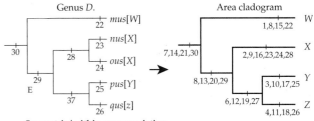

Sympatric/within-area speciation

BOX 12.3   *Primary and secondary Brooks parsimony analysis*   *(continued)*

Below, the diagram shows the effect of a *post-speciation dispersal* event of *D. nus* from area *X* to area *Z*, which now contains two species, the endemic *D. pus* and the newer arrival *D. nus* (which is now a *widespread species*). The resulting general area cladogram after duplicating area *Z* is:

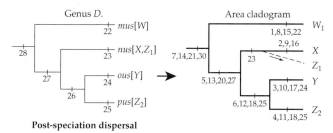

Post-speciation dispersal

Finally, the diagram below shows the extinction of *D. ous* from area *Y*, and the resulting general area cladogram:

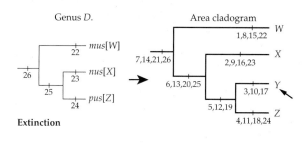

Extinction

---

cladogram is analyzed separately to develop a "taxon biogeography." The idea here is to establish a "cost" for each of several kinds of biogeographic "events"—the more likely the event is considered to be, the lower the cost assigned. The basic assumption used to build the cost matrix is that most speciation is allopatric and produced through vicariance. Thus, costs are usually assigned as follows: two kinds of events—vicariance and sympatric (or embedded allopatric) speciation events within a single area are assigned a cost of zero; whereas two other kinds of events—extinction and random dispersal—are assigned a cost of 1. The algorithm searches for the distribution of ancestral areas at each node on the tree that minimizes the overall cost of events. An example of a DIVA search for the ancestral area of a currently widespread and diverse taxon is shown in Figure 12.17.

While DIVA analyzes each taxon-area cladogram separately, results from each analysis can be evaluated collectively to identify any general, shared patterns. As such, there are cases in which either DIVA or BPA could be a method of choice to use in an "area biogeography" study. More generally, several recent studies have considered the relative performance of different methods on a given dataset (e.g., Crisci et al. 2003 compared ancestral areas analysis, reconciled trees, DIVA, component analysis, and panbiogeography to reconstruct the history of the southern beeches, *Nothofagus*, on Gondwanan continents).

Phylogeography, Again

In Chapter 11, we introduced the general approach of phylogeography, which has revolutionized "temporally shallow" historical biogeography by providing us with tools to reconstruct phylogenetic and population genetic structure across gene lineages within species and among closely related species. How-

FIGURE 12.16   A "worst-case scenario" of reticulate area relationships. Four Holarctic deciduous forest areas are recognized: WN = western Nearctic; EN = eastern Nearctic; WP = western Palearctic; EP = eastern Palearctic. These areas correspond, in order, to WNA, ENA, ER, and EA in Figure 12.24. Prior to time 1 (t_1), two ancestral areas connecting regions between current continents were recognized based on paleogeographic reconstructions. At t_1, each of these areas is fragmented by a vicariance event producing four separate areas. Subsequently, at t_2, geological events remove dispersal barriers between each of two sets of areas, but now, areas are connected within, rather than between, continents. Ronquist (1997) argued that DIVA is more appropriate than one of the hierarchical approaches to reconstructing biogeographic histories in such cases. (After Ronquist 1997; originally described in Enghoff 1996.)

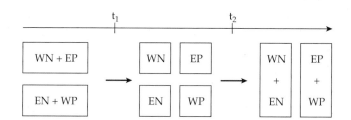

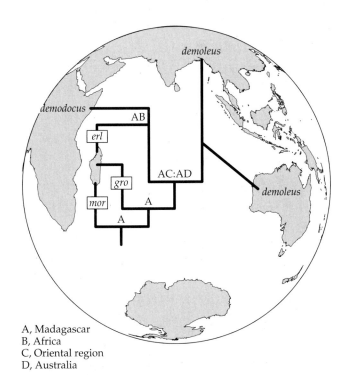

A, Madagascar
B, Africa
C, Oriental region
D, Australia

FIGURE 12.17 An example of a dispersal-vicariance analysis used to infer the ancestral area for lime swallowtails, *Papilio demoleus* group, including species: *demodecus, demoleus, erithonioides (erl), grosesmithi: (gro)*, and *morondavana (mor)*. Currently, populations of this group are distributed across landmasses bordering the Indian Ocean, but with highest species diversity on Madagascar. This DIVA analysis of a molecular phylogeny for the mtDNA cytochrome oxidase I and II genes reconstructed Madagascar as the most likely area of origin for the group, and suggested several independent dispersal events: one underlying speciation between a Madagascar species and one in Africa; the other between a different Madagascar species and one that is widespread in the Oriental region and Australia. This result was also produced through analysis of phylogenies based on morphology and two different nuclear DNA genes. (After Zakharov et al. 2004.)

ever, while we discussed some of the unique features that allow us to reconstruct lineage histories that trace back to relatively recent times—within the most recent thousands to several millions of years of Earth's history—we did not elaborate on the role of phylogeography in reconstructing biogeographic histories, and so pick up the discussion at that point.

RECONSTRUCTING SHALLOW BIOGEOGRAPHIC HISTORIES. The most revolutionary aspect of phylogeography is the way it has brought the scope of historical biogeography into a timeframe spanning the past several thousands to several millions of years, a period that experienced tremendous climatic oscillations with dramatic changes in the locations and composition of biotic assemblages (see Chapter 9). Not long ago, our best clues to how particular species and biotas responded to this dramatic climatic forcing came almost exclusively from fossils (e.g., pollen deposition in lakes, woodrat midden contents, and hard parts of animals and plants in sediments). While the importance of fossils has not faded, we now can employ phylogeographic analyses as a complementary—and in some situations, superior—form of discovery about the effect of Pleistocene paleoclimatic cycles on distributions of species and biotas. Moreover, we can address the challenging subject of the timeframes and geographic context of speciation during the past several millions of years, including an assessment of the importance of the Pleistocene on the formation of extant species (Klicka and Zink 1997, 1999; Johnson and Cicero 2004; Weir and Schluter 2004; Zink et al. 2004; Lovette 2005). In addition, we now can examine phylogeographic structure across two or more co-distributed species (using the approach called **comparative phylogeography**; Bermingham and Moritz 1998; Arbogast and Kenagy 2001) to assess the relative contributions of vicariance and dispersal to the assembly of continental, island, and marine biotas.

FIGURE 12.18 A mtDNA gene tree for the red-tailed chipmunk (*Neotamias ruficaudus*). This phylogeny sorts intraspecific genetic variation (mtDNA haplotypes labeled at tips of branches) into two well-supported clades (Eastern and Western) that delineate geographically distinct portions of the species' distribution (see Figure 12.19B, panel 1). Numbers along branches refer to several common measures of support for the robustness of a clade (see original reference for description); unlabelled branches represent outgroups used to root the tree. The key in the upper-left box refers to nested clades derived through a nested clade phylogeographic analysis (see Figure 12.19A and description in Box 12.4), with shading corresponding to clade distributions shown in Figure 12.19B, panels 2 and 3). (After Good and Sullivan 2001.)

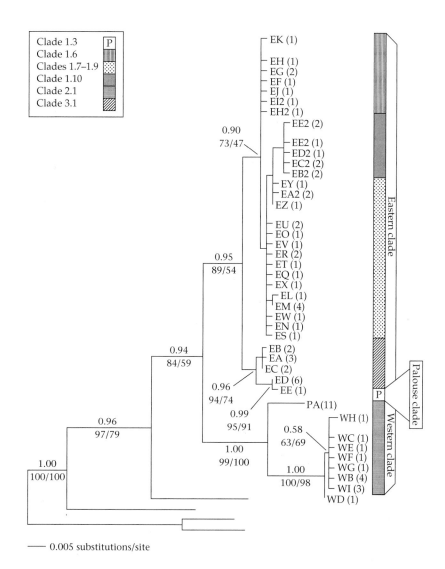

Earlier in this chapter, when we compared different approaches to addressing complex biogeographic histories, we emphasized the rapid accumulation of new and innovative methods. This point is no less true for phylogeography. Between the inception of phylogeography in 1987 (Avise et al. 1987) and its maturation into a widely recognized subdiscipline of biogeography (Avise 2000), phylogeography was biased toward the description of biogeographic patterns based on mtDNA gene trees in animals (and fewer but important cpDNA studies for plants). When gene trees for populations form discrete, reciprocally monophyletic clades that are genetically distinct and geographically disjunct, it is often quite clear that populations within each of the separate clades have had long histories of allopatric isolation and divergence (see, for example, the "coastal" vs. "inland" clades of the tailed frog in Figure 11.13; but see also Irwin [2002] for an interesting alternative view).

Now look at the phylogeny for the red-tailed chipmunk (*Neotamias ruficaudus*), a small mammal that occurs in the Pacific Northwest region of North America (Figure 12.18). Without undue controversy, we might infer through visual inspection of the tree that populations in the Western Clade and Eastern Clade have very likely experienced a history of allopatric isolation and divergence of sufficient duration to result in the clearly reciprocally monophyletic

(A)

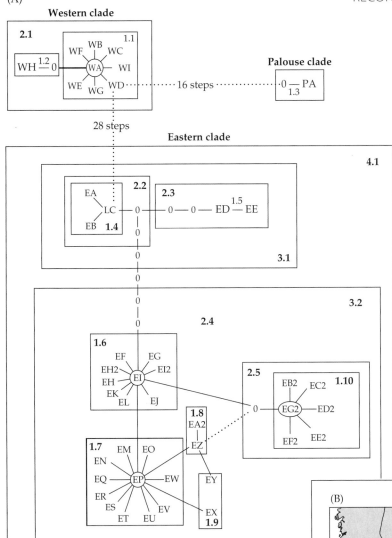

FIGURE 12.19 (A) A nested clade phylogeographic analysis created from a mtDNA haplotype network for the red-tailed chipmunk (*Neotamias ruficaudus*). The procedure for converting a haplotype network into a series of nested clades is described briefly in Box 12.4, and in more detail elsewhere (e.g., Avise 2000). Each branch connecting haplotypes represents a single mutation step; zeroes represent unsampled haplotypes, inserted in order to make connections between haplotypes that differ by more than a single mutation step. Compare with rooted phylogeny in Figure 12.18 and note that the Western and Palouse clades are too divergent from the Eastern clade and from each other to be nested into a single box by the nesting algorithm. (B) Geographic distributions for (1) the species; (2) sampling localities (with frequencies of nested clades at each location shown); and (3) nested clades (showing the largely complete separation of the Western and Palouse clades from those that are nested within the Eastern clade. (After Good and Sullivan 2001.)

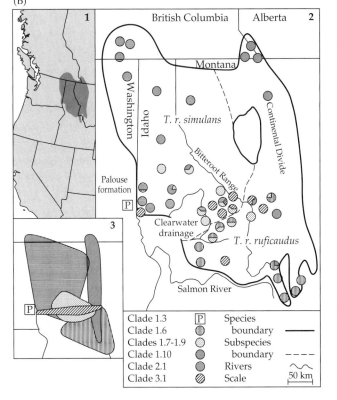

clades. Now, as we work our way up closer to the tips of the phylogeny, we still find several well-supported clades embedded within either the Western or Eastern clades, but the times of divergence and degree of separation between these clades becomes very shallow and their geographic separation tends to be less distinct (Figure 12.19B). We have two choices at this point. On one hand, we could conclude that little more can be inferred about geographic structure in this species once we get beyond the inference of historical isolation of the Western and Eastern clades. On the other hand, we can ask whether another approach might be available that would allow us to assess, first, whether there is a significant, non-random association of genetic variation and geography and, if there is, to determine what historical processes—allopatric isolation, recent range expansion, or simply a relationship between spatial proximity of isolated populations and rates of gene flow among them—might have generated the association. Increasingly, phylogeographers are finding innovative ways to be able to choose the second alternative within the emerging framework of **statistical phylogeography** (Box 12.4).

BOX 12.4 *Statistical phylogeography*

▓▓▓ The example given in this chapter for the red-tailed chipmunk (see Figures 12.18, 12.19) illustrates a situation that has motivated phylogeographers to call for—and develop—a rigorous statistical framework for phylogeographic research, and in many cases, to frame questions within explicitly stated alternative hypotheses about the geographic history of populations. The development of statistical phylogeography

includes two complementary, yet conceptually different, approaches (Knowles and Maddison 2002; Knowles 2004; Templeton 2004; Carstens et al. 2005). The first approach, developed in a number of papers over the past decade by Alan Templeton, Keith Crandall, and their co-workers, is called **nested clade analysis** (NCA; Templeton et al. 1995), or more recently, **nested clade phylogeographic analysis** (NCPA;

Templeton 2002). This method begins by building an *unrooted statistical phylogenetic network among haplotypes* (i.e., a haplotype network, which is similar to the gene tree phylogenies shown in Figures 11.13 and 12.18, but without a "root" and, therefore, no outgroup; see Figure 11.14). A statistical approach, such as that implemented in the program TCS (Clement et al. 2000), is used to determine the most likely relationships

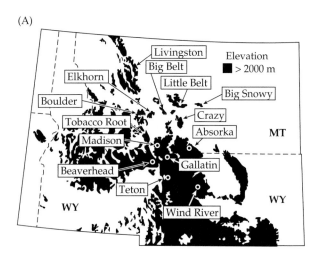

(A)

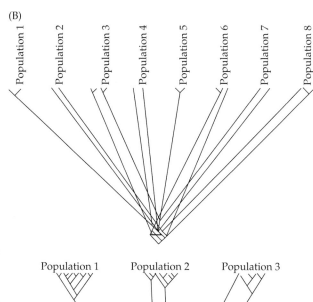

(B)

An example of an a priori approach to statistical phylogeography. (A) Sampling localities in the Northern Rocky Mountain montane forest 'sky island' archipelago for the grasshopper species *Melanoplus oregonensis*. Some of this area was glaciated during the late Pleistocene, and the species could have recolonized areas that were formerly under ice from either a single or multiple refugia. Alternative historical scenarios can be developed into two alternative hypotheses (B) to explain the distribution of genetic diversity within and among populations (see original reference for details). If populations were recolonized from a single ancestral source population, any of the populations might be expected to contain a set of genotypes drawn at random from the overall diversity within the species, as shown in the top tree. Alternatively, if post-Pleistocene recolonization came from more than one late-Pleistocene refugial population, ancestral lineage sorting between the refugial populations prior to establishment of post-Pleistocene geographic distributions would result in the pattern shown in the bottom figure. (After Knowles 2001.)

BOX 12.4 *(continued)*

among different haplotypes in the network. The next step is to convert this phylogenetic network into a hierarchical series of nested clades (see Figure 12.19), generally using nesting rules within the program GeoDis (Posada et al. 2000). For each nested clade (e.g., in Figure 12.19, clades in one perfectly nested series are labeled 4.1, 3.1, 2.2, and 1.4), the program addresses the "null hypothesis" of no statistically meaningful association between geography and the history of populations within a particular clade. Possible reasons underlying a statistically supported alternative to the null hypothesis of no association include allopatric fragmentation, restricted gene flow resulting from isolation by distance, and recent range expansions—all evaluated a posteriori using an inference key, with the most recent version in Templeton (2004).

A second approach to statistical phylogeography differs from NCPA by developing plausible alternative hypotheses a priori, and then using various coalescent-based methods (see Table below) to statistically address the probabilities of the alternative models (see Figure opposite page). Several phylogeographers who favor this approach also argue that a modeling approach should be used to determine the robustness of support for one or another of the alternative models (e.g., using the Mesquite program; Maddison and Maddison 2000). Another argument made by investigators favoring use of coalescent-based methods is that phylogeographers working with groups that have particularly shallow divergence times need to worry about the stochastic nature of the sorting of ancestral genetic polymorphism within and among recently diverged populations (see Figures 11.10, 11.15), and many methods, including those listed in the table below, do this.

The term "statistical phylogeography" has emerged only within the past few years and so its recognition as an approach distinct from an "older" form of phylogeography is relatively recent. Therefore, it is hard to say at this time what this newer approach will look like five or ten years from now, but we anticipate a rapid growth of innovative methods to address an historical biogeography of the very recent past and of rapidly evolving organisms and geographic structures (e.g., emerging infectious diseases that are transported by modern means of travel from "centers of origin" to far reaches of the Earth). ▌▐ ■

TABLE BOX 12.4 *Examples of programs and biogeographic utility of coalescent-based statistical phylogeography*

Program	Data type	Biogeographic utility	Source	Reference
FLUCTUATE	Sequence	Estimate exponential population expansion or decline	http://evolution.genetics washington.edu/lamarc.html	Kuhner et al. 1998
MIGRATE	Sequence, microsatellites	Estimate migration between subpopulations	http://evolution.genetics washington.edu/lamarc.html	Beerli and Felsenstein 1999
GENETREE	Sequence	Migration and growth rates in structured populations	http://www.stats.ox.ac.uk/ mathgen/griff/software.html	Griffiths and Tavaré 1994
BATWING	SNPs, microsatellites	Model population divergence, size, growth, and mutation rates	http://www/maths.abdn.ac.uk/ ijw/downloads/download.htm	Wilson et al. 2003
MDIV		Estimate divergence times and migration rates between two populations; test divergence with gene flow models	http://www.biom.cornell.edu/ Homepages/Rasmus_Nielsen/ files/html	Nielsen and Wakeley 2001
BEAST	Sequence	Estimate divergence dates, and population size and growth	http://evolve.zoo.ox.ac.uk/beast/	Drummond and Rambaut 2003
STRUCTURE	Microsatellites, RFLP's, SNPs	Infer population structure, testing models of admixture	http://pritch.bsd.uchicago.edu/	Pritchard et al. 2000
MESQUITE	Sequence	Test models of population divergence, including variance, fragmentation, and isolation by distance	http://mesquiteproject.org	Maddison and Maddison 2000

Source: Adapted from Knowles, 2003.

COMPARATIVE PHYLOGEOGRAPHY. Another aspect to phylogeography that has experienced an accelerating growth in popularity, although the term was beginning to be used nearly a decade ago (Zink 1996; Bermingham and Moritz 1998), is "comparative phylogeography," which describes the explicit comparison of phylogeographic structure across two or more co-distributed lineages (Arbogast and Kenagy 2001). Reasons to do so are essentially those that we described previously, and can be traced at least as far back as the 1970s and the American Museum group of vicariance biogeographers. As they asserted, if we are interested in revealing the general historical patterns in a biota—those that arise mainly through vicariance or geo-dispersal—we must examine more than a single co-distributed lineage for a signal of congruence. This is true whether the lineages are higher taxa, species, or gene trees. As such, one implementation of a comparative approach in phylogeography is really nothing more than co-opting the logical framework of cladistic biogeography (Figure 12.20) or phylogenetic biogeography II (Riddle and Hafner 2004).

While the goals of comparative phylogeography and phylogenetic biogeography II are broadly overlapping, each nevertheless brings a different, but complementary, set of strengths (and weaknesses) to an historical biogeographic study. The former approach offers strengths that derive from the sampling and analytical methods that make phylogeography a powerful means of examining relatively shallow and intraspecific histories, including the alternative histories of allopatric fragmentation versus range expansion. On the other hand, the latter approach offers the means to sort general from individualistic events in the vicariance, dispersal, and extinction histories of co-distributed species (through primary and secondary BPA and, in the near future, using PACT). Riddle and Hafner (2004) argued that the tree generated through BPA analysis could be treated as a set of hypotheses about events such as vicari-

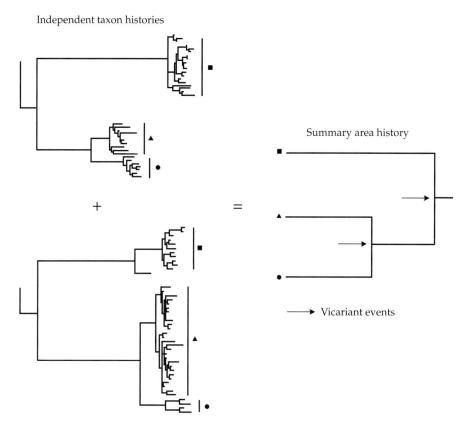

FIGURE 12.20 A depiction of the relationship between comparative phylogeography and historical biogeography, in which phylogroups with taxon phylogenies can be used in the same fashion as species or higher taxonomic units traditionally have, in order to postulate area histories and vicariant events (arrows). Each symbol in the taxon phylogenies depicts a reciprocally monophyletic set of alleles (haplotypes) occurring with, and among, a set of populations. (After Riddle and Hafner 2004.)

ance, jump dispersal, and range expansion, allowing an investigator to return from a phylogenetic biogeographic to phylogeographic framework in order to further test those hypotheses. On the other hand, comparative approaches in phylogeography are also being employed even when the phylogenetic tree is down-weighted in favor of one or more of the statistical population genetic approaches discussed above (Wares 2002; Lourie et al. 2005).

FINAL THOUGHTS ON PHYLOGEOGRAPHY WITHIN HISTORICAL BIOGEOGRAPHY. Phylogeography has enjoyed an explosive rate of growth over the past decade and continues to increase in popularity (19 hits in an ISI Web of Science search using topic terms "phylogeography or phylogeographic" for 1994, vs. 470 for 2004). Clearly, this approach is here to stay and has become a dominant force in modern historical biogeography. A major reason for the growing popularity of phylogeography is that it offers the potential to attract researchers from disparate disciplines that traditionally focused on a particular question using different approaches, perhaps without even acknowledging the overlap with other disciplines and approaches. For example, Wares (2002) demonstrated how a phylogeographic approach could be employed to address biotic and abiotic processes driving the assembly of a marine intertidal community in the Atlantic Ocean.

Because phylogeography lies temporally at the intersection between recent ecological and historical processes, it can also be used to address the relative roles of geological events, in situ diversification, colonization, and extinction in the biogeography of insular biotas (Heaney et al. 2005). Phylogeography also provides a valuable tool for conservation biologists, enabling them to objectively identify evolutionarily significant units (ESUs) and the geographic processes that underlie their histories and those of other populations, species, and biotas of high conservation priority.

The lesson for the current—and, perhaps more importantly, the next generation of—biogeographers seems clear. Regardless of whether they are most concerned with developing new theory in biogeography, or applying those lessons to conserving the distributions and diversities of native biotas, students should learn as much as they can about the rich tapestry that is modern historical biogeography. It is very apparent by now that historical biogeography is at the beginning of a major renaissance (e.g., Donoghue and Moore 2003; Riddle and Hafner 2004; Riddle 2005; Wiens and Donoghue 2005). Who can say what it will become in five, ten, or even twenty years? The only thing we can predict with some confidence is that students will enjoy a distinct advantage by having an understanding of the current array of methods and emerging approaches for reconstructing the phylogeography of lineages and biotas.

What Are We Learning about Biotic Histories?

This chapter began with an extended discussion of historical biogeography during the early- to mid-twentieth century. This period was dominated by a tradition of using one or more criteria (see Table 12.1) to "locate" a center of origin, generally in the Northern Hemisphere, and to propose scenarios for the dispersal of species away from that center, sometimes including untested notions of waves of derived species supplanting more primitive forms as they advanced out of the center. The death knell for this approach was two-pronged—Croizat's panbiogeography, with its emphasis on discovering the general patterns of distribution on one hand; and Hennig's phylogenetic methods, with its insistence on discovering monophyletic groups and the ancestor-descendant cladogenetic sequence, on the other hand. Brundin applied both methods in his classic study in the late 1960s and interpreted his results with the benefit of the recent revelations of plate tectonic theory. The 1970s saw the remolding of phylogenetics and track analysis into a form of biogeography that narrowed the field to a search for the general vicariant backbone shared by a set of taxon-area cladograms.

While most of today's historical biogeographers find this adherence to "vicariance only" unnecessarily and unrealistically narrow, vicariance biogeography provided a conceptual and methodological foundation for many of the approaches we use today, including those that incorporate methods to estimate dispersal, sympatric speciation, and extinction (see Table 12.2). Finally, from the arenas of molecular evolution and population biology, phylogeography emerged two decades ago and continues to mature into a remarkably popular aspect of modern historical biogeography.

Clearly, historical biogeography has experienced a series of important transformations over the past half century. Yet, until relatively recently, precious few data sets were available for addressing the history of terrestrial and marine biotas, and a good deal of the effort in historical biogeography focused on the "performance" of different approaches using exemplar data sets; most notable among these was Donn E. Rosen's (1978, 1979) poeciliid fish genera, *Heterandria* and *Xiphophorus*, from the uplands of Guatemala. Fortunately, all this changed with key technological advances of the past decade, including the increasing ease of obtaining DNA sequence and other forms of molecular data; the analytical power of sophisticated phylogenetic, population genetic, and biogeographic algorithms; and the availability of data from multiple co-distributed taxa, providing opportunities for exactly the kinds of comparative investigations required to sort general from individualistic biogeographic histories. These breakthroughs have greatly enhanced the analyses of biotic histories in both terrestrial and marine systems—covering a wide range of "deep" as well as "shallow" timeframes, and comparisons of biotas within, as well as among, the continents and island archipelagoes as well. The burgeoning number of publications from these studies is both encouraging and sometimes daunting, with the number and sophistication of publications increasing each year (e.g., see recent issues of these and other journals: Evolution, Journal of Biogeography, Molecular Ecology, Molecular Phylogenetics and Evolution, Biological Journal of the Linnean Society, Proceedings of the Royal Society of London, and Systematic Biology).

Earlier in this chapter and elsewhere in this book (Chapters 7 and 11) we featured a variety of examples of modern, molecular-based, biogeographic analyses of either single lineages or multiple co-distributed taxa (e.g., for the Hawaiian Archipelago). Here, we highlight a handful of intriguing studies that integrate information from a number of co-distributed taxa and demonstrate how modern historical biogeography is poised to produce synthetic and, in many cases, perhaps surprising insights about the histories of biotas.

Biotic Histories in Gondwanaland

From the beginning of modern historical biogeography, the plate tectonics model gave biogeographers one very clear exemplar system that should demonstrate a history of vicariance—the breakup of the continent of Gondwanaland. The timing and sequence of fragmentation of landmasses from the ancient Gondwanan continent is well known and provides for the construction of a geological area cladogram that offers explicit predictions about the topology of taxon-area cladograms for lineages that diversified in accordance with a vicariance model (see first Figure in Box 12.1). These lineages would have included the ratite and allied birds in the subclass Paleognathae, the chironomid midges studied by Brundin, the southern beeches (*Nothofagus*), and a number of lineages of fishes and reptiles.

What evidence could we use to reject vicariance in favor of dispersal for any of these Gondwanan lineages? First and most obvious, taxon-area cladograms that are incongruent with the geological area cladogram would provide a reason to reject vicariance in favor of dispersal. Second, as we discussed in Chapter 11, molecular data could be used to estimate the absolute and relative times of divergence, and would provide a strong argument against vicariance if an estimated divergence time was younger than the time of area fragmentation.

The results of a number of recent analyses are pointing to a surprising result—that transoceanic dispersal has played a far greater role in the biogeographic history of the Southern Hemisphere than had been predicted from the Gondwanan vicariance model. de Queiroz (2005) summarized many examples of disjunctions of a broad range of organisms, including primates, chameleons, frogs, and many genera of plants, distributed among landmasses across the Earth, many of Gondwanan origin (Figure 12.21). In each case, the disjunct distributions between sister taxa were interpreted as products of transoceanic dispersal, based on incongruence between molecular-based estimates of divergence times and geological estimates of the ages of fragmentation of landmasses. In another study, Sanmartín and Ronquist (2004) used a large data set, including 54 animal (insects, fish, reptiles, and mammals) cladograms and 19 plant taxon-area cladograms in a parsimony-based tree fitting analysis. Their analyses indicated that, overall, animal distributions are more

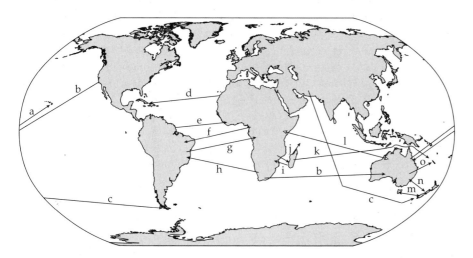

a *Scaevola* (Angiospermae: Goodeniaceae, three episodes of dispersal)
b *Lepidium* (Angiospermae: Brassicaceae)
c *Myosotis* (Angiospermae: Boraginaceae)
d *Tarentola* geckos from Africa to Cuba
e *Maschalocephalus* (Angiospermae: Rapateaceae)
f Monkeys (Platyrrhinii)
g Melastomes (Angiospermae: Melastomataceae)

h *Gossypium* (Angiospermae: Malvaceae)
i Chameleons, three episodes of dispersal
j Several frog genera
k *Acridocarpus* (Angospermae: Malpighiaceae)
l Baobab trees (Angiospermae: Bombacaceae)
m 200 plant species
n Many plant taxa
o *Nemuaron* (Angiospermae: Atherospermataceae)

FIGURE 12.21 Examples of trans-oceanic dispersal, derived mainly from recent molecular phylogenies with estimates of divergence time. The strongest case for dispersal rather than vicariance is made when the phylogeny suggests a divergence date between two lineages that is much younger than predicted from a geological area cladogram, such as that shown in Box 12.1 for Gondwana. Arrows on lines indicate direction of dispersal; a line with two filled arrows indicates bi-directional dispersal; and unfilled arrows indicate uncertainty about direction. (After de Queiroz 2005.)

FIGURE 12.22 Analysis of historical dispersal events between major land-masses of Gondwanaland, inferred from a parsimony-based analysis of (A) 54 animal, and (B) 19 plant cladograms. The width of the arrows is proportional to the frequency of a particular route (details of analysis provided in original reference). For the animals in (A), the thick arrows connecting Australia and southern South America via Antarctica, as well as the one connecting New Zealand and southern South America, are consistent with the vicariance model of area fragmentation (see first figure in Box 12.1), suggesting that "dispersal" between these areas occurred prior to the break-up of the ancient continent. However, one could argue the same thing for the high frequency of Madagascar and Africa dispersals, but many of these are now considered to have resulted from post-vicariance dispersal events (see discussion in text and Figure 12.23). For the plants in (B), the signal of transoceanic dispersal is stronger than for animals, particularly in the very high frequency of dispersal from Australia to New Zealand, clearly incongruent with the geological cladogram. Note also the weak connections between northern and southern South America, the latter having much stronger historical affinities with other southern landmasses. (After Sanmartín and Ronquist 2004.)

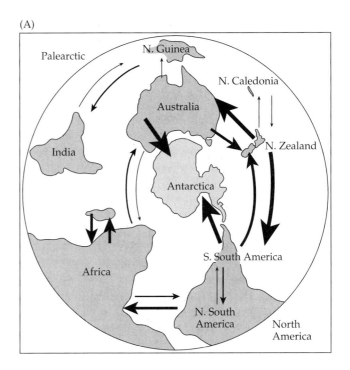

(A)

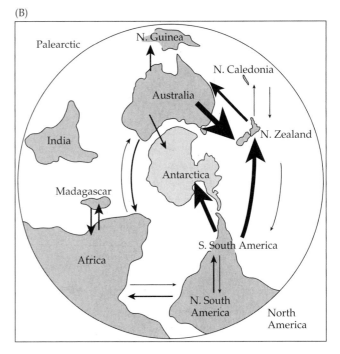

(B)

congruent with the fragmentation sequence of Gondwanaland than are those of plants (Figure 12.22). A dramatic case of incongruence in plants, for example, involves the modern flora of New Zealand, which may have originated in large part, if not in total, via long-distance dispersal following the near disappearance of exposed land in New Zealand during the Oligocene (37–23 million years B.P.; Pole 1994; Winkworth 2002). Generally, their results suggest that plants have dispersed more frequently and more recently than animals among landmasses in the Southern Hemisphere. But even in animals, with their better overall fit to a Gondwanan vicariance model (see Figure 12.22), several long-held presumptions about purely vicariant histories appear to be inconsistent with molecular estimates of divergence dates. For example, divergence dates between African and Malagasy chameleons (Raxworthy et al. 2002), frogs (Vences et al. 2003), plants (Renner 2004), primates (Yoder et al. 2004), and carnivores (Yoder et al. 2003; Figure 12.23) appear to be much younger than the geological estimate of about 120 million years B.P. for the separation of Africa from a Madascar-India landmass. This appears to be strong evidence for a history of multiple colonization events between Africa and Madagascar via sweepstakes dispersal, rather than vicariance.

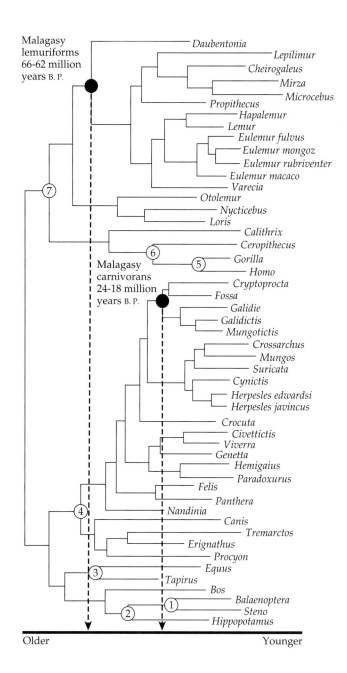

FIGURE 12.23 Molecular phylogeny comparing the ages of divergence of Malagasy primates (*Daubentonia*, the aye-aye; and a number of genera of lemurs) and carnivores (three genera of Malagasy 'mongooses'; *Fossa*, the Malagasy civet; and *Cryptoprocta*, the fossa). In each case, the Malagasy clade is monophyletic with a common ancestor at the black circle. Open circles with numbers are fossil-based calibration points used to estimate divergence times, which for the primates (66–62 million years B.P.) and carnivores (24–18 million years B.P.) post-date the geologically estimated time of separation of Africa and an ancestral Madagascar-India land mass (about 121 million years B.P.), and the separation of Madagascar from India (about 88 million years B.P.), suggesting colonization by ancestors of both clades by overwater "sweepstakes" dispersal. (After Yoder et al. 2003.)

We emphasize here the accumulating evidence of an important role for dispersal in the historical assembly of Southern Hemisphere biotas (McGlone 2005), but stress also that Sanmartín and Ronquist (2004) found congruence between some of their taxon-area cladograms and the geological cladogram, supporting vicariance as a component of biotic history as well. Interpreting the histories for any group of taxa can still be controversial (e.g., Briggs 2003; Sparks and Smith 2005).

We can mention two additional insights about the dynamic biogeography of Gondwanaland that have emerged from recent studies. The first provides substantial support for the reticulate nature of South America with Andean and southern parts of South America aligned historically with Australia and New Zealand, and northern (tropical) South America showing greater affinities to the Holarctic, and to some degree, Africa (see Sanmartín and Ronquist 2004; see Figure 12.22). This reticulated history formed the basis for Morrone (2002) subdividing South America into separate biogeographic regions (see Figure 10.16). Second, Gondwanaland appears to have played a surprisingly important role in the early diversification of a number of major groups of vertebrates previously thought to have originated in Laurasia, including neognathine birds (Cracraft 2001), ranid frogs (Bossuyt and Milinkovitch 2001; see Figure 7.11), placental mammals (Eizirik et al. 2001), and grasses (Bremer 2002).

Biotic Histories in the Holarctic

While the tectonic history of Gondwanaland can be summarized concisely into a geological area cladogram with a minimal number of area reticulations (first figure in Box 12.1), the geological history of connections and biotic interchange between Laurasian landmasses was much more complex throughout the Cenozoic. For example, although biogeographers have recognized only four broad areas of endemism for temperate deciduous forests—two in the Nearctic (eastern North America and western North America), and two in the Palearctic (Europe and eastern Asia; Figure 12.24A)—there has been a long history of debate about the historical sequence of connections between these areas. That these areas are likely to be highly reticulated is suggested by the inferred history of connections within and across the Holoarctic continents, summarized as follows:

1. Western and eastern Nearctic landmasses were separated by an epicontinental sea until the earliest Cenozoic epoch, the Paleocene (roughly 65 million years B.P.; see Chapter 8).

2. Multiple Beringian connections formed between the eastern Palearctic and western Nearctic landmasses during the Cenozoic (culminating in the important late-Pleistocene connections; see Chapter 9).

3. At least two Tertiary connections formed between the western Palearctic and eastern Nearctic across the North Atlantic (about 30 and 15 million years B.P., respectively).

The fossil evidence has previously been interpreted as demonstrating that ancient forests and taxa were widespread across Laurasia prior to its complex Cenozoic geological history, and that the current differences in species composition between areas is due primarily to extinction of ancestrally widespread taxa (Wolfe 1975; Tiffney 1985; Tiffney and Manchester 2001).

An increasing number of molecular phylogenetic data sets, many with estimates of divergence times, are becoming available for temperate deciduous forest plant and animal taxa distributed across the four recognized areas of

(A)

(B)

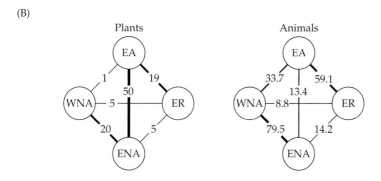

(C)

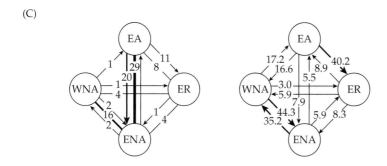

FIGURE 12.24 (A) Holarctic areas of endemism in plants and animals across eastern North America (ENA); western North America (WNA); Europe (ER); and Eastern Asia (EA). A comparison of disjunct patterns of distribution (B) and inferred ancestral areas and direction of movement (C), analyzed using animal data from Sanmartín et al. (2001). See text for discussion. (After Donoghue and Smith 2004.)

endemism. Donoghue and Smith (2004) used dispersal-vicariance analysis to compare taxon-area cladograms from 66 plant clades with the 57 animal clades analyzed by Sanmartín et al. (2001). As was the case in the Southern Hemisphere (Sanmartin and Ronquist 2004), the relative roles of dispersal and vicariance and patterns of dispersal among the Holarctic areas of endemism appear to differ between plants and animals, particularly in the historical relationship of eastern Asia and the Nearctic areas (as summarized in Figures 12.24B and 12.24C). That is, plants share a higher frequency of disjunct distributions of sister taxa between eastern Asia and eastern North America, and animals share more disjunction distributions between eastern Asia and western North America. Furthermore, again mirroring the Southern Hemisphere, there appears to be a higher frequency of more recent intercontinental dispersal events in plants than in animals in the Northern Hemisphere (but see Donoghue and Smith, 2004). Contrary to the "widespread ancient forest" model preferred by paleontologists, these and other studies support a history of multiple episodes of dispersal and vicariance between Palearctic and Nearctic areas during the Tertiary, primarily via a Beringian route (also supported in another study by estimated divergence dates on a molecular phylogeny for squirrels; Mercer and Roth 2003), but also to some degree via a Northern Atlantic route as well (see Figure 12.15).

Biotic Histories in, and Just Before, the Ice Ages

Phylogeographic—and particularly, comparative phylogeographic—studies have begun to reveal much about the responses of lineages and biotas to the dramatic climatic oscillations of the Pleistocene (Hewitt 2004). Yet, in a slightly expanded timeframe, many of the extant species and genera were also members of pre-Pleistocene, Pliocene, and Miocene biotas. We know that the Earth experienced dramatic geological and climatic changes during these epochs, including uplifting of mountains and plateaus, and closure of the Panamanian landbridge. Debate continues regarding the relative importance of Ice Ages versus earlier events on the origin of extant species and assembly of modern biotas (e.g., Johnson and Cicero 2004; Weir and Schluter 2004; Zink et al. 2004). Nevertheless, phylogeographic studies are clearly suggesting that the origination of many extant species and regional biotas date to pre-Pleistocene times, and are found in a wide range of biogeographic regions and biomes, including the tropical forests of northeastern Australia, central Africa, northern South America (Moritz et al. 2000), the Mexican Neovolcanic Plateau (Hulsey et al. 2003), the conifer forests of the Pacific Northwest in North America (Carstens et al. 2005), and the southwestern deserts of North America (Riddle et al. 2000; Zink et al. 2000; Riddle and Hafner, in press).

Within the Pleistocene, comparative phylogeography has provided interesting insights on the temporal cohesiveness of biotas across one or more glacial-interglacial climatic oscillations (reviewed by Hewitt 2004). For example, in the western Palearctic, biotas in Europe appear to have responded as cohesive subsets of taxa whose ranges retreated during glacial periods to one or more southern, unglaciated refugia (e.g. Iberian, Italian, Balkan), followed by northward range expansions following retreat of the glaciers (Figure 12.25). Often,

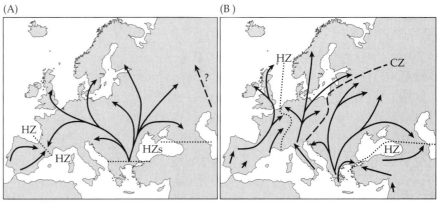

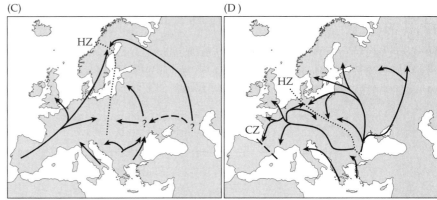

FIGURE 12.25 Four "paradigms" of postglacial colonization from Late Pleistocene southern refugia in the Palearctic as inferred from mtDNA phylogeographic studies. CZ and HZ are contact zones and hybrid zones, respectively, between lineages expanding from different refugia. The exemplars representing each of the four patterns here are for (A) the grasshopper (*Chorthippus parallelus*); (B) the hedgehog (*Erinaceus europaeus/concolor*); (C) the brown bear (*Ursus arctos*); and (D) the chub (*Leuciscus cephalus*). (After Hewitt 2004.)

separate "phylogroups" within a species can be recognized and assigned to a particular refugium, suggesting a history of isolation and divergence that extends deeper into the Pleistocene than just the latest of the 20 or so glacial-interglacial cycles. Farther to the east, Beringia served as a refugium for mammals, birds, plants, and invertebrates (Waltari et al. 2004), although the pattern of biotic responses appears to be more complex than has been the case in the western Palearctic (Hewitt 2004).

The Continuing Transformation of Historical Biogeography

As we close this chapter, and this unit of the book, we hope that it is clear that historical biogeography is rapidly blossoming into a productive and energetic discipline with the power to infer much about the geography of lineage and biotic diversification. We highlighted just a few studies demonstrating the tremendous progress being made in understanding the history of Earth's biotas, and we emphasized terrestrial systems, but recognize that much progress is being made in testing alternative hypotheses and elucidating the histories of marine biotas as well (e.g., Barber et al. 2000; Santini and Winterbottom 2002; Briggs 2003, 2004; Meyer et al. 2005). Although we discussed the fascinating biogeographic history of the Hawaiian Islands in some detail, we were unable to feature a growing number of other interesting studies of biogeographic and evolutionary experiments on oceanic archipelagoes (e.g., Cook et al. 2001; Emerson 2002; Heaney et al. 2005). Finally, we barely mentioned recent studies that are advancing the paleobiogeographies of long-extinct lineages, ranging from Paleozoic trilobites (Lieberman 2003, 2004) to Mesozoic dinosaurs (Upchurch et al. 2002).

Yet despite its great progress, especially in the past several decades, historical biogeography still has large hurdles to overcome if even more important advances are to be made. First, even though the large array of modern approaches illustrated throughout this chapter suggest that this is a discipline rich in theory and methods, historical biogeographers still are concerned about whether methods are sophisticated enough to unravel histories that are full of complexity, with reticulated biotas more often than not integrating multiple episodes of vicariance-driven speciation, dispersal, extinction, and sympatric speciation across timeframes spanning a few thousands to many millions of years of Earth's history. New methods continue to appear (e.g., Wojcicki and Brooks 2005), and what diverged to form distinct disciplines and methods are now merging into more synthetic approaches in which different methods are employed to address different questions at sequential stages in an analysis (Morrone and Crisci 1995; Althoff and Pellmyr 2002; Riddle and Hafner 2004, in press). Finally, along with continuing advances in methods, and growth in numbers and variety of lineages and biotas available for analyses, we are encouraged by recent calls for the re-integration of the historical biogeographic perspective into broader and more insightful ecological and evolutionary arenas (e.g., see Wiens and Donoghue 2004; summarized in Figure 15.36).

UNIT FIVE

Ecological Biogeography

CHAPTER 13

Island Biogeography

Patterns in Species Richness

I
SLANDS HAVE ALWAYS HAD A GREAT INFLUENCE on biogeography, far out of proportion to the tiny fraction of the Earth's surface that they cover. The reason for this is straightforward: Islands and other insular habitats, such as mountaintops, springs, lakes, and caves, are ideal subjects for natural experiments. They are well-defined, relatively simple, isolated, and numerous—often occurring in archipelagoes of tens or hundreds of islands.

Just as conditions can be varied in artificial manipulative experiments, islands can vary in a number of environmental features (e.g., area, isolation, or presence of predators and competitors). In this way, the effects of each factor on community structure can be assessed. Despite the limitations of such natural experiments, islands have an important advantage over artificial manipulations in that they were established sufficiently long ago that there has been time for evolutionary response to the variables. All one needs to do is to gather the data and interpret them correctly. This, however, has not been an easy task. Basic data on insular biotas are still being compiled, and their interpretation has often been the subject of controversy. Like any other scientific study, a natural experiment must be well designed. That is, out of all the islands we could study, we must prudently select a subsample of islands that minimizes or controls variation in all factors except those central to our working hypothesis.

Of course, the challenges of designing rigorous and insightful experiments, gathering and analyzing data, and debating and testing alternative methods and ideas are common to all scientific experiments—whether natural or manipulative. As we shall see, biogeographers have utilized both approaches to gain insights into the processes influencing insular communities and, in turn, have obtained important insights into the forces structuring mainland communities as well.

Historical Background

Since the early work of Johann Reinhold Forster and his colleagues during the eighteenth and early nineteenth centuries, studies of islands, mountaintops, and other isolated ecosystems have strongly influenced the development of biogeography. Forster contributed fundamental insights to what was to become island biogeography theory, including the very general tendencies for insular floras to have fewer species than those of the mainland, and for diversity of plants to increase with available resources (i.e., island area and variety of habitats). As we saw in Chapter 2, the influence of islands on biogeography, as well as on evolutionary biology and ecology, began in earnest in the early nineteenth century when various European nations undertook to explore, map, and study the world. Just as Johann Forster and Joseph Banks sailed with Captain James Cook in the late 1700s, other naturalists frequently accompanied voyages of exploration during the nineteenth century. The best of these naturalists—notably, Wallace, Darwin, and Hooker—not only described and collected specimens of what they found, but also noted patterns in nature and sought explanations for them. Some of the clearest patterns were apparent among the various islands of oceanic archipelagoes. Wallace's travels in the East Indies, Darwin's experiences in the Galápagos, and Hooker's explorations in the Southern Ocean had profound effects on the thinking of these scientists, and consequently, on the ideas about evolution and related areas of environmental biology that revolutionized scientific thought in the mid-1800s.

Another revolution—a much more modest one, but nevertheless a major shift in the direction of scientific thought—occurred in the mid-1900s with the integration of concepts from ecology, evolution, and biogeography. It is difficult to pinpoint the exact beginning of this endeavor, but islands again played a central role. One of the pioneers was David Lack (1947, 1976), who, early in his career, conducted a classic study of the evolution and ecology of Darwin's finches on the Galápagos Archipelago, and, shortly before his death, investigated the distribution and ecology of birds in the West Indies. As Lack had followed Darwin to the Galápagos, another ornithologist, Ernst Mayr (1942, 1963), followed Wallace to the East Indies and returned to make major contributions to the understanding of speciation and other aspects of the evolutionary process (see also Mayr and Diamond 2001). Another pioneer was G. E. Hutchinson (1958, 1959, 1967), who also traveled widely but studied lakes rather than terrestrial islands. In his 1959 paper, "Homage to Santa Rosalia, or why are there so many kinds of animals?" Hutchinson called attention to the problem of how to explain geographic variation in the diversity of species. This problem has remained a focus of research in ecological biogeography and community ecology right up to the present.

If any single contribution can be said to have triggered the recent revolution in ecological biogeography, however, it was Robert H. MacArthur and Edward O. Wilson's equilibrium theory of island biogeography (1963, 1967). This seminal work was completed when both men were young—still in their thirties. MacArthur had been a student of Hutchinson's at Yale. His doctoral dissertation (1958) was a classic study of competition and coexistence in several closely related species of warblers. After completing his degree, he did postdoctoral work in Britain with Lack and then held professorships at the University of Pennsylvania and at Princeton University. Wilson, who has spent his entire career at Harvard, began it as a systematist. Strongly influenced by Mayr, he had worked extensively on the origins and relationships of the ants on islands of the East Indies and South Pacific. He was also a coauthor of the classic paper on character displacement (Brown and Wilson 1956). Both men had extensive experience with islands: MacArthur in the montane islands of the southwestern United States, in the West Indies, and in small islands off the coasts of Maine and Panama; Wilson in the East Indies, Polynesia, and the Florida Keys. Both men went on to have illustrious careers. MacArthur died of cancer in 1972 at the age of 42, but had already produced many theoretical papers on population and community ecology that still motivate research in those fields today. Wilson continued to work on social insects, especially ants, but his interests have shifted from systematics and biogeography, to animal behavior (Wilson 1975), and most recently, to conservation of biodiversity.

MacArthur and Wilson's equilibrium theory represented a radical change in biogeographic thought. Prior to their work, investigators had focused on historical problems and idiosyncratic approaches. The primary questions of biogeography had always been those addressed in Chapter 2: Where did a particular taxonomic group of organisms originate, and how, as a result of subsequent dispersal, speciation, and extinction, did its diversity and distribution change? Obviously, these questions have a historical and phylogenetic focus, and they are ad hoc questions in the sense that they are normally applied to particular taxa or specific regions.

Prior to 1960, the dominant theme of island biogeography was what is sometimes referred to as the static theory of islands (Dexter 1978). Basically, this theory held that insular community structure was fixed in ecological time—that is, that species composition remained unchanged unless modified by long-term evolutionary processes. According to the static theory, insular community structure resulted from unique immigration and extinction events, and species number was determined by the limited number of niches available on each island (Lack's [1976] theory of ecological impoverishment). Either a species had already colonized the island in question, or it never would. Once it arrived, the species either found adequate resources for survival, or it failed to establish a population. Large islands had more species because they had more resources and a greater variety of niches which could support a greater diversity of species. Islands closer to the mainland had more species because they were within the (one-time) colonization abilities of more species.

Island biogeography was largely idiosyncratic, with no general explanations or rigorous models, mathematical or graphical, that would provide clear predictions and lead to tests among alternative explanations for observed patterns. All this was to change in a relatively short time with MacArthur and Wilson's theory. During the early decades of the twentieth century, models of dynamic equilibria had been developed to explain a variety of phenomena, ranging from chemical reactions and regulation of body temperature, to gene frequencies and demography. The central idea, or **paradigm,** in all these models was the concept of a dynamic equilibrium—that is, that opposing forces maintain constancy in some characteristic of a system despite continual changes, or turnover, in its other intrinsic properties. For example, the body temperature of a bird is regulated within narrow limits, despite changes in environmental temperature and internal heat production, by opposing mechanisms of heating and cooling. Given the early development of equilibrium models in other fields, rather than marveling over MacArthur and Wilson's contributions—which truly were revolutionary—we may wonder why ecological biogeography lagged so far behind. In fact, an early form of an equilibrium theory of island biogeography was developed by Eugene Gordon Munroe in 1948, but it was ignored by his contemporaries (Box 13.1). The following two decades witnessed a period of great growth in ecology, which had been a mathematically unsophisticated science. By the early 1960s, ecological biogeography was poised for a scientific revolution, to be spearheaded by one

BOX 13.1   *Independent discovery of the equilibrium theory of island biogeography*

■ ■ I A correlation of this kind [between number of species and logarithm of area of an island] is as interesting as it is unexpected, for it suggests the existence of an equilibrium value for the number of species in a given island, a value which acts as a limit to the size of the fauna. The processes which determine the equilibrium value for an island of given size must be, on the one hand, the extinction of species, and, on the other hand, the formation of new species within the island, and the immigration of new species from outside it.

The above quotation does not come from one of MacArthur and Wilson's (1963, 1967) two seminal publications on the equilibrium theory of island biogeography. It was written 15 years earlier by someone else. It appears on page 117 of Eugene G. Munroe's (1948) doctoral thesis on the distribution of butterflies in the West Indies.

The earlier and independent discovery of the equilibrium theory by Munroe is more than an amusing incident in the history of science. It warrants further examination for two reasons. First, in contrast to

some purported cases of prior discovery of important ideas, Munroe did not have just some vague, poorly articulated notion of species equilibrium. He clearly presented the empirical species-area relationship that stimulated his inductive discovery, investigated the generality of this pattern, and developed detailed verbal and mathematical models to explain it (Munroe 1953).

Second, given the striking similarity of the two models, it is worthwhile to ask why Munroe's discovery went unrecognized (but see Gilbert 1984) while MacArthur and Wilson's has been hailed as one of the great accomplishments of evolutionary ecology of the 1960s. Others have been given credit for developing precursors to the equilibrium theory (including Dammermann [1948—see Thornton 1992], and LaGreca and Sacchi [1957—see Vuilleumier 1975]), but Munroe's theory was conceptually and mathematically equivalent to MacArthur and Wilson's theory (see also Wilkinson 1993).

Mathematical models

Had Munroe stopped with his doctoral thesis, the similarities to MacArthur and Wil-

son's later work would have been striking enough. However, Munroe developed his equilibrium theory of island biography in a subsequent paper (1953), but only the abstract was published, and this appeared four years after the paper was presented at a meeting in 1949. This tantalizing abstract contains only three paragraphs. The first, quoted above, points out the generality of the semi-logarithmic species-area relationship, but then Munroe goes on to present a mathematical model:

The actual form of the curves is that of a shallow sigmoid, with the equation

$$F = k'A^* [iL/(i + kp)]$$

where F = number of species in the fauna at equilibrium; L = number of species in surrounding lands capable of immigrating into the island; i = the probability of any one species actually immigrating; p = the probability of extinction of a single pair of one species; and A = the area of the island, to which the population number of each species is assumed to be directly proportional. (Munroe 1953, p. 53)

MacArthur and Wilson (1967, p. 26, Figure

of the world's leading mathematical ecologists, Robert MacArthur, and an equally qualified naturalist and ecologist, Ed Wilson.

MacArthur and Wilson deliberately departed from the classic ad hoc, static and historical approach and asked radically new kinds of questions. They searched for general patterns in the distributions of diverse kinds of species, independent of their phylogenetic affinities, in the hope that such patterns would have general ecological explanations rather than idiosyncratic and historical ones. Although they recognized the importance of systematics and historical geology, MacArthur and Wilson were primarily interested in patterns that might be explained without invoking unique historical events. Their approach was to focus on variations in plant and animal distributions that appeared to be correlated with the functional attributes of contemporary organisms and with measurable characteristics of their present environments. They were more impressed, for example, with the fact that birds and bats are similar both in their ability to fly and in their wide distribution on oceanic islands, than with the many differences in morphological traits that reflect the long divergent evolutionary history of these two taxa.

Island Patterns

MacArthur and Wilson's theory was developed to explain two very general patterns in island biogeography: the tendencies for the number of species to

BOX 13.1 *(continued)*

11 and Equation 3-1) use a similar, but somewhat simpler, expression in their first mathematical model:

$$S = (IP)/(E + I)$$

where S = the equilibrium number of species; I = the initial immigration rate, if the island was empty of species; P = the number of species in the species pool available to colonize; and E = the extinction rate if P species were present on the island.

Conceptual advances and scientific progress

Why didn't Monroe promote his ideas?
Despite the remarkable similarity of their ideas, Munroe's prior discovery had no apparent impact and went virtually unrecognized to this day (but see Gilbert 1984), whereas MacArthur and Wilson's later, independent development of the same concept has been enormously influential. In one sense, it is not surprising that Munroe's concept of species equilibrium remained unknown to MacArthur, Wilson, and virtually all biogeographers and ecologists. Munroe's ideas were presented in five pages, one table, and one figure in a large unpublished

doctoral dissertation devoted primarily to the systematics and descriptive biogeography of Caribbean butterflies, and in a one-page abstract in a relatively obscure regional publication. And, it is not hard to understand why a young scientist with "competing interests and pressures" did not aggressively pursue ideas that apparently elicited interest from only a few colleagues. In the late 1940s, biogeography was dominated by descriptive and taxonomic approaches; this was not a propitious time for injecting mathematical theory and ecological concepts.

In another sense, however, it does seem surprising—especially with the clarity afforded by decades of hindsight—that Munroe did not make a greater effort to publicize his discovery. This is especially true considering that Munroe did not retire from productive science after receiving his degree, but went on to enjoy a distinguished career as a lepidopteran systematist. Munroe clearly devoted considerable time to developing his ideas on faunal equilibrium and recognized at least some of their important implications.

The lessons. It is unfortunate that Munroe

did not get more recognition for his discovery. One of the purposes of this essay is to rectify this situation. This case is strangely reminiscent of the independent discovery of the theory of evolution by natural selection by Wallace and Darwin. And like that story, it shows that, as important as new ideas are in the progress of science, they are often not the unique inspirations of genius that are portrayed in the textbooks. On one hand, scientific revolutions usually do depend on major conceptual innovations. On the other hand, in order for these insights to have an impact, they must be promoted cogently at a receptive stage in the development of a discipline. It is not sufficient to have a good idea; it is even more important to develop and publicize it. Munroe and MacArthur and Wilson had the same basic idea. Munroe, distracted by other interests and perhaps frustrated in his initial attempts to publish, allowed his idea to languish. MacArthur and Wilson vigorously pursued and advocated their idea, and had a major impact on their science.
(Excerpted and modified from Brown and Lomolino 1989.) ▮▮▮

---

increase with island area, and to decrease with island isolation. These patterns had been well known to biogeographers since the late 1700s (see Chapter 2). Perhaps the greatest inspiration for the equilibrium theory, however, came from the more recent observation that immigrations and extinctions were relatively frequent phenomena, even in ecological time. During the twentieth century, repeated biological surveys of Krakatau and other islands that had been cleared of life by volcanic eruptions revealed that immigrations and extinctions were recurrent processes, and that island communities exhibited substantial turnover as new colonists replaced extirpated species. MacArthur and Wilson's innovation was to recognize the common themes that underlie these observations (the species-area relationship, species-isolation relationship, and turnover) and to propose a single, unifying theory to account for them.

The Species-Area Relationship

> Theories, like islands, are often reached by stepping stones. The species-area curves are such stepping stones. (MacArthur and Wilson 1967, p. 8)

Schoener (1976) described the species-area relationship as "one of community ecology's few laws." Indeed, it is one of the most general, best-documented patterns in nature (Forster 1778; de Candolle 1855; Watson 1859; Jaccard 1902, 1908; Arrhenius 1921; Brenner 1921; Gleason 1922, 1926). Regardless of the tax-

FIGURE 13.1   The empirical relationship between number of species (S) and island area (A) for reptiles and amphibians of the West Indies, plotted from the original data of Darlington (1957). Note that both axes are logarithmic, and the points are well fitted by a straight line and the equation $S = cA^z$, where c and z are fitted values. Inset depicts this relationship in arithmetic space. (After MacArthur and Wilson 1967.)

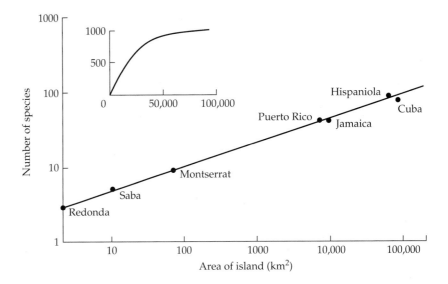

onomic group or type of ecosystem being considered, species number tends to increase with increasing area. This relationship, however, is not linear; richness increases less rapidly for larger islands (Figure 13.1).

Despite the long history of studies on this fundamental pattern, it wasn't until the 1920s that ecologists generalized it in mathematical form. In 1920, Arrhenius adapted the allometric equation (used for the scaling of morphology and metabolic processes as functions of body mass) to describe the species-area relationship. The Arrhenius equation, more commonly referred to as the power model, can simply be expressed as

$$S = cA^z$$

where S = species number (or richness); c is a fitted constant; A = island area; and z is another fitted parameter that represents the slope when both S and A are plotted on logarithmic scales. Typically, this relationship is linearized by taking the log of both sides of the equation:

$$log(S) = log(c) + z\ log(A)$$

The c and z values can now be easily estimated using simple linear regression of log-transformed data. In Box 13.2 we discuss the biological relevance and potential misinterpretations of c and z.

While the power model appears to be the most commonly used formula for the species-area relationship, a semi-logarithmic model also is frequently used, especially among plant ecologists. In 1922, Gleason used the following formula to study species-area relationships of plant communities:

$$S = d + k\ log(A)$$

As this formula implies, d represents the intercept, and k represents the slope of the line when S (richness) is plotted against the log of A (area).

Frank Preston, an engineer by profession and a naturalist by avocation, contributed to the mathematical development of ecology during the 1950s and 1960s. Preston (1962) noted that the species-area relationship of islands was a special case of the general multiplicative increase in the number of species with an increase in the area sampled of most ecosystems, either terrestrial or aquatic. He suggested that this was a consequence of what he termed the canonical lognormal distribution of the number of individuals among species

BOX 13.2 *Interpretations and comparisons of constants in the species-area relationship: An additional caution*

■■│ The species-area relationship is one of the most important and most frequently studied patterns in biogeography and, according to Schoener, "one of community ecology's few universal regularities" (1986, p. 560). Yet, the utility of the most common model describing this relationship—the power model ($S = cA^z$, where S is species richness, A is area, and c and z are fitted constants)—is debatable. The strongest controversy stems not from the "fit" provided by this model, but from the biological relevance of its exponent, z (Connor and McCoy 1979; Sugihara 1981; Abbott 1983).

Many misinterpretations can result from the frequent but unfortunate reference to the z and c values as the slope and intercept of the species-area relationship, respectively. The power function, however, "intercepts" at the origin (i.e., when $A = 0$, $S = 0$). Just as important, z values represent the slopes of the relationship between log

S and log A, not species richness and area. By themselves, z values do not indicate how rapidly S increases with A. For this, we require the values of both parameters in the power model, c and z (Gould 1979).

Although this is not a mathematically startling revelation, it is not rare to see, for example, a high z value equated with a rapid increase in S with increasing A. Yet, such an inference is valid only if c values for the archipelagoes and faunas under study are equal. On the contrary, c values vary considerably (often by an order of magnitude or more among archipelagoes or taxa), whereas z values tend to be conservative (typically ranging from 0.15 to 0.35; Gould 1979, his Table 1; Wright 1981, his Table 1). This fact led Gould (1979) to suggest that we draw inferences from comparisons of c values for archipelagoes with approximately equal z values (analogous to analysis of covariance for log-transformed data).

The effects of varying one of these parameters, c or z, while holding the other constant are illustrated in Figures A and B. Note that the species-area relationship (arithmetic scale) is strongly influenced by relatively modest variation in c, but comparatively insensitive to the variations in z, typical of natural communities. Furthermore, if these parameters vary simultaneously, as they surely do in nature, then the slope of the species-area relationship (again on an arithmetic scale) may actually be lower for studies reporting higher z values (Figure C of this box).

In summary, although the power model may continue to provide important insights into factors affecting species richness in isolated biotas, there remains considerable potential for misinterpretation. Studies comparing constants of the power model among archipelagoes, or comparisons with predicted values of these constants, should take measures to avoid the statistical problems discussed here and elsewhere (Connor and McCoy 1979; Gould 1979; Martin 1981).

(Excerpted and modified from Lomolino 1989.) ▐ ■

Effects of varying the values of c and z on the species-area relationship. The model in all cases is $S = cA^z$. (A) Effects of varying the value of c (z is held constant at 0.25). (B) Effects of varying the value of z (c is held constant at 1). (C) Effects of varying both c and z. Note that species richness increases more rapidly for the upper curve despite its substantially lower z value.

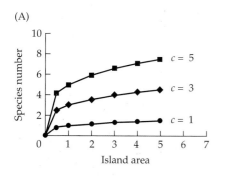

(A)

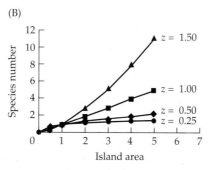

(B)

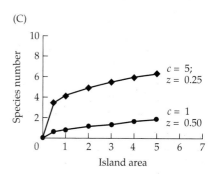

(C)

(see also Williams 1953, 1964). In any region, only a few species are extremely common, and most are moderately, or very, rare. Therefore, the distribution of the number of individuals among species, when plotted on a logarithmic abscissa (x-axis), is fairly well fitted by a normal, bell-shaped curve (Figure 13.2). Sometimes this curve is cut off on the left-hand side, and Preston suggested that this happens when the sample is so small that some of the rarest species are not observed. It is this effect that produces the species-area curve. As progressively larger areas are sampled, one obtains not only more individuals, but also more species, because some of the new individuals will be rep-

FIGURE 13.2 The relative abundances of species within a local biota often fit a lognormal distribution; in other words, the frequency distribution approximates a normal curve when abundance is plotted on a logarithmic scale. Note that often the left-hand tail of the distribution is cut off by what Preston called a veil line. Because the axis is logarithmic, this curve shows that every community contains more rare species than common ones. The data are from a bird census in Maryland (Preston 1957) and a count of moths at a light trap in England (Williams 1953).

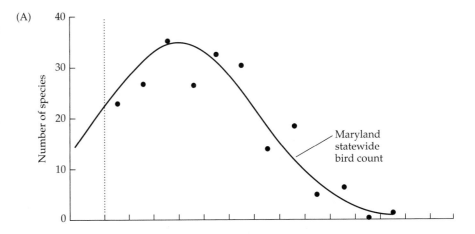

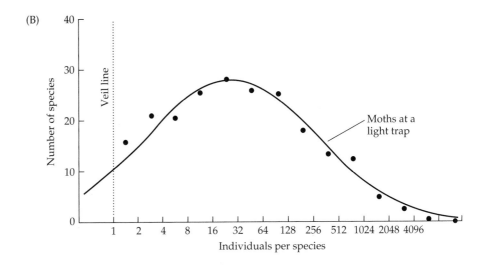

resentatives of rare species that had not yet been seen. Additionally, larger areas will tend to incorporate new kinds of habitats and therefore add specialized species that are restricted to those environments.

Preston pointed out that small, isolated islands have fewer species per unit of area and higher z values for the species-area curve than sample areas of comparable size within large regions of continuous habitat on continents (Figure 13.3; see Schoener 1974a; Sugihara 1981; Rosenzweig 1995). The reason for this should be intuitively apparent: small, isolated islands have fewer species than comparable areas on a continent, because if a species becomes too rare on an island, it is likely to become extinct; whereas on a continent its population can be sustained at low levels by the exchange of individuals between local areas. The effect of such extinctions is much more severe on small islands than on larger ones, resulting in the steeper slope of the species-area curve.

The Species-Isolation Relationship

Since the early 1800s, it has been well known that single, isolated islands far out in the ocean support fewer species than islands that are part of major archipelagoes or islands that are located nearer to continents. Assuming that the decline in species richness results from a decline in dispersal rates with isolation, the form of the species-isolation relationship should be a conse-

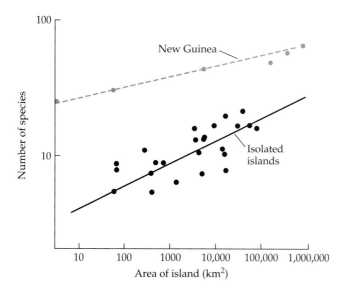

FIGURE 13.3 The slope of the species-area relationship in log-log space is much steeper for isolated islands than for sample areas of different sizes within a single large landmass. These data are for pomerine ants on the Moluccan and Melanesian islands (below) and in regions of increasing size on New Guinea (above). The difference between the two curves can be attributed to the greater likelihood of extinction without replacement by immigration of rare species on isolated islands. (After Wilson 1961.)

quence of dispersal curves for the pool of species (potential colonists from the mainland; see Chapter 6). Therefore, for a variety of taxa and ecosystems, species richness should decline as a negative exponential or sigmoidal function of isolation (Figure 13.4).

With appropriate transformations of one or both axes (i.e., richness and isolation), this relationship can be linearized to allow statistical analyses and comparisons among studies (e.g., $S = k_1 e^{-k_2(I)}$, or $S = k_1 e^{-k_2(I \cdot I)}$; where S = richness, k_1 and k_2 are fitted constants, and I = isolation). In practice, however, the species-isolation relationship often proves much less general than the species-area relationship. This difference may derive, in part, from the tendency for many studies of islands, such as those of a single archipelago, to include a broad range in island area but only a limited range in isolation. In addition,

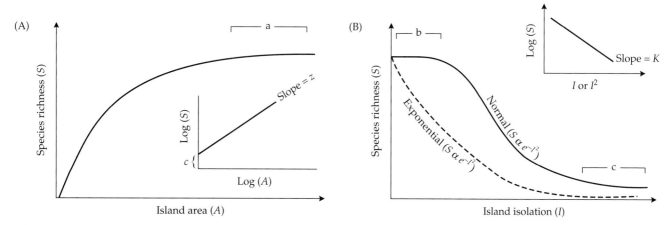

FIGURE 13.4 Two of the most common patterns in nature—the species-area (A) and the species-isolation (B) relationships. The log-transformed equivalents of these relationships are presented in the insets (S = species richness; A = island area; and c and z are fitted constants for the power model of the species-area relationship). Note that the species-area relationship will be difficult to detect (slope near 0) if biogeographic surveys are limited to the larger islands (i.e., region a; see also Figure 13.23). Similarly, the species-isolation relationship will be difficult to detect if surveys are restricted to the very near or very distant islands (regions b and c; see also Figures 6.19 and 13.27A).

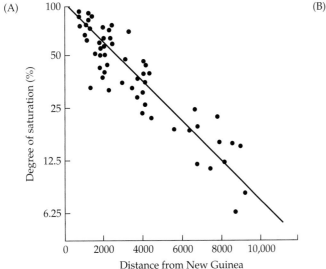

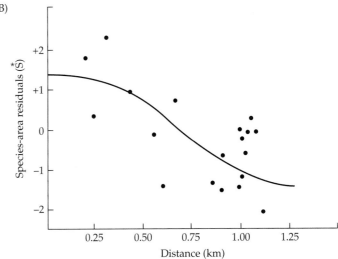

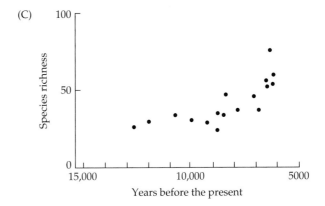

FIGURE 13.5 A sample of graphs illustrating the effects of isolation on the species richness of various insular biotas (see also Figure 13.30). (A) Resident land birds in the Moluccan and Melanesian archipelagoes. Here species richness is expressed as saturation, which is the species richness of isolated islands expressed as a percentage of that found on an island of equivalent size, but closer to New Guinea. (B) Nonvolant mammals of the Thousand Island region of the St. Lawrence River, New York. Here the ordinate equals the residuals about the species-area relationship [i.e., the difference between the observed insular species richness and that predicted for an island of its size based on the regression model $\hat{S} = 6.51(A^{0.305})$]. (C) Species richness of lizards on landbridge islands in the Gulf of California. Here, species richness is graphed as a function of time since isolation of these landbridge islands following rising sea levels of the most recent glacial recession. (A after Diamond 1972; B after Lomolino 1982; C after Wilcox 1978.)

while measures of island area are easily taken from maps, given the many possible immigration routes and sources, biologically relevant measurements of isolation are extremely challenging. Still, many studies report significant species-isolation relationships, especially when island isolation varies substantially and when the effects of island area are statistically "controlled" (e.g., by using correlation or regression analysis after accounting for the effects of area on species richness; Figure 13.5A and B).

Species Turnover

A third pattern that influenced MacArthur and Wilson was the rapidity of recolonization of the Krakatau Islands. These islands are located in the Sunda Straits between the Indonesian islands of Sumatra and Borneo (Figure 13.6). A violent volcanic eruption destroyed the original island of Krakatau in 1883, leaving several remnant islands devoid of life. Recolonization, apparently from the nearby large islands of Java and Sumatra, was rapid; by 1935, a tropical rain forest, supporting numerous species of plants, birds, and other organisms, was developing. Several scientific expeditions visited the remnant islands of Rakata and Sertung and inventoried the biota. Early censuses of the birds, which were particularly complete, are summarized in Table 13.1.

MacArthur and Wilson noted that the number of bird species on Rakata and Sertung increased rapidly until about 1920, but, after that, the total number of

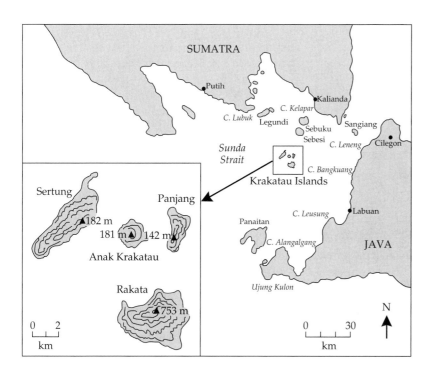

FIGURE 13.6 The Krakatau Islands are located in the Sunda Straits between the Indonesian Islands of Sumatra and Borneo. Rakata, Sertung, and Panjang are remnants of a much larger island that was destroyed by a massive volcanic explosion in 1883. Anak Krakatau is a relatively young island that emerged from a submarine volcanic vent in 1930. (After Whittaker and Jones 1994.)

TABLE 13.1 *Number of species of land and freshwater birds on Rakata and Sertung*

Number of species found

	Rakata			Sertung		
	Nonmigrant	Migrant	Total	Nonmigrant	Migrant	Total
1908	13	0	13	1	0	1
1919–1921	27	4	31	27	2	29
1932–1934	27	3	30	29	5	34

Number of extinctions and colonizations between censuses

	Rakata		Sertung	
	Extinctions	Colonizations	Extinctions	Colonizations
1908 to 1919–1921	2	20	0	28
1919–1921 to 1932–1934	5	4	2	7

Source: After MacArthur and Wilson 1967.

Note: The number of species increased from the census of 1883 to that of 1919–1921 and then remained relatively constant despite extinction of some species and colonization of others.

species remained relatively constant, despite changes in the composition of the avifauna. Species not only colonized rapidly prior to 1920, but also continued to immigrate afterwards. Some of these late arrivals became successful colonists, replacing about an equal number of species that became extinct. These offsetting colonizations and extinctions might simply have reflected successional changes in the avifauna in response to the development of a tropical forest and the concomitant elimination of open habitats, but they also suggested that continual turnover might be typical of insular biotas. Such turnover might be particularly high when, as is the case for birds on Rakata and Sertung, organisms need cross only modest barriers to reach small islands.

The Equilibrium Theory of Island Biogeography

MacArthur and Wilson produced a single theory to explain what they considered to be the three basic characteristics of insular biotas: the species-area relationship, the species-isolation relationship, and species turnover. They proposed that the number of species inhabiting an island represents a dynamic equilibrium between opposing rates of immigration and extinction. The equilibrium is termed dynamic because immigration and extinction are thought to be recurrent, opposing processes, maintaining a relatively stable species richness despite changes in species composition.

The equilibrium model can be presented graphically by plotting immigration and extinction rates as a function of the number of species present on an island (Figure 13.7). The number of species on the island (S) can range from zero to a maximum, P, the number in the pool of species that is available to colonize the island from a nearby continent or other source area. Now we can predict the shapes of the curves representing colonization and extinction rates. Let us start with an empty island. The immigration rate (defined as the rate of arrival of propagules of species not already present on the island) must decline from some maximum value when the island is empty, to zero when the island contains all the species in the pool and there are no more new species to arrive. As the number of resident species grows, there remain fewer new species on the mainland to colonize the island. Conversely, the extinction rate (defined as the rate of loss of existing insular species) should increase from zero when there are no species present on the island to become extinct, to some maximum value when all the species in the mainland pool are inhabiting the island. Simply put, as the island fills, the number of species that can suffer extinctions increases, and therefore, extinction rates should increase accordingly.

At some number of species between zero and P, the lines representing the immigration and extinction rates cross. At this point the two rates are exactly equal, resulting in an equilibrial number of species, $\hat{S}$, and an equilibrial rate of species turnover, $\hat{T}$. This point represents a stable equilibrium because if the number of species is perturbed from this value, it should (at least theoretically) always return. For example, suppose that a natural disaster, such as a hurricane, causes the extinction of several insular species, temporarily reducing the number of species from $\hat{S}$ to S' (see Figure 13.7). Then, the immigration rate will exceed the extinction rate, and the island will accumulate species until it has again reached $\hat{S}$. Similarly, if S is perturbed from $\hat{S}$ to a larger number, S'', then the extinction rate will be greater than the colonization rate, and species will be lost until $\hat{S}$ is restored.

Now let us incorporate the effects of island size and isolation into this model. MacArthur and Wilson assumed that the size of an island would affect only the extinction rate. Although they recognized that a large island would

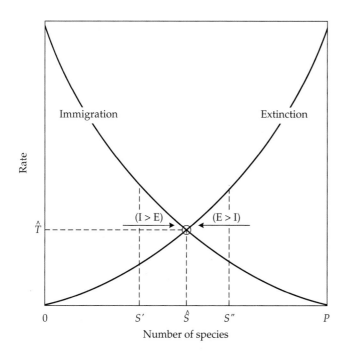

FIGURE 13.7   A simple model in which the number of species inhabiting an island represents an equilibrium between opposing rates of colonization and extinction. Note that the immigration rate declines and the extinction rate increases as the number of species increases from zero to P, the number in the mainland species pool. The point of intersection of the two curves represents a stable equilibrium, because if the number of species is displaced from $\hat{S}$ to either higher (S'') or lower (S') numbers, it will return to equilibrium (arrows).

provide a larger target for dispersing propagules than a small one, they reasoned that such an effect on immigration rate would be insignificant compared with the importance of island size in extinction. Population sizes of all species should decrease with decreasing island area and, as discussed in Chapter 7, the probability of extinction increases rapidly as a population gets very small. Consequently, for a source pool biota of many species, the extinction rate should be substantially greater for a small island than for a larger one. This can be shown in the graphic model by drawing two extinction rate curves; the one for a small island is always higher than the one for a large island (Figure 13.8). Examining the intersections of these extinction curves with the immigration curve, shows immediately that the small island is predicted to have a smaller equilibrium number of species and a higher equilibrium turnover rate than the large island.

MacArthur and Wilson used similar logic to show how immigration curves would be influenced by isolation. They assumed that the distance of an island from the source pool would affect only the immigration rate. No matter what the mechanism of dispersal, if a barrier exerts a filtering effect, then the probability of an organism crossing the barrier decreases as the width of the barrier increases. This effect of isolation by distance can be incorporated into the model by drawing two immigration curves; the one for an island near a source of species is always higher than the one for a more isolated island (see Figure 13.8). The intersections of these curves with either of the extinction curves predict that at equilibrium, near islands should have more species and higher rates of turnover than distant islands.

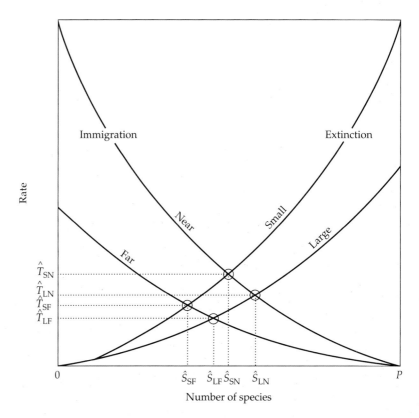

FIGURE 13.8 MacArthur and Wilson's (1963) equilibrium model of island biogeography, showing the effects of island size (different extinction rate curves) and isolation (different immigration rate curves) on the equilibrium number of species ($\hat{S}$) and rate of species turnover ($\hat{T}$). Intersections of the curves for islands of different combinations of size and distance can be used to predict the relative numbers of species and turnover rates at equilibrium.

By combining the effects of island size and isolation in a single graph (see Figure 13.8), one can see the predictions of the model. There are four intersections of immigration and extinction rate curves, one for each combination of island size, large (L) and small (S); and distance, near (N) and far (F). The number of species at equilibrium is predicted to be in the order $S_{SF} < S_{SN} \sim S_{LF} < S_{LN}$. (Note that whether a large far or small near island should have more species will depend on the exact shapes of the immigration and extinction curves.)

The model obviously explains the observations that motivated it: namely, that the number of species increases with area and decreases with isolation, and that there is a continual turnover of species. However, like any good theory, the model goes beyond what is already known to make additional predictions that can be tested only with new observations and experiments. Specifically, it predicts the following order of turnover rates at equilibrium: $T_{SN} > T_{SF} \sim T_{LN} > T_{LF}$. The model also predicts the relative rates at which islands of different sizes and degrees of isolation should return to equilibrium if the biota is perturbed. For example, a near island should return to equilibrium more rapidly than a distant island of the same size because it should have a higher immigration rate, but the same extinction rate.

Strengths and Weaknesses of the Theory

Like most important new ideas, MacArthur and Wilson's equilibrium theory elicited a mixed response from other scientists. It generated not only great interest and enthusiasm on the part of some investigators, but also severe skepticism and criticism from others. Certainly, the theory stimulated a new wave of research in ecological biogeography. Most of these studies involved islands or other isolated habitats and were designed specifically to evaluate or elaborate on MacArthur and Wilson's model. In a review of the equilibrium theory written only a decade after MacArthur and Wilson's (1963) first seminal paper, Simberloff (1974a) cited 121 references, and the pace of research has continued to accelerate. By 1987, the 1963 paper and the 1967 monograph had been cited in over 2000 publications, and the citations continue apace even in recent years, with over 3600 papers citing their monograph by 2002.

Several features of the equilibrium theory contributed to its favorable reception. Its elegantly simple graphical presentation made the essential elements of the model accessible to a wide audience, including persons with minimal mathematical training. Moreover, the equilibrium theory not only introduced stimulating new ideas that helped to bridge the gap between traditional biogeography and ecology, but presented them in the form of a model that made clear, testable predictions. Many mathematical models, especially those used in ecology, are not empirically operational—that is, they do not explicitly indicate what observations would be necessary to test the theories and reject the models. Such models can still be valuable, however, if they serve a heuristic function—if they cause one to think about a problem in new and more precise ways. MacArthur and Wilson's model certainly plays such a heuristic role, but it also indicates the kind of data that are necessary for a rigorous test. It predicts qualitative trends (increases or decreases) in numbers of species and turnover rates with island size and isolation. These predictions can be tested using simple lists of species inhabiting various archipelagoes at different times. Such lists can be compiled by original fieldwork, but the results of previous surveys of the biotas of many islands are also available in the published literature. The only other data required—the areas of the islands and distances to probable source areas—can be obtained from standard maps.

The simplicity of the theory, however, has also been the basis for much criticism. Some have argued that it is so simple as to be useless because it obscures, rather than clarifies, the patterns and processes that make island biogeography interesting (e.g., Sauer 1969; Lack 1970; Carlquist 1974; Gilbert 1980; Williamson 1989). In view of the research the theory has stimulated and the increased understanding of island distributions that has resulted, much of this criticism seems unwarranted. All models are intended to help investigators understand nature by presenting a simplified, abstracted concept of a more complex reality. They inevitably sacrifice a certain amount of precision for clarity and generality. Usually, new theories are presented initially in a very general, incomplete form and are corrected and refined as a result of subsequent empirical and theoretical research.

To their credit, MacArthur and Wilson themselves (1967) acknowledged some important weaknesses of their theory:

> First, we still know very little about the precise shape of the extinction and immigration curves, so that few numerical predictions can yet be made. Second, and more importantly, the model puts rather too simple an interpretation on the process by making an artificially clear-cut distinction between immigration and extinction ... The third difficulty

stems from the assumption that extinction and immigration curves have fairly regular shapes for different faunas and different islands and for different times on the same islands. When a new set of curves must be derived for a new situation, the model loses much of its virtue. Deviant cases do occur—for example the Krakatau flora. The extinction curves, furthermore, have a pronounced genetic component: rarity affects gene frequencies, and genetic as well as ecological causes of extinction act in concert.

Others were quick to echo these concerns and to identify additional short-comings of the theory. As you will see from the points listed below, most of these criticisms suggest that the theory is insufficient or incomplete, yet do not take issue with its basic tenets. That is, insular species richness results—at least in large part—from recurrent immigrations and extinctions.

Criticisms of MacArthur and Wilson's Theory include the following:

1. **Interspecific differences and interactions among species** The model assumes that the identities and characteristics of particular species can be ignored (this is sometimes referred to as "MacArthur's Paradox" given his earlier and extensive studies on niche differentiation among species). Immigration, extinction, and turnover are viewed as highly stochastic processes: new species immigrate and existing ones die out more or less at random, and only the approximate number of species remains the same. The implicit assumption that ecological processes—including interspecific interactions—do not determine *which* species can coexist on a particular island, is at least technically incorrect. Then again, the equilibrium model was developed to explain patterns in species richness, not *species composition*.

2. **Interdependence of immigration and extinction** The equilibrium theory treats immigration and extinction as independent processes. This is probably justified, given that immigration is defined as the arrival of propagules of new species and that secondary succession is not occurring. On the other hand, recruitment of additional individuals of species already present should tend to "rescue" declining populations on near islands from extinction. In a similar manner, area may affect immigration as well as extinction rates because larger islands may intercept more potential colonists (see the later sections on rescue and target area effects).

3. **Biogeographically meaningful measures of isolation** It may be difficult to identify the source of an island biota without careful investigation of the systematics and historical distribution of the species that are present. Indeed, the species inhabiting a single island may be derived from several sources, including over-water dispersal from continents and other islands, past connections with other landmasses, and endemic speciation within the island itself. If insular species are acquired by immigration from multiple sources, the theory can potentially be modified to deal with this complication. For example, MacArthur and Wilson considered stepping-stone colonization, in which a species disperses from one island to the next, down a chain of islands. In most cases, when measures of isolation are modified to better reflect the heterogeneity of dispersal barriers, currents, and ecological affinities of the focal species, the effects of isolation and immigrations become more clear (Taylor 1987; Lomolino et al. 1989; Lomolino 1994b; Lomolino and Perault 2000; Lomolino and Smith 2004; Walter 2004; see Figure 16.33). Put otherwise, the failure to

detect a significant effect of isolation on richness of insular communities may be more of a reflection of the difficulties in devising biogeographically meaningful measures of isolation than proof that insular community structure is not influenced by immigration.

4. **Biogeographically meaningful measures of island area** Total island area provides only a very general and indirect measure of the capacity of islands to support individuals and species. Even for the same island or archipelago, carrying capacity will vary substantially for different groups of species depending on their habitat requirements. In addition, although the extent of most habitat types increases with increasing island size, so usually does the diversity of habitats. Larger islands tend to have higher mountains, more aquatic habitats, and so on, as well as larger areas of most of the vegetation types found on small islands. Consequently, some of the increase in species diversity with island size may be owing to the addition of specialists whose habitat requirements are met only on large islands. In this case, a more elaborate model, which also incorporates specific habitat variables, should predict patterns of insular species diversity and distribution better than area alone (e.g., Power 1972; Johnson 1975). In a similar vein, others criticize the model for being so simplistic that it loses much of the spatial context of forces influencing biological diversity; to paraphrase Gertrude Stein (1937), there is no there there in modern biogeography. In a recent and provocative essay entitled "The mismeasure of islands," Hartmut Walter (2004) emphasizes the need to develop measures of area and isolation that better represent the ecological and biogeographic history of places (a concept he terms "eigenplace").

5. **The importance of speciation** If insular species are derived by speciation within the island itself, then a basic assumption of the model—that species richness is affected only by immigration and extinction—is clearly violated. Speciation, however, is likely to be particularly important on very large and isolated islands (Diamond and Gilpin 1983; Heaney 1986, 1991, 2000 ; Adler 1994; see Chapter 14). Again, although the original model did not do so, it can be modified to include the effects of speciation (see Figure 13.16). The key challenge here is to understand how all three fundamental processes—immigration, extinction, *and* speciation—are influenced by island characteristics (especially area and isolation) and how they combine to influence the structure of insular communities. One of Robert MacArthur's first students, Michael Rosenzweig, developed a very general model that went a long way toward achieving this goal (Rosenzweig 1995; see also Chapter 15).

6. **Disturbance in ecological to geological time scales** Many insular biotas may not be in equilibrium between opposing processes of immigration and extinction. Rather, the number of species may increase or decrease over evolutionary time, or with major environmental disturbances such as hurricanes and volcanic eruptions. This is particularly likely when immigration and extinction occur on approximately the same time scale as speciation, or the geological and climatic events that create, change, and destroy islands. Here, insular communities may never reach equilibrium (see later section on nonequilibrium biotas). An equilibrium theory may still be useful for interpreting these cases, however, because it may be instructive to consider these biotas as approaching a new equilibrium, or relaxing or rebounding toward the original equilibrium following a historical perturbation.

These criticisms and challenges are serious, and should be kept in mind as we proceed to discuss patterns in insular community structure in this and the following chapter. We will see many cases in which the model, in its original form, is inadequate to account for observed patterns (see Haila et al. 1982; Haila 1986; Case 1987; Williamson 1989). It will also be clear, however, that in following MacArthur and Wilson's general approach, in testing and sometimes rejecting their dynamic theory of island biogeography, we have learned a great deal about the ecological and historical processes that determine the diversity and distribution of organisms among islands and other isolated habitats. As we noted earlier, some biogeographers have drawn a parallel between the heuristic value of MacArthur and Wilson's equilibrium theory and that of the Hardy-Weinberg equilibrium model for gene frequencies (see Heaney 2000; Heaney and Vermeij 2004). The value of the latter model is not simply that it accurately predicts gene frequencies at equilibrium (a condition seldom achieved in natural populations), but that it identifies populations that deviate from equilibrium and the reasons for this (e.g., genetic drift and natural selection). As MacArthur and Wilson (1967:19–21) observed, "a perfect balance between immigration and extinction might never be reached … but to the extent that the assumption of a balance has enabled us to make certain valid new predictions, the equilibrium concept is useful." Again, given its paradigmatic influence on nearly all disciplines of biogeography and ecology, their theory served the field well. But over the four decades since MacArthur and Wilson first proposed the model, our understanding of the complexity of nature may have advanced beyond the conceptual envelope of the theory. Therefore, after reviewing some of the important tests of the model and related patterns in insular diversity, we conclude this chapter with a preview of some recent attempts to advance or replace this long-reigning paradigm of island biogeography.

Tests of the Model

Beginning in the late 1960s, numerous authors published papers purporting to test MacArthur and Wilson's theory. Often only one or two of its predictions—namely, covariation in number of species with area or isolation—were tested. Even these incomplete tests are valuable, however, because when the predictions are not borne out by the data, they indicate that the theory in its present form cannot account for the distributions of these biotas. On the other hand, when observations concur with the predictions, it is unwarranted to assume, as many authors have done, that this corroborates the theory. It is possible to obtain the right result for the wrong reason because alternative explanations, based on different assumptions, can generate some of the same predictions. Before accepting a particular model as the explanation for a set of observations, it is important not only to test all possible predictions of the model, but also to evaluate the validity of the underlying assumptions and to attempt to imagine and rule out alternative explanations.

Early studies that "tested" and purported to support the equilibrium model fell into three categories, which we describe below. Most common were analyses of data on insular distributions of various taxa at one point in time. Many of these simply confirmed for more groups of organisms and more archipelagoes the relationship between number of species, island size, and island isolation pointed out by MacArthur and Wilson (e.g., Hamilton et al. 1964; Hamilton and Armstrong 1965; Johnson et al. 1968; Johnson and Raven 1973). Others described similar patterns for other insular habitats, such as mountaintops (Vuilleumier 1970) and caves (Culver 1970; Vuilleumier 1973). Frequently, they

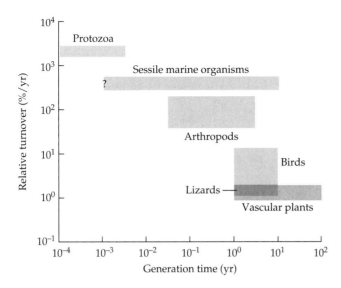

FIGURE 13.9 Relative turnover rates tend to be lower for organisms with longer generation times. (After Schoener 1983.)

advanced ad hoc explanations to explain differences among these relationships in such things as z values and goodness of fit of the regression equation to the data points. Although these studies supported the generality of the patterns, they cannot be viewed as rigorous tests of the theory. On the other hand, research on biotic turnover provided especially important insights into the assembly and dynamics of insular communities.

Biogeographers will continue to gain valuable insights not just from comparisons among archipelagoes, but also by comparing the community dynamics among functionally different groups of species (i.e., groups that differ in their abilities to immigrate to, or survive on, islands). For example, Thomas Schoener (1983) discovered some intriguing patterns in his review of 21 studies of turnover across a variety of organisms, ranging from protozoans and plants to terrestrial arthropods and vertebrates. Not only did turnover tend to be lower on larger islands, but it also decreased with the generation time of the organisms (Figure 13.9). As additional turnover studies are accumulated, reviews such as Schoener's will continue to provide important insights into the forces structuring isolated communities.

ESTIMATES OF TURNOVER ON LANDBRIDGE ISLANDS. In 1969, Jared M. Diamond reported a more direct test of one crucial prediction: continual turnover in species composition resulting from the balance of colonizations and extinctions. Diamond recensused the avifauna of the Channel Islands off the coast of southern California almost exactly 50 years after Howell (1917) had published a detailed account of the birds known to breed on each island at the turn of the century. Comparison of the two censuses revealed striking differences in species composition (Table 13.2), although there had been relatively little change in the number of species breeding on each island. In 1968, Diamond observed a number of species not known to breed there prior to 1917, and about an equal number of species present 50 years earlier that had apparently disappeared. From these observations, he concluded that at least 20 to 60 percent of the bird species on each island had turned over since 1917. He pointed

TABLE 13.2 *Turnover of breeding land bird species on the California Channel Islands between 1917 and 1968*

Island	Area (km²)	Distance to mainland (km)	Number of species 1917	Number of species 1968	Extinctions	Introductions (by humans)	Colonizations	Turnover (%)[a]
Los Coronados	2.6	13	11	11	4	0	4	36
San Nicholas	57	98	11	11	6	2	4	50
San Clemente	145	79	28	24	9	1	4	25
Santa Catalina	194	32	30	34	6	1	9	24
Santa Barbara	2.6	61	10	6	7	0	3	62
San Miguel	36	42	11	15	4	0	8	46
Santa Rosa	218	44	14	25	1	1	11	32
Santa Cruz	249	31	36	37	6	1	6	17
Anacapa	2.9	21	15	14	5	0	4	31

Source: Diamond 1969.

[a]Turnover rate, expressed as percentage of the resident species per 51 years, is calculated as 100 (extinctions + colonizations)/(1917 species + 1968 species – introductions).

out that actual turnover rates could well have been even higher; some species might have colonized and become extinct during the intervening 50 years. This methodological error is termed **cryptoturnover** and it, along with **pseudoturnover** (cases where species occurrences were missed in an earlier survey and therefore counted as immigrations [turnover] if detected in a later survey), may often confound our interpretations of community dynamics on islands (see Whittkaer 1998). Although there were not enough Channel Island communities to test rigorously for the relationship between turnover rate and either island size or isolation, Diamond noted that apparent turnover appeared to be greatest on the islands with the fewest species (i.e., those that were relatively small and/or isolated; see Table 13.2).

Diamond's study, widely cited as confirming or supporting the equilibrium model (e.g., MacArthur 1972), was also vigorously challenged. Lynch and Johnson (1974) pointed out that most of the thoroughly documented changes in the avifauna could be attributed directly to human influence, and some of the other apparent changes may have resulted from errors in conducting and interpreting the census. In particular, most of the extinctions involved the disappearance of large birds of prey, including the osprey (*Pandion haliaetus*), bald eagle (*Haliaeetus leucocephalus*), and peregrine falcon (*Falco peregrinus*), all of which were almost certainly eliminated by pesticide poisoning. Many colonizations were a result of the immigration of house sparrows (*Passer domesticus*) and starlings (*Sturnus vulgaris*), which colonized under their own power as they expanded their ranges after introduction into eastern North America from Europe. Turnover involving these latter two, and probably other species as well, had also been influenced by habitat changes caused by recent human activities. Lynch and Johnson concluded that, contrary to Diamond's claims, there was little evidence for natural turnover of breeding birds on the Channel Islands within the 50-year period (see also Walter 2000).

This debate, however, was far from resolved. Jones and Diamond (1976) surveyed the birds of the California Channel Islands—especially Santa Catalina—for several years in succession, and found year-to-year changes in the breeding status of several species. In a separate but related study, Diamond and May (1976) analyzed many years of careful records of birds breeding on the small Farnes Islands off the coast of Great Britain. Several species were reported to have been present only sporadically during the years for which data were available. In both of these studies, however, the "turnovers" (or pseudoturnovers) involved species that were migratory or highly nomadic. Their presence or absence on an island may have been more a matter of the arrival and departure of highly mobile individuals for which a few miles of water represent no significant barrier, than of the establishment and extinction of real resident populations. Terborgh and Faaborg (1973) reported apparent turnover in the avifauna of Mona, a small island west of Puerto Rico in the West Indies, as a result of comparisons between an early survey and their own subsequent census. Again, however, the island had been changed substantially by human activities, especially by the effects of introduced goats. Also in contrast to the theory, Lack (1976) reported the relative stasis of birds of Jamaica over more than 200 years of recorded history. On this large Caribbean island, with a resident land avifauna numbering 65 species, there have been just two extinctions and one colonization, and all of these can be attributed directly to human influence (see also Abbott 1983). Studies of turnover on some other isolated, oceanic islands such as those of the avifauna of the Revillagigedo Archipelago, located approximately 400 km south of Baja California, also reveal relative stability and thus seem consistent with Lack's theory of ecological impoverishment on islands (Walter 1998). However, MacArthur and Wilson's theory also predicts relatively low turnover rates on such isolated islands once they have accumulated their biotas.

KRAKATAU REVISITED: TURNOVER ON VOLCANICALLY ACTIVE ISLANDS.   While the species-area and species-isolation relationships fueled the development of the equilibrium theory, biotic surveys of the Krakatau Islands (see Figure 13.6) provided vital clues to the dynamics of insular biotas. Early surveys of the Krakatau communities between 1908 and 1934 (i.e., 25 to 41 years after the 1883 explosion cleared the islands of all life) revealed that they experienced frequent immigrations, extinctions, and turnover. Additional surveys conducted through 1990 continued to highlight some of the strengths of MacArthur and Wilson's theory, especially its emphasis on dynamic processes, but also revealed some of its shortcomings (Thornton et al. 1990; Bush and Whitaker 1991; Thornton et al. 1993; Whitaker and Jones 1994; Whittaker 1995, 1998; Thornton 1996; Whittaker et al. 2000; Thornton et al. 2002)

First, the now century-plus record confirms the dynamic nature of insular communities. Again, while MacArthur and Wilson's model is typically recognized as the "equilibrium theory," arguably its most salient and fundamental advance was its emphasis on community dynamics and recurrent processes. Equilibrium, of course, was an important outcome of these dynamic processes, and one that greatly simplified the derivation and predictions of the model.

Turnover remains relatively high for most groups of plants and animals of the Krakatau Islands. Moreover, the general trends in immigration and extinction rates appear consistent with the basic theory: immigration rate decreases and extinction rate increases with species richness (Figure 13.10). For a number of taxonomic groups—including seed plants, dragonflies, butterflies, reptiles, and resident land birds—the rate of species accumulation up to the year 1990 has slowed substantially, and species number appears to be approaching an equilibrium (Figure 13.11). The rate of approach toward an equilibrium

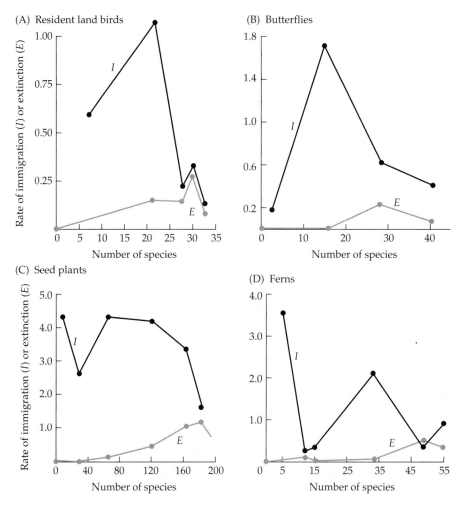

FIGURE 13.10   Immigration (*I*) and extinction (*E*) rates (as number of species/year) of animals and plants on Rakata, the largest of the Krakatau Islands, as a function of insular species richness (*S*). Consistent with MacArthur and Wilson's equilibrium theory, immigration rate declined and extinction rate increased as the islands accumulated species. Note that, at least for resident land birds, seed plants, and ferns, insular communities may have approached a dynamic equilibrium (*I* ≈ *E*). (After Thornton et al. 1993.)

varies considerably among taxa, and seems directly correlated with dispersal abilities. For example, while avian species richness on the Krakataus is only slightly below that expected based on island size, species richness of non-flying mammals is approximately one-tenth of the expected values (Figure 13.12). On Rakata, the largest island, sea-dispersed plants have accumulated much less rapidly than wind- or animal-dispersed plants, particularly during the last few decades of colonization (Figure 13.13).

While these trends confirm the basic tenets of the equilibrium theory, others reveal some important limitations of the theory. As MacArthur and Wilson (1967) anticipated, immigration and extinction curves may not be monotonic if major successional changes occur. Some species are strongly dependent on other pioneer species to create niches for them, or as key pollinators and agents of dispersal. This seems to be the case for butterflies, land birds, and ferns, whose accumulation rates increased substantially during forest development

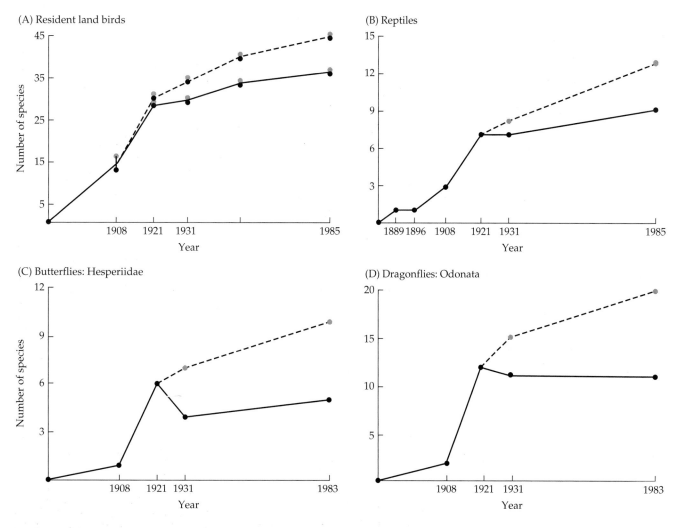

FIGURE 13.11 Colonization curves for various animal taxa in the Krakatau Islands: (A) resident land birds, (B) reptiles, (C) butterflies (Hesperiidae), and (D) dragonflies (Odonata). Solid lines indicate actual species numbers at the time the islands were surveyed. (After Thornton et al. 1990.)

(1908 to 1931; see Figure 13.12). Interpretations of these data are, however, plagued by our inability to study immigration directly and distinguish between rates of immigration and accumulation of species. Recall that MacArthur and Wilson (1967) criticized their own model for "making an artificially clear-cut distinction between immigration and extinction." Immigration is defined as "the process of arrival of a species on an island not occupied by that species" (MacArthur and Wilson 1967). This says nothing about the likelihood that an immigrant will establish a breeding population, and indeed, it is likely that most immigrants fail to do so. Biotic surveys such as those conducted on the Krakataus, while insightful, measure accumulation (or establishment) rates, not immigration rates. Although immigration rates should be unaffected by ecological succession, the chance that an immigrant will become established may be strongly influenced by successional changes. In short, the apparent non-monotonic immigration curves reported for the Krakataus' biota may be an artifact of the necessity to lump establishment (or "ecesis," sensu

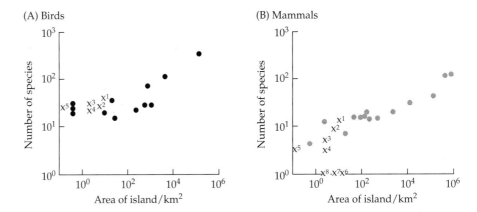

FIGURE 13.12   Species-area relationship for birds (A) and mammals (B) on the Krakatau Islands (x^1–x^8) and undisturbed oceanic islands of the Sunda Shelf (black circles). Superscripts: birds and volant mammals (bats): 1 = Rakata, 2 = Sertung, 3 = Panjang, 4 = Anak Krakatau (based on total land area), 5 = Anak Krakatau (based on vegetated area only); nonvolant mammals (rats): 6 = Rakata, 7 = Sertung, 8 = Panjang. Note that species richness of birds and bats on the Krakatau Islands may have rebounded to pre-eruption levels (see also Figure 13.11), whereas richness of nonvolant mammals remains relatively low in comparison to undisturbed islands of the same area. Anak Krakatau, which emerged from the sea in 1930, still lacks terrestrial mammals. (After Thornton et al. 1990.)

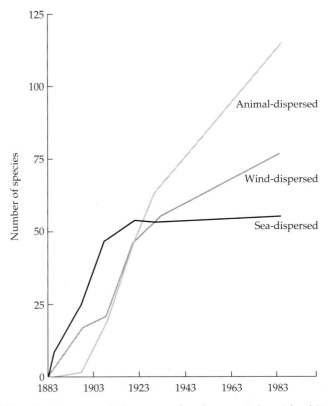

FIGURE 13.13   Species accumulation curves for plants on Rakata Island (Krakataus) differ markedly depending on the mode of dispersal. Sea- and wind-dispersed plants were first to arrive, whereas animal-dispersed plants did not accumulate until successional changes provided suitable habitat for their animal transporters. Following 1923, however, animal-dispersed plants accumulated much more rapidly than wind- and sea-dispersed forms. (After Bush and Whittaker 1991).

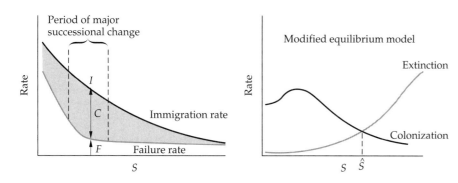

FIGURE 13.14 The effects of succession can be included in an equilibrium model by noting how it alters the likelihood that immigrants will survive and establish a breeding population. S = species richness; $\hat{S}$ = equilibrial species richness; I = immigration rate (number of propagules of new species arriving per unit of time); F = failure rate (number of propagules failing to establish a breeding population per unit of time); and C = colonization rate (= $I - F$). (After MacArthur and Wilson 1967.)

MacMahon 1987) together with immigration (Figure 13.14). Interestingly, in their attempt to include successional effects in their model, MacArthur and Wilson (1967:51) lumped failure to become established, together with extinction, not immigration. This problem is not simply one of semantics. If the equilibrium theory is to include important ecological processes such as succession, it must explicitly and independently address the effects of these processes on establishment as well as immigration and extinction.

Bush and Whitaker (1991) have voiced some other, more fundamental concerns; namely, that the equilibrium model does not include the effects of processes occurring at other time scales. If the approach to equilibrium is as slow as Bush and Whittaker predict—on the order of millennia—then disturbances such as hurricanes and volcanic eruptions, which occur on similar time scales, may prevent an equilibrium for some, perhaps many, taxa (Figure 13.15). Similarly, the equilibrium model does not include speciation, which, at least on very large and isolated islands, may rival or possibly exceed immigration rates (Diamond and Gilpin 1983; Heaney 1986, 1991; Adler 1994). With just a minor modification, however, the model can include the effects of spe-

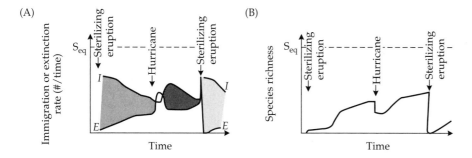

FIGURE 13.15 For some insular biotas, major disturbances such as volcanic eruptions and hurricanes may occur so frequently that the insular communities seldom achieve a dynamic equilibrium (S_{eq}) as envisioned by MacArthur and Wilson. (A) The shading indicates when immigration rate (I) exceeds extinction rate (E), and therefore, when species richness (B) should increase. (After Bush and Whittaker 1991; Whittaker 1995.)

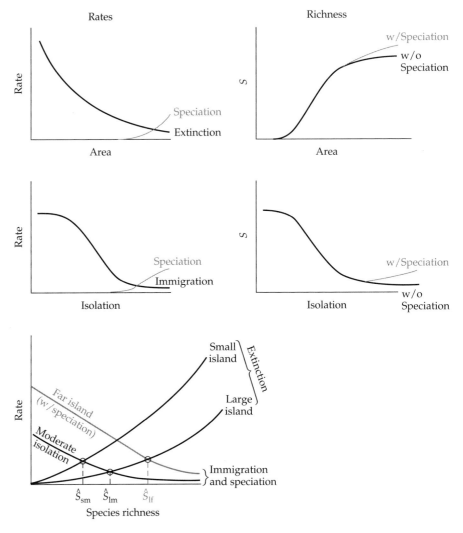

FIGURE 13.16 The effects of evolution on species richness can be incorporated into the equilibrium theory by noting that speciation will supplement immigration, and that speciation rates should be significant for only the largest and most isolated islands. As a result, the equilibrium number of species for distant islands may actually exceed that for less isolated islands. S = species richness; $\hat{S}$ = equilibrium species richness; s = small; l = large; m = moderately isolated; and f= far or distant islands. (See also Rosenzweig 1995.)

ciation as well as immigration (Figure 13.16; see also Rosenzweig 1995; Heaney 2000).

In summary, the large-scale and long-term natural experiments afforded by volcanic eruptions of the Krakatau, Long, and other archipelagoes in volcanically active regions continue to provide valuable insights into the dynamics and development of insular communities. Comparisons of patterns in community development across different islands and archipelagoes should prove especially valuable. For example, studies of bird communities on Long Island (Western Bismarck Archipelago, north of New Guinea), which was created by volcanic eruptions in the seventeenth century, reveal that its avifauna may have already achieved an equilibrium (or quasi-equilibrium, apparently in the interactive stage illustrated in Figure 13.19). Avian richness on Long Island has remained at approximately 50 species and not changed much over the past three decades, with three apparent extinctions replaced by three colonizing species (Schipper et al. 2001). In contrast, the much younger and smaller, Krakatau Islands, have fewer birds, and avian richness continues to increase (albeit at an apparently slowed pace in comparison to earlier stages of recolonization; see Figure 13.11). Again, these results seem entirely consistent with MacArthur and Wilson's theory.

Actually, there is a fascinating diversity of archipelagoes created by volcanic activity, each providing different, but complementary and important opportunities to explore the dynamics and assembly of isolated ecosystems. In addition to islands or archipelagoes that must be recolonized after sterilizing eruptions (e.g., the Karakatua Islands, San Benedicto Island of Mexico, and Volcano Island in the Phillipines), newly emergent islands such as Surtsey in the North Atlantic and Tuluman in the tropical western Pacific must be colonized de novo, often from great distances. Still other islands may emerge, or re-emerge from calderas submerged under marine waters (e.g., Anak Krakatau) or, in a truly intriguing case, within land-locked calderas that have filled with freshwater. Examples of the latter include Motmot Island in Lake Wisdom of Long Island Volcano Island in Lake Taal of the Phillipines, and Wizard Island in Crater Lake of North America, each of these providing invaluable opportunities to study community development within nested ecosystems (i.e., colonization of relatively small, volcanic islands, with freshwater lakes [also undergoing colonization], within a ring of terrestrial habitats [or barriers], all of this isolated by marine waters). Relatively recent studies by Ian Thornton and his colleagues (published in a special feature of the *Journal of Biogeography*, Volume 28[11/12]), provide a tantalizing case study of complexity of forces influencing assembly of these geologically and ecologically dynamic ecosystems.

As Schipper and his colleagues (Schippper et al. 2001) conclude, "natural experiments such as this are too rare not to be fully exploited by island ecologists." These insights may contribute to continued modifications and advancement of MacArthur and Wilson's theory, or, perhaps just as likely, lead to its ultimate replacement by a new and radically different paradigm. We return to this latter theme (i.e., the possible emergence of a new paradigm of island biogeography), in the last section of this chapter.

TURNOVER ON RECENTLY CREATED ANTHROPOGENIC ISLANDS. Human activities have fragmented and insularized once massive and continuous ecosystems across the globe. The effects of these activities on biodiversity are the focus of Chapters 16 and 17. Here, we consider the biogeographic dynamics of two anthropogenic archipelagoes created by the flooding of mountainous areas in the American tropics.

Early in the twentieth century, the Chagras River was dammed to create the Panama Canal and Gatun Lake, in turn flooding lowland areas and transforming forested hilltops into islands. Repeated biotic surveys have been conducted to trace the relaxation (decline toward a new equilibrium) of plant and animal communities on these islands, especially those on the largest island, Barro Colorado (area = 1600 ha). Surveys conducted during the 1970s and early 1980s revealed that about 45 of the estimated 108 original species of breeding birds had disappeared (see Willis 1974; Karr 1982, 1990; Wright 1985). More recent surveys conducted between 1994 and 1996 reveal that, consistent with the equilibrium theory and Brown's (1971) relaxation model, the rate of species loss over the past two decades has declined (Robinson 1999). However, species extinctions—especially for forest-interior birds and others with limited abilities, or propensities, for dispersal—continue to overwhelm colonizations. As a

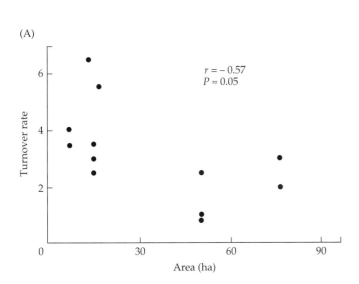

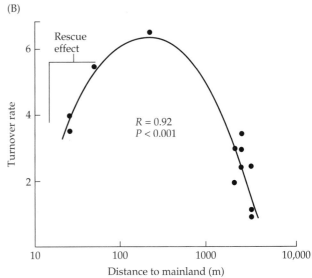

FIGURE 13.17 Consistent with the predictions of the equilibrium theory, turnover rates of birds on islands of Gatun Lake, Panama, tend to decrease with island area (A) and with isolation (B). Yet, because frequent immigrations can rescue otherwise dwindling populations from extinction, turnover rates may also be lower on near islands than on distant islands (e.g., the three near islands in B; see also Figure 13.21). (After Wright 1985.)

result, although it has been over eight decades since this anthropogenic island was formed, species richness of its avian community continues to decline.

Because intensive surveys were not conducted until well after the island was formed, and because extinction rates should be highest during the earliest stages of relaxation, additional extinctions may have gone unrecorded. The salient observations from these studies are entirely consistent with MacArthur and Wilson's theory: extinctions and immigrations are recurrent, and turnover is lowest on the largest and most isolated islands (Figure 13.17A). Numerous other studies report a similar inverse relationship between turnover rates and island area or isolation (Brown 1971; Diamond 1972; Wilcox 1978, 1980; Terborgh and Winter 1980; Nilsson and Nilsson 1982; Heaney 1984, 1986). Although the evidence is limited, it appears that species richness declines most rapidly during the early stages of relaxation, immediately following isolation (MacArthur and Wilson 1967; Terborgh 1974; Wilcox 1980; see also Lovejoy et al. 1986; Robinson et al. 1992; Kattan et al. 1994; Stouffer and Bierregaard 1995). As predicted by the equilibrium theory, relaxation rates then tend to slow as species richness approaches the new equilibrium.

More recent studies by John Terborgh and his colleagues suggest that relaxation on newly created islands may be surprisingly rapid (Terborgh et al. 1997a), again suggesting that relaxation is most rapid in its early stages (i.e., immediately after area is reduced, isolation increases, and when the differences between immigration and extinction rates are highest). Avian communities on relatively small (~1 ha) islands in Lago Guri, a Venezuelan lake created during the 1980s for a hydroelectric project, may have already achieved a new dynamic equilibrium just seven years after the islands were formed (Terborgh et al. 1997b). In contrast, avian species richness still appears to be declining on the larger islands, a result that is also consistent with the predictions of the equilibrium theory (see Lambert et al. 2003, for studies of rodent communities on these islands). Tracking the results of these and similar opportunistic "experiments" is certain to provide additional insights into the dynamic forces structuring insular communities.

EXPERIMENTAL DEFAUNATION. The most rigorous early tests of MacArthur and Wilson's theory were the defaunation experiments conducted by Wilson and

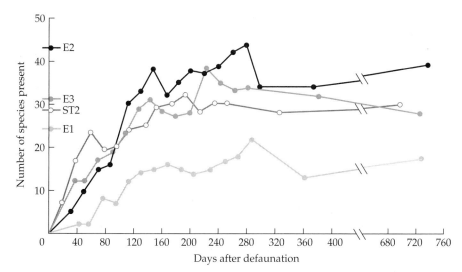

FIGURE 13.18 Recolonization by terrestrial arthropods of four small mangrove islands as a function of time since the fauna was removed. The initial number of species present is indicated along the vertical axis. Note that after defaunation the number of species increases rapidly, tends to overshoot the initial number, declines, and then increases gradually to approximately the initial number. Island E1, with a lower rate of colonization and a smaller number of species, was more isolated from a source of colonists than the other islands. (After Simberloff and Wilson 1970.)

his student Daniel Simberloff (Simberloff and Wilson 1969, 1970; Wilson and Simberloff 1969). This study has become justly famous as an example of the use of controlled, manipulative experimentation to test theoretical models in biogeography and ecology. The basic design was simple: all species of arthropods were eliminated from tiny islets of red mangrove (*Rhizophora mangle*) in the Florida Keys, and the subsequent changes were monitored closely. Simberloff and Wilson hired an exterminator, who used methyl bromide gas to kill all insects, spiders, mites, and other terrestrial animals while leaving the mangrove vegetation virtually undamaged. This was a drastic but effective perturbation. Recolonization, monitored by careful surveys, was surprisingly rapid (Figure 13.18). Within less than a year, all but the most distant island had recovered their initial number of species. In fact, the number of species increased rapidly and appeared to overshoot the initial number before declining and stabilizing close to the initial value. Furthermore, there was a great deal of turnover, even after the number of species had stopped changing significantly. Individual species colonized and disappeared, sometimes repeatedly, during the short-term study. The high turnover rates were not surprising given the proximity of these islands to the mainland (0.002 to 1.2 km), their small size (75 to 250 m^2), and the lack of intervening dry land.

Thus, Simberloff and Wilson's results strongly supported several predictions of the equilibrium model (see also Molles 1978; Rey 1981; Hockin 1982; Strong and Rey 1982). Although there were too few islands to test rigorously for the predicted relationships of initial colonization rate, equilibrium turnover rate, and equilibrium number of species to island size and isolation, the results were generally consistent with the predictions. For example, the most isolated island (E1 in Figure 13.18) had the fewest species and the lowest rate of recolonization. Simberloff and Wilson also suggested that developing communities may pass through three—and possibly, four—types of dynamic equilibria (Figure 13.19). First, species may tend to accumulate rapidly to reach a "non-interactive equilibrium," with most species occurring at relatively low population levels—perhaps too low to inhibit populations of other species. Populations may later increase to the point at which interspecific interactions can cause local extinctions of some species, resulting in a reduced, or "interactive" equilibrium. Simberloff and Wilson's studies also suggest that succession or ecological sorting may generate a subsequent increase in species

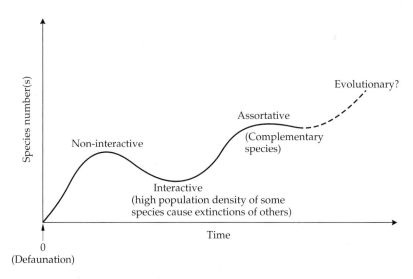

FIGURE 13.19 As an empty island accumulates species, its insular communities may pass through a series of equilibria reflecting demographic, ecological, and evolutionary processes. (After Simberloff and Wilson 1969, 1970.)

richness toward an "assortative" equilibrium (i.e., one with a combination of species more likely to coexist). Finally, on the largest and most isolated islands, species richness may continue to increase until extinction rates balance the combined effects of immigration plus speciation ("evolutionary" equilibrium). Thus, the graphical model of Figure 13.19 captures the salient features of MacArthur and Wilson's theory, while also incorporating the importance of ecological interactions among species, dynamics of species composition as well as species richness, and the influence of all three, fundamental biogeographic processes—immigration, extinction, and evolution.

Additional Patterns in Insular Species Richness

As we discussed earlier, the equilibrium theory assumes that extinction is affected only by area and that immigration is affected only by isolation. Here, we consider violations of these assumptions that may warrant modification of the original model.

THE RESCUE EFFECT. Another study of arthropods on isolated patches of vegetation points to a potentially important problem with MacArthur and Wilson's equilibrium model. James H. Brown and Astrid Kodric-Brown (1977) censused arthropods (mostly insects and spiders) on individual thistle (*Cirsium neomexicanum*) plants growing in desert shrubland in southeastern Arizona. Although the intervening habitat may have been suitable for some of the arthropod species, the thistle plants constituted isolated patches of favorable habitat. Brown and Kodric-Brown counted the individuals and species of arthropods on the plants at 5-day intervals. Their results confirmed several major predictions of the MacArthur-Wilson model. The number of individuals and species increased with plant size and decreased with increasing distance from the nearest plants. Although the arthropods did not maintain real populations on the plants, there was a dynamic equilibrium between the rates of arrival and disappearance. Defaunated thistles were reinhabited rapidly, those near other thistles more quickly than isolated plants; plants closely surrounded by other plants regained 94% of their original arthropod biota in 24

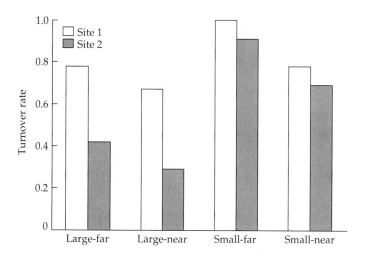

FIGURE 13.20 Turnover of arthropod species on individual thistle plants. Note that, although turnover rates are lower on larger "islands" of thistle plants, turnover also was lower on less-isolated plants (compare Large-far to Large-near islands, and Small-far to Small-near islands). The latter result was inconsistent with MacArthur and Wilson's equilibrium theory (see Figure 13.7), but is an important result that Brown and Kodric-Brown termed the rescue effect. (In order to study the relationships between turnover of "insular" populations, Brown and Kodric-Brown [1977] divided the plants into objective size and isolation categories on the basis of the number of flowers and the number of other plants in the immediate vicinity. Turnover rate was calculated as the number of species present only in the first census, plus the number of species present only in the second census, divided by the total number of species present in both censuses.)

hours, whereas isolated plants acquired only 67% of the initial number of species in the same period. The turnover of individuals and species was higher on small plants than on large ones.

All of these results were consistent with the predictions of the equilibrium theory. However, contrary to the model's prediction, turnover rates were lower on plants in close proximity to others than on isolated ones (Figure 13.20). This single exceptional result is important, because it suggests a problem with the model that may be as relevant for organisms on real islands as for arthropods on thistles. The most likely explanation for all of the results taken together is that there is an insular equilibrium maintained by opposing mechanisms of immigration and extinction, as envisioned by MacArthur and Wilson, but the factors affecting the arrival of new species are not independent of those influencing the extinction of species already present. Proximity to a source of immigrants increases the immigration rate of all species, and a continual influx of individuals belonging to species already present tends to prevent the disappearance of those species.

In the case of arthropods on thistles, this **rescue effect** is probably simply statistical: high rates of immigration reduce the probability that a species will temporarily be absent and, hence, recorded as a turnover. On real islands, however, immigrants may rescue populations from extinction by contributing to the breeding stock and by injecting new genetic variability to counteract the deleterious effects of inbreeding, which can be severe in small, isolated populations. In a separate study, Smith (1980) corroborated the rescue effect of immigrants on extinction rates, showing that populations of pikas (*Ochotona princeps*, rat-sized mammals similar in appearance to small rabbits) inhabiting isolated rockslides in North America had higher turnover rates than those near a source of colonists. Other researchers have subsequently reported rescue effects for a variety of organisms (e.g., Wright 1985, Laurance 1990; see Figure 13.17B) and it has become a fundamental feature of metapopulation theory (Gilpin and Hanski, 1991; Hanski and Gilpin 1997; Hanski, 1999).

The rescue effect was not anticipated by MacArthur and Wilson, but it is easy to modify their model slightly to incorporate it. Drawing different extinction rate curves, as well as different colonization rate curves, for near and far islands (Figure 13.21A,B) allows us to take into account the decrease in the extinction rate on near (vs. distant) islands that is due to the rescue effect. Note that when this is done, it may reverse the order of the equilibrium turnover

FIGURE 13.21 Two modifications of MacArthur and Wilson's equilibrium theory of island biogeography can be made to take observed patterns into account. The rescue effect (B) refers to the reduction in extinction (and therefore, turnover) rates on near islands because recruitment of immigrants can supplement otherwise dwindling populations. The target area effect (D) refers to the tendency for immigration rates to be higher on larger islands. I = immigration rate; E = extinction rate; T = turnover rate; S = species richness; P = richness of the source or "pool" biota; s = small island; L = large island; N = near island; and F = far island. (After Gotelli 1995.)

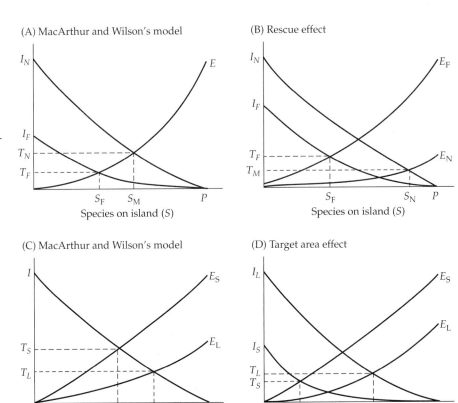

rates ($T_F > T_N$) from that predicted by the MacArthur-Wilson model, but the predicted relationship for the equilibrium number of species remains unchanged ($S_N > S_F$).

Brown and Kodric-Brown's study emphasizes the importance of testing all predictions of a model as well as critically evaluating its basic assumptions. Unless this is done, investigators risk misinterpreting data that are merely consistent with the model as corroborating evidence, and miss an opportunity to advance the theory.

THE TARGET AREA EFFECT.   Just as extinction rates may be influenced by island isolation, immigration rates may be influenced by island area. Again, this is a violation of one of the assumptions of the equilibrium model, but one that requires only a slight modification. Michael Gilpin and Jared Diamond (1976) suggested that at least part of the increase in species richness with increasing island area may derive from a tendency for immigration rates to be higher on larger islands. That is, larger islands may serve as more effective target areas for potential immigrants because they are more likely to be seen (by active immigrators) or encountered (by passive immigrators). In his studies on recolonization of defaunated *Spartina* islands, Jorge Rey (1981) found that immigration rates of arthropods were positively correlated with island area. In a quite different system, Hanski and Peltonen (1988) found that colonization rates of shrews on islets in a Finnish lake increased with island area. While this suggests a target area effect for active immigrators, it is important to note that colonization includes the survival and establishment of breeding populations, as well as immigration.

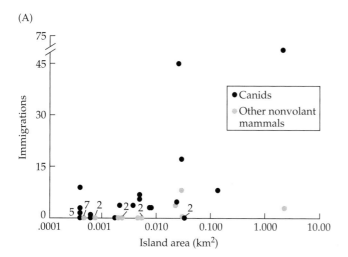

(A)

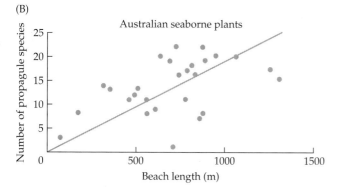

(B)

FIGURE 13.22   Two examples of the target area effect—the tendency for larger islands to attract or intercept more immigrants. (A) Nonvolant mammals dispersing across the snow-covered ice to islands of the St. Lawrence River, which forms the border between the northeastern United States and Canada. Immigration rates increase significantly with island area for the canids (primarily coyotes and red foxes) as well as other mammals (including raccoons, weasels, red squirrels, voles, deer mice, and shrews). (B) The diversity of seaborne plant propagules found along the beaches of islands in the vicinity of northeastern Australia increases with the length of the beachfront. (A after Lomolino 1990; B after Buckley and Knedlhans 1986.)

Lomolino (1990) was able to conduct a more direct test of the target area hypothesis by tracking the movements of terrestrial mammals across the ice-covered St. Lawrence River of North America during winter. The characteristics of tracks in the snow allowed the identification of species and the mapping of actual movements among islands and sites along the mainland. Immigration rates of both small and large mammals were significantly and positively correlated with island area (Figure 13.22A). In a very different set of studies of plant propagules drifting onto the beaches of oceanic islands, Buckley and Knedlhons (1986) also found that immigration rates increased with island size (Figure 13.22B). Thus, there is direct evidence, albeit limited, that at least some of the species-area relationship can be attributed to the target area effect. Again, this calls for a slight modification of the equilibrium model, one that alters relative turnover rates, but not predicted patterns in species richness (see Figure 13.21C,D; see also Hamilton et al. 1964; Connor and McCoy 1979; Coleman et al. 1982; McGuiness 1984).

THE SMALL ISLAND EFFECT.   In their 1967 monograph, MacArthur and Wilson remarked that the species-area relationships of very small islands may be "truly anomalous." They referred to Niering's (1963) study of higher plants on islands of the Kapingamarangi Atoll, Micronesia, which indicated that species richness appeared to be independent of area for the smaller islands (those of less than 3 acres; Figure 13.23). Other studies also reported a **small island effect** (see Wiens 1962; Whitehead and Jones 1969; Woodruffe 1985; Dunn and Loehle 1988; but see also Barrett et al. 2003) and, more recently, Lomolino and

FIGURE 13.23 The small island effect refers to the tendency of the species richness of some insular faunas to remain relatively low and independent of area for the smallest islands. This pattern was found among the higher plants of the Kapingamarangi Atoll, Micronesia. (After MacArthur and Wilson 1967; data from Niering 1963.)

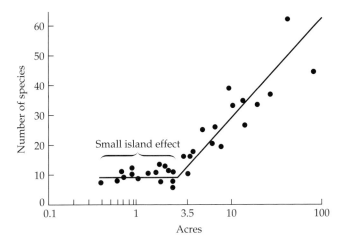

Weiser (2001) assessed the relative importance of the small island effect for just over 100 insular biotas. They found small island effects to be surprisingly common and that the range of the small island effect (i.e., the range over which species richness varies independently of island area) varies in a manner consistent with the resource requirements, immigration abilities, and degree of isolation of these biotas. The upper limit of the small island effect tended to be highest for species groups with relatively high resource requirements and low dispersal abilities, and for biotas of more isolated archipelagoes.

Perhaps the most important and most general inference from these studies is that the species-area relationship may be fundamentally different than the pattern we have been describing for well over a century. The relationship may actually be sigmoidal and, if so, future studies in island biogeography may well focus on three fundamentally different realms of the species-area relationship (Figure 13.24): (1) small islands where species richness varies independent of area, but with idiosyncratic differences among islands and with catastrophic events such as hurricanes (see also Barrett et al. 2003); (2) islands beyond the upper limit of small island effects where richness varies in a more deterministic and predictable manner with island area and associated, ecological factors; (3) islands large enough to provide the internal geographic isolation (large rivers, mountains, and other barriers within islands) necessary for in situ speciation.

Similarly, for reasons discussed in Chapter 6 (see Figure 6.19) and earlier in this chapter (see Figure 13.4B), the species-isolation relationship may also be sigmoidal—this time, a negative sigmoidal. Taken together, these patterns suggest that species richness tends to decline most rapidly across intermediate levels of area or isolation, presumably those levels that approximate the modal resource requirements or dispersal distances of the species pool. In contrast, both species-area and species-isolation relationships may be difficult to detect for extremes of island size and isolation, respectively (see Figures 13.4 and 13.24; see also Lomolino 1998, 2000a,b, 2002). Future studies on the potential sigmoidal nature of these relationships may prove extremely valuable because, if corroborated, they may change the way we study two of the most common patterns in nature.

Nonequilibrium Biotas

Comparing observed patterns of insular species diversity with the predictions of the MacArthur-Wilson model reveals many cases in which the diversity clearly does not achieve an equilibrium between contemporary rates of immi-

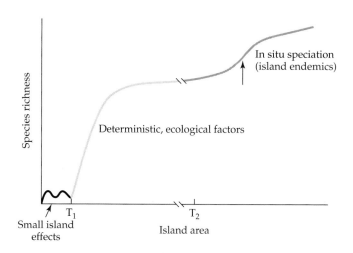

FIGURE 13.24 A general model for the species-area relationship (Lomolino 2000a,b) that includes the scale-dependent changes in the principal factors structuring insular communities including: (1) small island effects and idiosyncratic differences among islands and effects of hurricanes and other stochastic extinction forces—predominately on the small islands (i.e., those < the ecological threshold T_1); (2) more deterministic, ecological factors associated with habitat diversity, carrying capacity and extinction/immigration dynamics as envisioned by MacArthur and Wilson (1967)—on islands of intermediate size; and (3) in situ speciation—on the relatively large islands (i.e., those > the evolutionary threshold T_2).

gration and extinction. The number of species on these islands is not remaining approximately constant; instead, it is either increasing or decreasing steadily in response to major historical events. Ironically, as a result of the failure of some of its predictions, MacArthur and Wilson's approach to island biogeography has proved valuable for detecting and interpreting the dynamics of nonequilibrium biotas. In fact, many of the observed patterns can be explained in terms of the biota approaching a new equilibrium number of species following a historical perturbation. As we noted earlier, the heuristic value of MacArthur and Wilson's theory may be entirely analogous to that of the Hardy-Weinberg equilibrium model for gene frequencies (see Heaney 2000; Heaney and Vermeij 2004). In the latter case, the model's true value is not its prediction of a stable equilibrium, but its utility in identifying those nonequilibrial situations where factors beyond the model (in particular, genetic drift, natural selection and catastrophic disturbance) are important.

PLEISTOCENE REFUGIA. As pointed out in Chapter 9, changes in climate and sea level during the Pleistocene caused major shifts in the distributions of organisms on the continents, especially terrestrial and freshwater species. Their effects on certain insular habitats also were quite profound. In some cases, once isolated habitat islands were connected by habitat bridges, permitting a free interchange of biotas; in other instances, new islands or habitat islands were created by the intrusion of formidable barriers to dispersal.

The legacy of these Pleistocene perturbations is still apparent in many insular distributions. Many islands and patches of isolated habitat were formed by rising sea levels and climatic changes at the end of the Pleistocene. Sometimes the fragmentation of extensive areas of once continuous habitat left small islands oversaturated with species and, following the course of biotic relaxation discussed earlier in this chapter, diversity began to decrease as certain species were eliminated by extinction. In other cases, completely new insular habitats were created and began to acquire species by colonization and speciation. The extent to which contemporary biotas still show the effects of these historical changes depends largely on the kinds of barriers isolating these post-Pleistocene islands and on the ability of different kinds of organisms to disperse across them.

In one of the first attempts to test the applicability of the MacArthur-Wilson model to the distributions of organisms among habitat islands, Brown (1971b, 1978) studied the small non-flying mammals—and later, the birds—inhabiting isolated mountaintops in southwestern North America (see also Johnson 1975;

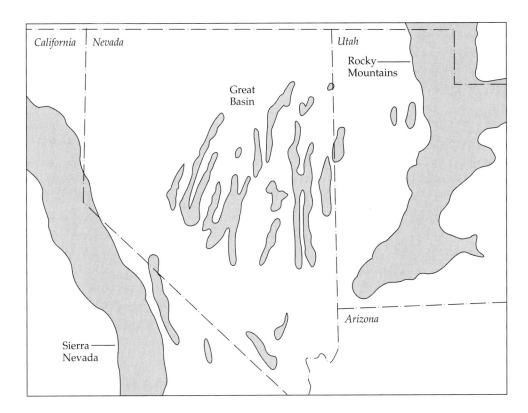

FIGURE 13.25 Isolated mountain ranges of the Great Basin in western North America are islands of cool, mesic forest habitat in a sea of sagebrush desert. The ranges shown, with peaks mostly higher than 3000 m, lie between two montane "mainlands": part of the Rocky Mountains to the east, and the Sierra Nevada Mountains to the west. The desert valleys between the mountains are readily crossed by birds, but they are virtually absolute barriers to dispersal of many small mammal species. (After Barbour and Brown 1994.)

Behle 1978; Grayson 1987a,b, 1993; Grayson and Livingston 1993; Grayson 2000; Grayson and Madson 2000). During glacial maxima, the climate of this region was relatively cool and wet, and forests formed an expansive "mainland" of habitats for boreal species. As we explained in Chapter 9, the moisture-laden westerlies shifted northward following the last glacial recession and the regional climate warmed and dried. As a result, the Great Basin, which lies at about 1500 m elevation between the Rocky Mountains to the east and the Sierra Nevada Mountains to the west (Figure 13.25), developed into a vast region of desert. The now isolated mountain ranges, some rising to more than 3000 m, formed islands of isolated coniferous forest and other mesic habitats surrounded by a sea of sagebrush desert.

These isolated mountaintops are now inhabited by a number of boreal (northern forest) mammal and bird species that are restricted to the cool, moist habitats of higher elevations. The patterns of diversity of these mammals and birds exhibit both similarities and differences when compared with each other and with the biotas of oceanic islands. As predicted by MacArthur and Wilson's theory, the number of boreal forest species increases with the size (area) of the mountain range (Figure 13.26). However, there is no detectable effect of isolation by distance on the diversity of either taxon. The Rocky Mountains and the Sierra Nevada Mountains (likely sources of colonists) support a much

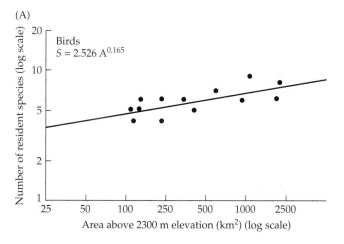

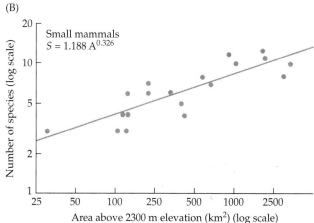

FIGURE 13.26 Species-area relationships for the boreal resident birds (A) and small terrestrial mammals (B) inhabiting the isolated mountain ranges of the Great Basin. Birds continually recolonize these mountains, whereas the mammals are relicts of more widespread populations that existed during the Pleistocene. (After Brown 1978.)

higher diversity of boreal forms than the isolated mountaintops do, including all of the species that are found on the isolated mountaintops.

Brown (1971) proposed that the present boreal mammal populations of these montane forest islands are relicts, vicariant remnants of once widespread distributions during the Pleistocene. This model is consistent with plant fossils, which show that as recently as 10,000 to 12,000 years ago, the climate of the Great Basin was cooler and wetter, and the vegetation zones were shifted several hundred meters below their present elevations (Wells and Berger 1967; Wells 1976, 1979; Thompson and Mead 1982). This shift would have connected several presently isolated habitats across the entire Great Basin, permitting all forest islands to be colonized by all species for which appropriate habitat bridges existed. At the end of the Pleistocene, however, the cool, mesic habitats shrank back to higher elevations, completely isolating on the mountaintops those boreal mammal species unable to disperse across the desert valleys (see Grayson 1993). After being isolated, some insular populations became extinct, reducing the diversity of the boreal biota—especially on small mountaintops— but relictual populations of other species have survived until the present.

Brown's model proposes that, in the absence of post-Pleistocene immigration, the mammalian faunas of the isolated mountaintops have been relaxing toward an equilibrium of zero species. Small islands have had high extinction rates and have lost most of their fauna in 10,000 years, whereas large islands still retain most of their original species. One additional observation supports this relaxation model. Late Pleistocene or more recent fossils of boreal species, including some forms still found on some of the larger isolated mountain ranges, have been found in the intermontane valleys and on several mountaintops, indicating that these species were indeed once present and have become extinct within the last 12,000 years (Grayson 1981, 1987a,b, 1993; Thompson and Mead 1982).

Paradoxically, birds also fail to exhibit significant species-isolation relationships; in this case, because they are excellent dispersers. In contrast to mammals, boreal bird species appear to be present on all mountain ranges where there are sufficient areas of suitable habitat (Johnson 1975; Behle 1978). Boreal birds have been observed flying across the desert valleys, which appear to pose no significant barriers to their colonization of even the most isolated mountaintops. A high immigration rate apparently continually replenishes bird populations on many small islands. Thus, the distribution of boreal birds may well represent an equilibrium between immigration and extinction. How-

(A)

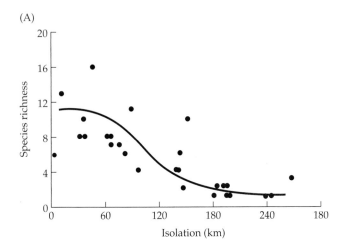

(B)

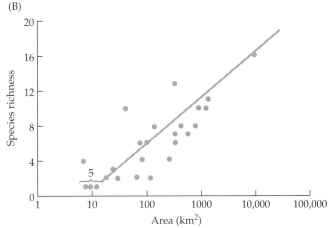

FIGURE 13.27 (A) Species-isolation and (B) species-area relationships for nonvolant forest mammals on 27 montane "islands" in the American Southwest (primarily Arizona, New Mexico, southern Colorado, and southern Utah). Because these mesic forests are isolated by "seas" of xeric woodlands and deserts, their biotic communities often exhibit biogeographic patterns similar to those of species on true islands. (After Lomolino et al. 1989.)

ever, even this is not quite the sort of equilibrium predicted by MacArthur and Wilson, because the immigration and extinction processes appear to be highly deterministic instead of stochastic. The rate of colonization is so high that habitats are almost completely saturated with those species that can live there. When, on rare occasions, an insular population does become extinct, it probably is replaced rapidly by individuals of the same (or ecologically similar) species.

Other studies also indicate that relaxation rates of supersaturated communities are inversely proportional to island area and tend to be most rapid during the period immediately following isolation (e.g., see Diamond 1972; Wilcox 1978, 1980; Terborgh et al. 1997). It may, however, be invalid to assume that all habitat islands are inhabited by fauna in perpetual states of relaxation. Even Brown's interpretation of the Great Basin mammals may need slight modification, because some species may occasionally disperse across xeric habitats (see Grayson and Livingston 1993; Lawlor 1998).

Still, it is likely that most species of forest mammals in the Great Basin, especially in the more xeric, southern reaches of this region, are completely isolated by the intervening deserts (see Lomolino and Davis 1997). In contrast, intermountain habitats in the American Southwest, which lie south of the Great Basin, may well represent filters, not perfect barriers, for the movements of many forest mammals (see Figure 9.22). The key difference between these two regions is that most mountain forests of the American Southwest are isolated by woodlands, not deserts. As a result, post-Pleistocene immigrations are likely for many, if not most, species, and mammalian communities may have approached a dynamic equilibrium as envisioned by MacArthur and Wilson. As illustrated in Figure 13.27, the species richness of forest mammals in this region is significantly correlated with both isolation and area. In addition, intermountain dispersal and post-Pleistocene colonization has been documented for a number of these species (Figure 13.28; Davis and Dunford 1987; Davis and Brown 1989; Lomolino et al. 1989; Davis and Callahan 1992; Brown and Davis 1995; Lomolino and Davis 1997).

Again, however, we caution against overgeneralizing from these studies. Patterson (1980, 1984, 1995) has provided convincing evidence that at least some of these communities, especially those on the more isolated mountains in southern New Mexico, may truly be Pleistocene relicts. Just as Brown observed in the Great Basin, some of the very distant islands of the American Southwest are surrounded by deserts as well as woodlands, making post-

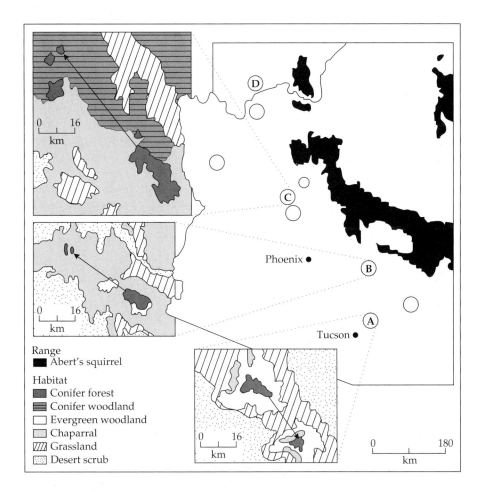

FIGURE 13.28 Relatively rapid intermountain dispersal by Abert's squirrel (*Sciurus aberti*) in Arizona. Open circles indicate sites where the squirrels were artificially introduced between 1940 and 1945. Lettered circles indicate introductions that were followed by colonization of nearby montane forests far removed from the natural range of this species. The time between introduction and apparent colonization was approximately 20, 30, 40, and 10 years for sites A, B, C, and D, respectively. The dispersal distances of 23, 29, 35, and 10 km, respectively, are roughly proportional to these times. (After Lomolino et al. 1989; data from D. E. Brown and R. Davis, pers. comm.)

Pleistocene immigration very unlikely, if not impossible, for many mammal species. Thus, mammalian communities on the more isolated mountains of this region still may indeed be relaxing from supersaturated levels of richness in the absence of contemporary immigrations (see also Lomolino and Davis 1997).

The take-home lesson is relatively simple: heterogeneous landscapes and species pools render simple measures of geographic isolation inadequate for assessing immigration potential. Immigration filters vary both within, as well as across, archipelagoes. Thus, species richness of mammals of the American Southwest is more strongly correlated with extent of intervening deserts versus the overall distance between montane islands and their potential sources (the latter of which included woodlands and other less-isolating habitats; see Lomolino et al. 1989; see also Figure 16.33). Also, as we noted above, comparisons across archipelagoes also pose special challenges. Great Basin birds

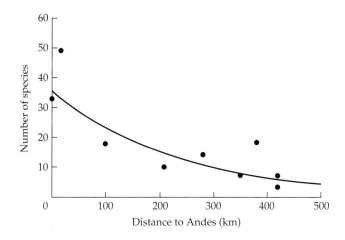

FIGURE 13.29 Much like that of their counterparts on oceanic islands, the species richness of birds in montane forest islands near the Andes in Venezuela, Colombia, and Ecuador decreases with isolation. (After Nores 1995.)

appear to be uninfluenced by isolation: that is, most islands may fall well within their immigration abilities. In other, more isolated regions, avian species richness declines significantly with isolation (e.g., montane birds of southern California and Baja California [Kratter 1992]; and the Andes of Venezuela, Colombia, Ecuador [Vuilleumier 1970], and Argentina [Nores 1995]; Figure 13.29).

LANDBRIDGE ISLANDS. Rising sea levels since the Pleistocene have created many new islands in coastal regions. The approximate 100 to 200 m rise in sea level that occurred between 18,000 and 10,000 years ago inundated many landbridges and created numerous continental islands. These are called landbridges or continental islands because, unlike oceanic islands, they were once part of the mainland, or at least connected to it by a bridge of terrestrial habitats.

Diamond (1972, 1975b) found that the influence of past connections to the mainland of New Guinea is apparent in the composition of the avifauna of the landbridge islands off its coast. Islands that are separated from the mainland by water less than 200 m deep support a greater number of species than oceanic islands of comparable size and distance from New Guinea (i.e., those that arose as undersea volcanoes and have never been connected to the mainland). Although the landbridge islands lack many bird species found in comparable habitats on New Guinea, they have several kinds of birds that are found on the mainland but never occur on the oceanic islands. Of course, there are still other species that occur on both landbridge and oceanic islands. Because these birds obviously have crossed water barriers to colonize the oceanic islands, investigators cannot be sure whether their populations on the continental islands are Pleistocene relicts or whether they have been replenished by subsequent immigration (see also Lawlor 1983, 1986).

POST-PLEISTOCENE DYNAMICS OF FRESHWATER FAUNAS. The present distribution of freshwater fishes in southwestern North America reflects a history of aquatic habitat connections during the cooler and wetter climate of the Pleistocene, followed by isolation and subsequent extinctions (Hubbs and Miller 1948; Miller 1948; Smith 1978). Hard as it may be to believe, only a few thousand years ago, Death Valley, in the most arid part of the North American desert, was almost completely filled by a large lake, which was supplied by a major system of permanent rivers and springs (see Figure 9.24). At least three genera of fishes inhabited this basin, and they have persisted as relictual populations in isolated springs. In the case of these fishes, there can be no doubt about the susceptibility of small, isolated populations to extinction. Many have disappeared within the last few years as humans have diverted water or introduced exotic species of competing or predatory fishes.

In most of the examples discussed so far, the biotas of insular habitats isolated since the Pleistocene are relictual. Once widespread habitats containing diverse biotas were diminished in size and became fragmented. As soon as these islands formed, they were oversaturated with species and, as a result, they have been losing species by extinction ever since. In contrast, some isolated habitats, as well as areas scoured by glaciers, begin with few, if any, species and remain undersaturated long after glaciers have receded (see Chapter 9).

An excellent example of the latter situation is provided by the glacier-formed lakes of northern North America and Eurasia. Many lakes, including those of northern Eurasia and the Great Lakes and the Finger Lakes of northeastern North America, were gouged out by the advancing continental ice sheets, and then filled with water as the glaciers retreated (see Figure 9.23). Some groups, such as certain algae, protozoans, and invertebrates that use cysts or other effective means of long-distance dispersal, may have achieved an equilibrium between rates of colonization and extinction in these habitats over a relatively short period of time. In contrast, other less vagile animals, such as fishes (Smith 1981) and some mollusks, may never achieve an equilibrial level of species richness. The fish faunas of many of these lakes appear to be undersaturated and still gradually increasing in diversity (Barbour and Brown 1974; see also Smith 1979; Browne 1981; Holland and Jain 1981; Tonn and Magnunson 1982; Brown and Dinsmore 1988; Hugueny 1989; Watters 1992; Elmberg et al. 1994; Oberdorff 1997).

Only a handful of fish species can survive for more than a few minutes out of water at any stage of their life cycle, so fishes can colonize new areas only when connections of aquatic habitat are present. They usually colonize lakes from rivers and streams, but most lotic waters contain only a few species that can also be successful in lentic environments. Many glacier-formed alpine lakes lack native fishes entirely because they have never been connected by suitable habitat bridges to waters inhabited by fishes. The fact that several introduced species thrive in some of these lakes is further evidence that the absence of fishes is a consequence of barriers to dispersal. Many of those glacier-formed lakes that do contain fishes still appear to be undersaturated with species. This is particularly true of the very large lakes, such as the Great Lakes in North America and Lake Baikal in the Soviet Union. There were not enough species in the rivers and streams draining these lakes to fill the niches available to fishes. Some taxa, such as the ciscoes or whitefishes (*Coregonus*) in the Great Lakes, have been diversifying by endemic speciation and adaptive radiation, but there has not been sufficient time since the Pleistocene to achieve an equilibrium between speciation and extinction.

This conclusion is supported by a comparison of species-area relationships for fishes inhabiting the glacier-formed lakes in temperate North America and the much older lakes of tropical Africa (Figure 13.30). As mentioned in Chapter 7, the relatively high diversity of Africa's Rift Valley lakes derives primarily from the spectacular endemic speciation and adaptive radiation of cichlids in the larger lakes (Fryer and Iles 1972; Greenwood 1974; Meyer 1993). Lake Victoria, the largest lake in Africa, now harbors over 300 endemic species of cich-

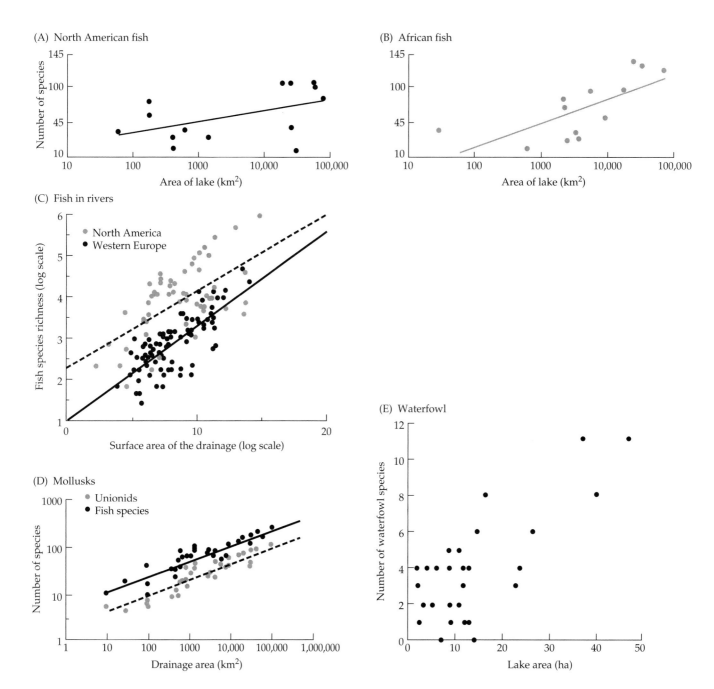

FIGURE 13.30 Species-area relationships for the fishes inhabiting lakes in northern North America recently formed by glaciers (A) and much older lakes of central and eastern Africa (B). The large African lakes have acquired diverse fish faunas by endemic speciation. Although such speciation is occurring in the North American lakes, it has not yet produced high species richness in the largest lakes. (C) Fish species richness in rivers as a function of total surface area of drainage basins in Western Europe and North America. The generally higher species richness of North American rivers may result from higher speciation rates or the tendency for many North American rivers to flow north-to-south, which thus has provided an opportunity for dispersal south of advancing glaciers and therefore higher persistence during periods of glaciation. (D and E) Just as species diversity of fish increases with area of lakes or catchment basins of rivers, diversity of dependent species (e.g., parasitic unionid mollusks of the Ohio River system, and waterfowl in boreal lakes of Finland and Sweden) also increases with surface area of their aquatic ecosystems. (A–B after Barbour and Brown 1974; C after Oberdorff 1997; D after Watters 1992; E after Elberg et al. 1994.)

lids, yet it appears to have dried up completely during the late Pleistocene (Johnson et al. 1996; Seehausen 2002). Following recolonization, speciation of these fishes must have proceeded at an extremely rapid rate indeed, possibly reaching a dynamic equilibrium before the relatively recent surge in extinctions due to introduced species and other anthropogenic activities. Finally, the relative time scale of underlying glacio-pluvial events holds some key insights for interpreting and predicting the biogeographic dynamics of insular systems. Recall from Chapter 9 that interglacial periods are relatively short compared with glacial periods (often just 10,000–30,000 years vs. 75,000–300,000 years, respectively). Thus, on a geological time scale, relaxation of relictual communities or accumulation of species in environments cleared by glaciers is likely to be a transient and incomplete phenomenon. Paleoecologists or biogeographers of future millennia may well reverse our original question and ask, To what extent do distribution patterns during glacial periods represent the legacy of the relatively short pulses of interglacial events?

Frontiers of Island Biogeography

The views expressed in Box 13.3 continue to be an accurate characterization of the exciting and dynamic status of island biogeography theory and the challenges facing any scientist who attempts to understand patterns in diversity of insular communities. If a new paradigm of island biogeography emerges in the near future, like MacArthur and Wilson's model, it will likely be dynamic, but, in this case, based on all three of the fundamental biogeographic processes (immigration, extinction, and speciation) and their combined but varying influences across a broad range of spatial and temporal scales (see the figure in Box 13.3, and Figures 13.24 and 13.31; see also Meadows 2001; Lomolino 2002). It is also our hope that any new theory of island biogeography be general enough to explain patterns of species richness for isolated ecosystems of the aquatic realm as well. Numerous studies reveal that lakes, rivers, wetlands, and coral reefs—or other island-like ecosystems of freshwater and marine systems—also exhibit properties of insular biotas: species-area and species-isolation relationships; biotic turnover; relaxation; endemicity; and the influence of speciation, disturbance, and other long-term processes (Browne 1981; Watters 1992; Elmberg et al. 1994; Oberdorff 1995; Johnson et al. 1996; Oberdorff et al. 1997; Vermeij 2004; see Figure 13.30).

While small-scale experimental studies such as those performed by Wilson and Simberloff (1970; discussed earlier in this chapter) on tiny mangrove islands may also provide important insights, island biogeographers are likely to continue to capitalize on natural experiments afforded by islands, especially given their perhaps unrivaled abilities to provide insights on the key, scale-dependent processes influencing biological diversity. Biogeographers must, of course, design these studies as carefully as other researchers design controlled, manipulative experiments. Again, given the abundance, diversity and tractability of islands, they will continue to serve as excellent subjects for invaluable scientific studies. Perhaps the most important insights will be generated by those studies that can strategically dissect or "deconstruct" (sensu Huston 1994; Marquet et al. 2004) diversity patterns by comparing those among functionally different groups of species or archipelagoes (i.e., those that differ with respect to factors influencing immigration, extinction, and speciation; see Schoener 1983; Ricklefs and Lovette 1998; Figure 13.9). In addition, whereas nearly all of the studies on insular "diversity" that we discussed in this chapter focused just on species richness, biogeographers have recently begun to explore analogous patterns in alternative measures of diversity (e.g.,

BOX 13.3 *Paradigms of island biogeography*

■■▮ We conclude our discussion on patterns in insular species richness with excerpts from two papers, including the introduction and the capstone to a Special Feature of the journal Global Ecology and Biogeography, which explored the potential for advancing or replacing the equilibrium theory of island biogeography.

A Call for a New Paradigm of Island Biogeography (Lomolino 2000a)

After 35 years … a growing consensus of ecologists now questions whether the Equilibrium Theory remains a useful paradigm for modern biogeography. We now have a much better appreciation of the complexity of nature and we study patterns that span a very broad range of spatial, temporal and ecological scales. The very reason for its heuristic appeal now appears to be the most constraining feature of the Equilibrium model (i.e., its simplicity). The Equilibrium Theory was developed to explain variation in species richness at the scale of an archipelago during ecological time … however, the Equilibrium Theory has some fundamental limitations that seem to make it inadequate as a modern theory of island biogeography. Three of the most important limitations are discussed below:

1. As we broaden the spatial and temporal scales of our studies, it becomes increasingly more clear that many systems are not only dynamic in species composition, but that they may seldom attain an equilibrium number of species. That is, many systems may still bear the imprint of speciation, major geological and climatic events, or anthropogenic disturbances that either modify the species pool or alter immigration and extinction rates before an equilibrium can be achieved.
2. The theory assumes that insular habitats and immigration filters (intervening landscapes or seascapes) are homogeneous within and among archipelagoes. On the contrary, much of the observed variation in insular species composition and richness within archipelagoes may result from predictable variation in habitat and other environmental conditions among islands (e.g., larger islands tend

to include a greater diversity of habitats). Similarly, differences in patterns of community structure among archipelagoes may be attributed to differences in immigration filters [e.g., bodies of water with and without strong currents (see Lomolino, 1994)]; intervening landscapes dominated by different habitats (see Aberg et al., 1995); or other factors that vary at geographic scales (e.g., clines in temperature and productivity along latitudinal gradients).
3. The theory is species neutral (i.e., it assumes that all species are independent and equivalent). Yet, as with many other complex systems, the structure and dynamics of insular communities is also influenced by ecological feedback (i.e., interactions among species and other system components). Because the Equilibrium Theory assumed that species are equivalent, it ignored the potential role of interspecific interactions. Moreover, this species-neutral model could not address patterns in species composition. Many of the most interesting patterns in biogeography concern not just how many, but *which* species inhabit islands.

This latter limitation is especially problematic for those attempting to conserve biological diversity on islands or isolated fragments of native ecosystems. There is no compelling reason to believe that rare and geographically restricted species are identical to the more common and wide-ranging ones. On the contrary, endangered species

are the oddballs, the species that are barely hanging on because of their relatively high resource requirements, limited dispersal abilities, and high sensitivity to human disturbance…

The task of developing a new, more comprehensive theory of island biogeography will be no less challenging for us now than it was when Eugene Gordon Munroe, then a graduate student at Cornell University, first proposed an equilibrium theory of island biogeography in 1948 (see Box 13.1; Munroe, 1948 and 1953; Brown and Lomolino, 1989). … It is now time to replace the Equilibrium Theory, a model that has reigned as paradigm for nearly 35 years. Rather than just dwell on its shortcomings, I will try to take a more constructive approach by outlining some of the desirable features of a future paradigm for the field.

- The model should be fundamentally simple: one based on the three fundamental biogeographic processes: immigration, extinction, and evolution (Figure A).

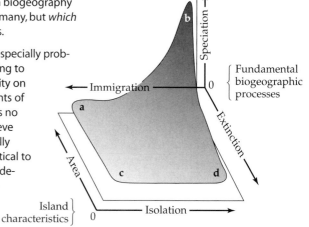

Figure A A model of island biogeography illustrating the variation in the three fundamental biogeographic processes (immigration, extinction, and evolution) as a function of characteristics of the islands in question. Immigration rates should increase with proximity to a source region and, equivalently, with immigration ability (ability to travel or be carried across a particular immigration filter) of the focal species. Extinction rates should decrease as island area increases, or increase with increasing resource requirements of the focal species. Speciation (within islands) should be most important where extinction and immigration are lowest and, therefore, increase with island area and isolation, but decrease with increasing resource requirements and vagility of the focal species. Shading indicates the relative levels of species richness, and the relative resistance to, and resilience of, insular biotas following disturbance. Community characteristics of islands within the four labeled regions of the shaded surface should be as follows: (a) moderate to relatively high richness, low endemicity and low turnover; (b) moderate to relatively high richness, high endemicity and low turnover; (c); moderate to low richness, low endemicity and high turnover; (d) depauperate islands.

BOX 13.3 *(continued)*

- The model should be hierarchical and include the system features (e.g., isolation, nature of immigration filters, area, geology, and climate) that influence immigration, extinction, and evolution at the spatial and temporal scales in question (Lomolino 1999).

- It should include potential feedback among system components (e.g., interspecific interactions) and interactions among biogeographic processes (e.g., the rescue effect [sensu Brown and Kodric-Brown, 1977], which refers to the tendency for relatively high immigration rates to reduce extinction rates on near islands).

- Finally, the model should be species-based. Species vary in many ways that affect immigration, extinction, and speciation, and many biogeographic patterns derive from, not despite, differences among species. For example, the form of the area species-isolation and species-area relationships may reflect very general patterns of variation in immigration abilities and resource requirements among species. ...

Judging from the surging interest in island biogeography over the past decade or so, our field may be primed for a second, modern revolution. Whatever their final form, the new models will likely build on the prescient insights of MacArthur, Wilson, and other charter members of the previous revolution. Accordingly, while we recommend a relaxation of the equilibrium assumption, we continue to agree with MacArthur and Wilson's more fundamental assumption that insular community structure is dynamic, resulting from recurrent extinctions and immigrations (plus, speciation).

Finally, if our comments here seem irreverent, that is terribly unfortunate because few scientists admire MacArthur and Wilson's contributions more than we do. In fact, we believe that, were he still alive, Robert MacArthur himself would be very disappointed to see that his model, largely unchanged, remained the paradigm of island biogeography for nearly 35 years. We feel strongly that the best way to honor MacArthur and Wilson's seminal contribution is to re-evaluate the equilibrium theory and replace it with one that offers a more versatile and comprehensive view of the fundamental processes influencing the structure of insular communities.

"Paraphrasing Picasso, MacArthur once observed 'a theory is a lie which makes you see the truth'. If the equilibrium theory forced us to pose new questions and to seek new evidence in order to better understand [insular] communities, its purpose has been fulfilled."

(Above quote from a paper by MacArthur's first graduate student, K. L. Crowell, 1986.)

Excerpted from *A call for a new paradigm of island biogeography*, Mark V. Lomolino, Global Ecology and Biogeography (2000) 9: 1–6.

Concluding remarks: Historical perspective and the future of island biogeography theory (Brown and Lomolino 2000.)

Most great ideas in science do not endure forever as laws or principles. Instead, they serve as stepping stones to newer theories and better understanding of the workings of nature. This is likely to be the role of MacArthur and Wilson's theory. Its many shortcomings have finally become apparent. Some of these, such as the nonequilibrial nature of landbridge islands and other historically interconnected insular systems, were pointed out decades ago (e.g., Brown, 1971; Diamond, 1972; Barbour & Brown, 1974; Heaney, 1986), but interpreted by the majority of biogeographers as "exceptions which prove the rule," rather than fundamental challenges to the theory. But the number of cases which could not adequately be explained continued to mount. Now, there are too many complications, exceptions, and violated assumptions to ignore.

By aiming to be more realistic and more general than MacArthur and Wilson's theory, such efforts might aspire to be stepping stones toward better understanding and new paradigms. Whether they will eventually be seen as having played such a role, will be for history to determine.

What does seem clear is that the discipline of biogeography is again, as in the 1960s and 1970s, in transition and ferment. The simple old paradigms are not holding up. This is true not only of MacArthur and Wilson's equilibrium theory, but also of the vicariance approach which sought to use phylogenetic reconstructions of lineages to interpret the histories of regions and their biotas (e.g., Rosen et al., 1979; Nelson & Platnick, 1981; Humphries & Parenti, 1986; Brown & Lomolino, 1998). For example, the phylogenies of several lineages inhabiting the Hawaiian Islands show not just one highly concordant pattern as predicted by vicariance theory, but a diversity of relationships reflecting differences in colonization sequences and ecological requirements among the clades, and differences in barriers to dispersal and modes of speciation within and among islands (Wagner & Funk, 1995; Brown & Lomolino, 1998). The papers in this special feature (of Global Ecology and Biogeography, Volume 9, Issue 1) are symptomatic of the discipline as a whole.

What is not so clear is where the current trends are heading. Does the future hold the promise of developing and testing theories that may apply across a broad range of taxonomic groups, types of ecosystems, and geographic regions? Or, does it hold the less exciting prospect of entering another stage of descriptive science, in which each situation is treated as a special case? Lacking a crystal ball, we can only note that currently, the discipline of biogeography itself is far from equilibrium. Its present unsteady state reflects not only the inadequacy of the traditional approaches and theories, but also, so far, the absence of accepted alternative frameworks to supply theoretical and empirical direction. ... While islands may not be such simple systems as MacArthur and Wilson's theory envisaged, they still represent some of nature's most invaluable experiments. They are replicated systems, which typically have limited biological diversities, relatively well-understood histories, and only moderate environmental complexities. Islands will continue to offer invaluable insights into how Earth history, phylogenetic history, and contemporary and past environments have interacted to shape biodiversity.

Excerpted from Concluding remarks: *Historical perspective and the future of island biogeography theory*, James H. Brown and Mark V. Lomolino, Global Ecology and Biogeography (2000) 9: 87–92. ▮▮▮

FIGURE 13.31 Whittaker's (1998, 2000) conceptualization of the major theories associated with extremes in dynamics and equilibrial characteristics of archipelagoes and their insular biotas (see also Vermeij 2004; Whittaker 2004). The dynamic equilibrial condition corresponds to MacArthur and Wilson's (1963, 1967) theory; the static equilibrial equates to David Lack's (e.g., 1969) ideas on island turnover of birds; the static nonequilibrium to James H. Brown's (1971b) work on nonequilibrium mountaintops; and the dynamic nonequilibrium to Bush and Whittaker's (1991, 1993) interpretations of Krakatau plant and butterfly data. Considering a single taxon, different positions in this diagram may correspond to different islands or archipelagos, while different taxa in the same island group may also correspond to different positions.

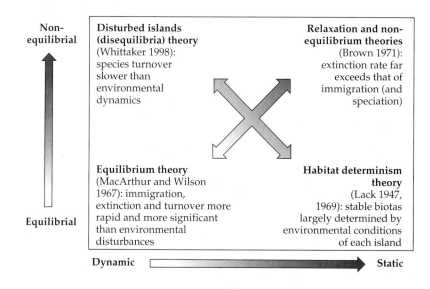

those based on morphological, functional, or phylogenetic variation; see Roy et al. 2004).

In summary, whatever form this new or revised paradigm of island biogeography takes, it would do well to draw on the full spectrum of insular patterns, sometimes referred to as the "island syndrome" (see Adler and Levins 1994). Indeed, islands exhibit many fascinating differences from their mainland counterparts, not just in terms of diversity of their communities, but also in their surprisingly high endemicity along with a suite of fascinating adaptations (physiological, morphological, behavioral, demographic, ecological, and evolutionary) to insular environments. Taken together, these new efforts to understand insular patterns as the products of all three fundamental biogeographic processes, across a much broader range of temporal and spatial scales, and including a more full range of patterns in ecological and evolutionary development of island biotas promise to provide a much more comprehensive understanding of the diversity and distinctness of isolated biotas. We explore many of these truly fascinating patterns in assembly and evolution of insular biotas in Chapter 14.

Island Biogeography

Assembly and Evolution of Insular Communities

ONE OF THE MOST GENERAL PATTERNS in biogeography is that, in comparison to mainland communities, islands tend to be species-poor. Islands are not only isolated, but also small, and thus contain a limited diversity of habitats and resources. As we learned in the previous chapter, the impoverishment of islands has been attributed to their relatively low rates of immigration and their limited ability to support populations. MacArthur and Wilson's (1967) equilibrium theory accounted for differences in species richness among islands as resulting from differences in immigration rates (as affected by isolation) and extinction rates (as affected by island area). Their model, however, included the simplifying assumptions that species are identical with respect to traits influencing their dispersal ability and their ability to survive on islands, and that insular community structure is not influenced by interactions among species.

Biogeographers often are concerned with species composition, that is, *which* species occur in a place, rather than just how many. The equilibrium model, while successfully predicting very general tendencies in species richness, cannot account for patterns in species composition. Furthermore, the structure of insular communities, especially on large and isolated islands, is strongly influenced by evolution in situ. Insular "novelties," including giant flightless birds, dwarf elephants, and assemblages dominated by an astounding diversity of taxa that are otherwise insignificant components of the mainland biota have attracted attention throughout the history of biogeography. Moreover, these novelties—and related patterns in the assembly of insular biotas—have provided invaluable insights into the forces influencing community structure and evolution of all biotas (Weiher and Keddy 1999; Gotelli 2004).

General explanations for these intriguing patterns must recognize that differences among the species matter, because their particular physiological, morphological, behavioral, and genetic characteristics ultimately combine to influence immigration, extinction, and evolution. It should not be necessary to point out that MacArthur and Wilson knew as well as any of their colleagues that species differ in ways directly related to their potential to colonize, survive, and evolve on islands. Both of these scientists were superb naturalists and ecologists, and both produced seminal papers exploring the ecological and evolutionary consequences of differences among species (e.g., see MacArthur 1958; Wilson 1959, 1961; MacArthur and MacArthur et. al. 1966; MacArthur and Levins 1967). The discrepancy between their previous, detailed ecological studies—which emphasized deterministic niche dynamics and differences among species—and their species-neutral equilibrium theory, is sometimes referred to as MacArthur's Paradox (Schoener 1983; Loreau and Mouquet 1999).

In this chapter, we focus on factors influencing the structure of insular communities in both ecological and evolutionary time. First, we discuss the selective nature of immigration and extinction, and then explore how these processes can contribute to **community assembly** (i.e., the development of nonrandom patterns in the structure of insular communities). After reviewing these and other ecological processes, we examine some truly fascinating evolutionary responses to insular environments.

Assembly of Insular Communities

Given that both species and islands differ with respect to factors influencing immigration and extinction, insular communities should represent nonrandom samples of their mainland source pools—samples biased in favor of the better immigrators and better survivors. On the other hand, under the null hypothesis that all species and islands are equivalent, we would expect insular communities to represent random draws from the mainland pool.

Biogeographers often use the terms **harmonic** and **disharmonic** to characterize insular biotas. Harmonic, or **balanced**, biotas are assemblages that are similar in composition to the source biota. They may have fewer species, but the proportions of species in each taxon or ecological category are roughly the same as on the mainland (Figure 14.1). Again, this is the null prediction. Where species and islands are not equivalent, we would predict some nonrandom patterns in species composition, not just differences between insular and mainland communities but among insular communities as well. Certain types of organisms should be overrepresented on islands, and this bias should increase as islands become either more isolated or smaller. As we shall see, while biogeographers have as yet failed to develop a comprehensive model of patterns in species composition, they have provided seemingly incontrovertible evidence for the selective nature of immigration and extinction—that is, for the nonrandom assembly of insular communities.

The Selective Nature of Immigration

If species vary in traits affecting their immigration potential, then they will exhibit different patterns of distribution among islands that vary in isolation. As we have seen in Chapter 6, dispersal abilities do vary—often dramatically—among species, even within a group of closely related species (Table 14.1).

To the extent that these interspecific differences are predictable, they will translate into predictable, nonrandom patterns of species distributions and community structure among islands. For example, groups with obvious adaptations for dispersing long distances over inhospitable habitats tend to be disproportionately common on islands. Birds, bats, and flying insects tend to be well represented on distant oceanic islands, whereas few, if any, native species

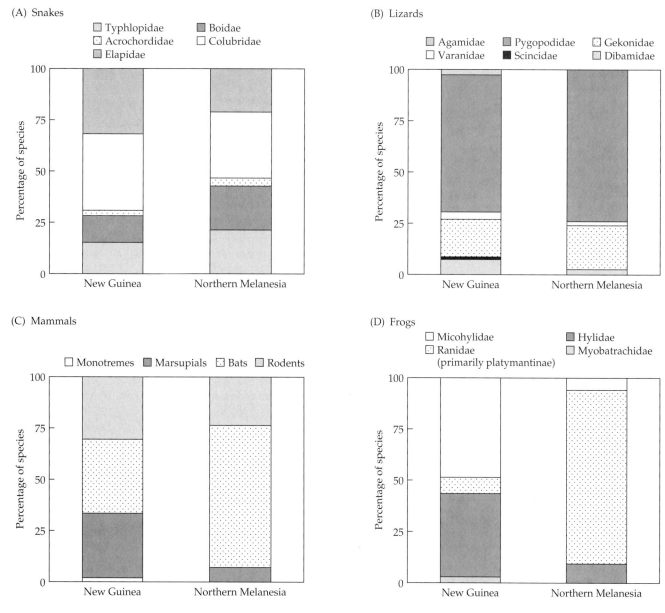

FIGURE 14.1 Species composition of terrestrial vertebrates on islands of Northern Melanesia illustrate how some biotas may be relatively harmonic (A and B), with insular communities similar to those of the mainland with respect to contributions (here, expressed as percentages of species within different families or orders) of component taxa, while others may be highly disharmonic (C and D). (After Mary and Diamond 2001.)

of nonvolant mammals, amphibians, and freshwater fish are present there (see Carlquist 1974; Nilsson and Nilsson 1978; J. Smith 1994). Similarly, the primary means of dispersal for plants inhabiting isolated oceanic archipelagoes is in the feathers, fur, or guts of birds and bats (see Whittaker and Jones 1994; Whittaker 1998). The predominance of avian-dispersed plants on isolated islands is probably a reflection of the relative efficiency of **zoochory** (transport by animals) rather than sheer number of propagules. That is, while more propagules may be dispersed passively by winds and water (**anemochory** and **hydrochory**, respectively), only a tiny fraction of those potential colonists will reach land. Indeed, the chance that a randomly dispersed propagule will reach an island decreases as an exponential function of its isolation. In contrast, avian dispersal to islands typically is a nonrandom, directed and, therefore, highly efficient phenomenon. Thus, plants and other avian-dispersed "hitchhikers" have a much higher probability of colonizing oceanic islands.

TABLE 14.1 *Some examples of documented long-distance dispersal by terrestrial animals*

Reptiles	
Freshwater turtles	Possible 200 miles (322 km) to Madagascar
Snakes	500 miles (805 km) or more to the Galápagos
Lizards	1000 miles (1609 km) to New Zealand; perhaps more than 1000 miles (1609 km) for geckos
Terrestrial Mammals	
Large mammals	Perhaps 30 miles (48 km) or more[a]
Small mammals	Possibly 200 miles (322 km) to Madagascar for civets and insectivores
Rodents	500 miles (805 km) to the Galápagos
Amphibians	Perhaps 500 miles (805 km) to the Seychelles; perhaps 1000 miles (1609 km) to New Zealand
Land mollusks	More than 2000 miles (3218 km) in Polynesia (to Juan Fernandez Island)
Insects and spiders	More than 2000 miles (3218 km)

Source: Wenner and Johnson 1980.

[a]Elephants have been estimated to swim up to 48 km (Johnson 1978). Also, a semiaquatic hippopotamus occurred on Madagascar.

In an analogous fashion, mammals can be divided into two groups—bats and nonvolant forms—on the basis of their over-water dispersal ability (e.g., see Lawlor 1986). Bats have naturally colonized such distant outposts as New Zealand, New Caledonia, and the Canary and Hawaiian Islands, and they are represented by 32 genera on the Greater and Lesser Antilles. In contrast, native nonvolant mammals are completely absent from New Zealand, New Caledonia, the Canaries, and the Hawaiian archipelago, and only 26 genera are known from the Antilles (all but 5 of which have become extinct since the late Pleistocene, owing, at least in part, to human activity; see Morgan and Woods 1986; Woods and Sergile 2001). Note that the disharmonic nature of mammalian communities across Melanesia illustrated in Figure 14.1C was primarily derived from the disproportionately high number of bats on these islands.

For those groups that lack flight or are otherwise incapable of being carried by winds, salt tolerance is an especially important influence on oceanic distributions (see Dunson and Mazotti 1989). Freshwater fishes and amphibians are so limited by their intolerance of salt water that their presence was often taken as an indication that an island was once connected to the mainland. However, it is now clear that nearly all species of freshwater fish found on isolated, oceanic islands were either introduced by humans or they are descendants of marine species that, over evolutionary time, developed abilities to invade brackish—and then, freshwater—systems. New Zealand, for example, has no native primary division freshwater fishes, but introduced trout thrive in its magnificent rivers. Among amphibians, the ranids and bufonids tend to be the dispersal champions (Meyers 1953). Crab-eating frogs (*Rana cancrivora*) and giant marine toads *(Bufo marina)* are especially well adapted for oceanic transport, with high salinity tolerance in both adults and tadpoles (Inger 1954; Udvardy 1969). Consequently, these species are much more widely distributed on oceanic islands than their freshwater counterparts (see also Neill 1958). Likewise, the disharmonic nature of frog communities across Melanesia (see Figure 14.1D) results primarily from the relatively high diversity of platymantines, which lay terrestrial eggs, forego the tadpole stage, and therefore have relatively high tolerance for salt water during dispersal.

Among the mollusks, slugs seem equivalent to freshwater fishes and hygrophylic amphibians in their intolerance of saline waters and their generally limited distributions on oceanic islands. In contrast, land snails adapted to arid conditions, and those small enough to be transported by the winds (so-called micro-snails), are often significant components of oceanic biotas (Solem 1959; Vagvolyi 1975; see also Jaenike 1978).

Among nonvolant mammals, relatively few insectivores and small rodents are found on islands along coastal and inland bodies of water, presumably because of their limited immigration abilities (Crowell 1986; Peltonen et al. 1989). In regions where waters freeze during at least part of the year, insular distributions are strongly influenced by the ability of species to remain active and withstand subfreezing temperatures during winter. In contrast to hibernators and other seasonally inactive species, those that remain active during winter are disproportionately common on islands. Absent, or surprisingly rare, are chipmunks, ground squirrels, woodchucks, jumping mice, and other species unable or unlikely to utilize the winter ice cover as a seasonal avenue for migration (see Tegelstrom and Hansson 1987; Lomolino 1988, 1993b; Peltonen et al. 1989). Even among winter-active mammals, larger species generally are more capable of withstanding winter conditions, and thus are often better at dispersing across ice to colonize islands in winter. Nonvolant mammals of the Kuril Islands, which stretch from Hokkaido—the northernmost island of Japan—to the southern tip of the Kamchatka Peninsula, provide a nice illustration of selection for immigration abilities. As we move away from the mainland, carnivores (primarily red fox, brown bear, and mustelids) become increasingly more important components of the mammalian fauna, presumably because their large size confers greater immigration abilities (Hoekstra and Fagan 1998).

Establishing Insular Populations

In Chapter 6, we tried to emphasize the distinction between immigration and establishment of an insular population. Obviously, both are requisite phases of successful colonization, yet they are often logistically difficult—if not impossible—to study as separate phenomena. For some, and perhaps many, groups of species, ability to establish breeding populations may not be correlated with immigration ability. Or, just as confounding, these traits might be positively correlated in some species groups and negatively correlated in others. While empirical studies addressing this question remain scarce, ecological theory suggests that the former may be the case; that is, good dispersers may, in general, be pre-adapted to establish populations on islands and in other isolated ecosystems. Species adapted to exploit disturbed or newly created environments often combine high dispersal abilities with broad ecological tolerances. Such species should therefore be pre-adapted for successfully colonizing islands. These species are often referred to as *r*-strategists. *K*-strategists, on the other hand, are better adapted for efficient utilization of limited resources, especially when population levels approach the carrying capacity of the environment (K). While most species probably fall between the idealistic extremes of an r/K continuum, biogeographers have long studied the characteristics of a prototypical colonizer (see MacArthur and Wilson 1967).

Empirical studies, while still very limited, tend to support most, but certainly not all, of MacArthur and Wilson's predictions. For example, Baur and Bengtsson (1987) found that, among land snails of the Baltic uplift archipelago, colonizing abilities were higher for species with a broader niche and for those capable of self-fertilization, a trait obviously advantageous for reproducing in newly colonized environments. Similarly, insular biotas of the more isolated, oceanic archipelagoes are often characterized by relatively high frequencies of self-fertilizing plants (Carlquist 1974; Barrett 1998) and parthenogenic lizards (Hanely et al. 1994).

Ebenhard's (1991) review of the available information on characteristics associated with colonization success in other animals is, for the most part, consistent with theoretical predictions. For example, colonization success among introduced birds and mammals was directly dependent on propagule size and clutch size (see Crowell 1973; Mehlop and Lynch 1978; Ebenhard 1989). Numerous introduction experiments with dung beetles (Scarabaeidae) across the globe revealed that colonization success tended to be higher for species with intermediate size, rapid development, high fecundity, and broad niches (see Doube et al. 1991; Hanski and Cambefort 1991; see also Schoener and Schoener 1983; Harrison 1989; Kotze et al. 2000).

The Selective Nature of Extinction

Despite the tremendous variation among insular environments, they all tend to be similar in that they provide limited resources, and as island size decreases, the amount of resources and their replenishment rate decreases. Thus, insular communities should be biased not just in favor of good dispersers, but also in favor of those species that require few resources to maintain their populations. Although this prediction is far from a rule, many studies are consistent with it. The record of extinctions since the Pleistocene, in particular (see Chapter 9), suggests that extinction has by no means been an entirely random process. As we discussed in Chapter 7, the fossil record is replete with evidence for species selection. During the last 10,000 years, some types of species have become locally extinct much more frequently than others, and their differential susceptibility to extinction appears to be related to their ecological characteristics. Of course, this is just what would be expected if, as suggested in earlier chapters, the ecological characteristics of species determine insular carrying capacities, and in turn, these equilibrium population densities largely determine the probability of extinction (e.g., Brown 1971b, 1978, 1981; Van Valen 1973a,b). In fact, it should be possible to predict the relative abilities of species to persist on isolated landbridge islands based on some easily measured ecological traits. Animals of large body size, carnivorous diet, and specialized habitat requirements should experience higher extinction rates than species that are smaller, herbivorous, or more generalized in habitat use.

This predicted pattern is precisely what is observed for terrestrial mammals on isolated, montane "islands" in western North America (Figure 14.2). Small herbivores that are habitat generalists make up the limited subset of species that are found on virtually all mountain ranges; whereas, carnivores and herbivores of larger body size or specialized habitat requirements have become extinct on many montane islands and persist on only the largest ones. This pattern is particularly impressive because it is repeated on two different, geographically isolated sets of mountaintop islands in North America: those of the Great Basin (Brown 1971b) and those farther south in the American Southwest (Brown 1978; Patterson 1984). In Chapter 9, we noted that vertebrate extinctions during the late Quaternary also tended to be biased in favor of large, carnivorous, or specialist species—that is, species with relatively high resource demands and low reproductive rates. The record of historical extinctions and lists of currently endangered species also bear witness to the vulnerability of large animals (see Chapter 17).

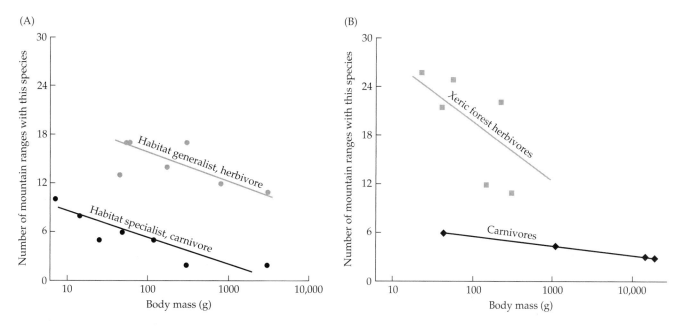

FIGURE 14.2 All else being equal, species with relatively low resource requirements tend to be disproportionately more common on islands. The small mammal communities of mountaintop forest "islands" of the Great Basin (A) and of New Mexico and Arizona (B) both tend to be dominated by small, generalist herbivores, presumably because—in comparison to mammals with higher resource requirements—they can maintain higher populations on the same area and, therefore, are less prone to extinction. (A after Brown 1971, 1978; B after Patterson 1984.)

Additional insights can be gained from studies of introductions of birds to hundreds of islands across the globe. Many ecologists have capitalized on these anthropogenic experiments to test some of the factors purported to influence the assembly of insular communities. Moulton and Pimm's studies of birds introduced to the Hawaiian Islands and Tahiti suggest that body size and demography strongly influenced the ability of these exotics to establish and maintain insular populations (Moulton and Pimm 1983, 1986, 1987; Pimm et al. 1988; Moulton 1993). In a similar study of recent extinctions of birds on 16 British islands, Pimm et al. (1988) observed that risk of extinction was highest for species that are migratory and tend to maintain relatively low and variable population densities. This study also revealed an interesting interaction between the effects of population size and body size. At population sizes above seven pairs, large-bodied species exhibited a relatively high risk of extinction, while the reverse was true for populations below seven pairs. Karr's (1982) study of extinctions of birds on Barro Colorado Island, which was isolated during the construction of the Panama Canal, also produced mixed results. Extinctions of forest birds were not significantly associated with population density, but species with more variable populations were more prone to extinction.

James's (1995) review of Recent fossils of Pacific Island birds indicated that avian richness on the Hawaiian Islands declined as much as 50 to 75% following the arrival of Polynesians some 1500 to 2000 years B.P. Again, larger species with greater resource requirements, including raptors (eagles, hawks, and owls) and water birds (ibises, rails, geese, and ducks), were much harder hit than the relatively small passerines (see also Steadman and Martin 2003). Extinctions also appear to have been selective with respect to trophic strategies. Extinctions were especially prevalent among predators and terrestrial omnivores and herbivores. Faring much better were the nectarivores, which now represent the most common native birds in Hawaiian forests (Figure 14.3; James 1995; see also Scott et al. 1986). Finally, James emphasized that the "extinction profile" may vary substantially among island ecosystems. For example, while prehistoric extinctions removed some 69 to 74% of the birds on Maui, Recent fossil evidence also indicates that avian richness on Madagascar

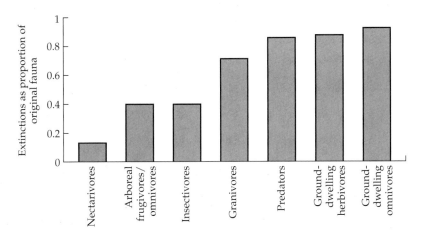

FIGURE 14.3 Prehistoric extinctions among birds on the Hawaiian Islands tended to be highly selective, being most prevalent among predators and ground-dwelling birds, while extinctions were relatively infrequent among nectarivores, which now represent the most common native birds on the islands. (After James 1995.)

declined by just 11 to 15% during the same period, and that extinctions there were even more skewed toward loss of the larger species (see also Dickerson and Robinson 1986; Burney et al. 1997).

In short, while many studies suggest that extinction is selective, there remains much debate over which characteristics are most closely associated with extinction. Many ecologists argue that insular survival is more strongly influenced by island attributes, including area and isolation, than by the intrinsic properties of the species (for example, see Richman et al. 1988; Tracy and George 1992, 1993; Diamond and Pimm 1993; Haila and Hanski 1993; Gotelli and Graves 1996; Terborgh et al. 1997, 2001; Boyett et al. 2000; Lambert et al. 2003). The geography of extinction has intrigued biogeographers throughout the distinguished history of the field, and it is likely to provide information critical to understanding and mitigating the current extinction crisis. We, therefore, explore this theme in depth in our discussions on biodiversity (Chapters 16 and 17).

Patterns Reflecting Differential Immigration and Extinction

NESTEDNESS OF INSULAR COMMUNITIES. MacArthur and Wilson's equilibrium model has sometimes been criticized for, among other things, its assumption that all species are equivalent. This criticism, however, seems a bit unfair, because their model attempted to explain patterns in species richness, not the composition of insular communities. In addition, MacArthur and Wilson's model implicitly included some effects of interspecific differences in immigration and extinction potential among species. If each species were identical with respect to these important characteristics, immigration and extinction rates would be linear functions of species richness. Each time a species colonized an island, the immigration rate would decline and the extinction rate would increase by constant amounts. However, because MacArthur and Wilson realized that the names of the species matter, they predicted that immigration and extinction curves should be concave (see Figures 13.7 and 13.8). Moreover, MacArthur and Wilson (1967:80) cited specific cases in which insular biotas tend to be biased in favor of species with superior dispersal and colonizing abilities. For example, they noted that the insects of Micronesia tend to

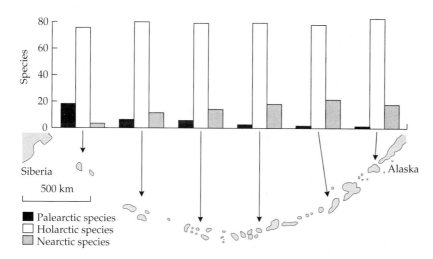

FIGURE 14.4 Much like the Lesser Sunda Islands (see Figure 6.14), the waters between Siberia and Alaska form an oceanic double filter, or transition zone, characterized here by the reciprocal attenuation of land plants on the Aleutian Islands. (After Williamson 1981.)

be relatively small, a trait that facilitates their transport by winds (MacArthur and Wilson 1967; after Gressitt 1954). Similarly, Wollaston's (1877) earlier work had demonstrated that insect colonists of Saint Helena Island in the Atlantic Ocean were predominately wood borers and other forms pre-adapted to rafting on logs or vegetation mats. As these and many other studies demonstrated, oceans act as filters, allowing the selective passage of certain types of species. Often, these oceanic filters form transition zones, characterized by differences in the attenuation rates of species groups that differ in dispersal ability (Figure 14.4; see also Figures 6.14 and 6.15).

Surprisingly, the selective nature of immigration is also evident for some islands that are quite near the coast. Simberloff and Wilson's (1969, 1970; Simberloff 1978) classic defaunation experiments on invertebrates of mangrove islands not only tested some critical predictions of the equilibrium theory, but also revealed that, as islands were colonized, they tended to converge on their original species composition. Leston's earlier (1957) comparisons of insect communities on isolated islands of the Atlantic and Pacific Oceans also revealed that these communities are convergent with one another and tend to form regular subsets of the mainland biota. Some taxonomic groups tend to be underrepresented on islands, while others are disproportionately common on islands in comparison with mainland areas. Just as important, these biases in species composition seem to increase with increasing isolation. This tendency for communities on species-poor islands to form regular subsets of those on species-rich islands is known as **nestedness**. Subsequent studies have confirmed that nestedness of insular biotas is an extremely common phenomenon (Wright 1996), one exhibited by most animals and plants and by an overwhelming variety of islands, mountaintops, and other isolated ecosystems (see Figure 14.5)—the result, highly non-random assembly of insular communities, this time caused by differential immigration abilities.

In his seminal monograph on biogeography, Darlington (1957) presented a graphical model for nestedness that he termed the **immigrant pattern** (Figure 14.6A). Interestingly, Darlington's model ignored the possibility that community assembly can also result from selection extinctions. His **relict pattern** pre-

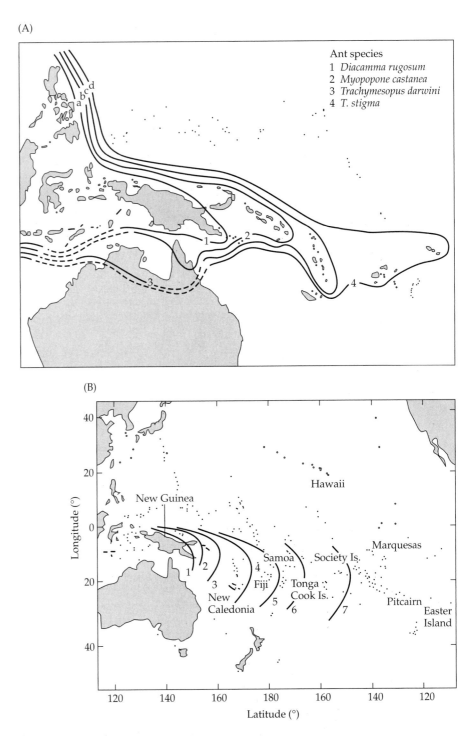

(A)

Ant species
1 *Diacamma rugosum*
2 *Myopopone castanea*
3 *Trachymesopus darwini*
4 *T. stigma*

(B)

FIGURE 14.5 The selective nature of immigration is revealed in the geographic nestedness of many insular biotas. Both ants (A) and birds (B) of the western Pacific exhibit analogous patterns of distributions extending eastward from their sources in New Guinea and Indonesia. Lines indicate the eastern range limits of taxa. Land and freshwater breeding birds: 1 = pelicans, storks, larks, pipits, birds of paradise, and nine others; 2 = cassowaries, quails, and pheasants; 3 = owls, rollers, hornbills, drongos, and six others; 4 = grebes, cormorants, ospreys, crows, and three others; 5 = hawks, falcons, turkeys, and wood swallows; 6 = ducks, thrushes, waxbills, and four others; 7 = barn owls, swallows, and starlings; and beyond 7 = herons, rails, pigeons, parrots, cuckoos, swifts, kingfishers, warblers, and flycatchers. (A after Wilson 1959; B from Williamson 1981, after Firth and Davidson 1945.)

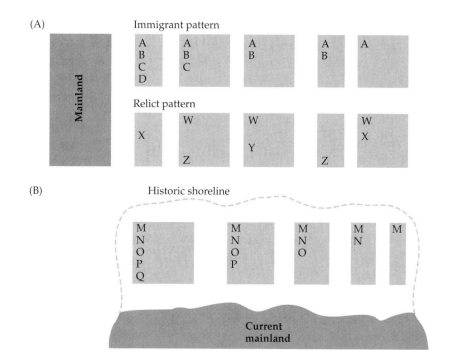

FIGURE 14.6 Nestedness of insular communities may result from species-selective immigration (the immigrant pattern, A) or extinction (B). Note that Darlington's "relict pattern" implies that the partial extinction of faunas during ecological relaxation might be random (A). Most recent studies, however, suggest that both immigration and extinction are selective processes. (A after Darlington 1957.)

dicted that, while smaller islands should have fewer species, their biotas should be random draws of the mainland pool. However, as we have discussed earlier, ability to survive on limited insular resources may differ dramatically and predictably among species, even within a closely related group. Therefore, regular patterns in species composition on islands may result from extinctions as well as immigrations. According to Brown's **relaxation model**, as the area of islands decreases, their communities should converge on a similar set of species (i.e., those with minimal resource requirements; see Figure 14.6B). Thus, across a great variety of archipelagoes, insular communities tend to form nested subsets when ordered by decreasing area or increasing isolation (or by species richness; Figure 14.7). Colonization studies of volcanically active islands such as Krakatoa, Long Island and its inland, freshwater island—Motmot (see discussion on Krakatoa in Chapter 13)— confirm the selective nature of immigration and colonization, and demonstrate how natural disturbances can set the stage for non-random assembly of insular communities. On the other hand, activities of modern human societies, including habitat destruction and fragmentation, make it clear that anthropogenic extinctions also can be highly selective and result in non-random collapse of native communities—a process known as **disassembly** (Fox 1987; Mikkelson 1993; Lomolino and Perault 2000; see Chapters 16 and 17).

Although nestedness had attracted a modest degree of attention throughout the 1960s and 1970s, it wasn't until 1984 that Bruce Patterson and Wirt Atmar developed the first statistically rigorous approach for analyzing nested subsets (Patterson and Atmar 1986). Their important paper triggered a flurry of studies that confirmed the generality of nestedness, provided alternative statistical approaches, and began to explore the causality of nestedness among taxonomic groups and across archipelagoes (see Kitchener et al. 1980; Lazell 1983; Schoener and Schoener 1983; Patterson 1987, 1990; Cutler 1991; Patterson and Brown 1991; Simberloff and Martin 1991; Wright and Reeves 1992; Atmar and Patterson 1993; Doak and Mills 1994; Cook 1995; Cook and Quinn 1995;

FIGURE 14.7 The distributions of five fish taxa among 28 springs in the Dalhousie Basin of South Australia illustrate the nested subset pattern. Here, nestedness is nearly perfect in that, with the exception of the taxon that is absent from spring G, the faunas of the more depauperate springs are regular subsets of those found in the richer springs. (After Brown and Kodric-Brown 1993.)

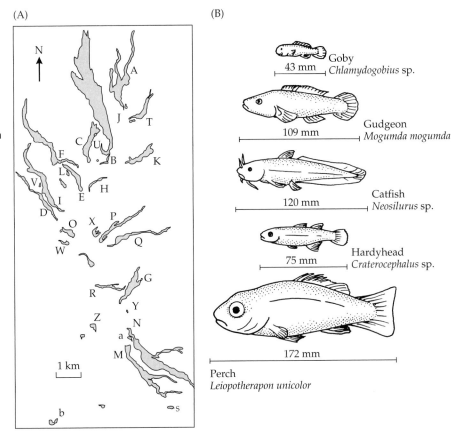

(C)

Spring (see map in A)

Species	A B C D E F G H I J K L M N O P Q R S T U V W X Y Z a b	No. of springs
Goby	X X X X X X X X X X X X X X X X X X X X X X X X X X X X	28
Gudgeon	X X X X X X X X X X X X X X X X X X X	19
Catfish	X X X X X X X X X X X X X X	14
Hardyhead	X X X X X X X X X	9
Perch	X X X X X X X	7
Number of species	5 5 5 5 5 5 5 4 4 4 4 3 3 3 3 2 2 2 2 2 1 1 1 1 1 1 1 1	

Kadmon 1995; Lomolino 1996; Wright et al. 1998; McClain and Pratt 1999; Fischer and Lindenmayer 2002; Loo et al. 2002; Hausdorf 2003). As this fascinating area of research continues to mature, it is certain to contribute substantially to our understanding of the forces structuring isolated communities.

DISTRIBUTIONS OF PARTICULAR SPECIES. Nestedness, while an intriguing and insightful measure of patterns among communities, provides only limited information on the distribution patterns of particular species. Again, significant nestedness indicates that species and islands differ with respect to factors influencing immigration and extinction. But how are these species actually distributed among islands that vary in both area and isolation?

A useful starting point for addressing this question may again be MacArthur and Wilson's equilibrium theory—specifically, its basic assumption that insular community structure results from the combined effects of recurrent immigrations and extinctions. Here, however, the terms *immigration* and *extinction* refer to the frequency of these events for a particular species. With the above assumption as a conceptual foundation, Levins, Hanski, and other scientists working during the past few decades, have developed an entirely new theory of population biology called metapopulation theory (see Gilpin and Hanski 1991; Hanski 1999; Hanski and Gagiotti 2004). Basically, metapopulation studies attempt to estimate the proportion of islands, or "patches," that must be occupied to ensure the survival of the interacting populations of a species—its metapopulation. Alternatively, given information on the actual proportion of patches occupied, metapopulation models can be used to estimate the time to extinction of the metapopulation.

In a similar fashion, a species-based model of assembly and insular distributions for a particular species can be derived based on the same fundamental assumption plus two additional ones: (1) that the probability or frequency of immigration decreases with increasing isolation, and (2) that the frequency of extinction decreases as island area increases (see Lomolino 1986, 1997). As you can see, these assumptions are essentially equivalent to those of the equilibrium theory. Given this, we predict that the distribution of the focal species should result from the interactive, or **compensatory effects** of immigration and extinction. The basic rule is that a focal species will occur on those islands where its immigration rate exceeds its extinction rate. Thus, the focal species is expected to occur on isolated islands, but only those islands that are large enough so that extinction rates are low enough to compensate for infrequent immigrations. Conversely, the species may be common on small islands if they are close enough to the mainland so that high immigration rates compensate for frequent extinctions. The overall result is that minimal area requirements to maintain populations of the focal species should increase as isolation increases. When this **insular distribution function** is plotted on a graph depicting island area and isolation, the intercept can be interpreted as a measure of resource requirements, while the slope is an inverse measure of the immigration ability of the focal species (Figure 14.8; see also Alatalo 1982; Hanski 1986; Peltonen et al. 1989; Peltonen and Hanski 1991; Thiollay 1998; Lomolino 2000; Ovaskainen and Hanski 2003).

This species-based approach to island biogeography is a relatively new and untested one. However, it complements the dynamic, ecosystem-level approach championed by MacArthur and Wilson. Here, however, the identities of the species are not lost. Indeed, many of the patterns community ecologists study result from, not in spite of, differences among species. For example, a high degree of nestedness implies that insular distributions vary in a very regular manner among species (see Figure 14.7; see also Patterson 1984). If species were equivalent, nestedness would be a rare phenomenon indeed.

Insular distribution functions, in addition to providing tools for exploring patterns in assembly of insular communities, highlight the need to study some critical questions. As MacArthur and Wilson (1967) lamented, we still know relatively little about interspecific differences in immigration and extinction. The general tendency for insular communities to be nested along gradients of the geographic template, including those of isolation and area, suggests some very regular patterns of variation in immigration abilities and resource requirements—but what are those patterns? If we were to draw histograms of the frequency distributions of immigration abilities, what form would they take? Do immigration abilities and resource requirements (possibly inferred from slopes and intercepts of insular distribution functions) covary among species? That is, do good immigrators require larger or smaller islands to

(A)

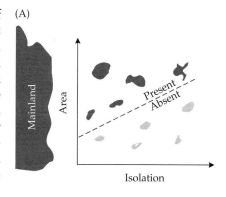

(B) Masked shrew (*sorex cinereus*)

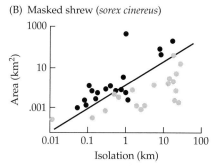

FIGURE 14.8 Insular distributions of species should be influenced by the combined effects of immigration and extinction. Species populations may persist on small islands if they are close enough to the mainland to experience frequent immigrations. Because immigration rates should decrease with isolation, the minimum area required to maintain insular populations of a focal species should increase with isolation. The predicted pattern (A) has been observed for a variety of organisms, including small mammals, fishes (see Figure 16.37), and butterflies (see Figure 16.35). (After Lomolino 1986, 1998.)

maintain their populations than do poor immigrants? These questions are challenging, but if resolved, they will certainly provide some fundamental insights into the assembly—and disassembly—of insular communities.

Patterns Reflecting Interspecific Interactions

Despite their heuristic value, MacArthur and Wilson's equilibrium theory and metapopulation theory ignore the potential importance of interspecific interactions. Nestedness models such as Figure 14.6 also ignore the potential importance of interactions such as competition, predation, and parasitism, which, if important, may reduce nestedness for many biotas. Modifying these general models to include the influence of one species on the distributions of others will certainly add a substantial degree of complexity (see Lomolino 2000). Yet, more than any other systems, islands provide compelling evidence for the importance of interspecific interactions.

Several biogeographic patterns suggest that the distributions and abundances of particular species are influenced not only by their own characteristics, but also by interactions with other species. All types of interactions, including amensalism, commensalism, predation, parasitism, and mutualism, are probably important, but most studies of insular ecology have emphasized competition. Three patterns that have been hypothesized to be caused by interspecific competition are listed below.

1. Ecologically similar species tend to exhibit mutually exclusive distributions, seldom if ever occurring together on the same island. A corollary of this pattern is the prediction that species that do coexist should tend to be more dissimilar in ecologically relevant traits than would be expected by chance.

2. In comparison to conspecific populations on the mainland, populations on species-poor islands often should exhibit relatively high densities.

3. Insular populations tend to exhibit ecological release, characterized by significantly broader niches and shifts to habitats, feeding strategies, activity periods, or other characteristics that would be considered atypical for the species on the mainland.

DISTRIBUTIONS, CHECKERBOARDS, AND INCIDENCE FUNCTIONS. In the absence of direct experiments on the competitive exclusion of one species by another, a pattern of mutually exclusive distributions of two or more species with similar resource requirements can serve as empirical evidence for competition. Archipelagoes consisting of many similar islands provide numerous examples of such patterns. Diamond (1975b) described several mutually exclusive distributions—which he called checkerboards—for pairs of congeneric bird species inhabiting the Bismarck Archipelago east of New Guinea. The flycatcher *Pachycephala melanura dahli*, for example, is found on 18 islands, and its congener *P. pectoralis* on 11 islands, but the two species never occur together on the same island (Figure 14.9).

Other negative associations resulting from interspecific competition may be more complex than simple checkerboards. Species that do not occur together on small islands may coexist on larger ones, suggesting that they competitively exclude each other where carrying capacities are low, but are able to subdivide the more plentiful resources of larger islands (see Schoener and Adler 1991). In still other cases, only certain combinations of three or more species may be able to coexist, whereas other combinations appear to be "forbidden," or at least, much less common, because of competitive exclusion. Diamond (1975b) cites another example from the Bismarck Archipelago—that

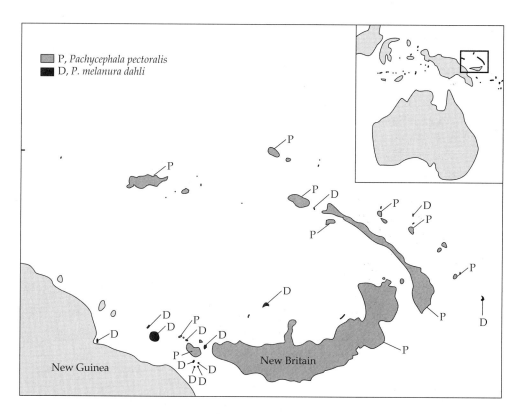

FIGURE 14.9 Mutually exclusive, or checkerboard, distributions of *Pachycephala* fly-catchers on the Bismarck Archipelago off New Guinea. There are two species, *P. pectoralis* and *P. melanura dahli*. Most islands have one of these species, no islands have both, and a few islands (especially the smallest) have neither. (After Diamond 1975b.)

of cuckoo doves in the genera *Macropygia* and *Reinwardtoena*. There are four species in this group, so there are fifteen possible combinations of species that would give biotas of from one to four species (Table 14.2). Only six of these combinations are actually observed, however, suggesting that certain sets of species are incompatible (see also McFarlane 1989; Gotelli and Graves 1996; Gotelli 2000, 2004; Gotelli and McCabe 2002). A recent meta-analysis of species distributions matrices confirms the predictions of Diamond's (1975a) general assembly rules: plant and animal communities exhibit fewer species combinations, more checkerboard patterns, and less species co-occurrence than expected by chance (Gotelli and McCabe, 2002).

Sometimes **diffuse competition** among many species, rather than interactions between just two closely related ones, has been implicated in determining insular distributions. Again, perhaps the best examples come from Diamond's (1974, 1975) studies of the birds of the Bismarck Archipelago. He noted that certain species, which he called **supertramps**, are usually found only on small and/or isolated islands containing few other bird species (Figure 14.10). Supertramps are also among the few species that have recolonized islands following a recent volcanic eruption. By plotting **incidence functions**—the proportion of islands inhabited by a given species as a function of the number of other species present—Diamond was able to view what he described as the "'*fingerprint' of the distributional strategy of a species*" and quantify some interesting distributional patterns (note incidence can be plotted against other variables such as island area and isolation to explore other bio-

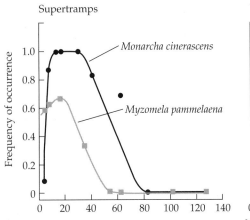

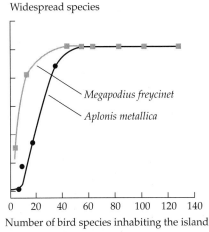

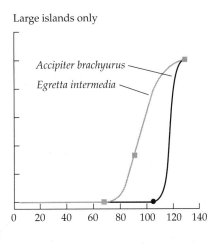

Number of bird species inhabiting the island

FIGURE 14.10   Incidence functions for various bird species in the Bismarck Archipelago. Note that some species, called supertramps, such as *Monarcha cinerascens* and *Myzomela pammelaena*, occur on most islands with few other bird species, but are absent from islands with diverse avifaunas; other species, such as *Aplonius metallica* and *Megapodius freycinet*, are found on all but the smallest islands; and still other species, including *Accipiter brachyurus* and *Egretta intermedia*, occur on only the largest islands with many other bird species. (After Diamond 1975b.)

geographic patterns and their causal processes). Using such graphs, supertramps, such as the flycatcher *Monarcha cinerascens* and the honeyeater *Myzomela pammelaena*, are readily distinguished as species that appear to be excellent colonists, but poor competitors in diverse bird communities (see Figure 14.10). Other species, including a starling (*Aplonis metallica*) and the incu-

TABLE 14.2   *Combinations of four cuckoo dove species that are present or absent in the Bismarck Archipelago*

	Observed combination		
Number of species	Species combinations	Number of islands inhabited by this species combination	Missing species combinations
1	A	3	N
	M	8	R
2	A, M	5	A, N
	A, R	4	M, N
	M, R	2	N, R
3	A, N, R	5	A, M, N
			A, M, R
			M, N, R
4	None		A, M, N, R

Source: After Diamond 1975b.

Note: A, *Macropygia amboinensis*; M, *M. mackinlayi*; N, *M. nigrirostris*; R, *Reinwardtoena* superspecies.

TABLE 14.3 *Coexistence of pairs of hummingbird species of varying size combinations on the islands of the Greater and Lesser Antilles*

	Both small	One small, one large	Both large
Observed species pairs	6	27	16
Species pairs expected from a random distribution	14.3	18.7	24.3

Source: Data from Brown and Bowers, 1984.

Note: The bill length of every other species was expressed as a ratio relative to the bill length of the smallest species, and the species with ratios less than or greater than 1.8 were categorized as small or large, respectively. Note that coexisting pairs of species tend to be of different sizes; the probability of this occurring by chance is less than 0.02.

bator bird (*Megapodius freycinet*), appear to be both capable colonists and good competitors, as they are found with relatively high frequency on all but the smallest islands. Still other species, especially birds with large body sizes, specialized diets, or restricted habitat requirements, such as hawks (e.g., *Accipiter brachyurus*) and herons (e.g., *Egretta intermedia*), are restricted to the largest islands, which presumably offer more resources, but also contain more bird species (see also Peltonen and Hanski 1991; Taylor 1991; Hanski 1992).

A similar approach involves testing the prediction that ecologically similar species coexist on islands less frequently than expected by chance. Several investigators have noted that islands contain relatively few congeners, and few species that are similar in ecologically important morphological traits, such as bill size in birds (e.g., Grant 1965, 1966b, 1968, 1986; MacArthur 1972; Diamond 1973, 1975b; Lack 1973, 1976; see review in Gotelli 2004). In analyzing these patterns, the critical question is whether the differences among co-occurring insular species are greater than would be expected if the island biotas were assembled at random from the mainland species pool. Some analyses that have addressed this problem have produced results that are consistent with the competition hypothesis. For example, in the case of both Darwin's finches in the Galápagos Islands (Grant and Abbott 1980; Hendrickson 1981; Case and Sidell 1983) and hummingbirds in the Antilles (Lack 1973, 1976; Case et al. 1983; Brown and Bowers 1985), it appears that species that coexist on the same island tend to be more different in bill and wing measurements than would be expected if the species associated at random (Table 14.3). Actually, the pattern is even more dramatic than Table 14.3 shows, because when species of similar size occur on the same island, they tend to be segregated by elevation and habitat.

Earlier, we mentioned Pimm and Moulton's studies on interspecific competition and success among birds introduced to Tahiti and the Hawaiian Islands (Figure 14.11; Moulton and Pimm 1983, 1986, 1987; Pimm et al. 1988; Moulton 1993). Later eco-morphological studies on birds introduced to Bermuda provided similar results (Lockwood and Moulton 1994; see also Simberloff and

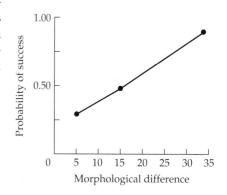

FIGURE 14.11 The probability that introduced species of birds will successfully establish an insular population is directly correlated with the degree to which they differ morphologically from the resident Hawaiian avifauna. Morphological difference is represented here by the difference in bill length between an introduced bird species and its most similar resident congener. (After Moulton and Pimm 1986.)

TABLE 14.4 *Diamond's assembly rules for insular communities*

1. *If one considers all the combinations that can be formed from a group of related species, only certain ones of these combinations exist in nature.*
2. *These permissible combinations resist invaders that would transform them into a forbidden combination.*
3. *A combination that is stable on a large or species-rich island may be unstable on a small or species-poor island.*
4. *On a small or species-poor island, a combination may resist invaders that would be incorporated on a larger or more species-rich island.*
5. *Some pairs of species never coexist, either by themselves or as part of a larger combination.*
6. *Some pairs of species that form an unstable combination by themselves may form part of a stable larger combination.*
7. *Conversely, some combinations that are composed entirely of stable subcombinations are themselves unstable.*

Source: Diamond 1975b.

Boecklen 1991; Moulton 1993). Invasion success was significantly lower when ecologically similar species were present, and surviving assemblages tended to be overdispersed in morphology.

Some patterns in insular community structure appear to be so regular that Diamond termed them **assembly rules** (Table 14.4). Each of these rules addresses the tendency of insular communities to represent nonrandom subsets of the mainland species pool. These rules have been the target of much criticism, yet they have served an important purpose—that of refocusing our attention from relatively simple patterns in species richness to other, more challenging, questions related to the forces influencing the organization of communities (see reviews by Gotelli and Graves 1996; Weiher and Keddy 1999; Gotelli 2004). Moreover, we may use insular distribution functions such as those in Figure 14.8 to put Diamond's assembly rules—and, in particular, checkerboard patterns—in a more explicit, geographic context. The resultant patterns (Figure 14.12) clearly illustrate how interspecific interactions and immigration abilities may restrict a species' *realized geographic range* to a subset of its *fundamental range* (i.e., the potential distribution of a species based solely on its environmental tolerances; see Chapter 3).

DENSITY COMPENSATION AND NICHE SHIFTS ON ISLANDS. If competition influences the organization of communities by limiting species to only part of their fundamental niches, one would expect populations to expand their niches and increase in density on islands where they interact with few other species. Both of these phenomena, niche expansion (or shifts) and increased densities, appear to be relatively common among insular populations. Crowell (1962) was among the first to focus attention on these patterns by comparing bird populations on Bermuda (900 km east of North Carolina in the Atlantic Ocean) with those of similar habitats on the North American mainland. He not only found niche shifts among these birds, but also observed that three species on Bermuda maintained a combined population density at least as great as that of the entire avian community on the mainland (see also Diamond 1970a, 1973, 1975; MacArthur et al. 1972, 1973; Yeaton 1974; Nilsson 1977; Kohn 1978; Faeth 1984; Martin 1992; Terborg et al. 1997; Terborgh et al. 2001; Gaston and Matter 2002; Lambert et al. 2003).

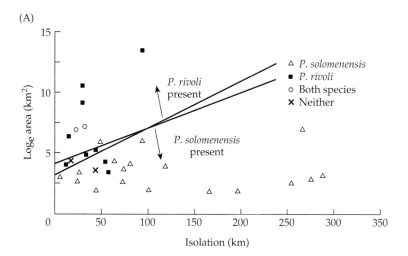

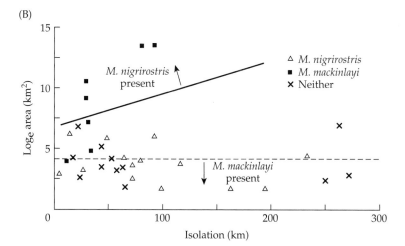

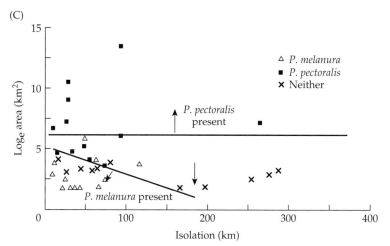

FIGURE 14.12 Insular distribution functions for three of Diamond's (1975) avian guilds of the Bismarck Archipelago reveal that species distributions are likely influenced by the combined effects of interspecific interactions and interspecific differences in immigration abilities and resource requirements. While the potential or fundamental ranges of these birds may overlap substantially, interspecific interactions appear to limit realized distributions of at least some of these species to a subset of the archipelago. Diamond's checkerboards, or exclusive distributions, appear to result from species segregating their realized distributions over different ranges of isolation and area; (A) fruit pigeon guild (*Ptillinopus rivoli* and *P. solomenensis*), (B) cuckoo dove guild (*Macropygia nigrirostris* and *M. mackinlayi*), and (C) gleaning flycatcher guild (*Pachcephala pectoralis* and *P. melanura*). (After Lomolino 2000.)

This latter phenomenon, termed **density compensation**, is often attributed to relatively low levels of interspecific competition, but it may just as easily result from the paucity of insular predators. Small rodents and shrews inhabiting islands along the coasts and within freshwater systems often exhibit relatively high densities along with substantial niche shifts (e.g., see Hatt et al.

(A) Densities of voles

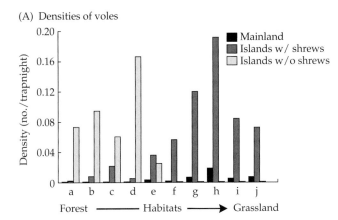

(B) Densities of short-tailed shrews

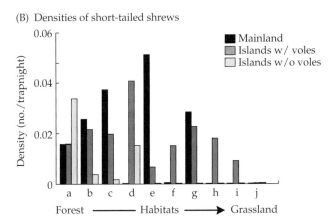

FIGURE 14.13 Insular populations often show evidence of niche shifts, exhibiting relatively high densities and occupying habitats considered atypical for the species on the mainland. Such shifts may result from the paucity of competitors or predators on islands. (A) In the Thousand Island region of the St. Lawrence River, meadow voles (*Microtus pennsylvanicus*) exhibit significant niche shifts on islands lacking one of their predators, the short-tailed shrew (*Blarina brevicauda*). (B) Conversely, on islands lacking meadow voles, the distribution of short-tailed shrews contracts to habitats where the availability of alternative prey items (primarily macroinvertebrates in forest soils) remains relatively high.

1928; Bishop and Delaney 1963; Grant 1971; Crowell and Pimm 1976; Lomolino 1984; Adler and Levins 1994). At least in some of these cases, predation is implicated as an important factor influencing the niche dynamics and densities of these mammals. For example, in many regions of North America, short-tailed shrews (*Blarina brevicauda*) often prey on small rodents such as meadow voles (*Microtus pennsylvanicus*) and deer mice (*Peromyscus* spp.), especially on the nestlings and juveniles. In the absence of these and other predators, voles and deer mice often exhibit significant ecological release, occupying habitats considered atypical for their species, often at relatively high densities (Figure 14.13A). In the absence of short-tailed shrews, meadow voles—typically grassland specialists—expand their niches to occupy all available habitats at relatively high densities. Experimental introductions of short-tailed shrews onto islands of the St. Lawrence River confirmed that these shrews prey heavily on juvenile voles and, if the shrew populations persist, can cause extinction of insular vole populations (Lomolino 1984). Interestingly, short-tailed shrews also exhibit insular niche shifts. On islands where voles are absent, they retreat from meadows to concentrate on more heavily forested sites where earthworms and other alternative prey items are more abundant (see Figure 14.13B).

The role of release from predation may actually be much more important than originally believed (see Polis and Hurd 1995; Terborgh et al. 2001; Lambert et al. 2003; Beauchamp 2004) . In an elegant field experiment, Schoener and Spiller (1987) tested whether excessive densities of spiders on small Bahamian islands resulted from predatory release. The primary predators of these spiders are lizards (*Anolis* spp. and *Ameiva* spp.). On islands lacking lizards, spider densities were about ten times as high as on islands where those predators were present. Schoener and Spiller tested the predatory release hypothesis by removing lizards from 84 m² enclosed plots and monitoring spider densities on experimental and control plots for just over a year.

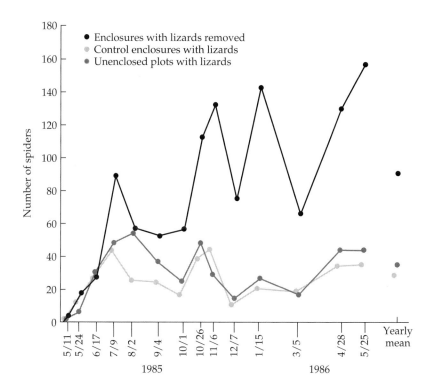

FIGURE 14.14 On Bahamian islands lacking predatory lizards, spiders often exhibit predatory release, reaching densities about ten times as high as on islands inhabited by these lizards. Experimental removal of lizards from enclosed plots demonstrated that population densities of spiders are strongly influenced by predation. (After Schoener and Spiller 1987.)

Spider density rapidly increased in the absence of lizards, reaching a level approximately 2.5 times that of the control plots by the end of the experiment (Figure 14.14).

Schoener and Spiller's study is just one in a long and distinguished series of studies on the ecology and evolution of West Indian lizards. No field biologist can visit a small Caribbean island without being impressed by the incredible abundance of *Anolis*. These small reptiles seem to be everywhere, from the ground to the tops of the tallest trees, from disturbed habitats along roadsides and in cities, to pristine native forests. Indeed, these lizards are much more abundant on most islands than they are anywhere on the tropical American mainland. E. E. Williams of Harvard University and his students T. W. Schoener, G. C. Gorman, J. Roughgarden, B. Lister, and R. Holt, and their students in turn, have studied the evolution, ecology, and biogeography of the Caribbean *Anolis* in great detail. Although many distributional patterns have been well documented, the underlying causal processes have proved more difficult to demonstrate convincingly.

Anolis is but one of several important lizard genera on the mainland of tropical America, but it is by far the most abundant genus of vertebrates on the Caribbean Islands. On the large islands of the Greater Antilles, a few colonizing species have given rise by endemic speciation and adaptive radiation to a diverse *Anolis* fauna. Hispaniola, the second largest and ecologically most diverse island, has at least 35 species, which were probably derived from four separate invasions (Williams 1976, 1983). These species occupy a variety of ecological niches. They range from tiny insectivores to large carnivores, with head and body lengths ranging from 40 mm to more than 200 mm. They are morphologically, physiologically, and behaviorally specialized for distinctive habitats and microenvironments—from sunny sites to deep shade, from open ground and rocks to grasslands and scrub forests, to different layers in the complex vegetation of mature tropical and montane forests (Figure 14.15). In

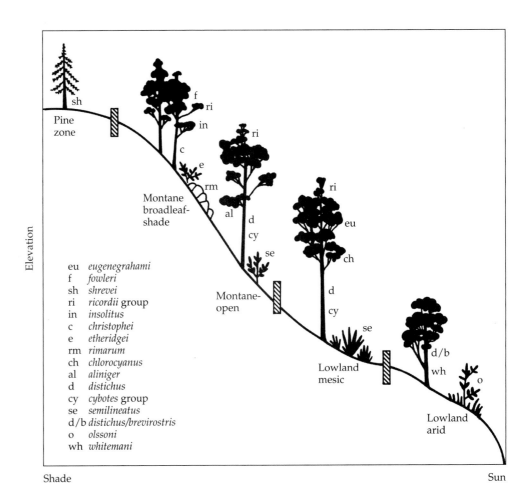

eu *eugenegrahami*
f   *fowleri*
sh *shrevei*
ri *ricordii* group
in *insolitus*
c   *christophei*
e   *etheridgei*
rm *rimarum*
ch *chlorocyanus*
al *aliniger*
d   *distichus*
cy *cybotes* group
se *semilineatus*
d/b *distichus/brevirostris*
o   *olssoni*
wh *whitemani*

Shade   Sun

FIGURE 14.15 Diagrammatic representation of the habitats occupied by different *Anolis* species on the northern part of Hispaniola. Their niches differ in elevation, vegetation type, perch height, and position on a gradient from sunlight to shade. The species indicated here—only a fraction of at least 35 species that inhabit the island—have been produced largely by speciation and adaptive radiation within the island. (After Williams 1983.)

contrast, the small islands of the Lesser Antilles each have only one or two generalized species (Roughgarden and Fuentes 1977; Roughgarden et al. 1983). When two species coexist on a small island, they differ in body size, prey size, and habitat; but when only one species is present, it is intermediate in size, takes a wide range of prey, and occupies virtually all habitats (Figure 14.16). Clearly, the fundamental niche of an *Anolis* species that has evolved in the absence of congeners is very broad. Some of the observed niche expansion of *Anolis* on small islands may represent behaviorally mediated ecological responses to the absence of competing species. Nevertheless, most of the niche dynamics in Caribbean species represent ecological and evolutionary adaptations to communities containing different numbers and kinds of other species, and these are reflected in morphological, physiological, and behavioral changes (see Roughgarden 1995; Losos and Queiroz 1997; Knox et al. 2001).

Variation in densities and niche characteristics of *Anolis* between islands and between habitats within islands has also been investigated, especially by Schoener (1968a, 1975), Lister (1976a,b), and Holt (1977). They have shown that when a species occurs on an island or in habitats where there are few

FIGURE 14.16 Body sizes of *Anolis* lizards of the Lesser Antilles. Note that all of these islands have either one or two species. When two species co-occur, they tend to be displaced in size, whereas when only one species is present, it tends to be of intermediate size. (After Roughgarden 1974; Roughgarden et al. 1983; Roughgarden and Fuentes 1997.)

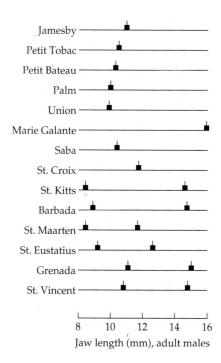

other lizards, increases in niche breadth often parallel increases in density (Figure 14.17). Again, it is tempting to attribute these patterns of density compensation and niche expansion to a release from competition with other *Anolis* species, but this assumption may be unwarranted. The small islands where the lizards show the most spectacular increases in density contain not only fewer *Anolis* species, but lower diversity of most taxa. Thus, *Anolis* populations are potentially released from competition not only with congeners, but also with insectivorous arthropods (e.g., spiders), birds, frogs, and lizards of other genera, and from predation by birds and other lizards. It is also possible that small islands support higher standing stocks of insect prey. Holt studied *Anolis* on islands located off the coast of Trinidad and concluded that direct competition with other *Anolis* or other lizard species could not account for the observed density changes. He suggested that the absence of predatory birds might be the most important of the various factors contributing to increased *Anolis* density and niche breadth on small islands.

Thus, while ecological release and high densities among insular populations are often attributed to reduced competition, they are more generally a consequence of the depauperate nature and reduced intensity of interspecific interactions, in general.

DENSITY OVERCOMPENSATION. One puzzling aspect of insular density compensation is that the total population densities of a few species inhabiting a small island may actually exceed the combined densities of a much greater number of species occupying similar habitats on the mainland. This phenomenon, called **density overcompensation** or **excess density compensation**, is fairly well documented, especially for birds on small oceanic islands. In fact, Crowell's (1962) studies, mentioned earlier in this chapter, revealed that just 10 species of small passerine birds on Bermuda together maintained a population not just equal to, but about 1.5 times greater than the combined densities of 20 to 30 species on the mainland. Subsequent studies by MacArthur et al. (1972) on the Pearl Islands south of Panama, by Diamond (1970b, 1975b) on the Bismarcks and other archipelagoes north and east of New Guinea, and by Emlen (1978) on the Bahamas, have described similar patterns. Case (1975) documented density overcompensation among lizards on the islands in the Gulf of California compared with the mainland of Baja California and Sonora; and Adler, Terborgh and their colleagues have reported the same phenomenon for mammals and birds on anthropogenic islands of Gatun Lake, Panama, and Lago Guri, Venezuela (Adler 1996; Terborgh et al. 1997; Lambert et al. 2001; Terborgh et al. 2001; Lambert et al. 2003).

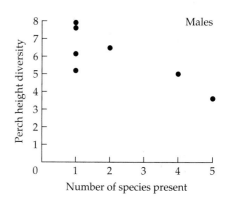

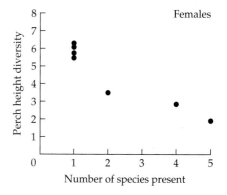

FIGURE 14.17 Relationship between the diversity of heights of perches used by males and females of *Anolis sangrei* and the number of other *Anolis* species occurring in the same habitat on the same island. Note that, when *A. sangrei* is the only species present, it uses a wide variety of perch heights, but the breadth of this niche dimension contracts and *A. sangrei* is excluded from arboreal habitats as it encounters an increasing number of coexisting congeneric species. (After Lister 1976a.)

Several explanations have been proposed for such overcompensation, and for density compensation as well:

1. Because vertebrates of large body size and higher trophic levels tend to be absent from small islands, the same resources can support substantially larger populations of smaller species. Note, however, that Crowell's (1962) study found that biomass, as well as density, was substantially higher among insular birds than among mainland birds.

2. Population densities of species—such as birds—reflect release from competition not only with missing bird species, but also with other taxa, such as mammals and amphibians, which use similar foods and other resources, but are even less common on oceanic islands than birds because they are poorer over-water dispersers.

3. Inflated population densities on islands reflect the absence or paucity of predators and parasites, which also tend to be poorly represented on oceanic islands (Grant 1966; MacArthur et al. 1972; Case 1975; George 1987; McLaughlin and Roughgarden 1989; Fallon et al. 2003).

4. Oceanic islands may be more productive, at least in terms of foods and other resources required by smaller vertebrates (i.e., small mammals, lizards and birds; see Case [1975]).

5. On oceanic islands, renewable food resources are harvested at rates closer to their maximum sustained yields than on mainlands, where intense competition can lead to overexploitation and lower productivity of resources.

6. Populations can become more finely adapted to their local environment, and hence, attain higher densities on isolated islands than on continents, where extensive gene flow among populations occupying different habitats tends to prevent specialization for more efficient use of local resources (Emlen 1978, 1979).

7. The surrounding waters may prevent population losses that would otherwise result from emigration of individuals into marginal habitats (the fence effect; see Krebs et al. 1969; MacArthur et al. 1972; Emlen 1979; Ostfeld 1994).

This is a long, but not all-inclusive list. As is often the case with the most general patterns in ecology and biogeography, these patterns may result from a combination of convergent mechanisms, rather than just one unique, overriding mechanism.

Patterns of combined densities of insular species, including compensation and overcompensation, may be summarized in a graphical model (Figure 14.18A; Wright 1980). We again emphasize that, like their mainland counterparts, insular communities often are Gleasonian, with species responding sometimes uniquely to one another as well as to environmental conditions. That is, while the overall pattern exhibited by a particular insular community may be density compensation, some of its species are likely to exhibit atypically low populations, while densities of others are compensatorily high. For example, Crowell (1962) found that common crows, house sparrows, and starlings were extremely rare in all but farmland habitats on Bermuda, while catbird, cardinal, and white-eyed vireo populations were 3 to 30 times higher in comparison to the same habitats on the mainland. Similarly, in a study of Caribbean insects, Janzen (1973) found that generalists, especially the Homoptera, tended to exhibit density compensation on islands, while other insects did not.

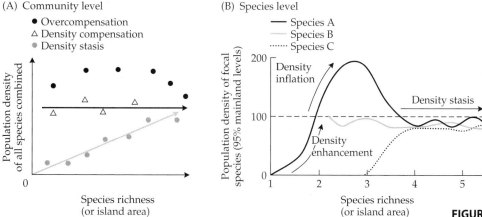

(A) Community level
- ● Overcompensation
- △ Density compensation
- ● Density stasis

Population density of all species combined

Species richness (or island area)

(B) Species level
— Species A
— Species B
⋯⋯ Species C

Population density of focal species (95% mainland levels)

Density inflation

Density enhancement

Density stasis

Species richness (or island area)

FIGURE 14.18 (A) Insular population densities (number of individuals per unit of area) may exhibit one of three qualitatively different responses to variations in species richness (or island area). If population densities are not influenced by competition, predation, and other interactions among species, then the densities of all species combined should increase in proportion to the total number of species (density stasis). On the other hand, if interspecific interactions do limit insular populations, they may exhibit density compensation or overcompensation on species-poor islands (i.e., populations of one or a few insular inhabitants may equal or exceed those of a much richer assemblage of species). (B) Population densities of individual species also may exhibit three qualitatively different responses to differences in species richness (or island area) among islands. Density enhancement (sensu Jaenike 1978): on very small, species-poor islands, population levels of a species may fall below those of its conspecifics on the mainland, but increase as island area increases and environmental conditions become more suitable. Density inflation: population levels may continue to increase, with increasing area, above their mainland levels, until interspecific interactions begin to regulate population densities of the focal species. Density stasis: at intermediate to high levels of species richness, many species may regulate one another's densities such that few, if any, exhibit any consistent trend of population density with species richness or island area. (A after Wright 1980.)

A species-based alternative to Wright's model (see Figure 14.18B) introduces an additional density pattern termed **density enhancement** (see Jaenike 1978). Population densities of some, and perhaps most, species on very small islands may indeed be much lower than their typical densities of the mainland, but increase with island area. Larger islands not only support more species—some of which may be mutualists or commensals—but also tend to provide less harsh, more stable, and otherwise more favorable habitats for particular species. Alternatively, population densities of at least some species may remain relatively stable over a broad range of species richness—a pattern that Williamson (1981) termed **density stasis**. In fact, the same species may exhibit all three patterns—density enhancement on tiny islands, density inflation on islands of intermediate size, and density stasis on larger, more species-rich islands (see Figure 14.18B, species A). Perhaps density inflation also varies systematically among species: highest for supertramps and lowest for those species restricted to the richest, and largest islands. We admit that these are largely untested predictions, but the pertinent data seem to be available, at least for some insular communities.

In summary, studies on the ecological responses of insular populations have been, and will certainly continue to be, fertile ground for research on the forces influencing community structure.

Evolutionary Trends on Islands

By virtue of their size, typically distinctive environmental character and, in particular, their isolation in both space and time, islands represent a fascinating collection of novel selective regimes—evolutionary arenas distinct from those of the mainland and each with a unique suite of environmental features. As a result, islands are rightly recognized as hotspots of endemicity and evolutionary marvels. As we observed in the previous chapter, despite the many challenges of reaching and establishing populations on oceanic islands, those species that are successful have indeed won an ecological and evolutionary sweepstake.

MacArthur and Wilson's (1967) model, at least in its traditional form, cannot explain patterns in endemicity on islands (but, see Rosenzweig 1995). In this case *speciation*, and not immigration is the primary means of increasing richness of endemic taxa. Both immigration and extinction oppose development of distinct biotas; speciation cannot occur in the face of immigration and significant gene flow from the mainland, nor can it occur if insular populations do not survive long enough for significant evolutionary divergence. For these reasons, the two geographic factors that are most strongly associated with insular richness per se (island area and island isolation), are also correlates of endemicity as well. Endemicity increases with island area because extinction rates are relatively low (i.e., persistence times are long) on large islands and because larger islands are more likely to include rivers, mountain chains, and other features that provide within-island isolation of local populations. Endemicity on islands should also be correlated with their isolation, but in this case the correlation should be opposite of that observed for richness per se. That is, because gene-flow decreases with isolation, endemicity should be highest for the most distant islands and archipelagoes (provided, again, that they are sufficiently large). In fact, most insular biotas exhibit just this pattern (i.e., endemicity increases with area and isolation; see Figure 14.19A and B).

Thus, the disharmonic nature of isolated biotas results not just from selective immigrations, but also from in situ speciation, which accentuates ecological disharmony and distinctiveness, especially on the more isolated islands (see Gillespie and Roderick 2002). For example, as we consider increasingly more isolated archipelagoes of the Pacific Ocean ranging eastward from New Guinea and Melansia, the relative dominance of ants attenuates to the point where they are totally absent from most archipelagoes east of Samoa (Wilson and Taylor 1975). Along this same gradient of increasing isolation, richness and endemicity of other anthropods such as the so-called micro-moths (families Pyralidae and Cosmopterigidae) increases (Munroe 1996; Gillespie and Roderick 2002). Also, as we might expect, endemicity varies with species traits, being highest for those whose relatively limited colonization abilities approximate isolation of the archipelago in question (see Figure 14.19C).

Perhaps the most general lesson to be gleaned from this discussion is that islands are not only hotspots of diversity per se, but they harbor a highly disproportionate number of endemic species. For example, if New Guinea is considered the world's largest island, then islands constitute roughly 3% of the Earth's land surface, but some 17% of today's avifauna are endemic to islands (and this number was certainly much higher before humans colonized oceanic islands, causing wholesale extinction of their native avifaunas; see Chapters 16 and 17). Similarly, approximately 14% of the world's plants are endemic to islands (Whittaker 1998). We see a similar pattern for invertebrates, where, for example, 9% of the world's snails are endemic to just eight isolated archipelagoes (Hawaiian Islands, Japan, Madagascar, New Caledonia, Madeira, Canary Islands, Mascarene Islands, and Rapa; Whittaker 1998).

As we implied above, these insular forms include a fascinating menagerie of evolutionary marvels, lineages that have achieved truly remarkable transformations of their ancestral forms, including plants and animals that have lost behavioral, morphological, and physiological defenses against predators, and those with highly restricted dispersal abilities and bizarre morphologies—at least by mainland standards. As we discuss near the end of this chapter, these evolutionary transformations may constitute the later stages of a predictable series of ecological and evolutionary events called **taxon cycles**—changes that are often punctuated by extinctions, either natural or anthropogenic.

Flightlessness and Reduced Dispersal Ability on Islands

> As with mariners shipwrecked near a coast, it would have been better for the good swimmers if they had been able to swim still further, whereas it would have been better for the bad swimmers if they had not been able to swim at all and had stuck to the wreck. (Darwin 1859, p. 177.)

(A)

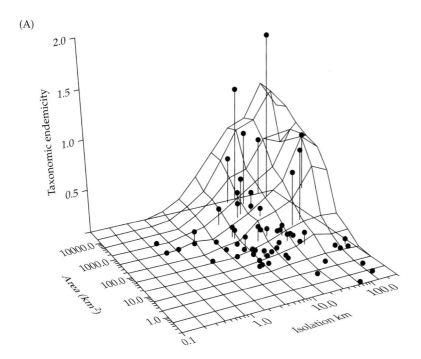

(B)

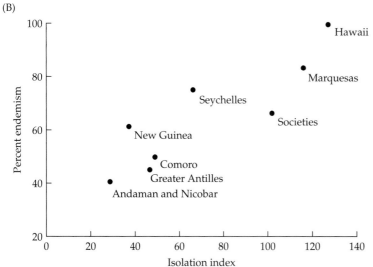

(C)

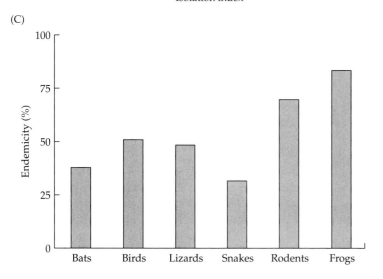

FIGURE 14.19   (A) Taxonomic endemicity of insular avifauna of Northern Melanesia increases with both island area and island isolation (this index of endemicity reflects endemicity at the levels of subspecies to genera, with increasing weight for endemicity at higher taxonomic levels). (B) Endemicity of spiders in the genus *Tetraghnatha* increases with isolation of islands in the Pacific Ocean (endemicity is calculated as percentage of species in this genus that are endemic to a particular island; isolation is calculated as the sum of the square root distances to the nearest larger island, nearest archipelago, and nearest continent). (C) Endemicity of North Melanesia vertebrates (as percent of species within each taxon) is highest for those groups with limited vagilities—in particular, rodents and frogs. (A after Mayr and Diamond 2001; B after Gillespie and Roderick 2002; C after Mayr and Diamond 2001.)

With this one metaphor, Charles Darwin explained a truly fascinating paradox of insular biotas: although insular communities are typically biased in favor of taxa with superior immigration abilities, many insular forms, especially those on the most isolated oceanic islands, have little or no ability to disperse to other islands. As many have remarked, there is no greater anomaly in nature than a bird that cannot fly (Darwin 1859). Yet, flightless birds and insects are relatively common on many oceanic islands, as are other animals and plants with little ability or propensity for dispersal. And herein lies the paradox: how could these relatively sedentary forms have colonized such remote ecosystems?

During Darwin's day, the extensionists may have used these observations to bolster their argument that ancient landbridges once connected even the most isolated archipelagoes with the continents (see also Wallace 1893). As we saw in Chapter 2, nothing "vexed" Darwin more than these post hoc scenarios of the Earth's dynamism. Instead, he offered an explanation embodied in the second part of the above quote. There was no doubt, at least in Darwin's mind, that these peculiar insular forms arrived via long-distance dispersal, but were then transformed by evolution.

Unfortunately, Darwin attributed some of these patterns to what was termed the "law of disuse." For example, he believed "that the nearly wingless condition of several birds, species which now inhabit several oceanic islands tenanted by no beast of prey, has been caused by disuse." He acknowledged, however, that the flightlessness of many insular forms, such as 200 of the 550 described beetles of Madeira Island, was "wholly, or mainly due to natural selection." The key point is that the selective forces operating on a potential immigrant may be entirely different from those operating on its insular descendants (see Carlquist 1974). In Darwin's metaphor, selection during immigration favors "good swimmers," but once they (or their descendants) have reached the "wreck" (island), selection then favors "bad swimmers"—those less likely to swim off or be carried away by winds and lost at sea.

BIRDS. Interestingly, flightlessness has never been documented in two of the major, or once dominant volant taxa—bats and pterosaurs. New Zealand's only endemic mammals, short-tailed bats (Mystacinidae) are clearly the most terrestrial of all bats, with incisors specialized for burrowing, and well-developed claws and wing membranes that roll up like furled sails (and, in the case of the extinct, greater short-tailed bat, *Mystacina robusta*, folded into pouch-like flaps of skin) to enable adept movement along the forest floor. Yet, these species still retained the ability, albeit somewhat limited, to fly.

On the other hand, derived insular flightlessness has likely evolved thousands of times in birds and is reported in at least ten orders of birds including kiwis (Apterygiformes), moas (Dinornithiformes), cormorants (Pelecaniformes), ibises (Ciconiiformes), ducks and geese (Anseriformes), kagus, rails and gallinules (Gruiformes), the dodo, solitaires and Fiji pigeon (Columbiformes), parrots (Psittaciformes), owls (Strigiformes), and wrens (Passerformes) (McNab 2002). On New Zealand, some 25 to 35% of the terrestrial and freshwater birds are—or were, in the case of extirpated forms—flightless. Similarly, 24% (20 species) of Hawaii's endemic birds were flightless. The long list of now extinct, flightless birds includes many flightless giants such as the dodo (*Raphus cucullatus*) of Mauritius, the solitaires (*R. solitarius* and *Pezophaps solitariaon*) of Réunion and Rodriguez, the elephant birds (Aepyornithidae) of Madagascar, and the 15 or so moas (*Pachyornis* spp.) of New Zealand (see review by McNab 1994a). Flightlessness is especially common among the rails, having evolved independently within at least 11 groups of rails across hundreds of islands (Steadman 1989, 1995; Diamond 1991). In fact, evolution of derived flightlessness may occur very rapidly, on the order of perhaps just a few centuries and, while this may seem astounding, paleontological evidence suggests that, before human colonization, most Pacific islands were inhabited by at least one species of flightless rail (see Olson 1973; Steadman and Olson 1985; Steadman 1986 and 1989; Silkas et al. 2001; Steadman and Martin 2003).

Currently, the most widely accepted explanation for the evolution of flightlessness in birds involves selective pressures associated with the absence of predators and with limited resources on islands (Diamond 1991; McNab 1994b, 2002; see also McCall et al. 1998). Under reduced pressure from predators, and for that matter, from competitors as well, insular populations would be likely to undergo ecological release. Evolutionary responses could include changes that would conserve energy, such as reduced size overall, or a reduction in metabolically expensive tissues, such as the otherwise large flight muscles. On the mainland, such selection to conserve energy is of course countered by the selective advantage of being able to escape ground-dwelling predators. On many islands, however, this is not the case, as "reduction in flight muscle brings great energy savings with little penalty" (Diamond 1991). If these selective factors were combined with natural, heritable variation in the mass of flight muscles, their relative size and ability to power flight would atrophy over generations. Thus, at least for the rails, flightlessness was probably achieved primarily by reduction in mass of flight muscle, with perhaps a modest reduction in overall mass as well.

Evolution of flightlessness in other birds, however, is often associated with an increase in body size (Morton 1978; Livezey 1993). This suggests an alternative path to the loss of flight: many insular birds may have simply outgrown their wings. In the absence of mammals and other large terrestrial vertebrates, selective pressures may have promoted increased body size overall, without a compensatory increase in flight muscle mass (see the following section on the evolution of insular body size). The result was moas, the dodo, solitaires, and elephant birds—avian giants no more capable of flight than we are. Indeed, these and other flightless insular birds may have taken over the large herbivore niches left vacant by mammals—shifting their diets, developing more drab, cryptic coloration, and increasing in size so as to eventually lose their power of flight (see Livezey 1993; Baker et al. 1995; Trewick 1996). A related, biogeographic pattern supports McNab's (1994b, 2002) energetics-based explanation for the evolution of flightlessness. As he observed, derived flightlessness is relatively common for birds on tropical islands, much less common for those of temperate regions, and absent for those of islands in higher latitudes. In the latter case, energetic demands and the need to migrate to warmer regions during winter, renders flightlessness untenable.

These evolutionary transformations of thousands of insular endemics provide poignant illustrations of both the marvels and the perils of island life. Flightless forms that became adapted and highly specialized for insular environments were also the first to suffer extinctions following colonization by humans and their ground-dwelling commensals (see Chapters 16 and 17). Such was the case of thousands of flightless rails and other insular birds (Steadman and Martin 2003). Similarly, of the three species of ground-dwelling, short-tailed bats discussed above, one *(M. tuberculata)* is critically endangered, and another—the poorest flyer *(M. robusta)*—is extinct.

INSECTS AND OTHER INVERTEBRATES. The flightless beetles of Madeira Island in the Atlantic Ocean are just one of the countless groups of insular insects whose powers of flight have been lost, or greatly reduced. In addition to Coleoptera,

FIGURE 14.20 Two examples of flight-lessness in insular insects: (A) *Scaptomyza frustulifera,* a drosophilid from Tristan da Cunha, and (B) the Stephens Island weta, *Deinacrida rugosa,* a flightless orthopteran from Stephens and Mana Islands, Cook Strait, New Zealand. The weta may be an ecological equivalent of rodents. It has been extirpated from the New Zealand mainland and many offshore islands (which lacked native nonvolant mammals) by introduced rats and other ground-dwelling mammals. (A after Williamson 1981; B after Collins and Thomas 1991.)

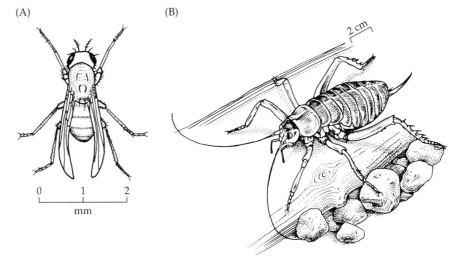

derived flightlessness has been reported in nearly all orders of insects across a variety of habitats including those at high altitudes, in deserts, and on islands (Wagner and Liebherr 1992; Roff 1994; Kotze et al. 2000). Insular forms that have lost the power of flight include butterflies (Lepidoptera), flies (Diptera), ants, bees and wasps (Hymenoptera), mayflies (Ephemeroptera), bugs (Homoptera), and crickets, grasshoppers and wetas (Orthoptera). Flightlessness and reduced dispersal abilities of insects appear to be most prevalent on the most isolated archipelagoes. For example, 40% of the native insects of Campbell Island (the southern most of New Zealand's sub-Antarctic islands) exhibit flightlessness, or at least some degree of wing reduction; 90% of Tristan de Cunha's (south Atlanctic) endemic beetles have reduced wings; and 94% of New Zealand's moths and butterflies have limited powers of flight (Williamson 1981; Gressitt, J. L. 1982; Howarth 1990).

While the evolution of flightlessness in insects of xeric habitats may have involved energy and water conservation (Roff 1986, 1990), it is likely that other factors were much more important for development of flightlessness on islands. Like many of the avian examples discussed above, the wetas (Orthoptera) of New Zealand represent another fascinating case of ecological release and evolutionary convergence on a mammalian niche (Figure 14.20). In the absence of ground-dwelling mammals, wetas have lost the power of flight, increasing in size and occupying the "geophilous" niche otherwise occupied by many rodents. Similarly, in the absence of predators along the surfaces of their wetland habitats in Madagascar, some mayflies (genus *Cheirogenesia*) have wings that are so reduced that they are incapable of flight and instead the adults use these oar-like appendages to skim the water surface while feeding (Ruffieux et al. 1998).

In contrast to birds, mammals, and other relatively large animals, insects perceive a much more coarse-grained environment. Insects may live out their entire life cycle in just a small patch of a bird's home range. Thus, site fidelity, or what Darlington (1943) termed **precinctiveness,** may be at a premium. Because more distant patches or habitats tend to be more dissimilar than those closer to their natal sites, selection may favor offspring or adults that limit their movements—in Darwin's terms, those that "stick to the wreck." Precinctiveness, however, applies at a much finer scale than in Darwin's metaphor: individuals do not have to be blown off the island to suffer lower fitness, only blown to a less favorable habitat.

Of course, such advantages are not unique to insects on islands. Many mainland landscapes also offer patchy, yet relatively stable, environments where energy conservation and site fidelity may increase fitness. Accordingly, flightlessness is quite common among insects on the mainland as well as on islands. Flightlessness, or reduced capacity and propensity for flight, tends to be more common among insects of montane environments than among those living at lower elevations (Figure 14.21; see Darlington 1943; Roff 1990; Barlow 1994; Hawkins and Lawton 1995). Again, the characteristics of biotas inhabiting these habitat "islands" seem to be convergent with those inhabiting true islands.

Land snails, while of course never achieving the power of flight, demonstrate sometimes astounding abilities to disperse great distances, especially the smaller species. As we mentioned in Chapter 6, ability to be dispersed by wind is inversely correlated with body size. As a result, oceanic islands tend to be colonized by a highly disproportionate number of micro-snails (Vagvolyi 1975; see also Zimmerman 1948). Just as we have seen for the micro-moths of isolated oceanic islands, the relative dominance of these snails may be accentuated by subsequent, in situ speciation. Such adaptive radiations, however, often include evolution of larger body sizes, with concomitant decreases in their dispersal abilities—once again, selection for species that "stick to the wreck."

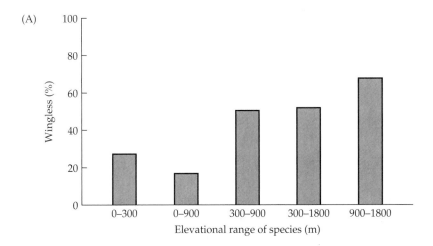

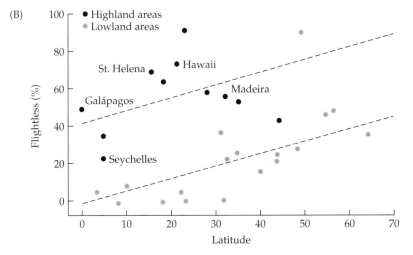

FIGURE 14.21 Flightlessness in insects is a common phenomenon among both insular and mainland forms, and tends to be more common in mountainous regions and at higher latitudes. (A) Wingless coleopterans of Madeira Island. (B) Flightless carabids on oceanic islands. (After Roff 1990.)

REDUCED DISPERSAL ABILITY OF INSULAR PLANTS. Reduced dispersal ability also is remarkably common in insular plants. In his comprehensive summary on this subject, Carlquist (1974: 429) wrote the following:

> While working with evolutionary phenomena in the Hawaiian Islands and other Polynesian islands, I observed types of fruits and seeds that seemed incongruously poorly adapted at dispersal. The floras of these islands must have arrived by long-distance dispersal, yet during evolution on the island areas various groups of plants have lost their dispersibility.

Many of the changes that result in reduced dispersal ability in plants are associated with shifts from adaptations for dispersing to and colonizing habitats along the beachhead, to traits better suited for living in interior, mesic forests. Precinctiveness appears to be just as important for insular plants as it is for insular insects. Accordingly, both seeds and fruits tend to become heavier, less buoyant, and less resistant to seawater, and structures that facilitate air lift or attachment to animals (spines, wings, and hooks) are greatly reduced (Figure 14.22; Carlquist 1974). Morphological changes associated with reduced dispersal ability in insular plants may evolve very rapidly, perhaps in less than a decade (see Cody and Overton 1996).

Many species of plants that are typically herbaceous on the mainland evolve woodiness and large size on isolated islands (Darwin 1859:392), resulting, for example, in sunflowers the size of oak trees (Figure 14.23). This pattern is as widespread as it is impressive, occurring on islands throughout Polynesia and Melanesia and on New Zealand, Juan Fernández, the Desventuradas, the Galápagos, California offshore islands, the Canary Islands, Madeira, the Cape Verde Islands, the Azores, the West Indies, Madagascar, the Comoros, and Saint Helena (see Carlquist 1974). The cause of such evolutionary novelties in insular plants is probably complex, but may stem from the disharmonic nature of insular biotas. In comparison to those of large trees and other woody plants, the propagules of small herbaceous plants are typically more easily transported by winds and water currents. Thus, isolated islands are colonized by a highly disproportionate number of small herbaceous species. Just as important, isolated islands often lack large herbivorous mammals, including browsers and folivores, which tend to feed most heavily on woody plants. Under such reduced herbivory (predatory) pressures, plants often undergo ecological release and sometimes evolve to occupy the large, woody plant niche left vacant by their mainland competitors. As we shall see in the following section, these types of responses to the depauperate and disharmonic nature of insular communities are common among animals as well as plants.

Finally, many cases of reduced dispersal ability among insular plants and insects may result from a type of ecological lottery that characterizes most islands. Given that islands are depauperate by nature, plants that depend on particular animals for dispersal (exozoochory), or animals that depend on larger animals for dispersal (**phoresy**), may find themselves stranded (see Whittaker et al. 1997). Luck may still be with them, however, if they can adapt and find a niche somewhere on the "wreck," and stick to it.

Evolution of Growth Form and Body Size on Islands

Flightlessness, of course, is not the only way in which a species can respond to limited resources and a paucity of interspecific interactions. As we have mentioned above, those species that adapt to these insular conditions often under-

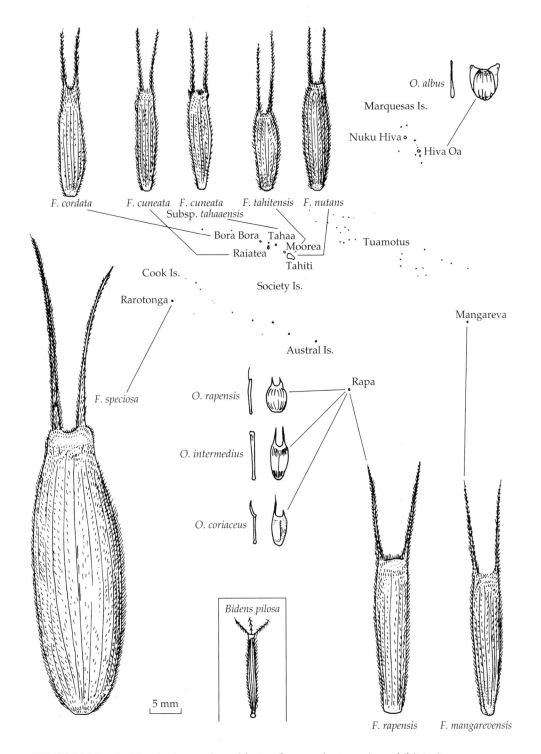

FIGURE 14.22   On islands, the seeds and fruits of many plant species exhibit traits reflecting reduced dispersal ability, including fruits and seeds that are heavier, less buoyant, and less resistant to seawater, as well as a reduction in spines, wings, and structures that would otherwise facilitate lift or attachment to animal dispersers. Fruits of insular forms of *Fitchia* and *Oparanthus* (Asteraceae) are shown; *Bidens pilosa,* which exhibits a more typical mainland morphology, is shown in the inset for comparison.

(A)   (B)

FIGURE 14.23 Evolution of tree-like stature and woodiness in insular plants include (A) the silversword (*Dubautia reticulata*) of the Hawaiian Islands and (B) the cactus tree (*Opuntia echios*) of the Galápagos. (A courtesy of Gerry Carr; B courtesy of A. Sinauer.)

go remarkable changes in body size (Figure 14.24). These remarkable and seemingly divergent changes, ranging from dwarfism in elephants to gigantism in earwigs, seem as chaotic as they are fantastic. But there is order to this variation. Gigantism and dwarfism may simply reflect differences in selective pressures among different islands, and predictable responses among different species.

Many insular species exhibit trends toward larger size. In the absence of competitors and predators, and with consequent increases in *intra*-specific competition, selective pressures shift so as to favor intraspecific competition and exploitation of a greater breadth of resources. Under these conditions, large size has a number of selective advantages:

1. Larger individuals can exploit a greater diversity of resources. Larger plants can shade and outcompete smaller ones for water and energy; larger predators can feed on large as well as small prey; larger squirrels and other granivores can crack large as well as small nuts.

2. Given that they can acquire sufficient resources to survive and reproduce, larger individuals can set more or higher quality seeds and fruits, or produce larger litters or clutches.

3. Although larger individuals require more resources, they also tend to dominate in territorial battles or other intraspecific interactions associated with competing for those resources.

Dwarfing in Giant Deer (*Megaceros*) during the Pleistocene

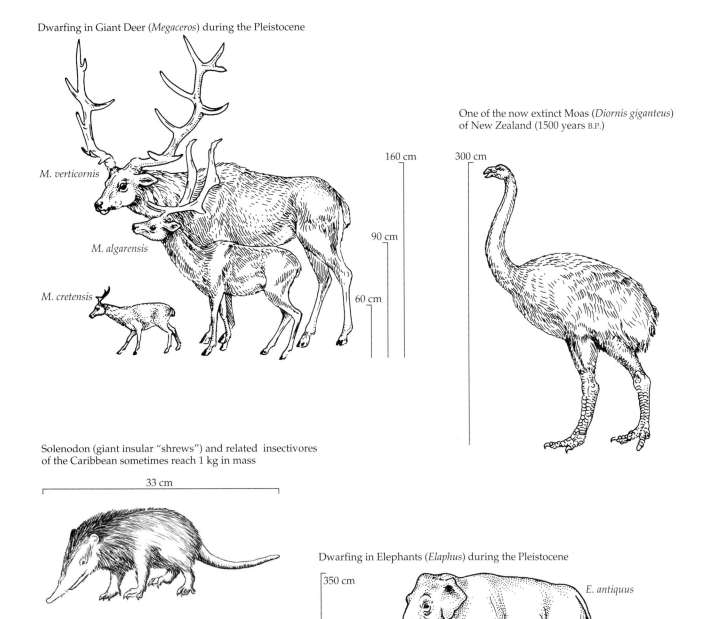

One of the now extinct Moas (*Diornis giganteus*) of New Zealand (1500 years B.P.)

M. verticornis

M. algarensis

M. cretensis

160 cm

90 cm

60 cm

300 cm

Solenodon (giant insular "shrews") and related insectivores of the Caribbean sometimes reach 1 kg in mass

33 cm

Dwarfing in Elephants (*Elaphus*) during the Pleistocene

350 cm

170 cm

75 cm

E. antiquus

E. mhaidriensis

E. falconeri

FIGURE 14.24 Insular forms of mammals and birds often exhibit markedly divergent morphological trends, ranging from dwarfism in typically large species such as elephants and ungulates, to gigantism in shrews, mice, and many birds.

4. Finally, larger individuals have relatively greater energy and water reserves and, therefore, greater ability to withstand shortages of energy, nutrients, and water.

Given all these advantages of being large, why are not all insular forms giants? Body size and growth form on islands—or for that matter, size, in general—represents a compromise between benefits and costs. Being small on islands also has its advantages:

1. Smaller individuals require fewer resources to survive and reproduce. This should be especially important on small islands where resources are in short supply.
2. Smaller individuals, because they require fewer resources, can be more specialized and more efficient at assimilating nutrients and energy.
3. Smaller individuals can exploit small shelters and refugia from predators (including herbivores), and avoid stressful environmental conditions, that larger forms cannot.

In addition to the above selective pressures—ecological release (promoting gigantism) and resource limitation (promoting dwarfism)—insular body size also may be influenced by selection for better immigrators: in Darwin's terms, selection in the good swimmers for the ability to swim still farther. If body size is somehow associated with immigration ability, then the founders of an insular population may represent a biased subset of the mainland population. For example, among active immigrators, larger individuals should have higher vagilities, and founding populations may be comprised of relatively large individuals. On the other hand, we might expect just the opposite for passive immigrators, where small individuals are more easily dispersed by wind or water. The effects of immigrant selection can persist, however, only if the trait in question—body size—has a genetic basis (see Falconer 1960; Roth and Klein 1986). Immigrations also must be frequent enough to counter the effects of genetic drift and opposing postcolonization selection, or these selective forces must be weak or convergent with the effects of immigrant selection.

Given that insular populations may be influenced by a combination of selective forces favoring gigantism or dwarfism, we might well expect trends in insular body size to seem random, lacking any recognizable pattern across taxonomic groups or trophic categories. As we will see, however, insular faunas often exhibit some remarkably consistent patterns of variation in body size, suggesting that at least one of the above selective forces varies in some systematic way among species.

WOODINESS AND "GIGANTISM" IN PLANTS. "Again, islands often possess trees or bushes belonging to orders which elsewhere include only herbaceous species." (Darwin 1859: 392) Darwin, along with a long and distinguished list of subsequent biogeographers and evolutionary biologists, marveled over the nature of island plants. On the most isolated archipelagoes such as the Galápagos and Hawaiian Islands, otherwise diminutive and herbaceous plants are often transformed into woody "giants." Whereas elsewhere across their extensive geographic range, *Opuntia* cacti typically are shrubs no greater than a meter or so tall; indeed, the island of Santa Cruz in the Galápagos boasts a 10 meter giant *Opuntia* cactus (*Opuntia echios*; see Figure 14.23B see Bohle et al. 1996; Grant 1998, 2001). Among Hawaii's most spectacular plants are the silverswords, woody relatives of sunflowers whose flowering stalks can tower up to 2 meters (see Figure 14.23A; see Baldwin and Robichaux 1995).

Woodiness, or arborescence, is essential for tree-form stature, and evolution of these features on islands has occurred repeatedly in several families of angiosperms (including Asteraceae, Boraginaceae, Campanulaceae, Lobeliaceae, Apiaceae, Brassicaceae and Euphorbiaceae; Carlquist 1965, 1974; Bohle et al. 1996). This phemonenon is particularly common in the sunflower family (Asteraceae), evolving independently in different tribes from numerous islands including St. Helena and those of the Canary, Juan Ferndandez, and Hawaiian Islands (Grant 1998). Woody giants have also evolved in the high elevation, habitat "islands," of the Andes and East African highlands (Carlquist 1974; Knox et al. 1993).

Evolution of woodiness on islands is typically associated with ecological shifts from herbaceous ancestors that occupied open, early successional habitats, to arborescent forms that have invaded and created a secondary, "forest' niche, left vacant by relatively poor dispersing trees which seldom colonize isolated archipelagos. In fact, Darwin (1859) speculated that the limited dispersal abilities and paucity of trees on islands was a key factor contributing to evolution of secondary woodiness on islands. Wallace (1878) also proposed an ecological explanation for this phenomenon, noting that woodiness also connotes greater longevity and a concomitant higher probability of reproductive success on isolated islands, which are notorious for their paucity of pollinators (see also Bohle et al. 1996). It appears that this may also account for the general tendency for insular plants to have relatively drab and otherwise inconspicuous flowers, and for their dependence on generalized pollinators, self fertilization, or wind pollination—especially in dioecious plants (Baker 1955; Carlquist 1974; Ehrendorfer 1979; Sakai et al. 1995; Barrett 1998). Carlquist (1974), on the other hand, downplayed these ecological explanations for insular woodiness in favor of a climatic one. Essentially, he hypothesized that insular climates tend to be relatively moderate, thus favoring species with a different suite of life history strategies. In MacArthur's terms, these plants shifted or evolved from r-selected strategies to k-selected ones (including increased size, slower growth rate, and increased longevity) once their populations approached and adapted to carrying capacities of their insular environments. As with most general patterns in biogeography, searches for *the one* overriding force driving these evolutionary patterns may well prove futile, as it is just, if not more, likely that the generality of these patterns results from a set of diverse factors, each being more or less important for different species and on different islands, but all part of the island syndrome and all with similar influences on the ecology and evolution of insular floras.

INSULAR MAMMALS. "Pygmy mammoth" is, of course, an oxymoron, but one familiar to most evolutionary biologists and, indeed, to many of the lay public as well. Paleontological studies have documented the ability of these now-extinct relatives of today's elephants to colonize and adapt to island environments. The fossil record reveals that these and other proboscidians were once widespread throughout Malaysia and on selected islands of the Mediterranean, the California Channel, and the Arctic Ocean (see Hooijer 1976; Sondaar 1977; Roth 1990; Lister 1993; Lister and Bahn 1994).

One of the most remarkable cases of insular dwarfism was described by Vartanyan and his colleagues (see Martin 1995; Vartanyan et al. 1995). Although the woolly mammoth (*Mammuthus primigenius*) was widespread across Siberia and northern Europe during the last Ice Age, its geographic range contracted rapidly during the onset of the current interglacial. Its demise may have resulted from the contraction of its primary habitat—steppe/grasslands—and hunting by aboriginal humans, who expanded their ranges northward as the glaciers receded. The mammoth's range continued to contract northward during the glacial recession (20,000 to 10,000 B.P.) until it survived on just one isolated refuge, Wrangel Island, which lies in the East Siberian Sea just north of the Arctic Circle. Here, provided with remnants of its primary habitat and some isolation from its human predators, mammoths persisted well into the current interglacial, perhaps as recently as 2000 B.P.

An additional key to the persistence of mammoths on Wrangel Island was a marked and relatively rapid development of dwarfism (see also Lister's [1989] example of rapid development of dwarfed forms of the red deer on Jersey Island during the previous interglacial). Wrangel Island mammoths decreased in size from 6 tons to just 2 tons over roughly 5000 years. These changes no doubt reduced the resource requirements of the animals, effectively increasing insular capacity and prolonging their persistence by many centuries, if not millennia.

This phenomenon, while truly remarkable, is not unique to mammoths. Other insular proboscidians also exhibited insular dwarfism, at least as remarkable as that exhibited by Wrangel Island mammoths. In fact, "mammoths" of the channel islands (*Mammuthus exilis*) are estimated to have reached just 10% of the mass of their mainland contemporaries, and insular elephants (*Elaphas falconeri*) of Sicily and Malta in the Mediterranean were even smaller, perhaps just 5% of the mass of those on the mainland (Figures 14.24 and 14.25C; see Sondaar 1977; Lister 1993; Lister and Bahn 1994). Interestingly, the size of Mediterranean island elephants appears to have been directly proportional to island area (see Heaney, 1978; see also Figure 14.26). Insular dwarfism is also reported for other large mammals, including ground sloths of the Caribbean, deer of the Mediterranean Islands and islands off the coast of Europe, and hippopotami on Madagascar and islands of the Mediterranean (Matthew and de Paulo Coutu 1959; Hooijer 1967; Hooijer 1975; Sondaar 1977; Lister 1989; Lister and Bahn 1994; White and MacPhee 2001; Anderson and Handley 2002).

While mammoths and some other large mammals tend to be dwarfed on islands, many small mammals exhibit gigantism on islands. In 1964, J. Bristol Foster published an important synthesis that revealed that these patterns were, in fact, features of a very general pattern, which later became known as the **island rule** (Van Valen 1973a,b). According to Foster (1963, 1964), different groups of mammals tend to exhibit different trends in insular body size. Carnivores—including canids and felids—exhibit dwarfism, while rodents tend to exhibit gigantism (Table 14.5). Rather than representing any taxonomic bias, however, the island rule can be more simply described as a graded trend from gigantism in smaller species to dwarfism in larger species (Heaney 1978; Lomolino 1985). This pattern is evident within, as well as among, taxonomic groups (Figure 14.25A), and therefore may reflect predictable changes in the relative importance of selective forces influencing body size on islands (Figure 14.25B).

Small mammals are relatively poor immigrators. In addition, they are more likely to be the smaller members of a guild of species—or prey—that escape predators by seeking the protection of small refugia (see McNab 1971; F. Smith 1992). In either case, these relatively small species tend to increase in size in the absence of larger competitors and predators. Thus, immigrant selection, ecological release, and increased pressures from *intra*-specific competition (i.e., selective forces favoring gigantism) should be most important for the smaller species of mammals. On the other hand, larger species, such as mammoths, deer, and wolves, require more energy to survive and reproduce. On the mainland, these species also tend to overcome prey and competitors, and to avoid

TABLE 14.5 *Taxonomic patterns in the relative sizes of insular mammal subspecies as compared with their relatives on the mainland*

	Number of subspecies		
	Smaller	**Same**	**Larger**
Marsupials	0	1	3
Insectivores	4	4	1
Lagomorphs	6	1	1
Rodents	6	3	60
Carnivores	13	1	1
Artiodactyls	9	2	0

Source: Foster 1964.

predators by outgrowing them. On islands, however, resource limitation and ecological release should promote dwarfism in these otherwise large species. The combined effects of these selective forces, which vary in intensity with the body size of the species, result in the graded trend in insular body size illustrated in Figure 14.25A.

Some biogeographers have scrutinized an additional feature of this pattern. If the island rule results from a balance between forces promoting gigantism in the smaller species and dwarfism in the larger species of insular mammals, where is the fulcrum? That is, is there an "optimal" insular body size—above which species tend to decrease, and below which they tend to increase in size? This size may be considered optimal or fundamental for a particular body plan (*bau plan*) and ecological strategy (e.g., large, running herbivorous mammal, or small, insectivorous, ground-dwelling lizard). As a result of the complexity of interspecific pressures in mainland communities, species will tend to diverge from this fundamental size over time. On islands, however, the paucity of interspecific interactions and the increased intensity of intraspecific pressures may promote convergent evolution to the fundamental size (i.e., relatively small species increasing, and relatively large species decreasing in size).

The hypothesized fundamental size can be estimated empirically by simple linear regression (estimating where the trend line intersects the horizontal line of Figure 4.24). Based on this analysis, the putative optimal size for nonvolant, terrestrial mammals on islands lies somewhere between 200 and 500 g, roughly the size of a squirrel (see Maurer et al. 1992). As implied above, fundamental sizes vary among different groups of mammals, ranging from just 26 g in shrews, 272 g in rodents, and 377 g in terrestrial carnivores, to 2.1 kg in rabbits and hares, and 6.9 kg in ungulates (Lomolino 2005). Brian Maurer and his colleagues also argued cogently that, in addition to bau plan and ecological strategies of the species, fundamental or optimal body size should also be a function of area of the land mass inhabited. Bolstering their argument, they reported that the median body size of land mammals ranges from 85 g for species in North America, to 216 g in Australia, and 231 g in Madagascar—a trend converging on the hypothetical optimal size for land mammals on smaller islands (Maurer et al. 1992).

Like most other biogeographic patterns, these observed trends in insular body size in mammals leave us with many intriguing questions. While the rel-

FIGURE 14.25 (A) The island rule refers to the graded trend from gigantism in small species of mammals (e.g., mice and voles) to dwarfism in large species of mammals (e.g., canids, ungulates, and elephants). (B) This pattern may result from the combined effects of ecological release and selection for better immigrants, which tends to select for gigantism in small mammals, and resource limitation, which tends to select for dwarfism, especially in larger species. Solid line represents trend (regression line) for mammals in general. (C) Body size trends for mammals of fragmented forests in northern Europe. Insular birds, insular turtles and tortoises, and insular snakes also exhibit graded trends from gigantism in the smaller species to dwarfism in the larger species. Body size trends of Australia's surviving large mammals ("time dwarfs"; sensu Flannery 1994) and extinct insular proboscideans of the late Pleistocene (elephants and mammoths) are consistent with the general trend for insular mammals, with degree of dwarfism increasing with body size of the species' mainland ancestors. (D) Insular bats (includins fruit bats and other megachiropterans) also tend to exhibit body size trends consistent with the island rule. (A, B, after Lomolino 1985; C after Flannery 1994; after Clegg and Owens 2002; Boback and Guyer 2003; Schmidt and Jensen 2003; N. Karraker, pers. comm. 2004; D after Krzanowski 1967.)

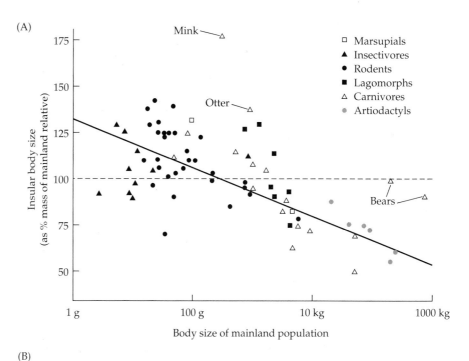

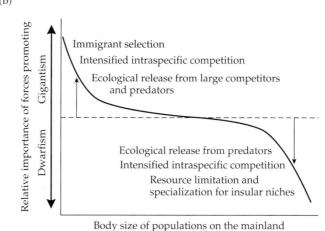

atively simple model shown in Figure 14.25B successfully accounts for the island rule, it remains largely untested. How much of the residual variation in Figure 14.25A can be accounted for by factors in addition to the body size of the ancestral species? The model predicts, for example, that the relative body sizes of insular mammals should increase with island area and isolation, and that they should be higher for herbivores than for carnivores. Consistent with the former prediction, the body sizes of tri-colored squirrels and fruit bats of Southeast Asia increase with island area (Figure 14.26; Heaney 1978; McNab 1994b; see also Krzanowski 1967). In addition, body size of these squirrels, and that of at least nine other nonvolant mammals tend to increase with island isolation (Figure 14.27; Lomolino 2005).

The potential influence of diet on the evolution of insular body size is suggested by the relative sizes of lagomorphs (rabbits and hares) and artiodactyls (deer and related species), which tend to fall above the general trend in Figure 14.25A, while insular canids and felids tend to fall below the general trend. Similarly, rodents exhibit stronger tendencies toward gigantism than do insec-

(C)

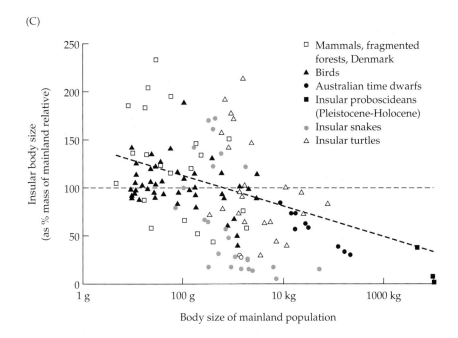

(D)

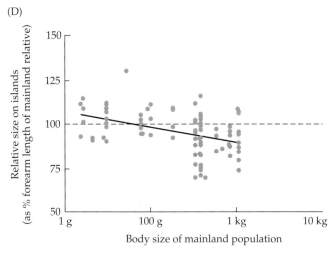

tivores (shrews and moles). On the other hand, insectivores (primarily, shrews and moles) seem anomalous in that, although they appear to be relatively large on islands, they fail to exhibit the expected graded trend (within the order) from gigantism to dwarfism. Bears (including black bears, *Ursus americanus*, and brown bears, *Ursus arctos*) also appear to be exceptional in that they exhibit either no, or only a modest degree of, dwarfism, despite their relatively large size and carnivorous habits (see rightmost triangles in Figure 14.25A; but see Gordon 1986). Further analyses of these exceptional cases, however, actually support the hypothesis that insular body size of large mammals is strongly influenced by resource limitation. After tracing the dynamics of body size of black bears inhabiting the Queen Charlotte Islands over the past 10,000 years, Gordon (1986) found that their ancestors were actually very large bears. Their insular populations subsequently underwent dwarfism, but not as rapidly or as marked as those that continued to inhabit the mainland through the warming periods of the Holocene (see also F. Smith et al. 1995). Again, the apparently exceptional cases (i.e., dwarfism, but relatively slow) may be explained by the

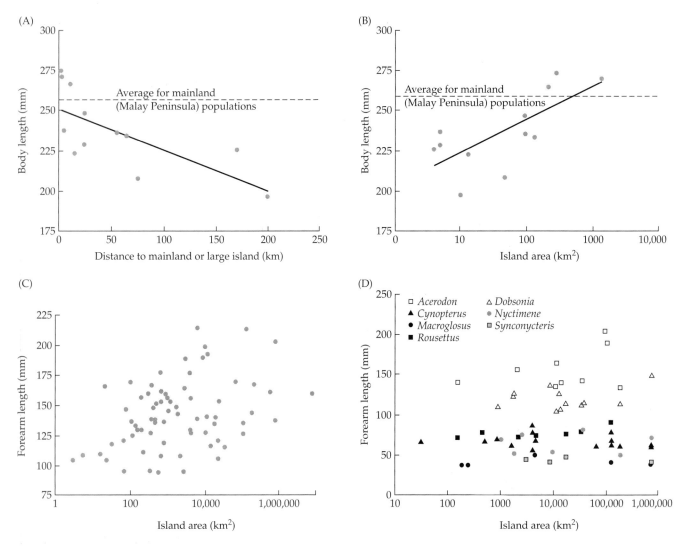

FIGURE 14.26 Mean body size of subspecies of the tri-colored squirrel (*Callosciurus prevosti*) as a function of (A) island isolation and (B) island area. Body size of fruit bats tends to vary with island area (megachiropteran bats of the genera *Pteropus* (C) and (D) *Acerodon, Cynopterus, Macroglosus, Rousettus, Dobsonia, Nyctimene,* and *Synconycteris*. (A, B after Heaney 1978, C, D after Krzanowski 1967; see also McNab 2002.)

special characteristics of these insular carnivores and the nature of their food resources. Despite the limited area of their insular environs, some of their key prey items—freshwater and marine fish (and possibly marine mammals as well)—are especially abundant, energy and nutrient rich resources on these islands (Hildebrand et al. 1999; see also McNab, 2002); resource limitation and resultant trends toward dwarfism are just not as intense for insular bears as it appears to be for carnivores that are more dependent on terrestrial prey. As we might expect, semi-aquatic species such as otters *(Lutra canadensis)* and mink *(Mustela vison)* also tend to be relatively large on islands (Lomolino 1983; after Goldman 1935; Cowan and Guiget 1956).

In summary, nonvolant mammals exhibit a very regular pattern in insular body size, but one with some potentially informative outliers and exceptions. Studies focusing on alternative taxa or those using different surrogates of body mass (e.g., linear measurements of skulls or teeth versus body length or actual mass) provide some equivocal results (e.g., see Meiri and Dayan 2003; Meiri et al. 2004a,b), although most recent studies continue to support the generality of the island rule in mammals and other vertebrates as well (see Case and Schwaner 1993; Ganem et al. 1995; Ventura and Lopez-Fuster 2000; Anderson and Handley 2002; Clegg and Owens 2002; and see following sec-

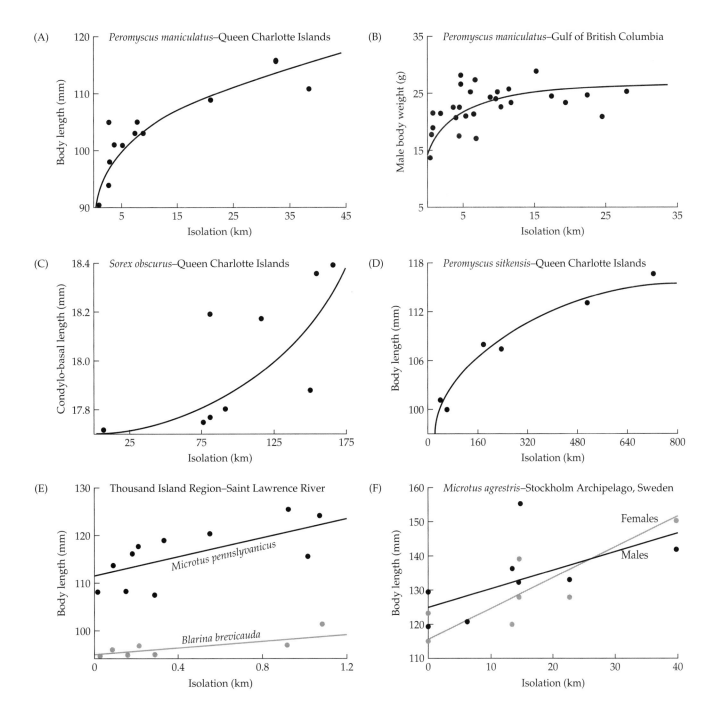

FIGURE 14.27 The body sizes of insular forms of a number of small mammals tend to increase with island isolation. (A–E after Lomolino 1983, 1984; F after Ebenhard 1988.)

tions). Paleontological research indicates that the pattern is not just a recent one, with similar patterns (i.e., gigantism in smaller species and dwarfism in larger species) being reported for insular mammals of the late Miocene and middle- to late-Pleistocene (Woods and Sergile 2001; Agusti and Anton 2002). Secondary patterns, including the relationships between body size and island age, area, isolation and latitude, and differences among groups with different energetic and trophic strategies provide especially important opportunities to explore not just the generality, but the underlying causes of body size variation on islands. Perhaps the most promising strategy for future studies will be to "deconstruct" (sensu Huston 1994; Marquet et al. 2004) the general pattern and focus on species groups with dramatically different resource requirements and vagilities. Bats, for example, have higher vagilities and much greater energy demands than nonvolant mammals of similar mass. Thus, bats should be less influenced by immigrant selection and more strongly affected by resource limitation. Consistent with this prediction, a review by Krzanowski, published some three decades ago, indicated that bats tend to be dwarfed on islands (74 cases of relatively small size vs. 32 cases of relatively large size; Krzanowski 1967). Even the bats (megachiropterans, including fruit bats and related species), however, exhibit the predicted, graded trend from gigantism in the smaller species to dwarfism in the larger species (see Figure 14.25 D). Finally, just like their insular counterparts, rodents on isolated mountaintops of Europe seem to exhibit gigantism (Zimmermann 1950; Carlquist 1974), but again, more rigorous tests are needed to determine whether the island rule applies to montane mammals. The generality of the island rule and its corollaries, including the size-isolation and size-area relationships, the effects of trophic strategies, relevance to Cope's rule, Bergmann's rule, latitudinal gradients in diversity and primary productivity, and parallel patterns for mountaintops and other isolated ecosystems, remain promising subjects for future studies (see Box 14.1).

Finally, these patterns may have relevance to current challenges of conserving the diversity and natural character of native mammals in nature reserves and other island-like systems (including zoos). While this is the focus of the final chapters of this book, it appears that the effects of anthropogenic insularization of native mammals may be very rapid indeed. For example, mammals in heavily fragmented forests of Europe appear to have undergone body size changes consistent with the island rule: in just 175 years of isolation, it appears that larger species are undergoing dwarfism, while smaller species are tending to increase in size, with the "optimal" size (i.e., that of mammals that are neither increasing nor decreasing in size) being about 280 g (see Figure 14.25C; Schmidt and Jensen, 2003). These results are limited to just one system and should therefore be taken as preliminary, but the possibility of anthropogenic changes in body mass of native species certainly warrants increased attention from biogeographers and conservation biologists (see Lomolino et al. 2001; Aponte et al. 2003).

OTHER INSULAR VERTEBRATES.   In comparison to other insular faunas, birds and reptiles represent unrivaled subjects for ecological and evolutionary studies. In comparison to mammals, birds and reptiles are more widely distributed on islands, and are generally easier to observe and collect. Some studies suggest that insular populations of birds and reptiles exhibit trends in body size that are not consistent with the island rule. The equivocal nature of these "patterns" may derive, paradoxically, from the wealth of information available for insular reptiles and birds. Measurements of body size have been taken from populations over a great variety of islands, which differ substantially in area, isolation, climate, and other factors that may influence the evolution of insular body size. Moreover, these islands are often much more isolated than those inhabited by mammals, thus making it more difficult to identify the source populations suitable for comparative studies. Yet, some general trends are emerging.

Until recently it appeared that, although birds exhibited a general trend toward increasing bill size on islands (Lack 1947; Grant 1965; Blondell 2000; but see Clegg and Owens 2002), they failed to exhibit trends in body size consistent with the island rule (at least when wing length was used as a surrogate of body size; Grant 1965). On the other hand, some relatively small birds such as wrens—and, possibly, fruit pigeons—tend to exhibit gigantism on islands

BOX 14.1   *Time dwarfing on the "island continent" of Australia*

■■▮ Evolutionary trends in the divergence of insular forms from their mainland ancestors in body size, such as those illustrated in Figure 14.25 and summarized in Table 14.5, are inferred from comparisons of the respective forms in a snapshot in time. Inferences regarding these evolutionary patterns are based on the assumption that insular and mainland forms shared a common ancestor, but then became isolated by some vicariant or dispersal event, after which the populations were free to evolve in isolation and respond to the different selective pressures of their respective environments (i.e., mainland or insular).

Occasionally, paleobiologists can piece together a more comprehensive picture of evolutionary trends in the morphological characteristics of a lineage through time. Perhaps most notable among these trends is Cope's rule, which refers to the tendency for body size to increase during the early evolutionary history of a lineage. In a recent and truly intriguing account of evolutionary trends in Australian mammals, Tim Flannery (1994) reported that, like their counterparts on the larger continents, Australian mammals tended to increase in body size during the past 15 million years. Flannery, however, also reported a startling reversal of this trend during the late Pleistocene. During the last 40,000 years, body sizes of most large Australian marsupials (i.e., > 5 kg) have actually decreased. This phenomenon, labeled time dwarfing by Flannery, appears to be a very general one: the very few apparent exceptions include just three extant species of wombats (*Lasiorhinus krefftii, L. latifrons,* and *Vombatus ursinus*) and possibly humans. Flannery also discovered that the degree of dwarfing of Australian marsupials during the late Pleistocene was greatest for the largest forms (see the figure on this page). Species less than 5 kg in mass failed to exhibit any detectable degree of dwarfism, while those ranging in size from 5 to 300 kg exhibited a graded trend in dwarfism much like the trend for insular mammals illustrated in Figure 14.22. In fact, when the data for time dwarfs are plotted on a similar graph (Figure 14.25C), these points fall along the general trend for insular mammals, suggesting evolutionary trends toward some optimal body

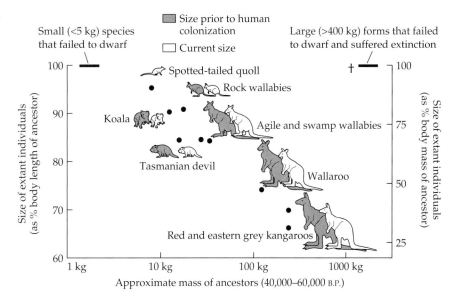

sizes for both insular and anthropogenically altered mammals. The largest of Australia's extant mammals—red and grey kangaroos (*Macropus rufus* and *M. giganteus*)—are quite small in comparison to their late-Pleistocene ancestors, perhaps just 70 to 75 percent of their body length and roughly half their mass. The largest of Australia's marsupials, including a diverse assemblage of species ranging from the hippo-like diprotodons, weighing in at nearly 2000 kg, to marsupial rhinos, marsupial lions, short-faced kangaroos, five wombats, and seven kangaroos, remained relatively unchanged in body size, but suffered extinction (see the figure).

Explanations for the time dwarfing of Australia's marsupials are the same as those proposed to account for the megafaunal extinctions during the Pleistocene: climatic change and overkill by humans (see Chapter 7). Proponents of the climatic change hypothesis argue that time dwarfing is a response to a presumed reduction in forage quality, which may have been associated with climatic change during the most recent glacial recession. Yet, the quality of forage for large herbivorous marsupials began to decline as early as 15 million years b.p., when most regions of Australia became increasingly arid. The body size of Australia's largest herbivores, the diprotodons, actually *increased* during this period. Further, opponents of the climatic

change hypothesis argue that this explanation ignores the fact that there were some 17 glacial cycles during the past 2 million years. Why should marsupials start to dwarf just during the last one, and why just those weighing 5 to 300 kg?

Flannery and many other paleoecologists have suggested that time dwarfing may have resulted from hunting by aboriginal humans, which appears to have concentrated on the larger species (those > 5 kg) and may have culled the largest individuals. The onset of time dwarfing in Australia is coincident with the arrival of humans, roughly 40,000 to 60,000 years B.P. According to this version of the overkill hypothesis, aboriginal hunting pressures either resulted in extinctions or, by decreasing the fitness of relatively large individuals, caused the dwarfism of surviving species.

The intriguing pattern of time dwarfing serves as another compelling case for the need to explore the deep history of biogeographic patterns. The existing assemblage of Australian mammals is in many ways not representative of those that occurred there prior to the arrival of humans. Inferences based solely on the limited diversity and relatively small size of the extant species may allow only an incomplete, and possibly misleading, view of the forces influencing the geographic variation of nature. ■■▮

(Williamson 1981; McNab 1994a,b), while larger birds such as rails, ducks, and ratites tend toward insular dwarfism (Wallace 1857; Greenway 1967; Lack 1974; Weller 1980). Cassowaries are large by most avian standards, but as Wallace noted in 1857, the New Guinea form *(Casuarius bennetti)* is small compared with its relative on the Australian mainland *(C. casuarinus;* body lengths of 52 vs. 65 inches, respectively). Emus *(Dromaeius novaehollandiae)* of the small islands of Bass Strait are much smaller than those of Tasmania and the Australian mainland (Greenway 1967). Thus, the trends suggest a general pattern.

Clegg and Owens (2002) have provided a much more clear and comprehensive assessment of morphological variation in insular birds, revealing that both bill length and body mass of a sample of 110 insular populations vary in a manner consistent with the island rule (i.e., a graded trend from gigantism in smaller birds to dwarfism in the larger birds, with a possible, "optimal" mass of approximately 125 g; see Figure 14.25C; see also Cassey and Blackburn [2004]). Similar to explanations discussed above, Clegg and Owens attribute these patterns to energetic constraints, interspecific interactions, and physiological optimization for insular environments. Interestingly, and just as we saw for some insular mammals (see Figure 14.27), body size of wrens of the British and Scottish Isles tends to increase with isolation (Berry 1964).

Insular reptiles show a diversity of patterns in body size evolution, but again a general trend may be emerging. Much as Foster (1964) first reported for mammals, different reptilian orders and families appear to exhibit different evolutionary tendencies. Gigantism, for example, is common in insular iguanids, herbivorous lizards, whiptails, tiger snakes, and possibly tortoises (Mertens 1934; Case 1978; Case and Bolger 1991; Case and Schwaner 1993; Petren and Case 1997; Caccone et al. 1999), while rattlesnakes tend to be dwarfed on islands (Schwaner and Sarre 1990). Body size of insular populations of the lizard *Uta stansburiana* increases as the number of resident competitor species decreases (Figure 14.28), suggesting that, just as has been postulated for insular rodents (see Figure 14.26), these lizards tend to exhibit gigantism in the absence of competitors (Soule 1966; see also Case and Bolger 1991). Other studies report different trends even in the same group of species (e.g., in snakes, see Boback 2003; Boback and Guyer 2003).

In fact, it is not clear whether the "giant" tortoises of some oceanic islands are actually giants, or dwarfs. This situation is reminiscent of the case of the Wrangel Island mammoths. Compared with almost all other mammals—except, of course, their Pleistocene ancestors—they seemed gigantic. Yet, we know that they underwent drammatic and relatively rapid dwarfism during the early Holocene. Now, consider the extant reptilian "mammoths." The land tortoises of the Galápagos in the Pacific and Aldabra Islands in the western

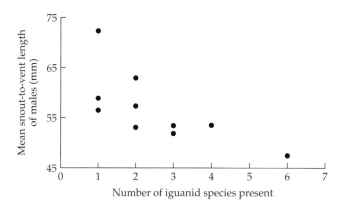

FIGURE 14.28 The lizard *Uta stansburiana* tends to exhibit increased body size on islands lacking its mainland competitors. (After Soulé 1966.)

Indian Ocean are indeed the largest extant tortoises—but are they larger than their ancestors? Recent studies indicate that Galápagos tortoises may have increased in size following colonization by a relatively small ancestor (Caccone et al. 1999). On the other hand, the "giant" tortoises of the Aldabras may actually be insular dwarfs. As recently as 500 years ago, tortoises occurred on at least 20 islands in the Indian Ocean—including Madagascar, which harbored two species. According to Arnold (1979), the supposed ancestor of the Aldabra Island tortoise is a Madagascan species that was at least 115 cm in length—larger than the biggest Aldabra species, which was 105 cm. One Madagascan species reached 122 cm in body length, rivaling the extant giant tortoises, while the two species of Rodriguez Island tortoise were just 85 cm and 42 cm. As Williamson (1981) observed, we cannot be certain whether large size evolved after the islands were colonized, or whether it simply made dispersal easier for the initial colonist from Madagascar.

Even monitor lizards—including the famous Komodo dragon—have proved to be a perplexing case study in insular evolution. The Komodo is, after all, the world's largest living lizard, measuring up to 3 m in length and tipping the scales at up to 150 kg (Auffenberg 1981). Even when considering just extant forms, however, the trend for monitors is equivocal, with some insular forms tending toward gigantism while others appear to be dwarfs (see also Gould and MacFadden 2004). Yet, all of these forms are tiny when compared with a monitor that inhabited Australia during the Pleistocene—a 7 m, 600 kg behemoth (Rich 1985). Until sound phylogenies are available, we must admit the possibility that the extant Indonesian "dragons" are actually insular dwarfs!

The above difficulties, while challenging, are probably not insurmountable. The influence of island characteristics on trends in insular body size can be statistically controlled, so that we can better focus on the influence of differences among species (e.g., trophic category, or size of presumed ancestor). In addition, phylogenetic analyses can be used to estimate the most likely ancestral state—in this case, the relative size of the immediate mainland ancestor (see Pianka 1995; Miles and Dunham 1996). Recent studies again reveal some interesting, emerging patterns within particular groups of reptiles. Nancy Karraker's (see Figure 14.25C) ongoing studies of insular turtles and tortoises indicate the expected trend from gigantism in the smaller species to dwarfism in the larger species, with an optimal size of approximately 20 to 25 cm (maximum carapace length; see also Aponte et al. 2003). Boback and Guyer's (2003) recent analysis of body size variation of insular snakes reveals a similar pattern, in this case with an optimal length of approximately 80 cm (see Figure 14.25C). Boback and Guyer also present an intriguing analysis of the influence of island area on species composition. While relatively large islands are inhabited by many species spanning a broad range of body sizes, as island area decreases, species richness decreases and species composition converges on those few snakes of the putative, optimal body size (i.e., those with lengths between 60 and 120 cm; Figure 14.29).

In summary, this continues to be an active area of research and it is an exciting time for any scientist interested in one of the most fundamental characteristics of life forms—body size. Patterns appear to be emerging, but many questions remain and many areas of future research will likely provide important insights into the forces influencing evolution of insular biotas. Perhaps most promising among these will be studies comparing patterns among taxa and among island archipelagoes, and those exploring the relationships between these patterns for insular biotas and related patterns including Cope's rule, Bergmann's rule, and geographic and temporal gradients in primary productivity and diversity. This research program will no doubt be challenging, as it

FIGURE 14.29 As we consider assemblages of insular snakes across a gradient of islands from the most to least isolated, species richness declines and species composition converges on snakes of intermediate size (from 80 to 120 cm long; see Figure 14.25). (After Boback and Guyer 2003.)

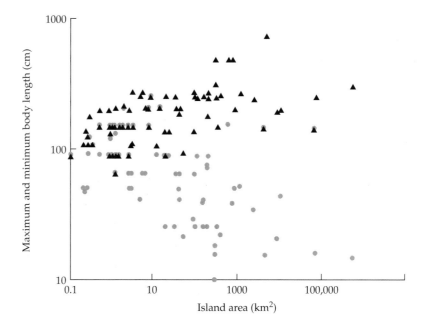

calls for broadening both the spatial and temporal context of our studies. We are, however, justifiably encouraged by recent insights, and by the enthusiasm and expertise of our colleagues and students. We are equally confident that biogeographers, ecologists, and evolutionary biologists will continue to provide some fundamental advances to our understanding of the assembly and evolution of insular biotas. It is appropriate, therefore, that we conclude this chapter with a theory, first advanced by one of the co-authors of the equilibrium theory, but one that predates that theory and, in many ways, is even more general.

The Taxon Cycle

Oceanic barriers severely limit the immigration of propagules, but those species that manage to gain a foothold have a high probability of success and, as George Gaylord Simpson observed, they win an ecological and evolutionary sweepstakes (see Chapter 6). This infrequent, but continual, establishment of colonists drives sequential phases of expansions and contractions of species' ranges that E. O. Wilson (1959, 1961) termed the **taxon cycle.** Wilson was, of course, one of the co-founders of the equilibrium theory of island biogeography, which was inspired in part by his concept of taxon cycles. In contrast to the equilibrium theory, however, species differences and interspecific interactions are not ignored but actually hypothesized to drive taxon cycles.

According to Wilson, insular species evolve through a series of stages from newly arrived and broadly distributed colonists—indistinguishable from their mainland relatives—to highly differentiated and ecologically specialized endemics, which ultimately become extinct (Table 14.6). This process is termed a "cycle" because, once insular populations begin to differentiate and adapt to island life, they appear to be doomed to extinction and replacement by new colonists from the mainland. As Ricklefs and Bermingham (2002) observed, taxon cycles are not expected to characterize all biotas, only those occurring on systems where isolation approximates dispersal abilities of the focal taxon; for less-isolated archipelagoes, immigration and gene flow will preclude ecological and genetic differentiation, whereas for the very distant archipelagoes, few islands will be colonized by the species.

TABLE 14.6 *Stages of the taxon cycle*

Stage I Initial expansion: The initial stage of colonization and establishment of populations of a species across an archipelago before its insular populations have differentiated from one another or from the source population on the mainland. In this stage, the species has a relatively continuous range across the archipelago, but exhibits little geographic variation. Such "invaders" are often broad-niched species from marginal habitats on the mainland, and are therefore preadapted for marginal habitats along beachfronts and other habitats occupying the island's periphery.

Stage II Ecological and evolutionary specialization: The insular populations have differentiated to the point at which they may represent endemic subspecies, or even species. The populations have invaded and become adapted to habitats within the island's interior, and often exhibit associated changes such as reduced dispersal ability, shifts in body size, and increased specialization. Some insular populations, however, have become extinct, so that the range of the taxon has contracted and its distribution is now spotty.

Stage III Initial contraction: Differentiation and range contraction have continued to the point that the taxon now comprises just a few relictual, endemic species whose populations are highly specialized and restricted to interior habitats.

Stage IV Single island endemics: The ranges of the relictual populations have contracted further, both within and among islands of the archipelago. As a result of their extreme specialization, perhaps hastened by competition with new invaders (stage I species), relictual populations disappear from the archipelago, presumably to be replaced by other, stage III species.

Source: After Wilson 1959 and 1961.

Whether fortuitously or by design, Wilson studied an insular fauna that met these criteria—ants of the islands of Melanesia, north and east of New Guinea in the western Pacific. By analyzing differences in distributions and ecological associations of ants in a snapshot of time, he inferred what appeared to be stages in the expansion and taxonomic differentiation of ants (see Figure 14.5A). He also analyzed the distributions of these species across habitats, within and among islands, from which he inferred successive changes in the niches of these ants as they evolved through the taxon cycle. Those species that are good over-water colonists typically occur in coastal or disturbed habitats in New Guinea; recently arrived, undifferentiated populations are found in similar habitats on the Melanesian Islands. As they differentiate in isolation, however, the ants also change their ecological requirements and expand into interior habitats, such as native forests (Figures 14.30 and 14.31). They are replaced by a new wave of colonists occupying the beaches and disturbed habitats. Highly differentiated endemic forms, representing the

FIGURE 14.30 Ecological changes accompanying the taxon cycle in the ants of Melanesia. Species of forest ancestry that have secondarily invaded marginal (usually disturbed or coastal) habitats in Southeast Asia (1) tend to be good dispersers and to colonize similar habitats on the islands of Melanesia (2). These populations become extinct fairly rapidly (3), or else they invade interior rain forest habitats (4), where they may undergo differentiation and adaptive radiation (5). Sometimes they give rise to forms that secondarily invade marginal habitats and disperse in stepping-stone fashion to more distant islands (6). (After Wilson 1959.)

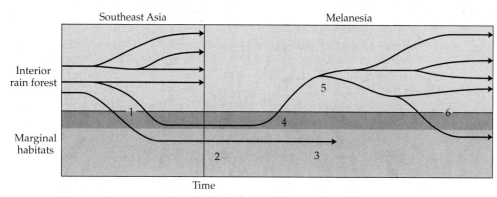

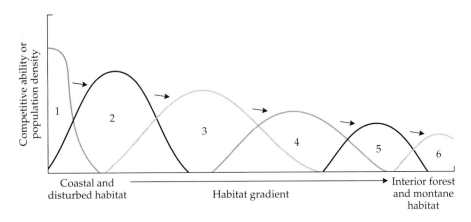

FIGURE 14.31 A graphic model of the processes that are thought to be involved in the insular taxon cycle. Numbers are as in Figure 14.30. The cycle is driven by the colonization of generalized species, which then evolve to become more specialized, sacrificing competitive ability and experiencing reduced population density within particular habitats. Pushed further along the habitat gradient by superior competitors, species in the terminal stages are forced to specialize even more, but evolutionary constraints prevent them from becoming well adapted to these new niches and, hence, from increasing in density and competitive ability. Consequently, they evolve into rare endemics and ultimately become extinct.

last stage of the taxon cycle, typically are restricted to just one island and to a narrow range of environments, usually rain forest or montane forest deep in the island's interior.

Thus, even before co-developing what would prove to be a paradigm of ecological biogeography—the equilibrium theory—Wilson presented a theory that could account for an even more diverse collection of patterns, including those associated with colonization, ecological release, niche shifts, character release and displacement, reduced dispersal ability, shifts in insular body size, range contraction, endemicity, and extinction. Later, Ricklefs and Cox (1972, 1978) reported ecological and evolutionary shifts in West Indian birds that showed successive stages of what they interpreted as a taxon cycle (Figure 14.32). Similarly, Roughgarden and his colleagues used the theory of taxon cycles to explain the evolution of body size in West Indian anoles (Figure 14.33; Roughgarden and Roughgarden et al. 1987; Pacala 1989; Roughgarden 1992; see also Erwin 1981; Miles and Dunham 1996).

More recently, Ricklefs and Bermingham (2002) have reviewed this subject and provided a much more explicit, causal explanation for taxon cycles. While immigration and the many factors influencing this fundamental process, explain the expansion phase, the contraction phase appears to result from subsequent selective pressures of insular environments—including interspecific interactions. For ants of Melansia and birds and *Anolis* lizards of the Lesser Antilles, ecological and evolutionary changes may be driven by changes in the balance of co-evolution between these species and their predators, competitors, and parasites (Ricklefs and Cox 1972; Ricklefs and Bermingham 2002; Ricklefs and Fallon 2002; Fallon et al. 2003). New colonists enter the expansion phase because they have escaped these mainland enemies, but new competitors begin to expand from the beachfront, and insular predators and parasites subsequently learn to exploit these new prey and hosts, eventually triggering a contraction phase.

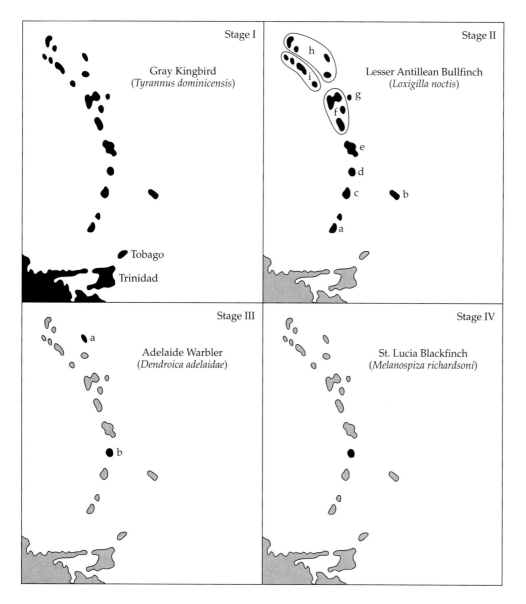

FIGURE 14.32 Stages of the taxon cycle as illustrated by birds of the West Indies. Stage I, a widespread, undifferentiated species that presumably has recently colonized from South America, represented here by the flycatcher *Tyrannus dominicensis*. Stage II, a widespread form with well-differentiated races (letters) on different islands, represented by the finch *Loxigilla noctis*. Stage III, a species with well-differentiated races on only a few islands, represented by the warbler *Dendroica adelaidae*. Stage IV, a narrowly endemic species, represented by the finch *Melanospiza richardsoni*, confined to Saint Lucia. (After Ricklefs and Cox 1972.)

Despite the initial attraction of the concept and its logical explanation, other researchers questioned both the generality and the above explanations for putative taxon cycles, instead offering alternative and seemingly more parsimonious explanations for observed patterns in distributions, niche shifts, and morphological differences (Losos 1992; Taper and Case 1992; Losos et al. 1993; see also Pielou 1979; Pregill and Olson 1981). In fact, Wilson's theory has fallen far short of achieving the status of a paradigm. For the past four decades, it has attracted surprisingly little attention from most biogeographers and ecologists, and many have dismissed it as an interesting, but failed, attempt. Such declarations, however, seem unfortunate and premature. In fact, a series of more recent studies by Ricklefs and his colleagues using phylogenetic analyses coupled with detailed information on the ecology and distributions of Lesser Antillean birds have confirmed most of the tenets of the theory, at least for this insular fauna (see Ricklefs and Bermingham 2002). In fact, Ricklefs and Bermingham (2002:359) assert that "alternating phases of expansion and con-

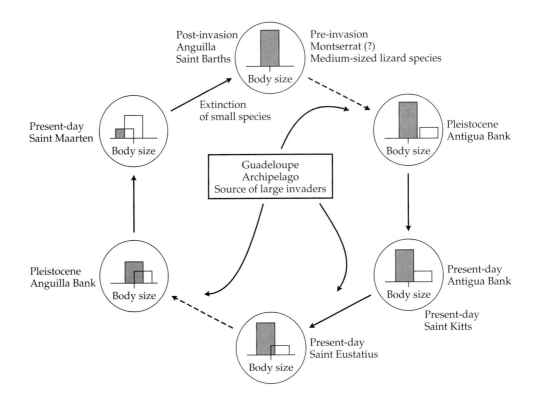

FIGURE 14.33 A proposed taxon cycle for the *Anolis* lizard species in the northern Lesser Antilles of the Caribbean. An island with just one resident species (gray bar) evolves a "characteristic" or "optimum" body size (top circle; bars indicate relative population density). This island is then invaded by a second and larger species (open bar), which then evolves toward the size of the original resident, which in turn, becomes smaller. The range of the initial resident begins to contract until it eventually becomes extinct. The "invader" now becomes the specialized resident, and its size and range continue to decrease as the island is again colonized by another, larger species, to reinitiate the cycle. (After Roughgarden et al. 1989.)

traction are nearly universal and that it is possible to study taxon cycles analytically in a wide variety of groups and regions." Briggs' (2003, 2004) recent research reveals that, consistent with this statement and the theory of taxon cycles, expansions and contractions of geographic ranges in the marine realm are correlated. Much earlier, Diamond (1977) drew parallels between what Wilson and others saw as taxon cycles and similar patterns of colonization and extinctions of native biotas and of human populations on islands. Therefore, given the possible generality of these patterns, the ability of Wilson's theory to integrate distributional, ecological, and evolutionary phenomena, and its potential relevance to our own species and to our attempts to conserve imperiled species (i.e., prolonging or possibly reversing the contraction phase of their own taxon cycles), the theory of taxon cycles merits far more attention and more rigorous assessment before we evaluate its true worth as pyrite or gold.

CHAPTER *15*

Areography, Ecogeographic Rules, and Diversity Gradients

*I*SLANDS, THE SUBJECT OF THE PREVIOUS two chapters, are often inhabited by evolutionary marvels and they provide fascinating and perhaps unrivaled insights into the forces influencing the ecology, evolution, and biogeography of biotas. On the other hand, islands cover just a small fraction of Earth's surface and their physical and environmental characteristics are, almost by definition, quite unlike that of most ecosystems (i.e., islands are isolated, relatively small, and characterized by "unusual" environments and biotic communities). It is important, therefore, to ask to what extent the patterns and processes known to occur on islands also hold true in geographically less isolated and biologically more diverse settings—the world's continents and ocean basins (see Crowell 1983).

Biogeography's first and most fundamental pattern, Buffon's Law (see Chapter 2), described differences in species composition among continental biotas—a pattern that Wallace, Darwin, and the following century of biogeographers ultimately attributed to the equally fundamental processes of evolution, dispersal (immigration), and extinction. Throughout the Age of European Exploration, equally distinguished naturalists and biogeographers such as Forster (1778) and von Humboldt (1808) not only documented the generality of Buffon's Law, but also discovered many related patterns in the geography of nature. Across their historic voyages, they were struck—not just by the complexity of nature, but also by its indefatigable geographic signal. The diversity of the biotas they studied, and the size and shape of the individuals in those biotas varied in a highly regular manner along gradients of latitude, elevation, and depth. While these clines in diversity and composition of ecological communities were traditionally viewed as the realm of "ecological biogeography," modern scientists now realize that they result from the combined effects of processes occurring over ecological—as well as, evolutionary—time (e.g., predation, competition, and dispersal; along with climate change, tectonic events, adaptation, and speciation). Thus, the ecological biogeography of continents and ocean basins is indeed a complex and challenging subject, but one which holds the promise of great rewards for anyone interested in the origins, distributions, and diversity of life.

In this, the final chapter before we turn to applying the lessons of biogeography for conserving biological diversity, we focus on two of the field's persistent themes (see Table 2.1):

- explaining differences in numbers (diversity) of species among geographic areas or along geographic gradients, and
- explaining geographic variation in the characteristics of individuals and populations.

Because species richness actually represents the overlap of species distributions, we first explore patterns in biogeography's fundamental unit—the geographic range. We then review the earliest accounts of so-called **ecogeographic rules**—very regular geographic gradients in body size, shape, and other characteristics of individuals and populations across the continents and ocean basins. We conclude this chapter with an overview of geographic gradients in diversity and, in particular, what has been referred to as one of ecology's oldest and most general, yet, at times, most perplexing patterns—the latitudinal gradient in species diversity.

While the complexity of all of these large-scale patterns—from geographic variation among individuals and populations, to variation in the size of geographic ranges and diversity among regional biotas—may seem overwhelming, each of these patterns should ultimately be explicable within the central framework of modern biogeography. Environmental factors vary in a very regular manner across the geographic template, and species respond to those environmental conditions by either dispersing, adapting in situ (and sometimes diversifying), or suffering extinction. Thus, the very regular geographic variation of the natural world is not, in retrospect, that remarkable. These large-scale patterns in biological diversity, however, do constitute some of science's most fascinating and challenging phenomena.

The Geographic Range: Areography and Macroecology

Given that one of the most fundamental units of biogeography is the geographic range, an equally fundamental question is, what determines species distributions (see Chapter 4)? This, of course, is a question biogeographers have been addressing since the origins of the field. Drawing on distributional information collected by naturalists over the past three centuries, modern biogeographers have made great strides in describing the patterns and underlying processes influencing the sizes, shapes, and location of geographic ranges. During the 1970s, this research program began to coalesce into what is recognized today as one of biogeography's most active and intriguing subdisciplines—**areography**. This field of research also became recognized as one of the most ambitious and challenging. As we have remarked elsewhere, use of experimental manipulations at spatial scales appropriate to study geographic ranges are rendered either logistically infeasible or unethical. Areographers are therefore limited to applying the comparative method, but they have shown some great ingenuity and creativity. In fact, their attempts to investigate the influences of processes operating over multiple spatial and temporal scales stimulated the development of a new approach to exploring patterns and processes in ecology and biogeography—**macroecology.**

Macroecology has been described as a top-down—and, perhaps more accurately, a multi-scale— approach to investigating the assembly and structure of biotas (Brown and Maurer 1987, 1989; Brown 1995; Gaston and Blackburn

1999, 2000; Blackburn and Gaston 2003). Because it is a potentially powerful statistical approach to understanding abundance and distributions of species, the macroecological approach has provided many important insights for both ecologists and areographers.

The essence of macroecology is that it tries to identify general patterns and to understand the underlying mechanisms by focusing on the statistical distributions of variables across spatial and temporal scales, and among large numbers of equivalent (but not identical) ecological *particles*. These particles can be almost anything: individual organisms within a local population or entire species, replicated sample areas or patches of habitat, or species within local communities or larger biotas (for examples, see Brown 1995; Gaston and Blackburn 2000; Blackburn and Gaston 2003). The premise of macroecology is that, although these particles may form seemingly random collections at one scale, their interactions and statistical properties (e.g., means, variances, frequency distributions) yield patterns that become emergent at larger spatial or longer temporal scales. Appropriately, Gaston and Blackburn (2000:16) attribute the early seeds of the macroecological approach to Robert MacArthur (1972), who observed that "we may reveal patterns in the whole that are not evident at all in its separate parts."

Macroecologists continue to discover some fascinating emergent patterns in the statistical distributions of abundances and geographic ranges, and to apply and advance the approach to explore general mechanistic processes involved across different scales of space, time, and biological complexity. As many of these studies focus on questions with only a minor spatial context, we refer our readers to general works by Brown (1995), Gaston and Blackburn (2000), and Blackburn and Gaston (2003) for comprehensive discussions of the subject; we instead limit our discussion below to patterns in geographic ranges.

Areography: Sizes, Shapes, and Overlaps of Ranges

SIZES OF RANGES. There is enormous variation in the sizes of geographic ranges of individual species. Among the smallest ranges are the natural distributions of the Soccoro isopod (*Thermosphaeroma thermophilum*) and the Devil's Hole pupfish (*Cyprinodon diabolis*), each of which occurs in a single freshwater spring with a surface area of less than 100 m^2. Among the largest ranges are those of several marine organisms, such as the blue whale (*Balaenoptera musculus*), whose ranges include most of the world's unfrozen oceans, an area on the order of 300,000,000 km^2. Among terrestrial organisms, species with very large native ranges include the peregrine falcon, barn owl, and osprey, all of which are widely distributed over all of the continents except Antarctica. Of course, modern *Homo sapiens* is now one of the most widely distributed species, and humans have carried several species of symbionts and exotics with them as they have spread over the entire Earth.

The year 1982 saw the publication of the English language edition of a fascinating little book entitled *Areography,* by the Argentine ecologist and biogeographer Eduardo Rapoport (Rapoport 1975). Working largely in isolation in Latin America, Rapoport showed that simple quantitative analyses of spatial distributions of organisms—such as the maps of geographic ranges in Hall's (1981) *Mammals of North America,* or Critchfield and Little's (1966) *Geographic Distribution of Pines of the World*—could reveal fascinating patterns and suggest hypotheses about the mechanisms that limit species' distributions and influence community diversity. Rapoport's book inspired a number of ecological biogeographers to undertake conceptually and methodologically similar studies on variation in geographic ranges. The result was a renewed emphasis on comparative biogeography that was similar in some ways to the resurgence of research on insular biogeography that was stimulated by MacArthur and Wilson's (1967) seminal monograph about 15 years earlier. Areography remains an active and stimulating research program, exploring all aspects of the structure of geographic ranges (see Gaston 2003).

The frequency distribution of range size. One of the most fundamental and interesting areographic patterns is the distinctively shaped frequency distribution of the sizes (areas) of geographic ranges, first documented by Willis (1922) and subsequently confirmed by Rapoport (1982) and others (Figure 15.1; see

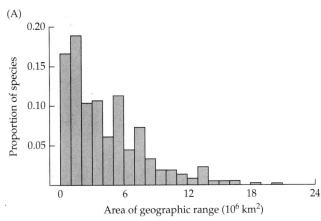

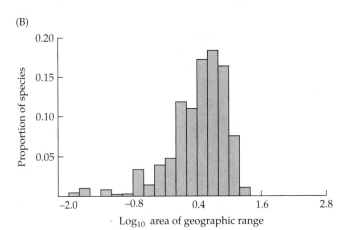

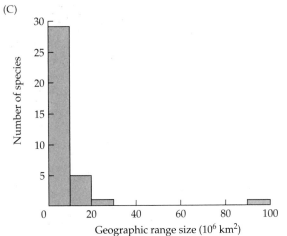

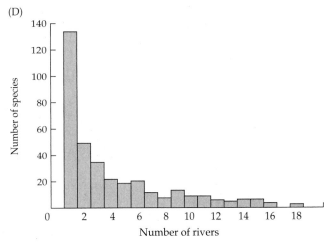

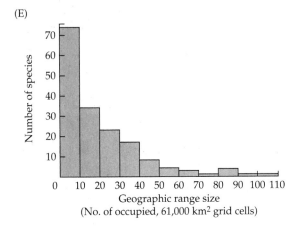

FIGURE 15.1 Frequency distribution plots of geographic range sizes reveal that, despite the substantial variation in size, geographic ranges of most species tend to be relatively small. Plotted here are frequency distributions of North American mammals on an arithmetic scale (A), the same data plotted on a log-transformed scale (B), and arithmetic scale plots for wild cats of the world (C), African freshwater fishes (D), and breeding ranges of wildfowl (E). (A–B after Brown 1995; C–E after Gaston 2003.)

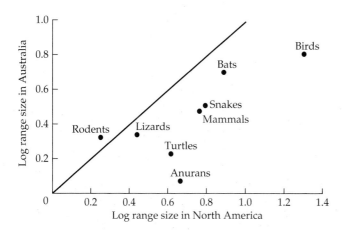

FIGURE 15.2 Geographic range size tends to be a direct function of, among other things, size of the continents. For example, the mean geographic range size (in 10^5 km^2) of North American vertebrates (except for rodents) exceeds those of Australian species (solid line is the line of equality). (From Gaston 2003; after Anderson and Marcus 1992.)

Gaston 1991, 1996; Pagel et al. 1991; Brown 1995; Brown et al. 1996; Gaston and Blackburn 2000; Gaston 2003). Two features of this frequency distribution are of particular interest: (1) it is unimodal, and (2) it is right skewed. That is, despite substantial variation in range size among species—often spanning three or more orders of magnitude—most species tend to have relatively small geographic ranges. The pattern appears to be a very general one, exhibited by terrestrial, freshwater, and marine assemblages, including modern forms, those from the paleontological record (Jablonski 1987; Roy 1994), and newly established invasive species as well (Gaston 2003; after Halloy 1999).

The consistent shape of these frequency distributions among such a diverse collection of organisms, ecosystems, and time periods, encourages the search for a very general explanation. Perhaps the most important clues to discovering underlying causal mechanisms, however, are not the similarities among these distribution functions, but their differences. For example, while geographic ranges for most terrestrial assemblages tend to occupy less than a third of the continents, the ranges of birds tend to be much larger than those of reptiles and amphibians (Figure 15.2; Anderson and Marcus 1992; Gaston 2003). Similarly, geographic ranges of mammals, while typically larger than those of lizards, turtles, and anurans, differ markedly between bats and non-flying mammals such as rodents; see Figure 15.2). Perhaps just as intriguing, geographic range sizes of the same taxa differ considerably and in a predictable manner among the continents, being smaller for assemblages from the smaller continents (note that most points in the graph of Figure 15.2 fall to the right of the line of equality, indicating that ranges of nearly all of these assemblages, except for rodents, tend to be greater on the larger continent—North America). Rapoport (1982) was perhaps the first to report this relationship, noting that the geographic ranges of terrestrial birds tend to vary with the continental size such that most species occupy approximately one-fourth of the continent's land surface.

Geographic gradients in range size.   Rapoport (1982) also studied patterns of variation of geographic range size along geographic gradients across particular continents. One pattern seemed so general that it later became known as

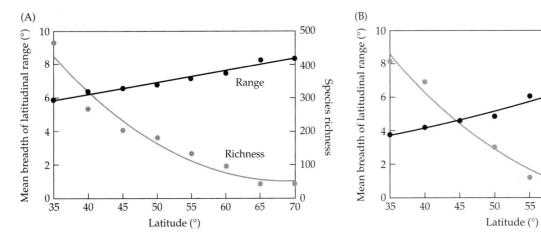

(A)

(B)

FIGURE 15.3 Two examples of Rapoport's rule in marine mollusk species along the Pacific Coast of North America (A), and in tree species in the continental United States and Canada (B). As species richness declines with increasing latitude, the remaining species tend to have geographic ranges that extend over a broader range of latitudes. This method of plotting Rapoport's rule (after Stevens 1992, 1989) depicts the average breadths of the latitudinal ranges of the species whose ranges are centered on 10° latitudinal bands. (After Brown 1995.)

Rapoport's rule—the tendency for geographic range size to increase with latitude (Figures 15.3 and 15.4; Stevens 1989; see also earlier report of this pattern by Lutz 1921). In subsequent studies, Stevens (1989, 1992) reported that the pattern is exhibited for many different kinds of organisms, including mammals, birds, mollusks, amphibians, reptiles, and trees, while other biogeographers reported the pattern for such diverse groups as fish, crayfish, amphipods, and beetles (see Gaston 2003:100–101). Rapoport's rule also has its analogue in elevational gradients in terrestrial environments and in depth gradients in the marine realm, but its generality remains uncertain (Figure 15.5; Stevens 1992, 1996; but see Rohde et al. 1993; Rohde 1996; Rohde and Heap 1996; Blackburn and Ruggiero 2001; Smith and Gaines 2003).

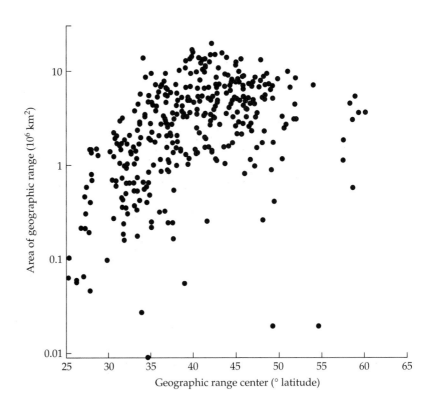

FIGURE 15.4 Variation in size of geographic range with latitude for North American breeding land birds. Although this method of plotting the data appears to show much more variation than in Figure 15.3, it also provides an example of Rapoport's rule: increasing range size with increasing latitude. (After Brown 1995.)

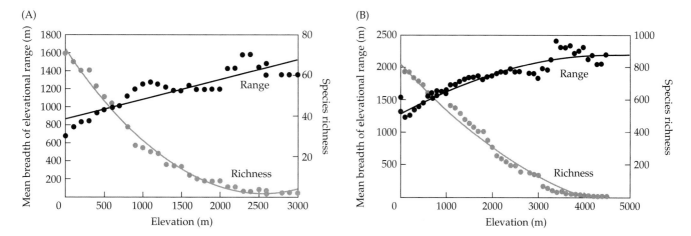

FIGURE 15.5 The elevational equivalent of Rapoport's rule: species tend to be distributed over a wider range of elevations as elevation increases. This pattern is seen both in tree species in Costa Rica (A) and in bird species in Venezuela (B). Compare with the latitude patterns in Figure 15.3. (After Stevens 1992.)

According to Stevens (1989), the pattern results from the effects of climatic regimes on ecological capacities of species across different regions. To be adapted to their seasonal and variable climatic regions, species occupying temperate regions should have relatively generalized niches and the capacity to disperse across climatic and topographic barriers (Janzen 1967). The cumulative effect of these selected pressures over space and time is that, in comparison to species in the tropics, those from temperate regions should be able to cross more barriers and, thus, have larger geographic ranges.

As with many large-scale patterns, however, the generality of Rapoport's rule has been challenged on both empirical and conceptual grounds. Many, if not most, species fail to exhibit the predicted pattern, or they exhibit it for some biogeographic regions and not for others (see Gaston 2003). At least one of the alternative explanations for the pattern—that based on a hypothesized link between range size, species diversity, and intense competition of the tropics (Brown 1995)—has been criticized for being circular (see also Smith and Gaines 2003). Are geographic ranges smaller because of competition from a diversity of species, or is it that more species can be packed into tropical regions because their ranges are smaller? Some biogeographers attribute trends consistent with the "rule" to a statistical artifact: the random placement of geographic (latitudinal) ranges between two cartographic extremes—the poles (the "mid-domain effect, Colwell and Hurt 1994; Colwell and Lees 2000; see also Willig and Lyons 1998; Lees et al. 1999; McCain 2003). According to this null hypothesis, we can think of ranges as being similar to a set of pencils randomly positioned within a long box; just by chance, many more of them will be located near the middle of the box ("the equator") then near either of its edges ("the poles"). While this thought experiment seems unrealistic in many respects, the mid-domain explanation can serve as a useful null hypothesis for this, and related, unimodal patterns in biogeography (e.g., the tendencies for species diversity and range size to peak at some intermediate elevation between sea level and mountain summits; Colwell and Hurtt 1994; see also Stevens 1992; Willig and Lyons 1998; Lees et al. 1999).

Cowlishaw and Hacker's (1997) study of diversity and distribution of geographic ranges in primates is an exemplary case study, albeit one leaving us

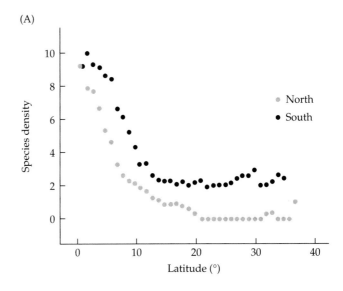

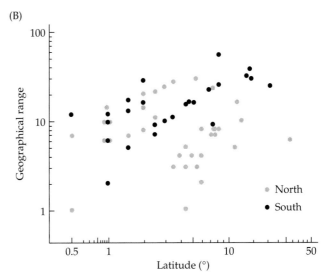

FIGURE 15.6 Across both the Northern and Southern Hemispheres, species density (number of species per 100,000 km²) of African primates (A) decreases, while the size of their geographic ranges (B)—although exhibiting much variation—tends to increase as we move from the equator toward the poles. (After Cowlishaw and Hacker 1997.)

with as many interesting questions as answers. As with most groups, diversity of primates tends to increase toward the equator, but they, as a group, fail to exhibit Rapoport's rule (Figure 15.6). These results did not support Brown's hypothesized causal relationship between species diversity and geographic range size. The actual patterns, however, are more complicated than this. When considered separately, geographic ranges of primates south of the equator follow the predicted pattern, whereas those in the Northern Hemisphere do not (see Figure 15.6A). As Cowlishaw and Hacker concluded, latitudinal breadth in geographic ranges may well be associated with climatic variation as hypothesized by Stevens (1989) but, because the correlation between climates and latitudes is far from perfect, latitudinal gradients in diversity can exist independently of gradients in range size (see also Rohde 1996; France 1998; Harcourt 2000).

In his review of this and related biogeographic patterns, Gaston (2003) concluded that the predicted pattern, while occurring in an impressive variety of species, is only exhibited in a minority of cases and its generality depends on the methods and the particular span of latitude used (see also Rohde et al. 1993; Rohde 1999). On the other hand, although the pattern may not be so universal as to merit the designation as a true "rule" of nature, the list of species exhibiting this pattern is quite long and continues to grow. We firmly believe that neither the cases consistent with the predicted pattern, nor the exceptional ones, should be ignored. On the contrary, anyone wishing to fully understand the factors and underlying processes influencing geographic range size, in general, should explore the full complexity of patterns—comparing the characteristics of the many groups and biogeographic regions that do exhibit the pattern, to those of the anomalous cases.

Geographic range size as a function of body size. In our above discussion of areography, we noted that the sizes of the geographic ranges of species within a continent exhibit a distinctively shaped frequency distribution (see Figure 15.1). What are the mechanisms that account for the preponderance of relatively small ranges, and might this be causally related to a similar macroecological pattern in body size? Within nearly all groups of organisms, the frequency distribution of body size takes on a highly skewed, unimodal form very similar to that of geographic range size (see Calder 1974; Peters 1983; Brown and Nicoletto 1991; Brown 1995; Gaston and Blackburn 2000; Blackburn and Gaston 2003). One relatively simple inference is that geographic range size is just a reflection of body size: species comprised of smaller individuals require less energy and less space and, therefore, have smaller geographic ranges. This, however, appears to be too simplistic. Fortunately, there is a more direct and more informative approach to investigating the relationship between body size and size of the geographic range: graphing geographic range size as a function of body size for an assemblage of species (Figure 15.7). Given the substantial variation in both of these variables, such plots are typically drawn on logarithmically transformed axes. Along with describing an interesting relationship between these important variables that links a fundamental property of biogeography (the geographic range) to one equally fundamental to physiology (body size), these graphs also illustrate an innovative contribution of the macroecological approach—constraint lines.

These and other macroecological plots often fail to suggest a tight, linear relationship between the variables of interest (in this case, range size and body size). The relationship, however, is far from random but resembles a cloud of points that is constrained to certain regions of the bivariate space. Macroecologists have thus convinced many of us not to focus just on general trends (i.e., trend lines through a cloud of points), but on the factors and processes constraining the extremes of those variables. Body size and range size plots for North American birds and mammals (see Figure 15.7) are typical of those for most taxa, including fish (Gotelli and Pyron 1991; Pyron 1999; Rosenfield 2002), amphibians (Murray et al. 1998), birds (Brown and Maurer 1987; Sant'ana and Diniz-Filho 1999), and reptiles (Bonfirm et al. 1998). The points tend to fall within a roughly triangular space, indicating that range size varies substantially among the smallest species, but varies less, and is typically larger, for the larger species. More to the point, the triangular shape of these plots

FIGURE 15.7 Relationship between area of geographic range and body size on logarithmic axes for North American terrestrial mammals (A) and birds (B). Note that, while there is much variation, it is far from random. In particular, there are few large species with small geographic ranges. (After Brown 1995.)

(A) Land mammals, North America

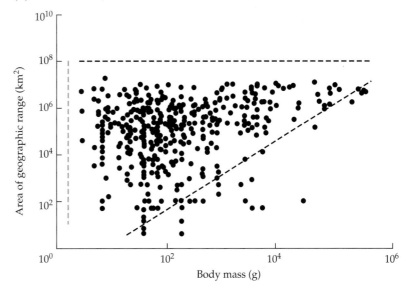

(B) Land birds, North America

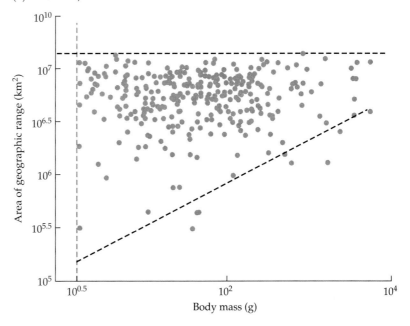

suggests at least three constraint lines. First, and perhaps easiest to overlook, there seems to be a constraint—perhaps a physiological one—on how small a species can be (the vertical dashed line in Figure 15.7).

The other apparent constraint lines in Figure 15.7 are no less interesting. The upper, horizontal constraint line in this figure is most likely a reflection of the influence of total size of the continent or ocean basin and so, at first glance, this seems trivial. Comparisons among continental biotas do, in fact, indicate that average geographic range size varies with total area of the continent. The relationship, however, differs among species groups and most species occupy a relatively small proportion of the available space. Thus, we still need to understand why this constraint line does not correspond to the total area available and, perhaps even more important, how and why these constraint lines vary among different groups of species.

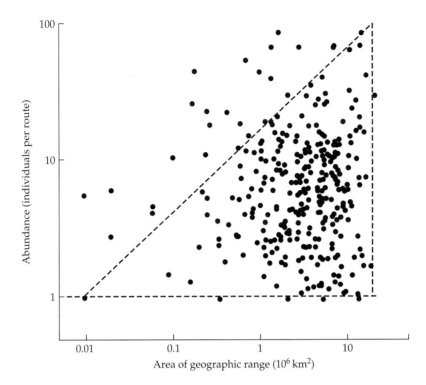

FIGURE 15.8 Relationship between area of geographic range and local population density for North American land birds. Note that, while there is extensive variation, it is far from random. In particular, there are few species with high abundances and small geographic ranges. (After Brown 1995.)

The remaining constraint line is perhaps the most intriguing one, and may have great relevance not only for ecology and biogeography, but for conservation of endangered species as well. Because population density varies inversely with body size (Brown and Maurer 1987; Damuth 1991; Silva and Downing 1995), and probability of extinction varies inversely with absolute population size (e.g., MacArthur and Wilson 1967; Leigh 1975; Gilpin and Hanski 1991), species with large body sizes tend to be rare and extinction-prone (Figure 15.8). Only if they have large geographic ranges are they likely to have sufficient numbers of individuals to persist for long periods. This phenomenon was clearly demonstrated by the many waves of anthropogenic extinctions that occurred during the latter part of the Pleistocene. Of the diversity of animals that existed then, it was the largest species (the Pleistocene megafauna) that suffered the highest extinction rates (Chapter 9). The diagonal constraint lines in Figure 15.7 can also serve as valuable tools for understanding recent and impending extinctions. It is not just the largest species that are in jeopardy—they certainly are—but more to the point, it is the species whose ranges are critically small *relative to their body size* that should receive the lion's share of conservation efforts (including species ranging in size from Kirtland's warblers, Morro Bay kangaroo rats and Marianas fruit bats, to lesser prairie chickens, black-footed ferrets, and giant pandas). Constraint lines should vary among taxa, bau plans, and tropic strategies, and they certainly merit further attention by macroecologists and biogeographers (for an overview of methods that can be used to calculate constraint lines, see Enquist et al. 1995; Garvey et al. 1998; Guo 1998; Cade et al. 1999; Gotelli and Entsminger 2001).

Temporal dynamics of range size. The fact that areographic patterns are so similar across so many different kinds of organisms—from butterflies and conifers to birds and mammals—encourages the search for unifying principles of biogeography. As Gaston (2003:81) concluded, "… species ranges must ultimately be a product of three processes: speciation, extinction, and the temporal dynamics of the range sizes of species between speciation and extinctions." We shall focus on the latter process (collapse of geographic ranges that may ultimately lead to extinctions) during our discussions of conservation biogeography in the next two chapters. Here, we turn our attention to a complementary phenomenon: the dynamics of geographic ranges during speciation and diversification of a lineage. Webb and Gaston (2000) have addressed this question by investigating the relationship between geographic range size and age of the species. They found that, for several monophyletic groups of birds with well-resolved phylogenies, range size tended to expand relatively rapidly at first, but then gradually declined with species age (Figure 15.9). Actually, the earlier stages in phylogeographic diversification are entirely analogous to those of range expansion in recently introduced, exotic species (Chapters 9 and 16); that is, once the invasive species becomes locally established, its individuals and populations expand rapidly until they encounter physical or biological barriers, which then slow, redirect, and eventually restrict their range

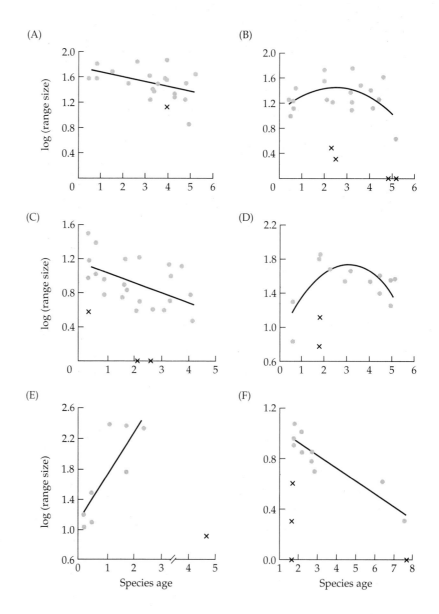

FIGURE 15.9 Geographic range size of species within monophyletic groups of birds tends to be correlated with species' age (abscissa in million years), increasing relatively rapidly (in evolutionary time) at first, but then declining over longer time periods. Each circle represents a different species of that monophyletic group, including (A) Old World reed warblers (*Acrocephalus* and *Hippolais*), (B) New World wood warblers *(Dendroica)*, (C) New World orioles *(Icterus)*, (D) storks, (E) gannets and boobies, and (F) albatrosses. The key outliers (marked by x) are those species that have suffered recent collapse as a result of anthropogenic factors. (From Gaston 2003; after Webb and Gaston 2000.)

expansion. As Wells and Gaston (2000) note, one important and foreboding set of exceptions to the general patterns illustrated in Figure 15.9 are species whose ranges have been drastically reduced by anthropogenic activities, and thus fall well below the general trend lines for species age as a function of range size (see Figure 15.9).

In the next two chapters, we discuss these and other emerging themes in the relatively newly articulated field of conservation biogeography. It is clear, however, from the foregoing discussion on geographic range size that species distributions and their dynamics over time result from properties of the species, environmental variation across the geographic template, and interactions between the two—in particular, the abilities of some species, especially our own, to modify their environments and alter the range dynamics of many other species.

SHAPES OF RANGES.   In addition to the impressive and sometimes regular variation in the sizes of geographic ranges, there are also interesting areographic patterns in their shapes. Rapoport (1982) noted that, despite the orders-of-magnitude variation in the areas of the ranges of North American mammals, one measure of shape—their perimeter-to-area ratio—remained relatively constant. That is, when he measured the length of a range boundary and the area encompassed within that boundary, he found that the ratio of the two variables was a nearly constant 10. This is surprising, because for similarly shaped geometric figures, perimeter-to-area ratios should decrease with range size.

This apparent invariance of perimeter-to-area ratios suggests that, when considered at the same spatial scale, large ranges have more convoluted boundaries than do small ranges. While this pattern might suggest something interesting about colonization-extinction dynamics, ecological limiting factors, or some combination of these, it is first important to evaluate an alternative, or null hypothesis; that is, it simply reflects an unintentional bias of mapmakers. We have an inherent tendency to draw maps with a fractal structure, including more detail about boundaries (and other features) for larger areas. In many cases, including Hall's (1981) treatise on North American mammals, small ranges are mapped at greater magnifications than are large ones. On the other hand, if the fractal inclinations of mapmakers could be taken into account, the periphery-to-area ratio could serve as a simple measure of range shape that warrants further study.

Brown and Maurer (1989) used an alternative and relatively simple technique to quantify both the shapes and the orientations of ranges (for other approaches to investigating variation in range shape, see also Maurer 1994; Ruggiero 2001; Fortin et al. 2005). In order to investigate whether range shape varied in any regular manner with range size, Brown and Mauer (1989) plotted maximum north-south extent across a species' geographic range as a function of maximum east-west extent. The result is the kind of graph shown in Figure 15.10B, in which a line of equality has been plotted for reference. In such a graph, ranges with equal dimensions, such as circles or squares, would fall along the diagonal line (with small ranges falling to the lower left and large ranges falling to the upper right of this graph). Ranges longer in a north-south direction would fall above the line of equality, whereas those longer in an east-west direction would fall below the line.

Although there is considerable scatter in such graphs (see Figure 15.11; Brown and Maurer 1989; Brown 1995), North American mammals, birds, and reptiles all show a consistent trend: small ranges tend to be oriented north-south, whereas large ranges tend to be oriented east-west. European mammals and birds present an interesting contrast, with most ranges oriented east-west.

FIGURE 15.10 A simple means of representing both the shape and size of geographic ranges is to measure their maximum extent along the cardinal directions, and then plot these measures for various species on one graph to investigate whether the shape of geographic ranges changes with their size. (A) This map illustrates the geographic range of four populations of the desert kangaroo rat (*Dipodomys deserti*). (B) The results are shown for four hypothetical species (*a*, *b*, *c*, and *d*); we have added highly simplified representations of range shapes to note whether they were relatively circular (small or large, for species *a* and *b*), or more elliptical ("stretched" north-south or east-west for species *c* and *d*).

(A)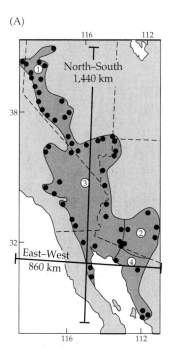

(B)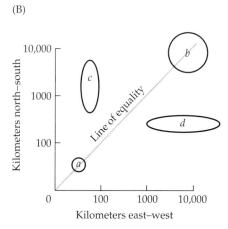

Brown and Maurer suggested that these patterns of range orientation reflect the physical geography of the continents. The east-west orientation of large ranges on both continents likely results from similar orientation of major climatic and vegetation zones (see Chapters 3 and 5). On the other hand, very small ranges (those of localized endemics) will appear roughly circular or symmetrical because of the cartographic artifact described above. Somewhat larger ranges, however, are likely constrained by large-scale, topographic barriers. As Brown and Mauer (1989) observed, most of the major topographic barriers in North America (e.g., the Atlantic and Pacific coastlines; the Mississippi River; the Rocky Mountains; and the Sierra Nevada, West Coast, and Appalachian Mountains) run roughly north and south. Accordingly, species whose distributions range over a small-to-intermediate portion of the continent should have ranges with a north-south "stretch." Species with larger ranges are, by definition, not constrained by these barriers and thus their ranges tend to be more circular; or, in those species with the largest ranges—stretched east-west, following the orientation of climatic and vegetation zones, or the northern coastlines of the continent.

Why then don't the species of Europe exhibit the same patterns? First, unlike North America, most of the major geographic barriers of Europe run east-west. Thus, dispersal and range expansion to the south is blocked, first by a broken string of mountain chains (including the Pyrenees, Alps, and Caucasus Mountains), and then by large bodies of water (the Mediterranean, Aegean, and Black Seas) further south. Northward dispersal is similarly limited by the Norwegian and Baltic Seas, and by the stresses associated with the subarctic and artic climates farther north. This explanation seems plausible, but it does not account for another important difference between European and North American avifauna (see Figure 15.11, compare B and C). Europe is distinguished by a surprising lack of birds with relatively small geographic ranges. In this case, the explanation is equally as plausible, but involves both the climatic dynamics of the Pleistocene as well as the differences in shapes between the continents. As climates changed and the glaciers expanded across

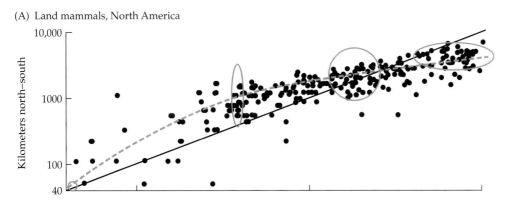

(A) Land mammals, North America

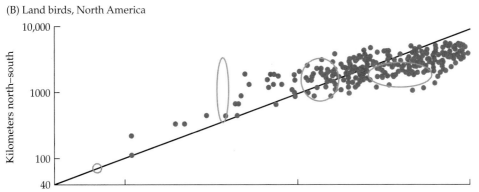

(B) Land birds, North America

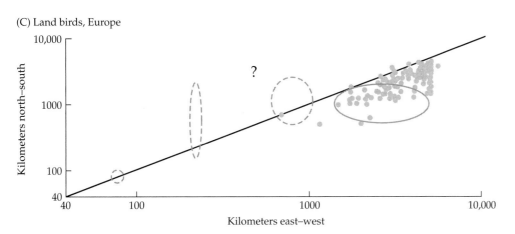

(C) Land birds, Europe

Kilometers east–west

FIGURE 15.11 The shapes of geographic ranges of North American mammals (A) and birds (B) changes in a consistent manner with range size from elliptical ranges stretched north-south for species with ranges of intermediate size, to circular and then east-west stretched ranges for species with relatively large ranges (solid line is the line of equality and denotes where the ranges are roughly circular; i.e., where the ranges extend from north-south equals that from east-west). In contrast, geographic ranges of European birds (C) almost all tend to be stretched east-west and to be relatively large.

these northern continents, native biotas either adapted by dispersing south of the glaciers, or they suffered extinctions. While the biogeographic dynamics were quite complex (Chapter 9), it is clear from any topographic map that this was much less challenging for species in North America. In Europe, however, only species capable of dispersing across its formidable geographic barriers (i.e., species with superior dispersal abilities and the largest geographic ranges) survived. On the other hand, geographically restricted species who, by definition, had limited dispersal abilities, could not escape the glaciers and severe swings in climatic conditions, and they perished.

OVERLAP OF RANGES. Knowledge of the extent to which the ranges of species overlap is central to understanding geographic patterns of species richness. It can also provide, at least, indirect evidence about the environmental (in particular, the biotic) factors that limit distributions and set range boundaries. In Chapter 4 we presented examples of pairs of closely related species whose ranges are adjacent but almost completely nonoverlapping, suggesting that their distributions are limited by interspecific competition. Such competitive exclusion has been clearly demonstrated by experimental studies on organisms as different as barnacles (Connell 1961) and chipmunks (Brown 1971). In some other organisms, such as desert rodents, these patterns of nonoverlapping distributions are so striking that they scarcely need statistical analyses to show that they are not due to chance (see Figure 4.26, and below). Such isolated examples, however, are hardly sufficient to show a causal relationship between competition and distributions. Indeed, we know of many examples of species that compete intensely, yet overlap in their geographic ranges and coexist in the same local habitats. And, of course, the majority of species with nonoverlapping geographic ranges are not closely related, do not use similar resources, and do not compete to any significant extent.

So, can we find any general patterns in the extent to which the geographic ranges of different species overlap? Specifically, do the ranges of close relatives—such as congeners—overlap, either more or less than one might expect on the basis of chance? There are good a priori reasons for predicting either outcome. On one hand, closely related forms are likely to occur in the same general region because they share a fairly recent common ancestor, possess similar dispersal capabilities, and have similar requirements for environmental conditions and resources. On the other hand, if species have originated by allopatric speciation, we might expect the vicariant pattern to be preserved, and if they are too similar in their requirements, we might expect competition to inhibit or preclude their coexistence.

Is it possible to determine which of these patterns predominates by quantitative analyses of the geographic ranges of well-studied taxa? It is simplest to consider the overlap of pairs of species. Pielou (1978) analyzed the distributions of algae along the coasts of North and South America. Within each of three groups—Rhodophyta (red algae), Phaeophyta (brown algae), and Chlorophyta (green algae)—on both Atlantic and Pacific coasts, she considered all pairs of species. Each pair was scored as to whether it was congeneric or more distantly related, and whether its distribution was overlapping or disjunct. Results for one group, the Rhodophyta on the Pacific coast, are shown in Table 15.1.

Statistical analyses of these data provide no grounds for rejecting a null hypothesis of independent (i.e., random) distributions of these species. The probability that this pattern could be due to chance is relatively high (P = 0.2). There is, however, a qualitative tendency for congeneric pairs to overlap more so than distantly related ones. This same trend is present in each of the six data sets Pielou examined. Because this, in itself, is statistically unlikely (P <

TABLE 15.1 *Patterns of overlap in the geographic ranges of red algae (Rhodophyta) on the Pacific coasts of North and South America*

Species pairs	Number of species pairs studied	Percentage with overlapping ranges
Congeneric	37	57%
More distantly related	2664	44%

Source: Pielou 1978.

TABLE 15.2 *Comparison of observed and expected frequencies of overlapping geographic ranges for pairs of desert rodent species in the same or different genus or of similar or different body size*

	Observed	Expected
Same genus	111	146.3
Different genus	785	749.7
Similar size	217	242.2
Different size	697	653.8

Source: After Kelt and Brown, 1999.

Note: Closely related and morphologically similar species overlap less frequently than expected by chance. In both cases the differences are statistically significant ($p < 0.05$).

0.02), she concluded that closely related species tend to overlap in their ranges, and that interspecific competition does not usually result in mutually exclusive geographic ranges.

Pielou's results were similar to those of Simberloff (1970), who used somewhat different methods to analyze the coexistence of congeneric pairs of bird and plant species on islands (see also Gotelli 2004). The fact that these two studies, performed independently using different techniques, gave similar results for widely divergent groups of organisms is significant. It suggests that closely related species often have similar dispersal capabilities and ecological requirements, and that these shared traits tend to result in overlapping distributions. No influence of either allopatric speciation or competitive exclusion is apparent at this level of analysis.

It would be premature, however, to conclude that the effects of vicariance and interspecific competition cannot be detected in the distributions of closely related species. In fact, studies of North American seed-eating desert rodents suggest that interspecific competition does, indeed, limit the overlap of both local and geographic ranges, as might be expected from our discussion in Chapter 4. Kelt and Brown (1999) found that congeneric and morphologically similar pairs of species overlapped in their geographic ranges less than expected by chance (Table 15.2). While quantitative analyses remain to be done for other kinds of organisms, many appear to exhibit similar patterns. In North America, these include chipmunks (*Eutamias*; see Figure 4.27), ground squirrels (*Spermophilus*, *Amospermophilus*), marmots (*Marmota*), prairie dogs (*Cynomys*), and pocket gophers (*Thomomys*, *Geomys*) among mammals; and jays (*Aphelocoma*, *Cyanocitta*), chickadees, and titmice (*Parus*), among birds. The latter genus of birds is especially interesting, because in Eurasia, the exact opposite pattern obtains: several species of *Parus* have broadly overlapping ranges and forage together in mixed-species flocks.

At least with respect to overlap of congeners then, desert rodents exemplify patterns that are just the opposite from those observed in Pielou and Simberloff's studies. How do we account for this discrepancy? We can suggest several hypotheses. First, compared with some other kinds of organisms, competition among desert rodents may be more frequent and more intense. In experimental studies, several species of desert rodents doubled or tripled their local populations when competitors were removed (e.g., Brown and Munger 1985; Valone and Brown 1995; Heske et al. 1997). The issue of scale may also be important. Species may compete strongly and exclude each other at small spatial scales, yet have broadly overlapping distributions at geographic scales. Many intertidal algae and land plant species show clearly segregated patterns of local elevational zonation (e.g., Pielou 1978; Yeaton 1981). As Connell demonstrated for barnacles (1961; see Figure 4.12), such zonation may reflect local competitive exclusion among species with extensively overlapping geographic ranges.

A second and related hypothesis is that intensity of competition is more closely correlated with taxonomic affinity in desert rodents than in some other organisms. Desert rodent species in the same genus tend to be similar in many characteristics that probably influence competitive interactions, such as diet, body size, and morphological specialization for locomotion, foraging, and predator avoidance (Kelt and Brown 1999). The term "guild" (see Root 1967) has been coined to describe sets of species that use similar resources in similar ways. By this criterion, most congeneric desert rodents are clearly members of the same guild.

A third hypothesis is that the nonoverlapping ranges of congeneric desert rodents reflect, in part, the enduring influence of allopatric speciation. Such a historical legacy would be difficult to disentangle from the influence of contemporary ecological interactions. Indeed, if speciation has occurred as a result of geographic isolation, we would expect the most closely related species to be most similar, to compete most intensely and, as a result, to maintain largely allopatric ranges. There is abundant evidence to support such a scenario in desert rodents. There are many species that are morphologically and ecologically nearly indistinguishable but recognizable as distinct species by clear differences in their chromosomal or molecular genetic characteristics (e.g., Patton et al. 1984; Sullivan et al. 1986; Lee et al. 1996). Such species are often called sibling species because of their close phylogenetic relationship, or cryptic species because of their great similarity. They often appear to have formed fairly recently, during the Pleistocene or late Tertiary, as a result of geographic isolation of populations due to geological and climatic changes. The predominant pattern is for pairs of such species to exhibit abutting but largely nonoverlapping distributions, even at very small scales. Thus, what was once thought to be a single species with a relatively continuous geographic range is often shown to consist of two or more very closely related species that come into close proximity, but rarely co-occur in the same habitat. Of course, competitive exclusion could play a major role by preventing overlap in geographic ranges, thus maintaining the biogeographic legacy of allopatric speciation events. In some other organisms, if speciation has occurred by sympatric mechanisms or if it has occurred at a much earlier time in the geological record (as is likely in the algae, for example), then the relationships between speciation, competition, and geographic isolation may not be so clear.

Geography of Abundance: Internal Structure of Geographic Ranges

Still further insights into the factors influencing geographic ranges can be obtained by investigating the relationships between population size and two biogeographic variables: range size and relative location of each population within the species' geographic range. First, by plotting average local abundance of species against the size of their geographic ranges (see Figure 15.8), macroecologists obtained a triangular-shaped plot, again suggesting a constrained relationship (Brown 1995; Blackburn et al. 1998; Gaston et al. 1998; Gaston and Blackburn 2000; Gaston 2003). The constraint lines include a trivial artifact (the absolute minimum abundance of 1), and a vertical constraint line that again corresponds to the maximum range size for the assemblage (equivalent to the dashed horizontal line in Figure 15.7). The third constraint line (the diagonal one) is certainly the most interesting, suggesting that maximal abundance increases with the size of the geographic range. This pattern supports the relationship between niche breadth, abundance, and distribution developed in Chapter 4. Species with less restrictive resource requirements and broad tolerances for abiotic and biotic limiting factors are able to be both locally abundant and widely distributed. Furthermore, in part for these reasons, these species are also likely to have a low probability of extinction, even in the face of major environmental change.

In the above examples, the particles of analysis are individual species within continental or oceanic biotas, and the emergent statistical patterns are the constrained plots of Figures 15.7 and 15.8. We can also examine a related, but more local scale relationship between abundance and geography, this time with populations serving as the particles and the general patterns emerging at the species level. In fact, we have already described this pattern in Chapter 4 (Figures 4.6, 4.7, and 4.17). Rapoport's (1982) analysis of aerial photographs revealed that the density of a species of a palm (*Copernicia alba*) decreased from sites near the center of its geographic range to those along its periphery. Subsequent studies have revealed that this pattern is very general indeed, and reported for a broad diversity of species including numerous plants, insects, mollusks, fish, birds, and mammals (Brown 1995; Gaston 2003). Not only are population densities lower near the edge of a range, but peripheral populations also tend to be more sparsely distributed and more variable over time. The overall result, and one especially relevant to conservation biology as well as biogeography, is that peripheral populations tend to be sinks, experiencing frequent extirpations and, thus, relying heavily on recolonization from more central (source) populations.

In summary, areographers and macroecologists have made great strides in discovering a diversity of patterns. They have capitalized on the great legacy of distributional information provided by centuries of biological surveys, and they have developed innovative and insightful approaches for investigating the forces that determine one of biogeography's most fundamental units—the geographic range.

Ecogeographic Rules

As we noted in the introduction to this chapter, one of biogeography's early and persistent themes is one that explains geographic variation in the characteristics of individuals and populations. While the diversity of such characteristics (including genetic, demographic, morphological, physiological, behavioral, and ecological) that could be studied seems overwhelming, some surprisingly regular patterns have emerged. Foremost among these are the so-called **ecogeographic rules**. As Ernst Mayr (1956) observed, the term "ecogeographic" is less than ideal because it seems to refer to geographic variation in ecological characteristics or processes (e.g., diversity or predation rates, respectively). The term, however, has been firmly established and is generally understood to refer to very regular geographic gradients in the characteristics of organisms (especially, but not limited to, their morphological traits) across continents and ocean basins.

Ecogeography of the Terrestrial Realm

BERGMANN'S RULE. Some of the oldest and best-known biogeographic patterns are regular geographic clines in body size of terrestrial animals, especially vertebrates. In 1847, Carl Bergmann noted that among closely related kinds of

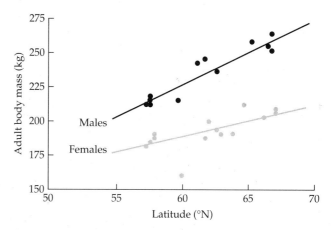

FIGURE 15.12 Body size of moose (*Alces alces*) in Sweden increases with latitude in a manner consistent with Bergmann's rule. (After Sand et al. 1995.)

mammals and birds, the largest species often occurred at higher latitudes (Figure 15.12). Bergmann realized that larger species have lower surface-to-volume ratios than their smaller relatives, which he suggested must be advantageous to endotherms in conserving body heat in colder climates.

The history of the study of latitudinal size variation in body size is typical of many questions in ecological biogeography. After Bergmann, many other studies also documented a positive correlation between body size and latitude among species or among populations of particular species. The pattern seemed to be sufficiently general that it came to be called **Bergmann's rule** (e.g., Allee et al. 1949; Rensch 1960; Mayr 1963). At first, the link between climate and body size appeared to be strengthened by studies showing that patterns of geographic size variation were often more closely correlated with environmental temperature than with latitude (Brown and Lee 1969; James 1970; Smith et al. 1995; Hadley 1997). But as more studies were reported, the picture became less clear. Enough exceptions to the rule were noted that some authors began to question its empirical generality and its presumed adaptive basis in temperature. In particular, McNab cogently argued that organisms do not live on a per gram (or per volume) basis (McNab 1971, 2002; see also Scholander 1955). Granted, larger organisms have less surface area per mass, but their total surface area, total heat loss, and total energy required to survive and reproduce exceed those of smaller organisms. It is difficult to understand why greater daily energy requirements would be adaptive in high-latitude environments, where temperatures and productivity are chronically low and often highly unpredictable. Indeed, as early as 1956, Ernst Mayr (p. 106) concluded that "since an increase in size results in an increased demand for food, size may actually decrease to the north where food is the limiting factor."

Although Bergmann's (1847) original description of the pattern was based on differences between species of endotherms (birds and mammals), Mayr (1956) and Rensch (1938) redefined it in a more general evolutionary context, restricting comparisons to those among populations within a species, but generalizing it to apply to ectotherms as well as endotherms. We believe this was a significant improvement as it focused attention closer to the levels at which natural selection operates (within and among populations) and included a more diverse assemblage of species groups, each capable of responding differently to selective pressures that vary with latitude (e.g., those associated with climatic regimes and diversity of competitors or predators). Both of these features enhance our abilities to use the comparative approach to evaluate—not just the generality of the pattern—but alternative explanations for the pattern as well. We, therefore, prefer to define Bergmann's rule in a similar manner, formally stated as the tendency for the average body mass of geographic populations of an animal species to increase with latitude. We, of course, realize that the pattern may not be so general as to apply to all taxa and functional groups of animals, or to all regions, ecosystems, and habitats. This more inclusive description of the pattern, however, provides more opportunity to use the comparative approach to assess underlying, mechanistic explanations for different patterns among functionally different groups of species.

Fortunately, a number of distinguished biogeographers have recently reassessed the empirical validity of Bergmann's rule, along with the various explanations for the pattern. First, it does appear that the pattern is a very general one, albeit far from universal. In Mayr's view (1956:105), Bergmann's—or any other ecogeographic rule—did not have to be a "law" without exception, just a general tendency exhibited by most of the species or races. Adopting this criterion (i.e., that a majority of species exhibit the tendency), Bergmann's rule appears to be valid for mammals (Ashton et al. 2000; Meiri and Dayan 2003; Meiri et al. 2004), birds (Ashton 2002a; Meiri and Dayan 2003; see also James 1970; Kendeigh 1976), salamanders (Ashton 2002b), turtles (Ashton and Feldman 2003), parasitic flatworms (Poulin 1996), and ants (Cushman et al. 1993; Kaspari and Vargo 1995; see also Ray 1960; Van Voohries 1996).

Even for these groups, however, there still remains some controversy as to the generality and, in particular, the explanation for the pattern. For example, among mammals, the pattern is equivocal for mustelids (weasels, and related species) and actually opposite of that expected for heteromyid rodents (Ashton et al. 2000). In addition, some studies indicate that the correlation between body size and latitude differs between large- and small-bodied species (Ashton 2002; Freckleton et al. 2003; Meiri and Dayan 2003; but see Ashton et al. 2000), for species that are more- or less-sedentary (e.g., birds; Ashton 2002a, 2004), and for different regions (e.g., tropical vs. temperature latitudes; McNab 1971; but see Ashton et al. 2000). Whether most of the species actually exhibit the pattern also depends on methodological issues. For example, in Ashton et al.'s (2000) review they found that 78 out of 110 (71%) species of mammals exhibited the pattern (i.e., body size of their geographic ranges increased with latitude), but the intra-specific trends were statistically significant for just 43 (39%) of the species—results that are similar to those McNab (1971) reported earlier as evidence against the generality of the rule. Ashton et al. (2000), however, attributed many of these nonsignificant trends to cases that were not sampled extensively. They found that, when limited to 33 species that were adequately studied, a clear majority (23 of those species, or 70%) exhibited significant, positive correlations between body mass and latitude.

A limited number of important studies have demonstrated that the pattern is dynamic, with body size varying over time as well as across latitudinal gradients. As we discussed in Chapter 9, species survived the repeated climatic cycles of the Pleistocene either by dispersing or by adapting to the highly altered, local conditions. The fact that mammals and other species that recolonized these regions since the glaciers receded (12,000 to 18,000 years B.P.) now exhibit the expected trends of body size with latitude, indicates relatively rapid adaptation and evolution (Blackburn and Hawkins 2004). For at least some of these species, their body size closely tracked the temporal changes in the thermal regime. Smith and her colleagues (Smith et al. 1995) found that over the past 20,000 years, body size of woodrats (North American rodents of the genus *Neotoma*) varied in a manner entirely consistent with the thermoreg-

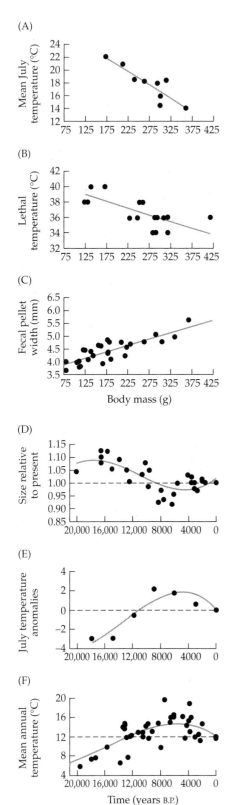

FIGURE 15.13 Bergmann's rule and adaptation to temperature in the bushy-tailed woodrat (*Neotoma cinerea*) over both geographic space and evolutionary time. (A) Geographic variation in body size among contemporary populations as a function of July temperature, showing that larger individuals inhabit cooler climates. (B) Upper lethal temperature as a function of body size, showing that larger individuals are more susceptible to heat stress. (C) Fecal pellet size as a function of body size, showing that larger animals produce larger feces, allowing body sizes of past populations to be estimated from fossilized fecal pellets. (D) Relative sizes of past populations, estimated from the sizes of radiocarbon-dated fecal pellets, over the last 20,000 years. (E, F) Two measures of environmental temperature over the last 20,000 years, showing a pattern that almost exactly mirrors the changes in body size. Therefore, over both space and time, there are adaptive shifts toward smaller sizes in warmer environments. (After Smith et al. 1995.)

ulatory explanations for Bergmann's rule: that of increasing in size as temperatures cooled, and decreasing in size as temperatures warmed (Figure 15.13). Similar patterns of geographic and temporal variation in body size were documented in pocket gophers (Hadley 1997). Recent studies also indicate that patterns consistent with Bergmann's rule can be established relatively rapidly. For example, just 20 years after it was introduced into the new world, the fly *Drosophila subobscura* not only occupied a broad geographic range, but exhibited a geographic cline in body size similar to that exhibited by populations in its native range—a cline entirely consistent with Bergmann's rule (Figure 15.14; Huey et al. 2000; see also Thurber and Peterson 1991).

For reasons discussed above, Bergmann's original explanation (based on gram-specific surface area) seems invalid (see Panteleev et al.'s [1998] report of Bergmann's rule exhibited by hibernating mammals; see also Kojola and Laitala 2001). This, of course, does not preclude the operation of other physiological or ecological mechanisms that may account for the general increase in body size as we move poleward. For example, lower critical temperatures (the temperatures below which endotherms must increase metabolic energy expenditures to maintain a constant body temperature) decrease with increasing body size. In addition, because energy and water stores increase more rapidly with body size than do requirements for these resources, larger animals can endure food and water shortages for longer periods than can smaller individuals (Calder 1974; Lindstedt and Boyce 1985; McNab 2002). This increased thermal independence of larger organisms may be particularly adaptive in the seasonal, often stressful and unpredictable climates of the high latitudes. Put simply, larger animals are more able to withstand environmentally harsh conditions including cold spells or long, severe winters (this explanation is variably known as the "thermal independence," "endurance," or "starvation resistance hypothesis"; see Arnett and Gotelli 2003; Blackburn and Hawkins 2004).

Patterns consistent with Bergmann's rule may also be associated with another very general geographic gradient—the latitudinal gradient in species richness, which we discuss later in this chapter. Low diversity in high latitudes means reduced interspecific interactions, including competition—a condition that often triggers ecological release and character release in some species (McNab 1971; see also Dayan et al. 1989; Dayan 1990; Iriarte et al. 1990; see related discussion on the island rule in Chapter 14). In species-rich environments, small size may allow some species to avoid intense competition by specializing on smaller prey, and avoid interference competition and predation by exploiting smaller refugia. When these interspecific pressures are released (e.g., in species-poor communities of the higher latitudes, or on iso-

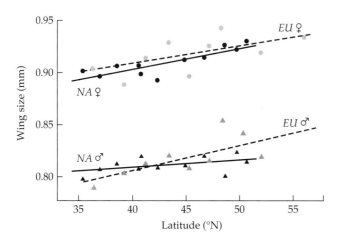

FIGURE 15.14 Within two decades after their introduction into North America (NA), body size of populations of the fruit fly, *Drosophila subobscura*, evolved to the point that the pattern has converged upon that of populations in their native range in Europe (EU), which is consistent with Bergmann's rule. (After Huey et al. 2000.)

lated islands), an increase in size in an otherwise small species may be favored because larger individuals tend to dominate intra-specific competitors, and they can exploit a broader range of resources (e.g., larger squirrels can crack large as well as small nuts; larger predators can take large as well as small prey). McNab's (1971) analysis of body size trends in nonvolant mammals appears consistent with this ecological explanation for patterns consistent with Bergmann's rule. For a number of guilds of mammals, trends toward increasing body size (whether toward the north or south, or along clines of longitude as well) appear to be associated with prey availability and the absence of a larger member of the guild (Figure 15.15; see also Fuentes and Jaksic 1979; King and Moors 1979; Erlinge 1987; Dayan et al. 1989; Dayan 1990; Iriarte 1990; Alacantra 1991).

In summary, Bergmann's rule may represent still another case in which a general pattern in nature results, not from a single mechanism, but from a combination of convergent forces, all of which vary in a predictable manner across the geographic template. Whatever unifying explanation emerges, it must account for the many exceptions to, and variations on, the rule—including animals that fail to exhibit the pattern (e.g., in mammals: Fuentes and Jaksic 1979; Erlinge 1987; Alacantra 1991; Thurber and Peterson 1991; Ochocinska and Taylor 2003; in reptiles: Ashton and Feldman 2003; in butterflies: Barlow 1994; Hawkins and Lawton 1995; in landsnails: Hausdorf 2003; and other arthropods: Blanckenhorn and Demont 2004), variation in trends across regions and body sizes (see McNab 1971; Hawkins and Lawton 1995; Ashton 2004), and related but conflicting clines in body size along elevational gradients (e.g., in mammals: Zimmerman 1950; Whitford 1979; Taylor et al. 1985; and in butterflies, birds, and moths: Roff 1990; Ashton 2002a; Brehm and Fiedler 2003; Hausdorf 2003).

We believe that future studies will continue to provide results that are qualitatively similar to those discussed above; that is, many groups of species will exhibit body size gradients consistent with Bergmann's rule, while others will not. The most important questions are not whether the pattern is universal—clearly, it is not. Instead, future studies of this and related patterns should focus on the exceptions and the differences among species groups. Such stud-

FIGURE 15.15 Geographic trends in body size of guilds (groups of species with similar feeding strategies) may be influenced by interspecific competition. For example, body sizes of smaller species of mustelids (A) and felids (B) sometimes increase along latitudinal gradients (north or south) beyond the range limits of a larger member of that guild. (Trend lines drawn here are simplified from actual data presented by McNab 1971.)

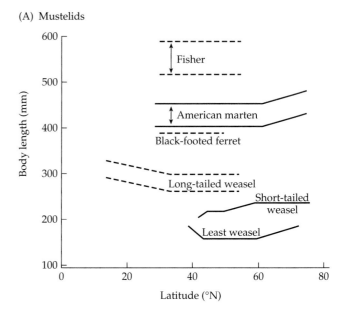

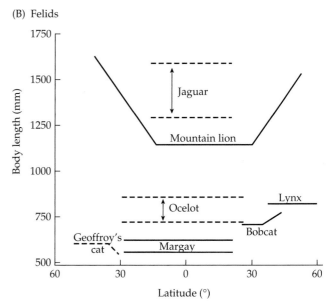

ies, in addition to those examining correlations of body size with other geographic gradients (e.g., elevation), and with factors more closely associated with causal explanations (e.g., temperature, productivity, and seasonality; and diversity of prey, competitors, or predators) are key to exploring and discriminating between causal mechanisms and developing a more integrative explanation for this intriguing pattern.

ALLEN'S RULE. Other ecogeographic rules have been coined to describe other widespread patterns of morphological variation. For example, Allen (1877) noted that among closely related endothermic vertebrates, those forms living in hotter environments (generally, those of the lower latitudes or lower elevations) often tend to have longer appendages. Again, the hypothesized mechanism invoked thermoregulatory advantages: that shorter appendages promote heat conservation in cold environments by reducing the total surface area for

(A)

(B)

FIGURE 15.16 Two classic examples of Allen's rule: variation in the relative size of the ears of hares (A, from left: *Lepus alleni*, *L. californicus*, *L. americanus*, and *L. articus*) and foxes (B, from left: *Fennecus zerda*, *Vulpes vulpes*, and *Alopex lagopus*). In each case, the species on the left occurs in a hot desert habitat, the one on the right in a cold tundra habitat, and the one (or two) in the middle in intermediate, temperate environments.

heat loss. On the other hand, relatively long appendages increase surface area and thus facilitate heat dissipation where ambient temperatures reach stressfully high levels. Physiological research has shown that the relatively long ears of desert rabbits and elephants, and the long and relatively thin legs of camels, do play important roles in heat dissipation (Schmidt-Nielsen 1963, 1964). In North America, hares of the genus *Lepus* show a dramatic reduction in the length of their ears with decreasing temperature, from the black-tailed and white-sided jackrabbits of the hot, arid Southwest to the varying and arctic hares of the perpetually cold north (Figure 15.16).

At present, however, the empirical evidence supporting Allen's rule still consists of a limited number of isolated cases (Hesse et al. 1951; Mayr 1963; Johnston and Selander 1971). The number of exceptions has led some authors to question the generality of this ecogeographic rule. For example, it is questionable whether North American rabbits of the genus *Sylvilagus* exhibit the expected pattern (Stevenson 1986). The presumed mechanistic basis of heat dissipation through enlarged appendages, can also be questioned. While long ears can be excellent heat dissipators, the short, well-insulated tails of rabbits and hares can play little role in dissipating heat. In fact, heat exchange can be readily regulated by changing insulation (adding to, or fluffing up, fur or feathers) and/or by altering blood flow to the extremities. Some relatively large, well-insulated endotherms use their long appendages for heat dissipation after prolonged exercise (e.g., the legs of caribou, the tails of beavers, the flippers of seals). Aside from thermoregulatory considerations, long ears may be advantageous in warm environments because they aid in sound detection (sound attenuates more rapidly at higher temperatures: Erulkar 1972; Harris 1996). However, the principal causal explanation for Allen's rule is fundamentally similar to that for Bergmann's rule, i.e., thermoregulation; but of the two, the latter has received surprisingly little attention, especially during the last two decades.

With advances in morphometrics and, in particular, the means to objectively describe relative shape, we believe there is a largely untapped opportunity to evaluate the generality of Allen's rule and, where proven valid, to assess casual explanations. We find claims that human populations exhibit morpho-geographic trends consistent with both Bergmann and Allen's rules tantalizing (see Newman 1953; Roberts 1953, 1978; Houghton 1990; Ruff 1994;

Katzmarzyk 1995; Bindon and Baker 1997), but equivocal. Clearly, more comprehensive studies of both of these patterns are in order if we wish to gain a better understanding of the factors influencing morphological variation of animals, including our own species.

GLOGER'S RULE. A few decades after Bergmann described the pattern of latitudinal variation in body size, Gloger (1883) noted that the coloration of related forms was often correlated with the humidity of their environments: that darker colors occurred in more humid environments. In this case, while some physiological mechanisms have been proposed, they have received little support. Dark colors might be thought to facilitate basking to absorb solar radiation in cold environments, and light colors to reflect solar radiation and prevent overheating in hot environments. But much radiant heat exchange occurs in the nonvisible wavelengths, and the thermally effective absorbance and reflectance of fur, feathers, scales, and other body surfaces is often not well correlated with their color. In fact, many animals tend to be white in cooler environments (e.g., polar bears, arctic foxes, arctic hares, and ptarmigans)—just the opposite of what we would expect based on this thermoregulatory argument. For some time, it was claimed that white or hollow fur may be advantageous in cold climates since the hairs might act as fiber optic conductors of ultraviolet light toward the skin. Detailed analyses of the optical quality of hairs in polar bears, however, clearly demonstrate that this is not the case (Koon 1998).

For polar bears and for species exhibiting marked variation in color among their populations, the most likely explanation is selection for crypsis to avoid visual detection by predators or prey. Humid climates are usually associated with dark soils and dense vegetation that casts deep shadows, whereas arid environments (hot or cold) often have light-colored soils and relatively sparse vegetation. Prey species that match the color of their environment should have an advantage in avoiding detection by visually hunting predators. Kettlewell's (1961) famous study of industrial melanism showed that moths matching the soot-darkened tree trunks and other surfaces in industrial areas were less likely to be discovered and eaten by birds. Less well publicized, but perhaps even more thorough and convincing are the studies of Sumner (1932), Blair (1943), and Dice (1947) on the adaptive basis of geographic variation in coat color in deer mice (*Peromyscus maniculatus*). They made three key observations:

1. The pattern of geographic variation is such that dorsal pelage of local and regional populations tends to closely match background coloration in the environment.

2. Most of this variation is heritable, controlled by both simple Mendelian genes and more complex polygenic systems.

3. Individuals that do not match their backgrounds are selectively captured by predators (owls) in controlled experiments.

Crypsis can also help predators avoid detection by their prey, potentially explaining why polar bears and some mustelids are white, and why lions are the color of dry grass. The extensive variation in pelage color of North American black bears also appears to be cryptic, with a higher frequency of light-colored forms in relatively dry habitats (Rounds 1987).

Although its label would suggest that Gloger's rule is a very general ecogeographic pattern, in retrospect it may be neither general nor geographic. First, compared to the great wealth of information on Bergmann's rule, there seems to be precious few studies of Gloger's rule and no comprehensive assessment of its generality. Perhaps just as relevant, Gloger's rule is not associated with any explicit geographic gradient such as latitude or elevation. Instead, it is based on visual characteristics of the background environment, which vary in no consistent manner with these geographic variables. This is not to say that these patterns of variation in color are not interesting or important—they certainly are, as they reflect the operation of natural selection across different local environments. Given the limited geographic context and spatial scale of these patterns, however, it is likely that biogeographers will continue to focus on the truly large-scale, ecogeographic patterns of the terrestrial and marine realms (discussed in the preceding and following sections, respectively).

Ecogeography of the Marine Realm

During the latter part of the nineteenth century, information on geographic variation in the marine realm also accumulated to the point where biogeographers began describing some interesting ecogeographic patterns. We review three such patterns for marine biotas here, including Jordan's rule, Thorson's rule, and gradients in a suite of morphological characteristics summarized by Vermeij (1978) in his classic book, *Biogeography and Adaptation: Patterns of Marine Life*.

JORDAN'S RULE. In 1892, the American ichthyologist David Starr Jordan observed that the number of vertebrae of marine fish increases along a gradient from the warm waters of the tropics to cooler waters of the high latitudes (Jordan 1892; Hubbs 1922). Like the other ecogeographic patterns described above, this one seemed so general that it became known as **Jordan's rule of vertebrae**.

While a comprehensive and systematic review of the generality of Jordan's rule may still be lacking, it is clear that all species do not exhibit the pattern, and the pattern often varies depending on the particular geographic region or the habitats of the populations studied (Figure 15.17; McDowall 2003). While the correlation of vertebrae number with latitude is often highly significant, it is clear that latitude per se cannot be the driving force for this pattern. Instead, ichthyologists and aquatic biogeographers attribute Jordan's rule to natural selection and adaptation of local populations to geographic variation water temperatures (Lindsey 1975; Lindsey and Arnason 1981; Billerbeck et al. 1997; McDowall 2003). This relationship may also be related to a body size gradient analogous to Bergmann's rule. As we have observed for many terrestrial species, body size of many fish also tends to increase with latitude. Regardless of the actual causal mechanism, number of vertebrae (which is highly variable, genetically controlled, and fixed during early development; Billerbeck et al. 1997) is greater in larger fish.

McDowall's (2003) relatively recent study provides a tantalizing account of the pattern for selected species of salmoniform fish, demonstrating how the pattern differs across regions and environments, and for species with different migration patterns (see Figure 15.17). This study provides an excellent example of how dissecting a large-scale geographic pattern among regions and functional groups can provide especially important insights into the causal mechanisms of that pattern (see the discussion on deconstruction of the latitudinal gradient later in this chapter).

THORSON'S RULE. Marine invertebrates such as mollusks and gastropods have two principal modes of development: (1) their larvae may become planktonic and disperse with the currents away from their natal range, or (2) they may

(A)

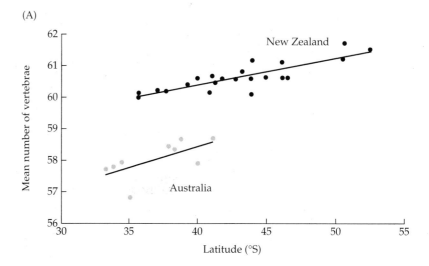

(B)

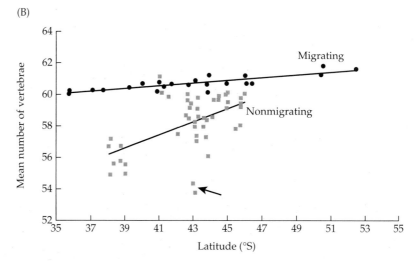

FIGURE 15.17 *Galaxias brevipinnis* is a salmoniform fish that clearly exhibits Jordan's rule: the tendency for vertebrae number to increase with latitude. Note, however, that the pattern differs not just among species, but also among populations of species from different regions (A), and for those from different environments or for those having different migration strategies (B) as well. (After McDowall 2003.)

undergo direct development to adults in their natal home range. It appears that, because of these differences in dispersal abilities and the subsequent effects on isolation and speciation in allopatry, invertebrates with direct development (and limited dispersal abilities) are more likely to become isolated and, thus, exhibit relatively high speciation rates (see Astorga et al. 2003). While the causal link between dispersal mode and speciation rates requires further investigation, it may explain an ecogeographic pattern first reported by Thorson in the 1930s: along a gradient of increasing latitude, the dominant mode of development for marine invertebrates switches from pelagic (indirect, planktotrophic) development in the tropics to direct development in the higher latitudes (Figure 15.18). In fact, the latter group is one of the few to exhibit a trend of increasing diversity with increasing latitude (see the discussion on the latitudinal gradient in species diversity later in this chapter).

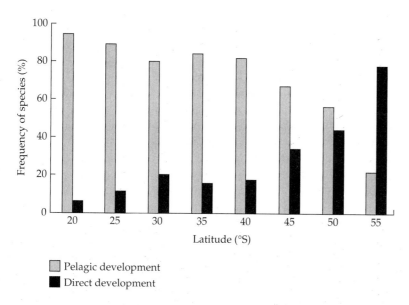

FIGURE 15.18 The principal mode of development in marine mollusks is strongly correlated with latitude, with direct development (i.e., lacking planktonic larvae) being the predominant mode in the higher latitudes. (From Marquet et al. 2004; after Gallardo and Penchaszadeh 2001.)

Thorson's rule appears to be a general one exhibited in a variety of invertebrates including mollusks, gastropods, and crustaceans (Thorson 1936, 1946, 1950; Mileikovsky 1971; Gallardo and Penchaszadeh 2001). However, the generality of this rule for other marine animals, differences in the trend among taxa and among regions, and its corollaries for latitudinal gradients in development and dispersal modes of other species groups (e.g., other marine invertebrates, marine vertebrates, or terrestrial animals) certainly merits more attention by biogeographers.

GRADIENTS IN PREDATION AND MORPHOLOGICAL DEFENSES. According to Geerat Vermeij (1978:108), "...antipredatory defenses are developed to a greater degree and are found in an increasing proportion of species along a gradient from high to low latitudes. With qualifications, such gradients have been shown to exist in all the groups for which data are available: gastropods, bivalves, barnacles, sponges, holothurians, and marine sessile photosynthesizers." Vermeij based this on his painstaking and prescient examination of the shells and life forms of many hundreds of specimens of these invertebrates. Shells from lower latitudes tend to be thicker, and more sculpted—often with a higher spire and more narrow aperture, which tends to be equipped with a thicker and more rugged lip and with a sturdier valve (Figure 15.19). In addition to comparing intact shells from a wide range of latitudes, Vermeij (1978; see also Vermeij 2004) also inspected shell fragments and damaged shells (i.e., those that were attacked by predators; see also Dietl and Kelly [2002]). By doing this, he was able to demonstrate that predation pressures also vary with latitude, being highest for the tropics where the abundance and diversity of predators (and most other types of species) is highest. Thus, Vermeij described what appears to be a general pattern and has simultaneously provided a convincing causal explanation for that pattern; in this case, linking gradients in morphology with those of interspecific interactions (predation) and species diversity.

FIGURE 15.19 According to Vermeij's (1978) predation hypothesis, the high diversity and selective pressures from predators in the tropical latitudes (or those from large vs. relatively small and isolated biotas) favor shell morphologies that are thicker and more sculpted, possessing a more narrow aperture (often equipped with a relatively thick and rugged lip, and a sturdy valve; compare the shells for the marine gastropods *Drupa aperta* (left) and *D. clathrata* (right), from low and high diversity environments, respectively). (Photo courtesy of G. Vermeij; see Vermeij 2005).

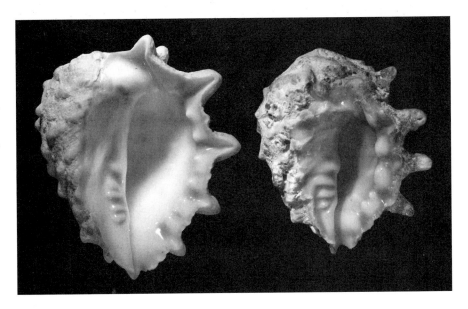

Vermeij's predation hypothesis has recently been tested in the terrestrial realm. In seed plants, the size of seeds appears to be a measure of predation pressure in perhaps an analogous fashion to shell morphology in the marine realm. Thus, the tendency for seed size to increase from the poles to the equator (Figure 15.20) appears consistent with Vermeij's hypothesis. It is not yet clear, however, whether the actual intensity of seed predation varies with latitude. Therefore, the seed size-latitudinal gradient may result from some other phenomenon, such as latitudinal variation in climatic conditions, plant growth form, growing period, or seed dispersal. Some of these factors, especially precipitation, temperature, and their effects on soil development, may explain another latitudinal gradient in plant morphology. Although highly variable, rooting depth tends to be relatively shallow in the tropics, and then increases along a cline toward the higher latitudes (Figure 15.21).

GEOGRAPHIC VARIATION IN LIFE HISTORY CHARACTERISTICS. As we saw above, interesting patterns of geographic variation are not confined to morphological traits. There is an extensive literature on geographic variation in life history characteristics such as clutch size and age ratios in birds and other egg-laying animals (e.g., Moreau 1944; Lack 1947; Skutch 1949; Cody 1966; Moll and Legler 1971; MacArthur 1972; Iverson et al. 1993; Baker 1995; Graves 1997), and parallel trends of geographic variation in litter size of mammals (Lord 1960; Cockburn et al. 1983; Glazier 1985; Millar 1989; Cockburn 1991; Reinhart 1997). Often, the observed patterns are related to latitude, with larger clutches or litters at higher latitudes being the general rule (e.g., Figures 7.7 and 15.22). Again, however, there is more consensus about the pattern than about its actual mechanism(s).

Several of the studies listed above suggest that clutch size is influenced by the combined effects of climate, food availability, and interspecific interactions. For example, Cody (1966) explained geographic variation in clutch size within the framework of *r*- and *K*-selection (essentially, selection favoring higher population growth rates and productivity vs. selection favoring more efficient utilization of resources, respectively; see also MacArthur and Wilson 1967). Cody reasoned that birds inhabiting seasonally variable mainland environments with short growing seasons should be subject to *r*-selection, producing rela-

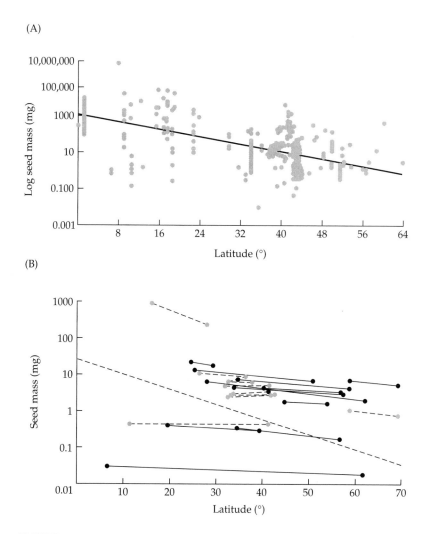

FIGURE 15.20 Seed size tends to increase from the poles to the equator, both among species (A) as well as within species (B). The two points shown for each species are the upper and lower latitudinal limits of the studies; the dashed line represents the slope of the global, multi-species relationship between latitude and seed mass. (After Moles and Westoby 2003.)

tively few, but large, clutches. Conversely, on temperate islands, where climates tend to be milder and less variable, selective pressures may shift toward more efficient use of available resources, and clutches of fewer—but higher-quality—offspring (i.e., *K*-selection; see Cody 1966; Isenmann 1982). Other studies have reported the same trend for insular salamanders (Anderson 1960) and for microtine rodents (Tamarin 1977; Gliwicz 1980; Ebenhard 1988). Similarly, in the mainland tropics, where the stable climates and the high diversity of competitors and predators put a premium on efficient resource use, specialization, and predator avoidance, clutch size should again be relatively low. Even some invertebrates such as estuarine (brackish water) copepods exhibit similar abilities to compensate for the relatively cool temperatures and short growing seasons of high latitude waters, with individuals from higher latitudes exhibiting more rapid rates of growth and development (Lonsdale and Levington 1985; see also Endler 1977; Vermeij 1978).

95% rooting depth (cm)

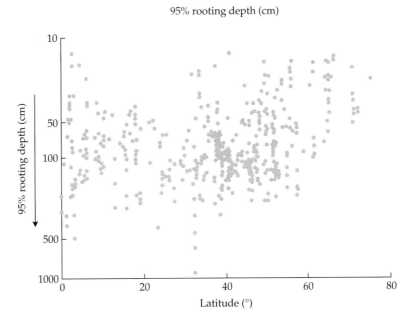

FIGURE 15.21 Rooting depth of plants in tropical regions tends to vary markedly, but then increase with increasing latitudes in temperate and subarctic regions (rooting depth calculated as the depth above which 95% of the roots are found). (After Schenk and Jackson 2002.)

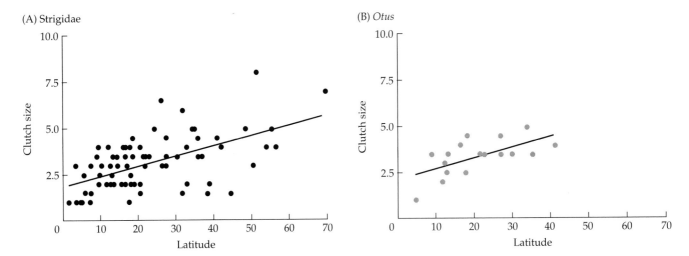

FIGURE 15.22 Clutch size as a function of latitude in owls. Such patterns have been found at three levels of relationship: (A) among the entire owl family, Strigidae, and (B) among the genus *Otus,* which contains the screech owls and their relatives; similar, though perhaps not statistically significant, trends have been found among populations of several widely distributed owl species. The strong positive relationships in these graphs are obvious, although the mechanisms underlying these typical patterns are much debated. (Courtesy of J.-L. Cartron and J. Kelly.)

TABLE 15.3 *Differences between clutch sizes (numbers of eggs per nest) of the same bird species on islands and nearby mainland localities of the same latitude*

For temperate islands off the coast of New Zealand [a]			For tropical islands in the Caribbean [b]		
Species or genus	**Average mainland clutch**	**Average island clutch**	**Species or genus**	**Average mainland clutch**	**Average island clutch**
Anas spp.	8 (3 sp.)	3.5 (1 sp.)	*Saltator albicollis*	2.0	2.5
Bowdleria punctata	3.1	2.5	*Tangara gyrola*	2.0	2.0
Gerygone spp.	4.5 (1 sp.)	4.0 (1 sp.)	*Habia rubica*	2.3	2.0
Petroica macrocephala	3.5	3.0	*Cacicus cela*	2.0	2.0
Miro australis	2.6	2.5	*Coereba flaveola*	2.5	2.5
Anthornis melanura	3.5	3.0	*Empidonax euleri*	2.0	3.0
Cyanoramphus novaezelandeae	6.5	4.0	*Elaenia flavogaster*	2.3	2.5

Source: MacArthur and Wilson 1967; Cody 1966.

Note: Clutch size averages smaller on the islands, as opposed to the mainland, in the temperate region but not in the tropical region.

[a]Species considered were either indigenous to the offshore islands and had a mainland relative in the same genus, or were subspecifically distinct on the island. The difference between the mainland and island means is 90% significant, by *t*-test.

[b]Species on isolated West Indian islands are compared to the mainland, and also the extreme southerly Lesser Antillean islands (isolated) are compared to Trinidad, which is very close to the mainland. The means are not significantly different.

While the *r/K* paradigm has been considered too simplistic to explain this suite of patterns in life history characteristics, Cody (1966) offered some interesting data in support of his hypothesis (Table 15.3). Other patterns of geographic variation in clutch and litter size, and growth rate may require somewhat different explanations. Why, for example, do clutch and litter sizes sometimes decrease from the tropical lowlands to the cool highlands of equatorial mountains (see Lack and Moreau 1965; Flux 1969)? Others have offered alternatives to Cody's explanation for geographic variation in clutch and litter size (e.g., see Lack 1947; Skutch 1949, 1967; Ashmole 1963; Charnov and Krebs 1974; Dunn et al. 2000). For the most part, these hypotheses are based on trade-offs associated with adapting to climatic fluctuations, risks of migration, available energy, seasonal pulses in productivity, risk of predation, and other factors. As is often the case, these hypotheses are not mutually exclusive, and some combination of them may be necessary to explain the observed geographic clines in clutch and litter size within any particular species as well as the pattern for terrestrial vertebrates, in general.

Diversity Gradients

Diversity Measures and Terminology

SPECIES RICHNESS AND DIVERSITY INDEXES. For many biogeographers, it is sufficient to be concerned with **species richness,** the number of species in a sample from a local area or geographic region. In many cases, the rare species may be as interesting and important as the common ones; in fact, many taxa of interest to both historical and ecological biogeographers are both highly restricted in their distributions and uncommon where they do occur (see Chapters 4, 10, and 11). To compare the similarity in species diversity or taxonomic composition of two regions, it is often most useful—as well as, most convenient—simply to compare lists of species, which can usually be obtained from systematic works, and faunal and floral surveys. Assuming that the samples are relatively complete, or that the sampling efforts for each region is at least comparable, species counts can provide simple and relatively unambiguous measures of the biological diversity and distinctiveness of different geographic areas.

While species richness is the most commonly used measure of community diversity, other measures also provide important information on the complexity of nature. In fact, many ecologists are interested primarily in the organization of natural communities and how *particular* species contribute disproportionately to that structure. To them, the relative abundances of species and other characteristics indicative of their ecological roles may be of more concern than the number of species present. Rare species may be important components of ecological systems (see Gaston 1994; Kunin and Gaston 1997), but ecologists often deemphasize them and focus on species that dominate the community in terms of abundance, biomass, energy use, cover, or some other estimate of "importance." Consequently, ecologists have adopted several measures, or indexes, of species diversity that give greater weight to the dominant species. Most of these measures of community complexity are based on the uncertainty principle, which, as applied by ecologists, asks the following question: given an individual is to be drawn randomly from a community or assemblage of species and individuals, what is the likelihood that we can correctly guess the identity of that randomly selective individual? Information theory (developed, in part, to decipher codes and other encrypted information) has proven very useful in this application, with the identity of a particular element or letter in a message being analogous to that of a species in a community (Shannon and Weaver 1949). The toughest codes to crack (i.e., those with the greatest uncertainty) are those that have the greatest number of elements (letters), and those whose elements are used in more equal frequencies. Similarly, most ecologists and biogeographers understand that local diversity increases with both the number of species in the community and with the evenness or degree to which population levels of these species are the same.

Many diversity indexes have been proposed, and several are used frequently in the ecological literature. These indexes are defined, and their uses, strengths, and weakness discussed, in several reviews (e.g., Pielou 1975; Whittaker 1977; Ludwig and Reynolds 1988; Magurran 1988; Rosenzweig 1995). Although a thorough understanding of these indexes is important for interpreting many ecological studies, a detailed treatment of them is not necessary here. For our purposes, it is sufficient to know that, in most natural communities, nearly all of the approximately 20 indexes commonly used to quantify species diversity are highly correlated with species richness. For this reason, we will use the term "species diversity" somewhat loosely, and synonymously with species richness.

SCALES OF DIVERSITY: ALPHA, BETA, GAMMA, AND DELTA. Like most other ecological patterns, species diversity varies with the spatial scale on which it is studied. The continuum of spatial scales is commonly divided into four convenient categories (Cody 1975; Whittaker 1975, 1977; Wilson and Schmida 1984; Magurran 1988; Ricklefs and Schluter 1993). **Alpha diversity** refers to the species richness of a local ecological community—that is, the number of species recorded within some standardized area, such as a hectare, a square kilometer, or some naturally delineated patch of habitat. **Beta diversity** refers to the change (or turnover) in species composition over a relatively small distance, often between recognizably different but adjacent habitats. An example of beta diversity would be the difference between two distinct communities on a mountain slope, such as that between lowland rain forest and montane evergreen forest in the tropics, or between deciduous and coniferous forests in a temperate region. Several methods have been used to measure beta diversity (see Wilson and Schmida 1984; Ricklefs and Schluter 1993), but their goals are the same: to express the overall turnover in species composition. **Gamma diversity** refers to the total species richness of large geographic areas ranging from a combination of local ecological communities to those of entire biomes, continents, and ocean basins. Thus, gamma diversity reflects the combined influences of alpha and beta diversity, so it will be highest in regions such as the Amazon basin where there are many species within local communities and a high turnover of species among habitats and across the landscapes. Finally, just as species composition of local communities can be compared to calculate measures of beta diversity, we can compare species lists for large geographic areas to calculate an analogous, large scale measure of dissimilarity, or **delta diversity**. Among these four measures, delta diversity may be the most fundamental to biogeography and evolution, as it describes the differences in biogeographic regions that were central to Buffon's Law (see Chapter 2) and challenged generations of scientists to explain how these large-scale differences in biotas developed. As we have thoroughly discussed diversification among biotas in Chapters 10–12, our focus here will be on geographic variation in alpha and gamma diversity (i.e., species richness at local to global scales).

The primary data available for numerous taxa and a variety of geographic regions are species counts for small areas. These counts usually are published either by systematists describing the results of biotic surveys or by ecologists summarizing samples of local habitats. An alternative and complementary method is to map the distributions of all species of a particular taxon, and then count the number of geographic ranges that overlap a particular point along some geographic gradient (e.g., latitude, elevation, or depth). Regardless of which of these methods is used to estimate species richness, in order to justify comparisons along geographic gradients, diversity should be calculated as **species density,** which is the number of species per standardized sample effort and area. This will avoid the potential confounding effects of differences in sampling protocol on species richness and, thus, enable more insightful investigations of the patterns and underlying processes influencing geographic gradients in species diversity.

RELATIONSHIPS BETWEEN LOCAL AND REGIONAL DIVERSITY. Some investigators have taken a multi-scale approach to questions about the assembly of continental biotas by investigating the relationship between local and regional diversity (e.g., Ricklefs 1987; Cornell 1993; Schluter and Ricklefs 1993; Caley and Schluter 1997). While this approach has much in common with both areography and macroecology, it differs in several important respects. It focuses on how local biotas are assembled from the regional species pool. It assumes that the composition of local communities is determined primarily by ecological interactions that usually occur on small spatial and temporal scales. These local interactions determine which of the species potentially able to colonize a

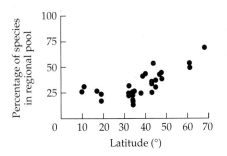

FIGURE 15.23 The relationship between beta diversity and latitude for North American terrestrial mammals. In this case, beta diversity is expressed as the percentage of the species in the regional species pool that occurs in small patches of relatively uniform habitat; low percentage values indicate high turnover of species among different habitat types, and thus, high beta diversity. (Data compiled by D. M. Kaufman.)

habitat are actually able to tolerate the abiotic environmental conditions and coexist with the other species that occur there. The composition of the pool of potential colonists, on the other hand, is assumed to reflect regional and historical processes such as barriers to dispersal and speciation and extinction events.

One line of research that is especially relevant to questions regarding spatial scales in diversity relies on plots of local diversity as a function of regional diversity. In Figure 15.23 we show the results of a somewhat similar analysis in which we plot the fraction of the species in the regional pool that coexist in local habitats. In that case, relative local diversity decreases as species diversity increases from high latitudes toward the equator. When the data are plotted as in Figure 15.24, somewhat different inferences are often drawn. For example, a leveling off in the slope of the relationship between local and regional diversity may be claimed to imply "saturation" of communities and limitation of community membership by local interactions, whereas a continuously increasing relationship may suggest that local communities are

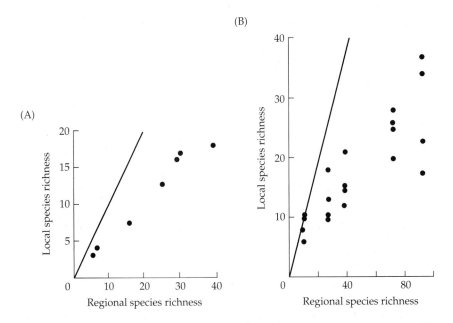

FIGURE 15.24 A plot of local versus regional species richness for (A) cynipine gall wasps on oaks in California and (B) Caribbean Island land birds. The straight line, drawn for reference, indicates values for which local diversity would equal regional diversity. Note that nearly all of the local communities contain fewer species than the regional species pool, and that the proportion of species that coexist in local habitats increases, but with a decreasing slope with regional diversity. While this pattern presumably represents the outcome of some combination of small-scale ecological interactions and larger-scale regional and historical processes, the exact nature of these processes is far from clear. (After Ricklefs 1987.)

"undersaturated" and are limited primarily by regional and historical processes (Ricklefs 1987; Cornell 1993; Schluter and Ricklefs 1993; but see Caley and Schluter 1997).

We applaud these efforts to characterize community patterns quantitatively, either by comparing real and null communities or by analyzing plots of local versus regional diversity. We are concerned, however, that interpreting the results in terms of a simplistic dichotomy between local or regional, or between ecological or historical phenomena, obscures more complex relationships between patterns and processes (see Crestwell et al. 1995; Griffiths 1999; Srivastava 1999; Gaston and Blackburn 2000; Witman et al. 2004). Several authors have assumed that "local" ecological processes, such as interspecific competition, do not influence the composition of the regional species pool. This is clearly incorrect. As pointed out in Chapter 4, the edges of nearly all geographic ranges are limited by local ecological niche variables that prevent populations of species from expanding into adjacent areas. If the species pool, by definition, consists of those species whose geographic ranges overlap at some local site, then local interactions can influence the composition of the pool. Put another way, no species would occur in the regional pool unless its local ecological relationships enabled it to occur within some local assemblages of the region. So, we expect the composition of the regional species pool to be strongly influenced, not just by regional and historical processes, but by local ecological interactions, including competition (see Colwell and Winkler 1984).

We also see problems in deciding when, or if, a local community is "saturated." A decreasing slope in the plot of local versus regional diversity is not sufficient to show saturation. Indeed, so long as the relationship is increasing at all, species obviously are being added to local communities, so they cannot be considered saturated. We question whether communities are ever truly saturated. This would imply that they could resist invasion of all exotics. At best, this is a virtually impossible proposition to test empirically. Furthermore, it represents a very rigid equilibrial view of ecology and biogeography. As pointed out in Chapter 13, views such as this are increasingly being challenged by nonequilibrial concepts, models, and theories that reflect the reality of a geographically complex and dynamic world.

The Latitudinal Gradient in Species Diversity

One of the most striking characteristics of life on Earth is the gradient of increasing species diversity from the poles to the equator. The earliest explorer-naturalists noticed that the tropics teem with life, the temperate zones have fewer kinds of animals and plants, and the Arctic and Antarctic are stark and barren by comparison (Forster 1778; Humboldt 1808 [see Hawkins 2001]; Wallace 1878; see Chapter 2). This pattern is a very general one in that it holds true not only for diversity of all organisms combined, but also for most major taxa (classes, orders, and families) of microbes, plants, and animals, and for those inhabiting most of the major types of ecosystems in the terrestrial and aquatic realms. In addition, the latitudinal gradient appears to be a very ancient pattern, although one varying not just among taxa, but also within taxa over time. Crame and Lidgard (1989) estimate that there has been a latitudinal gradient of land plant diversity for at least 110 million years. The fossil record for marine animals, such as brachiopods, corals, and single-celled foraminiferans and diatoms, also indicates that the latitudinal gradient in species diversity in the marine realm has been in existence for more than 250 million years in brachiopods and 70 million years in planktonic foraminifera (Stehli et al. 1969). More recent studies by Alistair Crame (Crame 2001, 2002, 2004) revealed

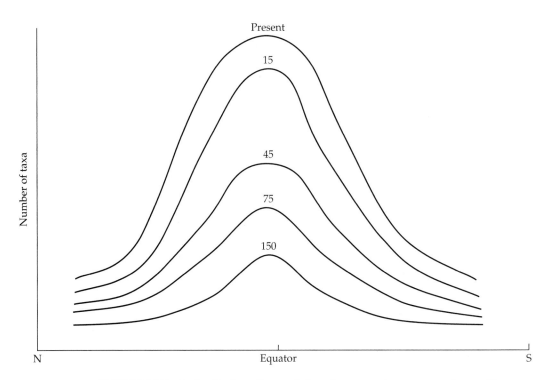

FIGURE 15.25 Latitudinal gradients of diversity are ancient patterns but, as illustrated here, those in the marine realm have actually intensified over time (from 150 million years B.P. to present), especially during the last 30 million years. (After Crame 2004.)

another interesting, dynamic feature of the pattern. The latitudinal gradient for marine bivalves has steadily intensified over time (Figure 15.25). According to Crame (2004), this may have resulted from global cooling following the climatic optimum of the Early Eocene (the so-called "mid-Eocene" sauna). This cooling appears to have been caused by tectonic events and associated changes in insolation and water currents. Among many other changes, this caused abrupt declines in mean oceanic temperatures by about 2° C at around 45 million years B.P., 5° C at around 37 million years B.P., and 4° C at around 14 million years B.P.; all this followed by more gradual cooling thereafter. As a result of simultaneous changes in oceanic currents, latitudinal gradients in water temperatures intensified, polar ice caps formed, and the tropical zone contracted as thermally defined marine provinces became more firmly established.

Perhaps just as remarkable as these long-term dynamics in diversity gradients is the observation that, at least for some taxa, latitudinal gradients in diversity may become established in an incredibly short amount of time. For example, Sax (2001) reports that within a span of decades to just a few centuries, naturalized (established) exotic species of Europe and North America now exhibit significant latitudinal trends in diversity (Figure 15.26).

NATURE AND COMPLEXITY OF THE PATTERNS. Beginning with classic studies by Alfred Fischer (1960) and George Simpson (1964), several authors have quantified latitudinal diversity gradients (for important reviews, see Fischer 1960; Pianka 1966; Stehli 1968; Stehli et al. 1969; Kiester 1971; Stehli and Wells 1971; Arnold 1972; Buzas 1972; Scriber 1973; Wilson 1974; Horn and Allen 1978; Sil-

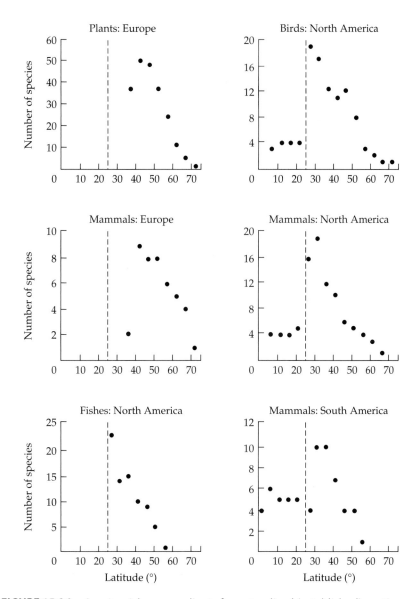

FIGURE 15.26 Species richness gradients for naturalized (established) exotic species of Europe, North America, and South America are similar in many respects to those of native biotas. For some groups of species, however, the gradient appears most prevalent outside of the tropics (dashed line indicates either the Tropic of Cancer or the Tropic of Capricorn). (After Sax 2001.)

verton 1985; Rohde 1992; Stuart and Rex 1994; Rosenzweig 1995; Roy et al. 1998; Willig 2000; Willig et al. 2003; Brown and Sax 2004; Crame 2004; Hillebrand 2004; Turner and Hawkins 2004). For groups of organisms whose distributions are well known, this can be done by counting the species in local areas of approximately equal size, and then drawing contour maps of species richness.

The diversity patterns for North American land birds, mammals, and plants (global biodiversity of vascular plants) are shown in Figures 15.27 and 15.28. The most striking feature that emerges from such maps is, of course, the rapid increase in diversity from the poles to the tropics. Interestingly, however,

(A)

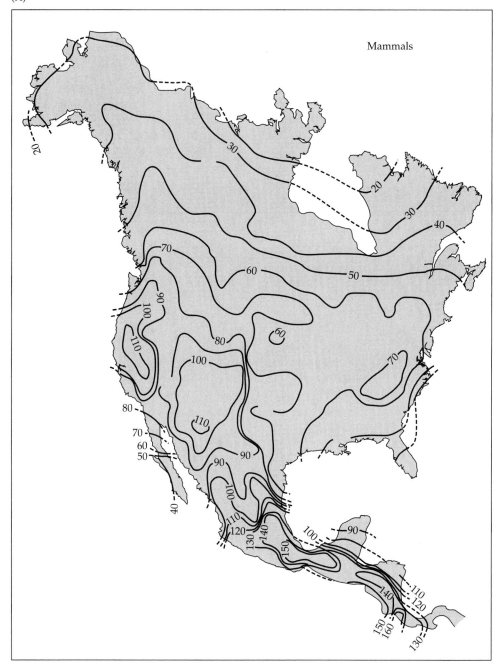

FIGURE 15.27 Geographic variation in the species richness of North American terrestrial mammals (A) and breeding land birds (B). These data were compiled by dividing the continent into grid squares, tallying the numbers of species whose geographic ranges overlap each square, and then interpolating and smoothing the data to produce the isopleths depicting diversity. Note the similar patterns in both groups, with numbers of species increasing toward the tropics; also shown are the high values in the diverse habitats in the mountainous regions of western North America. (After Simpson 1964 and Cook 1969.)

(B)

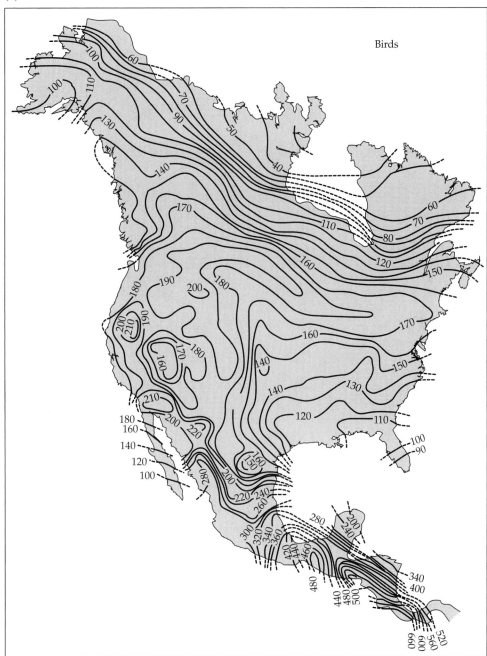

the rate of change of species richness with latitude is not constant, nor is it the same in different taxa. For example, not only are there more plants than birds, and more birds than mammals in every region, but bird species richness in North America increases about twelve-fold in the 60° of latitude shown, whereas mammalian diversity increases only eight-fold (see Figure 15.27). There are also differences among the different families of birds. Flycatchers (Tyrannidae) show a typical gradient of decreasing species richness from the tropics to the Arctic, but sandpipers (Scolopacidae) are one of the exceptional groups that exhibit the opposite pattern (Figure 15.29). In addition, it is clear

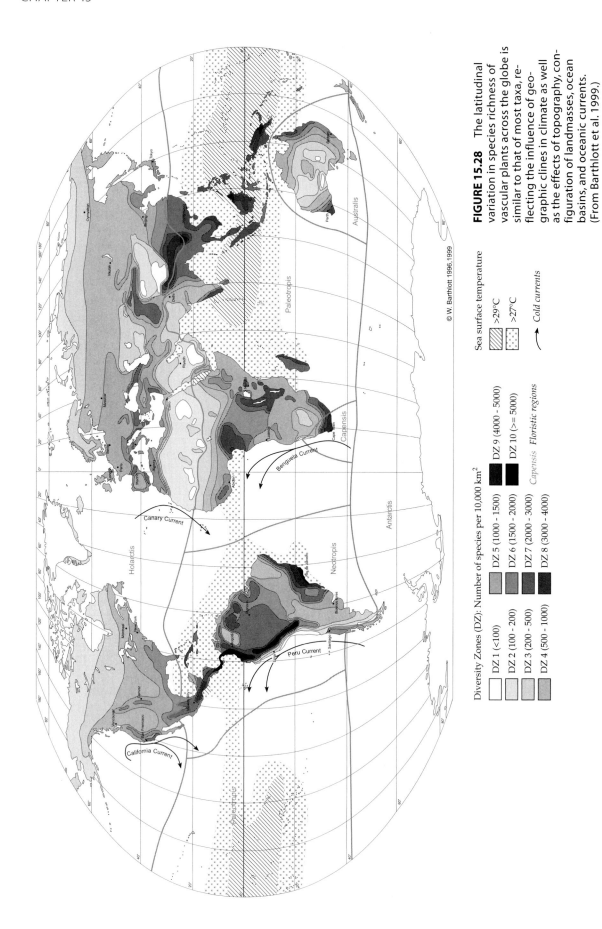

© W. Barthlott 1996,1999

FIGURE 15.28 The latitudinal variation in species richness of vascular plants across the globe is similar to that of most taxa, reflecting the influence of geographic clines in climate as well as the effects of topography, configuration of landmasses, ocean basins, and oceanic currents. (From Barthlott et al. 1999.)

Diversity Zones (DZ): Number of species per 10,000 km²

DZ 1 (<100)
DZ 2 (100 - 200)
DZ 3 (200 - 500)
DZ 4 (500 - 1000)

DZ 5 (1000 - 1500)
DZ 6 (1500 - 2000)
DZ 7 (2000 - 3000)
DZ 8 (3000 - 4000)

DZ 9 (4000 - 5000)
DZ 10 (>= 5000)

Capensis Floristic regions

Sea surface temperature

>29°C
>27°C

Cold currents

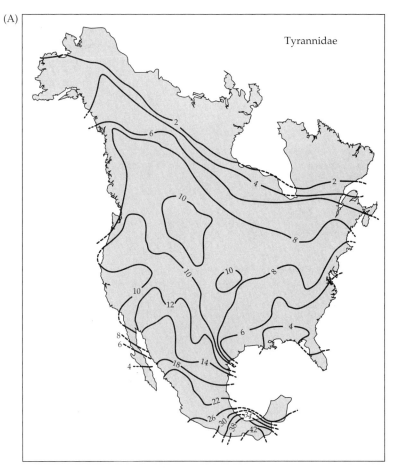

Tyrannidae

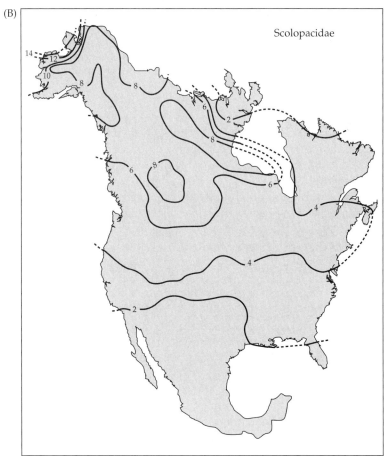

Scolopacidae

FIGURE 15.29 Geographic variation in the species richness of two families of birds in North America: (A) flycatchers (Tyrannidae) and (B) sandpipers (Scolopacidae). These maps, compiled from maps of the birds' breeding ranges, show contrasting patterns. The flycatchers, typical of most groups, increase in diversity toward the tropics; whereas, the greatest numbers of sandpiper species breed in tundra habitats at high latitudes. (After Cook 1969.)

(A) Pinaceae

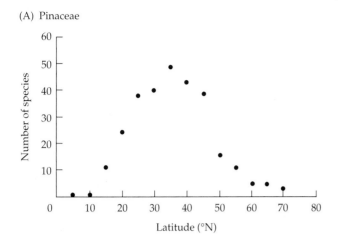

(B) Ichneumonidae

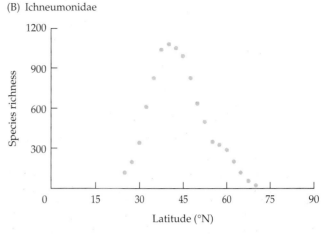

(C) Insectivorous mammals

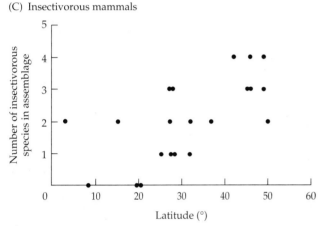

FIGURE 15.30 (A) One exceptional group that has been particularly well studied is the plant family Pinaceae—the conifers (pines, spruces, firs, and their relatives). These plants are restricted to the Northern Hemisphere and achieve their highest diversity at mid-latitudes in both North America and Eurasia. (B) The Ichneumonidae, a large family of parasitic wasps, are also most diverse in temperate regions, with the greatest species richness in North America occurring at about 40° N latitude. (C) The relationship between species richness of small insectivorous mammals (shrews and moles) and latitude is statistically significant, but opposite of that observed for most taxa, with diversity for this group peaking in the temperate latitudes. (D) Species richness of crayfish and amphipods tends to peak, not at the equator, but in the subtropical latitudes of the Nearctic Region, then species richness declines toward the higher latitudes. (A after Stevens and Enquist 1998; B after Owen and Owen 1974; Janzen 1981; see also Hespenheide 1978; C after Cotgreave and Stockley 1994; D after Stevens 1989.)

(D) Crayfish

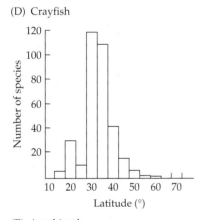

(E) Amphipods

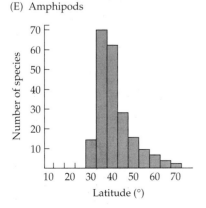

from these maps that other geographic factors, including topography and bathymetry, proximity to coastlines, configuration of landmasses and ocean basins, and climate and currents, also strongly influence global scale patterns of biological diversity.

Although many hundreds of studies have documented the generality of the latitudinal gradient in species diversity (see Hillebrand's [2004] meta-analysis of these gradients), there are, of course, exceptions. Some species groups, such as sawflies (Hymenoptera, Symphyta; Kouki 1994), aphids (Dixon et al. 1987), ichneumonid and braconid wasps (Janzen 1981; Gauld 1986; Gauld et al. 1992), willows (Myklestad and Birks 1993), marine mammals and their helminth parasites (Rohde 1992; but see Rohde 1999), communities of water-filled pitcher plants (Buckley et al. 2003), and birds of peatlands in Finland (Kouki 1999) and eastern deciduous forests of North America (Rabenold 1979), exhibit what appears to be a reversed trend, with species richness increasing as we move away from the equator. In nearly all cases, however, the trend is actually unimodal, or hump-shaped, with species richness peaking at intermediate latitudes and then declining in either direction (north or south) as we approach the range limits of that species group. This is, in fact, a corollary of one of biogeography's most general trends: diversity, and for that matter productivity and density, of nearly every species group peaks not at the edges of the groups' geographic distribution, but at locations (latitudes, longitudes, elevations, or depths) closer to the center of their combined

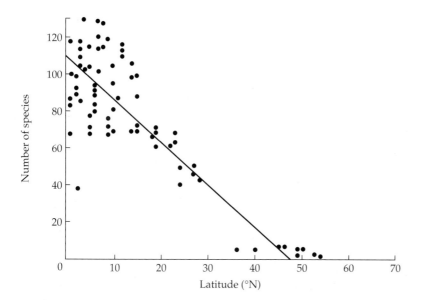

FIGURE 15.31 South American bats exhibit the classical latitudinal pattern in diversity, with richness varying markedly but peaking across a broad band of tropical latitudes. (After Willig and Selcer 1989.)

range (see also Colwell and Hurtt's [1994] discussion of the "mid-domain effect"). For most groups, the peak in species diversity occurs somewhere in the tropical latitudes, but for some species groups (namely those adapted to habitats and climates of the higher latitudes), the peak lies somewhere beyond the tropics. For example, coniferous trees (Pinaceae), freshwater crayfish and amphipods, insectivorous small mammals, and ichneumonid wasps (Ichneumonidae), are most diverse in the subtropical and temperate zones (Figure 15.30); while marine mammals and penguins (Spheniscidae) reach their greatest diversity at higher latitudes.

For every such exception, however, there are numerous groups that are confined, or nearly so, to the tropics; New World bats (Figure 15.31), Indo-Pacific giant clams (Tridacnidae), and palms (Arecaceae; Figure 15.32) are just a few examples. It is very important, however, that we not ignore the exceptions or discount them as mere nuisances and noise. Indeed, investigations as

FIGURE 15.32 Pantropical distribution of the palm family, Arecaceae. The primary factor limiting the distribution of these plants is temperature. Although a few palms can tolerate freezing, none can withstand extreme cold. Where palms are found outside the tropics, they invariably occur in areas with moderate climates, such as New Zealand and southeastern Australia. (After Good 1974; data from Moore 1973.)

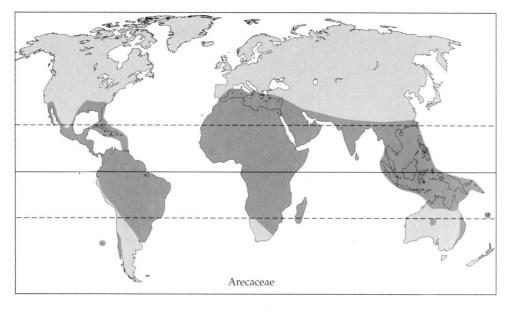

Arecaceae

to why these rare exceptions fail to exhibit the predicted pattern may reveal more about the underlying causal processes than studies of the many species that conform to the general pattern.

Latitudinal trends in species richness are also apparent from ecological studies of small areas of terrestrial habitat. Perhaps the most dramatic trend is in numbers of forest tree species. A single hectare of tropical rain forest in South or Central America may contain 40 to 100 different kinds of trees (Richards 1957; Anderson and Brown 1980; Hubbell and Foster 1986); whereas, a hectare of deciduous forest in eastern North America is likely to contain 10 to 30 species; and a hectare of coniferous forest in northern Canada, only 1 to 5 species (Braun 1950; Glenn-Lewin 1977; Latham and Ricklefs 1993; Scheiner and Rey-Benayas 1994). Because many insect species are specific herbivores and pollinators of individual plant species, it is expected that many insect taxa will show comparable patterns when the data become available (see Otte 1976; Erwin 1983, 1988; Cushman et al. 1993; Gotelli and Ellison 2002). When Fleming (1973) compiled data on numbers of mammal species in North American forest habitats, he found that 15 to 16 species occur in boreal coniferous forests in Alaska, 31 to 35 species inhabit deciduous forests in the eastern United States, and 70 species are found in both wet and dry tropical forests in Panama. For many, if not most, species groups, the latitudinal gradients in diversity can be attributed largely to the rapidly increasing numbers of species in the subtropics (see Figures 15.32 and 15.33), with richness being relatively high but variable across the tropics, then declining rapidly between 20° and 30° N or S latitude. For example, non-flying mammals of certain orders—such as Primates (monkeys), Edentata (armadillos, sloths, and anteaters), and families (especially of rodents) that are specialized for arboreal life or have insectivorous or frugivorous diets—contribute significantly to the diversity of tropical communities (Willig and Sandlin 1989; Willig and Selcer 1989; Kaufman 1995; Cowlishaw and Hacker 1997; Kaufman and Willig 1998), but have very few species whose ranges extend into the higher latitudes.

Clear latitudinal gradients of species diversity, qualitatively similar to those seen on the continents, also occur in the oceans. Figure 15.33B illustrates the pattern for planktonic foraminifera, which has also been reported for a diversity of other marine organisms (see also Figure 15.25; Stehli 1968; Stehli et al. 1969; Taylor and Taylor 1977; Turner 1981; Clarke 1992; Rex et al. 1997; Culver and Buzas 2000; Crame 2004.). Sanders (1968) found about five times as many species (100 vs. 20) of bottom-dwelling annelids and bivalve mollusks in the Bay of Bengal (20° N latitude) as in comparable areas off Cape Cod, Massachusetts (40° N latitude). There is even a pronounced latitudinal gradient of diversity in the abyssal depths (Figure 15.34; Stuart and Rex 1994). This gradient is of particular interest because, although latitudinal gradients in diversity are sometimes attributed to geographic clines in environmental conditions, the abiotic environment at such great depths is relatively constant throughout the world.

It was once thought that freshwater organisms were an exception to the latitudinal patterns of species richness observed in other habitats (Patrick 1961, 1966). Now, it appears that this is not the case (Clarke 1992; France 1992; Rex et al. 1993; Cotgreave and Stockley 1994; Oberdorff 1995). In a quantitative study of stream insects, Stout and Vandermeer (1975) usually found 30 to 60 species of these freshwater organisms at tropical American sites compared with 10 to 30 species in the temperate United States. The number of freshwater fish species clearly decreases with increasing latitude in North America (see Figure 10.15), and a similar trend holds true for lakes around the world (Barbour and Brown 1974; Amarasinghe and Welcomme 2002). Despite this general pattern, there is a great deal of variation owing largely to differences in the sizes and histories of individual bodies of water. The highest diversity of lacustrine fish species occurs in the Great Lakes of the African Rift Valley (Lakes Malawi, Tanganyika, and Victoria), where the family Cichlidae is represented by hundreds of species (Fryer and Iles 1972; Meyer 1993). With respect to rivers, species diversity of fish and invertebrates also appears to peak in the tropics, although again, the pattern is complex and varies markedly among species groups and biogeographic regions (Matthews and Matthews 2000; Vinson and Hawkins 2003). There seems to be little doubt, however, that the Amazon supports more fish species than any other river system in the world. This is true both for the entire river basin (gamma diver-

(A)

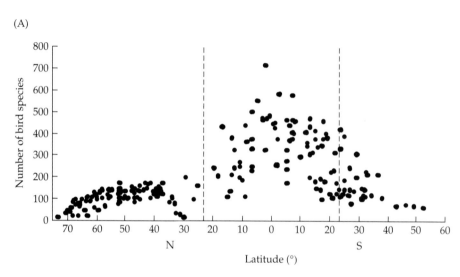

FIGURE 15.33 (A) Bird species exhibit the general latitudinal gradient in diversity, with species richness peaking in the tropics (delineated by the vertical dashed lines at the Tropics of Cancer and Capricorn). Birds also exhibit another relatively common pattern: species richness, albeit generally higher, is also highly variable across tropical latitudes (richness measured as number of bird species recorded in each of 206 randomly selected, 48,400 km^2 grid cells). (B) The worldwide pattern of species richness in planktonic foraminifera shows a latitudinal gradient that is fairly typical for marine organisms. Dots represent locations of samples used to draw the isopleths. Corals, mollusks, and fishes are other marine groups that are most diverse in tropical waters. (A after Turner and Hawkins 2004; B after Stehli 1968.)

(B)

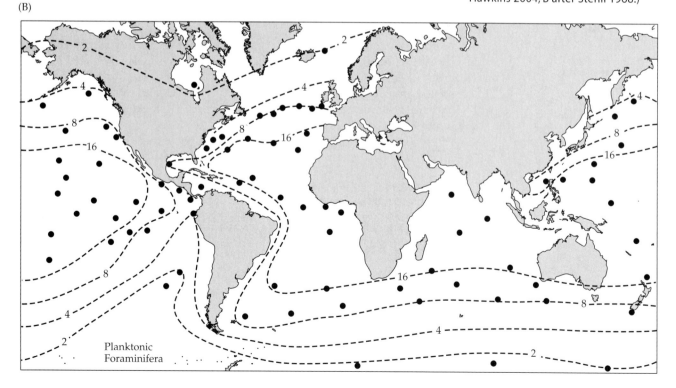

Planktonic Foraminifera

FIGURE 15.34 Species richness of prosobranch (gastropod) mollusks as a function of latitude in the abyssal depths of the North Atlantic Ocean. Samples were collected using a dragged sled trap, and species diversity was measured using Hurlbert's expected number of species (one of the diversity indexes). Note two things: first, there is considerable variation in diversity, even among samples collected at comparable latitudes; presumably, this reflects the spatial heterogeneity of abyssal environments. Second, despite this variation, there is a clear tendency for diversity to increase toward the equator. (After Stuart and Rex 1994.)

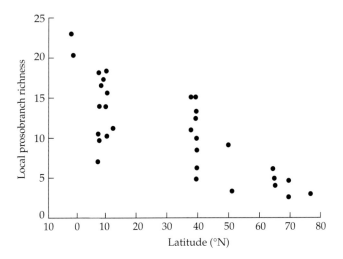

sity) and within small areas of relatively uniform habitat (alpha diversity). As documented in numerous scientific studies and television specials, the Amazon fish fauna includes a substantial component that moves into seasonally flooded forests to exploit the unusual habitats and food resources there, including even fruit.

As noted above, latitudinal gradients vary among taxa, realms and ecosystems, and over time as well. For most groups, the pattern is not a simple, symmetrical decline from the equator. In fact, diversity often declines at different rates in different hemispheres and just as often includes a broad, equatorial plateau where species richness remains high or, in some cases seems to vary erratically within the tropics even within the same taxonomic group (see Figures 15.28, 15.31 and 15.33). Recent compilations and analyses of geographic distributions and diversity of vascular plants and all 4740 species of terrestrial mammals provide clear illustrations of the generality and the complexity of latitudinal diversity gradients (Figures 15.28 and 15.35). As we have seen in the earlier maps generated by Simpson (Figures 15.27 and 15.29), these more recent and more comprehensive maps of species densities illustrate the very general tendency for diversity to be highest in the tropics. The pattern, however, differs substantially among geographic regions, for example with plant species density of the Western Hemisphere and that across Asia, Indonesia and Australia varying markedly within tropical latitudes (23° N to 23° S), while the gradient is much tighter across western landmasses of the Old World (Europe and Africa)—in this case peaking both at the equator and

FIGURE 15.35 Latitudinal gradients in species density (number per standardized area) may exhibit substantial variation, much of this due to variation among geographic regions (A) and among species groups comprising a relatively heterogeneous taxon (B–D; see also Figure 15.28). (E) Comparison of two methods of mapping hotspots of vascular plants including that by Meyers et al. (2000; mapping those sites with at least 0.5% of global plant diversity being endemic to that area, and with a loss of 70% or more of its primary vegetation) to that by Mutke and Barthlott (2005; identifying sites with at least 3,000 species per 10,000 km² and higher than average human impact according to Sanderson et al's [2003] human footprint index). A recent compilation and analysis of geographic distributions of all 4740 terrestrial mammals (F–K) also illustrates the generality and complexity of latitudinal diversity gradients (geographically restricted mammals refer to those in the lowest quartile of geographic range size for this taxon (plotted are species densities—number per 10,000 km² grid cell). (A–E from Mutke and Barthlott 2005; F–K courtesy of John Gittleman and Wes Sechrest, from Sechrest et al. in press.)

(A) All vascular plants

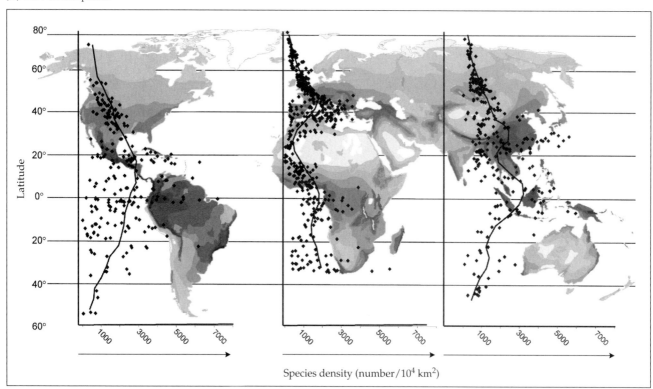

Species density (number/10^4 km^2)

(B) Gymnosperms

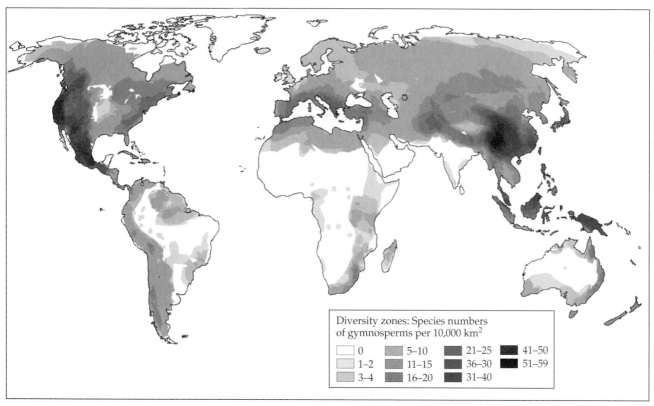

Diversity zones: Species numbers
of gymnosperms per 10,000 km^2

0   5–10   21–25   41–50
1–2   11–15   36–30   51–59
3–4   16–20   31–40

(C) Cacti

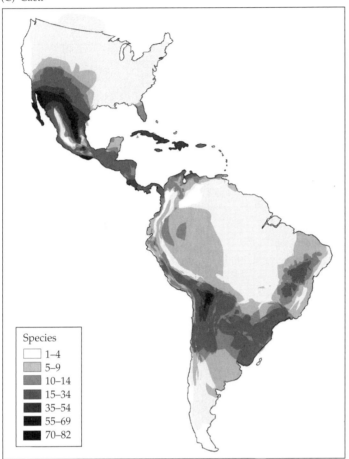

Species
- 1–4
- 5–9
- 10–14
- 15–34
- 35–54
- 55–69
- 70–82

(D) Mosses

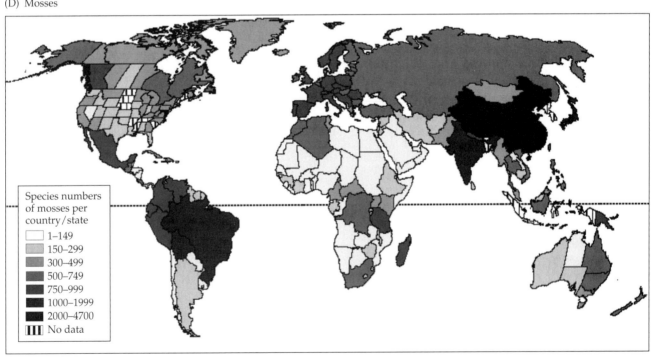

Species numbers
of mosses per
country / state
- 1–149
- 150–299
- 300–499
- 500–749
- 750–999
- 1000–1999
- 2000–4700
- No data

(E) Hotspots of vascular plant diversity

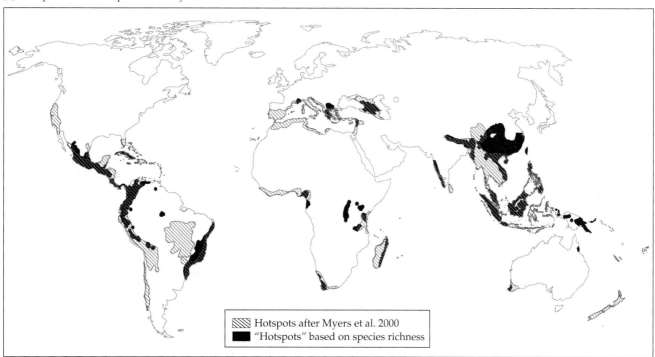

Hotspots after Myers et al. 2000
"Hotspots" based on species richness

(F) All terrestrial mammals

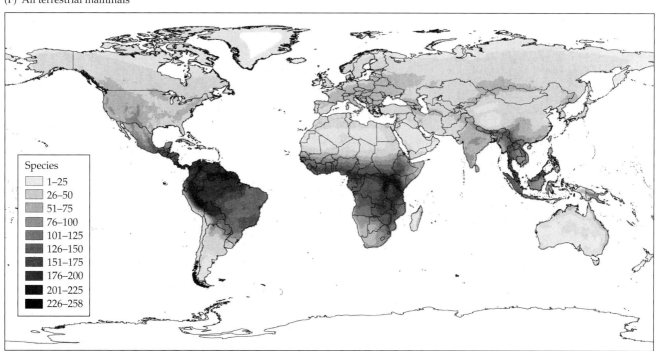

Species
1–25
26–50
51–75
76–100
101–125
126–150
151–175
176–200
201–225
226–258

(G) Bats

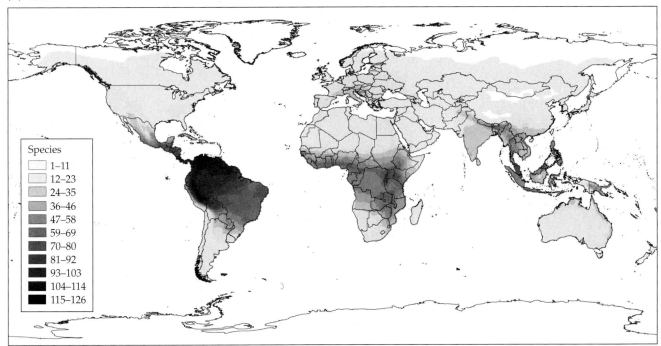

(H) Primates

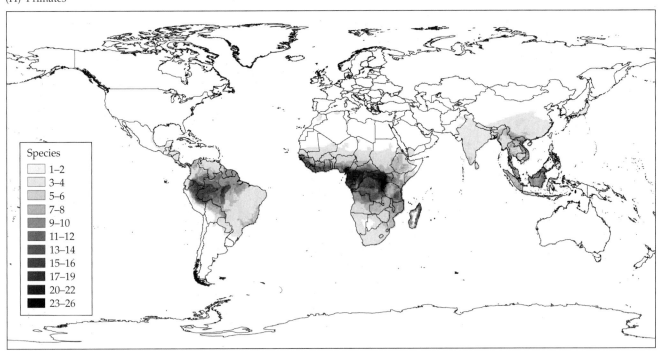

(I) Rodents

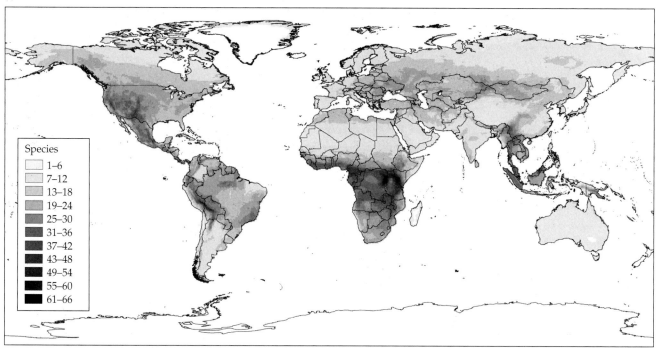

Species
	1–6
	7–12
	13–18
	19–24
	25–30
	31–36
	37–42
	43–48
	49–54
	55–60
	61–66

(J) Marsupials

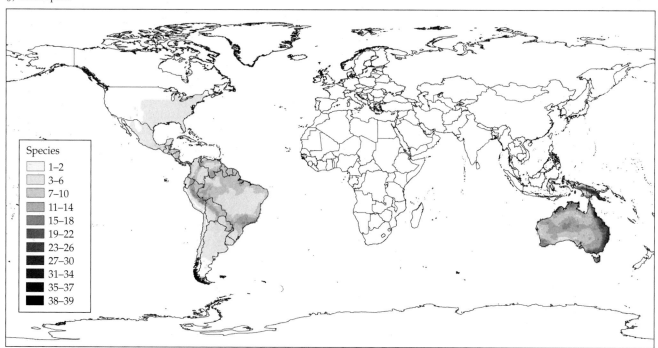

Species
	1–2
	3–6
	7–10
	11–14
	15–18
	19–22
	23–26
	27–30
	31–34
	35–37
	38–39

(K) Geographically restricted mammals

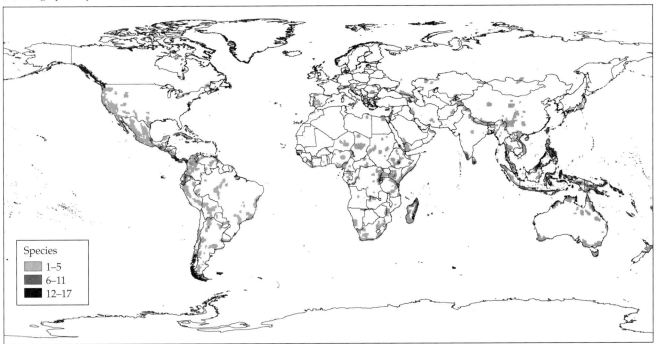

between 35° and 45° N latitude (Figure 15.35A). In addition to differences in environmental gradients across regions, another reason for the considerable variation in these latitudinal gradients in diversity is that these groups—vascular plants and terrestrial mammals—are comprised of heterogeneous collections of taxa with a diversity of environmental tolerances and, as a result, quite different geographic signatures (e.g., compare Figures 15.35B to C, and G through J to each other). Finally, a note of caution for those attempting to apply lessons from these general patterns for developing strategies for conserving rare and endangered species. Because these species are almost by definition atypical—exhibiting unusually limited ecological tolerances and geographic ranges—their diversities may exhibit highly irregular geographic signatures. That is, hotspots for species richness, per se, may differ markedly from sites with a high concentration of imperiled species (Figures 15.35E and 15.35K; see also Figures 16.11B, 16.9, 16.10, and 16.11).

PROCESSES AND CAUSAL EXPLANATIONS.   Latitudinal gradients in species diversity have intrigued ecologists, evolutionary biologists, and biogeographers for well over a century (see Chapter 2, and Brown and Sax 2004). Johann Reinhold Forster (1778), who sailed with Captain James Cook, reported that plants exhibit a latitudinal trend in diversity, and Forster attributed the pattern to the increased intensity of heat (i.e., sunlight and productivity) as one moves from the poles to the equator. A few decades later, Alexander von Humboldt (1807, whose work was translated into English in 1850; see Hawkins 2001:470) made similar observations based on his explorations the New World tropics, describing an increase in diversity of not just numbers, but other features of biological diversity as well: "Thus, the nearer we approach the tropics, the greater the increase in structure, grace of form, and mixture of colors, as also in perpetual youth and vigor of organic life." Later, an admirer of Humboldt—Charles Darwin (1839:29)—offered similar observations based on his own voyages and his studies of the reports of many other nineteenth century naturalists: "In England any person fond of natural history enjoys in his walks a great advantage, by always having something to attract his attention; but in these fertile lands [in Brazil] teeming with life, the attractions are so numerous, that he is scarcely able to walk at all."

In a work entitled *Tropical Nature and other Essays,* Alfred Russel Wallace (1878) demonstrated that he and his colleagues were not merely describing patterns in nature, but simultaneously developing causal explanations for those patterns. Wallace's explanation for the latitudinal gradient in species diversity was largely based on the increasing severity of climates and the resultant intense struggle for existence as one moves from the equator to the poles. In his words, "The equatorial zone, in short, exhibits to us the result of a comparatively continuous and unchecked development of organic forms; while in the temperate regions, there have been a series of periodical checks and extinctions of a more or less disastrous nature … The equatorial regions are then, as regards their past and present life history, a more ancient world than represented by the temperate zones, a world in which the laws which have governed the progressive development of life have operated with comparatively little check for countless ages, and have resulted in those infinitely varied and beautiful forms." (Wallace 1878:123.)

Our ever-increasing understanding of the geography of nature has in turn led to a proliferation of hypotheses attempting to account for latitudinal gradients in species richness. In his review published in 1966, Eric Pianka listed six explanations for the pattern, but by 1992, Klaus Rohde had listed nearly 30, with some particular to specific groups of species, ecosystems, or realms. Clearly, this pattern cannot be attributed to the influence of unique events in the histories of just a few lineages. Instead, it must reflect some pervasive way in which the geography of the Earth has influenced the diversification of living things (via evolution) and the spread, maintenance, or loss, of that diversity over time (via dispersal or extinction).

Fortunately, we believe the search for a general explanation for this very general pattern is much simpler than this. First, it is important to maintain the distinction between proximate factors and ultimate processes. Table 15.4 summarizes the diversity of explanations for this very general pattern in biological diversity. We note that this list, while quite long, is not exhaustive. On the other hand, many of these "alternative" hypotheses seem redundant, and most are not mutually exclusive. Null models (of which only two are listed in Table 15.4) may sometimes seem unrealistic, but they are very useful starting points for most investigations in ecology, evolution, and biogeography (see Gotelli and Graves 1996). Explanations based on biological interactions and ecological processes (see Table 15.4) certainly are more realistic, for the most part, and are based on empirical observations and sound ecological theory. A quick perusal of these explanations, however, reveals that many seem circular or, at best, just beg the question of what ultimately determines the latitudinal gradient in species diversity. Many explanations based on the influence of abiotic factors (see Table 15.4) also appear to fall short of identifying ultimate causality and, at least in some cases, focus on isolated factors that apply just to a small subset of all the groups that exhibit this very general pattern. We believe that the most promising explanations for this pattern are those that are more integrative and take into account the combination of factors that affect the fundamental processes that influence distributions of all species—evolution, dispersal, and extinction (see Table 15.4).

TABLE 15.4 *A diversity of explanations for latitudinal gradients in species richness*[a]

Hypothesis	Explanation	Source
Null models		
Mid-domain effect	Random placement of geographic ranges between two presumed hard boundaries (the polar regions) results in the highest overlap of ranges (highest richness) occurring midway between those boundaries (i.e., near the equator).	Colwell and Hurtt 1994
Neutral theory of biodiversity and biogeography	Even a set of species that are assumed to be identical with respect to their birth and death rates and dispersal abilities of their individuals, and whose population dynamics is similar to that of random walk, should be comprised of populations that vary over time—changing in population density and some suffering random extinction, while others experience mutations, genetic recombinations, and random speciation. The factors influencing population densities, random extinctions, and random speciation of these equivalent species may favor higher species richness in the tropics.	Hubbell 2001; Hubbell and Lake 2003; Turner and Hawkins 2004
Biological interactions and processes		
Competition	Competition, especially diffuse competition (that from many species of competitors), tends to hold each population in check, allowing more species to coexist in competitor-rich, tropical communities.	Dobzhansky 1950; Pianka 1966; Huston 1979
Predation	Predators (including herbivores and carnivores) tend to hold prey populations in check by switching and preying most heavily on the most abundant prey, thus preventing prey populations from increasing to levels at which they would exclude each other, especially in the tropics with its high diversity of predators.	Paine 1966; Pianka 1966; Harper 1969; Janzen 1970; Lubchenco 1978, 1980; Lubchenco and Menge 1978
Mutualism	Mutualists, by definition, promote the coexistence of their symbiots and, thus, the empirically observed high frequency of mutualism in the tropics promotes higher diversity of symbiots.	Dobzhansky 1950; Paine 1966; Janzen 1970; Menge and Sutherland 1976
Host diversity—parasitism	High diversity of hosts promotes high diversity of parasites. Therefore, because the topics are inhabited by more host species, diversity of parasites also is higher in the tropics.	Rohde 1989
Epiphyte load	Diversity of epiphytes is highest where diversity of their host trees are highest (i.e., in the lush forests of the tropics).	Strong 1977
Environmental patchiness (beta diversity)	Tropical habitats tend to be patchily distributed, thus providing more numerous patches to be colonized and occupied by different species.	McCoy and Connor 1980
Habitat (e.g., foliage height) diversity; biotic spatial heterogeneity	Diversity, in general, tends to be higher in more complex and heterogeneous habitats, and the tropics tend to be spatially and vertically more heterogeneous.	MacArthur et al. 1966; Cody 1968, 1974, 1975; MacArthur 1972; Huston 1979; Thiollay 1990
Niche width	Tropical species tend to be more specialized (i.e., have narrower niches) and, therefore, more species can be packed into tropical habitats.	Ben-Eliahu and Safriel 1982; Brown and Gibson 1983
Population growth rate	Tropical species tend to be comprised of populations with higher growth rates, which in turn provides more opportunities for ecological specialization.	Huston 1979
Productivity	Higher levels of primary productivity of tropical plant communities provides more energy, which in turn	Hutchison 1959; Connell and Orias 1964; MacArthur 1965,

TABLE 15.4 *(continued)*

Hypothesis	Explanation	Source
	supports higher populations and more diverse communities of consumers.	1969; Pianka 1966; Wright 1983, Currie and Paquin 1987; Tilman 1988; Currie 1991; Roy et al. 1998
Canopy/crown morphology (incidence of solar radiation)	Tropical forests receive more direct sunlight, which tends to favor trees with more shallow crowns, also allowing more light to penetrate below the canopy and support multiple layers of trees below and, thus, a greater diversity of trees overall.	Terborgh 1985
Geographic ranges (Rapoport's rule)	Tropical species tend to have smaller geographic ranges, thus allowing more species to coexist in tropical versus temperate regions.	Rapoport 1982; Pianka 1989; Stevens 1989

Abiotic/Environmental factors

Hypothesis	Explanation	Source
Environmental stability and predictability (over different time periods—see below)	Tropical environments tend to be more stable over both short and long time periods, thus their species are less subject to extinctions and more capable of adapting to—and specializing for—these more predictable environments.	Slobodkin and Sanders 1969; Janzen 1970
Tectonic dynamics	Despite significant changes in the development and drifting of Earth's plates, some areas that are currently in the tropics have been in tropical regions for much of the evolutionary history of their biota.	See Scotese 2004
Glacial fluxes	Glacial expansions and climatic fluxes of the Pleistocene (Chapter 9) caused extinctions of high-latitude species, and the subsequent inter-glacial period has been insufficient to re-establish diversity in these regions.	Fischer 1960; Pianka 1966; Simpson 1974,1975; Fischer and Arthur 1977; Stanley 1979
Annual stability (seasonality)	Tropical environments tend to be much less variable throughout the year, with little seasonality in temperature or precipitation. Thus, tropical species can become more specialized, allowing more species to coexist in the same amount of space.	Begon et al. 1986
Antiquity of the tropics (evolutionary time)	As indicated above, in comparison to those in higher latitudes, tropical regions tend to have occurred in tropical latitudes for longer periods of time. Thus, tropical regions have accumulated species over longer periods of time than those at higher latitudes.	Pianka 1966, 1988; Whittaker 1969
Antiquity of the tropics (ecological time)	Tropical species have had more time to disperse to, and colonize, a greater portion of suitable habitats than have their counterparts in temperate regions and higher latitudes (equilibrium levels of species richness between speciation, extinction, *and dispersal* are higher in the tropics).	Fisher 1960; Pianka 1966
Environmental harshness	Tropical environments tend to be more benign, and thus can be inhabited by more species.	Terborgh 1973b; Brown 1981, 1988; Thiery 982; Begon et al. 1986
Abiotic rarefaction (intermediate disturbance)	Tropical environments are intermediate with respect to frequency and intensity of storms and other abiotic disturbances that would otherwise reduce population densities. According to this modified form of the intermediate disturbance hypothesis, diversity should thus be higher in the tropics.	Dobzhansky 1950; Connell 1978

(continued on following page)

TABLE 15.4 *A diversity of explanations for latitudinal gradients in species richness[a]* *(continued)*

Hypothesis	Explanation	Source
Solar energy (species energy)	Tropical environments receive more intense solar energy, supporting higher productivity of plant communities, thus promoting higher diversity of these producers and dependent consumers.	Forster 1778; Willdenow 1805; Wright 1983; Turner 1986; Currie 1991; Roy et al. 1998
Heat (temperature and physiological reactions)	Higher environmental temperatures in the tropics promote higher metabolic rates and shorter generation times in tropical species (at least for the ectotherms), allowing more rapid speciation rates and greater overall accumulation of species.	Alekseev 1982; Rohde 1992
Aridity	Because precipitation tends to be higher in the tropics and because water is required by all life forms, species diversity (of terrestrial life forms) should be higher in the tropics.	Begon et al. 1986
Evapotranspiration	Plants require both water and energy in the form of sunlight. Tropical environments tend to be both wet and warm and, therefore, productivity of plant communities should be higher in the tropics, supporting a greater diversity of producers and consumers.	Wright 1983
Solar radiation and mutation rates (evolutionary speed)	The intensity of solar (ultraviolet) radiation in the tropics causes greater mutation rates, thus providing more raw material for selection and adaptive radiation.	Rensch 1959; Stehli et al. 1969; Rohde 1992; see also Bromham and Cardillo 2003
Area	Tropical regions tend to occupy a greater portion of Earth's surface. Because diversity tends to increase with area for nearly all types of ecosystems, diversity should be higher in the tropics.	Connor and McCoy 1979; Currie 1991; but see Chown and Gaston 2000
Fundamental processes and integrative hypotheses		
Speciation, extinction and area (equilibrium models)	As discussed above, tropical ecosystems tend to occupy a larger portion of the Earth's surface than those in higher latitudes. According to equilibrial models of MacArthur and Wilson (1963, 1967) and others, larger ecosystems should be characterized by relatively high immigration and speciation rates, but low extinction rates. Therefore, at equilibrium, larger ecosystems (i.e., the tropics) should be inhabited by more species.	Rosenzweig 1992, 1995, 1997; Hawkins and Porter 2001; Rohde 1997
Speciation, extinction, and dispersal (tropical niche conservatism)	In contrast to other regions, the tropics are larger and older and, therefore, should have higher speciation rates and lower extinction rates. In addition, because tropical climates tend to be relatively benign, predictable, and productive; selection favors species with relatively narrow niche breadths and limited dispersal abilities.	Wiens and Donoghue 2004

Source: Modified from Rohde 1992.

[a]This list, while a long one, is not necessarily an exhaustive one. In addition, these hypotheses are not necessarily mutually exclusive.

John Wiens, a distinguished ecologist, and Michael Donoghue, an equally distinguished historical biogeographer, recently presented a conceptually broad and potentially insightful explanation for latitudinal gradients in species richness (Wiens and Donoghue 2004). They quite cogently argue the need for new syntheses—in this case, one drawing on the principles and insights of ecology, evolution, and biogeography. Counter to some prevailing assumptions of historical biogeographers, Wiens and Donoghue argue that ecological interactions and climatic conditions *are* important, but primarily through their influence on evolutionary and biogeographic processes—again, these being speciation, dispersal, and extinction. Their tropical conservatism hypothesis combines the insights and arguments of numerous earlier biogeographers and ecologists, but its integrative nature is even more compelling. For reasons discussed above and summarized in Table 15.4, speciation rates tend to be higher in the tropics. In addition, because the tropics include a larger portion of the Earth's surface area, and because tropical ecosystems are more stable, more predictable, and less harsh, extinction rates should be lower in the tropics. For these same reasons (i.e., the relatively stable, predictable, and benign nature of tropical environments), tropical species tend to adapt over time by becoming more specialized. Put another way, species adapted to the variable and sometimes unpredictable nature of high-latitude environments must have broad niches, which allow them to adapt in situ—or disperse to—other environments during inclement periods.

Latitudinal gradients in speciation and extinction rates described above explain how the pattern is established, but niche conservatism and its influence on ecological interactions and dispersal can explain why the pattern is maintained, and why it has intensified over time (see Figure 15.25). As David Janzen (1967) remarked in one of his classic papers, "mountain passes *are* higher in the tropics"—not because the mountains are actually taller, but because tropical species tend to have relatively narrow niches and, therefore, more limited abilities or propensities to disperse across high montane habitats to colonize other lowland forests. Although a small fraction of these tropical species may eventually colonize regions in the higher latitudes, their dispersal (immigration) rate is insufficient to compensate for the relatively low speciation rate and high extinction rates of temperate and high latitude ecosystems.

There are at least two interesting corollaries of Wiens and Donoghue's hypothesis (Figure 15.36). First, geographic ranges of many animals and plants seem to be limited along their higher-latitude boundaries by climatic factors, suggesting that cold climate and niche conservatism prevent many tropical lineages from invading the temperate zone. Second, many species exhibiting the predicted gradient in species richness also exhibit a complementary phylogeographic pattern—"with an origin in the tropics and more recent dispersal to temperate regions" (Wiens and Donoghue 2004:642; see Figure 15.36).

The list of explanations in Table 15.4 include the fundamental biogeographic processes and the principle features of the Earth's geographic template that influence those processes. Perhaps, not surprisingly, the more integrative hypotheses that we favor are also some of the most recently developed explanations for

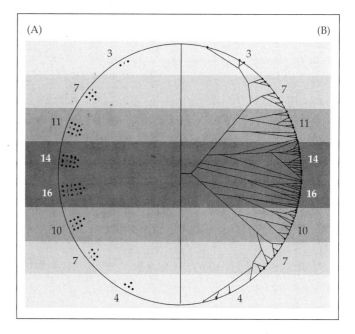

FIGURE 15.36 (A) Many earlier explanations for the latitudinal gradient in diversity were based on standard ecological approaches (see Table 15.4) and on correlations between species richness and an environmental variable (e.g., temperature or solar energy, represented here by intensity of shading). Here, each dot represents a different species, and the numbers along the Earth's surface represent species richness at that point. (B) In contrast, more integrative explanations for this general pattern are based on the history of lineages and of place, and how ecology, phylogeny, and adaptation have combined to determine the development and maintenance of biological diversity. Wiens and Donoghue's (2004) explanation is based, in part, on the tropical conservatism hypothesis (dots represent species, and lines connecting them represent both evolutionary relationships and simplified paths of dispersal). Because tropical climates are relatively benign, aseasonal, and predictable, their species tend to become ecologically specialized and limited in their abilities to disperse to other sites within the tropics or to those in the higher latitudes. Thus, because their species tend to be more isolated, and because the tropics tend to be larger and older than other biomes and regions, speciation rates and total number of species accumulated should be higher in the tropics. Wiens and Donoghue's hypothesis not only offers a synthetic explanation for this pattern, based on ultimate causes—speciation, extinction, and dispersal—but it also proposes other, testable predictions (e.g., that tropical lineages should, on average be relatively old, whereas those in temperate regions are often recently derived from the few clades that disperse from tropical regions). (After Wiens and Donoghue 2004.)

this long-studied pattern. We find this increased emphasis on more synthetic explanations—those that combine the insights of the great wealth of empirical information with sound theories from many disciplines—especially promising and characteristic of modern biogeography, in general. In short, while the true complexity of the latitudinal pattern along with the multitude of explanations has caused many heads to spin, we are optimistic that our colleagues and students will continue to make significant progress in understanding the true nature and ultimate causation of one of science's most intriguing and important patterns.

A META-ANALYSIS AND DECONSTRUCTION OF THE LATITUDINAL GRADIENT. Among the most recent and promising new directions in developing a more thorough understanding of this pattern and its general causes are rigorous statistical analyses across many taxa, and analyses that strategically "deconstruct" complex patterns into those exhibited by functionally different groups of species. Using the former of these approaches, Hillebrand (2004) reviewed nearly 581 studies of latitudinal gradients to statistically assess the generality of this pattern, and to explore how the pattern differed among species groups, types of ecosystems, and geographic regions. He applied the methods of meta-analysis (Gurevitch and Hedges 1993; Rosenberg 2000), which is a rigorous approach for analyzing the results of a collection of different studies and testing for the statistical relationships among variables that influence those relationships. Hillebrand's data included latitudinal gradients for a very diverse assortment of species and environments, including those varying in thermoregulatory strategies (440 ectotherms; 139 endotherms), body mass (from < 1 g to those > 50 kg), dispersal type (including 170 flyers, 119 with pelagic larvae, 73 seed dispersed, and 68 passive dispersers), trophic level (280 omnivores, 87 autotrophs, 65 herbivores, 58 carnivores, 34 suspension dwelling species, 41 microbivores, and 16 parasites), realm (305 terrestrial, 204 marine, and 69 freshwater), habitat (16 types), hemisphere (335 Northern, 180 Southern, and 64 both), and scale at which richness was measured (349 regional, and 222 local).

Despite the impressive heterogeneity of these data, Hillebrand's meta-analysis confirmed the generality of the pattern in the combined data set and in an overwhelming majority of the gradients studied. Just as important, he demonstrated that the strength (correlation coefficient) and the slope of the gradient varied in a systematic manner among the types of species and habitats studied. That is, *the* pattern is actually many patterns, with the same general signal, but varying to different degrees depending on the scale of the study, characteristics of the species group, and type of habitat or realm considered. For example, the strength and slope of the latitudinal gradients were greater when richness was measured at the regional scale, and for species groups comprised of larger species such as endotherms versus ectotherms, those at higher trophic levels, and for those from marine and terrestrial habitats versus those from freshwater ecosystems. Latitudinal gradients also differed significantly between the continents and between habitat types, but did not differ with latitudinal range covered or between hemispheres (indicating a general symmetry of gradients north and south of the equator).

Rather than just being viewed as the confounding complexity of this large and heterogeneous data set, these differences among functional groups may provide important clues into the causal mechanisms involved. In fact, Hillebrand's analysis is an exemplary case study in *deconstructing* a pattern to explore the underlying causal mechanisms for that pattern. According to Huston (1994:2), "… biological diversity can be broken down into components that have consistent and understandable behavior … [and] … various components of biological diversity are influenced by different processes, to the extent that one component may increase, while another decreases in response to the same change in conditions." The strategy and heuristic value of deconstructing general patterns is not new. In fact, Thorson's rule, describing different latitudinal gradients in marine invertebrates with direct versus indirect development (see Figure 15.18), is one of the earliest and clearest examples of deconstructing a complex pattern into different functional components (in this case, suggesting that latitudinal gradients are influenced by dispersal and its effects on isolation and speciation of marine lineages). Similarly, gradients in species diversity differ between mammals of different body size and feeding guilds (Andrews and O'Brien 2000), for carnivorous versus noncarnivorous marine gastropods (Figure 15.37), for marine bivalves that burrow into substratum versus those that live along its surface (infaunal vs. epifaunal species [Roy et al. 2004]), and for groups of plants with different growth forms (Bhattarai and Vetaas 2003).

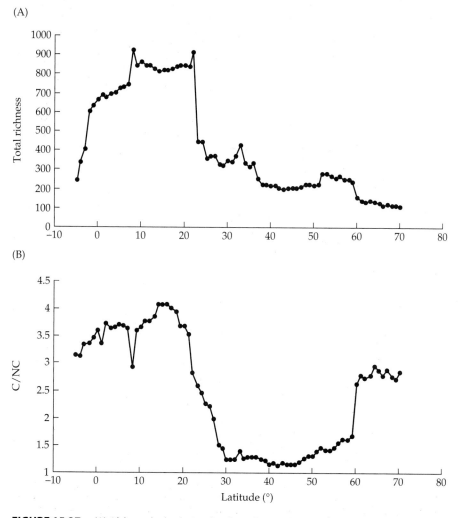

FIGURE 15.37 (A) Although the latitudinal gradient of species richness of marine gastropods in the Pacific Ocean follows the general trend, the pattern differs substantially for species with different feeding strategies. (B) In particular, the relative diversity of noncarnivores (measured as the ratio of carnivores to noncarnivores [C/NC]) is highest in the mid-latitudes, while that of carnivores is highest in the tropics and, to a lesser degree, in the waters of the higher latitudes. (From Roy et al. 2004; after Valentine et al. 2002.)

Marquet et al. (2004:192) use the term **deconstruction** in its etymological sense, meaning "a 'turning to the roots' of what is being measured, and to 'disaggregate' in order to make apparent what is hidden." They go on to list four reasons for deconstructing patterns in species richness:

1. To understand the underlying, causal mechanism
2. To reconcile seemingly disparate explanations for general patterns in diversity
3. To emphasize the importance of overcoming methodological challenges of analyzing heterogeneous data sets
4. To restate complex questions of biological diversity in a more tractable, comparative framework

One key feature of this approach is to strategically separate (deconstruct) heterogeneous data sets into smaller subsets that are internally homogeneous, but differ among each other with respect to putative causal factors and their associated processes. Rather than simply explore the pattern(s), subgroups are chosen based on the hypothesized, underlying causal explanations, selecting groups that differ markedly in their responses to the hypotheses' central factors or processes (see Roy et al. 2000). For example, to evaluate a hypothesis that is based on energy use, subgroups should be comprised of species that are similar in their trophic and energetic characteristics, with different subgroups representing disparate strategies for acquiring or possessing energy (e.g., endotherms vs. ectotherms, carnivores vs. herbivores, or large vs. small species). On the other hand, if we wish to test a hypothesis that latitudinal gradients are ultimately driven by differences in temperature, solar energy, or precipitation, then gradients for terrestrial and shallow water organisms might be compared to those inhabiting the abyssal depths, which are invariably cold, dark, and certainly not water limited. In still other cases, we can compare latitudinal gradients for the same species groups, but across regions with different environmental characteristics (e.g., those with very different climates, currents, and sea surface temperatures; see Figure 15.38) or different histories (geological age).

The foundation of this approach is the accumulated knowledge of centuries of studies by countless and dedicated naturalists, ecologists, and biogeographers. Modern biogeographers, however, have advanced the comparative method to where it is now a statistically rigorous and insightful tool—one holding, perhaps unrivaled potential for understanding the geography and diversity of the natural world.

LATITUDINAL GRADIENTS IN OTHER MEASURES OF DIVERSITY. As a final coda to this discussion of latitudinal gradients in species diversity, we must admit that we have been rather myopic in our discussion to date, focusing on just one—albeit a very important and, intuitively appealing, measure of diversity—species richness (or species density). But, as we indicated earlier, biological diversity is much more than this, as it includes the sum total of variation among life forms in all of their characteristics (genetic, physiological, morphological, and behavioral) as well as variation in all biological, ecological, and biogeographic entities and processes, and across all scales of the biosphere (from organelles and cells to biomes and biogeographic regions). Given that biogeographic studies often analyze combinations of communities, studies of gradients in species diversity may include not just measures of alpha and gamma diversity (i.e., richness and evenness of local to regional communities), but measures of beta and delta (between habitat/region) diversity as well. A comprehensive overview of all of

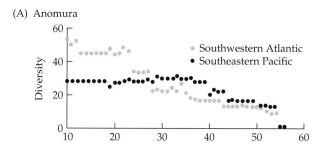

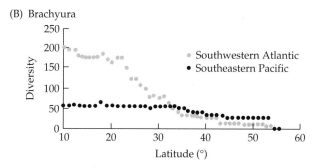

FIGURE 15.38 Latitudinal gradients in diversity of crustaceans differ markedly (A) among species groups (anomurans [false crabs and hermit crabs] and brachyurans [true crabs]), and (B) among geographic regions (southwestern Atlantic and southeastern Pacific). (From Marquet et al. 2004; after Astorga et al. 2003.)

these levels of variation would be far beyond our abilities to cogently distill the available information, so we offer here examples at just two scales: (1) measures of community diversity (i.e., other than species richness), and (2) morphological diversity with, and among, biotas.

Biogeographers of both the terrestrial and marine realms have begun to investigate geographic patterns in beta diversity. The conceptual basis for such studies was articulated by Robert MacArthur in 1965. He argued that evolution and diversification in the tropics acted across habitats, with new species being added to different habitats rather than packing them more tightly into one habitat. Thus, changes in diversity of an area or a latitudinal band comprised of a collection of habitats (i.e., gamma diversity) would be due to difference in richness among communities (i.e., beta diversity). Put another way, although localized or within-habitat diversity may vary little across latitudinal gradients, beta diversity should vary and be highest in the tropics. The difficulty is that most studies of latitudinal gradients are conducted at the gamma scale, combining results from a complexity of adjacent habitats, thus rendering MacArthur's hypothesis untestable. Fortunately, sufficient fine scale data is now available to test this hypothesis, at least for some systems. Clarke and Lidgard's (2000) study of North American bryozoans is an interesting case in point. Their studies indicate that beta-diversity is, indeed, correlated with latitude, peaking between 10 ° and 30° N and then steadily declining in the higher latitudes.

As we reported earlier, bats are one of the many groups of terrestrial organisms that exhibit latitudinal gradients in species richness (see Figure 15.31). Stevens and Willig (2002) analyzed latitudinal gradients in 14 measures of community diversity at both local and regional scales. Local species richness (alpha diversity) increased and became more variable along a gradient from the poles to the equator, whereas evenness of populations did not vary in any systematic manner along the same gradient (Figure 15.39). Latitudinal clines in gamma diversity (regional scale richness) were substantially steeper that those of alpha diversity, indicating that differences in species composition among local communities (i.e., beta diversity) is higher in the tropics (see Fig-

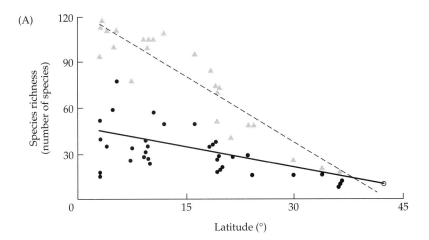

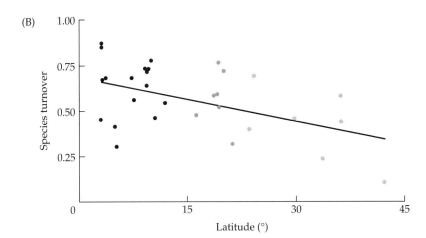

FIGURE 15.39 The latitudinal gradient in diversity is actually many gradients, each differing depending on the particular taxon, region, scale, and measure of diversity employed. (A) Regional species richness (gamma diversity: triangles and dashed line) of New World bats declines much more rapidly than local species richness (alpha diversity: circles and solid line), indicating that beta diversity (turnover in species composition among local communities) is highest in the tropics (B). These graphs also reveal that, across the tropical latitudes (23.5° S to 23.5° N), diversity tends to be high, but also highly variable and without an obvious latitudinal cline. Thus, the global-scale gradient in diversity of New World bats results primarily from differences between tropical communities and those of subtropical and temperate regions. (C) On the other hand, evenness of populations comprising these communities (Shannon's index of evenness) does not vary in any regular manner with latitude. (After Stevens and Willig 2002.)

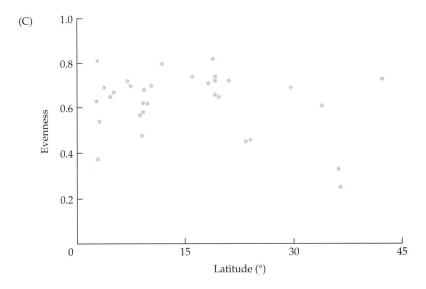

ure 15.39). Overall, Stevens and Willig's study indicates that the dramatic increase in gamma diversity toward the tropics is equally influenced by ecological and evolutionary processes which determine differentiation among local communities.

Latitudinal gradients in species richness of marine invertebrates were reported over three decades ago (Stehli 1968; Rex et al. 1993). Only recently, however, have marine biogeographers and ecologists begun to analyze gradients in other measures of diversity. Rex et al.'s (2000) reanalysis of diversity patterns of epibenthic invertebrates (including bivalves, gastropods, and isopods) from 37° S to 77° N indicates that species richness of the three groups combined, was strongly correlated with latitude. The correlation between latitude and evenness, however, was much steeper, in general, but varied among the three groups (being relatively strong for the isopods, weak for the gastropods, and not significant at all for the bivalves [see also Rex 2004]).

While these scientists and their colleagues will continue to investigate geographic variation in evenness, beta diversity, and related measures of community diversity, there are many other features of biological diversity that may vary, albeit in different and informative fashions, along geographic gradients. We strongly encourage our colleagues and their students to study other features of biological diversity, including geographic gradients in genetic, behavioral, physiological, and morphological measures of diversity.

In a groundbreaking study, Shepherd (1998) examined geographic clines in morphological diversity in 237 species of North American mammals. After accounting for the strong latitudinal gradient in species number, she found that, whereas diversity of body sizes increased with latitude, diversity in body shapes decreased. In their more recent studies of morphological diversity in the marine realm, Roy and his colleagues (Roy et al 2001; Roy et al. 2004) used principal components analysis to calculate a standardized measure of morphological diversity in strombid gastropods (conches and related species). As illustrated in the map of Figure 15.40, morphological diversity of these species exhibits some interesting geographic clines with both latitude and longitude, trends that are distinctly different from those of species richness.

FIGURE 15.40 Morphological diversity of strombid gastropods (conches and related species) exhibits a complex, but highly nonrandom pattern of geographic variation across gradients of both latitude and longitude (morphological diversity was measured using principal components analyses of body sizes and shapes of these species). (From Roy et al. 2004; after Roy et al. 2001.)

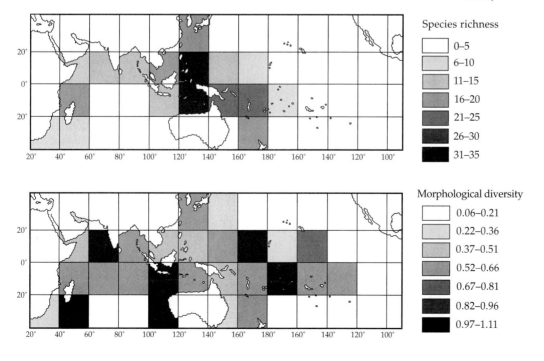

While our colleagues have only begun to compare geographic gradients in these and other alternative measures of biological diversity, we predict that Roy et al.'s results may be typical: many, if not most of the measures (taxonomic, genetic, ecological, functional, or morphological) of diversity will exhibit significant geographic clines, but the trends will differ among measures perhaps as much, if not more, than they differ among taxa (see Stevens et al. 2003). Again, comparative methods, and pattern deconstruction in particular (see the discussion on pattern deconstruction earlier in this chaper), will no doubt be key to deciphering this fascinating variety of diversity gradients and ultimately yielding a much more comprehensive understanding of the processes influencing the geography and complexity of nature.

Other Geographic Gradients in Species Richness

PENINSULAS. Maps illustrating diversity isoclines (see Figures 15.27, 15.29, 15.33, and 15.35) reveal the influence of geographic features in addition to latitude; in particular, the influence of proximity to coastlines and peninsulas and that of elevational variation along mountain chains. In the next section, we discuss patterns of variation in species richness along elevational gradients, while here, we consider the **peninsular effect**: the tendency for species richness to decrease from the continental interior toward the terminus of a peninsula (Figures 15.41 and 15.42). This pattern seems relatively general, being reported for plants, invertebrates, and vertebrates along a variety of peninsulas including those of Florida and Baha California in the Nearctic; the Iberian, Balkan, Scandinavian, Italian, and Korean Peninsulas in the Palearctic; and the Cape York Peninsula in Australia (Cook 1969; Taylor and Regal 1978; Seib 1980; Bussack and Hedges 1984; Due and Polis 1986; Brown 1987; Means and Simberloff 1987; Brown and Opler 1990; Martin and Gurrea 1990; Barbosa and Benzal 1996; Contoli 2000; Baquero and Telleria 2001; Johnson and Ward 2002; Choi 2004; but see Schwartz 1988). As Figures 15.41 and 15.42 indicate, the pat-

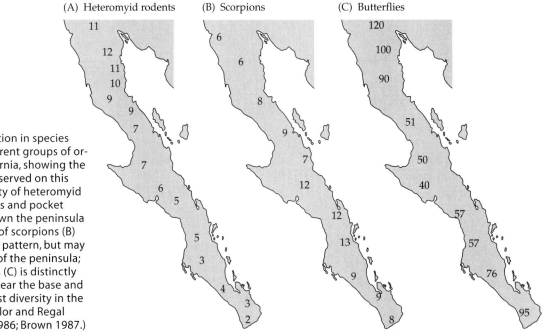

FIGURE 15.41 Variation in species richness of three different groups of organisms in Baja California, showing the variety of patterns observed on this peninsula. The diversity of heteromyid rodents (kangaroo rats and pocket mice, A) decreases down the peninsula from base to tip; that of scorpions (B) does not show a clear pattern, but may peak near the center of the peninsula; and that of butterflies (C) is distinctly bimodal, with peaks near the base and tip and with the lowest diversity in the center. (Data from Taylor and Regal 1978; Due and Polis 1986; Brown 1987.)

(A) Europe

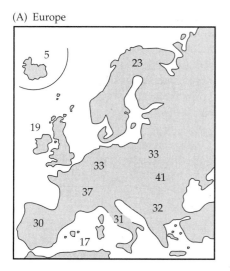

(B) Italy

FIGURE 15.42 Geographic variation in species richness of (A) native, nonvolant mammals in Europe and (B) rodents of Italy, clearly illustrates the influence of isolation on true islands and along peninsulas where increased isolation combines with the influences of maritime climates and reduced area and habitat diversity to limit diversity of these species. (A after Baquero and Telleria 2001; B after Contoli 2000.)

terns do differ somewhat among species groups and among regions. For the most part, however, richness within the peninsula tends to be substantially less than that of the continental interior and, in nearly all cases, richness is lowest at the terminus of the peninsula. It is also important to note that, for those peninsulas extending toward the lower latitudes (e.g., Cape York, Baha California, Florida, Italy, and Korea), the gradient in species richness actually runs counter to the otherwise very general latitudinal clines in richness (i.e., richness along these peninsulas decreases as we move closer to the equator).

Explanations for the peninsular effect are varied, but nearly all center on the island-like nature of peninsulas—especially their isolation, limited area, maritime climates, and limited habitat diversity. As we discussed in Chapter 13, relatively low species richness is expected for biotas inhabiting isolated and relatively small ecosystems because immigration rates will be relatively low, while extinction rates should be relatively high (Simpson 1964; MacArthur and Wilson 1967). Area affects not just extinction rates but also influences total amount of available resources, diversity of habitats, and the likelihood that the peninsula (or island) will contain topographic barriers that would promote allopatric speciation (e.g., mountain chains and large rivers; Taylor and Regal 1978; Lawlor 1983; Due and Polis 1986; Milne and Forman 1986; Brown 1987; Means and Simberloff 1987). That is, if the peninsula is sufficiently large and not subject to frequent catastrophes, then its isolation may favor evolutionary diversification and formation of peninsula endemics. In fact, endemicity is relatively common for biotas of large peninsulas, especially for lineages comprised of species with limited dispersal abilities (e.g., scorpions of Baha California, Due and Polis 1986; nonvolant mammals of European peninsulas, Baquero and Telleria 2001; ants of Baha California, Johnson and Ward 2002). Therefore, cases where biotas exhibit mid-peninsular peaks in richness (e.g., Figure 15.41C) likely result from the combined actions of two opposing gradients: (1) the general effects of peninsula size, isolation, and habitat diversity described above, and (2) the opposing effects of either those factors associated with latitudinal clines in diversity or those factors promoting the formation of peninsular endemics. We can extend this line of reasoning one step further by viewing the world's major isthmuses as conjoined peninsulas. As Janzen (1967) and others have noted, "the isolation provided by the peninsular situation" of isthmuses may lead "to differentiation and ultimately, speciation," promoting relatively high concentrations of endemics in these regions (e.g., the Isthmus of Panama; see Watson and Peterson 1999).

These intriguing complexities once again highlight the heuristic value of strategically exploring patterns, not just in species richness, but in composition and functional characteristics of component species. Each of the alternative hypotheses is based on the potential influence of geometry or related geographic factors on immigration, extinction, or speciation. Given that species vary in their abilities to disperse to, survive in, or eventually diversify in such environments, comparisons among species or functionally different groups of species represent powerful means of discriminating among alternative causal explanations for these patterns.

ELEVATION. Although the latitudinal gradient in species richness is one of biogeography's most studied patterns, it is probably not its most ancient one. As we stated in the first paragraph of this book, "even the earliest human societies were aware that as they expanded their search area they would encounter a greater number and greater diversity of plants and animals. Those societies living in mountainous regions could see that vegetation changes in an orderly manner as they moved from the lowlands to the summit." In fact, it was likely impossible for them not to see the striking transitions from deserts, woodlands, and forests to treeless zones and ice covered peaks—all of this knowledge vital to their survival and all of this clear from one vista or experienced first hand along an ascent of just a few hours (Lomolino 2001; see also Brown 2001).

Thus, knowledge of environmental variation along mountain slopes long predates the origins of biogeography. Accordingly, some of the earliest explanations for the origins of life on Earth were based on what we now call elevational gradients in biotic communities. As we discussed in Chapter 2, Linnaeus (1781) hypothesized that at the initial time of creation and during the biblical flood, all life forms survived in a montane region. There, along its slopes, could be found a compressed and very orderly succession of the Earth's biomes that could, he hypothesized, accommodate all of its species. Mountains and elevational gradients in climate and biotas also played a central role in Willdenow's (1805) explanation for the origins and differences among regional biotas: each surviving in different mountainous regions during periods of high sea level, and then descending down slopes, along with their habitats, as the waters receded. Willdenow also hypothesized that geography, climate, energy, and species diversity are interrelated:

> "Soil, situation, cold, heat, drought, and great moisture are all of powerful influence on vegetation ... [and] ... the warmer the climate, the greater must be the number of growing plants." (Willdenow 1805:348, 349.)

Research on montane biotas continued to play a prominent role in the development of biogeography, from Alexander von Humboldt's (1805) legion of detailed studies along Mount Chimborazo in the Ecuadorian Andes during the 1800s, to those of Charles Darwin (1839, 1859) in the Chilean Andes; from the accounts of Alfred Russel Wallace in Indonesia (1859, 1876), Joseph Dalton Hooker (1877), Asa Gray (1878), and C. H. Merriam (1890) in the North American Rockies during mid- to later decades of the nineteenth century, to those of Robert H. Whittaker (1960; Whittaker and Niering 1965) in the Santa Catalina, Great Smoky, and Siskiyou Mountains (southeastern and western North

America), and James H. Brown's studies of montane islands of the Great Basin region of North America during this past century (Brown 1971b, 1978).

Each of these distinguished scientists capitalized on the great heuristic value and natural experiments offered by elevation gradients along mountain slopes. For much of the early history of research on montane biogeography, they detailed the geographic variation in abiotic variables and discovered what appeared to be very general patterns in variation among biological communities along elevational transects. Originally, it appeared that diversity of these communities decreased monotonically from the lowest to the highest elevations (Whittaker 1960, 1977; Yoda 1967; Kikkawa and Williams 1971; Terborgh 1977; Heaney 1991; Heaney et al. 1989; Daniels 1992; Stenthouraskis 1992; Fernandes and Lara 1993; Patterson et al. 1996, 1998). Unfortunately, the apparent patterns may have been influenced to some degree by preferentially sampling some elevations (especially those along the foot hills or initial slopes) more than others, and by limiting their surveys to transects along mountain chains, but ignoring the lower elevations (e.g., the coastal plains).

As a result of more rigorous and more extensive surveys during the past two decades, a somewhat different pattern is now emerging. Biogeographic surveys of a variety plants, invertebrates, and vertebrates—especially those conducted along transects extending from sea level to mountain summits—indicate a unimodal, or hump-shaped, pattern (Whittaker 1960, 1977; Whittaker and Niering 1975; Dressler 1981; Brown 1988; Rosenzweif, 1992, 1995; Rahbek 1995, 1997; Fleishman et al. 1998; Hawkins 1999; Heaney 2001; Nor, S. Md. 2001; Rickart 2001; Sanchez-Cordero 2001; Sanders 2002; Sanders et al. 2003; Bhattarai et al. 2004; Lee et al. 2004). That is, species richness of most groups of terrestrial organisms increases gradually along the coastal plains, increasing more rapidly to peak as we ascend to the foothills or mid-elevations of mountains, and then declining again as we approach their ice-covered summits (Figures 15.43 and 15.44).

As you can imagine, over the long history of research on this pattern, there have been many hypothesized, causal explanations. The ultimate explanation for this pattern should be based on responses of species to the complex, but predictable, variation in the geographic template along elevational transects. As we described in Chapter 5, species typically respond to geographic clines in environmental conditions in an individualistic fashion, with very few species sharing precise ecological requirements or limits to their geographic ranges. The overlap among those ranges, however, exhibits some very general patterns along elevational gradients. This overlap, of course, is species richness.

Although some earlier explanations for elevational gradients in diversity were based on one overriding factor, it is clear that many factors and processes are involved. That is, species are responding to a combination of factors that vary in a highly regular manner as we ascend elevational gradients: some of these promoting immigration, survival, and speciation, and others having the opposite effect. For example, as elevation increases, air temperatures increase and air pressure (along with partial pressures of important gases including oxygen and carbon-dioxide) decrease. Because cooler air cannot hold as much moisture as warmer air, precipitation often increases as we move up a mountain slope, although at the highest elevations the moisture is frozen and generally unavailable to most organisms. Thus, the optimal combination of temperature, gases, and precipitation for most plants and other organisms most likely occurs at intermediate elevations.

It is also likely that the tight juxtaposition and concentration of climatic zones along elevational gradients contributes to the relatively high gamma diversity of these mountainous regions. Unlike the similar series of biomes

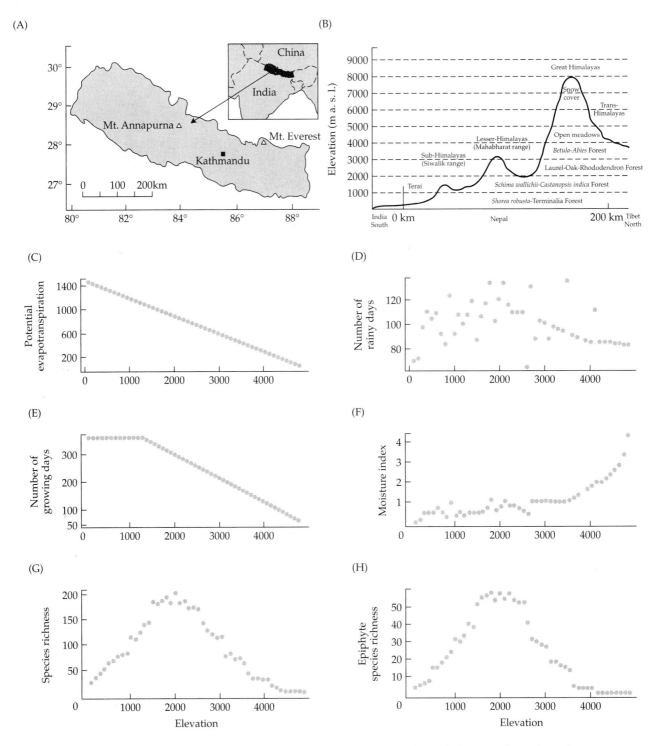

FIGURE 15.43 (A, B) Along a gradient through the central Himalayas from 100 to 4800 m above sea level, climatic conditions change in a predictable, albeit complex, fashion. The result is that the most favorable combination of environmental conditions for ferns (C–F) occurs at intermediate elevations, and species richness of ferns (G, all species combined; H, epiphytic species) peaks at approximately 2000 m above sea level. (After Bhattarai et al. 2004.)

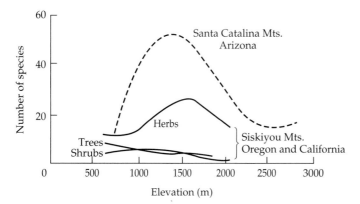

FIGURE 15.44 Species-elevation patterns vary among species groups and mountain chains, but most elevational gradients tend to exhibit a peak in richness at some intermediate elevation. Note that these surveys were not conducted below 500 m, so it is difficult to determine whether trees of the Siskiyou Mountains also exhibit a unimodal, or hump-shaped, cline in richness. (Data from Whittaker 1960; Whittaker and Niering 1965.)

that are encountered along extensive latitudinal gradients, geographic ranges of ecologically distinct species are in very close proximity and, therefore, more likely to overlap—or "spillover"— into other communities along the slopes of mountains (see Schmida and Wilson 1985; Pulliam 1988; Grytnes 2002, 2003). The tendency for elevational transects to reveal multiple peaks in species richness, each corresponding with an ecotone between principal climatic zones or habitats, seems entirely consistent with this explanation (Figure 15.45). As we approach the highest elevations, however, diversity should decline because of the increasingly harsh climates and the paucity, or absence, of habitats at higher elevations.

In addition to these climatic variables and their influences on montane habitats and landscapes, physiographic features also vary along elevational transects. Given the roughly conical shape of mountains, the area of any climatic or elevational zone actually decreases as we move from the lowlands to summit. In addition, communities at the higher elevations are more isolated, both from species assemblages occupying the lowlands and from those occurring in similar alpine communities, but in other mountains. Thus, as Robert MacArthur and other biogeographers noted, montane ecosystems can be viewed as archipelagoes comprised of ecosystems that vary in area and isolation—both of these factors affecting the rates of extinction and immigration, respectively. Based on this montane-island model alone, we would thus predict that species richness should be highest at the lowest elevations because lowland ecosystems occupy the largest area and are the least isolated. However, immigrations and extinctions are also influenced by the climatic factors

FIGURE 15.45 (A) Biotic interchange between adjacent communities along an elevational gradient may create a series of repeated minor peaks in species density at transition zones between those communities. The maximum peak should be located at the transition zone between the two most diverse communities. In this hypothetical example, zonal communities are labeled *a* through *e*, and illustrated by alternating bands of dark grey and white (transition zones in light grey). (B) The empirical pattern of floristic richness (number of species in each 100 m elevation belt) along the slopes of Mt. Hermon in Israel is consistent with this graphical model. (A after Lomolino 2001; B after Schmida and Wilson 1985.)

(A)

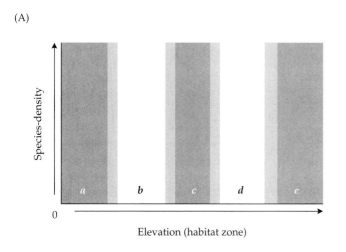

(B)

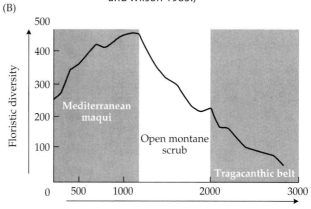

discussed above and, perhaps just as important, area and isolation also influence the other fundamental process—speciation. As we discussed in Chapters 13 and 14, speciation rates and endemicity increase with both area and isolation. Because these two physiographic features change in opposite directions along elevational gradients, however, speciation rates and richness of montane endemics should peak at some intermediate elevation (e.g., see Heaney 2001; Vetaas and Grytnes 2002; Bhattarai et al. 2004).

Again, the overall effect of geographic variation in all of these important environmental factors and their influence on immigration, speciation, and extinction is that species richness peaks at intermediate elevations along a montane transect.

DEPTH AND DIVERSITY IN AQUATIC ENVIRONMENTS. The marine realm is clearly one of the last frontiers for biogeographers, as well as for most other biologists. Until quite recently, the immense pressures, extreme cold, and complete darkness of deepwater environments limited most of our studies to the top few meters of the oceans. Fortunately, with advances in direct- and remote-controlled sampling techniques, we have been able to explore deeper into the oceans and begin to develop a better understanding of the geography of nature across all three dimensions of the marine realm. As we observed earlier, many of the geographic patterns of species richness found in the oceans are remarkably similar to those found on land. For example, like small islands, small lakes and seas contain fewer species than large ones, and the biotas of more-isolated bodies of water tend to be more distinct than those of less-isolated ones. Freshwater and marine communities exhibit latitudinal gradients in species richness quite similar to those exhibited by terrestrial biotas.

In addition, patterns of variation in species richness along gradients of depth in the marine realm appear analogous to those along gradients of elevation on land. For most marine macroinvertebrates and fish, species richness exhibits a unimodal, or hump-shaped, cline with depth, increasing as we descend below sea level and peaking somewhere along the continental shelf, then declining along the continental slope toward the abyssal plain (Figures 15.46 and 15.47; Vinogradova 1962; Sanders 1968; Slobodkin and Sanders1969; Rex 1981; MacPherson and Duarte 1994; Gappa 2000; Smith and Brown 2002). Freshwater communities of relatively deep lakes exhibit similar clines in species diversity with increasing depth. Some large lakes, including the Great Lakes and the Finger Lakes in northeastern North America, Lake Baikal in Siberia, and several of the large lakes of central Africa, are more than 100 m deep, but few organisms are found in the deepest waters. In Chapter 7, we described the great diversity of fishes, especially in the cichlids, that inhabit the Great Lakes of Africa's Rift Valley. Although Lake Malawi and Lake Tanganyika are 704 and 1470 m deep, respectively, their hundreds of species of fish are most diverse in the surface waters, while none are found below a depth of about 200 m. Their lower depths apparently are inhabited by only a few kinds of anaerobic invertebrates and bacteria.

The precise pattern and depth of peak diversity, of course, varies substantially depending on the particular species group or region studied. The general pattern along with its explanations, however, is qualitatively similar to those developed to explain diversity gradients on land. Just as environmental and physiographic factors (e.g., area, isolation, and slope) vary in a complex but predictable manner along gradients of latitude and elevation, light, nutrients, pressure, temperature, currents, and a seemingly endless list of chemicals, vary as we descend from sea level to the deepest reaches of the oceans. Again, the gradient is bracketed by two harsh boundaries—in this case, the

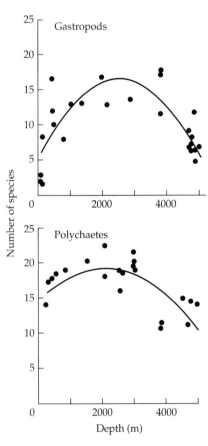

FIGURE 15.46 Variation in species richness with depth in two groups of benthic marine organisms: gastropod mollusks (above) and polychaete worms (below). These hump-shaped relationships suggest that diversity is low in shallow coastal waters, highest in the heterogeneous terrain of the continental slope, and low in the abyssal depths. This is probably the typical pattern for samples of small areas, but recent data suggest that the abyssal plain may have high diversity at larger scales. (After Rex 1981.)

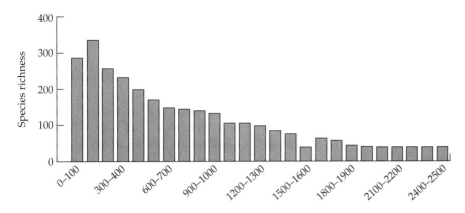

FIGURE 15.47 Species richness of pelagic marine fish in the northeast Pacific Ocean (40° to 50° N) shows a general tendency to decrease with depth, but peaks at around 200 m below sea level (although not shown here, species richness continues to decline at depths below 2500 m). The somewhat lower species richness, between 0 and 200 m below the surface, may result from several factors including more intense predation (due to higher visibility), relatively frequent physical disturbance (due to wave actions and storms), and highly variable temperatures and water chemistry of shallow environments. (After Smith and Brown 2002.)

terrestrial environment above, and the extreme pressures, darkness, and cold of the abyssal plains, below. Surface waters may at first seem to be favorable to many species, given their relatively high temperatures and insolation, but they also tend to be most subject to frequent and, sometimes intense, disturbance including erosion and rapid fluxes in temperature and water chemistry. In addition, a quick look at a bathymetric map reveals that deep water environments cover most of the Earth's surface (i.e., area, which tends to be positively correlated with species diversity, increases with depth in the marine realm). Thus, because of the complex and sometimes opposing clines of factors influencing species diversity, optimal conditions for most species occur somewhere between sea level and the continental slopes.

A Synthesis across Three Gradients: Latitude, Elevation, and Depth

We have considered each of three major geographic gradients in diversity separately, but they all represent similar responses of biotas to nonrandom variation in the geographic template. Because species compete with, prey on, and parasitize each other, or serve as mutualists and commensals for other species, they further modify and redefine each other's niche and, over time, they differentiate and diversify. One ultimate effect of all of this is that there is not one optimal niche or geographic distribution. Instead, the geographic ranges of biotas have diversified, just like the species themselves. Then again, while the overlap of their geographic ranges has never been perfect, it is far from random. After all, living organisms, especially those in well-defined taxonomic or ecological assemblages, share many resources and limiting factors. As we move across any geographic gradient, organisms will encounter a limited number of environments where the combination of abiotic factors is optimal for these species. Again, interspecific interactions will prevent all species from converging on a narrowly restricted niche, but species ranges should overlap most near these generally optimal sites.

We believe that it may be especially instructive—albeit unconventional—to view all three of the principal clines in species diversity discussed above (i.e., those along gradients of elevation, depth, and latitude) as three forms of the same phenomenon. The degree of overlap in species' ranges is, of course, what we have been referring to as species richness. For the same reasons discussed above, species richness should be relatively high near locations where the combination of environmental conditions are less extreme. Given the very general nature of spatial autocorrelation (including the tendency for environmental similarity to decrease with increasing distance between sites), as we move away from these locations toward either end of the gradient, conditions should become increasingly more harsh and, therefore, species richness should decline.

The patterns now emerging from research on gradients in species diversity with both elevation and depth reveal that species richness peaks in environments located intermediate between the ends of each gradient (i.e., between sea level and extremely cold, low pressure, ice-covered peaks of mountains; or the anoxic, dark, cold, and extreme high pressure of the abyssal depths). In both cases, intermediate does not mean "midway," and it is clear that, in both realms, diversity peaks closer to sea level than to the highest summits or the abyssal depths (i.e., for most species assemblages, species richness typically peaks along the foothills and initial slopes of mountain ranges in the terrestrial realm, and along the continental slopes in the marine realm).

We also believe that the latitudinal gradient in species richness can be viewed as another version of this same phenomenon—diversity peaking at some location intermediate between two relatively harsh and inhospitable ends of a continuum; in this case, the polar environments. The precise latitudes where conditions are optimal for a particular assemblage of species will certainly vary with the characteristics of those species; but once more, it will be located intermediate between the two extremes and, for most species groups, in the tropics. Again, if not for the tendency for interspecific interactions to promote diversification, the overlap among ranges would be much greater, and few, if any, diversity curves would peak outside the tropics; i.e., beyond where the combination of light, heat, temperature, precipitation, and other environmental factors appears to be optimal for most species.

This very general explanation may seem similar in some respects to Colwell and Hurt's mid-domain model, but their explanation was based on completely random placement of species' ranges. It should, however, be very clear from the above discussion and the many papers referenced in the previous sections, that causal explanations summarized here are almost entirely deterministic. That is, the causal mechanism is multifactorial and complex, but based on predictable responses of organisms to very regular patterns of variation in resources and limiting factors across the geographic template. Ultimately, these in some way influence immigrations, extinctions, and speciation and, therefore, species distributions and geographic variations in biological diversity.

One final coda to this chapter, and a prelude to the next: as general and ancient as these gradients in species diversity are, they are likely to change—perhaps markedly—during the next few centuries. In fact, it appears that these clines may have already been modified by various activities of human societies. Initially, our ancestral populations were largely restricted to lowland environments of the tropics. Even though they overlapped with the tropical hotspots of biodiversity, it wasn't until recently that our societies advanced to the point that they have become unrivaled ecosystem engineers and begun to attenuate the peak in the latitudinal diversity gradient.

During the past one hundred thousand years, human societies have spread across the lowlands of most of the terrestrial realm, but it wasn't until very recently that our abilities to modify native environments pressed up to higher elevations. Various anthropogenic threats, including fragmentation, habitat loss, pollution, and invasive species, have risen above the foothills and now threaten the diversity hotspots of subtropical and temperate mountain chains (see Lomolino 2001; Lee et al. 2004). Finally, the lessons of countless environmental studies over the past few decades have made it undeniably clear that anthropogenic threats (including pollution, introduced species, and diseases) are reaching further down into the marine realm to threaten many of its hotspots of diversity.

We believe, therefore, that this is a critical time and one when biogeography, in particular, can provide invaluable insights not just for understanding patterns in biological diversity, but for conserving them as well.

Conservation Biogeography and New Frontiers

CHAPTER 16

Biodiversity and the Geography of Extinctions

BIODIVERSITY IS, IN THE SIMPLEST TERMS, the variety of life. It encompasses the variation among species or other biological elements, including alleles and gene complexes, populations, guilds, communities, ecosystems, landscapes, and biogeographic regions. Biodiversity can be expressed as the variation within a given location, or the variation among elements across geographic units. This variation can include the number of different types of species or elements, their relative frequencies, the degree of variation among these elements, or variation in key processes such as dispersal, gene flow, interspecific interactions, or ecological succession.

Given such a broad and inclusive concept, our hope of conserving biodiversity may at first seem like an impossible dream. Yet, in more pragmatic terms, conserving biodiversity distills down to one simple, yet still challenging, goal: maintaining diversity of species. As Aldo Leopold put it, "The first requisite of intelligent tinkering is to save all the pieces." By "pieces," of course, he meant native species. With the loss of a species, biodiversity is diminished at all levels, from genetic and local scales to biogeographic and global ones. With each species extinction, associated ecological and evolutionary phenomena such as mass migrations of birds and butterflies, pollination of Hawaii's endemic plants by now-extinct honeycreepers, and a myriad of other interspecific interactions are also lost. Charles Elton's (1958) views were similar to Leopold's, but with a more explicit emphasis on the geographic dimensions of what we now term biodiversity. In his discussion of a "wilderness in retreat," Elton stated that "conservation should mean the keeping or putting in the landscape of the greatest possible ecological variety—in the world, every continent or island, and so far as practicable in every district." Given the increased emphasis on the geographic context of biodiversity, we might modify Leopold's initial statement and suggest that the goal of conservation biology is to preserve species distributions—the geography of nature—and, in so doing, preserve the diversity of species populations and the processes (ecological and evolutionary) that will enhance conservation of biodiversity for the long-term (see the discussion on conservation biogeography in Chapter 17).

Our goals in this chapter are three-fold. First, we will summarize the current status of biodiversity. Second, we will review some of the geographic dimensions of biodiversity—patterns we discussed in earlier chapters, especially those in species diversity and endemicity across geographic gradients. In the next section, we will explore the geography of extinction, or the tendency for extinctions to be nonrandomly distributed across the globe, focusing in particular on three phenomena that have profoundly altered the geographic template—species invasions, landscape transformations (including habitat loss and fragmentation), and climate change. In the last section of this chapter, we summarize patterns emerging from an important, but relatively new theme in biogeography and conservation—geographic range collapse.

The Biodiversity Crisis and the Linnaean Shortfall

Few fully informed and objective students of biological diversity question that the current rate loss of biodiversity far exceeds that of purely natural conditions. Leopold, Elton, and other ecologists and evolutionary biologists of this and the past two centuries repeatedly warned of the ongoing anthropogenic losses of native species and ecosystems. The existence of a biodiversity crisis is documented in the record of historical extinctions and the vulnerable and imperiled status of many extant species (Figure 16.1; World Conservation Monitoring Centre 1992; IUCN Redlist of Threatened Species 2003). Since the early 1600s, biologists have tallied extinctions of 129 species of birds and 83 species of mammals, including such wondrous forms as the great auk (*Alca impennis*), dodo (*Raphus cucullatus*), passenger pigeon (*Ectopistes migratorius*), Tasmanian tiger (*Thylacinus cynocephalus*), Cuban soledon (*Soledon cubanis*), and Stellar's seacow (*Hydrodamalis gigas*). Over the same time period, 21 species (including 100 subspecies or geographic races) of reptiles, and 7 species (64 subspecies) of amphibians have vanished, including one amphibian family and 38 genera (most historic extinctions of amphibians have occurred during the last three decades; see Duellman and Trueb 1986).

Aquatic species, including freshwater fishes, crayfishes, and mussels, are also included in this wave of historic extinctions. The rate of extinctions of freshwater fishes in North America has steadily increased over the past 100 years, rising to some 40 species or subspecies lost during the latter decades of the twentieth century (Williams and Miller 1990). In the Great Lakes of East Africa, the diversity of native fishes has plummeted just in the past few decades. In Lake Victoria alone, cichlid diversity has declined by as much as 200 species. Well over ten percent of the freshwater fish of many countries is now threatened, and global extinction rates of these species continue to rise (see Figure 16.1C,D). Worldwide, as much as 20% of all freshwater fish species (about 1800 species) are now extinct or in serious jeopardy.

Loss in diversity of mollusks has been almost equally severe. Since 1600, invertebrate biologists have recorded 191 extinctions of mollusks—including freshwater, marine, and terrestrial forms. The prospects for the near future are just as grave. For example, 7% of the 297 recognized species of mussels in the freshwater systems of the United States are presumed extinct, while another 65% are either endangered, threatened, or candidates for federal protection (Williams and Neves 1995).

This wave of extinctions has by no means been restricted to the animal kingdom. Worldwide, plant diversity has declined by over 600 species during the past 400 years (Smith et al. 1993). In the United States alone, 176 recog-

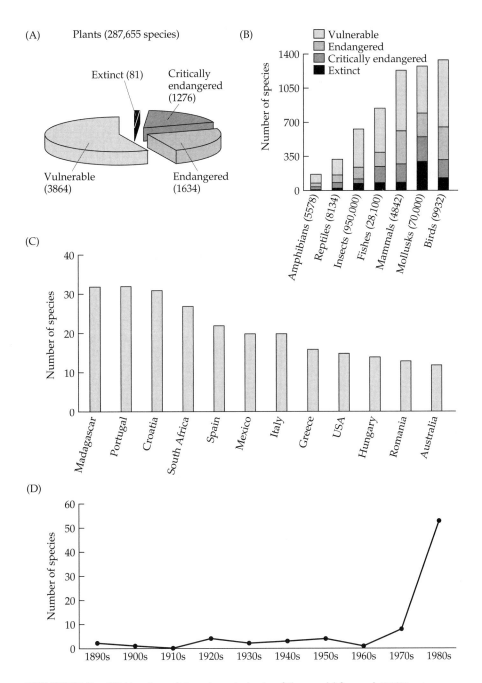

FIGURE 16.1 (A) Number of threatened plants of the world for each IUCN category. (B) Number of threatened species of selected groups of animals of the world for each IUCN category (total number of described species in parentheses). (C) Percent of native freshwater fish species that are threatened, and (D) chronologic pattern of global extinctions in freshwater fish species since 1890. (A–B after IUCN Red List of Threatened Species 2003; C–D after Loh 2000.)

nized species and subspecies are now extinct, while another 2465 species are officially listed as imperiled (Natural Heritage Databases 1995). The status of insular plants is especially grave. On the Hawaiian Islands, some 108 endemic plant taxa are now extinct, and another 175 are listed as endangered or vulnerable. On Saint Helena Island in the Atlantic Ocean, seven endemic plant

---

TABLE 16.1 *The Linnaean shortfall*

---

1. Eleven of the 80 extant cetaceans have been discovered in this century, one in 1991.

2. One of the largest shark species, "megamouth," was discovered in 1976.

3. Three new families of flowering plants were discovered in Mexico during the past decade.

4. Two new *phyla* were discovered during the 1990s, one (Loricefera) in the marine benthos and the other (Cycliophora) clinging to the mouthparts of lobsters.

5. Perhaps 90% of tropical forest insects remain unknown.

6. Almost 1.5 million fungi remain to be discovered.

7. About 4000 bacterial species have been described; a gram of soil may contain 4000 to 5000 species.

(Modified from Raven and Wilson 1992.)

species are extinct, and all of the remaining 46 endemic species of plants are endangered or threatened.

Even the fungi, an often neglected yet diverse and fascinating kingdom, have experienced a marked pulse of extinctions during recent decades. For example, between 1930 and 1990, the species diversity of European fungi dropped by 40% to 50% (Jaenike 1991; see also Webster 1997; Koune 2001; Moore et al. 2002). Today, some 20% of the approximate 8000 species of European fungi are threatened with extinction due to the degradation and loss of their habitats (Dahlberg and Croneborg 2003). The larger fungi (primarily mushrooms and puffballs) are especially vulnerable to both habitat loss and overharvesting, with 35% of their approximate 1000 species being threatened (Cherfas 1991).

As alarming as these statistics may seem, it is likely that they underestimate the actual decline in biodiversity. While scientists have now described over 1.7 million species, this represents only a fraction of the total species thought to occur on Earth. Estimates of total diversity vary, but typically range between 5 and 25 million species (UNEP World Conservation Monitoring Centre 2000). By almost all estimates, most of what is out there is unknown to us. This deficiency, sometimes called the **Linnaean shortfall**, represents not just a pressing challenge but also a glorious opportunity for field biologists. There are many species waiting to be discovered, some of them so distinct that once described they become the sole known representatives of new families, orders, or even phyla (Table 16.1; see also Staley 1997, and Fenchel and Finlay 2004 for reviews on the global diversity and conservation of microbes).

Given current trends in the specialization and training of taxonomists, however, the Linnaean shortfall is likely to remain with us for some time. While a great majority of undiscovered animals are thought to be insects, spiders, and other invertebrates, only 30% of today's taxonomists specialize in these groups (Figure 16.2). The geographic distribution of taxonomists—or, at least, their home institutions—is no more reassuring. Over three-fourths of today's taxonomists are trained in temperate areas of the Nearctic and Palearctic regions—which, as we have seen, are not the world's most species-rich areas. Similarly, a quick review of Table 16.1 reveals that the marine realm is still a great biological frontier. Less than 10% of the world's oceans have been adequately sampled for biological diversity, and hundreds if not thousands of small and even moderately rare species are easily missed (Culotta 1994). For example, the 295 km² reef system off the west coast of

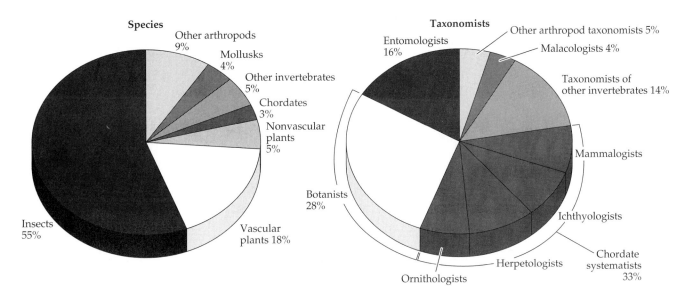

Species

Other arthropods 9%

Mollusks 4%

Other invertebrates 5%

Chordates 3%

Nonvascular plants 5%

Insects 55%

Vascular plants 18%

Taxonomists

Entomologists 16%

Other arthropod taxonomists 5%

Malacologists 4%

Taxonomists of other invertebrates 14%

Mammalogists

Ichthyologists

Chordate systematists 33%

Herpetologists

Ornithologists

Botanists 28%

FIGURE 16.2 A comparison between the relative number of species worldwide in each of the principal taxonomic groups and the number of taxonomists who specialize in them. (After Barrowclough 1992.)

New Caledonia (southwestern Pacific) is inhabited by at least 3000 species of mollusks—most of them less than 5 mm long (Vermeij 2004; after Bouchet et al. 2002).

One of the most serious downsides of the Linnaean shortfall is the likelihood that many of the undiscovered species will go extinct before they are known to science. E. O. Wilson (1992) calls these **Centinelan extinctions**, and they no doubt represent a significant component of the actual loss in biodiversity.

Whatever the extent of this unknown component, it is certain that we are witnessing a major surge in extinctions, one perhaps beginning to approach some of the mass extinctions of the fossil record. The avian extinction rate during the early Quaternary period is estimated to have been one species every 83 years. In recent years, this rate may have risen to one bird species every 4 years, and may have reached one species every 6 months by the year 2000 (Temple 1986). Overall, the global extinction rate of plants and animals combined may now be 100 species per year. One species—*Homo sapiens*– has transformed the geographic template and now dominates the Earth's ecosystems. We utilize some 20% to 40% of the total primary production of terrestrial ecosystems, and exploit— or overexploit—approximately 80% of marine fish production (Figure 16.3). Our activities account for over half of terrestrial nitrogen fixation, and we utilize over half of all accessible fresh surface water. There can be little doubt that this human domination of the Earth's life support systems has caused the current biodiversity crisis.

As scientists, we are called upon to identify the specific causes of the biodiversity crisis and objectively apply our knowledge to develop effective strategies to minimize the losses. The task is great, and it requires an integrative approach, one involving many disciplines and methods. Therefore, biogeography—being one of science's most holistic disciplines—has much to offer. More than any other scientific discipline, ours focuses on biological variation at local to global scales. In addition to studying patterns in the sizes, locations, and dynamics of geographic ranges, we study patterns in species diversity and endemicity across space and time. As we shall see below, the geographic signature of historical extinctions has allowed us to identify some of their principal causes and to develop some alternative strategies for abating the ongoing wave of extinctions.

FIGURE 16.3 Measures of human dominance and exploitation of several major components of the Earth's life support systems. Marine fisheries = percentage of major marine fisheries that are fully exploited, overexploited, or depleted by humans; nitrogen fixation = percentage of terrestrial nitrogen fixation that is human-caused; water use = percentage of accessible surface fresh water used by humans; land transformation = percentage of the land surface transformed by humans; bird extinctions = percentage of bird species on Earth that have become extinct in the past two millennia, almost all as a consequence of human activities; carbon dioxide concentration = percentage of the current atmospheric carbon dioxide concentration that results from human activities; and, plant invasions = percentage of plant species in Canada that humanity has introduced from elsewhere. (After Vitousek et al. 1997.)

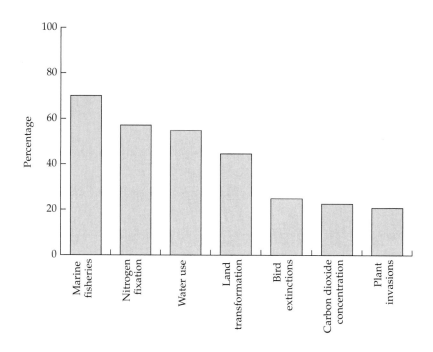

Geographic Variations in Biodiversity

Geographic patterns of biodiversity and dynamics of species distributions are central to both biogeography and conservation biology. We have discussed many of these patterns in earlier chapters of this book (Chapters 13 through 15). Three of the most relevant are listed below.

1. Species diversity tends to increase as we move toward the equator. Systems at lower latitudes tend to have both higher local, or alpha, diversity and higher between-system, or beta, diversity.

2. On islands, on mountaintops, and in other isolated systems, species diversity tends to increase with area and decrease with isolation.

3. Endemicity, or the relative number of unique species, tends to be higher for larger and more isolated regions.

Each of the above patterns derives from more fundamental patterns of variation in the characteristics of geographic ranges. It is clear from areographic studies that geographic ranges not only vary markedly among species, but vary in a systematic and predictable manner along geographic gradients. For example, as we move toward the equator, geographic ranges tend to decrease in size, allowing tighter packing of species and higher species richness. In addition, geographic ranges are not randomly distributed across the globe, but tend to concentrate and overlap in particular regions called **hotspots**. The term "hotspot" can refer either to the simple geographic co-occurrence of many species or, more specifically, to a site or region with an unusually high number of local endemics, also termed restricted-area species. It would seem that such intense hotspots of endemicity would be most relevant to conserving biodiversity.

Terrestrial Gradients and Hotspots of Biodiversity

While patterns of diversity and endemicity provide important clues for locating and ultimately protecting many rare and endangered species, two important questions remain: (1) What are the intensities and locations of hotspots for

TABLE 16.2 *The relative number of restricted-range species of birds varies dramatically among families*

Family[a]		Restricted-range species	Percent of family
Alcedinidae	Alcedinid kingfishers	20	84
Zosteropidae	White-eyes	75	78
Tytonidae	Barn and grass owls	10	59
Megapodidae	Megapodes	11	58
Rhinocryptidae	Tapaculos	16	57
Formicariidae	Ground antbirds	30	56
Scolopacidae	Sandpipers, curlews, etc.	7	8
Paridae	Titmice, chickadees, etc.	5	8
Anatidae	Ducks, swans and geese	11	7
Falconidae	Falcons and caracaras	6	6
Threskiornithidae	Ibises and spoonbills	2	6
Ardeidae	Herons, bitterns and egrets	3	5
Meropidae	Bee-eaters	1	4

Source: Bibby et al. 1992.

Note: The families listed above have significantly more (above the dashed line) or fewer (below the line) restricted range species than would be expected by chance. Restricted-range species include those with breeding ranges less than 50,000 km^2.

[a]Following taxonomy in Sibley and Monroe (1990).

a particular taxonomic group? (2) To what degree do different taxon-specific hotspots overlap?

Our task in conserving biodiversity would, of course, be easier if hotspots were few, intense, and restricted to the same locations for most taxa. Again, most of the Earth's diversity has yet to be catalogued, so we do not know to what extent this is the case. We have, however, developed reasonably reliable estimates of diversity for some groups, especially the more easily observable taxa such as birds. Under the assumption that extinctions are more likely for birds with smaller ranges, conservation of restricted-area species should be of paramount importance. Analyses conducted by the International Council for Bird Preservation during the previous decade (Bibby et al. 1992) indicated that 2609 land bird species, or 27% of all birds, have breeding ranges restricted to less than 50,000 km^2. Restricted ranges tend to be relatively common among some avian families, such as kingfishers, white-eyes, barn and grass owls, and megapodes, but rare in other families, especially those composed of larger, migratory, or nomadic species, such as ducks, falcons, herons, and egrets (Table 16.2).

The International Council for Bird Preservation refers to areas containing the breeding ranges of at least two restricted-range species as **endemic bird areas** (EBAs). As we might expect, EBAs are not randomly distributed, but tend to be concentrated in tropical regions (Figure 16.4). In addition, a highly disproportionate number of EBAs occur on islands. Although islands cover much less than one-tenth of the Earth's land area, nearly half of all EBAs are insular (Figure 16.5). Furthermore, whether insular or continental, the richness of EBAs increase with their area (Figure 16.6). Thus, we have answered our first question, at least for this one well-studied group of terrestrial vertebrates: *the intensity of avian hotspots is quite high.* The total area occupied by all EBAs

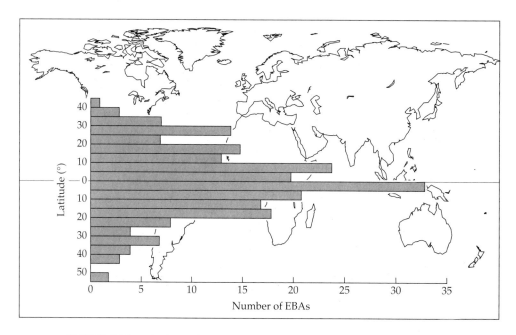

FIGURE 16.4 The latitudinal trend in the distribution of endemic bird areas (i.e., areas containing the breeding ranges of at least two restricted-range species) follows that of species richness in general. (After Bibby et al. 1992.)

combined is 6.5 million km². That means that protecting 4.5% of the world's total land area would provide breeding habitats for over 70% of globally threatened birds (Long et al. 1996; Johnson et al. 1998).

Now to our second question: To what degree do avian hotspots overlap with those of other animals and plants? The preliminary findings are equivocal, and depend on both the particular taxa and regions in question. Areas of endemism in Central America, for example, correspond very closely for birds, reptiles, and amphibians, but much less so when butterflies are included (Figure 16.7). On the African continent, areas of endemism are strikingly similar for amphibians and mammals, but birds and plants have additional hotspots exclusive of these groups (see Figure 16.8). Similar results have been obtained for other regions on a global scale for most major groups (Figure 16.9; see Pendergast et al. 1993; Vaisanen and Heliovaara 1994; Long et al. 1996; Williams et al. 1996; Dobson et al. 1997; Medail and Quezel 1999; Araujo 2002; Brooks et al.

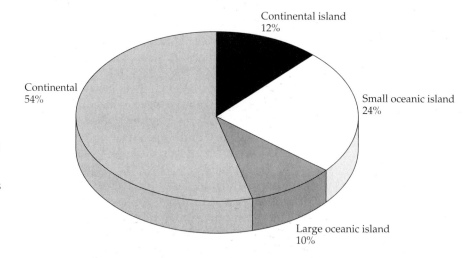

FIGURE 16.5 Endemic bird areas are not randomly distributed across the globe, but tend to be disproportionately common on islands. While islands cover less than 10 percent of the total land surface, nearly half of the world's EBAs are insular. (After Bibby et al. 1992.)

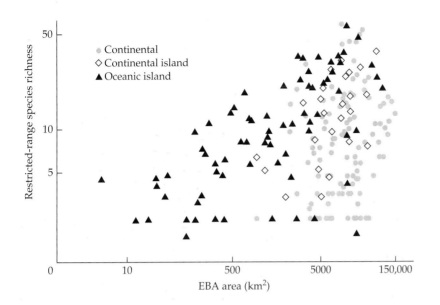

FIGURE 16.6 Consistent with the species-area relationship for most taxa, the number of restricted-range species increases with the area of endemic bird areas. The slope for the log-log relationship falls within the range observed for most insular systems (see Chapter 13). (After Bibby et al. 1992.)

2002). Species within each of the well-studied taxa tend to be nonrandomly distributed, with definite and intense hotspots of diversity and endemicity. Although hotspots for different taxa do overlap much more than would be expected by chance, the overlap is not perfect. In addition, hotspots of diver-

(A) Birds

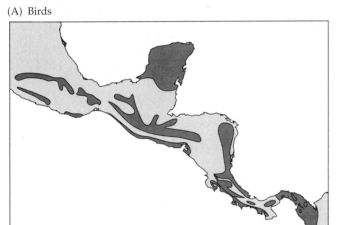

(B) Reptiles and amphibians

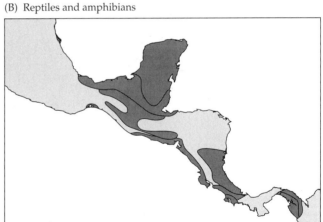

(C) Butterflies

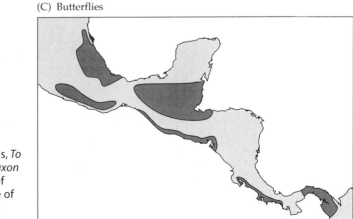

FIGURE 16.7 A critical question in conservation biology is, *To what degree do hotspots or areas of high endemism for one taxon overlap with those for other taxa?* In Central America, areas of endemism for birds show relatively high overlap with those of reptiles and amphibians, but considerably less overlap with areas of endemism of butterflies. (After Bibby et al. 1992.)

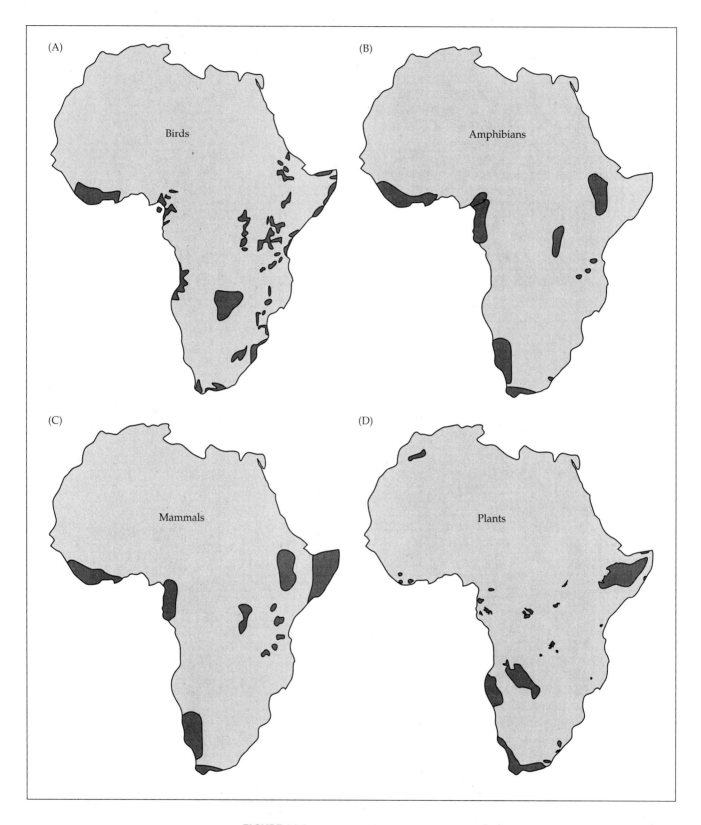

FIGURE 16.8 Just as we have seen for Central American faunas (Figure 16.7), overlap of areas of endemism in Africa are quite high for some groups (e.g., amphibians and mammals), but relatively low for others (e.g., plants and mammals). (A–C after Bibby et al. 1992; D after World Conservation Monitoring Centre 1992.)

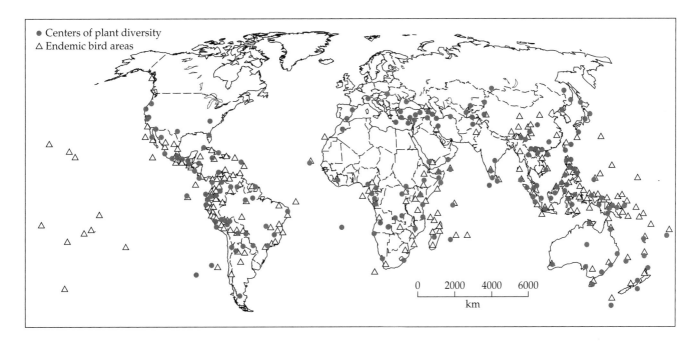

- • Centers of plant diversity
- △ Endemic bird areas

FIGURE 16.9 The global distribution of endemic bird areas and centers of plant diversity. (After World Conservation Monitoring Center 1992.)

sity may show little overlap with hotspots of endemicity. Still, hotspots of endemicity—those of paramount importance for conserving rare species—are fairly well known for some groups and, even if all were to be made reserves, only a small fraction of the Earth's land surface would need to be protected to maintain a great majority of these species.

Gradients and Hotspots in the Marine Realm

As this book has pointed out repeatedly, most of our knowledge as biogeographers comes from studies of terrestrial and, to a lesser degree, freshwater systems. In contrast, we know precious little about marine systems—not because of a lack of interest, but principally because of the logistic challenges associated with underwater—especially deep-water—studies. Just over 70% of the Earth's surface is covered by oceans, and nearly half of the world's marine waters are over 3000 m deep (World Conservation Monitoring Centre 1992). Coral reefs, mostly found in shallow coastal waters, are famous for their productivity and diversity. In contrast, deep-water benthic communities were until very recently thought to be relatively depauperate and inhabited largely by cosmopolitan species.

Technological advances in recent decades have provided some fascinating new insights into the marine realm. While most patterns in marine biodiversity are just emerging due to the early nature of research on deep-sea communities, marine faunas often exhibit biogeographic patterns similar to their mainland counterparts. For example, species diversity of marine biotas tends to increase as we move from the poles to the equator (Figures 15.25, 15.33, and 15.34; Stehli 1968; Buzas and Culver 1991; Rex et al. 1993; Rex et al. 2000; Crame 2004; see also Briggs 2004). As we saw in Chapter 15, marine species also tend to exhibit aerographic patterns that are similar to their counterparts in the terrestrial realm. For example, geographic ranges of marine fishes and invertebrates tend to increase along gradients from tropical to polar waters. In addition, diversity and geographic range size of at least some marine organisms tends to be strongly correlated not just with latitude, but with depth as well, with species richness increasing to reach a maximum at depths between 2000 and 3000 m

(A)

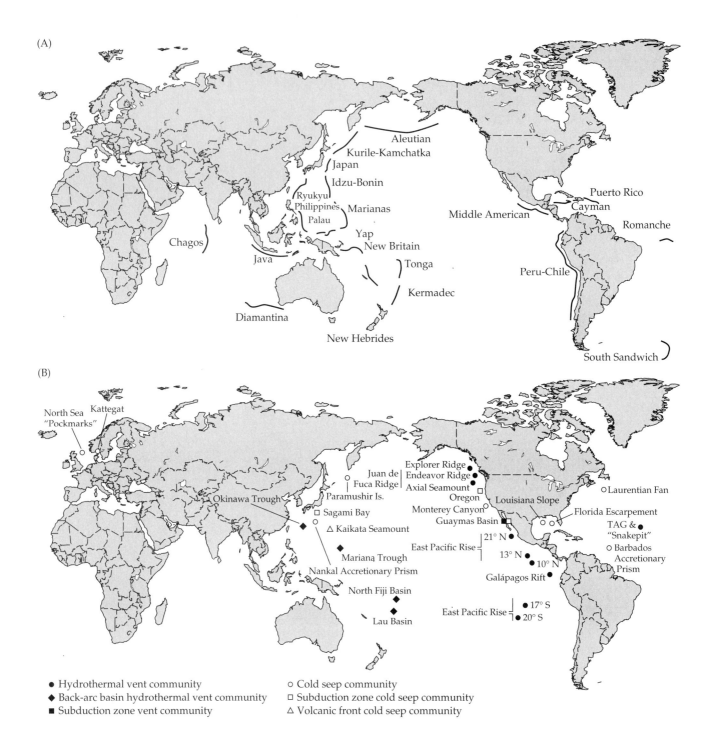

Aleutian
Kurile-Kamchatka
Japan
Idzu-Bonin
Ryukyu
Philippines
Palau
Marianas
Yap
New Britain
Chagos
Java
Tonga
Kermadec
Diamantina
New Hebrides
Puerto Rico
Cayman
Middle American
Romanche
Peru-Chile
South Sandwich

(B)

North Sea "Pockmarks"
Kattegat
Okinawa Trough
Paramushir Is.
Sagami Bay
Juan de Fuca Ridge
Kaikata Seamount
Explorer Ridge
Endeavor Ridge
Axial Seamount
Oregon
Monterey Canyon
Guaymas Basin
Louisiana Slope
Laurentian Fan
Florida Escarpement
TAG & "Snakepit"
Barbados Accretionary Prism
21° N
East Pacific Rise
13° N
10° N
Galápagos Rift
Mariana Trough
Nankal Accretionary Prism
North Fiji Basin
Lau Basin
East Pacific Rise
17° S
20° S

● Hydrothermal vent community
◆ Back-arc basin hydrothermal vent community
■ Subduction zone vent community
○ Cold seep community
□ Subduction zone cold seep community
△ Volcanic front cold seep community

(Figures 15.46 and 15.47; see Rex 1983; Gray 1997; Levin et al. 2001). Below these depths, diversity tends to decrease, but endemicity may increase.

One of the key reasons for the high endemicity of deep-water communities derives from the isolated nature of the deepest areas. Indeed, trench, seep, and hydrothermal vent communities are marine analogues of isolated oceanic islands. These hadal communities, defined as those below 6000 m, account for just 1% of the ocean's benthic area and are distributed among highly disjunct sites, including those of oceanic trenches (Figure 16.10A). Trenches harbor highly distinct communities with endemicities ranging between 50% and 90%.

(C)

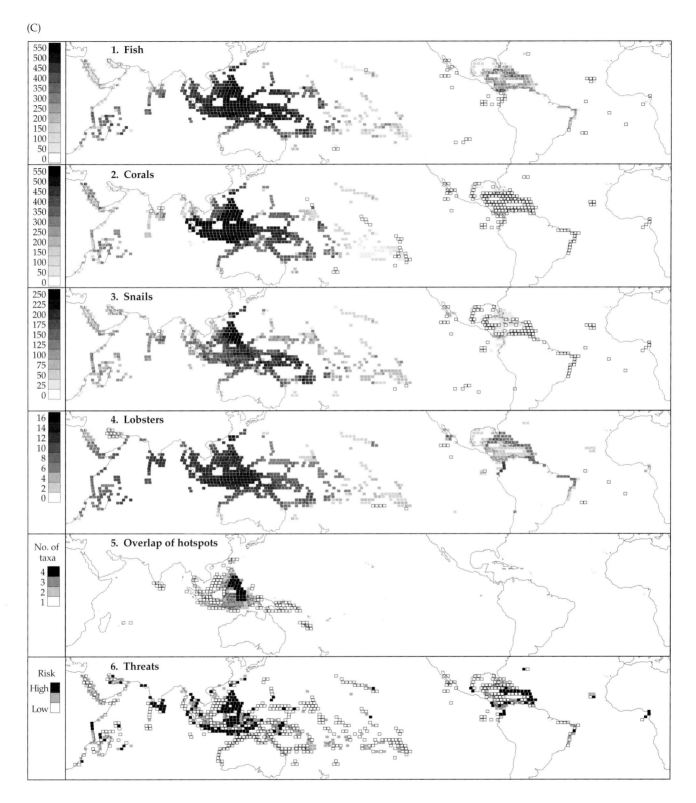

FIGURE 16.10 Global distributions of (A) principal oceanic trenches and (B) hydrothermal vents and cold seep communities. (C) Geographic variation in species richness of fish (1), corals (2), snails (3), and lobsters (4), while not identical, exhibit high spatial concordance. Overlap of the top 10 percent most species-rich sites among taxa (5) identifies the East Indies Triangle as a hotspot of marine diversity, in general (scale ranges from black cells which indicate sites that are hotspots for all four taxa [fish, corals, snails, and lobsters] to white cells for hotspots of just one taxon). Unfortunately, these same sites also tend to be "hotspots" for ongoing and future threats (6) by human activities. (A–B after World Conservation Monitoring Center 1992; C from Roberts et al. 2002.)

Recent studies indicate that biotas of a variety of shallow water marine environments including those of isolated coral reefs, kelp beds, and seagrass meadows, or those encircling the shores of isolated, oceanic islands (e.g., mudflats, intertidal rocks, sandy beaches, and mangroves) exhibit patterns of diversity and niche shifts in many ways similar to terrestrial islands. For example, species richness of mollusks, crabs, corals, and seagrasses tends to be much higher in the large, mainland-like environments and relatively low for the smaller or more isolated marine islands (Figure 15.19; see review by Vermeij 2004). The island syndrome for marine biotas also includes ecological shifts and intriguing trends in morphology of insular populations. For example, shells of gastropods on isolated marine islands often lack the heavy and complex antipredatory armor of the nearest, more widespread species— presumably in response to the paucity of predators on these systems (Vermeij 1978, 2001, 2004; Vermeij and Snyder 2002). Unlike terrestrial islands, however, even some of the most isolated shallow-water marine biotas exhibit relatively low endemicties, presumably because relatively high rates of dispersal prevent evolutionary divergence in these gastropods (Vermeij 2004).

In sharp contrast are the tectonically derived hydrothermal vent communities (see Figure 16.10B). In comparison to these shallow water marine islands, the diversity of hydrothermal vent communities appears to be relatively low, but their endemicity is extremely high. Although discovered only in 1977, surveys of vent communities have yielded one new phylum, over 14 new families, and 50 new genera—most of which are monospecific (Grassle 1989; Gage and Tyler 1991; see also Karl et al. 1996; Black et al. 1997; Fautin and Barber 1999; Ferrari and Markhaseva 2000; Sideleva 2002; Stuart et al. 2003; Crame 2004). Mollusks, polychaetes, and arthropods account for approximately 93% of the 375 species described from all vents worldwide; 90% are restricted to vent habitats. The high distinctness and endemicity of vent communities derives largely from three characteristics of hydrothermal vents: their isolation, their antiquity (vent habitats have existed throughout the Phanerozoic, providing ample time for evolutionary divergence), and the special adaptations required to live in a high-pressure, chemically reducing, and metal-rich environment where the food chain is based on chemosynthetic—rather than photosynthetic—producers (see Tunnicliffe and Fowler 1996; Luther et al. 2001; Dover 2002).

Recent global-scale analyses have identified intense hotspots of diversity and endemicity for marine taxa, including fish, corals, snails and lobsters (Roberts et al. 2002). As with terrestrial hotspots, these marine taxa exhibit high—albeit, far from complete—geographic concordance, with the most intense multi-taxa hotspots concentrated in Indonesia (see Figure 16.10C). While these are obvious targets for leveraging future conservation efforts, the most important areas of concordance are those where ongoing and future threats overlap with areas of high endemicity (see Figure 16.10C, part 5).

Thus, even though marine biotas, especially those of deep-water environments, may be less in jeopardy than their terrestrial biotas, they are not safely isolated from—or immune to—anthropogenic extinctions. The threats to shallow water marine environments, especially biodiversity hotspots such as the East Indies Triangle (Philippines to Malay Peninsula, and India to New Guinea) are ongoing. As Briggs (2004) states, "If we want to protect the place of highest diversity and also the process that produces that diversity, the choice becomes obvious." Hotspots such as the East Indies Triangle possess some of the highest species and generic diversity of the marine realm, and they function as centers of evolutionary radiation, thus supporting global diversity at a regional to global scale as well (Briggs 2004; see also Roberts et al. 2002). Identifying these and other hotspots in the Earth's last biological frontier has and will continue to represent one of the greatest challenges to our science. Protecting these hotspots, once identified, will be even more challenging as this will require conservation in the vast open environments that characterize the marine realm and, no less challenging, will require unprecedented cooperation among those who share, but do not own, these waters.

The Geography of Extinctions

Given that diversity varies in a nonrandom manner across geographic gradients and that extinction forces (many of them related to geographic expansions of our own species) move across and transform the geographic template in a highly nonrandom manner, we might expect that extinctions will also exhibit a definite geographic signature. Yet, only recently have we had the information and technology to tackle some truly intriguing questions that address this interface between biogeography and conservation biology. Do extinctions vary with latitude, elevation, depth, area, or isolation? Do different taxa exhibit the same geographic trends in susceptibility to anthropogenic extinctions? These are no longer just academic questions. Given the limited time, energy, and financial support available for preserving biodiversity and preventing extinctions, it is imperative that we direct our efforts *where* they will do the most good.

The Prehistoric Record of Extinctions

Studies of the fossil record provide some useful clues to some of these questions. The mass extinctions that occurred at the end of the Cretaceous period are marked by a 60% to 80% drop in species diversity at a global scale. David Raup and David Jablonski have searched for hotspots of local extinction among the rich fossil record of marine bivalves (Raup and Jablonski 1993). When all 340 genera were considered, they found that both the absolute number of end-Cretaceous extinctions and the proportion of local genera suffering extinctions were highest for the equatorial regions. The region with the highest extinction rate was the New World Tropics, which includes the site of the Chicxulub crater where the infamous asteroid impacted (see Sharpton et al. 1992). Raup and Jablonski warned, however, that their results were strongly influenced by the rudist bivalves (large, attached, or recumbent filter feeders) that were primarily restricted to tropical waters of the Jurassic and Cretaceous Periods. When rudists were excluded, the proportional loss of bivalve genera was uniform across the globe—that is, regardless of latitude, the proportion of local assemblages suffering extinctions was approximately constant (about 52%). Still, the total number of extinctions, an index perhaps more relevant to geographic clines in biodiversity, was highest in the richer tropical waters.

Other paleontological evidence sheds additional light on the geographic correlates of natural extinctions. Plate tectonics and continental drift may have played an important role in many of the extinction events recorded in the fossil record. The Great American Interchange is an important case in point (see Chapter 10). With the formation of the Central American landbridge approximately 3.5 million years B.P., the Nearctic and Neotropical regions were connected after a long period of isolation. The ensuing migrations and radiations of immigrants caused a wave of extinctions among native species, especially in South America. As we shall see, this lesson in paleobiogeography is especially relevant to the biodiversity crisis. Throughout the fossil record, whether due to the formation or breakup of Pangaea, the Great American Interchange, or the climate-driven range shifts associated with glacial cycles of the Pleistocene, exchange among biotas that evolved in isolation from one another has typically resulted in waves of extinctions among the native assemblages.

It is only in the most recent slice of this great record, the last 2 million years, that our own species has become a significant component of environmental change, resource depletion, biotic exchange, and resultant extinctions. Throughout this period, prehistoric human societies colonized almost all corners of the globe, preying on native species, transforming landscapes and introducing exotics. Throughout the world, as human societies colonized and developed advanced hunting and agricultural practices in new regions, native species suffered. Technological advances also spread among human populations, often with devastating effects on native biotas. The extinction of the Pleistocene megafauna is an important case in point. The mass extinction of mammoths, ground sloths, giant raptors such as teratorns, and other large mammals and birds was not synchronous across the globe, but almost invariably followed on the heels of human colonization, or more precisely, the spread of advanced hunting techniques and intensive agricultural practices, including fire (see Chapter 9). In Australia, the use of fire on a grand scale by Aborigines caused massive extinctions of native wildlife (Flannery 1994). In the Holarctic, hunting and gathering societies migrated across Beringia to colonize North America between 15,000 and 30,000 years B.P. Again, it wasn't until they began more intensive use of fire and developed advanced hunting skills that they had a significant effect on the diversity of the native biota. As we remarked in Chapter 9, Africa was the exception that proves the rule. Unlike other biotas that evolved in isolation from humans, Africa's rich wildlife had the opportunity to gradually adapt to the activities of aboriginal societies. It is only within the twentieth century that human population densities and their transformation of African landscapes have become so intense that they now threaten its megafauna along with many other native animals and plants (see the discussion on the biogeography of humanity in Chapter 18).

The Historical Record of Extinctions

Reviews of the historical record of extinctions have provided us with important clues for understanding the ongoing biodiversity crisis. Like the prehistoric extinctions of the Pleistocene megafauna, many historical extinctions occurred when waves of human colonization and development advanced over previously pristine hotspots of endemicity. The thousands of recorded extinctions have a definite geographic signature—one bearing witness to the high endemicity and fragility of insular communities. As Figure 16.11 reveals, a highly disproportionate number of the animal extinctions recorded since 1600 have occurred on islands (see Chapters 13 and 14). This insular bias applies to plants as well (see Figure 16.11A), and although just one in six species of plants are insular, one in three threatened species are endemic to islands (Schemske et al. 1994).

These insular extinctions have resulted from a combination of factors, nearly all related to human activities and the isolation and ecological naïveté of insular biotas. Many insular plants and animals, especially those endemic to more isolated and relatively depauperate islands, have been subjected to the effects of scores of exotic species introduced after humans colonized their islands. On Chiloe Island, located off the coast of Chile, Darwin found the native foxes—later named Darwin's fox (*Pseudalopex fulvipes*)—so naive that he was able to collect the type specimen by hitting it over the head with his geological hammer. On visiting the Galápagos, he was amazed that the avifauna was so tame that he could prod a hawk with the butt of his rifle.

659

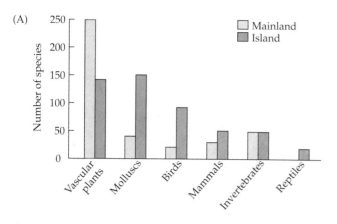

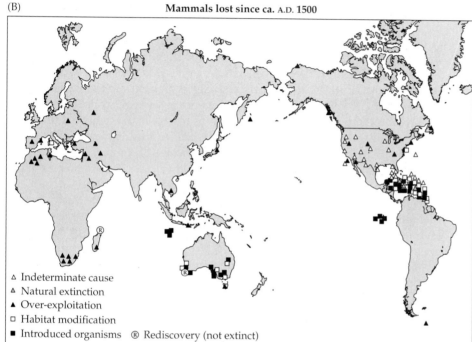

FIGURE 16.11 (A) Patterns in recorded extinctions (1600–1990) of plants, terrestrial animals, and mollusks. Extinctions have been far more common among animals inhabiting islands than among their mainland counterparts. (B) Extinctions of two well-documented groups of vertebrates, mammals, and birds, illustrate the strong geographic bias in extinctions. While particular distributions of historic extinctions vary among these two faunas, they both reveal the high susceptibility of insular biotas. In addition, the causes for historic extinctions in these two groups differ among groups as well as across geographic regions (e.g., extinctions of mammals on the continents, excluding Australia, were primarily due to over-exploitation, while those of most islands were caused by a combination of factors, including habitat modification and introduced species). (A after Reid and Miller 1989; World Conservation Monitoring Centre 1992; IUCN 2003; B and C after N. C. Heywood, pers. comm.)

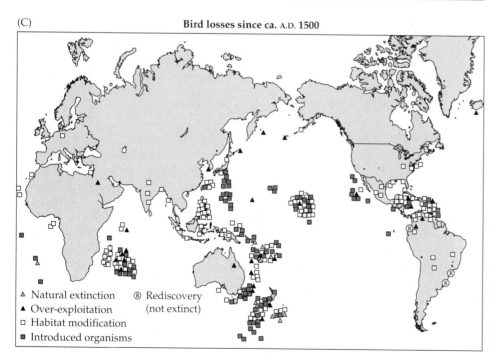

A gun here would be superfluous. What havoc the introduction of any beast of prey must cause in a country before the inhabitants have become adapted to the stranger's craft of power. (Darwin 1860.)

Of course, this was one of the many cases in which Darwin's words proved prescient. Two-thirds of the historical extinctions of insular birds have been caused either directly or indirectly by introduced mammals.

Throughout their history, most insular biotas evolved in the absence of large, ground-dwelling homeotherms. Many species of insular plants, for example, were able to flourish without evolving defenses against mammalian herbivores. In a very real sense, the biological novelty and fragility of insular biotas derives from their isolation. With the introduction of exotic species, whether planned or accidental, this "splendid isolation" was lost. Just as the formation of the Central American landbridge triggered the great faunal interchange and the subsequent extinctions of many species endemic to South America, anthropogenic introductions of exotic species onto islands effectively connected them to the mainland and triggered a wave of extinctions.

Actually, there have been multiple waves of insular extinctions. Europeans were not the only civilization to colonize islands and subsequently devastate the native biota. Subfossil evidence now indicates that many insular extinctions were caused by "native" islanders, centuries—and sometimes millennia—before Europeans ever set foot on the islands. These groups include the aboriginals of Malaysia, Indonesia, Australia, and Tasmania; the Micronesians, Melanesians, and Polynesians of the isolated archipelagoes across the Pacific; and the Amerindians of the West Indies (Morgan and Woods 1986; Woods and Sergile 2001; Steadman and Martin 2003). For example, Olson and James (1982a) estimate that the Polynesians may have caused the extinctions of nearly half of the native avifauna of Hawaii. In New Zealand, all 15 species of moas (immense flightless birds that were endemic to these islands), were extirpated by the Maori people and their introduced dogs and rats before European colonization (Halliday 1978; Worthy and Holdaway 2002). These extinctions were far from random, with the largest species falling first in a tragic, domino-like wave of extinctions. New Zealand's only surviving ratites are the much smaller kiwis (*Apteryx* spp.).

Because introduced species have played such an important role in historical extinctions, we now review the ecology and geography of species introductions and then consider the geographic nature and ecological effects of two other major threats to biodiversity—landscape transformation (including habitat loss and fragmentation), and global climate change.

The Ecology and Geography of Invasions

MAGNITUDE OF THE PROBLEM. The frequency and diversity of anthropogenic introductions of exotic species are staggering, and now appear to exceed natural invasion rates (Hodkinson and Thompson 1997). Species introductions, whether intentional or accidental, have touched and transformed all types of ecosystems, from oceanic islands and isolated mountaintops to tropical rain forests and the far reaches of the Antarctic (Williamson 1996; Vitousek et al. 1997b; Lonsdale 1999; Mack et al. 2000; Pimental et al. 2000; Table 16.3; Figure 16.12). More than any other events, including those associated with plate tectonics or glacial cycles, species introductions have had a demonstrable effect on the diversity and geography of the world's biota (McKinney and Lockwood 1999; Lockwood and McKinney 2001; Lockwood 2004). They have driven many hundreds of geographically restricted, native species to extinction and replaced them with a redundant set of species adapted to human-

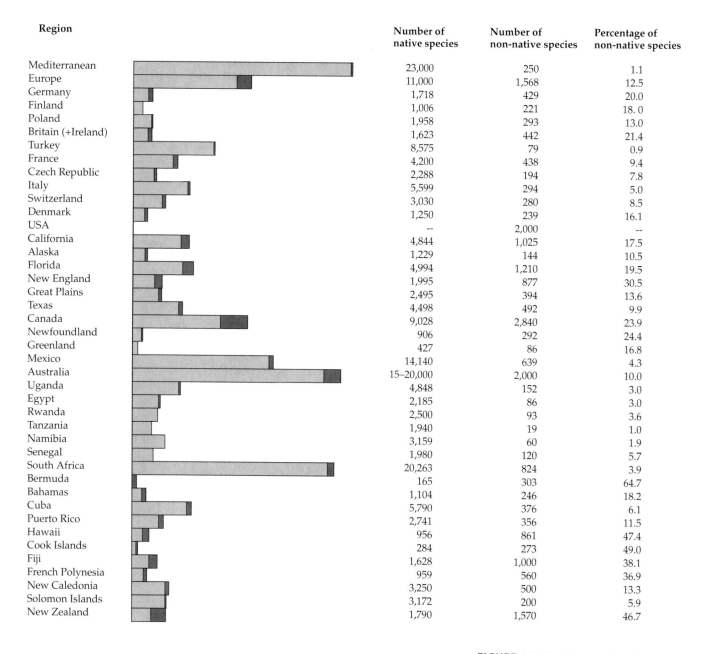

Region		Number of native species	Number of non-native species	Percentage of non-native species
Mediterranean		23,000	250	1.1
Europe		11,000	1,568	12.5
Germany		1,718	429	20.0
Finland		1,006	221	18.0
Poland		1,958	293	13.0
Britain (+Ireland)		1,623	442	21.4
Turkey		8,575	79	0.9
France		4,200	438	9.4
Czech Republic		2,288	194	7.8
Italy		5,599	294	5.0
Switzerland		3,030	280	8.5
Denmark		1,250	239	16.1
USA		--	2,000	--
California		4,844	1,025	17.5
Alaska		1,229	144	10.5
Florida		4,994	1,210	19.5
New England		1,995	877	30.5
Great Plains		2,495	394	13.6
Texas		4,498	492	9.9
Canada		9,028	2,840	23.9
Newfoundland		906	292	24.4
Greenland		427	86	16.8
Mexico		14,140	639	4.3
Australia		15–20,000	2,000	10.0
Uganda		4,848	152	3.0
Egypt		2,185	86	3.0
Rwanda		2,500	93	3.6
Tanzania		1,940	19	1.0
Namibia		3,159	60	1.9
Senegal		1,980	120	5.7
South Africa		20,263	824	3.9
Bermuda		165	303	64.7
Bahamas		1,104	246	18.2
Cuba		5,790	376	6.1
Puerto Rico		2,741	356	11.5
Hawaii		956	861	47.4
Cook Islands		284	273	49.0
Fiji		1,628	1,000	38.1
French Polynesia		959	560	36.9
New Caledonia		3,250	500	13.3
Solomon Islands		3,172	200	5.9
New Zealand		1,790	1,570	46.7

FIGURE 16.12 The number of introduced, or non-native, species of plants varies substantially among geographic regions and tends to be disproportionately high for islands such as Bermuda, Hawaii, the Cook Islands, and New Zealand, where exotics may comprise roughly half of the extant flora. (After Vitousek et al. 1996.)

modified habitats. As McKinney and Lockwood (1999) observed, the result is a homogenization of nature—a global scale process where relatively "few winners are replacing many losers in the next mass extinction." Thus, anthropogenic invasions have dramatically reduced both diversity of the world's species and the distinctness of regional biotas—the latter, of course, being the basis for biogeography's most fundamental pattern, Buffon's Law.

Although it may be impossible to estimate the total number and diversity of species introductions that have occurred even within historic times, we know that many were conducted as part of "naturalization" programs—deliberate attempts by Europeans to surround themselves with familiar species in exotic lands. While no doubt devastating to many native species, such deliberate programs do at least provide us with useful information on the magnitude, diversity, and geographic patterns of introductions. In 1988, Torbjorn

TABLE 16.3 *The number of introduced, or non-native, species of birds and freshwater fishes in selected regions*

Region	Breeding birds		Freshwater fishes	
	Native	Non-native	Native	Non-native
Europe	514	27		74
California			76	42
Alaska			55	1
Canada			177	9
Mexico			275	26
Australia		32	145	22
South Africa	900	14	107	20
Peru				12
Brazil	1635	2	517	76
Bermuda		6		
Bahamas	288	4		
Cuba		3		10
Puerto Rico	105	31	3	32
Hawaii	57	38	6	19
New Zealand	155	36	27	30
Japan	248	4		13

Source: Vitousek et al. 1996.

Note: The relative number of exotic species is especially high on islands such as Puerto Rico, Hawaii and New Zealand, where the number of non-native fishes may rival or exceed that of native species.

Ebenhard reviewed all available records of historical introductions of birds and mammals. Again, this has to be viewed as an incomplete record: an inestimable number of introductions went unreported. Yet, Ebenhard's survey is highly instructive. He found published reports documenting 788 mammalian introductions, including 118 species, 30 families, and 8 orders. The most frequently introduced mammals included common rabbits (*Oryctolagus cuniculus*), domestic cats (*Felis catus*), rats (including *Rattus norvegicus*, *R. rattus*, and *R. exulans*), house mice (*Mus musculus*), and domestic pigs, cattle, goats, and dogs. Not surprisingly, the pool of introduced species was not a random sample of the mainland faunas. Carnivores and artiodactyls (deer and related species of ungulates) constituted 19% and 31% of the introductions, respectively, whereas they represent less than 7% of the mainland species pool. Ebenhard also tallied 771 introductions of birds, including 212 species from 46 families and 16 orders. Again, species introductions were not random: waterfowl, gallinaceous birds, pigeons, and parrots were overrepresented in the sample of introduced species (representing 46% of the introduced species vs. just 12% of all avian species belonging to these orders).

Species introductions of mammals and birds also had a strong geographic bias (Figure 16.13). On the continents, the number of species introductions appears to be roughly a function of the area of the region, with more introductions in the Palearctic region than in any other continental region. Australia, however, appears to be an important exception to this species-area relationship, receiving many more introductions per area than any other continental landmass. The most striking feature of the geographic distribution of introductions is that, despite their isolation and relatively small area, oceanic islands have received the bulk of species introductions. Indeed, 60% of all mammal and bird introductions have occurred on oceanic islands. In another important monograph on species introductions in birds, Long (1981) catalogued 162 species introductions to the Hawaiian Islands, 133 to New Zealand, 56 to Tahiti, and well over 500 bird species introductions to some 90 other islands or archipelagoes across the globe. In addition to receiving a greater number of introductions, islands have also received a greater diversity of invasive species. More species of mammals and many more species of birds have been introduced onto oceanic islands than to the continents or islands of the continental shelves combined.

Introductions of exotic plants, although of greater magnitude, were similar in their geographic bias. Again, Australia and oceanic islands received a highly disproportionate number of introductions (Drake et al. 1989). It is estimated that there are now between 1500 and 2000 exotic plants in Australia (see also MacDonald et al. 1986). Nearly half the plants of New Zealand are invaders (1790 natives vs. 1570 exotics; Heywood 1989; Atkinson and Cameron 1993). Before human colonization, over 90% of the flowering plants of the Hawaiian Islands were endemic. Now, nearly half of Hawaii's flowering plants are invaders (over 850 out of about 2000 species).

Aquatic systems have also been significantly affected by introductions of exotic species (see Lodge 1993; Wonham et al. 2000). In the state of California alone, 48 out of 137 species of fish are exotics. Indeed, 46 of these are either known or suspected to have negatively affected native species. Of the 95 species that have bred in Arizona, 67 (71%) are not native to the state. Many exotic species established in freshwater systems of the United States subsequently invaded Canadian waters (Crossman 1991). In Africa, introduction of the Nile perch (*Lates niloticus*) into Lake Victoria appears to have been one of the primary causes of the extinction of as many as 200 endemic cichlids (see Baskin 1992; Goldschmidt et al. 1993).

Perhaps the most troubling feature of the history of species introductions is that most have occurred in those systems with the greatest endemicity and vulnerability to invading species: oceanic islands. The prospect for the future is no less troubling. The proportion of threatened terrestrial vertebrates that are affected by introduced species is much higher on oceanic islands (31%) and on the "island continent" of Australia (51%) than in other continental areas (0 to 10%; Table 16.4). Ironically, biological invasions also threaten the biological diversity of nature reserves. While we are wise to invest much time and energy in managing and protecting these reserves, most of them have already been significantly impacted by exotic species (MacDonald et al. 1989). On the continents, introduced species represent from 3% to 28% of all vascular plant species found in nature reserves. Again, the situation is much worse on islands, where between 31% and 64% of all vascular plant species in nature reserves are exotics. Similarly, while introduced birds and mammals tend to be minor components of mainland reserves (averaging 1% to 5% for most reserves), they represent significant, if not the dominant, elements of the biota in reserves on oceanic islands (20% and 81% of the total species for birds and mammals, respectively). In many reserves that include lakes and streams, well over half the species of freshwater fish are exotic.

Recent analyses of 1378 introduction events for 426 bird species provide some interesting and potentially important insights into the geography of human-assisted invasions and, in turn, the geography of extinctions as well. As Ebenhard (1988) observed earlier, most introductions of birds involve a limited diversity of taxa (mostly from just five families including the pheas-

(A)

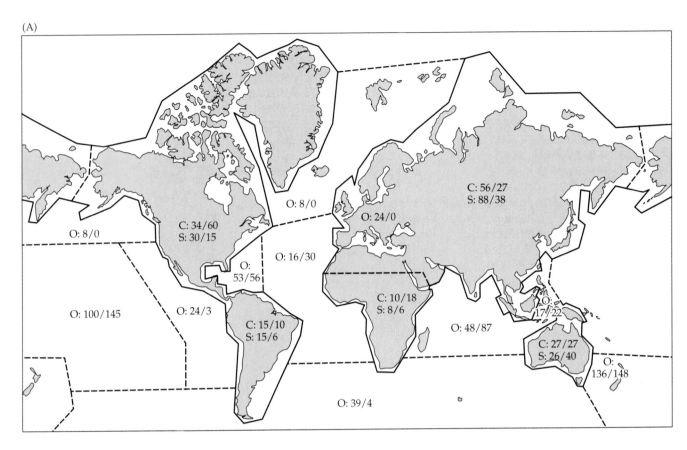

O: 8/0

C: 56/27
S: 88/38

O: 8/0

C: 34/60
S: 30/15

O: 24/0

O: 16/30

O:
53/56

C: 10/18
S: 8/6

O:
17/22

O: 100/145

O: 24/3

C: 15/10
S: 15/6

O: 48/87

C: 27/27
S: 26/40

O:
136/148

O: 39/4

(B) Mainlands

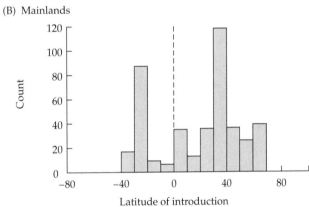

(D) Events

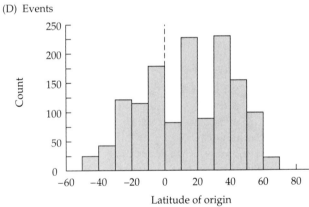

(C) Islands

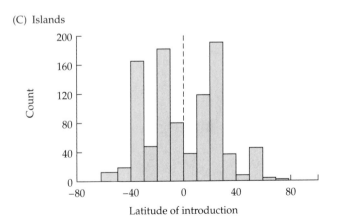

(E) Species

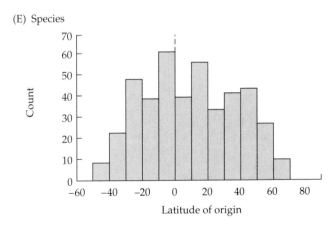

◀ **FIGURE 16.13** (A) The geographic pattern of recorded introductions of mammals and birds. C = continental; S = continental shelf islands; O = oceanic islands. The numbers before and after the slash represent numbers of mammal and bird species respectively. (B–E) The record of historic, human-assisted introductions of birds (including 1378 introduction events of 426 species) reveals strong geographic biases in locations of introductions (B and C), with a highly disproportionate number of introductions occurring on islands, and relatively few introductions to insular or mainland sites in the tropics or high latitudes. The source populations for these introductions (D and E) also tended to be geographically biased, with fewer taken from the tropical latitudes than would be expected based on their relatively high diversity. (A, after Ebenhard 1988b; B–E, after Blackburn and Duncan 2001.)

ants, passerines, parrots, ducks, geese, swans, and pigeons; Blackburn and Duncan 2001). Moreover, avian introductions were highly nonrandom with respect to both the sites of origin and sites of introduction (see Figure 16.13B–E). As Blackburn and Duncan (2001) observed, although islands cover just 3% of Earth's ice-free landsurface (i.e., landsurface where introductions are feasible; Mielke 1989), just over half of avian introductions were to islands. Similarly, although the greatest extent of ice-free land occurs in the tropics (Rosenzweig 1992), relatively few birds have been introduced to equatorial latitudes. These geographic biases of introduction largely reflect patterns of colonizations by Europeans during the eighteenth to twentieth centuries, with settlement and colonization especially concentrated in particular oceanic archipelagoes (especially Hawaii and New Zealand) and former British colonies (USA and Australia, in particular). Thus, the geography of extinction and endangerment may often reflect, not just the naïveté and vulnerability of insular biotas, but also the geographic dynamics of extinction forces— in this case, the geography of human populations. For this reason, we summarize the dynamic geography of our own species in the following chapter.

SUSCEPTIBILITY OF INSULAR BIOTAS TO INVASION. The insular biases in historic extinctions illustrated in Figure 16.11 are striking, even when we consider the geographic patterns in introduction sites summarized above (i.e., just over half of avian introductions were to islands, but well over 90% of historic extinctions of birds were of insular forms). Thus, islands are not just more frequently

TABLE 16.4 *Percentage of threatened terrestrial vertebrate species that are affected by introduced species*

Taxonomic group	Continental areas within biogeographic realms								All mainland areas	All insular areas
	Eurasia	N. America	Africa	Indo-Malaya	Oceania	Antarctica	S. America	Australia		
Mammals	16.7 (42)	3.4 (29)	8.0 (100)	12.7 (55)	0.0 (8)	0.0 (0)	10.0 (60)	64.4 (45)	19.4 (283)	11.5 (61)
Birds	4.2 (24)	13.3 (15)	2.5 (118)	0.0 (30)	0.0 (1)	0.0 (0)	4.2 (71)	27.3 (11)	5.2 (250)	38.2 (144)
Reptiles	5.9 (17)	16.7 (24)	25.0 (16)	4.3 (23)	14.3 (7)	0.0 (0)	14.3 (28)	22.2 (9)	15.5 (84)	32.9 (76)
Amphibians	0.0 (8)	6.3 (16)	0.0 (3)	0.0 (0)	0.0 (0)	0.0 (0)	0.0 (1)	0.0 (2)	3.3 (30)	30.8 (13)
Total for all groups considered	9.9 (91)	9.5 (84)	6.3 (237)	7.4 (108)	6.3 (16)	0.0 (0)	8.1 (160)	50.7 (67)	12.7 (647)	31.0 (294)

Source: MacDonald et al. 1989.

Note: The total number of threatened species in each realm is given in parenthesis.

targeted for human-assisted introductions, but in comparison to mainland biotas, insular species are more susceptible to "the stranger's craft of power." In retrospect, this may not be surprising given the special nature of island environments and their native communities. As we have seen, isolated oceanic islands tend to be inhabited by relatively depauperate communities that are often composed of novel—yet, ecologically naive—species. Wilson's taxon cycle theory (see Chapter 14) may be especially relevant to understanding the susceptibility of insular biotas to the effects of exotic species. According to this theory, one of the most striking patterns in insular biogeography is that insular populations may be evolutionary dead ends, doomed to eventual extinction. This results in the taxon cycle, the net movement of species from continents to islands over evolutionary time. Systematic studies show that insular biotas are ultimately derived from taxa that originated on larger landmasses. Even for those island species that are formed by endemic speciation within an island or an archipelago, as in the Hawaiian *Drosophila*, continents remain the ultimate source of insular forms. This appears to be a special case of the general tendency of organisms originating in large areas with diverse biotas to invade small areas with lower organic diversity whenever the opportunity arises, and ultimately to replace the native species (see Chapters 6 and 10; see also Elton 1958).

Even the exceptions to this pattern support the general rule. The region around Miami in southern Florida has been colonized successfully by several species from the Caribbean Islands, including about six *Anolis* species that apparently were imported either intentionally or accidentally by humans during the twentieth century. However, southernmost Florida is essentially an island—an isolated region of tropical habitat attached to temperate North America by a low-lying peninsula that was frequently inundated by seawater during the Pleistocene. Although the Caribbean species seem to thrive in the lush tropical gardens in the Miami region, they have not extended their ranges northward to invade more temperate habitats.

The extreme susceptibility of islands to invasion is dramatically illustrated by two case studies—the faunas of Hawaii and those of New Zealand. The avifauna of the Hawaiian archipelago is currently composed of 95 species of birds. Of these, 57 are native, while 38 (40%) are exotics that have been introduced since 1800. During the same period, 14 native species have become extinct and several others are rare and endangered. Of 94 species known to have been introduced prior to 1940, 53 became established (at least locally and temporarily), and only 41 failed completely. Most of the introduced species were deliberately released to augment the local avifauna with game birds and species of beautiful plumage and song. A strange mixture of species derived from different taxonomic groups and biogeographic provinces now coexists on the archipelago, including the mockingbird (*Mimus polyglottos*), cardinal (*Cardinalis cardinalis*), and California quail (*Callipepla californica*) from North America; the Indian mynah (*Acridotheres tristis*), lace-necked dove (*Streptopelia chinensis*), and Pekin robin (*Leiothrix lutea*—a babbler, not a thrush) from Asia; the house sparrow (*Passer domesticus*), skylark (*Alauda arvensis*), and ring-necked pheasant (*Phasianus colchicus*) from Europe; and the Brazilian cardinal (*Paroaria cristata*) from South America (Elton 1958). A quick review of Table 16.5 reveals that assemblages of the less vagile vertebrates, including non-volant mammals, amphibians, reptiles, and freshwater fishes, are dominated by exotic species.

The fate of the Hawaiian archipelago is by no means unique. Consider the birds and mammals of New Zealand (Table 16.6). The potential effect of these exotic species on the insular biota can be better appreciated when one realizes

TABLE 16.5 *Numbers of native, endemic, and introduced species of various taxonomic groups in the Hawaiian Islands*

Taxonomic Group	Native species	Endemic species	Introduced species
Flowering plants	970	883 (91%)	800 (45%)
Ferns and allies	143	105 (73%)	21 (13%)
Hepaticae (liverworts)	168	ca 112 (67%)	2 (0.01%)
Musci (mosses)	233	112 (48%)	3 (0.01%)
Lichens	678	268 (38%)	0 (0%)
Resident birds	57	44 (77%)	38 (42%)
Mammals (on land)	1	0 (0%)	18 (94%)
Reptiles (on land)	0	0	13 (100%)
Amphibians	0	0	4 (100%)
Freshwater fishes	6	6 (100%)	19 (76%)
Arthropods	6000–10,000	(98%)	ca 2000 (20–30%)
Mollusks	ca 1060	(99%)	9 (<1%)

Source: Loope and Mueller-Dombois 1989.

that the only native terrestrial mammals of New Zealand are two species of bats; all of the nonvolant mammals and over half of the bird species have been introduced (Wodzicki 1950; Elton 1958; Falla et al. 1966; Atkinson and Cameron 1993). It is little wonder that these exotics have had a tremendous impact on the native species of lizards and ground-nesting birds, which evolved in the complete absence of mammalian predators.

Given the magnitude and diversity of introductions, it should not be surprising that their effects on native species, while often negative, have involved many different mechanisms. Introduced species can directly compete with or prey on native species. Rats and mongooses are perhaps the most notorious of the predators introduced to oceanic islands. By preying on ground-nesting birds and reptiles, they have decimated many endemic species (Figures 16.14

TABLE 16.6 *Native and introduced land birds and mammals of New Zealand*

	Birds	Mammals	Total
Native			
Families	19	2	21
Species	39	2	41
Introduced			
Families	13	11	24
Species	26	26	52
Source of introduced species			
Europe	16	15	31
Australia	6	2	8
Asia	3	4	7
North America	1	3	4
Polynesia	0	2	2

FIGURE 16.14 The effect of introduced rats on the species diversity of native insular reptiles is revealed by consistent differences in species-area relationships for reptiles on islands off the coast of New Zealand, with and without rats. (After Whitaker 1978.)

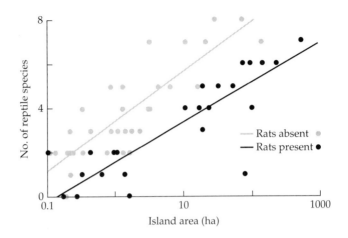

and 16.15). Overgrazing by pigs, goats, sheep, and other mammalian herbivores has affected many insular plants and has indirectly contributed to the demise of insular vertebrates by destroying their food supply or habitat. Other introduced species have served as vectors for pathogens, such as avian malaria and bird pox virus, Dutch elm (*Ophiostoma ulmi*) disease, and American chestnut (*Castanea dentata*) blight (see Jackson 1978; Berger 1981; von Broembsen 1989), which have caused range contractions and extinctions in many animals and plants. Hybridization of exotics with native species is another factor threatening biological diversity, especially in plants and fishes.

In most cases, however, native species have suffered from a combination of extinction forces. For example, four insular subspecies of the Galápagos tortoise were extirpated on islands that were infested with a combination of

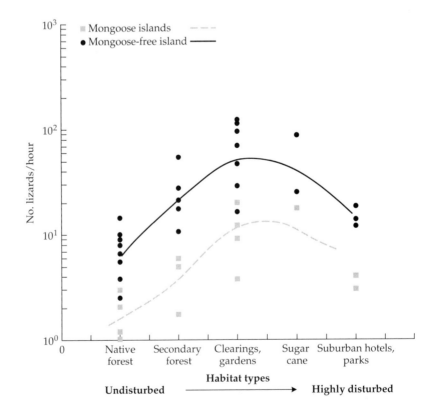

FIGURE 16.15 The effects of introduced mongooses on density and activity (as measured by number of lizards observed per hour) of native lizards inhabiting tropical islands of the Pacific Ocean. (After Case and Bolger 1991.)

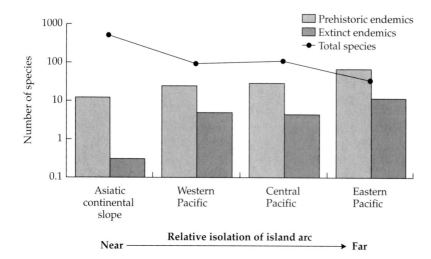

FIGURE 16.16 While the species richness of land birds decreases with the isolation of island arcs in the Pacific Ocean, endemicity and susceptibility to extinction increase. (These figures do not include prehistoric extinctions, which in some cases included as much as half of the native diversity of these islands prior to European colonization.) (After Greenway 1967.)

introduced mammals. Introduced rats preyed on their eggs, cats and dogs preyed on the young tortoises, and goats competed with the adult tortoises for vegetation. In still other cases, the effects of introduced species may be even more complex, involving indirect effects that cascade throughout the native community. Mongooses, which were introduced onto the Virgin Islands to control introduced rats, turned out to prey most heavily on native, ground-nesting birds and ground lizards (e.g., *Ameiva exsul*). The extirpation of these insectivores on some small islands was followed by outbreaks of insects, which in turn caused widespread damage to native plants.

Regardless of the specific causes of historical extinctions, they all speak to the susceptibility of insular biotas to invasion. These phenomena are consistent with theories of both general ecology and island biogeography. As Elton (1958) suggested, invasibility tends to be higher for disturbed and species-poor communities (see also Hobbs and Huenneke 1992). For example, the proportion of exotic birds and mammals tends to decrease with increasing richness of native species (Brown 1989). Greenway's (1967) studies of extinct and endangered birds of the Pacific Islands revealed that the avifauna of the more isolated archipelagoes were more prone to extinctions and harbored a higher proportion of threatened species (Figure 16.16). Within archipelagoes, avian extinction rates decreased with increasing area (Figure 16.17). This pattern appears to be a general phenomenon and one consistent with the equilibrium theory of island biogeography: extinction rates are higher on smaller and more isolated islands (see also Case and Bolger 1991; see also Burney et al. 2001).

Moreover, habitat alteration for agriculture and related anthropogenic disturbances tends to reduce the carrying capacity of islands for their native

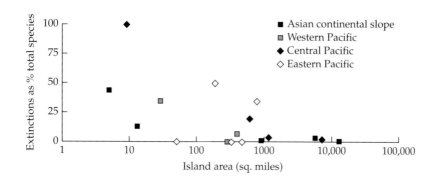

FIGURE 16.17 Extinction rates among land birds of the Pacific Islands tend to decrease with island area. (After Greenway 1967.)

FIGURE 16.18 The relationship between the number of introduced bird species and the number of extinct native bird species for several insular and continental regions. Invasion success appears to increase with the number of native species already extinct. Dashed line is line of equality; solid line is regression line. (After Case 1996.)

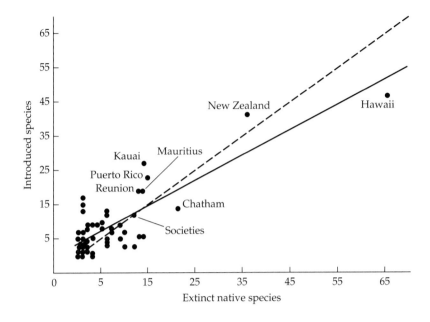

species while increasing the likelihood that introduced species can gain a foothold. This hypothesis is supported by Case's (1996) review of global patterns in the invasion success of exotic birds. Across a wide variety of archipelagoes and continental regions, invasion success increases with the number of native avian species already extirpated (Figure 16.18). In fact, Case found that the number of exotic species of birds established on an island was close to the number of species lost to extinctions.

While these results seem to support Elton's earlier contention that species-poor systems are more easily invaded, Case suggested that the invasion success of introduced birds was more directly linked to habitat disturbance. On islands, exotic species are largely restricted to disturbed sites, while native birds inhabit undisturbed refugia away from human activities. Thus, as humans and their commensals have converted native habitats into urban, agricultural, or otherwise disturbed sites, native birds have lost habitat and suffered extirpations, while exotic species have been provided with the habitat they require to establish and maintain populations. Of course, Case's and Elton's hypotheses are not mutually exclusive, and it is possible, if not likely, that habitat disturbance and reduced competition from native species combine to facilitate invasions by exotics. Although this debate will likely remain unresolved for some time, clearly once they have established populations on islands, exotic species have often devastated natives in both historic and recent times.

Geography of Endangerment

The historical record shows clearly that exotic species and habitat alteration have been important factors in causing extinctions, particularly among island endemics. Hunting, collection, and other forms of overexploitation have also taken a heavy toll, accounting for nearly a third of the avian extinctions and half of the mammalian extinctions worldwide since 1600 (Figure 16.19). Extinction patterns in mollusks, perhaps the hardest-hit group of animals, present a somewhat different picture. Most of their historical extinctions were associated with habitat destruction and pollution (60%), while introduced species and collecting accounted for a smaller, yet still significant, portion (24% and 14%, respectively).

(A) Causes of historical extinctions in animals

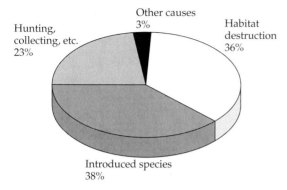

Hunting, collecting, etc. 23%

Other causes 3%

Habitat destruction 36%

Introduced species 38%

(B) Causes of endangerment in animals

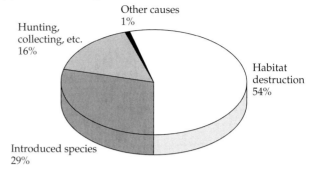

Other causes 1%

Hunting, collecting, etc. 16%

Habitat destruction 54%

Introduced species 29%

(C) Causes of historical extinctions in birds

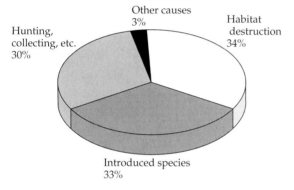

Other causes 3%

Hunting, collecting, etc. 30%

Habitat destruction 34%

Introduced species 33%

(D) Causes of endangerment in birds

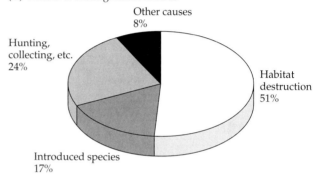

Other causes 8%

Hunting, collecting, etc. 24%

Habitat destruction 51%

Introduced species 17%

(E) Causes of historical extinctions in mammals

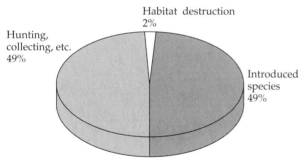

Habitat destruction 2%

Hunting, collecting, etc. 49%

Introduced species 49%

(F) Causes of endangerment in mammals

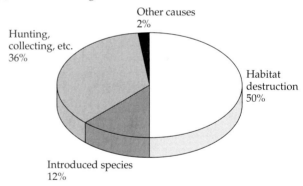

Other causes 2%

Hunting, collecting, etc. 36%

Habitat destruction 50%

Introduced species 12%

(G) Causes of historical extinctions in mollusks

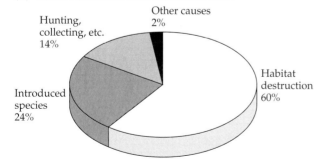

Other causes 2%

Hunting, collecting, etc. 14%

Introduced species 24%

Habitat destruction 60%

(H) Causes of endangerment in mollusks

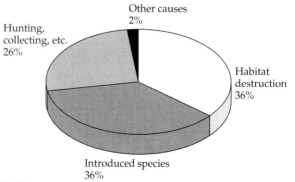

Other causes 2%

Hunting, collecting, etc. 26%

Habitat destruction 36%

Introduced species 36%

FIGURE 16.19 Causes of historical extinctions (1600–1980) and of current endangerment in selected animal taxa. (After Flather et al. 1994.)

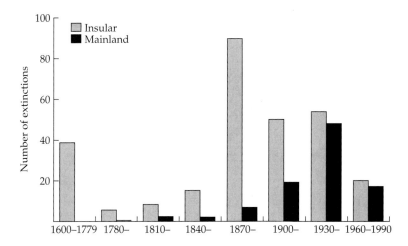

FIGURE 16.20 Historical trends in the relative numbers of insular and mainland extinctions (1600–1990) of animals worldwide. (After World Conservation Monitoring Centre 1992.)

While we have learned much by examining the patterns of past extinctions, it appears that the geographic signature of extinctions and their predominant causal forces are changing. Figure 16.20 reveals an important temporal-spatial trend in animal extinctions: the extinction front has expanded to include the mainland. Although nearly all animal extinctions recorded between 1600 and 1980 were insular, the number of extinctions of mainland species now rivals that of insular forms.

Ironically, the ongoing wave of extinctions may still be an insular phenomenon. Habitat destruction and fragmentation on the continents have transformed the geographic template and reduced once expansive stands of native habitats to ever-shrinking archipelagoes of habitat islands (Figure 16.21). Terrestrial biotas stranded in these remnants of native habitats may undergo a process of ecological relaxation similar to that described in Chapter 13. In contrast to the climatic and tectonic changes that have isolated habitats in the past, these anthropogenic habitat changes are occurring much more rapidly and may, therefore, take a heavier toll on the stranded species.

Whereas anthropogenic introductions and hunting were the primary causes of historical extinctions, habitat loss and fragmentation have now become the primary causes of endangerment in terrestrial animals (see Figure 16.19). In mollusks, species introductions and habitat alteration contribute equally to species endangerment. Patterns of endangerment in plants are similar to those reported for terrestrial animals (Figure 16.22). Habitat alteration, including development for housing, water control, mining, oil and gas, logging, off-road vehicles, and agriculture, represents the primary cause of endangerment for plant species listed by the U.S. Fish and Wildlife Service under the Endangered Species Act. The effects of introduced species and the effects of grazing and trampling by domestic livestock account for another 25% of species endangerment. Perhaps the most sobering finding is that natural causes account for just 1% of the endangerment of plants.

Habitat Loss and Fragmentation

MAGNITUDE OF THE PROBLEM.   According to many conservation biologists, habitat loss and fragmentation now represent the most serious threats to biological diversity (Wilcox and Murphy 1985; Laurance and Bierregaard 1996; Settele et al. 1996; Birdlife International 2000; Debinski and Holt 2000; Hilton-Taylor 2000; Bissonette and Storch 2002; Haila 2002; McGarigal and Cushman 2002). This is evident from the patterns of endangerment described above. Yet, we need to avoid overstating and simplifying what really is a very complex problem. The relative importance of different extinction forces varies markedly among taxa (see Figures 16.19 and 16.22) and among regions. Species introductions, hunting, pollution, climatic change, and other effects of human activities continue to pose serious threats to most endangered species. Again, almost every endangered species is threatened by a combination of these factors, many of them interacting with the effects of habitat loss and fragmentation. Even on oceanic islands such as Hawaii, however, where half the plants are already exotics, habitat destruction by humans and their domestic and feral livestock has become the primary threat to ferns and other endangered plants (Wagner 1995). Here, we shall first discuss the intensity and magnitude of habitat loss and fragmentation before examining the specific effects of these landscape dynamics on biological diversity.

The Earth's natural ecosystems are being transformed at an unprecedented rate. These include some of the world's most diverse systems, including the tropical rain forest biome, in which just 7% of the Earth's surface is home to over 50% of its species (Wilson 1988). A single rain forest tree can contain over 40 species of ants from 26 genera, approximately equal to the entire ant fauna of the British Isles. Yet, tropical rain forests are being destroyed at a rate of approximately 76,000 km^2 per year. Madagascar retains only 7% of its tropical forests, and only 1% of Brazil's Atlantic Coastal Forest remains uncut. Since 1819, the primary forest of Singapore has been all but completely removed, with less than 1% remaining (Corlett 1992). Similar patterns of deforestation have severely impacted tropical forest communities across the globe (Figure 16.23).

Loss of natural ecosystems has by no means been restricted to tropical rain forests, however (Figure 16.24). Prior to the impact of humans, grasslands covered approximately 40% of the Earth's surface (Clements and Shelford 1939). Now, just half of this amount remains (see World Conservation Monitoring Centre 1992). Pristine prairies once covered approximately 1 million km^2 of North America; now only 10% remains. Moreover, less than 1% of tallgrass prairie remains, nearly all of it on shallow, rocky soils or in tiny isolated reserves. The loss of wetlands has been no less severe. Estimates are that as much as 50% of the world's wetlands have been lost, most of this within the twentieth century, and that 65% of biologically significant tropical wetlands are currently threatened (Moyle and Leidy 1992). Freshwater systems are heavily affected by dams and other water control projects, which effectively fragment streams and rivers in most developed regions. In the marine realm, coral reefs are the aquatic equivalent of tropical rain forests in both diversity and productivity. Yet, many are threatened by a variety of human activities, including pollution, sedimentation, and overexploitation, all of which spread in a highly nonrandom fashion like a contagion across native landscapes and seascapes (see Figure 16.45).

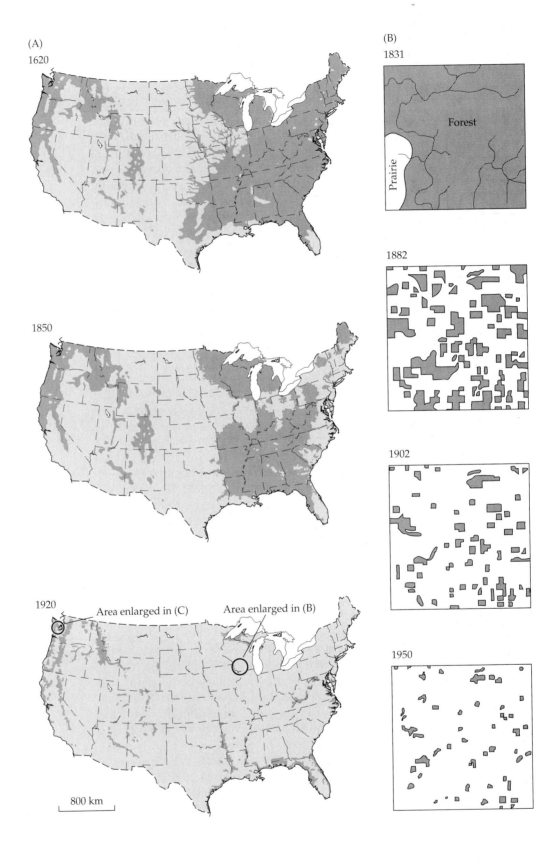

(A)
1620

1850

1920   Area enlarged in (C)   Area enlarged in (B)

800 km

(B)
1831   Forest   Prairie

1882

1902

1950

(C)
Original forest

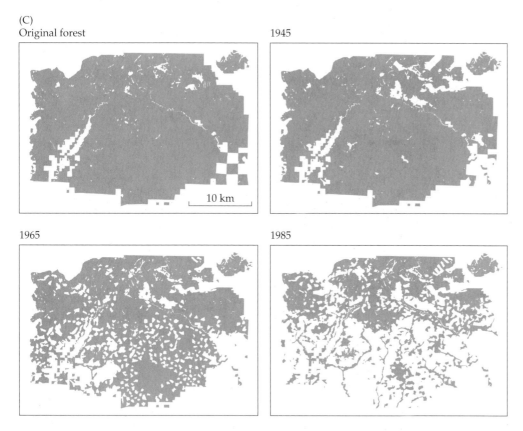

1945

1965

1985

10 km

FIGURE 16.21 The clearing of native North American forests (darker shaded areas) by European settlers has rapidly reduced once expansive stands of native forest habitat to archipelagoes of habitat islands. (A) United States. (B) Cadiz Township, Wisconsin. (C) Olympic National Forest, Washington. (A after Greely 1925, Williams 1990; B after Thomas 1956; C from D. R. Perault, unpublished report.)

ECOLOGICAL EFFECTS OF HABITAT LOSS AND FRAGMENTATION. Few of us can deny that modern civilizations are transforming the very nature of the Earth's land surface to a degree rivaling—and, perhaps exceeding—the climate-driven changes that characterized the glacial cycles of the Pleistocene. There are, however, a number of key differences. Anthropogenic transformations are occurring at a much more rapid pace, too rapid for many species to track their

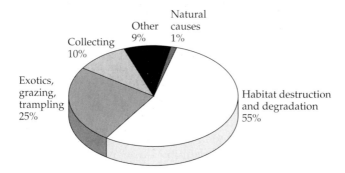

FIGURE 16.22 Causes of endangerment for plants of the United States. (After Schemske et al. 1994.)

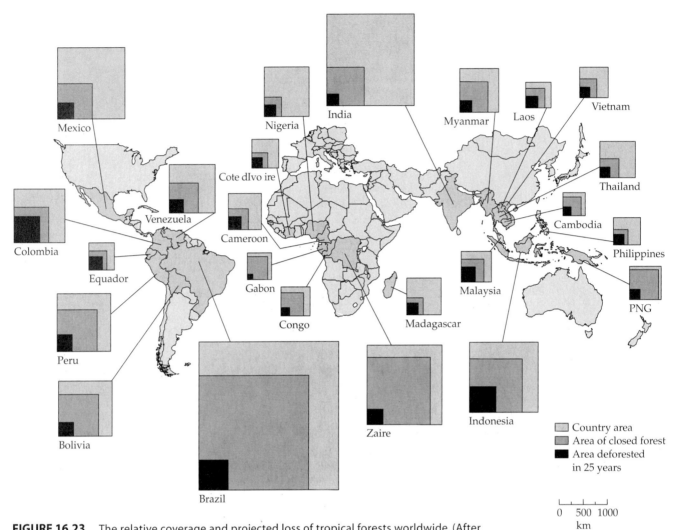

FIGURE 16.23 The relative coverage and projected loss of tropical forests worldwide. (After World Conservation Monitoring Centre 1992.)

optimal environments. Just as important, natural landscapes are not just being lost, but those that remain are being fragmented and degraded in quality.

The physical changes associated with habitat loss and fragmentation include the following (see reviews by Saunders et al. 1991; Harrison and Bruna 1999; Ricketts 2001; Villard 2002):

1. A reduction in the total area, resources, and productivity of native habitats

2. Increased isolation of remnant fragments and their local populations

3. Significant changes in the environmental characteristics of the fragments, including changes in solar radiation, wind, and water flux

Especially worrisome is the possibility that such chronic changes in local climate will trigger feedback effects, which in turn can accelerate the rate of habitat loss or prevent the regeneration of native ecosystems. Deforestation, for example, often leads to drier conditions in remnant patches of forest, which makes them susceptible to fires of unnaturally high intensity and frequency. In Borneo, deforestation—the slash-and-burn agriculture that replaced the forests—and seasonal droughts associated with El Niño events combined to

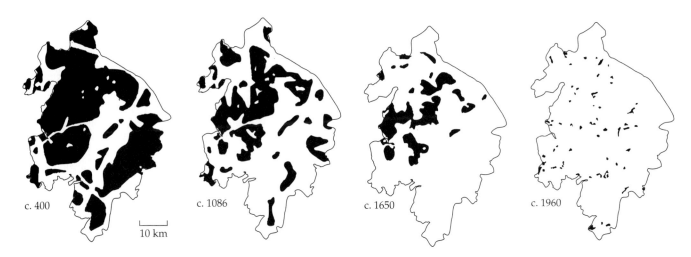

c. 400

10 km

c. 1086

c. 1650

c. 1960

FIGURE 16.24 Tropical rain forests are not the only habitats that have suffered extensive losses. This map shows the deforestation and fragmentation of temperate native forests in Warwickshire, England, between 400 and 1960 A.D. (After Wilcove et al. 1986.)

trigger massive forest fires, which destroyed a large portion of the island's remaining forests. Although fire is a natural and essential component of many ecosystems, it is now becoming a much more prevalent and serious threat to conserving the diversity of rain forests and other ecosystems not normally subjected to its effects.

As a result of these anthropogenic changes, natural systems may undergo a type of ecological relaxation similar to that described by Brown (1971) and Diamond (1972) for landbridge islands (see Chapter 13). In most cases, relaxation or biotic collapse after habitat fragmentation is expected to proceed in a sequence of stages (after Wilcove 1987):

Stage 1: Initial exclusion. Some species will be lost from the landscape simply because none of their populations were included in the remnant patches.

Stage 2: Extirpation due to lack of essential resources. Species vary tremendously in their resource requirements, many requiring very large areas and/or very rare resources. The likelihood that all of a species' resource requirements can be met decreases as the remaining area decreases.

Stage 3: Perils associated with small populations. Small populations are much more susceptible to a host of genetic, demographic, and stochastic problems (see Soulé 1987). As the total area of the remnant patches decreases, these problems become increasingly severe.

Stage 4: Deleterious effects of isolation. As discussed in Chapter 13, some populations may be rescued from extinction by migration and recruitment of individuals from other populations. The likelihood of such rescue effects decreases as isolation increases.

Stage 5: Ecological imbalances. While communities may not respond as superorganisms or units in the Clementsian sense (see Chapter 5), most species are strongly influenced by interactions with other species. Loss of one species during any of the above stages of relaxation may result in the subsequent loss of its predators, parasites, mutualists, or commensals. In addition, habitat disturbance and reductions in community diversity during the earlier stages of relaxation may facilitate the establishment of introduced species, triggering a cascade of subsequent extirpations.

Ecological imbalances and the cascading effects of initial losses in biological diversity are fairly well documented by the historical record and appear highly nonrandom with respect to species identities—a phenomenon sometimes termed **community disassembly** (see Chapter 14; Fox 1987; Mikkelson

1993; Lomolino and Perault 2000). For example, as a result of the fragmentation of native forests in North America, populations of brown-headed cowbirds (*Molothrus ater*) underwent dramatic increases in density and expanded their ranges into newly cleared areas and edges of remnant forests. Nest parasitism by cowbirds now takes a heavy toll on many Neotropical songbirds, especially those breeding within 500 m of the forest edge (see Brittingham and Temple 1983; O'Conner and Faaborg 1992). These songbirds suffer from indirect as well as direct effects of deforestation in both their breeding and wintering ranges. In the United States, deforestation has caused significant range contractions among large mammalian predators, including the cougar (*Puma concolor*), gray wolf (*Canis lupus*), and red wolf (*Canis rufus*). The loss of these top carnivores has resulted in compensatory increases in the densities of smaller carnivores (especially coyotes) and omnivores (raccoons, opossums, and skunks), which prey heavily on smaller animals—namely, songbirds and their eggs or nestlings. Fragmentation may also influence community structure by limiting the densities of native decomposers. In the fragmented tropical rain forests north of Manaus, Brazil, the diversity and density of carrion- and dung-feeding beetles was found to be markedly lower in clear-cuts and fragments than in adjacent stands of contiguous forest (Klein 1989; see also Bierregaard et al. 1992; Lomolino et al. 1995; Lomolino and Creighton 1996; Laurance and Gascon 1997). As a result, decomposition and recycling of dead organic material was greatly reduced.

Although ecological imbalances and secondary extinctions represent the final stage of relaxation, this stage may be of long duration. Modeling studies by David Tilman and his colleagues (Tilman et al. 1994) suggested that even moderate intensities of habitat destruction may well cause time-delayed, secondary extinctions even among currently dominant species. The cascading effects we have just discussed often take generations to move from one trophic level to another. Because any realistic scenario may involve many steps, today's habitat losses are likely to result in a substantial "extinction debt," a future ecological cost that may take many generations, possibly centuries, before it is fully realized.

Finally, even those species that somehow survive may exhibit morphological, physiological, ecological, and possibly evolutionary responses to fragmentation, changes consistent with those natives of true islands. Recall from Chapter 14 that the island syndrome includes a suite of ecological and evolutionary changes, among these being evolutionary trends in body size (i.e., the island rule; see Figure 14.25A). Schmidt and Jensen's (2003) study of body size trends in mammals inhabiting fragmented forests of northern Europe may be the first systematic and broad-scale study of mammals to report morphological trends in response to fragmentation that are entirely consistent with the island rule (i.e., small species tending toward gigantism and large species tending toward dwarfism; see Figure 14.25C; see also Fredickson and Hedrick [2002] for a report of similar trends, i.e., decreased body size, for captive and reintroduced Mexican wolves [*Canus lupus baileyi*]; for similar body size responses to fragmentation in skinks, see Sumner et al. 1999, and in African rainforest birds, see Smith et al. 1997).

Biogeography of Global Climate Change

Humanity's influence on the diversity and the geography of nature goes beyond our ability to transform landscapes. It is now clear that we can cause climatic changes at magnitudes perhaps rivaling those associated with the glacial cycles of the Pleistocene. If current trends continue, we will be heating up an interglacial (i.e., an already warm) period by perhaps as much as 6° C over the next century. Most ecologists and biogeographers agree that such a temperature change, if it does occur, will dramatically influence the diversity and distribution of life. Many plants and animals may already be signaling the early effects of global warming; but surely we can be more specific. Which species and which ecosystems will be most affected? As biogeographers, we should have some especially relevant insights for predicting geographic dynamics (e.g., shifts, expansions, and contractions of geographic ranges) provided, of course, reliable information on the magnitude and geographic profile of climate change is forthcoming (see Figure 16.25C; Stott and Kettleborough 2002; Zwiers 2002; see also Knutti et al. 2002). How much, and in what ways, will the geographic ranges of species expand or contract, and which and how many species will go extinct?

Magnitude and Geography of the Problem

The consensus among most meteorologists is that the already high and still increasing levels of carbon dioxide (CO_2), methane (CH_4), chlorofluorocarbons (CFCs), sulfur dioxide (SO_2), nitrous oxide (NO_2), and other anthropogenically derived greenhouse gases in the atmosphere have increased average global temperatures by approximately 0.7° C since 1860 (Figure 16.25). The term "greenhouse gas" refers to the fact that these gases allow solar radiation in the form of sunlight to penetrate the atmosphere, but retard the outward movement of infrared radiation from the Earth's surface, thereby trapping heat and increasing global surface temperatures. Human activities over the past two centuries have been responsible for an estimated 25% increase in atmospheric levels of CO_2 and a 100% increase in levels of methane (see Peters and Lovejoy 1992; Vitousek 1992; Karieva et al. 1993; IPCC 2001a,b). Nitrous oxides have also increased dramatically, especially during the past three or four decades. Of all the greenhouse gases, CFCs, which are solely anthropogenic and are the primary agents in the erosion of the protective ozone layer in the upper atmosphere, have increased most rapidly since their first detection during the early 1950s. Although these gases vary in their origin, concentration, and ability to absorb infrared radiation, they all are increasing exponentially, reflecting the exponential increases in human population densities and industrial technologies during the past two centuries. According to a consensus of scientists (see IPCC 2001; Schneider and Root 2002; Zwiers 2002), the combined effects of these anthropogenic increases in greenhouse gases has accounted for most of the global warming over the past 50 years.

While predicting changes in global climate is an exceptionally complex task, projected concentrations of greenhouse gases can serve as a useful yardstick. If the current trends continue over the next century, levels of these gases are likely to increase so as to effectively double their present capacity to trap heat in the lower atmosphere (referred to as "CO_2" doubling). Schneider (1993) has summarized the climatic changes predicted under a CO_2-doubling scenario, which include changes in a variety of environmental factors in addition to average global temperatures (Table 16.7; see also Schneider and Root 2002). For the moment, however, let's focus on temperature change. The projected climatic changes include a 1.5° to 6° C increase in average global temperatures over the next century, with a "best guess" mean of 2.5° C. It is important that we put this increase in context. Since the peak of the last glacial maximum, average global temperatures increased by approximately 4.5° C, most of this taking place over a period of about 5000 years (see Chapter 9). This change was sufficient to melt the great continental glaciers and cause sea levels to rise by approximately 100 m. Now contrast this rate of warming—

FIGURE 16.25 (A) Natural variation in atmospheric concentrations of carbon dioxide (solid line) over the past 160,000 years (based on ice core analysis). Recent anthropogenic increases in carbon dioxide concentrations are coincident with exponential growth of the human population (dashed line). (B) Increases in the concentrations of relatively stable greenhouse gases since the onset of the industrial age. (C) Map of predicted global change in surface air temperatures (°C) from 2000 to 2003, which is expected to have an uneven distribution with greater warming over land, especially at higher latitudes. (D) Over the past century and a half, the average global temperature has increased between 0.6 and 0.8°C, with most of the increase occurring over the past four decades. (A–B after Vitousek 1992; C after Zwiers 2002; D after Patz 2002.)

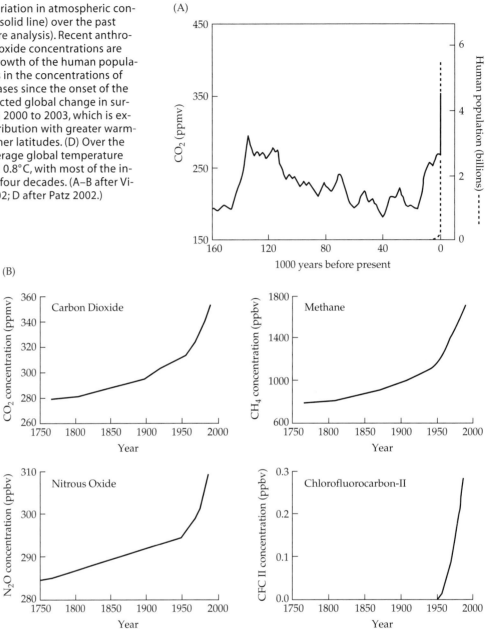

less than 1° C per thousand years—with the predicted increase in average global temperatures of 2.5° to 3.0° C in just one century. We know that even the relatively slow, natural warming of global climates had an enormous impact on biogeography and the biodiversity of late-Quaternary biotas. If the predicted rise in global temperatures actually occurs, its effects may be even more dramatic. Again, it is important to emphasize that the predicted global warming is occurring during an interglacial (i.e., warm period of the Pleistocene); thus, the predicted changes may result in one of the warmest periods of the past 2 million years. Just as important, and far different from the much earlier and simplistic assumptions that climate changes will be homogeneous across the globe, current models predict that changes in temperature and pre-

(C)

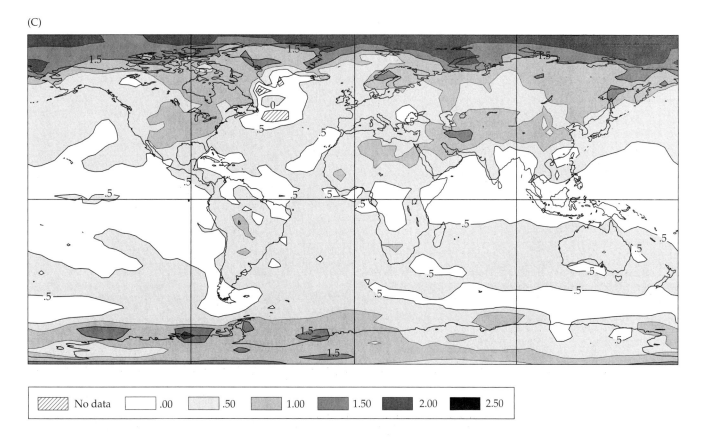

(D)

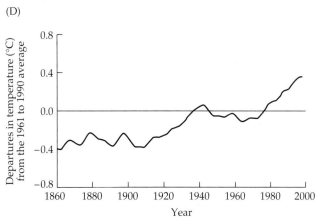

cipitation will be markedly uneven, with especially intense warming over land and at the higher latitudes (see Figure 16.25C).

It appears that the complex geographic profile of climate change may be characteristic of natural as well as anthropogenic changes. Reviews of biogeographic dynamics during the last glacial recession, however, reveal that species often shift their ranges individualistically, independent of other species (see Chapter 9). Even within the same taxonomic group, species differ markedly in their ability to adapt to new environmental conditions or to disperse to other sites when local conditions deteriorate. Thus, during the last glacial recession, species ranges shifted in different directions as well as at different rates, with some species migrating so slowly that their range expansion

TABLE 16.7 *Predicted changes in climatic conditions as a result of doubling in atmospheric concentrations of CO_2*

Degree of confidence[a]	Predicted change
	Temperature
*****	The lower atmosphere and Earth's surface warm
*****	The stratosphere cools
***	Near the Earth's surface, the global average warming lies between +1.5°C and +4.5°C, with a "best guess" of 2.5°C
***	The surface warming at high latitudes is greater than the global average in winter but smaller than in summer (in time-dependent simulations with a deep ocean, there is little warming over the high-latitude ocean)
***	The surface warming and its seasonal variation are least in the tropics
	Precipitation
****	The global average increases (as does that of evaporation); the larger the warming, the larger the increase
***	Increases at high latitudes throughout the year
***	Increases globally by 3–15% (as does evaporation)
**	Increases at mid-latitudes in winter
**	The zonal mean value increases in the tropics, although there are areas of decrease. Shifts in the main tropical rain bands differ from model to model, so there is little consistency between models in simulated regional changes.
**	Changes little in subtropical arid areas
	Soil Moisture
***	Increases in high latitudes in winter
**	Decreases over northern mid-latitude continents in summer
	Snow and Sea Ice
****	The area of sea ice and seasonal snow cover diminish

Source: Schneider 1993.

[a]The number of asterisks indicates the degree of confidence determined subjectively from the amount of agreement between models, and confidence in the representation of the relevant process in the model. Five stars indicate virtual certainties: one star indicates low confidence in the prediction.

continues today, some 5000 to 10,000 years following the last major climatic shifts (see Figure 9.19A). Then again, seldom throughout the entire Quaternary (the past 2 million years) have global temperatures increased with the rapidity of predicted and, apparently ongoing, anthropogenic changes in climate. Thus, if trends continue, biotic turnover and extinctions may exceed those of even the most severe episodes of the Pleistocene (see Chapter 9). Many species may not be able to track predicted geographic shifts in climate regimes, while others may suffer extinction when their optimal climate contracts or shifts to create unfavorable combinations of weather and soil conditions (see example below for Kirtland's warbler).

Effects on Species Distributions

If the projected changes in global climates do occur, they will no doubt alter the distributions of most terrestrial plants and animals. A starting point for predicting climate-driven change in the geographic range of a species is to define the species' climate space—the climatic dimensions of its niche, or the set of climatic conditions required by a particular species (Figure 16.26). Once we have defined the limits of a species' climate space, we can apply them to maps of projected climates to predict changes in the species' geographic range. But this, of course, is too simplistic. We know that species do not occupy all regions within their climate space. Much like the ecological concepts of real-

(A)

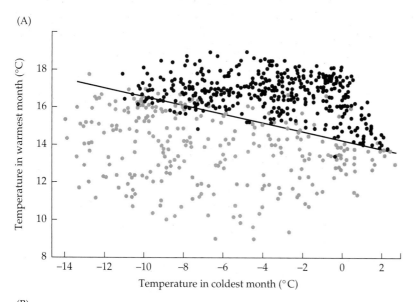

(B)

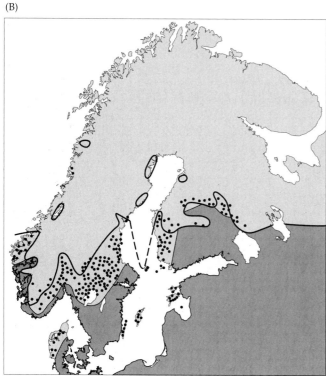

FIGURE 16.26 (A) A depiction of the climate space of the basswood tree (*Tilia cordata*), based on average temperatures during the warmest and coldest months. (Black dots indicate locations within the range of this species; gray dots depict those outside the range.) (B) The actual distribution (black dots). The range margin (solid line), as translated from the straight line in A, defines the marginal climate space for this species. (After Hengeveld 1992.)

ized and fundamental niches, the realized geographic range attained by a species will always be less than that defined by its climate space (i.e., its fundamental range; Chapter 4).

We can think of at least two reasons for this to be so. First, realized geographic ranges are influenced by species' interactions with all features of the geographic template, including spatial variation in abiotic factors (e.g., soil type and disturbance history) and interactions with other species. Reproduction, mortality, immigration, and emigration are influenced by other niche dimensions, both abiotic (requirements for soil, salinity, and nutrient conditions) and biotic (interactions with predators, competitors, parasites, and mutualists). Second, as we noted above, range shifts and expansions may be quite slow, and, at least in some—perhaps many—instances, species may never jump across all barriers to colonize otherwise suitable habitats. Thus, we need at least four types of information to predict the effects of climatic change on geographic ranges of native species.

1. The geographic profile and rate of climatic shifts, and the projected distribution of climatic zones across geographic regions. Note, that as we saw for the major climatic shifts of the Pleistocene (see Chapter 9), new climatic/edaphic zones are likely to be created while others have contracted and some vanished.

2. Estimates of the climate space for focal species, with special emphasis on geographic variation of factors typically included in climatic change models (e.g., average annual precipitation and mean monthly temperatures).

3. Predicted distributions of habitats and of landscape features (e.g., rivers, mountains, roadways, and other inhospitable habitats) that might serve as immigration barriers, filters, or corridors.

4. Estimates of the habitat affinities of the species, their capacities to disperse across future landscapes, and their abilities to influence distributions of other species.

Below we provide several examples of how such information can be used to predict climate-driven changes in the geographic ranges of organisms.

Effects on Terrestrial Biotas

Climate space is a well-established concept among plant ecologists. The environmental limits of this space can be determined by laboratory growth experiments or by noting environmental conditions that coincide with the geographic range limits—both past and present—of the focal species in question. In an early study, Zabinski and Davis (1989) used three climatic variables (mean January temperature, mean July temperature, and annual precipitation) to predict range changes in four tree species of deciduous forests in the Great Lakes region of North America. The northern distributional limits of all four species coincided strongly with the mean January isotherm of –15° C. On the other hand, the southern limits of yellow birch and hemlock coincided closely with mean July temperatures and the growth of all four species was strongly limited by annual precipitation. Zabinski and Davis then used the output of a global circulation model (GCM) to map the projected steady-state climate under the scenario of CO_2 doubling. They assumed that these tree species could migrate at a rate of 100 km per century. As Gates (1993) noted, these rates are probably much too high, perhaps by a factor of five or ten. In her earlier work on range shifts of trees during the last glacial recession, Davis (1981) estimated that dispersal rates ranged between 10 and 40 km per century (20 km/century for hemlock, beech, and maple: see Table 9.1). Given that projected landscapes will be even more fragmented and divided by roadways and other anthropogenic barriers to dispersal, future rates of tree dispersal should be much lower than those during the last glacial recession.

Nevertheless, Zabinski and Davis's (1989) results are informative. The differences between the current ranges of the four tree species and their projected ranges are presented in Figure 16.27. Even with the relatively high dispersal rate used in this exercise, the projected ranges after 100 years of migration are much less than the potential ranges based on the species' climate space. In addition, the projected realized ranges for all four species are much less than the extant ranges of these species.

Cascading ecological effects, with range dynamics of one species affecting that of others, may be no less important than the more direct effects of climatic change (Gates 1993). With reductions in the ranges of the now dominant deciduous tree species, the geographic ranges and cover of aspen and other early successional species are likely to increase, in turn triggering substantial increases in populations of browsing herbivores such as white-tailed deer. Overbrowsing by deer, compounded by a reduction in beech seed production, may in turn contribute to significant declines in birds and small mammals. Gates (1993) also suggests that fire may become a more prevalent component of these ecosystems, and that trees persisting in marginal environments will be stressed and likely to suffer from a higher incidence of insect pests and dis-

FIGURE 16.27 Present (left) and predicted (right) geographic ranges for four species of trees common to eastern regions of North America. The predicted ranges are based on the assumption of a doubling of current concentrations of CO_2; the dark gray area depicts the projected range in 2090 considering estimated species-specific rates of migration; the lighter gray area depicts the ultimate potential ranges provided enough time for species to migrate to all regions within their climate space. (After Gates 1993.)

Eastern Hemlock (*Tsuga canadensis*)

Beech (*Fagus grandifolia*)

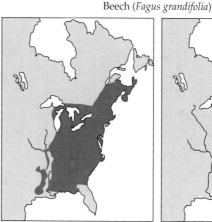

Yellow Birch (*Betula alleghaniensis*)

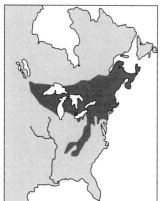

Sugar Maple (*Acer saccharum*)

400 km

ease. In fact, even with the high rates of dispersal assumed by Zabinski and Davis, they concluded that many hardwood species of the Great Lakes region would be threatened with extinction under this climatic scenario.

In another early, illustrative case study, Terry Root modeled the effects of climate on the winter ranges of songbirds in North America (Root 1988a,b,c 1993; Root and Schneider 1993, 1995; see also Repasky 1991; Root 1994, 2000; Stenseth et al. 2002). Both Hutchinson and MacArthur had hypothesized that, while the southern range limits of terrestrial animals of North America are largely set by interactions with other species, their northern range limits are more strongly influenced by abiotic factors, especially temperature. Following their lead, Root based her projections primarily on bioenergetic considerations and the locations of isotherms during the most stressful time of year: winter. She assumed that during winter, these diurnal birds must survive the night by tapping the energy that they accumulated and stored as fat and other metabolizable tissues during the day. The cooler the temperature, the more energy a bird must use for thermoregulation. Using known or estimated metabolic rates for a variety of species, Root was able to estimate the minimum winter temperature at which a bird of particular body mass would exhaust all of its energy supplies to survive through the night. This winter temperature or isotherm represents a critical dimension of the climate space of these birds. For most of the species she studied, it lies between 0 and –5° C. Root then compared the actual winter ranges of particular songbird species with minimum January isotherms across North America, and found that the northern range limits of 51 species of passerines were strongly associated with minimum winter temperatures (Figure 16.28). She also found that among these 51 species, larger birds had ranges that extended farther north, consistent with bioenergetic considerations (i.e., cold tolerance and energy stores in homeotherms generally increase with body size).

Using maps of projected shifts in winter isotherms under climatic change, Root's model can be applied to predict climate-induced shifts in the winter ranges of birds (see Root and Schneider 1995, 2002). But the shifting of winter isotherms is just one of the many factors—both abiotic and biotic—that can influence range dynamics. For example, Kirtland's warbler, which is narrowly restricted during the breeding season to jack pine (*Pinus banksiana*) habitats of

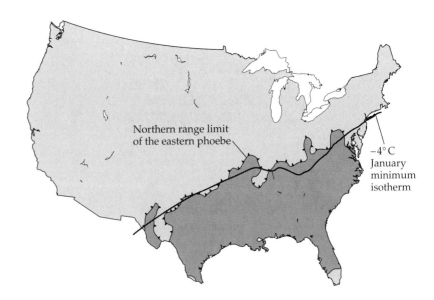

FIGURE 16.28 The northern limit of the winter range of the eastern phoebe (*Sayornis phoebe*) corresponds very closely with the –4° C minimum January isotherm (heavy solid line). Bioenergetic models also indicate that below this temperature, energy stores of these small birds are insufficient to maintain a constant body temperature during their nightly fasts. (After Root 1993.)

Michigan, is unlikely to survive even a small climatic change (Root 1993; Root and Schneider 1993; see Botkin et al. 1991). Jack pines require relatively well drained, sandy soils for survival—soils that do not occur directly north of the birds' current range. Therefore, even if Kirtland's warblers could potentially track their climate space as it shifts northward, they would probably be stranded because of their other niche requirements.

Root (1988c) also investigated spatial associations between the northern, eastern, and western range boundaries of songbird species and a variety of climatic conditions, as well as potential vegetation (Figure 16.29; southern range boundaries were not analyzed because they often extended beyond the region sampled in Root's study). Overall, range boundaries were most closely associated with potential vegetation, which is not surprising given that it reflects the combined variation in climatic and many other abiotic and biotic variables. The association of range boundaries with the climatic variables varied substantially. For example, northern range boundaries were strongly associated with average minimum January temperatures and mean length of the frost-free period. In contrast, eastern range boundaries were most closely associated with annual precipitation, while western range boundaries were associated with these two variables plus elevation. Elevational limits along western range boundaries are coincident with the Rocky Mountains, which run north to south and mark the range boundaries of many North American species (see also Harte and Shaw 1995).

Since these and other earlier studies, biogeographers, ecophysiologists, landscape ecologists, and many other scientists have studied responses to predicted climate change (e.g., see Abrahamson 1989; Peters and Lovejoy 1992; Gates 1993; Karieva et al. 1993; Harte et al. 1995; Parmesan et al. 1999; Root 2000; Parmesan and Yohe 2003; Root et al. 2003; Thomas et al. 2004). Here, we illustrate an alternative approach to predicting the ecological and biogeographic effects of global climate change. This particular approach is based on the patterns of biotic relaxation and disassembly described in Chapters 13 and 14. Using the fairly extensive data on the distributions of small boreal mammals inhabiting montane forests of the Great Basin (see Figure 13.25), McDonald and Brown (1992) were able to predict how many and which species should be lost from these forests if global climates warm by 3° C. Such a change would cause an upward shift in the lower elevational limits of woodlands and forests by approximately 500 m. Because there is less area at the

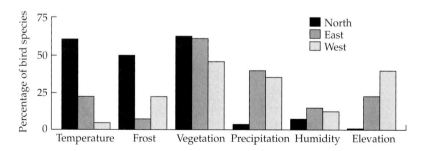

FIGURE 16.29   The percentage of North American bird species whose northern, eastern, and western range boundaries during winter were associated with various environmental factors. Temperature = average minimum January temperature; frost = mean length of frost-free period; vegetation = potential vegetation; precipitation = average annual precipitation; humidity = average relative humidity; elevation = elevation above sea level. Southern boundaries of winter ranges were not analyzed because they often extended beyond the study area. (After Root 1988c.)

(A)

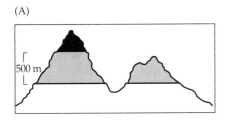

(B)

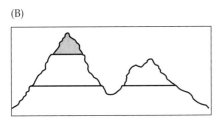

FIGURE 16.30 The approximate elevational boundaries of the vegetation types on the isolated mountain ranges of the Great Basin (A) today and (B) in the future after postulated climatic warming of approximately 3° C. Unshaded = desert shrub; gray = piñon-juniper woodland; black = mixed coniferous forest. An elevational shift of 500 m would decrease the area of woodland on all mountain ranges in the region and eliminate coniferous forest from some of them. (After McDonald and Brown 1992.)

higher elevations of a mountain, such elevational shifts would reduce the habitat area available to boreal mammals (Figure 16.30). McDonald and Brown used topographic maps and planimetry to predict the future area of woodlands and forests in each mountain range. Then, using the species-area relationships of extant communities, they were able to calculate how many species should be lost from each mountain range (Figure 16.31).

McDonald and Brown's results are sobering. They predicted a total of 45 extirpations across the 19 mountain ranges, with 4 ranges losing at least half their species. Assuming that the extinctions would be selective, McDonald and Brown were able to go one step further and predict not just how many, but which species would be lost from each mountain range. Selective extinction in these communities is thought to have produced the nested structure illustrated in Table 16.8. Given this structure, extirpations should be most likely for the populations along the diagonal running from the lower left to the upper right of the table—that is, those marked with an E. Of the 14 species considered in this study, only two were predicted to survive on all mountain ranges, whereas three were predicted to be lost from the entire system.

As McDonald and Brown (see Brown 1995) acknowledged, their predictions should be viewed with caution. In order to arrive at them, they had to make several assumptions about the magnitude of regional temperature change, elevation shifts in boreal habitats, and the responses of small mammals to these changes. They also assumed that the effects of increased fragmentation and isolation on these habitats and interactions among species were unimportant. Thus, while they provided a sobering thought experiment, McDonald and Brown's approach was probably too simplistic to reliably predict the biotic effects of global warming.

FIGURE 16.31 The species-area relationship can be used to predict changes in boreal mammal species richness among the isolated mountain ranges of the Great Basin as a result of climatic warming. Numbers of species were plotted as a function of the area above 2280 m elevation. Arrows show the changes in area and numbers of species predicted to be caused by climatic warming; the open circle at the base of each arrow indicates the present number of species in each mountain range, and the solid circle at the point of the arrow indicates the number of species predicted to remain after a 3° C increase in average temperature. (After McDonald and Brown 1992.)

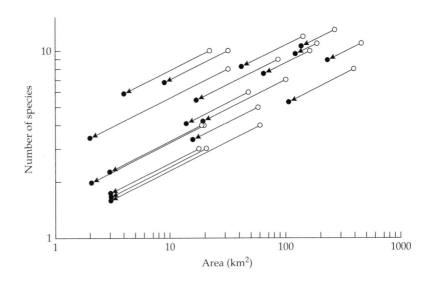

TABLE 16.8 *Distribution of 14 small boreal mammal species among 19 isolated mountain ranges in the Great Basin at present and after predicted extinctions due to the effects of global warming*

Species	Mountain Ranges																			Number of ranges occupied	
	4	**8**	**5**	**1**	**15**	**19**	**16**	**13**	**3**	**18**	**12**	**9**	**11**	**7**	**6**	**14**	**10**	**2**	**17**	**Present**	**Predicted**
Eutamias umbrinus	X	X	X	X	X	X	X	X	X	X	X	X	X	X	X	X	X			17	17
Neotoma cinerea	X	X	X	X	X	X	X	X	X	X	X	X		X		X	X	X	X	17	17
Eutamias dorsalis	X		X	X	X	X	X	X	X	X	X	X	E	E		E	E	X	X	17	14
Spermophilus lateralis	X	X	X	X	X		X	X	E		X	X	X	E		E			E	14	10
Microtus longicaudus	X	X	X	X	X	X	X	X	E	X	E		E			X				13	10
Sylivilagus nuttalii	X	X	X	X	X		E		E	E		E	E	E				E		12	5
Marmota flaviventris	X	X	X	X	X	X	E		E	E	X		E							11	7
Sorex vagrans	X	X	X	X	X	X	E	E		X	E									10	7
Sorex palustris	X	X	X	X	E	X			E				E							8	5
Mustella erminea	X			E	E	E	E			E										6	1
Ochotona princeps	X	E	E	E					E											5	1
Zapus princeps	E	E	E		E								E							5	1
Spermophilus beldingi	E	E																		2	0
Lepus townsendii		E			E															2	0
Present number of species	13	12	11	11	10	10	9	8	8	8	7	6	5	4	4	4	3	3	3		
Predicted number of species	11	8	9	10	8	7	5	5	3	6	4	4	3	2	2	2	2	2	2		

Source: From McDonald and Brown 1992.

Note: X = species now present and expected to persist; E = species now present and predicted to go extinct.

However, McDonald and Brown's approach, along with earlier models of Root and her colleagues, did serve to illustrate the multifactorial nature of ecology and biogeography. Just as important, they identified the geographically explicit information essential to predicting the effects of global climate change. Accurate predictions of the biogeographic responses to global climate change will no doubt depend on many advances, including better assessments of changes in landscapes and seascapes that may affect dispersal (see the discussion on changes in connectivity and isolation).

Effects on Freshwater and Marine Biotas

The geographic ranges of marine and freshwater species will also be affected by global warming if precipitation and evaporation potentials change according to current projections (see Table 16.7, see Figure 16.25). For example, Nash and Gleick (1991) estimate that a 4° C increase in average annual temperature will result in a 9% to 21% increase in mean annual runoff into streams of the Colorado Basin of western North America. The anticipated anthropogenic changes in climates over the next century may also cause seasonal shifts in periods of peak runoff and streamflow (e.g., see Gleick 1989). The predicted increased ratio of rain to snowfall and the anticipated elevation of the snowline may increase streamflow during fall and winter, while decreasing it during spring and summer. These shifts could cause changes in water quality and in the levels and circulation patterns of lakes and other freshwater systems. This, in turn, could result in significant and widespread changes in the composition and diversity of aquatic communities (see also Grimm 1993; Sanford 2002). Because of the limited size of freshwater systems compared with terrestrial ones, it is much more likely that freshwater species will become stranded and unable to persist by dispersing to other streams and lakes or across watersheds.

In contrast to freshwater systems, it may seem logical to assume that, because of their massive volume, marine systems will be buffered against most effects of anthropogenic changes in global and regional climates. Yet, marine communities actually may be severely affected by a combination of environmental changes associated with warming climates. It appears that global climate change will increase warming and evapotranspiration of surface waters. Ocean currents, including major ones such as the Gulf Stream and Humboldt, will then shift substantially. As Norse (1993) observed, the atmosphere and the ocean are two tightly linked parts of the same system, exchanging huge amounts of gases and heat. Oceanographers have identified several changes in marine environments that may directly or indirectly affect native species. The effects of CO_2 doubling on the marine environment are likely to include the following (see Titus 1990; Fields et al. 1993; Rahmstorf 1999; Sagarin et al. 1999; Smith et al. 1999):

1. Latitudinal and vertical shifts in water temperatures will occur. In general, warm waters will shift poleward, and to greater depths.

2. Currents and their effects on horizontal and vertical mixing of waters will be altered significantly.

3. Surface water temperatures will increase in many regions, but decrease in others, especially areas along the west coasts of continents, where upwelling may increase in magnitude.

4. Salinity of surface waters near the equator and at higher latitudes will decrease due to increased regional precipitation.

5. pH of surface waters is expected to drop by 0.3, equivalent to a doubling in acidity.

6. Sea levels are expected to rise by 0.5 to 2.0 m globally due to thermal expansion of seawater and additional ice melt.

While the precise magnitude and geographic profile of these disturbances are impossible to predict with any reasonable degree of certainty, they will dramatically affect species distributions and biodiversity in marine communities. For example, the predicted rise in global sea levels will substantially reduce the size of coastal wetlands and possibly coral reefs as well (Norse 1993). The projected rate of increase in sea levels is likely to overwhelm the ability of wetlands to re-form farther inland, and of reef growth to track the rising waters. As we learned in earlier chapters, decreases in habitat area translate into decreases in species diversity. In the United States alone, some 80 species of endangered plants and animals occur only in the narrow 3-m band above sea level (Reid and Trexler 1991). These and many other rare intertidal species worldwide are likely to perish unless they can migrate apace with rising sea levels. Not surprisingly, insular biotas will be especially threatened by the projected increase in sea levels. For example, the Maldives Archipelago, which includes 1190 small islands rising just 2 m above the surface of the Indian Ocean, would be submerged, thus destroying natural habitat for five endemic species of plants (Norse 1993).

As we also learned in earlier chapters, mixing of biotas often results in surges in the extinction rates of native species. This fact is especially relevant for predicting the effects of climatic change on marine communities. If surface waters warm substantially, the geographic ranges of cold-stenothermal species should contract, while marine species now restricted to tropical and subtropical regions may expand their ranges far into temperate and possibly subarctic regions (see Figure 9.25). If this mixing of long-isolated biotas occurs, it will result in the loss of many species due to competitive exclusion, predation, parasitism, or other interspecific interactions.

Studies of some recent transient shifts in ocean circulation may provide a sample of the ecological and biogeographic effects of more intense and persistent changes in global climates. El Niño phenomena are caused by fluctuations in the strength of trade winds. As a result, warm and nutrient-poor subtropical waters flow farther poleward along the western continental margins. The combined effects of increased water temperatures, reduced nutrient availability, decreased primary productivity, and the influx of subtropical invaders have severely affected marine communities. For example, the 1982–1983 El Niño event increased surface temperatures by as much as 2°C along the western coasts of North America, reduced primary productivity off the coast of California to its lowest recorded level, and triggered northward range expansions of at least 18 species of invertebrates and 38 vertebrates (McGowan 1985; Pearcy and Schoener 1987). Such effects no doubt cascaded through marine communities, affecting trophic levels from zooplankton to seabirds (Fields et al. 1993).

The high frequency and intensity of El Niño events during recent decades may actually be an early signal that global climate change is already affecting marine communities. Other potential signals are also apparent. In particular, the massive die-offs, or "bleaching," of coral reefs that occurred during the 1980s may be connected with increases in global temperatures. Coral reefs, the most productive and diverse systems in the marine realm, arise from a complex yet fragile mutualistic relationship between dinoflagellate algae and scleractinian corals. This mutualism is obligatory, and is easily disrupted by environmental disturbances, including relatively slight but abrupt changes in water temperature (see Glynn 1991). If current trends in levels of greenhouse gases continue, they may cause massive bleaching of coral reefs, drastically reducing their geographic ranges and triggering cascading effects on the overall diversity of the marine realm.

Changes in Connectivity and Isolation

How well can we now predict the effects of climatic change on the distributions of terrestrial and aquatic species? Let us review our list of requisite information for predicting the effects of climatic change on the geographic ranges of species (p. 684). It should be clear by now that the earlier modeling approaches discussed above, while offering many potential insights, are based on many untested assumptions. Climatologists, however, continue to make great progress in their ability to model and map climate change at global to regional levels (item 1 on our list). Climate space (item 2)—the climatic limits of a species' fundamental niche—has been estimated for a variety of species, both for plants and animals. Shifts in landscape features and habitats (item 3) must be estimated by integrating information from items 1 and 2 (i.e., climatic shifts and estimates of climate space for plants), which, in turn, constitute a major habitat component for many other organisms. Recent climatic shifts, such as El Niño events, along with the more prolonged shifts associated with

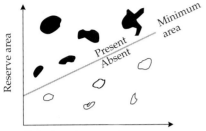

Isolation of prospective reserve

FIGURE 16.32 This species-based empirical model can serve as a spatially explicit tool for predicting reserve areas required to maintain populations of endangered species, or those of other high-priority species in isolated reserves. Solid symbols indicate species presence; open symbols indicate absence.

glacial cycles, can serve as valuable yardsticks for evaluating the assumptions used to model projected climates. They can also provide estimates of the abilities of some species to migrate across relatively heterogeneous and dynamic landscapes and seascapes (item 4).

Most approaches to modeling the biogeographic effects of climatic change, however, share one shortcoming: they do not explicitly address the effects of changes in habitat connectivity and isolation on the abilities of species to shift their ranges or survive within fragments of their current ranges. Real landscapes are heterogeneous mixtures of discontinuous habitats and topographic features that serve as dispersal barriers, filters, or corridors, depending on the focal species. Here again is a case in which biogeographic data may provide some valuable information. McDonald and Brown (1992) used such data to estimate the effects of climate-driven reductions in forest cover on boreal mammals of the Great Basin. Unfortunately, their approach cannot be used to assess the effects of changes in habitat isolation on species distributions. Such effects, however, can be predicted by using spatially explicit models, such as the mechanistic models developed by Hanski and others working in metapopulation theory (e.g., see Hanski and Thomas 1994; Hanski et al. 1995; Wahlberg et al. 1996). Empirically based, spatially explicit models such as the one illustrated in Figure 16.32 also can be used to estimate the effects of fragmentation on the distributions of endangered species regardless of whether the landscape changes result from climate change or from more direct activities (e.g., timber harvest and wetland conversion).

Metapopulation theory and the species-based model of island biogeography, presented in Chapter 14, provide some important insights for predicting the minimal area required to conserve populations of an endangered species. Both of these models are derivatives of MacArthur and Wilson's model, and both predict that, contrary to conventional wisdom, the minimum area or minimum reserve size required to maintain populations of a species increases with isolation (see Figure 16.32). Basically, because immigration rates decrease with isolation, populations on more isolated reserves must be larger to prevent extinctions from occurring before immigrants can rescue the population. These models also predict that minimum area requirements will be higher for less vagile species and for populations isolated by less hospitable matrix habitats. Distributions of hazel grouse *(Bonasa bonasia)* in isolated habitat fragments in Sweden (Aberg et al. 1995) and of small mammals on forested islands in the Great Lakes region of North America (Lomolino 1994) are consistent with the latter prediction (Figure 16.33).

Beyond these qualitative predictions, the species-based model summarized in Figure 16.32 can serve as a spatially explicit tool for designing and managing nature reserves or for predicting the effects of future anthropogenic disturbances. Given information on the distribution of an endangered species across a sample of islands, fragments, or prospective reserves, we can calculate the relationship between minimum area requirements and isolation—that is, the biogeographic space where the species should occur. The distribution of red squirrels *(Tamiasciurus hudsonicus)* among isolated montane forests of the American Southwest provides an illustrative case study. As Figure 16.34

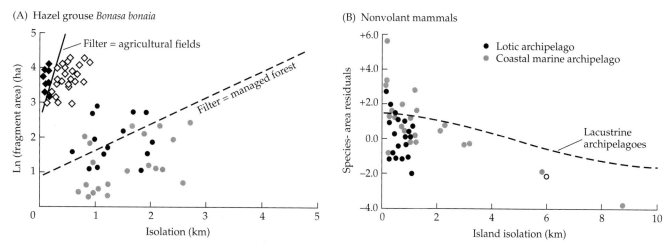

FIGURE 16.33 Species distribution patterns can reveal the influence of intervening habitats and immigration filters on distributions of focal species among isolated habitats and nature reserves. (A) Distributions of hazel grouse (*Bonasa bonasia*) in isolated forest fragments in Sweden. Although the grouse are widely distributed across forest fragments that are isolated by managed forests, in regions where the habitat matrix is dominated by agricultural fields, populations of grouse are restricted to the largest and least-isolated fragments. Black symbols indicate presence; gray symbols indicate absence. (B) Distributions of small mammals on forested islands in the Great Lakes region of North America. In comparison to that of lacustrine systems (archipelagoes in Lakes Huron and Michigan), the species richness of nonvolant mammals inhabiting forested islands of lotic and coastal marine systems (the Thousand Island region of the St. Lawrence River, and islands off the coast of Maine), declines much more rapidly with increasing isolation, presumably because immigrations (against currents, or across ice in winter) are more difficult in the latter systems. (A after Aberg et al. 1995; B after Lomolino 1994b.)

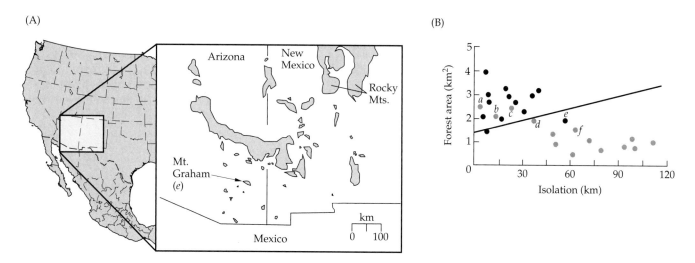

FIGURE 16.34 (A) Montane forest islands of the American Southwest. (B) Distribution of red squirrels (*Tamiasciurus hudsonicus*) among these forests (black symbols indicate presence; gray symbols indicate absence). Montane forest *a* appears to be a site with a high potential for natural colonization by red squirrels, while sites *b, c, d,* and possibly *f* may be good candidate sites for translocations and reestablishment of additional populations. The graph also indicates that the most tenuous and isolated population of red squirrels in this region inhabits site *e*, in the Pinaleno Mountains, the only forest inhabited by the endangered subspecies *T. h. grahmensis*.

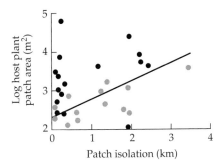

FIGURE 16.35 The area of cranberry patches required to maintain populations of the bog copper butterfly (*Lycaena epixanthe*) increases with patch isolation. (Black symbols indicate presence; gray symbols indicate absence.) (After Michaud 1995.)

reveals, the area requirements of red squirrel populations tend to increase with isolation. Montane site *a* appears to be one with a high potential for natural colonization by red squirrels, while sites *b*, *c*, and *d* may be good candidates for translocation and reestablishment of additional populations. Again, we caution against indiscriminately translocating species outside their native range without evaluating the potential effects of the focal species on the native biota. In addition to these concerns, it is important to note that the most tenuous and most isolated population of red squirrels in this region inhabits site *e*. This is Mount Graham, the only montane forest inhabited by the endangered subspecies *T. h. grahamensis*. Given this fact, we should modify our translocation strategy. If we wish to establish additional populations of this genetically distinct subspecies (Brett Riddle, pers. comm. 1995; Sullivan and Yates 1995), we should translocate a founding population to montane forests that are both large enough to promote population establishment and survival, and isolated enough such that intermountain migration and gene flow are unlikely. Thus, site *d*, and possibly site *f*, may be the optimal sites for establishing additional populations of the Mount Graham red squirrel.

The bog copper butterfly *(Lycaena epixanthe)* provides another illustrative example of the utility of spatially explicit tools in conservation biology. Populations of this species inhabit patchily distributed peatlands of New England. Joanne Michaud (1995) has shown that populations of *L. epixanthe* in southern Rhode Island are strongly limited by both the area and isolation of patches of their host plants (cranberries: *Vaccinium* spp.). As Figure 16.35 reveals, populations of this butterfly may persist in some relatively small cranberry patches if they are located near other occupied patches; whereas even some of the largest patches may be unoccupied if they are isolated from other populations. In a similar manner, the distribution of the endangered bay checkerspot butterfly of California (*Euphydryas editha bayensis*) reveals the importance of biogeographic variables (Harrison et al. 1988). In this case, however, the distribution of the butterflies appears to be most strongly associated with isolation, and populations occur on some of the smallest patches of their optimal habitat (serpentine grasslands) as long as those patches lie within 5 km of source populations (Figure 16.36). Prior to 1986, checkerspots occupied all but four patches of grasslands within this critical isolation distance. In 1986, they colonized two of those patches (labeled *A* and *B* in the figure), including the largest unoccupied patch within 5 km of a source population. Thus, even when populations do not exhibit the predicted relationship between mini-

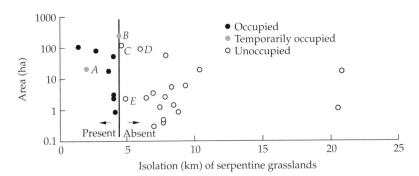

FIGURE 16.36 The distribution of the endangered bay checkerspot butterfly (*Euphydryas editha bayensis*) is strongly associated with the isolation of serpentine grassland patches, but not with area of those habitats. (After Harrison et al. 1988.)

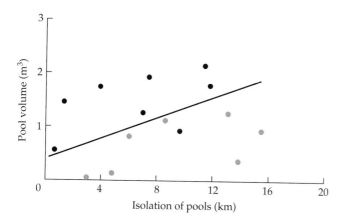

FIGURE 16.37 Just as for many terrestrial animals, the minimum habitat requirements (here, pool volume) of the endangered Ouachita Mountain shiner (*Lythurus snelsoni*) of eastern Oklahoma increase with the isolation of stream pools (black symbols indicate presence; gray symbols indicate absence). (After Taylor 1997.)

mum area and isolation, this spatially explicit approach still can serve to assist conservation strategies. For example, of the remaining unoccupied sites, those labeled *C*, *D*, and *E* would seem to be good candidates for establishing viable populations of this endangered butterfly.

This approach can also be illustrated with a hypothetical case study for aquatic systems: hypothetical in that we arbitrarily assume values for changes in habitat area and isolation due to climatic change, but real in that the projections are based on the actual, extant distribution of a rare fish species—the Ouachita Mountain shiner *(Lythrurus snelsoni)*—in stream pools in eastern Oklahoma (Figure 16.37). For simplicity, we have assumed that changes in area and isolation would be constant for all stream pools in this freshwater system. If regional precipitation decreases, then stream pools should become smaller and more isolated. Using arbitrarily chosen values for changes in volume (−0.5 m³) and isolation (+2 km), we can now predict the new distribution of the shiner. Our hypothetical changes in stream habitats would cause five of the eight extant populations of this threatened fish species to become extinct (Figure 16.38). Note also that, because population persistence should be influenced by the combined effects of immigration (isolation) and extinction (area), even the population now occupying the largest stream pool (labeled 1 in Figure 16.38B) would be extirpated, while those inhabiting smaller but less isolated pools (3 and 4) should survive.

FIGURE 16.38 Given accurate information on the distribution of a focal species and estimates of changes in the area and isolation of its principal habitat, the species-based model of Figure 16.32 can be used to predict the effects of anthropogenic climatic change. (A) A general model for predicting the effects of changes in the size and isolation of habitat patches for a hypothetical species. (B) An illustration of how the model can be used to predict changes in distribution and site occupancy, based on the actual current distribution of the Ouachita Mountain shiner *(Lythrurus snelsoni)*. This example arbitrarily assumes that a decline in regional precipitation would reduce pool volumes by 0.5 m³ and increase their isolation by 2 km. Solid symbols indicate presence, open symbols indicate absence, and arrows indicate changes in patch or pool characteristics. (Data from Taylor 1996.)

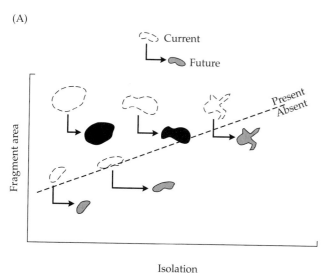

(A)

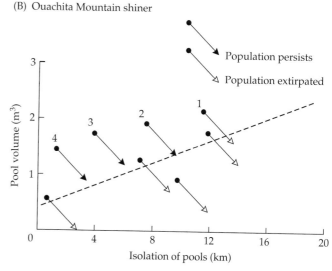

(B) Ouachita Mountain shiner

Again, we emphasize that this is just a hypothetical example used for illustrative purposes. Without more direct information on the dispersal abilities of focal species, these types of biogeographic models cannot be used to estimate rates of relaxation, only the final distributions after climates and landscapes (or seascapes) have stabilized. In fact, none of the approaches discussed in this chapter can stand alone.

SYNOPSIS OF ONGOING RESPONSES TO GLOBAL CLIMATE CHANGE. While it appears that we are experiencing just the earliest stage and relatively modest levels of the predicted changes in global climate (e.g., "just" 0.7° C of the predicted warming of 3° to 6°), hundreds of studies have now reported that biotic and biogeographic responses are not just likely, but ongoing (Price 2000; Patz 2002; Schneider and Root 2002; Root et al. 2003). Both plants and animals are exhibiting a myriad of predicted responses including range dynamics (shifts, contractions, and expansions in both natives and exotics; see Zavaleta and Royval 2002), extirpations, changes in local population densities, morphology, physiology, behavior, phenology (timing of important life history functions such as breeding or flowering), and higher scale changes in the structure and function of ecological communities resulting from altered distributions and interspecific interactions (e.g., from the loss of a mutualist or keystone predator, or the invasion of a potential predator, competitor, parasite, or disease; see Brown et al. 1997; Davis et al. 1998; Post et al. 1999; Rogers and Randolph 2000; Patz 2002; Sanford 2002; Zavaleta 2002). In an important meta-analysis (statistical analyses of the results from a combination of studies), Root et al. (2003) found that of 500 species (450 animals and 50 plants) studied from around the world, 440 (88%) exhibited significant responses to global climate change of the previous decade. Three-hundred and seventy species (74%) of the 500 species responded in a manner consistent with predictions based on their physiology and the local climate change they experienced (Root and Schneider 2002; Root et al. 2003). Among these observed responses are shifts, expansions, and contractions in geographic ranges of butterflies, birds, nonvolant mammals and plants; earlier spring (later fall) migrations in birds, bats, and amphibians; altered phenologies including earlier hatching in insects, frogs, birds, frogs, and newts; declines in densities of frogs and toads with increased temperatures and decreased precipitation; altered growth rates and adult body sizes in mammals, birds, fish, and amphibians; altered growth form in trees; reduced resistance to pests such as aphids in plants; and increased mortalities from road kills for mammals (see Schneider and Root 2002; Root et al. 2003).

Thus, the study of biotic responses to anthropogenic climate change is no longer an exercise in conjecture and hypothetical scenarios, but one that is already providing scientifically rigorous assessments of predictive models evaluated against realized responses. Yet, the challenge of predicting the ecological and biogeographic effects of the predicted full extent of global climate change may well pose a scientific problem of unrivaled complexity. It calls for spatially explicit models that span a range of scales from local to global levels, and include many potential interactions and feedback loops among model components, including atmospheric, hydrological, and nutrient cycles, and sometimes complex interactions among many species. Fortunately, we have learned a great deal over the past two centuries of studies on the dynamics and geography of nature, and we now possess an impressive arsenal of technological and analytical tools. While we need to continue to develop these tools, our greatest handicap lies in our limited knowledge of the diversity and distributions of the Earth's species, especially its rare and imperiled ones (a shortfall we will return to in the next chapter). On the other hand, with each newly described life form and each new map of a geographic range, we not only advance relevant theory, but we also enhance our ability to understand and perhaps mitigate the ongoing crisis in biological diversity.

Geographic Range Collapse

It is axiomatic that all extinctions— historic or prehistoric, natural or anthropogenic—are preceded by declines in geographic ranges. Throughout most of the history of biogeography and conservation biology, however, this process (i.e., geographic range collapse) has largely remained a mystery. When a species' range collapses, does the process exhibit a consistent or predictable pattern? For example, do populations in particular portions of the range (e.g., central vs. peripheral, poleward vs. equatorial, or high vs. low elevation) tend to fare better than others? Abilities to manage and analyze large, spatially explicit databases are now taken for granted, but these are relatively recent advances not available to most biogeographers and conservation biologists prior to the latter decades of the twentieth century. In addition, scientists tackling these questions are often plagued by a catch-22 of conservation biology: we need to know the most about the species that are most difficult to study— the rare, geographically restricted—and, likely atypical—species. Some very creative scientists, however, first tackled these challenges of studying range collapse by consulting classic biogeographic theory, along with lessons emerging from the then relatively new fields of areography and macroecology. Resultant papers did provide some interesting predictions and potentially important guidelines for conservation biologists. Later, enhanced data sets and geographic information systems enabled rigorous, empirical tests of the predicted patterns of geographic ranges. Before discussing these studies, however, it is important to emphasize that this is not just an interesting academic question, but also one with important implications for guiding conservation strategies. Efforts to reintroduce or translocate populations of endangered species, to locate nature reserves, establish conservation corridors, or to search for undiscovered remnants of a once-widespread species, would be greatly facilitated if we could develop a spatial search image for locations most likely to yield favorable results.

In fact, such a search image has been developed. Biogeographic and areographic theory suggests that reintroductions, translocations, and searches for remnant populations should be conducted near the core of a species' historical geographic range. After all, the range boundaries are just that, sites where environmental conditions are so poor that the species cannot maintain its populations without recruitment from other populations. Areas along the periphery of a species' range are often thought to represent sink habitats (sensu Pulliam 1988; see Chapter 4) for individuals emigrating out from more central source populations. In addition, aerographic studies indicate that, as we move from the center to the more peripheral areas of a species' geographic range, densities tend to decrease while variability in densities over time tends to increase (Figure 4.17; but see Blackburn et al. 1999; Sagarin and Gaines 2002). Thus, under the "melting range hypothesis," which assumes an equal and simultaneous decline in all populations across a species' geographic range, dwindling populations should implode, with the final populations persisting near the center of the species' historic range (see also Gaston 2003:167–177). These predictions of melting and imploding ranges are not new and were, in fact, anticipated by some of the most prominent figures in the history of biogeography, including Alfred Russel Wallace and Philip Darlington.

On a continent, the process of extinction will generally take effect on the circumference of the area of distribution, because it is there that the species comes into contact with such adverse conditions or competing forms as prevent it from advancing further. (A. R. Wallace 1876)

The actual limits of range will then be determined not by the limits of favorable ground but by a constantly fluctuating equilibrium between tendency to spread at the center of the range and tendency to recede at the margins. (P. J. Darlington, Jr. 1957: 548)

Conservation biologists of the latter part of the twentieth century followed the leads of Wallace and Darlington, often viewing the range periphery as the domain of the "living dead" and the "domain of zombies"—sites with little value for conserving endangered species (but see Hunter and Hutchinson 1994; Lawton 1995; Lessica and Allendorf 1995; Peterson 2001; Rodriguez 2002). Unfortunately, this may be another case in which generic prescriptions, even those based on very sound theory and well-documented empirical patterns, can prove misleading. Empirical analyses of geographic range collapse in rare and endangered species of vertebrates, invertebrates, and plants indicate that, contrary to the above predictions, relictual populations of these now geographically restricted species occur along the periphery, not near the core of the species' historical geographic ranges (Lomolino and Channell 1995,

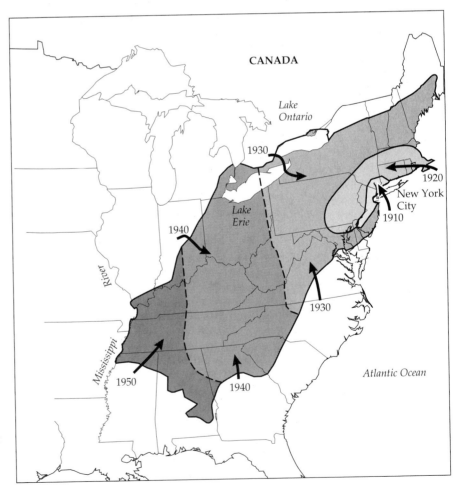

FIGURE 16.39 The American chestnut (*Castanea dentata*) was once one of the most common trees in the eastern deciduous forests of the United States. The chestnut blight, first introduced near New York City around 1910, spread rapidly to cause the decline and range collapse of the chestnut. The solid outline indicates the natural range of the American chestnut prior to 1910; dashed lines and shaded regions indicate the sequential collapse of chestnut populations between 1910 and 1950. (After Anderson 1974; Bell and Walker 1992.)

1998; Channell 1998; see also Safriel et al. 1994; Towns and Daughtery 1994; Laliberte and Ripple 2004). Two hundred-forty (98%) of the 245 species studied maintained at least one population along the periphery of their historical range, and 167 species (68%) maintained more populations along the range periphery than expected by chance. In fact, populations of 91 of these species were exclusively limited to the range periphery while only 5 species maintained populations exclusively within the range center. Among the species persisting along the range periphery are at least five "poster species" for the conservation movement: giant panda (*Ailuropoda melanoleuca*), red wolf (*Canis rufus*), black-footed ferret (*Mustela nigripes*), California condor (*Gymnogyps californianus*), and whooping crane (*Grus Americana*; see Figures 16.39 and 16.40). Analyses of range collapse in other taxa are concordant with these results (Figures 16.41, 16.42, and 16.43). The actual progression of range collapse is quite different from that predicted by earlier biogeographers and aerographers. Initially, geographic range collapse does begin along one of the peripheries of the historic range, but then it "spreads" to include central populations, with the

FIGURE 16.40 Patterns of range collapse in three species of terrestrial mammals: the giant panda (*Ailuropoda melanoleuca*), red wolf (*Canis rufus*), and black-footed ferret (*Mustela nigripes*). The locations of remnant populations of these endangered species are indicated in white; the dark gray polygons represent the historical ranges of the species. (After Lomolino and Channell 1995.)

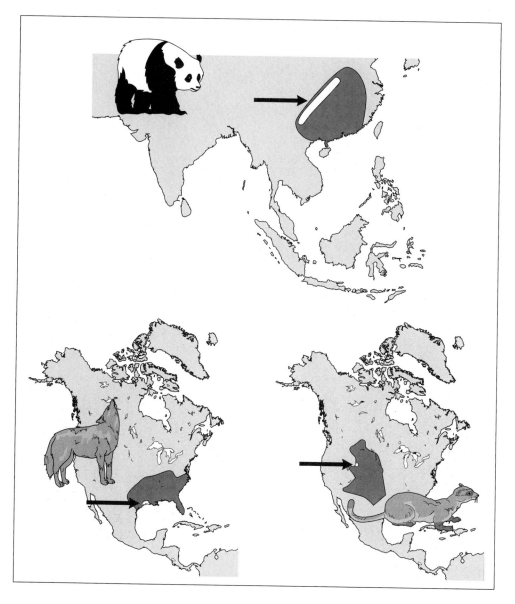

(A)

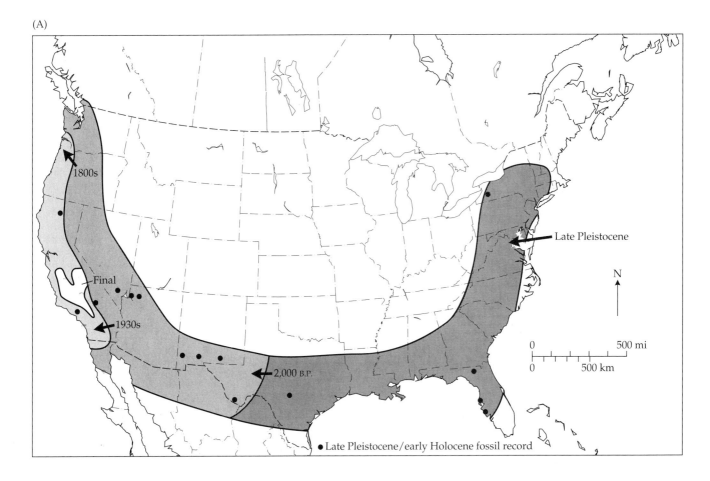

(B)

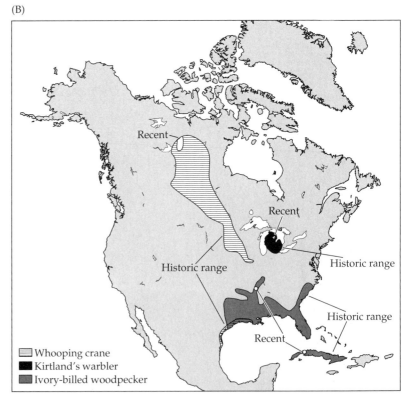

FIGURE 16.41 Patterns of range collapse in four endangered or extinct North American bird species: (A) California condor (*Gymnogyps californianus*), and (B) the ivory-billed woodpecker (*Campephilus principalis*), Kirtland's warbler (*Dendroica kirtlandii*), and whooping crane (*Grus americana*). Locations of the remnant or last known populations of these species are indicated in white; dark gray polygons represent the historical ranges of the species. (After Lomolino and Channell 1995; Channell and Lomolino 2000.)

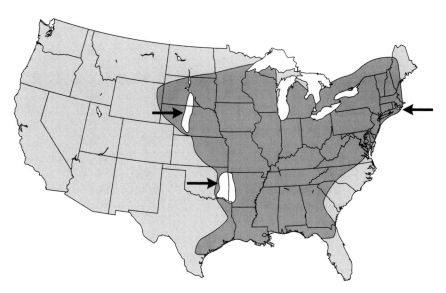

FIGURE 16.42 Range collapse of the American burying beetle (*Nicrophorus americanus*). The historical range of this species (ca. 1900–1940) is indicated by shading, while extant range is indicated by white polygons and arrows. (After Lomolino et al. 1995, and references therein.)

final, persisting populations being those at the periphery most distant from the initial onset of range collapse and extirpations (Channell and Lomolino 2000a,b).

Why do these patterns of range collapse appear so anomalous? The answer may well lie in the nature and geographic dynamics of the extinction forces. Even though the macroecological patterns of population density and variability appear to be very general ones, they may be overwhelmed by anthropogenic disturbance. Indeed, given that we are studying range collapse and extinction, these disturbances must (by definition) be significant enough to render the aerographic patterns irrelevant. And here is a second key distinction between the melting range hypothesis and actual processes causing range collapse and extinction: whereas the melting range hypothesis assumes a ubiquitous extinction force whose effects are nonvarying and simultaneous across space, nearly (if not all) known anthropogenic disturbances are spatially nonrandom and predictably dynamic—that is, they tend to spread across the landscape like a contagion from the beachheads to the interior, and from the lowlands to mountain peaks (Figures 16.44 and 16.45). Anthropogenic activities, including agriculture, fragmentation, introduced species, diseases, and advanced hunting technologies, tend to move across the landscape like fire across a burning leaf: regardless of the flashpoint, the last piece to burn will be along an edge. In these circumstances, isolation may be the key to persistence. Thus, remnant populations should occur in isolated regions, either on an island, a mountain range, or an otherwise isolated portion of the species' historical range.

An additional, albeit nonexclusive, explanation for persistence of at least some peripheral populations during the latter stages of range collapse derives from the relative number, genetic diversity, and local adaptations of these populations. In contrast to the relatively limited number of central populations, those along the periphery of a species' geographic range are more numerous and may, therefore, exhibit greater variation (physiological, behavioral, eco-

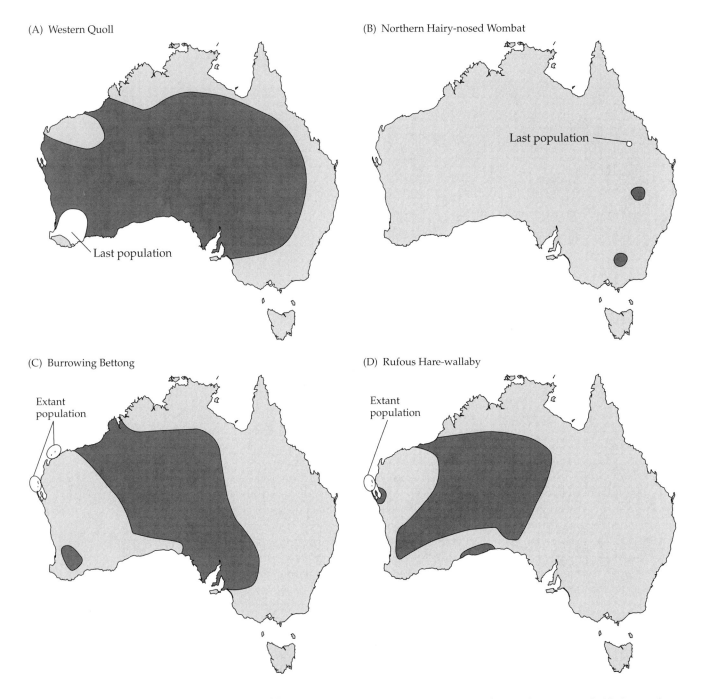

(A) Western Quoll

Last population

(B) Northern Hairy-nosed Wombat

Last population

(C) Burrowing Bettong

Extant population

(D) Rufous Hare-wallaby

Extant population

FIGURE 16.43 Range collapse in four species of Australian mammals (dark gray denotes historical range; white, current or final range). Note that remnant populations are typically found along the periphery of the species' historical ranges, especially on islands. (After Lomolino and Channell 1995.)

logical, and genetic) among populations. Perhaps just as important, these peripheral populations must adapt to a much greater range of environmental conditions in comparison to those prevailing near the center of the species' range. Therefore, it is more likely that one of the many and diverse peripheral populations will be pre-adapted to extinction forces that might threaten its

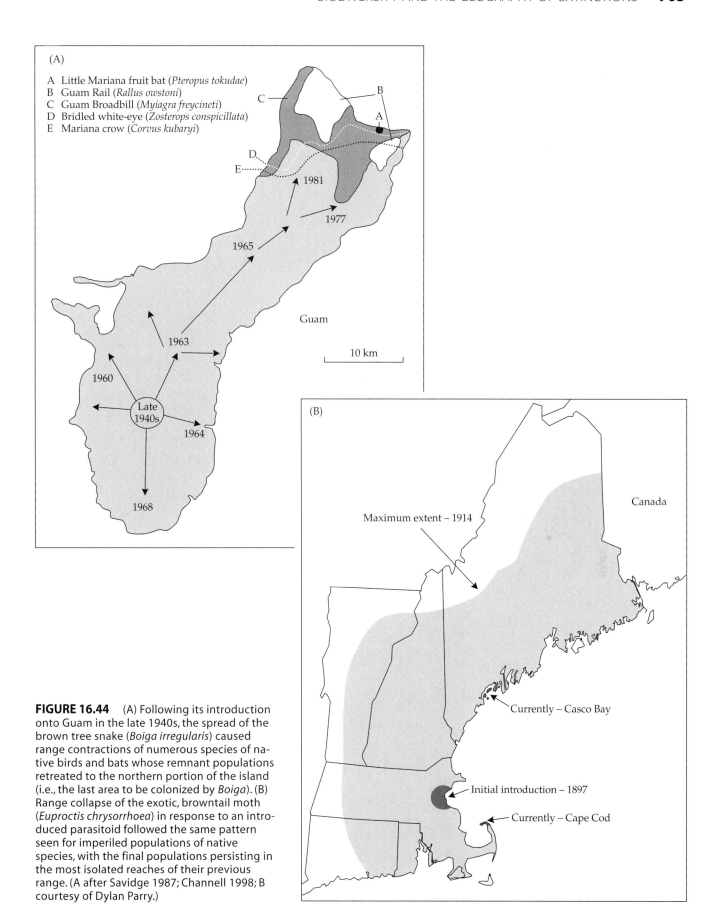

(A)

A Little Mariana fruit bat (*Pteropus tokudae*)
B Guam Rail (*Rallus owstoni*)
C Guam Broadbill (*Myiagra freycineti*)
D Bridled white-eye (*Zosterops conspicillata*)
E Mariana crow (*Corvus kubaryi*)

Guam

10 km

FIGURE 16.44 (A) Following its introduction onto Guam in the late 1940s, the spread of the brown tree snake (*Boiga irregularis*) caused range contractions of numerous species of native birds and bats whose remnant populations retreated to the northern portion of the island (i.e., the last area to be colonized by *Boiga*). (B) Range collapse of the exotic, browntail moth (*Euproctis chrysorrhoea*) in response to an introduced parasitoid followed the same pattern seen for imperiled populations of native species, with the final populations persisting in the most isolated reaches of their previous range. (A after Savidge 1987; Channell 1998; B courtesy of Dylan Parry.)

(B)

Canada

Maximum extent – 1914

Currently – Casco Bay

Initial introduction – 1897

Currently – Cape Cod

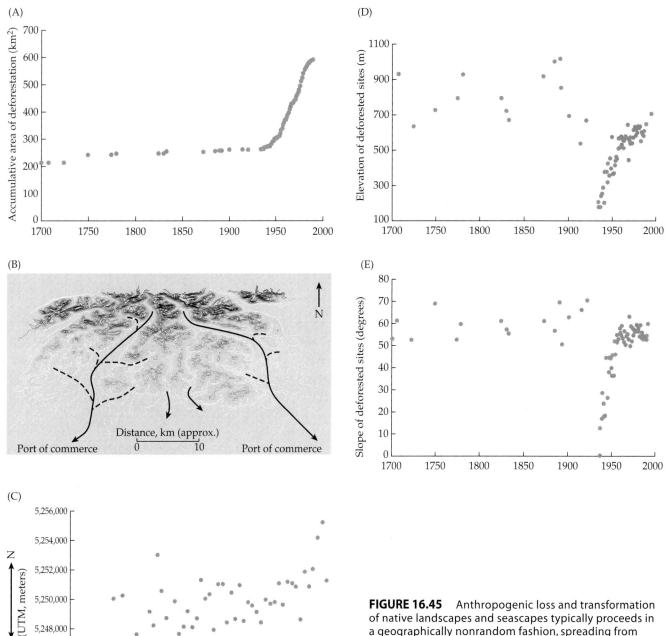

FIGURE 16.45 Anthropogenic loss and transformation of native landscapes and seascapes typically proceeds in a geographically nonrandom fashion, spreading from sites of human colonization toward the most isolated, elevated, and otherwise most remote sites. (A) Anthropogenic loss of old-growth temperate rainforests in the Olympic Peninsula (Washington) increased rapidly with advances in timber harvest technology during the middle decades of the twentieth century, and continued to accelerate until a moratorium banned timber harvest in this district of the U. S. National Forest. (B–E) At first, timber harvests were located in the lowlands close to the shipping ports (located to the south of this area), but they quickly spread northward first along the broad valleys and then to higher and steeper sites as extraction methods improved and the relative value of remaining old-growth trees increased. Arrows in B indicate paths of timber extraction (solid arrows = earliest harvests; dashed lines = subsequent harvests). (After Lomolino and Perault 2004; see also Hall et al. 1995.)

populations in the future. Geographic range collapse in the American burying beetle *(Nicrophorus americanus)* illustrates the potential role of both the contagion and peripheral diversity/pre-adaptation hypotheses (see Figure 16.42). The historic decline of this, the largest member of a guild of carcass-burying beetles, can be attributed to a combination of at least three factors: (1) anthropogenic decline and fragmentation of forested habitats (which provided optimal substrates for burying carcasses and rearing offspring), (2) range expansion of raccoons, skunks, and other carnivorous mammals that compete for carcasses, and (3) extinction of one of its key sources of carcasses—passenger pigeons (Lomolino et al. 1995; Lomolino and Creighton 1996; Rosenzweig and Lomolino 1997). The locations of the remaining populations of this species, including those along the western margins of its historic range and on an island off the east coast of Rhode Island (USA), are entirely consistent with the contagion hypothesis. Then again, western populations of this species occurred in the transition zone between the deciduous forests of the east and prairies of the Great Plains. Therefore, these populations may also have been pre-adapted to deforestation that extirpated nearly all of the populations to the east.

It is important to note that this series of studies represents some of the first articulations of empirical patterns in range collapse (see also Raup and Jablonski 1993), and we have no doubt that many surprises await those pursuing these and similar questions in dynamic biogeography. Because taxonomic groups vary in their vagilities and their susceptibilities to anthropogenic disturbance, we expect that patterns of collapse will vary among taxa. Even within the same taxonomic group, we expect to find different geographic trends in range collapse among regions. Because humans coevolved with the native plants and animals of Africa, for example, range collapse on this continent may fail to exhibit any noticeable directionality. In fact, Channell (1998) found that, while over 75% of the species of the Australian, North American, and European/Asian regions collapsed to the range periphery, only 58% of the African species exhibited peripheral collapse.

Patterns of range collapse for many insular species also appear anomalous, with relatively high persistence of interior ("central") populations. These patterns are, however, entirely consistent with the contagion hypothesis and patterns of human colonization and expansion across island habitats. Once we have established our populations along the beachfronts, subsequent populations spread from these coastal lowlands toward the interior and higher elevations. As a result, the ranges of native birds and snails of Hawaii and many other oceanic islands collapsed away from lower elevations and other areas of anthropogenic disturbance (J. M. Scott, pers. comm.), with many populations of native species persisting only at the higher elevations (see also Burney et al. 2001). In another illustrative case study, ranges of native birds on the island of Guam collapsed not along an elevational gradient, but with the well-documented expansion of the range of the introduced brown tree snake *(Boiga irregularis)*, which preyed heavily on native birds (Figure 16.44A). Similarly, range expansion of an exotic parasitoid (the generalist tachinid, *Compsilura concinnata)* introduced into the New England region of North America resulted in range collapse of one of its key target species, the exotic browntail moth *(Euproctis chrysorrhoea)*. The pattern of range collapse, this time for an exotic species, followed the pattern observed for most natives, with the final populations of browntail moths persisting in the most isolated reaches of their alien range (Figure 16.44B). Unfortunately, the parasitoid did not discriminate between exotics and natives, and simultaneously caused range collapse in giant silk moths (including *Hyalophora cecropia, Callosamia promethea, Antherea*

(A)

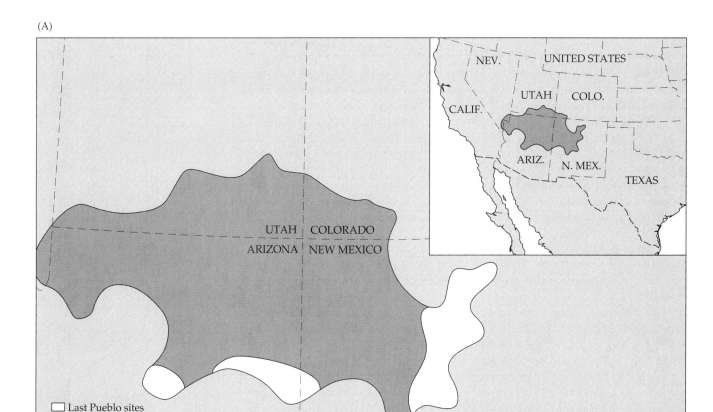

FIGURE 16.46   (A) Range collapse of the Anasazi of North America. As the shaded regions indicate, the last pueblo sites of these Native Americans were situated along the periphery of their historical range. (B) With the invasions of the Spanish conquistadores (arrows), the once vast range of the Incas collapsed to their final stronghold in the high and isolated sites of Machu Picchu and Vilcabamba. (A courtesy of National Geographic Society; B after Davies 1995.)

polyphemus, and *Actias luna*) and numerous other species of native, megainvertebrates (pers. comm., Dylan Parry).

In all these cases, whether populations persist along the edges of once expansive ranges, on offshore islands, or at high elevations, the common factor is that the final refugia were isolated and, therefore, the last to be affected by the spread of anthropogenic disturbances.

Finally, the putative cause of range collapse in these endangered species—the dynamic biogeography of human civilizations—may itself serve as a subject for parallel studies. As the ranges of aboriginal civilizations collapsed, did they exhibit spatial trends similar to those of other mammals? We have looked at only a few such maps, but the results are tantalizing. The range of the Anasazi of North America has collapsed toward the periphery of their historical range (Figure 16.46A). The Incas, whose expansive empire once encompassed much of western South America, present another interesting case of range collapse, one that seems similar to the patterns exhibited by birds of Guam and Hawaii. With the invasion of the Spanish Conquistadores, which spread southward from Ecuador, the Inca Empire collapsed to its final stronghold of Vilcabamba high in the Andes Mountains (see Figure 16.46B).

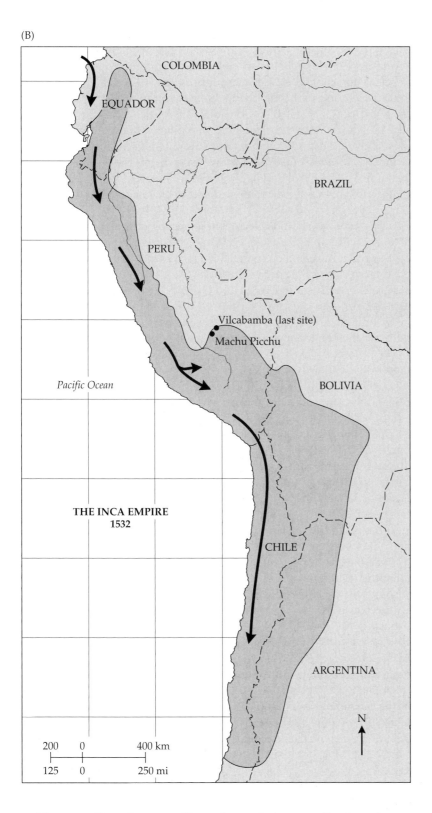

These studies of range collapse have obvious applications for conserving endangered species. In developing effective strategies, conservation biologists must consider the geographic dynamics of extinction forces as well as the ecological and biogeographic characteristics of the species in question. Just as important, these studies show that sites along the periphery of a species' historic range should no longer be dismissed as sites with little conservation value. Instead, reintroductions, translocations, locations of nature reserves and corridors, and surveys for relictual populations should be conducted wherever favorable sites can be located throughout the historical range of an endangered species, and perhaps outside the known range as well. Of course, conservation biologists must also consider the possibility that a species translocated outside its native range would negatively affect species native to that site (see above case study where a parasitoid was introduced to control browntail moths). It is also possible that we may apply the lessons of these and future studies of geographic range collapse in reverse—in this case developing a geographically explicit, range-wide strategy to accelerate and direct range collapse in exotics and otherwise problematic species.

In a practical sense, we should realize that geographic range boundaries are subjective summaries, open to interpretation or misinterpretation. For example, the last population of black-footed ferrets was not actually found just along the western edge of the species' historic range, but outside its described range (see Figure 16.40; see also Ferrari and Queiroz 1994; Blackburn and Gaston 1995). Moreover, sites outside a species' historic range may actually prove to be more favorable under the present regime of extinction forces. The plight of the American chestnut, discussed above, is an excellent case in point. The last field population of American chestnut trees is composed of individuals transplanted outside their historic range, and outside the range of the blight that wiped out other populations (Rosenzweig 1996). Conservation biologists of New Zealand, Guam, and Australia are using the same strategy to establish breeding populations of endangered birds and reptiles on offshore islands (Atkinson 1988; Towns and Ballantine 1993). The islands may be outside the historic ranges of these species, but they also lack the introduced cats, stoats (weasels), goats, and rats that apparently destroyed their conspecifics in less isolated areas. In an ironic twist of fate, it seems that islands, the site of a highly disproportionate number of prehistoric and historical extinctions, may now represent critical refugia for many endangered species. Even when extinction forces reach these islands, it is again the most isolated populations of imperiled species—typically those of the highest and steepest reaches of the island—that are the last to persist (e.g., see Burney et al. 2001).

Given that the fundamental unit of biogeography is the geographic range, studies on patterns of range dynamics would appear to be of fundamental importance to biogeography. The patterns of dispersal, range expansion, and invasions that we discussed in earlier chapters—and the patterns of range collapse that we have focused on here—all have obvious applications for conserving endangered species. Yet, as we observed earlier, it is only very recently that biogeographers have acquired adequate mapping and analytical technologies to investigate these patterns. Beyond the many exciting questions that await the biogeographers who study these patterns, the results of their work will have immediate applications for conserving the diversity and geography of nature. Since we believe biogeography has so much potential for conservation biologists, and because the major threats to biological diversity are directly associated with the dynamics of human populations (see Abbitt et al. 2000; Laliberte and Ripple 2004), we focus on these topics in our next chapter.

CHAPTER 17

Conservation Biogeography and the Biogeography of Humanity

THE TWO THEMES OF THIS CHAPTER—*Conservation Biogeography* and *Biogeography of Humanity*, while important and fascinating in their own right, are intricately related. Many, if not all of the major threats and potential solutions to the biodiversity crisis have a strong geographic context and they all have been, or will be, impacted by invasions and activities of human populations. As we discussed in previous chapters and those of Unit 5, nearly all patterns of biological diversity have geographic signatures. Indeed, the significant and nonrandom geographic variation of the natural world provided clues that proved fundamental to our understanding of the origins, spread, and diversification of life (see Chapter 2). It is also clear that the dynamics and diversity of human populations, along with associated anthropogenic threats to biodiversity (e.g., species introductions, habitat loss and fragmentation, climate change, and pollution) all have strong geographic signatures (see Chapter 16).

Conservation Biogeography and the Wallacean Shortfall

A Review: Fundamental Themes of Biogeography

At the foundation of all biogeographic patterns is the geographic template—the highly nonrandom spatial variation in environmental conditions (including soils, climatic conditions, water quality, and distributions of other species), which is both significant across all geographic gradients (e.g., those of latitude, area, isolation, and depth) and highly predictable for most biotas and regions of the terrestrial and aquatic realms. Species distributions, gradients in species richness, morphology, productivity, and all other biogeographic patterns result from differential responses of organisms to the geographic template. These "responses" include ecological interactions, adaptation, evolution, dispersal to more suitable environments or, if all this fails, extinction. Each of these processes, in turn, occur (or occurred) at particular times and in particular places. Each intermediate result served as a precursor to the next, and ecological and evolutionary phenomena occurring in one place affected the biotas of all other places, albeit to different degrees depending on their proximity. Alter the sequence of events or relative location and characteristics of those ecological and evolutionary arenas, and the processes and associated assemblages will be fundamentally altered as well. The asteroid that crashed into the Caribbean at the end of the Cretaceous Period caused the extinction of over half the species then in existence—including the dinosaurs—forever changing the character of life on Earth; despite 65 million years they have never "re-evolved."

To add one final layer of complexity to this synopsis of the geography and diversity of nature, we now know that the Earth's geographic template is highly dynamic across seasons, millennia, and geological periods (see Chapter 8). As we discussed in the previous chapter, we have begun to stir the mix and transform the geographic template by degrading, reducing, and fragmenting native ecosystems and, at the same time, "connecting" long-isolated biotas through many thousands of episodes of species introductions.

Even more fundamental than the concept of the geographic template and the importance of geographic context is the assertion that geographic variation is central to understanding the natural world. Life forms, indeed all features of nature from rocks and climates to continents, ocean basins, and entire planets, develop and diversify over both time and space. Few of us would deny the importance of any of these processes, or that they occur over particular spatial and temporal scales. Yet, we often fail to appreciate the linkage between temporal and spatial scales. Processes that require large spatial scales, including those fundamental to biogeography (immigration, evolution, and extinction) also require long time periods.

One of the esteemed founders of modern evolutionary theory—Theodosius Dobzhansky (1973)—once told us that "nothing in biology makes sense except in light of evolution." We certainly do not take issue with this, but instead offer our own observation that is even more general and possibly more strident. 'Little in ecology, evolution, and conservation biology makes sense unless viewed within a geographic context.' We hope that our reasoning and, in particular, the relevance of biogeography to ecology and evolutionary biology is now obvious from the many patterns and processes discussed in foregoing chapters of this book. The relevance and potential utility of biogeography for conserving biological diversity in particular may also be clear, especially from Chapter 16, and it appears to be appreciated by most of the world's leading scientists in both biogeography and conservation biology.

In his foreword to a book accounting the voyages of Alfred Russel Wallace, Edward O. Wilson (1999; Figure 17.1A) observed that "the vastness of the tropical archipelago [the islands of Malaysia and Indonesia] also provided the knowledge Wallace needed to conceive the biological discipline of biogeography, which has expanded during the late 20th century into a cornerstone of ecology and conservation biology." The linkages between these two fields—*conservation biology* and *biogeography*—have strengthened over time and quite recently have begun to crystallize into a new synthesis which we will discuss below.

A New Synthesis: Conservation Biogeography

Although Wilson's quote is a relatively recent one, the relevance of biogeography to both evolutionary biology and conservation biology was appreciated well before both of these fields were recognized as distinct and rigorous disciplines. As early as 1839, Charles Darwin was so amazed at the mass extinc-

(A)

(B)

(C)

(D)

FIGURE 17.1 Biogeographers were among the first to detect declines in distributions and diversity of native species. Perhaps most prominent among these early "conservation biologists" were (A) Edward O. Wilson, (B) Charles Darwin, (C) Alfred Russel Wallace, and (D) Charles Elton. (A courtesy of Rose Lincoln/Harvard News Office; D courtesy of the Zoology Department Library, University of Oxford.)

tions of the "Ice Age" that he remarked that "it is impossible to reflect on the state of the American continent without astonishment. Formerly it must have swarmed with great monsters: now we find mere pygmies compared with the antecedent, allied races." Perhaps it is not too surprising that Alfred Russel Wallace made similar observations in his seminal book on *The Geographic Distributions of Animals* (1876:150): "It is clear, therefore, that we are now in an altogether exceptional period of earth's history. We live in a zoologically impoverished world, from which all of the hugest, and fiercest, and strangest forms have recently disappeared ..." He returned to this theme of lost life-forms in his 1901 monograph on Darwinism (Wallace 1901:394): "So we find, both in Australia and South America, that in quite a recent period many of the largest and most specialized forms have become extinct, while only the smaller types have survived to our day ... a group progressing and reaching maximum size or complexity and then dying out, or leaving at most but a few pygmy representatives."

It is not clear whether Darwin and Wallace attributed the extinctions of the "hugest and strangest forms" to human activities, or to natural processes. Eventually, however, the anthropogenic nature of historic (and prehistoric) extinctions became more clear. In his classic book entitled *The Ecology of Invasions by Animals and Plants,* Charles Elton (1958:31) echoed observations of earlier biogeographers: "We must make no mistake; we are seeing one of the great historical convolutions of the world's flora and fauna." Later in the same book (1958:154–155), Elton described what he called "a wilderness in retreat," discussing both the causes for species extinctions (in particular, introduced species, habitat loss, and fragmentation) and potential applications of ecology and biogeography in the emerging field of conservation biology.

In retrospect, perhaps it should not be surprising that some of the first individuals to recognize the early waves of anthropogenic extinctions and their likely causes were biogeographers. Indeed, many of the most distinguished contributors to modern conservation biology were first trained as biogeographers (e.g., Wilson, who we first recognized for developing the taxon cycle theory and codeveloping the equilibrium theory of island biogeography). As biogeographers, we adopt holistic approaches to understanding the complexity and diversity of nature, and we view both the processes and resultant patterns in explicit temporal and geographic contexts. Fittingly, the theme of the second meeting of the International Biogeography Society, held in 2005, was *Conservation Biogeography.*

Emerging from the continuing and increasingly interdependent development of these two fields—biogeography and conservation—is the newly articulated subdiscipline, **conservation biogeography**. While we hesitate to introduce any additional jargon, especially in one of the last chapters of this book, we think the use of the term is well justified because it emphasizes two important points.

1. **Success in conserving biological diversity depends heavily on our understanding of the geography of nature.** As Wilson and his colleagues have amply demonstrated, biogeographers have provided many valuable insights for conserving biological diversity, tackling the many conservation challenges that require geographic information (e.g., what are the effects of reduced area and increased isolation of habitats; where are the hotspots of diversity and endemicity; where will an exotic species spread to in the future; where will native species spread to under climate change predictions; and where should we locate nature reserves and conservation corridors).

2. **In order to conserve "the hugest, and fiercest, and strangest forms" and the true nature of native species (including their distinct physiologies, morphologies, behaviors, and ecological interactions), we need to conserve their distributions (i.e., the geographic, ecological, and evolutionary context of nature).** The lessons from island biogeography and from seemingly unrelated fields of animal husbandry and horticulture are sobering ones. If we favor and select certain types of individuals (e.g., the tamest, or smallest, or those with the showiest flowers, or the smallest thorns), and if we attempt to "maintain" them only in small, ecologically simplified or otherwise altered environments such as nature reserves, zoos, and botanical gardens, their descendants are not likely to retain the diversity or natural character that we hope to conserve.

This latter point is a relatively new—but especially important—one, which we illustrate here with a case study of Earth's largest extant land animals, elephants. As you may remember, proboscidians (elephants, mammoths, mastodons, and their relatives) once dominated many terrestrial ecosystems, outgrowing even the largest predators, and diversifying and expanding their geographic ranges from their ancestral homeland in Africa, only to suffer extinctions during the latter part of the Pleistocene. Today's surviving species are now limited to isolated patches of reserves and other fragmented habitats across Africa and Asia. Save for these populations, the only other proboscidian to survive into the Holocene was the woolly mammoth (*Mammuthus primigenius*) of Wrangel Island, a 7300 km² steppe ecosystem located north of the Siberian mainland. Here, in isolation from many of the competitors and predators that inhabited the mainland (including expanding populations of our own species), woolly mammoths persisted for some 7000 years after they vanished from the mainland (Figure 17.2). Their new insular environs, however, were quite different from those that shaped the evolution of their ancestors. Now isolated on an ecologically simple, isolated, and relatively small ecosystem, they were no longer challenged by large predators (which on the mainland select for larger elephants), but instead had to cope with limited resources of the island. Consistent with the island rule (see Chapter 14), these—the largest land mammals—underwent insular dwarfism, in this case decreasing in size over just a few hundred generations to one-third of the mass of their mainland relatives. Woolly mammoths of Wrangel Island had

(A)

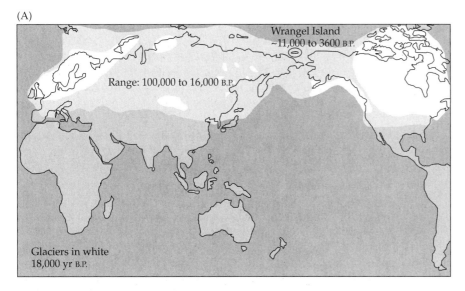

(B)

FIGURE 17.2 (A) The geographic range of the woolly mammoth (*Mammuthus primigenius*), included most of the ice-free zone of the Nearctic and Palearctic Regions during glacial periods of the late Pleistocene, but contracted to its final stronghold in Northern Siberia, Wrangel Island, during the early Holocene where it survived for some 7000 years after its disappearance from the mainland. (B) In its new island environs, the woolly mammoth underwent a remarkable evolutionary transformation, dwarfing to just one-third the mass of its mainland ancestors.

become one of the most remarkable oddities of nature, and a true oxymoron—pygmy mammoths.

The relevance of this account of Wrangel Island mammoths to the challenges of modern conservation biology may be obvious. When we alter the geographic and evolutionary character of native ecosystems—indeed, when we maintain species in small and isolated reserves lacking the challenges of surviving in the face of competitors, parasites, and predators—surviving generations of those species may well lose many of their most distinctive features. Again, we return to the illustrative case study of conserving extant elephants. Today's populations are now maintained in a complex collection of fragmented and isolated reserves, wildlife parks, and zoos. In addition to being relatively small, most of these systems lack large predators. In addition, the future of these populations may have been compromised by artificial selection for certain types of individuals, either during the initial stocking or subsequent culling operations. Some zoos and other artificial facilities may have selected for more docile, smaller, or otherwise less problematic individuals. Poaching or controlled hunts have also altered evolutionary pressures, in this case selecting *against* individuals with the largest tusks, which tend to be the largest individuals as well. The overall result of altering the geographic and evolutionary context for these species is ongoing, with some populations possibly exhibiting early stages of dwarfism (Dudley 1999) and others having lost one of their most distinguishing characteristics—the ability to grow tusks (Raubenheimer 2000). Elephants may well persist in zoos and reserves for centuries, but in what form?

The solution for conserving not just species richness, but also the natural character of these and other imperiled species, may seem relatively simple. As we emphasized in the second point above, we need to preserve the geographic and evolutionary context of native species. This, however, will not be easy given the continuing and increasing demands of human populations to further dominate native ecosystems. Even if we do control and redirect our activities to minimize threats to biodiversity, we still suffer from a great shortfall in our knowledge of the geography of nature, especially the distributions and geographic variation of the most imperiled species.

A Continuing Challenge: The Wallacean Shortfall

In the previous chapter, we discussed the Linnaean Shortfall—the gap between the number of described species (roughly 1.7 million) and the actual diversity of species in existence (estimated to be between 5 and 25 million). Field biologists continue to discover new species, or rediscover those long thought to be extinct (see Ferrari and Queiroz 1994; Patterson 1994; Pine 1994; Sunquist et al. 1994; Finlay et al. 1996; Morell 1996). However, even if we were able to heed Peter Raven and Ed Wilson's call to address this shortfall and "describe" and catalog most of what is out there (Raven and Wilson 1992), we suffer from an even more critical shortfall. We need more than just Latin binomials and descriptions of identifying characteristics of species. This, of course, is indisputably essential information, but to conserve biological diversity and imperiled species in particular, we need much more; we need to greatly expand our understanding of the geography of nature. This, after all, is what Wallace viewed as one of the most distinguished preoccupations of science—the relations of life-forms "to space and time, or in other words, their geographic and geological distributions and causes."

We define the **Wallacean Shortfall** as the paucity of information on geographic distributions of species (past and present), and on the geographic dynamics of extinction forces—namely, the geography of humans. In order to address this shortfall, we need to capitalize on the previous centuries of natural history and biogeography surveys, and on recent and continuing advances in technology and our abilities to map and analyze variation in the geographic template. We should expand and enhance biogeography surveys by enlisting what Daniel Simberloff, E. O. Wilson's graduate student, called the "army of unemployed and underemployed ecologists" and biogeographers to describe the geography and geographic dynamics of biodiversity. In particular, we need to focus our limited resources and personnel on surveys designed to determine distributions of imperiled species and ecosystems, to locate and assess the intensity of hotspots of diversity and endemicity in both the terrestrial and marine realms, and to develop methods to reliably predict both the impact and future geographic dynamics of key threats to biological diversity.

Perhaps the great wealth of biogeographic patterns we have discussed in this book, many of them documented for just a handful of terrestrial taxa, will prove to be equally applicable to all of the poorly studied groups—the undiscovered majority and, in particular, the imperiled and rare species. While typically optimistic, we strongly doubt that this is the case: biological diversity cannot be so simple. First, imperiled species are the oddballs; the exceptional ones whose requirements and characteristics are so specialized, or extreme or heavily impacted by various anthropogenic forces that they barely survive. Second, we may well learn a great deal more from these exceptional cases than from the majority of species that faithfully follow the general patterns and descriptions. Indeed, we hope that there will be many exceptions, each one telling us a bit more about the forces structuring the natural world and the great diversity of organisms that inhabit it.

It is just as clear, however, that we cannot afford to put conservation initiatives on hold until we develop more complete accounts of the geography of imperiled species and extinction forces. Species continue to go extinct. In fact, we should pursue both fronts: addressing the Wallacean Shortfall while continuing to apply lessons from the past two centuries of biogeographic studies to conserve biodiversity.

Applied Biogeography

Our purpose in this unit on conservation biogeography is not to present a gloom and doom scenario on the status of biodiversity. Yes, the challenge of conserving biodiversity in the face of expanding human populations is a great one, but we are guardedly optimistic that conservation biologists, aided by insights from biogeography and other relevant fields, can make some real progress toward understanding the problems and reducing the losses.

Ecological and Biogeographic Characteristics of Extinction-Prone Species

Given the limited resources allocated for conserving endangered species, it is essential that we focus our efforts where they will do the most good. Toward this end, we can learn a great deal from past extinction events. Reviews of records of prehistoric and historical extinctions reveal several common characteristics of extinction-prone species (see Diamond 1984; Marshall 1988; Kunin and Gaston 1997; Courtillot 1999; Sheehan 2001; see also Chapters 7 and 9):

1. Specialized habitat requirements and diets
2. Position in the higher trophic levels (e.g., top carnivores)
3. Small and variable populations
4. Large body size

FIGURE 17.3 Extinction rates of marine bivalves and gastropods living along the Atlantic and Gulf coastal plains of North America during the late Cretaceous were consistently higher for species with relatively small geographic ranges. (After Jablonski 1986.)

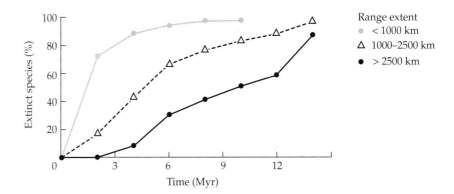

5. Long generation times (age to reproductive maturity) or low reproductive potentials

6. Poor dispersal abilities or limited opportunities for dispersal

7. Restricted geographic ranges

While each of these characteristics may represent useful clues for identifying endangered species, the latter two are fundamental to biogeography, and they are interrelated. That species with limited dispersal abilities and restricted ranges have been more prone to extinctions throughout the fossil record, as well as in recent times, should not be surprising. In the previous chapter, we discussed the relatively high vulnerability of insular biotas—groups of species that, by the very nature of their insular environment, have limited opportunities for dispersal and restricted geographic ranges. In contrast, there is less chance that natural disturbances such as floods, fires, and hurricanes will simultaneously threaten all populations of a broadly distributed species. Even in those cases in which environmental disturbances are pervasive, the more vagile species can (provided undivided landscapes and seascapes) adapt by shifting their geographic ranges.

The relationship between susceptibility to extinction and geographic range size is well evidenced by the fossil record. During periods of "normal"—or background—extinction, most of the above listed factors have influenced species' persistence, whereas during mass extinction events, range size has been especially important (Figure 17.3; Jablonski 1986, 1989; Rosenzweig 1996). It appears that the forces of mass extinction operate on such a grand scale that they often overwhelm the influence of traits that are honed by natural selection and operate at much finer scales (e.g., traits listed in points 1–5, above). As Jablonski (1991) put it, many species suffered extinctions not because they "were poorly adapted by the standards of background extinctions, but because they occurred in lineages lacking the environmental tolerances or geographic distributions necessary for surviving the mass extinction." Jablonski predicts that the ongoing extinction crisis will also "impinge most heavily on rare, geographically restricted species."

These extrapolations from the fossil record to the present are substantiated by studies of more recent extinctions and extirpations. You may remember Simberloff and Wilson's (1969, 1970) classic defaunation experiments, which recorded frequent extinctions of invertebrate populations on mangrove islands in Florida (see Chapter 13). Hanski (1982) re-examined their results and found that, as predicted, the more widespread invertebrate species were less likely to be extirpated from the islands (Figure 17.4). Hanski also found the same relationships for mollusks inhabiting ponds in England, and for

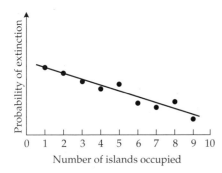

FIGURE 17.4 The probability of extirpation for insects inhabiting mangrove islands of coastal Florida decreases as the number of islands occupied by the species increases. (After Hanski 1982; based on analyses of Simberloff's 1976 data.)

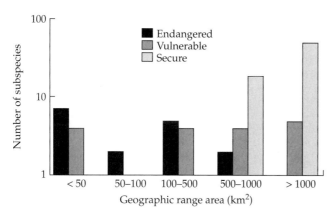

FIGURE 17.5 The relationship between geographic range size and endangerment in waterfowl (*Anseriformes*). "Endangered" refers to species listed as vulnerable or endangered based on three criteria: population size, rate of population decline, and estimated impact and frequency of disturbances. (After Mace 1993.)

leafhoppers in Finnish meadows: species that occurred across more sites (and presumably were better dispersers) were less prone to extirpation from any particular site. Not surprisingly, lists of species currently threatened with extinction are composed of a disproportionately high number of species with relatively small geographic ranges (Figure 17.5).

Then again, the tendency for extinction-prone species to have small geographic ranges, while very general, is far from the rule. It would be dangerous to assume that all species with broad geographic ranges were immune to extinction and, therefore, to ignore these species when setting priorities for conservation. Even abundant and widespread species can go extinct. The passenger pigeon (*Ectopistes migratorius*) and the American chestnut tree (*Castanea dentata*) provide two dramatic examples. Both of these species were broadly distributed across most of eastern North America, and both were dominant members of their communities, yet both suffered dramatic range collapse and extinction (or virtual extinction, in the case of the American chestnut) in less than a century (see Rosenzweig 1996; Rosenzweig and Lomolino 1997). The passenger pigeon was the most common bird in eastern North America during the eighteenth and early nineteenth centuries. Total population estimates vary, but ranged between 2 and 5 billion birds. If these estimates are accurate, then before its decline and ultimate extinction in 1914, one out of every two or three birds in eastern North America was a passenger pigeon. Similarly, before their decline, chestnut trees may have accounted for one-fourth of all trees in their range. Yet, deforestation and overhunting wiped out the passenger pigeon, and the introduced chestnut blight (*Endothia parasitica*) destroyed chestnut trees across all of their historical range (see Figure 16.39; see Roane et al. 1986).

Unfortunately, these are not the only exceptions to the rule. Some of the most broadly distributed species have suffered range reductions and regional or global extinction. This includes a long and diverse list of species from nearly all continents (see Chapter 16), along with a select number of species from the marine realm (e.g., Stellar's sea cow [*Hydrodamalis gigas*]). Species with broad ranges may have been better off before ecologically significant humans arrived, but anthropogenic disturbances became so powerful as to often render the initial biogeographic advantages of historically large ranges irrelevant. Species with critically restricted geographic ranges obviously constitute a highly disproportionate number of all endangered species. Then again, many of them enjoyed extensive ranges in the past, and each decade they are being joined by additional species experiencing the initial stages of geographic range collapse. Therefore, it seems essential that conservation biologists remember that the geography of nature is dynamic and that they focus not just on the species with critically restricted ranges, but also on the process that can sometimes—and with surprising rapidity—cause previously broad-ranging species to join the ranks of the imperiled. As Michael Soulé (1983) put it, "The extinction problem has little to do with the death rattle of its final actor. The curtain in the last act is but a punctuation mark—it is not interesting in itself. What biologists want to know about is the process of decline in range and numbers."

Designing Nature Reserves, Corridors, and Landscapes

The great majority of recent episodes of range collapse can ultimately be traced to waves of human colonization and advances in technology that overwhelmed hotspots of diversity and generated surges in extinction rates, both locally and globally. Thus, anthropogenic extinctions are important case studies in dynamic biogeography. As we have seen, historical extinctions have been especially severe on oceanic islands, and the ongoing wave of extinctions appears to be concentrated on the "archipelagoes" of fragmented remnant habitats. It is only logical, therefore, to believe that island biogeography may offer some important lessons for conserving biodiversity.

Not surprisingly, E. O. Wilson, the author of the taxon cycle theory and coauthor of the equilibrium theory of island biogeography, was one of the first and most distinguished scientists to make this argument. In a symposium volume dedicated to Robert MacArthur, Wilson and his colleague Ed Willis (1975), presented a set of recommendations and geometric "rules" for designing nature reserves and networks of reserves. These rules were based largely on the principles and patterns of island biogeography: extinction rates decrease with increasing area; immigration rates (and rescue effects) decrease with increasing isolation (Figure 17.6). Wilson and Willis urged planners and managers of nature reserves to follow four basic recommendations:

1. Individual reserves should be made as large as possible. Extinction rates are lower on larger reserves.

2. Unique habitats and biotas should be contained in multiple reserves. While each isolated population has a real risk of extinction, it is much less likely that extirpations will occur simultaneously among all the isolated populations.

3. Reserves should be located as close to one another as possible, or connected by corridors of favorable habitat. Given that the isolation is not too great, reserves can be recolonized by populations that persist in other reserves.

4. Reserves of fixed area should be as round in shape and as continuous as possible. This recommendation was based on the **peninsular effect** (the tendency for peninsulas to have relatively low diversities), but it also follows from edge effects (the potential negative effects of exotic species and disturbances that act along the edges of habitat fragments). The relative size of edges is minimized in circular reserves.

5. Conservation biologists should give highest priority to those biotas with the highest degrees of endemicity and vulnerability.

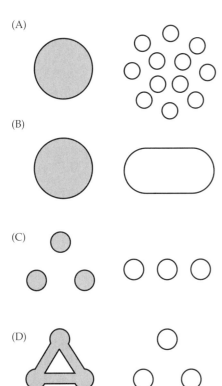

FIGURE 17.6 The application of biogeographic principles to the optimal design of nature reserves results in several recommendations. In each pair of figures, the configuration on the left is to be preferred over that on the right, even though both incorporate the same total area. (A) A continuous reserve is preferable to a fragmented one. (B) The ratio of area to edge should be maximized. (C) Distance between reserves should be minimized. (D) Dispersal corridors should be provided between isolated fragments. (After Wilson and Willis 1975.)

Jared Diamond (1975) and John Terborgh (1974) presented nearly identical sets of rules for designing nature reserves (see Figure 17.6).

Each of these recommendations seems perfectly sound, based on a wealth of empirical studies and biogeographic theory. Yet, these papers—or more specifically, their geometric rules for designing networks of nature reserves—set off one of the most contentious debates ever to visit the fields of biogeography and conservation biology. Unfortunately, they were taken too literally and often applied out of context. The most infamous case in point is what conservation biologists called the SLOSS debate. Basically, this debate focused on which was best, a Single Large reserve Or Several Small reserves (i.e., the two alternatives in Figure 17.6A). But "best" for what—conserving a maximum variety of species or preserving populations of a particular endangered species for as long as possible? Conservation biologists and biogeographers alike locked horns in the SLOSS debate for over a decade (Abele and Connor 1979; Gilpin and Diamond 1980; Simberloff and Abele 1982; Soulé and Simberloff 1986). The outcome was a stalemate, at best. Wilson and Willis's recommendations were sound, but neither they nor the geometric "rules" were meant to be generic prescriptions for all environmental problems.

More to the point, the optimal strategy for designing reserves and networks of reserves depends on the scale of the problem and the particulars of the species, communities, and landscapes in question. Where most species are ubiquitous, or where communities are strongly nested (i.e., when species-rich communities include all the species found in species-poor communities), one large reserve should indeed capture more species than several small reserves of the same total area (Patterson 1987). Yet the SLOSS debate considered just two of a great many possible combinations of reserves (see Lomolino 1994a; Rosenzweig 2004). It turns out that, for a variety of taxa and ecological systems, neither of the two SLOSS strategies is optimal (Figure 17.7). To include all focal species, either of the two SLOSS strategies would require much more area than the optimal combination of reserves (based on comprehensive biogeographic surveys and computer simulations of all possible combinations of reserves) and, in fact, they would tend to require more area than just a random combination of reserves. Yes, bigger is better, and many reserves are better than one or just a few, but on the question of whether one big reserve is better than many small ones of the same total size, biogeographic theory cannot provide a definitive answer.

Putting the SLOSS debate aside, biogeographers have developed a very useful set of tools for designing networks of nature reserves. Computer simulations are now readily available to reserve planners faced with the challenge of selecting reserves or prioritizing their efforts among many potential reserves. Given specific conservation objectives, these tools can be used to select reserves most likely to achieve particular goals. The critical limiting factor here is information on the distributions of the focal species among prospective reserves. Provided with these data, however, reserve planners can select sets of reserves that maximize the number of species included in a limited number or total area of reserves and, depending on the particular conservation goals, include those reserves that harbor the most endemics, or contribute most to redundancy (i.e., including multiple populations of target species to enhance metapopulation survival) or complementarity (i.e., including sets of reserves that are inhabited by species that occur on few, if any other reserves; see Pressey and Cowling 2001; Briers 2002; Cowling et al. 2003; Gaston and Rodrigues 2003).

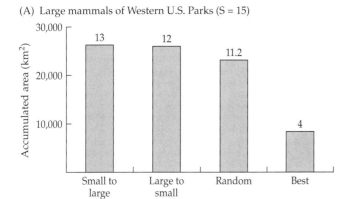

(A) Large mammals of Western U.S. Parks (S = 15)

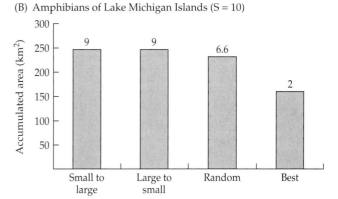

(B) Amphibians of Lake Michigan Islands (S = 10)

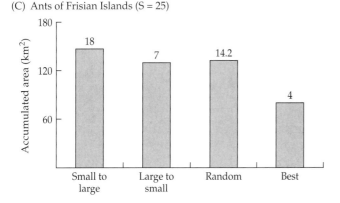

(C) Ants of Frisian Islands (S = 25)

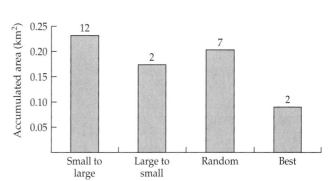

(D) Arctic and alpine plants of Adirondack Mountains (S = 18)

FIGURE 17.7 Given a large set of natural areas to be included in developing a reserve network, which ones should be selected? If the goal of reserve design is to include a maximal number of target species in a limited total area of protected lands, then the two accumulation strategies considered in the SLOSS debate (i.e., accumulating natural areas from the smallest to the largest, or largest to the smallest) perform poorly, requiring a much greater area and number of reserves (numbers above bars) to conserve the same total number of species (S = number of target species) than the optimal collection of reserves (Best, determined by computer simulations). In fact, the SLOSS strategies perform no better than if reserves were selected at random. (From Lomolino 1994a.)

Yet, we know that the inclusion of a species or even multiple populations of a particular target species in a reserve will not guarantee its preservation over a significant time period. Biotas are not static, but instead are strongly influenced by recurrent extinctions and immigrations. Much of the current emphasis on establishing conservation corridors stems from this observation. Indeed, Wilson and Willis included the role of immigration and corridors in their design principles (see their recommendation 3). Conservation planners have, therefore, been urged to select reserves that are clustered in space (Figure 17.6C), or to increase connectivity among reserves by establishing conservation corridors (Figure 17.6D), or by improving characteristics of the intervening landscape (see Laurance and Gascon 1997; Ricketts 2001).

From these suggestions, a rich literature on landscape design and the potential importance of corridors has developed (e.g., see Nicholls and Margules 1991; Soulé and Gilpin 1991; Newmark 1993; Machtans et al. 1996; Reed and Dunn 1996; Rosenberg et al. 1997; Beier and Noss 1998; Bennett 1999; Laurance and Laurance 1999; Lima and Gascon 1999; Estrada and Coates-Estrada 2001; Mech and Hallett 2001; Tewksbury et al. 2002). Although the term *corridor* is co-opted from George Gaylord Simpson's original conception, the contemporary concept of corridors is not consistent among today's scientists, and current applications of the concept for conservation is different than Simpson's (1940) original hypothesis. The term originally referred to a long-term route of range expansion over many generations and among continental or regional biotas (i.e., not jump dispersal of individuals over relatively short periods). Conservation biologists are applying the term "corridors" to phenomena occurring at a much more limited geographic scale, more abbreviated temporal scale, and to more restricted groups of species. On the other hand,

Simpson viewed corridors as nonselective routes for interchange among entire biotas; in Simpson's example, the Pre-Columbian landscape between Florida and New Mexico served as a corridor for terrestrial vertebrates of this region (Simpson, 1940). He used the term "filter-bridge" to refer to a migration route that was selective among species or species groups, thus resulting in an assemblage of successful colonists that was substantially different from a random subsample of the biotas of either pool of species. Most biogeographers and conservation biologists would agree with Simpson's observation that what is a barrier to some species is a migration route to others, and that what we refer to as "corridors," are species-selective. The important point is not so much what we call them, but that we test the hypothesis that purported "conservation corridors" facilitate dispersal (over the short- or long-term) among otherwise isolated populations and, as a result, increase their likelihood of persisting in the face of fragmentation of their native habitats.

It is not our purpose here to review all of the empirical and theoretical information on conservation corridors, but we do caution that along with their obvious potential benefits, corridors also may have some negative effects (see Hobbs 1992; Simberloff et al. 1992; but see also Beier and Noss 1998; Bennet 2003). For example, corridors may facilitate the spread of exotic species and disease or, by increasing dispersal and gene flow among populations, corridors may promote swamping of local adaptation. In a very real sense, corridors may violate Wilson and Willis's second recommendation by putting all our eggs in one basket and; as studies of geographic range collapse (see Chapter 16) have revealed, isolation may at times be key to persistence of at least some imperiled species. Fortunately, conservation biologists are well aware of these potential problems and are actively evaluating their relevance. Just as important, a variety of studies are assessing the extent to which particular species utilize corridors and the various geographic and ecological factors influencing dispersal along corridors. Many important questions can be addressed by field studies strategically designed to answer some key questions regarding optimal design of corridors (see Beier and Noss 1998; Laurance and Laurance 1999; Perault and Lomolino 2000). How wide should a corridor be? How long can a corridor be before its benefits begin to attenuate? To what degree are the benefits of corridors influenced by characteristics of the adjacent landscape, and to what degree do all these features (e.g., optimal width, length, quality, and characteristics of the surrounding matrix) vary among different species and different landscapes, waterways, or seascapes?

The latter feature, character of the adjacent landscape or "matrix," has justifiably been receiving increasing attention over recent years (e.g., see Bennett 1999; Lomolino and Perault 2000; Perault and Lomolino 2000; Ricketts 2001). A number of conservation biologists have made the sobering observation that if we had just reserves and corridors, we may still lose the majority of native species. Most individuals of a great majority of species live outside reserves, in the intervening matrix of habitats. If the habitat matrix is degraded or destroyed, all of the consequences of habitat loss and fragmentation listed in Chapter 16 on page 677 will ensue.

Biological Surveys and Biogeographic Monitoring

It should be clear that biogeographic theory can play an important role in conserving biological diversity. It can help us identify the key factors and processes to study, which sets of data we need, and how to use them. Yet, theory alone won't solve any of these problems. Without fundamental information on the ecology, systematics, and biogeography of particular species, even the most sophisticated theory will be rendered useless. Optimal strategies for selecting nature reserves can be used only when we know which species occur among the prospective sites. Similarly, McDonald and Brown's application of biogeographic theory (Chapter 16) would have been impossible without a good record of species distributions on the mountain ranges of the Great Basin. In short, while we should continue to look to theoretical advances in all relevant disciplines, a critical factor for conserving biological diversity is biogeographic data—the distributions of the focal species. But we need more than accurate lists of which species occur *somewhere* on Earth. As we stated in the beginning of this chapter, we need global biodiversity surveys to describe species distributions and identify hotspots of diversity, and we need a better understanding of the geographic progression of extinction forces (i.e., how they spread across, and transform, native landscapes and seascapes).

As we noted earlier, some systematists and conservation biologists suggest that a global biodiversity survey could be completed over the course of 50 years (Raven and Wilson 1992). While this may be a bit optimistic, efforts toward this goal would certainly go a long way toward reducing the great gaps in our knowledge. We are encouraged by the increasing number of biological surveys that are being undertaken in the meantime. Many of these, such as Conservation International's Rapid Assessment Program (RAP), have targeted their limited resources at the sites in greatest need of surveys and conservation: the tropical hotspots of diversity. RAP is essentially a SWAT team of some of the world's best field biologists, who can, within a span of weeks, conduct first-cut surveys to document some of the biological diversity of sites facing immediate danger. RAP selects sites based on four criteria: high total diversity, high endemicity, uniqueness of the ecosystem, and high risk of extinctions. Other efforts, such as BIOCLIM (Busby 1999), DOMAIN (Carpenter et al. 1993), FloraMap (www.floramap-ciat.org), WorldMap (www.nhm.ac.uk/science/projects/worldmap/), GARP (Stockwell and Peters 1999), and GAP (Gap Analysis Program, Scott et al. 1993, 1996, 2002), utilize remote sensing and GIS (Geographic Information Systems) technology along with extensive existing information on climate, soils, and species distributions and their ecological and climatic associations, to predict species distributions and locations of biodiversity hotspots (see also Sanchez-Cordero et al. 2004). Although each of these methods vary somewhat, the basic approach is to define the climatic and ecological limits of a species' predicted range, to limit the species' potential distribution to regions where the species is known to occur, then to project these limits on to GIS maps depicting the climate, soils, and other relevant features of focal landscapes. Distribution maps for each species can then be overlaid to create contours of species diversity and identify areas of conservation priority—either predicted occurrences of key, imperiled species or hotspots of diversity and endemicity. Such maps or GIS images can then be compared with those depicting the locations of national parks and other protected lands to identify gaps in the protection of biodiversity (i.e., hotspots not falling into a protected area; Figure 17.8). These gaps can then be given top priority by environmental planners and conservation biologists.

It is important to remember that these maps are predictions and not documented distributions. They will, therefore, include errors of both commission (i.e., predicting a species occurs in areas outside its actual range, or "over-prediction") and omission (i.e., predicting a species is absent from sites it actually inhabits, or "under-prediction"). Accuracy of these predictive maps can, however, be assessed using future surveys or existing data that was not used to generate the predictive models. In addition, once validated, the models can be used to monitor or predict the dynamics of species distributions over time, or to predict distributions of focal species or diversity hotspots at some future time (e.g., under alternative scenarios for global climate change or anthro-

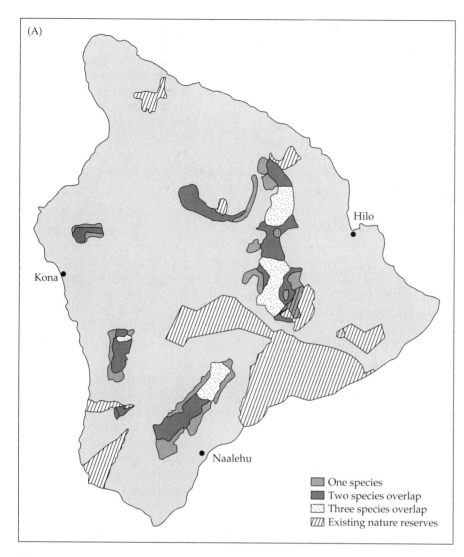

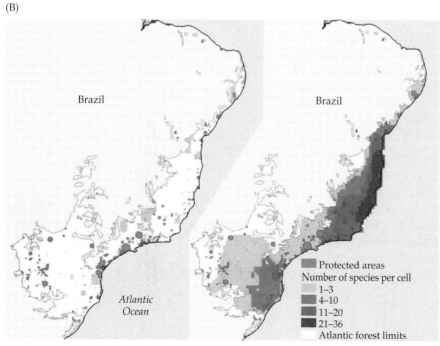

FIGURE 17.8 (A) The U.S. GAP Analysis Program's approach for identifying sites in need of protection is illustrated here in its first application: developing conservation strategies for the avifauna of the Island of Hawaii. Maps of species distributions are overlaid on one another to identify hotspots (i.e., sites with a significant number of priority species). Hotspots that are not located within existing nature reserves (hatched areas) are identified as "gaps" in the protection of biodiversity, and are thus targeted for future nature reserves. (B) A more recent global gap analysis conducted by Ana Rodrigues and her colleagues illustrates how information on habitat, climate, soils, landscape characteristics, and species distributions can be used to identify high priority areas to be included in a global network of nature reserves (i.e., areas that are currently unprotected, yet harbor relatively high numbers of geographically restricted species). The example shown here is that of Brazil's Atlantic Coast hotspot illustrating the species density (number of species per quarter-degree grid cell) of mammals, amphibians, turtles, and threatened birds whose geographic ranges fall completely outside, or have less than 5% of their range overlapping with, existing protected areas (maps on left and right, respectively). (A from Scott et al. 1993; B from Rodrigues et al. 2004.)

pogenic modifications of landscapes and seascapes). Each of these potential applications of advancing technologies again highlight the need for strategically designed biogeographic surveys to document species distributions—not just in a snapshot of time, but also over time periods when species' distributions changed in response to past and ongoing changes in the geographic template. While countless studies have contributed a wealth of information on distributions of thousands of species of plants and animals, these studies fall far short of what we believe is both needed and achievable in conservation biogeography. We recommend that a globally standardized program for biogeographic surveys and monitoring include at least the following.

- Use of internationally standardized methods for detecting occurrences and relative abundances for particular taxa (e.g., see Heyer et al. 1994; Wilson et al. 1996; Agosti 2000).

- Repetition of biological surveys at appropriate intervals, which will vary with characteristics of the focal species (e.g., their dispersal abilities and mean generation time) and on the rate of change in regional landscapes and seascapes. Quite often, biodiversity databases include the results of studies conducted decades earlier (i.e., during periods when landcover, water quality, and climates may have been substantially different than prevailing conditions).

- Development of globally standardized, geographic strategies for biological surveys (e.g., stratifying prospective survey sites among different regions, and randomizing them within these strata) to enable more rigorous tests of climatic and ecological associations of species and development of more accurate biogeographic models used to predict their distributions. Most of these models assume that the data used to predict climatic and ecological associations of the species and to evaluate the models' predictions were geographically unbiased. We know, however, that our surveys are conducted in a small and biased subsample of the world, often in those sites that are either easiest to access (i.e., near roads and waterways) or most attractive for various aesthetic reasons.

- Consultation of historic reconstructions and, when possible, repeated historical surveys of species distributions to describe the dynamics of species distributions in response to historical changes in native landscapes and seascapes. Reconstructions of the New England landscape prior to deforestation (see Foster 2002); recent, repeated surveys of the biota along the path of Lewis and Clark's explorations (Laliberte and Ripple 2003); and reconstructions of prehistoric land use patterns in the Amazon Basin (Moffat 2002), are exemplary cases of applying past surveys to current challenges of conserving biological diversity.

- Make the results of all of these surveys—past and future—available to all scientists in a standardized, digitized and geographically explicit format.

We are encouraged that a number of conservation organizations have developed survey programs that utilize many of the elements we list above. These are, of course, initiatives where biogeographers have and should continue to contribute information and expertise fundamental to understanding and conserving the geography of nature. We feature just one of these applications of biogeographic surveys and theory here—the exemplary case study of conserving biological diversity of the Philippines (Box 17.1).

BOX 17.1 *Case study in conservation biogeography—biological diversity of the Philippines*

"Science is built up with facts, as a house is with stones.

But a collection of facts is no more a science than a heap of stones is a house."

■■❚ The above quote from the nineteenth century mathematician, Jules Henri Poincare, is as applicable to conservation biogeography as it is to any field of science. We are optimistic that the many dedicated scientists and organizations tackling the biodiversity crisis will increase their efforts to conduct biogeographic surveys and lessen the Wallacean Shortfall. But even in our optimism, we must admit that it will take many decades to describe the geographic distributions of the majority of today's imperiled species. During that time, and unless we take action to conserve these species now, many of them will have gone extinct. Conservation biology is a science of urgency, but of all the actions we can take, which should receive the highest priority? The solution to this dilemma is that we continue to advance on three important and interdependent fronts:

1. We should increase our efforts to conduct biogeographic surveys to map species distributions and describe the geographic variation in diversity and endemicity;

2. We should apply relevant theory to interpret the results of these surveys; and

3. In lieu of more complete atlases on distributions of imperiled species, we should develop predictive, biogeographic models to identify likely hotspots of diversity and endemicity and focus future surveys and conservation initiatives on these sites.

It may appear that we have clouded the distinction between facts and theory, but this is both intentional and instructive. Even maps of geographic distributions—perhaps the most fundamental units of biogeography—are neither pure fact nor pure theory, but a combination of the two (see Chapter 4). As Thomas Kuhn (1970) observed, "Theory and facts are not categorically distinct. Scientific revolutions are driven by the dynamic tension between

Figure A Much of the Philippines was originally covered by a diversity of habitats (left and right), including lowland forests, montane forests, mossy forests, and others. (Left, courtesy of Paul D. Heideman; right, courtesy of Ben C. Tan.)

theory and empiricism." In biogeography, each distribution map is drawn based on documented locations of occurrence for that species (empirical "facts"), and on interpolations and extrapolations based on inferred (i.e., theoretical) associations of the species with ecological, climatic, and topographic features. These associations and, in turn, predicted distributions must be adjusted when biological surveys provide new distributional data.

The relevance of these lessons for conservation biologists is clearly illustrated in the exemplary case study of conservation in the Philippines, which has recently been summarized by Heaney (2004). The development of the Philippines Archipelago is quite complex, with some of the oldest geological units developing from tectonic activities which began about 35 million years ago. The three principal units, or plates, that now comprise the Philippines drifted and converged, eventually resulting in the emergence of islands from tectonic uplift and volcanic activities. The Philippines are now comprised of over 7000 islands ranging in size from hundreds of tiny islets to the

largest island— Luzon, which is just over 100,000 km^2 (Figure B). Given their tectonic and volcanic origins, these islands are topographically and ecologically complex, often with mountain ranges covered with an elevational sere of habitats ranging from lowland tropical forests to cloud and mossy forests at the highest elevations. Even during relatively recent periods— too short for substantial tectonic activities—the topographic and ecological characteristics of these islands, and indeed even their size, isolation, number and total area, changed dramatically with each of the 20 or so glacial/interglacial episodes and sea level dynamics of the Pleistocene (see Figure B; see also Chapter 9).

This complex history, along with the high insolation, high precipitation and relatively rich, volcanic soils, created a productive and dynamic evolutionary arena that allowed the archipelago's biota to diversify to the point that the Philippines rank among the world's richest hotspots of biological diversity. To put this in perspective, we can compare the diversity and

(continued on following page)

BOX 17.1 Case study in conservation biogeography—biological diversity of the Philippines (continued)

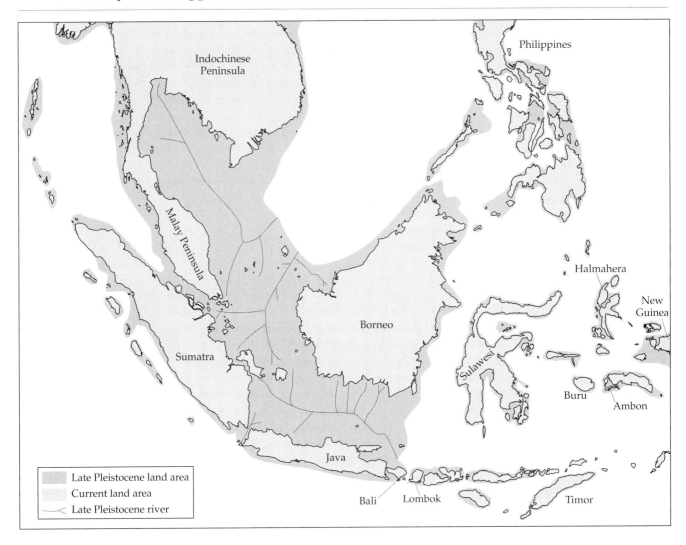

Figure B The Indo-Australian Region showing the configuration and locations of its major archipelagoes during recent times and during the previous glacial maximum when water levels were relatively low and many now-isolated islands and landmasses were connected. (After Heaney 2004.)

endemicity of the Philippines to an equal area of one of the world's most renown mega-diversity countries—Brazil. The total area of the Philippine Islands is 300,780 km², which is just under 3.5% of the total land area of Brazil. Using this percentage and figures for the number of vertebrates and endemic vertebrates of Brazil, we can estimate that a 300,780 km² region of Brazil will be inhabited by about 110 species of vertebrates, with around 28 (25%) of these endemic to that region. Yet, the total number of vertebrates in the Philippines is 952, with 542 (57%) of these endemic to these islands (Heaney 2004;

see also Heaney and Regalado 1998). Endemicity is high within each class of terrestrial vertebrates, but varies with dispersal abilities, being lowest for the best dispersers (i.e., breeding land birds, where endemicity = 44%), intermediate for land mammals and reptiles (i.e., with endemicities of 64% and 65%, respectively), and highest for amphibians (i.e., with endemicity = 73%), which are most limited by their inability to osmoregulate while dispersing in saltwater.

Unfortunately, this is another geographic area where a global hotspot of diversity and endemicity coincides with

one of human development. Prior to 1600, mature tropical forests covered about 90% of the land surface of the Philippines. By 1898, mature forest cover was reduced to 65%, and now just 8% of the original forests remain (Kummer 1991; Heaney and Regalado 1998; Environmental Science for Social Change 1999). Also, just as in other regions where human populations have advanced over hotspots of biological diversity, the geographic progression of deforestation in the Philippines was highly nonrandom. The first forests to be cleared were those that occupied the relatively broad and flat lowlands, but then deforestation advanced to the higher and steeper sites as lowland forests became more scarce and timber technologies advanced (see also Figure 16.45).

BOX 17.1 *(continued)*

Clearly, the Philippines are in dire need of continued and intensified biogeographic surveys, but they also require immediate conservation action. Given the limited time and financial resources, it is imperative that these efforts be applied *where* they will do the most good (i.e., on the islands and within the subregions of those islands that harbor the majority of the endemic and imperiled species). Fortunately, the Philippines have been attracting the attention of distinguished naturalists and biogeographers for well over a century (see Heaney and Regalado 1998). Granted, just as for any of the world's hotspots of biological diversity, comprehensive catalogs and atlases of the Philippines' imperiled plants and animals may be many decades away. However, biogeographers have used the available information to identify some very general geographic patterns and, thus, to develop a comprehensive plan for conservation in lieu of complete biogeographic surveys. The most general and relevant patterns are summarized by Heaney (2004):

1. Each deep-water island (or cluster of islands that were united during glacial periods) is a unique center of endemism (see Figure C).

2. Each isolated mountain range on the larger islands is a subcenter of endemism.

Heaney and his colleagues then applied these patterns to develop quantitative models that could be used to predict the locations of hotspots of endemicity, and to predict (i.e., based on area, isolation, and elevation of the island and its mountains) the actual number of endemic species (see Heaney and Regalado 1998; Heaney 2004). Subsequent surveys by these researchers not only confirmed the locations of these hotspots, but they typically yielded the same or similar numbers of endemic species as predicted by the model (Figure C). One of the few and most foreboding exceptions was the Island of Siquijor, which was predicted to be inhabited by two species of endemic land mammals,

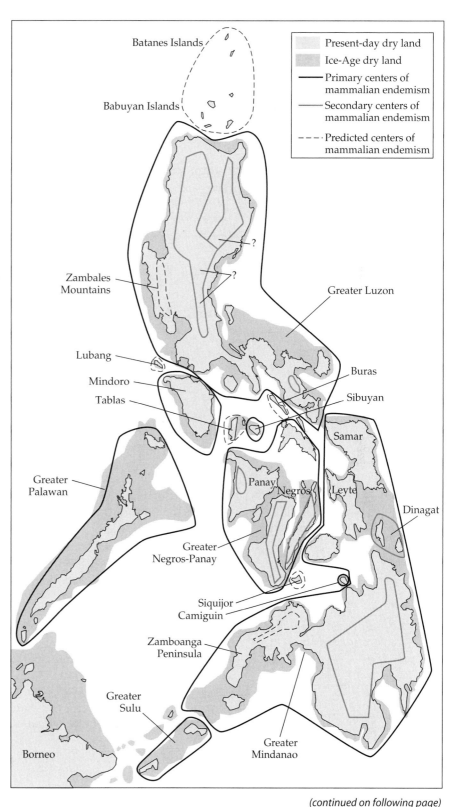

Figure C Locations of the major regions and subregions of endemicity of the Philippines' land mammals. (After Heaney 2004.)

(continued on following page)

---

but had none. Unfortunately, this island has been completely cleared of primary forests so that if it had any endemic land mammals, they likely went extinct before the naturalists could survey the island (an apparent case of Centinelan extinctions; see Chapter 16, p. 647).

Comparisons among patterns for different taxonomic groups revealed that, for the most part, hotspots of endemicity of plants, insects, and land vertebrates occur in the same regions and subregions, with just a few exceptions. In addition to being influenced by island area, isolation, and elevation, endemism of reptiles, amphib-

ians, and plants were also influenced by edaphic characteristics (endemicity of reptiles and amphibians being highest in areas of exposed limestone, and those of plants being highest in areas of ultra-basic soils; Ong et al. 2002). Thus, the accuracy and generality of the predictive, geographic models can be significantly enhanced with information from geological maps of the islands.

The success of this mixed strategy, where insights from both biological surveys and those from biogeographic theory are simultaneously applied to better focus conservation initiatives, is cause for opti-

mism. As Heaney (2004) concluded in a recent paper on conservation biogeography, the destruction of the Philippine forests continues at such a rate that we cannot afford to wait until we inventory the biota of each of its 7000 islands—indeed, there is no longer any need to wait. We can now concentrate our limited funds and personnel on the predicted and now verified centers and subcenters of endemicity and protect those sites now inhabited by most and, in the case of such well-studied groups as mammals (see Figure C), virtually all of the Philippines' endemic species. ∎∎∎

The Biogeography of Humanity

The biogeography of humanity is a subject that must intrigue nearly all of us, scientist and layperson alike. Where did our species originate, and how did we come to occupy and dominate nearly all points of land on Earth? These questions are as challenging as they are captivating in that they require a synthesis of information from many disciplines, ranging from genetics and evolutionary biology to comparative and functional anatomy, systematics, physical and cultural anthropology, archaeology, paleoecology, and paleoclimatology.

The geographic history of humanity provides biogeographers with an exceptional data base with which to test the generality of many biogeographic principles. Has our own species exhibited patterns of dispersal and modification similar to those of most other animals? Were the diversification and range expansion of our human ancestors strongly influenced by tectonic and climatic events? Were our colonizations of remote continental regions and oceanic islands the result of unique immigration events involving a limited number of founders, or did they depend on recurrent waves of immigration consistent with MacArthur and Wilson's equilibrium theory? Once they colonized remote oceanic islands, were human populations then locked into a taxon cycle, destined to undergo ecological and evolutionary changes that would eventually lead to their extinction, perhaps at the hands of a new wave of human colonists?

While the biogeographic dynamics of our own species is a compelling subject in its own right, it is also one of profound relevance to the prehistoric and historical waves of global extinctions. We may never be able to tease apart all of the potential causal factors, but we are certain that the great majority of these extinctions have been anthropogenic, or at least hastened by the activities of anatomically modern (and ecologically significant) man, *Homo sapiens sapiens*. Here, we present an overview of the origin and early expansion of our species (see the more extensive discussions in Howells 1973; Terrell 1986; Fagan 1990; Bell and Walker 1992; Clark 1992; Irwin 1992; Gamble 1994; Straus et al. 1996;

Kirch and Hunt 1997; Wenke 1999; Agusti and Anton 2002). We acknowledge that our account is far from complete, but we have tried to summarize the principal features and events of the geographic history of humanity, especially those most relevant to the late Holocene surge in extinction rates.

Human Origins and Colonization of the Old World

The origin of Primates, the order to which we and all other hominids belong, dates at least as far back as the early Tertiary Period, some 65 million years B.P. Over the next 20 to 30 million years, the primates diversified and expanded their ranges to include much of both the New and Old Worlds. These early prosimian primates comprised a collection of relatively primitive forms related to extant species, including galagos of tropical Africa, lemurs of Madagascar, lorises of Southeast Asia, and tarsiers of the Philippines and Indonesia. By 35 million years B.P. the prosimians gave rise to a second, more derived group of primates, the anthropoids, which includes the platyrrhines (New World monkeys) and the catarrhines (Old World monkeys and apes). Humans are descended from a subgroup of the catarrhines that is often referred to as the great apes, or hominoids, and includes gorillas, orangutans, chimpanzees, and hominids—the latter including australopithecines and members of the genus *Homo* (including *H. habilis*, *H. ergaster*, *H. erectus*, and *H. sapiens*).

It appears that the hominid lineage diverged from gorillas and chimpanzees between 5 and 7 million years B.P. in Africa. While these dates are, and will continue to be, the subject of much speculation and refinement, it is generally accepted that Africa was the site of human origins (Guilaine 1991; Larick and Ciochon 1996; Wenke 1999; Ambrose 2001). Fossils of the oldest known hominid, *Australopithecus afarensis*, found at sites in Ethiopia and Tanzania, date back to about 3.7 million years B.P. Over the next 1 to 2 million years, it appears that the evolutionary and cultural development of our ancestors was confined to savannas of eastern and southern Africa. The first representative of the genus *Homo* was *H. habilis*, which means "handy person." This was apparently the first hominid to make its own tools—relatively simple flaked stones, which it used to prepare animal carcasses and cut vegetation. The tools were rudimentary and unspecialized, but they mark a fundamental advance in our ability to exploit and modify our environment.

By approximately 1.4 million years B.P., a new—and, by most standards—more "advanced" hominid, *Homo erectus*, had evolved. In comparison to *H. habilis*, *H. erectus* was much larger (5 to 6 ft.), larger-brained, more dexterous, and more advanced in its ability to manufacture tools and presumably to communicate, at least in a rudimentary fashion, with conspecifics. *H. erectus* was able to develop effective strategies for hunting big game, and may have been able to capture and use fire. This latter advance cannot be overemphasized. With fire, *H. erectus* could greatly expand its niche and modify its environment to the extent that it could influence the geographic ranges of other species (Fagan 1990). Fire could be used for protection against predators, for hunting game, and for cooking smaller prey such as rodents and insects. Fire could also be used to render toxins, common to many vegetable foods, harmless. At first, *H. erectus* just captured natural fires, but eventually developed the ability to create as well as preserve fire. With this important tool, in combination with those fashioned from rock, bone, and wood, *H. erectus* could expand its range out of the savannas and woodlands and into a variety of other habitats (see Ambrose 2001; Balter 2004). In fact, because of their relatively large body size and more developed social systems, populations of *H. erectus* may have needed to expand their home ranges and colonize new habitats in order to meet their heightened energy demands.

By 1.5 million years B.P., and perhaps as early as 2 million years B.P., *H. erectus*—the apparent predecessor of *H. sapiens*—may have been capable of expanding its geographic range out of Africa and into Asia and Europe, but these migrations did not occur for many more hundreds of thousands of years. Other catarrhine primate groups (including the ancestors of the Asian great apes) had migrated out of Africa when the African and Arabian plates collided some 17 million years ago. The convergence of these plates closed the Tethys Sea and provided a landbridge to Eurasia across the Arabian Peninsula. The range expansions of hominids, however, did not immediately follow the formation of a terrestrial dispersal route out of Africa. Instead, these emigrations seem to have been strongly influenced by shifts in climate and habitat associated with the glacial cycles of the Pleistocene. With each cycle, the tropical savanna habitats of hominids—and many other mammals for that matter—split, shifted, and rejoined, setting the stage for the vicariant processes that often drive evolutionary and biogeographic change. Each wave of hominid expansion out of Africa, from *H. habilis* to *H. ergaster*, *H. erectus*, and finally *H. sapiens*, occurred in at least two stages and probably required repeated migrations of many populations.

First, hominid populations long adapted to mesic savannas and woodlands had to penetrate and eventually cross the great xeric barrier formed by the Sahara Desert. These expansions into the Sahara region most likely occurred during the relatively brief, but wet, interglacial periods. The second stage of hominid migration out of Africa was probably triggered by the return of glacial conditions that again brought a period of great aridity. Populations of *H. erectus* migrated out of the Sahara in all directions, retreating to the savannas and grassland to the south, but also emigrating northward and then eastward across the Sinai and into western Asia (Figure 17.9). Just as it had served as a landbridge earlier for the catarrhine primates, the Arabian Peninsula now

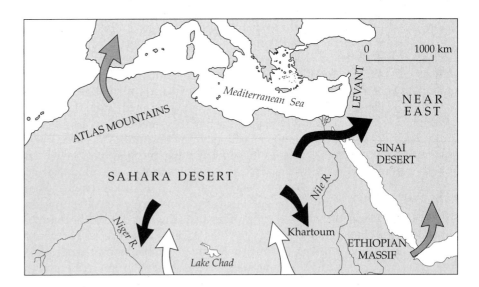

FIGURE 17.9   Stages in the colonization of the Old World by hominids. This two-stage migration was strongly associated with glacial cycles and was probably repeated during recurrent waves of hominid emigrations out of Africa during the Pleistocene. Open arrows = penetration of the Sahara Desert from the savanna "homeland" during interglacial periods; black arrows = emigration from the Sahara during relatively arid glacial maxima; and gray arrows = possible migration routes during periods of low sea level during glacial maxima. (After Fagan 1990.)

□ Distribution of *Homo erectus*

FIGURE 17.10 Estimated geographic range of *Homo erectus* between 600,000 and 500,000 B.P. (After Guilaine 1991.)

served as a migration route for hominids into Asia and Europe. Once each wave of hominids passed beyond this barrier, they rapidly expanded their geographic range to include most areas of Eurasia below 45° N latitude.

H. erectus first colonized the Arabian Peninsula around 800,000 B.P., then spread to establish populations from the eastern edge of Asia to western Europe by 500,000 B.P., and perhaps as early as 600,000 B.P. (Figure 17.10; see Larick and Ciochon 1996). Yet, with all its skills and advances, and despite its relatively extensive geographic range, *H. erectus* was soon to be overshadowed and replaced by another wave of hominid migrations, again originating in Africa. A few anthropologists have hypothesized that *H. sapiens* resulted from the independent, parallel evolution of isolated subspecies of *H. erectus* in Europe, Africa, and Asia. The general consensus, however, holds that *H. sapiens* evolved in the savannas of Africa prior to 300,000 B.P. as a descendant of the African form of *H. erectus*. Then, following in the footsteps of its primate and hominid ancestors, *H. sapiens* repeated the two-stage journey into the Sahara and then across the Sinai to colonize Eurasia. By 100,000 B.P., *H. sapiens* had expanded its range out of Africa and then, either by assimilation or by direct interaction, replaced *H. erectus* (see Gibbons 1996, 2001; Larick and Ciochon 1996; Mellars 1998; Balter 2001a,b).

In comparison to *H. erectus*, *H. sapiens* was empowered by a more advanced brain, more sophisticated language, and superior abilities to plan, organize, and coordinate group activities (Figure 17.11). It manufactured diverse tool kits, each with a variety of implements specialized for different local environments, from forests and open savannas to coastal habitats. It continued to

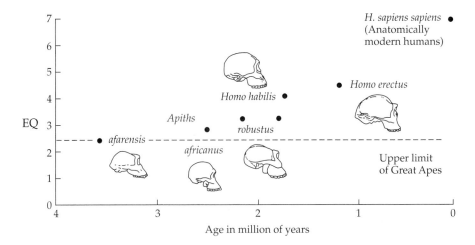

FIGURE 17.11   Relative brain size (as measured by encephalization quotient, or EQ) of hominoids has steadily increased over the past 3.5 million years. (After Gamble 1994.)

develop the ability to manufacture and use fire for protection, warmth, hunting, and preparing foods. It also built shelters and used caves for protection against the elements. In short, *H. sapiens* possessed an unrivaled ability to adapt to, modify, and eventually dominate a variety of environments. Once *H. sapiens* migrated out of Africa, its range expansion was extremely rapid, overtaking and eventually extending far beyond the ultimate range of *H. erectus*. The probable pathways and timing of the aboriginal migrations of *H. sapiens* across the globe are summarized in Figure 17.12.

Conquering the Cold: Expansion to the New World

The global expansion of *H. sapiens* beyond the range of its hominid predecessor posed two principal challenges: conquering the cold, and navigating the oceans. In many ways, *H. sapiens* was pre-adapted for meeting these challenges, or at least for rapidly developing the ability to master them. With its ability to create fire and to fashion tools for new purposes, along with its relatively advanced intelligence and ability to communicate, *H. sapiens* achieved an unrivaled ability to adapt to new environmental challenges (see Ambrose 2001). Yet, its expansion into Europe and the cooler climates of the higher latitudes was a relatively slow process, perhaps occurring over some 40,000 to 50,000 years after its first appearance in the Near East.

Again, geographic barriers and glacial cycles seem to have had a strong influence on this phase of range expansion by *H. sapiens*. Full glacial conditions persisted throughout Europe for most of the period between 75,000 and 40,000 B.P., except for a relatively brief period of warming around 50,000 B.P. As Fagan (1990) noted, it may not be just a coincidence that modern humans first appeared in Europe during this brief interglacial. Tools, shelters, and social organization associated with cold adaptation were developed rapidly after 40,000 B.P., allowing populations of *H. sapiens* to extend their ranges far into the higher latitudes of Eurasia and then into North America. During their migrations, they developed the projectile weapons and group strategies necessary for taking large game. They were then able to use the bones of mammoths and other large mammals to build shelters and fashion tools. Their cultural and ecological adjustments to the cold may have been augmented by morphological and physiological adjustments. Now populations of *H. sapiens*

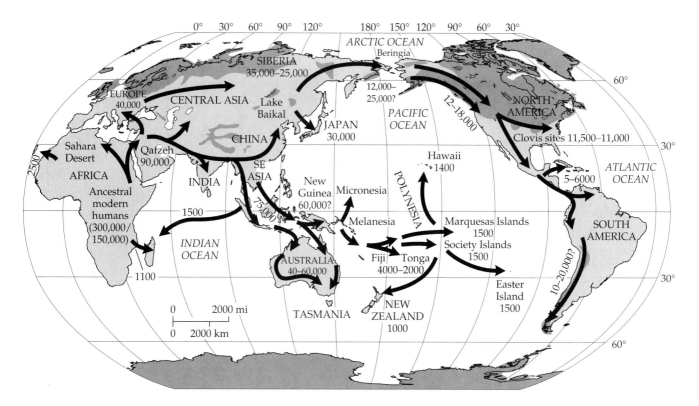

FIGURE 17.12   Colonization of the globe by anatomically modern humans (*Homo sapiens sapiens*). The exact dates of these migrations, given here in years before present, are the subject of continuing debate and frequent revision. Shading indicates extent of glaciers during the last glacial maximum (~18,000 years B.P.). (After Gamble 1994.)

could migrate along with the game and other species that tracked the retreat of the glaciers and shrub-steppe communities with each glacial cycle.

By 50,000 to 40,000 B.P., populations of *H. sapiens* had expanded northward to the edge of the glaciers in Europe and Siberia. Their colonization of northeastern Siberia between 40,000 and 30,000 B.P. set the stage for the great leap into the Western Hemisphere. Again, this major event in human colonization was strongly influenced by climatic cycles and may have required two fundamental steps: colonization of Alaska, and southward dispersal through the remainder of North America and into South America (Figure 17.13; see Batt and Pollard 1996; Gibbons 1996; Roosevelt et al. 1996). Between 25,000 and 18,000 B.P. glacial conditions prevailed, and with the lowering of sea levels by roughly 100 m, Beringia formed a 1500 km wide landbridge between the Palearctic and Nearctic regions. During this period, Beringian environments probably varied substantially, ranging from a relatively productive cold steppe that supported a diversity of large mammals, to a polar desert (see Chapter 9, and Hoffecker et al. 1993). Recent evidence suggests that *H. sapiens* may have migrated across Beringia along with their game species to colonize Alaska before 25,000 B.P. Following this first stage, subsequent expansion through the remainder of the Americas was probably blocked by the glaciers that persisted until about 16,000 B.P. It is possible that human populations migrated southward along the ice-free coasts or along the chain of nunataks between the Laurentian and Cordilleran glaciers prior to 16,000 B.P. Colonization of the New World prior to 15,000 B.P. is still controversial, however, and a

FIGURE 17.13 The colonization of North America by anatomically modern humans may have occurred in two stages and may have been closely associated with glacial cycles. During glacial maxima, lowered sea levels exposed Beringia, which served as a corridor for the dispersal of many terrestrial animals, including humans. Southward migration from Alaska, however, may have been blocked by continuous ice sheets that formed an east-west barrier across northern North America until glacial recession (after 16,000 B.P.). Recent claims of earlier colonization of North America by humans (before 25,000 B.P.) imply that intermittent corridors may have existed between the Cordilleran and Laurentian glaciers, or along a relatively thin strip of unglaciated land along the Pacific coast. (After Guilaine 1991.)

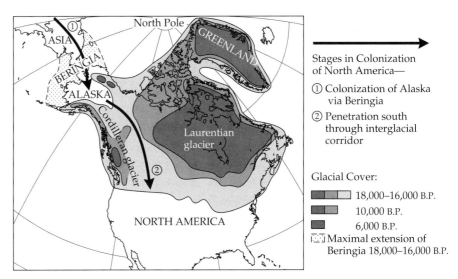

number of paleoecologists maintain that colonization occurred only after the last glaciation (see Batt and Pollard 1996; Josenhans et al. 1997; Marshall 2001a,b).

Either way, following their penetration south of the ice sheets, human populations expanded southward through North America and then South America with perhaps unprecedented rapidity (Mosiman and Martin 1975). Barring some taphonomic artifact, it appears that the dates of the earliest known sites of ecologically significant human occupation across the extent of the Americas (from North America to Tierra del Fuego in South America) span just a few thousand years. In contrast to the continents of the Eastern Hemisphere, where most barriers (e.g., the Sahara, the Mediterranean, and the Alps) run east to west, these features tend to run north to south in the Western Hemisphere, and thus may have actually facilitated human migrations southward from Alaska to the remainder of the Nearctic and Neotropical regions. Certainly, future paleontological studies will revise many of these dates, but the peopling of the Americas may continue to represent one of the most rapid and dramatic episodes of range expansion evidenced for any animal taxon. This rapid spread of ecologically significant hunting societies between 12,000 and 10,000 B.P. seems to be one of the prime causes of coincident megafaunal extinctions in North America (see Figure 9.30).

Conquering the Oceans: The Island Biogeography of Humanity

The last stage of human colonization represents one of the most important with respect to losses in biodiversity. Again, patterns in the colonization of isolated archipelagoes by *H. sapiens* provide an excellent opportunity to test the generality of many principles and hypotheses of island biogeographic theory. Moreover, because the prehistoric and historical waves of extinctions occurred most strongly on islands, this stage of human colonization is likely to shed some light on coincident losses in biological diversity.

Our key focus here is on the colonization of Indonesia, Australia, and the islands of the Pacific, all of which lie between 40° N and S latitudes. Yet, even in these now tropical and subtropical regions, human colonization was strongly influenced by the geographic barriers and glacial cycles of the Pleistocene (see Gibbons 2001a,b; Oppenheimer and Richards 2001). Populations of anatomically modern humans were established along the coasts of East Asia by about 100,000 to 80,000 B.P. Their migrations across much of what are now the islands of Indonesia paradoxically required only modest seafaring abilities. In fact, most of this area may have been colonized across ancient landbridges. During glacial maxima of the Pleistocene, this area comprised four main regions: Sunda, Wallacea, Sahul, and Oceania (Figure 17.14). As a result of the lowering of sea levels, the Asian mainland extended as a continuous landmass from the Malay Peninsula northeast to the southern reaches of the Philippines and eastward to Bali and Borneo—that is, to the western border of Wallacea (see Box 17.1). Thus, the colonization of most of what is present-day Indonesia probably took place by walking, and it probably occurred soon after *H. sapiens*

FIGURE 17.14 The lowering of sea levels during glacial maxima of the Pleistocene caused the exposure of continental shelves and the formation of dispersal routes across four regions of the eastern Pacific: Sunda, Wallacea, Sahul, and Oceania. (White areas = land exposed during glacial maxima; dark shading = deep water (> 200 m); possible dispersal routes are indicated by arrows). (After Fagan 1990; Guilaine 1991.)

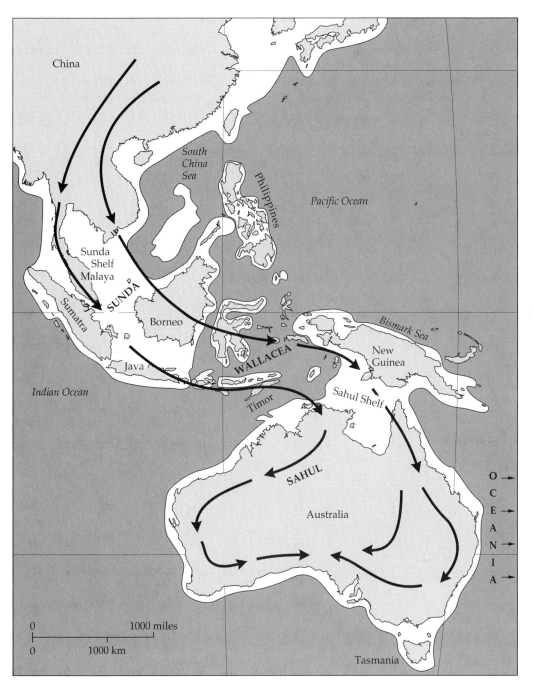

first colonized Southeast Asia, perhaps by 75,000 B.P. (Fagan 1990). The jump from Sunda to Sahul (the united landmasses of New Guinea, Australia, and Tasmania) required crossing the Wallacean archipelago, with a maximum inter-island distance of perhaps just 60 km. The crossing to New Guinea may have occurred as early as 70,000 B.P., and from there humans spread rapidly to occupy most of Australia and Tasmania by 60,000 to 40,000 B.P.

The final step in the peopling of the Pacific Islands was the most demanding because, even during periods of lowered sea levels, it required true navigation and seafaring skills. The archipelagoes closest to Sahul were colonized first, including the Bismarck Archipelago (colonized around 32,000 B.P.) and the Solomon Islands (28,000 to 20,000 B.P.; see Fagan 1990). Colonization of the more isolated archipelagoes, however, took thousands of years longer and required the discovery of new islands hundreds to thousands of kilometers from colonized sites. (Interestingly, flocks of native colonial seabirds may have provided important clues for locating some of the most remote oceanic islands.) These migrations by early seafaring people were indeed explorations, purposeful movements against the prevailing winds and currents, requiring sophisticated navigational skills and the development and construction of vessels specially designed for long voyages (Figure 17.15). These prehistoric

(A)

(B)

(C)

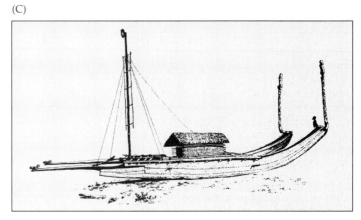

FIGURE 17.15 Colonization of the Pacific realm by humans required the development of sophisticated seaworthy sailing vessels many centuries before the Europeans began their "age of exploration." Shown here are (A) detailed plans for a voyaging canoe, (B) an artist's rendering of one of the Fijian vessels spotted by D'Urville's ship on its voyage between 1826 and 1829, and (C) a Tahitian double canoe sketched by Webber in 1777 during Cook's third voyage. (From Irwin 1992.)

Pacific voyagers carried with them essential resources, including domesticated plants and animals, along with the men and women required to establish viable populations. But the cultural shift from hunter-gatherers to sophisticated seafarers took millennia to evolve. Thus, Fiji wasn't colonized until about 4000 B.P., over 35,000 years after human populations were firmly established across Sahul. Once people had acquired the necessary technology, however, the colonization of what is now Polynesia was extremely rapid. The leap from Fiji to Tonga took approximately 2000 years, but in just another 500 to 600 years Polynesians reached the distant outposts of the Marquesas, Societies, Hawaiian, and Easter Islands (see Figure 17.12).

Yet, despite the Polynesians' mastery of the seas, it appears that some isolated islands were never colonized, perhaps never even visited, before the age of European expansions. These include many isolated islands of central and eastern Polynesia. Many other islands were apparently visited and perhaps briefly colonized by the Polynesians, but they failed to maintain populations for significant periods (Figure 17.16).

The implications of these distribution patterns with respect to the equilibrium theory of island biogeography are intriguing. In an insightful account of the colonization of the Pacific Islands, John Terrell (1986) analyzed patterns in

FIGURE 17.16 At the time of European discovery, many islands in the central and eastern Pacific were uninhabited (large dots). Crosses mark the islands that were colonized by Pacific Islanders prior to 1500 A.D.; those same islands were then abandoned in historic times, usually after extinctions of many native species. Question marks indicate those islands where pre-European colonization is suspected, but not clearly documented. (After Terrell 1986.)

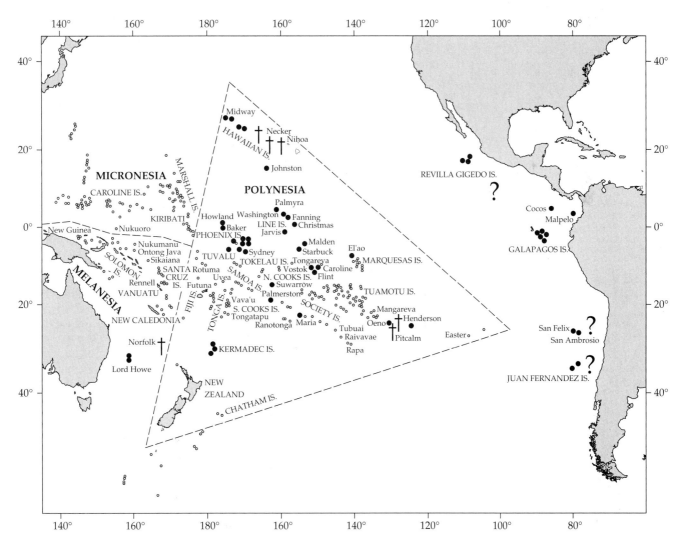

the distribution and diversity of humans using many of the principles and models of island biogeographic theory (see also Howells 1973; Terrell 1976; Terrell et al. 1977). He found, for example, that just as the theory predicts, humans first colonized the less isolated islands, and were likely to maintain populations on the more isolated islands only if they also were relatively large. That is, it appears that the minimum island area required to maintain populations of *Homo sapiens* increased with isolation (see the discussion of rescue effects and compensatory effects in Chapter 13). In the Atlantic, prehistoric civilizations occupied many small islands relatively close to continents (e.g., the Scilly Isles, and the Scottish and Channel Islands) by, if not many millennia before, the Bronze Age (5500 to 3000 B.P.). In contrast, the very isolated islands of the South Atlantic (e.g., Ascension, Saint Helena, Tristan da Cunha, and South Georgia) were uninhabited when Europeans first reached them. Other isolated islands of Oceania and elsewhere were colonized by repeated waves of colonists that either supplemented existing insular populations or replaced those that went extinct.

As observed for other insular faunas, persistence of human populations on islands appears to have been positively correlated with island area. For example, during glacial recession and the associated rise in sea levels during the early Holocene, the landbridge connecting Tasmania to the Australian mainland was fragmented. As a result, insular populations of humans were extirpated on all but the largest islands of Bass Strait. That is, just as Brown (1971) and Diamond (1972) reported for the relaxation of other insular biotas, human populations were subject to nonrandom extinctions following fragmentation of once expansive landmasses. All of these observations are of course consistent with the fundamental tenets of the equilibrium theory of island biogeography: immigrations and extinctions tend to be recurrent processes and are strongly influenced by island characteristics (i.e., isolation and area).

Terrell and his colleagues took this analogy one step further (see Terrell 1986). Not only does the similarity of languages and other cultural and morphological characteristics among insular populations decrease with isolation, but linguistic diversity (i.e., number of different cognate words) increases with island area and decreases with island isolation (Figure 17.17). Interestingly, and consistent with patterns of species diversity, linguistic diversity on the continents also exhibits a marked latitudinal gradient, increasing from the poles toward the equator (Mace and Pagel 1995). According to Terrell, the diversity of languages on islands can be explained by a linguistic version of MacArthur and Wilson's equilibrium theory. He hypothesized that linguistic diversity results from a balance between rates of borrowing ("immigration") of new words from other insular populations and the loss ("extinction") of shared, extant words. These rates are functions of both the existing number of cognate words on an island and its physical characteristics (isolation and area; see Figure 17.17). Similarly, the development and distribution of native linguistic groups in North America appears to have been strongly influenced by the distributions of physical barriers and ecogeographic zones of the late Pleistocene and early Holocene (Rogers et al. 1990; see also Fitzhugh 1997).

Finally, Jared Diamond (1977) suggested that insular human populations may undergo subsequent transformations much like those associated with Wilson's (1961) taxon cycles (see Chapter 14; see also Flannery 1994; Diamond 1997). Like taxon cycles, the colonization cycles of humans tend to be unidirectional, with most movements from larger landmasses to smaller ones, but not the reverse. After the initial colonists arrive, human populations often expand, not just geographically but ecologically as well, exploiting a broader range of resources and habitats. As population densities increase and resources become

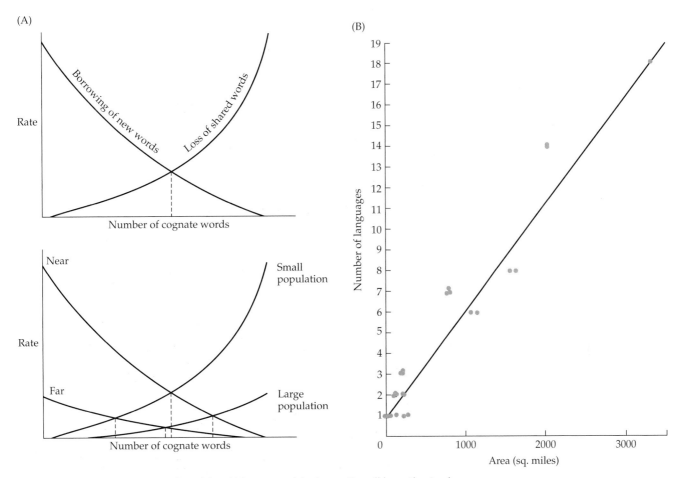

FIGURE 17.17 (A) Borrowing from island biogeographic theory, Terrell hypothesized that linguistic diversity (number of cognate words) results from a balance between the rate of borrowing of new words between insular societies (which should decrease with increasing distance between islands) and the rate of loss of shared words (which should be higher for smaller islands with smaller populations). Accordingly, linguistic diversity (number of cognate words and number of languages) on islands decreases with isolation and increases with area. (B) The relationship between island area and the number of languages spoken on a given island for the Solomon Islands supports this hypothesis. (After Terrell 1986.)

saturated, the initial founding population may enter a stage of local adaptation in which selective pressures switch from those favoring colonizing abilities (high dispersal ability, high reproductive rate, and generalized niches) to those favoring efficiency of resource use and competition with conspecifics. Increased competition from newly arriving human populations may then trigger a "retreating stage" in which the niche and distributional range of the initial colonists contracts. Finally, just as envisioned by Wilson in his taxon cycle hypothesis, Diamond's colonization cycle of man ends in extinction—that is, with the more ancient and specialized civilizations being replaced by waves of invading populations of generalists.

The biotas of the systems that were the last to be colonized by humans—oceanic islands—were the most severely affected by human activities (Figure 17.18). Studies by David Steadman and his colleagues reveal that the extinctions of as many as two-thirds of the native birds and bats of Oceania's islands were coincident with colonization by humans (Koopman and Steadman 1995;

(A)

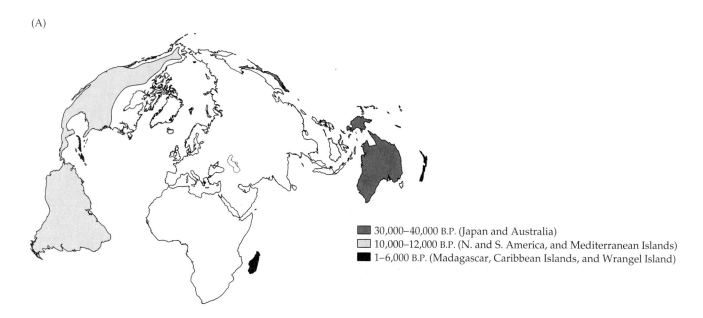

■ 30,000–40,000 B.P. (Japan and Australia)
□ 10,000–12,000 B.P. (N. and S. America, and Mediterranean Islands)
■ 1–6,000 B.P. (Madagascar, Caribbean Islands, and Wrangel Island)

FIGURE 17.18   (A) Waves of extinctions of large mammals and birds during the Pleistocene seem to be coincident with waves of colonization of environmentally significant humans (see Figures 17.12 and 9.30). (B) The relative number of late Pleistocene/early Holocene (40,000 to 10,000 B.P.) extinctions or extirpations of the mammalian megafauna (species heavier than 44 kg). Note that the continents long occupied by humans (Africa and Asia) were relatively immune to this wave of extinctions. (A after Martin 1984; B after Martin 1990.)

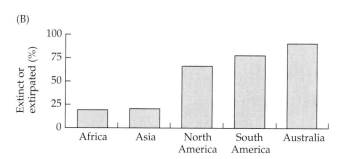

(B)

Steadman 1995; Burney et al. 2001; Steadman and Martin 2003). These extinctions included an estimated average loss of ten species (or populations) per island over 800 major islands, yielding 8000 extinctions or extirpations. Steadman (1993) suggests that human colonization of oceanic islands influenced insular avifauna more strongly than any tectonic, climatic, or biological event of the past 10,000 years (Figure 17.19).

Lessons from the Biogeography of Humanity

Before turning our attention to the frontiers of biogeography (Chapter 18), it is worth summarizing some of the general lessons of the foregoing discussion:

1. On a global scale, human expansions were not gradual and continual, but episodic. Long periods of geographic stasis were followed by rapid expansions once landscapes changed or hominids developed technologies necessary to cross persistent barriers.

2. Early expansions of hominids were strongly influenced by the same types of geographic barriers, filters, and corridors that influenced our mammalian relatives. Thus, the geographic expansions of our species were strongly influenced by tectonic events, and especially by the climatic cycles that were prevalent during the Pleistocene.

3. Our range expansions on continents and across archipelagoes were the result of recurrent bouts of colonization and expansion, often followed by

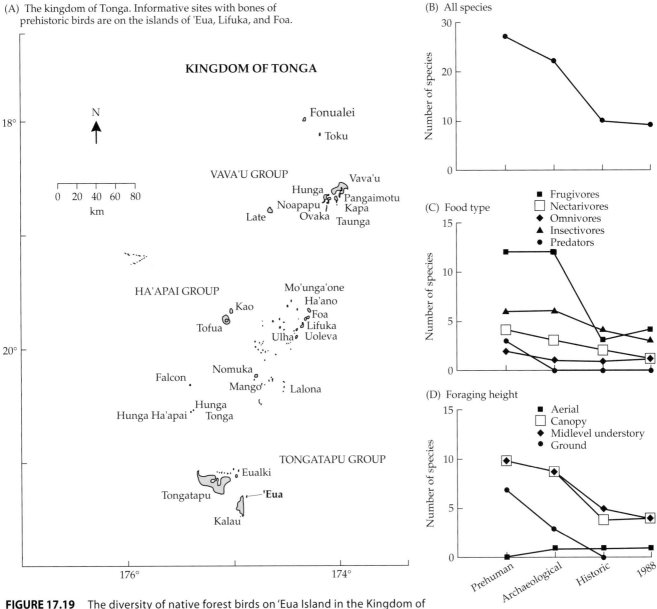

(A) The kingdom of Tonga. Informative sites with bones of prehistoric birds are on the islands of 'Eua, Lifuka, and Foa.

FIGURE 17.19 The diversity of native forest birds on 'Eua Island in the Kingdom of Tonga (A) has decreased dramatically since the island was colonized by Polynesians (B–D). Hardest hit were the frugivorous and ground-dwelling species, which were especially susceptible to predation by humans and their ground-dwelling commensals. (After Steadman 1995.)

collapse and extinction at the hands of other, "more advanced" peoples, technologies, and cultures.

4. Throughout our evolutionary history, the ability of humans to colonize distant sites and to modify—and, eventually dominate—a greater variety of ecosystems has continually increased.

As we discuss in the following and final chapter, we view the dynamic biogeography of humanity, both past and future, as one of the most intriguing and important frontiers of science. The origins, geographic expansions, diver-

BOX 17.2 *Mapping the ecological impact of human populations*

The Human Footprint

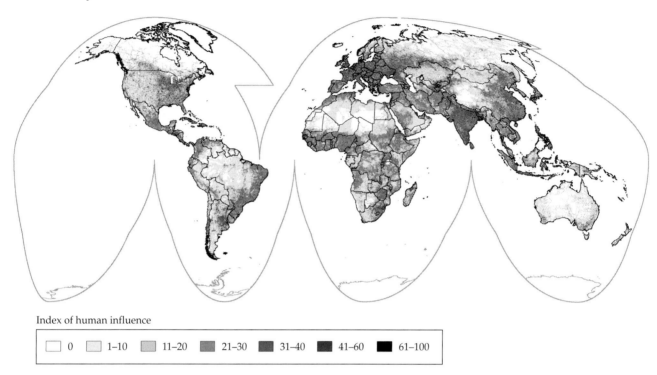

Index of human influence

| | 0 | | 1–10 | | 11–20 | | 21–30 | | 31–40 | | 41–60 | | 61–100 |

■■▮ The impact of modern human civilizations on native biotas, although pervasive, exhibits a strong geographic signature reflecting both our responses to the geographic template, and impacts on and opportunities for conserving biological diversity. The Wildlife Conservation Society's map of *The Human Footprint* (above) illustrates geographic variation in our impact on native landscapes, which is strongly influenced by proximity to coastlines, rivers, roads, mountains, and sources of power (primarily, fossil fuels and electricity). The index of human influence is based on quantification and integration of information on human population densities, anthropogenic land transformation, human access, and power infrastructure (see Sanderson et al. 2002).

Spatial analysis of this map was then used to identify the wildest and least impacted areas within each of 15 terrestrial biomes. The resultant map, *The Last of the Wild* (on facing page), illustrates the ten

sification and, for many once dominant civilizations, extinctions of human populations have been strongly shaped by the same characteristics of the geographic template that influenced the development and geographic dynamics of all species. It is just as clear that our range expansion and rise to global-scale ecological dominance is at the heart of the current wave of range collapse and extinctions. We remain, however, encouraged that our rapidly advancing understanding of the geography of nature, including that of current and future human populations (Box 17.2), will prove invaluable resources for solving one of history's greatest challenges—that of conserving the distributions, diversity, and natural character of Earth's imperiled species.

BOX 17.2   *(continued)*

The Last of the Wild

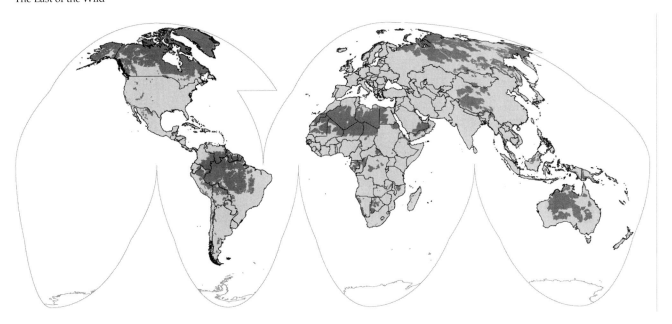

largest, contiguous areas within each biome that are least impacted by human activities (human influence factor of less than 10; source: Center for International Earth Science Information Network). These areas cover a relatively small fraction of Earth's land surface, but they represent sites of disproportionately high value for focusing our efforts to conserve terrestrial biotas.

Development and refinement of these maps continues as new information and new technologies become available. The next generations of these initiatives should certainly focus on both the aquatic as well as terrestrial realms, and should be developed at geographic scales fine enough to illustrate human impacts and conservation potential of small but biologically diverse ecosystems including lakes, rivers, estuaries, and oceanic islands. Such maps, combined with those depicting geographic variation in biological diversity and endemicity (e.g., Figures B and C in Box 17.1) and with dynamic models predicting anthropogenic changes in the future, seem essential to long-term planning and successful conservation of biological diversity. (Maps courtesy of the Wildlife Conservation Society and the Center for International Earth Science Information Network (CIESIN) at Columbia University.) ▮▮▮

CHAPTER 18

The Frontiers of Biogeography

RECENTLY, THE INTERNATIONAL BIOGEOGRAPHY SOCIETY (www.biogeography.org) commissioned a multi-authored volume—*The Frontiers of Biogeography: New Directions in the Geography of Nature*. In that volume, 34 distinguished biogeographers characterized the status of the field and showcased some of its most active and potentially insightful research programs. In the next two sections, we summarize these contributions and the emerging frontiers of the field (excerpted and modified from Lomolino and Heaney 2004).

From the Foundations to the Frontiers of Biogeography

As we have observed throughout this book—Chapter 2, in particular—the long and distinguished history of biogeography is tightly interwoven with that of ecology and evolutionary biology. In fact, biogeography draws widely not only from the biological sciences but also from such disparate fields as geology, meteorology, anthropology, and philosophy. The challenge of integrating these fields to develop a comprehensive understanding of the origins and maintenance of global patterns of biodiversity is one of the field's great attractions, but the isolation of people with such diverse interests into disparate disciplines has also represented a major hurdle to progress throughout the historical development of biogeography.

For centuries, the geography of nature served as a Rosetta stone, providing key insights into the many mysteries of what we now call biological diversity. Indeed, that the natural world varied over space and time was ancient knowledge, information essential to survival of hunting and gathering societies in their struggle for existence. But these ancient observations—that the types and numbers of organisms encountered varied with area searched, distance traveled, elevation of land, and depth of waters—did not coalesce into a true science until the Age of European Exploration. With the procession of broadly trained natural scientists—from Linnaeus, Forster, and Buffon during the eighteenth century to Darwin, Hooker, Sclater, Wallace, Wilson, MacArthur, Briggs, Diamond, and their colleagues of the nineteenth and twentieth centuries—our knowledge of the patterns and causal explanations for the geography of nature accumulated and matured into the science of biogeography.

This period of discovery marked the origins of the natural sciences in terrestrial and marine environments—geology, meteorology, systematics, paleontology, evolution, and ecology—all fields practiced by biogeographers (i.e., scientists studying the spatial variation in the Earth and its life forms, past and present). From these distinguished origins, each of these disciplines along with historical and ecological biogeography and the many subdivisions of those fields, evolved, diverged, and eventually flourished (or languished) as increasingly more distinct disciplines. With this growth and diversification of sciences and scientists came a presumed "need" for specialization and splintering, and the grand view—the ultimate synthesis across space and time—became murky and more elusive.

If this abbreviated history of natural science tells us anything, it is that biogeography is not only the foundation but also the frontier and future of many disciplines. The greatest strides that we can make in unlocking the mysteries and complexities of nature are those from syntheses and collaborations among scientists across the many descendant disciplines, long diverged but now reticulating within a strong spatial context—the *new biogeography*. From the earliest vicariance hypotheses of Joseph Dalton Hooker to the most recent methods of analyzing reticulating phylogenies and phylogeographies, geographic variation over time and space is the key. The unifying question of modern biogeography is how do life-forms from all kingdoms—from the tiniest unicellular organisms to the greatest beasts, and from ancient to current (and into the future)—vary across geographic gradients? The new emphasis is one of reintegration of divergent conceptual lineages, and new syntheses and bold collaborations among scientists that, in the past, had seldom read each other's work, let alone worked together.

The Frontiers

During the past three decades, biogeography has become an increasingly prominent cornerstone of modern ecology, evolution, and conservation biology. Modern biogeographers explore a great diversity of patterns in the geographic variation of nature from physiological, morphological, and genetic variation among individuals and populations, to differences in the diversity and composition of biotas along geographic gradients, and to the historical and evolutionary development of those patterns. It is difficult not to be enthusiastic about recent advances in biogeography and its great promise for answering some of science's greatest challenges (i.e., explaining the origins, spread, distribution, and diversification of life across the Earth). A tall order, granted, but our colleagues and our science have made great strides in recent decades and we are entirely confident that the next generation of biogeographers will not disappoint us.

The frontiers of biogeography are now being defined and driven by the push to develop more general, more complex, and more integrative theories of the distributions and diversity of life—theories based on the dynamics of the Earth, its environments, and its species. We believe that the best means for advancing the frontiers of our science is to foster novel collaborations among complementary research programs. The new series of syntheses—more complex, scale-variant and multifactorial views of how the natural world develops and diversifies—may be less appealing to many, but it is likely to be a much more realistic and more illuminating view of the complexity of nature.

This reintegration of biogeography will continue in earnest largely through the efforts of many broad-thinking scientists who no longer shy away from—but embrace the complexity of—nature, and foster collaborations and conceptual reticulations in modern biogeography. New challenges are emerging, and new approaches—and, in some cases, entirely new disciplines—are developing to meet those challenges. We see ten principal, emerging frontiers of this field, but we admit that this list may not be exhaustive. Indeed, we encourage our colleagues and students to redefine frontiers and surprise us with qualitatively novel contributions to this list.

1. Increasing attention to scale-dependence and to developing process-based linkage across scales of time, space, and biological complexity. As we have noted throughout this book, the influence of many, if not most, processes are emergent at particular scales of space and time, and not at others. Analyses conducted at just the local or global scales are likely to provide only a fragmentary view of the forces influencing geographic variation in biological diversity.

2. Advancing and developing more creative applications of the comparative approach, deconstructing complex patterns, and using resultant insights to develop comprehensive explanations for the diversity, geographic variation, and distribution of biotas.

3. Avoiding the siren's call to specialize and split into more isolated and independent subdisciplines. Biogeography's greatest distinction and strength is that it is the prototypic, interdisciplinary science. As our colleague, Stephanie Forrest observed, "The most effective way to do interdisciplinary science is through collaborations among disciplinary scientists."

4. Addressing the "Wallacean Shortfall" (i.e., the paucity of information on the geographic variation of nature), and intensifying our efforts to understand the biogeography of imperiled species as well as the geographically restricted, less accessible, or otherwise more elusive and novel biotas, including

   - biotas of remote recesses of the biosphere (e.g., those of the abyssal zones and deep soils)
   - a global scale, biogeography of freshwater communities
   - parasites, parasitoids, microbes, fungi, and bryophytes
   - genetically engineered and other anthropogenically modified species
   - invasive, and now "naturalized," species
   - species inhabiting, and in many ways surviving, only in anthropogenic ecosystems (especially zoos, botanical gardens, nature reserves, fragmented landscapes, and the matrix of anthropogenic habitats)

5. Utilizing the rapidly expanding databases (e.g., in marine paleontology, palynology, and integrated museum collections data) and increasingly powerful technology (e.g., those in Geographic Information Systems, molecular biology and geology, molecular, and geological) in the service of developing more integrated, conceptually based biogeography.

6. Developing an integrated theory differentiation and diversification of lineages and biotas which includes vicariance, extinction, geodispersal, and

reticulating phylogenies (i.e., developing a more comprehensive theory on how the production of living diversity is influenced by geographic circumstances and the dynamics of Earth's geographic template).

7. Developing a theory of immigration that considers the selective nature of immigration and its potential influence on the development, diversification, and distributions of biotas.

8. Developing a much more comprehensive understanding of the geography of extinction (i.e., how population persistence varies across the major geographic gradients of area, isolation, latitude, elevation, and depth; across geographic regions in terrestrial and marine realms, or with size of geographic ranges and relative position [e.g., center vs. periphery] across those ranges).

9. Advancing our understanding of the dynamic geography of our own species: how we, our commensals, associated diseases, fragmentation, and other anthropogenic, environmental disturbances have—and will—expand across and transform native ecosystems.

10. Finally, we should apply these approaches and resultant lessons to developing better strategies for conserving, not just isolated samples of imperiled life forms, but their geographic, evolutionary, and ecological context as well.

Three Research Programs

Here, we highlight three of the many active and exciting research programs in biogeography, each of which epitomizes the increasing relevance of the field and its renewed emphasis on creative applications of the comparative approach and on new collaborations among scientists from a diversity of disciplines.

I: The New Synthesis: Historical and Ecological Biogeography

Two major theoretical advances in biogeography during the late twentieth century—the vicariance model of Nelson, Platnick, and Rosen and the dynamic equilibrium model of MacArthur and Wilson—enriched biogeography with their explicitly stated and testable predictions, but also motivated a rift between "historical" and "ecological" approaches to deciphering biogeographic pattern and process. Our separation of largely historical and ecological topics into Units 4 and 5, respectively, is a clear indication that this rift persists. Fortunately, a growing number of distinguished and visionary scientists have taken up the challenge of reconciling this artificial dichotomy between historical and ecological approaches, and instead are developing a new, more integrative biogeography.

A careful reading of the relevant sections in Units 4 and 5 will, in fact, demonstrate that it is becoming increasingly difficult to separate historical from ecological concepts in emerging arenas of biogeographic research. For an excellent example, recall from Chapter 15 Wiens and Donoghue's (2005) proposition that the latitudinal gradient in species richness—which was often associated with one evolutionary or ecological process—might be better approached from a more integrative perspective that considers how environmental variability, adaptation, ecological niche conservatism, and the age and area of habitats, influence the three fundamental processes of biogeography—immigration, extinction, and evolution. As another example, several recent studies have demonstrated how molecular phylogeography and phylogenetics can be used in association with ecological niche modeling to reconstruct the distributional and evolutionary responses of lineages to paleoclimatic shifts (e.g., Hugall et al. 2002; Graham et al. 2004). We continue to be encouraged by these and other efforts to develop explicit connections between community phylogenetics, biogeography, and ecology (e.g., Wares 2002; Webb et al. 2002; Brooks 2004; Riddle and Hafner 2004; Riddle 2005) and, although no one can predict the precise nature of these innovations, we are confident they will provide fundamental insights for unraveling the temporal and spatial complexity of lineages, and for understanding the assembly (and disassembly) of biotas over ecological to evolutionary time scales.

Of course, the newly emerging synthesis between historical and ecological biogeography also bears directly on the future development of the two topics discussed below. As the geography and timing of events that mark the relatively recent expansion of *Homo sapiens* across the Earth has become increasingly better understood (e.g., Templeton 2002), the historical context for understanding the biogeography and ecology of our own species will increase in significance; and an expanding, molecular-based understanding of the geography of biological diversity and the historical and ecological processes that have shaped it has a critical role to play in the emerging field of conservation biogeography.

II: Biogeography of Homo sapiens

In Chapter 17, we summarized current information on the phylogenetic and geographic history of our own species, which has been studied extensively by many anthropologists, but few biogeographers (for notable case studies, see Diamond 1977, 1997; Terrell 1986).

Fortunately, this situation is changing in response to an inescapable reality. Biogeographers and ecologists can no longer pretend that nature is separate from humanity when the influences of humans are apparent in even the most remote and seemingly unspoiled places on Earth. Particles of industrial waste are being deposited in the polar icecaps and sinking to the ocean depths. It has become clear not only that humans are the single main cause of the drastic, recent decrease in global biodiversity, but also that active human intervention will be required to save endangered species and habitats (see Chapters 16 and 17). Respected scientists, once nearly universally unwilling to study anything "applied," are now doing research aimed at preserving biological diversity, wild habitats, and ecosystem processes. Some are even beginning to study the dynamic biogeography and ecology of such human-dominated systems as agricultural fields, suburban landscapes, and inner cities.

While these changes are praiseworthy, there is still a long way to go. Still largely missing is a perspective that considers *Homo sapiens* and all of its activities as an integral part of the system. Thus, for example, studies in "urban ecology" are still more likely to focus on the animals and plants living in the parks and streams than on the rats in the sewers, weeds in sidewalk cracks and vacant lots, and the people, pets, and houseplants living in the apartments in that urban environment. While there are increasing numbers of studies on the effects of agricultural, suburban, and urban areas on the surrounding lands and waters, there are few on the ecological resources and interactions that influence the human populations in these densely settled areas (Cohen 1995; Kates 1996).

We foresee a greatly increased emphasis on biogeography and ecology of human societies, driven both by compelling scientific curiosity and by environmental necessity. The dynamic biogeography of *Homo sapiens* throughout its four million-year history and, in more recent times, its rise to become the Earth's dominant life form are fascinating topics in their own right. To what degree have features of the geographic template that influenced countless other species also shaped the dispersal, distribution, and evolutionary development of our own species? Does diversity in morphology, physiology, languages, and customs among human populations on islands vary in a manner parallel to diversity among native species comprising other insular communities? Do these same traits in variation among human populations exhibit tendencies similar to the ecogeographic rules discussed in Chapter 15 (i.e., do they vary in any regular manner among continental regions, or with latitudinal position, elevation along montane slopes, proximity to coastlines or mountain slopes, or geographic range size of those populations)?

In addition to these intriguing questions in basic research, the dynamic biogeography of our species has undeniable insights for the applied sciences, especially conservation biology. As we remarked much earlier in this book, species are not just influenced by features of the geographic template, but at least in some cases, they can strongly alter those features and, in turn, the distributions and diversity of many other species as well. Certainly, there are very few, if any, ecosystems that are untouched, let alone untransformed by the activities of human societies ("primitive" or "advanced"). Therefore, the success of any attempt to understand the development and geographic signature of biological diversity—and to conserve it—should consider human societies as an integral part of that development. Tim Flannery (1994, 2001) and Jared Diamond (1997, 2005) have produced exemplary works using just this approach: insightful and provocative case studies that interweave the ecology and biogeography of humans with those of the ecosystems they have colonized (see also Vermeij 2004). As compelling as these works are, we believe this is still fertile ground for many novel contributions at the interface of what has, for all too long, been considered distinct fields—anthropology, archaeology, paleontology, sociology, ecology, and biogeography.

III: Conservation Biogeography

As we observed in the preceding chapter, conservation biogeography is a recently articulated discipline, albeit one whose origins trace deep into the early history of biogeography, and one that is intricately related to the biogeography of humanity. Because human populations advance across and transform the geographic template in a highly nonrandom manner, our effects on native communities—including thousands of extirpations and extinctions—should also have highly nonrandom geographic signatures. Again, the degree to which we can predict and mitigate these threats to biodiversity of native biotas depends largely on our abilities to understand, predict, and modify the dynamic biogeography of our own species. This is the first tenet of conservation biogeography (see Chapter 17, p. 712).

The second tenet, however, is just as fundamental: in order to conserve the true nature of native species (including their distinct physiologies, morphologies, behaviors, and ecological interactions), we need to conserve their distributions (i.e., the geographic, ecological, and evolutionary context of nature). While the Wallacean Shortfall remains significant, we applaud the efforts of visionary funding agencies and battalions of gifted and dedicated field biologists who are rapidly diminishing this gap and providing information essential to describing and ultimately conserving and restoring the distributions of native biotas. Thanks to continuing advances in GIS and related technologies, this information is being made available in an accelerating number of atlases and GIS catalogues that now describe geographic distributions for thousands of species. Unfortunately, we are often handicapped by a catch-22 of endangered species studies: we need to know the most about the species that are most difficult to study (i.e., those that are rare and restricted to critically limited, yet incompletely described, geographic ranges).

These and other critical applications of conservation biogeography have been discussed in two recent papers (Lomolino 2004; Whittaker et al. 2005; see also Sax et al. 2005), and it is clear from these papers and from numerous success stories that the promise of conservation biogeography far outweighs its significant challenges. In fact, the field is now recognized as a unifying linkage between biogeography and conservation biology, and this is reflected in the revised title of the journal *Diversity and Distributions*, now subtitled *A Journal of Conservation Biogeography*. Given its integrative nature and great promise, conservation biogeography is likely to continue to gain increased recognition as one of the most intriguing and important frontiers of science, and one that may well serve as a new paradigm for applying lessons from the geography of nature to conserving biological diversity.

Biogeography: Past, Present, and Future

It may be instructive to conclude by reflecting on the persistent themes in biogeography (i.e., those that have formed the conceptual framework of the field since long before Darwin and Wallace published their seminal works; see Table 2.1). It is a valuable lesson in humility to realize that, over the past two centuries, we have not added to this list, only modified the emphasis that we have placed on different themes at different periods.

Theme 1, classifying geographic regions based on their biotas, was once the dominant theme in biogeography, but now receives only passing attention from most of today's scientists. Either we are satisfied with the existing systems of geographic classification, or such systems are viewed to be trivial descriptive exercises. In either case, this seems unfortunate, as there remains much interesting and important work here, especially in classifying biogeographic regions of the marine realm and biogeographic classifications of the terrestrial realm based on lineages other than vascular plants and vertebrates.

In contrast, aided by advances in evolutionary biology and computer science, reconstructing the historical development of biotas (Theme 2) may have become the most popular theme of today's biogeography. Phylogenies have been developed for many species groups, which in turn have provided important insights for both historical and ecological biogeography. Yet, even here, some of the most interesting theories of the past—perhaps, most notably, the taxon cycle—seem to have been largely ignored. Few papers deal with this theory, yet it remains one of the most general explanations for a suite of ecological and evolutionary characteristics of insular biotas (see Ricklefs and Bermingham 2002).

It appears that we also have made significant progress in describing patterns of diversity across a great variety of regions, ecosystems, and geographic gradients (first part of Theme 3). Indeed, we may be on the verge of witnessing a new synthesis and the development of a much more comprehensive understanding of some of biogeography's earliest described patterns—the very regular patterns of variation in biological diversity from the poles to the equator and along other, large scale gradients across the marine and terrestrial realms. The second part of Theme 3, explaining differences in species composition among geographic areas, has received much less attention, yet it is no less fundamental to the field. Community convergence and the development of ecologically equivalent biotas, invasions of exotic species and their subsequent effects on—and integration into—local and regional assemblages, the geographic and ecological dynamics of novel assemblages of species through the fossil record, and the development of nested subsets of species are no less intriguing now than when these patterns were first described, but general explanations for most of them remain elusive.

Finally, Theme 4—explaining geographic variation in the characteristics of individuals—seems to have been dormant throughout most of the history of biogeography, with only brief episodes of activity evidenced by the works of distinguished biogeographers such as Bergmann, Allen, Jordan, and Vermeij. However, it now appears that this beast has awakened. With the earlier works of Rapoport (1982) and the more recent developments in the burgeoning field of macroecology, biogeographers and ecologists have now begun to explore the inner structure of geographic ranges and variation in morphology, demography, and other local-scale characteristics among geographic populations.

The new generation of biogeographers is marshaling the knowledge garnered from over 200 years of biological surveys, the collective intelligence of some of humanity's greatest minds, and an unparalleled tool kit of technological advances to explore geographic patterns in nature. This is an exciting time to be a biogeographer, and we have found it both challenging and immensely rewarding to update this textbook with the many advances in the field. We are equally confident that our colleagues and students will continue to challenge existing theories and contribute insights fundamental to understanding and conserving the geography of nature.

Glossary

abiotic   Pertaining to the nonliving components of an ecosystem, such as water, heat, solar radiation, and minerals.

abscissa   On a graph, the horizontal (x) axis or the horizontal coordinate of a point.

abyssal plain   The relatively flat floor of a deep ocean, mostly between 4 and 6 km beneath the surface.

abyssal zone   In deep bodies of water, the zone between 4 and 6 km through which solar radiation does not penetrate (aphotic zone) and in which temperature remains at or slightly below 4° C year-round.

acid rain   Precipitation with an extremely low pH. The acid condition is caused by the combination of water vapor in the atmosphere with chemicals such as hydrogen sulfide vapor released from the burning of fossil fuels, producing sulfuric acid.

active dispersal (vagility)   The movement of an organism from one point to another by its own motility, such as by active swimming, walking, or flying, rather than by being carried along by some other force; compare with *passive dispersal*.

actualism   The philosophical assumption that the physical processes now operating are timeless, and therefore that the fundamental laws of nature have remained unchanged; also called *uniformitarianism*.

adaptation   Any feature of an organism that substantially improves its ability to survive and leave more offspring over that of other ancestral forms or coexisting phenotypes.

adaptive radiation   The evolutionary divergence of a monophyletic taxon (from a single ancestral condition) into a number of very different forms and lifestyles (adaptive zones).

adaptive zone   A way of life, including such properties as ecological preference and mode of feeding, that has been adopted by a group of organisms.

adiabatic cooling   The decrease in air temperature as a result of a decrease in air pressure (not a loss of heat to the outside) as warm air rises and expands. The rate of cooling is about 1° C per 100 m for dry air and 0.6° C per 100 m for moist air.

aerial   Occurring in the air.

aerial plankton   A diverse collection of organisms that are so tiny and light that they are carried by strong winds high above the surface of the Earth and to places far removed from their natal ranges.

aestivation   A specialized type of animal behavior and physiology in which the organism lives through the summer in a dormant condition.

age and area hypothesis   According to Willis, a hypothesis stating that the greater the age of a taxon, the larger its distributional range.

albedo   The fraction of solar energy that is reflected from the surface of an object back into the atmosphere, or from the Earth back into space.

allele   One of two or more alternative forms of a gene located at a single point (locus) on a chromosome.

allelopathy   A type of interspecific interaction in which one species inhibits the growth of another by releasing chemicals into the soil.

Allen's rule   Among homeotherms, the ecogeographic (morphogeographic) trend for limbs and extremities to become shorter and more compact in colder climates than in warmer ones.

allochthonous   Having originated outside the area in which it now occurs.

allochthonous endemic   An endemic taxon that originated in a different location than where it is found today.

allometry   The manner in which the relative size of one part of an organism increases in relation to the size increase of the entire organism; also known as *scaling*.

allopatric   Occurring in geographically different places; i.e., ranges that are mutually exclusive.

allopatric speciation   The formation of new species that occurs when populations are geographically separated.

allopolyploid   A hybrid polyploid formed following the union of two gametes, usually from distantly related species, with nonhomologous chromosomes.

alluvial   A large, fan-shaped pile of sand, clay and other sediments gradually deposited by moving water along the shores of lakes and estuaries or a river bed that flows onto a flat plain at the foot of a mountain range.

alpha diversity   The species richness of a local ecological community, i.e., the number of species recorded within some standardized area, such as a hectare, a square kilometer, or some naturally delineated patch of habitat.

amphitropical   Occurring in subtropical or temperate areas on opposite sides of the tropics.

anagenesis   The process of evolution that produces entirely new levels of structural organization (grades) without branching events. Compare with *cladogenesis*.

ancestor   The individual or population that gave rise to some subsequent individuals or populations with different features.

anemochory   Passive dispersal of propagules by winds.

aneuploidy   The formation of a new chromosomal arrangement resulting in an increase or decrease of the chromosome number by one pair; often caused by an uneven meiotic division.

anthropochory   The intentional transport of disseminules by humans.

aphotic zone   The lower zone in a water column, usually below 50 to 100 m, in which the intensity of solar radiation is too low to permit photosynthesis by plants.

apomixis   Reproduction without the union of sexual cells (gametes).

apomorphy   In a transformation series, the derived character state.

apterous   Wingless; often used to contrast these forms with their primitive ancestors that had wings and the ability to fly.

aquatic   Living exclusively or for most of the time in water.

arboreal   Living predominantly or entirely in the canopies of trees.

arborescent   Treelike.

arctic   Pertaining to all nonforested areas north of the coniferous forests of the Northern Hemisphere, especially everything north of the Arctic Circle.

area biogeography   The primary goal of an analysis is to reconstruct the biogeographic history of a set of areas of endemism based on the affinities of taxa distributed across those areas.

area cladogram   A cladogram of relationships among areas, generated variously from a model of geological history (*geological area cladogram*) or taxon history (see *taxon-area cladogram*).

area of endemism   A geographic region containing two or more endemic taxa, defined and diagnosed variously (see Box 12. 1).

areography   The study of the structure of geographic ranges, including variation in their sizes, shapes, and overlap.

arid   Exceedingly dry; strictly defined as any region receiving less than 10 cm of annual precipitation.

assembly rules   Highly non-random patterns in the organization and structure of ecological communities resulting from selective immigrations, extinctions or interactions among species.

asthenosphere   A fluid, viscous zone of the upper mantle on which the continental and oceanic plates float (ride) and over which they move.

athenosphere   See *asthenosphere*.

austral   Pertaining to the temperate and subtemperate zones of the Southern Hemisphere.

Australasia   The continental fragments of the original Australian Plate, including Australia, New Zealand, New Guinea, Tasmania, Timor, New Caledonia, and several smaller islands.

Australian Region   The biogeographic region including Australia, New Zealand, and New Guinea, other nearby islands, and the Indonesian islands lying east of Wallace's line.

autapomorphy   In a transformation series, a derived character state present in a single taxon.

autochthonous   Having originated in the area in which it presently occurs.

autochthonous endemic   An endemic taxon that differentiated where it is found today.

autopolyploid   A polyploid possessing more than two sets of homologous chromosomes.

autosome   A chromosome that is not a sex chromosome; a somatic chromosome.

autotroph   An organism that uses carbon dioxide occurring in the environment as its primary source of cellular carbon.

avifauna   All the species of birds inhabiting a specified region.

bajada   A broad, sloping depositional surface at the base of a mountain range in deserts, resulting from the coalescing of alluvial fans.

balanced (biota)   See *harmonic (biota)*.

barrier   Any abiotic or biotic feature that totally or partially restricts the movement (flow) of genes or individuals from one population or locality to another.

basal meristem   A plant whose growth (mitotic) tissues are located in or near the soil layer, thus adapting this plant to continual grazing of its apical tissue.

basal metabolic rate   A standardized measure of the rate of energy and oxygen requirements of an organism (usually measured at rest and within a range of temperatures at which the organism is not under thermal stress), and indicative of the minimal amount of energy required to maintain vital functions under stress-free conditions.

basin-and-range topography   Landscapes characterized by many low, flat areas (basins) that are interrupted by isolated mountain ranges. During pluvial times, many of these basins filled with water.

bathyal zone   The deep sea; in particular, that portion within the aphotic zone but less than 4 km deep.

bathymetry   The depth and configuration of the bottom of a body of water.

Bejerinck's Law   "Everything is everywhere, but the environment selects. " A statement which holds that where the likelihood of dispersal and eventual colonization is relatively high (especially in microscopic organisms with high dispersal capacities), the geographic distributions of these species are largely limited by local environmental conditions and physiological tolerances of the species, not by dispersal.

Benioff zones   Zones of high earthquake activity located on the back sides of trenches. The earthquakes, which are caused by the subduction of a plate, are shallow near the trench and progressively deeper at greater distances.

benthic   Living at, in, or associated with structures on the bottom of a body of water.

Bergmann's rule   The tendency for the average body mass of geographic populations of an animal species to increase with latitude.

Beringia   The geographic area of western Alaska, the Aleutians, and eastern Siberia that was connected in the Cenozoic by a landbridge when the Bering Sea and adjacent shallow waters receded.

beta diversity   The dissimilarity (turnover) in species composition between local ecological communities.

biogeographic relicts   The narrowly endemic descendants of once widespread taxa.

biogeography   The science that attempts to document and understand spatial patterns of biological diversity. Traditionally defined as the study of distributions of organisms (both past and present), modern biogeography now includes studies of all patterns of geographic variation in life, from genes to entire communities and ecosystems; elements of biological diversity that vary across geographic gradients including those of area, isolation, latitude, depth and elevation.

biological species   A group of potentially interbreeding populations that are reproductively isolated from all other populations.

biomass   The total body mass of an organism, population, or community.

biome   A major type of natural vegetation that occurs wherever a particular set of climatic and soil conditions prevail, but that may contain different taxa in different regions; e.g., temperate grassland.

biosphere   Collectively, all the living things of the Earth and the areas they may inhabit.

biota   All species of plants, animals, and microbes inhabiting a specified region.

biotic   Pertaining to the components of an ecosystem that are living or came from a once-living form.

bipedal   Using two hindlimbs for locomotion, usually by hopping or jumping, such as a kangaroo or a kangaroo rat.

bipolar   Occurring at both poles, in the cold or sub-temperate zones.

bivoltine   Breeding twice per year.

boreal   Occurring in that portion of the temperate and sub-temperate zones of the Northern Hemisphere that characteristically contains coniferous (evergreen) forests and some types of deciduous forests.

bottleneck   In evolutionary biology, any stressful situation that greatly reduces the size of a population.

brackish   Having a salt concentration greater than fresh water (> 0.5‰) and less than seawater (35‰)

bradytelic   Term used by Simpson (1944) to refer to a slow rate of morphological change through time in a lineage with a fossil record.

breeding area   In migratory land animals, the area in which populations mate and produce offspring.

Brooks parsimony analysis (BPA)   A method in historical biogeography that uses parsimony analyses to construct general area cladograms from a set of taxon-area cladograms.

browser   An animal that feeds on plant materials, especially on woody parts of trees and shrubs.

Buffon's law   Named in honor of the eighteenth century biogeographer Georges-Louis Leclerc, Comte de Buffon (1707–1788) who observed that different regions, even those with similar environmental characteristics, are inhabited by different assemblages of species.

calcareous   In soil biology, pertaining to a soil whose horizons are rich in calcium carbonate and have a basic reaction.

calcification   The formation of a soil under continental climatic conditions of relatively low moisture and hot to cool temperatures, resulting in a soil rich in calcium carbonate ($CaCO_3$) because rainfall is not sufficient to leach calcium from the upper soil horizons.

caliche   A hard, often rocklike layer of calcium carbonate that forms in soils of arid regions at the level to which the leached calcium salts from the upper soil horizon are precipitated.

canonical distribution   A lognormal distribution of the number of individuals or species according to the mathematical formulation of Preston (1962).

carnivore   An animal that feeds mostly or entirely on animal prey.

carrying capacity (*K*)   An estimate of the population size or biomass of a particular species that can be supported by a given habitat or environment.

catastrophic death assemblages   Fossil deposits characterized by large numbers of individuals or species preserved at the same time and informative about biological and ecological characterstics of lineages and biotas.

catastrophic extinction   See *mass extinction*.

census population size   The total number of individuals in a population. Compare with *effective population size*.

Centinelan extinctions   The loss of species in some unstudied area and time period before they became known to science.

chaparral   A type of sclerophyllous scrub vegetation occurring in the southwestern region of North America with a Mediterranean climate.

character   A heritable trait with different *character states* that can be recognized by systematists and used to infer evolutionary relationships and classify organisms into taxa.

character displacement   The divergence of a feature of two similar species where their ranges overlap so that each uses different resources.

character state   One of several alternative forms of a character; e.g., the ancestral form or one of several derived forms.

chromosome   A connected sequence of genetic material on long-stranded deoxyribonucleic acid (DNA) wrapped with proteins.

chronospecies   A recognizable stage along a sequence from ancestral to descendant species through non-branching (anagenetic) evolutionary change. See *phyletic speciation*.

circumboreal   Occurring in the temperate or subtemperate zones of the New and Old World portions of the Northern Hemisphere.

clade   Any monophyletic evolutionary branch in a phylogeny, using derived characters to support genealogical relationships.

cladistic biogeography   See *vicariance biogeography*.

cladistics   The method of reconstructing the evolutionary history (phylogeny) of a taxon by identifying the branching sequence of differentiation through analysis of shared derived character states. Also called *phylogenetic systematics*. Compare with *evolutionary systematics* and *numerical phenetics*.

cladogenesis   The process of evolution that produces a series of branching events.

cladogram A line diagram derived from a cladistic analysis showing the hypothesized branching sequence (genealogy) of a monophyletic taxon and using shared derived character states (synapomorphies) to determine when each branch diverged.

classical species concept See *morphological species concept.*

climax Pertaining to a community that perpetuates itself under the prevailing climatic and soil conditions; therefore, the last stage in secondary succession.

cline A change in one or several heritable characteristics of populations along a geographic transect, attributable to changes in the frequencies of certain alleles and often correlated with a gradual change in the environment.

coalescence In a gene tree, the point in time (absolute time or scaled to generations) at which two allelic lineages diverged from an ancestral lineage.

coenocline A graphical method developed by Robert H. Whittaker to illustrate changes in local abundances or densities of a group of species along an environmental or geographic gradient.

coevolution The simultaneous, interdependent evolution of two unrelated species that have strong ecological interactions, such as a flower and its pollinator or a predator and its prey.

coexistence Living together in the same local community.

cohesion See *cohesion species concept.*

cohesion species concept A species concept that defines species as "…the most inclusive population of individuals having the potential for phenotypic cohesion through intrinsic cohesion mechanisms [genetic and/or demographic exchangeability]" (Templeton 1989, p. 12).

collision zones Regions where two tectonic plates converge, resulting either in the subduction of one of the plates beneath the other or, if both are of equal buoyancy (as in convergence of two continental plates), uplift and formation of montains.

colonization The immigration of a species into new habitat followed by the successful establishment of a population.

commensalism An interspecific relationship in which one species draws benefits from the association and the other is unaffected.

community An assemblage of organisms that live in a particular habitat and interact with one another.

community assembly The development of nonrandom patterns in the structure of ecological communities.

community ecology The study of interactions among co-occurring organisms living in a particular habitat.

comparative phylogeography In phylogeography, the analysis of geographic pattern across multiple co-distributed species or groups of species.

compensatory effects Distributions of populations of a particular, focal species tend to result from the interactive (rather than just the independent) effects of immigration and extinction, such that its populations will occur on isolated islands if their extinction rates on those islands are compensatorily low (i.e., the island is relatively large), and its populations will occur on relatively small islands if their immigration rates are compensatorily high (i.e., the islands are near the mainland or other source of immigrations).

competition Any interaction that is mutually detrimental to both participants. Interspecific competition occurs among species that share requirements for limited resources.

competitive exclusion The principle that when two species with similar resource requirements co-occur, one eventually outcompetes and causes the extinction of the other.

component analysis An approach in historical biogeography that searches for shared cladogenetic events among a set of partially but not completely congruent taxon-area cladograms by using one or more assumptions to develop rules about how to remove the incongruent parts of the cladograms.

congeners Species belonging to the same genus.

consensus area cladogram Summarizes the shared cladogenetic events among a set of taxon-area cladograms.

conservation biogeography A newly articulated discipline that applies lessons from biogeographic theory and patterns to conserve biological diversity, and which emphasizes the need to conserve the geographic, ecological, and evolutionary context of nature.

continental drift A model, first proposed by Alfred Wegener, stating that the continents were once united and have since become independent structures that have been displaced over the surface of the globe.

continental island An island that was formed as part of a continent and that has a nucleus of continental (sialic) rocks.

contour maps Maps that use isoclines ("contours" of similar levels of a variable) to illustrate changes in characteristics (e.g., population densities of a species or diversity of ecological communities) across a geographic area.

convergent evolution The development of two or more species with strong superficial resemblances from totally unlike and unrelated ancestors.

Cope's rule A trend in directional evolution (orthogenesis) toward increased body size.

coprolite Fossil excrement.

Coriolis effect A physical consequence of the law of conservation of angular momentum whereby, as a result of the Earth's rotation, a moving object appears to veer to the right in the Northern Hemisphere and to the left in the Southern Hemisphere.

corridor A dispersal route that permits the direct spread of many or most taxa from one region to another.

cosmopolitan Occurring essentially worldwide, as on all habitable landmasses or in all major oceanic regions.

craton The stable crustal nucleus of a continent or continental island, which is older than 600 million years; also called *Precambrian shield.*

crust The outermost rock layer of the Earth, covering the mantle.

cryptic species Species within a genus that are morphologically so similar that they cannot be visually distinguished by superficial features. See *sibling species.*

cryptoturnover Biotic turnover (immigrations and extinctions) of species that may go unrecorded because they occur between survey periods.

Curie point The temperature at which remnant magnetism develops in cooling minerals; e.g., 680° C for hemitite and 580° C for magnetite.

cyclical vicariance model (Also known as the "speciation-pump model.") A model that attributes the high diversity of tropical communities in South America to repeated fragmentation (and allopatric speciation) and reconnection of tropical forests caused by the 20 or so glacial/interglacial cycles of the Pleistocene.

deciduous   In plants, having leaves that are shed for at least one season, usually in response to the onset of cold or drought.

declination   Dipping of magnetic needles, which orient toward Earth's magnetic poles, which lie deep beneath its crust. Thus, declination can be used to estimate latitude, either under existing conditions or in magnetically active rocks that crystallized in ancient periods, but now drifted to distant positions.

decomposer   An organism (usually a bacterium or fungus) capable of metabolically breaking down organic materials into simple organic and inorganic compounds and releasing them into the ecosystem.

deconstruction (pattern deconstruction)   An attempt to understand underlying, causal mechanisms for patterns of biodiversity by examining the differences in patterns among functionally different groups of species, or differences among patterns for similar species across different habitats, ecosystems or different regions (see Huston 1994).

deductive reasoning   A method of analysis in which one reasons from general constructs to specific cases.

defaunation   The elimination of animal life from a particular area.

dehiscence   The act of splitting along a natural line to discharge the contents, such as that of an anther to release pollen grains or of a capsule to release seeds.

deletion   A form of mutation in which one or more nucleotides are eliminated from a DNA sequence.

delta diversity   The dissimilarity (turnover) in species composition between large, geographic areas.

density compensation   In island biogeography, an increase in the density of a species inhabiting an island habitat when one or more taxonomically similar competitors are absent.

density enhancement   The tendency for densities of populations inhabiting relatively small, species-poor islands to increase with island area.

density inflation   The tendency for densities of populations inhabiting islands of intermediate size and species richness to increase beyond the level of this species on the mainland. The accumulated effects of density inflation over many species is referred to as "density overcompensation" (see below).

density overcompensation   In island biogeography, a case in which the total densities of a few species inhabiting a small island exceed the combined densities of a much greater number of species of the same taxon occupying similar habitats on a large island or continent.

density stasis   On islands of intermediate to relatively large area and high levels of species richness, the combined effects of many species interactions may regulate one another's densities such that few, if any, exhibit any consistent trend of population density with species richness or island area.

desert   A general term for an extremely dry habitat, especially one where water is unavailable for plant growth most of the year; in particular, a habitat with long periods of water stress and sparse coverage by plants, often with perennials covering less than 10% of the total area.

desertification   The degradation of land in arid, semi arid and dry sub-humid areas into desert.

deterministic   Determined or controlled by some regulatory force, such as natural selection. Compare with *stochastic*.

detritivore   An organism that feeds solely on detritus.

detritus   Freshly dead or partially decomposed organic matter.

dew point   The temperature at which air becomes saturated with water vapor and it condenses to form fog or other forms of precipitation.

diadromous   Referring to an aquatic animal that must migrate between fresh water and seawater to complete its life cycle, such as certain lampreys and eels.

diapause   An arrested state of development in the life cycle, especially of many insects, during which the organism has reduced metabolism and is more resistant to stressful environmental conditions, such as cold, heat, or drought.

diaspore   Any part or stage in the life cycle of an organism that is adapted for dispersal.

diffuse competition   A type of competition in which one species is negatively affected by numerous other species that collectively cause a significant depletion of shared resources.

diffusion   A form of range expansion that is accomplished over generations by individuals spreading out from the margins of the species' range.

dimorphic   Having two distinct forms in a population.

dioecious   Having individuals with only male or only female reproductive systems.

diploid   Having two sets of chromosomes ($2N$).

disassembly (community disassembly)   The non-random loss of particular species from ecological communities, typically associated with anthropogenic disturbances and resultant, selective extinctions.

discontinuous   A common pattern of variation in trait characteristics, where some individuals are very similar to one another but separated by other groups of individuals by distinct gaps.

disharmonic (biota)   A biota that is not a random subset of the mainland or source biota, but one biased in favor of species with superior immigration abilities or abilities to survive on islands.

disjunct   A taxon whose range is geographically isolated from that of its closest relatives.

disjunction   A discontinuous range of a monophyletic taxon in which at least two closely related populations are separated by a wide geographic distance.

dispersal   The movement of organisms away from their point of origin.

dispersal biogeography   A form of historical biogeography that attempts to account for present-day distributions based on the assumption that they resulted from differences in the dispersal abilities of individual lineages.

dispersal-vicariance analysis (DIVA)   An event-based approach in historical biogeography that reconstructs ancestral areas at each node in a taxon-area cladogram and infers events based on an *a priori* assignment of the "cost" of vicariance, dispersal, and extinction events.

dispersion The spatial distribution of individual organisms within a local population.

disruptive selection Selection that favors extreme and eliminates intermediate phenotypes.

disseminule Any part of a plant that is used for dispersal; occasionally restricted to include only seeds and seed-bearing structures. See also *diaspore*.

distance-decay The tendency for some property (e.g., species richness) to decline with increasing isolation, or for similarity in characteristics of two or more sites to decrease with increasing distance between those sites.

doldrums A narrow equatorial zone characterized by long periods of calm or light shifting winds, caused by the upward movement of air masses from this region of high atmospheric pressure to higher latitudes with relatively lower pressure.

dominant A species having great influence on the composition and structure of a community by virtue of its abundance, size, or aggressive behavior.

dot maps Maps depicting the geographic range of a species by place dots at points of documented occurrence of its individuals and populations.

driftless area A possible glacial refugium located in what is present day southern Wisconsin and adjacent areas of Illinois and Iowa.

eccentricity The degree of ellipticity in Earth's orbit about the sun, which varies in a cyclical manner over a 100,000 year period.

ecogeographic (morphogeographic) rules A regular change in characteristics of organisms (in particular, but not limited to, their morphological characteristics) along geographic gradients (e.g., Bergmann's, Allen's, Jordan's, and Thorson's rules).

ecological biogeography The study of distributions and geographic variation of extant biotas, with special emphasis on the influences of interactions between organisms and their abiotic and biotic environments.

ecological niche See *niche*.

ecological release The tendency for populations inhabiting species-poor environments such as small or isolated islands to occur in a broader range of habitats, feed on a broader variety of prey, or otherwise occupy a broader range of their fundamental niche than they do in species-rich communities.

ecological speciation The development of reproductive isolation between two incipient species as a result of divergent natural selection.

ecological time The period during which a population can interact with its environment and respond to environmental fluctuations without undergoing substantial evolutionary modification; compare *evolutionary time.*

ecology The study of the abundance and distribution of organisms and of the relationships between organisms and their biotic and abiotic environments.

ecoregions As defined by Robert G. Bailey, a geographic grouping of landscapes, each comprised of a mosaic of ecosystems.

ecosystem engineer A species that significantly alters the functioning and environmental characteristics of an ecosystem.

ecosystem geography The study of the distribution of ecosystems and the processes that have differentiated them in space and time.

ecosystem The set of biotic and abiotic components in a given environment.

ecotone A zone of transition between two habitats or communities.

ectoparasite A parasite that lives on the exterior of its host, such as a louse.

ectotherm An animal whose body temperature is determined largel.

edaphic Pertaining to soil.

edge effects In conservation biology, the potential negative effects of exotic species and disturbances that act along the edges of habitat fragments.

effective population size The number of breeding individuals in a population. Compare with *census population size.*

El Niño Southern Oscillation (ENSO) An approximate 5- to 7-year cycle of regional climatic changes that is caused by variation in sea surface temperatures and oceanic currents in the tropical regions of the Pacific (similar events occur in the tropical Atlantic).

emigration Dispersal of organisms away from a region of interest.

endemic Pertaining to a taxon that is restricted to the geographic area specified, such as a continent, lake, biome, or island.

endemic bird areas Areas containing the ranges of at least two restricted-range species of birds (i.e., those whose breeding ranges are less than 50,000 km^2).

endotherm An animal whose body temperature is maintained largely by its own metabolic heat production; compare with *ectotherm.*

entropy A measure of the unavailable energy in a closed thermodynamic system.

epeiric sea A large but relatively shallow body of salt water that lies over a part of a continent.

epicenter The point on the Earth's surface directly above the origin (focus) of an earthquake.

epicontinental sea See *epeiric sea.*

epifaunal Living on a substrate, such as a barnacle or coral.

epipelagic Living in open water, mostly within the upper 100 m.

epiphyllous Living on a leaf.

epiphyll Thin layers of mosses, lichens, and algae that grow along the surfaces of trees in rainforests.

epiphyte A plant that usually lives on another plant (is not rooted in soil) and derives its moisture and nutrients from atmospheric precipitation and whatever materials are released by the organisms in the immediate vicinity.

equatorial countercurrent A small current running west to east along the equator (i.e., opposite the major oceanic gyres) in the eastern region of the Pacific Ocean (see Figure 3.5).

equilibrium A condition of balance between opposing forces, such as birth and death rates or immigration and extinction rates.

equilibrium theory of island biogeography The theory proposed by MacArthur and Wilson stating that the number of species on an island results from a dynamic equilibrium between the opposing rates of immigration and extinction.

equilibrium turnover rate   The change in species composition per unit of time when immigration equals extinction.

equinox   Either of two times (March 21 and September 22) in a year when the sun passes the equator so that day and night are the same length everywhere on Earth.

establishment   The successful start or founding of a population.

estuary   A body of water where freshwater from rivers mixes with saltwater from the sea.

Ethiopian Region   The portion of Africa south of the Sahara Desert plus Madagascar and other nearby islands.

euryhaline   Having a tolerance to an extremely wide range of salt concentrations.

eurythermal   Having a tolerance to a broad range of temperatures.

eurytopic   Having a tolerance to an extremely wide range of habitats and environmental conditions.

eustatic (eustasis)   Global fluctuations in sea level resulting from the freezing or melting of great masses of sea ice, thus decreasing or increasing the global volume of liquid water, respectively.

eutherian   A placental mammal.

eutrophic lake   A lake that is rich in dissolved nutrients and highly productive, but that is usually shallow and seasonally deficient in oxygen.

evapotranspiration   The sum total of water lost through evaporation from land and transpiration by plants.

evolution   In the strictest sense, any irreversible change in the genetic composition of a population.

evolutionarily significant units   Defined in several ways, but one that is popular is as "historically isolated sets of populations for which a stringent and qualitative criterion is reciprocal monophyly for mitochondrial DNA (mtDNA) combined with significant divergence in frequencies of nuclear alleles" (Moritz et al. 1995, p. 249).

evolutionary biogeography   An earlier approach in historical biogeography motivated by the search for centers of origin and dispersal over a stable geographic template.

evolutionary species concept   A species concept that recognizes a species as "…an entity composed of organisms that maintains its identity from other such entities through time and over space and that has its own independent evolutionary fate and historical tendencies" (Wiley and Mayden 2000, p. 73).

evolutionary systematics   A method of reconstructing the evolutionary history (phylogeny) of a taxon by analyzing the evolution of major features along with the distribution of both shared primitive and shared derived characteristics. Compare with *cladistics* and *numerical systematics*.

evolutionary time   The period during which a population can evolve and become adapted to an environment by means of genetic changes. Compare *ecological time.*

excess density compensation   See *density overcompensation.*

exoskeleton   An external skeleton of an animal, such as that of a clam or insect.

exploitative competition   A negative interaction between species or conspecifics in which individuals use up resources and make them unavailable to others.

exponential growth   In ecology, a population that increases in proportion to its population size and, therefore, at a continually accelerating rate.

extant   Living at the present time.

extinct   No longer living.

extratropical   Occurring outside the tropics.

facultative relationship   An interaction between two organisms that is not essential to the survival of either.

Family   A taxonomic category above the level of genus and below the level of order.

fault   In geology, a weakness in the Earth's crust along which there can be crustal motion and displacement.

filter   A geographic or ecological barrier blocks the passage of some forms, but not others.

filter feeder   An aquatic animal that feeds on plankton or other minute organic particles by using one of a variety of filtering mechanisms.

first law of thermodynamics   Energy is neither created nor destroyed, but can be converted from one form to another.

fitness   The ability of a genotype to leave offspring in the next generation or succeeding generations as compared with that of other genotypes.

floristic belts   A series of plant communities that are characterized by different growth forms or physiognomies that occur in predictable series (e.g., deserts, savanna, dry woodlands, coniferous forests and tundra) along elevational and latitudinal gradients.

flyway   An established air route used by vast numbers of migratory birds.

food chain   A diagram or list of species that describes predator and prey relationships and the transfer of energy and nutrients within an ecological community.

forest   Any of a variety of vegetation types dominated by trees and usually having a fairly well developed or closed canopy when the trees have leaves.

fossil   A remnant, impression, or other trace of a living organism from the past.

fossorial   Referring to an animal that lives in and forages on plants from a burrow.

founder effect   Genetic drift that occurs when a newly isolated population is founded, such as on an island, by one or a few colonists (founders). The features of the new population may be markedly different from those of the ancestral population because the gene pool of the founders may be a biased and small sample of the source population.

fragment   In plate tectonics, a portion of a former landmass.

fresh water   In the strictest sense, water that has a salt concentration of less than 0.5‰.

frugivore   An animal that feeds mainly on juicy fruits.

fundamental geographic range   The theoretical distribution that populations of a particular species may achieve based solely on its physiological and abiotic tolerances, assuming that species interactions are unimportant and opportunities of dispersal are unlimited.

fundamental niche   The total range of environmental conditions in which a species can survive and reproduce.

fynbos   A type of sclerophyllous scrub vegetation occurring in the region of South Africa with a Mediterranean climate.

gamete   One of two cells, usually from different parents, that fuse to form a zygote.

gamma diversity   The total species richness of large geographic areas ranging from a combination of local ecological communities to those of entire biomes, continents and ocean basins.

Gaussian fuction   A frequency distribution that takes the approximate form of a normal distribution, or bell-shaped curve.

gene   The small unit of a DNA molecule that codes for a specific protein to produce one of the chemical, physiological, or structural attributes of an organism.

gene conversion   A process by which damaged genes can be repaired.

gene flow   The movement of alleles within a population or between populations caused by the dispersal of gametes or offspring.

gene frequency (allelic frequency)   The proportions of gene forms (alleles) in a population.

gene tree   The phylogeny of a particular gene or group of tightly linked genes embedded within the phylogeny of a population or species.

genealogical concordance concept   A species concept that proposes that "…population subdivisions concordantly identified by multiple independent genetic traits should constitute the population units worthy of recognition as phylogenetic taxa" (Avise and Ball 1990, p. 52).

genealogy   The study of the exact sequence of descent from an ancestor to all derived forms.

general area cladogram   A cladogram summarizing vicariance (and depending on approach, also dispersal, sympatric speciation, and extinction) events from a set of taxon-area cladograms.

generalized track   In panbiogeography, a line drawn on a map representing the coincident distributions of numerous disjunct taxa.

genetic drift   Changes in gene frequency within a population caused solely by chance—i.e., which individuals happen to mate and leave offspring in the next generation—without any influence of natural selection.

genome   A full set of chromosomes.

genotype   The total genetic message found in a cell or an individual.

genus   A taxonomic category for classifying species derived from a common ancestor; a level below that of family and tribe.

geo-dispersal   The concordant dispersal of several species out of an area of endemism following erosion of a dispersal barrier with subsequent re-emergence of a barrier and a new round of speciation.

geographic coordinate systems   A means of using two geographic variables (e.g., latitude and longitude, or easting and northings [UTM system]) to locate points on a map.

geographic isolation   Spatial separation of two potentially interbreeding populations; allopatry.

geographic speciation   See *allopatric speciation*.

geographic template   The highly non-random, spatial variation of environmental conditions that forms the foundation for all biogeographic patterns.

glacial-interglacial cycles   The 20 or so repeated, global-scale climatic changes that occurred during the Pleistocene Epoch and profoundly affected biogeographic patterns of Earth's biotas.

glacio-pluvial   The alternating, "ice" to "rain" (glacial to interglacial) stages of the Pleistocene Epoch.

gleization   The formation of a soil under moist and cool or cold conditions, resulting in an acidic soil with a large amount of organic matter and iron present in a reduced state (FeO).

Gondwanaland   The southern half of the supercontinent Pangaea, consisting of all the southern continental landmasses and India, which were united for at least 1 billion years but broke up during the late Mesozoic.

granivore   An animal that feeds mainly on dry seeds and fruits; a subtype of herbivore.

grassland   Any of a variety of vegetation types composed mostly of grasses and other herbaceous plants (forbs) but few if any trees and shrubs.

Great American Interchange   Term used by the paleontologist George Gaylord Simpson to describe the waves of biotic exchange of terrestrial organisms between Nearctic and Neotropical Regions (North and South America) following the emergence of the Central American landbridge at roughly 3.5 million years B.P.

greenhouse effect   The retention of heat in the atmosphere when clouds (water vapor) and carbon dioxide absorb the infrared (heat) radiation reradiated from the Earth rather than permitting the heat to escape.

grouping rule   Only synapomorphies provide evidence of common ancestry relationships; symplesiomorphies, autapomorphies and homoplasies are uninformative for doing so.

guide fossil   A fossil used to date the age of a sedimentary stratum; also called an *index fossil*.

guild   Groups of species that exploit the same class of environmental resources in a similar manner (e.g., desert granivores or foliage-gleaning birds).

guyot   A flat-topped submarine volcano.

habitat island   A geographically isolated patch of habitat, such as a pond, mountaintop, or cave, that can be studied in the same ways as oceanic islands for patterns of colonization and extinction.

habitat selection   The preference of an organism for a particular habitat type.

hadal   Of or pertaining to the deepest zones of the ocean, below 6 km, which have nearly constant environments with year-round temperatures near 4° C and no light penetration; e.g., in the oceanic trenches.

half-life   The amount of time needed for half of the radioactive material in a rock to decay to a stable element.

halomorphic soil   Soil that is characterized by very high concentrations of sodium, chlorides, and sulfates and which forms in estuaries and salt marshes, and in arid inland basins where shallow water accumulates and evaporates, leaving behind high concentrations of salts.

halophytic   "Salt-loving" plant species that grow in areas of soils with high salt concentrations, including a variety of taxonomic and functional groups, each of with special adaptations for dealing with the problem of maintaining osmotic and ionic balance in these environments (e.g., the ability to excrete salts or store them in special cells and tissues).

haploid   Having one set of chromosomes (*N*).

haplotype   A *haploid* allele for a gene or group of tightly linked genes.

harmonic (biota)   A biota that is similar to a random subset of its source pool. See *balanced* (*biota*).

herbivore   An animal that feeds mostly or entirely on plants.

hermaphroditic   Having both male and female reproductive structures in the same individual.

heterogeneity   The state of being mixed in composition, as in genetic or environmental heterogeneity.

heterotroph   An organism that uses organic carbon (compounds made by living organisms) as its source of cellular carbon.

historical biogeography   The study of the development of lineages and biotas including their origin, dispersal and extinction.

histosols   Soils that are characterized by a substantial layer of organic matter (>30%) of more than 40 cm either extending down from the surface or taken cumulatively within the upper 80 cm of the soil. These soils are formed because the production of organic debris exceeds its decay.

Holarctic region   The extratropical zone of the Northern Hemisphere, which includes both the Neartic and Palearctic regions.

homeostasis   The maintenance of a constant internal state despite fluctuations in the external environment.

homeotherm   An animal that maintains a fairly constant body temperature.

homologies   Character states shared between two or more taxa because it was inherited from a common ancestor.

homologous chromosomes   Two chromosomes (in a diploid organism) that have essentially the same gene sequence and that are similar enough to pair during meiosis and mitosis.

homology   A character shared by a group of organisms or taxa due to inheritance from a common ancestor.

homoplasies   Character states shared between two or more taxa because of convergent or parallel evolution.

homoplasy   A character that appears similar among a group of organisms or taxa, but the similarity is due to parallel or convergent evolution rather than inheritance from a common ancestor.

horizon   In soils, the major stratifications or zones, each of which has particular structural and chemical characteristics.

horizontal gene transfer   The movement of a piece of DNA or a gene from one species, and incorporation into the genome of a new species, often mediated through an intermediate vector (e.g., bacterium or parasite).

horotelic   Term used by Simpson (1944) to refer to a moderate rate of morphological change through time in a lineage with a fossil record.

horse latitudes   The zones of dry descending air between 30° and 40° N and S latitude, where many deserts of the world are located.

hotspot   In plate tectonics, a stationary weak point in the upper mantle that discharges magma (molten rock) as a plate passes over it, producing a narrow chain of islands. In biodiversity studies, an area with a relatively high number of species or high number of endemic species.

humus   A brown or black organic substance in soils consisting of partially or wholly decayed vegetable or animal matter that provides nutrients for plants and increases the ability of soil to retain water.

hybridization   The production of offspring by parents of two different species or dissimilar populations or genotypes.

hydrochory   Passive dispersal of propagules by water.

hyperosmotic   Referring to an environment in which water will diffuse from an organism because the external solution has a salt concentration higher than its internal concentration.

hypersaline   Having a higher salt concentration than normal seawater.

hypothetico-deductive reasoning   A method of analysis in which one starts with a new, tentative hypothesis and then tests the predictions and assumptions following from it one by one in an attempt to falsify it.

ichthyofauna   All the species of fishes inhabiting a specified region.

immigration   The arrival of new individuals to an isolated site.

immigrant pattern   The tendency for islands along a gradient of increasing isolation to be inhabited by assemblages of species that represent highly non-random subsets of less-isolated communities, in this case biased in favor of species with superior immigration abilities.

incidence functions   A method of exploring community assembly by graphing the proportion of islands inhabited by a given species as a function of the number of other species present.

included niche   A niche of a specialized species characterized by a narrow range of conditions and lying entirely within the larger fundamental niche of a more generalized species.

index fossil   See *guide fossil*.

inductive reasoning   A method of analysis in which one uses specific observations to derive a general principle.

infaunal   Living within a substrate, such as clams that bury themselves in sand or mud and that feed by means of a long siphon.

ingroup   The focal monophyletic group in a cladistic analysis.

insectivore   An animal that feeds mainly on insects.

insertion   A form of mutation in which one or more nucleotides are inserted into a DNA sequence.

insular distribution function   A method of assessing the relative importance of immigration and extinction as processes influencing community assembly by using a bivariate plot to illustrate the distribution of populations of a focal species on islands of varying isolation and area.

interference competition   A negative interaction between species in which aggressive dominance or active inhibition is used to deny other individuals access to resources.

interglacial   A phase during the Quaternary, when glacial ice sheets retreated and the climate became more equable.

interglacials   The relatively warm and wet periods that alternated with the glacial periods of the Pleistocene Epoch.

intertidal zone   The zone above the low tide mark and below the high tide mark of a body of water; the littoral zone.

inversion   A form of mutation in which the orientation of a portion of a chromosome becomes reversed with respect to its former orientation.

island rule   A graded trend in insular vertebrates from gigantism in the smaller species to dwarfism in the larger species.

isolating mechanism   Any structural, physiological, ecological, or behavioral mechanism that blocks or strongly interferes with hybridization or gene exchange between two populations.

isostasy   In geology, the term that describes how continental blocks float on a viscous layer of the mantle. While doing this the block may rise or fall to achieve an equilibrium.

isotherm   A line on a map connecting all locations with the same mean temperature.

Jordan's rule (law of vertebrae)   The tendency for the number of vertebrae in marine fish to increase along a gradient from the warm waters of the tropics to cooler waters of the high latitudes.

jump dispersal   Long-distance dispersal that is accomplished by movement of individuals within a relatively short period.

karyotype   The morphological appearance and characteristic number of a set of chromosomes in a cell, and generally conserved across a population or species.

kettle lake   Lakes that formed during the onset of an interglacial period when large blocks of ice separated from the melting glaciers and formed deep, persistent depressions that later filled with runoff from ice-melt and precipitation.

key innovation   Evolution of a new trait that confers a strong adaptive advantage and is often association with an adaptive radiation.

keystone species   A species whose activities have a disproportionate effect on the structure and function of ecological communities.

kriging   A statistical means of estimating the value of a particular variable at an unsampled location by calculating a moving-average from data collected at other locations.

K-selection   Selection favoring a more efficient utilization of resources, which is more pronounced when the species is at or near carrying capacity (*K*).

K-strategists   Species whose life-history and ecological characteristics are adapted to selective pressures associated with populations near carrying capacity (*K*) of an environment.

laterite   In tropical soils, a hard, rocklike layer composed principally of ferric oxide, produced when this compound accumulates in high concentration in the soil.

laterization   The formation of a soil under conditions of abundant moisture, warm temperatures, and high decomposer activity, resulting in a soil from which bases and silica have been removed (leached), leaving behind a clay rich in ferric and aluminum oxides.

Laurasia   The northern half of the supercontinent Pangaea, including North America, Europe, and parts of Asia.

leaching   In soil science, the removal of soluble substances by water.

lentic   Referring to standing freshwater habitats, such as ponds and lakes.

leptokurtic   A mathematical distribution characterized by a sharply peaked curve with a long tail.

liana   A climbing woody or herbaceous vine that is especially common in wet tropical forests.

Liebig's law of the minimum   An early ecological generalization, now discredited or greatly modified, which stated that abundance and distribution are limited by the single factor in the shortest supply. See also *limiting factor*.

life zones   The characteristic changes in vegetation composition and form that occur along an elevational or latitudinal gradient.

limiting factor   The resource or environmental parameter that most limits the abundance and distribution of a population.

limnetic zone   The offshore waters of a lake that receive light sufficient to support photosynthesis of their plant (macrophyte and phytoplankton) communities.

limnologist   A scientist who studies the physical, chemical and biological characteristics of freshwater systems, especially lakes, rivers and ponds.

lineage sorting   In a gene tree, the loss of ancestral polymorphism within a population or species as a consequence of stochastic effects influencing the probability of an allele either being retained or lost during genetic transmission across generations.

Linnaean shortfall   The disparity between the number of described species and the total number of species in existence.

lithosphere   The Earth's crust, exclusive of water (hydrosphere) and living organisms (biosphere).

littoral zone   The marginal zone of the sea; i.e., the intertidal zone. In fresh water, the shallow zone that may contain rooted plants.

living fossils   An extant taxon that has changed very little morphologically from ancestral taxa in the fossil record.

log-transformed   Referring to the change of values in a data set by taking the logarithm of each.

long-distance (jump) dispersal The movement of an organism across inhospitable environments to colonize a favorable distant habitat.

lotic Referring to running freshwater habitats, such as brooks and rapids.

macchia A type of sclerophyllous scrub vegetation occurring in regions of the Old World with a Mediterranean climate.

macroecology A top-down and multi-scale approach to investigating the assembly and structure of biotas, which identifies general patterns and underlying mechanisms by focusing on the statistical distributions of variables across spatial and temporal scales, and among large numbers of equivalent (but not identical) ecological *particles* (e.g., particles can include individual organisms within a local population or entire species, replicated sample areas or patches of habitat, or species within local communities or larger biotas).

macroevolution A general term for evolution above the population level.

magma The molten rock material under the Earth's crust, from which igneous rock is formed by cooling.

magnetic reversals In geology, episodes during which the direction of Earth's magnetic field has been reversed, occurring approximately twice every million years.

magnetic stripes In geology, long alternating stripes of normally and reversely magnetized basaltic rock on the ocean floor.

malacofauna All the species of mollusks inhabiting a specified region.

mangrove swamp A wetland found in tropical climates and along coastal regions that is dominated by mangrove trees and shrubs, particularly red mangroves (*Rhizophora*), black mangroves (*Avicennia*) and white mangroves (*Laguncularia*).

mantle The second and thickest layer of the Earth.

mantle drag One of the forces responsible for continental drift, in this case resulting from the tendency of the crust to ride the gyres of circulating mantle beneath it much like boxes on a conveyor belt.

maquis A type of sclerophyllous scrub vegetation occurring in portions of the Mediterranean region with its characteristic climate.

marginal population A population that has difficulty in surviving as a result of limiting abiotic or biotic factors.

marine Living in salt water.

maritime climate In general, a coastal or island environment with little or no freezing, much cloud cover and fog, and less variance in temperature, and thus a milder year-round climate, than nearby inland or mainland localities.

marsh (moor) A wetland that is dominated by herbaceous vegetation.

mass extinction A major episode of extinction for many taxa, occurring fairly suddenly in the fossil record.

matorral A type of sclerophyllous scrub vegetation occurring in the region of Chile with a Mediterranean climate.

Mediterranean climate A semiarid climate characterized by mild, rainy winters and hot, dry summers.

megafauna A general term for the large terrestrial vertebrates inhabiting a specified region.

mesic Relatively moist and equable.

mesophytes Land plants that grow in environments with an average ("mesic") supply of water.

Messinian crisis The sudden drainage of the Mediterranean Sea in the Cenozoic.

metabolism The sum total of the positive and negative chemical reactions in an organism or a cell that provide the energy and chemical substances necessary for its existence.

metamorphosis A major change in the form of an individual animal during development; e.g., a caterpillar becomes a moth or a butterfly and a tadpole becomes a frog or a toad.

metapopulation A set or constellation of local populations of a particular species that are linked by dispersal among those populations.

metatherian A marsupial mammal.

microallopatric Refers to spatially small forms of allopatric divergence, such as might occur among different volcanoes on a single oceanic island.

microclimate The fine-scale environmental regime.

microcosm A small community that represents in miniature the components and processes of a larger ecosystem.

microevolution Evolution below the species level, within and among populations (e.g., changes in gene diversity and gene frequencies resulting from mutation, genetic drift, natural selection, and gene flow).

microhabitat The fine-scale environment that often determines the presence or absence of each kind of organism.

microphyllous Having small, narrow leaves; characteristic of many plants in very dry habitats.

microvicariance Vicariance at a relatively small spatial scale (e.g., on a single island, or between small peripheral versus large core populations).

midden A solid mass of collected organic debris left by an animal such as pack rats (Rodentia).

midoceanic ridge In plate tectonics, a submarine mountain chain, within which seafloor spreading of the oceanic plates occurs.

migration Dispersal from and return to and from an area, typically a breeding site, by an individual or its immediate descendants.

Milankovitch cycles Named in honor of their discoverer, Serbian astrophysicist Milutin Milankovitch, the cyclical changes in the characteristics of Earth's orbit about the sun.

mimicry The marked resemblance of an organism to another organism or a background (e.g., a leaf, tree bark, or sand) to deceive predators or prey by "disappearing" (crypsis) or by causing the predator or prey to confuse the mimic with something it is not.

mobility The ability to move or be moved.

molecular phylogenetics (systematics) A subdiscipline of phylogenetic systematics that employs characters derived from DNA sequences, protein electrophoresis, protein immunology, etc.

monoecious In botany, having separate male and female flowers on the same plant.

monomorphic Having only one form in a population.

monophyletic   A group, or clade, of organisms that includes an ancestral taxon and all of its descendant taxa.

monospecific   Having only one species in the genus.

monotypic   Having only one species in the taxon.

monsoon   Predictable and typically intense summer rains that occur in tropical and subtropical regions where heat and rising air masses over the land surface draws cool, moisture laiden air in from the adjacent oceans. As this air rises over land, it cools, reaches its dew point, and results in heavy precipitation.

monsoon forest   Tree-dominated ecosystems, sometimes called the "layperson's jungle," that develop in areas with intense seasonal rainfall (summer monsoons) and are dominated by trees with large leaves and a luxuriant undergrowth of shrubs and small trees.

morphogeographic rule   See *ecogeographic rule.*

morphological species concept   A species concept that proposes that "species are the smallest groups that are consistently and persistently distinct, and distinguishable by ordinary means" (Cronquist 1978, p. 15).

motility   The ability to move under one's own power, as by wings or a flagellum.

mutation   Any change in the genetic information that results in either the alteration in a gene (e.g., point mutation, insertion, or deletion) or a major modification in the karyotype (chromosomal mutation), neither of which is usually reversible in the strictest sense.

mutualism   An interspecific relationship in which both species receive positive benefits from their interaction.

mycorrhiza   A symbiotic relationship between a fungus and a plant root that benefits the plant by providing a source of useful nitrogen, which is manufactured by the fungus.

natural selection   The process of eliminating from a population through differential survival and reproduction those individuals with inferior fitness.

neap tide   A less than average tide occurring at the first and third quarters of the moon.

Nearctic region   The extratropical region of North America.

nectarivore   An animal that feeds on plant nectar.

nekton   Organisms that are free-swimming in the upper zone of open water and strong enough to move against currents.

neoendemic   An endemic that evolved in fairly recent times. Compare with *paleoendemic.*

Neotropical Region   The region from southern Mexico and the West Indies to southern South America.

neritic zone   The shallow water adjoining a seacoast, especially the zone over a continental shelf.

nested clade analysis (NCA)   A method in phylogeography that converts a statistical haplotype network into a hierarchically nested set of clades based on measures of geographical distribution of clades and nested clades relative to one another.

nested clade phylogeographic analysis (NCPA)   See *nested clade analysis.*

nestedness   The tendency for ecological communities from species-poor sites to form proper subsets (include only those species) found in richer sites.

net primary productivity   The rate of formation of plant tissue in an ecological community that represents the rate of energy made available for consumption by herbivores (equal to gross primary productivity minus total respiration of all plants in that community).

niche   The total requirements of a population or species for resources and physical conditions.

niche breadth   The range of resources and physical conditions used by a particular population relative to that of other populations.

niche expansion   An increase in the range of habitats or resources used by a population, which may occur when a potential competitor, predator or other interacting species is absent.

nomadic   Having no fixed pattern of migration.

nonvolant   An organism that is unable to fly (e.g., rodents versus bats and birds).

null hypothesis   A statistical hypothesis stating what would be expected by chance alone, which can be tested in order to determine whether an observation could be a result of chance or is instead the result of some directing force.

null phenetics   The method of classifying a taxon by using quantitative measures of a large number of characters to assess its overall character similarity to other related taxa. Compare with *evolutionary systematics* and *cladistics* (*phylogenetic systematics*).

nunatak   An area in a glaciated region that was not covered by an ice sheet; hence, a refugium.

obligate relationship   An interaction between species in which at least one of the species cannot survive or reproduce without the other; e.g., many host-parasite relationships, in which a single species of host is required.

obliquity   The tilt of the Earth on its axis, which varies between 22.1° to 24.5° over a 41,000-year period.

oceanic   In marine ecology, of or pertaining to open ocean with very deep water.

oceanic island   An island that was formed de novo from the floor of the ocean through volcanic activity and that has never been attached to a continent.

oceanographer   A scientist who studies the physical, chemical and biological characteristics of marine systems.

oligotrophic lake   A deep lake with low primary productivity.

omnivore   An animal that feeds on both plants and animals.

order   A taxonomic category above the level of family and below the level of class.

ordinate   On a graph, the vertical (y) axis or the vertical coordinate of a point.

Oriental Region   The tropical zone of Southeast Asia eastward to the margin of the continental shelf (Wallace's line).

orogeny   The process of mountain building resulting from the upward thrust of Earth's crust due to volcanic or tectonic activity.

orthogenesis   The supposed intrinsic tendency of organisms to evolve steadily in a particular direction, e.g., to become larger or smaller.

osmoconformer   An organism that does not osmoregulate, but instead has internal salt concentrations in osmotic balance with its environment.

osmoregulation   The process of maintaining homeostasis by maintaining a constant internal concentration of body fluids in changing external solutions.

outbreak area   In organisms that irrupt, the area into which populations expand during peak population densities.

outcrossing   Having gametes that are exchanged between different genotypes.

outgroup   A taxon related to the ingroup, used in a cladistic analysis to infer primitive and derived character states in a transformation series.

outline maps   A map depicting the geographic limits (range boundary) of a particular population or species.

overkill hypothesis   A theory that attributes the mass extinctions of many megafaunal species during the late-Pleistocene and early Holocene to hunting and other forms of over-exploitation by humans.

overturn   Vertical mixing of the water column in a lake caused by temperature changes over the seasons.

Pacifica   A hypothesized ancient continent that may have existed somewhere in the Pacific basin.

pagility   The ability of an organism to be transported by passive dispersal, i.e., where the principal dispersal force is from some force external to the organism (e.g., wind, water currents or other organisms).

Palearctic region   The region of extratropical climates in Eurasia and in the coastal area of northernmost Africa.

paleocirculation   The ocean currents of the past.

paleoclimatology   The study of past climates, as elucidated mainly through the analysis of fossils.

paleoecology   The branch of paleontology that attempts to reconstruct the structure of and the processes affecting ancient populations and communities.

paleoendemic   An endemic that evolved in the distant past. Compare with *neoendemic*.

paleoflora   All the species of plants inhabiting a specified region in the past.

paleomagnetism   The magnetism or magnetically induced orientation of microstructures that has existed in a rock since its origin.

paleontology   The field of study devoted to describing, analyzing, and explaining the fossil record.

paleotropical   Occurring in the Old World tropics and subtropics; i.e., in Africa, Madagascar, India, and Southeast Asia.

panbiogeography   An approach in historical biogeography, developed by Croizat, that attempts to reconstruct the events leading to observed distributions of taxa by drawing lines on a map (tracks) connecting known distributions of related taxa. Unlike cladistic vicariance biogeography, panbiogeography does not require phylogenies of the focal taxa.

Pangaea   In plate tectonics, the supercontinent of the Permian that was composed of essentially all the present continents and major continental islands.

panmixis   The condition whereby interbreeding within a large population is totally random.

Panthalassa   The great global ocean that existed during the Permian coincident with the supercontinent Pangaea.

pantropical   Occurring in all major tropical areas around the world.

paradigm   A unifying principle, pattern or theory, such as MacArthur and Wilson's equilibrium theory of island biogeography, or Darwin's theory of natural selection.

parallel evolution   The evolution of species with strong resemblances from fairly closely related ancestors; hence the evolution of two or more taxa in the same direction from related ancestors.

paralogy-free subtrees   An approach in historical biogeography that modifies component analysis by removing geographic paralogy (repetition of geographic areas resulting from sympatric or embedded allopatric speciation) prior to analysis of congruence among taxon-area cladograms.

parámo   Tropical alpine vegetation, characteristic of high mountains at the equator, that has a low and compact perennial cover as a response to the perpetually wet, cold, and cloudy environment.

parapatric   Having contiguous but narrowly overlapping distributions.

parapatric speciation   A mode of speciation in which differentiation occurs when two populations have contiguous but narrowly overlapping ranges, often representing two distinct habitat types.

paraphyletic   In phylogenetic systematics, referring to taxa that include an ancestral taxon and some, but not all, of its descendant taxa; an artificial taxon.

parasitism   An interspecific relationship in which one species (the parasite) draws nutrition from or is somehow dependent for survival on the other species (the host), which is negatively affected by the interaction.

parsimony   In phylogenetic analysis, provides a "rule" stating that the tree requiring the fewest number of character state changes is preferred over more complex trees.

parsimony analysis for comparing trees (PACT)   A method in historical biogeography conceptually similar to primary and secondary BPA but that employs a different algorithm to incorporate all taxon-area cladogram area relationships into a general area cladogram.

parthenogenesis   The development of eggs without fertilization by a male gamete.

passive dispersal (vagility)   The movement of an organism from one location to another by means of a stronger force, such as wind or water or via a larger animal.

pattern   Nonrandom, repetitive organization.

pedogenic regimes   The soil-forming processes; e.g., laterization, podzolization, calcification, and gleization.

pelagic   Occurring in open water and away from the bottom.

peninsular effect   The hypothesized tendency for species richness to decrease along a gradient from the axis to the most distal point of a peninsula.

peripheral population   Any population of a species that occurs along or near the edges of its geographic range, around either the perimeter or the elevational limit. Not synonymous with *marginal population*.

permanence theory The view, widely held before continental drift was accepted, that the distribution pattern of ocean basins and continents remained relatively constant over the history of the Earth.

perturbation Any event that greatly upsets the equilibrium of or alters the state or direction of change in a system.

phenetics The study of the overall similarities of organisms. Compare with *phylogeny*.

phenotype The expression of the genetic message of an individual in its morphology, physiology, and behavior.

phoresy The phenomenon where relatively small animals depend on larger animals for dispersal.

photic (euphotic) zone The uppermost zone in a water column where solar radiation is adequate to permit photosynthesis by plants.

photoautotroph An organism that uses light as the energy source and carbon dioxide as the carbon source for its basic metabolism; hence, an organism that is photosynthetic.

photosynthesis The chemical process of using pigments to capture sunlight and then using that energy, along with water and carbon dioxide, to make organic compounds (sugars), releasing oxygen in the process.

phyletic gradualism An evolutionary process whereby a species is gradually transformed over time into a different organism; compare with *punctuated equilibrium*.

phyletic speciation The transformation of one ancestral species into a single descendant species through non-branching (anagenetic) evolutionary change.

PhyloCode A proposed new system of taxonomic classification.

phylogenetic biogeography I An earlier approach in historical biogeography that used Hennig's phylogenetic methods, but also incorporated his progression rule under the assumption of a history of sequential dispersal and speciation events as a lineage expanded out of a center of origin.

phylogenetic biogeography II An approach in historical biogeography that searches for shared cladogenetic events among a set of taxon-area cladograms, but also explains incongruent events in terms of post-vicariance dispersal, peripheral isolates speciation, sympatric speciation, and extinction.

phylogenetic systematics See *cladistics*. Compare with *evolutionary systematics* and *numerical phenetics*.

phylogeny The evolutionary relationships between an ancestor and all of its known descendants.

phylogeography An approach in biogeography that studies the geographic distributions of genealogical lineages, within species and among similar species, and attempts to differentiate between historical and ongoing processes leading to the development of observed patterns.

phylogroup A term used in phylogeography to define a population or species whose alleles within a given gene tree form a monophyletic lineage with respect to other populations or species.

physiognomy The external aspect of a landscape; e.g., the topography and other physical characteristics of a land form and its vegetation.

physiological ecology The study of how the physiological characteristics of organisms relate to the abundances and distributions of those organisms in their natural habitats.

phytogeography The study of the distribution of plants.

phytoplankton The collection of small or microscopic plants that float or drift in great numbers in freshwater or marine environments, especially at or near the surface, and serve as food for zooplankton.

phytosociology The quantitative study of the composition of plant communities and how these relate to environmental factors.

placental A mammal that has a placenta connecting the mother to the fetus.

plankton Small organisms (especially tiny plants, small invertebrates, and juvenile stages of larger animals) that inhabit water and are transported mainly by water currents and wave action rather than by individual locomotion.

planktotrophic Referring to aquatic larvae that have no long-term storage of nutrients and so must feed on small organisms in the plankton during their development.

plate In plate tectonics, a portion of the Earth's upper surface, about 100 km thick, that moves over the asthenosphere during seafloor spreading.

plate tectonics The study of the origin, movement, and destruction of plates and how these events have been involved in the evolution of the Earth's crust.

plesiomorphy In a transformation series, the primitive character state.

plunge pool A relatively deep, roughly circular lake which formed as glacial meltwaters flowed over the surface of a glacier and then plunged off its edge to carve a basin in the Earth some 2 to 3 km below.

pluvial Referring to periods of high rainfall and water runoff.

podzolization The formation of a soil under conditions of adequate moisture and low decomposer activity, resulting in a soil in which the bases, humic acids, colloids, and ferric and aluminum oxides have been removed (leached) from the upper horizon.

poikilotherm An animal with a relatively variable body temperature, often ectothermic; i.e., relying on the external environment to control its body temperature.

pollination The transfer of pollen grains to a receptive stigma, usually by wind or flower-visiting animals.

polymorphic Having several distinct forms in a population.

polyphagous Feeding on a variety of different kinds of food.

polyphyletic A group of organisms that does not include their common ancestor; an artificial taxon.

polyploid Any organism or cell that has three or more sets of chromosomes.

population bottleneck Reduction in effective population size, often through the founding of a new population via jump dispersal of one or a few individuals across a barrier; a general result is loss of genetic variation.

Precambrian shield See *craton*.

precession Cyclical changes, or "wandering" of Earth's orientation where the axis of the North Pole shifts from one "north star" (presently Polaris of Ursa Minor) to another (Vega of Lyra) with a periodicity of approximately 22,000 years.

precinctiveness   The tendency for organisms of some species to exhibit relatively strong site fidelity and remain in their natal range.

predation   The act of feeding on other organisms; an interspecific interaction that has negative effects on the species that is consumed or used.

predator   In a predatory relationship, the species that consumes other organisms.

predator mediated coexistence   Coexistence of two intense competitors which results from the actions of a predator that preys most heavily on the more abundant or otherwise dominant competitor, therefore preventing populations of either competitor from increasing to the point where it can exclude the other.

prey   In a predatory relationship, the species that is consumed or used by another organism.

prezygotic isolating mechanism   An attribute that serves as a form of reproductive isolation between separate *biological species* prior to the formation of a viable zygote.

primary BPA   See *Brooks parsimony analysis*.

primary consumer   An organism that consumes plants.

primary division freshwater fish   Any fish that is totally intolerant of salt water.

primary production   The production of biomass by green plants.

primary succession   The gradual transformation of bare rock or another sterile substrate into a soil that supports a living ecological community.

profundal zone   The deepwater zone of lakes where light is insufficient to support photosynthesis.

progression rule   The idea that a primitive form remains in the center of origin, whereas progressively more derived forms are found at progressively greater distances.

projections (geographic projections)   A means of using light projected through plastic models of the Earth (traditionally), or the use of various mathematical formulae to transform spatial data from the curved, 3-dimensional surface of the Earth to a two-dimensional map.

propagule   Any part of an organism, or group of organisms, or stage in the life cycle that can reproduce the species and thus establish a new population.

provincialism   The coincident occurrence of large numbers of well-differentiated endemic forms in an area; regional or provincial distinctiveness.

pseudo-congruence   Superficially congruent taxon-area relationships that in fact have developed at different times in response to different vicariance or dispersal events.

pseudoturnover   An overestimate of biotic turnover resulting from failure to detect a species that was actually present during a biological survey, and was therefore either recorded as an immigration or an extinction in subsequent surveys.

punctuated equilibrium   The hypothesis that evolution occurs during periods of rapid differentiation (often accompanying speciation), which are followed by long periods in which few if any characters evolve. Compare with *phyletic gradualism*.

quadrupedal   Having four limbs for locomotion.

radiation   In evolutionary biology, a general term for the diversification of a group, implying that many new species have been produced.

rafting   The transport over water of living organisms on large floating mats of debris.

rain-green forest   The dominant vegetation of tropical-dry forests characterized by trees that leaf out during the first heavy rains following the dry season.

rain shadow   More generally, a "precipitation shadow," which is a relatively dry region typically found along the leeward side of a mountain range where the relative humidity of the descending air decreases as it becomes adiabatically warmed.

range maps   Maps depicting the geographic distribution of particular species (includes dot, contour and outline maps).

Rapoport's rule   The tendency for geographic range size to increase with latitude.

realized geographic range   The actual distribution of populations of a species, which is restricted to only a subset of the theoretical, *fundamental geographic range* as a result of interactions with other species and limited opportunities for dispersal.

realized niche   The actual environmental conditions in which a species survives and reproduces in nature; a subset of the fundamental niche.

reciprocally monophyletic   Two populations or species are reciprocally monophyletic if, for a given gene tree, the alleles within each population have coalesced to a monophyletic lineage with respect to the other population.

recombination   The exchange of genetic material between chromosomes, resulting in new combinations of genes on the chromosomes of offspring derived from each of the parental chromosomes.

reconciled tree   A method in historical biogeography that reconstructs historical events (sympatric speciation, dispersal, vicariance, extinction) in order to "reconcile" phylogenetic and geographic incongruence between co-distributed taxa.

Red Queen hypothesis   A hypothesis that states that a species must continually evolve in order to keep pace with an environment that is perpetually changing because all other species are also evolving, altering the availability of resources and the nature of biotic interactions.

refugium   An area in which climate and vegetation type have remained relatively unchanged while areas surrounding it have changed markedly, and which has thus served as a refuge for species requiring the specific habitat it contains.

reinforcement   At a contact zone between differentiated populations, the process of selection for prezygotic mechanisms that completes the reproductive isolation between two new *biological species*.

relative apomorphy rule   Homologous characters found within members of a monophyletic group that are also found in the sister group are plesiomorphic, while homologous characters found only in the ingroup are apomorphic.

relative rate test   A method used in molecular systematics to determine whether two clades are evolving at a similar rate relative to a more distantly related third clade.

relaxation model   An explanation for non-random assembly of ecological communities inhabiting ecosystems that have decreased in size over time (due to natural or anthropogenic changes) which holds that as the area of these ecosystems decreased, their communities lost the most resource-intensive species and began to converge on a similar set of species, i.e., those with relatively low resource requirements.

relict   A surviving taxon from a group that was once widespread (biogeographic relict) or diverse (taxonomic relict).

relict pattern   In island biogeography, Philip Darlington's prediction that ecological communities from relatively small ecosystems would represent a random subset of those found on the mainland or source pool.

remant magnetism   The property of rocks containing iron and titanium oxides to become magnetized as they solidify and cool, thus recording the magnetic fields in their crystalline structure. Such rocks can later serve as a "fossil compasses" indicating both the direction and declination of the magnetic fields at the sites and periods the rocks originally solidified.

reproductive isolation   Inability of individuals from different populations to produce viable offspring.

rescue effect   The tendency for extinction rates to be relatively low on less-isolated islands because their relatively high immigration rate tends to supplement declining populations before they suffer extinction.

resident   A species that lives year-round in a particular habitat or location.

respiration (community respiration)   In ecology, the rate of chemical breakdown of organic material in an ecological community.

reticulate evolution   The formation of new species by the hybridization of dissimilar populations; e.g., interspecific hybridization.

reticulate speciation   A new species that arises through hybridization between two separate parental species.

ridge push   One of the forces responsible for continental drift, in this case resulting from the rise of hot magma toward and then outward from Earth's surface.

rift zone   A region where one or more continental plates are separating, thus forming a low-lying, tectonically active area (spreading zone). Examples include those that in the past created the Red Sea and the great, deep lakes of the Baikal rift zone and the East African Rift Valley.

*r***-selection**   Selection that favors high population growth rate ("*r*"), which is more prominent when population size is far below carrying capacity.

*r***-strategists**   Species whose life-history and ecological characteristics are adapted to selective pressures associated with relatively low population levels.

saline   Having high concentrations of salts and especially ions of chloride and sulfate.

salt marsh   A wetland dominated by herbaceous vegetation and characterized by its relatively high salinity and salt adapted inhabitats.

scaling   See *allometry.*

scavenger   An animal that feeds on carrion.

sclerophyllous   Having tough, thick, evergreen leaves.

scrub   Any of a wide variety of vegetation types dominated by low shrubs; in exceedingly dry locations, scrub vegetation has few or no trees and widely spaced low shrubs, but in areas of fairly high rainfall, scrub has trees and grades into either woodland or forest.

seafloor spreading   In plate tectonics, the process of adding crustal material at a midoceanic ridge and thus displacing older rocks, usually on both sides, away from their point of origin.

seamount   A peaked submarine volcano.

second law of thermodynamics   As energy is converted into different forms, its capacity to perform useful work diminishes, and the disorder (entropy) of the system increases.

secondary BPA   A method in historical biogeography that duplicates areas on a primary BPA tree only as much as required to remove all homoplasy; the resulting tree includes area relationships that are the results of vicariance, post-speciation dispersal, peripheral isolates speciation, and extinction.

secondary consumer   An animal that consumes herbivorous animals (i.e., a predator).

secondary division freshwater fish   A fish that prefers fresh water but can live for short periods in salt water.

secondary succession   A series of changes in the vegetational composition of an environment in response to disturbance, involving the gradual and regular replacement of species and ending, at least hypothetically, with the return to a stable state (climax).

secular migration   A means of geographic range expansion which is so slow (i.e., over many generations) that it is often accompanied by substantial evolutionary changes in the populations en route, i.e., as they expand to colonize new regions.

selective sweep   Rapid differentiation of populations resulting from the fixation of a new mutation with a strong positive fitness effect.

self-incompatible   Requiring two individuals to exchange gametes in order to produce offspring.

semiaquatic   Living partly in or adjacent to water.

semiarid   Having a fairly dry climate with low precipitation, usually 25 to 60 cm per year, and a high evapotranspiration rate, so that potential loss of water to the environment exceeds the input.

semidesert   A semiarid habitat characterized by low vegetation; e.g., small, widely spaced shrubs.

seral stage   One of the stages in the ecological succession of communities.

sere   A series of stages in community transformation during ecological succession.

serpentine   A rock or soil type rich in magnesium but deficient in calcium.

sessile   Remaining fixed in the same spot throughout adulthood; e.g., most plants and certain benthic aquatic invertebrates.

shrubland   Any of a wide variety of vegetation types dominated by shrubs, which may form a fairly solid cover; e.g., sclerophyllous scrub (chaparral), or sparse cover; e.g., desert scrub.

sial   Rock rich in silica-aluminum; the principal component of continental rocks.

sibling species   Different species that are difficult to delineate with morphological traits (also called *cryptic species*).

siliceous   Containing silica.

sima   Rock rich in silica-magnesium; the principal component of basalt and of oceanic plates.

similarity index   An estimate of the similarity or relatedness of two communities, biota, or taxa.

sink habitats   Environments where the birth rate of a particular species is exceeded by its death rate and, therefore, the presence of this species indicates that most individuals are derived from dispersal from other ("source") habitats.

sister group (clade)   In a phylogeny, the group or clade most closely related to the focal group, and therefore the most useful outgroup for rooting the phylogeny.

sister taxa   In cladistics, the two taxa that are most closely (and therefore most recently) related.

slab pull   One of the forces responsible for continental drift, in this case resulting from the tendency for portions of a plate to adhere to the leading, subducting edge of that plate.

small island effect   The tendency for species richness to vary independent of island area on relatively small islands.

solstice   Either of two times in a year (June 22, December 22) when the sun reaches its highest latitude (23.5° N and S, respectively).

source habitats   Environments in which the birth rate of a particular species exceeds its death rate and, therefore, results in a surplus of individuals which contributes to dispersal to other ("sink") habitats.

speciation   The process in which two or more contemporaneous species evolve from a single ancestral population.

speciation-pump model   See *cyclical vicariance model*.

species   The fundamental taxonomic category for organisms; variously defined and diagnosed using different species concepts (see Table 7.1 and definitions).

species-area relationship   A plot of the numbers of species of a particular taxon against the area of islands or other ecosystems. The relationship is often linearized by log-transforming one or both variables.

species composition   The types of species that constitute a given sample or community.

species concept   Any one of some 22 criteria and approaches (see Table 7.1) used to delineate separate species.

species density   The number of species per standardized sample area.

species pool   All the organisms present in neighboring source areas that are theoretically available to colonize a particular habitat or island.

species richness   A relatively simple, but important measure of diversity representing the number of species in an ecological community.

species selection   An analogue of natural selection at the species level, in which some species with certain characteristics increase while others decrease or become extinct.

spreading zone   A region where two or more continental plates are separating, often associated with the formation of a new ocean basin (sea-floor spreading) or rift valleys on the continents.

spring tide   A greater than average tide occurring during the new and full moons.

stasipatric speciation   See *parapatric speciation*.

statistical phylogeography   The framing of phylogeographic investigations within a rigorous statistical framework, either through development of alternative hypotheses that are evaluated with coalescence-based statistics; or through a nested clade phylogeographic analysis (see *nested clade analysis*).

stenohaline   Having a tolerance for only a narrow range of salt concentrations.

stenothermal   Having a tolerance for only a narrow range of temperatures.

stochastic   Random, expected (statistically) by chance alone; compare with *deterministic*.

stratigraphy   The branch of geology dealing with the sequence of deposition of rocks and fossils as well as their composition, origin, and distribution.

subduction   In plate tectonics, the movement of one plate beneath another, leading to the heating and subsequent remelting of the lower plate.

subduction zone   A region where one relatively dense tectonic plate (oceanic plate, composed sima) slides beneath a less dense (continental, sial) plate.

subfamily   A taxonomic category used for grouping genera within a family.

sublittoral zone   The coastal marine zone below the intertidal zone and therefore below the point at which the sea bottom is periodically exposed to the atmosphere.

subspecies   A taxonomic category used by some systematists to designate a genetically distinct set of populations with a discrete range.

substitution   A form of DNA mutation in which one nucleotide base is replaced by one of the three other nucleotide bases.

subterranean   Living underground.

supercontinent   An ancient landmass formed by the collision and connection of most, if not all, of the present global landmasses (e.g., Pangaea).

superfamily   A taxonomic category used for grouping families within a suborder.

superpáramo   A vegetation type of high elevations in the equatorial Andes mountains.

superspecies   A group of closely related species.

supertramps   Species that are relatively common on relatively small and isolated islands (i.e., those with few ecologically similar species), but are absent from islands with more diverse communities.

swamp (marl)   A wetland dominated by woody vegetation.

sweepstakes route   A severe barrier that results in the partly stochastic dispersal of some elements of a biota, and the establishment of a disharmonic biota.

symbiosis   A long-term interspecific relationship in which two unrelated and unlike organisms live together in close association so that each receives some adaptive benefit.

sympatric   In the strictest sense, living in the same local community, close enough to interact; in the more general sense, having broadly overlapping geographic ranges.

sympatric speciation   The differentiation of two reproductively isolated species from one initial population within the same local area; hence, speciation that occurs under conditions in which much gene flow potentially could or actually does occur. Compare with *allopatric speciation; parapatric speciation.*

symplesiomorphy   In a transformation series, a character state shared by taxa in a focal clade, but also shared by other clades at more basal nodes in a phylogeny and therefore not useful in diagnosing the focal group as being monophyletic. See *grouping rule.*

synapomorphy   In a transformation series, a derived character state shared by taxa due to inheritance from a common ancestor and therefore useful in diagnosing a monophyletic group or clade. See *grouping rule.*

systematics   The study of the evolutionary relationships between organisms. Includes *phylogenetics* and *taxonomy.*

tachytelic   Term used by Simpson (1944) to refer to a rapid rate of morphological change through time in a lineage with a fossil record.

taphonomy   A subdiscipline of paleontology concerned with the processes by which remains of living things become fossilized, and the ways in which these processes can bias the fossil record or cause problems of interpretation.

taxon (pl. **taxa**)   A convenient and general term for any taxonomic category; e.g., a species, genus, family, or order.

taxon-area cladogram   An area cladogram generated by replacing the taxa on a taxon cladogram with their areas of occurrence.

taxon biogeography   The primary goal of an analysis is to reconstruct the biogeographic history of a single taxon.

taxon cycle   A series of ecological and evolutionary changes in insular populations from their colonization of early successional sites along the beachfronts, to their expansion and increased specialization for interior habitats, and their eventual extinction and replacement by subsequent waves of colonists and their descendenats.

taxonomic relicts   The sole survivors of once diverse taxonomic groups.

taxonomy   In the strictest sense, the study of the names of organisms, but often used for entire process of classification. See *systematics.*

tectonic   Referring to any process involved in the production or deformation of the Earth's crust.

terrane   An accumulation of marine basin sediments which was scraped off of the upper surface of an oceanic plate while it was being subducted beneath a continental plate.

terrestrial   Living on land.

tetrapod   Any four-legged vertebrate, including amphibians, reptiles, and mammals.

thermocline   In a water column, the subsurface zone in which the temperature drops sharply.

thorn scrub   A relatively xeric ecosystem typically receiving less than 30 cm of annual rainfall, experiencing a 6-month dry season, and dominated by sparse woody vegetation.

three-item statement analysis   An approach in historical biogeography that modifies component analysis by converting taxon-area cladograms to subsets of three-area relationships prior to analysis of congruence.

tillites   Glacial rock deposits.

timberline   The upper elevational limit of trees on mountains.

time dwarfing   The tendency for some initially large animals (in particular, large marsupials of Australia; Flannery 1994) to decrease in size over time, especially following significant reduction and isolation of their native habitats.

tokogenetic   Reticulate (netlike or interwoven) relationships between individuals within a sexual species.

total evidence   An approach in systematics that combines characters drawn from different sources (e.g., morphological and molecular characters) into a single data set for phylogenetic analyses.

trace fossils   Taxa known only from their activities (e.g., tracks or burrows).

track   In panbiogeography, a line drawn on a map connecting the geographically isolated ranges of the species in a taxon that are closest relatives (vicariants).

trade winds   Winds blowing toward the equator between the horse latitudes and the doldrums in the Northern and Southern Hemispheres.

transformation series   For a character, two or more *character states*, assumed to be products of evolutionary changes and therefore used by systematists to infer evolutionary relationships and classify organisms into taxa.

transform fault   A fault in an oceanic plate, perpendicular to the midoceanic ridge, that divides the plate into smaller units.

transform zone   (Also known as a "strike-slip fault.") A region where two tectonic plates slide against eachother but in opposite directions.

translocation   A kind of chromosomal mutation in which a segment of one chromosome becomes attached to a different chromosome.

transpiration   The loss of water vapor from plants through pores called stomates.

transposable elements   Nucleotide sequences that promote their own movement between chromosomal sites.

trench   In plate tectonics, an exceedingly deep cut in the ocean floor where the subduction of an oceanic plate is occurring.

triple junction   In plate tectonics, a point at which three oceanic plates meet; the position of the junction shifts because each of the plates drifts at a different rate.

trophic level   A functional, ecological classification of organism based on their primary source of energy (including primary producers = plants; primary consumers = herbivores; secondary consumers = predators; tertiary and in some cases, higher order consumers; and decomposers).

trophic status   The position or role of an organism in the nutritional structure of a community; e.g., primary producer, herbivore, or top carnivore.

tropical alpine scrubland   Vegetative communities found above the timberline in equatorial regions and dominated by relatively sparse, low-growing vegetation including tussock grasses and bizarre, erect rosette perennials with thick stems.

Tropical Convergence Zone   The zone along tropical regions of the Earth's surface with most direct sunlight and most intense heating, which is also associated with rising air masses and high precipitation, and shifts with the seasons between the Tropics of Cancer and Capricorn.

turnover   The rate of replacement of species in a particular area as some taxa become extinct but others immigrate from outside.

undersaturated   Having fewer taxa of a particular kind than expected on the basis of an equilibrium between colonization, speciation, and extinction.

unequal crossover   An asymmetric exchange of genetic material between a chromosome pair during recombination.

uniformitarianism   See *actualism*.

upwelling   The vertical movement of deep water, containing dissolved nutrients from the ocean bottom, to the surface.

vagility   The ability to move actively from one place to another.

vicariance biogeography   An approach in historical biogeography that attempts to reconstruct the historical events that led to observed distributional patterns based largely on the assumption that these patterns resulted from the splitting (vicariance) of areas and not long-distance dispersal. Compare *dispersal biogeography*.

vicariant events   Tectonic, eustatic, climatic, or oceanographic events that result in the geographic isolation of previously connected populations.

vicariants   Two disjunct and phylogenetically related-species that are assumed to have been created when the initial range of their ancestor was split by some historical event.

Viking funeral ship   A term coined by McKenna for a landmass, such as a fragment, containing fossils that were laid down when the land was in one location, but that were transported to a completely different locality via continental drift.

waif   In dispersal biogeography, a diaspore or any type of individual that is carried passively by waves or air currents to a distant place; e.g., most colonizers of oceanic island beaches.

Wallace's line   The most famous biogeographic line, running between Borneo and Celebes and between Bali and Lombok, which marks the boundary between the Oriental and Australian Regions.

Wallacean Shortfall   The paucity of information on geographic distributions of species (past and present), and on the geographic dynamics of extinction forces, especially the dynamic geography of humans, their commensals and other anthropogenic threats.

Wegenerism   The general idea of continental drift according to Alfred Wegener.

westerlies   Prevailing winds in the temperate regions (30° to 60°) of both the Northern and Southern Hemispheres which, as a result of the Coriolis effect, have a strong west-to-east component.

wintering area   In migratory land animals, the area where populations spend the cold season and feed, but do not breed.

woodland   Any of a variety of vegetation types consisting of small, widely spaced trees with or without substantial undergrowth.

xerophytes   Land plants that grow in relatively dry ("xeric") environments.

zonal soils   Soils that have distinctive characteristics and are formed by the actions of climate and organisms on the so-called "typical" rocks (sandstone, shale, granite, gneiss and slate).

zoochory   Transport of propagules by animals.

zoogeography   The study of the distributions of animals.

zooplankton   The collection of small or microscopic animals that float or drift in great numbers in fresh or salt water, especially at or near the surface, and serve as food for fish and other larger organisms.

Bibliography

Abatzopoulos, T. J. 2002. *Artemia: Basic and Applied Biology.* Kluwer Academic Publishers.

Abbitt, R. J. F., J. M. Scott and D. S. Wilcove. 2000. The geography of vulnerability: Incorporating species geography and human development patterns into conservation planning. *Biological Conservation* 96: 169–175.

Abbott, I. 1978. Factors determining the number of land bird species on islands around southwestern Australia. *Oecologia* 33: 221–233.

Abbott, I. 1983. The meaning of z in species/area regressions and the study of species turnover in island biogeography. *Oikos* 41: 385–390.

Abbott, I., L. K. Abbott and P. R. Grant. 1977. Comparative ecology of Galapagos ground finches (*Geospiza* Gould): Evaluation of the importance of floristic diversity and interspecific competition. *Ecological Monographs* 47: 151–184.

Abele, L. G. and E. F. Connor. 1979. Application of island biogeography theory to refuge design: Making the right decision for the wrong reasons. In R. M. Linn (ed.), *Proceedings of the First Conference on Scientific Research in the National Parks, New Orleans, LA,* 89–94. Washington, DC: Department of the Interior (National Park Service).

Abele, L. G. and W. Kim. 1989. The decapod crustaceans of the Panama canal. *Smithsonian Contributions to Zoology* 482: 1–50. Washington, DC: Smithsonian Institution Press.

Aberg, J., G. Jansson, J. E. Swenson and P. Angelstam. 1995. The effect of matrix on the occurrence of hazel grouse (*Bonasa bonasia*) in isolated habitat fragments. *Oecologia* 103: 265–269.

Abrahamson, D. E. 1989. *The Challenge of Global Warming.* Washington, DC: Island Press.

Ackerly, D. D. and M. J. Donoghue. 1995. Phylogeny and ecology reconsidered. *Journal of Ecology* 83: 730–733.

Adams, C. C. 1902. Southeastern United States as a center of geographical distribution of flora and fauna. *Biological Bulletin* 3: 115–131.

Adams, C. C. 1909. *An Ecological Survey of Isle Royale, Lake Superior.* Lansing, MI: Michigan Biological Survey.

Adams, C. G. and D. V. Ager. 1967. *Aspects of Tethyan Biogeography.* The Systematics Association Publications, no. 7, Wetteren, Universa.

Adams, J., M. Maslin and E. Thomas. 1999. Sudden climate transition during the Quaternary. *Progress in Physical Geography* 23: 1–36.

Adler, G. H. 1994. Avifaunal diversity and endemism on tropical Indian Ocean islands. *Journal of Biogeography* 21: 85–95.

Adler, G. H. 1996. The island syndrome in isolated populations of a tropical forest rodent. *Oecologia* 108: 694–700.

Adler, G. H. and R. Levins. 1994. The island syndrome in rodent populations. *The Quarterly Review of Biology* 69: 473–490.

Agassiz, L. 1840. Etudes sur les Glaciers/Studies of the glaciers. Neuchatel.

Agosti, D. 2000. Standard Methods for Measuring and Monitoring Biodiversity: Ants. Washington, DC: Smithsonian Institution Press.

Agusti, J. and M. Anton. 2002. *Mammoths, Sabertooths and Hominids: 65 Million Years of Mammalian Evolution in Europe.* New York: Columbia University Press.

Alacantra, M. 1991. Geographical variation in body size of the wood mouse *Apodemus sylvaticus* L. *Mammalian Reviews* 21: 143–150.

Alatalo, R. V. 1982. Bird species distributions in the Galapagos and other archipelagoes: Competition or chance? *Ecology* 63: 881–887.

Albrecht, F. O. 1967. Polymorphisme phasaire et biologie des acridiens migrateurs. Paris: Masson.

Allard, G. O. and V. J. Hurst. 1969. Brazil-Gabon geologic link supports continental drift. *Science* 163: 528–532.

Allee, W. C., O. Park, A. E. Emerson, T. Park and K. P. Schmidt. 1949. *Principles of Animal Ecology.* Philadelphia: W. B. Saunders.

Allen, J. A. 1878. The influence of physical conditions in the genesis of species. *Radical Review* 1: 108–140.

Alley, R. B. and P. U. Clark. 1999. The deglaciation of the Northern Hemisphere: A global perspective. *Annual Review of Earth and Planetary Sciences* 27: 149–182.

Alroy, J. 1999. Putting North America's end-Pleistocene extinction in context. In R. MacPhee (ed.), *Extinctions in Near Time,* 105–143. New York: Kluwer Academic/Plenum Publishers.

Alroy, J. 2001. A multispecies overkill simulation of the End-Pleistocene megafaunal mass extinction. *Science* 292: 1893–1896.

Althoff, D. M. and O. Pellmyr. 2002. Examining genetic structure in a bogus yucca moth: A sequential approach to phylogeography. *Evolution* 56: 1632–1643.

Alvarez, L. W., W. Alvarez, F. Asaro and H. V. Michel. 1980. Extraterrestrial cause for the Cretaceous-Tertiary extinction. *Science* 208: 1095–1108.

Alvarez, W., E. G. Kauffman, F. Surlyk, L. W. Alvarez, F. Asaro and H. V. Michel. 1984. Impact theory of mass extinctions and the invertebrate fossil record. *Science* 223: 1135–1141.

Amadon, D. 1950. The Hawaiian honeycreepers (Aves, Drepaniidae). *Bulletin of the American Museum of Natural History* 95: 153–262.

Amarasinghe, U. S. and R. L. Welcomme. 2002. An analysis of fish species richness in natural lakes. *Environmental Biology of Fishes* 65: 327–339.

Ambrose, S. H. 2001. Paleolithic technology and human evolution. *Science* 291: 1748–1753.

Amiran, D. H. K. and A. W. Wilson (eds.). 1973. *Coastal Deserts: Their Natural and Human Environments.* Tucson, AZ: University of Arizona Press.

Anderson, A. B. and W. W. Brown. 1980. On the number of tree species in Amazonian forests. *Biotropica* 12: 235–237.

Anderson, I. W. 1974. The chestnut pollen decline as a time horizon in lake sediments in eastern North America. *Canadian Journal of Earth Sciences* 11: 678–685.

Anderson, P. K. 1960. Ecology and evolution in island populations of salamanders in San Francisco Bay region. *Ecological Monographs* 30: 359–385.

Anderson, R. P. and C. O. Handley. 2002. Dwarfism in insular sloths: Biogeography, selection and evolutionary rate. *Evolution* 56: 1045–1058.

Anderson, S. and L. S. Marcus. 1992. Areography of Australian tetrapods. *Australian Journal of Zoology* 40: 627–651.

Andrewartha, H. G. and L. C. Birch. 1954. *The Distribution and Abundance of Animals.* Chicago: University of Chicago Press.

Andrews, P. and E. M. O'Brien. 2000. Climate, vegetation and predictable gradients in mammal species richness in southern Africa. *Journal of Zoology* (London) 251: 205–231.

Aponte, C., G. R. Barreto and J. Terborgh. 2003. Consequences of habitat fragmentation of age structure and life history in a tortoise population. *Biotropica* 35: 550–555.

Araujo, M. B. 2002. Biodiversity hotspots and zones of ecological transition. *Conservation Biology* 16: 1662–1663.

Arbogast, B. S. and G. J. Kenagy. 2001. Comparative phylogeography as an integrative approach to historical biogeography. *Journal of Biogeography* 28: 819–825.

Arbogast, B. S., S. V. Edwards, J. Wakeley, P. Beerli and J. B. Slowinski. 2002. Estimating divergence times from molecular data on phylogenetic and population genetic timescales. *Annual Review of Ecology and Systematics* 33: 707–740.

Arnett, A. E. and N. J. Gotelli. 2003. Bergmann's rule in larval ant lions: Testing the starvation resistance hypothesis. *Ecological Entomology* 28: 645–650.

Arnold, E. N. 1979. Indian Ocean giant tortoises: Their systematics and island adaptations. *Philosophical Transactions of the Royal Society of London*, Series B 286: 127–145.

Arnold, M. L. and S. K. Emms. 1998. Paradigm lost: Natural hybridization and evolutionary innovations. In D. J. Howard and S. H. Berlocher (eds.), *Endless Forms: Species and Speciation*, 379–389. New York: Oxford University Press.

Arnold, S. J. 1972. Species densities of predators and their prey. *American Naturalist* 106: 220–236.

Arrhenius, O. 1921. Species and area. *Journal of Ecology* 4: 68–73.

Ashmole, N. P. 1963. The regulation of numbers of tropical ocean birds. *Ibis* 103b: 458–473.

Ashton, K. G. 2002b. Do amphibians follow Bergmann's rule? *Canadian Journal of Zoology* 80: 708–716.

Ashton, K. G. 2002a. Patterns of within-species body size variation in birds: Strong evidence for Bergmann's rule. *Global Ecology and Biogeography* 11: 505–523.

Ashton, K. G. 2004. Sensitivity of intraspecific latitudinal clines of body size for tetrapods to sampling, latitude and body size. *Integrative Comparative Biology* 44: 403–412.

Ashton, K. G. and C. R. Feldman. 2003. Bergmann's rule in non–avian reptiles: Turtles follow it, lizards and snakes reverse it. *Evolution* 57: 1151–1163.

Ashton, K. G., M. C. Tracy and A. de Queiroz. 2000. Is Bergmann's rule valid for mammals? *American Naturalist* 156: 390–415.

Astorga, A., M. Fernandez, E. E. Boschi and N. Lagos. 2003. Two oceans, two taxa and one mode of development: Latitudinal diversity patterns of South American crabs and test for possible causal processes. *Ecology Letters* 6: 420–427.

Athias-Binche, F. 1993. Dispersal in varying environments: The case of phoretic ceropodid mites. *Canadian Journal of Zoology-Revue Canadienne de Zoologie* 71: 1793–1798.

Atkinson, I. A. E. 1988. Presidential address: Opportunities for ecological restoration. *New Zealand Journal of Ecology* 11: 1–12.

Atkinson, I. A. E. and E. K. Cameron. 1993. Human influence on terrestrial biota and biotic communities of New Zealand. *Trends in Ecology and Evolution* 8: 447–451.

Atmar, W. and B. D. Patterson. 1993. The measure of order and disorder in the distribution of species in fragmented habitats. *Oecologia* 96: 373–382.

Auffenberg, W. 1971. A new fossil tortoise, with remarks on the origin of South American testudines. *Copeia* (1): 106–117.

Auffenberg, W. 1981. *The Behavioral Ecology of the Komodo Monitor.* Gainesville: University of Florida Presses.

Avise, J. C. 1994. *Molecular Markers, Natural History and Evolution.* New York: Chapman and Hall.

Avise, J. C. 2000. *Phylogeography: The History and Formation of Species.* Cambridge, MA: Harvard University Press.

Avise, J. C. 2004. *Molecular Markers, Natural History and Evolution.* 2nd edition. Sunderland, MA: Sinauer Associates.

Avise, J. C. and R. M. Ball, Jr. 1990. Principles of genealogical concordance in species concepts and biological taxonomy. In D. Futuyma and J. Antonovics (eds.), *Oxford Surveys in Evolutionary Biology*, 45–67. Oxford: Oxford University Press.

Avise, J. C. and W. S. Nelson. 1989. Molecular genetic relationships of the extinct dusky seaside sparrow. *Science* 243: 646–648.

Avise, J. C. and D. Walker. 1998. Pleistocene phylogeographic effects on avian populations and the speciation process. *Proceedings of the Royal Society of London*, Series B-Biological Sciences 265: 457–463.

Avise, J. C. and D. E. Walker. 1999. Species realities and numbers in sexual vertebrates: Perspectives from an asexually transmitted genome. *Proceedings of the National Academy of Sciences, USA* 96: 992–995.

Avise, J. C., D. Walker and G. C. Johns. 1998. Speciation durations and Pleistocene effects on vertebrate phylogeography. *Proceedings of the Royal Society of London*, Series B-Biological Sciences 265: 1707–1712.

Avise, J. C., J. Arnold, R. M. Ball, E. Bermingham, T. Lamb, J. E. Neigel, C. A. Reeb and N. C. Saunders. 1987. Intraspecific phylogeography—The mitochondrial-DNA bridge between population-genetics and systematics. *Annual Review of Ecology and Systematics* 18: 489–522.

Axelrod, D. I. 1967. Quaternary extinctions of large mammals. *University of California Publications in Geological Science* 74: 1–42.

Azzaroli, A. 1982. Insularity and its effects on terrestrial vertebrates: Evolutionary and biogeographic aspects. In E. M. Gallitelli (ed.), *Paleontology, Essentials of Historical Geology*, 193–213. Mucchi, Modena, Italy: S.T.E.M.

Backus, R. H. 1986. Biogeographic boundaries in the open ocean. In *Pelagic Biogeography*, 9–14. UNESCO Technical Papers in Marine Science, 49.

Bahre, C. J. 1995. Human impacts on the grasslands of Southwestern Arizona. In M. P. McClaran and T. R. Van Devender (eds.), *The Desert Grassland*, 230–264. Tucson, AZ: University of Arizona Press.

Bailey, E. 1963. *Charles Lyell.* New York: Doubleday and Company.

Bailey, R. G. 1996. *Ecosystem Geography.* New York: Springer.

Bak, P. 1996. How Nature Works: The Science of Self-Organized Criticality. New York: Copernicus.

Baker, A. J., C. H. Daugherty, R. Colbourne and J. L. McLennan. 1995. Flightless brown kiwis of New Zealand possess extremely subdivided population structure and cryptic species like small mammals. *Proceedings of the National Academy of Sciences, USA* 92: 8254.

Baker, H. A. and E. G. H. Oliver. 1967. *Ericas in Southern Africa.* Capetown/Johannesburg: Purnell & Sons.

Baker, H. G. 1955. Self-compatibility and establishment after 'long-distance' dispersal. *Evolution* 9: 347–349.

Baker, H. G. and G. L. Stebbins (eds.). 1965. *The Genetics of Colonizing Species.* New York: Academic Press.

Baker, M. 1995. Environmental component of latitudinal clutch-size variation in house sparrows (*Passer domesticus*). *The Auk* 112: 249–252.

Baker, R. H. 1968. Habitats and distribution. In J. A. King (ed.), *Biology of Peromyscus*, 98–126. American Society of Mammalogists Special Publication no. 2.

Baker, R. R. 1978. *The Evolutionary Ecology of Animal Migration.* London: Hodder & Stoughton.

Baldwin, B., P. J. Coney and W. R. Dickinson. 1974. Dilemma of a Cretaceous time scale and rates of sea-floor spreading. *Geology* 2: 267–270.

Baldwin, B. G. and R. H. Robichaux. 1995. Historical biogeography and ecology of the Hawaiian silversword alliance (*Asteraceae*): New molecular phylogenetic perspectives. In W. L. Wagner and V. A. Funk (eds.), *Hawaiian Biogeography: Evolution on a Hot Spot Archipelago.* Washington, DC: Smithsonian Institution Press.

Balter, M. 2001a. Anthropologists duel over modern human origins. *Science* 291: 1782–1783.

Balter, M. 2001b. In search of the first Europeans. *Science* 291: 1722–1725.

Balter, M. 2004. Earliest signs of human-controlled fire uncovered in Israel. *Science* 304: 663–663.

Banfield, A. W. F. 1954. The role of ice in the distribution of mammals. *Journal of Mammalogy* 35: 104–107.

Banfield, A. W. F. 1961. A revision of the reindeer and caribou, genus *Rangifer*. *Bulletin of the National Museum of Canada* 177: 1–137.

Banfield, A. W. F. 1974. *The Mammals of Canada*. Toronto: University of Toronto Press.

Banks, H. P. 1975. Early vascular land plants: Proof and conjecture. *Bioscience* 25: 730–737.

Baquero, R. A. and J. L. Telleria. 2001. Species richness, rarity and endemicity of European mammals: A biogeographical approach. *Biodiversity and Conservation* 10: 29–44.

Barber, P. H., S. R. Palumbi, M. V. Erdmann and M. K. Moosa. 2000. Biogeography—A marine Wallace's line? *Nature* 406: 692–693.

Barbosa, A. and J. Benzal. 1996. Diversity and abundance of small mammals in Iberia: Peninsular effect or habitat suitability? *Zeitschrift fur Saugetierkunde* 61: 236–241.

Barbour, C. D. and J. H. Brown. 1974. Fish species diversity in lakes. *American Naturalist* 108: 473–489.

Barker, F. K., G. F. Barrowclough and J. G. Groth. 2002. A phylogenetic hypothesis for passerine birds: Taxonomic and biogeographic implications of an analysis of nuclear DNA sequence data. *Proceedings of the Royal Society of London*, Series B—Biological Sciences 269: 295–308.

Barlow, N. D. 1994. Size distributions of butterfly species and the effect of latitude on species sizes. *Oikos* 71: 326–332.

Barnes, R. K. and K. H. Mann (eds.). 1980. *Fundamentals of Aquatic Ecosystems*. Oxford: Blackwell Scientific Publications.

Barnosky, A. D., P. L. Koch, R. S. Feranec, S. L. Wing and A. B. Shabel. 2004. Assessing the causes of Late Pleistocene extinctions on the continents. *Science* 306: 70–75.

Barraclough, T. G. and A. P. Vogler. 2000. Detecting the geographical pattern of speciation from species-level phylogenies. *American Naturalist* 155: 419–434.

Barrett, K., D. A. Wait and W. B. Anderson. 2003. Small island biogeography in the Gulf of California: Lizards, the subsidized island biogeography hypothesis and the small island effect. *Journal of Biogeography* 30: 1575–1581.

Barrett, S. C. H. 1998. The reproductive biology and genetics of island plants. In P. R. Grant (ed.), *Evolution on Islands*, 18–34. New York: Oxford University Press.

Barron, E. J., C. G. A. Harrison, J. L. Sloan and W. W. Hay. 1981. Paleogeography, 180 million years ago to the present. *Ecologae Geologicae Helveticae* 74: 443–470.

Barrowclough, G. F. 1992. Biodiversity and conservation biology. In N. Eldredge (ed.), *Systematics, Ecology and the Biodiversity Crisis*, 121–143. New York: Columbia University Press.

Barton, N. H. and B. Charlesworth. 1984. Genetic revolutions, founder effects and speciation. *Annual Review of Ecology and Systematics* 15: 133–164.

Baskin, Y. 1992. Africa's troubled waters: Fish introductions and a changing physical profile muddy Lake Victoria's future. *BioScience* 42: 476–481.

Bates, J. M., S. J. Hackett and J. Cracraft. 1998. Area-relationships in the Neotropical lowlands: An hypothesis based on raw distributions of passerine birds. *Journal of Biogeography* 25: 783–793.

Batt, C. M. and A. M. Pollard. 1996. Radiocarbon calibration and the peopling of North America. *American Chemical Society Symposium Series* 625: 415–433.

Baum, D. A. and M. J. Donoghue. 1995. Choosing among alternative phylogenetic species concepts. *Systematic Botany* 20: 560–573.

Baur, B. and J. Bengtsson. 1987. Colonizing ability in land snails on the Baltic uplift archipelagoes. *Journal of Biogeography* 14: 329–341.

Bazzaz, F. A. 1996. Plants in Changing Environments: Linking Physiological, Population and Community Ecology. Cambridge: Cambridge University Press.

Beadle, N. C. W. 1966. Soil phosphate and its role in molding segments of the Australian flora and vegetation, with special reference to xeromorphy and sclerophylly. *Ecology* 47: 991–1007.

Beadle, N. C. W. 1981. *The Vegetation of Australia*. Stuttgart: Gustav Fischer Verlag.

Beard, C. 2002. East of Eden at the Paleocene/Eocene boundary. *Science* 295: 2028–2029.

Beauchamp, G. 2004. Reduced flocking by birds on islands with relaxed predation. *Proceedings of the Royal Society of London*, Series B-Biological Sciences 271: 1039–1042.

Beckwith, S. L. 1954. Ecological succession on abandoned farm lands and its relationship to wildlife management. *Ecological Monographs* 24: 349–376.

Beerli, P. and J. Felsenstein. 1999. Maximum-likelihood estimation of migration rates and effective population numbers in two populations using a coalescent approach. *Genetics* 152: 763–773.

Beerli, P. and J. Felsenstein. 2001. Maximum likelihood estimation of a migration matrix and effective population sizes in n subpopulations by using a coalescent approach. *Proceedings of the National Academy of Sciences, USA* 98: 4563–4568.

Begon, M., J. L. Harper and C. R. Townsend. 1986. *Ecology: Individuals, Populations and Communities*. Oxford: Blackwell Scientific Publications.

Behle, W. H. 1978. Avian biogeography of the Great Basin and Intermontane Region. *Great Basin Naturalist Memoirs* 2: 55–80.

Behrensmeyer, A. K., J. D. Damuth, W. A. DiMichele, R. Potts, H. Sues and S. L. Wing. 1992. *Terrestrial Ecosystems Through Time: Evolutionary Paleoecology of Terrestrial Plants and Animals*. Chicago: University of Chicago Press.

Beier, P. and R. F. Noss. 1998. Do habitat corridors really provide connectivity? *Conservation Biology* 12: 1241–1252.

Beilmann, A. P. and L. G. Brenner. 1951. The recent intrusion of forests in the Ozarks. *Annals of the Missouri Botanical Garden* 38: 261–282.

Bell, M. and M. J. C. Walker. 1992. *Late Quaternary Environmental Change: Physical and Human Perspectives*. New York: John Wiley and Sons.

Benioff, H. 1954. Orogenesis and deep crustal structure: Additional evidence from seismology. *Bulletin of the Geological Society of America* 65: 385–400.

Bennett, A. F. 1999. *Linkages in the Landscape. The Role of Corridors and Connectivity in Wildlife Conservation*. Gland, Switzerland and Cambridge: IUCN.

Benson, L. and R. S. Thompson. 1987. The physical record of lakes in the Great Basin. In W. F. Ruddiman and H. E. Wright, Jr. (eds.), *Geology of North America Volume K-3: North America and Adjacent Oceans during the Last Deglaciation*, 241–260. Boulder, CO: Geological Society of America.

Benton, M. J. 1995. Diversification and extinction in the history of life. *Science* 268: 52–58.

Benton, M. J. and R. J. Twitchett. 2003. How to kill (almost) all life: The end-Permian extinction event. *Trends in Ecology and Evolution* 18: 358–365.

Bercovici, D. 2002. The generation of plate tectonics from mantle convection. *Earth and Planetary Science Letters* 205: 107–121.

Berger, A. J. 1981. *Hawaiian Birdlife*. 2nd edition. Honolulu: University of Hawaii Press.

Bergmann, C. 1847. Über die Verhältnisse der Wärmeökonomie der Thiere zu ihren Grösse. *Göttinger Studien* 1: 595–708.

Berlocher, S. H. 1998. Origins: A brief history of research on speciation. In D. J. Howard and S. H. Berlocher (eds.), *Endless Forms: Species and Speciation*. Oxford: Oxford University Press.

Bermingham, E. and J. C. Avise. 1986. Molecular zoogeography of freshwater fishes in the southeastern United States. *Genetics* 113: 939–965.

Bermingham, E. and A. P. Martin. 1998. Comparative mtDNA phylogeography of neotropical freshwater fishes: Testing shared history to infer the evolutionary landscape of lower Central America. *Molecular Ecology* 7: 499–517.

Bermingham, E. and C. Moritz. 1998. Comparative phylogeography: Concepts and applications. *Molecular Ecology* 7: 367–369.

Bernabo, J. C. and T. Webb. 1977. Changing patterns in the Holocene pollen record from northeastern North America: A mapped summary. *Quaternary Research* 8: 64–96.

Bernatchez, L. and C. C. Wilson. 1998. Comparative phylogeography of nearctic and palearctic fishes. *Molecular Ecology* 7: 431–452.

Berra, T. M. 2001. *Freshwater Fish Distribution.* Oxford: Elsevier Science and Technology.

Berry, R. J. 1964. The evolution of an island population of the house mouse. *Evolution* 18: 468–483.

Berry, W. B. N. 1973. Silurian-Early Devonian graptolites. In A. Hallam (ed.), *Atlas of Palaeobiogeography*, 81–87. Amsterdam: Elsevier.

Bertness, M. D. 1984. Habitat and community modification by an introduced herbivorous snail. *Ecology* 65: 370–381.

Bertness, M. D. 1985. Fiddler crab regulation of *Spartina alterniflora* production on a New England salt marsh. *Ecology* 66: 1042–1055.

Bertness, M. D. 1989. Competitive and facilitative interactions and acorn barnacle populations in a sheltered habitat. *Ecology* 70: 257–268.

Bertness, M. D. 1991. Interspecific interactions among high marsh perennials in a New England salt marsh. *Ecology* 72: 125–137.

Bertness, M. D. and S. W. Shumway. 1993. Competition and facilitation in marsh plants. *American Naturalist* 142: 718–724.

Betancourt, J. L. 2004. Advances in arid lands paleobiogeography: The rodent midden record in the Americas. In M. V. Lomolino and L. R. Heaney (eds.), *Frontiers of Biogeography—New Directions in the Geography of Nature.* London: Cambridge University Press.

Betancourt, J. L., T. R. Van Devender and P. S. Martin. 1990. *Packrat Middens: The Last 40,000 Years of Biotic Exchange.* Tucson, AZ: University of Arizona Press.

Beven, S., E. F. Conner and K. Beven. 1984. Avian biogeography in the Amazon basin and the biological model of diversification. *Journal of Biogeography* 11: 383–399.

Bhattarai, K. R. and O. R. Vetaas. 2003. Variation in plant species richness of different life forms along a subtropical elevation gradient in the Himalayas, east Nepal. *Global Ecology and Biogeography* 12: 327–340.

Bhattarai, K. R., O. R. Vetaas and J. A. Grytnes. 2004. Fern species richness along a central Himalayan elevational gradient, Nepal. *Journal of Biogeography* 31: 389–400.

Bibby, C. J., N. J. Collar, M. J. Crosby, M. F. Heath, Ch. Imboden, T. H. Johnson, A. J. Long, A. J. Stattersfield and S. J. Thirgood. 1992. *Putting Biodiversity on the Map: Priority Areas for Global Conservation.* Cambridge: International Council for Bird Preservation.

Bierhorst, D. W. 1971. *Morphology of Vascular Plants.* New York: Macmillan.

Bierregaard, R. O., Jr., T. E. Lovejoy, V. Kapos, A. A. dos Santos and R. W. Hutchings. 1992. The biological dynamics of tropical rainforest fragments: A prospective comparison of fragments and continuous forest. *Bioscience* 42: 859–866.

Biju-Duval, B. and L. Montadert (eds.). 1977. Structural history of the Mediterranean basins. International Symposium of the 25th Plenary Congress Assembly of the International Commission for the Scientific Exploration of the Mediterranean. Paris: Société des Éditions Technip.

Billerbeck, J. M., G. Ortí and D. O. Conover 1997. Latitudinal variation in vertebrate number has a genetic basis in the Atlantic silverside, *Menidia menidia. Canadian Journal of Fisheries and Aquatic Sciences* 54: 1796–1801.

Bilton, D. T., J. R. Freeland and B. Okamura. 2001. Dispersal in freshwater invertebrates. *Annual Review of Ecology and Systematics* 32: 159–181.

Bindon, J. R. and P. T. Baker. 1997. Bergmann's rule and the thrifty genotype. *American Journal of Physical Anthropology* 104: 201–210.

Biological Sciences Curriculum Study. 1963. *Biological Science: Molecules to Man.* Boston: Houghton Mifflin.

BirdLife International. 2000. *Threatened Birds of the World.* Barcelona and Cambridge: Lynx Editions and BirdLife International.

Birge, E. A. and C. Juday. 1911. The inland lakes of Wisconsin. *Bulletin of the Wisconsin Geological Natural History Survey* 22: 1–259.

Birks, H. J. B. 1987. Recent methodological developments in quantitative descriptive biogeography. *Annales Zoologici Fennici* 24: 165–178.

Birks, H. J. B. 1989. Holocene isochrone maps and patterns in tree-spreading in the British Isles. *Journal of Biogeography* 16: 503–540.

Bisby, F. A. 1995. Characterization of biodiversity. In V. H. Heywood (ed.), *Global Biodiversity Assessment*, 21–106. Cambridge: Cambridge University Press.

Bishop, I. R. and M. J. Delaney. 1963. The ecological distribution of small mammals in the Channel Islands. *Mammalia* 27: 99.

Bissonette, J. A. and I. Storch. 2002. Fragmentation: Is the message clear? *Conservation Ecology* 6(2): 14. Available from: http://www.consecol.org/vol6/iss2/art14 (accessed 2004).

Black, M. B., K. M. Halanych, P. A. Y. Maas, W. R. Hoeh, J. Hashimoto, D. Desbruyeres, R. A. Lutz and R. C. Vrijenhoek. 1997. Molecular systematics of vestimentiferan tubeworms from hydrothermal vents and cold-water seeps. *Marine Biology* 130: 141–149.

Blackburn, T. M. and R. P. Duncan. 2001. Establishment patterns of exotic birds are constrained by nonrandom patterns in introductions. *Journal of Biogeography* 28: 927–939.

Blackburn, T. M. and K. J. Gaston. 1994. Animal body size distributions change as more species are described. *Proceedings of the Royal Society of London*, Series B 257: 293–297.

Blackburn, T. M. and K. J. Gaston. 1995. What determines the probability of discovering a species? A study of South American oscine passerine birds. *Journal of Biogeography* 22: 7–14.

Blackburn, T. M. and K. J. Gaston. (eds.). 2003. *Macroecology: Concepts and Consequences.* Oxford: Blackwell Scientific Publications.

Blackburn, T. M. and B. Hawkins. 2004. Bergmann's rule and the mammal fauna of northern North America. *Ecography* 27: 715–724.

Blackburn, T. M. and A. Ruggiero. 2001. Latitude, elevation and body mass variation in Andean passerine birds. *Global Ecology and Biogeography* 10: 245–259.

Blackburn, T. M., K. J. Gaston and N. Loder. 1999. Geographic gradients in body size: A clarification of Bergmann's rule. *Diversity and Distributions* 5: 165–174.

Blackburn, T. M., K. J. Gaston, J. J. D. Greenwood and R. D. Gregory. 1998. The anatomy of the interspecific abundance-range size relationship for British avifauna. II. Temporal dynamics. *Ecology Letters* 1: 47–55.

Blackburn, T. M., K. J. Gaston, R. M. Quinn and R. D. Gregory. 1999. Do local abundance of British birds change with proximity to range edge? *Journal of Biogeography* 26: 493–505.

Blair, W. F. 1943. Activities of the Chihuahuan deer-mouse in relation to light intensity. *Journal of Wildlife Management* 7: 92–97.

Blanckenhorn, W. U. and M. Demont. 2004. Bergmann and converse Bergmann latitudinal clines in arthropods: Two ends of a continuum? *Integrative Comparative Biology* 44: 413–424.

Blondel, J. 2000. Evolution and ecology of birds on islands: Trends and prospects. *Life and Environment* 50: 205–220.

Boback, S. M. and C. Guyer. 2003. Empirical evidence for an optimal body size in snakes. *Evolution* 57: 345–351.

Bock, C. E. and L. W. Lepthian. 1976. Synchronous eruptions of boreal seed-eating birds. *American Naturalist* 110: 559–571.

Bohle, U. R., H. H. Hilger and W. F. Martin. 1996. Island colonization and evolution of the insular woody habit in Echium L. (Boraginaceae). *Proceedings of the National Academy of Sciences, USA* 93: 11740–11745.

Bohning-Gaese, K., L. I. Gonzalez-Guzman and J. H. Brown. 1998. Constraints on dispersal and the evolution of the avifauna of the Northern Hemisphere. *Evolutionary Ecology* 12: 767–783.

Bonfirm, F. S., J. A. Diniz-Filho and R. P. Bastos. 1998. Spatial patterns and the macroecology of South American viperid snakes. *Revista Brasileira de Biologia* 58: 97–103.

Borodin, A. M., A. G. Bannikov and V. E. Sokolov (eds.). 1984. *Red Data Book of the USSR: Rare and Endangered Species of Animals and Plants.* 2nd edition. Moscow: Forest Industry Publishers.

Bossuyt, F. and M. C. Milinkovitch. 2001. Amphibians as indicators of early Tertiary "out-of-India" dispersal of vertebrates. *Science* 292: 93–95.

Botkin, D. P., D. A. Woodby and R. A. Nisbet. 1991. Kirtland's warbler habitat: A possible early indicator of climate warming. *Biological Conservation* 56: 63–78.

Bouchard, P. and D. R. Brooks. 2004. Effect of vagility potential on dispersal and speciation in rainforest insects. *Journal of Evolutionary Biology* 17: 994–1006.

Boucher, D. H. 1985. The Biology of Mutualism: Ecology and Evolution. London: Croom Helm.

Boucher, D. H., S. James and K. Kesler. 1984. The ecology of mutualism. *Annual Review of Ecology and Systematics* 13: 315–347.

Bouchet, P., P. Lozouet, P. Maestrati and V. Heros. 2002. Assessing the magnitude of species richness in tropical marine environments: Exceptionally high numbers of mollusks at a New Caledonia site. *Biological Journal of the Linnean Society* 75: 421–436.

Boucot, A. J. and J. G. Johnson. 1973. Silurian brachiopods. In A. Hallam (ed.), *Atlas of Paleobiogeography,* 59–65. Amsterdam: Elsevier.

Bourliere, F. 1973. The comparative ecology of rain forest mammals in Africa and tropical America: Some introductory remarks. In B. J. Meggers, E. S. Ayensu and W. D. Ducksworth (eds.), *Tropical Forest Ecosystems in Africa and South America: A Comparative Review,* 279–292. Washington, DC: Smithsonian Institution Press.

Bowen G. J., W. C. Clyde, P. L. Koch, S. Ting, J. Alroy, T. Tsubamoto, Y. Wang and Y. Wang. 2002. Mammalian dispersal at the Paleocene/Eocene boundary. *Science* 295: 2062–2065.

Bowers, M. 1982. Insular biogeography of mammals in the Great Salt Lake. *Great Basin Naturalist* 42: 589–596.

Bowers, M. A. and J. H. Brown. 1982. Body size and coexistence in desert rodents: Chance or community structure. *Ecology* 63: 391–400.

Bowers, M. A. and C. H. Lowe. 1986. Plant-form gradients on Sonoran Desert bajadas. *Oikos* 46: 284–291.

Boyd, E. M. and S. A. Nunneley. 1964. Banding records substantiating the changed status of 10 species of birds since 1900 in the Connecticut Valley. *Bird-Banding* 35: 1–8.

Boyett, W. D., M. E. Endries and G. H. Adler. 2000. Colonization and extinction dynamics of opossums on small islands in Panama. *Canadian Journal of Zoology* 78: 1972–1979.

Braba, F., L. Frechilla and G. Orizaola. 1996. Effect of introduced fish on amphibian assemblages in mountain lakes of Northern Spain. *The Herpetological Journal* 6: 145.

Bradai, M. N., J. P. Quignard, A. Bouain, O. Jarboui, A. Ouonnes-Ghorbel, L. Ben Abdallah, J. Zaouali and S. Ben Salem. 2004. Autochthonous and exotic fish species of the Tunisian coasts: Inventory and biogeography. *Cybium* 28: 315–328.

Braun, E. L. 1950. *Deciduous Forests of Eastern North America.* New York: Hafner Press.

Brehm, G. and K. Fiedler. 2003. Bergmann's rule does not apply to geometrid moths along an elevational gradient in Andean montane rain forest. *Global Ecology and Biogeography* 12: 7–14.

Bremer, K. 1992. Ancestral areas—A cladistic reinterpretation of the center of origin concept. *Systematic Biology* 41: 436–445.

Bremer, K. 1993. Intercontinental relationships of African and South American Asteraceae: A cladistic biogeographic analysis. In P. Goldblatt (ed.), *Biological Relationships Between Africa and South America,* 105–135. New Haven, CT: Yale University Press.

Bremer, K. 1995. Ancestral areas: Optimization and probability. *Systematic Biology* 44: 255–259.

Bremer, K. 2002. Gondwanan evolution of the grass alliance of families (Poales). *Evolution* 56: 1374–1387.

Brenner, W. 1921. Vaxtgeografiska studien: Barosunds skargard. I. Allman del och floran. *Acta Societatis pro Fauna et Flora Fennica* 49: 1–151.

Briers, R. A. 2002. Incorporating connectivity into reserve selection procedures. *Biological Conservation* 103: 77–83.

Briggs, D. and S. M. Walters. 1984. *Plant Variation and Evolution.* Cambridge: Cambridge University Press.

Briggs, J. C. 1968. Panama sea-level canal. *Science* 162: 511–513.

Briggs, J. C. 1970. *Tropical Shelf Zoogeography.* Publications of the California Academy of Sciences: San Francisco.

Briggs, J. C. 1974. *Marine Zoogeography.* New York: McGraw-Hill.

Briggs, J. C. 1987. *Biogeography and Plate Tectonics.* Amsterdam: Elsevier.

Briggs, J. C. 1994. The genesis of Central America: Biology vs. geophysics. *Global Ecology and Biogeography Letters* 4: 169–172.

Briggs, J. C. 1995. *Global Biogeography.* New York: Elsevier.

Briggs, J. C. 2003. Marine centres of origin as evolutionary engines. *Journal of Biogeography* 30: 1–18.

Briggs, J. C. 2003a. The biogeographic and tectonic history of India. *Journal of Biogeography* 30: 381–388.

Briggs, J. C. 2003b. Fishes and birds: Gondwana life rafts reconsidered. *Systematic Biology* 52: 548–553.

Briggs, J. C. 2004. A marine center of origin: Reality and conservation. In M. V. Lomolino and L. R. Heaney (eds.), *Frontiers of Biogeography: New Directions in the Geography of Nature,* Chapter 13. Cambridge University Press.

Briggs, J. C. 2004. Older species: A rejuvenation on coral reefs? *Journal of Biogeography* 31: 525–530.

Briggs, J. C. 2004. The ultimate expanding earth hypothesis. *Journal of Biogeography* 31: 855–857.

Brittingham, M. C. and S. A. Temple. 1983. Have cowbirds caused forest songbirds to decline? *BioScience* 33: 31–35.

Bromham, L. and M. Cardillo. 2003. Testing the link between the latitudinal gradient in species richness and rates of molecular evolution. *Journal of Evolutionary Biology* 16: 200–207.

Brook, B. W. and D. M. J. S. Bowman. 2002. Explaining the Pleistocene megafaunal extinctions: Models, chronologies and assumptions. *Proceedings of the National Academy of Sciences, USA* 99: 14624–14627.

Brooks, D. R. 1985. Historical ecology: A new approach to studying the evolution of ecological associations. *Annals of the Missouri Botanical Garden* 72: 660–680.

Brooks, D. R. 1988. Macroevolutionary comparisons of host and parasite phylogenies. *Annual Review of Ecology and Systematics* 19: 235–259.

Brooks, D. R. 2004. Reticulations in historical biogeography: The triumph of time over space in evolution. In M. V. Lomolino and L. R. Heaney (eds.), *Frontiers of Biogeography,* 123–144. Sunderland, MA: Sinauer Associates.

Brooks, D. R. and S. M. Bandoni. 1988. Coevolution and relics. *Systematic Zoology* 37: 19–33.

Brooks, D. R. and D. A. McLennan. 1991. *Phylogeny, Ecology and Behavior.* Chicago: University of Chicago Press.

Brooks, D. R. and D. A. McLennan. 2002. *The Nature of Diversity: An Evolutionary Voyage of Discovery.* Chicago: University of Chicago Press.

Brooks, T. M., R. A. Mittermeier, C. G. Mittermeier, G. A. B. da Fonseca, A. B. Rylands, W. R. Konstant, P. Flick, J. Pilgrim, S. Oldfield, G. Magin and C. Hilton-Taylor. 2002. Habitat loss and extinction in the hotspots of biodiversity. *Conservation Biology* 16: 909–923.

Brower, L. P. 1977. Monarch migration. *Natural History* 86(6): 40–53.

Brower, L. P. and S. B. Malcolm. 1991. Animal migrations: Endangered phenomena. *American Zoologist* 31: 265–276.

Brown, D. E. and R. Davis. 1995. One hundred years of vicissitude: Terrestrial bird and mammal distribution changes in the American Southwest. In *Biodiversity and Management of the Madrean Archipelago: The Sky Islands of Southwestern United States and Northwestern Mexico,* 231–244. USDA Forest Service, General Technical Report RM-GTR-264.

Brown, D. E., F. Reichenbacher and S. E. Franson. 1998. *A Classification of North American Biotic Communities.* Salt Lake City: University of Utah Press.

Brown, J. H. 1968. Adaptation to environmental temperature in two species of woodrats, *Neotoma cinerea* and *N. albingula. University of Michigan Museum of Zoology Miscellaneous Publications* 135: 1–48.

Brown, J. H. 1971a. Mechanisms of competitive exclusion between two species of chipmunks (*Eutamias*). *Ecology* 52: 306–311.

Brown, J. H. 1971b. Mammals on mountaintops: Nonequilibrium insular biogeography. *American Naturalist* 105: 467–478.

Brown, J. H. 1971c. The desert pupfish. *Scientific American* 225(5): 104–110.

Brown, J. H. 1975. Geographical ecology of desert rodents. In M. L. Cody and J. M. Diamond (eds.), *Ecology and Evolution of Communities*, 315–341. Cambridge, MA: Belknap Press.

Brown, J. H. 1978. The theory of insular biogeography and the distribution of boreal birds and mammals. *Great Basin Naturalist Memoirs* 2: 209–227.

Brown, J. H. 1981. Two decades of homage to Santa Rosalia: Toward a general theory of diversity. *American Zoologist* 21: 877–888.

Brown, J. H. 1984. On the relationship between abundance and distribution of species. *American Naturalist* 124: 255–279.

Brown, J. H. 1986. Two decades of interaction between the MacArthur-Wilson model and the complexities of mammalian distributions. *Biological Journal of the Linnean Society* 28: 231–251.

Brown, J. H. 1988. Species diversity. In A. Myers and R. S. Giller (eds.), *Analytical Biogeography*, 57–89. London: Chapman and Hall.

Brown, J. H. 1989. Patterns, modes and extents of invasions in vertebrates. In J. A. Drake et al. (eds.), *Biological Invasions: A Global Perspective*, 85–110. New York: John Wiley and Sons.

Brown, J. H. 1995. *Macroecology*. Chicago: University of Chicago Press.

Brown, J. H. 1999. The legacy of Robert MacArthur: From geographical ecology to macroecology. *Journal of Mammalogy* 80: 333–344.

Brown, J. H. 2001. Mammals of mountainsides: Elevational patterns of diversity. *Global Ecology and Biogeography* 10: 101–109.

Brown, J. H. and M. A. Bowers. 1984. Patterns and processes in three guilds of terrestrial vertebrates. In D. R. Strong, D. Simberloff, L. G. Abele and A. B. Thistle (eds.), *Ecological Communities: Concepts, Issues and the Evidence*, 282–296. Princeton, NJ: Princeton University Press.

Brown, J. H. and M. A. Bowers. 1985. On the relationship between morphology and ecology: Community organization in hummingbirds. *Auk* 102: 251–269.

Brown, J. H. and C. R. Feldmeth. 1971. Evolution in constant and fluctuating environments: Thermal tolerances of desert pupfish (*Cyprinodon*). *Evolution* 25: 390–398.

Brown, J. H. and A. C. Gibson. 1983. *Biogeography*. St. Louis, MO: Mosby.

Brown, J. H. and A. Kodric-Brown. 1977. Turnover rates in insular biogeography: Effect of immigration on extinction. *Ecology* 58: 445–449.

Brown, J. H. and A. Kodric-Brown. 1993. Highly structured fish communities in Australian desert springs. *Ecology* 74: 1847–1855.

Brown, J. H. and A. K. Lee. 1969. Bergmann's rule and climatic adaptation in woodrats (*Neotoma*). *Evolution* 23: 329–338.

Brown, J. H. and M. V. Lomolino. 1989. On the nature of scientific revolutions: Independent discovery of the equilibrium theory of island biogeography. *Ecology* 70: 1954–1957.

Brown, J. H. and M. V. Lomolino. 2000. Concluding remarks: Historical perspective and the future of island biogeography theory. *Global Ecology and Biogeography* 9: 87–92.

Brown, J. H. and B. A. Maurer. 1986. Body size, ecological dominance and Cope's rule. *Nature* 324: 248–250.

Brown, J. H. and B. A. Maurer. 1987. Evolution of species assemblages: Effects of energetic constraints and species dynamics on the diversification of the North American avifauna. *American Naturalist* 130: 1–17.

Brown, J. H. and B. A. Maurer. 1989. Macroecology: The division of food and space among species on continents. *Science* 243: 1145–1150.

Brown, J. H. and J. C. Munger. 1985. Experimental manipulation of a desert rodent community: Food addition and species removal. *Ecology* 66: 1545–1563.

Brown, J. H. and P. F. Nicoletto. 1991. Spatial scaling of species assemblages: Body masses of North American land mammals. *American Naturalist* 138: 1478–1512.

Brown, J. H. and D. F. Sax. 2004. Gradients in species diversity: Why are there so many species in the tropics? In M. V. Lomolino, D. F. Sax and J. H. Brown (eds.), *Foundations of Biogeography*, 1145–1154. Chicago: University of Chicago Press.

Brown, J. H., D. W. Mehlman and G. C. Stevens. 1995. Spatial variation in abundance. *Ecology* 76: 2028–2043.

Brown, J. H., G. C. Stevens and D. M. Kaufman. 1996. The geographic range: Size, shape, boundaries and internal structure. *Annual Review of Ecology and Systematics* 27: 597–623.

Brown, J. H., T. J. Valone and C. G. Curtin. 1997. Reorganization of an arid ecosystem in response to recent climate change. *Proceedings of the National Academy of Sciences, USA* 94: 9729–9733.

Brown, J. M. and D. S. Wilson. 1992. Local specialization of phoretic mites on sympatric carrion beetle hosts. *Ecology* 73: 463–478.

Brown, J. W. 1987. The peninsular effect in Baja California: An entomological assessment. *Journal of Biogeography* 14: 359–365.

Brown, J. W. and P. A. Opler. 1990. Patterns of butterfly species density in peninsular Florida. *Journal of Biogeography* 17: 615–622.

Brown, K. S. Jr. 1982. Paleoecology and regional patterns of evolution in neotropical forest butterflies. In G. T. Prance (ed.), *Biological Differentiation in the Tropics*, 255–308. New York: Columbia University Press.

Brown, M. and J. J. Dinsmore. 1988. Habitat islands and the equilibrium theory of island biogeography: Testing some predictions. *Oecologia* 75: 426–429.

Brown, W. H., E. D. Merrill and H. S. Yates. 1919. The revegetation of Volcano Island, Luzon, Philippine Islands, since the eruption of Taal Volcano in 1911. *Philippine Journal of Science*, Sect. C, vol. 12.

Brown, W. L. and E. O. Wilson. 1956. Character displacement. *Systematic Zoology* 5: 49–64.

Brown, W. M. 1983. Evolution of animal mitochondrial DNA. In M. Nei and R. K. Koehn (eds.), *Evolution of Genes and Proteins*, 62–88. Sunderland, MA: Sinauer Associates.

Brown, W. M., M. George, Jr. and A. C. Wilson. 1979. Rapid evolution of animal mitochondrial DNA. *Proceedings of the National Academy of Sciences, USA* 76: 1967–1971.

Browne, R. A. 1981. Lakes as islands: Biogeographic distribution, turnover rates and species composition in the lakes of central New York. *Journal of Biogeography* 8: 75–83.

Browne, R. A. and G. H. MacDonald. 1982. Biogeography of the brine shrimp, *Artemia*: Distribution of parthenogenetic and sexual populations. *Journal of Biogeography* 9: 331–338.

Bruhnes, B. 1906. Recherches sur la direction d'aimentation des roches volcaniques (1). *Journal Physique*, 4e Sér., 5: 705–724.

Brundin, L. 1965. On the real nature of transantarctic relationships. *Evolution* 19: 496–505.

Brundin, L. 1966. Transantarctic relationships and their significance, as evidence by chironomid midges. *Kungliga Svenska Vetenskapsakademiens Handlingar*, 4th ser., II, no. 1: 1–472.

Brundin, L. 1967. Insects and the problem of austral disjunctive distribution. *Annual Review of Entomology* 12: 149–168.

Brundin, L. 1972. Phylogenetics and biogeography. *Systematic Zoology* 21: 69–79.

Brundin, L. 1988. Phylogenetic biogeography. In A. A. Myers and P. S. Gillers (eds.), *Analytic Biogeography: An Integrated Approach to the Study of Animal and Plant Distributions* 343–369. London: Chapman and Hall.

Bryant, P. J. 2004. *Biodiversity and Conservation*. Available from: http://darwin.bio.udi.edu/~sustain/bio65/titlepage.htm/ (accessed 2004).

Buckley, H. L., T. E. Miller, A. M. Ellison and N. J. Gotelli. 2003. Reverse latitudinal trends in species richness of pitcher-plant food webs. *Ecology Letters* 6: 825–829.

Buckley, R. C. and S. B. Knedlhans. 1986. Beachcomber biogeography: Interception of dispersing propagules by islands. *Journal of Biogeography* 13: 69–70.

Buffet, B. A. 2000. Earth's core and the geodynamo. *Science* 288: 2007–2012.

Buffon, G. L. L., Comte de. 1761. *Histoire Naturelle, Generale et Particuliere*, vol. 9. Paris: Imprimerie Royale.

Bullard, E. C., J. E. Everett and A. G. Smith. 1965. Fit of the continents around the Atlantic. In P. M. S. Blackett, E. C. Bullard and S. K. Runcorn (eds.), *A Symposium on Continental Drift*, 41–75. *Philosophical Transactions of the Royal Society of London*, Series A 248.

Bullock, J. M. and R. Kenward. 2001. *Dispersal Ecology*. Malden, MA. Blackwell Publishing.

Bunt, J. S. 1975. Primary productivity of marine ecosystems. In H. Lieth and R. H. Whitaker (eds.), *Primary Productivity of the Biosphere*, 169–202. Washington, DC: National Academy of Sciences.

Bureau de Recherches Géologiques et Minières. 1980a. Colloque C5. Géologie des chaines alpines issues de la Téthys. *Mémoirs*, no. 115.

Bureau de Recherches Géologiques et Minières. 1980b. Colloque C6. Géologie de l'Europe du Précambrien aux bassins sedimentaires post-hercyniens. *Mémoirs*, no. 108.

Burke, K., J. F. Dewey and W. S. F. Kidd. 1977. World distribution of sutures: The sites of former oceans. *Tectonophysics* 40: 69–99.

Burney, D. A., H. F. James, F. V. Grady, J.-G. Rafamantanantsoa, Ramilisonina, H. T Wright and J. B. Cowart. 1997. Environmental change, extinction and human activity: Evidence from caves in NW Madagascar. *Journal of Biogeography* 24: 755–768.

Burney, D. A., H. F. James, L. P. Burney, S. L. Olson, W. Kikuchi, W. L. Wagner, M. Burney, D. McCloskey, D. Kikuchi, F. V. Grady, R. Gage II and R. Nishek. 2001. Fossil evidence for a diverse biota from Kaua'I and its transformation since human arrival. *Ecological Monographs* 71: 615–641.

Burns, K. J., S. J. Hackett and N. K. Klein. 2002. Phylogenetic relationships and morphological diversity in Darwin's finches and their relatives. *Evolution* 56: 1240–1252.

Burr, B. M. and R. L. Mayden. 1992. Phylogenetics and North American freshwater fishes. In R. L. Mayden (ed.), *Systematics, Historical Ecology and North American Freshwater Fishes*. Stanford, CA: Stanford University Press.

Busack, S. D. and S. B. Hedges. 1984. Is the peninsular effect a red herring? *American Naturalist* 123: 266–275.

Busby, J. R. 1991. BIOCLIM—A bioclimatic analysis and predictive system. In C. R. Margules and M. P. Austin (eds.), *Nature Conservation: Cost Effective Biological Surveys and Data Analysis*, 64–68. Canberra: CSIRO.

Bush, A. O., J. C. Fernandez, G. W. Esch and J. R. Seed. 2001. *Parasitism—The diversity and ecology of animal parasites*. London: Cambridge University Press.

Bush, G. L. 1975. Modes of animal speciation. *Annual Review of Ecology and Systematics* 6: 339–364.

Bush, M. B. 1994. Amazonian speciation: A necessarily complex model. *Journal of Biogeography* 21: 5–17.

Bush, M. B. and R. J. Whittaker. 1991. Krakatau: Colonization patterns and hierarchies. *Journal of Biogeography* 18: 341–356.

Bush, M. B. and R. J. Whittaker. 1993. Non-equilibrium in island theory and Krakatau. *Journal of Biogeography* 20: 453–458.

Bussing, W. A. 1985. Patterns of distribution of the Central American Icthyofauna. In G. G. Stehli and S. D. Webb (eds.), *The Great American Interchange*, 453–473. New York: Plenum Press.

Butler, R. 1995. When did India hit Asia? *Nature* 373: 20–21.

Butlin, R. 1998. What do hybrid zones in general and the *Chorthippus parallelus* zone in particular, tell us about speciation? In D. J. Howard and S. H. Berlocher (eds.), *Endless Forms: Species and Speciation* 367–378. Oxford: Oxford University Press.

Buzas, M. A. 1972. Patterns of species diversity and their explanation. *Taxon* 21: 275–286.

Buzas, M. A. and S. J. Culver. 1991. Species diversity and dispersal of benthic foraminifera. *BioScience* 41(7): 483–489.

Cabrera, A. L. and A. Willink. 1973. *Biogeografia de America Latina*. Washington, DC: Programa Regional de Desarrollo Cientifico y Tecnologico, Departamento Asuntos Cientificos, Secretario General de la Organizacion de los Estados Americanos.

Caccone, A., J. P. Gibbs, V. Ketmaier, E. Suatoni and J. R. Powell. 1999. Origin and evolutionary relationships of giant Galapagos tortoises. *Proceedings of the National Academy of Sciences, USA* 96: 13223–13228.

Caccone, A., G. Gentile, J. P. Gibbs, T. H. Fritts, H. L. Snell, J. Betts and J. R. Powell. 2002. Phylogeography and history of giant Galapagos tortoises. *Evolution* 56: 2052–2066.

Cade, B. S., J. W. Terrell and R. L. Schroeder. 1999. Estimating effects of limiting factors with regression quantiles. *Ecology* 80: 311–323.

Cain, S. A. 1944. *Foundations of Plant Geography*. New York: Harper and Brothers.

Calder, W. A. 1974. Consequences of body size for avian energetics. In R. A. Paynter, Jr. (ed.), *Avian Energetics*, 86–151. Cambridge, MA: Nuttal Orinithology Club, Publ. 15.

Calder, W. A. III. 1984. *Size, Function and Life History*. Cambridge: Harvard University Press.

Caley, M. J. and D. Schluter. 1997. The relationship between local and regional diversity. *Ecology* 78: 70–80.

Calvert, A. J., E. W. Sawyer, W. J. Davis and J. N. Ludden. 1995. Archaean subduction inferred from seismic images of a mantle suture in the Superior Province. *Nature* 375: 670–674.

Candolle, A. P. de. 1820. *Essai Elementaire de Geographie Botanique*. De l'imprimerie de F. G. Levrault.

Candolle, A. P. de. 1855. *Géographie Botanique Raisonnée*. 2 vols. Paris: Masson.

Carey, S. W. 1955. The orocline concept of geotectonics, part I. *Proceedings of the Royal Society of Tasmania*, Papers 89: 255–288.

Carey, S. W. 1958. A tectonic approach to continental drift. In *Continental Drift: A Symposium*, 177–355. Hobart: University of Tasmania.

Carey, S. W. 1996. "Earth, Universe, Cosmos." Produced and distributed by Geology Department, University of Tasmania. GPO Box 252-79 Hobart, Tasmania, Australia.

Carlquist, S. 1965. *Island Life*. Garden City, NY: Natural History Press.

Carlquist, S. 1974. *Island Biology*. New York: Columbia University Press.

Carlquist, S. 1981. Chance dispersal. *American Scientist* 69: 509–515.

Carothers, S. W. and R. R. Johnson. 1974. Population structure and social organization of Southwestern riparian birds. *American Zoologist* 14: 97–108.

Carpenter, F. M. 1977. Geological history and evolution of the insects. In D. White (ed.), *Proceedings of the 15th International Congress of Entomology, Washington*, 63–70. College Park, MD: Entomological Society of America.

Carpenter, G., A. N. Gillison and J. Winter. 1993. DOMAIN: A flexible modeling procedure for mapping potential distributions of animals and plants. *Biodiversity and Conservation* 2: 667–680.

Carpenter, S. R. 1988. *Complex Interactions in Lake Communities*. New York: Springer-Verlag.

Carpenter, S. R. 2002. Ecological futures: Building an ecology of the long now. *Ecology* 83: 2069–2083.

Carpenter, S. R., J. F. Kitchell and J. R. Hodgson. 1985. Cascading tropic interactions and lake ecosystem productivity. *BioScience* 35: 635–639.

Carpenter, S. R., J. F. Kitchell, J. R. Hodgson, P. A. Cochran, J. J. Elser, M. M. Elser, D. M. Lodge, D. Kretchmer, X. He and C. von Ende. 1987. Regulation of lake primary productivity by food web structure. *Ecology* 68: 1867–1876.

Carroll, R. L. 1988. *Vertebrate Paleontology and Evolution*. New York: W.H. Freeman and Company.

Carson, H. L. 1971. Speciation and the founder principle. *University of Missouri, Stadler Symposium* 3: 51–70.

Carson, H. L. 1981. Microevolution in insular ecosystems. In D. Mueller-Dombois, K. W. Bridges and H. L. Carson (eds.), *Island Ecosystems: Biological Organization in Selected Hawaiian Communities*, 471–482. Stroudsburg, PA: Hutchinson Ross.

Carson, H. L. and K. Y. Kaneshiro. 1976. *Drosophila* of Hawaii: Systematics and ecological genetics. *Annual Review of Ecology and Systematics* 7: 311–346.

Carson, H. L. and A. R. Templeton. 1984. Genetic revolutions in relation to speciation phenomena: The founding of new populations. *Annual Review of Ecology and Systematics* 15: 97–131.

Carson, H. L., D. E. Hardy, H. T. Spieth and W. S. Stone. 1970. The evolutionary biology of the Hawaiian Drosophilidae. In M. K. Hecht and W. C. Steere (eds.), *Essays in Evolution and Genetics in Honor of Theodosius Dobzhansky*, 437–543. New York: Appleton-Century-Crofts.

Carstens, B. C., J. D. Degenhardt, A. L. Stevenson and J. Sullivan. 2005. Accounting for coalescent stochasticity in testing phylogeographical hypotheses: Modeling Pleistocene population structure in the Idaho giant salamander *Dicamptodon aterrimus*. *Molecular Ecology* 14: 255–265.

Case, T. J. 1975. Species numbers, density compensation and the colonizing ability of lizards on islands in the Gulf of California. *Ecology* 56: 3–18.

Case, T. J. 1978. A general explanation for insular body size trends in terrestrial vertebrates. *Ecology* 59: 1–18.

Case, T. J. 1987. Testing theories of island biogeography. *American Scientist* 75: 402–411.

Case, T. J. 1996. Global patterns in the establishment and distribution of exotic birds. *Biological Conservation* 76: 69–96.

Case, T. J. and D. T. Bolger. 1991. The role of interspecific competition in the biogeography of island lizards. *Trends in Ecology and Evolution* 6(4): 135–139.

Case, T. J. and T. D. Schwaner. 1993. Island/mainland body size differences in Australian varanid lizards. *Oecologia* 94: 102–109.

Case, T. J. and R. Sidell. 1983. Pattern and chance in the structure of model and natural communities. *Evolution* 37: 832–849.

Case, T. J., D. T. Bolger and K. Petren. 1994. Invasion and competitive displacement among house geckos in the tropical Pacific. *Ecology* 75: 464–477.

Case, T. J., J. Faaborg and R. Sidell. 1983. The role of body size in the assembly of West Indian bird communities. *Evolution* 37: 1062–1074.

Cassey, P. and T. M. Blackburn. 2004. Body size trends in a Holocene island bird assemblage. *Ecography* 27: 59–67.

Caswell, H. 1978. Predator-mediated coexistence: A nonequilibrium model. *American Naturalist* 112: 127–154.

Caughley, G. 1987. The distribution of eutherian body weights. *Oecologia* 74: 319–320.

Censky, E. J., K. Hodge and J. Dudley. 1998. Over-water dispersal of lizards due to hurricanes. *Nature* 395: 556–557.

Channell, R. 1998. *A Geography of Extinction: Patterns in the Contraction of Geographic Ranges*. Ph.D. dissertation, University of Oklahoma, Norman, OK.

Channell, R. and M. V. Lomolino. 2000a. Dynamic biogeography and conservation of endangered species. *Nature* 403: 84–86.

Channell, R. and M. V. Lomolino. 2000b. Trajectories toward extinction: Dynamics of geographic range collapse. *Journal of Biogeography* 27: 169–179.

Chappell, M. A. 1978. Behavioral factors in the altitudinal zonation of chipmunks (*Eutamias*). *Ecology* 59: 565–579.

Charnov, E. L. and J. R. Krebs. 1974. On clutch size and fitness. *Ibis* 116: 217–219.

Cheatham, A. H. and J. E. Hazel. 1969. Binary (presence-absence) similarity coefficients. *Journal of Paleontology* 43: 1130–1136.

Cherfas, J. 1991. Disappearing mushrooms: Another mass extinction? *Science* 254: 1458.

Chester, R. H. 1969. Destruction of Pacific corals by the sea star *Acanthaster planci*. *Science* 165: 280–283.

Choi, S. 2004. Trends in butterfly species richness in response to the peninsular effect in South Korea. *Journal of Biogeography* 31: 587–592.

Choquenot, D. and D. M. J. S. Bowman. 1999. Marsupial megafauna, Aborigines and the overkill hypothesis: Application of predator-prey models to the question of Pleistocene extinction in Australia. *Global Ecology and Biogeography* 7: 167–180.

Chown, S. L. and K. J. Gaston. 2000. Areas, cradles and museums: The latitudinal gradient in species richness. *Trends in Ecology and Evolution* 15: 311–315.

Clagg, H. B. 1966. Trapping of air-borne insects in the Atlantic-Antarctic area. *Pacific Insects* 8: 455–466.

Clark, G. 1992. *Space, Time and Man: A Prehistorian's View*. New York: Cambridge University Press.

Clarke, A. 1992. Is there a latitudinal cline in the sea? *Trends in Ecology and Evolution* 7: 286–287.

Clausen, J., D. D. Keck and W. M. Hiesey. 1940. *Experimental Studies on the Nature of Species*. I. *Effect of Varied Environments on Western North American Plants*. Carnegie Institute Publication 520. Washington, DC: Carnegie Institute.

Clausen, J., D. D. Keck and W. M. Hiesey. 1947. Heredity of geographically and ecologically isolated races. *American Naturalist* 81: 114–133.

Clausen, J., D. D. Keck and W. M. Hiesey. 1948. *Experimental Studies on the Nature of Species*. III: *Environmental Responses of Climatic Races of* Achillea. Carnegie Institute Publication 581: 1–129. Washington, DC: Carnegie Institute.

Clegg, J. A., M. Almond and P. H. S. Stubbs. 1954. The remnent magnetism of some sedimentary rocks in Britain. *Philosophical Magazine* 45: 583–598.

Clegg, S. M. and I. P. F. Owens. 2002. The "island rule" in birds: Medium body size and its ecological explanation. *Proceedings of the Royal Society of London,* Series B 269: 1359–1365.

Clement, M., D. Posada and K. A. Crandall. 2000. TCS: A computer program to estimate gene genealogies. *Molecular Ecology* 9: 1657–1660.

Clements, F. E. 1916. *Plant Succession: An Analysis of the Development of Vegetation*. Carnegie Institute Publication no. 242. Washington, DC: Carnegie Institute.

Clements, F. E. and V. E. Shelford. 1939. *Bio-Ecology*. New York: John Wiley & Sons.

CLIMAP project members. 1976. The surface of the ice-age earth. *Science* 191: 1131–1137.

Clobert, J. 2001. *Dispersal*. New York: Oxford University Press.

Cockburn, A. 1991. *An Introduction to Evolutionary Ecology*. London: Blackwell Scientific Publications.

Cockburn, A., A. K. Lee and R. W. Martin. 1983. Macrogeographic variation in litter size in *Antechinus* spp. (Marsupialia: Dasyuridae). *Evolution* 37: 86–95.

Cody, M. L. 1966. The consistency of intra- and inter-continental grassland bird species counts. *American Naturalist* 100: 371–376.

Cody, M. L. 1968. On the methods of resource division in grassland bird communities. *American Naturalist* 102: 107–137.

Cody, M. L. 1973. Parallel evolution and bird niches. In F. di Castri and H. A. Mooney (eds.), *Mediterranean-Type Ecosystems: Origin and Structure*, 307–338. Ecological Studies 7. New York: Springer-Verlag.

Cody, M. L. 1974. *Competition and the Structure of Bird Communities*. Monographs in Population Biology, no. 7. Princeton, NJ: Princeton University Press.

Cody, M. L. 1975. Towards a theory of continental species diversity. In M. L. Cody and J. M. Diamond (eds.), *Ecology and Evolution of Communities*, 214–257. Cambridge, MA: Belknap Press.

Cody, M. L. 1980. Evolution of habitat use: Geographic perspectives. In R. Nöhring (ed.), *Acta XVII Congressus Internationalis Ornithologici*, vol. 2, 1013–1018. Berlin: Verlag der Deutschen Ornithologen-Gesellschaft.

Cody, M. L. 1985. *Habitat Selection in Birds*. Orlando, FL: Academic Press.

Cody, M. L. and J. M. Diamond (eds.). 1975. *Ecology and Evolution of Communities*. Cambridge, MA: Belknap Press.

Cody, M. L. and H. A. Mooney. 1978. Convergence vs. nonconvergence in Mediterranean-climate ecosystems. *Annual Review of Ecology and Systematics* 9: 265–321.

Cody, M. L. and J. M. Overton. 1996. Short-term evolution of reduced dispersal ability in island plant populations. *Journal of Ecology* 84: 53–61.

Cogger, H. G. 1975. Sea snakes of Australia and New Guinea. In W. A. Dunson (ed.), *The Biology of Sea Snakes*, 59–139. Baltimore, MD: University Park Press.

Cohan, F. M. 2001. Bacterial species and speciation. *Systematic Biology* 50: 513–524.

Cohen, J. E. 1995. *How Many People Can the Earth Support?* New York: W. W. Norton.

Cole, K. L. 1982. Late Quaternary zonation of vegetation in the eastern Grand Canyon. *Science* 217: 1142–1145.

Coleman, B. D., M. A. Mares, M. R. Willig and Y.-H. Hsieh. 1982. Randomness, area and species richness. *Ecology* 63: 1121–1133.

Coleman, R. G. 1993. *Geologic Evolution of the Red Sea*. New York: Oxford University Press.

Colinvaux, P. A. 1981. Historical ecology in Beringia: The southland bridge coast at St. Paul Island. *Quaternary Research* 16: 18–36.

Colinvaux, P. A. 1989. Ice-age Amazonia revisited. *Nature* 340: 188–189.

Colinvaux, P. A. 1996. Low-down on a landbridge. *Nature* 382: 21–23.

Colinvaux, P. A. and P. E. De Oliveira. 2002. Paleoecology and climate of the Amazon Basin during the last glacial cycle. *Journal of Quaternary Science* 15: 347–356.

Colinvaux, P. A., P. E. De Oliveira and M. B. Bush. 2000. Amazonian and Neotropical plant communities on glacial time-scales: The failure of the aridity hypothesis. *Quaternary Science Reviews* 19: 141–169.

Colinvaux, P. A., P. E. De Oliveira, J. E. Moreno, M. C. Miller and M. B. Bush. 1996. A long pollen record from lowland Amazonia: Forest and cooling in glacial times. *Science* 274: 86–88.

Collins, N. M. and J. A. Thomas. 1991. *The Conservation of Insects and Their Habitats*. London: Academic Press.

Colwell, R. K. 1973. Competition and coexistence in a simple tropical community. *American Naturalist* 107: 737–760.

Colwell, R. K. 1979. The geographical ecology of hummingbird flower mites in relation to their host plants and carriers. In J. S. Rodrigues (ed.), *Recent Advances in Acarology*, vol. 2, 461–468. New York: Academic Press.

Colwell, R. K. and G. C. Hurtt. 1994. Non-biological gradients in species richness and a spurious Rapoport effect. *American Naturalist* 144: 570–595.

Colwell, R. K. and D. C. Lees. 2000. The mid-domain effect: Geometric constraints on the geography of species richness. *Trends in Ecology and Evolution* 15: 70–76.

Colwell, R. K. and D. W. Winkler. 1984. A null model for null models in biogeography. In D. R. Strong, D. Simberloff, L. G. Abele and A. B. Thistle (eds.), *Ecological Communities: Conceptual Issues and the Evidence*, 344–359. Princeton, NJ: Princeton University Press.

Coney, P. J. 1982. Plate tectonic constraints on biogeographic connections between North and South America. *Annals of the Missouri Botanical Garden* 69: 432–443.

Coney, P. J., D. L. Jones and I. W. H. Monger. 1980. Cordilleran suspect terrains. *Nature* 288: 329–333.

Connell, J. H. 1961. The influence of interspecific competition and other factors on the distribution of the barnacle *Chthamalus stellatus*. *Ecology* 42: 710–723.

Connell, J. H. 1975. Some mechanisms producing structure in natural communities: A model and evidence from field experiments. In M. L. Cody and J. M. Diamond (eds.), *Ecology and Evolution of Communities*, 460–490. Cambridge, MA: Belknap Press.

Connell, J. H. 1978. Diversity in tropical rain forests and coral reefs. *Science* 199: 1301–1310.

Connell, J. H. and E. Orias. 1964. The ecological regulation of species diversity. *American Naturalist* 98: 399–414.

Connell, T. H. and R. O. Slatyer. 1977. Mechanisms of succession in natural communities and their role in community stability and organization. *American Naturalist* 111: 1119–1144.

Connor, E. F. and E. D. McCoy. 1975. The statistics and biology of the species-area relationship. *American Naturalist* 113: 791–833.

Connor, E. F. and D. Simberloff. 1978. Species number and compositional similarity of the Galápagos flora and avifauna. *Ecological Monographs* 48: 219–248.

Conrad, C. P. and C. Lithgow-Berteloni. 2002. How mantle slabs drive plate tectonics. *Science* 298: 207–209.

Conran, J. G. 1995. Family distributions in the Liliiflorae and their biogeographical implications. *Journal of Biogeography* 22: 1023–1034.

Contoli, L. 2000. Rodents of Italy species richness maps and forma Italiae. *Hystrix* 11: 39–46.

Cook, J. A., A. L. Bidlack, C. J. Conroy, J. R. Demboski, M. A. Fleming, A. M. Runck, K. D. Stone and S. O. MacDonald. 2001. A phylogeographic perspective on endemism in the Alexander Archipelago of southeast Alaska. *Biological Conservation* 97: 215–227.

Cook, R. E. 1969. Variation in species density of North American birds. *Systematic Zoology* 18: 63–84.

Cook, R. R. 1995. The relationship between nested subsets, habitat subdivision and species diversity. *Oecologia* 101: 204–210.

Cook, R. R. and J. F. Quinn. 1995. The influence of colonization in nested species subsets. *Oecologia* 102: 413–424.

Cooke, H. B. S. 1972. The fossil mammal fauna of Africa. In A. Keast, F. C. Erk and B. Glass (eds.), *Evolution, Mammals and Southern Continents*, 89–139. Albany: State University of New York Press.

Cooper, A., C. Lalueza-Fox, S. Anderson, A. Rambaut, J. Austin and R. Ward. 2001. Complete mitochondrial genome sequences of two extinct moas clarify ratite evolution. *Nature* 409: 704–707.

Cooper, W. S. 1913. The climax forest of Isle Royale, Lake Superior and its development. *Botanical Gazette* 15: 1–44, 115–140, 189–235.

Corlett, R. T. 1992. The ecological transformation of Singapore, 1819–1990. *Journal of Biogeography* 19: 411–420.

Cornelius, J. M. and J. F. Reynolds. 1991. On determining the statistical significance of discontinuities with ordered ecological data. *Ecology* 72: 2057–2070.

Cornell, H. V. 1993. Unsaturated patterns in species assemblages: The role of regional processes in setting local species richness. In R. E. Ricklefs and D. Schluter (eds.), *Species Diversity in Ecological Communities: Historical and Geographical Perspectives*, 243–252. Chicago: University of Chicago Press.

Cotgreave, P. and P. Stockley. 1994. Body size, insectivory and abundance in assemblages of small mammals. *Oikos* 71: 89–96.

Courtillot, V. 1999. *Evolutionary Catastrophes: The Science of Mass Extinction*. Cambridge: Cambridge University Press.

Cowan, I. McT. and C. J. Guiget. 1956. *The Mammals of British Columbia*. British Columbia Provincial Museum Handbook No. 11.

Cowles, H. C. 1889. The ecological relations of the vegetation of the sand dunes of Lake Michigan. *Botanical Gazette* 27: 95–117, 167–202, 281–308, 361–391.

Cowles, H. C. 1901. The physiographic ecology of Chicago and vicinity: A study of the origin, development and classification of plant societies. *Botanical Gazette* 31: 73–108, 145–182.

Cowling, R. M., R. L. Pressey, R. Sims-Castley, A. le Roux, E. Baard, C. J. Burgers and G. Palmer. 2003. The expert or the algorithm? Comparison of priority conservation areas in the Cape Floristic Region identified by park managers and reserve selection software. *Biological Conservation* 112: 147–167.

Cowlishaw, G. and J. E. Hacker. 1997. Distribution, diversity and latitude in African primates. *American Naturalist* 150: 505–512.

Cox, C. B. 1973a. The distribution of Triassic terrestrial tetrapod families. In D. H. Tarling and S. K. Runcorn (eds.), *Implications of Continental Drift to the Earth Sciences*, vol. 1, 369–371. New York: Academic Press.

Cox, C. B. 1973b. Triassic tetrapods. In A. Hallam (ed.), *Atlas of Palaeobiogeography*, 213–223. Amsterdam: Elsevier.

Cox, C. B. 1990. New geological theories and old biogeographical problems. *Journal of Biogeography* 17: 117–130.

Cox, C. B. and P. D. Moore. 1985. *Biogeography: An Ecological and Evolutionary Approach*. Boston: Blackwell Scientific Publications.

Cox, G. W., L. C. Contreras and A. V. Milewski. 1994. Role of fossorial animals in community structure and energetics of Mediterranean ecosystems. In M. T. Kilinin de Arroyo, P. H. Zedler and M. D. Fox (eds.), *Ecology of Convergent Ecosystems: Mediterranean-Climate Ecosystems of Chile, California and Australia*, 383–398. New York: Springer-Verlag.

Coyne, J. A. and H. A. Orr. 2004. *Speciation*. Sunderland, MA: Sinauer Associates.

Cracraft, J. 1980. Avian phylogeny and intercontinental biogeographic patterns. In R. Nöhring (ed.), *Acta XVII Congressus Internationalis Ornithologici*, vol. 2, 1302–1308. Berlin: Verlag der Deutschen Ornithologen-Gesellschaft.

Cracraft, J. 1981. Toward a phylogenetic classification of the Recent birds of the world (class Aves). *Auk* 98: 681–714.

Cracraft, J. 1982. Geographic differentiation, cladistics and vicariance biogeography: Reconstructing the tempo and mode of evolution. *American Zoologist* 22: 411–424.

Cracraft, J. 1988. Deep-history biogeography: Retrieving the historical pattern of evolving continental biotas. *Systematic Zoology* 37: 221–236.

Cracraft, J. 1989. Speciation and its ontology: The empirical consequences of alternative species concepts for understanding patterns and processes of differentiation. In D. Otte and J. A. Endler (eds.), *Speciation and Its Consequences*, 27–59. Sunderland, MA: Sinauer Associates.

Cracraft, J. 2001. Avian evolution, Gondwana biogeography and the Cretaceous-Tertiary mass extinction event. *Proceedings of the Royal Society of London,* Series B-Biological Sciences 268: 459–469.

Cracraft, J. and M. J. Donoghue. 2004. *Assembling the Tree of Life.* New York: Oxford University Press.

Cracraft, J. and R. O. Prum. 1988. Patterns and processes of diversification: Speciation and historical congruence in some Neotropical birds. *Evolution* 42: 603–620.

Crame, J. A. 1993. Latitudinal range fluctuations in the marine realm through geological time. *Trends in Ecology and Evolution* 8: 162–166.

Crame, J. A. 2001. Taxonomic diversity gradients through geologic time. *Diversity and Distributions* 7: 175–189.

Crame, J. A. 2002. Evolution of taxonomic diversity gradients in the marine realm: A comparison of Late Jurassic and Recent bivalve faunas. *Paleobiology* 28: 184–207.

Crame, J. A. 2004. Pattern and process in marine biogeography: A view from the poles. In M. V. Lomolino and L. R. Heaney (eds.), *Frontiers of Biogeography I: New Directions in the Geography of Nature*, 272–292 Sunderland, MA: Sinauer Associates.

Crame, J.A. and B. R. Rosen. 2002. Cenozoic palaeogeography and the rise of modern biodiversity patterns. In J. A. Crame and A. W. Owen (eds.), *Palaeobiogeography and Biodiversity Change: the Ordovician and Mesozoic-Cenozoic Radiations.* Geological Society, London, Special Publications, No. 194, 153–168.

Crandall, K. A. and A. R. Templeton. 1996. Applications of intraspecific phylogenetics. In P. H. Harvey, A. J. Leigh Brown, J. Maynard Smith and S. Nee (eds.), *New Uses for New Phylogenies*, 81–99. New York: Oxford University Press.

Crane, P. R. and S. Lidgard. 1989. Angiosperm diversification and paleolatitudinal gradients in Cretaceous floristic diversity. *Science* 246: 675–678.

Craw, R. C. 1982. Phylogenetics, areas, geology and the biogeography of Croizat: A radical view. *Systematic Zoology* 31: 304–316.

Craw, R. C. 1988. Panbiogeography: Method and synthesis in biogeography. In A. A. Myers and P. S. Giller (eds.), *Analytical Biogeography: An Integrated Approach to the Study of Animal and Plant Distributions*, 405–435. London: Chapman and Hall.

Craw, R. C., J. R. Grehan and M. J. Heads. 1999. *Panbiogeography: Tracking the History of Life.* New York: Oxford University Press.

Craycraft, J. and R. O. Prum. 1988. Patterns and processes of diversification: Speciation and historical congruence in some neotropical birds. *Evolution* 42: 603–620.

Creer, K. M., E. Irving and S. K. Runcorn. 1954. The direction of the geomagnetic field in remote epochs in Great Britain. *Journal of Geomagnetism and Geoelectricity* 6: 163–168.

Creer, K. M., E. Irving and S. K. Runcorn. 1957. Geophysical interpretation of palaeomagnetic directions from Great Britain. *Philosophical Transactions of the Royal Society of London*, Series A 250: 144–156.

Cressie, N. 1991. *Statistics for Spatial Data.* New York: John Wiley and Sons.

Crestwell, J. E., V. M. Vidal-Martinez and N. J. Crichton. 1995. The investigation of saturation in the species richness of communities: Some comments on methodology. *Oikos* 72: 301–304.

Crisci, J. V., L. Katinas and P. Posadas. 2003. *Historical Biogeography: An Introduction.* Cambridge, MA: Harvard University Press.

Crisci, J. V., M. M. Cigliano, J. J. Morrone and S. Roigjunent. 1991. Historical biogeography of Southwestern South America. *Systematic Zoology* 40: 152–171.

Critchfield, W. B. and E. J. Little. 1966. *Geographic Distributions of Pines of the World.* USDA Forest Service, Miscellaneous Publication 991.

Croizat, L. 1952. *Manual of Phytogeography.* The Hague: Dr. W. Junk.

Croizat, L. 1958. *Panbiogeography.* 2 vol. Caracas: Published by the author.

Croizat, L. 1960. *Principia Botanica.* Caracas: Published by the author.

Croizat, L. 1964. *Space, Time, Form: The Biological Synthesis.* Caracas: Published by the author.

Croizat, L. 1982. Vicariance/vicariism, panbiogeography, "vicariance biogeography," etc: A clarification. *Systematic Zoology* 31: 291–304.

Croizat, L., G. J. Nelson and D. E. Rosen. 1974. Centers of origin and related concepts. *Systematic Zoology* 23: 265–287.

Cronquist, A. 1978. Once again, what is a species? In L. V. Knutson (ed.), *BioSystematics in Agriculture*, 3–20. Allenheld Osmum, Montclair.

Crosby, G. T. 1972. Spread of the cattle egret in the Western Hemisphere. *Bird-Banding* 43: 205–212.

Crossman, E. J. 1991. Introduced freshwater fishes: A review of the North American perspective with emphasis on Canada. *Canadian Journal of Fisheries and Aquatic Sciences* 48 (suppl. 1): 46–57.

Crowell, K. L. 1962. Reduced interspecific competition among the birds of Bermuda. *Ecology* 43: 75–88.

Crowell, K. L. 1973. Experimental zoogeography: Introductions of mice to small islands. *American Naturalist* 107: 535–558.

Crowell, K. L. 1983. Islands—insights or artifacts? Population dynamics and habitat utilization in insular rodents. *Oikos* 41: 442–454.

Crowell, K. L. 1986. A comparison of relict vs. equilibrium models for insular mammals of the Gulf of Maine. *Biological Journal of the Linnean Society* 28: 37–64.

Crowell, K. L. and S. L. Pimm. 1976. Competition and niche shifts of mice introduced onto small islands. *Oikos* 27: 251–258.

Cruden, R. W. 1966. Birds as agents of long-distance dispersal for disjunct plant groups of the temperate Western Hemisphere. *Evolution* 20: 517–532.

Cuellar, O. 1977. Animal parthenogenesis. *Science* 197: 837–843.

Culotta, E. 1994. Is marine diversity at risk? *Science* 263: 918–920.

Culver, D. C. 1970. Analysis of simple cave communities. I. Caves as islands. *Evolution* 29: 463–474.

Culver, S. J. and M. A. Buzas. 2000. Global latitudinal species diversity gradient in deep-sea benthic foraminifera. *Deep-Sea Research I* 47: 259–275.

Cumber, R. A. 1953. Some aspects of the biology and ecology of bumblebees bearing upon the yields of red clover seed in New Zealand. *New Zealand Journal of Science and Technology* 34: 227–240.

Cunningham, C. W. and T. M. Collins. 1994. Developing model systems for molecular biogeography: Vicariance and interchange in marine invertebrates. In B. Schierwater, B. Streit, G. P. Wagner and R. DeSalle (eds.), *Molecular Ecology and Evolution: Approaches and Applications.* Basel, Switzerland: Birkhauser.

Currie, D. J. 1991. Energy and large-scale patterns of animal- and plant-species richness. *American Naturalist* 137: 27–49.

Currie, D. J. and V. Paquin. 1987. Large-scale biogeographical patterns of species richness of trees. *Nature* 329: 326–327.

Curtis, J. T. 1959. *The Vegetation of Wisconsin.* Madison: University of Wisconsin Press.

Cushman, J. H., J. H. Lawton and B. F. J. Manly. 1993. Latitudinal patterns in European ant assemblages: Variation in species richness and body size. *Oecologia* 95: 30–37.

Cutler, A. 1991. Nested faunas and extinction in fragmented habitats. *Conservation Biology* 5: 496–505.

Dahlberg, A. and H. Croneborg. 2003. *Thirty-three Threatened Fungi of Europe.* European Council for Conservation of Fungi, Sweden.

Dalrymple, G. B., E. A. Silver and E. D. Jackson. 1973. Origin of the Hawaiian Islands. *American Scientist* 61: 294–308.

Dammermann, K. W. 1948. The fauna of Krakatau, 1883–1933. *Koninklijke Nederlandsche Akademie Wetenschappen Verhandelingen* 44: 1–594.

Damuth, J. 1991. Of size and abundance. *Nature* 351: 268–269.

Dana, J. D. 1853. On an isothermal oceanic chart, illustrating geographic distribution of marine animals. *The American Journal of Science and Arts,* 2nd Series, 66: 153–167, 325–327, 391–392.

Daniels, R. J. R. 1992. Geographic distribution patterns of amphibians in the western Ghats, India. *Journal of Biogeography* 19: 521–529.

Dansereau, P. M. 1957. *Biogeography: An Ecological Perspective.* New York: Ronald Press.

Darlington, P. J. Jr. 1938. The origin of the fauna of the Greater Antilles, with discussion of dispersal of animals over water and through the air. *Quarterly Review of Biology* 13: 274–300.

Darlington, P. J. Jr. 1943. Caribidae of mountains and islands: Data on the evolution of isolated faunas and on atrophy of wings. *Ecological Monographs* 13: 37–61.

Darlington, P. J. Jr. 1957. *Zoogeography: The Geographical Distribution of Animals.* New York: John Wiley & Sons.

Darlington, P. J. Jr. 1959a. Darwin and zoogeography. *Proceedings of the American Philosophical Society* 103: 307–319.

Darlington, P. J. Jr. 1959b. Area, climate and evolution. *Evolution* 13: 488–510.

Darlington, P. J. Jr. 1965. *Biogeography of the Southern End of the World: Distribution and History of Far Southern Life and Land, with an Assessment of Continental Drift.* Cambridge, MA: Harvard University Press.

Darwin, C. 1839. *Journal of the Researches into the Geology and Natural History of Various Countries Visited by H.M.S. Beagle, under the Command of Captain Fitzroy, R.N. from 1832 to 1836.* London: Henry Colburn.

Darwin, C. 1859. *On the Origin of Species by Means of Natural Selection or the Preservation of Favored Races in the Struggle for Life.* London: John Murray.

Darwin, C. 1860. *The Voyage of the Beagle.* New Jersey: Doubleday.

Daubenmire, R. 1978. *Plant Geography with Special Reference to North America.* New York: Academic Press.

Davies, N. 1995. *The Incas.* Niwat, CO: University Press of Colorado.

Davis, A. J., L. S. Jenkinson, J. H. Lawton, B. Shorrocks and S. Wood. 1998. Global warming, population dynamics and community structure in a model insect assemblage. In R. Harrington and N. E. Stork (eds.), *Insects in a Changing Environment,* Chapter 18. San Diego, CA: Academic Press.

Davis, A. L. V., C. H. Scholtz and S. L. Chown. 1999. Species turnover, community boundaries and biogeographical composition of dung beetle assemblages across an altitudinal gradient in South Africa. *Journal of Biogeography* 26: 1039–1055.

Davis, M. B. 1969. Palynology and environmental history during the Quaternary period. *American Scientist* 57: 317–332.

Davis, M. B. 1976. Pleistocene geography of temperate deciduous forests. *Geoscience and Man* 13: 13–26.

Davis, M. B. 1981. Quaternary history and the stability of forest communities. In D. C. West, H. H. Shugart and D. B. Botkin (eds.), *Forest Succession: Concepts and Application,* 132–153. New York: Springer-Verlag.

Davis, M. B. 1986. Climatic stability, time lags and community disequilibrium. In J. M. Diamond and T. Case (eds.), *Community Ecology,* 269–284. New York: Harper and Row.

Davis, M. B. 1994. Ecology and paleoecology begin to merge. *Trends in Ecology and Evolution* 9: 357–358.

Davis, R. 1996. The Pinalenos as an island in a montane archipelago. In C. A. Istock and R. S. Hoffmann (eds.), *Storm over a Mountain Archipelago: Conservation Biology and the Mt. Graham Affair,* 123–134. Tuscon: University of Arizona Press.

Davis, R. and D. E. Brown. 1989. Role of post-Pleistocene dispersal in determining the modern distribution of Albert's squirrel. *Great Basin Naturalist* 49: 425–434.

Davis, R. and J. R. Callahan. 1992. Post-Pleistocene dispersal in the Mexican vole (*Microtus mexicanus*): An example of an apparent trend in the distribution of southwestern mammals. *Great Basin Naturalist* 52: 262–268.

Davis, R. and C. Dunford. 1987. An example of contemporary colonization of montane islands by small, nonflying mammals in the American Southwest. *American Naturalist* 129: 398–406.

Dawson, W. R. and C. Carey. 1976. Seasonal acclimation to temperature in cardueline finches. *Journal of Comparative Physiology* 112: 317–333.

Dayan, T. 1990. Feline canines: Community-wise character displacement in the small cats of Israel. *American Naturalist* 136: 39–60.

Dayan, T., E. Tchernov, Y. Yom-Tov and D. Simberloff. 1989. Ecological character displacement in Saharo-Arabian *Vulpes:* Outfoxing Bergmann's rule. *Oikos* 55: 263–272.

Dayton, L. 2001. Mass extinctions pinned on Ice Age hunters. *Science* 292: 1819.

Dayton, P. K. 1971. Competition, disturbance and community organization: The provision and subsequent utilization of space in a rocky intertidal community. *Ecological Monographs* 41: 351–389.

Debinski, D. M. and R. D. Holt. 2000. A survey and overview of habitat fragmentation experiments. *Conservation Biology* 14: 342–355.

DeGraaf, R. M. and J. H. Rappole. 1995. *Neotropical Migratory Birds: Natural History, Distribution and Population Change.* Ithaca, NY: Comstock Press (Cornell University Press).

Dehant, V., K. C. Creager, S. Karato and S. Zatman. 2003. *Earth's Core: Dynamics, Structure and Rotation.* Washington, DC: American Geophysical Union.

Delany, M. J. 1972. The ecology of small rodents in tropical Africa. *Mammalian Reviews* 2: 1–41.

Delcourt, H. R. and P. A. Delcourt. 1994. Postglacial rise and decline of *Ostrya virginiana* (Mill.) K. Koch and *Carpinus caroliniana* Walt. in eastern North America: Predictable responses of forest species to cyclic changes in seasonality of climate. *Journal of Biogeography* 21: 137–150.

Delcourt, P. A. and H. R. Delcourt. 1977. The Tunica Hills, Louisiana-Mississippi: Late glacial locality for spruce and deciduous forest species. *Quaternary Research* 7: 218–237.

Delcourt, P. A. and H. R. Delcourt. 1987. Late-quatenary dynamics of temperate forests: Applications of paleoecology to issues of global environmental change. *Quaternary Science Review* 6: 129–146.

Delcourt, P. A. and H. R. Delcourt. 1996. Quaternary paleoecology of the lower Mississippi Valley. *Engineering Geology* 45: 219–242.

Dell, B., G. Muir and M. Palmer. 1980. Lizard assemblages and reserve size and structure in the Western Australian wheatbelt—Some implications for conservation. *Biological Conservation* 17: 25–61.

Den Boer, P. J. 1977. *Dispersal Power and Survival: Carabids in a Cultivated Countryside.* Wageningen, The Netherlands: H. Veenman and Zonen B. V.

Deplus, C. 2001. Plate tectonics: Indian Ocean actively deforms. *Science* 292: 1850–1851.

de Queiroz, K. 1998. The general lineage concept of species, species criteria and the process of speciation. In D. J. Howard and S. H. Berlocher (eds.), *Endless Forms: Species and Speciation,* 57–75. New York: Oxford University Press.

de Queiroz, K. 2005. The resurrection of oceanic dispersal in historical biogeography. *Trends in Ecology and Evolution* 20: 68–73.

DeSalle, R., T. Freedman, E. M. Prager and A. C. Wilson. 1987. Tempo and mode of sequence evolution in mitochondrial DNA of Hawaiian *Drosophila. Journal of Molecular Evolution* 26: 157–164.

Desbruyeres, D., Almeida, A., Biscoito, M., Comtet, T., Khripounoff, A., LeBris, N., Sarradin, P.M., Segonzac, M. 2000. A review of the distribution of hydrothermal vent communities along the northern Mid-Atlantic Ridge: Dispersal vs. environmental controls. *Hydrobiologia* 440(1–3): 201–216.

Desmond, A. 1965. How many people have ever lived on earth? In L. K. Y. Ng and S. Mudd (eds.), *The Population Crisis: Implications and Plans for Action.* Bloomington: Indiana University Press.

de Vries, A. L. 1971. Freezing resistance in fishes. In W. S. Hoar and D. J. Randall (eds.), *Fish Physiology,* vol. 6, 157–190. New York: Academic Press.

de Wet, J. M. J. 1979. Origins of polyploids. In W. H. Lewis (ed.), *Polyploidy: Biological Relevance,* 3–15. New York: Plenum Press.

Dewey, J. F. 1977. Suture zone complexities: A review. *Tectonophysics* 40: 53–67.

Dexter, R. W. 1978. Some historical notes on Louis Agassiz's lectures on zoogeography. *Journal of Biogeography* 5: 207–209.

D'Hondt, S. and M. A. Arthur. 1996. Late Cretaceous oceans and the cool tropic paradox. *Science* 271: 1838–1841.

Diamond, J. M. 1969. Avifaunal equilibria and species turnover rates on the Channel Islands of California. *Proceedings of the National Academy of Sciences, USA* 64: 57–63.

Diamond, J. M. 1970a. Ecological consequences of island colonization by Southwest Pacific birds. I. Types of niche shifts. *Proceedings of the National Academy of Sciences, USA* 67: 529–536.

Diamond, J. M. 1970b. Ecological consequences of island colonization by Southwest Pacific birds. II. The effect of species diversity on total population density. *Proceedings of the National Academy of Sciences, USA* 67: 1715–1721.

Diamond, J. M. 1972. Biogeographic kinetics: Estimation of relaxation times for avifaunas of Southwest Pacific islands. *Proceedings of the National Academy of Sciences, USA* 69: 3199–3203.

Diamond, J. M. 1973. Distributional ecology of New Guinea birds. *Science* 179: 759–769.

Diamond, J. M. 1974. Colonization of exploded volcanic islands by birds: The supertramp strategy. *Science* 184: 803–806.

Diamond, J. M. 1975a. The island dilemma: Lessons of modern biogeographic studies for the design of natural reserves. *Biological Conservation* 7: 129–146.

Diamond, J. M. 1975b. Assembly of species communities. In M. L. Cody and J. M. Diamond (eds.), *Ecology and Evolution of Communities*, 342–444. Cambridge, MA: Belknap Press.

Diamond, J. M. 1977. Colonization cycles in man and beast. *World Archaeology* 8: 249–261.

Diamond, J. M. 1984. "Normal" extinctions of isolated populations. In M. N. Nitecki (ed.), *Extinctions*, 191–246. Chicago: University of Chicago Press.

Diamond, J. M. 1991a. Did Komodo dragons evolve to eat pygmy elephants? *Nature* 326: 832.

Diamond, J. M. 1991b. A new species of rail from the Solomon Islands and convergent evolution of insular flightlessness. *Auk* 108: 461–470.

Diamond, J. M. 1997. Guns, Germs and Steel: The Fates of Human Societies. New York: W. W. Norton.

Diamond, J. M. 2005. *Collapse: How Societies Chose to Fail or Succeed*. New York: Viking Press.

Diamond, J. M. and T. J. Case (eds.). 1986. *Community Ecology*. New York: Harper and Row.

Diamond, J. M. and M. E. Gilpin. 1983. Biogeographic umbilici and the origin of the Philippine avifauna. *Oikos* 41: 307–321.

Diamond, J. M. and R. M. May. 1976. Island biogeography and the design of natural preserves. In R. M. May (ed.), *Theoretical Ecology: Principles and Applications*, 163–186. Philadelphia: W. B. Saunders Co.

Diamond, J. M. and S. L. Pimm. 1993. Survival times of bird populations: A reply. *American Naturalist* 142: 1030–1035.

Dice, L. R. 1947. Effectiveness of selection by owls of deer mice (*Peromyscus maniculatus*) which contrast in color with their background. *Contributions, Laboratory Vertebrate Biology, University of Michigan* 50: 1–15.

Dice, L. R. and P. M. Blossom. 1937. *Studies of Mammalian Ecology in Southwestern North America with Special Attention to the Colors of Desert Mammals*. Carnegie Institute Publication 485: 1–29. Washington, DC: Carnegie Institute.

Dickerson, J. E. Jr. and J. V. Robinson. 1986. The controlled assembly of microcosmic communities: The selective extinction hypothesis. *Oecologia* 71: 12–17.

Dickman, C. R. 1996. The impact of exotic generalist predators on the native fauna of Australia. *Wildlife Biology* 2: 185–195.

Dietl, G. P. and P. H. Kelley. 2002. The fossil record of predator-prey arms races: Coevolution and escalation hypotheses. In M. Kowalewski and P. H. Kelley (eds.), *The Fossil Record of Predation: The Paleontological Society Papers* 8: 353–374.

Dietz, R. S. 1961. Continental and ocean basin evolution by spreading of the sea floor. *Nature* 190: 854–857.

Dixon, A. F. G., P. Kindlmann, J. Leps and J. Holman. 1987. Why are there so few species of aphids, especially in the tropics. *American Naturalist* 129: 580–592.

Dixon, J. R. 1979. Origin and distribution of reptiles of low-land tropical rain forest of South America. In W. E. Duellman (ed.), *The South American Herpetofauna: Its Origin, Evolution and Dispersal*, 217–240. Monograph no. 7. Lawrence: Museum of Natural History, University of Kansas.

Doak, D. F. and L. S. Mills. 1994. A useful role for theory in conservation. *Ecology* 75: 615–626.

Dobson, A. P. 1996. *Conservation and Biodiversity*. New York: Scientific American Library.

Dobson, A. P., J. P. Rodriguez, W. M. Roberts and D. S. Wilcove. 1997. Geographic distribution of endangered species in the United States. *Science* 275: 550–553.

Dobzhansky, Th. 1937. *Genetics and the Origin of Species*. New York: Columbia University Press.

Dobzhansky, Th. 1950. Evolution in the tropics. *American Scientist* 38: 209–221.

Dobzhansky, Th. 1951. *Genetics and the Origin of Species*. 3rd edition. New York: Columbia University Press.

Dobzhansky, Th. 1973. Nothing in biology makes sense except in the light of evolution. *The American Biology Teacher* 35: 125–129.

Docters van Leeuwen, W. M. 1936. Krakatau, 1883–1933. *Ann. Jard. Bot. Buitenzorg* 46–47: 1–506.

Dodd, A. P. 1959. The biological control of prickly pear in Australia. In A. Keast (ed.), *Biogeography and Ecology in Australia*, 565–577. Monographiae Biologicae 8. The Hague: Dr. W. Junk.

Dominey, W. J. 1984. Effects of sexual selection and life history on speciation: Species flocks in African cichlids and Hawaiian *Drosophila*. In A. A. Echelle and I. Kornfield (eds.), *Evolution of Fish Species Flocks*, 231–249. Orono: University of Maine at Orono Press.

Donoghue, M. J. and B. R. Moore. 2003. Toward an integrative historical biogeography. *Integrative and Comparative Biology* 43: 261–270.

Donoghue, M. J. and S. A. Smith. 2004. Patterns in the assembly of temperate forests around the Northern Hemisphere. *Philosophical Transactions of the Royal Society of London*, Series B 359: 1633–1644.

Dorst, J. 1962. *The Migrations of Birds*. Boston: Houghton Mifflin.

Doube, B. M., A. MacWueen, T. J. Ridsill-Smith and T. A. Weir. 1991. Native and introduced dung beetles in Australia. In I. Hanski and Y. Camberfort (eds.), *Dung Beetle Ecology*, 255–278. Princeton, NJ: Princeton University Press.

Dover, C. L. 2002. Variation in community structure within hydrothermal vent mussel beds of the East Pacific Rise. *Marine Ecology Progress Series* 253: 550–66.

Doyle, J. A. 1978. Origin of angiosperms. *Annual Review of Ecology and Systematics* 9: 365–392.

Drake, J. A., H. A. Mooney, F. di Castri, R. H. Groves, F. J. Kruger, M. Rejmanek and M. Williamson (eds.). 1989. *Biological Invasions: A Global Perspective*. New York: John Wiley & Sons.

Drude, O. 1887. *Atlas der Pflanzenverbreitung*. 5 Abt. Gotha, Berghaus Physikalischer Atlas.

Drummond, A. J. and A. Rambaut. 2003. BEAST. Version 1.0. Available from: http://evolve.zoo.ox.ac.uk/beast/.

Drury, W. H. and I. C. T. Nisbet. 1973. Succession. *Journal of the Arnold Arboretum* 54: 331–368.

Dudley, J. P. 1999. Coevolutionary implications of an endemic Pleistocene megaherbivore fauna for insular floras of the California Channel Islands. *Conservation Biology* 13: 209–210.

Due, A. D. and G. A. Polis. 1986. Trends in scorpion diversity along the Baja California Peninsula. *American Naturalist* 128: 460–468.

Duellman, W. E. and L. Trueb. 1986. *Biology of Amphibians*. London, New York: McGraw-Hill.

Dunmire, W. W. 1960. An altitudinal survey of reproduction in *Peromyscus marliculatus*. *Ecology* 41: 174–182.

Dunn, C. P. and C. Loehle. 1988. Species-area parameter estimation: Testing the null model of lack of relationships. *Journal of Biogeography* 15: 721–728.

Dunn, P. O., K. J. Thusius, K. Kimber and D. W. Winkler. 2000. Geographic and ecological variation in clutch size of tree swallows. *The Auk* 117: 215–221.

Dunne, J. A., R. J Williams and N. D.Martinez. 2002. Food-web structure and network theory: The role of connectance and size. *Proceedings of the National Academy of Sciences, USA* 99: 12917–12923.

Dunson, W. A. (ed.) 1975. *The Biology of Sea Snakes*. Baltimore, MD: University Park Press.

Dunson, W. A. and F. J. Mazotti. 1989. Salinity as a limiting factor in the distribution of reptiles in Florida Bay: A theory for the estuarine origin of marine snakes and turtles. *Bulletin of Marine Science* 44: 229–244.

Durham, J. W. and F. S. MacNeil. 1967. Cenozoic migrations of marine invertebrates through the Bering Strait region. In D. M. Hopkins (ed.), *The Bering Land Bridge*, 326–349. Stanford, CA: Stanford University Press.

D'Urville, M. J. D. 1833. Voyage de la corvette L'Atrolable execute pendant les annees 1826–1827–1828–1829 sans le commandement de M. Jules Dummont D'Urville, capitaine de vaisseau. *Atlas Historique Paris, J. Tartu.*

du Toit, A. L. 1927. *A Geological Comparison of South America with South Africa*. Publication no. 381, 1–157. Washington, DC: Carnegie Institute.

du Toit, A. L. 1937. *Our Wandering Continents*. Edinburgh: Oliver & Boyd.

Duxbury, A. 2000. *An Introduction to the World's Oceans.* 3rd edition. Berkshire: McGraw-Hill.

Dyer, R. A. 1975. *The Genera of Southern African Flowering Plants*, vol. 1. Pretoria: Department of Agricultural Technical Services, Botanical Research Institute.

Ebach, M. C. and C. J. Humphries. 2002. Cladistic biogeography and the art of discovery. *Journal of Biogeography* 29: 427–444.

Ebenhard, T. 1988a. *Demography and Island Colonization of Experimentally Introduced and Natural Vole Populations*. Ph.D. dissertation, Uppsala University, Sweden.

Ebenhard, T. 1988b. Introduced birds and mammals and their ecological effects. *Swedish Wildlife Research* 13: 1–107.

Ebenhard, T. 1989. Bank vole (*Clethrionomys glareolus*) propagules of different sizes and island colonization. *Journal of Biogeography* 16: 173–180.

Ebenhard, T. 1991. Colonization of metapopulations: A review of theory and observations. *Biological Journal of the Linnean Society* 42: 105–121.

Echelle, A. A. and T. E. Dowling. 1992. Mitochondrial-DNA variation and evolution of the Death Valley pupfishes (Cyprinodon, Cyprinodontidae). *Evolution* 46: 193–206.

Echelle, A. A. and I. Kornfield (eds.). 1984. *Evolution of Fish Species Flocks*. Orono: University of Maine Press.

Edwards, P. J., R. M. May and N. R. Webb (eds.). 1994. *Large-Scale Ecology and Conservation Biology*. London: Blackwell Scientific Publications.

Edwards, S. V. and P. Beerli. 2000. Perspective: Gene divergence, population divergence and the variance in coalescence time in phylogeography studies. *Evolution* 54: 1839–1854.

Egyed, L. 1956. The change of Earth's dimensions determined from paleographical data. *Geofisica Pura e Applicata* 33: 42–48.

Egyed, L. 1957. A new dynamic conception of the internal constitution of the Earth. *Geologische Rundschau* 46: 101–121.

Ehrendorfer, F. 1979. Reproductive biology in island plants. In D. Bramwell (ed.), *Plants and Islands*, 293–306. London: Academic Press.

Ehrlich, P. R. 1961. Intrinsic barriers to dispersal in the checkerspot butterfly. *Science* 134: 108–109.

Ehrlich, P. R. 1964. Some axioms of taxonomy. *Systematic Zoology* 13: 109–123.

Ehrlich, P. R. 1965. The population biology of the butterfly *Euphydryas editha*. II. The structure of the Jaspar Ridge colony. *Evolution* 19: 327–336.

Ehrlich, P. R. and R. W. Holm. 1963. *The Process of Evolution*. New York: McGraw-Hill.

Ehrlich, P. R. and P. H. Raven. 1965. Butterflies and plants: A study in coevolution. *Evolution* 18: 586–608.

Ehrlich, P. R., R. R. White, M. C. Singer, S. W. McKechnie and L. E. Gilbert. 1975. Checkerspot butterflies: A historical perspective. *Science* 188: 221–228.

Eisenberg, J. F. 1981. *The Mammalian Radiations: An Analysis of Trends in Evolution, Adaptation and Behavior*. Chicago: University of Chicago Press.

Eisenberg, J. F. and E. Gould. 1970. The Tenrecs: A study in mammalian behavior and evolution. *Smithsonian Contributions to Zoology* 27: 1–137.

Eizirik, E., W. J. Murphy and S. J. O'Brien. 2001. Molecular dating and biogeography of the early placental mammal radiation. *Journal of Heredity* 92: 212–219.

Ekman, S. 1953. *Zoogeography of the Sea*. London: Sidgwick & Jackson.

Eldredge, N. and I. Cracraft. 1980. *Phylogenetic Patterns and the Evolutionary Process*. New York: Columbia University Press.

Eldredge, N. and S. J. Gould. 1972. Punctuated equilibria: An alternative to phyletic gradualism. In T. J. M. Schopf (ed.), *Models in Paleobiology*, 82–115. San Francisco: Freeman, Cooper & Co.

Elena, S. F., V. S. Cooper and R. E. Lenski. 1996. Punctuated evolution caused by selection of rare beneficial mutations. *Science* 272: 1802–1804.

Elias, S. A. 1997. *The Ice-Age History of Southwestern Parks*. Washington, DC: Smithsonian Institution Press.

Elias, S. A., S. K. Short and C. H. Nelson. 1996. Life and times of the Bering landbridge. *Nature* 382: 60–63.

Elliot, D. H., E. H. Colbert, W. J. Breed, I. A. Jensen and T. S. Powell. 1970. Triassic tetrapods from Antarctica: Evidence for continental drift. *Science* 169: 197–201.

Elliot, G. F. 1951. On the geographical distribution of terebratelloid brachiopods. *Annals and Magazines of Natural History,* Series 12. 4: 305–334.

Ellson, J. A. 1969. Late Quaternary marine submergence of Quebec. *Review of Geography, Montreal* 23: 247–258.

Elmberg, J., P. Nummi, H. Paysa and K. Sjoberg. 1994. Relationship between species number, lake size and resource diversity in assemblages of breeding waterfowl. *Journal of Biogeography* 21: 75–84.

Elton, C. 1927. *Animal Ecology*. New York: Macmillan.

Elton, C. S. 1958. *The Ecology of Invasions by Animals and Plants*. London: Methuen & Co.

Elton, C. S. 1966. *The Patterns of Animal Communities*. London: Methuen & Co.

Embleton, B. J. J. 1973. The palaeolatitude of Australia through Phanerozoic time. *Journal of the Geological Society of Australia* 19: 475–482.

Emerson, B. C. 2002. Evolution on oceanic islands: Molecular phylogenetic approaches to understanding pattern and process. *Molecular Ecology* 11: 951–966.

Emlen, J. T. 1978. Density anomalies and regulation mechanisms in land bird populations on the Florida peninsula. *American Naturalist* 112: 265–286.

Emlen, J. T. 1979. Land bird densities on Baja California islands. *Auk* 96: 152–167.

Emslie, S. D. and G. S. Morgan. 1994. A catastrophic death assemblage and paleoclimatic implications of Pliocene seabirds of Florida. *Science* 264: 684–685.

Enckell, P. H., S. A. Bengtson and B. Wiman. 1987. Serf and waif colonization, distribution and dispersal of invertebrate species in Faroe Island settlement areas. *Journal of Biogeography* 14: 89–104.

Endler, J. A. 1977. *Geographic Variation, Speciation and Clines*. Monographs in Population Biology, no. 10. Princeton, NJ: Princeton University Press.

Endler, J. A. 1982. Problems in distinguishing historical from ecological factors in biogeography. *American Zoologist* 22: 441–452.

Engel, S. R., K. M. Hogan, J. F. Taylor and S. K. Davis. 1998. Molecular systematics and paleobiogeography of the South American sigmodontine rodents. *Molecular Biology and Evolution* 15: 35–49.

Enghoff, H. 1996. Widespread taxa, sympatry, dispersal and an algorithm for resolved area cladograms. *Cladistics* 12: 349–364.

Enquist, B. J., M. A. Jordan and J. H. Brown. 1995. Connections between ecology, biogeography and paleobiology: Relationship between local abundance and geographic distribution in fossil and recent molluscs. *Evolutionary Ecology* 9: 586–604.

Environmental Science for Social Change. 1999. *Decline of Philippine Forests.* Map. Makati: Bookmark.

Erickson, R. O. 1945. The *Clematis fremontii* var. *riehlii* population in the Ozarks. *Annals of the Missouri Botanical Garden* 32: 413–460.

Erlinge, S. 1987. Why do European stoats *Mustela erminea* not follow Bergmann's rule? *Holarctic Ecology* 10: 33–39.

Erulkar, S. D. 1972. Comparative aspects of spatial localization of sound. *Physiological Reviews* 52: 237–360.

Erwin, D. H. 1993. *The Great Paleozoic Crisis: Life and Death in the Permian.* New York: Columbia University Press.

Erwin, T. C. 1981. Taxon pulses, vicariance and dispersal: An evolutionary synthesis illustrated by carabid beetles. In G. Nelson and D. E. Rosen (eds.), *Vicariance Biogeography: A Critique,* 159–196. New York: Columbia University Press.

Erwin, T. L. 1983. Tropical forest canopies: The last biological frontier. *Bulletin of the Entomological Society of America* 19: 14–19.

Erwin, T. L. 1988. The tropical forest canopy: The heart of biotic diversity. In E. O. Wilson (ed.), *Biodiversity,* 123–129. Washington, DC: National Academy Press.

Estes, J. A. and D. O. Duggins. 1995. Sea otters and kelp forests in Alaska: Generality and variation in a community ecology paradigm. *Ecological Monographs* 65: 75–100.

Estes, R. and A. Baez. 1985. Herptofaunas of North and South America during the Late Cretaceous and Cenozoic: Evidence for interchange? In F. G. Stehli and S. D. Webb (eds.), *The Great American Interchange,* 140–200. New York: Plenum Press.

Estes, R. and O. A. Reig. 1973. The early fossil record of frogs: A review of the evidence. In J. L. Vial (ed.), *Evolutionary Biology of the Anurans,* 11–63. Columbia: University of Missouri Press.

Estrada, A. and R. Coates-Estrada. 2001. Bat species richness in live fences and in corridors of residual rain forest vegetation at Los Tuxtlas, Mexico. *Ecography* 24: 94–102.

Facelli, J. and S. T. A. Pickett. 1990. Markovian chains and the role of history in succession. *Trends in Ecology and Evolution* 5: 27–30.

Faeth, S. H. 1984. Density compensation in vertebrates and invertebrates: A review and an experiment. In D. R. Strong, D. Simberloff, L. G. Abele and A. B. Thistle (eds.), *Ecological Communities: Conceptual Issues and the Evidence,* 491–509. Princeton, NJ: Princeton University Press.

Fagan, B. M. 1990. *The Journey from Eden: The Peopling of Our World.* London: Thames and Hudson.

Falconer, D. S. 1960. *Introduction to Quantitative Genetics.* Edinburgh: Oliver & Boyd.

Falla, R. A., R. B. Sibson and E. G. Turbott. 1966. *A Field Guide to the Birds of New Zealand and Outlying Islands.* London: William Collins Sons & Co.

Fallaw, W. C. 1979. Trans-North Atlantic similarity among Mesozoic and Cenozoic invertebrates correlated with widening of the ocean basin. *Geology* 7: 398–400.

Fallon, S. M., E. Bermingham and R. E. Ricklefs. 2003. Island and taxon effects in parasitism revisited: Avian malaria in the Lesser Antilles. *Evolution* 57: 606–615.

Farrell, T. M. 1991. Models and mechanisms of succession: An example from a rocky intertidal community. *Ecological Monographs* 61: 95–113.

Fautin, D. G. and B. R. Barber. 1999. *Maractis rimicarivora,* a new genus and species of sea anemone (Cnidaria: Anthozoa: Actiniaria: Actinostolidae) from an Atlantic hydrothermal vent. *Proceedings of the Biological Society of Washington* 112: 624.

Feduccia, A. 1980. *The Age of Birds.* Cambridge, MA: Harvard University Press.

Feduccia, A. 1995. Explosive evolution in Tertiary birds and mammals. *Science* 267: 637–638.

Feduccia, A. 1996. *The Origin and Evolution of Birds.* New Haven, CT: Yale University Press.

Felsenstein, J. 1985. Phylogenies and the comparative method. *American Naturalist* 125: 1–15.

Fenchel, T. 2003. Biogeography of bacteria. *Science* 301: 925–926.

Fenchel, T. and B. J. Finlay. 2004. The ubiquity of small species: Patterns of local and global diversity. *BioScience* 54(8): 777–784

Fernandes, G. W. and A. C. F. Lara. 1993. Diversity of Indonesian gall-forming herbivores along altitudinal gradients. *Biodiversity Letters* 1: 186–192.

Ferrari, F. D. and E. L. Markhaseva. 2000. *Grievella shanki,* a new genus and species of scolecitrichid calanoid copepod (Crustacea) from a hydrothermal vent along the southern East Pacific Rise. *Proceedings of the Biological Society of Washington* 113: 1079–1088.

Ferrari, S. F. and H. L. Queiroz. 1994. Two new Brazilian primates discovered, endangered. *Oryx* 28: 31–36.

Fichman, M. 1977. Wallace: Zoogeography and the problem of land-bridges. *Journal of the History of Biology* 10: 45–63.

Fiedel, S. J. 1999. Older than we thought: Implications of corrected dates for paleo-indians. *American Antiquity* 64: 95–116.

Fields, P. A., J. B. Graham, R. H. Rosenblatt and G. N. Somero. 1993. Effects of expected global climate change on marine faunas. *Trends in Ecology and Evolution* 8: 361–366.

Figuerola, J. and A. J. Green. 2002. Dispersal of aquatic organisms by waterbirds: A review of past research and priorities for future studies. *Freshwater Biology* 47: 483–494.

Findley, J. S. 1969. Biogeography of southwestern boreal and desert mammals. In J. K. Jones Jr. (ed.), *Contributions in Mammalogy* 113–128. Lawrence: University of Kansas, Museum of Natural History Publication no. 51.

Findley, I. S. and C. Jones. 1964. Seasonal distribution of the hoary bat. *Journal of Mammalogy* 45: 461–470.

Finerty, J. P. 1980. *The Population Ecology of Cycles in Small Mammals: Mathematical Theory and Biological Fact.* New Haven, CT: Yale University Press.

Finlay, B. J. 2002. Global dispersal of free-living microbial eukaryote species. *Science* 296: 1061–1063.

Finlay, B. J., J. O. Corliss, G. Esteban and T. Fenchel. 1996. Biodiversity at the microbial level: The number of free-living ciliates in the biosphere. *Quarterly Review of Biology* 71: 221–237.

Firth, R. and J. W. Davidson. 1945. *Pacific Islands,* vol. 1. General Survey. Naval Intelligence Division, London, 20.3.5.

Fischer, A. G. 1960. Latitudinal variation in organic diversity. *Evolution* 14: 64–81.

Fischer, A. G. 1984. The two Phanerozoic supercycles. In W. A. Berggren and J. A. Van Couvering (eds.), *Catastrophes and Earth History,* 129–150. Princeton, NJ: Princeton University Press.

Fischer, A. G. and M. A. Arthur. 1977. Secular variations in the pelagic realm. *Society for Economic Paleontology and Mineralogy Special Publications* 25: 19–50.

Fischer, J. and D. B. Lindenmayer. 2002. Treating the nestedness temperature calculator as a "black box" can lead to false conclusions. *Oikos* 99: 193–199.

Fitzhugh, W. W. 1997. Biogeographical archaeology in the Eastern North American Arctic. *Human Ecology* 25: 385–418.

Flannery, T. 2001. *The Eternal Frontier: An Ecological History of North America and Its Peoples.* New York: Grove Press.

Flannery, T. F. 1994. *The Future Eaters: An Ecological History of the Australasian Lands and People.* Kew, Victoria, N.S.W., Australia: Reed International Books.

Flather, C. H., L. A. Joyce and C. A. Bloomgarden. 1994. *Species Endangerment Patterns in the United States.* USDA Forest Service, General Technical Report RM-241.

Fleishman, E., G. T. Austin and A. D. Weiss. 1998. An empirical test of Rapoport's rule: Elevational gradients in montane butterfly communities. *Ecology* 79: 2482–2493.

Fleming, T. H. 1973. Numbers of mammal species in North and Central American forest communities. *Ecology* 54: 555–563.

Fleming, T. H. and A. Estrada (eds.). 1993. *Frugivory and Seed Dispersal: Ecological and Evolutionary Aspects.* Dordrecht, The Netherlands: Kluwer Academic.

Flenley, J. R. 1979a. *The Equatorial Rain Forest: A Geological History.* London: Butterworth.

Flenley, J. R. 1979b. The Late Quaternary vegetational history of the equatorial mountains. *Progress in Physical Geography* 3: 488–509.

Flessa, K. W. 1975. Area, continental drift and mammalian diversity. *Paleobiology* 1: 189–194.

Flessa, K. W. 1980. Biological effects of plate tectonics and continental drift. *Bioscience* 30: 518–523.

Flessa, K. W. 1981. The regulation of mammalian faunal similarity among the continents. *Journal of Biogeography* 8: 427–438.

Flessa, K. W. and J. J. Sepkoski Jr. 1978. On the relationship between Phanerozoic diversity and changes in habitable area. *Paleobiology* 4: 359–366.

Flessa, K. W., S. G. Barnett, D. B. Cornue, M. A. Lomaga, N. Lombardi, J. M. Miyazaki and A. S. Murer. 1979. Geologic implications of the relationship between mammalian faunal similarity and geographic distance. *Geology* 7: 15–18.

Flint, R. F. 1971. *Glacial and Quaternary Geology*. New York: John Wiley & Sons.

Flohn, H. 1969. *Climate and Weather*. New York: World University Library, McGraw-Hill.

Flux, J. E. C. 1969. Current work on the African hare, *Lepus capensis* L. in Kenya. *Journal of Reproduction and Fertilization* 6 (suppl.): 225–227.

Foote, M. 2003. Origination and extinction through the Phanerozoic: A new approach. *Journal of Geology* 111: 125–148.

Forbes, E. 1843. Report on the Mollusca and Radiata of the Aegean Sea. *Reports of the British Association of Science 1843 (1844)*, 130–193.

Forbes, E. 1856. Map of the distribution of marine life. In A. K. Johnston (ed.), *The Physical Atlas of Natural Phenomena*. Philadelphia: Lea and Blanchard.

Forbes, E. 1859. *The Natural History of European Seas*. London: John Van Voorst.

Forman, R. T. T. (ed.). 1979. *Pine Barrens: Ecosystem and Landscape*. New York: Academic Press.

Forster, J. R. 1778. *Observations Made during a Voyage Round the World, on Physical Geography, Natural History and Ethic Philosophy*. London: G. Robinson.

Fortin, M. J., T. H. Keith, B. A. Maurer, M. L. Taper, D. M. Kaufman and T. M. Blackburn. 2005. Species' geographic ranges and distributional limits: Pattern analysis and statistical issues. *Oikos* 108: 7–17.

Foster, D. R. 2002. Insights from historical geography to ecology and conservation: Lessons from the New England landscape. *Journal of Biogeography* 29: 1269–1278.

Foster, J. B. 1963. *The Evolution of Native Land Mammals of the Queen Charlotte Islands and the Problem of Insularity*. Ph.D. dissertation, University of British Columbia, Vancouver.

Foster, J. B. 1964. Evolution of mammals on islands. *Nature* 202: 234–235.

Fox, B. J. 1987. Species assembly and evolution of community structure. *Evolutionary Ecology* 1: 201–213.

France, R. 1992. The North American latitudinal gradients in species richness and geographic range of freshwater crayfish and amphipods. *American Naturalist* 139: 342–354.

France, R. L. 1998. Refining latitudinal gradient analyses: Do we live in a climatically symmetrical world? *Global Ecology and Biogeography Letters* 7: 295–296.

Fratini, S. M. Vannini, S. Cannicci and C. D. Schubart. 2005. Tree-climbing mangrove crabs: A case of convergent evolution. *Evolutionary Ecology Research* 7: 219–233.

Freckleton, R. P., P. H. Harvey and M. Pagel. 2003. Bergmann's rule and body size in mammals. *American Naturalist* 161: 821–825.

Fredrickson, R. and P. Hedrick. 2002. Body size in endangered Mexican wolves: Effects of inbreeding and cross-lineage matings. *Animal Conservation* 5: 39–43.

Free, J. B. 1970. *Insect Pollination of Crops*. New York: Academic Press.

Friis, E. M., W. G. Chaloner and P. R. Crane (eds.). 1987. *The Origins of Angiosperms and Their Biological Consequences*. Cambridge: Cambridge University Press.

Fritts, H. C. 1976. *Tree Rings and Climate*. New York: Academic Press.

Fryer, G. and T. D. Iles. 1972. *The Cichlid Fishes of the Great Lakes of Africa: Their Biology and Evolution*. Edinburgh: Oliver & Boyd.

Fuentes, E. R. 1976. Ecological convergence of lizard communities in Chile and California. *Ecology* 57: 3–17.

Fuentes, E. R. and F. M. Jaksic. 1979. Latitudinal size variation of Chilean foxes: Tests of alternative hypotheses. *Ecology* 60: 43–47.

Funk, V. A. 2004. Revolutions in historical biogeography. In M. V. Lomolino, D. F. Sax and J. H. Brown (eds.), *Foundations of Biogeography: Classic Papers with Commentaries* 647–657. Chicago: University of Chicago Press.

Funk, V. A. and W. L. Wagner. 1995. Biogeographic patterns in the Hawaiian Islands. In W. L. Wagner and V. A. Funk (eds.), *Hawaiian Biogeography: Evolution on a Hot Spot Archipelago* 379–419. Washington, DC: Smithsonian Institution Press.

Futuyma, D. J. 1998. *Evolutionary Biology*. 3rd edition. Sunderland, MA: Sinauer Associates.

Gadow, H. F. 1913. *The Wanderings of Animals*. Cambridge: Cambridge University Press.

Gaffney, E. S. 1975. A phylogeny and classification of the higher categories of turtles. *Bulletin of the American Museum of Natural History* 155: 387–436.

Gage, J. G. and P. A. Tyler. 1991. *Deep-Sea Biology: A Natural History of Organisms at the Deep-Sea Floor*. Cambridge: Cambridge University Press.

Gaines, S. D. and M. D. Bertness. 1992. Dispersal of juveniles and variable recruitment in sessile marine species. *Nature* 360: 579–580.

Galil, B. S. 2000. A sea under siege—Alien species in the Mediterranean. *Biological Invasions* 2: 177–186.

Gallardo, C. S. and P. E. Penchaszadeh. 2001. Hatching mode and latitude in marine gastropods: Revisiting Thorson's paradigm in the Southern Hemisphere. *Marine Biology* 138: 547–552.

Gamble, C. 1994. *Timewalkers: The Prehistory of Global Colonization*. Cambridge, MA: Harvard University Press.

Ganem, G., L. Granjon, K. Ba and J.-M. Duplantier. 1995. Body size variability and water balance: A comparison between mainland and island populations of *Mastomys huberti* (Rodentia: Muridae) in Senegal. *Experientia* 51: 402–410.

Gappa, J. L. 2000. Species richness of marine Bryozoa in the continental shelf and slope of Argentina (south-west Atlantic). *Diversity and Distributions* 6: 15–27.

Gardner, S. L. 1991. Phyletic coevolution between subterranean rodents of the genus *Ctenomys* (Rodentia, Hystricognathi) and nematodes of the genus *Paras*. *Zoological Journal of the Linnean Society* 102: 169–201.

Garvey, J. E., E. A. Marschall and R. A. Wright. 1998. From star charts to stoneflies: Detecting relationships in continuous bivariate data. *Ecology* 79: 442–447.

Gascon, C., J. R. Malcolm, J. L. Patton, M. N. F. da Silva, J. P. Bogart, S. C. Lougheed, C. A. Peres, S. Neckel and P. T. Boag. 2000. Riverine barriers and the geographic distribution of Amazonian species. *Proceedings of the National Academy of Science, USA* 97: 13672–13677.

Gaston, K. J. 1991. How large is a species' geographic range? *Oikos* 61: 434–438.

Gaston, K. J. 1994. *Rarity*. London: Chapman and Hall.

Gaston, K. J. 1996. Species-range-size distributions: Patterns, mechanisms and implications. *Trends in Ecology and Evolution* 11: 197–201.

Gaston, K. J. 2003. *The Structure and Dynamics of Geographic Ranges*. Oxford: Oxford University Press.

Gaston, K. J. and T. M. Blackburn. 1996. The tropics as a museum of biological diversity: Analysis of the New World avifauna. *Proceedings of the Royal Society of London*, Series B-Biological Sciences 263: 63–68.

Gaston, K. J. and T. M. Blackburn 2000. *Pattern and Process in Macroecology*. Oxford: Blackwell Scientific Publications.

Gaston, K. J. and R. David. 1994. Hot spots across Europe. *Biodiversity Letters* 2: 108–116.

Gaston, K. J. and S. F. Matter. 2002. Individuals-area relationships: Comment. *Ecology* 83: 288–293.

Gaston, K. J. and R. M. May. 1992. Taxonomy of taxonomists. *Nature* 356: 281–282.

Gaston, K. J. and A. S. L. Rodgrigues. 2003. Reserve selection in regions with poor biological data. *Conservation Biology* 17: 188–195.

Gaston, K. J., T. M. Blackburn and J. I. Spicer. 1998. Rapoport's rule: Time for an epitaph? *Trends in Ecology and Evolution* 13: 70–74.

Gaston, K. J., T. M. Blackburn, R. D. Gregory and J. J. D. Greenwood. 1998. The anatomy of the interspecific abundance-range size relationship for British avifauna: I. Spatial patterns. *Ecology Letters* 1: 38–46.

Gates, D. M. 1993. *Climate Change and Its Biological Consequences.* Sunderland, MA: Sinauer Associates.

Gauld, I. D. 1986. Latitudinal gradients in ichneumonid species-richness in Australia. *Ecological Entomology* 11: 155–161.

Gauld, I. D., K. J. Gaston and D. H. Janzen. 1992. Plant allelochemicals, tritrophic interactions and the anomalous diversity of tropical parasitoids: The "nasty" host hypothesis. *Oikos* 65: 353–357.

Gause, G. F. 1934. *The Struggle for Existence.* Baltimore, MD: Williams & Wilkins.

Gayet, M., J. C. Rage, T. Sempere and P. Y. Gagner. 1992. Mode of interchanges of continental vertebrates between North and South America during the late Cretaceous and Paleocene. *Bulletin de la Societe Geologique de France* 163: 781–791.

Genoways, H. H. and J. H. Brown (eds.). 1993. *Biology of the Heteromyidae.* American Society of Mammalogists, Special Publication, no. 10.

George, T. L. 1987. Greater land bird densities on island vs. mainland: Relation to nest predation level. *Ecology* 68: 1393–1400.

Ghebreab, W. 1998. Tectonics of the Red Sea region reassessed. *Earth-Science Reviews* 45: 1–43.

Gibbons, A. 1996a. Did Neandertals lose an evolutionary arms race? *Science* 272: 1586–1587.

Gibbons, A. 1996b. The peopling of the Americas. *Science* 274: 31–33.

Gibbons, A. 2001a. The peopling of the Pacific. *Science* 291: 1735–1737.

Gibbons, A. 2001b. The riddle of coexistence. *Science* 291: 1725–1729.

Gilbert, F. S. 1980. The equilibrium theory of island biogeography: Fact or fiction? *Journal of Biogeography* 7: 209–235.

Gilbert, L. E. 1984. The biology of butterfly communities. In R. I. Vane-Sright and P. R. Ackery (eds.), *The Biology of Butterflies,* 41–53. London: Academic Press.

Gili, J. M. 2002. Variability of the keystone species concept in marine ecosystems. *Trends in Ecology and Evolution* 17: 499.

Gillespie, R. G. and G. K. Roderick. 2002. Arthropods on islands: Colonization, speciation and conservation. *Annual Review of Entomology* 47: 595–632.

Gilpin, M. E. and J. M. Diamond. 1976. Calculations of immigration and extinction curves from the species-area distance relation. *Proceedings of the National Academy of Sciences, USA* 73: 4130–4134.

Gilpin, M. E. and J. M. Diamond. 1980. Subdivisions of nature reserves and the maintenance of species diversity. *Nature* 285: 567–568.

Gilpin, M. E. and J. M. Diamond. 1981. Immigration and extinction probabilities for individual species: Relation to incidence functions and species colonization curves. *Proceedings of the National Academy of Sciences, USA* 78: 392–396.

Gilpin, M. E. and I. Hanski. 1991. *Metapopulation Dynamics: Empirical and Theoretical Investigations.* London: Academic Press.

Gingerich, P. D. 1976a. Paleontology and phylogeny: Patterns of evolution at the species level in Early Tertiary mammals. *American Journal of Science* 276: 1–28.

Gingerich, P. D. 1976b. Cranial anatomy and evolution of early Tertiary Plesiadapidae. *University of Michigan Papers in Paleontology* no. 15, 1–140.

Gittleman, J. L. 1985. Carnivore body size: Ecological and taxonomic correlates. *Oecologia* 67: 540–554.

Glaessner, M. F. 1969. Decapoda. In R. C. Moore (ed.), *Treatise on Invertebrate Paleontology*, Part R, *Arthropoda*, vol. 2, 400–532. Boulder, CO: Geological Society of America.

Glazier, D. S. 1985. Energetics of litter size in five species of *Peromyscus* with generalizations for other mammals. *Journal of Mammalogy* 66: 629–642.

Gleason, H. A. 1917. The structure and development of the plant association. *Bulletin of the Torrey Botanical Club* 53: 7–26.

Gleason, H. A. 1922. On the relation between species and area. *Ecology* 3: 158–162.

Gleason, H. A. 1926. The individualistic concept of plant association. *Bulletin of the Torrey Botanical Club* 53: 7–26.

Gleick, P. H. 1989. Climate change, hydrology and water resources. *Reviews of Geophysics* 27(3): 329–344.

Glenn-Lewin, D. C. 1977. Species diversity in North American temperate forests. *Vegetatio* 33: 153–162.

Glick, P. A. 1939. *The Distribution of Insects, Spiders and Mites in the Air.* Washington, DC: U.S. Department of Agriculture.

Gliwicz, J. 1980. Island populations of rodents: Their organization and functioning. *Biological Reviews* 55: 109–138.

Gloger, C. L. 1883. Das Abandern der Vogel durch Einfluss des Klimas. Breslau: A. Schulz.

Glynn, P. W. 1991. Coral reef bleaching in the 1980s and possible connections with global warming. *Trends in Ecology and Evolution* 6(6): 175–179.

Godfrey, G. K. and P. Crowcroft. 1960. *The Life of the Mole.* London: Museum Press.

Goguel, J. 1962. *Tectonics.* English translation from the French ed. of 1952, by Hans E. Thalmann. San Francisco: W.H. Freeman.

Gofas, S. and A. Zenetos. 2003. Exotic molluscs in the Mediterranean basin: Current status and perspectives. *Oceanography and Marine Biology* 41: 237–277.

Goin, F. J. and A. A. Carlini. 1995. An early tertiary microbiotherid marsupial from Antarctica. *Journal of Vertebrate Paleontology* 15: 205–207.

Golani, D. 1993. The sandy shore of the Red Sea—Launching pad for Lessepsian (Suez Canal) migrant fish colonizers of the eastern Mediterranean. *Journal of Biogeography* 20: 579–585.

Golani, D. and A. Ben-Tuvia. 1990. Two Red Sea flatheads, immigrants in the Mediterranean (Pisces: Platycephalidae). *Cybium* 14: 57–61.

Goldblatt, P. 1978. An analysis of the flora of southern Africa: Its characteristics, relationships and origins. *Annals of the Missouri Botanical Garden* 65: 369–436.

Goldman, E. A. 1935. New American mustelids of the genera Martes, Gulo and Lutra. *Proceedings of the Biological Society of Washington* 48: 175–186.

Goldschmidt, T., F. Witte and J. Wanink. 1993. Cascading effects of the introduced Nile perch on the detritivorous/planktivorous species in the sublittoral areas of Lake Victoria. *Conservation Biology* 7: 686–700.

Good, J. M. and J. Sullivan. 2001. Phylogeography of the red-tailed chipmunk (*Tamias ruficaudus*), a northern Rocky Mountain endemic. *Molecular Ecology* 10: 2683–2695.

Good, R. 1974. *The Geography of the Flowering Plants.* 3rd edition. White Plains, NY: Longman. [First published in 1947.]

Goode, G. B. 1896. *Biobliograph of the Published Writings of Philip Lutley Sclater, FRS Secretary of the Zoological Society of London.* Washington, DC: Government Printing Office.

Goodman, C. S. and B. C. Couglin. 2000. The evolution of evo-devo biology. *Proceedings of the National Academy of Sciences, USA* 97: 4424–4425.

Gordon, K. R. 1986. Insular evolutionary body size trends in *Ursus.* *Journal of Mammalogy* 67: 395–399.

Gotelli, N. J. 1991. Metapopulation models: The rescue effect, the propagule rain, the core-satellite hypothesis. *American Naturalist* 138: 768–776.

Gotelli, N. J. 2000. Null model analysis of species co-occurrence patterns. *Ecology* 81: 2606–2621.

Gotelli, N. J. 2004. Part 7: Assembly rules. In M. V. Lomolino, D. F. Sax and James H. Brown, *Foundations of Biogeography,* 1027–1144. Chicago: Chicago University Press.

Gotelli, N. J. and A. M. Ellison. 2002. Biogeography at a regional scale: Determinants of ant species density in New England bogs and forests. *Ecology* 83: 1604–1609.

Gotelli, N. J. and G. L. Entsminger. 2001. *EcoSim: Null models software for ecology.* Version 7.0. Acquired Intelligence Inc. & Kesey-Bear. Available from: http://homepages.together.net/~gentsmin/ecosim.htm.

Gotelli, N. J. and G. R. Graves. 1996. *Null Models in Ecology*. Washington, DC: Smithsonian Institution Press.

Gotelli, N. J. and D. J. McCabe. 2002. Species co-occurrence: A meta–analysis of Diamond's (1975) assembly rules model. *Ecology* 83: 2091–2096.

Gotelli, N. J. and M. Pyron. 1991. Life history variation in North American freshwater minnows: Effects of latitude and phylogeny. *Oikos* 62: 30–40.

Gould, G. C. and B. J. MacFadden. 2004. Gigantism, dwarfism and Cope's rule: "Nothing in evolution makes sense without phylogeny." *Bulletin of the American Museum of Natural History* 285: 219–237.

Gould, S. J. 1965. Is uniformitarianism necessary? *American Journal of Science* 263: 223–228.

Gould, S. J. 1979. An allometric interpretation of species-area curves: The meaning of the coefficients. *American Naturalist* 114: 335–343.

Gould, S. J. and C. B. Calloway. 1980. Clams and brachiopods: Ships that pass in the night. *Paleobiology* 6: 383–396.

Grace, J. B. and R. G. Wetzel. 1981. Habitat partitioning and competitive displacement in cattails (*Typha*): Experimental field studies. *American Naturalist* 118: 463–474.

Graham, M. H., P. K. Dayton and M. A. Hixon. 2002. Paradigms in ecology: Past, present and future. *Ecology* 83: 1479–1480.

Graham, R. W. 1986. Response of mammalian communities to environmental changes during the Late Quaternary. In J. M. Diamond and T. J. Case (eds.), *Community Ecology*, 300–313. New York: Harper and Row.

Graham, R. W., E. L. Lundelius Jr., M. A. Graham, E. K. Schroeder, R. S. Toomey III, E. Anderson, A. D. Barnosky, J. A. Burns, C. S. Churcher, C. K. Grayson, R. D. Guthrie, C. R. Harrington, G. T. Jefferson, L. D. Martin, H. G. McDonald, R. E. Morlan, H. A. Semken Jr., S. D. Webb, L. Werdelin and M. C. Wilson. 1996. Spatial response of mammals to late-Quaternary environmental fluctuations. *Science* 272: 1601–1606.

Grande, L. 1985. The use of paleontology in systematics and biogeography and a time control refinement for historical biogeography. *Paleobiology* 11: 234–243.

Grant, B. R. and P. R. Grant. 1989. *Evolutionary Dynamics of a Natural Population: The Large Cactus Finch of the Galápagos*. Chicago: University of Chicago Press.

Grant, B. R. and P. R. Grant. 1998. Hybridization and speciation in Darwin's finches: The role of sexual imprinting on a culturally transmitted trait. In D. J. Howard and S. H. Berlocher (eds.), *Endless Forms: Species and Speciation* 404–422. Oxford: Oxford University Press.

Grant, P. R. 1965. The adaptive significance of some size trends in island birds. *Evolution* 19: 355–367.

Grant, P. R. 1966a. The density of land birds on the Tres Marias Island in Mexico. I. Numbers and biomass. *Canadian Journal of Zoology* 44: 391–400.

Grant, P. R. 1966b. Ecological compatibility of bird species on islands. *American Naturalist* 100: 451–462.

Grant, P. R. 1968. Bill size, body size and the ecological adaptations of bird species to competitive situations on islands. *Systematic Zoology* 17: 319–333.

Grant, P. R. 1971. The habitat preference of *Microtus pennsylvanicus* and its relevance to the distribution of this species on islands. *Journal of Mammalogy* 52: 351–361.

Grant, P. R. 1986. *Ecology and Evolution of Darwin's Finches*. Princeton, NJ: Princeton University Press.

Grant, P. R. 1998. *Evolution on Islands*. New York: Oxford University Press.

Grant, P. R. 2001. Reconstructing the evolution of birds on islands: 100 years of research. *Oikos* 92: 385–403.

Grant, P. R. and I. Abbott. 1980. Interspecific competition, island biogeography and null hypotheses. *Evolution* 34: 332–341.

Grant, P. R. and B. R. Grant. 1992. Hybridization of bird species. *Science* 256: 93–197.

Grant, V. 1981. *Plant Speciation*. New York: Columbia University Press.

Grassle, J. F. 1986. The ecology of deep-sea hydrothermal vent communities. *Advances in Marine Biology*, vol. 23. New York: Academic Press.

Grassle, J. F. 1989. Species diversity in deep-sea communities. *Trends in Ecology and Evolution* 4(1): 12–15.

Grassle, J. F. 1991. Deep-sea benthic biodiversity. *BioScience* 41: 464–469.

Grassle, J. F. and N. J. Maciolek. 1992. Deep-sea richness: Regional and local diversity estimates from quantitative bottom samples. *American Naturalist* 139: 313–341.

Graur, D. and W. Martin. 2004. Reading the entrails of chickens: molecular timescales of evolution and the illusion of precision. *Trends in Genetics* 20: 80–86.

Graves, G. R. 1997. Geographic clines of age rations of black-throated blue warblers (*Dendroica caerulescens*). *Ecology* 78: 2524–2531.

Gray, A. 1878. Forest geography and archaeology. *American Journal of Science and Arts* 16, 3rd series: 85–94, 183–196.

Gray, J. and A. J. Boucot (eds.). 1979. *Historical Biogeography, Plate Tectonics and the Changing Environment*. Corvallis: Oregon State University Press.

Gray, J., A. J. Boucot and W. B. N. Berry (eds.). 1981. *Communities of the Past*. Stroudsburg, PA: Hutchinson Ross.

Gray, J. S. 1997. Gradients in marine biodiversity. In R. F. G. Ormond, J. D. Gage and M. V. Angel (eds.), *Marine Biodiversity: Pattern and Process*, 18–34. Cambridge: Cambridge University Press.

Grayson, D. K. 1977. Pleistocene avifaunas and the overkill hypothesis. *Science* 195: 691–692.

Grayson, D. K. 1981. A mid-Holocene record for the heather vole, *Phenacomys* cf. *intermedius*, in the central Great Basin and its biogeographic significance. *Journal of Mammalogy* 62: 115–121.

Grayson, D. K. 1987a. An analysis of the chorology of late Pleistocene extinctions in North America. *Quaternary Research* 28: 281–289.

Grayson, D. K. 1987b. The biogeographic history of small mammals in the Great Basin: Observations on the last 20,000 years. *Journal of Mammalogy* 68: 359–375.

Grayson, D. K. 1993. *The Desert's Past: A Natural Prehistory of the Great Basin*. Washington, DC: Smithsonian Institution Press.

Grayson, D. K. 2000. Mammalian responses to Middle Holocene climatic change in the Great Basin of the western United States. *Journal of Biogeography* 27: 181–192.

Grayson, D. K. 2001. The archaeological record of human impacts on animal populations. *Journal of World Prehistory* 15: 1–68.

Grayson, D. K. and S. D. Livingston. 1993. Missing mammals on Great Basin Mountains: Holocene extinctions and inadequate knowledge. *Conservation Biology* 7: 527–532.

Grayson, D. K. and Madson, D. B. 2000. Biogeographic implications of recent low-elevation recolonization by Neotoma cinerea in the Great Basin. *Journal of Mammalogy* 81: 1100–1105.

Grayson, D. K. and D. J. Meltzer. 2002. Clovis hunting and large mammal extinction: A critical review of the evidence. *Journal of World Prehistory* 16: 313–359.

Grayson, D. K. and D. J. Meltzer. 2003. A requiem for North American overkill. *Journal of Archaeological Science* 30: 585–593.

Greely, W. B. 1925. The relation of geography to timber supply. *Economic Geography* 1: 1–11.

Greenway, J. C. Jr. 1967. *Extinct and Vanishing Birds of the World*. New York: Dover Publications.

Greenwood, P. H. 1974. The cichlid fishes of Lake Victoria, East Africa: The biology and evolution of a species flock. *Bulletin of the British Museum of Natural History* 6 (suppl.): 1–134.

Greenwood, P. H. 1984. African cichlids and evolutionary theories. In A. A. Echelle and I. Kornfield (eds.), *Evolution of Fish Species Flocks* 141–154. Orono: University of Maine Press.

Gressitt, J. L. 1954. *Insects of Micronesia*. Hawaii: Bishop Museum.

Gressitt, J. L. (ed.). 1963. *Pacific Basin Biogeography*. Honolulu: Bishop Museum Press.

Gressitt, J. L. 1982. Pacific-Asian biogeography with examples from the Coleoptera. *Entomological Genetics* 8: 1–11.

Griffiths, D. 1999. On investigating local-regional species richness relationships. *Journal of Animal Ecology* 68: 1051–1055.

Griffiths, M. 1978. *The Biology of the Monotremes*. New York: Academic Press.

Griffiths, R. C. and S. Tavaré. 1994. Simulating probability distributions in the coalescent. *Theoretical Population Biology* 46: 131–159.

Grimm, N. B. 1993. Effects of climate change for stream communities. In P. M. Karieva, J. M. Kingsolver and R. B. Huey (eds.), *Biotic Interactions and Global Change*, 293–314. Sunderland, MA: Sinauer Associates.

Grinnell, J. 1917. The niche-relationship of the California thrasher. *Auk* 34: 427–433.

Grinnell, J. 1922. The role of the "accidental." *Auk* 39: 373–380.

Grytnes, J. A. 2002. Species richness and altitude, a comparison between simulation models and interpolated plant species richness along the Himalayan altitudinal gradient. *American Naturalist* 159: 294–304.

Grytnes, J. A. 2003. Ecological interpretations of the mid-domain effect. *Ecology Letters* 6: 883–888.

Guilaine, J. 1991. *Prehistory: The World of Early Man*. New York: Facts on File Publishers.

Guilday, J. E. 1967. Differential extinction during Late Pleistocene and Recent times. In P. S. Martin and H. E. Wright Jr. (eds.), *Pleistocene Extinctions: The Search for a Cause*, 121–140. New Haven, CT: Yale University Press.

Guo, Q., J. H. Brown and B. J. Enquist. 1998. Using constraint lines to characterize plant performance. *Oikos* 83: 237–245.

Guthrie, R. D. 1990. *Frozen Fauna of the Mammoth Steppe: The Story of Blue Babe*. Chicago: University of Chicago Press.

Hackett, S. J. 1993. Phylogenetic and biogeographic relationships in the Neotropical genus *Gymnopithys* (Formicariidae). *Wilson Bulletin* 105: 301–315.

Hackett, S. J. and K. V. Rosenberg. 1990. A comparison of phenotypic and genetic differentiation in South American antwrens (Formicariidae). *Auk* 107: 473–489.

Haddrath, O. and A. J. Baker. 2001. Complete mitochondrial DNA genome sequences of extinct birds: Ratite phylogenetics and the vicariance biogeography hypothesis. *Proceedings of the Royal Society of London*, Series B 268: 939–945.

Hadley, E. A. 1997. Evolutionary and ecological response of pocket gopher (*Thomomys talpoides*) to late-Holocene climatic change. *Biological Journal of the Linnean Society* 60: 277–296.

Haeckel, E. 1876. (later editions in 1907, 1911) The history of creation, or, The development of the Earth and its inhabitants by the action of natural causes: A popular exposition of the doctrine of evolution in general and of that of Darwin, Goethe and Lamarck in particular. New York: D. Appleton.

Haeckel, E. H. P. A. 1866. *Generelle Morphologie der Organismen*. Berlin: Reimer.

Haffer, J. 1969. Speciation in Amazonian forest birds. *Science* 165: 131–137.

Haffer, J. 1974. *Avian Speciation in Tropical South America, with a Systematic Survey of the Toucans (Ramphastidae) and Jacamars (Galbulidae)*. Cambridge, MA: Nuttall Ornithological Club.

Haffer, J. 1978. Distribution of Amazon forest birds. *Bonner Zoologische Beitrage* 29: 38–78.

Haffer, J. 1981. Aspects of Neotropical bird speciation during the Cenozoic. In G. Nelson and D. E. Rosen (eds.), *Vicariance Biogeography: A Critique*, 371–391. New York: Columbia University Press.

Hafner, M. S. and S. A. Nadler. 1991. Cospeciation in host-parasite assemblages: Analysis of rates of evolution and timing of cospeciation events. *Systematic Zoology* 39: 192–204.

Hafner, M. S. and R. D. M. Page. 1995. Molecular phylogenies and host-parasite cospeciation: Gophers and lice as a model system. *Philosophical Transactions of the Royal Society of London*, Series B 349: 77–83.

Hagan, J. M. III and D. W. Johnston (eds.). 1992. *Ecology and Conservation of Neotropical Migrant Landbirds*. Washington, DC: Smithsonian Institution Press.

Haila, Y. 1986. On the semiotic dimension of ecological theory: The case of island biogeography. *Biology and Philosophy* 1: 377–387.

Haila, Y. 2002. A conceptual genealogy of fragmentation research: From island biogeography to landscape ecology. *Ecological Applications* 12: 321–334.

Haila, Y. and I. Hanski. 1993. Birds breeding on small British islands and extinction risks. *American Naturalist* 142: 1025–1029.

Haila, Y., I. Hanski, O. Jarvinen and E. Ranta. 1982. Insular biogeography: A Northern European perspective. *Acta Oecologica* 3: 303–318.

Hall, B. G. 2004. *Phylogenetic Trees Made Easy*. 2nd edition. Sunderland, MA: Sinauer Associates

Hall, C. A. S., T. Hanqin, Y. Qi, G. Pontius and J. Cornell, J. 1995. Modelling spatial and temporal patterns of tropical land use change. *Journal of Biogeography* 22: 753–757.

Hall, E. R. 1981. *The Mammals of North America*. 2nd edition. 2 vols. New York: John Wiley & Sons.

Hall, R. 1998. The plate tectonics of Cenozoic SE Asia and the distribution of land and sea. In R. Hall and J. D. Holloway (eds.), *Biogeography and Geological Evolution of SE Asia*, 99–131. Leiden: Backhuys Publishers.

Hallam, A. 1967. The bearing of certain palaeogeographic data on continental drift. *Palaeogeography, Palaeoclimatology and Palaeoecology* 3: 201–224.

Hallam, A. 1973a. Provinciality, diversity and extinction of Mesozoic marine invertebrates in relation to plate movements. In D. H. Tarling and S. K. Runcorn (eds.), *Implications of Continental Drift to the Earth Sciences*, vol. 1, 287–294. New York: Academic Press.

Hallam, A. (ed.). 1973b. *Atlas of Palaeobiogeography*. Amsterdam: Elsevier.

Hallam, A. 1983. Plate tectonics and evolution. In D. S. Bendall (ed.), *Evolution From Molecules to Men*, 367–386. Cambridge: Cambridge University Press.

Hallam, A. 1994. *Outline of Phanerozoic Biogeography*. Oxford Biogeography Series, 10. Oxford: Oxford University Press.

Hallam, A. and B. W. Sellwood. 1976. Middle Mesozoic sedimentation in relation to tectonics in the British area. *Journal of Geology* 84: 301–321.

Halliday, T. 1978. *Vanishing Birds: Their Natural History and Conservation*. New York: Holt, Rinehart and Winston.

Halloy, S. R. P. 1999. The dynamic contribution of new crops to the agricultural economy: Is it predictable? In J. Janick (ed.), *Perspectives on New Crops and New Uses*, 53–59. Alexandria, VA: ASHS Press.

Hamilton, T. H. and N. E. Armstrong. 1965. Environmental determination of insular variation in bird species abundance in the Gulf of Guinea. *Nature* 207: 148–151.

Hamilton, T. H., R. H. Barth Jr. and I. Rubinoff. 1964. The environmental control of insular variation in bird species abundance. *Proceedings of the National Academy of Sciences, USA* 52: 132–140.

Hanely, K. A., D. T. Bolger and T. J. Case. 1994. Comparative ecology of sexual and asexual gecko species (Lepidodactylus) in French Polynesia. *Evolutionary Ecology* 8: 438–454.

Hansen, T. A. 1980. Influence of larval dispersal and geographic distribution on species longevity in neogastropods. *Paleobiology* 6: 193–207.

Hanski, I. 1982. Dynamics of regional distribution: The core and satellite species hypothesis. *Oikos* 38: 210–221.

Hanski, I. 1986. Population dynamics of shrews on small islands accord with the equilibrium model. *Biological Journal of the Linnean Society* 28: 23–36.

Hanski, I. 1992. Inferences from ecological incidence functions. *American Naturalist* 139: 657–662.

Hanski, I. 1999. *Metapopulation Ecology*. Oxford: Oxford University Press.

Hanski, I. and Y. Cambefort. 1991. Spatial processes. In I. Hanski and Y. Camberfort (eds.), *Dung Beetle Ecology*, 283–304. Princeton, NJ: Princeton University Press.

Hanski, I. and O. E. Gaggiotti. 2004. *Ecology, Genetics and Evolution of Metapopulation*. Burlington, MA: Elsevier.

Hanski, I. and M. E. Gilpin. 1997. *Metapopulation Biology: Ecology, Genetics and Evolution*. San Diego, CA: Academic Press.

Hanski, I. and A. Peltonen. 1988. Island colonization and peninsulas. *Oikos* 51: 105–106.

Hanski, I. and C. D. Thomas. 1994. Metapopulation dynamics and conservation: A spatially explicit model applied to butterflies. *Biological Conservation* 68: 767–786.

Hanski, I., T. Pakkala, M. Kuussaari and G. Lei. 1995. Metapopulation persistence of an endangered butterfly in a fragmented landscape. *Oikos* 72: 21–28.

Harcourt, A. H. 2000. Latitude and latitudinal extent: A global analysis of the Rapoport effect in tropical mammalian taxon: primates. *Journal of Biogeography* 27: 1169–1182.

Hardin, G. 1960. The competitive exclusion principle. *Science* 131: 1292–1297.

Hardisty, M. V. 1979. *Biology of the Cyclostomes.* London: Chapman and Hall.

Harold, A. S. and R. D. Mooi. 1994. Areas of endemism—Definition and recognition criteria. *Systematic Biology* 43: 261–266.

Harper, J. L. 1969. The role of predation in vegetational diversity. *Brookhaven Symposia in Biology* 22: 48–62.

Harris, C. M. 1996. Absorption of sound in air vs. humidity and temperature. *Journal of Acoustic Society of America* 40: 148–159.

Harris, P., D. Peschken and J. Milroy. 1969. The status of biological control of the weed *Hypencum perforatum* in British Columbia. *Canadian Entomologist* 101: 1–15.

Harris, S. A. 2002. Global heat budget, plate tectonics and climatic change. *Geografiska Annaler* 84A: 1–9.

Harris, V. T. 1952. An experimental study of habitat selection by prairie and forest races of the deer mouse, *Peromyscus maniculatus. Contributions to the Laboratory of Vertebrate Biology, University of Michigan* 56: 1–53.

Harrison, R. G. 1998. Linking evolutionary pattern and process: The relevance of species concepts for the study of speciation. In D. J. Howard and S. H. Berlocher (eds.), *Endless Forms: Species and Speciation* 19–31. New York: Oxford University Press.

Harrison, S. 1989. Long-distance dispersal and colonization in the Bay checkerspot butterfly, *Euphydryas editha bayensis. Ecology* 70: 1236–1243.

Harrison, S. and E. Bruna. 1999. Habitat fragmentation and large-scale conservation: What do we know for sure? *Ecography* 22: 225–232.

Harrison, S., D. D. Murphy and P. R. Ehrlich. 1988. Distribution of the bay checkerspot butterfly, *Euphydryas editha bayensis:* Evidence for a metapopulation model. *American Naturalist* 132: 360–382.

Harte, J. and R. Shaw. 1995. Shifting dominance within a montane vegetation community: Results of a climate-warming experiment. *Science* 267: 876–879.

Harte, J., M. Torn, F. R. Chang, B. Feifarek, A. Kinzig, R. Shaw and K. Shen. 1995. Global warming and soil microclimate: Results from a meadow-arming experiment. *Ecological Applications* 5: 132–150.

Hartnett, D. C., K. R. Hickman and L. E. Fischer Walter. 1996. Effects of bison grazing, fire and topography on floristic diversity in tallgrass prairie. *Journal of Range Management* 49: 413–420.

Harvey, P. H. and M. D. Pagel. 1991. *The Comparative Method in Evolutionary Biology.* Oxford: Oxford University Press.

Hastings, J. R. and R. M. Turner. 1965. *The Changing Mile.* Tucson, AZ: University of Arizona Press.

Hastings, J. R., R. M. Turner and D. K. Warren. 1972. *An Atlas of Some Plant Distributions in the Sonoran Desert.* Tucson, AZ: University of Arizona Institute of Atmospheric Physics, Technical Reports of the Meteorology of Arid Lands.

Hatt, R. T. 1928. The relation of the meadow mouse *Microtus pennsylvanicus p.* to the biota of a Lake Champlain island. *Ecology* 9: 88–93.

Hausdorf, B. 1998. Weighted ancestral area analysis and a solution of the redundant distribution problem. *Systematic Biology* 47: 445–456.

Hausdorf, B. 2003. Latitudinal and altitudinal body size variation among north-west European land snail species. *Global Ecology and Biogeography* 12: 389–394.

Hausdorf, B. and C. Hennig. 2003. Nestedness of north-west European land snail ranges as a consequence of differential immigration from Pleistocene glacial refuges. *Oecologia* 135: 102–109.

Hawkins, A. F. A. 1999. Altitudinal and latitudinal distribution of east Malagasy forest bird communities. *Journal of Biogeography* 26: 447–458.

Hawkins, B. A. 1994. *Parasitoid Community Ecology.* Oxford: Oxford University Press.

Hawkins, B. A. 1995. Latitudinal body-size gradients for the bees of the eastern United States. *Ecological Entomology* 20: 195–198.

Hawkins, B. A. 2001. Ecology's oldest pattern? *Trends in Ecology and Evolution* 16: 470.

Hawkins, B. A. and J. H. Lawton. 1995. Latitudinal gradients in butterfly body sizes: Is there a general pattern? *Oecologia* 102: 31–36.

Hawkins, L. K. and P. F. Nicoletto. 1992. Kangaroo rat burrow structure and the spatial organization of ground-dwelling animals in a semiarid grassland. *Journal of Arid Environments* 23: 199–208.

Hay, O. P. 1927. *The Pleistocene of the Western Region of North America and its Vertebrate Animals.* Washington, DC: Carnegie Institute.

Heads, M. 1984. *Principia Botanica:* Croizat's contribution to botany. *Tuatara* 27(1): 8–13

Heaney, L. R. 1978. Island area and body size of insular mammals: Evidence from the tri-colored squirrel (*Callosciurus prevosti*) of Southwest Africa. *Evolution* 32: 29–44.

Heaney, L. R. 1984. Mammalian species richness on islands of the Sunda Shelf, Southwest Asia. *Oecologia* 61: 11–17.

Heaney, L. R. 1985. Zoogeographic evidence for middle and late Pleistocene land bridges to the Philippine Islands. *Modern Quaternary Research in Southeast Asia* 9: 127–143.

Heaney, L. R. 1991. An analysis of patterns of distribution and species richness among Philippine fruit bats (Pteropodidae). *Bulletin of the American Museum of Natural History* 206: 145–167.

Heaney, L. R. 1991. Distribution and ecology of small mammals along an elevational transect in southeastern Luzon, Philippines. *Journal of Mammalogy* 72: 458–469.

Heaney, L. R. 2000. Dynamic disequilibrium: A long-term, large-scale perspective on the equilibrium model of island biogeography. *Global Ecology and Biogeography* 9: 59–74.

Heaney, L. R. 2001. Small mammal diversity along elevational gradients in the Philippines: An assessment of patterns and hypotheses. *Global Ecology and Biogeography* 10: 15–40.

Heaney, L. R. 2004. Conservation biogeography in oceanic archipelagoes. In M. V. Lomolino and L. R. Heaney (eds.), *Frontiers of Biogeography: New Directions in the Geography of Nature,*345–360. Cambridge: Cambridge University Press.

Heaney, L. R. and J. C. Regalado, Jr. 1998. *Vanishing Treasures of the Philippine Rain Forest.* Chicago: The Field Museum.

Heaney, L. R. and E. A. Rickart. 1990. Correlations of clades and clines: Geographic, elevational and phylogenetic distribution patterns among Philippine mammals. In G. Peters and R. Hutterer (eds.), *Vertebrates in the Tropics,* 321–332. Bonn: Museum Alexander Koenig.

Heaney, L. R. and G. Vermeij. 2004. Diversification. In M. V. Lomolino, D. F. Sax and J. H. Brown (eds.), *Foundations of Biogeography,* 779–788. Chicago: University of Chicago Press.

Heaney, L. R., E. A. Rickart, R. B. Utzurrum and J. S. F. Klompen. 1989. Elevational zonation of mammals in the central Philippines. *Journal of Tropical Ecology* 5: 259–280.

Heatwole, H. 1987. *Sea Snakes.* Kensington, NSW: New South Wales University Press.

Hedges, S. B. 1996. Historical biogeography of West Indian vertebrates. *Annual Reviews of Ecology and Systematics* 17: 163–196.

Hedges, S. B. 2001. Biogeography of the West Indies: An overview. In C. A. Woods and F. E. Sergile (eds.), *Biogeography of the West Indies: Patterns and Perspectives* (2nd edition), 15-34. London: CRC Press.

Hedgpeth, J. W. 1957. Classification of marine environments. In J. W. Hedgpeth (ed.), *Treatise on Marine Ecology and Paleoecology,* vol. 1, *Ecology,* 17–18. Memoirs of the Geological Society of America, 67.

Hemmingsen, A. M. 1960. *Energy Metabolism as Related to Body Size and Respiratory Surfaces and Its Evolution.* Reports of the Steno Memorial Hospital and the Nordisk Insulin Laboratorium, Copenhagen, vol. 9(2).

Hendrickson, J. A. Jr. 1981. Community-wide character displacement reexamined. *Evolution* 35: 794–809.

Hengeveld, H. G. 1990. Global climate change: Implications for air-temperature and water supply in Canada. *Transactions of the American Fisheries Society* 119: 176–182.

Hengeveld, R. 1989. *Dynamics of Biological Invasions*. London: Chapman and Hall.

Hengeveld, R. 1992. *Dynamic Biogeography*. Cambridge: Cambridge University Press.

Hennig, W. 1950. Grundzüge einer theorie der phylogenetischen Systematik. Berlin: Deutscher Zentralverlag.

Hennig, W. 1966. *Phylogenetic Systematics*. 3rd edition. Trans. D. D. Davis and R. Zanderl. Urbana: University of Illinois Press.

Henstock, T. J. and A. Levander. 2003. Structure and seismotectonics of the Mendocino Triple Junction. *Journal of Geophysical Research,* 108 (BF-ESE 12) :1–17.

Herbertson, A. J. 1905. The major natural regions: An essay in systematic geography. *Geography Journal* 25: 300–312.

Hershler, R., H. P. Liu and M. Mulvey. 1999. Phylogenetic relationships within the aquatic snail genus Tryonia: Implications for biogeography of the North American southwest. *Molecular Phylogenetics and Evolution* 13: 377–391.

Heske, E. J., D. L. Rosenblatt and D. W. Sugg. 1997. Population dynamics of small mammals in an oak woodland-grassland-chaparral habitat mosaic. *Southwestern Naturalist* 42: 1–12.

Hespenheide, H. A. 1978. Are there fewer parasitoids in the tropics? *American Naturalist* 112: 766-769.

Hess, H. H. 1962. History of ocean basins. In A. E. J. Engel, H. L. James and B. F. Leonard (eds.), *Petrological Studies: A Volume in Honor of A. F. Buddington*, 599–620. New York: Geological Society of America.

Hesse, R., W. C. Allee and K. P. Schmidt. 1951. *Ecological Animal Geography*. 2nd edition. New York: John Wiley & Sons.

Hewitt, G. M. 1996. Some genetic consequences of ice ages and their role in divergence and speciation. *Biological Journal of the Linnean Society* 58: 247–276.

Hewitt, G. M. 1999. Post-glacial re-colonization of European biota. *Biological Journal of the Linnean Society* 68: 82–108.

Hewitt, G. M. 2004. Genetic consequences of climatic oscillations in the Quaternary. *Philosophical Transactions of the Royal Society of London,* Series B-Biological Sciences 359: 183–195.

Heyer, W. R. and L. R. Maxson. 1982. Distributions, relationships and zoogeography of lowland frogs/the *Leptodactylus* complex in South America, with special reference to Amazonia. In G. T. Prance (ed.), *Biological Differentiation in the Tropics*. 375–388. New York: Columbia University Press.

Heyer, W. R., M. A. Donelly, R. W. McDiarmid, L. C. Hayek and M. S. Foster. 1994. *Standard Methods for Measuring and Monitoring Biodiversity: Amphibians*. Washington, DC: Smithsonian Institution Press.

Heywood, V. H. 1989. Patterns, extents and modes of invasions by terrestrial plants. In J. A. Drake et al. (eds.), *Biological Invasions: A Global Perspective*, 31–60. New York: John Wiley and Sons.

Hildebrand, G. V., C. C. Schwartz, C. T. Robbins, M. E. Jacoby, T. A. Hanley, S. M. Arthur and C. Servheen. 1999. The importance of meat, particularly salmon, to body size, population productivity and conservation of North American brown bears. *Canadian Journal of Zoology* 77: 132–138.

Hildebrand, S. F. 1939. The Panama Canal as a passageway for fishes, with lists and remarks of the fishes and invertebrates observed. *Zoologica* 24: 15–46.

Hilgard, E. W. 1860. *Report on the Geology and Agriculture of State of Mississippi*. Jackson, Mississippi: Mississippi Printers.

Hillebrand, H. 2004. On the generality of the latitudinal diversity gradient. *American Naturalist* 163: 192–211.

Hillis, D. M. and C. Moritz. 1990. *Molecular Systematics*. Sunderland, MA: Sinauer Associates.

Hillis, D. M., C. Moritz and B. K. Mable. 1996. *Molecular Systematics*. 2nd edition. Sunderland, MA: Sinauer Associates.

Hilton-Taylor, C. 2000. *IUCN Red List of Threatened Species*, The World Conservation Union. Available from: http://www.redlist.org/ (accessed 2004).

Hintikka, V. 1963. Ueber das grossklima einiger pflanzenareale, durch zwei limakoordinatensystemen dargestellt. *Annales Botanici Societatis Zoologicae Botanicae Fennicae "Vanamo"* 34: 1–65.

Hnatiuk, S. H. 1979. A survey of germination of seeds of some vascular plants found on Aldabra Atoll. *Journal of Biogeography* 6: 105–114.

Hobbs, R. J. 1992. The role of corridors in conservation: Solution or bandwagon. *Trends in Ecology and Evolution* 7(11): 389–391.

Hobbs, R. J. and L. F. Huenneke. 1992. Disturbances, diversity and invasion: Implications for conservation. *Conservation Biology* 6: 324–337.

Hobson, J. A. 2005. Using stable isotopes to trace long-distance dispersal in birds and other taxa. *Diversity and Distribution* 11: 157–164.

Hobson, K A. 2002. Incredible journeys. *Science* 295: 981–983.

Hocker, H. W. Jr. 1956. Certain aspects of climate as related to the distribution of loblolly pine. *Ecology* 37: 824–834.

Hockin, D. C. 1982. Experimental insular zoogeography: Some tests of the equilibrium theory using meiobenthic harpacticoid copepods. *Journal of Biogeography* 9: 487–497.

Hocutt, C. H. and E. O. Wiley (eds.). 1986. *The Zoogeography of North American Freshwater Fishes*. New York: John Wiley & Sons.

Hocutt, C. H., R. F. Denoncourt and J. R. Stauffer Jr. 1978. Fishes of the Greenbrier River, West Virginia, with drainage history of the Central Appalachians. *Journal of Biogeography* 5: 59–80.

Hodkinson, D. J. and K. Thompson. 1997. Plant dispersal: The role of man. *Journal of Applied Ecology* 34: 1484–1496.

Hoekstra, H. E. and W. F. Fagan. 1998. Body size, dispersal ability and compositional disharmony: The carnivore-dominated fauna of the Kuril Islands. *Diversity and Distributions* 4: 135–149.

Hoekstra, H. E. and M. W. Nachman. 2003. Different genes underlie adaptive melanism in different populations of rock pocket mice. *Molecular Ecology* 12: 1185–1194.

Hoffecker, J. F., W. R. Powers and T. Goebel. 1993. The colonization of Beringia and the peopling of the New World. *Science* 259: 46–52

Hoffmann, R. S. 1971. Relationships of certain Holarctic shrews, genus *Sorex*. *Zeitschrift fur Saugetierkunde* 36: 193–200.

Hoffmann, R. S. 1981. Different voles for different holes: Environmental restrictions on refugial survival of mammals. In G. E. Scudder and J. L. Reveal (eds.), *Evolution Today: Proceedings of Systematic and Evolutionary Biology*, 24–45. Pittsburgh: Hunt Institute for Botanical Documentation, Carnegie-Mellon University.

Hoffmann, R. S., J. W. Koeppl and C. F. Nadler. 1979. The relationships of the amphiberingian marmots (Mammalia: Sciuridae). *Occasional Papers, Museum of Natural History, University of Kansas* 83: 1–56.

Holdaway, R. N. 1996. Arrival of rats in New Zealand. *Nature* 384: 225–226.

Holdaway, R. N. 1999. A spatio-temporal model for the invasion of the New Zealand archipelago by the Pacific rat *Rattus exulans*. *Journal of the Royal Society of New Zealand* 29: 91–105.

Holdaway, R. N. and C. Jacomb. 2000. Rapid Extinction of the Moas (Aves: Dinornithiformes): Model, Test and Implications. *Science* 287: 2250–2253.

Holdgate, M. W. 1960. The fauna of the mid-Atlantic islands. *Proceedings of the Royal Society of London*, Series B 152: 550–567.

Holdridge, L. R. 1947. Determination of world plant formations from simple climatic data. *Science* 105: 367–368.

Holland, M. M., P. G. Risser and R. J. Naiman (eds.) 1991. *Ecotones: The Role of Landscape Boundaries in the Management and Restoration of Changing Environments*. New York: Chapman and Hall.

Holland, R. F. and S. K. Jain. 1981. Insular biogeography of vernal pools in the central valley of California. *American Naturalist* 117: 24–37.

Holling, C. S. 1965. The functional response of predators to prey density and its role in mimicry and population regulation. *Memoirs of the Entomological Society of Canada* 45: 1–60.

Holloway, T. D. and N. Jardine. 1968. Two approaches to zoogeography: A study based on the distributions of butterflies, birds and bats in the Indo-Australian area. *Proceedings of the Linnean Society of London* 179: 153–188.

Holt, R. D. 1977. Predation, apparent competition and the structure of prey communities. *Theoretical Population Biology* 12: 197–229.

Holt, R. D., J. P. Grover and D. Tilman. 1994. Simple rules for interspecific dominance in systems with exploitative and apparent competition. *American Naturalist* 144: 741–771.

Hooijer, D. A. 1976. Observations on the pygmy mammoths of the Channel Islands, California. Athlon, Festschrift Loris Russel, Royal Ontario Museum, 220–225.

Hooker, J. D. 1853. *The Botany of the Antarctic Voyage of H.M.S. Discovery Ships "Erebus" and "Terror" in the Years 1839–1843.* London: Lovell Reeve.

Hooker, J. D. 1867. *Lecture on Insular Floras.* London. Delivered before the British Association for the Advancement of Science at Nottingham, August 27, 1866.

Hooker, J. D. 1877. *Journey to America.* (Unpublished bound volume archived at Royal Botanical Gardens, Kew.)

Hooper, E. T. 1942. An effect on the *Peromyscus maniculatus* Rassenkreis of land utilization in Michigan. *Journal of Mammalogy* 23: 193–196.

Hopkins, D. M. and P. A. Smith. 1981. Dated wood from Alaska and the Yukon: Implications for forest refugia in Beringia. *Quaternary Research* 15: 217–249.

Hopkins, D. M., J. W. Mathews Jr., C. E. Schweger and S. B. Young. 1982. *Paleoecology of Beringia.* New York: Academic Press.

Horai, S., K. Hayasaka, R. Kondo, K. Tsugane and N. Takahata. 1995. Recent African origin of modern humans revealed by complete sequences of hominoid mitochondrial DNAs. *Proceedings of the National Academy of Sciences, USA* 92: 532–536.

Horn, H. S. 1974. The ecology of secondary succession. *Annual Review of Ecology and Systematics* 5: 25–37.

Horn, H. S. 1975. Markovian processes in forest succession. In M. L. Cody and J. M. Diamond (eds.), *Ecology and Evolution of Communities*, 196–211. Cambridge, MA: Belknap Press.

Horn, H. S. 1981. Succession. In R. M. May (ed.), *Theoretical Ecology*, 253–271. Oxford: Blackwell Scientific Publications.

Horn, M. H. and L. G. Allen. 1978. A distributional analysis of California coastal marine fishes. *Journal of Biogeography* 5: 23–42.

Horner, E. 1954. Arboreal adaptations of *Peromyscus* with special reference to use of the tail. *Contributions, Laboratory of Vertebrate Biology, University of Michigan* 61: 1–84.

Hostetler, S. W., P. J. Bartlein, P. U. Clarke, E. E. Small and A. M. Solomon. 2000. Simulated influence of Lake Agassiz on the climate of central North America 11,000 years ago. *Nature* 405: 334–337.

Houghton, P. 1990. The adaptive significance of Polynesian body form. *Annals of Human Biology* 17: 19–32.

Howard, D. J. and S. H. Berlocher. 1998. *Endless Forms: Species and Speciation.* Oxford: Oxford University Press.

Howarth, F. G. 1990. Hawaiian terrestrial arthropods: An overview. *Bishop Museum Occasional Papers* 30: 30: 4–26.

Howe, H. F. 1985. Gomphothere fruits: A critique. *The American Naturalist* 125: 853–865.

Howell, A. B. 1917. *Birds of the Islands off the Coast of Southern California.* Pacific Coast Avifauna, no. 12. Hollywood, CA: Cooper Ornithological Club.

Howell, N., I. Kubacka and D. A. Mackey. 1998. How rapidly does the human mitochondrial genome evolve? *American Journal of Human Genetics* 59: 501–509.

Howells, W. 1973. *The Pacific Islanders.* London: Weidenfeld and Nicolson.

Hsu, K. J. 1972. When the Mediterranean dried up. *Scientific American* 227: 26–36.

Hsu, S. B. 1981. Predator-mediated coexistence and extinction. *Mathematical Biosciences* 54: 231–248.

Hubbell, S. P. 1995. Towards a theory of biodiversity and biogeography on continuous landscapes. In G. R. Carmichael, G. E. Folk and J. L. Schnoor (eds.), *Preparing for Global Change: A Midwestern Perspective*, 171–199. Amsterdam: SPB Academic Publishing.

Hubbell, S. P. and R. B. Foster. 1986. Biological chance, history and the structure of tropical rainforest tree communities. In J. M. Diamond and T. J. Case (eds.), *Community Ecology*, 314–329. New York: Harper and Row.

Hubbs, C. L. 1922. Variations in the number of vertebrae and other meristic characters of fishes correlated with the temperature of water during development. *American Naturalist* 56: 360–372.

Hubbs, C. L. and R. R. Miller. 1948. The zoological evidence: Correlation between fish distribution and hydrographic history in the desert basins of western United States, with emphasis on glacial and postglacial times. In *The Great Basin*, 17–166. *Bulletin of the University of Utah, Biology*: 107.

Huey, R. B., G. W. Gilchrist, M. L. Carlson, D. Berrigan and L. Serra. 2000. Rapid evolution of a geographic cline in an introduced fly. *Science* 287: 308–310.

Huffaker, C. B. and C. E. Kennett. 1959. A ten-year study of vegetational changes associated with biological control of Klamath weed. *Journal of Range Management* 12: 69–82.

Hugueny, B. 1989. West African rivers as biogeographic islands: Species richness of fish communities. *Oecologia* 79: 236–243.

Hultén, E. 1937. *Outline of the History of Arctic and Boreal Biota during the Quaternary Period.* Stockholm: Bokforlags Aktiebolaget Thule.

Humboldt, A. von. 1805. Essai sur la geographie des plantes accompagne d'un tableau physique des regions equinoxiales, fonde sur des mesures executees, depuis le dixieme degre de latitude boreale jusqu'au dixieme degre de latitude australe, pendant les annees 1799, 1800, 1801, 1802 et 1803. Paris: Levrault Schoell.

Humboldt, A. von. 1808. Ansichten der Natur mit Wissenschaftlichen Erlauterungen. Tubingen, Germany: J. G. Cotta.

Humphrey, S. R. 1974. Zoogeography of the nine-banded armadillo (*Dasyopus novemcinctus*) in the United States. *BioScience* 24: 457–462.

Humphries, C. J. 1981. Biogeographical methods and the southern beeches (Fagaceae: *Nothofagus*). In V. A. Funk and D. R. Brooks (eds.), *Advances in Cladistics: Proceedings of the First Meeting of the Willi Hennig Society*, 177–207. New York: New York Botanical Survey.

Humphries, C. J. 1993. Cladistic biogeography. In P. Forey, C. Humphries, I. Kitching, R. Scotland, D. Siebert and D. Williams (eds.), *Cladistics: A Practical Course in Systematics*, 137–159. The Systematics Association, Publication no. 10. Oxford: Clarendon Press.

Humphries, C. J. and L. R. Parenti. 1986. *Cladistic Biogeography.* Oxford: Clarendon Press.

Humphries, C. J., P. Y. Ladiges, M. Roos and M. Zandee. 1988. Cladistic biogeography. In A. A. Myers and P. S. Giller (eds.), *Analytical Biogeography: An Integrated Approach to the Study of Animal and Plant Distributions* 371–404. London: Chapman and Hall.

Humphries, J. M. 1984. Genetics of speciation in pupfishes from Laguna Chichancanab, Mexico. In A. A. Echelle and I. Kornfield (eds.), *Evolution of Fish Species Flocks*, 129–140. Orono: University of Maine Press.

Humphries, J. M. and R. R. Miller. 1981. A remarkable species flock of pupfishes, genus *Cyprinodon*, from Yucatan, Mexico. *Copeia* 1981: 53–64.

Hunter, M. I. and A. Hutchinson. 1994. The virtues and shortcomings of Parochialism: Conserving species that are locally rare, but globally common. *Conservation Biology* 8: 1163–1165.

Huntley, B. and H. J. B. Birks. 1983. *An Atlas of Past and Present Pollen Maps for Europe: 0–12,000 Years Ago.* Cambridge: Cambridge University Press.

Hurley, P. M. 1968. The confirmation of continental drift. *Scientific American* 218(4): 52–62.

Hurley, P. M. and J. R. Rand. 1969. Pre-drift continental nuclei. *Science* 164: 1229–1242.

Huston M. A. 1994. *Biological Diversity: The Coexistence of Species on Changing Landscapes.* Cambridge: Cambridge University Press.

Hutchinson, G. E. 1957. *A Treatise on Limnology*, vol. 1. New York: John Wiley and Sons.

Hutchinson, G. E. 1958. Concluding remarks. *Cold Spring Harbor Symposia on Quantitative Biology* 22: 415–427.

Hutchinson, G. E. 1959. Homage to Santa Rosalia, or why are there so many kinds of animals? *American Naturalist* 93: 145–159.

Hutchinson, G. E. 1961. The paradox of the plankton. *American Naturalist* 95: 137–145.

Hutchinson, G. E. 1967. *A Treatise on Limnology*, vol. 2. New York: John Wiley and Sons.

Hutchinson, G. E. 1975. *A Treatise on Limnology*, vol. 3. New York: John Wiley and Sons.

Hutchinson, G. E. 1978. *An Introduction to Population Ecology*. New Haven, CT: Yale University Press.

Hutchinson, G. E. 1993. *A Treatise on Limnology*, vol. 4. *The Zoobenthos*. New York: John Wiley and Sons.

Hutchinson, G. E. and R. H. MacArthur. 1959. A theoretical ecological model of size distributions among species of animals. *American Naturalist* 93: 117–125.

Hutton, J. 1795. *Theory of the Earth with Proofs and Illustrations*. Edinburgh.

Huxley, J. S. 1932. *Problems of Relative Growth*. New York: Dial.

Hyde, W. T., T. J. Crowley, S. K. Baum and W. R. Peltier. 2000. Neo-proterozoic "snowball Earth" simulations with a coupled climate/ice-sheet model. *Nature* 405: 425–429.

Inger, R. F. 1954. Systematics and zoogeography of Philippine Amphibia. *Fieldiana Zoology* 33: 181–531.

Inouye, R. S., N. J. Huntly, D. Tilman, J. R. Tester, M. A. Stillwell and K. C. Zinnel. 1987. Old field succession on a Minnesota sand plain. *Ecology* 68: 12–26.

Intergovernmental Panel on Climate Change (IPCC). 2001a. *Climate Change 2001: The Scientific Basis*. J. T. Houghton, Y. Ding, D. J. Griggs, M. Noguer, P. J. van der Linden and D. Xiaosu. Contribution of working group I to the third assessment report of IPCC. Cambridge: Cambridge University Press.

Intergovernmental Panel on Climate Change (IPCC). 2001b. *Climate Change 2001: Impacts, Applications and Vulnerability*. J. J. McCarthy, O. F. Canziani, N. A. Leary, D. J. Dokken and K. S. White. Contribution of working group II to the third assessment report of IPCC. Cambridge: Cambridge University Press.

International Union for the Conservation of Nature and Natural Resources (IUCN). 2003. *Red List of Threatened Species*. Cambridge: IUCN.

Iriarte, J. A., W. L. Franklin, W. E. Johnson and K. H. Redford. 1990. Biogeographic variation of food habits and body size of the American puma. *Oecologia* 85: 185–190.

Irving, E. 1956. Paleomagnetic and paleoclimatological aspects of polar wandering. *Pure and Applied Geophysics* 33: 23–41.

Irving, E. 1959. Paleomagnetic pole positions. *Journal of the Royal Astronomy Society Geophysics* 2: 51–77.

Irwin, G. 1992. *The Prehistoric Exploration and Colonisation of the Pacific*. Cambridge: Cambridge University Press.

Isenmann, P. 1982. The influence of insularity on fecundity in tits (Aves, Paridae) in Corsica. *Acta Oecologia* 3: 295–301.

Iturralde-Vinent, M. A. and R. D. E. MacPhee. 1999. Paleogeography of the Caribbean region: Implications for Cenozoic biogeography. *Bulletin of the American Museum of Natural History* 238: 1–95.

Ivantsoff, W., P. Unmack, B. Saeed and L. E. L. M. Crowley. 1991. A redfinned blue-eye, a new species and genus of the family Pseudomugilidae from central western Queensland. *Fishes of Sahul* 6: 277–282.

Iverson, J. B., C. P. Balgooyen, K. K. Byrd and K. K. Lyddan. 1993. Latitudinal variation in egg and clutch size in turtles. *Canadian Journal of Zoology* 71: 2448–2461.

Jablonski, D. 1982. Evolutionary rates and modes in Late Cretaceous gastropods: Role of larval ecology. In B. Mamet and M. J. Copeland (eds.), *Proceedings of the Third North American Paleontological Convention, Toronto*.

Jablonski, D. 1986. Background and mass extinctions: The alternation of macroevolutionary regimes. *Science* 231: 129–133.

Jablonski, D. 1987. Heritability at the species level: Analysis of geographic ranges of Cretaceous mollusks. *Science* 238: 360–363.

Jablonski, D. 1989. The biology of mass extinction: A paleontological view. *Philosophical Transactions of the Royal Society of London*, Series B 325: 357–368.

Jablonski, D. 1991. Extinctions: A paleontological perspective. *Science* 253: 754–757.

Jablonski, D. 2000. Micro- and macroevolution: Scale and hierarchy in evolutionary biology and paleobiology. *Paleobiology* 26: 15–52.

Jablonski, D. 2002. Survival without recovery after mass extinctions. *Proceedings of the National Academy of Sciences, USA* 99: 8139–8144.

Jablonski, D. and D. J. Bottjer. 1990. Onshore-offshore trends in marine invertebrate evolution. In R. M. Ross and W. D. Allmon (eds.), *Causes of Evolution: A Paleontological Perspective*, 21–75. Chicago: University of Chicago Press.

Jablonski, D. and D. J. Bottjer. 1991. Environmental patterns in the origins of higher taxa: The post-Paleozoic fossil record. *Science* 252: 1831–1833.

Jablonski, D. and R. A. Lutz. 1980. Molluscan shell morphology: Ecological and paleontological applications. In D. C. Rhoads and R. A. Lutz (eds.), *Skeletal Growth of Aquatic Organisms*, 323–377. New York: Plenum Press.

Jablonski, D., J. J. Sepkoski Jr., D. J. Bottjer and P. M. Sheehan. 1983. Onshore-offshore patterns in the evolution of Phanerozoic shelf communities. *Science* 222: 1123–1125.

Jaccard, P. 1902. Etude comparative de la distribution florale dans une portion des Alpes et du Jura. *Bulletin de la Societe Vaudoise de la Science Naturelle* 37: 547–579.

Jaccard, P. 1908. Nouvelles recherches sur la distribution florale. *Bulletin de la Societe Vaudoise de la Science Naturelle* 44: 223–276.

Jackson, E. D., E. A. Silver and G. B. Dalrymple. 1972. Hawaiian-Emperor chain and its relation to Cenozoic circumpacific tectonics. *Bulletin of the Geological Society of America* 83: 601–618.

Jackson, H. H. T. 1919. An apparent effect of winter inactivity upon the distribution of mammals. *Journal of Mammalogy* 1: 58–64.

Jackson, J. A. 1978. Alleviating problems of competition, predation, parasitism and disease in endangered birds: A review. In S. A. Temple (ed.), *Endangered Birds: Management Techniques for Preserving Threatened Species*, 75–112. Madison: University of Wisconsin Press.

Jackson, J. B. C. 1974. Biogeographic consequences of eurytopy and stenotopy among marine bivalves and their evolutionary significance. *American Naturalist* 108: 541–560.

Jackson, J. B. C. and A. H. Cheetham. 1999. Tempo and mode of speciation in the sea. *Trends in Ecology and Evolution* 14: 72–77.

Jackson, S. T. 2004. Quaternary biogeography: Linking biotic responses to environmental variability across timescales. In, M. V. Lomolino and L. R. Heaney (eds.), *Frontiers of Biogeography–New Directions in the Geography of Nature*. London: Cambridge University Press.

Jackson, S. T. and D. R. Whitehead. 1991. Holocene vegetation patterns in the Adirondack Mountains. *Ecology* 72: 641–653.

Jackson, S. T., J. T. Overpeck, T. Webb III, S. E. Keattch and K. H. Anderson. 1997. Mapped plant-macrofossil and pollen records of Late Quaternary vegetation change in eastern North America. *Quaternary Science Reviews* 16: 1–70.

Jaenike, J. 1978. Effect of island area on *Drosophila* population densities. *Oecologia* 36: 327–332.

Jaenike, J. 1991. Mass extinction of European fungi. *Trends in Ecology and Evolution* 6(6): 174–175.

Jaksic, F.M. and E. R. Fuentes. 1980. Why are native herbs in the Chilean matorral more abundant beneath shrubs: Microclimate or grazing? *Journal of Ecology* 68: 665–669.

James, F. C. 1970. Geographic size variation in birds and its relationship to climate. *Ecology* 51: 365–390.

James, F. C., R. F. Johnston, N. O. Wamer, G. J. Niemi and W. J. Boecklen. 1984. The Grinellian niche of the wood thrush. *American Naturalist* 124: 17–47.

James, H. F. 1995. Prehistoric extinctions and ecological changes on oceanic islands. In P. M. Vitousek, L. L. Loope and H. Andersen (eds.), *Islands: Biological Diversity and Ecosystem Function*, 87–102. New York: Springer Verlag.

Jannasch, H. W. and C. O. Wirsen. 1980. Chemosynthetic primary production at East Pacific sea floor spreading center. *Bioscience* 29: 592–598.

Janzen, D. H. 1966. Coevolution of mutualism between ants and acacias in Central America. *Evolution* 20: 249–275.

Janzen, D. H. 1967. Why mountain passes are higher in the tropics. *American Naturalist* 101: 233–249.

Janzen, D. H. 1970. Herbivores and the number of tree species in tropical forests. *American Naturalist* 104: 501–528.

Janzen, D. H. 1973. Sweep samples of tropical foliage insects: Effects of seasons, vegetation types, elevation, time of day and insularity. *Ecology* 54: 687–708.

Janzen, D. H. 1981. The peak in North American ichneumonid species richness lies between 38° and 42° N. *Ecology* 62: 532–557.

Janzen, D. H. 1985. The natural history of mutualisms. In D. H. Boucher (ed.), *The Biology of Mutualisms*, 40–99. London: Croom Helm.

Janzen, D. H. and P. S. Martin. 1982. Neotropical anachronisms: The fruits the gomphotheres ate. *Science* 215: 19–27.

Jarrard, R. D. and D. A. Clague. 1977. Implications of Pacific island and seamount ages for the origin of volcanic chains. *Review of Geophysics and Space Physics* 15: 57–76.

Järvinen, O. and L. Sammalisto. 1976. Regional trends in the avifauna of Finnish peatland bogs. *Annales Zoologica Fennici* 13: 31–43.

Järvinen, O., J. Kouki and U. Häyrinen. 1987. Reversed latitudinal gradients in total density and species richness of birds on Finish mires. *Ornis Fennici* 64: 67–73.

Johannsen, K. 2003. Evolution in Littorina: Ecology matters. *Journal of Sea Research* 49: 107–117.

Johansson, M. E. and P. A. Keddy. 1991. Intensity and asymmetry of competition between two plant pairs of different degrees of similarity: An experimental study on two guilds of wetland plants. *Oikos* 60: 27–34.

Johnson, C. N. 2002. Determinants of loss of the Late Quaternary 'megafauna' extinctions: Life history and ecology, but not body size. *Proceedings of the Royal Society of London*, Series B 269: 2221–2227.

Johnson, D. D. P., S. I. Hay and D. J. Rogers. 1998. Contemporary environmental correlates of endemic bird areas derived from meteorological satellite sensors. *Proceedings of the Royal Society of London*, Series B 265: 951–959.

Johnson, D. L. 1978. The origin of island mammoths and the Quaternary land bridge history of the Northern Channel Islands, California. *Quaternary Research* 10: 204–225.

Johnson, D. L. 1980. Problems in the land vertebrate zoogeography of certain islands and the swimming powers of elephants. *Journal of Biogeography* 7: 383–398.

Johnson, D. L. 1981. More comments on the Northern Channel Island mammoths. *Quaternary Research* 15: 105–106.

Johnson, G. A. L. 1973. Closing of the Carboniferous sea in Western Europe. In D. H. Tarling and S. K. Runcorn (eds.), *Implications of Continental Drift to the Earth Sciences*, vol. 2, 843–850. New York: Academic Press.

Johnson, L. A. S. and B. G. Briggs. 1975. On the Proteaceae: The evolution and classification of a southern family. *Botanical Journal of the Linnean Society* 70(2): 83–182.

Johnson, L. E. and J. T. Carlton. 1996. Post-establishment spread in large-scale invasions: Dispersal mechanisms of the zebra mussel *Dreissena polymorpha*. *Ecology* 77: 1688–1690.

Johnson, M. P. and P. H. Raven. 1973. Species number and endemism: The Galápagos Archipelago revisited. *Science* 179: 893–895.

Johnson, M. P., L. G. Mason and P. H. Raven. 1968. Ecological parameters and species diversity. *American Naturalist* 102: 297–306.

Johnson, N. K. 1975. Controls of the number of bird species on montane islands in the Great Basin. *Evolution* 29: 545–567.

Johnson, N. K. and C. Cicero. 2004. New mitochondrial DNA data affirm the importance of Pleistocene speciation in North American birds. *Evolution* 58: 1122–1130.

Johnson, R. A. and P. S. Ward. 2002. Biogeography and endemism of ants (Hymenoptera: Formicidae) in Baja California, Mexico: A first overview. *Journal of Biogeography* 29: 1009–1026.

Johnson, T. C., C. A. Scholz, M. R. Talbot, K. Kelts, R. D. Ricketts, G. Ngobi, K. Beuning, I. Ssemmanda and J. W. McGill. 1996. Late Pleistocene desiccation of Lake Victoria and rapid evolution of cichlid fishes. *Science* 273: 1091–1093.

Johnston, M. C. 1963. Past and present grasslands of southern Texas and northeastern Mexico. *Ecology* 44: 456–466.

Johnston, R. F. and W. J. Klitz. 1977. Variation and evolution in a granivorous bird: The house sparrow. In J. Pinowski and S. C. Kendeigh (eds.), *Granivorous Birds in Ecosystems*, 15–51. Cambridge: Cambridge University Press.

Johnston, R. F. and R. K. Selander. 1964. House sparrows: Rapid evolution of races in North America. *Science* 144: 548–550.

Johnston, R. F. and R. K. Selander. 1971. Evolution in the house sparrow. II. Adaptive differentiation in North American populations. *Evolution* 25: 1–28.

Jolivet, L. and C. Faccenna. 2000. Mediterranean extension and the Africa-Eurasia collision. *Tectonics* 19: 1095–1107.

Jones, C. G. and J. H. Lawton. 1994. *Linking Species and Ecosystems*. London: Chapman and Hall.

Jones, D. L., A. Cox, P. Coney and M. Beck. 1982. The growth of western North America. *Scientific American* 247(5): 70–84.

Jones, H. L. and J. M. Diamond. 1976. Short-time-base studies of turnover in breeding bird populations on the California Channel Islands. *Condor* 78: 526–549.

Jones, J. K. Jr. and H. H. Genoways. 1970. Chiropteran systematics. In B. H. Slaughter and D. W. Walton (eds.), *About Bats*, 3–21. Dallas: Southern Methodist University Press.

Jones, J. R. E. 1949. A further study of calcareous streams in the "Black Mountain" district of South Wales. *Journal of Animal Ecology* 18: 142–159.

Jones, M. B. 2003. *Migrations and Dispersal of Marine Organisms*. Dordrecht: Kluwer Academic.

Jordan, D. S. 1891. *Temperature and Vertebrae: A Study in Evolution*. New York: Wilder-Quarter Century Books.

Jordan, D. S. 1908. The law of geminate species. *American Naturalist* 42: 73–80.

Jordan, P. 1971. *The Expanding Earth: Some Consequences of Dirac's Gravitational Hypothesis*. New York: Pergamon Press.

Kadmon, R. 1995. Nested species subsets and geographic isolation: A case study. *Ecology* 76: 458–465.

Karieva, P. M., J. G. Kingsolver and R. B. Huey (eds.). 1993. *Biotic Interactions and Global Change*. Sunderland, MA: Sinauer Associates.

Karl, D. M., C. O. Wirsen and H. W. Jannasch. 1980. Deep-sea primary production at the Galápagos hydrothermal vents. *Science* 207: 1345–1347.

Karl, S. A., S. Schultz, D. Desbryires, R. C. Vrijenhoek and R. A. Lutz. 1996. Molecular analysis of gene flow in the hydrothermal-vent clam, *Calypogena magnifica*. *Molecular Marine Biology and Biotechnology* 5: 193–202.

Karr, J. R. 1982 Avian extinction on Barro Colorado Island, Panama: A reassessment. *American Naturalist* 119: 220–239.

Karr, J. R. 1990. Avian survival rates and the extinction process on Barro Colorado Island, Panama. *Conservation Biology* 4: 391–397.

Kaspari, M. and E. Vargo. 1995. Does colony size buffer environmental variation? Bergmann's rule and social insects. *American Naturalist* 145: 610–632

Kates, R. W. 1996. Population, technology and the human environment: A thread through time. *Daedalus* 125: 43–71.

Katsman, C. A., P. C. F. Van der Vaart, H. A. Dijkstra and W. P. M. de Ruijter. 2003. Stability of multilayer ocean vortices: A parameter study including realistic Gulf Stream and Agulhas Rings source. *Journal of Physical Oceanography* 33(6): 1197–1218.

Kattan, G. H., H. Alvarez-Lopez and M. Giraldo. 1994. Forest fragmentation and bird extinctions: San Antonio eighty years later. *Conservation Biology* 8: 138–146.

Katzmarzyk, P. T. and W. R. Leonard. 1995. Body mass, surface area and climate. *Human Biology Council Program and Abstracts* 132.

Kaufman, D. M. 1995. Diversity of New World mammals: Universality of the latitudinal gradient of species and bauplans. *Journal of Mammalogy* 76: 322–334.

Kaufman, D. M. and M. R. Willig. 1998. Latitudinal patterns of mammalian species richness in the New World: The effects of sampling method and faunal group. *Journal of Biogeography* 25: 795–805.

Kaufman, L. and P. Ochumba. 1993. Evolutionary and conservation biology of cichlid fishes as revealed by faunal remnants in northern Lake Victoria. *Conservation Biology* 7: 719–730.

Keast, A. 1971. Adaptive evolution and shifts in niche occupation in island birds. In W. L. Stern (ed.), *Adaptive Aspects of Insular Evolution*, 39–53. Pullman: Washington State University Press.

Keast, A. 1972a. Continental drift and the biota of the mammals on southern continents. In A. Keast, F. C. Erk and B. Glass (eds.), *Evolution, Mammals and Southern Continents*, 23–87. Albany: State University of New York Press.

Keast, A. 1972b. Australian mammals: Zoogeography and evolution. In A. Keast, F. C. Erk and B. Glass (eds.), *Evolution, Mammals and Southern Continents*, 195–246. Albany: State University of New York Press.

Keast, A. 1972c. Comparisons of contemporary mammal faunas of southern continents. In A. Keast, F. C. Erk and B. Glass (eds.), *Evolution, Mammals and Southern Continents*, 433–501. Albany: State University of New York Press.

Keast, A. 1977. Zoogeography and phylogeny: The theoretical background and methodology to the analysis of mammal and bird fauna. In H. J. Frith and J. H. Calaby (eds.), *Proceedings of the 16th International Ornithological Congress, Australian Academy of Sciences* 246–312.

Keddy, P. A. 1982. Population ecology on an environmental gradient: *Cakile edentula* on a sand dune. *Oecologia* 52: 348–355.

Keddy, P. A. and P. MacLelan. 1990. Centrifugal organization in forests. *Oikos* 59: 75–84.

Kelt, D. A. and J. H. Brown. 1999. Community structure and assembly rules: Confronting conceptual and statistical issues with data on desert rodents. In E. Weiher and P. A. Keddy (eds.), *Ecological Assembly Rules—Perspectives, Advances, Retreats*, 75–107 (Ch. 3). Cambridge: Cambridge University Press.

Kelt, D. A. and D. Van Vuren. 1999. On the relationship between body size and home range area: Consequences of energetic constraints in mammals. *Ecology* 80:337–400.

Kelt, D. A. and D. H. Van Vuren. 2001. The ecology and macroecology of mammalian home range area. *American Naturalist* 157: 637–645.

Kelt, D. A., J. H. Brown, G. Shenbrot and J. H. Brown. 1999. Patterns in the structure of Asian and North American desert small mammal communities. *Journal of Biogeography* 26: 825–841.

Kennedy, W. J. 1977. Ammonite evolution. In A. Hallam (ed.), *Patterns of Evolution as Illustrated by the Fossil Record*, 251–304. Amsterdam: Elsevier.

Keogh, J. S., R. Shine and S. Donnellan. 1998. Phylogenetic relationships of terrestrial Australo-Papuan elapid snakes (subfamily Hydrophiinae) based on cytochrome b and 16S rRNA sequences. *Molecular Phylogenetics and Evolution* 10: 67–81.

Kerr, R. A. 1995. Earth's surface may move itself. *Science* 269: 1214–1216.

Kerr, R. A. 2000. An appealing snowball Earth that's still hard to swallow. *Science* 287: 1734–1736.

Kerr, R. A. 2003. Megafauna died from big kill, not big chill. *Science* 300: 885.

Kettlewell, H. B. D. 1961. The phenomenon of industrial melanisms in Lepidoptera. *Annual Review of Ecology and Systematics* 6: 245–262.

Kidwell, S. M. and S. M. Holland. 2002. The quality of the fossil record: implications for evolutionary analyses. *Annual Review of Ecology and Systematics* 33: 561–588.

Kiester, A. R. 1971. Species density of North American amphibians and reptiles. *Systematic Zoology* 20: 127–137.

Kikkawa, J. and K. Pearse. 1969. Geographical distribution of land birds in Australia: A numerical analysis. *Australian Journal of Zoology* 17: 821–840.

Kikkawa, J. and E. E. Williams. 1971. Altitudinal distribution of land birds in New Guinea. *Search* 2: 64–69.

King, C. M. and P. J. Moors. 1979. On co-existence, foraging strategy and the biogeography of weasels and stoats (*M. nivalis* and *M. erminea*) in Britain. *Oecologia* 39: 129–150.

Kirch, P. V. and T. L. Hunt. 1997. *Historical Ecology in the Pacific Islands: Prehistoric Environmental and Landscape Change.* New Haven, CT: Yale University Press.

Kitchener, D. J., A. Chapman, J. Dell, B. G. Muir and M. Palmer. 1980. Lizard assemblage and reserve size and structure in the Western Australian wheatbelt—some implications for conservation. *Biological Conservation* 17: 25–61.

Klein, B. C. 1989. Effects of forest fragmentation on dung and carrion beetle communities in Central Amazon. *Ecology* 70: 1715–1725.

Klicka, J. and R. M. Zink. 1997. The importance of recent ice ages in speciation: A failed paradigm. *Science* 277: 1666–1669.

Klicka, J. and R. M. Zink. 1999. Pleistocene effects on North American songbird evolution. *Proceedings of the Royal Society of London*, Series B-Biological Sciences 266: 695–700.

Knoll, A. H. 1986. Patterns of change in plant communities through geological time. In J. Diamond and T. J. Case (eds.), *Community Ecology*, 126–141. New York: Harper and Row.

Knowles, L. L. 2001. Did the Pleistocene glaciations promote divergence? Tests of explicit refugial models in montane grasshoppers. *Molecular Ecology* 10: 691–701.

Knowles, L. L. 2003. The burgeoning field of statistical phylogeography. *Journal of Evolutionary Biology* 17: 1–10.

Knowles, L. L. and W. P. Maddison. 2002. Statistical phylogeography. *Molecular Ecology* 11: 2623–2635.

Knowlton, N. 1993. Sibling species in the sea. *Annual Review of Ecology and Systematics* 24: 189–216.

Knowlton, N. and L. A. Weigt. 1998. New dates and new rates for divergence across the Isthmus of Panama. *Proceedings of the Royal Society of London*, Series B 265: 2257–2263.

Knox, A. K., J. B. Losos and C. Schneider. 2001. Adaptive radiation vs. intraspecific differentiation: Morphological variation in Caribbean Anolis lizards. *Journal of Evolutionary Biology* 14: 904–909.

Knox, E. B., S. R. Downie and J. D. Palmer. 1993. Chloroplast genome rearrangements and the evolution of giant lobelias from herbaceous ancestors. *Molecular Biology and Evolution* 10: 414–430.

Knutti, R. T., F. Stocker, F. Joos and G. K. Palmer. 2002. Constraints on radiative forcing and future climate change from observations and climate model ensembles. *Nature* 416: 719–723.

Kocher, T. D., J. A. Conroy, K. R. McKaye and J. R. Stauffer. 1993. Similar morphologies of cichlid fish in Lakes Tanganyika and Malawi are due to convergence. *Molecular Phylogenetics and Evolution* 2: 158–165.

Kodric-Brown, A. and J. H. Brown. 1979. Competition between distantly related taxa and the co-evolution of plants and pollinators. *American Zoologist* 19: 1115–1127.

Kohn, A. J. 1978. Ecological shift and release in an isolated population: *Conus miliaris* at Easter Island. *Ecological Monographs* 48: 323–336.

Kojola, I. and H.-M. Laitala. 2001. Body size variation of brown bear in Finland. *Annales Zoologici Fennici* 38: 173–178.

Koopman, K. F. and J. K. Jones. 1970. Classification of bats. In B. H. Slaughter and D. W. Walton (eds.), *About Bats*, 22–28. Dallas: Southern Methodist University Press.

Koopman, K. F. and D. W. Steadman. 1995. Extinction and biogeography of bats on 'Eua, Kingdom of Tonga. *American Museum Novitates* 3125: 1–13.

Kornfield, I. and P. F. Smith. 2000. African cichlid fishes: Model systems for evolutionary biology. *Annual Review of Ecology and Systematics* 31: 163–196.

Korobytsina, K. V., D. F. Nadler, N. N. Vorontsov and R. S. Hoffmann. 1974. Chromosomes of the Siberian snow sheep, *Ovis nivicola* and implications concerning the origin of amphiberingian wild sheep. *Quaternary Research* 4: 235–245.

Kotze, D. J., J. Niemela and M. Nieminen. 2000. Colonization success of carabid beetles on Baltic Islands. *Journal of Biogeography* 27: 807–819.

Kouki, J. 1999. Latitudinal gradients in species richness in northern areas: Some exceptional patters. *Ecological Bulletins* 47: 30–37.

Kouki, J., P. Niemelä and M. Viitasaari. 1994. Reversed latitudinal gradient in species richness of sawflies (Hymenoptera, Symphyta). *Annales Zoologica Fennici* 31: 83–88.

Koune, J.-P. 2001. *Threatened Mushrooms in Europe.* Strasbourg: Council of Europe Publishers.

Krebs, C. J. 1978. Ecology: *The Experimental Analysis of Distribution and Abundance.* New York: Harper & Row.

Krebs, C. J., B. L. Keller and R. H. Tamarin. 1969. *Microtus* population biology: Demographic changes in fluctuating populations of *M. ochrogaster* and *M. pennsylvanicus* in southern Indiana. *Ecology* 50: 587–607.

Kremp, G. O. W. 1992. Earth expansion theory vs. statical assumption. In S. Chatterjee and N. Hotton III (eds.), *New Concepts in Global Tectonics*, 297–309. Lubbock, TX: Texas Tech University Press.

Krzanowski, A. 1967. The magnitude of islands and the size of bats (Chiroptera). *Acta Zoologica Cracoviensia* 15, XI: 281–348.

Kuch, M., N. Rohland, J. L. Betancourt, C. Latorre, S. Steppan and H. N. Poinar. Molecular analysis of an 11,700-year-old rodent midden from the Atacama Desert, Chile. *Molecular Ecology* 11: 913–924.

Kuhn, T. 1970. *The Structure of Scientific Revolutions.* Chicago: University of Chicago Press.

Kuhner, M. K., J. Yamato and J. Felsenstein. 1998. Maximum likelihood estimation of population growth rates based on the coalescent. *Genetics* 149: 429–434.

Kumar, S. and S. B. Hedges. 1998. A molecular timescale for vertebrate evolution. *Nature* 392: 917–920.

Kummer, D. M. 1992. *Deforestation in the Postwar Philippines.* Chicago: University of Chicago Press.

Kunin, W. E. and K. J. Gaston. 1997. *The Biology of Rarity: Causes and Consequences of Rare-Common Differences.* New York: Chapman and Hall.

Kurtén, B. and E. Anderson. 1980. *Pleistocene Mammals of North America.* New York: Columbia University Press.

Lack, D. 1947. *Darwin's Finches.* Cambridge: Cambridge University Press.

Lack, D. 1970. Island birds. *Biotropica* 2: 29–31.

Lack, D. 1973. The numbers of species of hummingbirds in the West Indies. *Evolution* 27: 326–337.

Lack, D. 1974. *Evolution Illustrated by Waterfowl.* Oxford: Blackwell Scientific Publications.

Lack, D. 1976. *Island Biology Illustrated by the Land Birds of Jamaica.* Studies in Ecology, vol. 3. Berkeley: University of California Press.

Lack, D. and R. E. Moreau. 1965. Clutch size in tropical passerine birds of forest and savanna. *L'Oiseau* 35: 76–89.

La Greca, M. and C. F. Sacchi. 1957. Problemi del popamento animale nelle piccole isole mediterranee. *Ann. Inst. Mus. Zool. Univ. Napoli* 9: 1–189.

Lai, D. Y. and P. L. Richardson. 1977. Distribution and movement of Gulf Stream rings. *Journal of Physical Oceanography* 7: 670–683.

Laliberte, A. S. and W. J. Ripple. 2003. Wildlife encounters by Lewis and Clark: A spatial analysis between Native Americans and wildlife. *Bioscience* 53: 1006–1115

Laliberte, A. S. and W. J. Ripple. 2004. Range contractions of North American carnivores and ungulates. *BioScience* 54: 123–138.

La Marche, V. C. 1973. Holocene climatic variations inferred from treeline fluctuations in the White Mountains, California. *Quaternary Research* 3: 632–660.

La Marche, V. C. 1978. Tree-ring evidence of past climatic variability. *Nature* 276: 334–338.

Lambert, T. D., G. H. Adler, C. M. Riveros, L. Lopez, R. Ascanio and J. Terborgh. 2003. Rodents on tropical land-bridge islands. *Journal of Zoology* (London) 260: 179–187.

Lanner, R. M. and T. R. Van Devender. 1998. The recent history of pinyon pines in the American Southwest. In D. M. Richardson (ed.), *Ecology and Biogeography of Pinus,* 171-182. Cambridge: Cambridge University Press.

Larick, R. and R. Ciochon. 1996. The first Asians: A cave in China yields evidence of the earliest migration out of Africa. *Archaeology* 49: 51–53.

Larson, H. K. 1995. A review of the Australian endemic gobiid fish genus Chlamydogobius, with description of five new species. *The Beagle, Records of the Museums and Art Galleries of the Northern Territory* 12: 19–51.

Latham, R. E. and R. E. Ricklefs. 1993. Global patterns of tree species diversity in moist forests: Energy-diversity theory does not account for variation in species richness. *Oikos* 67: 325–333.

Latorre, C., J. L. Betancourt, K. A. Rylander and J. A. Quade. 2002. Vegetation invasions into Absolute Desert: A 45,000-year rodent midden record from the Calama-Salar de Atacama Basins, Chile. *Geological Society of America Bulletin* 114: 349–366.

Laurance, S. G. and W. F. Laurance. 1999. Tropical wildlife corridors: Use of linear rainforest remnants by arboreal mammals. *Biological Conservation* 91: 231–239.

Laurance, W. F. 1990. Comparative responses of five arboreal marsupials to tropical forest fragmentation. *Journal of Mammalogy* 71: 641–653.

Laurance, W. F. and R. O. Bierregaard. 1996. Fragmented tropical forests. *Bulletin of the Ecological Society of America* 77: 34–36.

Laurance, W. F. and C. Gascon. 1997. How to creatively fragment a landscape. *Conservation Biology* 11: 577–580.

Lawlor, T. E. 1983. The mammals. In T. J. Case and M. L. Cody (eds.), *Island Biogeography of the Sea of Cortez,* 265–289, 482–500. Berkeley: University of California Press.

Lawlor, T. E. 1986. Comparative biogeography of mammals on islands. *Biological Journal of the Linnean Society* 28: 99–125.

Lawlor, T. E. 1998. Biogeography of great mammals: Paradigm lost? *Journal of Mammalogy* 79: 1111–1130.

Lawton, J. H. 1995. Population dynamic principles. In J. H. Lawton and R. M. May (eds.), *Extinction Rates,* 147–163. Oxford: Oxford University Press.

Lawton, J. H., S. Nee, A. J. Letcher and P. H. Harvey. 1994. Animal distributions: Patterns and processes. In P. J. Edwards, R. M. May and N. R. Webb (eds.), *Large-Scale Ecology and Conservation Biology,* 41–58. London: Blackwell Scientific Publications.

Lazell, J. D. Jr. 1983. Biogeography of the herptofauna of the British Virgin Islands, with description of a new anole (Sauria: Iguanidae). In A. G. J. Rhodin and K. Miyata (eds.), *Advances in Herpetology and Evolutionary Biology,* 99–117. Cambridge, MA: Museum of Comparative Zoology.

Lee, P.-F., T.-S. Ding, Fu-H. Hsu, G. Shu. 2004. Breeding bird species richness in Taiwan: Distribution on gradients in elevation, primary productivity and urbanization. *Journal of Biogeography* 31: 307–314.

Lee, T. E. Jr., B. R. Riddle and P. L. Lee. 1996. Speciation in the desert pocket mouse (*Chaetodipus penicillatus* Woodhouse). *Journal of Mammalogy* 77: 58–68.

Lees, D. C., C. Kremen and L. Andriamampianina. 1999. A null model for species richness gradients: Bounded range overlap of butterflies and other rainforest endemics in Madagascar. *Biological Journal of the Linnean Society* 67: 529–584.

Leibold, M. A. 1996. A graphical model of keystone predators in food webs: Trophic regulation of abundance, incidence and diversity patterns in communities. *American Naturalist* 147: 784–812.

Leigh, E. 1975. Population fluctuations and community structure. In W. H. Van Dobben and R. H. Lowe-McConnel (eds.), *Unifying Concepts in Ecology,* 67–88. The Hague: Dr. W. Junk.

Leigh, E. G. 1981. The average lifetime of a population in a varying environment. *Journal of Theoretical Biology* 90: 213–239.

Leith, H. 1956. Ein Beitrag zur Frage der korrelation zwischen mittleren klimawerten und vegetationsformationen. *Berichte Deutsche Botanische Gesellschaft* 69: 169–176.

Lesica, P. and F. W. Allendorf. 1995. When are peripheral populations valuable for conservation? *Conservation Biology* 9: 753–760.

Leston, D. 1957. Spread potential and colonization of the islands. *Systematic Zoology* 6: 41–46.

Leverington, D. and J. Teller. 2003. Paleotopographic reconstructions of the eastern outlets of glacial Lake Agassiz. *Canadian Journal of Earth Sciences* 40: 1259–1278.

Leverington, D., J. Mann and J. Teller. 2002. Changes in the bathymetry and volume of Glacial Lake Agassiz between 9200 and 7700 ^{14}C yr B.P. *Quaternary Research* 244–252.

Levin, D. A. (ed.). 1979. *Hybridization: An Evolutionary Perspective.* Stroudsburg, PA: Dowden, Hutchinson, & Ross.

Lewis, W. H. (ed.). 1979. *Polyploidy: Biological Relevance.* New York: Plenum Press.

Lewontin, R. C. and L. C. Birch. 1966. Hybridization as a source of variation for adaptation to new environments. *Evolution* 20: 315–336.

Li, W. H. 1979. *Molecular Evoolution.* Sunderland, MA: Sinauer Associates.

Lidicker, W. Z. Jr. 1988. Solving the enigma of microtine "cycles." *Journal of Mammalogy* 69: 225–235.

Lidicker, W. Z. Jr. 1992. *Animal Dispersal: Small Mammals as a Model.* New York: Chapman and Hall.

Lieberman, B. S. 2000. *Paleobiogeography: Using Fossils to Study Global Change, Plate Tectonics and Evolution.* New York: Kluwer Academic/Plenum Publishers.

Lieberman, B. S. 2002. Phylogenetic biogeography with and without the fossil record: Gauging the effects of extinction and paleontological incompleteness. *Palaeogeography Palaeoclimatology Palaeoecology* 178: 39–52.

Lieberman, B. S. 2003. Paleobiogeography: The relevance of fossils to biogeography. *Annual Review of Ecology Evolution and Systematics* 34: 51–69.

Lieberman, B. S. 2004. Range expansion, extinction and biogeographic congruence: A deep time perspective. In M. V. Lomolino and L. R. Heaney (eds.), *Frontiers of Biogeography,* 111–124. Sunderland, MA: Sinauer Associates.

Lieth, H. 1973. Primary production: Terrestrial ecosystems. *Human Ecology* 1: 303–332.

Lillegraven, J. A., Z. Kielan-Jaworowska and W. A. Clemens. 1979. *Mesozoic Mammals: The First Two-Thirds of Mammalian History.* Berkeley: University of California Press.

Lima, M. G. de and C. Gascon. 1999. The conservation value of linear forest remnants in central Amazonia. *Biological Conservation* 91: 241–247.

Linder, H. P. 2001. On areas of endemism, with an example from the African Restionaceae. *Systematic Biology* 50: 892–912.

Lindeman, R. 1942. The trophic-dynamic aspect of ecology. *Ecology* 23: 399–418

Lindsey, C. C. 1975. Peomerism, the widespread tendency among related fish species for vertebral number to be correlated with maximum body length. *Journal of the Fisheries Research Board of Canada* 28: 2453–2469.

Lindsey, C. C. and A. N. Arnason. 1981. A model for responses for vertebral number in fish to environmental influences during development. *Canadian Journal of Fisheries and Aquatic Sciences* 38: 334–347.

Lindstedt, S. L. and M. S. Boyce. 1985. Seasonality, body size and survival time in mammals. *American Naturalist* 125: 873–878.

Linnaeus, C. 1781. On the increase of the habitable earth. *Amonitates Academicae* 2: 17–27.

Lister, A. and P. Bahn. 1994. *Mammoths.* New York: MacMillan Press.

Lister, A. M. 1993. Mammoths in miniature. *Nature* 362: 188–189.

Lister, A. M. 1989. Rapid dwarfing of red deer on Jersey in the last interglacial. *Nature* 342: 539–542.

Lister, B. C. 1976a. The nature of niche expansion in West Indian *Anolis* lizards. I. Ecological consequences of reduced competition. *Evolution* 30: 659–676.

Lister, B. C. 1976b. The nature of niche expansion in West Indian *Anolis* lizards. II. Evolutionary consequences. *Evolution* 30: 677–692.

Lithgow-Bertelloni, C. and M. A. Richards. 1995. Cenozoic plate driving forces. *Geophysical Research Letters* 22: 1317–1320.

Liu, H. P., R. Hershler and K. Clift. 2003. Mitochondrial DNA sequences reveal extensive cryptic diversity within a western American springsnail. *Molecular Ecology* 12: 2771–2782.

Livezey, B. C. 1992. Morphological corollaries and ecological implications of flightlessness in the kakapo (Psittaciformes: *Strigops habroptilus*). *Journal of Morphology* 213: 105–145.

Livezey, B. C. 1993. An ecomorphological review of the dodo (*Raphus cucullatus*) and solitaire (*Pezophaps solitaria*), flightless Columbiformes of the Macarene Islands. *Journal of Zoology* (London) 230: 247.

Lockwood, J. L. 2004. How do biological invasions alter diversity patterns? A biogeographic perspective. In M. V. Lomolino and L. R. Heaney (eds.), *Frontiers of Biogeography,* Chapter 15. Sunderland, MA: Sinauer Associates.

Lockwood, J. L. and M. L. McKinney. 2001. *Biotic Homogenization.* New York: Kluwer Academic/Plenum Publishers.

Lockwood, J. L. and M. P. Moulton. 1994. Ecomorphological pattern in Bermuda birds: The influence of competition and implications for nature preserves. *Evolutionary Ecology* 8: 53–60.

Lodge, D. M. 1993. Biological invasions: Lessons for ecology. *Trends in Ecology and Evolution* 8: 133–137.

Loh, J. 2000. *Living Planet Report.* Gland, Switzerland: World Wildlife Fund.

Lomolino, M. V. 1982. Species-area and species-distance relationships of terrestrial mammals in the Thousand Island Region. *Oecologia* 54: 72–75.

Lomolino, M. V. 1983. *Island Biogeography, Immigrant Selection and Mammalian Body Size on Islands.* Ph.D. dissertation, Department of Biology, State University of New York at Binghamton.

Lomolino, M. V. 1984. Immigrant selection, predatory exclusion and the distributions of *Microtus pennsylvanicus* and *Blarina brevicauda* on islands. *American Naturalist* 123: 468–483.

Lomolino, M. V. 1985. Body size of mammals on islands: The island rule re-examined. *American Naturalist* 125: 310–316.

Lomolino, M. V. 1986. Mammalian community structure on islands: Immigration, extinction and interactive effects. *Biological Journal of the Linnean Society* 28: 1–21.

Lomolino, M. V. 1988. Winter immigration abilities and insular community structure of mammals in temperate archipelagoes. In J. F. Downhower (ed.), *Biogeography of the Island Region of Western Lake Erie,* 185–196. Columbus: Ohio State University Press.

Lomolino, M. V. 1989. Bioenergetics of cross-ice movements of *Microtus pennsylvanicus, Peromyscus leucopus* and *Blarina brevicauda. Holarctic Ecology* 12: 213–218.

Lomolino, M. V. 1989. Interpretation and comparisons of constants in the species-area relationship: An additional caution. *American Naturalist* 133: 71–75.

Lomolino, M. V. 1990. The target area hypothesis: The influence of island area on immigration rates of non-volant mammals. *Oikos* 57: 297–300.

Lomolino, M. V. 1993a. Matching of granivorous mammals of the Great Basin and Sonoran deserts on a species-for-species basis. *Journal of Mammalogy* 74: 863–867.

Lomolino, M. V. 1993b. Winter filtering, immigrant selection and species composition of insular mammals of Lake Huron. *Ecography* 16: 24–30.

Lomolino, M. V. 1994a. An evaluation of alternative strategies for building networks of nature reserves. *Biological Conservation* 69: 243–249.

Lomolino, M. V. 1994b. Species richness patterns of mammals inhabiting nearshore archipelagoes: Area, isolation and immigration filters. *Journal of Mammalogy* 75: 39–49.

Lomolino, M. V. 1996. Investigating causality of nestedness of insular communities: Selective immigrations or extinctions? *Journal of Biogeography* 23: 699–703.

Lomolino, M.V. 1999. A species-based, hierarchical model of island biogeography. In E. A. Weiher and P. A. Keddy (eds.), *The Search for Assembly Rules in Ecological Communities.* New York: Cambridge University Press.

Lomolino, M. V. 2000a. A call for a new paradigm of island biogeography. *Global Ecology and Biogeography* 9: 1–6.

Lomolino, M. V. 2000b. A species-based theory of insular zoogeography. *Global Ecology and Biogeography* 9: 39–58.

Lomolino, M. V. 2001. Elevation gradients of species-density: Historical and prospective views. *Global Ecology and Biogeography* 10: 3–12.

Lomolino, M. V. 2002. "There are areas too small and areas too large to show clear diversity patterns ..." R. H. MacArthur (1972:191). *Journal of Biogeography* 29: 555–557.

Lomolino, M. V. and R. Channell. 1995. Splendid isolation: Patterns of range collapse in endangered mammals. *Journal of Mammalogy* 76: 335–347.

Lomolino, M. V. and R. Channell. 1998. Range collapse, reintroductions and biogeographic guidelines for conservation. *Conservation Biology* 12: 481–484.

Lomolino, M. V. and J. C. Creighton. 1996. Habitat selection and breeding success of the endangered American burying beetle (*Nicrophorus americanus*). *Biological Conservation* 77: 235–241.

Lomolino, M. V. and R. Davis. 1997. Biogeographic scale and biodiversity of mountain forest mammals of western North America. *Global Ecology and Biogeography Letters* 6: 57–76.

Lomolino, M. V. and D. R. Perault. 2000. Assembly and disassembly of mammal communities in a fragmented temperate rainforest. *Ecology* 81: 1517–1532.

Lomolino, M. V. and D. R. Perault. 2004. Geographic gradients of deforestation and mammalian communities in a fragmented, temperate rainforest landscape. *Global Ecology and Biogeography* 13: 55–64.

Lomolino, M. V. and G. A. Smith. 2003. Terrestrial vertebrate communities at black-tailed prairie dog (*Cynomys ludovicianus*) towns. *Biological Conservation* 115: 89–100.

Lomolino, M. V. and G. A. Smith. 2004. Prairie dog towns as islands: Applications of island biogeography and landscape ecology for conserving non-volant terrestrial vertebrates. *Global Ecology and Biogeography* 12: 275–286.

Lomolino, M. V. and M. D. Weiser. 2001. Towards a more general species-area relationship: Diversity on all islands, great and small. *Journal of Biogeography* 28: 431–445.

Lomolino, M. V., J. H. Brown and R. Davis. 1989. Island biogeography of montane forest mammals in the American Southwest. *Ecology* 70: 180–194.

Lomolino, M. V., R. Channell, D. R. Perault and G. A. Smith. 2001. Downsizing nature: Anthropogenic dwarfing of species and ecosystems. In M. McKinney and J. Lockwood (eds.), *Biotic Homogenization: The Loss of Diversity Through Invasion and Extinction*. 33–56. New York: Kluwer Academic/Plenum Publishers.

Lomolino, M. V., J. C. Creighton, G. D. Schnell and D. L. Certain. 1995. Ecology and conservation of the endangered American burying beetle (*Nicrophorus americanus*). *Conservation Biology* 9: 605–614.

Long, A. J., M. J. Crosby, A. J. Stattersfield and D. C. Wege. 1996. Towards a global map of biodiversity: Patterns in the distribution of restricted-range birds. *Global Ecology and Biogeography* 5: 281–305.

Long, J. 1981. *Introduced Birds of the World*. London: David and Charles.

Lonsdale, D. J. and J. S. Levington. 1985. Latitudinal differentiation in copepod growth: An adaptation to temperature. *Ecology* 66: 1397–1407.

Lonsdale, W. M. 1999. Global patterns of plant invasions and the concept of invisibility. *Ecology* 80: 1522–1536.

Loo, S. E., R. MacNally and P. Quinn. 2002. An experimental examination of colonization as a generator of biotic nestedness. *Oecologia* 132: 118–124.

Loope, L. L. and D. Mueller-Dombois. 1989. Characteristics of invaded islands, with special reference to Hawaii. In J. A. Drake et al. (eds.), *Biological Invasions: A Global Perspective*, 257–280. New York: John Wiley and Sons.

Lord, R. D. Jr. 1960. Litter size and latitude in North American mammals. *American Midland Naturalist* 64: 488–499.

Loreau, M. and N. Mouquet. 1999. Immigration and the maintenance of local species diversity. *American Naturalist* 154: 427–440.

Losos, J. B. 1992. A critical comparison of the taxon cycle and character displacement models of size evolution in *Anolis* lizards in the lesser Antilles. *Copeia* 1991: 279–288.

Losos, J. B. and K. de Queiroz. 1997. Evolutionary consequences of ecological release in Caribbean *Anolis* lizards. *Biological Journal of the Linnean Society* 61: 459–483.

Losos, J. B. and H. W. Greene. 1988. Ecological and evolutionary implications of diet in monitor lizards. *Biological Journal of the Linnean Society* 35: 379–407.

Losos, J. B., J. C. Marks and T. W. Schoener. 1993. Habitat use and ecological interactions of an introduced and a native species of *Anolis* lizard on Grand Cayman, with a review of the outcomes of anole introductions. *Oecologia* 95: 525–532.

Lourie, S. A., D. M. Green and A. C. J. Vincent. 2005. Dispersal, habitat preferences and comparative phylogeography of Southeast Asian seahorses (*Syngnathidae: Hippocampus*). *Molecular Ecology* 14: 1073–1094.

Lovejoy, T. E., R. O. Birregaard, Jr., A. B. Rylands, J. R. Malcolm, C. E. Quintela, L. H. Harper, K. S. Brown, Jr., A. H. Powell, G. V. N. Powell, H. O. R. Schubart and M. B. Hays. 1986. Edge and other effects of isolation on Amazon forest fragments. In M. E. Soulé (ed.), *Conservation Biology: The Science of Scarcity and Diversity*, 257–285. Sunderland, MA: Sinauer Associates.

Lovette, I. J. 2005. Glacial cycles and the tempo of avian speciation. *Trends in Ecology and Evolution* 20: 57–59.

Lowther, P. E. and C. L. Cink. 1992. *Passer domesticus*: House sparrow. The Birds of North America. *American Ornithologists' Union*, no. 12.

Lubchenco, J. 1978. Plant species diversity in a marine intertidal community: Importance of herbivore food preferences and algal competitive abilities. *American Naturalist* 112: 23–39.

Lubchenco, J. 1980. Algal zonation in the New England rocky intertidal community: An experimental analysis. *Ecology* 61: 333–344.

Lubchenco, J. and B. A. Menge. 1978. Community development and persistence in a low rocky intertidal zone. *Ecological Monographs* 48: 67–94.

Lubick, N. 2002. Snowball fights. *Nature* 417: 12–13.

Ludwig, J. A. and J. F. Reynolds. 1988. *Statistical Ecology: A Primer on Methods and Computing*. New York: John Wiley and Sons.

Lundberg, J. G. and B. Chernoff. 1992. A Miocene fossil of the Amazonian fish *Arapaima* (Teleostei, Arapaimidae) from the Magdalena river region of Colombia: Biogeographic and evolutionary implications. *Biotropica* 24: 2–14.

Lundberg, J. G., A. Machado-Allison and R. F. Kay. 1986. Miocene characid fishes from Columbia: Evolutionary stasis and extirpation. *Science* 234: 208–209.

Luther, G. W. III, T. F. Rozan, M. Taillefert, D. B. Nuzzio, C. D. Meo, T. M. Shank, R. A. Lutz and S. C. Cary. 2001. Chemical speciation drives hydrothermal vent ecology. *Nature* 410: 813–816.

Lyell, C. 1834 (3rd of 6 editions). *Principles of Geology, Being an Attempt to Explain the Former Changes of the Earth's Surface, by Reference to Causes Now in Operation*. London: John Murray.

Lyell, C. 1969. *Principles of Geology*. New York. Johnson Reprint Company. [First published in 1830.]

Lynch, J. D. 1979. The amphibians of the lowland tropical forests. In W. E. Duellman (ed.), *The South American Herpetofauna: Its Origin, Evolution and Dispersal*, 189–215. Monograph no. 7. Lawrence: Museum of Natural History, University of Kansas.

Lynch, J. D. 1988. Refugia. In A. A. Myers and P. S. Giller (eds.), *Analytical Biogeography: An Integrated Approach to the Study of Animal and Plant Distributions*. 314–342. London: Chapman and Hall.

Lynch, J. D. and N. V. Johnson. 1974. Turnover and equilibria in insular avifaunas, with special reference to the California Channel Islands. *Condor* 76: 370–384.

MacArthur, R. H. 1958. Population ecology of some warblers of northeastern coniferous forests. *Ecology* 39: 599–619.

MacArthur, R. H. 1965. Patterns of species diversity. *Biological Review* 40: 510–533.

MacArthur, R. H. 1972. *Geographical Ecology: Patterns in the Distributions of Species*. New York: Harper & Row.

MacArthur, R. H. and T. H. Connell. 1966. *The Biology of Populations*. New York: John Wiley and Sons.

MacArthur, R. H. and R. Levins. 1967. The limiting similarity, convergence and divergence of coexisting species. *American Naturalist* 101: 377–385.

MacArthur, R. H. and J. W. MacArthur. 1961. On bird species diversity. *Ecology* 42: 594–598.

MacArthur, R. H. and E. O. Wilson. 1963. An equilibrium theory of insular zoogeography. *Evolution* 17: 373–387.

MacArthur, R. H. and E. O. Wilson. 1967. *The Theory of Island Biogeography*. Monographs in Population Biology, no. 1. Princeton, NJ: Princeton University Press.

MacArthur, R. H., J. M. Diamond and J. Karr. 1972. Density compensation in island faunas. *Ecology* 53: 330–342.

MacArthur, R. H., H. Recher and M. Cody. 1966. On the relation between habitat selection and species diversity. *American Naturalist* 100: 319–332.

MacArthur, R. H., J. MacArthur, D. MacArthur and A. MacArthur. 1973. The effect of island area on population densities. *Ecology* 54: 657–658.

MacDonald, I. A. W., F. J. Kruger and A. A. Ferrar. 1986. *The Ecology and Management of Invasions in Southern Africa*. Cape Town: Oxford University Press.

MacDonald, I. A. W., L. L. Loope, M. B. Usher and O. Hamann. 1989. Wildlife conservation and the invasion of nature reserves by introduced species: A global perspective. In J. A. Drake et al. (eds.), *Biological Invasions: A Global Perspective*, 215–256. New York: John Wiley and Sons.

Mace, G. M. 1993. An investigation into methods for categorizing the conservation status of species. In P. J. Edwards, R. M. May and N. R. Webb (eds.), *Large-Scale Ecology and Conservation Biology*, 293–312. London: Blackwell Scientific Publications.

Mace, R. and M. Pagel. 1995. A latitudinal gradient in the density of human languages in North America. *Proceedings of the Royal Society of London*, Series B 261: 117–121.

Machtans, C. S., M. Villard and S. J. Hannon. 1996. Use of riparian buffer strips as movement corridors by forest birds. *Conservation Biology* 10: 1366–1279.

Mack, R. N., D. Simberloff, W. M. Lonsdale, H Evans, M. Clout and F. A. Bazzaz. 2000. Biotic invasions: Causes, epidemiology, global consequences and control. *Ecological Applications* 10: 689–710.

MacMahon, J. A. 1987. Disturbed lands and ecological theory: An essay about a mutualistic association. In W. R. Jordan, M. E. Gilpin and J. D. Aber, (eds.), *Restoration Ecology: A Synthetic Approach to Ecological Research*. 221–237. New York: Cambridge University Press.

MacPhee, R. D. E. 1999. *Extinctions in Near Time: Causes, Contexts and Consequences*. New York: Kluwer Academic/Plenum Publishers.

MacPhee, R. D. E. and P. A. Marx. 1997. The 40,000-year plague: Humans, hyperdisease and first-contact extinctions. In S. Goodman and B. D. Patterson (eds.), *Human Impact and Natural Change in Madagascar*, 169–217. Washington, DC: Smithsonian Institution Press.

MacPherson, E. and C. M. Duarte. 1994. Patterns in species richness, size and latitudinal range of East Atlantic fishes. *Ecography* 17: 242–248.

Maddison, W. P. and D. R. Maddison. 2000. *Mesquite: A modular programming system for evolutionary analysis*. Available from: http://mesquiteproject.org. (Accessed 8 June 2005.)

Magurran, A. E. 1988. *Ecological Diversity and Its Measurement*. Princeton, NJ: Princeton University Press.

Maloney, B. K. 1980. Pollen analytical evidence for early forest clearance in North Sumatra. *Nature* 287: 324–326.

Manly, B. F. J. 1991. *Randomization and Monte Carlo Methods in Biology*. London: Chapman and Hall.

Mares, M. A. 1976. Convergent evolution in desert rodents: Multivariate analysis and zoogeographic implications. *Paleobiology* 2: 39–63.

Mares, M. A. 1993a. Desert rodents, seed consumption and convergence. *BioScience* 43: 373–379.

Mares, M. A. 1993b. Heteromyids and their ecological counterparts: A pandesertiv view of rodent ecology and evolution. In H. H. Genoways and J. H. Brown (eds.), *Biology of the Heteromyidae*, 652–719. American Society of Mammalogists, Special Publication no. 10.

Mares, M. A. and T. E. Lacher Jr. 1987. Ecological, morphological and behavioral convergence in rock-dwelling mammals. In H. H. Genoways (ed.), *Current Mammalogy*, vol. 1, 307–347. New York: Plenum Press.

Markgraf, V., M. McGlone and G. Hope. 1995. Neogene paleoenvironmental and paleoclimatic change in southern temperate ecosystems: A southern perspective. *Trends in Ecology and Evolution* 10: 143–147.

Marks, G. and W. K. Beatty. 1976. *Epidemics*. New York: Charles Scribner's Sons.

Marquet, P. A., M. Fernández, S. A. Navarrete and C. Valdovinos. 2004. Diversity emerging: Towards a deconstruction of biodiversity patterns. In, M. V. Lomolino and L. R. Heaney, *Frontiers of Biogeography: New Directions in the Geography of Nature*. Sunderland, MA: Sinauer Associates.

Marshall, E. 2001a. Clovis first. *Science* 291: 1732.

Marshall, E. 2001b. Pre-clovis sites fight for acceptance. *Science* 291: 1730.

Marshall, L. G. 1979. Evolution of metatherian and eutherian (mammalian) characters: A review based on cladistic methodology. *Zoological Journal of the Linnean Society* 66: 369–410.

Marshall, L. G. 1988. Extinction. In A. A. Meyers and P. S. Giller (eds.), *Analytical Biogeography: An Integrated Approach to the Study of Animal and Plant Distributions*, 219–254. New York: Chapman and Hall.

Marshall, L. G. and R. S. Corruccini. 1978. Variation, evolutionary rates and allometry in dwarfing lineages. *Paleobiology* 4: 101–119.

Marshall, L. G. and J. G. Lundberg. 1996. Miocene deposits in the Amazonian Foreland Basin. *Science* 273: 123–124.

Martens, K., B. Godderis and G. Coulter. 1994. *Speciation in Ancient Lakes*. Stuttgart: E. Schweizerbart'sche Verlagsbuchhandlund.

Martin, A. R. 1990. *The Illustrated Encyclopedia of Whales and Dolphins*. New York: Portland House.

Martin, J. and P. Gurrea. 1990. The peninsular effect in Iberian butterflies (Lepidoptera: Papilionoidea and Hesperioidea). *Journal of Biogeography* 17: 85–96.

Martin, J. L. 1992. Niche expansion in an insular bird community: An autecological perspective. *Journal of Biogeography* 19: 375–381.

Martin, P. S. 1967. Prehistoric overkill. In P. S. Martin and H. E. Wright Jr. (eds.), *Pleistocene Extinctions: The Search for a Cause*, 75–120. New Haven, CT: Yale University Press.

Martin, P. S. 1973. The discovery of America. *Science* 180: 969–974.

Martin, P. S. 1984. Prehistoric overkill. In P. S. Martin and R. G. Klein (eds.), *Quaternary Extinctions: A Prehistoric Revolution*, 354–403. Tucson, AZ: University of Arizona Press.

Martin, P. S. 1990. 40,000 years of extinction on the "planet of doom." *Palaeogeography, Palaeoclimatology, Palaeoecology* 82: 187–201.

Martin, P. S. 1995. Mammoth extinction: Two continents and Wrangel Island. *Radiocarbon* 37: 1–6.

Martin, P. S. and R. G. Klein (eds.). 1984. *Quaternary Exctinctions: A Prehistoric Revolution*. Tucson, AZ: University of Arizona Press.

Martin, P. S. and H. E. Wright Jr. (eds.). 1967. *Pleistocene Extinctions: The Search for a Cause*. New Haven, CT: Yale University Press.

Martin, T. E. 1981. Species-area slopes and coefficients: A caution on their interpretation. *American Naturalist* 118: 823–837.

Mason, H. L. 1954. Migration and evolution of plants. *Madroño* 12: 161–192.

Mast, A. R. and R. Nyffeler. 2003. Using a null model to recognize significant co-occurrence prior to identifying candidate areas of endemism. *Systematic Biology* 52: 271–280.

Matthew, W. D. 1915. Climate and evolution. *Annals of the New York Academy of Sciences* 24: 171–318.

Matthews, E. and W. J. Matthews. 2000. Geographic, terrestrial and aquatic factors: Which most influence the structure of stream fish assemblages in the midwestern United States? *Ecology of Freshwater Fish* 9: 9–21.

Maurer, B. A. 1994. *Geographic Population Analysis: Tools for Analysis of Biodiversity*. London: Blackwell Scientific Publications.

Maurer, B. A., J. H. Brown and R. D. Rusler. 1992. The micro and macro of body size evolution. *Evolution* 46: 939–953.

Maxlow, J. 1996. "Global expansion tectonics: small earth modelling of an exponentially expanding earth" Terrella Consultants, 29 Cecil Street, Glen Forrest, Western Australia.

Maxson, L. R. and J. D. Roberts. 1984. Albumin and Australian frogs—Molecular-data a challenge to speciation model. *Science*, 225: 957–958.

May, R. M. 1988. How many species are there on earth? *Science* 241: 1441–1449.

Mayden, R. L. 1987a. Historical ecology and North American highland fishes: A research program in community ecology. In W. J. Matthews and D. C. Heins (eds.), *Community and Evolutionary Ecology of North American Stream Fishes*, 210–222. Norman: University of Oklahoma Press.

Mayden, R. L. 1987b. Pleistocene glaciation and historical biogeography of North American Central Highland fishes. In W. C. Johnson (ed.), *Quaternary Environments of Kansas*, 141–151. Lawrence: Kansas Geological Survey.

Mayden, R. L. 1992a. Explorations into the past and the dawn of systematics and historical ecology. In R. L. Mayden (ed.), *Systematics, Historical Ecology and North American Freshwater Fishes*. 3–17. Stanford, CA: Stanford University Press.

Mayden, R. L. (ed.) 1992b. *Systematics, Historical Ecology and North American Freshwater Fishes*. Stanford, CA: Stanford University Press.

Mayden, R. L. 1997. A hierarchy of species concepts: The denouement in the saga of the species problem. In M. F. Clardige, H. A. Dawah and M. R. Wilson (eds.), *Species: The Units of Biodiversity*, 439. London: Chapman and Hall.

Mayden, R. L. and R. M. Wood. 1995. Systematics, species concepts and the evolutionarily significant unit in biodiversity and conservation biology. *American Fisheries Society Symposium* 17: 58–113.

Mayr, E. 1942. *Systematics and the Origin of Species*. New York: Columbia University Press.

Mayr, E. 1944a. Wallace's Line in the light of recent zoogeographic studies. *Quarterly Review of Biology* 19: 1–14.

Mayr, E. 1944b. The birds of Timor and Sunda. *Bulletin of the American Museum of Natural History* 83: 127–194.

Mayr, E. 1956. Geographical character gradients and climatic adaptation. *Evolution* 10: 105–108.

Mayr, E. 1963. *Animal Species and Evolution*. Cambridge, MA: Harvard University Press.

Mayr, E. 1965a. The nature of colonization in birds. In H. G. Baker and G. L. Stebbins (eds.), *The Genetics of Colonizing Species*, 3047. New York: Academic Press.

Mayr, E. 1965b. Avifauna: Turnover on islands. *Science* 150: 1587–1588.

Mayr, E. 1969. *Principles of Systematic Zoology*. New York: McGraw-Hill.

Mayr, E. 1970. *Populations, Species and Evolution*. Cambridge, MA: Harvard University Press.

Mayr, E. 1974. Cladistic analysis or cladistic classification? *Z. Zool. Syst. Evolut.-forsch.* 12: 94–128.

Mayr, E. 1982. *The Growth of Biological Thought*. Cambridge, MA: Harvard University Press.

Mayr, E. and P. D. Ashlock. 1991. *Principles of Systematic Zoology*. New York: McGraw-Hill.

Mayr, E. and J. M. Diamond. 2001. *The Birds of Northern Melanesia: Speciation, Ecology and Biogeography*. New York: Oxford University Press.

McAtee, W. L. 1947. Distribution of seeds by birds. *American Midland Naturalist* 38: 214–223.

McAuliffe, J. R. 1994. Landscape evolution, soil formation and ecological processes in Sonoran Desert bajadas. *Ecological Monographs* 64: 111–148.

McCain, C. M. 2003. North American desert rodents: A test of the mid–domain effect in species richness. *Journal of Mammalogy* 84: 967–980.

McCall, R. A., S. Nee and P. H. Harvey. 1998. The role of wing length in the evolution of avian flightlessness. *Evolutionary Ecology* 12: 569–580.

McCarthy, D. 2003. The trans-Pacific zipper effect: Disjunct sister taxa and matching geological outlines that link the Pacific margins. *Journal of Biogeography* 30: 1545–1561.

McClure, H. E. 1974. *Migration and Survival of the Birds of Asia*. Bangkok: Applied Scientific Research Corporation of Thailand.

McCook, L. J. 1994. Understanding ecological succession: Causal models and theories, a review. *Vegetatio* 110: 115–147.

McCook, L. J. and A. R. O. Chapman. 1997. Patterns and variations in natural succession following massive ice-scour of a rocky intertidal seashore. *Journal of Experimental Marine Biology and Ecology* 214: 121–147.

McCoy, E. D., S. S. Bell and K. Walters. 1986. Identifying biotic boundaries along environmental gradients. *Ecology* 67: 749–759.

McDonald, K. A. and J. H. Brown. 1992. Using montane mammals to model extinctions due to global change. *Conservation Biology* 6: 409–415.

McDowall, R. M. 2003. Variation in vertebral number in galaxiid fishes (Teleostei: Galaxiidae): A legacy of life history, latitude and length. *Environmental Biology of Fishes* 66: 361–381.

McDowell, S. B. 1969. Notes on the Australian sea snake *Ephalophis greyi* M. Smith (Serpentes: Elapidae, Hydrophiinae) and the origin and classification of sea snakes. *Zoological Journal of the Linnean Society* 48: 333–349.

McDowell, S. B. 1972. The genera of sea snakes of the *Hydrophis* group (Serpentes: Elapidae). *Transactions of the Zoological Society of London* 32: 189–247.

McDowell, S. B. 1974. Additional notes on the rare and primitive sea snake, *Ephalophis greyi*. *Journal of Herpetology* 8: 123–128.

McElhinny, M. W. 1973a. *Paleomagnetism and Plate Tectonics*. Cambridge: Cambridge University Press.

McElhinny, M. W. 1973b. Paleomagnetism and plate tectonics of eastern Asia. In P. J. Coleman (ed.), *The Western Pacific: Island Arcs, Marginal Seas, Geochemistry*, 407–414. New York: Crane, Russak & Co.

McFarlane, D. A. 1989. Patterns of species co-occurrence in the Antillean bat fauna. *Mammalia* 53: 59–66.

McGarigal, K. and S. A. Cushman. 2002. Comparative evaluation of experimental approaches to the study of habitat fragmentation studies. *Ecological Applications* 12(2): 335–345.

McGlone, M. S. 2005. Goodby Gondwana. *Journal of Biogeography*. 32: 739–740.

McGowan, J. A. 1985. El Niño effects in the Eastern Subarctic Pacific Ocean. In W. S. Wooster and D. L. Fluharty (eds.), *El Niño North*, 166–184. Seattle: Washington Sea Grant, University of Washington Press.

McGuinness, K. A., 1984. Equations and explanations in the study of species-area curves. *Biological Review* 59: 423–440.

McIntosh, R. P. 1967. The continuum concept of vegetation. *Botanical Review* 33: 130–187.

McIntosh, R. P. 1981. Succession in ecological theory. In D. C. West, H. H. Shugart and D. B. Botkin (eds.), *Forest Succession: Concepts and Applications*, 10–23. New York: Springer-Verlag.

McIntosh, R. P. 1999. The succession of succession: A lexical chronology. *Bulletin of the Ecological Society of America* 80: 256–264.

McKenna, M. C. 1972a. Eocene final separation of the Eurasian and Greenland-North American landmasses. *24th International Geological Congress, Section 7*: 275–281.

McKenna, M. C. 1972b. Was Europe connected directly to North America prior to the middle Eocene? In Th. Dobzkansky, M. K. Hecht and W. C. Steere (eds.), *Evolutionary Biology* 6: 179–188. New York: Appleton-Century-Crofts.

McKenna, M. C. 1973. Sweepstakes, filters, corridors, Noah's Arks and beached Viking funeral ships in paleogeography. In D. H. Tarling and S. K. Runcorn (eds.), *Implications of Continental Drift to the Earth Sciences*, vol. 1, 295–308. New York: Academic Press.

McKenna, M. C. and S. K. Bell. 1997. *Classification of Mammals above the Species Level*. New York: Columbia University Press.

McKenzie, D. P. and W. J. Morgan. 1969. The evolution of triple junctions. *Nature* 226: 239–243.

McKinney, H. L. 1972. *Wallace and Natural Selection*. New Haven, CT: Yale University Press.

McKinney, M. L. and J. L. Lockwood. 1999. Biotic homogenization: A few winners replacing many losers in the next mass extinction. *Trends in Ecology and Evolution* 14: 450–453.

McKitrick, M. C. and R. M. Zink. 1988. Species Concepts in Ornithology. *Condor* 90: 1–14.

McLain, D. K. and A. E. Pratt. 1999. Nestedness of coral reef fish across a set of fringing reefs. *Oikos* 85: 53–67.

McLaughlin, J. F. and J. Roughgarden. 1989. Avian predation on *Anolis* lizards in the northeastern Caribbean: An inter-island contrast. *Ecology* 70: 617–628.

McLaughlin, S. P. 1986. Floristic analysis of the southwestern United States. *Great Basin Naturalist* 46: 46–65.

McLaughlin, S. P. 1989. Natural floristic area of the western United States. *Journal of Biogeography* 16: 239–248.

McLaughlin, S. P. 1992. Are floristic areas hierarchically arranged? *Journal of Biogeography* 19: 21–32.

McMillan, W. O. and S. R. Palumbi. 1995. Concordant evolutionary patterns among Indo-West pacific butterflyfishes. *Proceedings of the Royal Society of London, Series B-Biological Sciences* 260: 229–236.

McNab, B. K. 1963. Bioenergetics and the determination of home range size. *American Naturalist* 97: 113–140.

McNab, B. K. 1971. On the ecological significance of Bergmann's rule. *Ecology* 52: 845–854.

McNab, B. K. 1994a. Energy conservation and the evolution of flightlessness in birds. *American Naturalist* 144: 628–642.

McNab, B. K. 1994b. Resource use and the survival of land and freshwater vertebrates on oceanic islands. *American Naturalist* 144: 643–660.

McNab, B. K. 2002. Minimizing energy expenditure facilitates vertebrate persistence on oceanic islands. *Ecology Letters* 5: 693–704.

McNab, B. K. 2002. *The Physiological Ecology of Vertebrates: A View from Energetics*. Ithaca, NY: Cornell University Press.

McNeill, W. H. 1976. *Plagues and Peoples*. Garden City, NY: Anchor Press.

McPhail, J. D. 1994. Speciation and the evolution of reproductive isolation in the sticklebacks (*Gasterosteus*) of southwestern British Columbia. In M. A. Bell and S. A. Foster (eds.), *Evolutionary Biology of the Threespine Stickleback*, 399–437. Oxford: Oxford University Press.

McPhail, J. D. and C. C. Lindsey. 1986. Zoogeography of the freshwater fishes of Cascadia (the Columbia River north of the Stikine). In C. H. Hocutt and E. O. Wiley (eds.), *Zoogeography of North American Freshwater Fishes*, 615–637. New York: John Wiley and Sons, Inc.

Meadows, M. E. 2001. Biogeography: Does theory meet practice? *Progress in Physical Geography* 25: 134–142.

Means, D. B. and D. Simberloff. 1987. The peninsula effect: Habitat-correlated species decline in Florida's herptofauna. *Journal of Biogeography* 14: 551–568.

Mech, S. G. and J. G. Hallett. 2001. Evaluating the effectiveness of corridors: A genetic approach. *Conservation Biology* 15: 467–474.

Medail, F. and P. Quezel. 1999. Biodiversity hotspots in the Mediterranean Basin: Setting global conservation priorities. *Conservation Biology* 13: 1510–1513.

Meiri, S. and T. Dayan. 2003. On the validity of Bergmann's rule. *Journal of Biogeography* 30: 331–351.

Meiri, S., T. Dayan and D. Simberloff. 2004a. Body size of insular carnivores: Little support for the island rule. *American Naturalist* 163: 469–479.

Meiri, S., T. Dayan and D. Simberloff. 2004b. Carnivores, biases and Bergmann's rule. *Biological Journal of the Linnean Society* 81: 579–598.

Melhop, P. and J. F. Lynch. 1978. Population characteristics of *Peromyscus leucopus* introduced to islands inhabited by *Microtus pennsylvanicus*. *Oikos* 31: 17–26.

Mellars, P. 1998. The fate of Neanderthals. *Nature* 395: 539–540.

Melville, R. 1981. Vicariance plant distributions and paleogeography of the Pacific region. In G. Nelson and D. E. Rosen (eds.), *Vicariance Biogeography: A Critique*, 238–274. New York: Columbia University Press.

Menge, B. A. and J. Lubchenco. 2001. Essays on ecological classics— On the genesis of "community development and persistence in a low rocky intertidal zone." *Bulletin of the Ecological Society of America* 82: 124–125.

Menge, B. A. and J. P. Sutherland. 1976. Species diversity gradients: Synthesis of the roles of predation, competition and temporal heterogeneity. *American Naturalist* 110: 351–369.

Menge, B. A., E. L. Berlow, C. A. Blanchette, S. A. Navarrete and S. B. Yamada. 1994. The keystone species concept: Variation in interaction strength in a rocky intertidal habitat. *Ecological Monographs* 64: 249–286.

Mercer, J. M. and V. L. Roth. 2003. The effects of Cenozoic global change on squirrel phylogeny. *Science* 299: 1568–1572.

Merriam, C. H. 1890. Results of a biological survey of the San Francisco Mountain region and the desert of the Little Colorado, Arizona. *North American Fauna* 3: 1–136.

Merriam, C. H. 1894. Laws of temperature control of the geographic distribution of terrestrial animals and plants. *National Geographic* 6: 229–238.

Mertens, R. 1934. Die Inseleidenchsen des Golfes von Salerno. *Senckenbergiana Biologica* 42: 31–40.

Meschede, M. and W. Frisch. 1998. A plate-tectonic model for the Mesozoic and Early Cenozoic history of the Caribbean plate. *Tectonophysics* 296: 269–291

Metcalf, H. and J. F. Collins. 1911. The control of the chestnut bark disease. *Farmer's Bulletin of the U. S. Deptartment of Agriculture* 467: 1–24.

Metcalfe, I. 1999. *Gondwana Dispersion and Asian Accretion*. Rotterdam, Netherlands: A. A. Balkema.

Meyer, A. 1993. Phylogenetic relationships and evolutionary processes in East African cichlid fishes. *Trends in Ecology and Evolution* 8: 279–284.

Meyer, A., T. D. Kocher, P. Basasibwaki and A. C. Wilson. 1990. Monophyletic origin of Lake Victoria cichlid fishes suggested by mitochondrial DNA sequences. *Nature* 347: 550–553.

Meyer, C. P., J. B. Geller and G. Paulay. 2005. Fine scale endemism on coral reefs: archipelagic differentiation in turbinid gastropods. *Evolution* 59: 113–125.

Meyers, G. S. 1953. Ability of amphibians to cross sea barriers, with especial reference to Pacific zoogeography. *Proceedings of the Seventh Pacific Science Congress* 4: 19–17.

Meyers, N. 1980. *The Sinking Ark*. Oxford: Pergamon Press.

Michaud, J. M. 1995. *Biogeography of the Bog Copper Butterfly (Lycaena epixanthe) in Southern Rhode Island Peatlands: A Metapopulation Perspective*. Master's Thesis, University of Rhode Island.

Mielke, H. W. 1989. *Patterns of Life: Biogeography of a Changing World*. Boston: Unwin Hyman.

Mikkelson, G. M. 1993. How do food webs fall apart? A study of changes in tropic structure during relaxation on habitat fragments. *Okios* 67: 539–547.

Mileikovsky, S. A. 1971. Types of larval development in marine bottom invertebrates, their distribution and ecological significance: A re-evaluation. *Marine Biology* 10: 193–213.

Miles, D. B. and A. E. Dunham. 1996. The paradox of the phylogeny: Character displacement of analyses of body size in island *Anolis*. *Evolution* 50: 594–603.

Miles, J. 1987. Vegetation and succession: Past and present perceptions. In A. J. Gray, M. J. Crawley and P. J. Edwards (eds.), *Colonisation, Succession and Stability*, 1–30. Oxford: Blackwell Scientific Publications.

Millar, J. S. 1989. Reproduction and development. In G. L. Kirkland Jr. and J. N. Layne (eds.), *Advances in the Study of Peromyscus (Rodentia)*, 169–232. Lubbock, TX: Texas Tech University Press.

Miller, A. I. 1989. Spatio-temporal transitions in Paleozoic Bivalvia: An analysis of North America fossil assemblages. *Historical Biology* 1: 251–273.

Miller, B., G. Ceballos and R. Reading. 1994. The prairie dog and biotic diversity. *Conservation Biology* 8: 677–681.

Miller, B., R. Reading, J. Hoogland, T. Clark, G. Ceballos, R. List, S. Forrest, L. Hanebury, P. Manzano, J. Pacheco and D. Uresk. 2000. The role of prairie dogs as a keystone species: Response to Stapp. *Conservation Biology* 14: 318–321.

Miller, G. H., J. W. Magee, B. J. Johnson, M. L. Fogel, N. A. Spooner, M. T. McCulloch and L. K. Ayliffe. 1999. Pleistocene extinction of *Genyornis newtoni*: Human impact on Australian megafaun. *Science* 283: 205–208.

Miller, R. R. 1948. The cyprinodont fishes of the Death Valley system of eastern California and southwestern Nevada. *University of Michigan Museum of Zoology Miscellaneous Publications* 42: 1–80.

Miller, R. R. 1961a. Man and the changing fish fauna of the American Southwest. *Papers of the Michigan Academy of Science, Arts and Letters* 46: 365–404.

Miller, R. R. 1961b. Speciation rates in some freshwater fishes of western North America. In W. F. Blair (ed.), *Vertebrate Speciation*, 537–560. Austin, TX: University of Texas Press.

Miller, R. R. 1966. Geographical distribution of Central American freshwater fishes. *Copeia* (4): 773–802.

Miller, R. S. 1964. Ecology and distribution of pocket gophers (Geomyidae) in Colorado. *Ecology* 45: 256–272.

Miller, R. S. 1967. Pattern and process in competition. *Advances in Ecological Research* 4: 1–74.

Milne, B. T. and R. T. Forman. 1986. Peninsulas in Maine: Woody plant diversity, distance and enviromental patterns. *Ecology* 67: 967–974.

Mitrovica, J. X. and B. L. A. Vermeersen. 2002. *Ice Sheets, Sea Level and Dynamic Earth*. Washington, DC: American Geophysical Union.

Moffat, A. S. 2002. Brazilian ecosystems. South American landscapes: Ancient and modern. *Science* 296: 1959–1960.

Moilanen, A., A. T. Smith and I. Hanski. 1998. Long-term dynamics in a metapopulation of the American pika. *American Naturalist* 152: 530–542.

Moles, A. T. and W. Westoby. 2003. Latitude, seed predation and seed mass. *Journal of Biogeography* 30: 105–128.

Moll, E. O. and J. M. Legler. 1971. The life history of a Neotropical slider turtle, *Pseudomys scripta* (Schoepff), in Panama. *Bulletin of the Los Angeles County Museum of Natural History of Science*, no. 11.

Molles, M. C., Jr. 1978. Fish species diversity on model and natural reef patches: Experimental insular biogeography. *Ecological Monographs* 48: 289–305.

Monk, C. D. 1968. Successional and environmental relationships of the forest vegetation of north central Florida. *American Midland Naturalist* 79: 441–457.

Mooney, H. A. (ed.). 1977. *Convergent Evolution in Chile and California: Mediterranean Climate Ecosystems*. US/IBP Synthesis Series 5. Stroudsburg, PA: Dowden, Hutchinson, & Ross.

Mooney, H. A., J. Kummerow, A. W. Johnson, D. J. Parsons, D. Kelley, A. Hoffman, R. I. Hays, J. Giliberto and C. Chu. 1977. The producers: Their resources and adaptive responses. In H. A. Mooney (ed.), *Convergent Evolution in Chile and California*, 85–143. Stroudsburg, PA: Dowden, Hutchinson and Ross.

Moore, D., M. M. Nauta, S. Evans and M. Rotheroe. 2001. *Fungal Conservation: Issues and Solutions*. Cambridge: Cambridge University Press.

Moore, H. E. Jr. 1973. Palms in the tropical forest ecosystems of Africa and South America. In B. J. Meggers, E. S. Ayensu and W. D. Duckworth (eds.), *Tropical Forest Ecosystems in Africa and South America: A Comparative Review*, 63–88. Washington, DC: Smithsonian Institution Press.

Moore, W. S. 1995. Inferring phylogenies from mtdna variation—Mitochondrial-gene trees vs. nuclear-gene trees. *Evolution* 49: 718–726.

Mora, C., P. M. Chittaro, P. F. Sale, J. P. Kritzer and S. A. Ludsin. 2003. Patterns and processes in reef fish diversity. *Nature* 421: 933–936.

Moreau, R. E. 1944. Clutch size: A comparative study with special reference to African birds. *Ibis* 86: 286–347.

Morell, V. 1996. New mammals discovered by biology's new explorers. *Science* 273: 1491.

Morgan, G. S. and C. A. Woods. 1986. Extinction and zoogeography of West Indian land mammals. *Biological Journal of the Linnean Society* 28: 167–203.

Morgan, W. J. 1972a. Deep mantle convection plumes and plate motion. *Bulletin of the American Association of Petroleum Geologists* 56: 203–213.

Morgan, W. J. 1972b. Plate motions and deep mantle convection. *Memoirs of the Geological Society of America* 132: 7–22.

Morin, P. J and S. P. Lawler. 1995. Food web architecture and population dydnamics: Theory and empirical evidence. *Annual Review of Ecology and Systematics* 26: 505–530.

Moritz, C., S. Lavery and R. Slade. 1995. Using allele frequency and phylogeny to define units for conservation and management. *American Fisheries Society Symposium* 17: 249–262.

Moritz, C., J. L. Patton, C. J. Schneider and T. B. Smith. 2000. Diversification of rainforest faunas: An integrated molecular approach. *Annual Review of Ecology and Systematics* 31: 533–563.

Morrone, J. J. 1994. On the identification of areas of endemism. *Systematic Biology* 43: 438–441.

Morrone, J. J. 2002. Biogeographical regions under track and cladistic scrutiny. *Journal of Biogeography* 29: 149–152.

Morrone, J. J. and J. V. Crisci. 1995. Historical biogeography: Introduction to methods. *Annual Review of Ecology and Systematics* 26: 373–401.

Morton, E. S. 1978. Avian arboreal folivores: Why not? In G. G. Montgomery (ed.), *The Ecology of Arboreal Folivores*, 123–130. Washington, DC: Smithsonian Institution Press.

Mosiman, J. E. and P. S. Martin. 1975. Simulating overkill by paleoindians. *American Scientist* 63: 304–313.

Moulton, M. P. 1985. Morphological similarity and co-existence of congeners: An experimental test with introduced birds. *Oikos* 44: 301–305.

Moulton, M. P. 1993. The all-or-none pattern in introduced Hawaiian passeriforms: The role of competition sustained. *American Naturalist* 141: 105–119.

Moulton, M. P. and J. L. Lockwood. 1992. Morphological dispersion of introduced Hawaiian finches: Evidence for competition and the Narcissus effect. *Evolutionary Ecology* 6: 45–55.

Moulton, M. P. and S. L. Pimm. 1983. The introduced Hawaiian avifauna: Biogeographical evidence for competition. *American Naturalist* 121: 669–690.

Moulton, M. P. and S. L. Pimm. 1986. The extent of competition in shaping an introduced avifauna. In J. M. Diamond and T. Case (eds.), *Community Ecology*, 80–97. New York: Harper and Row.

Moulton, M. P. and S. L. Pimm. 1987. Morphological assortment and introduced Hawaiian passerines. *Evolutionary Ecology* 1: 113–124.

Moyle, P. B. and R. A. Leidy. 1992. Loss of biodiversity in aquatic ecosystems: Evidence from fish faunas. In P. L. Fiedler and S. K. Jain (eds.), *Conservation Biology: The Theory and Practice of Nature Conservation, Preservation and Management*, 127–170. New York: Chapman and Hall.

Muller, J. 1970. Palynological evidence on early differentiation of angiosperms. *Biological Reviews of the Cambridge Philosophical Society* 45: 417–450.

Muller, P. 1974. *Aspects of Zoogeography*. The Hague: Dr. W. Junk.

Muller, R. A. and G. J. MacDonald. 1995. Glacial cycles and orbital inclination. *Nature* 377: 107–108.

Munroe, E. 1996. Distributional patterns of Lepidoptera in the Pacific Islands. In A. Keast and S. E. Miller (eds.), *The Origin and Evolution of Pacific Island Biotas. New Guinea to Eastern Polynesia: Patterns and Processes*, 275–295. Amsterdam: SPB Academic.

Munroe, E. G. 1948. *The Geographical Distribution of Butterflies in the West Indies*. Ph.D. dissertation, Cornell University, Ithaca, NY.

Munroe, E. G. 1953. The size of island faunas. In *Proceedings of the Seventh Pacific Science Congress of the Pacific Science Association*, vol. IV, Zoology, 52–53. Auckland, New Zealand: Whitcome and Tombs.

Murdoch, W. W. and A. Oaten. 1975. Predation and population stability. *Advances in Ecological Research* 9: 1–131.

Murray, B. R., C. R. Fonesca and M. Westoby. 1998. The macroecology of Australian frogs. *Journal of Animal Ecology* 67: 567–579.

Murray, J. 1895. A summary of scientific results. In *Challenger* Report Summary. 2 volumes. London: Nell and Company.

Murray, M. 2003. Overkill and sustainable use. *Science* 299: 1851–1853.

Mutke, J. and W. Barthlott. 2005. Patterns of vascular plant diversity at continental to global scales. In I. Friis and H. Balsey (eds.), *Plant Diversity and Complexity patterns: Local, Regional, and Global Dimensions.* 521–537. The Royal Danish Academy of Sciences and Letters, Copenhagen.

Myers, A. A. and P. S. Giller (eds.). 1989. *Analytical Biogeography.* London: Chapman and Hall.

Myklestad, A. 1993. The distribution of *Salix* species in Fennoscandia—A numerical analysis. *Ecography* 16: 329–344.

Myklestad, A. and H. J. B. Birks. 1993. A numerical analysis of the distribution patterns of Salix L. species in Europe. *Journal of Biogeography* 20: 1–32.

Nachman, M. W., H. E. Hoekstra and S. L. D'Agostino. 2003. The genetic basis of adaptive melanism in pocket mice. *Proceedings of the National Academy of Sciences, USA* 100: 5268–5273.

Nadler, C. F., N. N. Vorontosov, R. S. Hoffmann, I. I. Formichova and C. F. Nadler Jr. 1973. Zoogeography of transferins in arctic and long-tailed ground squirrel populations. *Comparative Biochemistry and Physiology* 44B: 33–40.

Nadler, C. F., M. Zhurkevich, R. S. Hoffmann, A. I. Kozlovskii, L. Deutsch and D. F. Nadler Jr. 1978. Biochemical relatedness of some Holarctic voles of the genera *Microtus, Arvicola* and *Clethrionomys* (Rodentia: Arvicolae). *Canadian Journal of Zoology* 56: 1564–1575.

Nalepa, T. F. and D. W. Schlosser. 1993. *Zebra Mussels: Biology, Impacts and Control.* Boca Raton, FL: CRC Press.

Nash, L. L. and P. H. Gleick. 1991. Sensitivity of streamflow in the Colorado Basin to climate changes. *Journal of Hydrology* 125: 221–241.

Nason, J. D., J. L. Hamrick and T. H. Fleming. 2002. Historical vicariance and postglacial colonization effects on the evolution of genetic structure in Lophocereus, a Sonoran Desert columnar cactus. *Evolution* 56: 2214–2226.

Nathan, R. 2005. Long-distance dispersal research: Building a network of yellow brick roads. *Diversity and Distribution* 11: 125–130.

Navarro, A. and N. H. Barton. 2003. Accumulating postzygotic isolation genes in parapatry: A new twist on chromosomal speciation. *Evolution* 57: 447–459.

Nei, M. 1987. *Molecular Evolutionary Genetics.* New York: Columbia University Press.

Neill, W. T. 1958. The occurrence of amphibians and reptiles in salt-water areas and the bibliography. *Bulletin of Marine Science of the Gulf and Caribbean* 8: 1–97.

Neill, W. T. 1969. *The Geography of Life.* New York: Columbia University Press.

Neilsen, R. and J. Wakeley. 2001. Distinguishing migration from isolation: A Markov chain Monte Carlo approach. *Genetics* 158: 885–896.

Nelson, G. and P. Y. Ladiges. 1992. *TAS and TAX: MSDos computer programs for cladistics.* New York and Melbourne: published by the authors.

Nelson, G. and P. Y. Ladiges. 1995. *TASS (Three Area Subtrees).* New York and Melbourne: published by the authors.

Nelson, G. and P. Y. Ladiges. 1996. Paralogy in cladistic biogeography and analysis of paralogy-free subtrees. *American Museum Novitates* 3167: 1–58.

Nelson, G. J. 1969a. The problem of historical biogeography. *Systematic Zoology* 18: 243–246.

Nelson, G. J. 1969b. Gill arches and the phylogeny of fishes, with notes on the classification of vertebrates. *Bulletin of the American Museum of Natural History* 141: 475–552.

Nelson, G. J. 1978. From Candolle to Croizat: Comments on the history of biogeography. *Journal of the History of Biology* 11: 269–305.

Nelson, G. J. and N. Platnick. 1981. *Systematics and Biogeography: Cladistics and Vicariance.* New York: Columbia University Press.

Nelson, G. J. and D. E. Rosen (eds.). 1981. *Vicariance Biogeography: A Critique.* New York: Columbia University Press.

Neumayr, M. 1887. Uber klimatische Zonen wahrend der Jura-und Kreidzeit. *Koniglische Akademie der Wissenschaft Wien Denkschrift* 47: 277–310.

Nevo, E. and H. Bar-El. 1976. Hybridization and speciation in fossorial mole rats. *Evolution* 30: 831–840.

Nevo, E., M. Corti, G. Heth, A. Beiles and S. Simpson. 1988. Chromosomal polymorphisms in subterranean mole rats: Origins and evolutionary significance. *Biological Journal of the Linnean Society* 33: 309–322.

Newman, M. T. 1953. The application of ecological rules to the racial anthropology of the aboriginal New World. *American Anthropology* 55: 311–327.

Newman, T. W. 1985. Human-wildlife competition and the passenger pigeon: Population growth from system destabilization. *Human Ecology* 13: 389–410.

Newmark, W. D. 1993. The role and design of wildlife corridors with examples from Tanzania. *Ambio* 22: 500–504.

Nicholls, A. O. and C. R. Margules. 1991. The design of studies to demonstrate the biological importance of corridors. In D. A. Saunders and R. J. Hobbs (eds.), *The Role of Corridors in Nature Conservation,* 49–61. Chipping Norton, NSW, Australia: Surrey Beatty.

Nielson, M., K. Lohman and J. Sullivan. 2001. Phylogeography of the tailed frog (*Ascaphus truei*): Implications for the biogeography of the Pacific Northwest. *Evolution* 55: 147–160.

Niering, W. A. 1963. Terrestrial ecology of Kapingamarangi Atoll, Caroline Islands. *Ecological Monographs* 33: 131–160.

Niethammer, G. 1958. Tiergeographie (Bericht uber die Jahren 1950–56). *Fortschr. Zool.* 11: 35–141.

Niklas, K. J., B. H. Tiffney and A. H. Knoll. 1983. Apparent changes in the diversity of fossil plants. In M. C. Hecht, W. C. Steere and B. Wallace (eds.), *Evolutionary Biology,* 12: 1–89. New York: Plenum Press.

Nilsson, S. G. 1977. Density compensation and competition among birds on small islands in a south Swedish lake. *Oikos* 28: 170–176.

Nilsson, S. G. and I. N. Nilsson. 1978. Species richness and dispersal of vascular plants to islands in Lake Mockeln, Southern Sweden. *Ecology* 59: 473–480.

Nobel, P. S. 1978. Surface temperarures of cacti: Influences of environmental and morphological factors. *Ecology* 59: 986–996.

Nobel, P. S. 1980a. Morphology, nurse plants and minimum apical temperatures for young *Carnegiea gigantea. Botanical Gazette* 141: 188–191.

Nobel, P. S. 1980b. Morphology, surface temperatures and northern limits of columnar cacti in the Sonoran Desert. *Ecology* 61: 1–7.

Nor, S. Md. 2001. Elevational diversity patterns of small mammals on Mount Kinabalu, Sabah, Malaysia. *Global Ecology and Biogeography* 10: 41–62.

Nores, M. 1995. Insular biogeography of birds on mountain-tops in north Argentina. *Journal of Biogeography* 22: 61–70.

Nores, M. 1999. An alternative hypothesis for the origin of Amazonian bird diversity. *Journal of Biogeography* 26: 475–485.

Norse, E. A. 1993. *Global Marine Biological Diversity: A Strategy for Building Conservation into Decision Making.* Washington, DC: Island Press.

Nur, A. and Z. Ben-Avraham. 1977. Lost Pacifica continent. *Nature* 270: 41–43.

Oberdorff, T., J. Guegan and B. Hugeny. 1995. Global scale patterns in fish species diversity in rivers. *Ecography* 18: 345–352.

Oberdorff, T., B. Hugueny and J. F. Gue'gan. 1997. Is there an influence of historical events on contemporary fish species richness in rivers? Comparison between Western Europe and North America. *Journal of Biogeography* 24: 461–467.

Ochocinska, D. and J. R. E. Taylor. 2003. Bergmann's rule in shrews: Geographical variation of body size in Palearctic *Sorex* species. *Biological Journal of the Linnean Society* 78: 365–381.

Ochumba, P. B. O., M. Gophen and L. S. Kaufman. 1993. Changes in oxygen availability in the Kenyan portion of Lake Victoria: Effects on fisheries and biodiversity. Abstract. In *People, Fisheries, Biodiversity and the Future of Lake Victoria*. Report 93–3. Boston: New England Aquarium.

O'Conner, R. J. and J. Faaborg. 1992. The relative abundance of the brown-headed cowbird (*Molothrus ater*) in relation to exterior and interior edges in forests of Missouri. *Transactions of the Missouri Academy of Science* 26: 1–9.

Odening, W. R., B. R. Strain and W. C. Oechel. 1974. The effects of decreasing water potential on net CO_2 exchange of intact desert shrubs. *Ecology* 55: 1086–1095.

Odum, E. P. 1969. The strategy of ecosystem development. *Science* 164: 262–270.

Odum, E. P. 1971. *Fundamentals of Ecology*. 3rd edition. Philadelphia: W. B. Saunders.

Odum, H. T. 1957. Trophic structure and productivity of Silver Springs, Florida. *Ecological Monographs* 27: 55–112.

Olifers, N., M., V. Vieira and C. E. V. Grelle. 2004. Geographic range and body size in Neotropical marsupials. *Global Ecology and Biogeography* 13: 439–444.

Olson, S. L. 1973. Evolution of the rails of the South Atlantic Islands. *Smithsonian Contributions to Zoology* 152: 1–43.

Olson, S. L. 1976. Oligocene fossils bearing on the origins of the Totidae and Momotidae (Aves: Coraciiformes). *Smithsonian Contributions to Paleobiology* 27: 111–119.

Olson, S. L. and H. F. James. 1982a. Fossil birds from the Hawaiian Islands: Evidence for wholesale extinction by man before Western contact. *Science* 217: 633–635.

Olson, S. L. and H. F. James. 1982b. Promodromus of the fossil avifauna of the Hawaiian Islands. *Smithsonian Contributions to Zoology* 365.

Olson, S. L. and H. F. James. 1984. The role of Polynesians in the extinction of the avifauna of the Hawaiian Islands. In P. S. Martin and R. G. Klein (eds.), *Quarternary Extinctions*, 768–780. Tucson, AZ: University of Arizona Press.

Omi, P. N., L. C. Wensel and J. L. Murphy. 1979. An application of multivariate statistics to land-use planning: Classifying land units into homogeneous zones. *Forest Science* 25: 399–414.

Ong, P., L. E. Afuang and R. G. Rosell-Ambal (eds.). 2002. *Philippine Biodiversity Conservation Priorities: A Second Iteration of the National Biodiversity Strategy and Action Plan*. Philippine Department of the Environment and Natural Resources, Quezon City. xviii + 113 pp.

Oppenheimer, S. J. and M. Richards. 2001. Polynesian origins: Slow boat to Melanesia? *Nature* 410: 166–167.

Oreskes, N. (ed.). 2001. *Plate Tectonics: An Insider's History of the Modern Theory of the Earth*. Cambridge, MA: Westview Press.

Ortelius, A. 1596. *Thesaurus Geographicus*.

Orti, G. and A. Meyer. 1997. The radiation of characiform fishes and the limits of resolution of mitochondrial ribosomal DNA sequences. *Systematic Biology* 46: 75–100.

Ortmann, A. E. 1896. *Grundzuge der marinen tiergeographie*. Jena: G. Fisher.

Ospovat, D. 1977. Lyell's theory of climate. *Journal of the History of Biology* 10: 317–339.

Ostfeld, R. S. 1994. The fence effect reconsidered. *Oikos* 70: 340–348.

Otte, D. 1976. Species richness patterns of New World desert grasshoppers in relation to plant diversity. *Journal of Biogeography* 3: 197–209.

Otte, D. and J. A. Endler (eds.). 1989. *Speciation and Its Consequences*. Sunderland, MA: Sinauer Associates.

Ovaskainen, O. and I. Hanski. 2003. The species-area relationship derived from species-specific incidence functions. *Ecology Letters* 6: 903–909.

Overpeck, J. T., R. S. Webb and T. Webb III. 1992. Mapping eastern North American vegetation changes of the past 18 ka: No-analogs and the future. *Geology* 20: 1071–1074.

Owen, D. F. and J. Owen. 1974. Species diversity in temperate and tropical Ichneumonidae. *Nature* 249: 583–584.

Owen, H. G. 1992. Has the earth increased in size? In S. Chatterjee and N. Hotton III (eds.), *New Concepts in Global Tectonics*, 289–296. Lubbock, TX: Texas Tech University Press.

Owen-Smith, N. 1987. Pleistocene extinctions: The pivotal role of megaherbivores. *Paleobiology* 13: 351–362.

Owen-Smith, N. 1989. Megafaunal extinctions: The conservation message from 11,000 years B.P. *Conservation Biology* 3: 405–412.

Owen-Smith, R. N. 1988. *Megaherbivores: The Influence of Very Large Body Size on Ecology*. Cambridge: Cambridge University Press.

Pacala, S. and J. Roughgarden. 1985. Population experiments with the *Anolis* lizards of St. Maarten and St. Eustatius. *Ecology* 66: 128–141.

Paetkau, D., G. F. Shields and C. Strobeck. 1998. Gene flow between insular, coastal and interior populations of brown bears in Alaska. *Molecular Ecology* 7: 1283–1292.

Page, R. D. M. 1990. Tracks and trees in the antipodes: A reply. *Systematic Zoology* 39: 288–299.

Page, R. D. M. 1991. Clocks, clades and cospeciation: Comparing rates of evolution and timing of cospeciation events in host-parasite assemblages. *Systematic Zoology* 40: 188–198.

Page, R. D. M. 1993a. Genes, organisms and areas: The problem of multiple lineages. *Systematic Biology* 42: 77–84.

Page, R. D. M. 1993b. Parasites, phylogeny and cospeciation. *International Journal for Parasitology* 23: 499–506.

Page, R. D. M. 1994. Parallel phylogenies: Reconstructing the history of host-parasite assemblages. *Cladistics* 10: 155–173.

Page, R. D. M. 2003. *Treemap 2.02. Program and User's Manual*. Division of Environmental and Evolutionary Biology, Institute of Biomedical and Life Sciences, University of Glasgow, Glasgow, U.K.

Pagel, M. D., R. M. May and A. R. Collie. 1991. Ecological aspects of the geographical distribution and diversity of mammalian species. *American Naturalist* 137: 791–815.

Paine, R. T. 1966. Food web complexity and species diversity. *American Naturalist* 100: 65–76.

Paine, R. T. 1974. Intertidal community structure: Experimental studies on the relationship between a dominant competitor and its principal predator. *Oecologia* 15: 93–120.

Paine, R. T. 1995. A conversation on refining the concept of keystone species. *Conservation Biology* 9: 962–965.

Palmer, M. A., R. F. Ambrose and N. L. Poff. 1997. Ecological theory and community restoration ecology. *Restoration Ecology* 5: 291–300.

Palumbi, S. R. 1997. Molecular biogeography of the Pacific. *Coral Reefs* 16: S47–S52.

Panteleev, P. A., M. Le Berre, A. N. Terekhina and R. Ramouz. 1998. Application of Bergmann's rule to hibernating animals: An example of the genus *Marmota*. *Russian Journal of Ecology* 29: 192–196.

Parenti, L. R. 1981. Discussion [of C. Patterson, *Methods of Paleobiogeography*]. In G. Nelson and D. E. Rosen (eds.), *Vicariance Biogeography: A Critique*, 490–497. New York: Columbia University Press.

Parker, V. T. 2002. Conceptual problems and scale limitations of defining ecological communities: A critique of the CI concept (community of individuals). *Perspectives in Plant Ecology, Evolution and Systematics* 4: 80–96.

Parmesan, C. 1996. Climate and species range. *Nature* 382: 765–766.

Parmesan, C. and G. Yohe. 2003. A globally coherent fingerprint of climate change impacts across natural systems. *Nature* 421: 37–42.

Parmesan, C., N. Ryrholm, C. Stefanescu, J. K. Hill, C. D. Thomas, H. Descimon, B. Huntley, L. Kaila, J. Kullberg, T. Tammaru, W. J. Tennent, J. A. Thomas and M. Warren. 1999. Poleward shifts in geographic ranges of butterfly species associated with regional warming. *Nature* 399: 579–583.

Parsons, T. J. and M. M. Holland. 1998. Mitochondrial mutation rate revisited: Hot spots and polymorphism—Response, *Nature Genetics* 18: 110.

Pascual, R., M. Archer, E. O. Jaureguizar, J. L. Prado, H. Godthelp and S. J. Hand. 1992. First discovery of monotremes in South America. *Nature* 356: 704–706.

Paterson, H. E. H. 1982. Perspective on speciation by reinforcement. *South African Journal of Science* 78: 53–57.

Patrick, R. 1961. A study of the numbers and kinds of species found in rivers in eastern United States. *Proceedings of the Academy of Natural Sciences of Philadelphia* 13: 215–258.

Patrick, R. 1966. The Catherwood Foundation Peruvian Amazon Expedition: Limnological and systematic studies. *Monographs of the Academy of Natural Sciences of Philadelphia* 14: 1–495.

Patterson, B. D. 1980. Montane mammalian biogeography in New Mexico. *Southwestern Naturalist* 25: 33–40.

Patterson, B. D. 1981a. Morphological shifts of some isolated populations of *Eutamias* (Rodentia: Sciuridae) in different congeneric assemblages. *Evolution* 35: 53–66.

Patterson, B. D. 1984. Mammalian extinction and biogeography in the southern Rocky Mountains. In M. H. Nitecki (ed.), *Extinctions*, 247–294. Chicago: University of Chicago Press.

Patterson, B. D. 1987. The principle of nested subsets and its implications for biological conservation. *Conservation Biology* 1: 323–334.

Patterson, B. D. 1990. On the temporal development of nested subset patterns of species composition. *Oikos* 59: 330–342.

Patterson, B. D. 1994. Accumulating knowledge on the dimensions of biodiversity: Systematic perspectives on Neotropical mammals. *Biodiversity Letters* 2: 79–86.

Patterson, B. D. 1995. Local extinctions and the biogeographic dynamics of boreal mammals in the southwest. In C. A. Istock and R. S. Hoffman (eds.), *Storm over a Mountain Island: Conservation Biology and the Mount Graham Affair*. Tucson, AZ: University of Arizona Press.

Patterson, B. D. and W. Atmar. 1986. Nested subsets and the structure of insular mammalian faunas and archipelagoes. *Biological Journal of the Linnean Society* 28: 65–82.

Patterson, B. D. and J. H. Brown. 1991. Regionally nested patterns of species composition in granivorous rodent assemblages. *Journal of Biogeography* 18: 395–402.

Patterson, B. D. and R. Pascual. 1972. The fossil mammal fauna of South America. In A. Keast, F. C. Erk and B. Glass (eds.), *Evolution, Mammals and Southern Continents*, 247–309. Albany: State University of New York Press.

Patterson, B. D., V. Pacheco and S. Solari. 1996. Distribution of bats along an elevational gradient in the Andes of south-east Peru. *Journal of the Zoological Society of London* 240: 637–658.

Patterson, B. D., D. F. Stotz, S. Solari, J. W. Fitzpatrick and V. Pacheco. 1998. Contrasting patterns of elevational zonation for birds and mammals in the Andes of southeastern Peru. *Journal of Biogeography* 25: 583–607.

Patterson, C. 1981. Methods of paleobiogeography. In G. Nelson and D. E. Rosen (eds.), *Vicariance Biogeography: A Critique*, 446–489. New York: Columbia University Press.

Patton, J. L. 1969. Chromosomal evolution in the pocket mouse, *Perognathus goldmani* Osgood. *Evolution* 23: 645–662.

Patton, J. L. 1972. Patterns of geographic variation in karyotype in the pocket gopher, *Thomomys bottae* (Eydoux and Gervais). *Evolution* 25: 574–586.

Patton, J. L. 1985. Population structure and the genetics of speciation in pocket gophers, genus *Thomomys*. *Acta Zoologica Fennica* 170: 109–114.

Patton, J. L. and M. N. F. da Silva. 1998. Rivers, refuges and ridges: The geography of speciation of Amazonian mammals. In D. J. Howard and S. H. Berlocher (eds.), *Endless Forms: Species and Speciation*, 202–213. Oxford: Oxford University Press.

Patton, J. L., B. Berlin and E. A. Berlin. 1982. Aboriginal perspectives of a mammal community in amazonian Perú: Knowledge and utilization patterns among the Aguaruna Jivaro. In M. A. Mares and H. H. Genoways (eds.), *Mammalian Biology in South America*, vol. 6, 111–128. Linesville, PA: University of Pittsburgh.

Patton, J. L., M. N. F. da Silva and J. R. Malcom. 1994. Gene genealogy and differentiation among arboreal spiny rats (Rodentia: Echymidae) of the Amazon Basin: A test of the riverine barrier hypothesis. *Evolution* 48: 1314–1323.

Patton, J. L., M. F. Smith, R. D. Price and R. A. Hellenthal. 1984. Genetics of hybridization between the pocket gophers *Thomomys bottae* and *Thomomys townsendii* in northeastern California. *Great Basin Naturalist* 44: 431–440.

Patz, J. A. 2002. A human disease indicator for the effects of recent global climate change. *Proceedings of the National Academy of Sciences, USA* 99: 12506–12508.

Pearcy, W. G. and A. Schoener. 1987. Changes in the marine biota coincident with the 1982–1983 El Niño in the northeastern subarctic Pacific Ocean. *Journal of Geophysical Research* 92, C13: 14417–14428.

Pearson, S. and J. L. Betancourt.2002. Understanding arid environments using fossil rodent middens. *Journal of Arid Environments* 50: 499–511.

Pearson, T. G. (ed.). 1936. *Birds of America*. Garden City, NY: Garden City Publishing Co.

Peattie, D. C. 1922. The Atlantic coastal plain element in the flora of the Great Lakes. *Rhodora* 24: 57–70, 80–88.

Peltonen, A. and I. Hanski. 1991. Patterns of island occupancy explained by colonization and extinction rates in shrews. *Ecology* 72: 1698–1708.

Peltonen, A., S. Peltonen, P. Vilpas and A. Beloff. 1989. Distributional ecology of shrews in three archipelagoes in Finland. *Annales Zoologici Fennici* 26: 381–387.

Pendergast, J. R., R. M. Quinn, J. H. Lawton, B. C. Eversham and D. W. Gibbons. 1993. Rare species, the coincidence of diversity hot spots and conservation strategies. *Nature* 365: 335–337.

Peng, C. H., J. Guiot, E. VanCamp and R. Cheddar. 1995. Temporal and spatial variations of terrestrial biomes and carbon storage since 13,000 yr B.P. in Europe: Reconstruction from pollen data and statistical models. *Water, Air and Soil Pollution* 82: 375–390.

Perault, D. R. 1998. *Landscape Heterogeneity and the Role of Corridors in Determining the Spatial Structure of Insular Mammal Populations*. Ph.D. dissertation, University of Oklahoma, Norman, OK.

Perault, D. R. and M. V. Lomolino. 2000. Corridors and mammal community structure across a fragmented, old-growth forest landscape. *Ecological Monographs* 70: 401–422.

Pesole, G., C. Gissi, A. De Chirico and C. Saccone. 1999. Nucleotide substitution rate of mammalian mitochondrial genomes. *Journal of Molecular Evolution* 48: 427–434.

Peters, R. H. 1983. *The Ecological Implications of Body Size*. Cambridge: Cambridge University Press.

Peters, R. L. and T. E. Lovejoy. 1992. *Global Warming and Biological Diversity*. New Haven, CT: Yale University Press.

Peterson, A. T. 2001. Endangered species and peripheral populations: Cause for reflection. *Endangered Species Update* 18: 30–31.

Peterson, C. H. 1991. Intertidal zonation of marine invertebrates in sand and mud. *American Scientist* 79: 236–249.

Peterson, R. L. 1955. *North American Moose*. Toronto: University of Toronto Press.

Petren, K. and T. J. Case. 1997. A phylogenetic analysis of body size evolution and biogeography in chuckwallas (*Sauromalus*) and other iguanines. *Evolution* 51: 206–219.

Petren, K., D. T. Bolger and T. J. Case. 1993. Mechanisms in the competitive success of an invading sexual gecko over an asexual native. *Science* 259: 354–358.

Petuch, E. J. 1995. Molluscan diversity in the late Neogene of Florida: Evidence for a two-staged mass extinction. *Science* 270: 275–277.

Pianka, E. R. 1966. Latitudinal gradients in species diversity: A review of concepts. *American Naturalist* 100: 33–46.

Pianka, E. R. 1967. On lizard species diversity: North American flatland deserts. *Ecology* 48: 331–351.

Pianka, E. R. 1978. *Evolutionary Ecology*. 2nd edition. New York: Harper & Row.

Pianka, E. R. 1986. *Ecology and Natural History of Desert Lizards*. Princeton, NJ: Princeton University Press.

Pianka, E. R. 1988. *Evolutionary Ecology*. New York: Harper and Row.

Pianka, E. R. 1995. Evolution of body size: Varanid lizards as a model system. *American Naturalist* 146: 398–414.

Pickett, S. T. A. and P. S. White. 1985. *The Ecology of Natural Disturbance and Patchiness*. New York: Academic Press.

Pielou, E. C. 1975. *Ecological Diversity.* New York: John Wiley and Sons.

Pielou, E. C. 1977a. *Mathematical Ecology.* New York: John Wiley and Sons.

Pielou, E. C. 1977b. The latitudinal spans of seaweed species and their patterns of overlap. *Journal of Biogeography* 4: 299–311.

Pielou, E. C. 1978. Latitudinal overlap of seaweed species: Evidence for quasi-sympatric speciation. *Journal of Biogeography* 5: 227–238.

Pielou, E. C. 1979. *Biogeography.* New York: John Wiley and Sons.

Pielou, E. C. 1991. *After the Ice Age.* Chicago: University of Chicago Press.

Pijl, L. van den. 1972. *Principles of Dispersal in Higher Plants.* 2nd edition. New York: Springer-Verlag.

Pimentel, D., L. Lach, R. Zuniga and D. Morrison. 2000. Environmental and economic costs of nonindigenous species in the United States. *BioScience* 50: 53–65.

Pimm, S. L. 1991. *The Balance of Nature? Ecological Issues in the Conservation of Species and Communities.* Chicago: University of Chicago Press.

Pimm, S. L., H. L. Jones and J. M. Diamond. 1988. On the risk of extinction. *American Naturalist* 132: 757–785.

Pine, R. H. 1994. New mammals not so seldom. *Nature* 368: 593.

Platnick, N. I. 1991. On areas of endemism. *Australian Systematic Botany* 4(unnumbered).

Platnick, N. I. and G. Nelson. 1978. A method of analysis for historical biogeography. *Systematic Zoology* 27: 1–16.

Platt, I. 1969. What we must do. *Science* 166: 1115–1121.

Platt, W. J. 1975. The colonization and formation of equilibrium plant species associations on badger disturbances in a tall-grass prairie. *Ecological Monographs* 45: 285–305.

Poinar, G. O. and D. A. Grimaldi. 1990. Fossil and extant macrochelid mites (Ascari: Macroelidae) phoretic on Drosophilid flies (Diptera: Drosophilidae). *Journal of the New York Entomological Society* 98: 88.

Poinar, G. O., G. M. Thomas and B. Lighthart. 1990. Bioassay to determine the effect of commercial preparations of bacillus-thuringiensis on entomogeneous rhabditoid nematodes. *Agriculture Ecosystems and Environment* 30: 195–202.

Poinar, H. N., M. Hofreiter, W. G. Spaulding, P. S. Martin, B. A. Stankiewicz, H. Bland, R. P. Evershed, G. Possnert and S. Paabo. 1998. Molecular coproscopy: Dung and diet of the extinct ground sloth *Nothrotheriops shastensis. Science* 281: 402–406.

Pole, M. 1994. The New Zealand flora—Entirely long-distance dispersal? *Journal of Biogeography* 21: 625–635.

Polis, G. A. and D. S. Hurd, 1995. Extraordinarily high spider densities on islands: Flow of energy from the marine to terrestrial food webs and the absence of predation. *Proceedings of the National Academy of Sciences, USA* 92: 4382–4386.

Popper, K. R. 1968a. *The Logic of Scientific Discovery.* 2nd edition. New York: Harper & Row.

Popper, K. R. 1968b. *Conjectures and Refutations.* New York: Harper & Row.

Por, F. D. 1971. One hundred years of Suez Canal: A century of Lessepsian migration: Retrospect and viewpoints. *Systematic Zoology* 20: 138–159.

Por, F. D. 1975. Pleistocene pulsation and preadaptation of biotas in Mediterranean seas: Consequences for Lessepsian migration. *Systematic Zoology* 24: 72–78.

Por, F. D. 1978. *Lessepsian Migration: The Influx of Red Sea Biota into the Mediterranean by Way of the Suez Canal.* Ecological Studies 23. Berlin: Springer-Verlag.

Por, F. D. 1990. Lessepsian migration: An appraisal and new data. *Bulletin, Institut Oceanographique (Monaco)* Special 7: 1–10.

Porter, J. W. 1972. Ecology and species diversity of coral reefs on opposite sides of the Isthmus of Panama. *Bulletin of the Biological Society of Washington* 2: 89–116.

Porter, J. W. 1974. Community structure and coral reefs on opposite sides of the Isthmus of Panama. *Science* 186: 543–545.

Posada, D. and K. A. Crandall. 1998. MODELTEST: Testing the model of DNA substitution. *Bioinformatics* 14: 817–818.

Posada, D., K. A. Crandall and A. R. Templeton. 2000. GeoDis: A program for the cladistic nested analysis of the geographical distribution of genetic haplotypes. *Molecular Ecology* 9: 487–488.

Post, E., R. O. Peterson, N. C. Stenseth and B. E. McLaren. 1999. Ecosystem consequences of wolf behavioral responses to climate. *Nature* 401: 905–907.

Poulin, R. 1995. The evolution of body size in Monogenea: The role of host size and latitude. *Canadian Journal of Zoology* 74: 726–732.

Poulson, T. L. and W. B. White. 1969. The cave environment. *Science* 165: 971–981.

Power, D. M. 1972. Numbers of bird species on the California Islands. *Evolution* 26: 451–463.

Power, M. E., D. Tilman, J. A. Estes, B. A. Menge, W. J. Bond, L. S. Mills, G. Daily, J. C. Castilla, J. Lubchenco and R. T. Paine. 1996. Challenges in the quest for keystones. *Biosience* 46: 609–620.

Prance, G. T. 1973. Phytogeographic support for the theory of Pleistocene forest refugia in the Amazon basin, based on evidence from distribution patterns in Caryocaraceae, Chrysobalanaceae, Dichapetalaceae and Lecythidaceae. *Acta Amazonica* 3(3): 5–28.

Prance, G. T. 1978. The origin and evolution of the Amazon flora. *Interciencia* 3: 207–222.

Prance, G. T. (ed.). 1982. *The Biological Model of Diversification in the Tropics.* New York: Columbia University Press.

Pregill, G. K. and S. L. Olson. 1981. Zoogeography of West Indian vertebrates in relation to Pleistocene climatic cycles. *Annual Review of Ecology and Systematics* 12: 75–98.

Pressey, R. L. and R. M. Cowling. 2001. Reserve selection algorithms in the real world. *Conservation Biology* 15: 275–277.

Preston. F. W. 1957. Analysis of Maryland statewide bird counts. *Maryland Birdlife [Bulletin of the Maryland Ornithological Society]* 13: 63–65.

Preston, F. W. 1962a. The canonical distribution of commonness and rarity: part I. *Ecology* 43: 185–215.

Preston, F. W. 1962b. The canonical distribution of commonness and rarity: part II. *Ecology* 43: 410–432.

Price, J. P. and D. A. Clague. 2002. How old is the Hawaiian biota? Geology and phylogeny suggest recent divergence. *Proceedings of the Royal Society of London,* Series B 269: 2429–2435.

Price, N. J. 2001. *Major Impacts and Plate Tectonics.* New York: Routledge.

Price, P. W. 1980. *Evolutionary Biology of Parasites.* Monographs in Population Biology, no. 15. Princeton, NJ: Princeton University Press.

Price, P. W. and H. Roininen. 1993. Adaptive radiation in gall induction. In M. Wagner and K. F. Raffa (eds.), *Sawfly Life History Adaptation to Woody Plants,* 229–257. San Diego: Academic Press.

Primack, R. 1998. *Essentials of Conservation Biology.* 2nd edition. Sunderland, MA: Sinauer Associates.

Pritchard, J. K., M. Stephens and P. Donnelly. 2000. Inferences of population structure using multilocus genotype data. *Genetics* 155: 945–959.

Pruvot, G. 1896. *Essai sur les fonds et la faune de la Manche occidentale (cotes de Bretagne) compares a ceux du golfe du Lion.* Paris: Schleicher Frères.

Pulliam, H. R. 1988. Sources, sinks and population regulation. *American Naturalist* 132: 652–661.

Purves, W. K., D. Sadava, G. H. Orians, and H. C. Heller. 2004. *Life: The Science of Biology,* 7th edition. Sunderland, MA: Sinauer Associates.

Putnam, R. J. 1994. *Community Ecology.* New York: Chapman and Hall.

Pyron, M. 1999. Relationships between geographical range size, body size, local abundance and habitat breadth in North American suckers and sunfishes. *Journal of Biogeography* 26: 549–558.

Rabenold, K. N. 1979. A reversed latitudinal diversity gradient in avian communities of eastern deciduous forests. *American Naturalist* 114: 275–286.

Rabinowitz, D., S. Cairns and T. Dillon. 1986. Seven forms of rarity and their frequency in the flora of the British Isles. In M. E. Soulé (ed.), *Conservation Biology: The Science of Scarcity and Diversity*, 182–204. Sunderland, MA: Sinauer Associates.

Raffi, S., S. M. Stanley and R. Marasti. 1985. Biogeographic patterns and Plio-Pleistocene extinction of Bivalvia in the Mediterranean and southern North Sea. *Paleobiology* 11: 368–388.

Rahbek, C. 1995. The elevational gradient of species richness: A uniform pattern? *Ecography* 18: 200–205.

Rahbek, C. 1997. The relationship among area, elevation and regional species richness in Neotropical birds. *American Naturalist* 149: 875–902.

Rahmstorf, S. 1999. Shifting seas in the greenhouse? *Nature* 3999: 523–524.

Rai, S. N., D. V. Ramana and A. Manglik. 2002. *Dynamics of Earth's Fluid System*. Rotterdam, Netherlands: A. A. Balkema., Netherlands.

Raikow, R. J. 1976. The origin and evolution of the Hawaiian honeycreepers (Drepaniidae). *Living Bird* 15: 95–117.

Rapoport, E. H. 1975. *Areografia: Estrategias Geograficas de las Especies*. Fondo de Cultura Economica.

Rapoport, E. H. 1982. *Areography: Geographical Strategies of Species*. New York: Pergamon Press.

Rapoport, E. H., M. E. Diaz Betancourt and I. R. Lopez Moreno. 1983. *Aspectos de la Ecologia Urbana en la Ciudad de Mexico: Flora y de las calles y baldios*. Mexico, D.F.: Editorial Limusa.

Rass, T. S. 1986. Vicariance icthyogeography of the Atlantic ocean pelagial. In *Pelagic Biogeography*, 237–241. UNESCO Technical Papers in Marine Science, 49.

Raubenheimer, E.J. Koedoe. 2000. Development of the tusk and tusklessness in African elephant (*Loxodonta africana*). *Koedoe* 43: 57–64.

Raunkiaer, C. 1934. *The Life Forms of Plants and Statistical Plant Geography*. Oxford: Clarendon Press.

Raup, D. M. 1972. Taxonomic diversity during the Phanerozoic. *Science* 177: 1065–1071.

Raup, D. M. 1976. Species diversity in the Phanerozoic: An interpretation. *Paleobiology* 2: 289–297.

Raup, D. M. 1979. Size of the Permo-Triassic bottleneck and its evolutionary implications. *Science* 206: 217–218.

Raup, D. M. 1994. The role of extinction in evolution. *Proceedings of the National Academy of Sciences, USA* 91: 6758–6763.

Raup, D. M. and D. Jablonski. 1993. Geography of end-Cretaceous marine bivalve extinctions. *Science* 260: 971–973.

Raup, D. M. and J. J. Sepkoski Jr. 1982. Mass extinctions in the marine fossil record. *Science* 215: 1501–1503.

Raup, D. M., S. J. Gould, T. J. M. Schopf and D. Simberloff. 1973. Stochastic models of phylogeny and the evolution of diversity. *Journal of Geology* 81: 525–542.

Raven, P. H. and E. O. Wilson. 1992. A fifty-year plan for biodiversity surveys. *Science* 258: 1099–1100.

Rawlinson, P. A., R. A. Zann, S. van Balen and I. W. B. Thornton. 1992. Colonization of the Krakatau islands by vertebrates. *Geojournal* 28: 225–31.

Raxworthy, C. J., M. R. J. Forstner and R. A. Nussbaum. 2002. Chameleon radiation by oceanic dispersal. *Nature* 415: 784–786.

Ray, C. 1960. The application of Bergmann's and Allen's rule to the poikilotherms. *Journal of Morphology* 106: 85–109.

Ray, C., M. Gilpin and A. T. Smith. 1991. The effect of conspecific attraction on metapopulation dynamics. *Biological Journal of the Linnean Society* 42: 123–134.

Real, L. A. and J. H. Brown (eds.). 1991. *Foundations of Ecology: Classic Papers with Commentaries*. Chicago: University of Chicago Press.

Reed, D. F. and C. P. Dunn. 1996. Corridors for wildlife. *Science* 271: 132–133.

Reichman, O. J. 1985. Impact of pocket gopher burrows on overlying vegetation. *Journal of Mammalogy* 66: 720–725.

Reichman, O. J. and S. C. Smith. 1985. Impact of pocket gopher burrows on overlying vegetation. *Journal of Mammalogy* 66: 720–725.

Reichman, O. J., D. T. Wicklow and C. Rebar. 1985. Ecological and mycological characteristics of caches in the mounds of *Dipodomys spectabilis*. *Journal of Mammalogy* 66: 643–651.

Reid, W. V. and R. K. Miller. 1989. *Keeping Options Alive—The Scientific Basis for Conserving Biodiversity*. Washington, DC: World Resources Institute.

Reid, W. V. and M. C. Trexler. 1991. *Drowning the National Heritage: Climate Change and U.S. Coastal Biodiversity*. Washington, DC: World Resources Institute.

Reig, O. A. 1989. Karyotypic repatterning as one triggering factor in cases of explosive speciation. In A. Fondevila (ed.), *Evolutionary Biology of Transient Unstable Populations*, 246–289. Berlin: Springer-Verlag.

Reinhardt, K. 1997. Breeding success of southern hemisphere skuas Catharacta spp.: The influence of latitude. *Ardea* 85: 73–82.

Renner, S. S. 2004. Multiple Miocene Melastomataceae dispersal between Madagascar, Africa and India. *Philosophical Transactions of the Royal Society of London,* Series B 359: 1485–1494.

Rensch, B. 1960. *Evolution above the Species Level*. New York: Columbia University Press.

Repasky, R. R. 1991. Temperature and the northern distribution of wintering birds. *Ecology* 72: 2274–2285.

Rex, M. A. 1981. Community structure in the deep-sea benthos. *Annual Review of Ecology and Systematics* 12: 331–354.

Rex, M. A. 1983. Geographic patterns of species diversity in the deep-sea benthos. In G. T. Rowe (ed.), *Deep Sea Biology*, vol. 8, *The Sea*, 452–472. New York: John Wiley and Sons.

Rex, M. A., R. J. Etter and C. T. Stuart. 1997. Large-scale patterns of species diversity in deep-sea benthos. In R.F.G. Ormond, J. D. Gage, and M. V. Angel, (eds.), *Marine Biodiversity*. 94–121. Cambridge: Cambridge University Press.

Rex, M. A., C. T. Stuart and G. Goyne. 2000. Latitudinal gradients of species richness in the deep-sea benthos of the North Atlantic. *Proceedings of the National Academy of Sciences, USA* 97: 4082–4085.

Rex, M. A., C. T. Stuart, R. R. Hessler, J. A. Allen, H. L. Sanders and G. D. F. Wilson. 1993. Global-scale patterns of species diversity in the deep sea benthos. *Nature* 365: 636–639.

Rey, J. R. 1981. Ecological biogeography of arthropods on *Spartina* islands in northwest Florida. *Ecological Monographs* 51: 237–265.

Rhode, D. 2001. Packrat middens as a tool for reconstructing historic ecosystems. In D. Egan and E. Howell (eds.), *Historical Ecology Handbook: A Restorationist's Guide to Reference Ecosystems*, 257–293. Covelo, CA: Island Press.

Rice, D. W. and A. A. Wolman. 1971. *The Life History and Ecology of the Gray Whale (Eschrichtius robustus)*. American Society of Mammalogists, Special Publication.

Rich, T. H. 1985. *Megalania prisca* (Owen, 1859), the giant goanna. In P. V. Rich and F. F. van Tets (eds.), *Kadimakara: Extinct Vertebrates of Australia*, 152–155. Lildale, Victoria: Pioneer Design Studio.

Richards, P. W. 1957. *The Tropical Rainforest*. Cambridge: Cambridge University Press.

Richards, P. W. 1996 *The Tropical Rainforest: An Ecological Study*. 2nd edition. Cambridge: Cambridge University Press.

Richman, A. D., T. J. Case and T. D. Schwaner. 1988. Natural and unnatural extinction rates of reptiles on islands. *American Naturalist* 131: 611–630.

Richter-Dyn, N. and N. S. Goel. 1971. On the extinction of a colonizing species. *Theoretical Population Biology* 3: 406–433.

Rickart, E. A. 2001. Elevational diversity gradients, biogeography and the structure of montane mammal communities in the intermountain region of North America. *Global Ecology and Biogeography* 10: 77–100.

Ricketts, J. H. 2001. The matrix matters: Effective isolation in fragmented landscapes. *American Naturalist* 158: 87–99.

Ricklefs, R. E. 1987. Community diversity: Relative roles of local and regional processes. *Science* 235: 167–171.

Ricklefs, R. E. and E. Bermingham. 2002. The concept of the taxon cycle in biogeography. *Global Ecology and Biogeography* 11: 353–361.

Ricklefs, R. E. and G. W. Cox. 1972. Taxon cycles of the West Indian avifauna. *American Naturalist* 106: 295–219.

Ricklefs, R. E. and G. W. Cox. 1978. Stage of taxon cycle, habitat distribution and population density in the avifauna of the West Indies. *American Naturalist* 112: 875–895.

Ricklefs, R. E. and S. M. Fallon. 2002. Diversification and host switching in avian malaria parasites. *Proceedings of the Royal Society of London,* Series B 269: 885–892.

Ricklefs, R. E. and I. J. Lovette. 1999. The role of island area per se and habitat diversity in the species-area relationships of four Lesser Antilliean faunal groups. *Journal of Animal Ecology* 68: 1142–1160.

Ricklefs, R. E. and D. Schluter (eds.). 1993. *Species Diversity in Ecological Communities: Historical and Geographical Perspectives.* Chicago: University of Chicago Press.

Riddle, B. R. 1995. Molecular biogeography in the pocket mice (*Perognathus and Chaetodipus*) and grasshopper mice (*Onychomys*): The late Cenozoic development of a North America aridlands rodent guild. *Journal of Mammalogy* 76: 283–301.

Riddle, B. R. 1996. The molecular phylogenetic bridge between deep and shallow history in continental biotas. *Trends in Ecology and Evolution* 11: 207–211.

Riddle, B. R. 2005. Is biogeography emerging from its identity crisis? *Journal of Biogeography* 32: 185–186.

Riddle, B. R. and D. J. Hafner. 2004. The past and future roles of phylogeography in historical biogeography. In M. V. Lomolino and L. R. Heaney (eds.), *Frontiers of Biogeography,* 93–110. Sunderland, MA: Sinauer Associates.

Riddle, B. R. and D. J. Hafner. in press. A step-wise approach to integrating phylogeographic and phylogenetic biogeographic perspectives on the history of a core North American warm deserts biota. *Journal of Arid Environments.*

Riddle, B. R., D. J. Hafner, L. F. Alexander and J. R. Jaeger. 2000. Cryptic vicariance in the historical assembly of a Baja California Peninsular Desert biota. *Proceedings of the National Academy of Sciences, USA* 97: 14438–14443.

Ridley, H. N. 1930. *The Dispersal of Plants throughout the World.* Ashford, England: L. Reeve & Co.

Rieppel, O. 2002. A case of dispersing chameleons. *Nature* 415: 744–745.

Rising, J. D. and J. C. Avise. 1993. Application of genealogical-concordance principles to the taxonomy and evolutionary history of the sharp-tailed sparrow (Ammodramus-Caudacutus). *Auk* 110: 844–856.

Roane, M. K., G. J. Griffin and J. R. Elkins. 1986. *Chestnut Blight, other Endothia Diseases and the Genus Endothia.* St. Paul, MN: American Phytopathological Society.

Roberts, C. M., C. J. McClean, J. E. N. Veron, J. D. Hawkins, G. R. Allen, D. E. McAllister, C. G. Mittermeier, F. W. Schueler, M. Spalding, F. Wells, C. Vynne and T. B. Werner. 2002. Marine biodiversity hotspots and conservation priorities for tropical reefs. *Science* 295: 1280–1288.

Roberts, D. F. 1953. Body weight, race and climate. *American Journal of Physical Anthropology* 11: 533–558.

Roberts, D. F. 1978. *Climate and Human Variability.* 2nd edition. Menlo Park, CA: C. A Cummings.

Roberts, R. G., T. F. Flannery, L. K. Ayliffe, H. Yoshida, J. M. Olley, G. J. Prideaux, G. M. Laslett, A. Baynes, M. A. Smith, R. Jones and B. L. Smith. 2001. New ages for last Australian megafauna: Continent-wide extinction about 46,000 years ago. *Science* 292: 1888–1892.

Robinove, C. J. 1979. *Integrated Terrain Mapping with Digital Landsat Images in Queensland, Australia.* Professional Paper 1102. Washington, DC: U.S. Geological Survey. 39 pp.

Robinson, E. and J. F. Lewis. 1971. *Field Guide to Aspects of the Geology of Jamaica. An International Field Institute Guidebook to the Caribbean Island-Arc System,* 2–29. American Geological Institute (Jamaica section).

Robinson, G. R., R. D. Holt, M. S. Gaines, S. P. Hamburg, M. L. Johnson, H. S. Fitch, E. A. Martinko. 1992. Diverse and contrasting effects of habitat fragmentation. *Science* 257: 524–526.

Robinson, W. D. 1999. Long-term changes in the avifauna of Barro Colorado Island, Panama, a tropical forest isolate. *Conservation Biology* 13: 85–97.

Roca, A. L., N. Georgiadis and S. J. O'Brien. 2005. Cytonuclear genomic dissociation in African elephant species. *Nature Genetics* 37: 96–100.

Rodrigues, A. S. L., H. R. Akcakaya, S. J. Andelman, M. I. Bakarr, L. Boitani, T. M. Brooks, J. S. Chanson, L. D. C. Fishpool, G. A. B. Da Fonesca, K. J. Baston, M. Hoffman, P. A. Marquet, J. D. Pilgrim, R. L. Pressey, J. Schipper, W. Sechrest, S. N. Stuart, L. G. Underhill, R. W. Waller, M. E. J. Watts and X Yan. 2004. Global gap analysis: priority regions for expanding the global Protected-Area Network. *BioScience* 54: 1092–1100.

Rodriguez, J. P. 2002. Range contraction in declining North American bird populations. *Ecological Applications* 12: 238–248.

Roff, D. A. 1986. The evolution of wing dimorphism in insects. *Evolution* 40: 1009–1020.

Roff, D. A. 1990. The evolution of flightlessness in insects. *Ecological Monographs* 60: 389–421.

Roff, D. A. 1994. The evolution of flightlessness, is history important? *Evolutionary Ecology* 8: 639–657.

Rogers, A. R. and H. Harpending. 1992. Population-growth makes waves in the distribution of pairwise genetic-differences. *Molecular Biology and Evolution* 9: 552–569.

Rogers, D. J. and S. E. Randolph. 2000. The global spread of malaria in a future, warmer world. *Science* 289: 1763–1766.

Rogers, R. A., L. D. Martin and T. D. Nicklas. 1990. Ice-age geography and the distribution of native North American languages. *Journal of Biogeography* 17: 131–143.

Rogers, R. A., L. A. Rogers, R. S. Hoffmann and L. D. Martin. 1991. Native American biological diversity and the biogeographic influence of Ice Age refugia. *Journal of Biogeography* 18: 623–630.

Rohde, K. 1992. Latitudinal gradients in species diversity: The search for the primary cause. *Oikos* 65: 514–527.

Rohde, K. 1996. Rapoport's rule is a local phenomenon and cannot explain latitudinal gradients in species diversity. *Biodiversity Letters* 3: 10–13.

Rohde, K. 1997. The larger area of the tropics does not explain latitudinal gradients in species diversity. *Oikos* 79: 169–172.

Rohde, K. 1999. Latitudinal gradients in species diversity and Rapoport's rule revisited: A review of recent work and what can parasites teach us about the causes of the gradients? *Ecography* 22: 593–613.

Rohde, K. and M. Heap. 1996. Latitudinal ranges of teleost fish in the Atlantic and Indo-Pacific oceans. *American Naturalist* 147: 659–665.

Rohde, K., M. Heap and D. Heap. 1993. Rapoport's rule does not apply to marine teleosts and cannot explain latitudinal gradients in species richness. *American Naturalist* 142: 1–16.

Romer, A. S. 1966. *Vertebrate Paleontology.* Chicago: Chicago University Press.

Ronquist, F. 1997. Dispersal-vicariance analysis: A new approach to the quantification of historical biogeography. *Systematic Biology* 46: 195–203.

Ronquist, F. 2002. Parsimony analysis of coevolving species associations. In R. D. M. Page (ed.), *Cospeciation,* 22–64. Chicago: University of Chicago Press.

Ronquist, F. and S. Nylin. 1990. Process and pattern in the evolution of species association. *Systematic Zoology* 39: 323–344.

Roosevelt, A. C., M. Lima deCosta, C. Lopes Machado, M. Michab, N. Mercier, H. Vallada, J. Feathers, W. Barnett, M. Imazio da Silveira and K. Schick. 1996. Paleoindian cave dwellers in the Amazon: The peopling of the Americas. *Science* 272: 373–384.

Root, R. B. 1967. The niche exploitation pattern of the blue-grey gnat-catcher. *Ecological Monographs* 37: 317–350.

Root, T. 1988a. *Atlas of Wintering North American Birds: An Analysis of Christmas Bird Count Data.* Chicago: University of Chicago Press.

Root, T. 1988b. Energy constraints on avian distributions and abundances. *Ecology* 69: 330–339.

Root, T. 1988c. Environmental factors associated with avian distributional boundaries. *Journal of Biogeography* 15: 489–505.

Root, T. L. 1993. Effects of global climate change on North American bird communities. In P. M. Karieva, J. M. Kingsolver and R. B. Huey (eds.), *Biotic Interactions and Global Change,* 280–292. Sunderland, MA: Sinauer Associates.

Root, T. L. 1994. Scientific/philosophical challenges of global change research: A case study of climatic changes in birds. *Proceedings of the American Philosophical Society* 138: 377–384.

Root, T. L. 2000. Ecology: Possible consequences of rapid global change. In G. Ernst (ed.), *Earth Systems,* 315–324. Cambridge: Cambridge University Press.

Root, T. L. and S. H. Schneider. 1993. Can large-scale climatic models be linked with multiscale ecological studies? *Conservation Biology* 7: 256–270.

Root, T. L. and S. H. Schneider. 1995. Ecology and climate: Research strategies and implications. *Science* 269: 334–431.

Root, T. L., J. T. Price, K. R. Hall, S. H. Schnieder, C. Rosenzweig and J. A. Pounds. 2003. Fingerprints of global warming on wild animals and plants. *Nature* 421: 57–60.

Rosen, B. R. and A. B. Smith. 1988. Tectonics from fossils? Analysis of reef coral and sea urchin distributions from late Cretaceous to Recent, using a new method. In M. G. Audley-Charles and A. Hallam (eds.), *Gondwana and Tethys,* 275–306. Special publication. Oxford: Geological Society of London.

Rosen, D. E. 1975. A vicariance model of Caribbean biogeography. *Systematic Zoology* 24: 431–464.

Rosen, D. E. 1978. Vicariant patterns and historical explanations in biogeography. *Systematic Zoology* 27: 159–188.

Rosen, D. E. 1979. Fishes from the uplands and intermontane basins of Guatemala: Revisionary studies and comparative geography. *Bulletin of the American Museum of Natural History* 162: 267–376.

Rosen, D. E., G. Nelson and C. Patterson. 1979. Foreword. In W. Hennig (ed.), *Phylogenetic Systematics,* 3rd edition, vii-xiii. Urbana, IL: University of Illinois Press.

Rosenberg, D. K., B. R. Noon and E. C. Meslow. 1997. Biological corridors: Form, function and efficacy. *BioScience* 47: 677–687.

Rosenfield, J. A. 2002. Pattern and process in the geographical ranges of freshwater fishes. *Global Ecology and Biogeography* 11: 323–332

Rosenzweig, M. L. 1966. Community structure in sympatric Carnivora. *Journal of Mammalogy* 47: 602–612.

Rosenzweig, M. L. 1968. Net primary productivity of terrestrial communities: Predictions from climatological data. *American Naturalist* 102: 67–74.

Rosenzweig, M. L. 1975. On continental steady states of species diversity. In M. L. Cody and J. M. Diamond (eds.), *Ecology and Evolution in Communities,* 121–141. Cambridge, MA: Belknap Press.

Rosenzweig, M. L. 1978. Competitive speciation. *Biological Journal of the Linnaean Society* 10: 275–289.

Rosenzweig, M. L. 1992. Species diversity gradients: We know more and less than we thought. *Journal of Mammalogy* 73: 715–730.

Rosenzweig, M. L. 1995. *Species Diversity in Space and Time.* New York: Cambridge University Press.

Rosenzweig, M. L. 1997. Species diversity and latitudes: Listening to area's signal. *Oikos* 80: 172–176.

Rosenzweig, M. L. 2004. Applying species-area relationships to the conservation of species diversity. In M. V. Lomolino and L. R. Heaney (eds.), *Frontiers of Biogeography: Advances in the Geography of Nature,* 325–344. Sunderland, MA: Sinauer Associates.

Rosenzweig, M. L. and M. V. Lomolino. 1997. Rarity and community ecology: Who gets the tiny bits of the broken stick. In W. E. Kunin and K. J. Gaston (eds.), *The Biology of Rarity: Causes and Consequences of Rare-Common Differences,* 63–90. London: Chapman and Hall.

Roth, V. L. 1990. Insular dwarf elephants: A case study in body mass estimation and ecological inference. In J. Damuth and B. J. MacFadden (eds.), *Body Size in Mammalian Paleobiology: Estimation and Biological Implications,* 151–179. New York: Cambridge University Press.

Roth, V. L. and M. S. Klein. 1986. Maternal effects of body size of large insular *Peromyscus maniculatus:* Evidence from embryo transfer experiments. *Journal of Mammalogy* 67: 37–45.

Roughgarden, J. 1974. Niche width: Biogeographic patterns among *Anolis* lizard populations. *American Naturalist* 108: 429–442.

Roughgarden, J. 1992. Comments on the paper by Losos: Character displacement vs. taxon loop. *Copeia* 1992: 288–295.

Roughgarden, J. 1995. Anolis *Lizards of the Caribbean: Ecology, Evolution and Plate Tectonics.* Oxford: Oxford University Press.

Roughgarden, J. and M. Feldman. 1975. Species packing and predation pressure. *Ecology* 56: 489–492.

Roughgarden, J. and E. R. Fuentes. 1977. The environmental determinants of size in solitary populations of West Indian *Anolis* lizards. *Oikos* 29: 44–51.

Roughgarden, J. and S. Pacala. 1989. Taxon cycle among *Anolis* lizard populations: Review of the evidence. In D. Otte and J. Endler (eds.), *Speciation and Its Consequences,* 403–442. Sunderland, MA: Sinauer Associates.

Roughgarden, J., S. D. Gaines and S. Pacala. 1987. Supply side ecology: The role of physical transport processes. In P. Giller and J. Gee (eds.), *Organization of Communities: Past and Present,* 491–518. Oxford: Blackwell Scientific Publications.

Roughgarden, J., D. Heckel and E. R. Fuentes. 1983. Coevolutionary theory and the biogeography and community structure of *Anolis.* In R. B. Huey, E. R. Pianka and T. W. Schoener (eds.), *Lizard Ecology: Studies of a Model Organism,* 371–410. Cambridge, MA: Harvard University Press.

Roughgarden, J., R. M. May and S. A. Levin. 1989. *Perspectives in Ecological Theory.* Princeton, NJ: Princeton University Press.

Rounds, R. C. 1987. Distribution and analysis of colourmorphs of the black bear (*Ursus americanus*). *Journal of Biogeography* 14: 521–538.

Rowe, J. S. 1980. The common denominator in land classification in Canada: An ecological approach to mapping. *Forestry Chronicle* 56: 19–20.

Roy, K. 1994. Effects of the Mesozoic Marine Revolution on the taxonomic, morphologic and biogeographic evolution of a group: Aporrhaid gastropods during the Mesozoic. *Paleobiology* 20: 274–296.

Roy, K., D. P. Balch and M. E. Hellberg. 2001. Spatial patterns of morphological diversity across the Indo-Pacific: Analysis using strombid gastropods. *Proceedings of the Royal Society of London,* Series B 268: 1–6.

Roy, K., D. Jablonski and K. K. Martien. 2000. Invariant size-frequency distributions along a latitudinal gradient in marine bivalves. *Proceedings of the National Academy of Sciences, USA* 97: 13150–12155.

Roy, K., D. Jablonski and J. W. Valentine. 1994. Eastern Pacific molluscan provinces and latitudinal diversity gradient: No evidence for Rapoport's rule. *Proceedings of the National Academy of Sciences, USA* 91: 8871–8874.

Roy, K., D. Jablonski and J. W. Valentine. 2001. Dissecting latitudinal diversity gradients: Functional groups and clades of marine bivalves. *Proceedings of the Royal Society of London,* Series B 267: 293–299.

Roy, K., D. Jablonski and J. W. Valentine. 2004. Beyond species richness: Biogeographic patterns and biodiversity dynamics using other metrics of diversity. In M. V. Lomolino and L. R. Heaney, *Frontiers of Biogeography: New Directions in the Geography of Nature,* 151–170. Sunderland, MA: Sinauer Associates.

Roy, K., D. Jablonski, J. W. Valentine and G. Rosenberg. 1998. Marine latitudinal gradients: Tests of causal hypotheses. *Proceedings of the National Academy of Sciences, USA* 95: 3699–3702.

Rubinoff, I. 1968. Central American sea-level canal: Possible biological effects. *Science* 161: 857–861.

Rubinoff, R. W. and I. Rubinoff. 1971. Geographic and reproductive isolation in Atlantic and Pacific populations of Panamanian *Bathygobius. Evolution* 25: 88–97.

Ruff, C. B. 1994. Morphological adaptation to climate in modern and fossil hominids. *Yearbook of Physical Anthropology* 37: 65–107.

Ruggiero, A. 2001. Size and shape of the geographical ranges of Andean passerine birds: Spatial patterns in environmental resistance and anisotropy. *Journal of Biogeography* 28: 1281–1294.

Runcorn, S. K. 1956. Paleomagnetic comparisons between Europe and North America. *Proceedings of the Geological Association of Canada* 8: 77–85.

Runcorn, S. K. 1962. Paleomagnetic evidence for continental drift and its geophysical cause. In S. K. Runcorn (ed.), *Continental Drift,* 1–40. International Geophysics Series 3. New York: Academic Press.

Rutherford, S. and S. D'Hondt. 2000. Early onset and tropical forcing of 100,000-year Pleistocene glacial cycles. *Nature* 408: 72–75.

Safriel, U. N., S. Volis and S. Kark. 1994. Core and peripheral populations and global climate change. *Israel Journal of Plants Sciences* 42: 331–345.

Sagarin, R. D. and S. D. Gaines. 2002. The 'abundant centre' distribution: To what extent is it a biogeographical rule? *Ecology Letters* 5: 137–147.

Sagarin, R. D., J. P. Barry, S. E. Gilman and C. H. Baxter. 1999. Climate-related change in an intertidal community over short and long time scales. *Ecological Monographs* 69: 465–490.

Sakai, A., W. L. Wagner, D. M. Ferguson and D. R. Herbst. 1995. Origins of dioecy in the Hawaiian flora. *Ecology* 76: 2517–2529.

Sanchez-Cordero, V. 2001. Elevation gradients of diversity for rodents and bats on Oaxaca, Mexico. *Global Ecology and Biogeography* 10: 63–76.

Sanchez-Cordero, V., M. Munguía and A. T. Peterson. 2004. GIS-based predictive biogeography in the context of conservation. In M. V. Lomolino and L. R. Heaney, *Frontiers of Biogeography: Advances in the Geography of Nature,* (eds.) 311–324. Sunderland, MA: Sinauer Associates.

Sand, H., G. Cederlund and Kjell Danell. 1995. Geographical and latitudinal variation in growth patterns and adult body size of Swedish moose (*Alces alces*). *Oecologia* 102: 433–442.

Sanders, H. L. 1968. Marine benthic diversity: A comparative study. *American Naturalist* 102: 243–282.

Sanders, H. L. 1969. Benthic marine diversity and the stability-time hypothesis. *Brookhaven Symp. Biol.* 22: 71–81.

Sanders, J. J. 2002. Elevational gradients in ant species richness: Area, geometry and Rapoport's rule. *Ecography* 25: 25–32.

Sanders, N. J., J. Moss and D. Wagner. 2003. Patterns of ant species richness along elevational gradients in an arid ecosystem. *Global Ecology and Biogeography* 12: 93–102.

Sanderson, E. W., M. Jaiteh, M. A. Levy, K. H. Redford, A. V. Wannebo and G. Woolmer. 2002. The human footprint and the last of the wild. *Bioscience* 52: 891–904.

Sanderson, M. J. 2002. Estimating absolute rates of molecular evolution and divergence times: A penalized likelihood approach. *Molecular Biology and Evolution* 19: 101–109.

Sanford, E. 2002. Community responses to climate change: Links between temperature and keystone predation in a rocky intertidal community. In S. H. Schnedier and T. L. Root (eds.), *Wildlife Responses to Climate Change: North American Case Studies,* 165–200. Washington, DC: Island Press.

Sanmartin, I. 2003. Dispersal vs. vicariance in the Mediterranean: Historical biogeography of the Palearctic Pachydeminae (Coleoptera, Scarabaeoidea). *Journal of Biogeography* 30: 1883–1897.

Sanmartin, I. and F. Ronquist. 2002. New solutions to old problems: Widespread taxa, redundant distributions and missing areas in event-based biogeography. *Animal Biodiversity and Conservation* 25: 75–93.

Sanmartín, I. and F. Ronquist. 2004. Southern Hemisphere biogeography inferred by event-based models: Plant vs. animal patterns. *Systematic Biology* 53: 216–243.

Sanmartin, I., H. Enghoff and F. Ronquist. 2001. Patterns of animal dispersal, vicariance and diversification in the Holarctic. *Biological Journal of the Linnean Society* 73: 345–390.

Sant'ana, C. E. R. and J. A. Diniz-Filho. 1999. Macroecologigia de corujas (Aves, Strigiformes) de America do Sul. *Ararajuba* 7: 3–11.

Santini, F. and R. Winterbottom. 2002. Historical biogeography of Indo-western Pacific coral reef biota: Is the Indonesian region a centre of origin? *Journal of Biogeography* 29: 189–205.

Sarich, V. M. and A. C. Wilson. 1967. Immunological time scale for hominid evolution. *Science* 158: 1200–1203.

Sauer, J. 1969. Oceanic islands and biogeographic theory: A review. *Geographical Review* 59: 582–593.

Sauer, J. D. 1988. *Plant Migration: The Dynamics of Geographic Patterning in Seed Plants.* Berkeley: University of California Press.

Saunders, D. A., R. J. Hobbs and C. R. Margules. 1991. Biological consequences of ecosystem fragmentation: A review. *Conservation Biology* 5: 18–32.

Savage, J. M. 1973. The geographic distribution of frogs: Patterns and predictions. In J. L. Vial (ed.), *Evolutionary Biology of the Anurans,* 351–445. Columbia: University of Missouri Press.

Savidge, J. A. 1987. Extinction of an island forest avifauna by an introduced snake. *Ecology* 68: 660–668.

Savilov, A. I. 1961. The distribution of the ecological forms of the by-the-wind sailor, *Velella lata,* Ch. and Eys. and the Portugese man-of-war *Physalia utriculus* (La Martiniere) Esch., in the North Pacific. *Transactions Institute Okenologyi, Akademis Nauk.* SSSR 45: 223–239.

Sax, D. F., J. J. Stachowicz and S. D. Gaines. 2005. *Species Invasions: Insights into Ecology, Evolution and Biogeography.* Sunderland, MA: Sinauer Associates.

Scanlon, J. D. and M. S. Y. Lee. 2004. Phylogeny of Australasian venomous snakes (*Colubroidea, Elapidae, Hydrophiinae*) based on phenotypic and molecular evidence. *Zoologica Scripta* 33: 335–366.

Schall, J. J. and E. R. Pianka. 1978. Geographical trends in numbers of species. *Science* 201: 679–686.

Scheiner, S. M. and J. M. Rey-Benayas. 1994. Global patterns of plant diversity. *Evolutionary Ecology* 8: 331–347.

Schemske, D. W., B. C. Husband, M. H. Ruckelshaus, C. Goodwillie, I. M. Parker and J. G. Bishop. 1994. Evaluating approaches to the conservation of rare and endangered plants. *Ecology* 75: 584–606.

Schenk, H. J. and R. B. Jackson. 2002. The global biogeography of roots. *Ecological Monographs* 72: 311–328.

Schimper, A. F. W. 1898. *Pflanzengeographie auf physiologischer Grundlage.* 1st edition. Jena.

Schimper, A. F. W. 1903. *Plant-Geography upon a Physiological Basis.* Trans. W. R. Fisher. Oxford: Clarendon Press.

Schipper, C., M. Shanahan, S. Cook and I. W. B. Thornton. 2001. Colonization of an island volcano, Long Island, Papau New Guinea and an emergent island, Motmot, in its calder lake. III. Colonization by birds. *Global Ecology and Biogeography* 28: 1339–1352.

Schlinger, E. I. 1974. Continental drift, *Nothofagus* and some ecologically associated insects. *Annual Review of Entomology* 19: 323–343.

Schluter, D. 1986. Tests of similarity and convergence of finch communities. *Ecology* 67: 1073–1086.

Schluter, D. 1988. The evolution of finch communities on islands and continents: Kenya vs. Galápagos. *Ecological Monographs* 58: 229–249.

Schluter, D. 1993. Adaptive radiation in sticklebacks: Size, shape and habitat use efficiency. *Ecology* 74: 699–709.

Schluter, D. 1996. Ecological speciation in postglacial fishes. *Philosophical Transactions of the Royal Society of London* 351: 807–814.

Schluter, D. 2000. *The Ecology of Adaptive Radiation.* Oxford: Oxford University Press.

Schluter, D. and P. R. Grant. 1984. The distribution of *Geospiza difficilis* in relation to *G. fuliginosa* in the Galápagos Islands: Tests of three hypotheses. *Evolution* 36: 1213–1226.

Schluter, D., T. D. Price and P. R. Grant. 1985. Ecological character displacement in Darwin's finches. *Science* 227: 1056–1059.

Schmida, A. and M. V. Wilson. 1985. Biological determinants of species diversity. *Journal of Biogeography* 12: 1–20.

Schmidt, N. M. and P. M. Jensen. 2003. Changes in mammalian body length over 175 years—Adaptations to a fragmented landscape? *Conservation Ecology* 7: 6.

Schmidt-Nielsen, K. 1963. Osmotic regulation in higher vertebrates. *Harvey Lectures,* ser. 58: 53–93.

Schmidt-Nielsen, K. 1964. *Desert Animals: Physiological Problems of Heat and Water.* New York: Oxford University Press.

Schmidt-Nielsen, K. 1984. *Scaling: Why Is Animal Size So Important?* New York: Cambridge University Press.

Schmitt, R. J. 1987. Indirect interactions between prey: Apparent competition, predator aggregation and habitat segregation. *Ecology* 68: 1887–1897.

Schneider, S. H. 1993. Scenarios of global warming. In P. M. Karieva, J. G. Kingsolver and R. B. Huey (eds.), *Biotic Interactions and Global Change,* 9–23. Sunderland, MA: Sinauer Associates.

Schneider, S. H. and T. L. Root. 2002. *Wildlife Responses to Climate Change: North American Case Studies*. Washington, DC: Island Press.

Schoener, T. W. 1968a. The *Anolis* lizards of Bimini: Resource partitioning in a complex fauna. *Ecology* 49: 704–726.

Schoener, T. W. 1968b. Sizes of feeding territories among birds. *Ecology* 49: 123–141.

Schoener, T. W. 1970. Size patterns in West Indian *Anolis* lizards. II. Correlations with the sizes of particular sympatric species: Displacement and divergence. *American Naturalist* 104: 155–174.

Schoener, T. W. 1974. The species-area relationship within archipelagos: Models and evidence from island land birds. In H. J. Firth and J. H. Calaby (eds.), *Proceedings of the 16th International Ornithological Congress*, 629–642. Canberra: Australian Academy of Science.

Schoener, T. W. 1975. Presence and absence of habitat shift in some widespread lizard species. *Ecological Monographs* 45: 233–258.

Schoener, T. W. 1976. The species-area relationship within archipelagoes: Models and evidence from island birds. *Proceedings of the XVI International Ornithological Congress* 6: 629–642.

Schoener, T. W. 1983. Rate of species turnover decreases from lower to higher organisms: A review of the data. *Oikos* 41: 372–377.

Schoener, T. W. 1986. Patterns in terrestrial vertebrate vs. arthropod communities: Do systematic differences in regulation exist? In J. Diamond and T. J. Case (eds.), 556–586. *Community Ecology*. New York: Harper and Row.

Schoener, T. W. 1988. The ecological niche. In J. M. Cherrett (ed.), *Ecological Concepts*, 79–113. Oxford: Blackwell Scientific Publications.

Schoener, T. W. 1989. Food webs from the small to the large. *Ecology* 70: 1559–1589.

Schoener, T. W. and G. H. Adler. 1991. Greater resolution of distributional complementarities by controlling for habitat affinities: A study of Bahamian lizards. *American Naturalist* 137: 669–692.

Schoener, T. W. and A. Schoener. 1983. Distribution of vertebrates on some very small islands. I. Occurrence sequences of individual species. *Journal of Animal Ecology* 52: 209–235.

Schoener, T. W. and D. A. Spiller. 1987. High population persistence in a system with high turnover. *Nature* 330: 474–477.

Scholander, P. F. 1955. Evolution of climatic adaptation in homeotherms. *Evolution* 9: 15–26.

Schopf, J. M. 1970a. Gondwana paleobotany. *Antarctic Journal of the U.S.* 5: 62–66.

Schopf, J. M. 1970b. Relation of floras of the Southern Hemisphere to continental drift. *Taxon* 19: 657–674.

Schopf, J. M. 1974. Permo-Triassic extinctions: Relation to sea-floor spreading. *Journal of Geology* 82: 129–144.

Schopf, J. M. 1975. Precambrian paleobiology: Problems and perspectives. *Annual Review of Earth and Planetary Sciences* 3: 213–249.

Schopf, J. M. 1976. Morphologic interpretation of fertile structures in glossopterid gymnosperms. *Review of Palaeobotany and Palynology* 21: 25–64.

Schouw, F. 1823. Grunzüge einer algemeinen Pflanzengeographie. Berlin.

Schuchert, C. 1932. Gondwana landbridges. *Bulletin of the Geological Society of America* 43: 875–916.

Schuster, M. F. and S. McDaniel. 1973. A vegetative analysis of a Black Belt prairie relict site near Aliceville, Alabama. *Journal of the Mississippi Academy of Science* 19: 153–159.

Schwaner, T. D. and S. D. Sarre. 1990. Body size and sexual dimorphism in mainland and island tiger snakes. *Journal of Herpetology* 24: 320–322.

Schwartz, M. W. 1988. Species diversity patterns in woody flora of three North American peninsulas. *Journal of Biogeography* 15: 759–774.

Schwarzbach, M. 1980. *Alfred Wegener: The Father of the Continental Drift*. Madison, WI: Science Technology.

Sclater, J. G. and C. Tapscott. 1979. The history of the Atlantic. *Scientific American* 240: 156–174.

Sclater, P. L. 1858. On the general geographical distribution of the members of the class Aves. *Journal of the Linnean Society, Zoology* 2: 130–145.

Sclater, P. L. 1897. On the distribution of marine mammals. *Proceedings of the Zoological Society of London* 41: 347–359.

Scotese, C. R. 2004. Cenozoic and Mesozoic paleogeography: Changing terrestrial biogeographic pathways. In M. V. Lomolino and L. R. Heaney (eds.), *Frontiers of Biogeography: New Directions in the Geography of Nature*, Chapter 1. Cambridge: Cambridge University Press.

Scotese, C. R., L. M. Gahagan and R. L. Larson. 1988. Plate tectonic reconstructions of the Cretaceous and Cenozoic ocean basins. *Tectonophysics* 155: 27–48.

Scott, J. M., T. H. Tear and F. W. Frank (eds.). 1996. *GAP Analysis: A Landscape Approach to Biodiversity Planning*. American Society of Photogrammetry and Remote Sensing. Maryland USA.

Scott, J. M., S. Mountainspring, F. L. Ramsey and C. B. Kepler. 1986. Forest bird communities of the Hawaiian Islands: Their dynamics, ecology and conservation. *Studies in Avian Biology* 9: 1–431.

Scott, J. M., P. J. Heglund, M. L. Morrison, J. B. Haufler, M. G. Raphael, W. A. Wall and F. B. Samson. (eds.) 2002. *Predicting Species Occurrences: Issues of Accuracy and Scale*. Washington, DC: Island Press.

Scott, M. J., F. Davis, B. Csuiti, B. Butterfield, C. Groves, H. Anderson, S. Caicco, F. D'Erchia, T. C. Edwards Jr., J. Ulliman and R. G. Wright. 1993. GAP analysis: A geographical approach to protection of biological diversity. *Wildlife Monographs* 123: 1–41.

Scriber, J. M. 1973. Latitudinal gradients in larval feeding specialization of the world Papilionidae (Lepidoptera). *Psyche* 80: 355–373.

Seastedt, T. R., O. J. Reichman and T. C. Todd. 1986. Microarthropods and nematodes in kangaroo rat burrows. *Southwestern Naturalist* 31: 114–116.

Sechrest, W. Brooks, T., Boitani, L. and J. L. Gittleman. In press. Global biogeography of mammals. *PLoS Biology*.

Seehausen, O. 2002. Patterns in fish radiation are compatible with Pleistocene desiccation of Lake Victoria and 14,600 year history for its cichlid species flock. *Proceedings of the Royal Society of London: Biological Sciences* 269: 491–497.

Selander, R. K., M. H. Smith, S. Y. Yang, W. E. Johnson and J. B. Gentry. 1971. Biochemical polymorphism and systematics in the genus *Peromyscus*. II. Genic heterozygosity and genetic similarity among populations of the old-field mouse (*Peromyscus polionotus*). *Studies in Genetics VI, University of Texas Publication*, 7103: 49–90.

Sepkoski, J. J. Jr. 1976. Species diversity in the Phanerozoic: Species-area effects. *Paleobiology* 2: 298–303.

Sepkoski, J. J. Jr. 1984. A kinetic model of Phanerozoic taxonomic diversity. III. Post-Paleozoic families and mass extinctions. *Paleobiology* 10: 246–267.

Sepkoski, J. J. Jr., R. K. Bambach, D. M. Raup and J. W. Valentine. 1981. Phanerozoic marine diversity and the fossil record. *Nature* 293: 435–437.

Settele, J., C. R. Margules, P. Poschlod and K. Henle, editors. 1996. *Species Survival in Fragmented Landscapes*. Dordrecht, The Netherlands: Kluwer Academic.

Shane, C. K. and E. J. Cushing. 1991. *Quaternary Landscapes*. Minneapolis: University of Minnesota Press.

Shannon, C. E. and W. Weaver. 1949 (1964). *The Mathematical Theory of Communication*. Urbana, IL: University of Illinois Press.

Sharpton, V. L., G. B. Dalrymple, L. E. Marin, G. Ryder and B. C. Schuraytz. 1992. New links between the Chicxulub impact structure and the Cretaceous/Tertiary boundary. *Nature* 359: 819.

Sheehan, P. M. 2001. The late Ordovician mass extinction. *Annual Review of Earth and Planetary Sciences* 29: 331–363.

Shefer, S. A. Abelson, O. Mokady and E. Geffen. 2004. Red to Mediterranean Sea bioinvasion: Natural drift through the Suez Canal, or anthropogenic transport? *Molecular Ecology* 13: 2333–2343.

Shepherd, U. L. 1998. A comparison of species diversity and morphological diversity across the North American latitudinal gradient. *Journal of Biogeography* 25: 19–29.

Shields, O., 1990. Plate tectonics on an expanding Earth. In *Critical Aspects of the Plate Tectonics Theory; Volume II, Alternative Theories*. Athens, Greece: Theophrastus.

Shipley, B. and P. A. Keddy. 1987. The individualistic and community-unit concepts as falsifiable hypothesis. *Vegetatio* 69: 47–55.

Short, J. and A. Smith. 1994. Mammal decline and recovery in Australia. *Journal of Mammalogy* 75: 288–297.

Shreve, F. 1921. Conditions indirectly affecting vertical distribution on desert mountains. *Ecology* 3: 269–274.

Shreve, F. 1942. The desert vegetation of North America. *The Botanical Review* 8: 195–246.

Shrode, J. B. 1975. Developmental temperature tolerance of a Death Valley pupfish (*Cyprirodon nevadensis*). *Physiological Zoology* 48: 378–389.

Shuntov, V. P. 1974. *Sea Birds and the Biological Structure of the Ocean.* Washington, DC: National Information Service, U. S. Department of Commerce.

Sibley, C. G. 1970. A comparative study of the egg-white proteins of passerine birds. *Bulletin of the Peabody Museum of Natural History, Yale University* 32: 1–131.

Sibley, C. G. and J. E. Ahlquist. 1985. The Phylogeny and Classification of the Australo-Papuan Passerine Birds. *Emu* 85: 1–14.

Sibley, C. G. and J. E. Ahlquist. 1990. *Phylogeny and Classification of Birds.* New Haven, CT: Yale University Press.

Sibley, C. G. and B. L. Monroe. 1990. *Distribution and Taxonomy of Birds of the World.* New Haven, CT: Yale University Press.

Sideleva, V. G. 2002. Ichthyofauna of the underwater hydrothermal vent (Frolikha Bay, Lake Baikal) with a description of a new species of neocottus (Abyssocottidae). *Journal of Ichthyology* 42: 219–222.

Silkas, B., S. L. Olson and R. C. Fleischer. 2001. Rapid, independent evaluation of flightlessness in four species of Pacific Island rails (Rallidae): An analysis based on mitochondrial sequence data. *Journal of Avian Biology* 32: 5–14.

Silva, M. and J. A. Downing. 1995. The allometric scaling of density and body size: A non-linear relationship for terrestrial mammals. *American Naturalist* 141: 704–727.

Silva, M. N. F. and J. L. Patton. 1993. Amazonian phylogeography: MtDNA sequence variation in arboreal echimyid rodents (Caviomorpha). *Molecular Phylogenetics and Evolution* 2: 243–255.

Silvertown, J. 1985. History of a latitudinal diversity gradient: Woody plants in Europe. *Journal of Biogeography* 12: 519–525.

Simberloff, D. S. 1970. Taxonomic diversity of island biotas. *Evolution* 24: 23–47.

Simberloff, D. S. 1974a. Equilibrium theory of island biogeography and ecology. *Annual Review of Ecology and Systematics* 5: 161–182.

Simberloff, D. S. 1974b. Permo-Triassic extinctions: Effects of area on biotic equilibrium. *Journal of Geology* 82: 267–274.

Simberloff, D. S. 1976. Trophic structure determination and equilibrium in an arthropod community. *Ecology* 57: 395–398.

Simberloff, D. S. 1978. Using island biogeographic distributions to determine if colonization is stochastic. *American Naturalist* 112: 713–726.

Simberloff, D. S. 1988. Contribution of population and community biology to conservation science. *Annual Review of Ecology and Systematics* 19: 473–511.

Simberloff, D. S. and L. G. Abele. 1982. Refuge design and island biogeographic theory: Effects of fragmentation. *American Naturalist* 120: 41–50.

Simberloff, D. S. and W. Boecklen. 1991. Patterns of extinction in the introduced Hawaiian avifauna: A re-examination of the role of competition. *American Naturalist* 138: 300–327.

Simberloff, D. S. and J. Martin. 1991. Nestedness of insular avifaunas: Simple summary statistics masking complex species patterns. *Ornis Fennica* 68: 178–192.

Simberloff, D. S. and E. O. Wilson. 1969. Experimental zoogeography of islands: The colonization of empty islands. *Ecology* 50: 278–296.

Simberloff, D. S. and E. O. Wilson. 1970. Experimental zoogeography of islands: A two-year record of colonization. *Ecology* 51: 934–937.

Simberloff, D. S., J. Cox and D. W. Mehlman. 1992. Movement corridors: Conservation bargains or poor investments? *Conservation Biology* 6: 493–504.

Simberloff, D. S., K. L. Heck, E. D. McCoy and E. F. Conner. 1981a. There have been no statistical tests of cladistic biogeographical hypothesis. In G. Nelson and D. E. Rosen (eds.), *Vicariance Biogeography: A Critique,* 40–64. New York: Columbia University Press.

Simberloff, D. S., K. L. Heck, E. D. McCoy and E. F. Conner. 1981b. Response. In G. Nelson and D. E. Rosen (eds.), *Vicariance Biogeography: A Critique,* 85–93. New York: Columbia University Press.

Simmons, N. B. and J. H. Geisler. 1998. Phylogenetic relationships of *Icaronycteris, Archaeonycteris, Hassianycteris* and *Palaeochiropteryx* to extant bat lineages, with comments on the evolution of echolocation and foraging strategies in Microchiroptera. *Bulletin of the American Musuem of Natural History* 4: 1–182.

Simpson, B. B. 1974. Glacial migrations of plants: Island biogeographical evidence. *Science* 185: 697–700.

Simpson, B. B. 1975. Pleistocene changes in the flora of the high tropical Andes. *Paleobiology* 1: 273–294.

Simpson, G. G. 1936. Data on the relationships of local and continental mammal faunas. *Journal of Paleontology* 10: 410–414.

Simpson, G. G. 1940. Mammals and land bridges. *Journal of the Washington Academy of Science* 30: 137–163.

Simpson, G. G. 1943a. Mammals and the nature of continents. *American Journal of Science* 241: 1–31.

Simpson, G. G. 1943b. Turtles and the origin of the fauna of Latin America. *American Journal of Science* 241: 413–429.

Simpson, G. G. 1944. *Tempo and Mode in Evolution.* New York: Columbia University Press.

Simpson, G. G. 1945. The principles of classification and a classification of mammals. *Bulletin of the American Museum of Natural History* 85: i–xvi, 1–350.

Simpson, G. G. 1952a. Periodicity in vertebrate evolution. *Journal of Paleontology* 26: 359–370.

Simpson, G. G. 1952b. Probabilities of dispersal in geologic time. *Bulletin of the American Museum of Natural History* 99: 163–176.

Simpson, G. G. 1953. *The Major Features of Evolution.* New York: Columbia University Press.

Simpson, G. G. 1956. Zoogeography of West Indian land mammals. *American Museum Novitates* no. 1759.

Simpson, G. G. 1961a. *Principles of Animal Taxonomy.* New York: Columbia University Press.

Simpson, G. G. 1961b. Historical zoogeography of Australian mammals. *Evolution* 15: 413–446.

Simpson, G. G. 1964. Species density of North American Recent mammals. *Systematic Zoology* 13: 57–73.

Simpson, G. G. 1965. *The Geography of Evolution.* Philadelphia: Chilton Book Co.

Simpson, G. G. 1969. South American mammals. In E. J. Fittkau, J. Illies, H. Klinge, G. H. Schwabe and H. Sioli (eds.), *Biogeography and Ecology in South America,* vol. 2, Monographiae Biologicae 20: 879–909. The Hague: Dr. W. Junk.

Simpson, G. G. 1970. Uniformitarianism: An inquiry into principle, theory and method in geohistory and biohistory. In M. K. Hecht and W. C. Steere (eds.), *Essays in Evolution and Genetics in Honor of Theodosius Dobzhansky,* 43–96. New York: Appleton-Century-Crofts.

Simpson, G. G. 1978. Early mammals in South America: Fact, controversy and mystery. *Proceedings of the American Philosophical Society* 122: 318–328.

Simpson, G. G. 1980a. *Why and How: Some Problems and Methods in Historical Biology.* Oxford: Pergamon Press.

Simpson, G. G. 1980b. *Splendid Isolation: The Curious History of Mammals in South America.* New Haven, CT: Yale University Press.

Sinclair, W. A. 1964. Comparisons of recent declines of white ash, oaks and sugar maple in northeastern woodlands. *Cornell Plantations* 20: 62–67.

Sites, J. W. and J. C. Marshall. 2003. Delimiting species: A Renaissance issue in systematic biology. *Trends in Ecology and Evolution* 18: 462–470.

Skelton, P. and A. Smith. 2002. *Cladistics: A Practical Primer on CD-ROM.* Cambridge: Cambridge University Press.

Skutch, A. F. 1949. Do tropical birds rear as many young as they can nourish? *Ibis* 91: 430–455.

Skutch, A. F. 1967. Adaptive limitation of the reproductive rate of birds. *Ibis* 109: 579–599.

Slatkin, M. 1973. Gene flow and selection in a cline. *Genetics* 75: 733–756.

Slaughter, B. H. 1967. Animal ranges as a cue to Late-Pleistocene extinctions. In P. S. Marion and H. E. Wright Jr. (eds.), *Pleistocene Extinctions: The Search for a Cause*, 155–167. New Haven, CT: Yale University Press.

Slobodkin, L. B. and H. L. Sanders. 1969. On the contribution of environmental predictability to species diversity. In G. M. Woodwell and H. H. Smith (eds.), *Diversity and Stability in Ecological Systems*, 82–93. Brookhaven Symp. Biol. 22.

Smith, A. T. 1974. The distribution and dispersal of pikas: Consequences of insular population structure. *Ecology* 55: 1112–1119.

Smith, A. T. 1980. Temporal changes in insular populations of the pika (*Ochotona princeps*). *Ecology* 61: 8–13.

Smith, C. H. 1983. A system of world mammal faunal regions I: Logical and statistical derivation of regions. *Journal of Biogeography* 10: 455–466.

Smith, F. A. 1992. Evolution of body size among woodrats from Baja California, Mexico. *Functional Ecology* 6: 265–273.

Smith, F. A., J. L. Betancourt and J. H. Brown. 1995. Evolution of body size in the woodrat over the past 25,000 years of climate change. *Science* 270: 2012–2014.

Smith, F. D. M., R. M. May, R. Pellew, T. H. Johnson and K. R. Walter. 1993. How much do we know about the current extinction rate? *Trends in Ecology and Evolution* 8(10): 375–377.

Smith, G. A. and M. V. Lomolino. 2004. Black-tailed prairie dogs and the structure of avian communities on the shortgrass plains. *Oecologia.* 138 [4]:592–602.

Smith, G. B. 1979. Relationships of eastern Gulf of Mexico reef-fish communities to the species equilibrium theory of insular biogeography. *Journal of Biogeography* 6: 49–61.

Smith, G. R. 1978. Biogeography of intermountain fishes. *Great Basin Naturalist Memoirs* 2: 17–42.

Smith, G. R. 1981. Late Cenozoic freshwater fishes of North America. *Annual Review of Ecology and Systematics* 12: 163–193.

Smith, J. M. B. 1994. Patterns of disseminule dispersal by drift in the northwest Coral Sea. *New Zealand Journal of Botany* 30: 57–67.

Smith, J. M. B., H. Heatwole, M. Jones and B. M. Waterhouse. 1990. Drift disseminules on cays of the Swain Reefs, Great Barrier Reef, Australia. *Journal of Biogeography* 17: 5–17.

Smith, K. F. and J. H. Brown. 2002. Patterns of diversity, depth range and body size among pelagic fishes along a gradient in depth. *Global Ecology and Biogeography* 11: 313–322.

Smith, K. F. and S. D. Gaines. 2003. Rapoport's bathymetric rule and the latitudinal species diversity gradient for Northeast Pacific fishes and Northwest Atlantic gastropods: Evidence against a causal link. *Journal of Biogeography* 30: 1153–1159.

Smith, M. H. and J. T. McGinnis. 1968. Relationships of latitude, altitude and body size and mean annual production of offspring in *Permoyscus*. *Research on Population Ecology* 10: 115–126.

Smith, R. C., D. G. Ainley, K. Baker, E. Domack, S. D. Emslie, W. R. Fraser, J. Kennet, A. Leventer, E. Mosley-Thompson, S. E. Stammerjohn and M. Vernet. 1999. Marine ecosystem sensitivity to historical climate change in the Antarctic Peninsula. *BioScience* 49: 393–404.

Smith, R. L. 1974. *Ecology and Field Biology*, 2nd edition. New York: Harper & Row Publishers, Inc.

Smith, T. B., R. K. Wayne, D. J. Birman and M. W. Bruford. 1997. A role for ecotones in generating rainforest biodiversity. *Science* 276: 1855–1857.

Snead, R. E. 1980. *World Atlas of Geomorphic Features*. Huntington, NY: Krieger.

Sneath, P. H. A. and R. R. Sokal. 1973. *Numerical Taxonomy*. San Francisco: W. H. Freeman.

Snelgrove, P. V. R., J. F. Grassle and R. F. Petrecca. 1992. The role of food patches in maintaining deep-sea diversity: Field experiments with hydrodynamically unbiased colonization trays. *Limnology and Oceanography* 37: 1543–1550.

Snelgrove, P. V. R., J. F. Grassle and R. F. Petrecca. 1996. Experimental evidence for aging food patches as a factor contributing to deep-sea macrofaunal diversity. *Limnology and Oceanography* 41: 605–614.

Solbrig, O. T. 1972. The floristic disjunctions between the "Monte" in Argentina and the "Sonoran Desert" in Mexico and the United States. *Annals of the Missouri Botanical Garden* 59: 218–223.

Solem, A. 1959. Zoogeography of the land and freshwater mollusca of the New Herbrides. *Fieldiana Zoology* 43: 239–343.

Solem, A. 1981. Land-snail biogeography: A true snail's pace of change. In G. Nelson and D. E. Rosen (eds.), *Vicariance Biogeography: A Critique*, 197–221. New York: Columbia University Press.

Soltis, D. E., M. A. Gitzendanner, D. D. Strenge and P. S. Soltis. 1997. Chloroplast DNA intraspecific phylogeography of plants from the Pacific Northwest of North America. *Plant Systematics and Evolution*, 206: 353–373.

Soltz, D. L. and R. J. Naiman. 1978. *The Natural History of Native Fishes in the Death Valley System*. Los Angeles, CA: Natural History Museum of Los Angeles County.

Sondaar, P. Y. 1977. Insularity and its effect on mammal evolution. In M. K. Hecht, P. C. Goody and B. M. Hecht (eds.), *Major Patterns in Vertebrate Evolution*, 671–707. New York: Plenum Press.

Soulé, M. E. 1966. Trends in insular radiation of a lizard. *American Midland Naturalist* 100: 47–64.

Soulé, M. E. 1983. What do we really know about extinction? In C. M. Schonewald-Cox, S. M. Chambers, B. MacBryde and W. L. Thomas (eds.), *Genetics and Conservation: A Reference for Managing Wild Animal and Plant Populations*, 111–124. London: Benjamin/Cummings Publishing Company.

Soulé, M. E. 1987. *Viable Populations for Conservation*. New York: Cambridge University Press.

Soulé, M. E. and M. E. Gilpin. 1991. The theory of wildlife corridor capability. In D. A. Saunders and R. J. Hobbs (eds.), *The Role of Corridors in Nature Conservation*, 3–8. Chipping Norton, NSW, Australia: Surrey Beatty.

Soulé, M. E. and D. S. Simberloff. 1986. What do genetics and ecology tell us about the design of nature reserves? *Biological Conservation* 35: 19–40.

Southern California Coastal Water Research Project. 1973. *The Ecology of the Southern California Bight: Implications for Water Quality Management*. Technical Report 104: 1–531. El Segundo, CA.

Souza, W. P., S. C. Schroeter and S. D. Gaines. 1981. Latitudinal variation in intertidal algal community structure. *Oecologia* 48(3): 297–303.

Sparks, J. S. and W. L. Smith. 2005. Freshwater fishes, dispersal ability and nonevidence: "Gondwana Life Rafts" to the rescue. *Systematic Biology* 54: 158–165.

Spaulding, W. G. and L. J. Graumlich. 1986. The plast pluvial climatic episodes in the deserts of Southwestern North America. *Nature* 320: 441–444.

Spencer, A. W. and H. W. Steinhoff. 1968. An explanation of geographical variation in litter size. *Journal of Mammalogy* 49: 281–286.

Spiller, D. A., J. B. Losos and T. W. Schoener. 1998. Impact of a catastrophic hurricane on island populations. *Science* 281: 695–696.

Spolsky, C. and T. Uzzell. 1986. Evolutionary history of the hybridogenetic hybrid frog *Rana esculenta* as deduced from mtDNA analyses. *Molecular Biology and Evolution* 3: 44–56.

Srivastava, D. 1999. Using local-regional richness plots to test for species saturation: Pitfalls and potentials. *Journal of Animal Ecology* 68: 1–16.

Staley, J. T. 1997. Biodiversity: Are microbial species threatened? *Current Opinion in Biotechnology* 8: 340–345.

Stanley, S. M. 1975. A theory of evolution above the species level. *Proceedings of the National Acadedmy of Sciences, USA* 72: 646–650.

Stanley, S. M. 1979. *Macroevolution: Pattern and Process*. San Francisco: W. H. Freeman.

Stanley, S. M. 1984. Marine mass extinctions: A dominant role for temperature. In R. H. Nitecki (ed.), *Extinctions*, 69–117. Chicago: University of Chicago Press.

Stanley, S. M. 1987. *Extinction.* New York: Scientific American Books, Inc.

Stapp, P. 1998. A reevaluation of the role of prairie dogs in Great Plains grasslands. *Conservation Biology* 12: 1253–1259.

Steadman, D. W. 1986. Two new species of rails (Aves: Rallidae) from Mangaia, southern Cook Islands. *Pacific Science* 40: 27–43.

Steadman, D. W. 1987. California condor associated with spruce-jack pine woodland in the late-Pleistocene of New York. *Quaternary Research* 28: 415–426.

Steadman, D. W. 1989. Extinction of birds in eastern Polynesia: A review of the record and comparisons with other Pacific islands groups. *Journal of Archaeological Science* 16: 177–205.

Steadman, D. W. 1993. Biogeography of Tongan birds before and after human contact. *Proceedings of the National Academy of Sciences* 90: 818–822.

Steadman, D. W. 1995. Prehistoric extinctions of Pacific island birds: Biodiversity meets zooarchaeology. *Science* 267: 1123–1131.

Steadman, D. W. and P. S. Martin. 1984. Extinctions of birds in the late Pleistocene of North America. In P. S. Martin and R. J. Klein (eds.), *Quaternary Extinctions*, 466–477. Tucson, AZ: University of Arizona Press.

Steadman, D. W. and P. S. Martin. 2003. The late Quaternary extinction and future resurrection of birds on Pacific islands. *Earth-Science Reviews* 61: 133–147.

Steadman, D. W. and S. L. Olson. 1985. Bird remains from an archaeological site on Henderson Island, South Pacific. *Acta XX Congresus Internationalis Ornithologici* 2: 361–382.

Stebbins, G. L. 1971a. Adaptive radiation of reproductive characteristics in angiosperms. 2. Seeds and seedlings. *Annual Review of Ecology and Systematics* 2: 237–260.

Stebbins, G. L. 1971b. *Chromosomal Evolution in Higher Plants.* London: Edward Arnold.

Steenbergh, W. F. and C. H. Lowe. 1976. Ecology of the saguaro. I. The role of freezing weather in a warm desert plant population. In *Research in the Parks: National Park Service Symposium*, ser. no. 1: 49–92. Washington, DC: U.S. Government Printing Office.

Steenbergh, W. F. and C. H. Lowe. 1977. Ecology of the saguaro. II. Reproduction, germination, establishment, growth and survival of the young plant. *National Park Service Scientific Monographs*, ser. no. 8. Washington, DC: U.S. Government Printing Office.

Stehli, F. G. 1968. Taxonomic diversity gradients in pole locations: The recent model. In E. T. Drake (ed.), *Evolution and Environment*, 163–227. Peabody Museum Centennial Symposium. New Haven, CT: Yale Unversity Press.

Stehli, F. G. and S. D. Webb (eds.). 1985. *The Great American Biotic Interchange.* New York: Plenum Press.

Stehli, F. G. and J. W. Wells. 1971. Diversity and age patterns in hermatypic corals. *Systematic Zoology* 20: 115–126.

Stehli, F. G., R. G. Douglas and N. D. Newell. 1969. Generation and maintenance of gradients in taxonomic diversity. *Science* 164: 947–949.

Stein, G. 1937. *Everybody's Autobiography.* New York: Random House.

Steinauer, E. M. and S. L. Collins. 1996. Prairie ecology: The tallgrass prairie. In F. B. Samson and F. L. Knopf (eds.), *Prairie Conservation*, 39–52. Washington, DC: Island Press.

Stenseth, N. C. and R. A. Ims. 1993. *Biology of Lemmings.* London: Academic Press.

Stenseth, N. C. and W. Z. Lidicker, Jr. 1992. *Animal Dispersal: Small Mammals as a Model.* New York: Chapman and Hall.

Stenseth, N. C., A. Mysterud, G. Ottersen, J. W. Hurrell, K. Chan and M. Lima. 2002. Ecological effects of climate fluctuations. *Science* 297: 1292–1295.

Stenthouarskis, S. 1992. Altitudinal effect on species richness of Oniscidea (Crustacea; Isopoda) on three mountains in Greece. *Global Ecology and Biogeography Letters* 2: 157–164.

Sterba, G. 1966. *Freshwater Fishes of the World.* London: Studio Vista.

Stephens, D. W. and J. Gardner. 1999. Brine shrimp in Great Salt Lake, Utah. *U.S. Geological Survey*, Utah District.

Stevens, G. C. 1989. The latitudinal gradients in geographic range: How so many species coexist in the tropics. *American Naturalist* 132: 240–256.

Stevens, G. C. 1992. The elevational gradient in altitudinal range: An extension of Rapoport's latitudinal rule to altitude. *American Naturalist* 140: 893–911.

Stevens, G. C. 1996. Extending Rapoport's rule to Pacific marine fishes. *Journal of Biogeography* 23: 149–154.

Stevens, G. C. and B. J. Enquist. 1998. Macroecological limits to the abundance and distribution of *Pinus*. In D. M. Richardson (ed.), *Ecology and Biogeography of the Genus* Pinus, 183–190. Cambridge: Cambridge University Press.

Stevens, G. C. and J. F. Fox. 1991. The causes of treeline. *Annual Review of Ecology and Systematics* 22: 177–191.

Stevens, R. D. and M. R. Willig. 2002. Geographical ecology at the community level: Perspectives on the diversity of new world bats. *Ecology* 83: 545–560.

Stevens, R. D., S. B. Cox, R. E. Strauss and M. R. Willig. 2003. Patterns of functional diversity across an extensive environmental gradient: Vertebrate consumers, hidden treatments and latitudinal trends. *Ecology Letters* 6: 1099–1108.

Stevenson, R. D. 1986. Allen's rule in North American rabbits (*Sylvilagus*) and hares (*Lepus*) is an exception, not a rule. *Journal of Mammalogy* 67: 312–316.

Stockwell, D. R. B. and D. Peters. 1999. The GARP modeling system: Problems and solutions to automated spatial prediction. *International Journal of Geographical Information Science* 13: 143–158.

Stodart, E. and I. Parer. 1988. Colonization of Australia by the rabbit, *Oryctolagus cunicularis*. Project Report no. 6, *CSIRO Division of Wildlife and Ecology*.

Stone, L., T. Dayan and D. Simberloff. 1996. Community-wide assembly patterns unmasked: The importance of species' differing geographical ranges. *American Naturalist* 148: 997–1015.

Stoneking, M., S. T. Sherry, A. J. Redd and L. Vigilant. 1992. New approaches to dating suggest a recent age for the human mtDNA ancestor. *Philisophical Transactions of the Royal Society of London*, Series B 337: 167–175.

Storey, B. C. 1995. The role of mantle plumes in continental breakup: Case histories from Gondwanaland. *Nature* 377: 301–308.

Stott, P. A. and J. A. Kettleborough. 2002. Origins and estimates of uncertainty in predictions of twenty-first century temperature rise. *Nature* 416: 723–726.

Stouffer, P. C. and R. O. Birregaard, Jr. 1995. Use of Amazonian forest fragments by understory insectivorous birds. *Ecology* 76: 2429–2445.

Stout, J. and J. Vandermeer. 1975. Comparison of species richness for stream-inhabiting insects in tropical and mid-latitude streams. *American Naturalist* 109: 263–280.

Strahler, A. N. and A. H. Strahler. 1973. *Environmental Geoscience.* Santa Barbara: Hamilton.

Strahler, A. N. 1975. *Physical Geography*, 4th edition. New York: John Wiley and Sons.

Strahler, A. N. 1998. *Plate Tectonics.* Cambridge, MA: Geo-Books.

Straus, L. G., B. V. Eriksen, J. M. Erlandson and D. R. Yesner (eds.). 1996. *Humans at the End of the Ice-age: The Archaeology of the Pleistocene-Holocene Transition.* New York: Plenum Press.

Strauss, S. Y. 1991. Indirect effects in community ecology: Their definition, study and importance. *Trends in Ecology and Evolution* 6: 206–210.

Strayer, D. L. 1991. Projected distribution of zebra mussel, *Dreissena polymorpha*, in North America. *Canadian Journal of Fisheries and Aquatic Sciences* 48: 1389–1395.

Strecker, U., C. G. Meyer, C. Sturmbauer and H. Wilkens. 1996. Genetic divergence and speciation in an extremely young species flock in Mexico formed by the genus *Cyprinodon* (Cyprinodontidae, Teleostei). *Molecular Phylogenetics and Evolution* 6: 143–149.

Strong, D. R. Jr. 1974. Rapid asymptotic species accumulation in phytophagous insect communities: The pests of cacao. *Science* 185: 1064–1066.

Strong, D. R. Jr. and J. R. Rey. 1982. Testing for MacArthur-Wilson equilibrium with the arthropods of the miniature *Spartina* archipelago at Oyster Bay, Florida. *American Zoologist* 22: 355–360.

Strong, D. R., E. D. McCoy and J. R. Rey. 1977. Time and the number of herbivore species: The pests of sugarcane. *Ecology* 58: 167–175.

Strong, D. R., D. Simberloff, L. G. Abele and A. B. Thistel (eds). 1984. *Ecological Communities: Conceptual Issues and the Evidence.* Princeton, NJ: Princeton University Press.

Stuart, C. T. and M. A. Rex. 1994. The relationship between developmental pattern and species diversity in deep-sea prosobranch snails. In C. M. Young and K. J. Eckelbarger (eds.), *Reproduction, Larval Biology and Recruitment of the Deep-Sea Benthos,* 119–136. New York: Columbia University Press.

Stuart, C. T., M. A. Rex and R. J. Etter. 2003. Large-scale spatial and temporal patterns of deep-sea benthic species diversity. In P. A. Tyler (ed.), *Ecosystems of the World-Ecosystems of the Deep Oceans,* 295–311. Amsterdam: Elsevier.

Stute, M., M. Forster, H. Frischkorn, A. Serejo, J. F. Clark, P. Schlosser, W. S. Broeker and G. Bonani. 1995. Cooling of tropical Brazil: 5°C during the last glacial maximum. *Science* 269: 379–380.

Stutz, H. C. 1978. Explosive evolution of perennial *Atriplex* in western America. In K. T. Harper and J. L. Reveal (eds.), *Intermountain Biogeography: A Symposium,* 161–168. Great Basin Naturalist Memoirs. Provo, UT: Brigham Young University.

Sugihara, G. 1981. $S = CAz, z = 1/4$: A reply to Connor and McCoy. *American Naturalist* 117: 790–793.

Sullivan, R. M. and T. L. Yates. 1995. Population genetics and conservation biology of relict populations of red squirrels. In C. A. Istock and R. S. Hoffman (eds.), *Storm over a Mountain Island: Conservation Biology and the Mt. Graham Affair,* 193–208. Tucson, AZ: University of Arizona Press.

Sumner, F. B. 1932. Genetic, distributional and evolutionary studies of the subspecies of deer mice (*Permoscus*). *Bibliographia Genetica* 9: 1–106.

Sumner, J., C. Moritz and R. Shine. 1999. Shrinking forest shrinks skink: Morphological change in response to rainforest fragmentation in the prickly forest skink (*Gnypetoscincus queenslandiae*)— A review. *Biological Conservation* 91: 159– 167.

Sunquist, M., C. Leh, F. Sunquist, D. M. Mills and R. Rajaratnam. 1994. Rediscovery of the Bornean bay cat. *Oryx* 28: 67–69.

Swisher, C. C. III, J. M. Grajales-Nishimura, A. Montanari, S. V. Margolis, P. Claeys, W. Alvarez, P. Renne, E. Cedillo-Pardo, F. J.-M. R. Maurrasse, G. H. Curtis, J. Smit and M. O. McWilliams. 1992. Coeval 40AR/39AR ages of 65.0 million years ago from Chicxulub crater melt rock and Cretaceous-Tertiary boundary tektites. *Science* 257: 954–958.

Swofford, D. L. 2003. *PAUP*: Phylogenetic Analysis Using Parsimony (*and Other Methods), Version 4.* Sunderland, MA: Sinauer Associates.

Szumik, C. A. and Goloboff. P. A. 2004. Areas of endemism: An improved optimality criterion. *Systematic Biology* 53: 968–977.

Taitt, M. J. and C. J. Krebs. 1985. Population dynamics and cycles. In R. H. Tamarin (ed.), *Biology of New World* Microtus, 567–620. American Society of Mammalogists, Special Publication no. 8.

Takhtajan, A. 1986. *Floristic Regions of the World.* Berkeley: University of California Press.

Tamarin, R. H. 1977. Dispersal in island and mainland voles. *Ecology* 58: 1044–1054.

Tamura, K. and M. Nei. 1993. Estimation of the number of nucleotide substitutions in the control region of mitochondrial DNA in humans and chimpanzees. *Molecular Biology and Evolution* 10: 512–526.

Taper, M. L. and T. J. Case. 1992. Models of character displacement and the theoretical robustness of taxon cycles. *Evolution* 46: 317–333.

Tarling, D. H. 1962. Tentative correlation of Samoan and Hawaiian Islands using "reversals" of magnetism. *Nature* 196: 882–883.

Tarr, C. L. and R. C. Fleischer. 1995. Evolutionary relationships of the Hawaiian honeycreepers (Aves, Drepanidinae). In W. L. Wagner and V. A. Funk (eds.), *Hawaiian Biogeography: Evolution on a Hot Spot Archipelago.* Washington, DC: Smithsonian Institution Press.

Taulman, J. F. and L. W. Robbins. 1996. Recent range expansion and distributional limits of the nine-banded armadillo (*Dasypus novemcinctus*) in the United States. *Journal of Biogeography* 23: 635–648.

Taylor, B. 1991. Investigating species incidence over habitat fragments of different areas: A look at error estimation. *Biological Journal of the Linnean Society* 42: 477–491.

Taylor, C. M. 1996. Abundance and distribution within a guild of benthic stream fishes: Local processes and regional patterns. *Freshwater Biology* 36: 385–396.

Taylor, C. M. 1997. Fish species richness and incidence patterns in isolated and connected stream pools: Effects of pool volume and spatial position. *Oecologia* 110: 560–566.

Taylor, C. M. and N. J. Gotelli. 1994. The macroecology of *Cyprinella*: Correlates of phylogeny, body-size and geographical range. *American Naturalist* 144: 549–569.

Taylor, D. W. 1996. *Flowering Plant Origin, Evolution and Phylogeny.* New York: Chapman and Hall.

Taylor, F. B. 1910. Bearing of the Tertiary mountain belt on the origin of the earth's plan. *Geological Society of America Bulletin* 21: 179–226.

Taylor, F. B. 1928. Sliding continents and tidal and rotational forces. In W. A. Van Der Gracht and J. M. Waterschoot (eds.), *Theory of Continental Drift,* 158-177. Tulsa, OK: American Association of Petroleum Geologists.

Taylor, J. D. and C. N. Taylor. 1977. Latitudinal distribution of predatory gastropods on the eastern Atlantic shelf. *Journal of Biogeography* 4: 73–81.

Taylor, J. M., S. C. Smith and J. H. Calaby. 1985. Altitudinal distribution and body size among New Guinean *Rattus* (Rodentia, Muridae). *Journal of Mammalogy* 66: 353–358.

Taylor, R. J. 1987. The geometry of colonization: 1. Islands. *Oikos* 48: 225–231.

Taylor, R. J. and P. J. Regal. 1978. The peninsular effect on species diversity and the biogeography of Baja California. *American Naturalist* 112: 583–593.

Teal, J. M. 1957. Community metabolism in a temperate cold spring. *Ecological Monographs* 27: 283–302.

Teal, J. M. 1962. Energy flow in the salt marsh ecosystem of Georgia. *Ecology* 43: 614–624.

Tegelstrom, H. and L. Hansson. 1987. Evidence for long distance dispersal in the common shrew (*Sorex araneus*). *Sonderdunk aus Zeitschrift fur Saugetierkunde* 52: 52–54.

Teller, J. T., D. W. Leverington and J. D. Mann. 2002. Freshwater outbursts to the oceans from glacial Lake Agassiz and their role in climate change during the last deglaciation. *Quaternary Science Reviews* 21: 879–887.

Temple, S. 1986. The problem of avian extinctions. *Current Ornithology* 3: 453–485.

Templeton, A. R. 1980a. The theory of speciation via the founder principle. *Genetics* 94: 1011–1038.

Templeton, A. R. 1980b. Modes of speciation and inferences based on genetic distances. *Evolution* 34: 719–729.

Templeton, A. R. 1981. Mechanisms of speciation: A population genetic approach. *Annual Review of Ecology and Systematics* 12: 23–48.

Templeton, A. R. 1998. Nested clade analyses of phylogeographic data: Testing hypotheses about gene flow and population history. *Molecular Ecology* 7: 381–397.

Templeton, A. R. 2002. Out of Africa again and again. *Nature* 416: 45–51.

Templeton, A. R. 2004. Statistical phylogeography: Methods of evaluating and minimizing inference errors. *Molecular Ecology* 13: 789–809.

Templeton, A. R., E. Routman and C. A. Phillips. 1995. Separating population-structure from population history—A cladistic-analysis of the geographical-distribution of mitochondrial-DNA haplotypes in the tiger salamander, *Ambystoma-Tigrinum. Genetics* 140: 767–782.

Terborgh, J. 1973a. Chance, habitat and dispersal in the distribution of birds in the West Indies. *Evolution* 27: 338–349.

Terborgh, J. 1973b. On the notion of favorableness in plant ecology. *American Naturalist* 107: 481–501.

Terborgh, J. 1974. Preservation of natural diversity: The problem of extinction prone species. *Bioscience* 24: 715–722.

Terborgh, J. 1977. Bird species diversity on an Andean elevational gradient. *Ecology* 58: 1007–1019.

Terborgh, J. 1986. Keystone plant resources in tropical forest. In M. E. Soulé (ed.), *Conservation Biology: The Science of Scarcity and Diversity*, 330–344. Sunderland, MA: Sinauer Associates.

Terborgh, J. 1992. *Diversity and the Tropical Rain Forest*. New York: Scientific American Library.

Terborgh, J. and J. Faaborg. 1973. Turnover and ecological release in the avifauna of Mona Island, Puerto Rico. *Auk* 90: 759–779.

Terborgh, J. and B. Winter. 1980. Some causes of extinction. In M. E. Soulé and B. A. Wilcox, (eds.), *Conservation Biology: An Ecological-Evolutionary Perspective*, 119–134. Sunderland, MA: Sinauer Associates.

Terborgh, J., L. Lopez and J. Tellos. 1997. Bird communities in transition: The Lago Guri Islands. *Ecology* 78: 1494–1501.

Terborgh, J., L. Lopez, V. P. Nunez, M. Rao, G. Shahababuddin, G. Orihuela, M. Riveros, R. Ascanio, G. H. Adler, T. D. Lambert and L. Balbas. 2001. Ecological meltdown in predator-free forest fragments. *Science* 294: 1923–1926.

Terrell, J. 1976. Island biogeography and man in Melanesia. *Archaeology and Physical Anthropology in Oceania* 11: 1–17.

Terrell, J. 1986. *Prehistory in the Pacific Islands: A Study of Variation in Language, Customs and Human Biology*. Cambridge: Cambridge University Press.

Terrell, J., M. Miller and D. Roe. 1977. *Human Biogeography*. World Archaeology, Vol. 8. Henley-on-Thames: Routledge and Kegan Paul.

Tewksbury, J. J., D. J. Levey, N. M. Haddad, S. Sargent, J. L. Orrock, A. Weldon, B. J. Danielson, J. Brinkerhoff, E. I. Damschen and P. Townsend. 2002. Corridors affect plants, animals and their interactions in fragmented landscapes. *Proceedings of the National Academy of Science, USA* 99: 12923–12926.

Thiollay, J. 1998. Distribution patterns and insular biogeography of South Asian raptor communities. *Journal of Biogeography* 25: 57–72.

Thomas, C. D., A. Cameron, R. E. Green, M. Bakkenes, L. J. Beaumont, Y. C. Collingham, B. F. N. Erasmus, M. F. De Siqueira, A. Grainger, L. Hannah, L. Hughes, B. Huntley, A. S. Van Jaarsveld, G. F. Midgley, L. Miles, M. A. Ortega-Huerta, A. T. Peterson, O. L. Phillips and S. E. Williams. 2004. Extinction risk from climate change. *Nature* 427: 145–148.

Thomas, W. L. 1956. *Man's Role in Changing the Face of the Earth*. Chicago: Univeristy of Chicago Press.

Thompson, D. B. 1990. Different spatial scales of adaptation in the climbing behavior of *Permyscus maniculatus*: Geographic variation, natural selection and gene flow. *Evolution* 44: 952–965.

Thompson, D. Q., R. L. Stuckey and E. B. Thompson. 1987. *Spread, Impact and Control of Purple Loosestrife* (Lythrum saclicaria) *in North American Wetlands*. Washington, DC: U.S. Fish and Wildlife Service, Research 2.

Thompson, R. S. and J. I. Mead. 1982. Late Quaternary environments and biogeography in the Great Basin. *Quaternary Research* 17: 39–55.

Thornton, I. 1996. *Krakatau: The Destruction and Reassembly of an Island Ecosystem*. Cambridge, MA: Harvard University Press.

Thornton, I. W. B. 1992. K. W. Dammerman: Forerunner of island theory? *Global Ecology and Biogeography Letters* 2: 145–148.

Thornton, I. W. B. 2001. Colonization of an island volcano, Long Island, Papau New Guinea and an emergent island, Motmot, in its calder lake. I. General introduction. *Global Ecology and Biogeography* 28: 1299–1310.

Thornton, I. W. B., R. A. Zann and S. van Balen. 1993. Colonization of Rakata (Krakatau Is.) by non-migrant land birds from 1883 to 1992 and implications for the value of island biogeography theory. *Journal of Biogeography* 20: 441–452.

Thornton, I. W. B., T. R. New, R. A. Zann and P. A. Rawlinson. 1990. Colonization of the Krakatau Islands by animals: Perspective from the 1980s. *Philosophical Transactions of the Royal Society of London*, Series B 328: 131–165.

Thornton, I. W. B., D. Runciman, S. Cook, L. F. Lumsden, T. Partomihardjo, N. Natasha, J. Junichi and S. A. Ward. 2002. How important are stepping stones in colonization of Krakatau? *Biological Journal of the Linnean Society* 77: 275–317.

Thorson, G. 1946. Reproduction and larval development of Danish marine bottom invertebrates, with special reference to the planktonic larvae in the Sound (Oresund). *Meddelelser fra Kommissionen for Danmarks Fiskeri– og Havundersogelser, Serie Plankton* 4: 1–523.

Thorson, G. 1950. Reproductive and larval ecology of marine bottom invertebrates. *Biological Reviews* 25: 1–45.

Thorson, G. 1936. The larval development, growth and metabolism of Arctic marine bottom invertebrates compared with those of other seas. *Meddelelser om Gronland* 100: 1–155.

Thrower, N. J. W. and D. E. Bradbury (eds.). 1977. *Chile-California Mediterranean Scrub Atlas: A Comparative Analysis*. Stroudsburg, PA: Dowden, Hutchinson & Ross.

Thurber, J. M. and R. O. Peterson. 1991. Changes in body size associated with range expansion in the coyote (*Canis latrans*). *Journal of Mammalogy* 72: 750–755.

Tiffney, B. H. 1985. Perspectives on the origin of the floristic similarity between eastern Asia and eastern North America. *Journal of the Arnold Arboretum* 66: 73–94.

Tiffney, B. H. and S. R. Manchester. 2001. The use of geological and paleontological evidence in evaluating plant phylogeographic hypotheses in the Northern Hemisphere Tertiary. *International Journal of Plant Science* 162: S41–S52.

Tilman, D. 1988. *Plant Strategies and the Dynamics and Structure of Plant Communities*. Princeton, NJ: Princeton University Press.

Tilman, D., R. M. May, C. L. Lehman and R. M. A. Nowak. 1994. Habitat destruction and the extinction debt. *Nature* 371: 65–66.

Titus, J. G. 1990. Greenhouse effects, sea level rise and land use. *Land Use Policy* 7: 138–153.

Tivy, J. 1993. *Biogeography: A Study of Plants in the Ecosphere*. New York: John Wiley and Sons.

Tonn, W. M. and J. J. Magnunson. 1982. Patterns in the species composition and richness of fish assemblages in northern Wisconsin lakes. *Ecology* 63: 1149–1166.

Towns, D. R. and W. J. Balantine. 1993. Conservation and restoration of New Zealand island ecosystems. *Trends in Ecology and Evolution* 8: 452–457.

Towns, D. R. and C. H. Daugherty 1994. Patterns of range contractions and extinctions in the New Zealand herpetofauana following human colonization. *New Zealand Journal of Zoology* 21: 325–339.

Tracy, C. R. and T. L. George. 1992. On the determinants of extinction. *American Naturalist* 139: 102–122.

Tracy, C. R. and T. L. George. 1993. Extinction probabilities of British island birds: A reply. *American Naturalist* 142: 1036–1037.

Trewick, S. A. 1996. Morphology and evolution of two takahe: Flightless rails of New Zealand. *Journal of Zoology* (London) 238: 221.

Tull, D. S. and K. Bohning-Gaese. 1993. Patterns of drilling predation on gastropods of the family Turritellidae in the Gulf of California. *Paleobiology* 19: 476–486.

Tunnicliffe, V. and C. M. R. Fowler. 1996. Influence of sea-floor spreading on the global hydrothermal vent fauna. *Nature* 379: 531–533.

Turner, J. R. G. 1981. Adaptation and evolution in *Heliconius*: A defense of neo-Darwinism. *Annual Review of Ecology and Systematics* 12: 99–121.

Turner, R. M., J. E. Bowers and T. L. Burgess. 1995. *Sonoran Desert Plants: An Ecological Atlas*. Tucson, AZ: University of Arizona Press.

Turrill, W. B. 1953. *Pioneer Plant Geography: The Phytogeography Researches of Sir Joseph Dalton Hooker*. The Hague: Martinus Nijhoff.

Udvardy, M. D. F. 1969. *Dynamic Zoogeography*. New York: Van Nostrand Reinhold.

Upchurch, P., C. A. Hunn and D. B. Norman. 2002. An analysis of dinosaurian biogeography: Evidence for the existence of vicariance and dispersal patterns caused by geological events. *Proceedings of the Royal Society of London*, Series B 269: 613–621.

Upton, G. and B. Fingleton. 1990. *Spatial Data Analysis by Example*. New York: John Wiley and Sons.

Urquhart, F. A. 1960. *The Monarch Butterfly*. Toronto: University of Toronto Press.

Usher, M. B. 1979. Markovian approaches to ecological succession. *Journal of Animal Ecology* 48: 413–426.

Vagvolygi, J. 1975. Body size, aerial dispersal and origin of the Pacific land snail fauna. *Systematic Zoology* 24: 465–488.

Vaisanen, R. and K. Heliovaara. 1994. Hot-spots of insect diversity in northern Europe. *Annales Zoologici Fennici* 31: 71–81.

Valentine, J. W. 1961. Paleoecologic molluscan geography of the Californian Pleistocene. *University of California Publications in Geological Science* 34: 309–442.

Valentine, J. W. 1966. Numerical analysis of marine molluscan ranges on the extratropical northeastern Pacific shelf. *Limnology and Oceanography* 11: 198–211.

Valentine, J. W. 1989. Phanerozoic marine faunas and the stability of the Earth system. *Global and Planetary Change* 1: 137–155.

Valentine, J. W. and D. Jablonski. 1982. Larval strategies and patterns of brachiopod diversity in space and time. *Geological Society of America Abstracts* 14: 241.

Valentine, J. W. and D. Jablonski. 1993. Fossil communities: Compositional variation at many time scales. In R. E. Ricklefs and D. Schluter (eds.), *Species Diversity in Ecological Communities: Historical and Geographical Perspectives*, 341–349. Chicago: University of Chicago Press.

Valentine, J. W., K. Roy and D. Jablonski. 2002. Carnivore-noncarnivore ratios in northeastern Pacific marine gastropods. *Marine Ecology-Progress Series* 228: 153–163.

Valone, T. J. and J. H. Brown. 1995. Effects of competition, colonization and extinction on rodent species diversity. *Science* 267: 880–883.

Van Den Bosch, F., R. Hengeveld and J. A. J. Metz. 1992. Analyzing the velocity of animal range expansion. *Journal of Biogeography* 19: 135–150.

Van der Hammen, T. 1982. Paleoecology of tropical South America. In G. T. Prance (ed.), *Biological Diversification in the Tropics*, 60–66. New York: Columbia University Press.

Van der Hammen, T., T. A. Wijmstra and W. H. Zagwijn. 1971. The floral record of the late Cenozoic of Europe. In K. K. Turekian (ed.), *The Late Cenozoic Glacial Ages*, 391–424. New Haven, CT: Yale University Press.

Van Devender, T. R. 1977. Holocene woodlands in the southwestern deserts. *Science* 198: 189–192.

Van Devender, T. R. and W. G. Spaulding. 1979. Development of vegetation and climate in the southwestern United States. *Science* 204: 701–710.

Vanni, M. J. and D. L. Findlay. 1990. Trophic cascades and phytoplankton community structure. *Ecology* 71: 921–937.

Van Riper, C., S. G. Van Riper, M. L. Goff and M. Laird. 1986. The epizootiology and ecological significance of malaria in Hawaiian islands. *Ecological Monographs* 56: 327–344.

Van Valen, L. 1970. Late Pleistocene extinctions. In *Proceedings of the North American Pelontologist Conference* 469–485. Lawrence, KS: Allen Press.

Van Valen, L. 1973a. Body size and the number of plants and animals. *Evolution* 27: 27–35.

Van Valen, L. 1973b. Pattern and the balance of nature. *Evolutionary Theory* 1: 31–49.

Van Valen, L. 1973c. A new evolutionary law. *Evolutionary Theory* 1: 1–30.

Van Valkenburgh, B. and C. M. Janis. 1993. Historical diversity patterns in North American large herbivores and carnivores. In R. E. Ricklefs and D. Schluter (eds.), *Species Diversity in Ecological Communities*, 330–340. Chicago: University of Chicago Press.

van Veller, M. G. P. and D. R. Brooks. 2001. When simplicity is not parsimonious: A priori and a posteriori approaches in historical biogeography. *Journal of Biogeography* 28: 1–11.

van Veller, M. G. P., D. R. Brooks and M. Zandee. 2003. Cladistic and phylogenetic biogeography: The art and science of discovery. *Journal of Biogeography* 30: 319–329.

van Veller, M. G. P., M. Zandee and D. J. Kornet. 1999. Two requirements for obtaining valid common patterns under different assumptions in vicariance biogeography. *Cladistics* 15: 393–406.

Van Voorhies, W. A. 1996. Bergmann size clines: A simple explanation for their occurrence in ectotherms. *Evolution* 50: 1259–1264.

Vanzolini, P. E. and W. R. Heyer. 1985. The American herptofauna and the interchange. In F. G. Stehli and S. D. Webb (eds.), *The Great American Interchange*, 475–483. New York: Plenum Press.

Vanzolini, P. E. and E. E. Williams. 1970. South American anoles: The geographic differentiation and evolution of the *Anolis chtysolepis* species group (Sauria, Iguanidae). *Arq. Zool.* (São Paulo) 19: 1–298.

Varela-Romero, A., G. Ruiz-Campos, L. M. Yépiz-Velázquez and J. Alaníz-García. 1998. Distribution, habitat and conservation status of desert pupfish (*Cyprinodon macularius*) in the Lower Colorado River Basin, Mexico. *Reviews in Fish Biology and Fisheries* 12, no. 2 (2002): 157–165.

Varley, G. C. 1970. The concept of energy flow applied to a woodland community. In A. Watson (ed.), *Animal Populations in Relation to Their Food Resources*, 389–405. London: Blackwell Scientific Publications.

Vartanyan, S. L., V. E. Garutt and A. V. Sherr. 1993. Holocene dwarf mammoths from Wrangel Island in the Siberian Arctic. *Nature* 362: 337–340.

Vartanyan, S. L., K. A. Arslanov, T. V. Tertychnaya and S. B. Chernov. 1995. Radiocarbon dating evidence for mammoths on Wrangel Island, Arctic Ocean, until 2000 B.C. *Radiocarbon* 37: 1–6.

Vaughan, T. A. 1967. Two parapatric species of pocket gophers. *Evolution* 21: 148–158.

Vaughan, T. A. and R. M. Hansen. 1964. Experiments on interspecific competition between two species of pocket gophers. *American Midland Naturalist* 72: 444–452.

Vences, M., D. R. Vieites, F.Glaw, H. Brinkmann, J. Kosuch, M. Veith and A. Meyer. 2003. Multiple overseas dispersal in amphibians. *Proceedings of the Royal Society of London,* Series B 270: 2435–2442.

Ventura, J. and M. J. Lopez–Fuster. 2000. Morphometric analysis of the black rat, *Rattus rattus*, from Congreso Island (Chafarinas Archipelago, Spain). *Orsis* 15: 91–102.

Vermeij, G. J. 1974. Marine faunal dominance and molluscan shell form. *Evolution* 28: 656–664.

Vermeij, G. J. 1978. *Biogeography and Adaptation: Patterns of Marine Life*. Cambridge, MA: Harvard University Press.

Vermeij, G. J. 1991a. Anatomy of an invasion: The trans-Arctic interchange. *Paleobiology* 17: 281–307.

Vermeij, G. J. 1991b. When biotas meet: Understanding biotic interchange. *Science* 253: 1099–1103.

Vermeij, G. J. 2001. Distribution, history and taxonomy of the Thais clade (Gastropoda: Muricidae) in the Neogene of tropical America. *Journal of Paleontology* 75: 797–805.

Vermeij, G. J. 2004. Island life: A view from the sea. In M. V. Lomolino and L. R. Heaney (eds.), *Frontiers of Biogeography: New Directions in the Geography of Nature*, 239–254. Sunderland, MA: Sinauer Associates.

Vermeij, G. J. 2004. *Nature: An Economic History*. Princeton, NJ: Princeton University Press.

Vermeij, G. J. and M. A. Snyder. 2002. Leucoznoia and related genera of fasciolariid gastropods: Shell-based taxonomy and relationships. *Proceedings of the Natural Academy of Sciences of Philadelphia* 152: 23–44.

Vetaas, O. R., J.-A. Grytnes. 2002. Distribution of vascular plant species richness and endemic richness along the Himalayan elevational gradient in Nepal. *Global Ecology and Biogeography* 11: 291–301.

Villard, M.-A. 2002. Habitat fragmentation: Major conservation issue or intellectual attractor? *Ecological Applications* 12(2): 319–320.

Vine, F. J. and D. H. Matthews. 1963. Magnetic anomalies over a young oceanic ridge. *Nature* 199: 947–949.

Vinogradova, N. G. 1962. Vertical zonation in the distribution of deep-sea benthic fauna in the ocean. *Deep-Sea Research* 8: 245–250.

Vitousek, P. M. 1992. Global environmental change: An introduction. *Annual Review of Ecology and Systematics* 23: 1–14.

Vitousek, P. M., L. L. Loope and H. Andersen. 1995. *Islands: Biological Diversity and Ecosystem Function*. Berlin: Springer Verlag.

Vitousek, P. M., C. M. D'Antonio, L. L. Loope and R. Westbrooks. 1996. Biological invasions as a global environmental challenge. *American Scientist* 84: 468–478.

Vitousek, P. M., P. R. Ehrlich, A. H. Ehrlich and P. A. Matson. 1986. Human appropriation of the products of photosynthesis. *BioScience* 36: 368.

Vitousek, P. M., H. A. Mooney, J. Lubchenco and J. M. Melillo. 1997. Human domination of Earth's ecosystems. *Science* 277: 494–499.

Vitousek, P. M., C. M. D'Antonio, L. L. Loope, M. Rejmanek and R. Westbrooks. 1997b. Introduced species: A significant component of human-caused global change. *New Zealand Journal of Ecology* 21: 1–16.

Voelker, G. 2002. Systematics and historical biogeography of wagtails: Dispersal versus vicariance revisited. *The Condor* 104: 725–739.

von Broembsen, S. L. 1989. Invasions of natural ecosystems by plant pathogens. In J. A. Drake et al. (eds.), *Biological Invasions: A Global Perspective*, 77–84. New York: John Wiley and Sons.

Von Holle, B., H.R. Delcourt and D. Simberloff. 2003. The importance of biological inertia in plant community resistance to invasion. *Journal of Vegetation Science* 14: 425–432.

Voris, H. K. 1977. A phylogeny of the sea snakes (Hydrophiidae). *Fieldiana Zoology* 70(4): 79–166.

Vrba, E. S. 1992. Mammals as a key to evolutionary theory. *Journal of Mammalogy* 73: 1–28.

Vuilleumier, F. 1970. Insular biogeography in continental species. I. The Northern Andes of South America. *American Naturalist* 104: 373–388.

Vuilleumier, F. 1973. Insular biogeography in continental regions. II. Cave faunas from Tesin, southern Switzerland. *Systematic Zoology* 22: 64–76.

Vuilleumier, F. 1975. Zoogeography. In D. S. Farner and J. R. King (eds.), *Avian Biology*, vol. 5, 421–469. New York: Academic Press.

Vuilleumier, F. 1984. Faunal turnover and development of fossil avifaunas in South America. *Evolution* 38: 1384–1396.

Vuilleumier, F. 1985. Fossil and recent avifaunas and the Interamerican interchange. In F. G. Stehli and S. D. Webb (eds.), *The Great American Biotic Interchange*, 387–424. New York: Plenum Press.

Vuilleumier, F. and D. Simberloff. 1980. Ecology versus history as determinants of patchy and insular distributions in high Andean birds. In M. K. Hecht, W. C. Steere and B. Wallace (eds.), *Evolutionary Biology* 12, 235–379. New York: Plenum Press.

Wagner, D. L. and J. K. Liebherr. 1992. Flightlessness in insects. *Trends in Ecology and Evolution* 7: 216–220.

Wagner, W. H. Jr. 1995. Evolution of Hawaiian ferns and fern allies in relation to their conservation status. *Pacific Science* 49: 31–41.

Wagner, W. L. and V. A. Funk (eds.). 1995. *Hawaiian Biogeography: Evolution on a Hot Spot Archipelago*. Washington, DC: Smithsonian Institution Press.

Wahlberg, N., A. Moilanen and I. Hanski. 1996. Predicting the occurrence of endangered species in fragmented landscapes. *Science* 273: 1536–1538.

Wahrman, R., R. Goitein and E. Nevo. 1969. Mole rat *Spalax*: Evolutionary significance of chromosome variation. *Science* 164: 82–84.

Wainright, P. C., D. R. Bellwood and M. W. Westneat. 2002. Ecomorphology of locomotion in labrid fishes. *Environmental Biology of Fishes* 65: 47–62.

Wake, D. B. 1966. Comparative osteology and evolution of the lungless salamanders, family Plethodontidae. *Memoirs of the Southern California Academy of Sciences* 4: 1–111.

Waleter, H. S. 1998. Driving forces of island biodiversity: An appraisal of two theories. *Physical Geography* 19: 351–377.

Walker, L. R. and R. Moral. 2003. *Primary succession and ecosystem rehabilitation*. London: Cambridge University Press.

Wallace, A. R. 1857. On the natural history of the Aru Islands. *Annals and Magazine of Natural History*, Supplement to Volume 20, December.

Wallace, A. R. 1860. On the zoological geography of the Malay Archipelago. *Journal of the Linnaean Society of London* 4: 172–184.

Wallace, A. R. 1869. *The Malay Archipelago: The Land of the Orangutan and the Bird of Paradise*. New York: Harper.

Wallace, A. R. 1876. *The Geographical Distribution of Animals*. 2 vols. London: Macmillan.

Wallace, A. R. 1878. *Tropical Nature and Other Essays*. New York: Macmillan.

Wallace, A. R. 1880. *Island Life, or the Phenomena and Causes of Insular Faunas and Floras*. London: Macmillan.

Wallace, A. R. 1893. Mr. H. O. Forbes's discoveries in the Chatam Islands. *Nature* (May 11, 1893): 27.

Wallace, A. R. 1901. *Darwinism: An Exposition of the Theory of Natural Selection, with Some of Its Applications*. Macmillan, London.

Waloff, Z. 1966. The upsurges and recessions of the desert locust plague: An historical survey. *Anti-Locust Memoirs* 8: 1–111.

Waltari, E., J. R. Demboski, D. R. Klein and J. A. Cook. 2004. A molecular perspective on the historical biogeography of the northern high latitudes. *Journal of Mammalogy* 85: 591–600.

Walter, H. S. 2000. Stability and change in the bird communities of the Channel Islands. In D. R. Browne, K. L. Mithcell and H. W. Chaney (eds.), *Proceedings of the 5th California Islands Symposium*, 307–314. U. S. Department of the Interior, Mineral Management Service (OCS Study MMS 99–0038).

Walter, H. S. 2004. The mismeasure of islands: Biogeographic theory and the conservation of nature. *Journal of Biogeography* 31: 177–197.

Ward, 2001. Sudden productivity collapse associated with the Triassic-Jurassic boundary mass extinction. *Science* 292: 1148–1151.

Wares, J. P. 2002. Community genetics in the Northwestern Atlantic intertidal. *Molecular Ecology* 11: 1131–1144.

Warming, E. 1895. *Plantesamfund*. Copenhagen.

Warming, E. 1905. *Oecology of Plants: An Introduction to the Study of Plant Communities*. Oxford: Oxford University Press.

Warner, R. E. 1969. The role of introduced diseases in the extinction of the endemic Hawaiian avifauna. *Condor* 70: 101–120.

Waser, N. M., L. Chittka, M. V. Price, N. M. Williams and J. Ollerton. 1996. Generalization in pollination systems and why it matters. *Ecology* 77: 1043–1060.

Waters, J. M., A. Lopez and G. P. Wallis. 2000. Molecular phylogenetics and biogeography of galaxiid fishes (Osteichthys: Galaxiida): Dispersal, vicariance and position of Lexidogalaxias salamandroides. *Systematic Biology* 49: 777–795.

Watson, D. M. and A. T. Peterson. 1999. Determinants of diversity in a naturally fragmented landscape: Humid montane forest avifaunas of Mesoamerica. *Ecography* 22: 582–589.

Watson, H. C. 1859. *Cybele Britannica, or British Plants and their Geographical Relations*. London: Longman and Company.

Watters, G. T. 1992. Unionids, fishes and the species-area curve. *Journal of Biogeography* 19: 481–490.

Watts, D. 1979. The new biogeography and its niche in physical geography. *Geography* 78: 324–337.

Wayne, R. K., B. Vanvalkenburgh and S. J. Obrien. 1991. Molecular distance and divergence time in carnivores and primates. *Molecular Biology and Evolution* 8: 297–319.

Webb, S. D. 1985. Late Cenozoic mammal dispersals between the Americas. In F. G. Stehli and S. D. Webb (eds.), *The Great American Biotic Interchange*, 357–386. New York: Plenum Press.

Webb, S. D. 1991. Ecogeography and the great American interchange. *Paleobiology* 17: 266–280.

Webb, S. D. and A. D. Barnosky. 1989. Faunal dynamics of Pleistocene mammals. *Annual Reviews of Earth and Planetary Sciences* 17: 413–438.

Webb, S. D. and L. G. Marshall. 1982. Historical biogeography of recent South American land mammals. In M. A. Mares and H. H. Genoways (eds.), *Mammalian Biology in South America*, 39–52. Special Publication Series 6, Pymatuning Laboratory of Ecology, University of Pittsburgh.

Webb, T. III. 1987. The appearance and disappearance of major vegetational assemblages: Long-term vegetational dynamics in eastern North America. *Vegetatio* 69: 177–187.

Webb, T. J. and K. J. Gaston. 2000. Geographic range size and evolutionary age in birds. *Proceedings of the Royal Society of London*, Series B-Biological Sciences 267: 1843–1850.

Weber, F. R., T. D. Hamilton, D. M. Hopkins, C. A. Repenning and H. Haas. 1981. Canyon Creek: A Late Pleistocene vertebrate locality in interior Alaska. *Quaternary Research* 16: 167–180.

Webster, J. 1997. Fungal biodiversity. *Biodiversity and Conservation* 6: 657.

Wecker, S. C. 1963. The role of early experience in habitat selection by the prairie deer-mouse, *Peromyscus maniculatus bairdi. Ecological Monographs* 33: 307–325.

Wecker, S. C. 1964. Habitat selection. *Scientific American* 211: 109–116.

Wegener, A. 1912a. Die Entstehung der Kontinente. *Petermanns Geogr. Mitt.* 58: 185–195, 253–256, 305–308.

Wegener, A. 1912b. Die Entstehung der Kontinente. *Geol. Rundsch.* 3: 276–292.

Wegener, A. 1915. *Die Entstehung der Kontinente und Ozeane.* Braunschweig: Vieweg. [Other editions 1920, 1922, 1924, 1929, 1936.]

Wegener, A. 1966. *The Origin of Continents and Oceans.* New York: Dover Publications. [Translation of 1929 edition by J. Biram.]

Weigert, R. G. and D. F. Owen. 1971. Trophic structure, available resources and populations densities in terrestrial ecosystems versus aquatic ecosystems. *Journal of Theoretical Biology* 30: 69–81.

Weiher, E. and P. Keddy (eds.). 1999. *Ecological Assembly Rules: Perspectives, Advances, Retreats.* Cambridge: Cambridge University Press.

Weir, J. T. and D. Schluter. 2004. Ice sheets promote speciation in boreal birds. *Proceedings of the Royal Society of London* Series B-Biological Sciences 271: 1881–1887.

Weller, M. W. 1980. *The Island Waterfowl.* Ames: Iowa State University Press.

Wells, P. V. 1976. Macrofossil analysis of wood rat (*Neotoma*) middens as a key to the Quaternary vegetational history of arid America. *Quaternary Research* 6: 223–248.

Wells, P. V. 1978. Postglacial origin of the present Chihuahuan Desert less than 11,500 years ago. In R. H. Wauer and D. H. Riskind (eds.), *Transactions of the Symposium on the Biological Resources of the Chihuahuan Desert Region, United States and Mexico,* 67–83. U.S. Department of Interior, National Park Service Transactions and Proceedings, ser. no. 3.

Wells, P. V. 1979. An equable glaciopluvial in the West: Pleniglacial evidence of increased precipitation on a gradient from the Great Basin to the Sonoran and Chihuahuan Deserts. *Quaternary Research* 12: 311–325.

Wells, P. V. 1983. Paleobiogeography of montane islands in the Great Basin since the last glaciopluvial. *Ecological Monographs* 53: 341–382.

Wells, P. V. and R. Berger. 1967. Late Pleistocene history of coniferous woodland in the Mohave Desert. *Science* 155: 1640–1647.

Wells, P. V. and C. D. Jorgensen. 1964. Pleistocene wood rat middens and climatic change in Mojave Desert: A record of juniper woodlands. *Science* 143: 1171–1174.

Wenke, R. J. 1999. *Patterns in Prehistory: Humankind's First Three Million Years* (4th edition). New York: Oxford University Press.

Wenner, A. M. and D. L. Johnson. 1980. Land vertebrates on the California Channel Islands: Sweepstakes or bridges? In D. M. Power (ed.), *The California Islands: Proceedings of a Multi-Disciplinary Symposium,* 497–530. Santa Barbara, CA: Museum of Natural History.

Werger, M. J. A. and A. C. van Bruggen (eds.). 1978. *Biogeography and Ecology of Southern Africa.* 2 vols. Monographiae Biologicae 31. The Hague: Dr. W. Junk.

Werner, P. A. 1975 A seed trap for determining patterns of seed deposition in terrestrial plants. *Canadian Journal of Botany* 58: 810–813.

West, R. G. 1977. *Pleistocene Geology and Biology.* 2nd edition. London: Longman Group.

Westing, A. H. 1966. Sugar maple decline: An evaluation. *Economic Botany* 20: 196–212.

Wetzel, R. G. 1975. *Limnology.* Philadelphia: W. B. Saunders.

Wheeler, Q. D. and R. Meier. 2000. *Species Concepts and Phylogenetic Theory: A Debate.* New York: Columbia University Press.

Whitaker, A. H. 1978. The effects of rodents on reptiles and amphibians. In P. R. Dingwall and I. A. Atkinson (eds.), *The Ecology and Control of Rodents in New Zealand's Nature Reserves,* 75–88. Department of Lands and Survey Information Series no. 4.

White, J. L. and B. C. Harvey. 2001. Effects of an introduced piscivorous fish on native benthic fishes in a coastal river. *Freshwater Biology* 46, no. 7 (2001): 987–995.

White, J. L. and R. D. E. MacPhee. 2001. The sloths of the West Indies: A systematic and phylogenetic review. In C. A. Woods and F. E. Sergile (eds.), *Biogeography of the West Indies: Patterns and Perspectives,* 2nd Edition, 201–236. London: CRC Press.

White, M. J. D. 1973. *Animal Cytology and Evolution.* 3rd edition. Cambridge: Cambridge University Press.

White, M. J. D. 1978. *Modes of Speciation.* San Francisco: W. H. Freeman.

White, T. C. R. 1976. Weather, food and plagues of locusts. *Oecologia* 22: 119–134.

Whitehead, D. R. and C. E. Jones. 1969. Small islands and the equilibrium theory of insular biogeography. *Evolution* 23: 171–179.

Whitford, P. C. 1979. An explanation of altitudinal deviations from Bergmann's rule as applied to birds. *The Biologist* 61: 1–10.

Whittaker, A. H. 1978. The effects of rodents on reptiles and amphibians. In P. R. Dingwall and I. A. Atkinson (eds.), *The Ecology and Control of Rodents in New Zealand's Nature Reserves,* 75–88. Department of Lands and Survey Information Series no. 4.

Whittaker, R. H. 1956. Vegetation of the Great Smoky Mountains. *Ecological Monographs* 22: 1–44.

Whittaker, R. H. 1960. Vegetation of the Siskiyou Mountains, Oregon and California. *Ecological Monographs* 30: 279–338.

Whittaker, R. H. 1967. Gradient analysis of vegetation. *Biological Review* 42: 207–264.

Whittaker, R. H. 1975. *Communities and Ecosystems.* 2nd edition. New York: Macmillan.

Whittaker, R. H. 1977. Evolution of species diversity in land communities. In M. K. Hecht, W. C. Steere and B. Wallace (eds.), *Evolutionary Biology* 10, 1–67. New York: Plenum Press.

Whittaker, R. H. and G. E. Likens. 1973. Carbon in the biota. In G. M. Woodwell and E. V. Pecan (eds.), *Carbon and the Biosphere,* 281–300. Conf. 72501. Springfield, VA: National Technical Information Service.

Whittaker, R. H. and W. A. Niering. 1965. Vegetation of the Santa Catalina Mountains, Arizona: A gradient analysis of the south slope. *Ecology* 46: 429–452.

Whittaker, R. H. and W. A. Niering. 1968. Vegetation of the Santa Catalina Mountains, Arizona. IV. Limestone and acid soils. *Journal of Ecology* 56: 523–544.

Whittaker, R. H. and W. A. Niering. 1975. Vegetation of the Santa Catalina Mountains, Arizona. V. Biomass, production and diversity along the elevation gradient. *Ecology* 56: 771–790.

Whittaker, R. H., S. H. Jones and T. Partomihardjo. 1997. The rebuilding of an isolated rain forest assemblage: How disharmonic is the flora of Krakatau? *Biodiversity and Conservation* 6: 1671–1696.

Whittaker, R. J. 1995. Disturbed island ecology. *Trends in Ecology and Evolution* 10: 421–425.

Whittaker, R. J. 1998. *Island Biogeography: Ecology, Evolution and Conservation.* New York: Oxford University Press.

Whittaker, R. J. 2004. Dynamic hypotheses of richness on islands and continents. In M. V. Lomolino and L. R. Heaney (eds.), *Frontiers of Biogeography: New Directions in the Geography of Nature,* 211–232. Sunderland, MA: Sinauer Associates.

Whittaker, R. J. and S. H. Jones. 1994. The role of frugivorous bats and birds in the rebuilding of a tropical forest ecosystem, Krakatau, Indonesia. *Journal of Biogeography* 21: 245–258.

Whittaker, R. J., R. Field and T. Partomihardjo. 2000. How to go extinct: Lessons from the lost plants of Krakatau. *Journal of Biogeography* 27: 1049–1064.

Whitaker, R. J., D. W. Grogan and J. W. Taylor. 2003. Geographic barriers isolate endemic populations of hyperthermophilic Archaea. *Science* 301: 976–978.

Whittaker, R. J., M. B. Araujo, P. Jepson, R. J. Ladle, J. E. M. Watson and K. J. Willis. 2005. Conservation biogeography: Assessment and prospect. *Diversity and Distributions* 11: 3–23.

Whittier, T. R. and T. M. Kincaid. 1999. Introduced fish in northeastern USA lakes: Regional extent, dominance and effect on native species richness. *Transactions of the American Fisheries Society* 128: 769–784.

Wiens, H. J. 1962. *Atoll Environment and Ecology*. New Haven, CT: Yale University Press.

Wiens, J. J. and M. J. Donoghue. 2004. Historical biogeography, ecology and species richness. *Trends in Ecology and Evolution* 19: 639–644.

Wilcove, D. S. 1987. From fragmentation to extinction. *Natural Areas Journal* 7: 23–29.

Wilcove, D. S., C. H. McLellan and A. P. Dobson. 1986. Habitat fragmentation in the temperate zone. In M. E. Soulé (ed.), *Conservation Biology: The Science of Scarcity and Diversity*, 237–256. Sunderland, MA: Sinauer Associates.

Wilcox, B. A. 1978. Supersaturated island faunas: A species-age relationship for lizards on post-Pleistocene land-bridge islands. *Science* 199: 996–998.

Wilcox, B. A. 1980. Insular ecology and conservation. In M. E. Soulé and B. A. Wilcox, (eds.), *Conservation Biology: An Ecological-Evolutionary Perspective*, 95–117. Sunderland, MA: Sinauer Associates.

Wilcox, B. A. and D. D. Murphy. 1985. Conservation strategy: The effects of fragmentation on extinction. *American Naturalist* 125: 879–887.

Wildenow, K. L. 1792. *Nomenclator botanicus sistens plantas omnes in Caroli a Linne Speciebus plantarum*. Halae Magdeb: Typis Io. Christ. Hendelii.

Wiley, E. O. 1981. *Phylogenetics: The Theory and Practice of Phylogenetic Systematics*. New York: Wiley-Interscience.

Wiley, E. O. 1988. Vicariance biogeography. *Annual Review of Ecology and Systematics* 19: 513–542.

Wiley, E. O. and R. L. Mayden. 1985. Species and speciation in phylogenetic systematics, with examples from the North American fish fauna. *Annals of the Missouri Botanical Garden* 72: 596–635.

Wiley, E. O. and R. L. Mayden. 2000. The evolutionary species concept. In Q. D. Wheeler and R. Meier (eds.), *Species Concepts and Phylogenetic Theory: A Debate*, 70–89. New York: Columbia University Press.

Wiley, E. O., D. J. Siegel-Causey, D. R. Brooks and V. A. Funk. 1991. *The Compleat Cladist: A Primer of Phylogenetic Procedures*. Lawrence, KS: Museum of Natural History, University of Kansas.

Wilkinson, D. M. 1993. Equilibrium island biogeography: Its independent invention and the marketing of scientific theories. *Global Ecology and Biogeography Letters* 3: 65–66.

Wilkinson, D. M. 1998. Mycorrhizal fungi and Quaternary plant migrations. *Global Ecology and Biogeography* 7: 137–140.

Wilkinson, D. M. 2001. What is the upper size limit for cosmopolitan distribution in free-living microorganisms? *Journal of Biogeography* 28: 285–291.

Willdenow, K. L. 1792. (trans. English 1805). *Grundriss de Kräuterkunde zu Vorlesungen (Principles of Botany)*. Haude und Spener, Berlin.

Williams, C. B. 1953. The relative abundance of different species in a wild animal population. *Journal of Animal Ecology* 22: 14–31.

Williams, C. B. 1964. *Patterns in the Balance of Nature and Related Problems in Quantitative Ecology*. New York: Academic Press.

Williams, E. E. 1976. West Indian anoles: A taxonomic and evolutionary summary. I. Introduction and a species list. *Breviora* no. 440.

Williams, E. E. 1983. Ecomorphs, faunas, island size and diverse end points in island radiations of *Anolis*. In R. B. Huey, E. R. Pianka and T. W. Schoener (eds.), *Lizard Ecology: Studies on a Model Organism*, 326–370. Cambridge, MA: Belknap Press.

Williams, J. D. and R. J. Neves. 1995. Freshwater mussels: A neglected and declining aquatic resource. In E. T. Laroe (ed.), *Our Living Resources: A Report to the Nation on the Distribution, Abundance and Health of U.S. Plants, Animals and Ecosystems*, 177–179. Washington, DC: U.S. Department of the Interior, National Biological Service.

Williams, J. E. and R. R. Miller. 1990. Conservation status of North American fish fauna in fresh water. *Journal of Fish Biology* 37: 79–85.

Williams, J. W., B. N. Shuman, T. Webb III, P. J. Bartlein and P. Leduc. 2003. Late Quaternary vegetation dynamics in North America: Scaling from taxa to biomes. *Ecological Monographs* 74: 309–334.

Williams, M. 1990. Clearing of the forests. In M. P. Conzen (ed.), *The Making of the American Landscape*, 146–168. Boston: Unwin Hyman.

Williams, P., D. Gibbons, C. Margules, A. Rebelo, C. Humphries and R. Pressey. 1996. A comparison of richness hotspots, rarity hotspots and complementary areas for conserving diversity of British birds. *Conservation Biology* 10: 155–174.

Williamson, M. 1981. *Island Populations*. Oxford: Oxford University Press.

Williamson, M. 1989. The equilibrium theory today: True but trivial. *Journal of Biogeography* 16: 3–4.

Williamson, M. 1996. *Biological Invasions*. London: Chapman and Hall.

Willig, M. R. 2000. Latitude, Trends with. In S. Levin (ed.), *Encyclopedia of Biodiversity*, 701–714. San Diego, CA: Academic Press.

Willig, M. R. and S. K. Lyons. 1998. An analytical model of latitudinal gradients of species richness with an empirical test for marsupials and bats in the New World. *Oikos* 81: 93–98.

Willig, M. R. and E. R. Sandlin. 1989. Gradients of species density and species turnover in New World bats: A comparison of quadrat and band methologies. In M. A. Mares and D. J. Schmidly (eds.), *Latin American Mammals: Their Conservation, Ecology and Evolution*, 81–96. Norman, OK: University of Oklahoma Press.

Willig, M. R. and K. W. Selcer. 1989. Bat species density gradient in the New World: A statistical assessment. *Journal of Biogeography* 16: 189–195.

Willig, M. R., D. M. Kaufman and R.D. Stevens. 2003. Latitudinal gradients of biodiversity: Pattern, process, scale and synthesis. *Annual Review of Ecology, Evolution and Systematics* 34: 272–309.

Willis, B. 1932. Isthmian links. *Bulletin of the Geological Society of America* 43: 917–952.

Willis, E. O. 1974. Populations and local extinctions of birds on Barro Colorado Island, Panama. *Ecological Monographs* 44: 153–169.

Willis, J. C. 1922. *Age and Area*. Cambridge: Cambridge University Press.

Willis, K. J. and R. J. Whittaker. 2000. The refugial debate. *Science* 287: 1406–1407.

Wilson, D. E., F. R. Cole, J. D. Nichols, R. Rudram and M. S. Foster. 1996. *Standard Methods for Measuring and Monitoring Biodiversity: Mammals*. Washington, DC: Smithsonian Institution Press.

Wilson, D. S. 1975. The adequacy of body size as a niche difference. *American Naturalist* 109: 769–784.

Wilson, E. O. 1959. Adaptive shift and dispersal in a tropical ant fauna. *Evolution* 13: 122–144.

Wilson, E. O. 1961. The nature of the taxon cycle in the Melanesian ant fauna. *American Naturalist* 95: 169–193.

Wilson, E. O. 1975. *Sociobiology*. Cambridge, MA: Belknap Press.

Wilson, E. O. (ed.) 1988. *Biodiversity*. Washington, DC: National Academy Press.

Wilson, E. O. 1992. *The Diversity of Life*. Cambridge, MA: Belknap Press.

Wilson, E. O. and D. S. Simberloff. 1969. Experimental zoogeography of islands: Defaunation and monitoring techniques. *Ecology* 50: 267–278.

Wilson, E. O. and R. W. Taylor. 1967. Ants of Polynesia (Hymenoptera, Formicidae). Honolulu: Department of Entomology, Bishop Museum.

Wilson, E. O. and E. O. Willis. 1975. Applied biogeography. In M. L. Cody and J. M. Diamond (eds.), *Ecology and Evolution of Communities*, 522–534. Cambridge, MA: Belknap Press.

Wilson, I., M. Weale and D. Balding. 2003. Inferences from DNA data: Population histories, evolutionary processes and forensic math probabilities. *Journal of the Royal Statistical Society* Series A 166: 155–188.

Wilson, J. T. 1963a. A possible origin of the Hawaiian Islands. *Canadian Journal of Physics* 41: 863–870.

Wilson, J. T. 1963b. Evidence from islands on the spreading of the ocean floors. *Nature* 197: 536–538.

Wilson, J. W. III. 1974. Analytical zoogeography of North American mammals. *Evolution* 28: 124–140.

Wilson, M. V. and A. Schmida. 1984. Measuring beta diversity with presence-absence data. *Journal of Ecology* 72: 1055–1064.

Winchester, S. 2001. *The Map That Changed the World: William Smith and the Birth of Modern Geology.* New York: Harper Collins.

Windley, B. F. 1977. *The Evolving Continents.* London: John Wiley and Sons.

Winemiller, K. O. 1990. Spatial and temporal variation in tropical fish trophic networks. *Ecological Monographs* 60: 331–367.

Winemiller, K. O., E. R. Pianka, L. J. Vitt. and A. Joern. 2001. Food web laws or niche theory? Six independent empirical tests. *American Naturalist* 158: 193–200.

Winfield, I. J. and C. Hollingworth. 2001. Nonindigenous fishes introduced into inland waters of the United States (American Fisheries Society, Special Publication 27) *Fish and Fisheries* 2: 172–173.

Winkworth, R. C., S. J. Wagstaff, D. Glenny and P. J. Lockhart. 2002. Plant dispersal N.E.W.S. from New Zealand. *Trends in Ecology and Evolution* 17: 514–520.

Wisheu, I. C. and P. A. Keddy. 1989. Species richness: Standing crop relationships along 4 lakeshore gradients: Constraints on the general model. *Canadian Journal of Botany* 67: 1609–1617.

Witman, J. D., R. J. Etter and F. Smith. 2004. The relationship between regional and local species diversity in marine benthic communities: A global perspective. *Proceedings of the National Academy of Sciences, USA* 101: 15664–15669.

Wodzicki, K. A. 1950. Introduced mammals of New Zealand: An ecological and economic survey. *Bulletin Department of Scientific and Industrial Research* 98: 1–255.

Wojcicki, M. and D. R. Brooks. 2004. Escaping the matrix: A new algorithm for phylogenetic comparative studies of co-evolution. *Cladistics* 20: 341–361.

Wojcicki, M. and D. R. Brooks. 2005. PACT: An efficient and powerful algorithm for generating area cladograms. *Journal of Biogeography* 32: 755–774.

Wolfe, J. A. 1975. Some aspects of plant geography in the Northern Hemisphere during the late Cretaceous and Tertiary. *Annals of the Missouri Botanical Garden* 62: 264–279.

Wolfe, J. A. 1979. *Temperature Parameters of Humid to Mesic Forests of Eastern Asia and Relation to Forests of Other Regions of the Northern Hemisphere and Australasia.* U.S. Geological Survey Professional Paper no. 1106. Washington, DC: U.S. Government Printing Office.

Wollaston, T. V. 1854. *Insecta Maderensia.* London: John Van Voorst.

Wollaston, T. V. 1877. *Coleoptera Sanctae-Helenae.* London: John Van Voorst.

Wonham, M. J., J. T. Carlton, G. M. Ruiz and L. D. Smith. 2000. Fish and ships: Relating dispersal frequency to success in biological invasions. *Marine Biology* 136: 1111–1121.

Woodburne, M. O. and J. A. Case. 1996. Dispersal, vicariance and the late Cretaceous to early Tertiary land mammal biogeography from South America to Australia. *Journal of Mammalian Evolution* 3: 121–161.

Woodburne, M. O. and W. J. Zinmeister. 1982. Fossil land mammals from Antarctica. *Science* 218: 284–286.

Woodring, W. P. 1966. The Panama canal landbridge as a sea barrier. *Proceedings of the American Philosophical Society* 110: 425–433.

Woodroffe, C. D. 1986. Vascular plant species-area relationship on Nui Atoll, Tuvalu, Central Pacific: A reassessment of the small island effect. *Australian Journal of Ecology* 11: 21–31.

Woods, C. A. and F. E. Sergile. 2001. Biogeography of the West Indies: Patterns and Perspectives, 2nd edition. London: CRC Press.

Woods, K. D. and M. D. Davis. 1989. Paleoecology of range limits: Beech in the upper peninsula of Michigan. *Ecology* 70: 681–696.

Wootton, J. T. 1992. Indirect effects, prey susceptibility and habitat selection: Impacts of birds on limpets and algae. *Ecology* 73: 981–991.

World Conservation Monitoring Centre. 2000. *Global Diversity: Earth's Living Resources in the 21st Century.* B.Groombridge and M. D. Jenkins (eds.) World Conservation Press, Cambridge, UK.

World Conservation Monitoring Centre. 2000. Freshwater biodiversity. *World Conservation Monitoring Centre Biodiversity Series No. 8:* PP-PP.

World Conservation Monitoring Centre. 1998. Freshwater biodiversity: A preliminary global assessment. *World Conservation Monitoring Centre Biodiversity Series* no. 8: 1–12.

World Conservation Monitoring Centre. 1992. *Global Biodiversity: Status of the Earth's Living Resources.* New York: Chapman and Hall.

Worthy, T. H. and R. N. Holdaway. 2002. *The Lost World of the Moa: Prehistoric Life of New Zealand.* Bloomington, IN: Indiana University Press.

Wright, D. H. 1983. Species-energy theory: An extension of species-area theory. *Oikos* 41: 496–506.

Wright, D. H. 1987. Estimating human effects on global extinction. *International Journal of Biometereology* 31: 293.

Wright, D. H. and J. H. Reeves. 1992. On the meaning and measurement of nestedness of species assemblages. *Oecologia* 92: 416–428.

Wright, D. H., B. D. Patterson, G. M. Mikkelson, A. Cutler and W. Atmar. 1996. A comparative analysis of nested subset patterns of species composition. *Oecologia* 113: 1–20.

Wright, H. E. 1976. Ice retreat and revegetation of the western Great Lakes area. In W. C. Mahaney (ed.), *Quaternary Stratigraphy of North America*, 119–132. Stroudsburg, PA: Dowden, Hutchison and Ross.

Wright, S. 1978. *Variability Within and Among Natural Populations.* Chicago: University of Chicago Press.

Wright, S. J. 1980. Density compensation in island avifaunas. *Oecologia* 45: 385–389.

Wright, S. J. 1981. Inter-archipelago vertebrate distributions: The slope of the species-area relation. *American Naturalist* 118: 726–748.

Wright, S. J. 1985. How isolation affects rates of turnover of species on islands. *Oikos* 44: 331–340.

Wulff, E. V. 1943. *An Introduction to Historical Plant Geography.* Trans. E. Brissenden. Waltham, MA: Chronica Botanica.

Wyatt, R. E. 1992. *Ecology and Evolution of Plant Reproduction.* New York: Chapman and Hall.

Xu, X., Z. H. Zhou, X. L. Wang, X. W. Kuang, F. C. Zhang and X. K. Du. 2003. Four-winged dinosaurs from China. *Nature* 421: 335–340.

Yeaton, R. I. 1974. An ecological analysis of chaparral and pine forest bird communities on Santa Cruz Island and mainland California. *Ecology* 55: 959–973.

Yeaton, R. I. 1981. Seedling morphology and the altitudinal distribution of pines in the Sierra Nevada of central California: A hypothesis. *Madroño* 28: 67–77.

Yeaton, R. I., R. W. Yeaton and J. E. Horenstein. 1981. The altitudinal replacement of digger pine by ponderosa pine on the western slopes of the Sierra Nevada. *Bulletin of the Torrey Botanical Club* 107: 487–495.

Yoda, K. 1967. A preliminary survey of the forest vegetation of eastern Nepal. II. General description, structure and floristic composition of sample plots chosen from different vegetation zones. *Journal of the College of Art and Science. Chi'ba University National Science* 5: 99–140.

Yoder, A. D. and Z. Yang. 2004. Divergence dates for Malagasy lemurs estimated from multiple gene loci: Geological and evolutionary context. *Molecular Ecology* 13: 757–773.

Yoder, A. D., M. M. Burns, S. Zehr, T. Delefosse, G. Veron, S. M. Goodman and J. J. Flynn. 2003. Single origin of Malagasy Carnivora from an African ancestor. *Nature* 421: 734–737.

Zabinski, C. and M. B. Davis. 1989. Hard times ahead for Great Lakes forests: A climate threshhold model predicts responses to CO_2-induced climate change. In J. B. Smith and D. Tirpak (eds.), *The Potential Effects of Global Climate Change on the United States*, Appendix D. Washington, DC: U.S. Environmental Protection Agency.

Zakharov, E. V., C. R. Smith, D. C. Lees, A. Cameron, R. I. Vane-Wright and F. A. H. Sperling. 2004. Independent gene phylogenies and morphology demonstrate a Malagasy origin for a wide-ranging group of swallowtail butterflies. *Evolution* 58: 2763–2782.

Zandee, M. and M. C. Roos. 1987. Component-compatibility in historical biogeography. *Cladistics* 3: 305–332.

Zaret, T. M. 1980. *Predation and Freshwater Communities*. New Haven, CT: Yale University Press.

Zaret, T. M. and R. T. Paine. 1973. Species introduction in a tropical lake. *Science* 182: 449–455.

Zavaleta, E. S. and J. L. Royval. 2002. Climate change and the susceptibility of U.S. ecosystems to biological invasions: Two cases of expected range expansion. In S. H. Schnedier and T. L. Root (eds.), *Wildlife Responses to Climate Change: North American Case Studies*, 277–242. Washington, DC: Island Press.

Zeh, D. W. and J. A. Zeh. 1992. Failed predation or transportation? Causes and consequences of phoretic behavior in the pseudoscorpion *Dinocheirus arizonensis* (Pseudoscorpionida: Chernetidae). *Journal of Insect Behavior* 5: 37–50.

Zeh, D. W., J. A. Zeh and G. Tavakilian. 1992. Sexual selection and sexual dimorphism in the harlequin beetle *Acrocinus longimanus*. *Biotropica* 24: 86–96.

Zimmerman E. C. 1948. *The Insects of Hawaii*, vol. 1. Honolulu: University of Hawaii Press.

Zimmerman, K. 1950. Die randformen der mitteleuropaischen Wuhlmause. In A. Jordans and F. Peus (eds.), *Syllegomena Biologica Festschrift*, 454–471. Leipzig: Kleinschmidt Verlag.

Zink, R. M. 1996. Comparative phylogeography in North American birds. *Evolution* 50: 308–317.

Zink, R. M. and J. C. Avise. 1990. Patterns of mitochondrial-DNA and allozyme evolution in the avian genus Ammodramus. *Systematic Zoology* 39: 148–161.

Zink, R. M., R. C. Blackwell-Rago and F. Ronquist. 2000. The shifting roles of dispersal and vicariance in biogeography. *Proceedings of the Royal Society of London*, Series B 267: 497–503.

Zink, R. M., J. Klicka and B. R. Barber. 2004. The tempo of avian diversification during the Quaternary. *Philosophical Transactions of the Royal Society of London*, Series B-Biological Sciences 359: 215–219.

Zink, R. M., A. E. Kessen, T. V. Line and R. C. Blackwell-Rago. 2001. Comparative phylogeography of some aridland bird species. *Condor* 103: 1–10.

Zuckerkandl, E. and L. Pauling. 1965. Evolutionary divergence and convergence in proteins. In V. Bryson and H. J. Vogel (eds.), *Evolving Genes and Proteins*, 97–166. New York: Academic Press.

Zwiers, F. W. 2002. The 20-year forecast. *Nature* 416: 690–691.

Index

Numbers in *italic* refer to information in an illustration, caption, or table.

About the authors

Mark V. Lomolino is a Professor in the Department of Environmental and Forest Biology at the SUNY College of Environmental Science and Forestry. His research and teaching focus on biogeography and conservation of biological diversity. He is a cofounder and past President of the International Biogeography Society. Dr. Lomolino received the American Society of Mammalogists Award, serves on the editorial advisory board for *Biological Conservation*, and is a coeditor of two books—*Foundations of Biogeography* (University of Chicago Press) and *Frontiers of Biogeography: New Directions in the Geography of Nature*, recently published by Sinauer Associates.

Brett R. Riddle is a Professor in the Department of Biological Sciences at the University of Nevada, Las Vegas. His research focuses primarily on the history of biodiversity in western North America, with ongoing projects including: historical assembly of the warm desert biotas; phylogeography of Great Basin montane island biotas; and molecular systematics and biogeography of diverse North American rodent groups. He is a cofounder and current President of the International Biogeography Society, and an editor of the *Journal of Biogeography*.

James H. Brown is a member of the U.S. National Academy of Science and a Distinguished Professor of Biology at the University of New Mexico. He is a past President of the International Biogeography Society, the American Society of Mammalogists, the American Society of Naturalists, and the Ecological Society of America. His research interests include: community ecology and biogeography, with special projects on granivory in desert ecosystems; biogeography of insular habitats; and structure of dynamics of geographic-scale assemblages of many species.

About the book

Editor: Andrew D. Sinauer

Project Editors: Kathaleen Emerson and David McIntyre

Production Manager: Christopher Small

Electronic Book Production: Janice Holabird

Illustration Program: Precision Graphics, Inc., Joanne Delphia, and Nancy Haver

Copy Editor: Suzanne Lain

Book and Cover Design: Jefferson Johnson

Book Manufacturer: Courier Companies, Inc.